APPLIED ELECTRONIC INSTRUMENTATION AND MEASUREMENT

DAVID BUCHLA
Yuba College

WAYNE McLACHLAN
Grass Valley Group
A Tektronix, Inc. Company

Merrill, an imprint of
Macmillan Publishing Company
New York

Collier Macmillan Canada, Inc.
Toronto

Maxwell Macmillan International Publishing Group
New York Oxford Singapore Sydney

Cover photos provided by Tektronix, Inc., Hewlett-Packard, and David Buchla.
Editor: Dave Garza
Developmental Editor: Carol Hinklin Robison
Production Editor: Rex Davidson
Art Coordinator: Vincent A. Smith
Text Designer: Debra A. Fargo
Cover Designer: Russ Maselli
Production Buyer: Pamela D. Bennett

This book was set in Times Roman by Bi-Comp, Inc. and was printed and bound by R.R. Donnelley & Sons Company. The cover was printed by Lehigh Press, Inc.

Printed in the United States of America

Macmillan Publishing Company
866 Third Avenue
New York, NY 10022

Macmillan Publishing Company is part of the
Maxwell Communication Group of Companies.

Maxwell Macmillan Canada, Inc.
1200 Eglinton Avenue East, Suite 200
Don Mills, Ontario M3C 3N1

Library of Congress Cataloging-in-Publication Data

Buchla, David.
Applied electronic instrumentation and measurement/by David M. Buchla and Wayne C. McLachlan.
p. cm.
Includes index.
ISBN 0-675-21162-X
1. Electronic instruments. 2. Electronic measurements.
I. McLachlan, Wayne C. II. Title.
TK7878.4.B79 1992
621.381'54--dc20

91-11690
CIP

Printing: 1 2 3 4 5 6 7 8 9 Year: 2 3 4 5

This book is dedicated to our wives, Lorraine and Donna.
They are beyond measure!

David and Wayne

MERRILL'S INTERNATIONAL SERIES IN ENGINEERING TECHNOLOGY

ADAMSON	*Applied Pascal for Technology,* 0-675-20771-1 *The Complete Dictionary of Electronics,* 0-02-300820-2 *Microcomputer Repair,* 0-02-300825-3 *Structured BASIC Applied to Technology,* 0-675-20772-X *Structured C for Technology,* 0-675-20993-5 *Structured C for Technology* (w/disks), 0-675-21289-8
ANTONAKOS	*The 68000 Microprocessor: Hardware and Software Principles and Applications,* 0-675-21043-7
ASSER/ STIGLIANO/ BAHRENBURG	*Microcomputer Servicing: Practical Systems and Troubleshooting,* 0-675-20907-2 *Microcomputer Theory and Servicing,* 0-675-20659-6 *Lab Manual to accompany Microcomputer Theory and Servicing,* 0-675-21109-3
ASTON	*Principles of Biomedical Instrumentation and Measurement,* 0-675-20943-9
BATESON	*Introduction to Control System Technology, Third Edition,* 0-675-21010-0
BEACH/ JUSTICE	*DC/AC Circuit Essentials,* 0-675-20193-4
BERLIN	*Experiments in Electronic Devices to accompany Floyd's Electronic Devices and Electronic Devices: Electron Flow Version, Third Edition,* 0-02-308422-7 *The Illustrated Electronics Dictionary,* 0-675-20451-8
BERLIN/ GETZ	*Experiments in Instrumentation and Measurement,* 0-675-20450-X *Fundamentals of Operational Amplifiers and Linear Integrated Circuits,* 0-675-21002-X *Principles of Electronic Instrumentation and Measurement,* 0-675-20449-6
BERUBE	*Electronic Devices and Circuits Using MICRO-CAP II,* 0-02-309160-6
BOGART	*Electronic Devices and Circuits, Second Edition,* 0-675-21150-6
BOGART/ BROWN	*Experiments in Electronic Devices and Circuits, Second Edition,* 0-675-21151-4
BOYLESTAD	*DC/AC: The Basics,* 0-675-20918-8 *Introductory Circuit Analysis, Sixth Edition,* 0-675-21181-6
BOYLESTAD/ KOUSOUROU	*Experiments in Circuit Analysis, Sixth Edition,* 0-675-21182-4 *Experiments in DC/AC Basics,* 0-675-21131-X
BREY	*Microprocessors and Peripherals: Hardware, Software, Interfacing, and Applications, Second Edition,* 0-675-20884-X *The Intel Microprocessors-8086/8088, 80186, 80286, 80386, and 80486—Architecture, Programming, and Interfacing, Second Edition,* 0-675-21309-6
BROBERG	*Lab Manual to accompany Electronic Communication Techniques, Second Edition,* 0-675-21257-X

BUCHLA
Digital Experiments: Emphasizing Systems and Design, Second Edition, 0-675-21180-8
Experiments in Electric Circuits Fundamentals, Second Edition, 0-675-21409-2
Experiments in Electronics Fundamentals: Circuits, Devices, and Applications, Second Edition, 0-675-21407-6

BUCHLA/ McLACHLAN
Applied Electronic Instrumentation and Measurement, 0-675-21162-X

CICCARELLI
Circuit Modeling: Exercises and Software, Second Edition, 0-675-21152-2

COOPER
Introduction to VersaCAD, 0-675-21164-6

COX
Digital Experiments: Emphasizing Troubleshooting, Second Edition, 0-675-21196-4

CROFT
Getting a Job: Resume Writing, Job Application Letters, and Interview Strategies, 0-675-20917-X

DAVIS
Technical Mathematics, 0-675-20338-4
Technical Mathematics with Calculus, 0-675-20965-X
Study Guide to Accompany Technical Mathematics, 0-675-20966-8
Study Guide to Accompany Technical Mathematics with Calculus, 0-675-20964-1

DELKER
Experiments in 8085 Microprocessor Programming and Interfacing, 0-675-20663-4

FLOYD
Digital Fundamentals, Fourth Edition, 0-675-21217-0
Electric Circuits Fundamentals, Second Edition, 0-675-21408-4
Electronic Devices, Third Edition, 0-675-22170-6
Electronic Devices: Electron Flow Version, 0-02-338540-5
Electronics Fundamentals: Circuits, Devices, and Applications, Second Edition, 0-675-21310-X
Fundamentals of Linear Circuits, 0-02-338481-6
Principles of Electric Circuits, Electron Flow Version, Second Edition, 0-675-21292-8
Principles of Electric Circuits, Third Edition, 0-675-21062-3

FULLER
Robotics: Introduction, Programming, and Projects, 0-675-21078-X

GAONKAR
Microprocessor Architecture, Programming, and Applications with the 8085/8080A, Second Edition, 0-675-20675-8
The Z80 Microprocessor: Architecture, Interfacing, Programming, and Design, 0-675-20540-9

GILLIES
Instrumentation and Measurements for Electronic Technicians, 0-675-20432-1

GOETSCH
Industrial Supervision: In the Age of High Technology, 0-675-22137-4

GOETSCH/ RICKMAN
Computer-Aided Drafting with AutoCAD, 0-675-20915-3

GOODY
Programming and Interfacing the 8086/8088 Microprocessor, 0-675-21312-6

HUBERT
Electric Machines: Theory, Operation, Applications, Adjustment, and Control, 0-675-21136-0

HUMPHRIES
Motors and Controls, 0-675-20235-3

HUTCHINS
Introduction to Quality: Management, Assurance and Control, 0-675-20896-3

KEOWN
PSpice and Circuit Analysis, 0-675-22135-8

KEYSER
Materials Science in Engineering, Fourth Edition, 0-675-20401-1

KIRKPATRICK
The AutoCAD Book: Drawing, Modeling and Applications, Second Edition, 0-675-22288-5
Industrial Blueprint Reading and Sketching, 0-675-20617-0

KRAUT
Fluid Mechanics for Technicians, 0-675-21330-4

KULATHINAL
Transform Analysis and Electronic Networks with Applications, 0-675-20765-7

LAMIT/ LLOYD
Drafting for Electronics, 0-675-20200-0

LAMIT/ WAHLER/ HIGGINS
Workbook in Drafting for Electronics, 0-675-20417-8

LAMIT/ PAIGE	*Computer-Aided Design and Drafting,* 0-675-20475-5
LAVIANA	*Basic Computer Numerical Control Programming, Second Edition,* 0-675-21298-7
MacKENZIE	*The 8051 Microcontroller,* 0-02-373650-X
MARUGGI	*Technical Graphics: Electronics Worktext, Second Edition,* 0-675-21378-9 *The Technology of Drafting,* 0-675-20762-2 *Workbook for the Technology of Drafting,* 0-675-21234-0
McCALLA	*Digital Logic and Computer Design,* 0-675-21170-0
McINTYRE	*Study Guide to accompany Electronic Devices, Third Edition* and *Electronic Devices, Electron Flow Version,* 0-02-379296-5 *Study Guide to accompany Electronics Fundamentals, Second Edition,* 0-675-21406-8
MILLER	*The 68000 Microprocessor Family: Architecture, Programming, and Applications, Second Edition,* 0-02-381560-4
MONACO	*Essential Mathematics for Electronics Technicians,* 0-675-21172-7 *Introduction to Microwave Technology,* 0-675-21030-5 *Laboratory Activities in Microwave Technology,* 0-675-21031-3 *Preparing for the FCC General Radiotelephone Operator's License Examination,* 0-675-21313-4 *Student Resource Manual to accompany Essential Mathematics for Electronics Technicians,* 0-675-21173-5
MONSSEN	*PSpice with Circuit Analysis,* 0-675-21376-2
MOTT	*Applied Fluid Mechanics, Third Edition,* 0-675-21026-7 *Machine Elements in Mechanical Design, Second Edition,* 0-675-22289-3
NASHELSKY/ BOYLESTAD	*BASIC Applied to Circuit Analysis,* 0-675-20161-6
PANARES	*A Handbook of English for Technical Students,* 0-675-20650-2
PFEIFFER	*Proposal Writing: The Art of Friendly Persuasion,* 0-675-20988-9 *Technical Writing: A Practical Approach,* 0-675-21221-9
POND	*Introduction to Engineering Technology,* 0-675-21003-8
QUINN	*The 6800 Microprocessor,* 0-675-20515-8
REIS	*Digital Electronics Through Project Analysis,* 0-675-21141-7 *Electronic Project Design and Fabrication, Second Edition,* 0-02-399230-1 *Laboratory Manual for Digital Electronics Through Project Analysis,* 0-675-21254-5
ROLLE	*Thermodynamics and Heat Power, Third Edition,* 0-675-21016-X
ROSENBLATT/ FRIEDMAN	*Direct and Alternating Current Machinery, Second Edition,* 0-675-20160-8
ROZE	*Technical Communication: The Practical Craft,* 0-675-20641-3
SCHOENBECK	*Electronic Communications: Modulation and Transmission, Second Edition,* 0-675-21311-8
SCHWARTZ	*Survey of Electronics, Third Edition,* 0-675-20162-4
SELL	*Basic Technical Drawing,* 0-675-21001-1
SMITH	*Statistical Process Control and Quality Improvement,* 0-675-21160-3
SORAK	*Linear Integrated Circuits: Laboratory Experiments,* 0-675-20661-8
SPIEGEL/ LIMBRUNNER	*Applied Statics and Strength of Materials,* 0-675-21123-9
STANLEY, B.H.	*Experiments in Electric Circuits, Third Edition,* 0-675-21088-7
STANLEY, W.D.	*Operational Amplifiers with Linear Integrated Circuits, Second Edition,* 0-675-20660-X

SUBBARAO	*16/32-Bit Microprocessors: 68000/68010/68020 Software, Hardware, and Design Applications,* 0-675-21119-0
TOCCI	*Electronic Devices: Conventional Flow Version, Third Edition,* 0-675-20063-6 *Fundamentals of Pulse and Digital Circuits, Third Edition,* 0-675-20033-4 *Introduction to Electric Circuit Analysis, Second Edition,* 0-675-20002-4
TOCCI/ OLIVER	*Fundamentals of Electronic Devices, Fourth Edition,* 0-675-21259-6
WEBB	*Programmable Logic Controllers: Principles and Applications, Second Edition,* 0-02-424970-X
WEBB/ GRESHOCK	*Industrial Control Electronics,* 0-675-20897-1
WEISMAN	*Basic Technical Writing, Sixth Edition,* 0-675-21256-1
WOLANSKY/ AKERS	*Modern Hydraulics: The Basics at Work,* 0-675-20987-0
WOLF	*Statics and Strength of Materials: A Parallel Approach,* 0-675-20622-7

PREFACE

The goal of *Applied Electronic Instrumentation and Measurement* is to teach principles of measurement using state-of-the art electronic instruments and instrumentation systems. It is intended to support courses in instrumentation at the college and technical school level for students who have had an introductory course in basic electronics and who are familiar with circuits in general. Because of the wide application of electronic instrumentation in engineering, science, and industry, we feel it is important to provide more material than is typical for a one-semester instrumentation course so that instructors can select from topics for their particular program.

Applied Electronic Instrumentation and Measurement is uniquc in that it presents a systematic view of measurement technology, including data and data analysis, instrumentation circuits, specific instruments, and a variety of measurement systems in a single introductory volume. Our guiding philosophy is that for a user to understand the limitations of a particular measurement, he or she should be aware of everything that affects the data from the input transducer to the final data analysis. Accordingly, we have considered electronic measurements from the systems view and devoted a significant part of the book to measurement problems associated with systems. We have made a considerable effort to ensure that the book is current by including the latest instruments and system concepts, such as testing methods for surface-mounted devices, boundary scan methods, the VXI bus, and more.

The text is divided into four sections in support of instrumentation and measurement technology.

Section One includes an overview of instruments, data, standards, and calibration; it includes introductory material on measurement systems. In addition, Section One contains descriptions of waveforms, linear and non-linear graphs, data analysis (including statistical analysis and linear regression), and other important topics necessary for reporting measured results. The chapter on standards and calibration includes a discussion of the new voltage and resistance standards adopted for the United States.

Section Two is concerned with circuits used in instrumentation and the measurements of circuits in general. For many readers, much of the circuit coverage will be review material; however, the emphasis is different from that of many devices texts. For example, the diode equation is applied to temperature sensors and again to logarithmic amplifiers, both important topics in instrumentation. Circuits used in instruments and instrumentation systems, such as analog switches, automatic gain control, A/D and D/A conversion, and bus systems are emphasized. An introduction to the transistor curve tracer is included in Chapter 4.

In addition to discussing instrumentation circuits, this section underscores how to make measurements of circuit parameters, such as the input impedance of an amplifier. Students who have taken an introductory electronics circuits course should have a basic familiarity with the operation of the oscilloscope and DMM. For that reason, circuit measurements in the second section are limited to applications in which these two instruments are the only ones necessary. Students who are familiar with this material can omit it with no loss of continuity; references are scattered throughout the text to specific circuits contained in this section. We feel this procedure is most efficient for providing the reader with the information needed to understand the instruments and systems in the later sections.

Section Three covers instruments such as meters, oscilloscopes, function generators, logic analyzers, spectrum analyzers, and other linear and digital instruments. The emphasis is on understanding a measurement, as well as applications of specific instruments. There are many practical examples, such as the relationship of sampling-to-frequency effects, probes, instrument loading, and discussion of newer instruments, such as arbitrary waveform generators and digital sampling oscilloscopes. The spectrum analyzer is discussed with relation to communication systems, an application where it is the dominant instrument. For this reason, we have included a discussion on modulation methods and modulation measurements, a topic normally reserved for communication books.

Section Four is concerned with problems associated with electronic measurements in systems and considers several selected systems in more depth. Coverage includes chapters of importance to nearly all instrumentation systems—namely, noise, transducers, and data acquisition. In particular, we tried to offer practical suggestions for understanding and solving problems associated with noise. The coverage on transducers includes suggestions for choosing and calibrating transducers and describes common sensing methods for those used in electronic systems.

Section Four also includes three chapters on representative systems: automatic test systems, video systems, and biomedical systems. We selected these topics because of their importance in the field of instrumentation. For instance, video systems are frequently used to acquire information about inaccessible areas; they also find application ranging from hospital operating rooms to robotic vision. Unfortunately, few "how to" resources exist for making specific video measurements. We have, therefore, included very specific procedures for making the most important video measurements with enough introductory information to explain the purpose of each measurement.

Each chapter begins with a set of objectives and a related short historical note. It is hoped that the instructor and student will find these historical notes as interesting as we did in researching them. Chapters contain key words in boldface near where they are first defined. Many examples and specific applications are included throughout. Chapters conclude with a Summary and set of Questions and Problems to test students' understanding of the key points. A complete Glossary follows the text.

An instructor's manual is available that has answers to Questions and Problems as well as several complete laboratory exercises selected from a number that we have prepared. These selected exercises are designed for the first portion of the book and may be reproduced by adopters at no charge. In addition, the instructor's manual contains multiple-choice quizzes for each chapter.

ACKNOWLEDGMENTS

There are many persons who contributed their expertise to help make this text a reality, but one person deserves special recognition. We are particularly grateful for the many helpful discussions over the last several years with our friend Dr. Richard Bliss. His contributions include writing Chapter 17 on biomedical instrumentation as well as complete reviews of each of the other chapters. His practical knowledge is obtained both from designing and using biomedical instruments for many years, and he has a gift for writing that will be evident to the reader.

There are many others who supported our efforts, including our colleges and many students who read drafts of this manuscript as it developed. Their suggestions were very helpful and we thank them. We also sincerely appreciate the constructive suggestions and help from the following people:

Jerry Allgood	Grass Valley Group
Dick Angus	Hewlett-Packard
Alex Avtgis	Wentworth Institute of Technology
Richard Bannister	Grass Valley Group
Nancy Boles	Yuba College
Eldred S. Bliss	Eldred Bliss and Associates
Bill Burns	Greenville Technical College
Dale Cigoy	Keithley Instruments
Mark Corrao	Vermont Technical College
David Delker	Kansas State University
Earl J. Farley	Texas Technical University
Joseph Farren	University of Dayton
Barbara Fields	Tektronix
Frank Gergelyi	Metropolitan Technical Institute
G.V. Guard	Greater New Haven Technical College
Vince Hamm	Hewlett-Packard
Bonnie Hansen	Yuba College
Steve Klein	Yuba College
Harvey Laabs	North Dakota State College of Science
Michael Maxwell	St. Agnes Medical Center, Fresno
Wally Miller	John Fluke Manufacturing Co.
Walter V. Morgan	Lawrence Livermore National Laboratory

Mike Parker	Grass Valley Group
Robert Pease	National Semiconductor
John Petrella	George Washington University
Phil Postel	Yuba College
Lee Rosenthal	Fairleigh Dickenson University
Blaine Russell	Yuba College
Don Schrader	Yuba College
Chandra Sekhar	Purdue University-Calumet
Norm Sprangle	Humbolt State University
Martha Spalding	Measurements Group, Inc.
Gary Snyder	Bently Nevada Co.
Lauren Syda	Yuba College
Barry N. Taylor	National Institute of Standards and Technology
Mike Thorson	Grass Valley Group
Bill White	Nashville State Technical Institute
Ulrich Zeisler	Salt Lake Community College

We would also like to express our appreciation to the staff at Merrill Publishing for their help and support in bringing this project to completion. Thanks to Dave Garza, Carol Robison, Rex Davidson, Vince Smith, Debra Fargo, and Linda Thompson. Finally, a special thank you to our wives, Lorraine and Donna, who have given us encouragement and support for this project.

CONTENTS

Section Three

BASIC INSTRUMENTS 279

Section One

MEASUREMENT AND DATA ANALYSIS

Chapter 1

Measurement Principles

OBJECTIVES

This chapter presents an overview of measurement systems and principles that will be addressed in greater detail in the remainder of this book. Measurement is the process of getting information about a physical quantity such as temperature, pressure, or voltage and presenting it in a useful form. An instrument is a device that forms an extension of our human senses by allowing us to determine the magnitude of the measured quantity. Instrumentation systems are concerned with the entire measurement of physical quantities from the first operation to the end result.

Compared to nonelectronic instrumentation, electronic instruments offer major advantages for measurements because of their high speed, great sensitivity, and ability to process and store information. They are vital to all fields of science, ranging from the study of the structure of atoms to the distances in astronomy that we cannot imagine. When an electronic instrument is used to measure a physical quantity, the quantity is generally first converted into an electrical signal. The electrical signal is then transmitted to the instrument and processed. The result is usually displayed, stored, or used for control purposes.

When you complete this chapter, you should be able to

1. Describe the function of the elements of a measurement system, including the transducer, transmission system, instrument, and user.
2. Define basic vocabulary words of measuring systems, including transfer function, impulse response, frequency response, and dynamic range.
3. Compare the time, frequency, and data domains for signals.
4. Given the equation for a sinusoidal waveform, find the amplitude, period, frequency, and phase shift.
5. Convert between peak, peak-to-peak, rms, and average values for a sinusoidal waveform.
6. Compare coax, twisted pair, flat cable, and fiber optics transmission paths.
7. Explain the principles for drawing block diagrams, logic diagrams, and schematic drawings.

HISTORICAL NOTE

Jean Baptiste Fourier (1768–1830), a French mathematician, was the nineteenth child of Joseph and Edmée Fourier. He showed an early interest in mathematics but was unable to realize his dream of becoming an officer in the army. Fourier established a scientific reputation for mathematical work and for his studies of ancient Egypt, including extensive writing gathered from an Egyptian expedition. His most famous scientific achievement was for his introduction of new and powerful mathematical techniques developed for solving problems in heat diffusion. The famous Fourier series and Fourier transform are but two of a number of contributions he made to mathematics. These techniques have wide application to the physical sciences and engineering.

1–1 MEASUREMENT SYSTEMS

Measurement is the process of associating a number with a quantity by comparing the quantity to a standard. When it is necessary to measure objectively a quantity such as temperature, rotational speed, or sound level from an industrial process or a scientific experiment, a measurement system is required. Although measurements can be made strictly by mechanical or other means, electronic measuring systems offer substantial advantages for many measurements. These include very high speed, high reliability, the capability to take data in the range of picoseconds to years, very large dynamic range, and the ability to send, process, and store information by electronic means.

There are a great variety of measurement systems. Electronic measuring systems can be very complex but usually contain the common elements in the simplified measuring system shown in Figure 1–1. These elements are a transducer to convert the physical quantity into an electrical signal, a transmission path, signal-conditioning elements, electronic processing to convert the signal into a suitable form, and recording or display devices for the user. The term **instrument** generally refers to all of these elements, but it is also common practice to refer to a specific electronic device within the measuring system (such as an oscilloscope or digital voltmeter) as an instrument.

The **measurand** is the physical quantity to be measured. It can be a quantity such as temperature, light intensity, or nuclear radiation. Transducers convert the measurand into an electrical signal. Transducers can include thermocouples, strain gauges, photocells, phonograph cartridges, and microphones, among others. Often, the desired signal from the transducer or sensor is a very low level signal; for these cases a means of separating the signal from noise or interference is important. When the signal delivered by a transducer and transmission line is a low-level signal, it generally requires amplification or other conditioning prior to processing. This can include filtering, such as in high-pass cable compensation networks, amplification, buffering, linearization, decoding, or detecting the signal. The signal is then processed electronically and may be converted to digital form (A/D conversion) for further processing. Finally, it is converted to a form

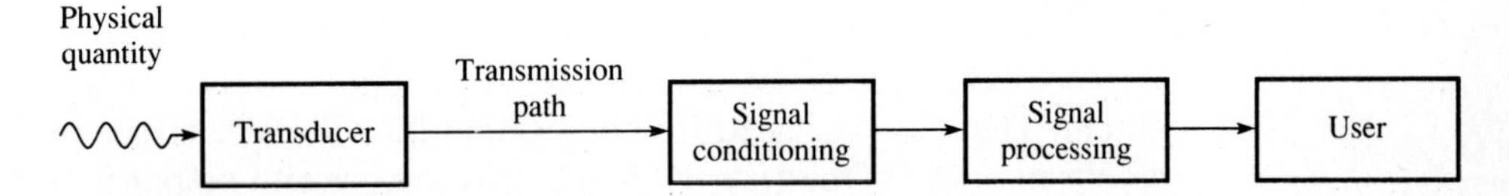

FIGURE 1–1
Simplified measuring system.

useful to the user. The user can be a human observer, a control mechanism, a computer, or a recording system (such as a tape recorder to log a performance of a system). The user can also be part of an automatic control system, in which the data is acted on in some manner to control a process. For human observers, the output from the instrument is normally made into some form of visual display but occasionally is made into an audible form (usually for warnings for certain conditions). Visual displays include a number or a set of numbers on a CRT (cathode-ray tube), a pointer on a scale, a graph or graphical display, a pictorial representation (such as an X ray), or printed information.

Electronic telemetry systems are sometimes used to send data from remote sensors to the instrument or user. Data is acquired at the remote location and sent via the telemetry system. The basic measurement system may include the telemetry system illustrated in Figure 1–2. The desired signal is acquired by the transducer and encoded into a signal suitable for the transmitter. It can then be used to modulate a high-frequency radio signal, which is fed to the transmission system. Typically, telemetry systems are designed to handle multiple channels of information. Such a system requires that the signals be **multiplexed**—that is, several signals are sent over the same link to the receiver and measuring instrument. The transmission link can be very high frequency (VHF), microwave radio, satellite, fiber optics, infrared laser, or simply a conducting path such as a telephone link, twisted-pair wires, a coaxial cable, or an electric power line to a receiver. The receiver decodes the original signal and converts it into an appropriate format for the user.

FIGURE 1–2
Simplified measuring system with telemetry.

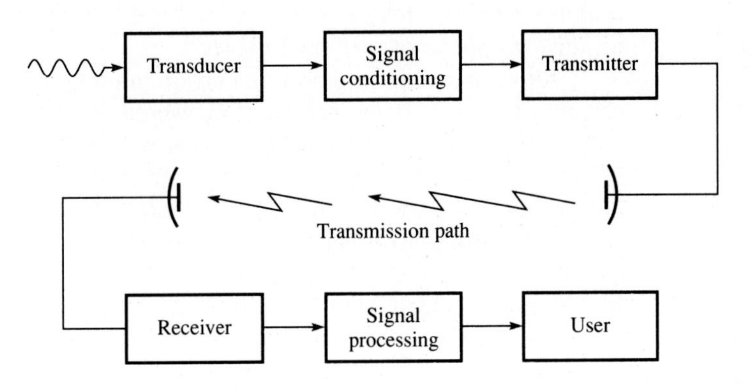

SYSTEM RESPONSE

Each component of a measuring system contributes to the overall response of the system. The observer needs to be aware of how the measurement system affects the observations. One way to characterize the system response is by the system transfer function. The **transfer function** of a system or subsystem is the *ratio* of the output signal to the input signal. It always includes the magnitude and may include the phase of the signals. If the input and output signals are voltages, as in a linear amplifier, then the transfer function is the ratio of the output voltage to the input voltage. Actually, the transfer function can represent the ratio of any two physical quantities. For example, the transfer function of a digital temperature-measuring system is the displayed reading divided by the actual input temperature. If the input and output quantities are measured in the same units, the transfer function is a dimensionless quantity. Examples of transfer functions are shown in Figure 1–3.

The transfer function can be characterized by the **frequency response** of the system. If the ratio of the output voltage to input voltage for a linear amplifier is plotted as a function of frequency, the resulting curve indicates the frequency response of the amplifier. An example is given in Figure 1–4. A high- or low-frequency signal passing through this system does not cause the same response as a signal in the middle of the band. In addition to the frequency response, a finite time delay is associated with the transfer function, causing a lag in the output response. This delay can be important in high-speed measurements.

Another way to characterize a system is by **transient response.** Transient response is the response time of a system to an abrupt input change. For example, if you could obtain an ideal square wave from a generator, it would rise in zero time. Of course, no signal generator can produce an ideal waveform like this. Neither can any real system *respond* instantaneously to this signal. Transient response is important in medical imaging systems, where an abrupt change in contrast in a video signal may occur. The time required to reproduce the change is a measure of the picture quality. Another type of transient signal is an impulse. Ideally, an **impulse** is a pulse with zero width yet with a finite area. Although it is not possible to generate a true impulse, very short, fast pulses are useful for testing purposes. The impulse response of a limited-bandwidth system is an envelope with a finite rise and exponential decay characteristic. One application of impulse response testing is that of fast cable systems. Figure 1–5 illustrates the impulse response of a coaxial cable.

All systems are limited to a certain minimum and maximum values that can be recorded accurately. The **input threshold** of a system is the smallest detectable value of the measured quantity starting near the zero value of the variable. The input threshold is limited by the system's **sensitivity,** which is the ratio of the output response to the input cause under static conditions. For example, the sensitivity of a meter can be stated in units of microamps required for a stated deflection of the meter. The sensitivity of measurement systems is generally limited by noise, which is inherent in all measuring systems. **Noise** is any signal that does not convey useful information and can include frequencies ranging from power-line to radio frequencies. The source of noise is typically electrical in

FIGURE 1–3
Examples of system transfer functions. The transfer function is the *ratio* of the output signal to the input signal.

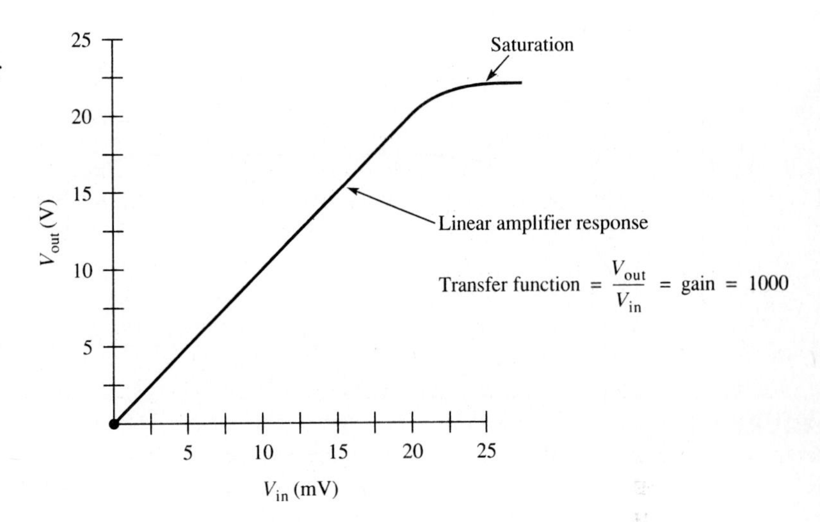

(a) Transfer function of a linear amplifier.

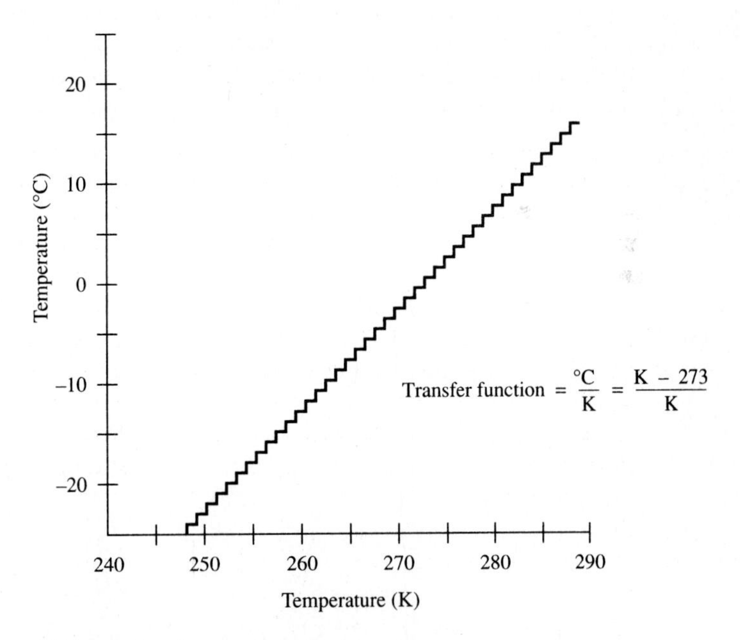

(b) Transfer function of a digital thermometer.

nature and includes several effects (see Chapter 12). Other factors can contribute to the minimum observable signal, including the sensitivity of the transducer or measuring instrument, the stability and reproducibility of the system, and even the data-reduction process. The maximum signal is limited by the transducer response or limitations of the instrument, including saturation or breakdown. The

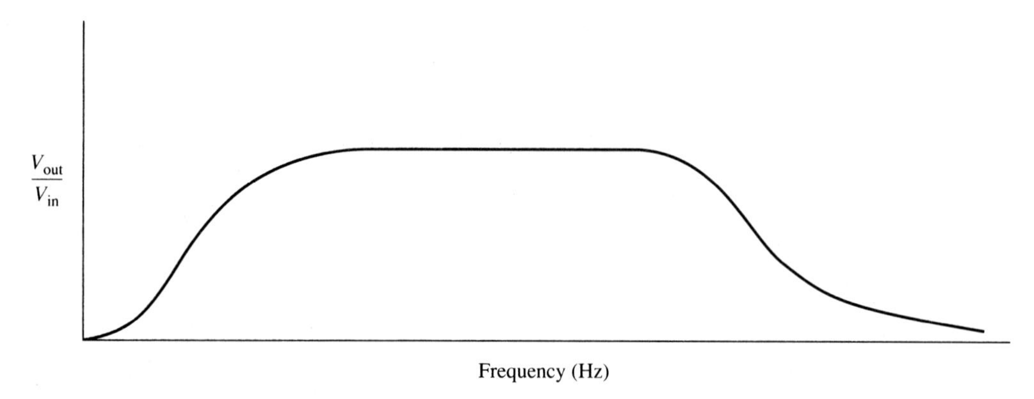

FIGURE 1–4
Example of a system transfer function plotted against frequency.

difference between the maximum and minimum signal to which a given system can respond with a stated degree of accuracy is the **dynamic range** of that system. Note that dynamic range is dependent on the frequency response that limits the maximum signal outside the passband. Frequency response refers to how the amplitude or phase characteristic of a system changes as frequency is varied.

FIGURE 1–5
Impulse response of coaxial cable.

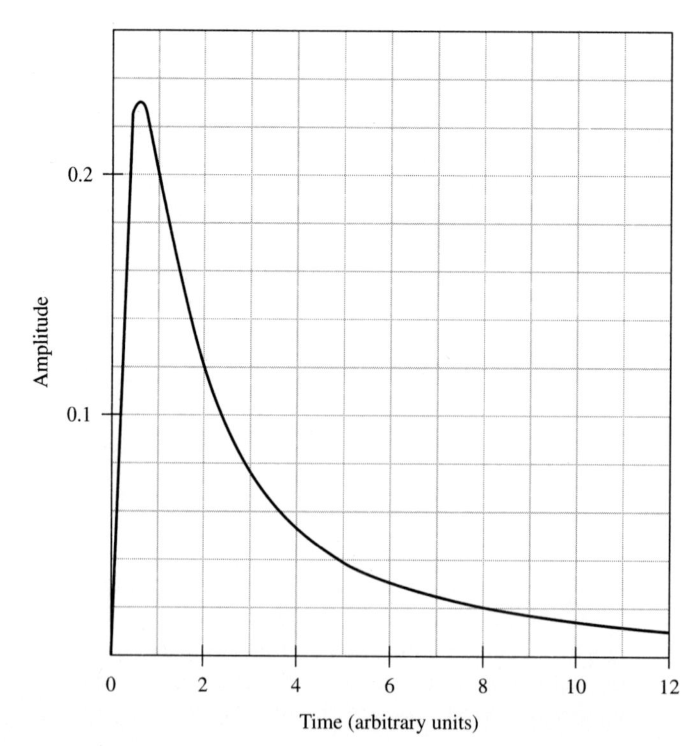

Another important parameter that characterizes measurement systems is resolution. **Resolution** is the minimum discernible *change* in the measurand that can be detected. The definition of resolution does not require the property to begin near the threshold level. For example, the resolution of a time-varying quantity is determined by the ability of the system to resolve time; it may also be set by the ability of the system to detect a small change in the amplitude of the measured quantity. The resolution of a system is affected by all the parts of the system but is ultimately determined by the weakest link in the system. An analog oscilloscope, for example, can generally resolve in the vertical direction 1 part in 500, whereas the same oscilloscope may resolve only 1 part in 300 in the horizontal direction. This might limit the resolution of the system to that of the oscilloscope. It is important that the user is aware of such limitations in any measuring system.

As previously mentioned, noise limits measurement systems. Noise can be induced in measurement systems because of electric or magnetic fields or mechanical coupling, or it can occur inside the system because of thermal noise due to random vibrations of electrons and other sources. Noise affects both the input threshold of the system and the resolution. Chapter 12 discusses the sources of noise and methods for dealing with noise problems in measurement systems.

1–2 CHARACTERISTICS OF SIGNALS

In a general sense, a **signal** is any physical quantity that varies with time (or other independent variable) and carries information. By this definition, a signal can be represented by audio, visual, or other physical means; however, in electronics work, it usually refers to information that is carried by electrical waves, either in a conductor or as an electromagnetic field.

Signals can be classified as either continuous or discrete. A **continuous** signal changes smoothly, without interruption. A **discrete** signal changes in definite steps, or in a quantized manner. The terms *continuous* and *discrete* can be applied either to the value (amplitude) or to the time characteristics of a signal. In nature, most signals can take on a continuous range of values within certain limits (although the limits can extend to infinity). We refer to such signals as **analog signals.** Signals that take on a limited number of discrete values are referred to as **digital signals;** they are discontinuous. As an example, consider a light dimmer that uses a rheostat; the light can be varied continuously within limits, and the result is an analog output. On the other hand, if a rotary switch is used, only certain positions can be selected, and a digital output results.

Signals can also be classified as either continuous-time or discrete-time. A **continuous-time signal** exists at all instances in time. A **discrete-time signal** is defined only at selected instants of time; frequently they are equally spaced intervals. Many instruments sample the signals at their input terminals only at certain times, in effect producing discrete-time signals. Others (frequency counters, for example) accumulate the signal for an interval of time and then show a discrete-time signal. An analog signal is converted to a digital signal by two steps: (1)

sampling, in which the signal is converted to a discrete-time signal, and (2) **quantizing,** in which the samples are given a numeric value (see Section 6–5).

PERIODIC SIGNALS

Repeating waveforms are said to be **periodic.** The period T represents the time for a periodic wave to complete one cycle. A **cycle** is the complete sequence of values that a waveform exhibits before an identical pattern occurs. The period can be measured between any two corresponding points on successive cycles.

The most basic and important periodic waveform is the sinusoidal wave. Sinusoidal waves can be defined in terms of either the sine function or the cosine function. Both the sine and cosine functions have the *shape* of a sinusoidal wave. (See Figure 1–6.) The term *sine wave* usually implies the trigonometric function, whereas the term *sinusoidal wave* means a waveform with the shape of a sine wave. The motion of a spring or pendulum, the vibration of a tuning fork, and ocean waves have cyclic motion called **simple harmonic motion.** A graph of simple harmonic motion as a function of time produces a sinusoidal wave. It is generated as the natural waveform from ac generators, lasers, and radio waves.

The sinusoidal curve can also be generated by plotting the projection of the endpoint of a rotating vector that is turning with uniform circular motion, as

FIGURE 1–6
Comparison of the sine and cosine functions.

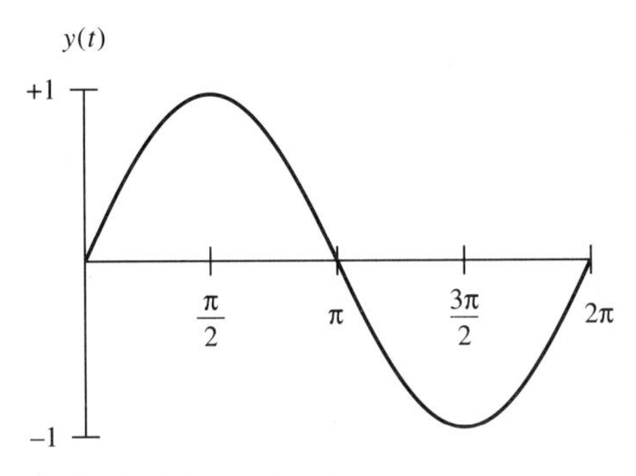

(a) Graph of the sine function.

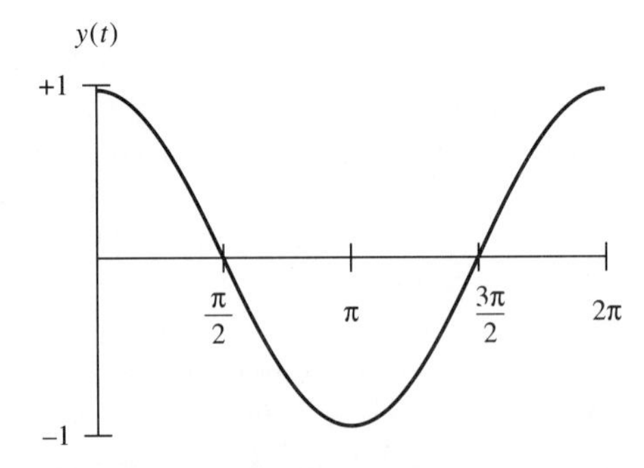

(b) Graph of the cosine function.

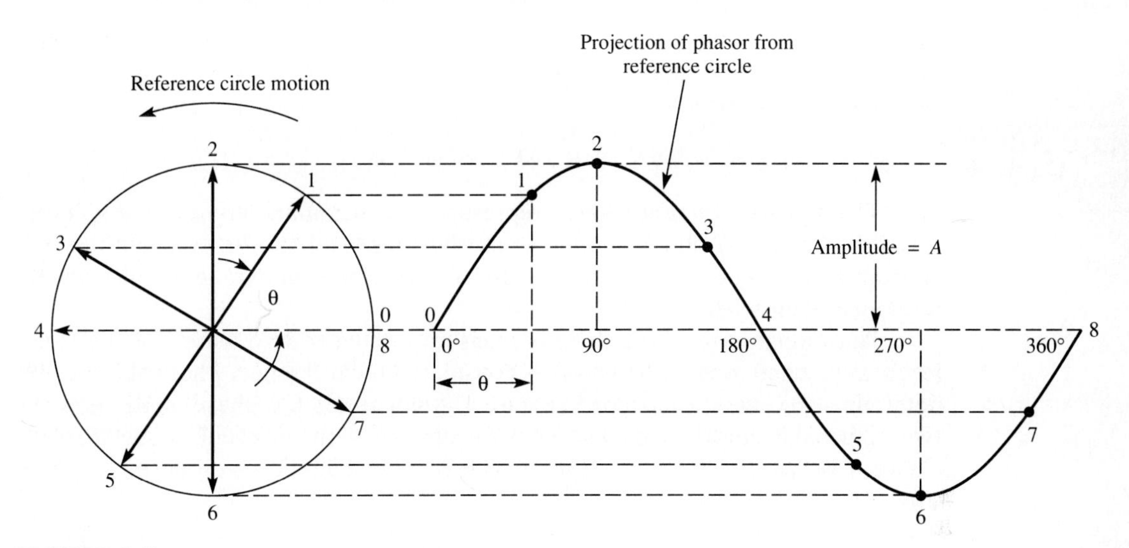

FIGURE 1–7
Generation of sinusoidal waveform.

illustrated in Figure 1–7. Successive revolutions of the point generate a periodic curve, which can be expressed mathematically as

EQUATION 1–1

$$y(t) = A \sin(\omega t \pm \phi)$$

where $y(t)$ = vertical displacement of a point on the curve from the horizontal axis

A = amplitude of the wave or maximum displacement from the horizontal axis

ω = angular velocity of the rotating vector, rad/s

t = time to a point on the curve, s

ϕ = phase angle, rad

The phase angle is simply a fraction of a cycle by which a waveform is shifted from a reference waveform of the same frequency. It is positive if the curve begins before $t = 0$ and is negative if the curve starts after $t = 0$.

The equation for a sinusoidal waveform was shown using the sine function. The cosine wave has the same shape as the sine wave but is shifted by 1/4 cycle. There is a trigonometric identity that states

$$\sin \phi = \cos\left(\phi - \frac{\pi}{2}\right)$$

Substituting this identity into Equation 1–1 allows us to write the equation for a sinusoidal waveform in terms of the cosine function:

EQUATION 1–2

$$y(t) = A \cos\left(\omega t \pm \phi - \frac{\pi}{2}\right)$$

For a sinusoidal voltage, we can substitute $v(t)$ for the vertical displacement and V_p for the amplitude into Equation 1–1. The equation for a sinusoidal voltage can therefore be written

EQUATION 1–3

$$v(t) = V_p \sin(\omega t \pm \phi)$$

The dependent variable, $v(t)$, represents the instantaneous value of voltage. The independent variable, t, represents the time, and the other quantities are constants. The phase shift can be either positive or negative depending on the reference point where $t = 0$.

Phase angle (or phase shift) is measured between two waves of the same frequency, where one of the waves is considered the reference. Phase shift can be determined by one oscilloscope. Figure 1–8 illustrates the phase angle between two sinusoidal curves such as might be observed on an oscilloscope display. Curve A is the reference wave, so we can write its equation with no phase shift. That is

$$v_A(t) = V_A \sin \omega t$$

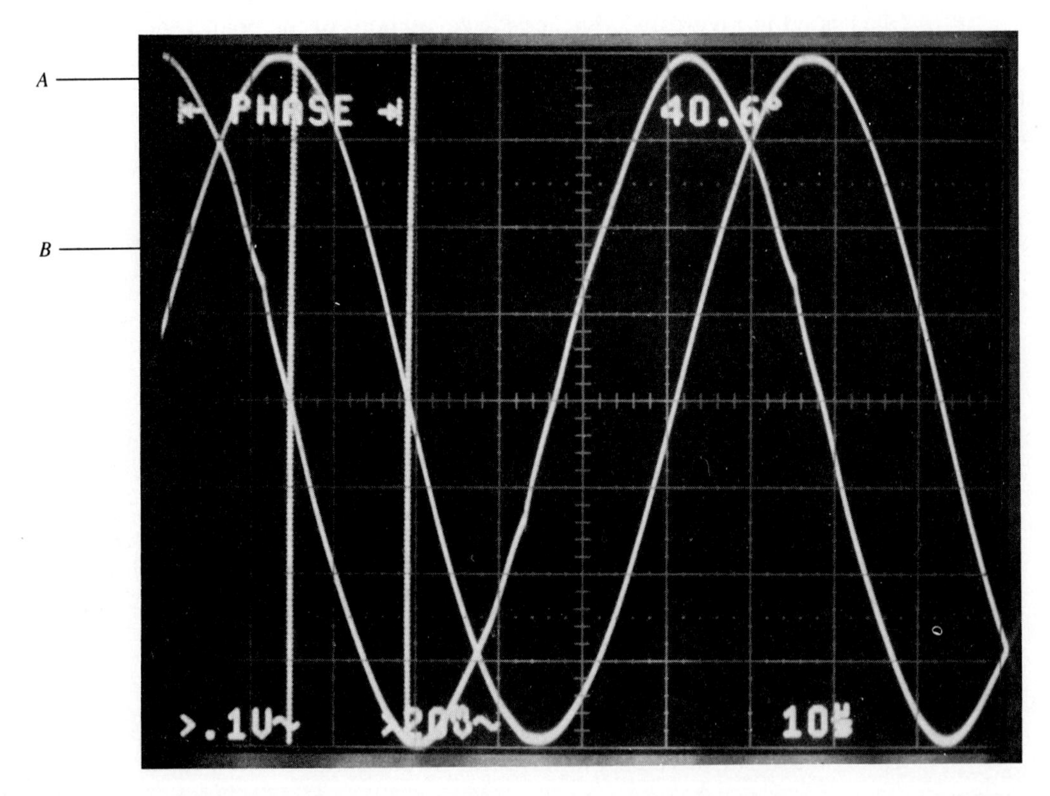

FIGURE 1–8
Phase difference between sinusoidal waveforms.

Curve B starts later than A; therefore A leads B. We see that A leads B by 40.6°. (Conversely, we could say B lags A by 40.6°.) The equation for curve B is therefore displaced from A by 40.6°:

$$v_B(t) = V_B \sin(\omega t - 40.6°)$$

The phase angle is shown as a negative number because curve B starts later than the reference.

EXAMPLE 1–1 A sinusoidal waveform has a peak amplitude of 5.0 V and an angular velocity of 27 rad/s. It leads a reference wave by 30°. Write the equation for the instantaneous voltage.

SOLUTION It is useful to express the phase angle and ωt in the same angular units. Start by converting the phase angle to radian measure:

$$\phi = \frac{30°}{57.3°/\text{rad}} = 0.524 \text{ rad}$$

Substituting into Equation 1–1:

$$v(t) = V_p \sin(\omega t \pm \phi)$$

$$v(t) = (5.0 \text{ V})\sin(27t + 0.524 \text{ rad})$$

Notice that the sign of the phase angle is positive because it leads the reference wave.

FREQUENCY AND PERIOD

When the rotating vector has made one complete cycle, it has rotated through 2π rad. The number of complete cycles generated per second is called the frequency. Dividing the angular velocity (ω) of the rotating vector (in radians per second) by the number of radians in one cycle (2π) gives the number of cycles per second:

EQUATION 1–4

$$f = \frac{\omega \text{ (rad/s)}}{2\pi \text{ (rad/cycle)}}$$

Prior to 1960 the unit of frequency was cycles per second (cps), but it was renamed the hertz (abbreviated Hz) in honor of Heinrich Hertz, a German physicist who demonstrated radio waves.[1] One cycle per second is equal to 1 Hz. Since the frequency of a periodic wave is the number of cycles in 1 s and the period is the time for one cycle, it follows that the reciprocal of the period is the frequency (f) and vice versa. That is

$$T = \frac{1}{f} \quad \text{and} \quad f = \frac{1}{T}$$

[1] An interesting account of Hertz's accomplishments can be found in the May 1988 issue of *IEEE Spectrum*. The old definition was more descriptive of the meaning of frequency.

For example, if a signal repeats every 10 ms, then its period is 10 ms and its frequency is 1/10 ms = 0.1 kHz.

EXAMPLE 1–2 What is the frequency of the sinusoidal wave for Example 1–1?

SOLUTION

$$f = \frac{\omega}{2\pi} = \frac{27}{2\pi} = 4.30 \text{ Hz}$$

PEAK, AVERAGE, AND EFFECTIVE VALUES OF A SINUSOIDAL WAVE

The amplitude of a sinusoidal wave is maximum displacement from the horizontal axis, as shown in Figure 1–7. For a sinusoidal *voltage* waveform, the amplitude is called the **peak voltage,** V_p. When making voltage measurements with an oscilloscope, it is easier to measure the peak-to-peak voltage, V_{pp}. The peak-to-peak voltage is the difference between the most positive excursion and the most negative excursion and is twice the peak value.

During one cycle, a sinusoidal waveform has equal positive and negative excursions. It follows that the average value of a sinusoidal waveform must be zero. However, the term **average value** is generally used to mean the average over a cycle without regard to the sign. That is, the average is usually computed by converting all negative values to positive values and then averaging. (This is equivalent to taking the average of a full-wave rectified sinusoidal wave.) By this definition, the average value is related to the peak value of a sinusoidal wave by

EQUATION 1–5

$$V_{ave} = 0.637V_p$$

The average value is useful in certain practical problems. For example if a rectified sine wave is used to deposit material in an electroplating operation, the quantity of material deposited is related to the average current.

If you apply a dc voltage to a resistor, a steady amount of power is dissipated in the resistor, which can be found by application of the power law:

EQUATION 1–6

$$P = IV$$

where V = dc voltage across the resistor (volts)
I = dc current in the resistor (amperes)
P = power dissipated (watts)

A sinusoidal waveform is different. It transfers maximum power at the peak excursions of the curve and no power at all at the instant the voltage crosses zero. In order to compare ac and dc voltages and currents, ac voltages and currents can be defined in terms of the equivalent heating value of dc. A reasonably accurate value of the effective value can be determined by taking a number of equally spaced instantaneous values of voltage along one cycle of a sinusoidal curve, computing the average of the values squared, and extracting the square root of the result. This result is called the **effective,** or **rms** (for root-mean-square), **value** of voltage. The rms voltage of a sinusoidal waveform is related to the peak voltage by the equation

EQUATION 1–7

$$V_{rms} = 0.707\,V_p$$

Likewise, the effective or rms current is

EQUATION 1–8

$$I_{rms} = 0.707\,I_p$$

For an alternating current, the effective power is

EQUATION 1–9

$$P_{rms} = (0.707\,I_p)(0.707\,V_p) = 0.5\,I_p V_p$$

where P_{rms} = effective power dissipated in a resistive circuit, W

Illustrations of the definitions for peak, peak-to-peak, average, and rms values for sinusoidal waves are shown in Figure 1–9. The relationship among these values is important to understand because measuring instruments respond differently or display different values depending on the particular instrument.

EXAMPLE 1–3 A voltage waveform is described by the equation $v(t) = 5.2 \sin(7540t)$ volts.

(a) From this expression, compute the peak, rms, and average voltage, the angular frequency, the frequency, and the period.

(b) If the waveform is across a 10 Ω resistor, what power is dissipated in the resistance?

SOLUTION (a) The peak voltage is the same as the amplitude of the waveform or 5.2 V. The rms voltage is

$$V_{rms} = 0.707\,V_p = 0.707(5.2\text{ V}) = 3.68\text{ V}$$

The average voltage is

$$V_{ave} = 0.637\,V_p = 0.637(5.2\text{ V}) = 3.31\text{ V}$$

The angular frequency is 7540 rad/s. Therefore, the frequency is $7540/2\pi = 1200$ Hz. The period is

$$T = \frac{1}{f} = \frac{1}{1200} = 833\ \mu\text{s}$$

(b) The power dissipated is found using the rms value of the voltage:

$$P = I_{rms} V_{rms} = \left(\frac{V_{rms}}{R}\right) V_{rms} = \frac{(V_{rms})^2}{R}$$

$$P = \frac{(3.68\text{ V})^2}{10\ \Omega} = 1.35\text{ W}$$

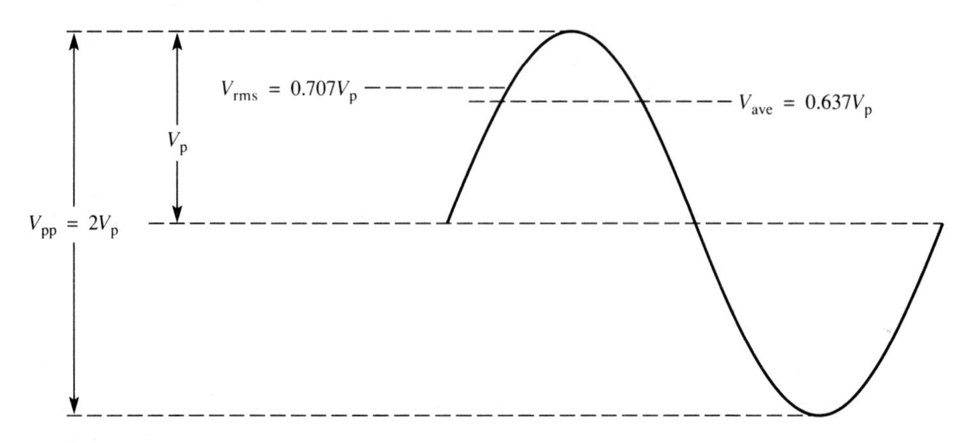

FIGURE 1–9
Definitions for a sinusoidal voltage waveform.

OTHER PERIODIC WAVES

Other important periodic waveshapes are used extensively in the measurement field because many practical electronic circuits generate periodic waves. Periodic waves are generated by circuits called oscillators. Most oscillators are designed to produce a particular waveform—either a sinusoidal wave or nonsinusoidal waves such as the square, rectangular, triangle, and sawtooth waves. These waveforms are illustrated in Figure 1–10.

A general-purpose instrument for generating various waveforms is called a **function generator.** Square waves from a function generator, for example, are

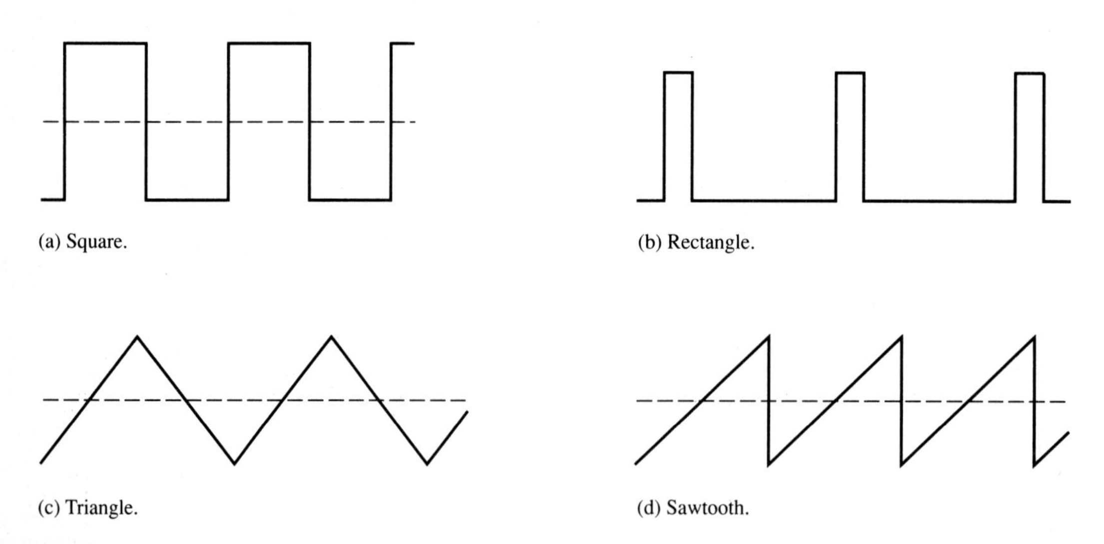

FIGURE 1–10
Examples of periodic, nonsinusoidal waveforms.

useful for testing amplifiers and compensation networks. The sawtooth wave is useful in electronic music systems because it can easily be altered with electronic filters to change tone color. It is also used to move the beam across an oscilloscope or to provide a varying timing-ramp control in level triggered circuits.

TIME-DOMAIN SIGNALS

The signals we have considered vary with time, and it is natural to assume time is the independent variable. Some instruments, such as an oscilloscope, record signals as a function of time. In such a case, time is the independent variable. The values assigned to the independent variable are called the **domain.** Signals in which voltage, current, resistance, or other quantities vary as a function of time are called **time-domain** signals.

FREQUENCY-DOMAIN SIGNALS

Sometimes it is useful to view a signal where frequency is represented on the horizontal axis and the signal amplitude (usually in logarithmic form) is plotted along the vertical axis. Since frequency is the independent variable, the instrument yields **frequency-domain** signals; the plot of amplitude versus frequency is called a **spectrum.** An instrument used to view the spectrum of a signal is a spectrum analyzer. These instruments are extremely useful in radio-frequency (rf) measurements for analyzing the frequency response of a circuit, testing for harmonic distortion, checking the percent modulation from transmitters, and in many other applications.

You have seen how the sinusoidal wave can be defined in terms of three basic parameters. These are the amplitude, frequency, and phase angle. A continuous sinusoidal wave can be shown as a time-varying signal defined by these three parameters. A sinusoidal wave can also be shown as a single line on a frequency spectrum. The frequency-domain representation gives information about the amplitude and frequency present in the waveform but does not show the phase angle. These two representations of a sinusoidal wave are compared in Figure 1–11. The height of the line on the spectrum is the amplitude of the sinusoidal wave.

FOURIER SERIES

Most periodic waves are not simple sine or cosine waves but are complex waveforms. The lowest single-frequency component in a periodic wave is called the **fundamental** frequency; it represents the repetition frequency of the waveform. Jean Fourier, a French mathematician interested in problems of heat conduction, found that all periodic waves can be represented by a series of sinusoidal waves containing a dc offset, the fundamental frequency, and a series of integer multiples of the fundamental called **harmonics.** This famous mathematical series is appropriately called the **Fourier series.**[2] It is generally written as a series of

---

[2] Although Fourier's work was significant and he was awarded a prize, his colleagues were uneasy about it. The famous mathematician Legrange argued in the French Academy of Science that Fourier's claim was impossible. For further information, see *Scientific American,* June 1989, p. 86.

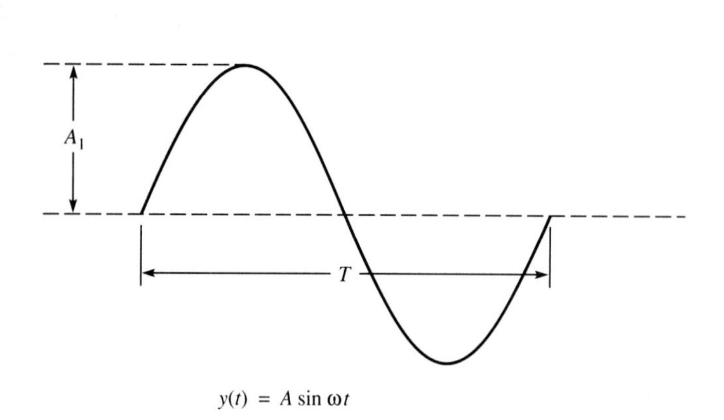

(a) Time domain.

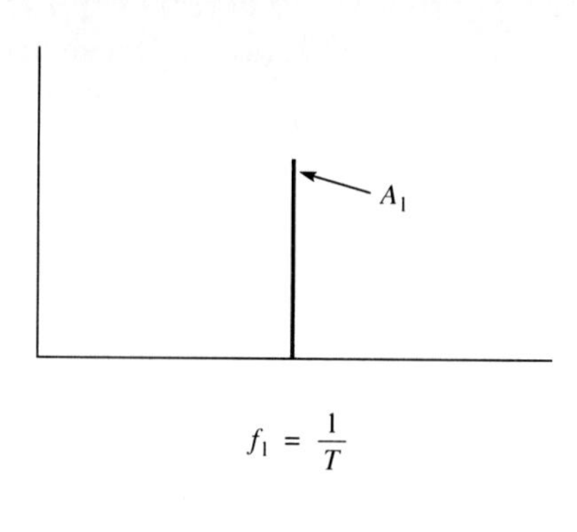

(b) Frequency domain.

FIGURE 1–11
Comparison of the time-domain and frequency-domain representations of a sinusoidal voltage waveform.

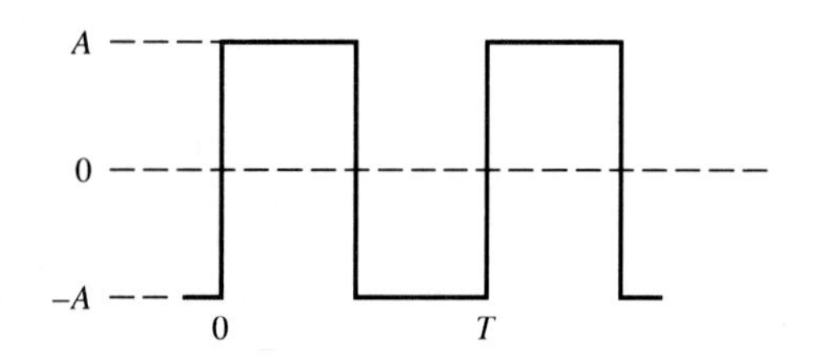

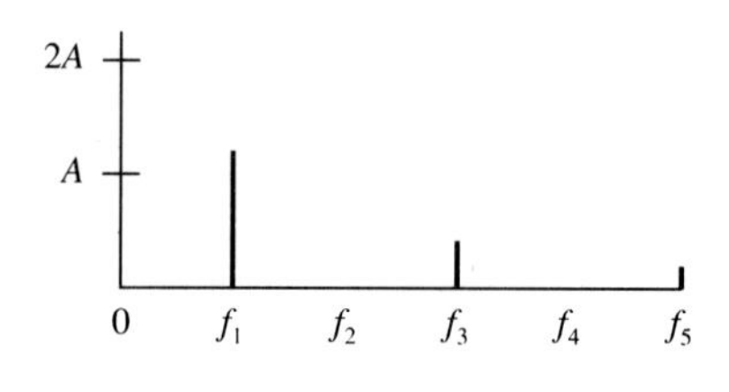

(a) Square.

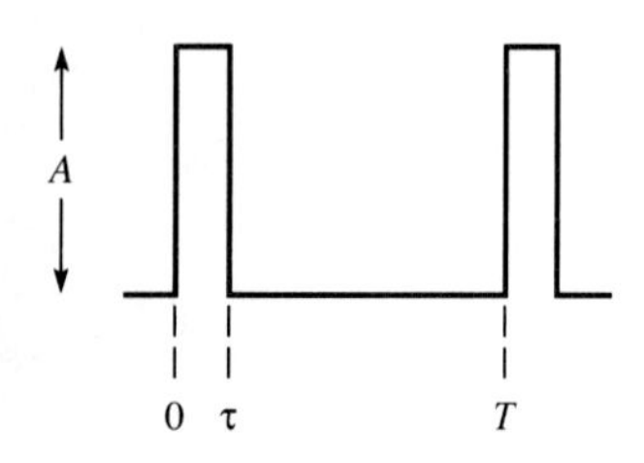

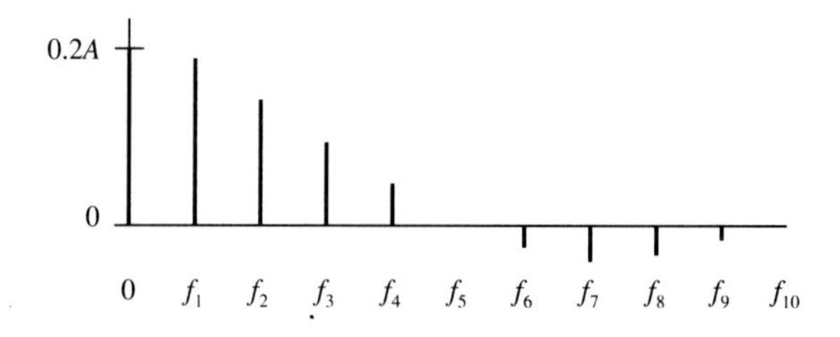

Note: Example shown is for $\frac{\tau}{T} = \frac{1}{5}$.

(b) Rectangle.

FIGURE 1–12
Examples of time- and frequency-domain representations of some periodic, nonsinusoidal waves.

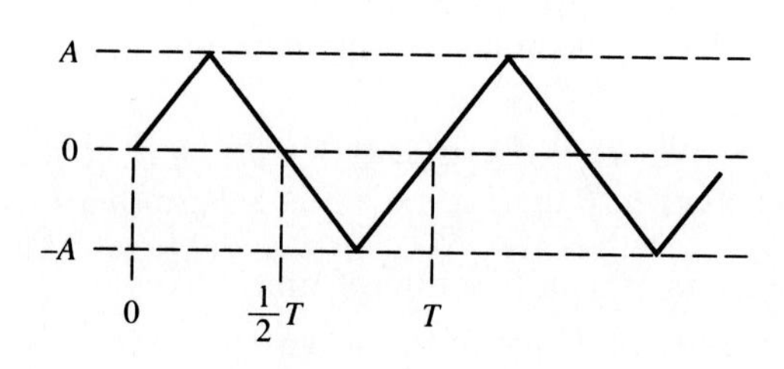

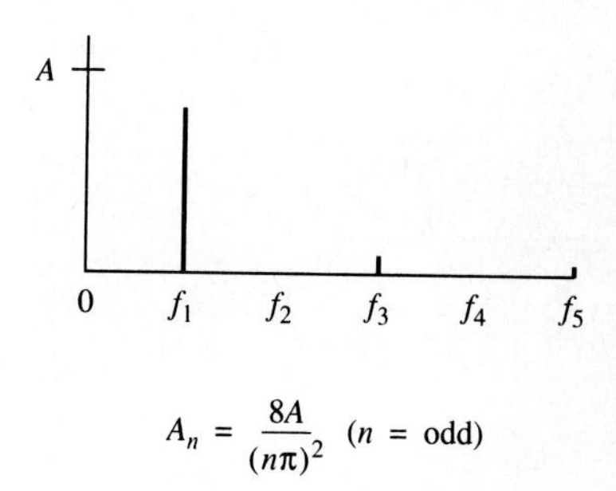

$$y(t) = \frac{8A}{\pi^2} \sum_{n = \text{odd}}^{\infty} \left[\frac{1}{n^2} \sin\left(\frac{n\pi}{2}\right)\right] \sin n\omega t$$

$$A_n = \frac{8A}{(n\pi)^2} \quad (n = \text{odd})$$

(c) Triangle.

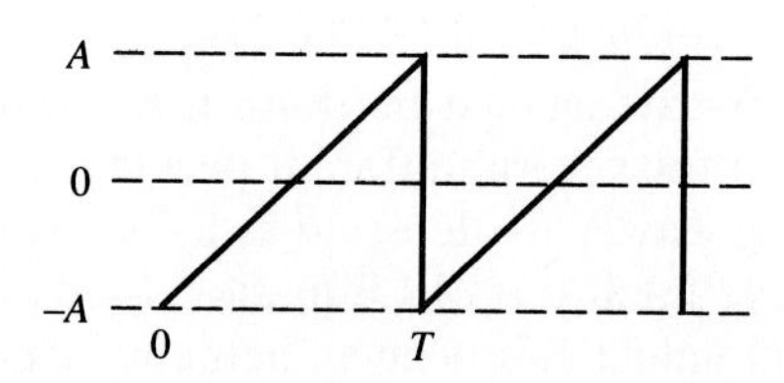

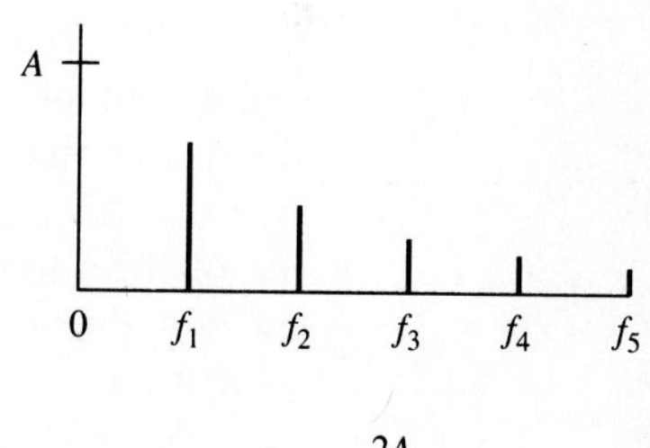

$$y(t) = \frac{2A}{\pi}\left(\sin \omega t - \frac{1}{2}\sin 2\omega t + \frac{1}{3}\sin 3\omega t - \ldots\right)$$

$$A_n = \frac{2A}{n\pi}$$

(d) Sawtooth.

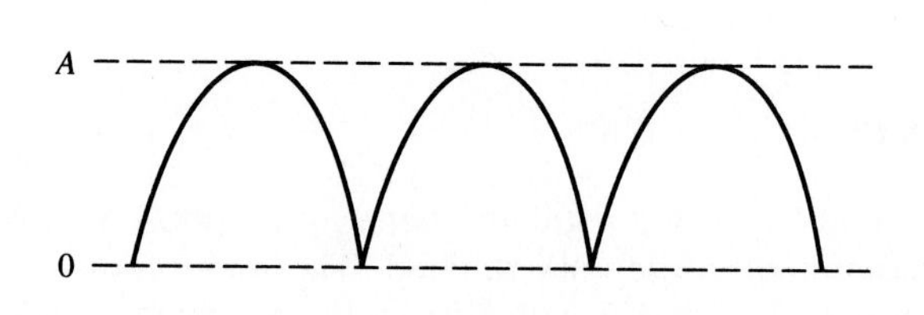

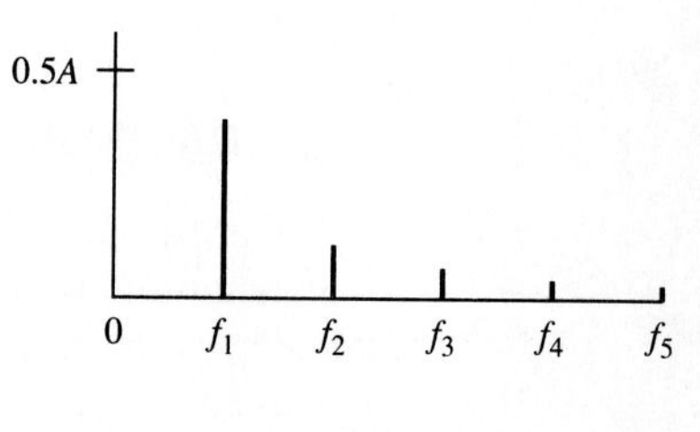

$$y(t) = \frac{2A}{\pi} - \frac{4A}{\pi} \sum_{n = 1}^{\infty} \frac{\cos n\omega t}{(4n^2 - 1)}$$

$$A_n = \frac{4A}{\pi} \frac{1}{(4n^2 - 1)}$$

(e) Full-wave rectified sinusoid.

FIGURE 1–12 *(continued)*

sine and cosine terms, with each term having an amplitude coefficient associated with it:

EQUATION 1–10

$$y(t) = A_0 + A_1\cos(\omega t) + A_2\cos(2\omega t) + \cdots + A_n\cos(n\omega t) + B_1\sin(\omega t) + B_2\sin(2\omega t) + \cdots + B_n\sin(n\omega t)$$

where $y(t)$ = y displacement as a function of time
A_0 = dc component (offset) of the wave
A_1, B_1 = amplitude coefficients of the terms that represent the fundamental
A_2, B_2 = amplitude coefficients of the terms that represent the second harmonic
$A_n B_n$ = amplitude coefficients of the terms that represent the nth harmonic

The frequency spectrum is often shown as an amplitude spectrum with voltage or power plotted on the y-axis against hertz on the x-axis. Figure 1–12 illustrates the amplitude spectrum for several different periodic waveforms. Notice that all spectrums for periodic waves are depicted as lines located at harmonics of the fundamental frequency. Each term in Equation 1–10 contributes one line on the spectrum. These individual frequencies actually occur and can be measured.

Nonperiodic signals such as speech or other transient waveforms can be represented by a spectrum; however, the spectrum is no longer a series of lines. Transient waveforms are computed by another method called the *Fourier transform*. The spectrum of a transient waveform generally contains all frequencies rather than just the harmonically related components of a periodic waveform. A representative Fourier pair of signals for a nonrepetitive pulse is shown in Figure 1–13.

DATA DOMAIN

Sometimes it is useful to view data still another way. Suppose you wish to see the instructions in a computer as they are fetched. The time required to execute an instruction is not constant, so it is not useful to view the data as a function of time; instead, it can be viewed as a function of when the instruction is fetched. Since the data is viewed as a series of data points, it is encoded in the **data domain**. Logic analyzers are instruments that can present information in the data domain.

1–3 TRANSDUCERS

A **transducer** is a device that receives energy from a measurand and responds to the data by converting it into some usable form for a measuring system. The simple mercury thermometer is a transducer that converts temperature into a displacement of the liquid mercury. The output signal from a transducer used in

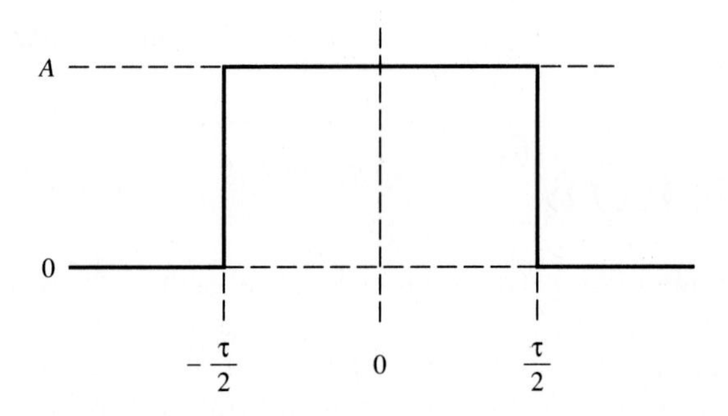

(a) Time-domain representation.

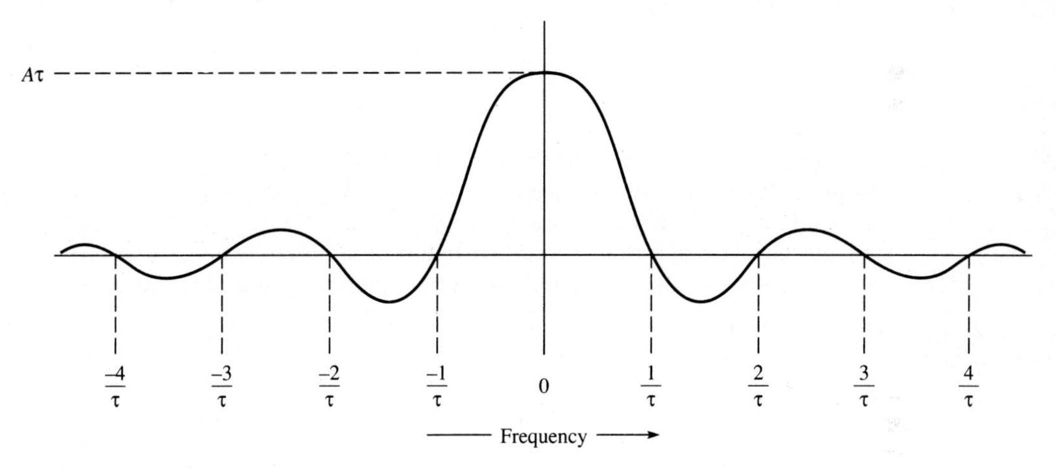

(b) Frequency domain representation.

FIGURE 1–13
Nonrepetitive signals produce a continuous spectrum, as illustrated here by the spectrum of a pulse.

electronic measuring systems is usually electrical for direct use by the instrument. There are an enormous variety of transducers found in all major industries and all the sciences for measuring quantities such as temperature, pressure, altitude, moisture content, humidity, velocity, liquid level, flow rate, and a host of other things.

The output of a transducer depends on the type of transducer and its sensitivity. For electronic measuring systems, the output is typically dc or ac voltage, resistance, current, inductance, capacitance, or a change in frequency. One simple resistive transducer is the common potentiometer. The resistance changes as a function of the rotational position of the control knob. The sensitivity of a transducer is the ratio of the output signal to the measured input parameter. A higher sensitivity implies greater response of the transducer (volts, current, resistance, etc.) to the parameter being measured (temperature, radiation, light intensity). Transducers are covered in Chapter 13.

1–4 THE TRANSMISSION PATH

The basic purpose of a **transmission line** is to transfer electrical signals from one point to another. In an instrumentation system, a transmission line is typically used between the transducer and the signal-conditioning circuits. At low frequencies and short distances, the transmission line does not present any particular problems; however, when low-level signals are transmitted over a distance, particularly at high frequencies, a number of adverse effects may be introduced. These include

1. Attenuation of the signal
2. Decrease in the high-frequency response of the system
3. Noise pickup

The electrical properties of transmission lines are determined by their physical characteristics, which also relate to the capacitance and inductance per unit length of the line. These characteristics are distributed along the length of the transmission line and are called **distributed parameters**. The distributed parameters of a transmission line include its series resistance and inductance, its parallel capacitance, and the conductance of the dielectric. (The dielectric is the insulating material between the conductors.) These parameters contribute to the attenuation of the line specified by the manufacturer in units of decibels per 100 ft at a specified frequency. At frequencies above about 1 MHz, the inductance and capacitance dominate the distributed parameters and determine the **characteristic impedance** of the line, an ac quantity that is basically independent of the signal frequency or the length of the line. The characteristic impedance is the complex ratio of the voltage to current of a signal traveling away from a generator or source of voltage and represents the opposition to the flow of current by the transmission line. For a transmission line, the characteristic impedance is a purely resistive quantity given by

EQUATION 1–11

$$Z_0 = \sqrt{\frac{L}{C}}$$

where Z_0 = characteristic impedance of a transmission line, Ω
L = inductance per unit length of the line, H
C = capacitance per unit length of the line, F

Another important parameter for transmission lines is the **propagation delay** time. The propagation delay time depends on the velocity of electromagnetic waves in the line and its length. For cable, the propagation delay time per unit length is given by

EQUATION 1–12

$$T_D = \sqrt{LC}$$

where T_D = delay time of a transmission line per unit length, s
L = inductance per unit length of the cable, H
C = capacitance per unit length of the cable, F

The velocity of electromagnetic waves is the distance divided by the time:

EQUATION 1–13

$$V_P = \frac{D}{\sqrt{LC}}$$

where V_P = propagation velocity for the cable, m/s
D = length of the line, m
L = inductance per unit length of the cable, H
C = capacitance per unit length of the cable, F

The voltage that exists at any point on a transmission line can be thought of as the sum of two waves. One of the waves, called the **incident wave**, travels from the source to the load. The other wave, called the **reflected wave**, travels from the load back to the source. The superposition of the two waves gives the voltage distribution along the line at any instant in time.

When the load impedance is equal to the characteristic impedance of the line, the incident wave is entirely absorbed and there is no reflected wave. Ideally, the source impedance, the transmission-line impedance, and the load impedance should all be the same to avoid reflections from discontinuities. In practice, small discontinuities (at connectors, circuit boards, or components) are present that prevent the ideal impedance match. At high frequencies, it is important to minimize whatever mismatch is present in the system. For frequencies above 100 MHz or pulse rise times faster than about 3 ns, connectors and other interface points can create serious impedance mismatches as shown in Figure 1–14. Mismatches can be observed and measured by a technique known as *time-domain reflectometry*. With large mismatches, more energy from the incident wave is reflected, and the distortion is increased.

It is frequently necessary to locate transducers a considerable distance from the signal-conditioning circuits. This can be necessitated because multiple locations are being monitored or because the transducer's environment is not suitable for the instrument. If the signal from the transducer is a low-frequency signal, the transmission line can be as simple as a pair of wires. For high-frequency signals, the characteristics of the transmission line can greatly affect measurement. When a signal has a wavelength that is much longer than the transmission-line delay time, ordinary wiring can be used. However, when the wavelength is shorter than about 1/100 of the transmission-line delay time, the signal can be seriously degraded by the transmission line. Special care must be taken to avoid distorting pulse signals when the rise time of a pulse approaches the delay time of the transmission line. When this occurs, the path must be treated as a transmission line with matching impedances and care must be taken to avoid **cross talk**, a form of interference caused by unwanted signal coupling between conductors. Figure 1–15 illustrates the decision threshold point where impedance-controlled interconnections need to be used for pulse transmission. Notice that controlled impe-

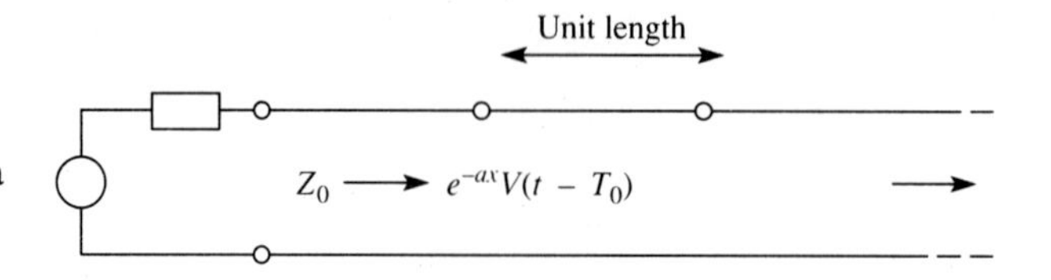

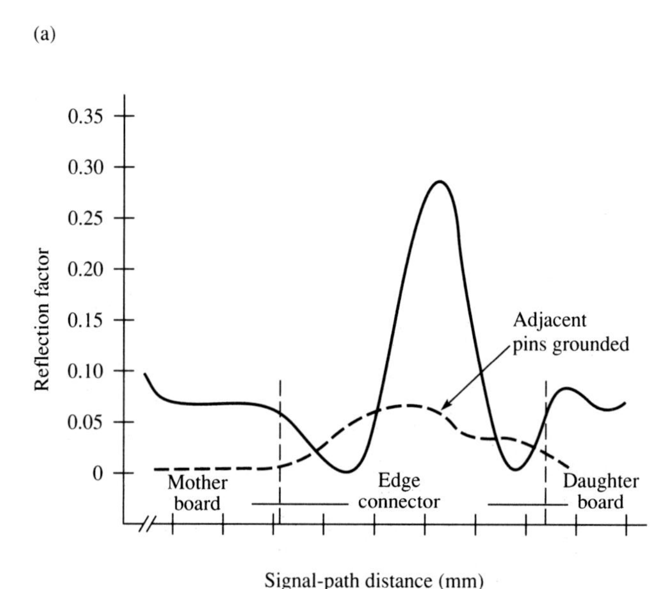

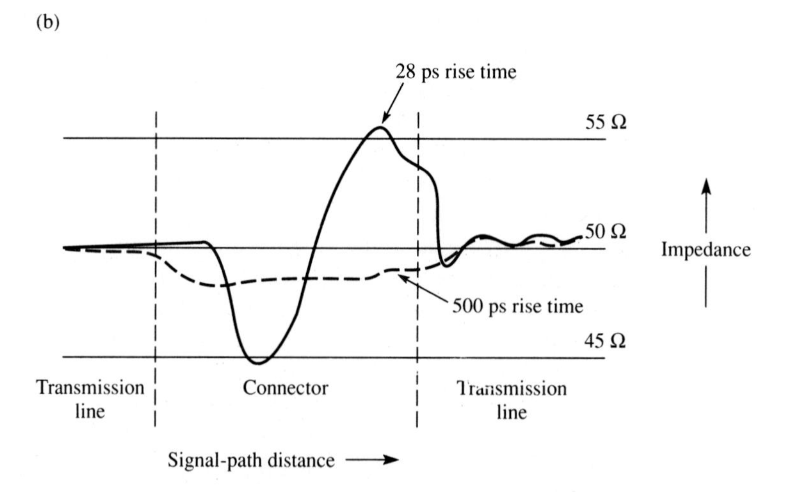

FIGURE 1–14
Effect of discontinuity on fast rising signals. Take the characteristic impedance of a transmission line (a) into consideration when interconnecting high-speed digital circuits. For example, nearly 30% reflectivity is exhibited by an edge connector to a 0.15 ns pulse; however, it can be reduced to 5% by proper capacitor-impedance compensation (b). In addition, note the much greater discontinuity effect a 28 ps rise-time signal has over that of a 500 ps rise-time signal when the signals encounter a connector in a transmission line (c). (Reprinted with permission from Electronic Design, vol. 28, no. 13, June 21, 1980. © 1980 VNU Business Publications, Inc.)

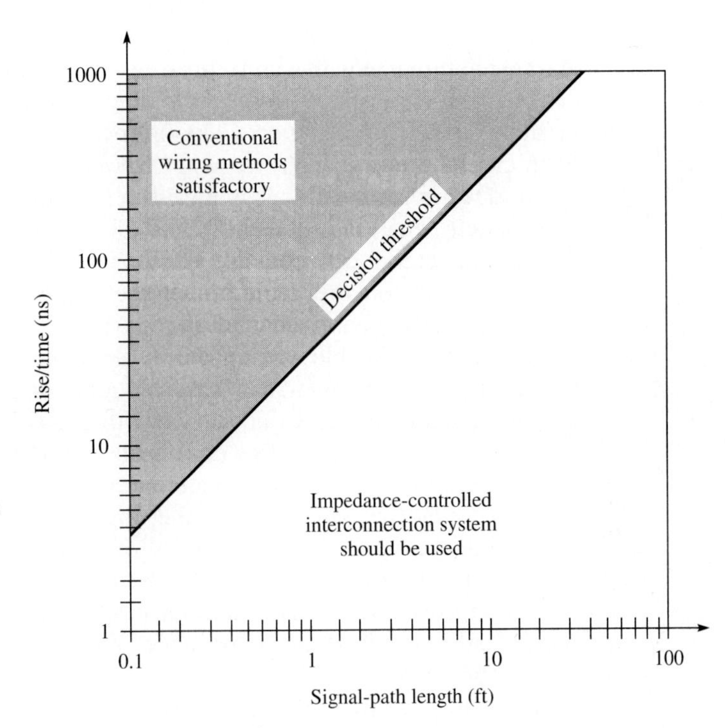

FIGURE 1–15
Threshold where impedance control interfacing is required. The decision-threshold line separates the rise-time/signal-path-length relationship into two regions: One does and the other does not require controlled-impedance interfacing. Thus, at rise times of 3.5 ns or less and signal-path lengths of about 0.1 ft, which are normal distances found within the confines of PC boards, controlled-impedance interconnecting is needed to avoid discontinuity problems. (Reprinted with permission from Electronic Design, vol. 28, no. 13, June 21, 1980. © 1980 VNU Business Publications, Inc.)

dance systems are a function of the rise time of the signal and the path length. For example, a transmission path as short as 1 foot long carrying a signal that rises in less than 40 ns requires an impedance-controlled interconnection.

In addition to matching cables in systems with transducers, the proper matching and terminating of cables is important when interconnecting general-purpose instruments. Improper termination can result in different signal levels than anticipated. In audio or video systems, a level shift due to improper termination can significantly affect the signal quality and any measurement that was made on that signal. If the termination is not purely resistive, the frequency response can be affected also. In addition, reflections can be introduced when the source and load impedance of different pieces of equipment are not the same. For example, a frequency counter that was improperly terminated could respond to a reflected pulse rather than the signal of interest, producing erroneous data.

TYPES OF TRANSMISSION LINES

There are several types of transmission lines used in instrumentation systems. These include coaxial cable (or *coax*), twisted-pair wiring, flat cable, and fiber optics cable. Signals from remote sensors can also be sent over the 60 Hz ac power-line wiring using a frequency-modulated (FM) radio-frequency signal. Other types of transmission lines are important at frequencies above those normally associated with instrumentation systems.

Coaxial cable is one of the most popular transmission methods. It is a broadband transmission line with a concentric center and outer conductors separated by an air or foam dielectric. The outer conductor forms a grounded shield for the inner signal-carrying conductor. Coax offers important advantages because of its uniform characteristics, high-frequency response, and grounded shields, which tend to reduce noise pickup. It is useful in limiting cross talk. Depending on its size and length, coax is useful for frequencies up to about 1 GHz.

The characteristic impedance of coaxial cable is independent of its length and depends only on the geometry of the conductors and the dielectric core material. Typically, coax is manufactured with 50, 75, 93 or 150 Ω characteristic impedance and is available with solid or braided outer conductor. Fifty ohm cable is standard for test equipment and high-frequency work (over 30 MHz). This cable is easy for manufacturers to construct with a stable mechanical structure and uniform impedance. Seventy-five ohm cable is widely used for video and broadcast work because it has minimum attenuation.

Many times the uniformity of the impedance is more important than the actual impedance itself. The uniformity of a coaxial cable can be affected by connectors or by physical damage to the cable itself. An improperly installed connector appears as a sharp discontinuity in the path. In addition, a sharp bend in the cable can also affect the impedance, providing a point for reflections and erratic performance of the line.

The characteristic impedance for a coaxial cable can be computed from its geometry and dielectric constant of the dielectric using the equation

EQUATION 1–14

$$Z_0 = \frac{138}{\sqrt{\epsilon_r}} \log\left(\frac{R}{r}\right)$$

where ϵ_r = dielectric constant of the dielectric
R = radius of outer conductor
r = radius of inner conductor

Dielectric constants for common dielectrics are listed in Table 1–1.

EXAMPLE 1–4 Find the characteristic impedance of a coaxial cable with a 22 gage solid inner conductor and 3.70 mm diameter outer conductor. The dielectric is cellular polyethylene. Number 22 wire has a diameter of 0.643 mm.

SOLUTION We must find the radius of the inner and outer conductors: R = 1.85 mm and r = 0.322 mm. Substituting the given values into Equation 1–14 yields

$$Z_0 = \frac{138}{\sqrt{1.64}} \log\left(\frac{1.85}{0.322}\right) = 82\ \Omega$$

For high-speed pulse measurements, the impulse response of a coaxial cable can be simulated by a method known as the *tau model*. This model enables an

TABLE 1–1
Dielectric constants for common coaxial core materials.

MATERIAL	CONSTANT
Air	1.0
Cellular polyethylene	1.64
Foamed polyethylene	1.55
Magnesium oxide	3.6
Polyethylene	2.3
Polystyrene	2.5
Polyurethane	7.0

experimenter to determine the effect on a given signal due to the finite transient response of the cable. Conversely, given the response function of a cable and the measured output signal, an approximation of the input signal can be determined by a mathematical process known as deconvolution.

Tau is measured in units of time and is related to the time-response function for unequalized coaxial cable. The equation for the response function is

EQUATION 1–15

$$y(t) = \sqrt{\frac{\tau}{\pi}}\frac{e^{-\tau/t}}{t^{1.5}}$$

where $y(t)$ = amplitude
τ = tau of cable, ns
t = time, ns

The tau of a coaxial cable depends on its length and type. As an example, 58 ft of 7/8 in. diameter, high-speed RF-19 cable has a tau of 1.0 ns. The response of this cable to an impulse is shown in Figure 1–16. Notice that although the peak response is less than 1 ns, a significant portion of the response occurs after 5 ns. The important point to keep in mind is that each element of a high-speed system, especially transducers and cables, affects the overall response of that system and hence the measurement accuracy.

For lower frequencies, twisted-pair wiring is widely used.[3] **Twisted-pair** wiring normally consists of AWG-22 or AWG-24 wires, twisted about 30 times per foot. It is less expensive than coax and is available in a variety of configurations and sizes, including shielding, high-temperature insulation, multiple-conductor paired cable, and so forth. It can help reduce cross talk and, when used with a differential amplifier, can reduce common mode interference. The propagation velocity of twisted-pair wiring is approximately 50% of coax. The rate that data can be sent over twisted-pair wiring varies inversely with the distance—as the distance increases, the data rate is reduced due to the limited bandwidth of longer wires.

[3] Although twisted-pair wiring is normally used below 1 MHz, it is being installed in some digital systems with bit rates as high as 10 million bits per second (10 Mb/s) in local area networks (LANs) for transmission over relatively short distances (typically less than 500 ft).

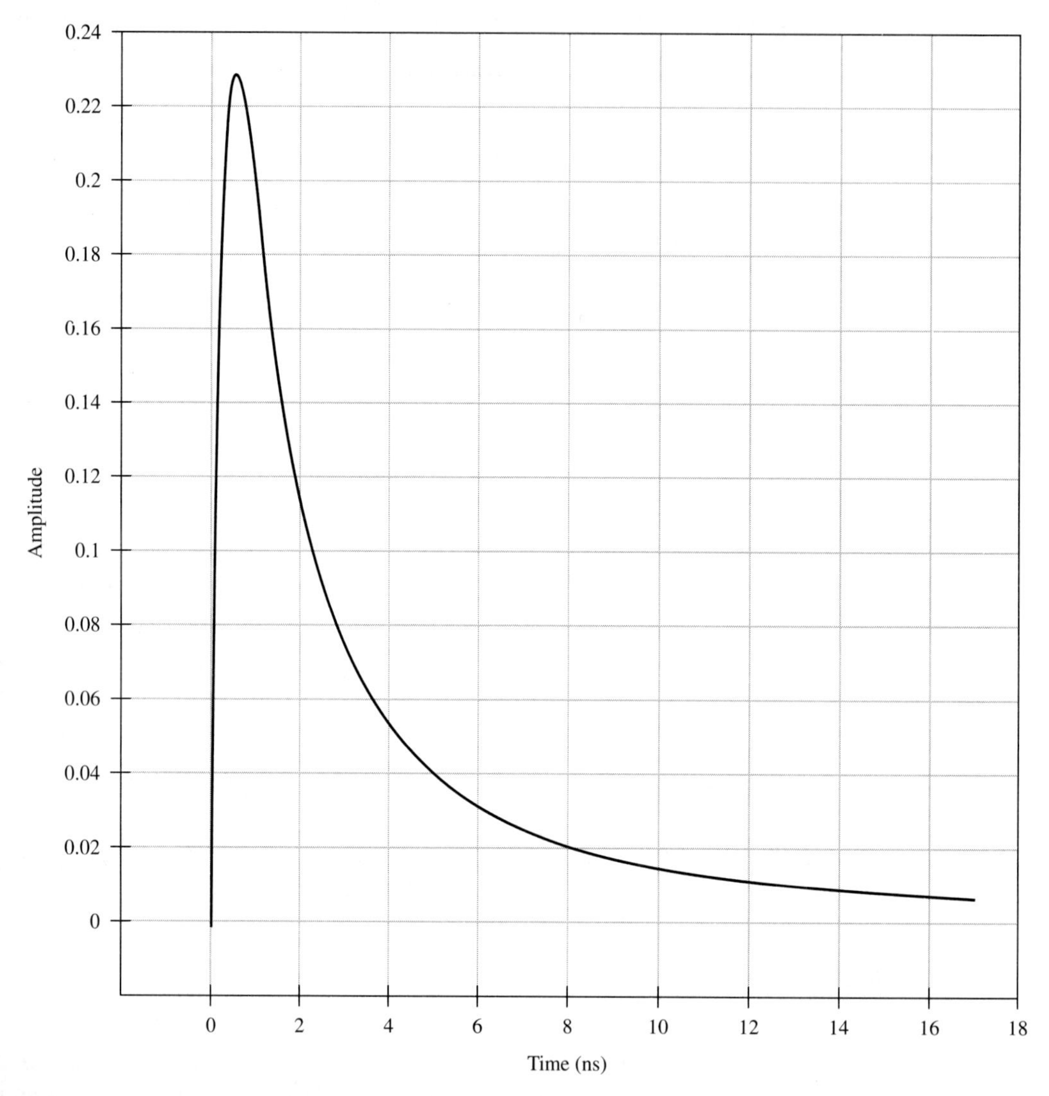

FIGURE 1–16
Response of a cable to an impulse. The cable shown has a tau of 1 ns.

Ordinary open wiring that connects two points can present an impedance of several thousand ohms. In addition to the noise benefits cited, twisted-pair wiring can control the characteristic impedance. Typically it has impedances from 70 to 140 Ω, which can be computed from the following equation:

EQUATION 1–16

$$Z_0 = \frac{120}{\sqrt{\epsilon_r}} \ln\left(\frac{2D}{d}\right)$$

where ϵ_r = dielectric constant of the insulation
D = center-to-center distance between the conductors
d = diameter of the conductors

EXAMPLE 1–5 A 22 gage twisted-pair wire has a polyethylene insulation with a dielectric constant of 1.64. The diameter of 22 gage wire is 0.643 mm, and the center-to-center spacing of the conductors is 1.46 mm. Compute the characteristic impedance.

SOLUTION Use Equation 1–16:

$$Z_0 = \frac{120}{\sqrt{\epsilon_r}} \ln\left(\frac{2D}{d}\right)$$

$$= \frac{120}{\sqrt{1.64}} \ln\left[\frac{(2)(1.46)}{0.643}\right]$$

$$= 142\ \Omega$$

73.17

Another means of transmitting a signal from one point to another is with flat cable. **Flat cable** is available as ribbons of side-by-side conductors. One advantage of ribbon cable is that it can easily be fitted with connectors using mass production methods. For cross talk control, flat cable can be connected with signal grounds used in every other position. For even higher protection, flat cable is available as ribbon coaxial cable. The impedance of flat cable ranges from 50 Ω to 200 Ω.

There are a number of other special-purpose cables available for instrumentation and other applications. Examples include cables particularly designed for special instrumentation and control applications, computers, thermocouples, telephone systems, and TV cameras, to name a few.

FIBER OPTICS

In the last two decades, many advances have occurred in optical-fiber technology, making this an increasingly popular type of transmission line for both telecommunications systems and for shorter distances required in instrumentation applications. A **fiber-optic cable** is made from thin strands of plastic or glass that carry light instead of electricity. Although they are generally more expensive than conductive lines, fiber-optic transmission lines offer significant advantages. Advantages include very high bit rates (greater than 1 Gb/s) and immunity from electromagnetic pickup and cross talk as well as low attenuation for signals. Other advantages are gained when data needs to be sent through severe environments such as nuclear radiation, chemical or explosive vapors, electrical noise, salt water, high humidity and dust, and so forth.

Fiber-optic links are often selected in instrumentation and control systems because they can solve electrical isolation and interference problems typical in industrial environments. Applications include object-sensing, counting, and positioning systems, connections between controllers and production equipment, electrical speed-control links, and connections from computer-control systems and the units they control. Many off-the-shelf systems are available with a complete link, including the transmitter, cable assembly, and receiver with electronic inputs and outputs that are compatible with logic systems.

1–5 INSTRUMENTS

Instruments can be considered as either *stimulus* instruments or *measuring* (response) instruments. A **measuring instrument** is a device that converts a quantity to be measured into some usable form for interpretation. Measuring instruments receive signals from the signal conditioner or cable and provide an output in some form for the user. Examples of measuring instruments include digital multimeters (DMMs), frequency counters, and oscilloscopes. **Stimulus instruments** are sources for test signals. They can be function generators, power supplies, oscillators, or pulse generators, for example. Stimulus instruments are used in measuring systems for testing, troubleshooting, or supplying power or signals needed by a transducer. Some instruments have been combined into multifunctional instruments, which can simplify a system and provide accuracy while at the same time reducing the power requirement compared to a collection of separate instruments. Plug-in modular instruments offer versatility by allowing the implementation of custom modules.

Electronic instruments, as contrasted with other types of instruments, offer many advantages for measuring physical phenomena. Among these are high speed, operation over large distances, internal signal processing, sensitivity, versatility, and reliability. All measurements, regardless of type, require a source of energy. For electronic measurements, the energy can be supplied from a circuit under test, an instrument, or from both. Despite the great variety of electronic measuring instruments, they can be divided into three categories, depending on the source of energy for the measurement:

1. Passive instruments, such as voltmeters, absorb energy from the circuit under test. When a passive instrument is connected to a circuit, it changes the circuit impedances. This effect is called **loading** the circuit.
2. Active instruments, such as ohmmeters, supply energy to the circuit under test. These instruments can, in some cases, supply too much current and destroy sensitive components such as meter movements. Active instruments are frequently used with the power removed from the circuit under test.
3. Balancing instruments, such as certain bridge circuits and potentiometer circuits, neither absorb nor supply energy to a circuit under test; instead they operate by detecting a null condition between the circuit and the instrument. This type of instrument does not violate the premise that all measurements require a source of energy because the energy for the measurement is supplied by the circuit under test. An advantage of these instruments is that the balanced condition results in almost no loading effect.

Many measurement and stimulus instruments contain internal microprocessors to make possible functions such as data analysis or automated measurements. Some instruments can be programmed for a series of automatic setups that can be recalled later. Instruments with built-in microprocessors that have the ability to perform one or more functions that would otherwise be performed by a person have been marketed by manufacturers as **intelligent** instruments. This definition includes instruments of limited "intelligence," such as instruments that can perform self-diagnostics (error checking) and self-calibration when they are turned on. Instruments with more intelligence can be programmed for decision making, such as comparing incoming signals with a reference and taking some action based on the comparison or performing a series of tests under control of a computer and sending the processed data back to the computer for analysis or storage. These instruments are frequently interconnected to form a complete test system. Many intelligent instruments have software-driven controls that have replaced a multitude of hardware front-panel controls. Software-driven keys (sometimes called *soft keys*) change function depending on a menu of options selected by the user to simplify instrument setup. Some intelligent instruments allow the user to store and recall various front-panel setups to allow quick cycling through a series of measurements or perform certain measurements automatically. Even power supplies are available that are endowed with some intelligence, allowing them to be programmed remotely, and that can read back the actual current or voltage out as well as other functions.

1–6 THE USER

Measurement instruments need to detect the measurand and provide an output. The output of an instrument can go to any of several users. The user can be a human, in which case the output is usually a visual display such as a meter scale or CRT; it can be audio, as in speech or an alarm. The output can be used directly in automatic control or be sent to a computer for processing; in either case the output is usually a digital signal. The advent of microprocessors has increased the applications of digital control systems, where the system itself is the end user of the data; one example is radar tracking and control of a missile. Other possible end users of an instrument are recording devices such as chart recorders or magnetic tape, which store the information for use at some later time.

1–7 BLOCK DIAGRAMS

Circuits and measurement systems can be depicted with various levels of abstraction. The simplest representation is that of the block diagram, drawn as a series of interconnected boxes. Block diagrams are used to show the flow of information and the basic functional relationships in a system. A block can represent a complete device (such as a digital multimeter) or it can represent a specific functional

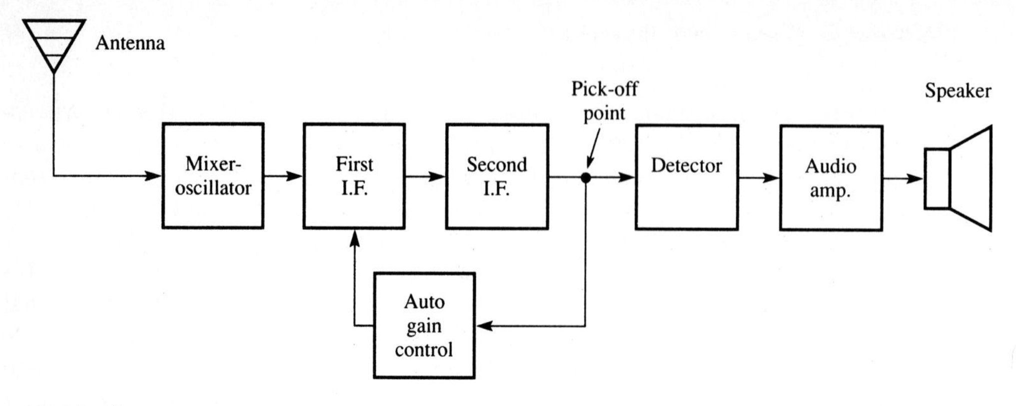

(a) AM radio.

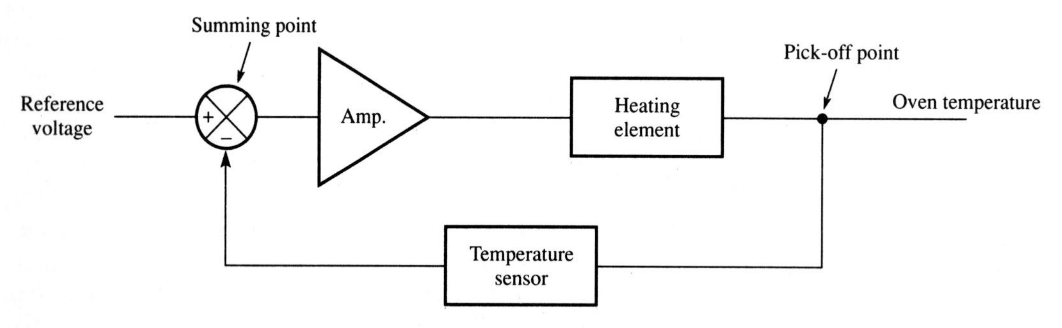

(b) Control system for oven.

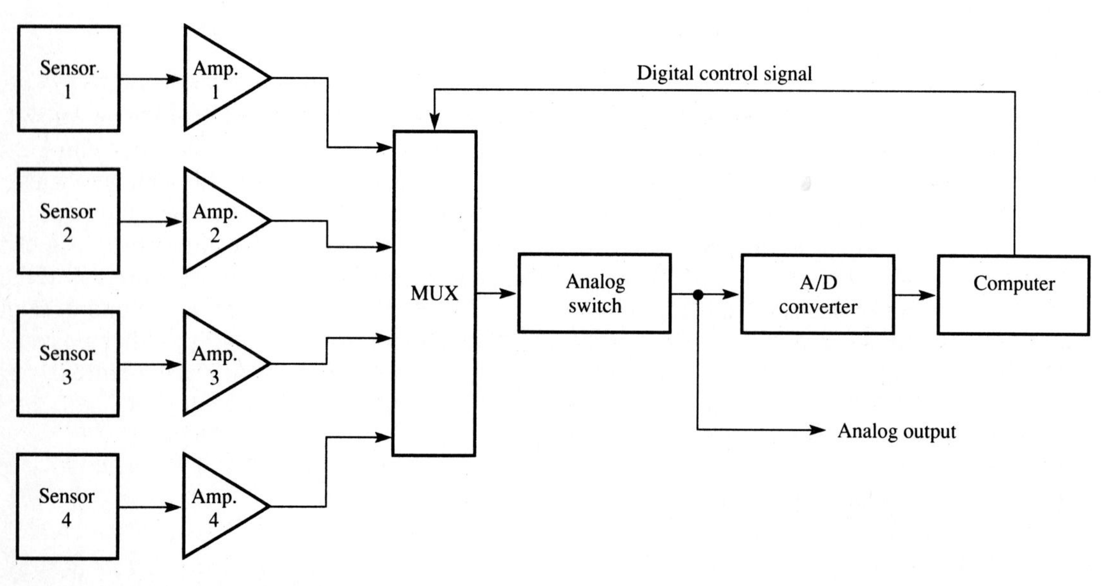

(c) Four-channel data-acquisition system.

FIGURE 1–17
Examples of block diagrams.

unit that occurs in a system (such as a mixer). Connecting lines on block diagrams are shown with arrows to indicate the signal paths.

The basic symbol used for block diagrams is a rectangle; however, other shapes and symbols are sometimes used. A triangle is frequently used to represent an amplifier. Other schematic symbols, such as those for resistors, capacitors, and antennas, may be used when they clarify the overall system or the signal flow. Each block is labeled with its function, or the function of the block may be shown symbolically.

When it is necessary to show a portion of signal removed from some point, a pick-off point is drawn, represented by a dot at the junction. Conversely, when two signals are either added or subtracted at some point, they can be brought into a common block, or a summing point can be shown. A summing point is drawn as a circle. If a variable goes through the summing point without a change in sign, a plus (+) sign is shown with the variable; if it is reversed, a minus (−) sign is placed beside the variable. Examples of block diagrams for several systems are illustrated in Figure 1–17.

SUMMARY

1. The basic elements of an instrumentation system include a transducer, a transmission path, signal-conditioning elements, an instrument, and the user.
2. The transfer function of a system or subsystem is the ratio of the output signal to the input signal.
3. Measurement systems are characterized by the sensitivity, dynamic range, and resolution of the measurement.
4. Electrical signals can be classified as analog signals or discrete-time signals. Digital signals are a special case of discrete-time signals.
5. Periodic waveforms repeat at a specific time interval, called the period of the waveform. The most basic and important periodic waveform is the sinusoidal wave.
6. A sinusoidal waveform can be defined by its amplitude, frequency, and phase angle. The general equation for a sinusoidal waveform is

$$y(t) = A \sin(\omega t \pm \phi)$$

7. Signals can be viewed in the time domain or the frequency domain. A frequency-domain picture of a signal is called a spectrum.
8. Periodic waveforms can be represented by a series of sinusoidal waves containing a dc offset, the fundamental frequency, and a series of integer multiples of the fundamental frequency.
9. Transducers are used in electronic measuring systems to convert energy from a measurand into an electrical signal suitable for the measuring system.
10. Typical transmission lines consist of coax, twisted-pair, or flat cable. The characteristic impedance of a transmission line is an ac quantity that is independent of the signal frequency or length of the line.
11. Measuring instruments receive signals and provide an output in some form for the user. Stimulus instruments provide a source of signals for test.
12. Block diagrams are used to show the flow of information in a circuit.

QUESTIONS AND PROBLEMS

1. (a) What is the function of a transducer in a measuring system?
   (b) Name some transducers that you might find in an automobile.
2. Compare the term *transient response* with the term *frequency response* of a system.
3. Figure 1–18 shows the linear operating region for three sensors: A, B, and C.
   (a) Which sensor has the largest dynamic range? Explain your answer.
   (b) Which sensor has the greatest sensitivity? Explain your answer.

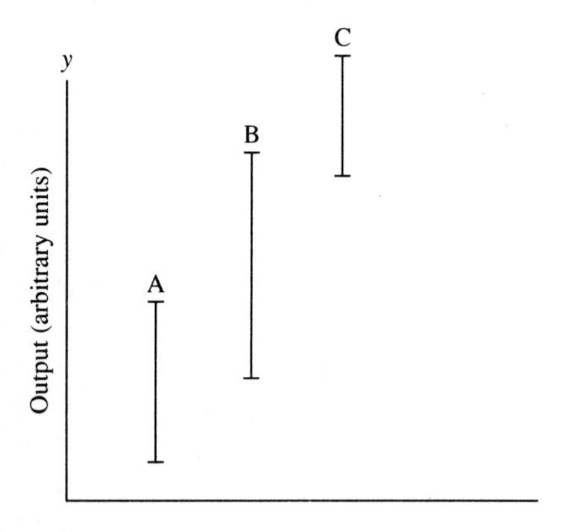

FIGURE 1–18

4. Assume the measurement units for the output in Figure 1–18 are volts and the transducers respond to input current. What are the measurement units for the transfer function?
5. (a) Write the equation for the instantaneous voltage of a sinusoidal waveform that has a peak voltage of 25 V and a frequency of 200 kHz. Assume the waveform lags the reference by 40°.
   (b) Find the instantaneous voltage at a time of 1.45 μs.
6. Find the average and rms voltage for the waveform in Problem 5.
7. The average value of a sinusoidal waveform is 1.5 A. Find the rms, peak, and peak-to-peak values of the current.
8. An alternating current in a resistor has a peak value of 2 A when the voltage has a peak value of 12 V. What is the rms power dissipated in the resistor?
9. (a) Compare the time domain with the frequency domain. Give an example of an instrument that is used in the time domain and one that is used in the frequency domain.
   (b) What is meant by the data domain?
10. Compare the spectrum of a periodic signal to that of a nonperiodic signal.
11. A coaxial cable uses solid polyethylene dielectric with a dielectric constant of 2.3. The outer conductor has a diameter of 7.4 mm and the inner conductor has a diameter of 2.1 mm. Compute the characteristic impedance of the cable.
12. A 16 gage twisted-pair wire uses PVC insulation with a dielectric constant of 4.5. The diameter of the conductor for 16 gage wire is 1.29 mm. The thickness of the shield is 0.58 mm. Assuming there is no air spacing between the wires, what is the characteristic impedance?
13. What are some advantages of fiber-optic cable as compared to conductive wiring?
14. Compare *measuring* instruments with *stimulus* instruments and give examples of each.
15. What is meant by an intelligent instrument?
16. What advantages do electronic measuring instruments have over nonelectronic instruments?
17. Compare a pick-off point with a summing point on a block diagram.

Chapter 2

Data and Data Analysis

OBJECTIVES

Two types of information about physical phenomena can be obtained. The first type is not associated with numbers and is called qualitative, or descriptive, information. The second type is numerical and gives information about the magnitude or intensity of the phenomena. It is called quantitative, or metric, information. The information we obtain using measuring instruments is called data. It is typically quantitative but can be qualitative.

Making a physical measurement is more than connecting an instrument and reading the display. The user must understand the type of measurement to be made and recognize the limitations of the instrumentation system, such as its sensitivity, linearity, and usable range. Measured data always contains error. Knowing the limitations of the measuring system, interpreting data, drawing conclusions, and reporting results are basic to the study of instrumentation.

When you complete this chapter, you should be able to

1. Define terminology related to data, including accuracy, precision, sensitivity, resolution, threshold, and error.
2. Explain the difference between the random and systematic error.
3. Determine the number of significant digits in a measurement.
4. Compute the average, average deviation, and standard deviation of a set of data.
5. Correctly plot both linear and logarithmic data on graphs.
6. Find the equation for straight line plots on linear, semilogarithmic, and logarithmic graph paper.
7. Find the best-fit straight line to data using the least squares method (linear regression).

HISTORICAL NOTE

When a technician wants to know the chance that experimental measurements are correct, the science of probability is used. The famous bell-shaped normal curve was first recognized by Abraham de Moivre 300 years ago. Carl Gauss, the famous nineteenth-century mathematician and scientist, wrote the equation for the curve that now bears his name. The Gaussian curve describes the expected variations in a large group of measurements and is widely used in experimental work to state the significance of measured quantities.

2–1 THE NATURE OF DATA

Information you gather is called **data.** Data can be a factual statement of physical phenomena. For example, the statement "the copper was removed by the chemical reaction with ferric chloride" is descriptive data. When data is purely descriptive, it is said to be **qualitative data.** When a quantity is measured, we associate numerical values with it, and the information is more useful in a scientific way because more information is present. Information about the magnitude or intensity of a physical phenomenon is called **quantitative data.** Recall that the quantity that is being measured is called the measurand. Instrumentation extends the human senses by allowing a numerical value or values to be associated with the measurand.

Numerical data can be categorized in several ways. It can be an isolated value or can be dependent on time or location. Values recorded directly from an experiment or observation are called **empirical data.** Prior to processing, empirical data is often referred to as **raw,** or **unprocessed, data,** whereas data that has been analyzed is called **processed data.** Data can also be generated by theoretical calculations. Frequently, theoretical data is compared to measured or processed data as a test of the theoretical model.

After data is collected, it may need to be processed either by applying mathematical computations to it or by arranging it in some meaningful manner. This procedure is called *data processing* or *data reduction.* Data may be entered into a computer for reduction, or, in some cases, the measurement instrument may perform the data reduction within the instrument. For example, a digital oscilloscope may present the rms value of a voltage as a displayed numeric value. As part of the process of data reduction, obvious errors or discrepancies should be looked for; sometimes statistical processing is applied to indicate the nature of experimental precision.

After the data is reduced, it is analyzed. **Data analysis** is the process of trying to make results from a measurement meaningful and to resolve any differences due to variations in the data. The data-analysis step should consider the consistency of the data, experimental errors and limitations, approximations in the data-reduction process, and other factors that could affect the interpretation of the data. The combination of these effects is used to support a conclusion. After analysis, the data serves as the basis of a report.

2–2 ERROR, ACCURACY, AND PRECISION

Data measured with test equipment is not perfect; rather, the accuracy of the data depends on the accuracy of the test equipment and the conditions under which the measurement was made. In order to interpret data from an experiment, we need to have an understanding of the nature of errors. Experimental error should not be thought of as a mistake. All measurements that do not involve counting are approximations of the true value. **Error** is the difference between the true or best accepted value of some quantity and the measured value. A measurement is said to be accurate if the error is small—**accuracy** refers to a comparison of the measured and accepted, or "true," value. It is important for the user of an instrument to know what confidence can be placed in it. Instrument manufacturers generally quote accuracy specifications in their literature, but the user needs to be cautioned to understand the specific conditions for which an accuracy figure is stated. The number of digits used to describe a measured quantity is not always representative of the true accuracy of the measurement.

Two other terms associated with the quality of a measurement are precision and resolution. **Precision** is a measure of the repeatability of a series of data points taken in the measurement of some quantity. The precision of an instrument depends on both its resolution and its stability. Recall that resolution was defined in Section 1–1 as the minimum discernible change in the measurand that can be detected. **Stability** refers to freedom from random variations in the result. A precise measurement requires both stability and high resolution. Precision is a measure of the dispersion of a set of data, not a measure of the accuracy of the data. It is possible to have a precision instrument that provides readings that are not scattered but that are not accurate because of a systematic error. However, it is not possible to have an accurate instrument unless it is also precise.

The resolution of a measurement is not a constant for a given instrument but may be changed by the measurand or the test conditions. For example, a nonlinear meter scale has a higher resolution at one end than at the other due to the spacing of the scale divisions. Likewise, noise induced in a system can affect the ability to resolve a very small change in voltage or resistance. Temperature changes can also affect measurements because of the effect on resistance, capacitance, dimensions of mechanical parts, drift, and so forth.

SIGNIFICANT DIGITS

When a measurement contains approximate numbers, the digits known to be correct are called **significant digits.** The number of significant digits in a measurement depends on the precision of the measurement. Many measuring instruments provide more digits than are significant, leaving it to the user to determine what is significant. In some cases, this is done because the instrument has more than one range and displays the maximum number of significant digits on the highest range. If the instrument is set to a lower range, the instrument may show the same number of digits despite the fact that the rightmost digits are not significant. This can occur when the resolution of the instrument does not change as the range is changed. The user needs to be aware of the resolution of an instrument to be able to determine correctly the number of significant digits.

When reporting a measured value, the least significant uncertain digit may be retained, but other uncertain digits should be discarded. To find the number of significant digits in a number, ignore the decimal place and count the number of digits from left to right, starting with the first nonzero digit and ending with the last digit to the right. All digits counted are significant except zeros to the right end of the number. A zero on the right end of a number is significant if it is to the right of the decimal place; otherwise it is uncertain. For example, 43.00 contains four significant digits, but the whole number 4300 may contain two, three, or four significant digits. In the absence of other information, the significance of the right-hand zeros is uncertain. To avoid confusion, a number should be reported using scientific notation. For example, the number 4.30×10^3 contains three significant figures and the number 4.300×10^3 contains four significant figures.

The rules for determining if a reported digit is significant are as follows:

1. Nonzero digits are always considered to be significant.
2. Zeros to the left of the first nonzero digit are never significant.
3. Zeros between nonzero digits are always significant.
4. Zeros at the right end of a number *and* the right of the decimal are significant.
5. Zeros at the right end of a whole number are uncertain. Whole numbers should be reported in scientific notation to clarify the significant figures.

EXAMPLE 2–1 Underline the significant digits in each of the following measurements:

(a) 40.0 (b) 0.3040 (c) 1.20×10^5 (d) 120,000 (e) 0.00502

SOLUTION

(a) 40.0 has three significant digits; see Rule 4.

(b) 0.3040 has four significant digits; see Rules 2 and 3.

(c) 1.20×10^5 has three significant digits; see Rule 4.

(d) 120,000 has at least two significant digits. Although this numeral represents the same numeral as in (c), zeros in this example are uncertain; see Rule 5. This is not a recommended method for reporting a measured quantity.

(e) 0.00502 has three significant digits; see Rules 2 and 3.

ROUNDING NUMBERS

Since measurements always involve approximate numbers, they should be shown only with those digits that are significant plus no more than one uncertain digit. The number of digits shown is indicative of the precision of the measurement. For this reason, you should **round** a number by dropping one or more digits to the right. The rules for rounding are as follows:

1. If the digit dropped is greater than 5, increase the last retained digit by 1.
2. If the digit dropped is less than 5, do not change the last retained digit.

3. If the digit dropped is 5, increase the last retained digit *if* it makes it even, otherwise do not. This is called the *round-even* rule.

EXAMPLE 2–2 Round each of the following numbers to three significant figures:

(a) 123.52 (b) 122.52 (c) 10.071 (d) 6.3948 (e) 29.961

SOLUTION

(a) 124; see Rule 3.
(b) 122; see Rule 3.
(c) 10.1; see Rule 1.
(d) 6.39; see Rule 2.
(e) 30.0; see Rule 2.

SYSTEMATIC AND RANDOM ERROR

There are two classes of errors that affect measurements: **systematic errors** and **random errors.** Systematic errors consistently appear in a measurement in the same direction. These could be caused by inaccurate calibration, mismatched impedances, response-time error, nonlinearities, equipment malfunction, environmental change, and loading effects. Systematic errors are often unknown to the observer and may arise from a source that was not considered in the measurement. Sometimes a systematic error occurs because of the misuse of an instrument outside its design range, such as when a voltmeter is used to measure a frequency beyond its specifications. (This is also called an applicational error.) Another common type of systematic error is *loading error*. Whenever an instrument is connected to a circuit, it becomes part of the circuit being measured and changes the circuit to some extent. Measurements in high-impedance circuits can be significantly affected if this is not taken into account. Another possible systematic error is *calibration error*. For example, a frequency counter uses an internal oscillator to count an unknown frequency for a specific amount of time. If this oscillator runs too slowly, then the counter waits too long, giving a result that is consistently too high. This produces a systematic error for all readings made with that counter. Other systematic errors can occur because the calibration was performed under different environmental conditions than those present when the instrument is in service. These might include temperature, humidity, atmospheric pressure, vibration, magnetic or electrostatic fields, and so forth.

The best way to detect the presence of a systematic error is to repeat the measurement with a completely different technique using different instruments. If the two measurements agree, greater confidence can be placed in the correctness of the measurement.

Random errors (also called *accidental errors*) tend to vary in both directions from the true value by chance. These errors are unpredictable and occur because of a number of factors that determine the outcome of a measurement. Random errors are generally small and may be caused by electrical noise, interference, vibration, gain variation of amplifiers, leakage currents, drift, observational error, or other environmental factors. The best way to reduce random errors is to make repeated measurements and use statistical techniques to determine the uncertainty of the final result.

2–3 STATISTICS

Suppose you need to measure the upper cutoff frequency of an amplifier using an oscilloscope. You adjust the input frequency until the output amplitude drops to 70.7% of the midband amplitude and determine the frequency by counting the time for one cycle. You might decide to repeat the procedure several times to ensure that you have made the best possible measurement. Even though you have used the same procedure and equipment, the results are usually not identical because of random errors. The question arises, What is the best number to report and what is the precision of the result? The answer to this and other questions pertaining to observational data is given by a branch of mathematics called statistics.

Statistics is concerned with methods for handling data and drawing conclusions from it. There are two parts to the science of statistics. The first is called **descriptive statistics** and deals with collecting, processing, and analysis of data in a way that makes it comprehensible. The second is called **statistical inference** and has the goal of interpreting the data and drawing conclusions from it. The two parts are closely related. The drawing of inferences about data is meaningful only if we have previously collected, processed, and analyzed it.

As we have seen, all measurements contain error, and a set of measurements of the same quantity will have a distribution of values. To describe data and attempt to decide what is the important information from it, we can characterize its distribution with a set of numbers. The number most commonly used for describing the location of the center of the data is commonly called the sample **mean** (or average). The symbol $\bar{x}$ is used for the sample mean. The sample mean is found by summing all of the observed values (i = 1 to n) and dividing the result by the number of values. This is written

EQUATION 2–1

$$\bar{x} = \frac{x_1 + x_2 + x_3 + \cdots + x_n}{n} = \frac{\sum_{i=1}^{n} x_i}{n}$$

where $\bar{x}$ = mean value for sampled data
x_i = value of the ith data point
n = total number of data points

THE NORMAL DISTRIBUTION

If a moderate-size group of measurements is tabulated, it can be difficult to see the significance of the data. Table 2–1 illustrates a set of 60 data points for measurement of signal-to-noise ratio taken over several hours. The data is rearranged in

Table 2–2 from the smallest to the largest value. This may help clarify certain features of the data but has too many data points to show the overall pattern. If, instead, we tally the number of occurrences for each measurement that lie within some small interval, a **frequency distribution** of the results is obtained. The data from Table 2–2 is listed as a frequency distribution in Table 2–3. The frequency distribution sacrifices some details of the data but gives an overall representation of the results in a form that is easy to digest.

Generally, error that is strictly random leads to a **normal distribution** (also called a Gaussian distribution); data points have an equal chance of being above or below the true value. The normal distribution curve is important because many

TABLE 2–1
Signal-to-noise-ratio measurements (decibels).

20.15	19.81	19.43	20.80	21.90	19.57
20.29	19.84	19.00	20.51	18.56	20.02
19.66	20.63	18.91	21.15	20.87	19.18
19.43	19.26	19.57	19.62	20.02	19.39
20.89	19.27	18.69	18.82	19.54	19.57
21.84	18.80	18.48	20.26	19.36	20.38
21.12	19.63	18.91	19.36	19.59	18.81
20.17	20.20	21.37	19.37	20.49	19.44
19.59	18.89	20.22	20.71	20.05	19.72
20.51	20.22	19.86	20.06	21.68	20.96

TABLE 2–2
Signal-to-noise-ratio measurements (decibels) in increasing order.

18.48	19.18	19.54	19.81	20.22	20.80
18.56	19.26	19.57	19.84	20.22	20.87
18.69	19.27	19.57	19.86	20.26	20.89
18.80	19.36	19.57	20.02	20.29	20.96
18.81	19.36	19.59	20.02	20.38	21.12
18.82	19.37	19.59	20.05	20.49	21.15
18.89	19.39	19.62	20.06	20.51	21.37
18.91	19.43	19.63	20.15	20.51	21.68
18.91	19.43	19.66	20.17	20.63	21.84
19.00	19.44	19.72	20.20	20.71	21.90

TABLE 2–3
Frequency distribution of signal-to-noise-ratio measurements (decibels).

CLASS INTERVAL	FREQUENCY
18.00–18.49	1
18.50–18.99	8
19.00–19.49	11
19.50–19.99	13
20.00–20.49	13
20.50–20.99	8
21.00–21.49	3
21.50–21.99	3

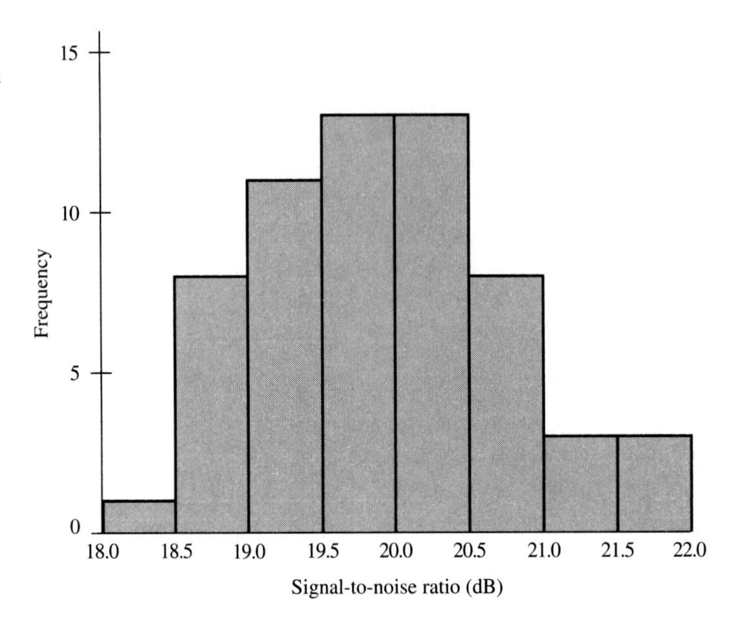

FIGURE 2–1
Frequency distribution graph.

measurements are normally distributed or nearly so. The frequency distribution of the data from Table 2–3 is plotted as a histogram in Figure 2–1. The histogram has the same information as the tabulated frequency distribution; however, it is easier to interpret. If we make the measurement interval smaller and add many more observations, the frequency distribution curve will approach the famous bell-shaped curve called the **normal curve,** and the sample mean will approach the true value, as shown in Figure 2–2. The normal curve extends indefinitely in both directions and is symmetrical about a vertical line drawn through the highest point. The vertical line represents the mean value of the measurement.

There are some very useful relationships obtained from the normal distribution that allow us to make assumptions about the precision of the measurements. The most useful relationship is the **standard deviation,** which is a measure of the

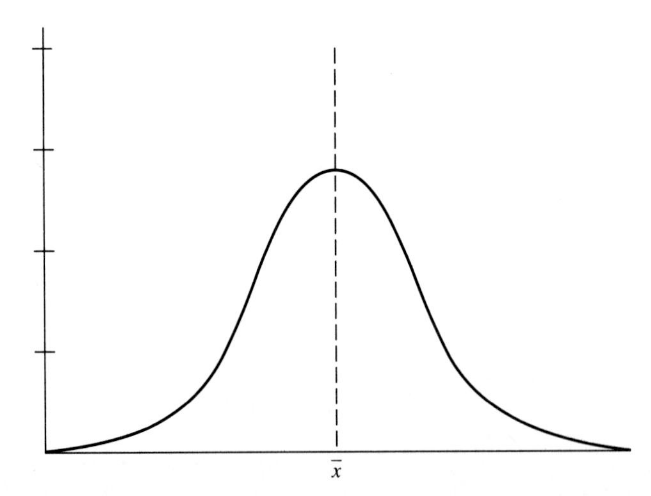

FIGURE 2–2
Bell-shaped normal distribution curve.

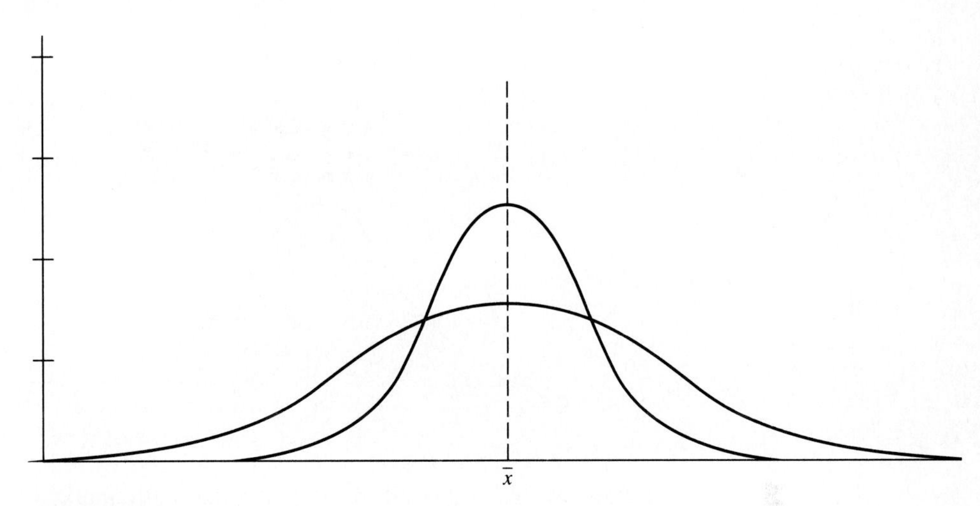

FIGURE 2–3
Two normal distribution curves with the same average but different standard deviations.

scatter of the data and permits us to assign a confidence range for the data. Standard deviation is abbreviated with the lowercase Greek letter sigma, $\hat{\sigma}$, shown with a circumflex to indicate an approximate value based on the measurements. If the data for the normal curve is tightly grouped about the mean, then the curve is narrow and the precision is high. As a result the standard deviation is small. On the other hand, if the data has large variations, then the normal curve is flatter, indicating that the results are less precise. As a consequence, the standard deviation is larger. Even though two different normal curves can have exactly the same mean, the standard deviations can be quite different, and the faith we place in the data is affected. This is illustrated in Figure 2–3.

One reason the standard deviation is useful is that we can readily obtain an idea of the precision of a measurement. Assuming a normal distribution of data, if we calculate the standard deviation for a group of measurements and look at the scatter, we find that about 68% of our measurements will fall within 1σ from the mean and 95% will fall within 2σ from the mean, as illustrated in Figure 2–4. Nearly all (about 99.7%) the values will fall within 3σ from the mean.

To find the standard deviation, we need to compute the mean of a set of measurements and then find how much each value differs from the mean. The amount each value differs from the mean is called the **deviation.** The standard deviation is the root-mean-square value of the deviations; however, for samples, the root-mean-square definition is modified slightly. To find the sample standard deviation of a set of n measurements, the following steps are taken:

1. Find the mean, $\bar{x}$
2. Determine the deviation of each of the n values, $x_i - \bar{x}$
3. Square each of the deviations
4. Find the sum of the squared deviations
5. Divide the sum by $n - 1$
6. Take the square root of the result found in Step 5

FIGURE 2–4
The relationship of σ on the normal distribution curve.

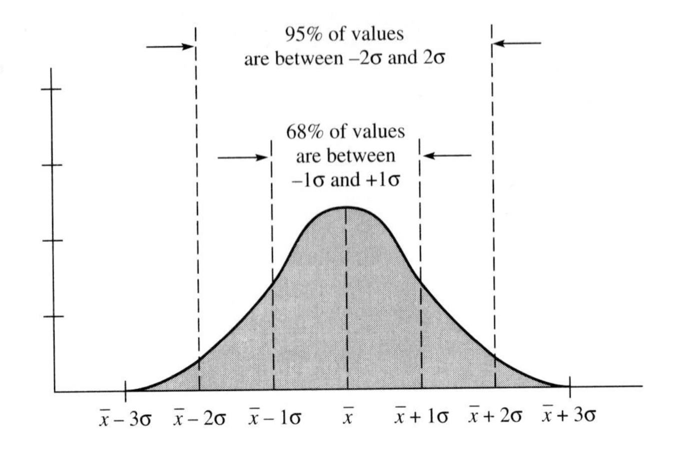

These steps are summarized in the following mathematical formula for the sample standard deviation:

EQUATION 2–2

$$\hat{\sigma} = \sqrt{\frac{\sum_{i=1}^{n} (x_i - \bar{x})^2}{n - 1}}$$

where $\hat{\sigma}$ = sample standard deviation
x_i = value of the ith data point
n = total number of data points

EXAMPLE 2–3 The collector-emitter breakdown voltage of nine similar transistors is measured on a curve tracer and the following values are observed: 96 V, 82 V, 99 V, 106 V, 83 V, 91 V, 108 V, 103 V, and 97 V. Determine the sample standard deviation.

SOLUTION The steps are shown in tabulated form:

i	x_i	$x_i - \bar{x}$	$(x_i - \bar{x})^2$
1	96	−0.1	0.0
2	82	−14.1	198.8
3	99	2.9	8.4
4	106	9.9	98.0
5	83	−13.1	171.6
6	91	−5.1	26.0
7	108	11.9	141.6
8	103	6.9	47.6
9	97	0.9	0.8
	$\bar{x} = 96.1$		$\sum(x_i - \bar{x})^2 = 692.8$

$$\hat{\sigma} = \sqrt{\frac{\sum_{i=1}^{n} (x_i - \bar{x})^2}{n - 1}} = \sqrt{\frac{692.8}{8}} = 9.3$$

Example 2–3 is tedious to compute. It is usually easier to compute the sample standard deviation using the following computing formula, which gives the same result:

EQUATION 2–3

$$\hat{\sigma} = \sqrt{\frac{n\left(\sum_{i=1}^{n} x_i^2\right) - \left(\sum_{i=1}^{n} x_i\right)^2}{n(n-1)}}$$

EXAMPLE 2–4 Repeat Example 2–3 using Equation 2–3.

SOLUTION

i	x_i	x_i^2
1	96	9216
2	82	6724
3	99	9801
4	106	11234
5	83	6889
6	91	8281
7	108	11664
8	103	10609
9	97	9409

$(\sum x_i)^2 = 748{,}225 \quad \sum x_i^2 = 83{,}827$

$$\hat{\sigma} = \sqrt{\frac{n\left(\sum_{i=1}^{n} x_i^2\right) - \left(\sum_{i=1}^{n} x_i\right)^2}{n(n-1)}}$$

$$= \sqrt{\frac{9 \times 83{,}827 - 748{,}225}{9 \times 8}} = 9.3$$

The result using the computing formula is exactly the same as before. This formula is frequently used by electronic calculators to compute standard deviation.

2–4 LOGARITHMIC SCALES

DECIBEL POWER RATIOS

When it is necessary to compare signals, such as the input and output of an amplifier, the ratio can be a very large number (1 million or more is not uncommon). One method of representing the ratio of two numbers is to use a logarithmic ratio named the bel, in honor of Alexander Graham Bell. The bel originated in the telephone industry because communication engineers needed a method for describing the logarithmic attenuation of their signals in long-distance cables. To calculate power in bels

EQUATION 2–4

$$\text{bel} = \log \frac{P_2}{P_1}$$

where P_1 and P_2 are the powers being compared and P_1 is the referenced power. Both P_1 and P_2 can be expressed in any consistent units, but if the waveforms are not alike, rms values must be used.

In practice, a more convenient unit is the decibel (dB). A decibel is one-tenth of a bel and is defined as the logarithmic ratio of two powers. Stated mathematically, this is

EQUATION 2–5

$$\text{dB} = 10 \log \frac{P_2}{P_1}$$

The bel and decibel are dimensionless quantities because they are ratios. The decibel is not defined in terms of an absolute quantity, which means that any two powers with the same ratio are the same number of decibels. For example, the power ratio between 500 W and 1 W is 500 : 1, which represents 27 dB. This is exactly the same number of decibels between 100 mW and 0.2 mW (500 : 1). If the power ratio is such that P_2 is greater than P_1, the ratio is greater than 1. This implies a power gain. When the power ratio is less than 1, there is a power loss, or attenuation. The decibel ratio is positive for power gains and negative for power losses.

One important power ratio is 2 : 1. The dB equivalent of a 2 : 1 power ratio is

$$\text{dB} = 10 \log \frac{2}{1}$$
$$= 3.01 \text{ dB}$$

This result is usually rounded to 3 dB. Since 3 dB represents a doubling of power, 6 dB represents another doubling of the original power, for a ratio of 4 : 1. Thus, 9 dB represents an 8 : 1 ratio of power, and so forth. Note that if the ratio is reversed, the decibel result remains the same except for the sign:

$$\text{dB} = 10 \log \frac{1}{2}$$
$$= -3.01 \text{ dB}$$

The negative result indicates that P_2 is less than P_1. This ratio is the defining power ratio for specifying the cutoff frequency of instruments, amplifiers, filters, and the like.

Another useful ratio to keep in mind is 10 : 1. Since the log of 10 is 1, 10 dB equals a power ratio of 10 : 1. One can use this quickly to estimate the overall gain (or attenuation) in certain situations. For example, if a signal is attenuated by 23 dB, it can be represented by two 10 dB attenuators and a 3 dB attenuator. Two 10 dB attenuators yield a factor of 100 and 3 dB represents another factor of 2, for an overall attenuation of 200 : 1.

EXAMPLE 2–5 The input power to an amplifier is 1 mW when the output power is 150 mW. What is the gain expressed in decibels?

SOLUTION

$$dB = 10 \log \frac{P_2}{P_1}$$

$$= 10 \log \frac{150 \text{ mW}}{1 \text{ mW}}$$

$$= 21.8 \text{ dB}$$

EXAMPLE 2–6 A passive network has an attenuation of 15 dB. If the input power is 150 mW, what is the power from the network?

SOLUTION

$$-15 \text{ dB} = 10 \log \frac{P_2}{150 \text{ mW}}$$

$$-1.5 \text{ dB} = \log \frac{P_2}{150 \text{ mW}}$$

$$P_2 = 150 \text{ mW} \times 10^{-1.5}$$

$$= 4.74 \text{ mW}$$

Decibels are useful in simplifying calculations when linear elements such as amplifiers or attenuators are connected together in a system. The ordinary gain of several cascaded stages is the product of the individual gains. Because the decibel is a logarithmic unit, the decibel gain is computed by adding the decibel gains and losses in the system.

EXAMPLE 2–7 An antenna with a gain of 10 dB is connected to a preamplifier with a gain of 14 dB. The transmission line attenuates the signal by −3 dB; it is then connected to an amplifier with a gain of 35 dB. What is the overall gain of the system? Express the answer as a decibel gain and as an ordinary gain.

SOLUTION The overall gain is the algebraic sum of the individual decibel gains and losses. Therefore, expressed in decibels, the gain is

$$10 \text{ dB} + 14 \text{ dB} - 3 \text{ dB} + 35 \text{ dB} = 56 \text{ dB} \quad \text{gain}$$

To convert the decibel gain to ordinary gain

$$dB = 10 \log \frac{P_2}{P_1}$$

$$56 \text{ dB} = 10 \log \frac{P_2}{P_1}$$

$$5.6 = \log \frac{P_2}{P_1}$$

$$\frac{P_2}{P_1} = 10^{5.6} = 398{,}000$$

A unit similar to the decibel is the dBm. A dBm is a means of expressing power levels in decibels when P_1 (the reference) is understood to be 1 mW. The dBm is defined as

EQUATION 2–6

$$\text{dBm} = 10 \log \frac{P_2}{1 \text{ mW}}$$

In instrumentation, the dBm is commonly used to specify the output level of signal generators (see Chapter 9). It is also used in telecommunications to simplify the computation of power levels. Thus a power level of 15 dBm means that the power is 15 dB above a reference of 1 mW.

EXAMPLE 2–8 Compute the power level of a signal that is 15 dBm.

SOLUTION

$$\text{dBm} = 10 \log \frac{P_2}{1 \text{ mW}}$$

$$15 = 10 \log \frac{P_2}{1 \text{ mW}}$$

$$1.5 = \log \frac{P_2}{1 \text{ mW}}$$

$$31.6 = \frac{P_2}{1 \text{ mW}}$$

$$P_2 = 31.6 \text{ mW}$$

A unit that is very similar to the dBm is the volume unit (VU), which is used in broadcasting. A volume unit, like a dBm, is referenced to 1 mW, but the reference is a sinusoidal waveform measured in a 600 Ω load. The volume unit is defined as

EQUATION 2–7

$$\text{VU} = 10 \log \frac{P_2}{1 \text{ mW}}$$

Compare this definition to that of the dBm. Note that 1 VU is numerically equal to 1 dB. VU meters are designed to respond to complex broadcast waveforms such as speech and have a response that is close to the response of the human ear. The VU meter reads between the average and the peak of complex waveforms.

Specialized reference levels are appropriate to various fields such as noise measurements, acoustical power measurements, and so forth. Some of the terms associated with these measurements are

dBW = dB referenced to 1 W
dBk = dB referenced to 1 kW
dBrap = dB referenced to 10^{-16} W (acoustical limit of human hearing)
dBc = dB referenced to a carrier level

These units are similar in form to the decibel and simply represent a different reference power level for measurement. The dBW, for example, is familiar to many amateur radio operators because it is used to describe output transmitter power. The dBc is used in specifying the noise level in signal generators.

DECIBEL VOLTAGE RATIOS

The decibel was previously defined by the equation

$$\text{dB} = 10 \log \frac{P_2}{P_1}$$

Recall from basic electronics that the power law states $P_1 = V_1^2/R_1$ and $P_2 = V_2^2/R_2$. Substituting the power law for P_1 and P_2 yields

EQUATION 2–8

$$\text{dB} = 10 \log \left(\frac{V_2^2/R_2}{V_1^2/R_1}\right)$$

where R_1, R_2 = resistances in which P_1 and P_2 are developed
V_1, V_2 = voltages across the resistances R_1 and R_2

If the resistances are equal, they cancel and we obtain

EQUATION 2–9

$$\text{dB} = 10 \log \left(\frac{V_2^2}{V_1^2}\right)$$

$$= 20 \log \left(\frac{V_2}{V_1}\right)$$

Equation 2–9 gives the decibel voltage gain. It was originally derived from the decibel power equation and is used when both the input and load impedance are the same (as in telephone systems). Both the decibel voltage gain equation and decibel power gain equation give the same results if the input and load resistance are the same. However, it has become common practice to apply the decibel voltage equation to cases where the impedances are *not* the same. When the impedances are not equal, the two equations do not give the same result. When this is done, it is a good idea to accompany the measurement with a specific statement indicating that the measurement was a voltage gain and giving the procedure followed.

EXAMPLE 2–9 An amplifier with an input impedance of 200 kΩ drives a load impedance of 16 Ω. If the input voltage is 100 μV and the output voltage is 18 V, calculate the decibel power gain and the decibel voltage gain.

SOLUTION The input and output power are

$$P_1 = \frac{V_1^2}{R_1} = \frac{(100\ \mu V)^2}{200\ \text{k}\Omega} = 5 \times 10^{-14}\ \text{W}$$

$$P_2 = \frac{V_2^2}{R_2} = \frac{(18\ \text{V})^2}{16\ \Omega} = 20.25\ \text{W}$$

Substituting for the power in Equation 2–5 gives

$$\text{Power gain:}\quad \text{dB} = 10 \log \frac{P_2}{P_1} = 10 \log \left(\frac{20.25 \text{ W}}{5 \times 10^{-14} \text{ W}}\right) = 146 \text{ dB}$$

Equation 2–9 gives the voltage gain in decibels:

$$\text{Voltage gain:}\quad \text{dB} = 10 \log \left(\frac{V_2^2}{V_1^2}\right) = 20 \log \left(\frac{18 \text{ V}}{100\ \mu\text{V}}\right) = 105 \text{ dB}$$

A specialized unit referenced to 1 mV in 75 Ω is called the decibel millivolt (dBmV). It is used for measurement of signal strength in community antenna television (CATV) systems. The mathematical definition of the decibel millivolt is

EQUATION 2–10

$$\text{dBmV} = 20 \log \left(\frac{V_2}{1 \text{ mV}}\right)$$

EXAMPLE 2–10 Determine the voltage level across a resistance of 75 Ω for a signal of +55 dBmV.

SOLUTION

$$\text{dBmV} = 20 \log \frac{V_2}{1 \text{ mV}}$$

$$55 \text{ dBmV} = 20 \log \frac{V_2}{1 \text{ mV}}$$

$$V_2 = 1 \text{ mV} \times 10^{2.75}$$

$$V_2 = 562 \text{ mV}$$

2–5 GRAPHING DATA

Graphs are useful for presenting data in a concise form. The purpose of a graph is to show the relationship between two or more variables. One of the variables is called the *dependent variable* and the other variable is called the *independent variable*. Think of the independent variable as the controlling variable and the dependent variable as the responding variable. For example, the current in a circuit (dependent) responds to a change in the voltage (independent). *The dependent variable should always be plotted on the y-axis and the independent variable should be plotted on the x-axis.*

There are certain important elements for constructing clear linear graphs:

1. Prepare a table of data containing an ordered list of the data pairs.
2. Observe the range of the data and choose a scale for each axis that allows all the pertinent data to be shown over as large a portion of the graph as possible. The selection of a scale should be done in such a way as to include all the pertinent data within the boundaries of the graph. The scale values should be

selected using simple multiples (such as 1, 2, 5, or 10) along each axis to allow the user to interpret the data easily. As a rule, it is best to include the zero point on the *y*-axis except in cases where the data is far removed from zero. In these cases warn the reader by showing a broken line to indicate that not all the true vertical scale is shown.

3. Labels should be shown on each axis, including the measurement units. The reader should be able to determine the value of any data point directly from the graph. Supplementary information may be shown in a location where there is no data. Include a title for the graph.
4. All pertinent data points should be placed on the graph (except that obviously erroneous data should be discarded and remeasured). If more than one curve is put on the same paper, the data points should be coded with different symbols (dots, circles, triangles, etc). Open symbols are preferred, with a small dot in the center representing the precise location of the data point.
5. Normally, for data representing a continuous phenomenon, a smooth curve should be shown that indicates the trend of the data. It is not necessary for the curve to touch every data point. An exception is for calibration curves or other data representing a discontinuous variable. In these cases, connect all data points with straight-line segments.

LINEAR GRAPHS

Linear graph paper has evenly spaced lines for both the *x*- and *y*-axis in a rectangular grid arrangement. Four-quadrant paper is designed for plotting both positive and negative values of the *x* and *y* variables. Such a system is referred to as a Cartesian coordinate system. The intersection of the axes is called the origin. Positive *x* values are plotted to the right of the origin and negative *x* values are plotted to the left of the origin. Likewise, positive *y* values are plotted above the origin and negative *y* values are plotted below the origin. An example of four-quadrant graph paper is shown in Figure 2–5.

Data that consists of positive values of both variables is plotted on single-quadrant paper. The origin is normally drawn in the lower left corner, with a convenient scale chosen to show all data points on the paper. Sometimes the range of the data is relatively small, but the actual values are large. In these cases, it is not necessary to include the origin in the graph. Examples of linear graphs are given in Figure 2–6.

POLAR GRAPHS

Linear graph paper lends itself to plotting points that are given in rectangular (x, y) coordinates. By contrast, data written in polar form $(M \angle \theta)$ can be plotted directly on polar graph paper. Polar graph paper is made with equally spaced concentric circles from the origin; major divisions are indicated by darker circles. Radial lines extend from the origin and are marked in degrees with respect to a pole or origin. Polar graph paper is selected when it is necessary to indicate directional characteristics, such as antenna or microphone patterns. Data points are usually con-

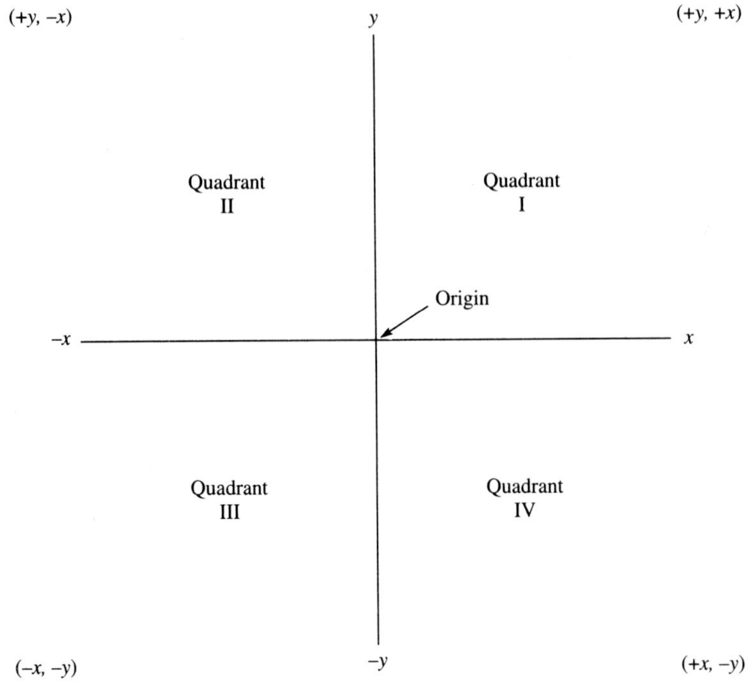

FIGURE 2–5
Rectangular four-quadrant graph paper.

nected in a smooth curve. An example of a polar graph is the microphone pattern shown in Figure 2–7.

LOGARITHMIC GRAPHS

Sometimes the range of values to be plotted is extremely large. For example, the frequency response of an amplifier may extend from 20 Hz to more than 20,000 Hz. If a linear scale is chosen for the graph that allows the largest data point to fit on the paper, then values at the low end are compressed and the reader cannot discriminate between data points. On the other hand, a logarithmic (log for short) scale is one in which distances along the scale are spaced logarithmically. A log scale allows data encompassing a large dynamic range to have good resolution for very small values.

Semilogarithmic (semilog) graph paper has one axis drawn with a logarithmic scale and the other axis drawn with a linear scale. Either the x-axis or the y-axis can be the log scale. In the case of the frequency response of an amplifier, the x-axis is made into a log scale to expand the frequency scale. A plot of an amplifier's frequency response drawn on semilog paper is shown in Figure 2–8. The frequency response curve plotted on semilog paper is called a *Bode plot.* Notice that the paper repeats in cycles along the x-axis. Each cycle represents a decade, or power of ten. Since there are three cycles shown, the scale represents 3 decades, or a factor of 1000 from one end to the other ($10 \times 10 \times 10$). There is no zero

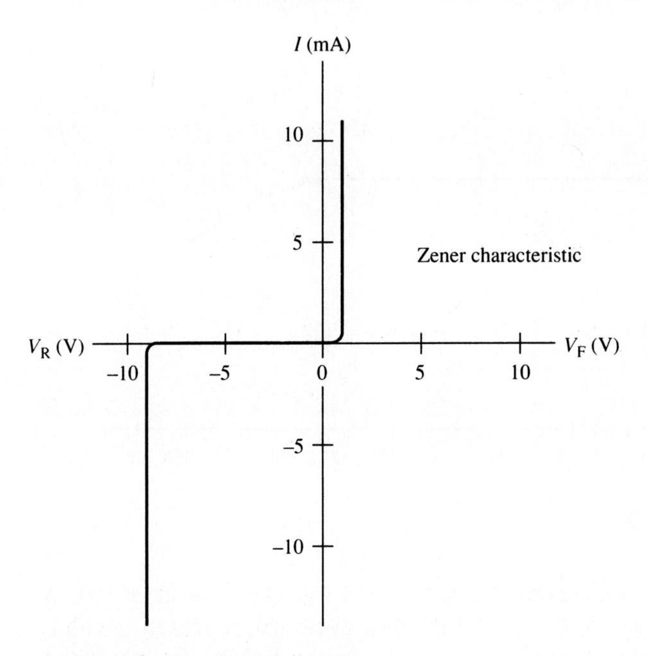

(a) Four-quadrant rectangular graph paper.

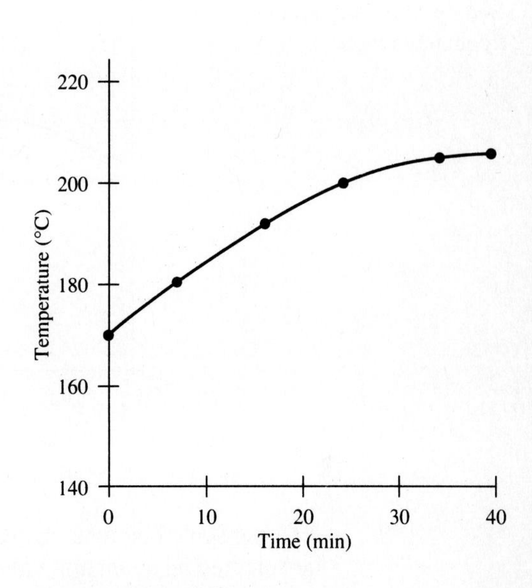

(b) Linear graph paper (first quadrant) note y-axis scale does not begin at 0.

FIGURE 2–6
Examples of linear graphs.

FIGURE 2–7
Directional microphone response. The microphone shown is a supercardioid that has some rear pickup.

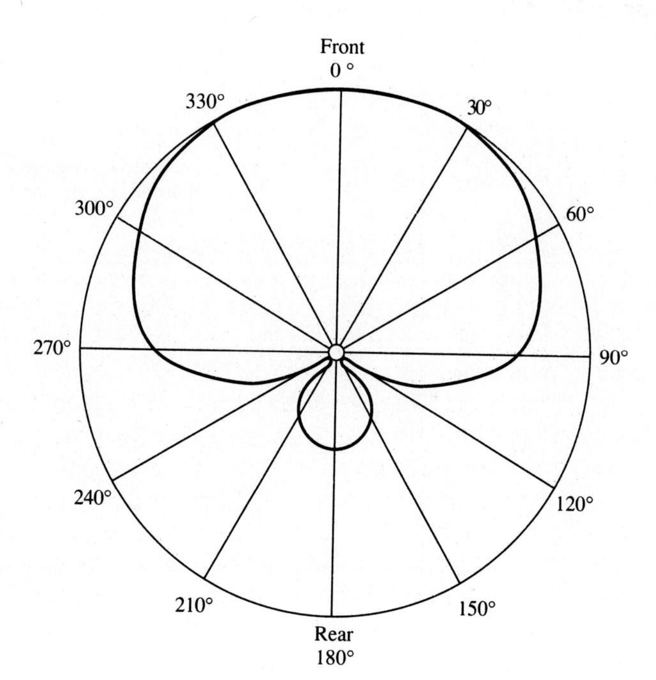

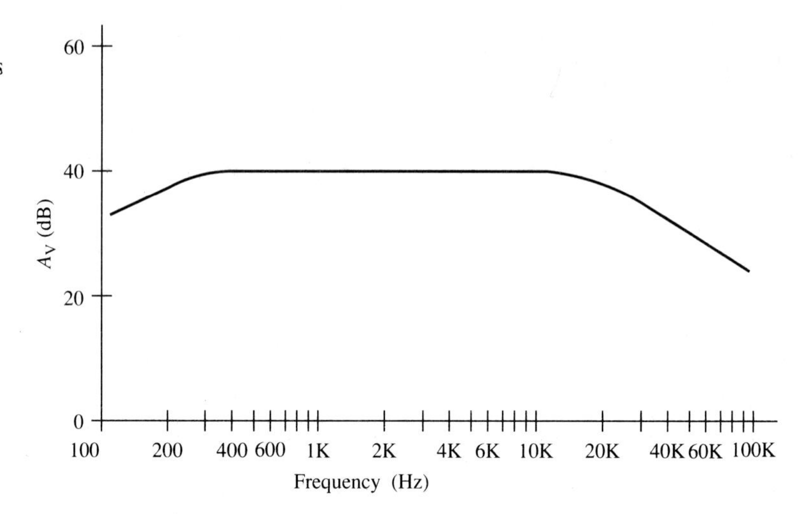

FIGURE 2–8
Bode plot of an amplifier's frequency response.

on a log scale because log (0) is undefined (negative infinity). Any power of 10 can be selected as a starting value, and the end of the first cycle will represent a value that is 10 times greater than the starting value. In this case, the *y*-axis is a normal linear scale.

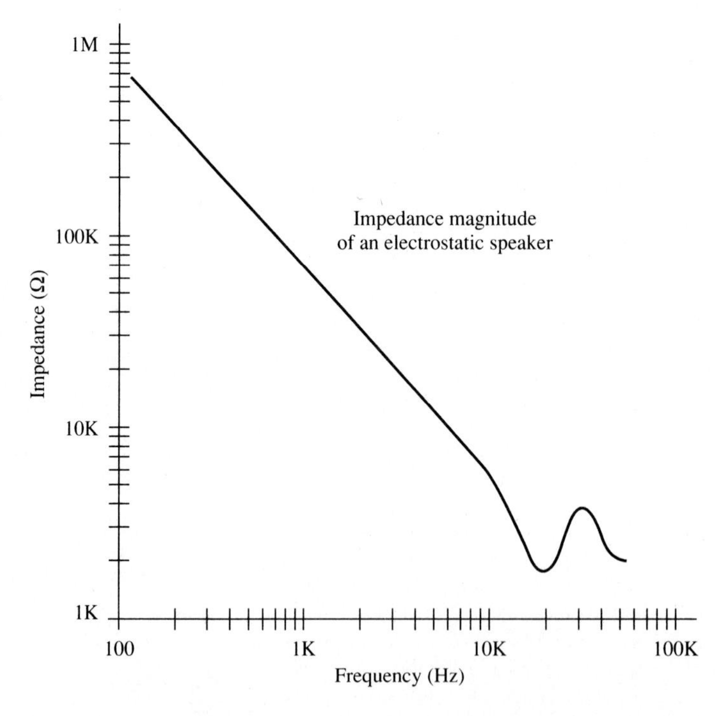

FIGURE 2–9
Example of a log-log plot using 3-by-3 log-log paper.

When the range of values is very large for both variables, log-log graph paper is useful. Log-log paper has a log scale on both the horizontal and vertical axes. The paper is named according to the number of repeating cycles on each axis. Thus 3-by-5 log-log paper has three cycles on the x-axis and five cycles on the y-axis. Figure 2–9 illustrates a graph of speaker impedance for an electrostatic loudspeaker as a function of frequency plotted on 3-by-3 log-log paper.

2–6 INTERPRETATION OF GRAPHS

Sometimes it is useful to determine the equation of a set of data from a graph of the empirical data. The resulting equation is called an **empirical equation.** The first objective in determining an empirical equation is to find the form of the equation. We will examine one method (the slope-intercept method) for determining an empirical equation for data that produces a straight line on various types of graph paper. A second method, called the method of least squares (linear regression) is discussed in Section 2–7.

A STRAIGHT LINE ON LINEAR GRAPH PAPER

The most basic relationship between variables is that of a straight line. Recall from elementary algebra that the general equation for a straight line can be written in slope-intercept form as

EQUATION 2–11

$$y = mx + b$$

where y = the dependent variable
x = the independent variable
m = the slope of the line
b = the value of the y-intercept

If the data points indicate that there is a straight-line relationship, a line is drawn representing the best-fit straight line. The equation for this line can be found by measuring the slope and the y-intercept. The slope is measured by determining the ratio of the vertical change in y, abbreviated Δy, to the horizontal change in x, abbreviated Δx. Because the slope, $\Delta y/\Delta x$, is constant, it can be measured between any two points that are separated by a reasonable distance. The variables are substituted for x and y, giving the straight-line equation in standard form.

EXAMPLE 2–11 A field-effect transistor has a set of curves that show its current-voltage characteristics. If one of the curves is expanded near the region of $V_{GS} = 0$, the resulting curve is a straight line, as shown in Figure 2–10. This graph shows the drain current as a function of drain-source voltage. Determine the equation for the line.

FIGURE 2–10
Determining the slope and y-axis intercept of a straight line.

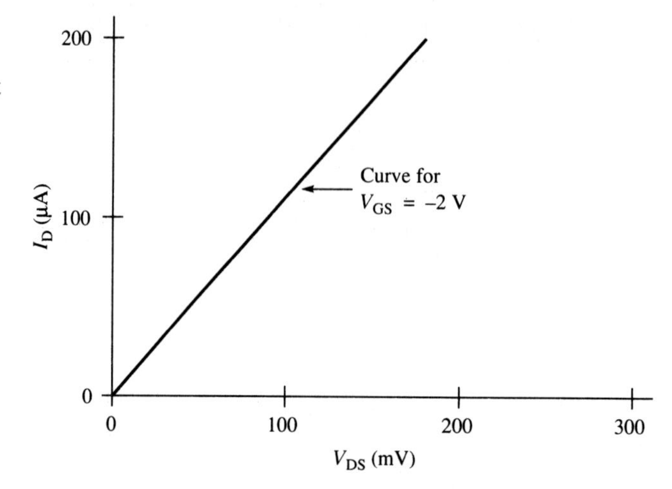

(a) Curve for an n-channel JFET when $V_{GS} = -2$ V (enlarged around $V_{DS} = 0$)

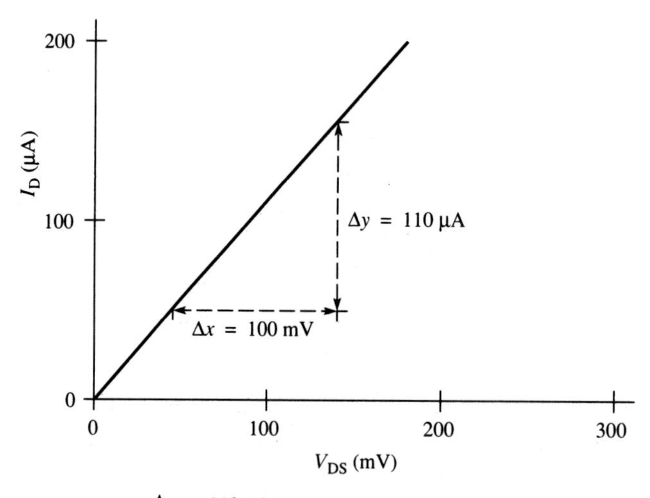

$$m = \text{slope} = \frac{\Delta y}{\Delta x} = \frac{110\ \mu\text{A}}{100\ \text{mV}} = 1.1\ \text{mS} = 1.1 \times 10^{-3}\ \text{S}$$

b = y-axis intercept = 0

(b) Finding the slope and intercept.

SOLUTION The slope is determined by finding the ratio of $\Delta y/\Delta x$, as shown. The y-axis intercept is 0. Substituting into Equation 2–11 for a straight line gives

$$y = mx + b$$
$$I_D = (1.1 \times 10^{-3})V_{DS}$$

Notice that the units for the slope are siemens, the reciprocal of ohms.

Sometimes a linear relationship is obscured if an inverse relationship is present. The graph of capacitive reactance, X_C, plotted against frequency is such a case. Recall from basic electronics that

$$X_C = \frac{1}{2\pi fC}$$

where X_C = capacitive reactance, Ω
f = frequency, Hz
C = capacitance, F

Figure 2–11 illustrates a graph of this equation for a 0.01 μF capacitor. The capacitive reactance clearly indicates an inverse relationship to frequency but

FIGURE 2–11

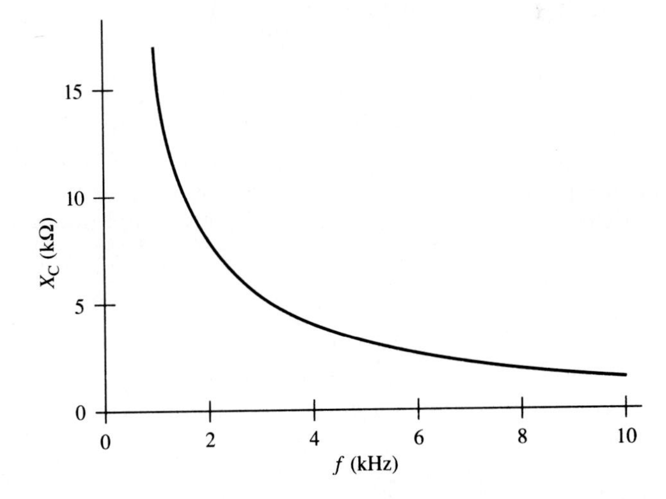

(a) Capacitive reactance as a function of frequency for a 0.01 μF capacitor.

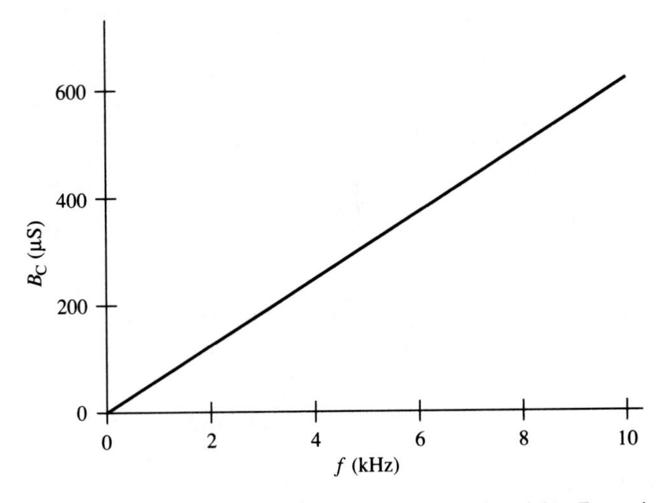

(b) Capacitive susceptance as a function of frequency for a 0.01 μF capacitor.

doesn't immediately show that the equation for the line is linear. By taking the reciprocal of X_C values, we obtain the capacitive susceptance, B_C. Plotting the capacitive susceptance as a function of frequency clearly shows the linear nature of the relationship.

STRAIGHT LINES ON SEMILOG GRAPH PAPER

There are two ways in which semilog paper is used; the log scale can be on the y-axis or the log scale can be on the x-axis. The data determines which graph should be used. Data that covers a large dynamic range in the dependent variable (y) but not in the independent variable (x) should be plotted with the y-axis being the log scale and the x-axis being the linear scale. If the data produces a straight line when the y-axis is the log scale, the form of the equation is

EQUATION 2–12

$$y = b10^{mx}$$

An example showing a comparison of the same data plotted on both linear and semilog plots is given in Figure 2–12. The y-intercept, b, is read directly as 3.0. The slope, m, is found to be 2.0, as illustrated. Notice that the vertical distance of the triangle represents $\Delta \log y$. It is necessary to find the difference in the log of the two vertical points ($\log y_2 - \log y_1$) and divide by Δx ($x_2 - x_1$)

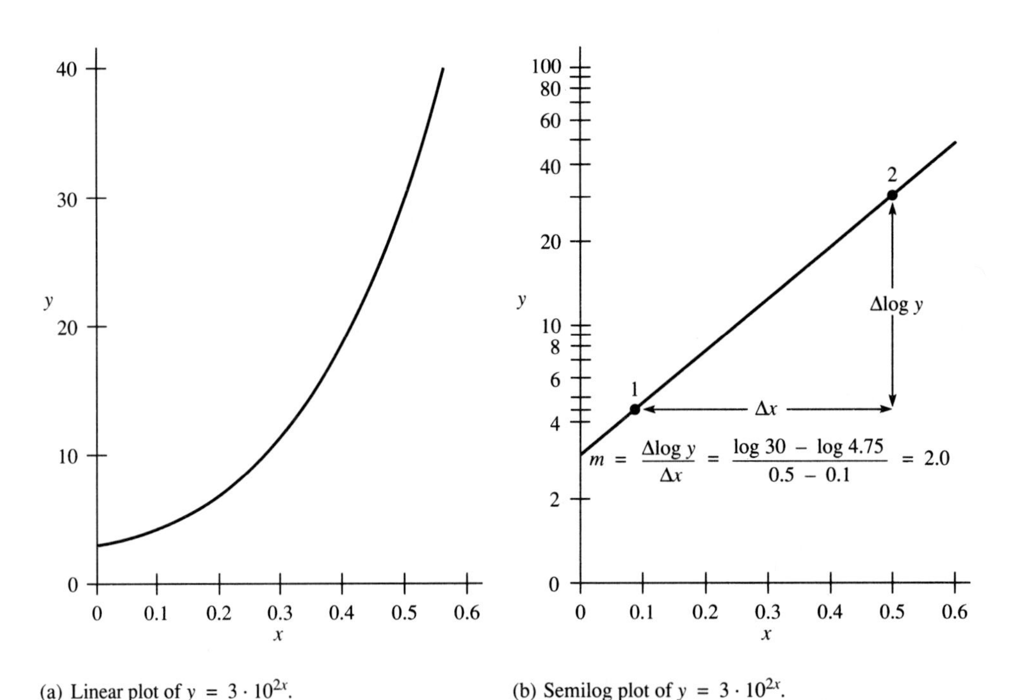

(a) Linear plot of $y = 3 \cdot 10^{2x}$.

(b) Semilog plot of $y = 3 \cdot 10^{2x}$.

FIGURE 2–12
Comparison of linear and semilog plots of an equation with the form $y = b10^{mx}$.

FIGURE 2–13
Comparison of linear and semilog plots of an equation with the form $y = m \log x + b$.

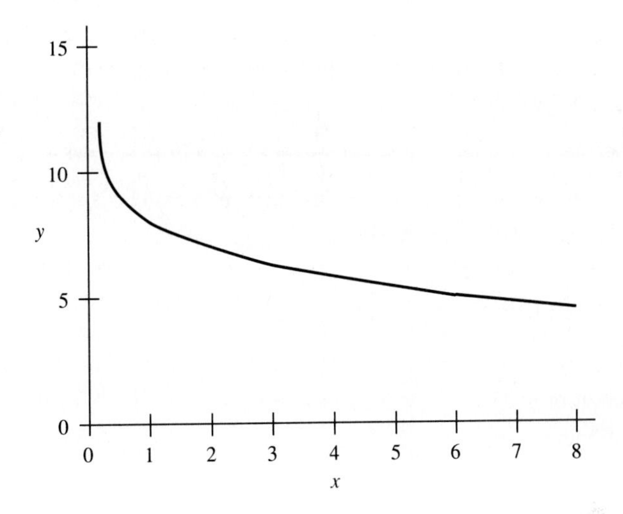

(a) Linear plot of $y = -4.0 \log x + 8.0$.

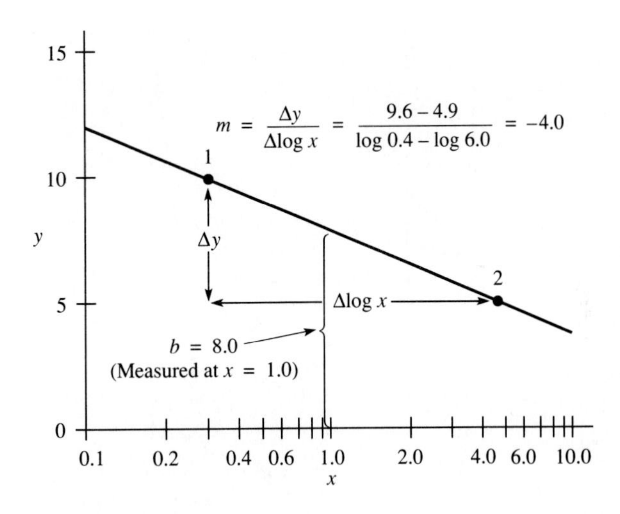

(b) Semilog plot of $y = -4.0 \log x + 8.0$.

to compute the slope. Substituting into Equation 2–12 gives the equation $y = 3 \times 10^{2x}$.

The form of the equation for a straight line on semilog paper when the x-axis is the log scale is

EQUATION 2–13

$$y = m \log x + b$$

Again, an example showing a comparison of the same data plotted on both linear and semilog plots is shown in Figure 2–13. In this case, the horizontal distance of the triangle represents $\Delta \log x$. The slope is found by dividing Δy $(y_2 - y_1)$ by the horizontal distance along the triangle $(\log x_2 - \log x_1)$. The y-

intercept, b, is found on the graph at the point where $x = 1$. Notice that at this point, the first term of the equation is zero (since $\log 1 = 0$) and the intercept can be read directly.

EXAMPLE 2–12 The data for the forward bias curve of a diode is shown plotted in Figure 2–14. Determine the equation for the curve.

SOLUTION See Figure 2–14.

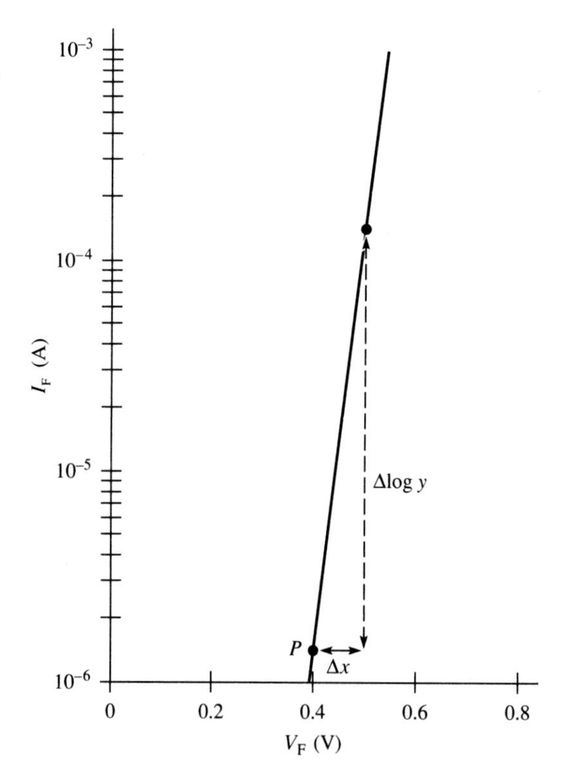

FIGURE 2–14 Finding the equation of a forward-biased diode.

Note that the equation has the form $y = b10^{mx}$.

$$\text{slope} = m = \frac{\Delta \log y}{\Delta x} = \frac{\log 1.5 \times 10^{-4} - \log 1.5 \times 10^{-6}}{0.52 - 0.40} = \frac{2.0}{0.12} = 16.7$$

The y-intercept can be found indirectly. Substitute known values of point P into the general equation and solve for b.

$$y = b10^{mx}$$
$$1.5 \times 10^{-6} = b10^{(16.7)(0.4)}$$
$$b = \frac{1.5 \times 10^{-6}}{4.64 \times 10^{6}} = 3.2 \times 10^{-13}$$

Therefore, the equation is $y = 3.2 \times 10^{-13} \cdot 10^{16.7x}$.

STRAIGHT LINES ON LOG-LOG PAPER

If data indicates a straight-line relationship on log-log graph paper, the form of the equation is

EQUATION 2–14

$$y = bx^m$$

This equation is a power function. The intercept, b, can be found directly on the graph at the point where $x = 1$ (since at this point $y = b$). The slope is $(\Delta \log y)/(\Delta \log x)$. If the length of the graph paper's x-cycle is equal to the length of the graph paper's y-cycle (the usual case), the slope can be found by direct measurement with a ruler. A comparison of the same data plotted on both linear and log-log plots is shown in Figure 2–15. The intercept, b, is found to be 2 by direct reading. Two methods of finding the slope, m, are shown. The first uses the ratio of $(\Delta \log y)/(\Delta \log x)$ and is illustrated between points 1 and 2. The method shown between points 3 and 4 is a direct measurement with a ruler and finding the ratio of $\Delta y/\Delta x$. Both methods give the result 2/3. Therefore the equation for the line is determined to be

$$y = 2x^{2/3}$$

The decision as to whether a particular set of data should be plotted on logarithmic paper depends primarily on the range of the data to be plotted. If the

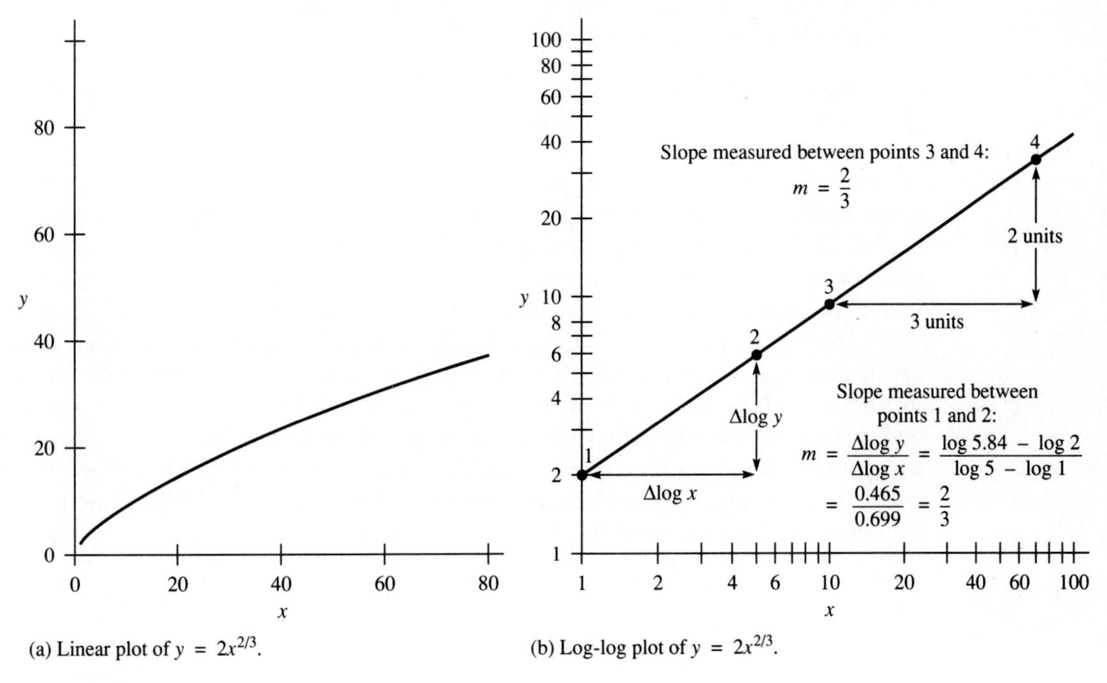

(a) Linear plot of $y = 2x^{2/3}$.

(b) Log-log plot of $y = 2x^{2/3}$.

FIGURE 2–15
Comparison of linear and log-log plots of an equation with the form $y = bx^m$.

TABLE 2–4
Comparison of equations for straight lines plotted on different types of paper.

PAPER	EQUATION FOR STRAIGHT LINE
Linear	$y = mx + b$
Semilog $\begin{cases} x = \text{linear} \\ y = \text{logarithmic} \end{cases}$	$y = b10^{mx}$
Semilog $\begin{cases} x = \text{logarithmic} \\ y = \text{linear} \end{cases}$	$y = m \log x + b$
Log-log	$y = bx^m$

range of data is very large, semilog or logarithmic paper is a good choice. A second reason for selecting logarithmic paper is to show a straight-line relationship in the data. If a straight line results from plotting a set of data on logarithmic paper, the equation for the line can be determined as previously discussed. As a summary, the equations for straight lines plotted on various types of paper are given in Table 2–4.

RATE OF CHANGE OF CONTINUOUSLY VARYING QUANTITIES

Many times it is difficult to express the relationship of two variables in a simplified form. Electronic measurements often are dependent on a number of variables, including frequency effects, environmental conditions, noise, and nonlinear responses, to name a few. Even when a simplifying relationship between two variables cannot be found, a measurement of the slope at various points may be useful. For example, data taken with a displacement transducer indicates the position of some point as a function of time. If the velocity of the point is needed, it can be found by measuring the slope of the position data as it changes with time.

The measurement of slope is useful for many electronic devices. One interesting device is a tunnel diode. A tunnel diode's characteristic *I-V* curve is shown in Figure 2–16(a). The slope of this curve is rather unique for electronic devices and even shows a region where the resistance is negative! To find the slope at some point on the curve, a line is drawn tangent to the point, as shown in Figure 2–16(b). The slope of this line is determined by finding $\Delta I/\Delta V$. The slope represents the conductance g of the device at the tangent point:

EQUATION 2–15

$$\text{Slope} = \frac{\Delta I}{\Delta V} = g$$

where g = conductance in siemens

FIGURE 2–16
Example of measuring the slope of a complex nonlinear curve.

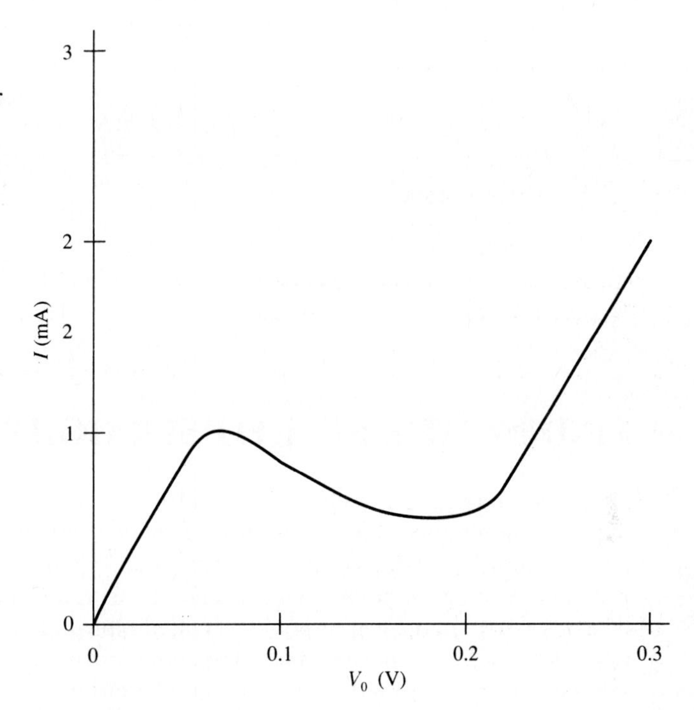

(a) *I-V* curve for a tunnel diode.

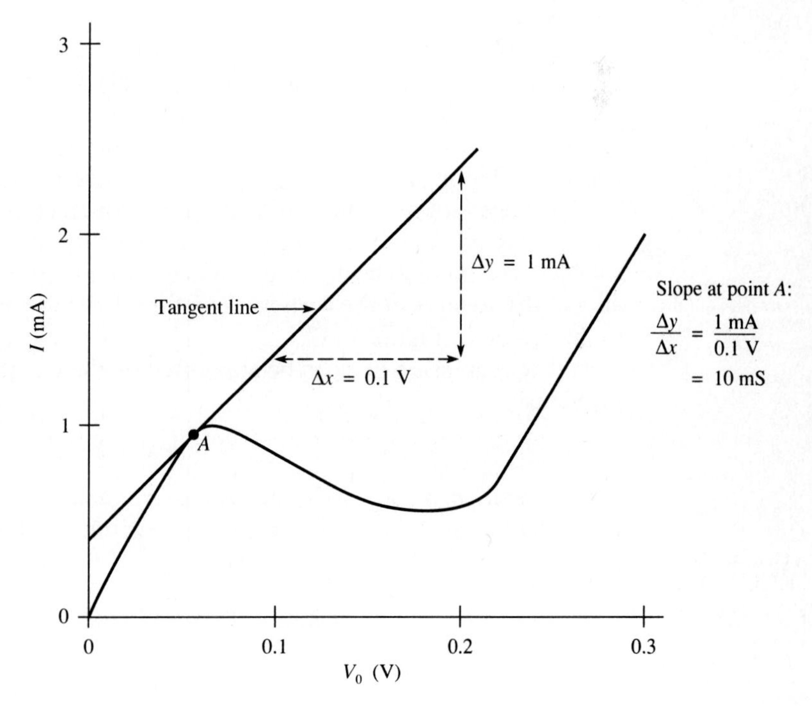

(b) Measurement of the slope at point *A*.

The reciprocal of the conductance is the resistance measured at the tangent point. The resistance is

EQUATION 2–16

$$r = \frac{1}{g} = \frac{\Delta V}{\Delta I}$$

where r = resistance, Ω

It is apparent that the resistance of the device depends on the point that is measured. This resistance depends on the value of the independent variable, so it is called **dynamic,** or **ac, resistance.**

2–7 FINDING THE BEST-FIT STRAIGHT LINE

Empirical data seldom falls in a perfectly straight line due to experimental error and the inherent limitation of all measuring instruments. If we can find the best-fit line to empirical data, we can determine an equation for the line and use it to predict the values of points that were not measured. Finding corresponding values of x or y between measured points is called **interpolation**, and finding corresponding values of x or y beyond measured points is called **extrapolation**. There is an infinite variety of curved lines that could be fitted to experimental data; however, we restrict our discussion to straight lines. Finding the best-fit line is known as finding the regression line. When the line is straight, the procedure is termed **linear regression.**

A straight line often gives a good fit to scattered data and is the easiest line to fit. However, for many measurement problems, the data is scattered and the "best" line to fit seems arbitrary. There are many lines that might fit a particular set of data in a satisfactory manner, and each person attempting to do so might draw a different line. This problem was studied by the French mathematician Adrien Legendre near the beginning of the nineteenth century. He reasoned that data points far from the average value should carry more "weight" as far as fitting the line. The method he devised is called the method of **least squares,** whereby the sum of the squares of the vertical deviations from the line are minimized. The idea is illustrated in Figure 2–17.

The regression line can be expressed by the equation for a straight line given previously:

$$y = mx + b$$

The slope m and y-axis intercept b are found by applying statistical functions to analyzing the data. The slope and y-axis intercept are found as follows:

EQUATION 2–17

$$m = \frac{n\Sigma x_i y_i - \Sigma x_i \Sigma y_i}{n\Sigma x_i^2 - (\Sigma x_i)^2}$$

EQUATION 2–18

$$b = \frac{\Sigma y_i - m\Sigma x_i}{n} \qquad (i \text{ varies from 1 to } n)$$

FIGURE 2–17
Example of fitting a line to a set of scattered data.

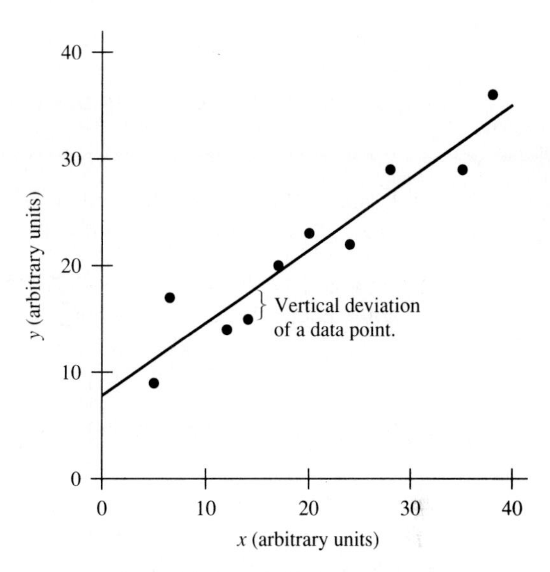

Deviations are measured from regression line to data point in the vertical direction.

where m = slope of the regression line
b = y-axis intercept of regression line
n = number of data points
x_i = x value of ith point
y_i = y value of ith point

The equation shown for finding the y-intercept b requires that the slope m be determined first. Many calculators are equipped to find the required summations for linear regression equations. In addition, some programmable calculators have the preceding equations programmed so the user need only enter the data.

After solving for the linear regression line for a set of data, it is useful to know how well the line fits the observed data. The result is not very meaningful if you don't have confidence in it. A graph can help determine if the data appears to fit but it isn't very quantitative. A quantitative measure of goodness of fit is called the **correlation coefficient,** R. It can be computed using the equation

EQUATION 2–19

$$R = \frac{n\Sigma x_i y_i - \Sigma x_i \Sigma y_i}{\{[n\Sigma x_i^2 - (\Sigma x_i)^2]\,[n\Sigma y_i^2 - (\Sigma y_i)^2]\}^{1/2}}$$

The correlation coefficient is a dimensionless quantity that varies between -1 and $+1$. Negative values occur when the regression line has a negative slope; positive values occur when the regression line has a positive slope. Perfectly correlated data has a correlation coefficient of either $+1$ or -1. Data that exhibits no correlation has a correlation coefficient of 0. The closer the value is to either $+1$ or -1, the better the fit.

R^2 is also frequently used to determine the goodness of fit. R^2 indicates the percent variation described by the line; the range is between 0 and +1. Many calculators have built-in routines to compute R^2.

EXAMPLE 2–13 The data for Figure 2–17 is listed in Table 2–5. Compute the equation for the regression line and find the correlation coefficient.

TABLE 2–5
Data for Figure 2–17.

i	x_i	y_i
1	5.	9.
2	6.	17.
3	12.	14.
4	14.	15.
5	17.	20.
6	20.	23.
7	24.	22.
8	28.	29.
9	35.	29.
10	38.	36.

SOLUTION Table 2–6 shows the data from Table 2–5 with the statistical information expanded.

TABLE 2–6
Data for Figure 2–17.

i	x_i	y_i	x_iy_i	x_i^2	y_i^2
1	5.	9.	45	25	81
2	6.	17.	102	36	289
3	12.	14.	168	144	196
4	14.	15.	210	196	225
5	17.	20.	340	289	400
6	20.	23.	460	400	529
7	24.	22.	528	576	484
8	28.	29.	812	784	841
9	35.	29.	1015	1225	841
10	38.	36.	1368	1444	1296

From data in Table 2–6, we find

$$n = 10 \qquad \Sigma x_i y_i = 5048$$
$$\Sigma x_i = 199 \qquad \Sigma x_i^2 = 5119 \qquad (\Sigma x_i)^2 = 39{,}601$$
$$\Sigma y_i = 214 \qquad \Sigma y_i^2 = 5182 \qquad (\Sigma y_i)^2 = 45{,}796$$

Substituting into Equations 2–17 and 2–18:

$$m = \frac{n\Sigma x_i y_i - \Sigma x_i \Sigma y_i}{n\Sigma x_i^2 - (\Sigma x_i)^2} = \frac{(10)(5048) - (199)(214)}{(10)(5119) - 39{,}601} = 0.681$$

$$b = \frac{\Sigma y_i - m\Sigma x_i}{n} = \frac{(214) - (0.681)(199)}{10} = 7.85$$

Therefore, the equation for the regression line is

$$y = 0.681x + 7.85$$

To find the correlation coefficient, substitute into Equation 2–19:

$$R = \frac{n\Sigma x_i y_i - \Sigma x_i \Sigma y_i}{\{[n\Sigma x_i^2 - (\Sigma x_i)^2][n\Sigma y_i^2 - (\Sigma y_i)^2]\}^{1/2}}$$

$$= \frac{(10)(5048) - (199)(214)}{\{[(10)(5119) - 39{,}601][(10)(5182) - 45{,}796]\}^{1/2}} = 0.94$$

The positive value of the correlation coefficient indicates the slope of the equation is positive. Since the result is close to 1.0, there is a high degree of linear correlation.

SUMMARY

1. All measured data contains experimental error. Error is the difference between the true, or accepted, value and the measured value.
2. The number of significant digits that should be retained for measured data depends on the precision of the measurement.
3. Error can be classified as random or systematic. Random errors vary by chance above and below the true value. Systematic errors consistently cause the measured data to be either higher or lower than the true value.
4. The standard deviation is a statistical measure of the dispersion of a set of data from the mean. If the standard deviation is small, the data is closely bunched about the mean.
5. The decibel is a logarithmic ratio of two powers, defined as

$$\text{dB} = 10 \log \frac{P_2}{P_1}$$

6. A graph is a pictorial representation of the relationship between a dependent and independent variable. To construct a graph, the data should be tabulated, a scale should be selected that covers the range of data, both axes should be labeled, all pertinent data should be included, and a smooth curve should be drawn that indicates the trend of the data.
7. Different types of graph paper are used to show data. These include linear, semilog, log-log, and polar graph paper.
8. A straight-line plot on linear, semilog, or log-log graph paper can be analyzed to determine the equation for the line. Curved lines can be analyzed for their instantaneous slope.
9. The best-fitting straight line to scattered data can be found by a technique called linear regression. The coefficient of correlation is a measure of how well the data fits the linear regression line.

QUESTIONS AND PROBLEMS

1. Explain the difference between the terms *accuracy* and *precision*.
2. Determine the number of significant figures in each of the following numbers.
   (a) 0.0500
   (b) 0.0001
   (c) 100.50
   (d) 1.60×10^{-5}

(e) 1.05×10^3
(f) 0.00807
(g) 450
(h) 2.914
(i) 10.0

3. The measurement 22,000 has three significant digits. Show how to indicate this number such that the number of significant figures is unambiguous.
4. Round the following numbers to three significant digits.
   (a) 1.297
   (b) 1.205
   (c) 1.0085
   (d) 1.155×10^4
   (e) 1.894
5. The data in Table 2–7 represents a set of measurements of the cutoff frequencies of 25 different amplifiers. Reorganize the data into a frequency distribution table with a class interval of 1 kHz. Then plot the data in a histogram.

TABLE 2–7
Cutoff frequency for 25 amplifiers (kHz).

98.4	101.1	98.6	100.2	100.4
99.5	102.2	100.0	98.3	98.7
103.5	102.9	99.6	102.0	97.1
100.4	101.1	101.9	101.8	101.3
99.0	100.9	99.4	98.7	97.2

6. Compute the standard deviation of the data in Table 2–7.
7. What type of error (random or systematic) is likely from the following problems?
   (a) Electrical noise is present with the desired signal.
   (b) An analog meter does not point to zero when no signal is applied.
   (c) Measurements are made with an instrument before it has had the required warm-up time.
   (d) An instrument is used in an environment in which the temperature is higher than the manufacturer's rating.
   (e) A 1 V signal is measured on the 100 V range of an instrument.
   (f) A measurement is made with an uncalibrated instrument.
   (g) An electronic counter that uses separate start and stop lines to measure a time interval has a delay cable that is not accounted for connected to the stop line.
8. An amplifier has a gain of 36 dB. If the input signal is 0.5 mW, what is the output-signal power?
9. An amplifier has an input power of 50 μW and an output power of 1 W. Calculate the decibel power gain.
10. If the amplifier in Problem 9 has an input impedance of 1.0 kΩ and an output impedance of 50 Ω, what is the decibel voltage gain?
11. What power level is represented by 45 dBm?
12. Compute the voltage across a 300 Ω line at 20 dBm.
13. Find an equation for the straight line shown in Figure 2–18.

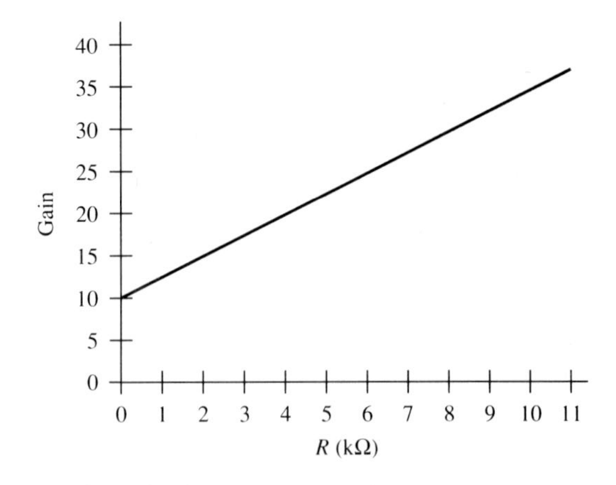

FIGURE 2–18
Gain of an amplifier as a function of a variable resistor's setting.

14. Find an equation for the line shown on log-log paper in Figure 2–19.

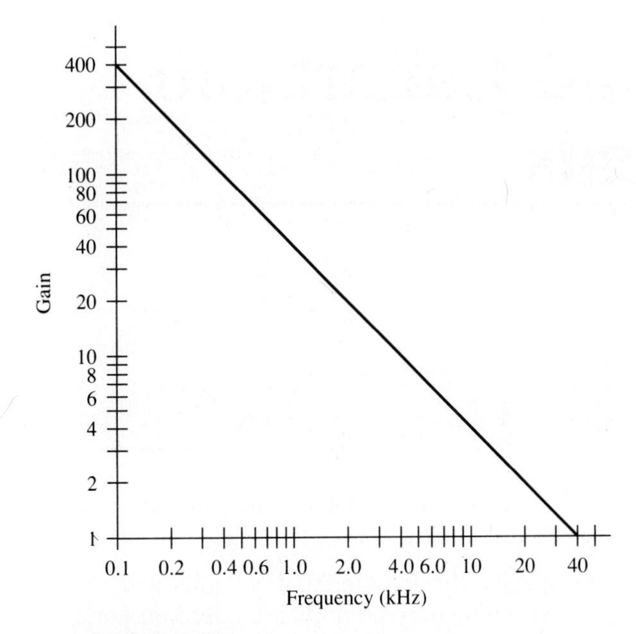

FIGURE 2–19
Frequency response of a low-pass active filter.

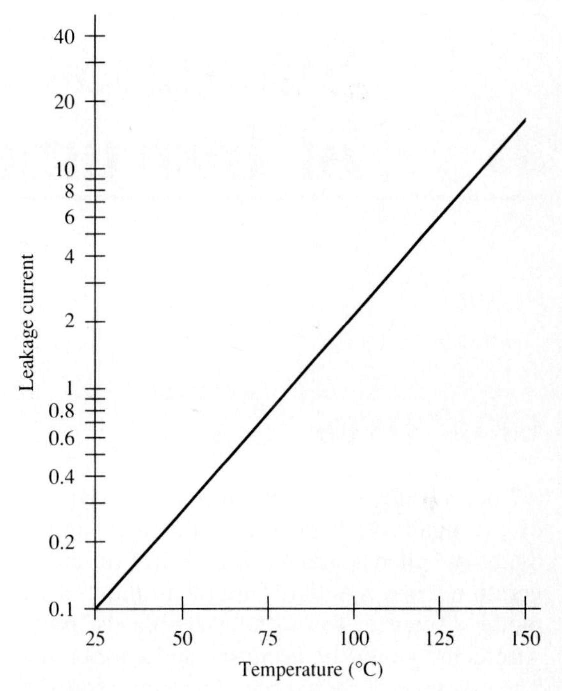

FIGURE 2–20
Leakage current as a function of temperature.

15. What are the criteria for determining whether a group of data points should be connected by line segments or by a smooth curve?
16. What is meant by 3-by-4 log-log paper?
17. What type of graph paper should be selected to show a transmitting antenna's radiation pattern?
18. Graph the equation $3x + 2y = 10$ on linear graph paper.
19. The leakage current for a certain integrated circuit increases as a function of temperature, as shown in Figure 2–20. Determine an equation for the curve.
20. An investigation of a JFET (junction field-effect transistor) was done in which the drain power supply (V_{DD}) was increased while monitoring the drain current (I_D). The gate voltage was held at 0 V. The data is shown as follows. Plot the data and answer the questions.
    (a) What is the slope ($\Delta I/\Delta V$) at V_{DD} = 5 V?
    (b) What is the slope at V_{DD} = 10 V?
    (c) What is the slope at V_{DD} = 15 V?

V_{DD}	I_D
1.00 V	1.50 mA
2.00 V	2.95 mA
3.00 V	4.20 mA
4.00 V	5.70 mA
5.00 V	7.25 mA
6.00 V	8.20 mA
7.00 V	9.30 mA
8.00 V	10.2 mA
9.00 V	10.8 mA
10.0 V	11.3 mA
11.0 V	11.5 mA
12.0 V	11.8 mA
13.0 V	11.9 mA
14.0 V	12.0 mA
15.0 V	12.0 mA

Chapter 3

Standards and Calibration of Instruments

OBJECTIVES

All measurements are based on a comparison of the quantity to be measured to a known quantity called a standard. It is not an exaggeration to say standards are the foundation of modern science, for without standards, measurements would be arbitrary and capricious. The advance of technology has increased the demand for more accuracy, wider range, and greater diversity in measurement standards.

Scientific and engineering work requires that measurements from various countries have a common definition using a common system of measurements. Prior to adoption of SI units, two measurement systems prevailed in the industrialized countries—the metric system and the customary, or English, system. Although the customary system had its origins in the length of a human foot and had other inconsistencies, it became precisely defined for standardization work in the United States and Britain. Although the customary measuring system is still used for many measurements, the scientific and engineering communities have universally adopted SI units.

When you complete this chapter, you should be able to

1. Specify the fundamental SI units and explain how derived units are based on these fundamental units.
2. List the metric prefixes and their meanings.
3. Describe the mission of the national and international organizations responsible for uniform measurement standards.
4. Explain the advantages of intrinsic standards and how they differ from prototype standards.
5. Compare international standards, primary standards, secondary standards, and working standards.
6. Describe standards for time and frequency, voltage, resistance, capacitance, inductance, temperature, and photometry.
7. Discuss the rationale for instrument calibration intervals and the benefit of record keeping.

HISTORICAL NOTE

The National Institute of Standards and Technology (NIST), formerly known as the National Bureau of Standards (NBS), was established in 1901, replacing the Office of Standard Weights and Measures. At the time, state laws for standard weights and measures were antiquated. Some laws were virtually unchanged since colonial times when states adopted standards of their own choosing. The NIST took over custody of basic standards for weights and measures but—more importantly—began a broad program of physical research to determine precise physical standards and construct new standards when necessary. Early in its history, it developed instruments to measure electrical current and other properties of electricity. It also contributed to the data in many areas other than fundamental measurement standards, including data on physical constants, properties of materials, and tests and procedures, and has contributed to national codes such as the National Electrical Safety Code.

3–1 UNITS OF MEASUREMENT

The measurement of a quantity is done with specified measurement units. Measurement **units** are the reference quantities that have been selected as a basis for comparison for most measurable quantities. The definitions for various units are modified and improved from time to time as necessary. The original "old" metric system was proposed by the French Academy of Science in 1790. In 1875, at a conference called by the French, the *Convention du Mètre* was signed by 18 countries to set up and maintain a body of weights and measures in Paris. Through a governing authority, the *Conférence Général des Poids et Mesures* (CGPM), delegates from member countries adopted necessary procedures for the improvement of the metric system. By international agreement, metric units have been adopted for use in all scientific and engineering work. The dominant system in use today is called the International System of Units (*Le Système International d'Unités,* abbreviated SI) which was defined at the Eleventh Session of the CGPM in 1960. The SI system was simplified from the original system and was based on six **fundamental units** (sometimes called base units) and two **supplementary units.** In 1971, the General Conference added a standard of quantity called the mole to the fundamental units. This addition was made because the number of atoms or molecules in a substance is frequently more relevant to chemists than the mass of substance. A mole is defined as the number of atoms in 12 grams of carbon 12. All other SI units can be formed by combining the fundamental and supplementary units so are called **derived units.** Table 3–1 lists the fundamental units and Table 3–2 lists the supplementary units. Definitions of the fundamental units are given in Table 3–3.

In addition to the fundamental and supplementary units, 27 derived units were adopted by the CGPM. Table 3–4 lists these 27 derived units and some additional useful electrical units. An important advantage of the derived units was that of coherence. Coherence means that derived units can be expressed directly in terms of products or quotients of fundamental units without numerical factors.

TABLE 3–1
SI fundamental units.

QUANTITY	BASE UNIT	ABBREVIATION
Length	meter	m
Mass	kilogram	kg
Time	second	s
Electric current	ampere	A
Temperature	kelvin	K
Luminous intensity	candela	cd
Amount of substance	mole	mol

TABLE 3–2
SI supplementary units.

QUANTITY	BASE UNIT	ABBREVIATION
Plane angle	radian	r
Solid angle	steradian	sr

For example, the unit of force is the newton. From physics, force is mass times acceleration. In fundamental SI units the newton is defined as the force that produces an acceleration of one meter per second squared to a mass of one kilogram. That is, $1\ N = 1\ kg \cdot m \cdot s^{-2}$. The definition does not include arbitrary numerical constants; hence it is coherent.

TABLE 3–3
Definitions of the fundamental units.

THE FUNDAMENTAL UNITS
Length (meter): The meter is the length of the path traveled by light in vacuum during a time interval of 1/(299,792,458) second (1983).
Mass (kilogram): The mass of the international prototype kilogram (1889).
Time (second): The second is the duration of 9,192,631,770 periods of the radiation corresponding to the transition between two hyperfine levels of the ground state of the cesium 133 atom (1967).
Electric current (ampere): The ampere is that constant current which, if maintained in two straight parallel conductors of infinite length, of negligible circular cross section, and placed 1 meter apart in a vacuum, would produce between these conductors a force equal to 2×10^{-7} newton per meter of length (1948).
Temperature (kelvin): The kelvin unit of thermodynamic temperature is the fraction 1/273.16 of the thermodynamic temperature of the triple point of water (1967).
Luminous intensity (candela): The candela is the luminous intensity, in a given direction, of a source that emits monochromatic radiation of frequency 540×10^{12} hertz and that has a radiant intensity in that direction of 1/683 watt per steradian (1979).
Amount of substance (mole): The mole is the amount of substance of a system that contains as many elementary entities as there are atoms in 0.012 kilogram of carbon 12 (1971).

TABLE 3–4
Derived units adopted by the CGPM.

QUANTITY	NAME OF UNIT	UNIT SYMBOL
Area	square meter	m^2
Volume	cubic meter	m^3
Frequency	hertz	Hz (s^{-1})
Density	kilogram per cubic meter	kg/m^3
Speed	meter per second	m/s
Angular velocity	radian per second	rad/s
Acceleration	meter per second squared	m/s^2
Angular acceleration	radian per second squared	rad/s^2
Force	newton	N ($kg \cdot m/s^2$)
Pressure	newton per square meter	N/m^2
Viscosity (dynamic)	newton second per square meter	$N \cdot s/m^2$
Viscosity (kinematic)	meter squared per second	m^2/s
Work, energy, quantity of heat	joule	J ($N \cdot m$)
Power	watt	W (J/S)
Quantity of electricity	coulomb	C ($A \cdot s$)
Electric tension, potential difference, electromotive force	volt	V (J/C)
Electric field strength	volt per meter	V/m
Resistance	ohm	Ω (V/A)
Capacitance	farad	F ($A \cdot s/V$)
Magnetic flux	weber	Wb ($V \cdot s$)
Inductance	henry	H ($V \cdot s/A$)
Magnetic flux density	tesla	T (Wb/m^2)
Magnetic field strength	ampere per meter	A/m
Magnetomotive force	gilbert	Gb (A)
Luminous flux	lumen	lm ($cd \cdot sr$)
Luminance	candela per square meter	cd/m^2
Illumination	lux	lx (lm/m^2)

ADDITIONAL USEFUL ELECTRICAL UNITS

QUANTITY	NAME OF UNIT	UNIT SYMBOL
Conductance	siemen	S
Gain/attenuation	decibel	dB

EXAMPLE 3–1 Lenz's law states

EQUATION 3–1

$$v = L\frac{di}{dt}$$

where v = instantaneous voltage, V
L = inductance, H
di = incremental change in current
dt = incremental change in time

Starting with Lenz's law and the definitions of basic units in Table 3–4, show that the unit for inductance, the henry (H), can be written using fundamental units as

$$\mathrm{H} = \frac{\mathrm{kg \cdot m^2}}{\mathrm{A^2 \cdot s^2}}$$

SOLUTION Rewriting Lenz's law:

$$L = v\frac{dt}{di}$$

Substituting the units for each quantity yields

$$\mathrm{H} = \left(\frac{\mathrm{J}}{\mathrm{C}}\right)\frac{\mathrm{s}}{\mathrm{A}} = \left(\frac{\mathrm{N \cdot m}}{\mathrm{A \cdot s}}\right)\frac{\mathrm{s}}{\mathrm{A}}$$

The newton, N, is a unit of force. It is shown in Table 3–4 as

$$\mathrm{N} = \frac{\mathrm{kg \cdot m}}{\mathrm{s^2}}$$

Therefore

$$\mathrm{H} = \left(\frac{\mathrm{kg \cdot m^2}}{\mathrm{A \cdot s^3}}\right)\frac{\mathrm{s}}{\mathrm{A}} = \frac{\mathrm{kg \cdot m^2}}{\mathrm{A^2 \cdot s^2}}$$

Except for the ampere,[1] all electrical units are derived units. For example, the unit for electrical resistance is the ohm, which can be defined in terms of the fundamental units for length and time. Derived units may be given names in terms of algebraic combinations of the fundamental units but are more frequently given unique names honoring people who worked in the field. The ohm was named in honor of Georg Simon Ohm (1787–1854), a German physicist who studied electricity.

[1] Note that the ampere is not a fully independent unit because it uses the second in its definition (coulomb/second).

TABLE 3–5
Names and abbreviations of metric prefixes.

MULTIPLYING FACTOR	PREFIX	SYMBOL
1,000,000,000,000 = 10^{12}	tera	T
1,000,000,000 = 10^{9}	giga	G
1,000,000 = 10^{6}	mega	M
1,000 = 10^{3}	kilo	k
100 = 10^{2}	hecto	h
10 = 10^{1}	deca	da
0.1 = 10^{-1}	deci	d
0.01 = 10^{-2}	centi	c
0.001 = 10^{-3}	milli	m
0.000001 = 10^{-6}	micro	μ
0.000000001 = 10^{-9}	nano	n
0.000000000001 = 10^{-12}	pico	p
0.000000000000001 = 10^{-15}	femto	f
0.000000000000000001 = 10^{-18}	atto	a

METRIC PREFIXES

To express very large and very small quantities, a prefix can be used in conjunction with the measurement unit. An important advantage of the metric system is that metric prefixes are based on the decimal system and stand for powers of ten. The names and abbreviations of the metric prefixes, together with the powers of ten they represent, are shown in Table 3–5. The prefixes can be used in conjunction with all measurement units; however, with electrical measurements, prefixes with an exponent that is a multiple of three are preferred. Numbers expressed with an exponent or metric prefix that is a multiple of three are said to be expressed in **engineering notation.** Engineering notation is used for marking many electronic components and is also an aid to computations.

For historical reasons, the kilogram is the only fundamental unit defined with a metric prefix. Names of decimal multiples and submultiples of the kilogram are formed by attaching prefixes to the word *gram*. As an example, it is *not* correct to write a *milligram* as a *microkilogram*.

3–2 STANDARDS ORGANIZATIONS

All measurements are based on comparison to some known quantity or reference. The reference is called a **standard.**[2] The earliest standards were for weights and measures. One very early standard, called the Egyptian royal cubit, was based on

[2] The word *standard* is also used to mean the documents and codes that control human endeavor. For example, if manufacturers agree to build a particular electronic interface to a standard, they are adopting a common basis for production. For this text, we are concerned only with measurement standards.

the length of the pharaoh's forearm. Such standards were devised primarily as a means of uniform dealing in commerce. When the metric system was adopted, the standard of length, the meter, was defined as one ten-millionth of the distance along the meridian line between the North Pole and the equator passing through Dunkirk, France, and Barcelona, Spain. The Dunkirk-Barcelona distance was surveyed by the French and the distance from the North Pole to the equator was extrapolated from the survey. A standard bar made of platinum-iridium was constructed as a standard of comparison based on the original definition. As measurements improved, there was a need to verify measurements independently. Before any new standard can be considered for adoption, it must be stable and reproducible. In 1927, the International Conference on Weights and Measures adopted a definition of length that could be reproduced in laboratories throughout the world. It was based on a specific number of wavelengths of red light from the emission of cadmium. Over the years, the definition has been modified to increase the degree of precision.

INTERNATIONAL STANDARDS ORGANIZATIONS

The need for a clear structure for international cooperation to assure the uniformity of measurements has long been recognized. The exchange of scientific information as well as international trade, communication, navigation, and a host of other activities require a high degree of cooperation between nations. Currently, the CGPM consists of nearly 50 member nations, including all the major industrialized nations, which have mutually agreed to accept the standards adopted by the CGPM. Every four years, delegates from the member nations meet to discuss and adopt improvements of the SI units. The general organization of the CGPM is shown in Figure 3–1.

Eighteen delegates from the CGPM are elected to serve on an overseeing group called the *Comité International des Poids et Mesures* (CIPM) that meets annually to take up matters that have been referred to it. CIPM members serve as chairpersons for various committees composed of international metrology experts. These committees, occasionally helped by working groups, examine advances in the field of physics and metrology that have a direct influence on standards. CIPM members also direct the activities of the laboratory arm of the CGPM known as the *Bureau International des Poids et Mesures* (BIPM). The BIPM maintains permanent laboratories in Sèvres, France (a suburb of Paris), and includes a staff of about 60 people. The principal task of the BIPM is to ensure worldwide uniformity of physical measurements. Activities of the BIPM include research for improving measurement standards and techniques, maintaining reference standards for various physical quantities with good long-term stability, and participating in international comparisons and carrying out calibrations. In addition, the BIPM serves as an international authority on matters concerning scientific metrology, including verification of national standards when required. The CIPM and the BIPM work closely with many other international organizations to provide a common base, both as a formal agreement and as a physical laboratory, for the world's complex measurement network.

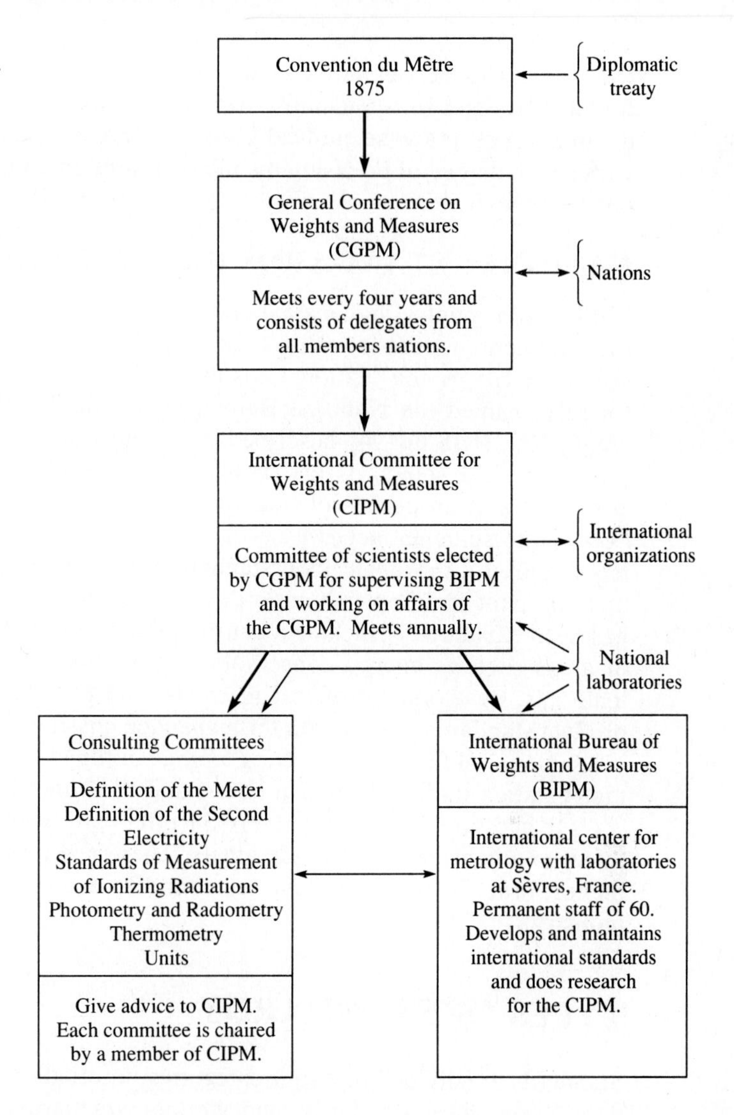

FIGURE 3–1
International organization of the *Convention du Mètre*.

The search for more accurate ways to measure physical quantities is a never-ending pursuit. Standards continue to be redefined through scientific advances in the field of metrology, particularly by the BIPM and the various national laboratories. The goal for standardization is to find standards based on constants in nature, preferably related to atomic behavior, that can be reproduced in technical facilities anywhere in the world. All the fundamental SI units have been defined in such terms except for the unit for mass. The kilogram is still defined by a prototype quantity of platinum-iridium known as the international kilogram. Work at the BIPM and national laboratories has lead to various improvements in the accepted definitions of the fundamental units. For example, the definition of the meter was changed from a prototype standard to a definition based on a certain orange spectral line of light in 1960. Then, at the seventeenth conference of the CGPM in October 1983, the definition of the meter was redefined as the length of path traveled by light in a vacuum during a time interval of 1/(299,792,458) s. This definition is clear and scientifically sound. More importantly, it can be reproduced with a high degree of precision by any of several methods in laboratories throughout the world.

NATIONAL STANDARDS ORGANIZATIONS

Within each country, national laboratories provide a central authority for uniform measurement standards within their respective countries. In the United States, that authority is the National Institute of Standards and Technology (NIST), formerly named the National Bureau of Standards (NBS). The legislation that created the NBS in 1901 described its intended function as the custodian of national physical standards and assigned it the mission of determining physical constants and properties of materials as well as measurements that cannot be obtained to sufficient accuracy elsewhere. The need for ever-increasing accuracy has required a thorough understanding of the underlying physical principles of measurement and implementation of new technology. The mission of the NIST has been expanded to include a considerable amount of related scientific research in areas such as energy conservation, environmental protection, utilization of materials, development of test methods, and coordination of activities such as product standards. Along with this expanded mission, the NIST publishes a significant amount of literature covering a range of topics, including product standards, mathematics studies, technical notes, and consumer information.

In addition to research, the NIST offers physical measurement services (for a fee) to organizations and individuals. These services include calibrations of the highest order available, special tests, and measurement assurance programs for industry.

3–3 TYPES OF STANDARDS

Standards are divided into four categories: international standards, primary standards, secondary standards, and working standards. All standards need to be rigorously traceable through documentation to accepted international standards or definitions representing the fundamental units. Traceability implies that local reference standards have been compared to higher-echelon standards through an unbroken chain of comparisons to national standards maintained by the NIST. Other important attributes of all standards include long-term stability, accuracy, and insensitivity to environmental conditions.

INTERNATIONAL STANDARDS

Except for the kilogram, all the fundamental units are defined in terms of nature. **Intrinsic standards** are those that can be realized directly from their definitions rather than comparing them to an artifact maintained at a laboratory. The advantage to intrinsic standards is that they can be constituted in specialized laboratories throughout the world, thus ensuring that the units remain stable, reproducible, and indestructible. Intrinsic (or "absolute") standards are tied to the basic laws of physics rather than an artifact. Rules for specifying the practical realization of various intrinsic standards were adopted by the CIPM in 1983. Advances in making standards based on their physical definitions has recently improved, particularly in electrical standards, and will ultimately improve the measurement accuracy at all levels.

When a physical object is used as a standard, as with the kilogram, it is called a **prototype standard**. Although it would be useful to have a mass standard that is defined by nature, no one has been able to propose a standard with the required precision. It is conceivable that, in the future, actual counting of atoms in a specified crystalline structure could be employed as a reproducible mass standard. The international kilogram is maintained by the BIPM along with a series of working prototype standards of the kilogram that are used for comparison with national prototype standards. The surest method of protecting the invariability of the international prototype kilogram is to make the least possible use of the international kilogram. The BIPM has only rarely compared the international kilogram with its own working prototype standards; a series of measurements begun in 1988 is only the third use of the international prototype in nearly 100 years.

The BIPM maintains a number of other standards for fundamental and derived units that can be used as calibration sources for the national laboratories. These standards are used for routine calibrations as needed by national laboratories for international comparisons. They are not available for use by the public.

NATIONAL PRIMARY STANDARDS

A **primary standard** is one that does not require any other reference for calibration. National Reference Standards are primary standards that are maintained and controlled by a national measurements laboratory such as the National Institute of Standards and Technology for the United States or the National Physical Laboratory for England. For example, a primary standard for time is generated from a cesium atomic beam–controlled oscillator. NIST uses such a device as the national reference standard.

The U.S. copy of the international kilogram (number 20) is kept in a vault at the NIST laboratory at Gaithersburg, Maryland. It is considered a primary standard for the United States but in fact is a very accurate copy of the international kilogram (number 1) retained by the BIPM in Sèvres, France. The U.S. prototype is used no more than once a year to compare with high-precision copies that are accurate to 1 part in 100 million. Other primary reference standards of the fundamental units can be reproduced from their definition in nature.

SECONDARY STANDARDS

The difference between a primary standard and a secondary standard is that a primary standard does not require any other reference for calibration, but a **sec-**

ondary standard must be compared periodically to a primary standard, depending on the accuracy required. Secondary standards are often used as **transfer standards,** to move a unit of measurement from a primary standard to one that can be used in a laboratory as a working standard. Secondary standards, with calibrations directly traceable to NIST, are retained by many laboratories and industries. An example of a secondary standard is the rubidium frequency standard. Although it is typically more than 100 times more accurate than an ordinary quartz crystal oscillator, the rubidium gas cell is subject to very small degrees of drift and must be periodically tested against a primary standard.

WORKING STANDARDS

Typical electronic measuring equipment cannot approach the precision of primary standards. Except for the most exacting scientific work, it is seldom necessary to calibrate electronic equipment to these standards. A **working standard** is a device used to maintain a unit of measurement for routine calibration and certification work on test equipment. Working standards require periodic calibration and certification to assure that they are accurate. They are normally not used for routine measurements in an organization but instead are reserved for calibration service. Calibration standards must have the attributes of accuracy and stability. They must be used only for the measurement range for which they are intended.

A common industrial working standard for length is the gage block. A gage block is a precision machined block with highly polished parallel faces used to check the reading of a micrometer. If the micrometer reading is a different value than that marked on the gage block, a recalibration is called for. For electrical measurements, common working standards include quartz crystal oscillators, standard resistors, capacitors, inductors, Weston cells, solid-state voltage references, and time comparators.

Working standards form the basis of the quality of measurement work for an organization. A record should be kept for all working standards. This record includes calibrations with higher-echelon standards, comparisons made with other working standards, adjustments or repairs that could give an indication of possible deterioration of the standard, and other pertinent information. The calibration interval depends on the delicacy of the standard, required accuracy, frequency of use, and the experience level of personnel who use the standard.

3–4 FREQUENCY AND TIME STANDARDS

Throughout history, time has been measured by observation of some periodic event. For many centuries, it was based on the earth's rotation, and the second was defined as 1/86,400 of a mean solar day, a definition that stood into the twentieth century. Unfortunately, the earth's rotation can vary as much as 3 s in one year—an amount that can easily be detected by precise time standards.

There is no fundamental difference between a time standard and a frequency standard, since both are derived from the second. There are natural oscillators that occur as a result of atomic transitions from one energy state to another that are observable as spectral lines. These extremely high frequencies were first used by the NIST in 1948 when a clock based on the transition observed in ammonia molecules was used to construct the first atomic clock. Atomic clocks use an atomic resonance to stabilize a quartz crystal oscillator. In 1967, the Thirteenth General Conference adopted a particularly suitable hyperfine transition in the cesium 133 isotope to define the second. Currently, even more accurate cesium standards have been constructed, with accuracies of better than 1 part in 10^{13}. These precisions are among the best of any type of standard.

TIME STANDARDS

There are two aspects to the measurement of time. One is the need to establish a time interval that defines the second; the other is the requirement to establish the time of day (epoch time). Epoch time is based on very stable atomic clocks and is called coordinated universal time (UTC). It is coordinated with time signals broadcast from stations around the world and differs from local time by an exact integer number of hours. UTC is actually the local time near Greenwich, England, and is equivalent to the familiar Greenwich Mean Time, which is 5 h later than Eastern Standard Time.

A second time scale, called universal time-1 (UT_1) (also called astronomical time) is a nonuniform time scale based on the earth's rotation and is inferred from astronomical observations. UT_1 and UTC are kept in synchronism by making time-step adjustments of exactly 1 second into UTC time whenever needed as determined by the International Time Bureau. (This is normally about once or twice a year and may be either the addition or deletion of 1 s.) A number of time and frequency broadcast stations transmit UTC throughout the world. In the United States, UTC is broadcast from NIST radio stations WWV and WWVB in Boulder, Colorado, WWVH in Kekaha, Hawaii, and the Geostationary Operational Environmental Satellite (GEOS) system, which uses two operational satellites of the National Oceanic and Atmospheric Administration (NOAA) operating at 468.8 MHz. As a service to navigators and others using UT_1, an audio code is included in broadcasts from WWV and GEOS to provide the current time difference between UTC and UT_1.

The NIST radio stations give time signals by voice announcements every minute with audible "ticks" broadcast every second (except the 29th and 59th second of every minute). In addition, time signals are encoded on a 100 Hz subcarrier in BCD (binary-coded decimal) code. Time signals can generally be received with an accuracy of about 1 ms. Time signals are also disseminated by the Naval Observatory in Washington, D.C., via telephone. These signals are also available on COMP U SERVE for personal computers. The signals are sent as standard telecommunication signals using frequency-shift keying (FSK). It is possible to measure and correct for loop delay to produce time accuracy of ±3 ms for users such as broadcast stations, industrial users, and others who need precise time. Figure 3–2 shows a master clock system that can be programmed to dial a time service automatically for verification.

FIGURE 3–2
A master clock system for broadcast stations and other applications requiring precision time. This system can automatically dial a standard time service and calibrate itself (courtesy of Leitch Video).

In addition to UTC time, the NIST stations transmit other information that can be used for calibration and testing by government and industry. All the stations transmit highly accurate radio frequency carriers (within 1 part in 10^{11}) on several shortwave bands, although the received frequency can be different due to changes in propagating conditions. WWV and WWVH broadcast on 2.5, 5, 10, and 15 MHz, and WWV is also on 20 MHz. Both stations also transmit precise audio frequencies (440, 500, and 600 Hz), musical pitches, time intervals, time signals, and UT_1 corrections as well as official announcements, which include marine storm warnings. WWVB transmits on very low frequency (VLF) at 60 kHz and broadcasts only time code. Transmissions from WWV and WWVH are also available via telephone lines without the radio-frequency carrier. Figure 3–3 shows the hourly broadcast formats of WWV and WWVH.[3]

Although the transmitted frequencies of WWV and WWVB are held to within 1 part in 10^{11}, the received error can be as high as 1 part in 10^7 for WWV (1 cycle in 10 MHz) when averaged for 10 min. Propagation effects are smaller on the received signals from WWVB because of the lower transmission frequency than from WWV and WWVH. Frequency comparisons with WWVB can be as high as 1 part in 10^{11}, if appropriate long-term averaging techniques are employed. For this reason, most master clock systems rely on either WWVB or the GEOS system for a time standard.

Time and frequency comparisons between stations at various points in the world have been made using atomic clocks that have been carried between locations and by a new satellite timing system called the global positioning system (GPS), based on the common observation of signals from a GPS satellite. Time differences between world timing centers in Boulder, Colorado; Braunschweig, Germany; and Tokyo, Japan, have been measured at less than 3 μs during comparisons over a number of months. The GPS allows the frequency difference between the best standards in the world to be compared to within a few parts in 10^{14} using several days for acquiring the data. A drawback to GPS is that it does not provide 24 h coverage and does not have complete coverage for a number of areas.

FREQUENCY STANDARDS

As mentioned previously, radio signals from the NIST high-frequency stations are transmitted with very high quality time and frequency standards; however, the received frequency of these stations can be in error due to atmospheric effects. Integrating over time improves the accuracy of the received high-frequency radio-frequency standards. A frequency standard that avoids these problems receives low-frequency (100 kHz) LORAN-C signals that are propagated by ground waves. LORAN (an acronym for long-range radio navigation) uses cesium atomic frequency reference oscillators that are traceable to NIST. Each LORAN-C transmitter maintains three cesium atomic standards that have a long-term stability of 1 part in 10^{12}. It can be received throughout the northern hemisphere (except portions of central Russia and parts of India). A LORAN-C receiver, designed as a frequency standard, is shown in Figure 3–4. The receiver locks onto the LORAN-C signal, enabling the receiver to have the same long term frequency stability as the LORAN-C transmitter. The output of the receiver is a precise frequency standard that can be selected from 0.01 Hz to 10 MHz.

Another frequency standard is transmitted by network television stations as part of color video signals. The standard is a 3.58 MHz[4] "color-burst" signal

[3] Further information about the services can be obtained from Special Publications #432, *NBS Time and Frequency Dissemination Services*, and #559, *Time and Frequency Users Manual*, available from the U.S. Government Printing Office.

[4] The frequency is actually 3.579545 MHz.

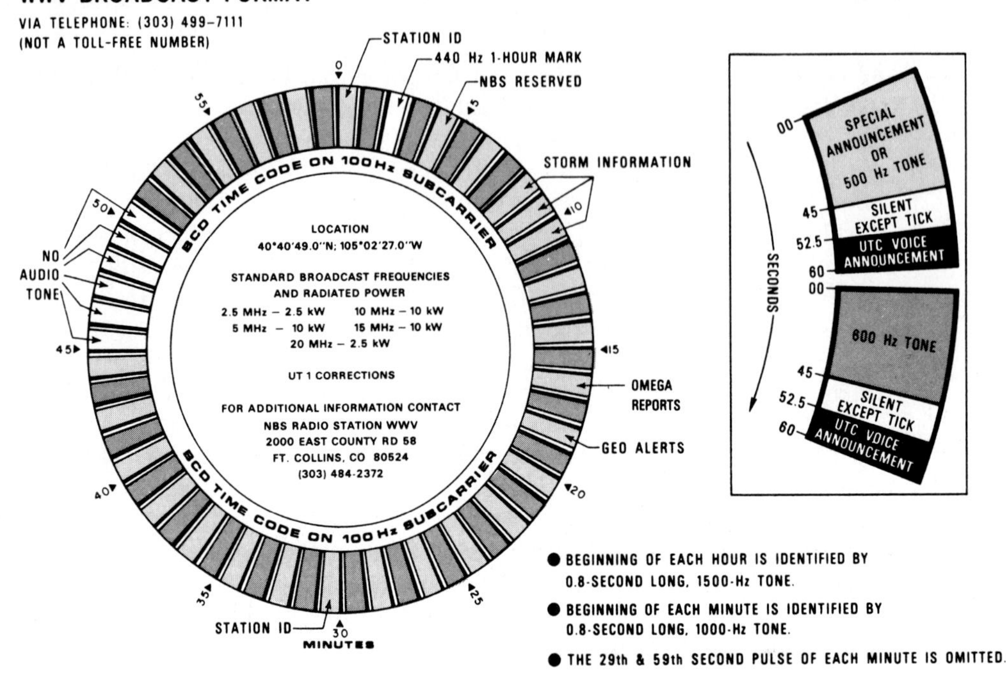

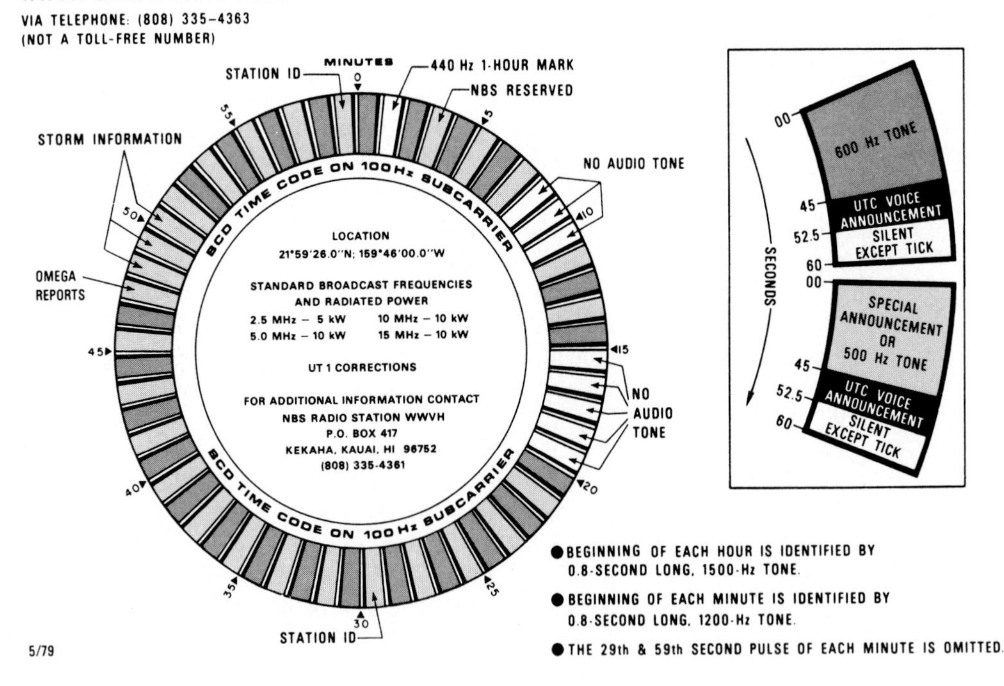

FIGURE 3–3
The hourly broadcast schedules of WWV and WWVH (courtesy of NIST).

FIGURE 3–4
The Stanford Research Systems FS-700 LORAN-C frequency standard. This instrument can provide precision frequencies with a long-term stability of a cesium clock (1 part in 10^{12}) (courtesy of Stanford Research Systems).

generated by national network broadcasters using an extremely stable rubidium or cesium oscillator. It is sent in the television signal after the horizontal synchronization pulse to provide a phase-lock reference for the 3.58 MHz crystal oscillator contained in all color television receivers. This means that the oscillator in any television set is locked to the network frequency standard when the set is receiving a network show. For network programs originating from New York or Los Angeles, the NIST monitors the signal and publishes monthly bulletins of the difference between the network oscillators and the NIST frequency standard. Thus, for a modest investment, a user can calibrate an oscillator against a standard that is traceable to the NIST frequency standard.

Several techniques are available for using the 3.58 MHz oscillator. The most economical is a color-bar comparator. The comparator uses a test oscillator as an input signal to synthesize a 3.58 MHz output signal using a phase-locked loop. For example, the input test oscillator can be set to 1 MHz to synthesize the 3.58 MHz output. This synthesized 3.58 MHz signal is then gated and sent to the TV chroma circuit, where it is displayed on the television screen as a vertical color bar. The color bar moves across the screen and changes color at a rate that depends on the difference in frequency between the test oscillator and the network's 3.58 MHz signal. By adjusting the test oscillator while observing the color bar, the test oscillator can be calibrated with extremely high accuracy. A color photo of the signal seen on a modified TV set is shown on the back cover.

Another technique involving the 3.58 MHz signal uses a digital offset computer to compare the user's oscillator with the TV color signal. The display on the TV screen is a digital number representing the frequency difference between the user's oscillator and the TV color signal.

3–5 VOLTAGE AND RESISTANCE STANDARDS

Although the fundamental SI electrical unit is the ampere, a current standard is difficult to realize and maintain in a laboratory. In fact, the absolute ampere can be accurately maintained for only a few minutes in the laboratory. Consequently, the volt and ohm are the units commonly used as departure points for electrical units. The ohm and volt are specified by the SI definition in terms of the ampere; however, it is difficult to realize the definition directly. For this reason, the ohm and volt are maintained by practical laboratory representations.

Recent technological advances have made practical new standards of voltage and resistance based on fundamental physical measurements instead of material objects. The impact of these developments is significant and soon may make electrical standards simpler and less expensive. The new advances are described in the next paragraphs.

VOLTAGE STANDARDS

Prior to the early 1970s, the national laboratories maintained the volt with banks of standard cells composed of two dissimilar electrodes immersed in an electrolytic solution. The Weston cell, a stable wet cell composed of cadmium sulfate electrolyte, generates a voltage of slightly greater than 1 V and is basically the only standard cell still in use. Weston cells are designed for long-term stability and precise voltage comparisons. They are not intended to supply current to circuits. Although Weston cells are very stable, they do drift with time, and there are further difficulties with transporting and maintaining them, including careful control of temperature. Figure 3–5 illustrates a saturated Weston cell.

In 1962, Brian Josephson, a graduate student at the University of Cambridge, predicted an effect that bears his name. He predicted that the tunneling of electron pairs through an extremely thin insulator (1 to 3 nm) sandwiched between two superconductors[5] would allow the dc current to continue through the insulator with no resistance, just as in the superconducting material. Another aspect of the Josephson effect is that a very high frequency ac voltage appears across the insulator. Conversely, if a Josephson junction is exposed to microwave radiation, steps appear in the current-voltage curve at certain discrete voltage levels. The voltage levels are dependent on the frequency and two fundamental physical constants: e, the charge on an electron, and h, Planck's constant. Thus a Josephson junction could be used to convert a microwave frequency into a voltage. The constant of proportionality between frequency and voltage is given by $2e/h$.

[5] A superconductor is a material that loses all resistance to electric current below some threshold temperature.

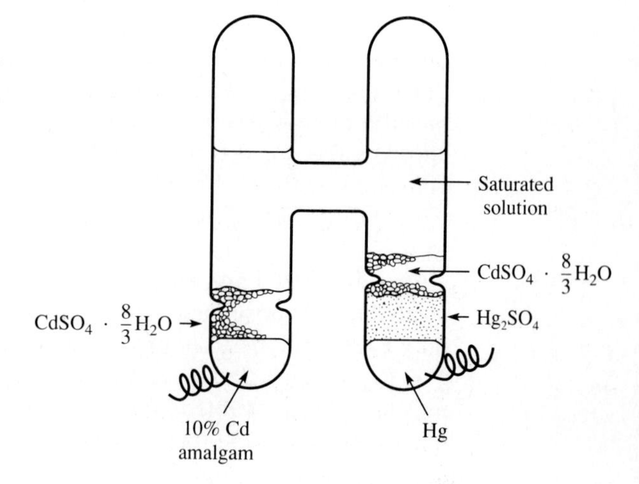

FIGURE 3–5
Cross section of a saturated cadmium standard cell.

The voltage of the nth level of a Josephson junction voltage standard is given by

EQUATION 3–2

$$V_n = \left(\frac{h}{2e}\right)nf$$

where V_n = voltage of the nth level, V
n = step number
e = the charge on an electron, 1.602×10^{-19} C
h = Planck's constant, 6.626×10^{-34} J/Hz
f = frequency, Hz

EXAMPLE 3–2 Compute the voltage of the 250th step of a Josephson junction that is radiated with a frequency of 96 GHz.

SOLUTION

$$V_{250} = \left(\frac{6.626 \times 10^{-34}\ \text{J/Hz}}{2 \times 1.602 \times 10^{-19}\ \text{C}}\right)(250)(96\ \text{GHz}) = 49.6\ \text{mV}$$

The voltage levels from a single Josephson junction are very small and hence are difficult to measure with high precision. By using a step such as the 250th step and connecting junctions in series, the voltage level can be increased. Josephson junctions can then be used to check on the stability of Weston cells, which are used as a secondary standard to maintain reference voltage measurements using Josephson junctions. The BIPM (since 1976) and the national standards laboratories of most countries have used this technique to define the national standard of voltage and maintain it constant with time. The advantage of the Josephson junction method is that it can be made more precise, has no decay, as in a standard cell, and has significantly less electrical noise. Prior to January 1, 1990, laboratories in various countries used slightly different values of the constant $2e/h$, making international voltage comparisons inconsistent by a few parts per million. After that date, the differences were resolved through adjustments made by CIPM based on detailed analysis of the experimental evidence. The internationally accepted value for the constant $2e/h$ is 483,597.9 GHz/V, which is believed to be consistent with the SI definition to within 0.4 ppm.

At the NIST, many Josephson junctions have been connected in series to form a 1 V standard, which, since February 1987, has been substituted for Weston cells in industrial and military standards laboratories. The NIST has also demonstrated a 10 V Josephson junction standard composed of 14,184 junctions, which may eventually be used as a commercial voltage standard.

Secondary standards for use by instrument manufacturers and other companies must be traceable to national standards. Typically, saturated cells have been used as secondary voltage standards kept in temperature- and humidity-controlled rooms. Solid-state standards are replacing saturated cells, and NIST is offering a certification service for the 10 V output of a solid-state reference.

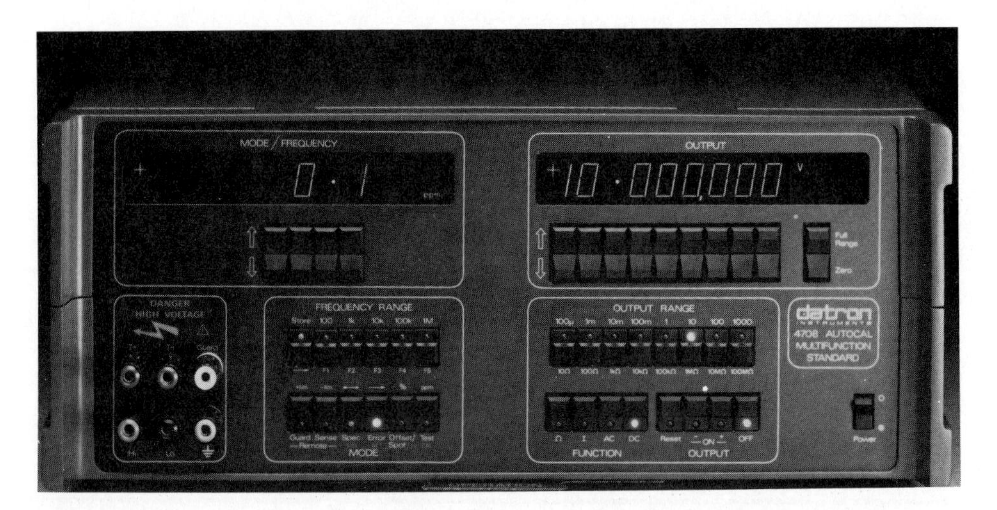

FIGURE 3–6
Datron 4708 Autocal Multifunction Standard. The Datron 4708 is a programmable calibrator that can be used as a working standard for dc and ac voltage, dc and ac current, and resistance. It is capable of calibrating 7½- and 8½-digit DMMs. Key specifications are

Total 1 y uncertainty ±5°C; uncertainties to better than 4.7 ppm (Vdc) and 40 ppm (Vac).
Fully variable Vac from 10 Hz to 1 MHz, providing 1100 V at 33 kHz or 750 V at 100 kHz.
IEEE-488 compatible, making it capable of operating in computer-controlled systems.

(Courtesy of Datron.)

Working voltage standards include specially designed calibration instruments that provide high-accuracy dc and ac voltages as well as internal standard resistors. Figure 3–6 illustrates a multifunctional calibrator used to calibrate digital voltmeters (DVMs) and digital multimeters (DMMs). It can serve as a working calibration station for dc and ac voltage as well as resistance. For less stringent applications, an accurate DVM or DMM can serve as the working standard for an organization.

The internal reference voltage within a voltage calibrator, DVM, or DMM used as a dc working standard must contain a stable and accurate dc reference. The reference voltage is usually derived from a special integrated circuit (IC) designed specifically for the purpose of providing a dc reference voltage. The reference IC has an internal zener diode (described in Section 4–2) as a voltage standard. In addition the reference IC contains temperature-compensating networks and is designed for low noise outputs and fast recovery after an off period. The reference IC requires a burn-in period to "age" the components and increase the stability. In high-precision applications, temperature stabilization of the reference IC itself is required. Using special selection and other techniques, the temperature coefficient (TC) of a reference IC can be made as low as 1 ppm/°C.

AC voltage standards are less accurate than dc standards. The traditional method for making high accuracy measurements of ac voltages is to convert the ac to dc and compare the result to dc standards. The most widely employed conversion technique uses a thermoresistive element through which the ac to be measured flows. A sensing element measures the temperature of the wire with high precision. Thermoelements have been constructed which can convert with an accuracy of 10 ppm.

RESISTANCE STANDARDS

In 1980, Klaus von Klitzing of the Max Plank Institute in Grenoble observed a startling new phenomenon, for which he won the 1985 Nobel prize. The effect was a new aspect of a phenomenon known for many years as the Hall effect. The **Hall effect** was first observed by Edwin Hall in 1879 when he noticed that a conductor in the presence of a perpendicular magnetic field produced a voltage drop perpendicular to the flow of current. A diagram of the Hall effect is shown in Figure 3–7. The voltage measured perpendicular to the current is called the Hall voltage. Dividing the Hall voltage by the current in the conductor gives a resistance called the Hall resistance.

Klaus von Klitzing was working with semiconductors cooled to temperatures near absolute zero. He discovered that when very high magnetic fields were present, the Hall resistance for an extremely thin flat-plate semiconductor exhibited discrete levels of resistance. The Hall voltage and resistance did not change linearly when the magnetic field increased but rather exhibited regions of constant resistance termed Hall plateaus. The effect is called the **quantum Hall effect** (QHE) because of the discrete resistance levels. Remarkably, the Hall resistance can be obtained by multiplying an integer number by Planck's constant and divid-

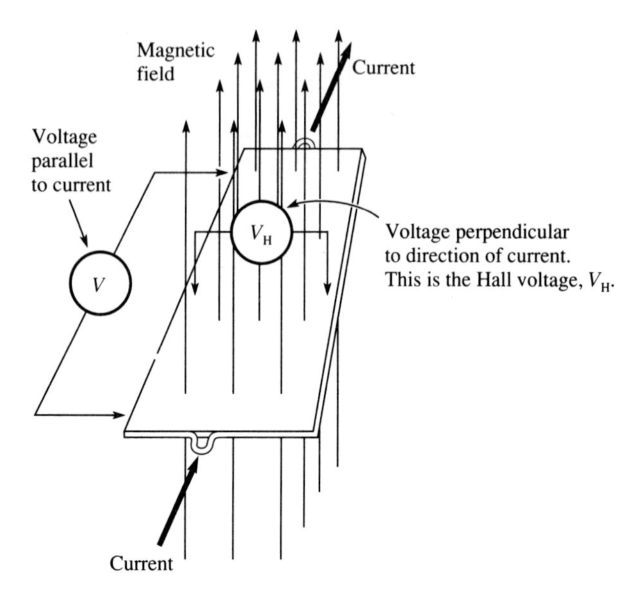

FIGURE 3–7
Hall effect. When a current flows in a flat plate situated in a perpendicular magnetic field, a voltage is induced perpendicular to the current and the magnetic field.

ing by the square of the charge on an electron. That is

EQUATION 3–3

$$R_H = \left(\frac{h}{e^2}\right)n$$

where R_H = Hall resistance
h = Planck's constant, 6.626×10^{-34} J/Hz
e = the charge on an electron, 1.602×10^{-19} C
n = an integer

In the past, most national laboratories represented the ohm with precision wirewound resistors; however, new definitions of the ohm, based on the QHE, were made on January 1, 1990. The uncertainty is 0.2 ppm from the SI definition. The ohm has been maintained by the BIPM, the NIST, and other national laboratories using a group of 1 Ω resistors whose mean value is known by comparing the impedance of a calculatable capacitor (which is known very accurately) to the resistance of the resistors. The comparison is done with a complex series of bridges. The new QHE standard provides an invariant standard of resistance that is highly uniform and reproducible at laboratories throughout the world.

There are different types of working standard resistors, depending on whether they are designed for use as dc standard resistors or for ac use. Standard resistors designed for dc and low audio frequencies are generally wirewound types with a low temperature coefficient and contained in an enclosure to avoid changes in humidity. Standard resistors for ac work need to be designed to minimize reactance losses as frequency is increased. The observed error due to reactance losses is greatest in higher-value resistors as the frequency is increased. For calibrating resistance-measuring devices, precision decade resistors are convenient. An example of a precision decade resistor is shown in Figure 3–8.

FIGURE 3–8
Precision decade resistor (courtesy of GenRad).

3–6 CAPACITANCE AND INDUCTANCE STANDARDS

CAPACITANCE STANDARDS

The NIST primary standard for capacitance is a specially constructed 1 pF calculatable capacitor based on theoretical work by A. M. Thompson and D. G. Lampard of the Australian National Measurements Laboratory (NML). The value of the calculatable capacitor can be computed from the length and the permittivity of free space, which in turn depends on the square of the speed of light. The practical realization of the calculatable capacitor consists of four equal cylindrical rods with their axes parallel and produces a capacitance of about 2 pF per meter. NIST has constructed five 10 pF calculatable capacitors, which have been calibrated against the primary standard. The accuracy is known to within 0.02 ppm, the highest resolution of any electrical unit. The average capacitance of the group is used as the basis for the U.S. legal farad.

The usefulness of such a tiny value of capacitance would be extremely limited were it not for a method of comparing it with larger practical values of capacitance. The method that was developed uses a transformer comparator bridge developed at NML and NIST. With a transformer comparator bridge, capacitors 100,000,000 times larger than the reference can be compared without a significant loss of accuracy. This is comparable to measuring the circumference of the earth using a 1 ft ruler without making a detectable error!

Working standard capacitors are available commercially with accuracies and long-term stabilities on the order of a few parts per million. Working standard capacitors include fixed precision capacitors with values ranging from 10 pF to 1 μF, precision variable air capacitors, and decade capacitors. Smaller-value working standards are usually constructed using air dielectric capacitors in a sealed environment. Larger-value standards are typically fixed silver-mica capacitors in a sealed case. Fixed working standard capacitors can have accura-

FIGURE 3–9
Reference standard capacitor (courtesy of GenRad).

FIGURE 3–10
Standard inductor (courtesy of GenRad).

cies of better than ±0.01%. Standard capacitors are designed for extremely low random fluctuations in value. An example of a standard capacitor is shown in Figure 3–9.

INDUCTANCE STANDARDS

In theory, a calculatable inductor could be constructed to maintain the value of inductance; however, NIST has found it unnecessary to maintain a standard inductor, because improvements in the calculatable capacitor gave it a decided advantage over a calculatable inductor. NIST calibrates precision inductors using special bridges that compare inductance to reference standards for capacitance and resistance. The bridge method can be used for values ranging from 1 μH to 10 H at audio frequencies.

Working standard inductors are available commercially in sealed containers with values ranging from about 100 μH to 10 H and accuracies of ±0.1%. An example of a commercial standard inductor is shown in Figure 3–10.

3–7 TEMPERATURE AND PHOTOMETRIC STANDARDS

TEMPERATURE STANDARDS

Temperature is a measure of the average kinetic energy of the particles of a substance. A gas thermometer is a good way to think of the problem of creating a temperature scale. Imagine a gas in a closed container at a temperature of T_A. Further, let's assume the gas is *ideal*—meaning it is made up of particles with no mass and it never changes states. If the temperature is lowered to T_B, the gas pressure will decrease. In fact, if we plot the temperature versus the pressure for our ideal gas, we find a linear relationship between pressure and temperature. This relationship, known as the ideal gas law, is given as

EQUATION 3–4

$$PV = nRT$$

where P = pressure, A
V = volume, l
n = number of moles of gas
R = the universal gas constant
T = temperature, K

For each degree of temperature loss, we find that the pressure decreases a certain amount. As this continues, there is a point where the pressure of our ideal gas drops to 0. This implies that no colder temperature can exist and we have reached a point where atoms and molecules lose all their kinetic energy. This point is said to be at a temperature of *absolute zero*. Figure 3–11 illustrates this behavior for an ideal gas.

The imaginary gas thermometer just described is not practical for constructing a high-precision thermometer using a real gas; however, it illustrates the idea

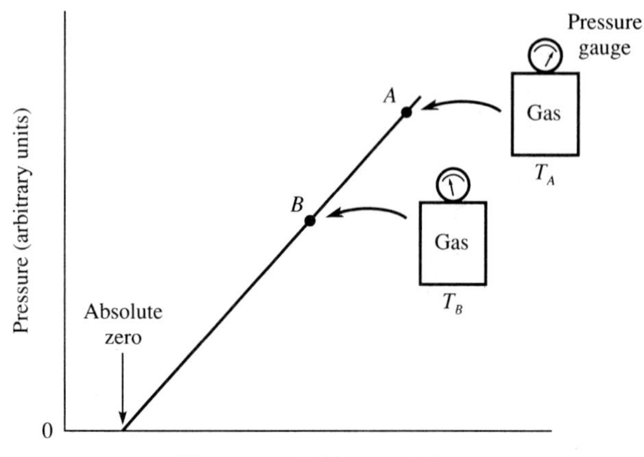

FIGURE 3–11
Pressure plotted against temperature for a gas at constant volume.

behind a temperature scale. The fundamental SI unit of temperature is the kelvin. It is an absolute scale, meaning there are no negative temperatures and the coldest possible temperature is 0 K. Temperatures on this scale are defined by the triple point of water, which is fixed at exactly 273.16 K. (This is the temperature of ice, water, and water vapor in equilibrium.) The problem for measurement of temperature is to set some practical fixed temperature points that can be used to provide reference temperatures for calibration of temperature scales. The SI temperature scale is defined by 11 reproducible temperature points (called "defining fixed points"), moving from the triple point of hydrogen (13.81 K) to the freezing point of gold (1337.57 K). The platinum resistance thermometer, a platinum 10% rhodium/platinum thermocouple, and the monochromatic optical pyrometer are standard instruments used to interpolate between the fixed points. Work at the BIPM and cooperating national laboratories resulted in a change of the international temperature scale to a more accurate, reproducible, and simpler scale. The new scale was adopted on January 1, 1990, and is known as the International Temperature Scale of 1990.

PHOTOMETRIC STANDARDS

Photometry deals with the measurement of visible radiated energy. The standard unit is the candela, which has roots in an earlier unit, the standard candle. The practical realization of a standard of luminous intensity was difficult because of uncertainty associated with measuring the temperature of a radiating cavity. As a result, the definition of the fundamental unit has undergone several changes. The 1979 definition, adopted by the Sixteenth CGPM, is given in Table 3–3. Luminous-intensity standard lamps are calibrated at either a set current or a specified color temperature. The uncertainty is approximately 2% relative to the SI definition of luminous intensity.

As part of photometric standards, the NIST helped develop standards for testing optical fibers. The tests measure the loss or bandwidth (or both) of optical fibers—two parameters of great interest in optical-fiber systems. Other photometric related standards are maintained for optical filters, atomic and plasma standards, and laser power and energy measurements.

3–8 CALIBRATION

Calibration refers to the comparison of a measurement instrument to a standard or other instrument of known accuracy in order to bring the instrument into substantial agreement with an established standard. The established standard is normally of at least four times greater accuracy than the instrument being calibrated or the average of multiple standards if the four-times criterion is not feasible. To say that an instrument is calibrated means that it indicates measurements within specified limits of error for that instrument. To calibrate an instrument, the person doing the calibration should consider the limits of accuracy of the instrument to be calibrated, the methods that will be used to ensure that the required accuracy can be obtained, and the accuracy of the standard or instrument used as a reference. The accuracy of an instrument is dependent on three major contributing factors:

1. The time since it was last calibrated
2. The difference between the calibration temperature and the operating temperature
3. The uncertainty of the standard used for calibration

In most organizations, calibration is the responsibility of a separate group. That group is responsible for maintaining working standards, keeping records of instruments for periodic calibration and certification, and being knowledgeable of the latest calibration procedures. The calibration facility is usually located in a separate standards laboratory, which is equipped to make the highest-accuracy measurements needed by the organization. The standards laboratory should have environmental controls to assure consistency because environmental factors can affect the accuracy of certain standards and the performance of instruments. Environmental factors that are controlled usually include temperature, humidity, vibration, dust, radio-frequency interference, grounding, and line voltage regulation.

In general, instruments should be recalibrated at specified intervals or after any repair work to ensure they are within accuracy specifications. Many instruments have accuracy specifications based on hours of operation—typically 1000 to 2000 hours. The recalibration interval is dependent on various factors, including the accuracy and confidence level required, the average daily use of the instrument, and the predicted or actual long-term drift of the instrument. Depending on these factors, the recalibration interval may be as short as 30 days for stringent requirements; more typically it is 6 months to one year or more. Other instruments, such as passive resistance boxes, may have longer recalibration intervals—frequently up to two years.

The long-term drift of instruments depends on the class of instrument (oscilloscope, digital instruments, meters), the specific model, and the type of use it receives. Advances in the manufacture of precision components having very low long-term drift and improved stable reference standards make it possible to predict instrument long-term drift, allowing extended recalibration intervals. Some instruments rarely require adjustment—only verification. The recommended calibration interval can be extended on instruments with demonstrated long-term stability; however, the documentation required to extend the interval may make it economically infeasible. A performance check of such instruments may be all that is required to assure that they are still within specifications.

The calibration procedure and instrument performance checks are generally specified by the instrument manufacturer in the manuals that are provided with the instrument. Sometimes calibration consists only of a performance check of the instrument to assure that it is operating within specified accuracy limits. Normally, if an instrument is operating well within its specified accuracy limits, adjustments to optimize its performance are unnecessary and should not be made. Frequent unnecessary adjustments may also interfere with the records for an instrument and accelerate progressive drift in values. An exception to this procedure can be made if the parameter is very near the tolerance limit and there is reason to believe it may be out of tolerance before the next calibration. The decision to adjust instruments that are close to the specified limits depends on the type of instrument and on the operating practice and philosophy of the laboratory doing the recalibration. The key factor is that the calibration laboratory is, in effect, guaranteeing that the instrument is calibrated and will remain calibrated until the next calibration due date (barring a breakdown). The technician making optimizing adjustments to an instrument needs to be aware that the adjustment of one parameter often has an interaction effect on another parameter. For example, if the power supply is "tweaked," it can have an effect on all other parameters and require a complete calibration check of the instrument.

DEFIBRILLATOR CALIBRATION

INDICATED	DELIVERED
WS	WS
WS	WS
WS	WS
WS	WS
WS	WS
MAX.	

DATE ________ BY ________

BE447

FIGURE 3–12
Calibration tags.

REPAIR/PERFORMANCE TEST RECORD

JOB #__________ PARTS P.O. #__________ ASSET #__________

MODEL NO.__________ COMPANY__________ S/N __________ PRIORITY__________

RECEIVED FROM__________ DATE RECEIVED__________ SERVICE REQUESTED__________

☐ SERVICE IN-HOUSE ☐ SEND TO FACTORY: DATE OUT__________ DATE IN__________

REPAIR OR SPECIAL INSTRUCTIONS__________

ACTY CODE	FAIL CODE	PART DESCRIPTION	PART NUMBER	* MFR	PART STAT	TOTAL QTY	PART ORDERED	PART RCVD	HOURS	DATE	** INITIAL

ACTION TAKEN:__________

PERFORMANCE LIMITATION:__________

CALIBRATION DATA:

DATE CALIBRATED __________ BY __________ DATE DUE __________ CYCLE (***) __________ MO.

STDS USED: MODEL # __________ S/N __________ MFR __________ DUE __________ CYCLE ( ) __________ MO.

MODEL # __________ S/N __________ MFR __________ DUE __________ CYCLE ( ) __________ MO.

MODEL # __________ S/N __________ MFR __________ DUE __________ CYCLE ( ) __________ MO.

MODEL # __________ S/N __________ MFR __________ DUE __________ CYCLE ( ) __________ MO.

MODEL # __________ S/N __________ MFR __________ DUE __________ CYCLE ( ) __________ MO.

MODEL # __________ S/N __________ MFR __________ DUE __________ CYCLE ( ) __________ MO.

MODEL # __________ S/N __________ MFR __________ DUE __________ CYCLE ( ) __________ MO.

TEMPERATURE __________ °F HUMIDITY __________ % RH

THIS SERVICE COMPLETED BY__________ DATE__________ TOTAL HRS __________

*IF DIFFERENT THAN EQUIPMENT OEM. **IF MORE THAN ONE PERSON ON THIS ITEM. ***R OR C ONLY

GVG FORM 1561R4 8-87

FIGURE 3–13
Calibration record (courtesy of Grass Valley Group).

CALIBRATION RECORDS

After an instrument's performance has been checked or it has been recalibrated, certain records should be completed. A label or other coding is attached to the instrument to indicate the instrument identification, date, person who performed the calibration, and the date when the instrument is to be rechecked. Different labels are used to indicate certain specific information about the calibration. A label may indicate conditions that affect the use of the instrument, such as a performance limitation, calibration deviation, that the instrument is traceable to NIST, or that no calibration is necessary. Figure 3–12 shows typical calibration tags that are attached to an instrument.

In addition to the instrument label, a file record should be kept for each instrument belonging to an organization, listed by model and serial number. It should include the previously cited information plus any service work performed on the instrument. Figure 3–13 illustrates a typical record used for each instrument. This record is valuable when troubleshooting an instrument, since it may point out likely problem locations. A similar record is often kept in a computer database to keep track of parts used on a particular instrument model, technician time, accounting charges, and the like. The computer record also simplifies the procedure for assuring that all instruments requiring calibration are recalled by the standards laboratory at the proper time interval. Careful monitoring of the calibration history of a model or class of instruments, including records on those that are out of tolerance and the repair history, can point to the need for corrective action, such as changing the recalibration interval, changing the recalibration procedure, derating the calibration, changing the specifications, or even removing an instrument from service.

SUMMARY

1. The SI system contains seven fundamental units (sometimes called base units) and two supplementary units.
2. Metric prefixes are based on the decimal system and represent powers of 10.
3. All measurements are based on comparison to a standard; all standards must be stable and reproducible.
4. An international system for measurement standards is headed by a governing authority called the Conférence Général des Poids et Mesures (CGPM). Delegates from member countries meet regularly to improve the metric system.
5. National laboratories provide a central authority for uniform measurement standards within their respective countries. In the United States, that authority is the National Institute of Standards and Technology (NIST).
6. The SI definition of one second is based on a hyperfine transition in the cesium 133 isotope. Epoch time is measured by coordinated universal time (UTC) and is broadcast from various stations around the world.
7. The BIPM and most national standards laboratories use Josephson junctions to check on the stability of Weston cells, which are used as secondary standards to maintain a reference voltage.
8. The present international resistance stan-

dard is a group of 1 Ω resistors whose mean value is known by comparing it to the impedance of a standard capacitor.

9. The international temperature scale is defined by 11 fixed points. Interpolation between points is done by the platinum resistance thermometer, thermocouple, and pyrometer.
10. Instrument accuracy is defined by the standards used for calibration and the inherent capability of the instrument.
11. The appropriate recalibration interval for instruments depends on the accuracy and confidence level required and on the predicted or actual long-term drift of the instrument. Records are an important part of any calibration operation.

QUESTIONS AND PROBLEMS

1. (a) What is the advantage of intrinsic international standards compared to prototype standards?
   (b) Cite an example of a fundamental unit maintained by an intrinsic standard and one maintained by a prototype standard.
2. What are two advantages of the metric system over the customary system for measurements?
3. What is the difference between a fundamental unit and a derived unit?
4. (a) How is an ampere defined?
   (b) Why do you think the ampere was selected for the fundamental electrical unit rather than the coulomb?
5. Beginning with Ohm's law, prove that the fundamental units for the ohm are

$$\Omega = \frac{\text{kg} \cdot \text{m}^2}{\text{A}^2 \cdot \text{s}^3}$$

6. What is the difference between an international, primary, secondary, and working standard?
7. What is the laboratory arm of the CGPM? What is its mission?
8. What is the Josephson effect? Why is it important as a standard?
9. Explain the difference between UTC and UT_1.
10. What disadvantage results from sending time signals by high-frequency radio?
11. How is network television used for a frequency standard?
12. Compute the voltage of the 200th step of a Josephson junction that is radiated with a frequency of 90 GHz.
13. Why doesn't the NIST maintain a calculatable inductor standard?
14. What is meant by absolute zero?
15. What factors determine the calibration interval suitable for an instrument?
16. What benefits accrue from keeping records of instrument service and recalibration?
17. Suppose you are working in the calibration facility of a large company and you want to start a database of instrument calibration and repair records for all instruments the company owns. List the reasons that justify the advantages of such a database.

Section Two

MEASUREMENT CIRCUITS

Chapter 4

Diode and Transistor Circuits

OBJECTIVES

Diode and transistor circuits are widely used in instrumentation systems and are the basic building blocks of integrated circuits. Even in this age of integrated circuits, discrete diodes and transistors are frequently used in instruments and instrumentation systems. In addition, diodes and transistors are used as light and temperature sensors. Measurements of diode and transistor parameters are also necessary in many troubleshooting problems. An instrument used for measuring diode and transistor characteristics is the transistor curve tracer.

This chapter includes information on diode and transistor parameters, the measurement of these parameters, and their application to instrument circuits. It also introduces the transistor curve tracer as a basic measuring instrument for diodes and transistors.

When you complete this chapter, you should be able to

1. Given the characteristic curve for a diode, determine the static and dynamic forward resistance.
2. Explain how the characteristic curve for a diode is affected by a change in temperature.
3. Describe the important specifications for various types of diodes, including rectifiers, signal diodes, and zeners.
4. Explain the operation of diode circuits, including rectifier circuits, clipping and clamping circuits, and diode transmission gates.
5. Given the characteristic curves for a bipolar transistor, determine β_{dc} and β_{ac}.
6. Show how a bipolar transistor's collector circuit can be represented by an equivalent Thevenin circuit and draw a dc load line for the equivalent circuit.
7. Solve for the dc and ac parameters for common emitter amplifiers using voltage-divider and emitter bias.
8. Compute the differential gain, the common-mode gain, and the common-mode rejection ratio of a simple differential amplifier.
9. Describe the basic characteristics of JFETs and MOSFETs, typical bias circuits, and applications.
10. Describe the operation of a transistor curve tracer by citing examples of common measurements that can be made.

HISTORICAL NOTE

The transistor had its beginnings at Bell Labs in 1947. It was invented by Walter Brattain, John Bardeen, and William Shockley. Their lab notebook entry for December 23, 1947, described measured parameters, including a voltage gain of 100, and included a simple schematic of their one transistor amplifier. The entry went on to state, "This circuit was actually spoken over and by switching the device in and out a distinct gain in speech level could be heard and seen on the scope presentation with no noticeable change in quality. By measurements at a fixed frequency in, it was determined that the power gain was on the order of 18 or greater. Various people witnessed this test. . . ."

In 1956 Brattain, Bardeen, and Shockley were awarded the Nobel Prize in Physics for their development of the transistor.

4–1 DIODE VOLT-AMPERE CHARACTERISTIC

A diode is a one-way valve that allows current to flow only when it is forward-biased. The ideal, or "perfect," diode has zero resistance in one direction and infinite resistance in the other direction. If we plot a curve of the current versus voltage for the ideal diode, we get a 90° break between turn-off and turn-on, as shown in Figure 4–1(a). Actual diodes differ from this curve in several ways. First, the minimum turn-on voltage is determined by the barrier potential (0.3 V for germanium and 0.7 V for silicon). In addition, the forward-biased diode has some resistance. There are two ways of considering the resistance of the diode. When a dc voltage is applied to a diode, a corresponding current flows. Dividing the voltage by the current gives the dc, or **static,** resistance, as stated by Ohm's law:

EQUATION 4–1

$$R_F = \frac{V_F}{I_F}$$

where V_F = voltage drop across the forward-biased diode
I_F = current through the forward-biased diode

If a changing, or ac, voltage is applied to a diode, a corresponding change in current occurs. The resistance can be found by dividing the change in voltage (ΔV_F) by the corresponding change in current (ΔI_F). This resistance, called the ac, **dynamic,** or **differential** resistance of the diode, is dependent on the current in the diode, as shown in Figure 4–1(b). It is shown as R_f, where the lowercase subscript indicates an ac parameter. That is

EQUATION 4–2

$$R_f = \frac{\Delta V_F}{\Delta I_F}$$

From the curve in Figure 4–1(b), it can be seen that both the ac and dc resistance depend on the current in the diode. For most applications, however, it is sufficiently accurate to approximate the diode as having a linear volt-ampere characteristic above the forward voltage drop, as shown in Figure 4–1(c). The

FIGURE 4–1
Diode characteristic curves.

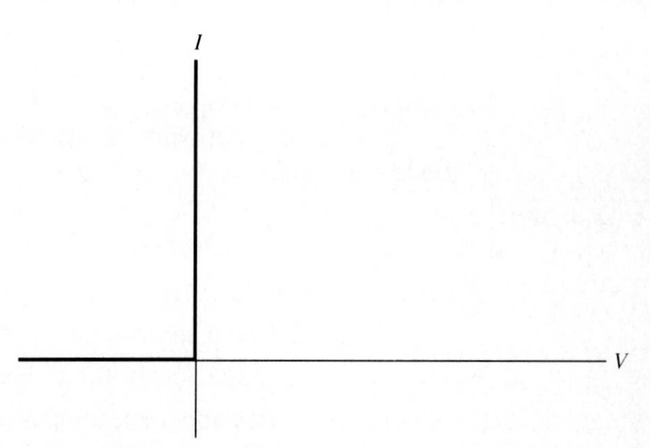

(a) Ideal diode curve.

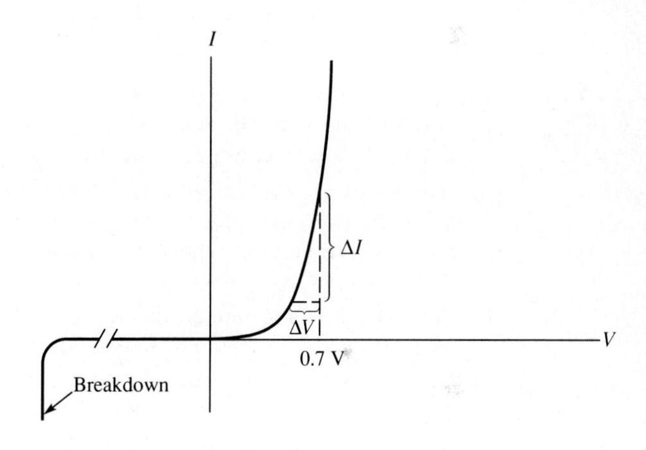

(b) Actual diode curve (silicon diode).

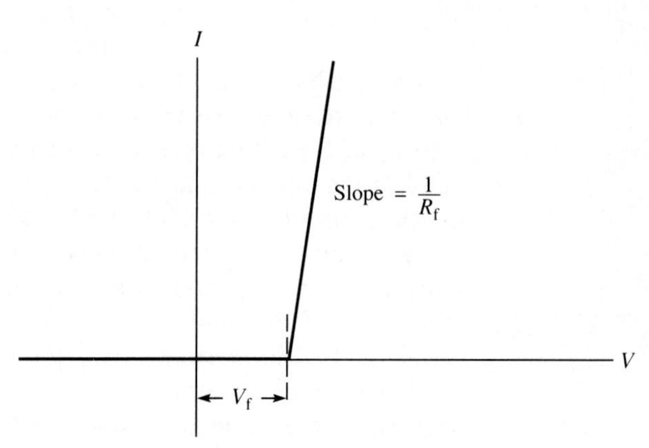

(c) Diode approximation.

forward voltage drop, V_f, is the voltage required to overcome the potential barrier. It is equivalent to having a small battery in series that opposes the forward bias. In many cases, the ideal diode (perfect one-way conductor) is sufficiently accurate to approximate circuit performance.

For more exact work, the diode can be theoretically modeled with the ideal diode equation:

EQUATION 4–3

$$I = I_S(e^{V/\eta V_T} - 1)$$

where I = diode current, A

I_S = reverse saturation current, A

V = voltage across the diode, V

V_T = volt-equivalent temperature, V (approximately 26 mV at room temperature)

η = emission coefficient, a value depending on the semiconductor

In practice, η varies with the voltage across the diode but is about 1 for silicon diodes in which the voltage across the diode is greater than approximately 0.5 V. At lower voltages, η is closer to 2. The saturation current, I_S, is a very tiny current in the reverse-biased diode. Notice that the ideal diode equation predicts *both* the forward and reverse voltage and current. Forward voltage and current are positive values, whereas reverse voltage and current are negative quantities. Although the ideal diode equation predicts that I_S is the maximum current that can flow in a reverse-biased diode, in real diodes the current is much larger due to leakage current. Reverse-biased diodes have a very high resistance (frequently over 10 MΩ). The reverse-biased diode also has a tiny current called *reverse leakage current* (in the nanoamp range), which is almost never of any importance until breakdown occurs. Except for zener diodes, diodes should not normally be subjected to voltages large enough to cause breakdown.

A graph of the "knee" region, indicating the forward voltage drop across the diode as a function of diode current assuming $\eta = 1$, is shown in Figure 4–2. The logarithmic characteristic of the diode forward-bias curve is clearly evident. This characteristic extends over an enormous range of currents—typically on the order of seven decades. Because of this, diodes are ideally suited for applications in which the logarithm or antilogarithm of a voltage is required. In instrumentation applications, logarithmic circuits are useful in signal compression, measurement of quantities with large dynamic range, conversion of quantities to decibel form, linearization, and certain computations (multiplication, division, roots, and powers) with analog quantities. Depending on the application, temperature compensation of the diode may be necessary. A number of integrated circuits are available that have compensating circuits.

The diode equation includes the volt-equivalent temperature, V_T, which has important implications for temperature measurements. The volt-equivalent temperature can be found from the equation

EQUATION 4–4

$$V_T = \frac{kT}{q}$$

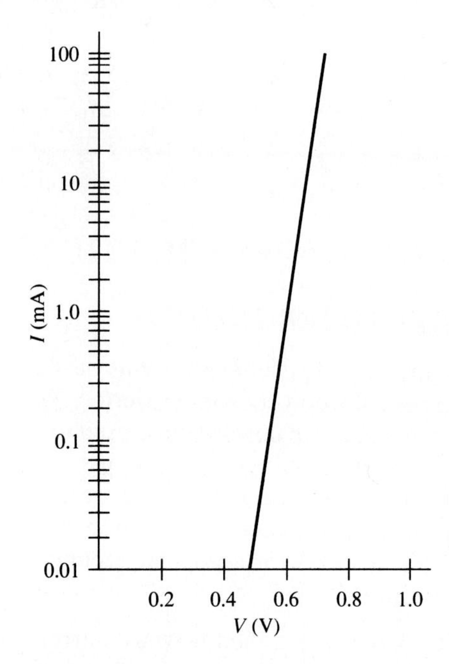

FIGURE 4–2
Diode forward-bias characteristic (for $I_S = 0.1$ pA, $\eta = 1.0$, $V_T = 26$ mV).

where k = Boltzmann's constant, 1.38×10^{-23} J/K
T = temperature, K
q = charge on an electron, 1.60×10^{-19} C

Notice that the volt-equivalent temperature is directly proportional to the kelvin temperature. An even stronger dependency with temperature is exhibited by the reverse saturation current, I_S. At a constant voltage drop across the diode, the diode current due to I_S is approximately doubled for each 10°C increase in temperature. The net effect of V_T and I_S is to shift the I-V curve to the left by approximately 20 mV for each 10°C increase in temperature. This dependency is exploited in certain solid-state temperature sensors (see Section 13–5). As mentioned earlier, temperature dependence can also be a problem in some cases, creating the need for special temperature-compensation techniques.

EXAMPLE 4–1 For the diode curve shown in Figure 4–2, determine the static and dynamic resistance at a current of 1.0 mA.

SOLUTION The static resistance can be found by dividing the forward voltage drop by the current at the point of interest. That is

$$R_F = \frac{V_D}{I_D} = \frac{0.6\text{ V}}{1.0\text{ mA}} = 600\ \Omega$$

The dynamic resistance is determined by finding the change in voltage over the change in current about the point of interest. We will select points located at $I_D = 0.5$ mA,

$V_D = 0.58$ V and $I_D = 2.3$ mA, $V_D = 0.62$ V. Then

$$R_f = \frac{\Delta V_D}{\Delta I_D} = \frac{0.62 - 0.58 \text{ V}}{2.3 - 0.5 \text{ mA}} = 22 \ \Omega$$

4–2 DIODE SPECIFICATIONS

RECTIFIER DIODES

Almost all instruments require a dc source of power. Portable instruments can obtain dc power from batteries, but other instruments use a power supply to change ac into dc. A device used to convert alternating current into direct current is called a **rectifier.** A diode used for this purpose is called a *rectifier diode* (or simply a rectifier). It is one of the simplest and most common applications of diodes. The major specifications for rectifier diodes follow. As an example, the specifications for a popular rectifier diode, the 1N4007, are given along with the definition.

1. Average rectified forward current (I_0) is the forward current averaged over a full cycle of operation with a half-sine-wave operation at 60 Hz and a conduction angle of 180° (1N4007 specification is 1 A).
2. Surge current (I_{FSM}) is the maximum (peak) safe current for a given number of cycles. The surge current is the initial charging current that flows due to uncharged capacitors in the power supply circuit. Current should never be allowed to exceed the surge current rating (1N4007 specification is 30 A).
3. Peak inverse voltage (PIV) (sometimes called peak reverse voltage, V_{RM}) is the maximum reverse voltage that can be applied across the diode before the onset of avalanche breakdown. Values vary from 50 V to a maximum of 1000 V for some diodes. Several diodes can be connected in series to obtain higher values of PIV (1N4007 specification is 1000 V).
4. Forward voltage drop (V_F) is the dc forward voltage drop across the diode when the specified I_F is flowing through the diode (1N4007 specification is 1.1 V with 1 A of forward current).

SIGNAL DIODES

The most widely used diodes are small-signal, general-purpose devices used in various applications such as switching, clipping and clamping, and demodulator circuits. Signal diodes have relatively low forward current limits. The important specifications for signal diodes follow. As an example, the specifications for a popular signal diode, the 1N914, are given along with the definition.

1. Peak inverse voltage (PIV) is the maximum reverse voltage that can be applied before avalanche breakdown is reached (1N914 specification is 75 V).
2. Forward current (I_F) is the value of current that flows through the diode in the forward direction (1N914 specification is 75 mA).

3. Forward voltage drop (V_F) is approximately 0.7 V (silicon) or 0.3 V (germanium) (1N914 specification is 1 V maximum).
4. Reverse recovery time (t_{rr}) is the switching time to turn off a forward-biased diode. It is specified as the time required from the application of a reverse voltage until the current has dropped to some specified value. This characteristic is important in high-speed applications (1N914 specification is 8 ns maximum).

ZENER DIODES

Zener diodes are heavily doped diodes designed to exploit the breakdown characteristic of the reverse-biased *pn* junction. This breakdown is controlled by the design of the diode and does not destroy the device if the power rating of the diode is not exceeded. Zener diodes are normally reverse-biased into the breakdown region in order to establish and maintain a constant voltage drop for applications such as a power supply reference voltage. The key specification for a zener diode is the nominal reference voltage (V_Z), defined as the voltage at which the *pn* junction breaks down. A zener characteristic curve and symbol are shown in Figure 4–3 along with a simple circuit showing the zener used as a reference. As an example, the series 1N746A through 1N759A is a popular series of zener diodes. This series offers zener voltages from 3.3 V to 12 V with a power dissipation of 400 mW steady state to 75°C. The maximum reverse current depends on the particular diode and conditions.

For critical applications, discrete zener diodes by themselves are not very high quality references for several reasons. First, zeners have poor temperature characteristics—the breakdown voltage changes as a function of temperature. (The temperature coefficient of a zener is given as the average change in output voltage as a function of temperature compared to its value at 25°C.) A second problem with a discrete zener is that the typical 10% tolerance specification is not good enough for many applications. In addition, zeners are typically high-impedance sources.

Instruments such as digital voltmeters require a very stable reference voltage to maintain accuracy. In order to improve on the specifications of discrete zeners, manufacturers have designed special integrated circuits for references. Some of these special reference circuits use internal zener diodes as a reference but have superior specifications to discrete zeners. Temperature coefficients as low as 1 ppm/°C are available with better long-term stability than discrete zeners.

SPECIAL-PURPOSE DIODES

Many special-purpose diodes have been developed that are important in instrumentation. Light-emitting diodes (LEDs) use compounds such as gallium arsenide to release light energy when forward-biased. The color of the light depends on the type of diode. For example, a typical small LED is the TIL209A, which produces red light and is useful as a visual indicator. Important specifications for this LED are maximum reverse voltage (V_{RM}), which is 3 V; the continuous forward current (V_{RM}), which is 40 mA (maximum); and the static forward voltage, which is 2 V

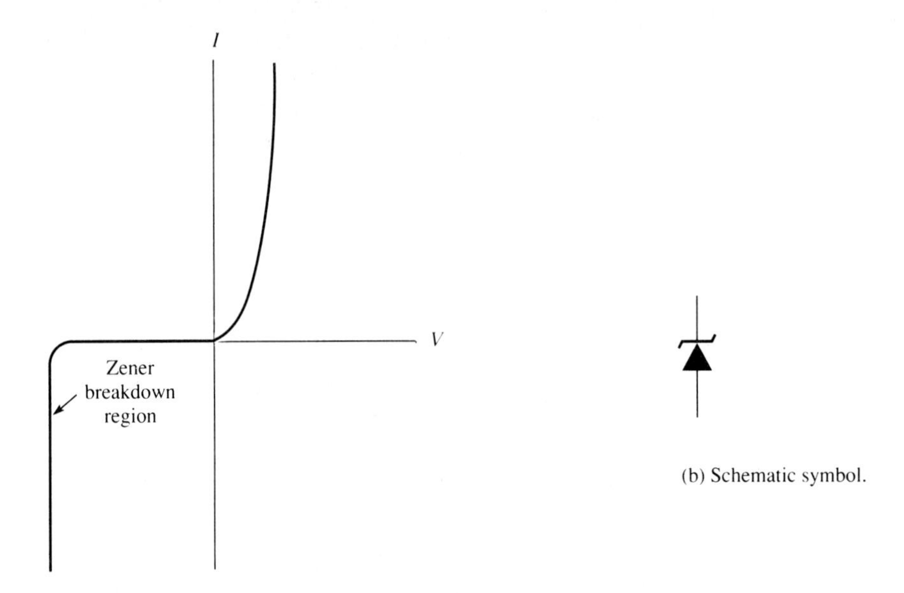

(a) Characteristic curve.

(b) Schematic symbol.

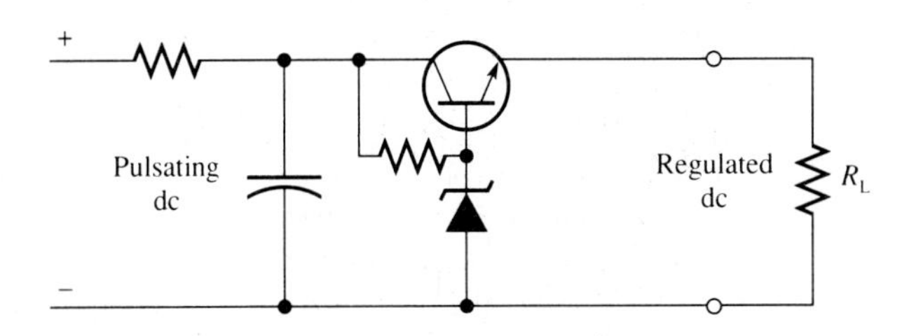

(c) Zener diode used as reference in simple regulator.

FIGURE 4–3
Zener diode characteristics and application.

(maximum). The spectral response indicates the color of light emitted by the diode. It is shown for reference in Figure 4–4. LEDs can be fabricated in a variety of shapes, including popular seven-segment displays. Seven-segment displays are frequently used as output transducers for digital multimeters, electrometers, and other instruments. In addition to display applications, LEDs can be used where a relatively monochromatic source of light is required, such as in a simple spectrophotometer or as a fiber-optic light source.

A photodiode is another special-purpose diode with application to instrumentation circuits. Photodiodes are transducers used to produce a linear current in response to light illumination. Light energy creates electron-hole pairs in the region near the junction, which, under reverse-bias conditions, cause a current to flow that is proportional to the incident light. This characteristic is useful for

FIGURE 4–4

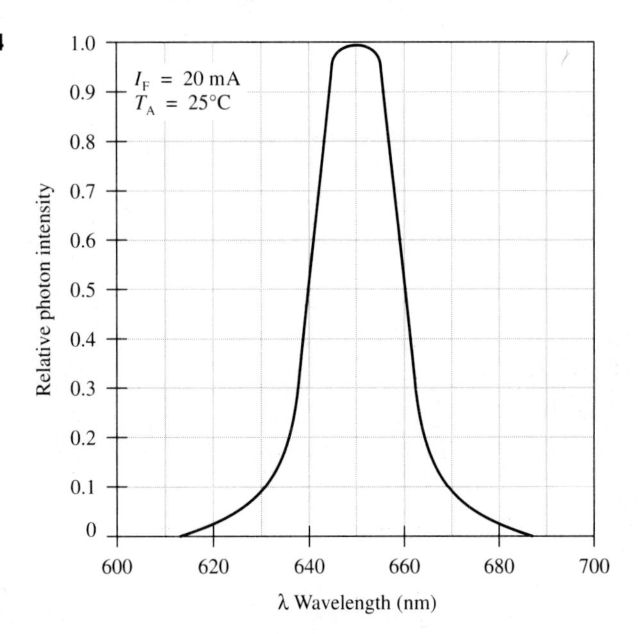

instrumentation transducers that detect light energy. Photodiodes also have applications as nuclear radiation detectors using a two-step process that first converts the radiation to light. Section 13–8 discusses photodiode transducers further.

Varactor (tuning) diodes make use of the decrease in capacitance that occurs when the reverse-bias voltage on a *pn* diode is increased. This characteristic makes the varactor diode a voltage-controlled capacitor useful in microwave oscillators and in electronic tuning applications.

Tunnel diodes are heavily doped diodes with a negative resistance region in the forward characteristic that is unstable. By exploiting this negative resistance region, tunnel diodes have applications as microwave oscillators and in certain high-speed switching circuits. Another special-purpose diode is the PIN diode used in microwave frequencies. The PIN diode is so named because of a center region of intrinsic (nondoped) material sandwiched between a *p*- and an *n*-layer. The impedance of a PIN diode is controlled by dc bias applied to it. It has applications in high-speed switching, modulators, and attenuators.

DIODE TESTING

Diodes can be tested for all of the characteristics described in the previous sections. Tests may be made to assure quality, predict circuit performance, improve a circuit, or verify that the diode is functional.

The simplest functional test of a diode is to check the forward (R_F) and reverse (R_R) static resistance using an ohmmeter. The test is illustrated in Figure 4–5. It is necessary to use a low-resistance scale to measure the forward resistance due to low output current from the meter on the higher-resistance scales. Auto-ranging meters have special instructions for down-ranging the meter; some

FIGURE 4–5
Diode ohmmeter tests.

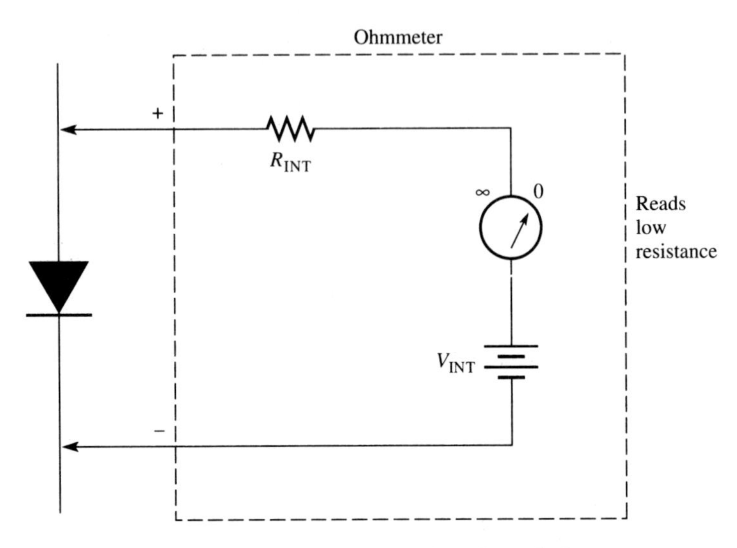

(a) Measuring R_F with an ohmmeter.

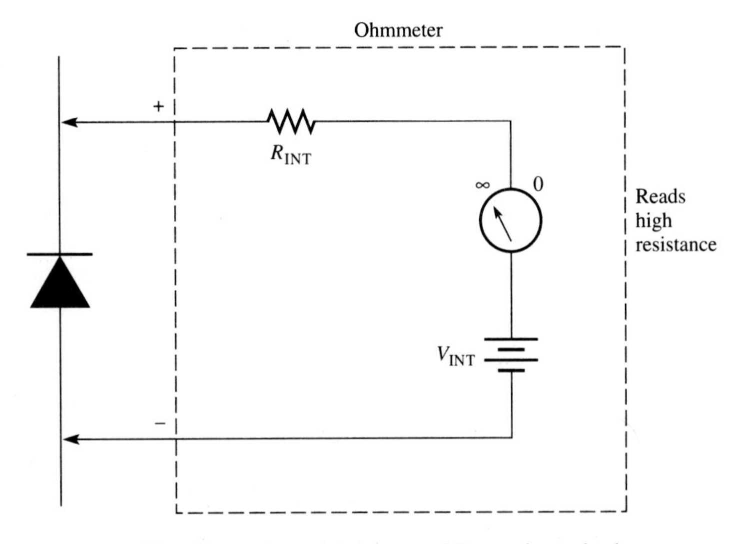

(b) Measuring R_R with an ohmmeter.

meters have special diode-testing provisions, as described in Section 7–6. The forward resistance depends on the type of diode but should be between a few ohms and several hundred ohms. Germanium diodes have a smaller forward resistance than silicon diodes. The reverse resistance should be extremely high; silicon diodes can be so high that they read open. Although the ohmmeter test is a relative test, for a good diode it should indicate a much higher resistance in one direction than the other. If a diode has low resistance in both directions or very high resistance in both directions, it is defective.

Another test of a diode is the breakdown voltage. Breakdown voltage occurs when the applied reverse voltage is large enough to cause a high reverse current to flow, as illustrated in the breakdown region on the characteristic curve shown in Figure 4–1(b) for a rectifier or signal diode and Figure 4–3(a) for a zener diode. The breakdown voltage can be as low as 2 V for zener diodes to several hundred volts for typical rectifier diodes. For zener diodes, the slope of the curve in the breakdown region is important. To measure the breakdown point, a series-limiting resistor is used and an increasing reverse bias is applied while monitoring reverse leakage current. Prior to breakdown, the reverse leakage current is very small, but it will increase linearly with the applied voltage. Breakdown is imminent when the reverse current no longer increases in direct proportion to the reverse voltage. The high electric field in the depletion region causes charge carriers crossing the region to have sufficient energy to break covalent bonds, freeing even more charge carriers, a condition known as **avalanching.** A current-limiting resistor should be used to prevent the avalanche current from destroying the diode. If other diodes of the same type are being tested, the breakdown voltage can be predicted by observing the reverse leakage current as the voltage is increased.

The characteristic curve for a diode can be displayed on a transistor curve tracer (discussed in Section 4–14) or on an oscilloscope (discussed in Chapter 8). The curve tracer can measure all the important voltage and current specifications that were previously listed. Forward dc and ac resistance as well as reverse current (as low as 1 nA) can also be measured. The characteristic curve can also be viewed on an oscilloscope operating in X-Y mode, as shown in Figure 4–6. The voltage across the diode is monitored by Channel 1, and the current through the diode develops a voltage that is monitored by Channel 2. Channel 2 should be inverted to display the curve in the proper orientation.

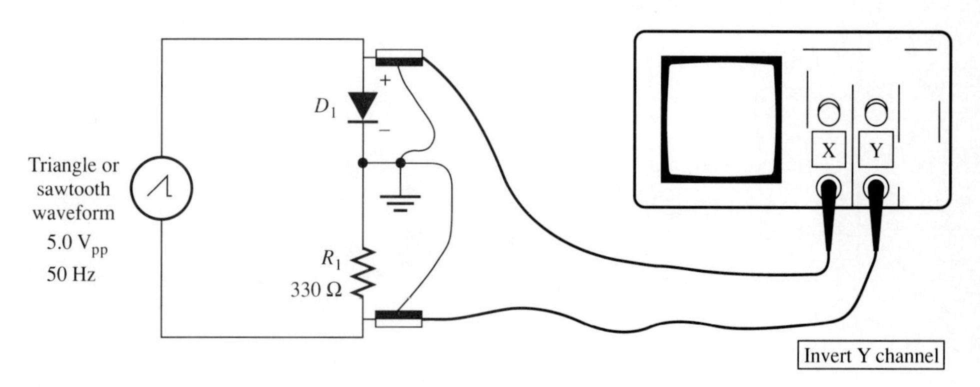

Note: Low side of signal generator must be isolated from ground.

FIGURE 4–6
Displaying the characteristic curve of a diode on an oscilloscope.

4–3 DIODE CIRCUITS

DIODE RECTIFIER CIRCUITS

Rectifier circuits are applied in instrumentation when it is necessary to convert ac to dc. Almost all instruments must convert ac power line voltage to dc voltage suitable for active circuits. Another common application is converting ac into dc for measuring instruments such as ac voltmeters.

A simple circuit for rectification is shown in Figure 4–7(a). The diode allows current to flow only during the positive half-cycle—hence the name half-wave rectifier. In this circuit, the transformer is used to change the line voltage as required. A filter capacitor is shown to smooth out the variations in the output. Half-wave rectifiers are seldom used as they require considerably more filtering than other types and deliver less power.

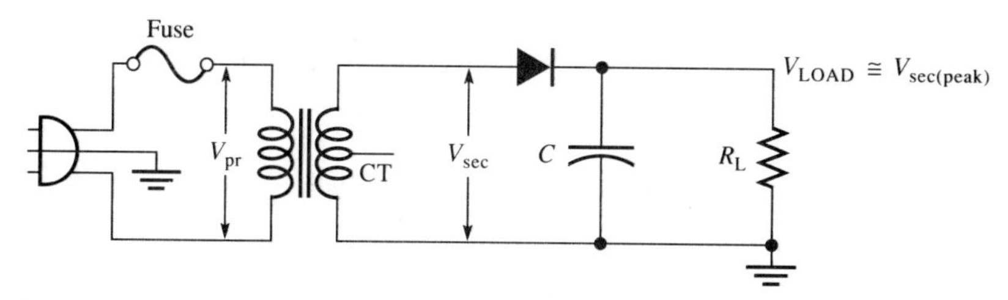

(a) Half-wave rectifier.

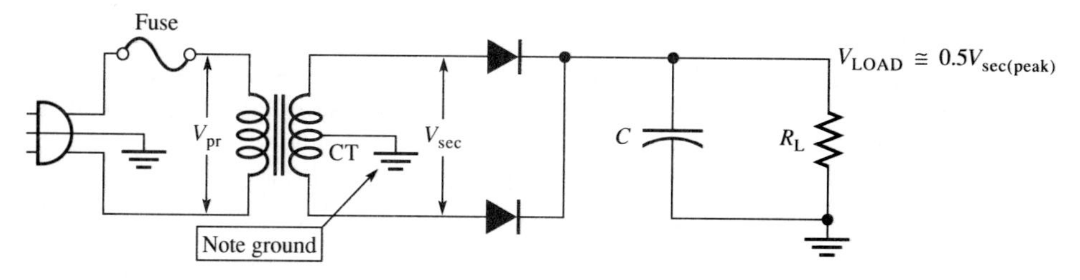

(b) Full-wave rectifier.

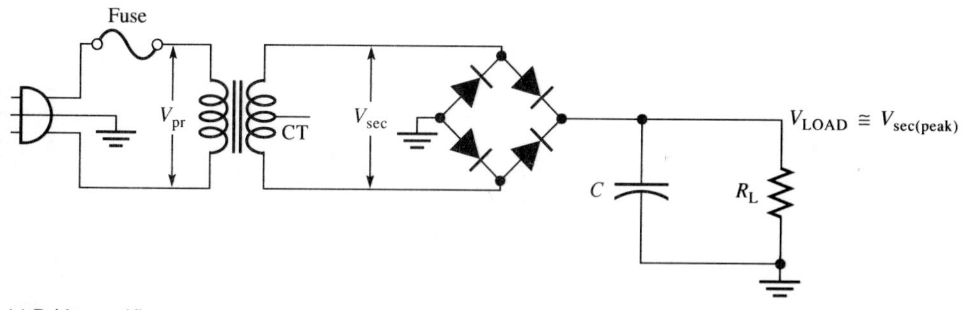

(c) Bridge rectifier.

FIGURE 4–7
Power supply rectifier circuits.

A better circuit is the full-wave circuit shown in Figure 4–7(b). A full-wave circuit uses a center-tapped transformer and is essentially equivalent to two half-wave circuits connected back-to-back. It makes use of both halves of the ac voltage and hence is easier to filter. In addition, each diode carries only one-half the load current, so the current rating of the diodes can be one-half the total load current. Notice that the half-wave rectifier uses one side of the transformer as the ground reference, whereas the full-wave rectifier uses the center-tap as the reference. It is therefore possible to obtain a higher voltage with a particular transformer using the half-wave circuit. It is also possible to connect two transformers with their primaries in parallel and their secondaries in series (with the proper polarities) to obtain a higher voltage.

A third arrangement of rectifiers is the bridge circuit shown in Figure 4–7(c). In the bridge arrangement, two diodes are connected in series with the load for each half-cycle. When the cycle reverses, the other two diodes are connected in series with the load. The output waveform has the same shape as the full-wave circuit, but it is not necessary to use a center-tapped transformer. Each pair of diodes in the bridge circuit conducts only half the time, so each diode handles only one-half the load current. Bridge rectifiers are the most common type used and are available in encapsulated form, containing all four diodes in a single package.

In power supply applications, the output of the rectifier must be smoothed with filtering and regulators. Each of the rectifiers in Figure 4–7 is shown with a capacitor input filter. The capacitor can charge to the peak voltage when the diodes conduct. The discharge time constant of the capacitor is designed to be long enough to cause the voltage to drop only slightly from the peak value during the time the diodes do not conduct. The peak voltage for the full-wave rectifier is one-half the secondary voltage because the reference ground is the center-tap of the transformer instead of one side of the transformer, as in the half-wave and bridge rectifiers.

The rectifier circuits discussed so far produce unregulated dc voltage. By themselves, these circuits are generally inadequate sources for most applications requiring a stable source of voltage. A high degree of regulation can easily be obtained by adding an integrated circuit (IC) regulator to the output of an unregulated supply. The IC regulator circuit uses negative feedback to compare the output voltage with an internal reference voltage; the result is a stable output over a wide range of loads. Fixed three-terminal IC regulators are typically within 5% of the nominal voltage—they are available with a limited number of fixed outputs ranging from 5 V to 24 V and can provide up to 1 A output. Popular fixed IC regulators include the 78xx series (positive regulators) and 79xx series (negative regulators). The last two digits (xx) represent the output voltage—for example, the 7805 provides +5.0 V. The 7805 regulator can supply 1 A with 80 dB of ripple rejection. Integrated circuit regulators such as this can be extended to higher currents by adding external pass transistors. Adjustable regulators allow the power supply to be varied over a wide range of values with a minimum of external components. Adjustable regulators are available with extra features, such as adjustable short-circuit current limiting. An example of an adjustable IC regulator is the popular LM317, adjustable from 1.2 V to 37 V.

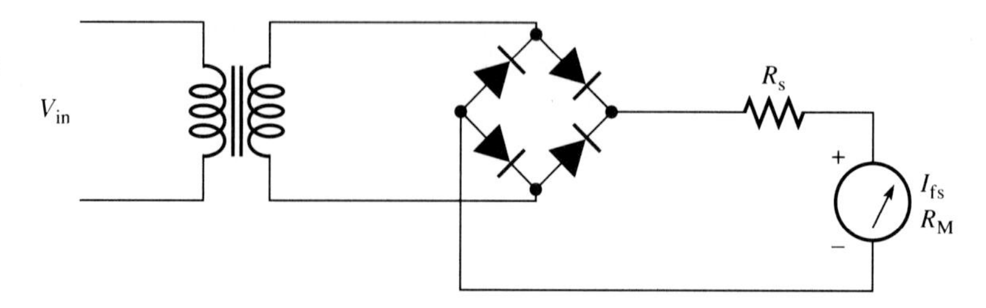

FIGURE 4–8
Bridge rectifier used in an analog ac meter.

Rectifier circuits are also used in measuring circuits to convert ac to dc. The full-wave bridge circuit, shown in Figure 4–8, is popular for ac voltmeters. The basic meter responds to the *average* value of the rectified waveform. To read rms voltage, the meter scale is calibrated accordingly. The knee region of the diode characteristic curve causes problems with nonlinearity at the lowest ranges of ac voltmeters, an effect that is greater for bridge rectifiers because of the two diodes in series, as previously mentioned. Meters and meter circuits are discussed in more detail in Chapter 7.

DIODE CLIPPERS

Diode-clipping circuits (sometimes called limiters) are used to prevent a waveform from exceeding some particular voltage, either positive or negative. Consider the negative series-clipping circuit shown in Figure 4–9(a). Whenever the waveform is greater than 0.7 V, the diode is forward-biased and the output waveform is approximately the same as the input waveform. However, when the input voltage is less than 0.7 V, the diode is reverse-biased and appears to be an open. The entire source voltage is then developed across the diode. The portion of the output waveform less than +0.7 V has been clipped. The circuit can be changed to a positive series-clipping circuit by simply reversing the diode. The clipping level can be shifted higher or lower by adding a dc level to the load resistor, as shown in Figure 4–9(b). This addition has the disadvantage of R_L not being connected to circuit ground.

Another clipping circuit, called a shunt clipper, is shown in Figure 4–9(c). Instead of taking the output across the resistor, it is taken across the diode. The series resistor should be much smaller than the load resistor R_L and much larger than the diode's forward resistance R_F for good clipping action. When the source voltage exceeds +0.7 V, the diode is forward-biased and the output is clipped above this level, causing positive clipping action. The shunt clipper can be biased by adding a series dc source, as shown in Figure 4–9(d). This causes the clipping level to be shifted higher or lower by the amount of the bias but leaves one side of R_L connected to circuit ground. As before, the circuit can be changed to a negative clipping circuit by reversing the diode.

A common instrumentation application of diode-clipping circuits is in the waveshaping circuits found in many function generators. Surprisingly, a sinusoidal waveform is not an easy waveform to generate over a large range of frequencies; however, it is easy to generate a triangular waveform and then convert it to a

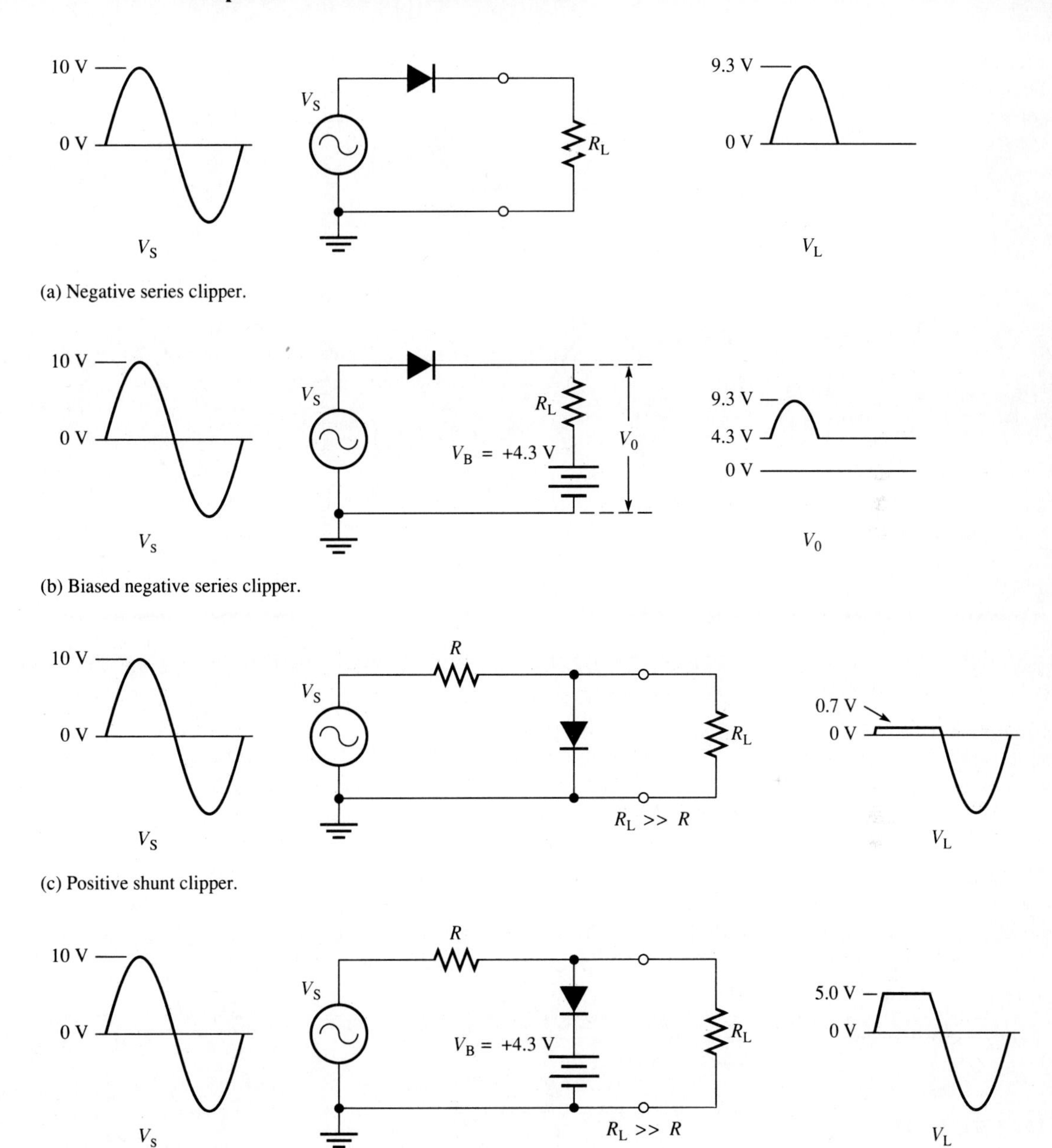

FIGURE 4–9
Diode-clipping circuits.

sinusoidal waveform with diodes. The basic circuit is shown in Figure 4–10. R_1 and R_2 form a voltage divider that sets the clipping points. The output waveform becomes nonlinear near the peak due to diode conduction. Many function generators use this technique to generate a high-quality sine wave.

FIGURE 4–10
Diode-clipping circuit used for waveshaping.

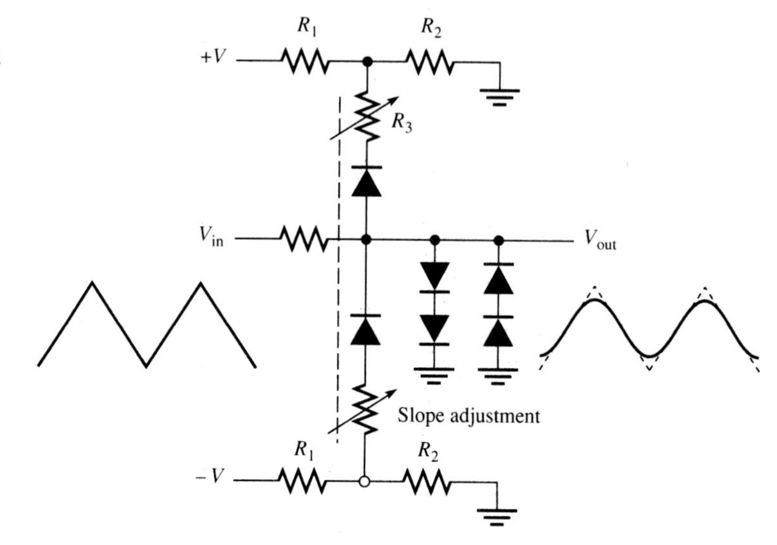

EXAMPLE 4–2 Determine the clipping level and output waveform for the biased shunt clipper shown in Figure 4–11.

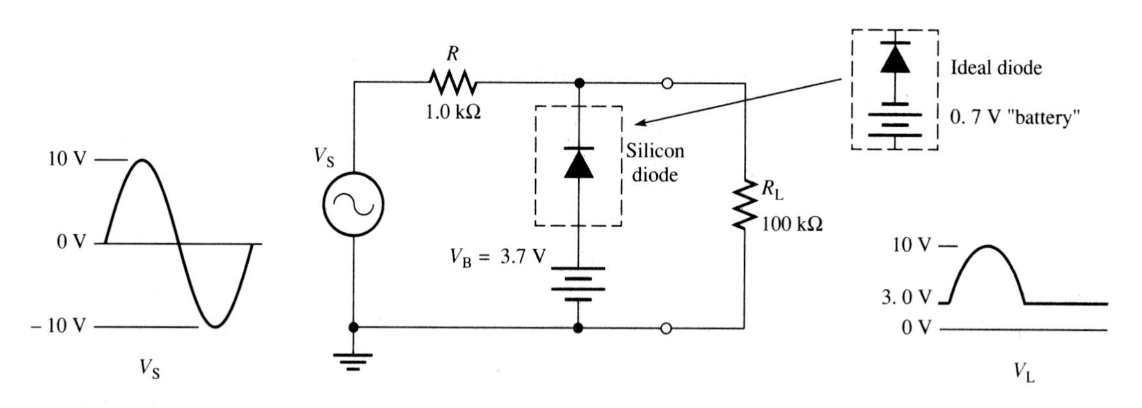

FIGURE 4–11

SOLUTION Think of the silicon diode as an equivalent ideal diode in series with a 0.7 V "battery," as shown in the dotted box to the upper right. This is the voltage that must be overcome to forward-bias the ideal diode. This diode battery appears to be in series with V_B, so we can add their voltages *algebraically*. The sum is 3.0 V, which is the clipping level. Observe that the waveform must be less than +3.0 V in order for the diode to conduct. The output waveform is clipped *below* this level, as shown.

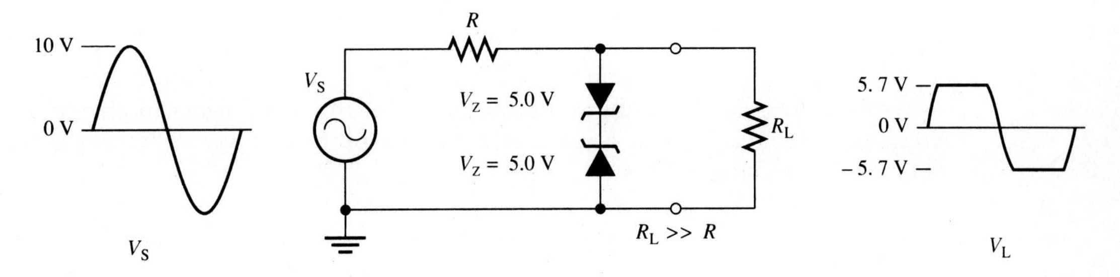

FIGURE 4–12
Double-ended zener clipper.

Another type of clipping circuit uses zener diodes to limit the output waveform between the zener forward and breakdown voltages. For symmetrical limiting, two back-to-back zeners can be connected in the form of a double-ended clipper. Zener clipping is illustrated in Figure 4–12. When the input signal rises above or below $V_F + V_Z$, both diodes conduct and the output is clipped.

DIODE CLAMPERS

Diode clampers are used to shift the average dc level of a waveform. When a signal has passed through a capacitor, the dc component of the signal is blocked. A clamping circuit can restore the dc level. For this reason these circuits are sometimes called *dc restorers*. Diode clamping action is illustrated in Figure 4–13 for both positive and negative clamping circuits. The diode causes the series

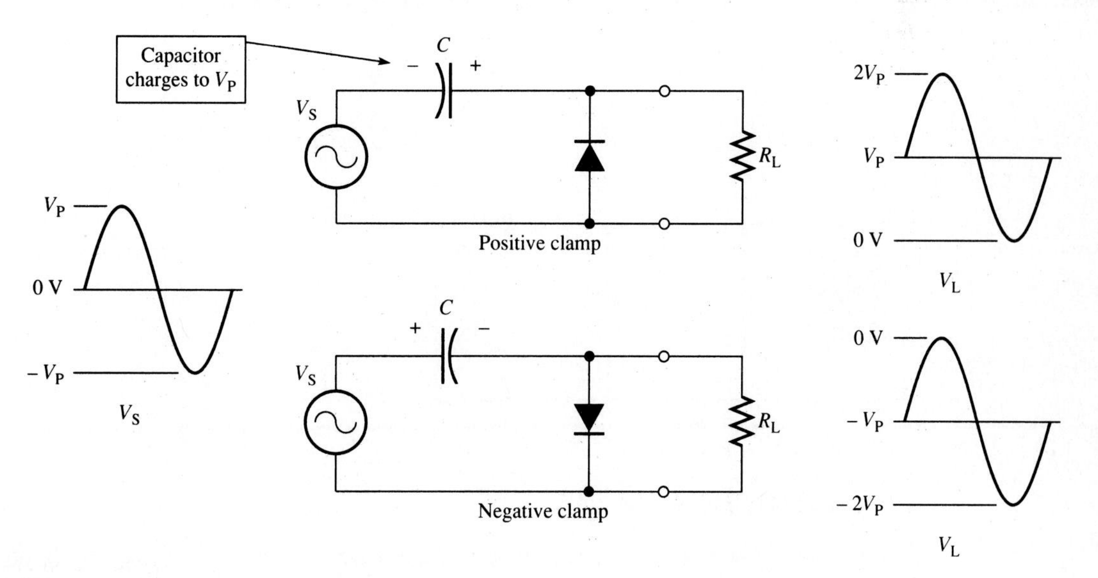

FIGURE 4–13
Diode-clamping circuits using ideal diodes.

capacitor to have a low-resistance charging path and a high-resistance discharge path through R_L. As long as the RC time constant is long compared to the period of the waveform, the capacitor will be charged to the peak value of the input waveform. This action requires several cycles of the input signal to charge the capacitor. The output load resistor sees the sum of the dc level on the capacitor and the input voltage.

EXAMPLE 4–3 The circuit shown in Figure 4–14 is called a *voltage doubler*. Explain how it works.

FIGURE 4–14

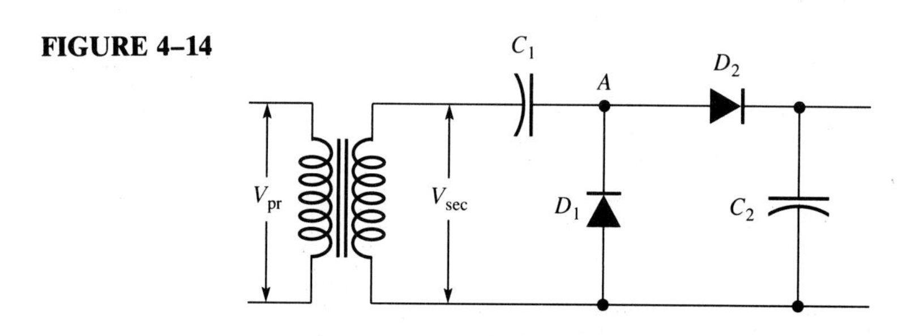

SOLUTION Diode D_1 and capacitor C_1 form a clamp circuit that will charge capacitor C_1 to a dc level nearly equal to $V_{sec(peak)}$. This action causes the ac voltage at point A to go from 0 V to approximately $2V_{sec(peak)}$, as shown in Figure 4–15. Diode D_2 is then forward-biased and charges capacitor C_2 to a voltage of nearly $2V_{sec(peak)}$. The output is taken across C_2 and is approximately twice the peak secondary voltage.

FIGURE 4–15

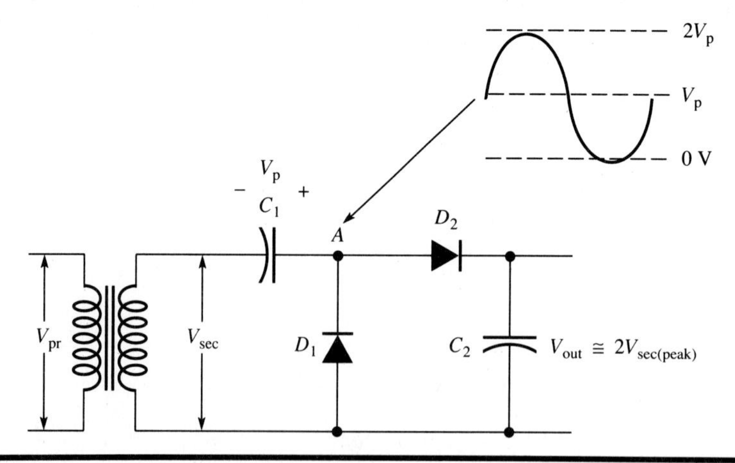

TRANSMISSION GATES

Transmission gates are essentially electronic analog switches and find many applications in instrumentation and control systems. In data-acquisition systems, it is frequently necessary to select only one of several sources for a measurement system input. Other applications include sample-and-hold and chopper circuits. Transmission gates can be made from discrete diodes and are also available in integrated circuit form using transistors (usually FETs (field-effect transistors)). FET analog switches are discussed in Section 4–13.

Diodes can be used as switches because of their low forward and high reverse resistance. By using Schottky diodes, the switching time can be less than 1 ns. The circuit for a diode transmission (or sampling) gate is shown in Figure 4–16. During selected times, the output is a duplicate of the input; otherwise it is zero. The diodes are turned on by a control signal, which has a greater amplitude than the signal voltage, and all four diodes conduct. If the diodes are thought of as ideal, then no voltage is dropped across them. The output node is at the same potential as the generator node, causing the output voltage to follow the generator. If the control voltage is reversed, the diodes are turned off and no current can flow in the load; hence the output is zero.

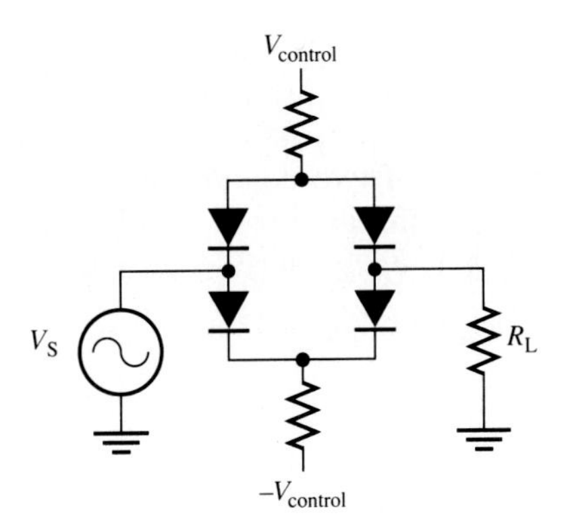

(a) Circuit.

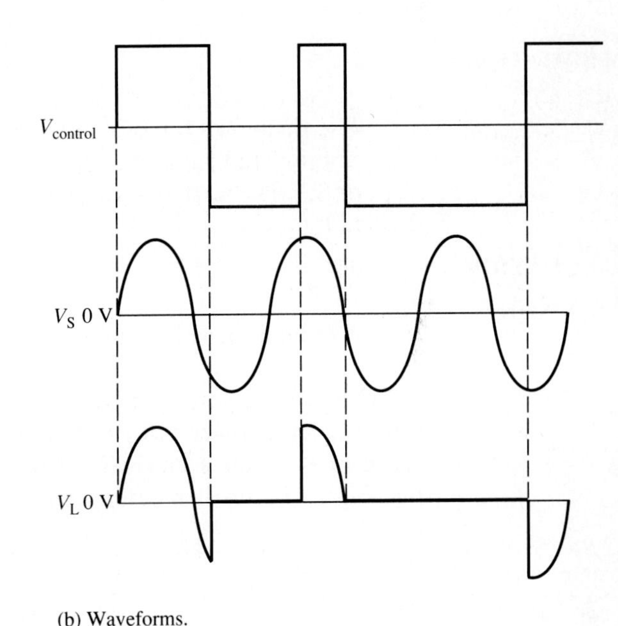

(b) Waveforms.

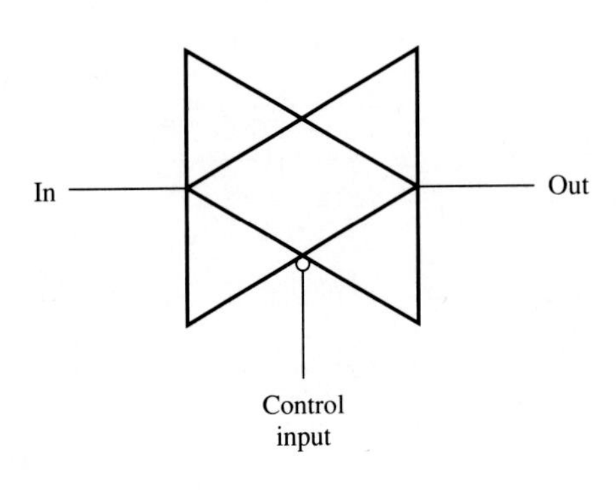

(c) Symbol.

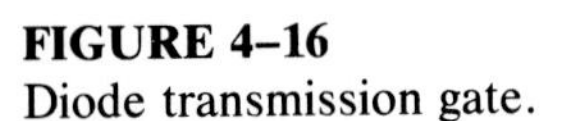

FIGURE 4–16
Diode transmission gate.

4–4 BIPOLAR JUNCTION TRANSISTORS

The bipolar junction transistor is a current amplifier; a small base current controls a larger current in the collector, an effect often referred to as *transistor action.* The ratio of the collector current, I_C, to the base current, I_B, is an important quantity called the dc β (beta), or h_{FE}, of a transistor. This quantity is frequently measured on a transistor curve tracer as a quick indicator of whether a transistor is functional. As a reminder, these definitions are shown in Figure 4–17 for both *npn* and *pnp* transistors.

The concept of β is valid only if the transistor is operating in its linear, or *active,* region. If the transistor has no base current, then both the emitter and collector diodes are reverse-biased and the transistor is said to be *cut off.* If the transistor has so much base current that both the base and collector diodes are forward-biased, the transistor is said to be *saturated,* and collector current is limited only by external resistance. The equations for β are invalid in the cutoff and saturation regions. When a transistor is operating in its active region, the ratio of I_C to I_B is written

EQUATION 4–5

$$\beta_{dc} = \frac{I_C}{I_B}$$

The definition for β in Equation 4–5 is the ratio of two dc currents and is often abbreviated β_{dc} to distinguish it from a second definition for β, which is the ac β, or β_{ac}. The ac β of a transistor is determined by a *change* in collector current divided by a *change* in base current:

EQUATION 4–6

$$\beta_{ac} = \frac{\Delta I_C}{\Delta I_B}$$

Although there is a difference in definition between β_{dc} and β_{ac}, the difference is usually not important (except at high frequencies). Typical small-signal transistors have values of β_{dc} ranging from 50 to 200 or more. Power transistors typically have lower values. Although β_{dc} is an easy parameter to measure, the range of β_{dc} values can vary widely for different batches of the same type of transistor. To further complicate matters, β_{dc} increases with temperature. For

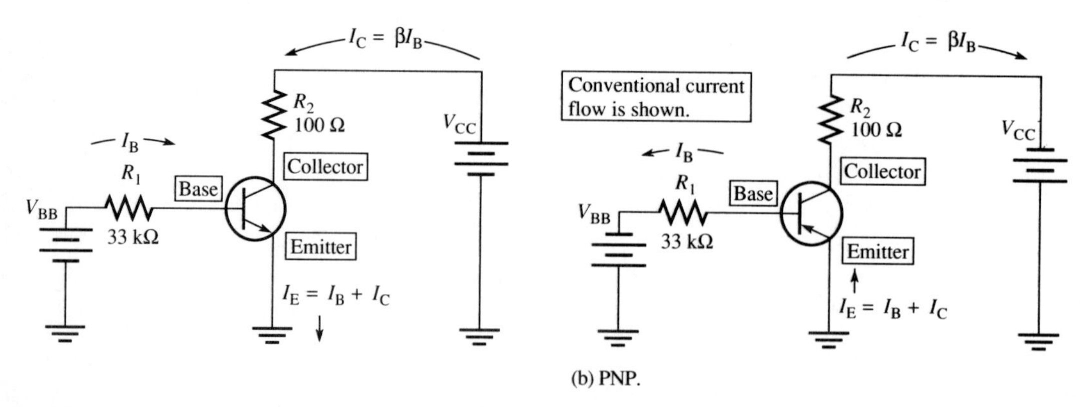

FIGURE 4–17
Definitions of I_B, I_C, and I_E.

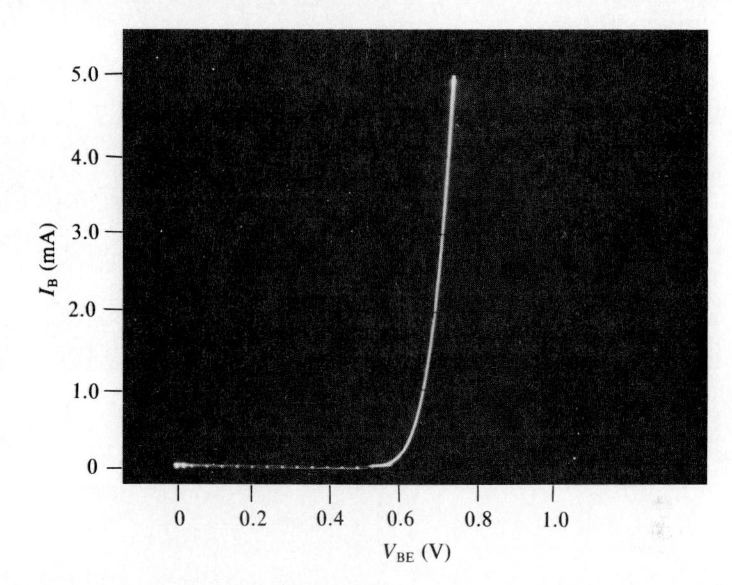

FIGURE 4–18
Input characteristics for a CE *npn* transistor.

these reasons, good circuit design must not depend too heavily on a specific value of β_{dc}. Observe that $I_E = I_B + I_C$. Since I_B is much smaller than I_C, the emitter current is approximately equal to the collector current—that is, $I_C \cong I_E$.

A bipolar transistor is a two-port device—that is, it has a pair of terminals to which the input signal is applied and a pair of terminals from which the output signal is extracted. Since it is a three-terminal device, one of the terminals must be common to the applied signal. Most circuits have the emitter as the common terminal, so we begin with this configuration. In the active region, a transistor's input circuit is a forward-biased diode (the base-emitter diode). As you might expect, the input characteristics of a normally operating transistor can be described by a diode curve with the base-emitter voltage as a function of the base current, as shown in Figure 4–18.

The characteristics of a diode can be fully represented by a single volt-ampere curve. However, it is not possible to represent a transistor's output characteristics with a single curve because it is a three-terminal device, with the output terminals controlled by the input terminals. The output characteristics must be described with a family of curves that depend on which of the three terminals is common to both the input and output circuits. Consider the curve shown for a common-emitter *npn* transistor in Figure 4–19(a). This curve is generated by holding a constant current in the base-emitter circuit while observing the collector current. Above the knee of the curve, an increase in collector-emitter voltage (V_{CE}) has little effect on collector current. In this region of the curve, the output circuit is often modeled as a Norton circuit—in this case, a current source with a large parallel resistance looking back into the collector. Now let the base-emitter current increase, and a new output curve is generated, as shown in Figure 4–19(b). Continuing in this manner, the family of curves in Figure 4–19(c) is generated, representing the output characteristics for the transistor. Since these curves represent the collector current as a function of the collector-emitter voltage, they are referred to as the **collector characteristics.**

FIGURE 4–19
Collector characteristics for a CE *npn* transistor. These photographs were taken with a transistor curve tracer, described in Section 4–14.

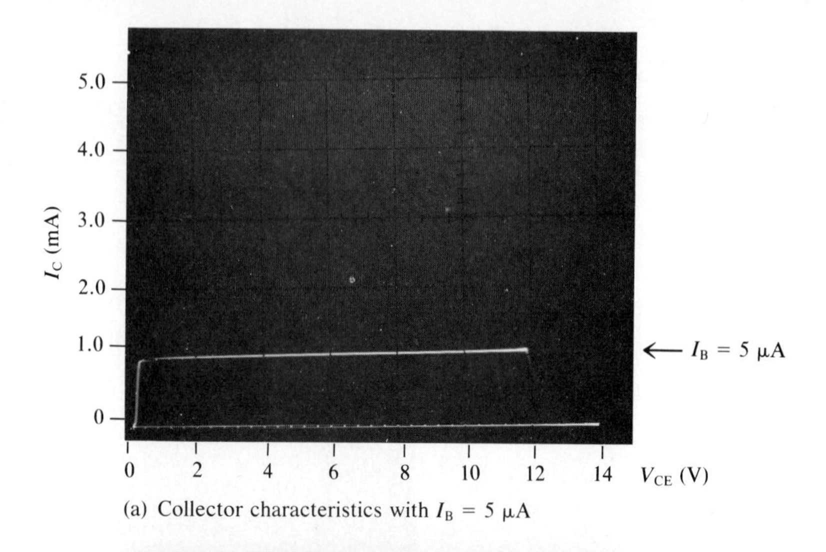

(a) Collector characteristics with $I_B = 5\ \mu A$

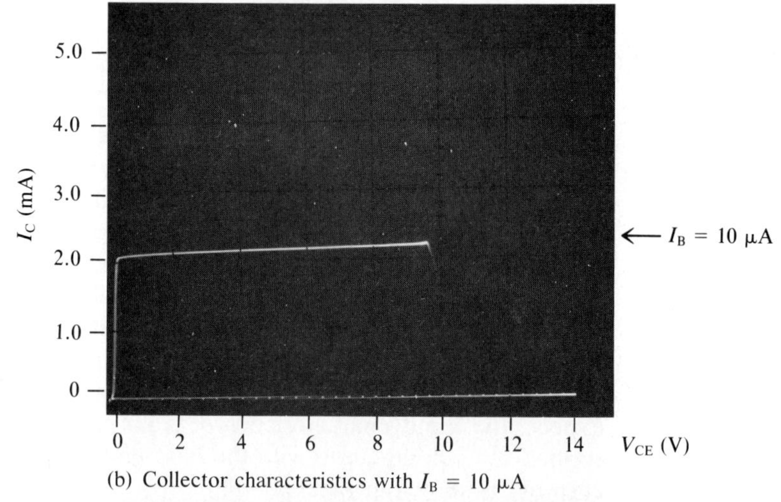

(b) Collector characteristics with $I_B = 10\ \mu A$

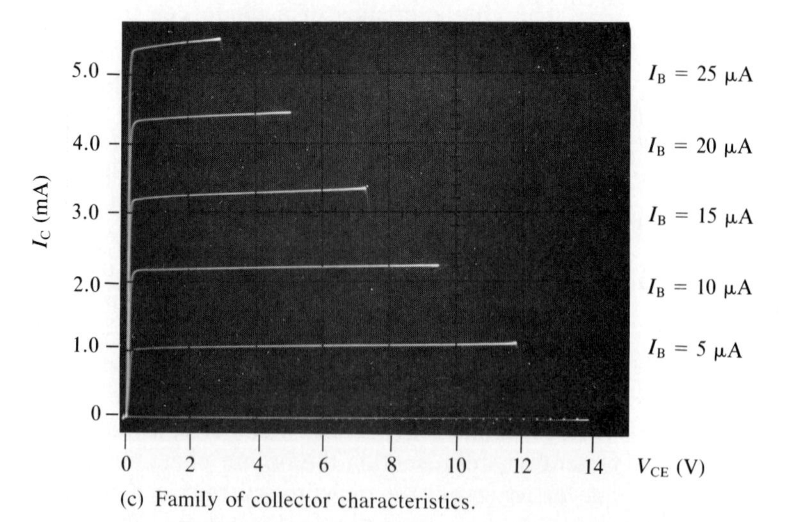

(c) Family of collector characteristics.

EXAMPLE 4–4 At base current of 20 μA, determine the β_{dc} and the collector current for the transistor described by the characteristic curves of Figure 4–19.

SOLUTION At 20 μA of base current, the collector current is approximately 4.3 mA. The β_{dc} is

$$\beta_{dc} = \frac{I_C}{I_B} = \frac{4.3 \text{ mA}}{20 \ \mu\text{A}} = 215$$

4–5 THE dc LOAD LINE

Imagine a series circuit consisting of a battery and a resistor driving a "black-box" load. Recall from basic electronics that the driving circuit is a Thevenin circuit, which is shown in Figure 4–20. From Kirchhoff's voltage law

$$V_{TH} = IR_{TH} + V_{bb}$$

The equation for current and voltage in the box is seen to be a linear (straight-line) equation with two unknowns (V_{bb} and I). To find the line that represents the solution to all possible quantities that could be in the box, put a short (zero resistance) in the box and note that $V_{bb} = 0$ and the current that flows is given by Ohm's law:

$$I_{short} = \frac{V_{TH}}{R_{TH}}$$

To find a second point on the line, put an open (infinite resistance) into the box and note that the current is zero. Therefore, V_{TH} will appear across the terminals of the box. Plotting these two points on the graph of Figure 4–21 and connecting the points with a straight line gives the load line for the circuit. The load line is determined by V_{TH} and R_{TH} and has a slope of $-1/R_{TH}$. *Any load* placed in the box will have a current and voltage point located somewhere on this line.

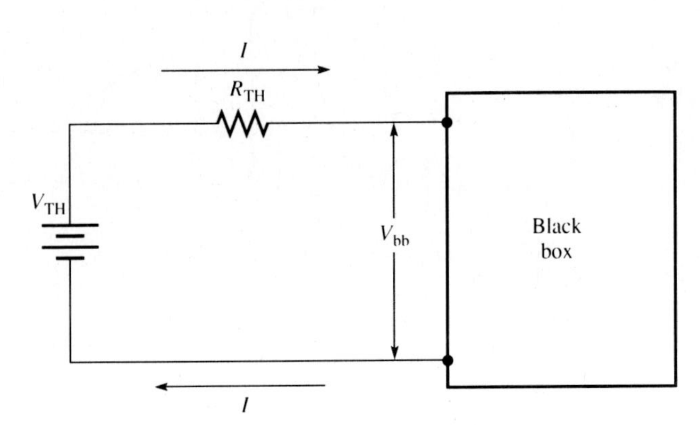

FIGURE 4–20
A Thevenin circuit driving a black box.

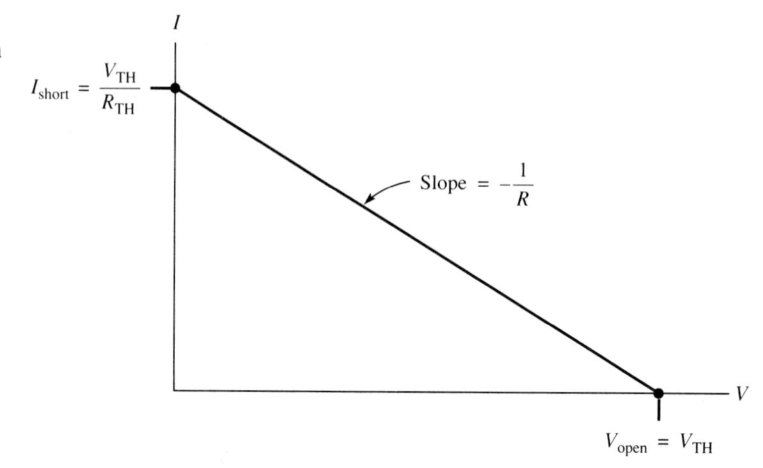

FIGURE 4–21
Load line for a Thevenin circuit.

Now imagine a transistor is placed in the box, with the collector and emitter connected to the terminals of the box as shown in Figure 4–22(a). The transistor's characteristic curves intercept the load line at the operating point for a given base current. Assume we set a base current of 20 μA. The current into the box and the voltage across the box can be found by locating the intersection of the 20 μA base current characteristic curve with the load line. If we change the base current, we can see the effect on the collector current and collector-emitter voltage by observing the new intersection corresponding to the appropriate base current curve with the load line. The collector current (the same as I_{bb}) can be read on the y-axis and the collector-emitter voltage (the same as V_{bb}) can be read on the x-axis, as shown in Figure 4–22(b). Load lines are a useful conceptual tool to show the relationships between various circuit parameters and for analyzing the dc operating point. The load line for a particular operation can even be set up and drawn electronically on a transistor curve tracer, an instrument similar to an oscilloscope that can

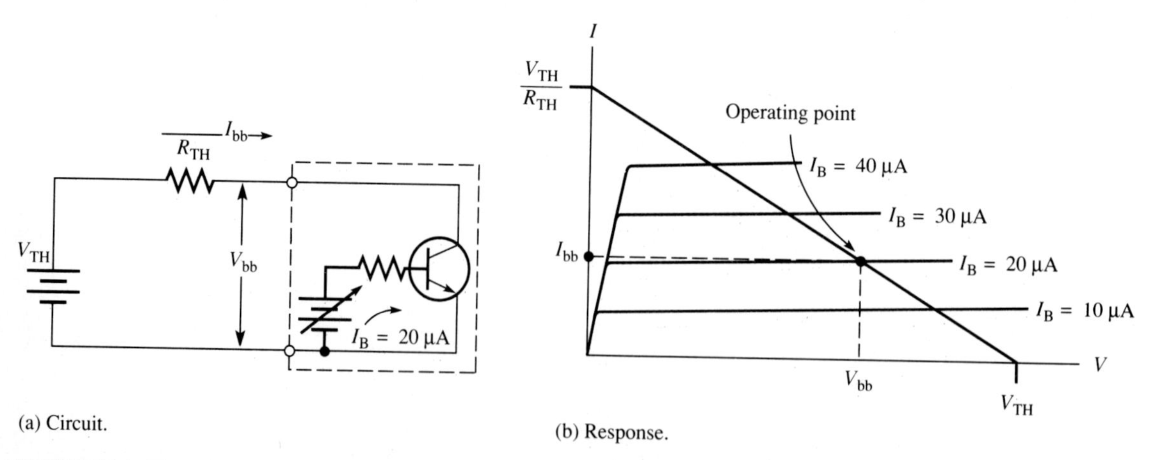

(a) Circuit.

(b) Response.

FIGURE 4–22
Transistor characteristics superimposed on Thevenin circuit load line.

display the characteristic curves of semiconductors (see Section 4–14). Unfortunately, a load line drawn for one circuit is not directly applicable to a similar circuit with different components because of variations in β_{dc} and other circuit parameters, restricting the usefulness of load-line analysis as a tool. Instead, purely algebraic methods are used, which are entirely adequate for practical applications.

EXAMPLE 4–5 Draw the dc load line for the transistor circuit shown in Figure 4–23.

FIGURE 4–23

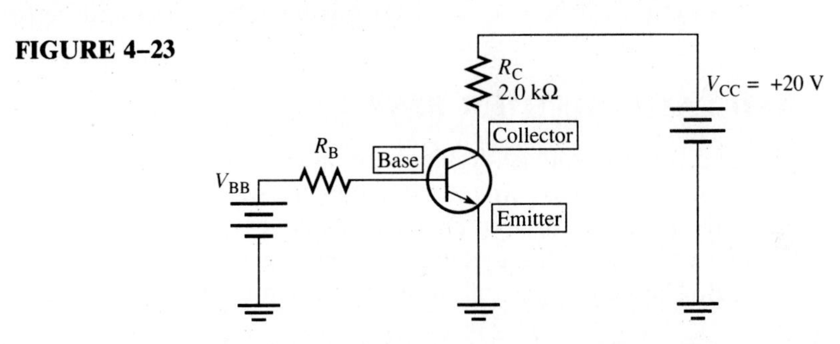

SOLUTION The current that will flow when the transistor's collector-emitter is represented by a short is called the saturation current. From Ohm's law, this current is

$$I_{SAT} = \frac{V_{CC}}{R_C} = \frac{20\text{ V}}{2.0\text{ k}\Omega} = 10\text{ mA}$$

When the transistor is represented by an open (no base current), V_{CC} appears across the collector-emitter terminals. The load line is then represented by a line joining these two points, as shown in Figure 4–24.

FIGURE 4–24

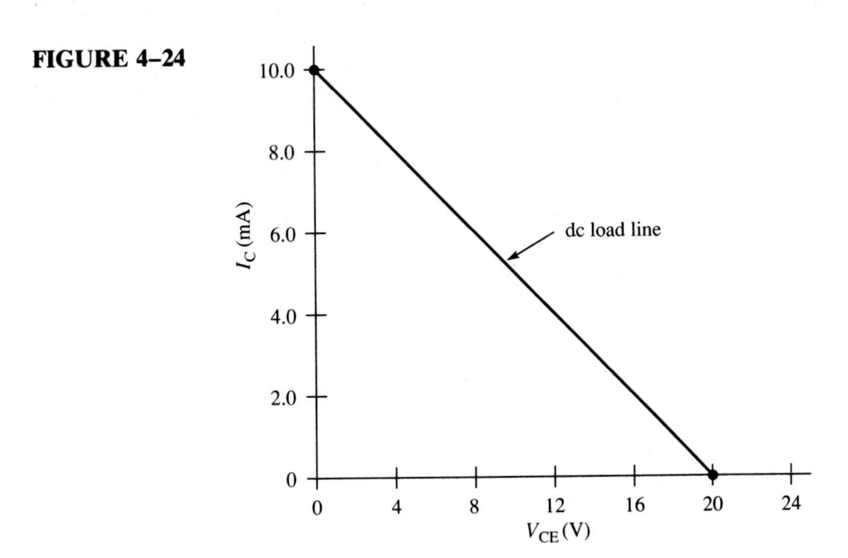

4–6 BIPOLAR TRANSISTOR BIASING

Biasing refers to establishing the dc parameters that set the operating point of a transistor. This operating point is called the quiescent point, or *Q*-point, and is located on the load line at the intersection of I_C, V_{CE}, and I_B. For stable operation, the *Q*-point should be near the middle of the dc load line. This allows the ac signal to vary both above and below the *Q*-point. In addition, the bias circuit must provide forward bias for the base-emitter junction and reverse bias for the base-collector junction. The bias must be stable so that changes in temperature or input signal or even a different transistor has no effect on circuit performance.

VOLTAGE-DIVIDER BIAS

The two-resistor voltage divider with an emitter resistor shown in Figure 4–25 is the most widely used circuit for achieving stable bias. The voltage divider provides the required forward bias to the base-emitter, and the emitter-resistor contributes to bias stability by tending to oppose any effect that changes the collector current. The dc analysis of circuit operation is straightforward:

1. Use the voltage-divider rule to find the base voltage.
2. Subtract 0.7 V (forward diode drop) from the base voltage to find the emitter voltage. (Add 0.7 V for *pnp* transistors.)
3. Apply Ohm's law to the emitter resistor to find the emitter current. The emitter current is approximately equal to the collector current.
4. Use Ohm's law to compute the drop across the collector-resistor.

FIGURE 4–25
Voltage-divider bias.

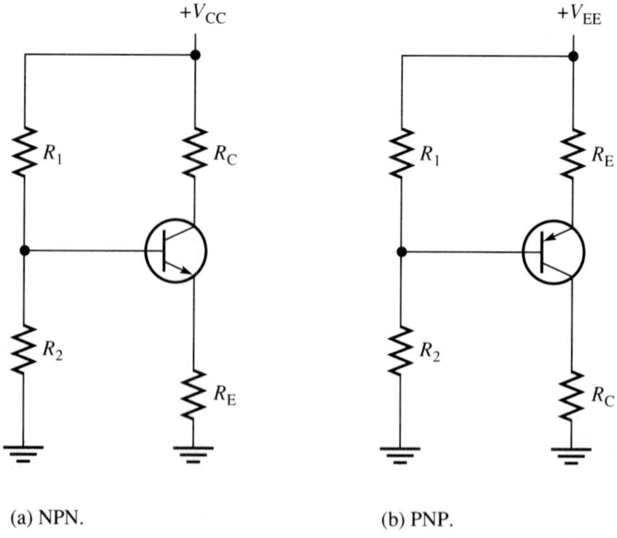

EXAMPLE 4–6 Compute the base, emitter, and collector voltages for the voltage-divider bias circuit shown in Figure 4–26.

FIGURE 4–26

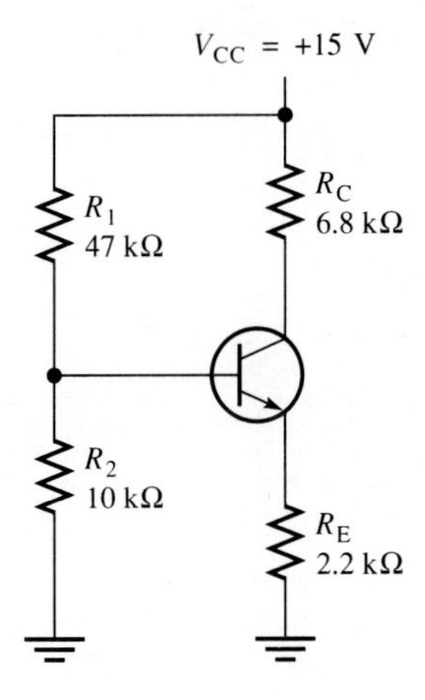

SOLUTION The base voltage is found by applying the voltage-divider rule to resistors R_1 and R_2:

$$V_B = V_{CC}\left(\frac{R_2}{R_1 + R_2}\right) = 15\left(\frac{10\text{ k}\Omega}{10\text{ k}\Omega + 47\text{ k}\Omega}\right) = 2.63\text{ V}$$

The emitter voltage is

$$V_E = V_B - 0.70\text{ V} = 2.63\text{ V} - 0.70\text{ V} = 1.93\text{ V}$$

The emitter current is found by applying Ohm's law to R_E:

$$I_E = \frac{V_E}{R_E} = \frac{1.93\text{ V}}{2.2\text{ k}\Omega} = 0.88\text{ mA}$$

The collector current is nearly the same as the emitter current. Therefore

$$I_C \cong 0.88\text{ mA}$$
$$V_{RC} = I_C R_C = (0.88\text{ mA})(6.8\text{ k}\Omega) = 5.96\text{ V}$$
$$V_C = V_{CC} - V_{RC} = 15.0\text{ V} - 5.96\text{ V} = 9.04\text{ V}$$

EMITTER BIAS

Emitter bias requires a positive and negative supply voltage, as shown in Figure 4–27. A rule of thumb for this circuit is that the dc emitter voltage is approximately −1.0 V for *npn* transistors and +1.0 V for *pnp* transistors. Using this rule, the voltage across the emitter-resistor can be found immediately. A more precise method to find the dc voltages is to write Kirchhoff's voltage law around the base-

FIGURE 4–27
Emitter bias.

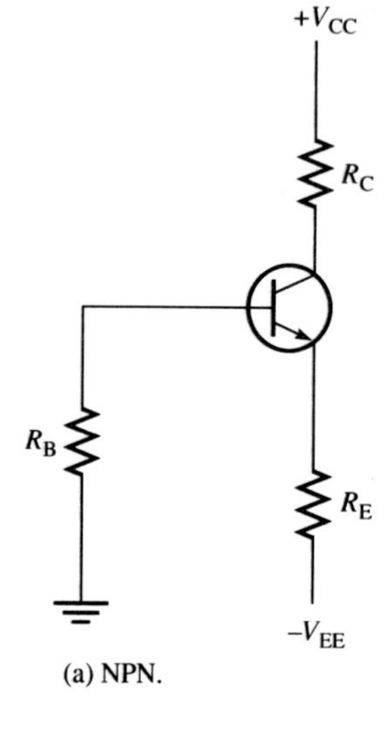

(a) NPN.

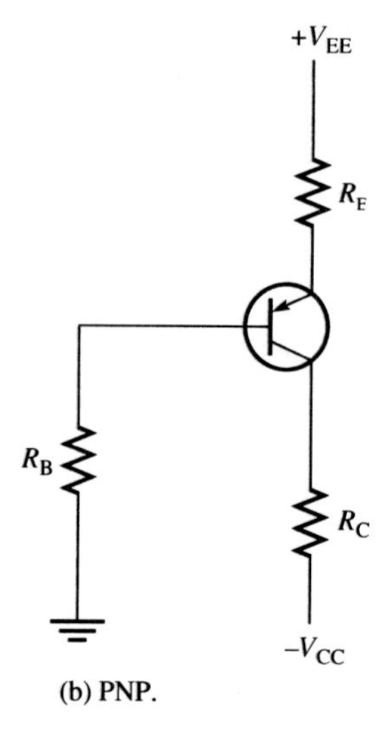

(b) PNP.

emitter path. For the *npn* circuit, the equation is

$$I_B R_B + V_{BE} + I_E R_E - V_{EE} = 0$$

Substituting for I_B

$$I_B \cong \frac{I_E}{\beta_{dc}}$$

$$\frac{I_E}{\beta_{dc}} R_B + I_E R_E = V_{EE} - V_{BE}$$

The emitter current is

EQUATION 4–7

$$I_E = \frac{V_{EE} - V_{BE}}{(R_B/\beta_{dc}) + R_E}$$

The emitter current is approximately equal to the collector current. Applying Ohm's law to the collector-resistor:

$$I_C \cong I_E$$

$$V_{RC} = I_C R_C \cong I_E R_C$$

EQUATION 4–8

$$V_C = V_{CC} - I_E R_C$$

Equation 4–7 was developed for the *npn* circuit. The equation for the *pnp* circuit is the same except the voltage V_{EB} is used for the *pnp* transistor instead of V_{BE}.

EXAMPLE 4–7 Compute the base, emitter, and collector voltages for the emitter-bias circuit shown in Figure 4–28.

FIGURE 4–28

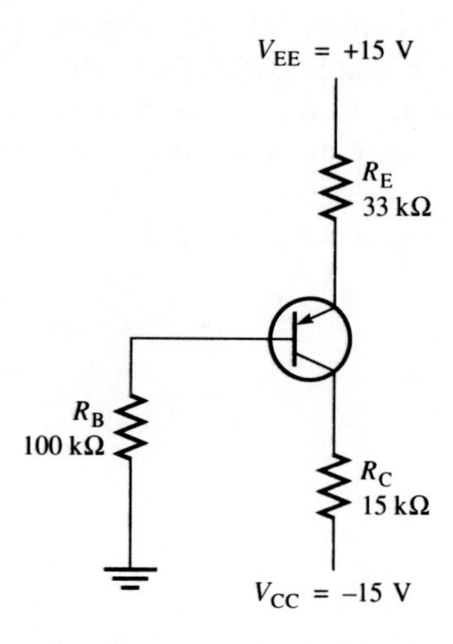

SOLUTION The circuit uses a *pnp* transistor. Assuming a β_{dc} of 100, we can find the emitter current by applying Equation 4–7 (using V_{EB} because of the *pnp* transistor):

$$I_E = \frac{V_{EE} - V_{EB}}{\left(\frac{R_B}{\beta_{dc}}\right) + R_E} = \frac{15\text{ V} - 0.7\text{ V}}{(100\text{ k}\Omega/100) + 33\text{ k}\Omega} = 0.42\text{ mA}$$

The voltage drop across the emitter-resistor is found by applying Ohm's law:

$$V_{RE} = (0.42\text{ mA})(33\text{ k}\Omega) = 13.9\text{ V}$$

The emitter voltage is found by subtracting the drop across R_E from V_{EE}:

$$V_E = V_{EE} - V_{RE} = 15\text{ V} - 13.9\text{ V} = 1.1\text{ V}$$

The base voltage is 0.7 V less than the emitter (*pnp* transistor):

$$V_B = V_E - V_{EB} = 1.1\text{ V} - 0.7\text{ V} = 0.4\text{ V}$$

The collector current is approximately equal to the emitter current:

$$I_C \cong I_E = 0.42\text{ mA}$$

When we apply Equation 4–8 to the *pnp* transistor to find the collector voltage, the emitter current is negative due to the direction of current:

$$\begin{aligned} V_C &= V_{CC} - I_E R_C \\ &= -15\text{ V} - (-0.42\text{ mA})(15\text{ k}\Omega) \\ &= -8.7\text{ V} \end{aligned}$$

4–7 THE COMMON-EMITTER AMPLIFIER

COUPLING AND BYPASS CAPACITORS

A capacitor represents an open to dc voltages. For ac, the reactance depends on the frequency, but for simplification, we will consider the larger-value capacitors shown as a short to ac. Consider the common-emitter (CE) amplifier shown in Figure 4–29(a). The dc bias is set up by a voltage divider, and the quiescent currents and voltages are shown. An ac signal can now be introduced to the base of the transistor through a capacitor and the output signal taken from the collector

FIGURE 4–29
Common-emitter amplifiers.

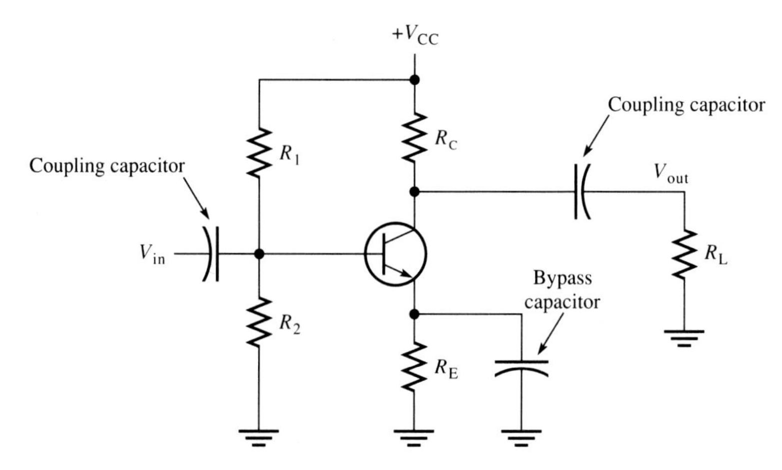

(a) CE amplifier with bypassed emitter resistance.

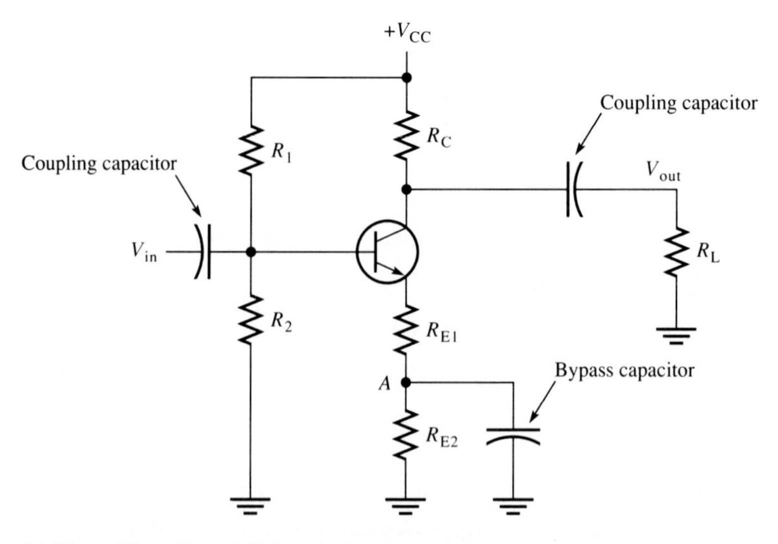

(b) CE amplifier with partially bypassed emitter resistance.

through another capacitor. These capacitors isolate the dc bias and couple the ac signal to and from the transistor. They are, therefore, called *coupling* capacitors.

Sometimes it is useful to place some point in a circuit at ac ground but not affect its dc potential. This can be accomplished with a capacitor called a *bypass* capacitor. A bypass capacitor is shown in Figure 4–29(a) across the emitter resistor that puts the emitter at ac ground. For this reason, the equivalent ac circuit is different than the dc circuit used to calculate the bias. Figure 4–29(b) shows a partially bypassed emitter-resistor. In this case, R_{E2} is bypassed, placing point *A* at ac ground. The dc bias stability is determined by both resistors, whereas R_{E1} provides gain stability for the ac signal.

ac EMITTER RESISTANCE

Recall that diodes have an ac resistance that is determined by the forward current in the diode (refer to Section 4–2). The base-emitter diode of a transistor also has an equivalent ac resistance r_e that appears to be in series with the emitter. This resistance is an ac parameter that is dependent on the dc emitter current, I_E. The value of r_e is found by

EQUATION 4–9

$$r_e = \frac{25 \text{ mV}}{I_E}$$

where r_e = equivalent ac resistance of the emitter diode, Ω
I_E = dc emitter current, mA

This equation is derived from the theoretical model (illustrated in Figure 4–2) of the *I-V* curve for a diode but applied to the base-emitter diode of a transistor.

THE ac EQUIVALENT CIRCUIT

When an ac signal is applied to a transistor amplifier, the capacitors appear to be shorted (except at very low frequencies). Figure 4–30(a) shows a CE amplifier with coupling capacitors and a bypass capacitor. Using the superposition theorem, the dc supply is replaced with a short, placing V_{CC} at ac ground. This causes the load resistor R_L to be in parallel with the collector resistor R_C and the bias resistors R_1 and R_2 to be in parallel with each other. In addition, the equivalent ac circuit contains r_e in the emitter. The ac equivalent circuit is shown in Figure 4–30(b). Observe that the emitter bypass capacitor has eliminated R_{E2} from the ac circuit, but not R_{E1}. The ac resistance of the emitter, r_e, is computed from Equation 4–9.

VOLTAGE GAIN

Voltage gain is the ratio of the output voltage to the input voltage. As you can see from the CE ac equivalent circuit, the input voltage of the CE amplifier is applied across r_e and the unbypassed resistance of the emitter circuit (R_{E1}). From Ohm's

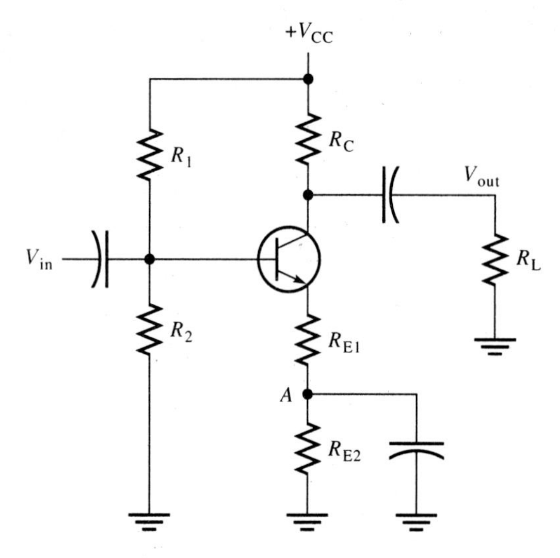

(a) CE amplifier.

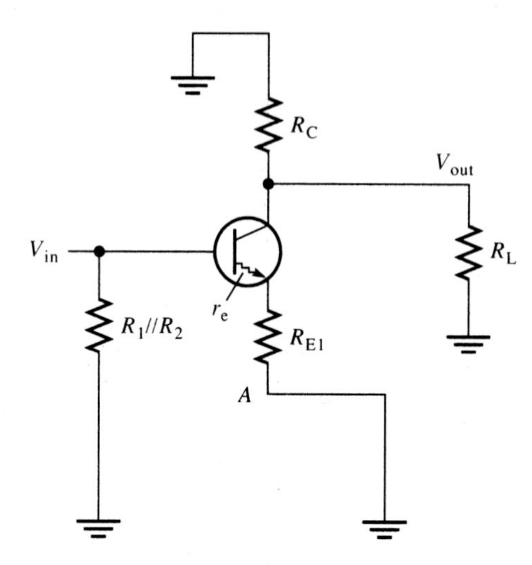

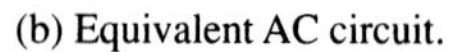

(b) Equivalent AC circuit.

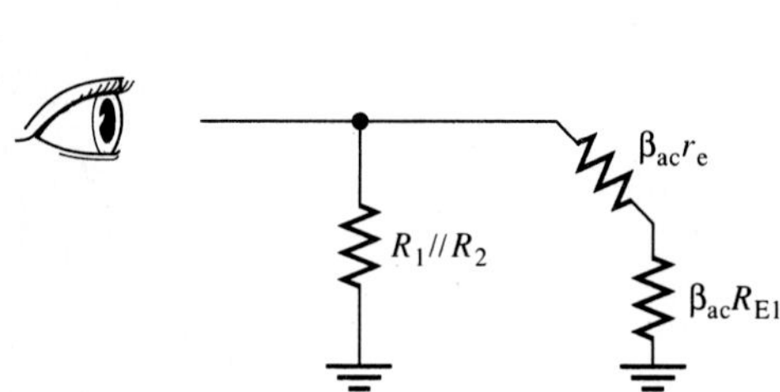

(c) "Looking" in at base to obtain input impedance.

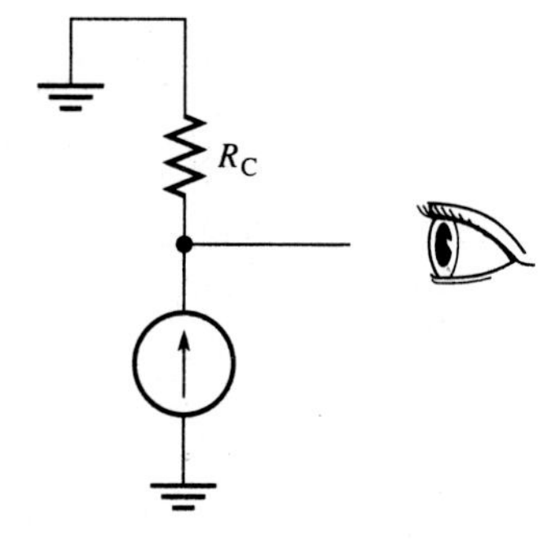

(d) "Looking" back from load to obtain output impedance.

FIGURE 4–30
CE amplifier and equivalent circuits.

law, the input voltage is

$$v_{\text{in}} = i_e(r_c + R_{\text{E1}})$$

The output voltage is taken across the ac resistance from the collector to ground. Looking from the transistor's collector, R_L appears to be in parallel with R_C. The output voltage, as given by Ohm's law, is

$$v_{\text{out}} = -i_c R_C \| R_L$$

The minus sign indicates phase inversion. That is, a positive signal on the input is turned into a negative-going signal on the output. Taking the ratio of output to input voltage gives

EQUATION 4–10

$$A_V = \frac{v_{\text{out}}}{v_{\text{in}}} = \frac{-i_c R_C \| R_L}{i_e(r_e + R_{\text{E1}})} \cong -\frac{R_C \| R_L}{r_e + R_{\text{E1}}}$$

The collector and emitter currents are nearly identical, so they cancel. The voltage gain is the ratio of the ac collector resistance to the ac emitter resistance. This result means that you can easily calculate the voltage gain of a CE amplifier by computing the ratio of ac collector resistance to ac emitter resistance. The unbypassed resistance in the emitter circuit is a form of negative feedback called *emitter degeneration*. Although the voltage gain is reduced, it is less dependent on the ac emitter resistance, resulting in predictable, stable gain. The gain is stable because it is less dependent on the transistor or variations between different transistors. The negative feedback also reduces the distortion caused by a variable ac emitter resistance.

INPUT AND OUTPUT IMPEDANCE

The input impedance is the ac resistance seen by the input signal. Remember that the V_{CC} supply looks like an ac ground, causing the input voltage divider resistors to be in parallel with each other and in parallel with the ac resistance looking into the transistor's base. Figure 4–30(c) shows the situation. Notice that the base resistance is the ac resistance in the emitter circuit (including r_e) times β_{ac}. That is

EQUATION 4–11

$$z_{in} = R_1 \| R_2 \| \beta_{ac}(r_e + R_{E1})$$

The output impedance is the Thevenin equivalent resistance looking back from the load, as shown in Figure 4–30(d). It "sees" the collector resistor in parallel with the transistor's collector, a current source. Since a current source looks like an open circuit, the output impedance is simply the collector resistor:

$$z_{out} = R_C$$

EXAMPLE 4–8 Compute the voltage gain for common-emitter amplifier shown in Figure 4–31.

FIGURE 4–31

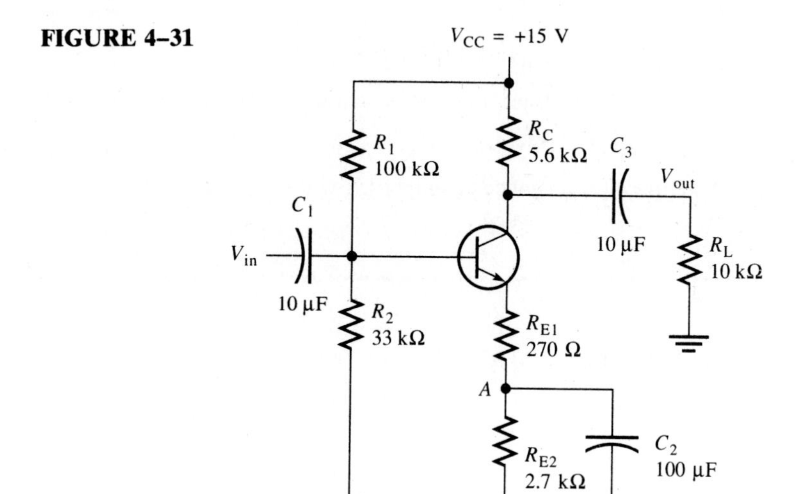

SOLUTION First find the dc parameters for the amplifier. The base voltage is found by applying the voltage-divider rule to R_1 and R_2:

$$V_B = V_{CC}\left(\frac{R_2}{R_1 + R_2}\right) = 15\text{ V}\left(\frac{33\text{ k}\Omega}{100\text{ k}\Omega + 33\text{ k}\Omega}\right) = 3.72\text{ V}$$

The emitter voltage is one diode drop less than the base voltage:

$$V_E = V_B - V_{BE} = 3.72\text{ V} - 0.70\text{ V} = 3.02\text{ V}$$

The emitter current is found by applying Ohm's law to R_{E1} and R_{E2}:

$$I_E = \frac{V_E}{R_{E1} + R_{E2}} = \frac{3.02\text{ V}}{270\ \Omega + 2.7\text{ k}\Omega} = 1.02\text{ mA}$$

The collector current is approximately the same as the emitter current, so the dc drop across R_C can be found if desired:

$$I_C \cong I_E = 1.02\text{ mA}$$
$$V_{RC} = I_E R_C = (1.02\text{ mA})(5.6\text{ k}\Omega) = 5.71\text{ V}$$

Now compute the ac parameters. The ac emitter resistance, r_e, is found by applying Equation 4–8:

$$r_e = \frac{25\text{ mV}}{I_E} = \frac{25\text{ mV}}{1.02\text{ mA}} = 24.5\ \Omega$$

The gain can be found by applying Equation 4–10:

$$A_V = \frac{R_C \| R_L}{r_e + R_{E1}} = -\frac{5.6\text{ k}\Omega \| 10\text{ k}\Omega}{24.5\ \Omega + 270\ \Omega} = 12.2$$

EXAMPLE 4–9 Compute the input and output impedance of the CE amplifier shown in Figure 4–31. Assume the β_{ac} for the transistor is 100.

SOLUTION The input impedance is given by Equation 4–11:

$$z_{in} = R_1 \| R_2 \| \beta_{ac}(r_e + R_{E1}) = 100\text{ k}\Omega \| 33\text{ k}\Omega \| 100(24.5\ \Omega + 270\ \Omega)$$
$$= 13.5\text{ k}\Omega$$

The output impedance is simply the collector resistor R_C:

$$z_{out} = R_C = 5.6\text{ k}\Omega$$

4–8 THE COMMON-COLLECTOR AMPLIFIER

In the common-collector (CC) amplifier (also called the emitter-follower), the input signal is applied to the base and the output signal is taken from the emitter, as shown in Figure 4–32(a). The collector is tied directly to V_{CC}, an ac ground. The

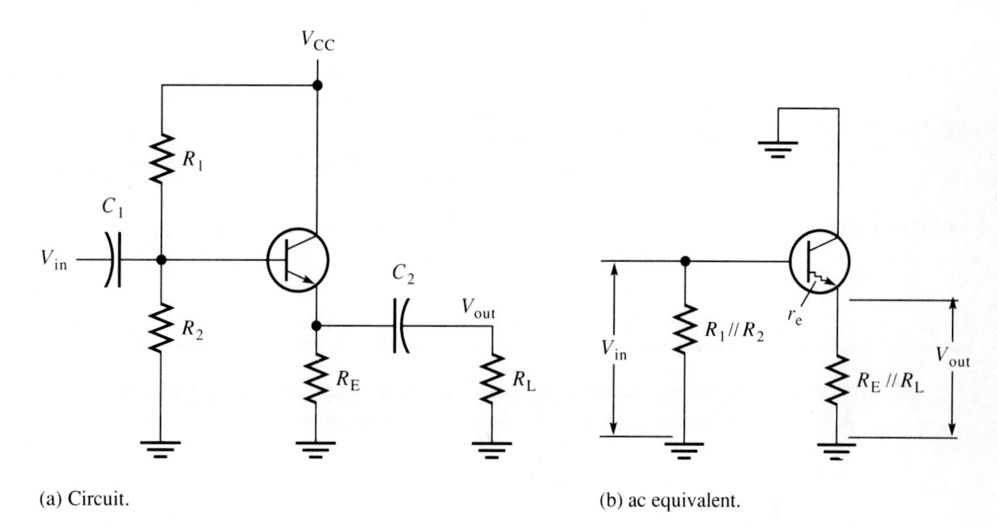

(a) Circuit.

(b) ac equivalent.

FIGURE 4–32
CC amplifier.

ac output signal is almost a perfect duplicate of the input voltage waveform (hence the alternate name emitter-follower). Although this means that the voltage gain is approximately 1, the current gain is not; hence, the common-collector amplifier can deliver increased signal power to a load. The CC amplifier is characterized by high input impedance and low output impedance. This is the most important characteristic of the CC amplifier.

To examine the characteristics of a CC amplifier, we can draw the equivalent ac circuit shown in Figure 4–32(b). The ac emitter resistance is found, as mentioned earlier, from the equation

$$r_e = \frac{25 \text{ mV}}{I_E}$$

The voltage gain is the ratio of the output voltage divided by the input voltage. The input voltage is applied across r_e *and* the ac emitter resistance, whereas the output voltage is taken only across the ac emitter resistance. Thus the voltage gain is based on the voltage-divider equation:

EQUATION 4–12

$$A_v = \frac{V_{out}}{V_{in}} = \frac{i_e(R_E \| R_L)}{i_e(r_e + R_E \| R_L)} = \frac{R_E \| R_L}{r_e + R_E \| R_L}$$

The input impedance includes the bias resistors (in parallel) and the impedance looking into the transistor's base circuit. The impedance of the emitter circuit is increased by the β_{ac} of the transistor looking from the base because of the transistor's current gain. The total input impedance seen by the ac signal is

EQUATION 4–13

$$z_{in} = R_1 \| R_2 \| [\beta_{ac}(r_e + R_E \| R_L)]$$

The power gain can be found by dividing the ac power delivered to the load resistor by the input power. The output power is V_{out}^2/R_L. The input power is V_{in}^2/z_{in}. Since the voltage gain is approximately 1, the power gain can be expressed as a ratio of z_{in} to R_L:

EQUATION 4–14

$$A_p = \frac{\left(\frac{V_{out}^2}{R_L}\right)}{\left(\frac{V_{in}^2}{z_{in}}\right)} = A_v^2\left(\frac{z_{in}}{R_L}\right) = \frac{z_{in}}{R_L}$$

EXAMPLE 4–10 Compute the input impedance and the voltage and power gain of the CC amplifier shown in Figure 4–33. Assume the β_{ac} for the transistor is 100.

FIGURE 4–33

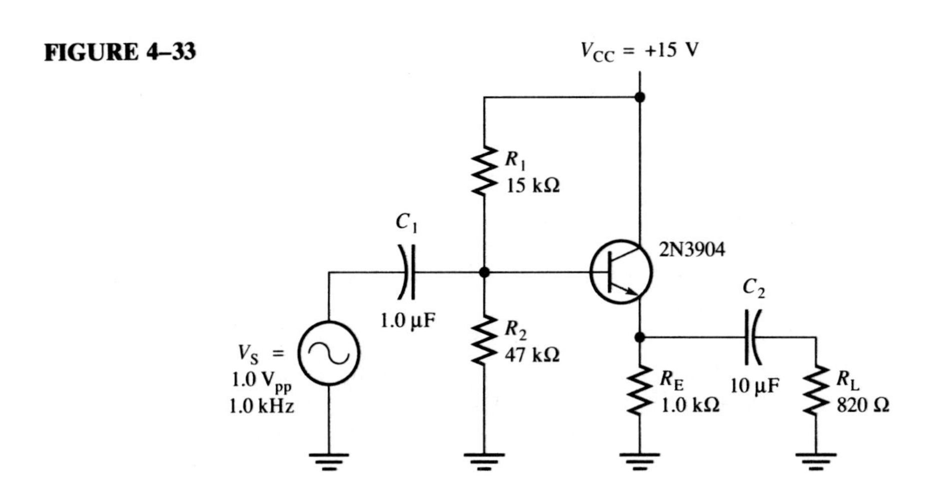

SOLUTION Start by finding the dc parameters. The base voltage is

$$V_B = V_{CC}\left(\frac{R_2}{R_1 + R_2}\right) = +15\text{ V}\left(\frac{47\text{ k}\Omega}{47\text{ k}\Omega + 15\text{ k}\Omega}\right) = 11.4\text{ V}$$

The emitter voltage is

$$V_E = V_B - V_{BE} = 11.4\text{ V} - 0.7\text{ V} = 10.7\text{ V}$$

The emitter current is found by applying Ohm's law to R_E:

$$I_E = \frac{V_E}{R_E} = \frac{10.7\text{ V}}{1.0\text{ k}\Omega} = 10.7\text{ mA}$$

Now solve for the ac emitter resistance, r_e:

$$r_e = \frac{25\text{ mV}}{I_E} = \frac{25\text{ mV}}{10.7\text{ mA}} = 2.34\ \Omega$$

The input impedance is given by Equation 4–13:

$$\begin{aligned} z_{in} &= R_1 \| R_2 \| \beta_{ac}(r_e + R_E \| R_L) \\ &= 15\ \text{k}\Omega \| 47\ \text{k}\Omega \| 100(2.34\ \Omega + 1.0\ \text{k}\Omega \| 820\ \Omega) \\ &= 9.1\ \text{k}\Omega \end{aligned}$$

The voltage gain is determined by applying Equation 4–12:

$$A_v = \frac{R_E \| R_L}{r_e + R_E \| R_L} = \frac{1.0\ \text{k}\Omega \| 820\ \Omega}{2.4\ \Omega + 1.0\ \text{k}\Omega \| 820\ \Omega} = 0.995$$

The power gain is found by applying Equation 4–14:

$$A_p = \frac{z_{in}}{R_L} = \frac{9.1\ \text{k}\Omega}{820\ \Omega} = 11.1$$

4–9 THE DIFFERENTIAL AMPLIFIER

A differential amplifier is one of the basic building blocks of operational amplifiers and is vital in many instrumentation systems. In this section, we look at the basic transistor differential amplifier; in Section 5–6 we study the operational amplifier version of this important amplifier.

A common instrumentation problem occurs when a weak signal is contaminated with interfering electrical signals. The desired signal may be a low-level signal from a transducer that is transmitted to an instrumentation system for amplification and processing. The desired signal is applied between the conductors (commonly twisted-pair wires), as illustrated in Figure 4–34. This difference signal is called the **normal-mode,** or **differential-mode,** signal. Interference causes both lines to change voltages together with respect to a common ground. This type of signal is said to be a **common-mode** signal, since it induces the same signal in each conductor. The great advantage of the differential amplifier is that it preferentially amplifies the differential mode signal compared to the common-mode

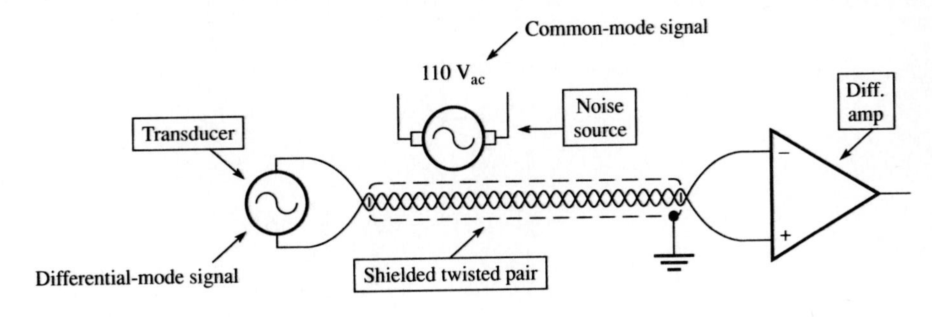

FIGURE 4–34
Common- and differential-mode signals.

signal. Notice that a differential amplifier cannot distinguish normal-mode noise from normal-mode signals, but almost all electrical interference is induced in common-mode.

Assume a line from a transducer is contaminated with common-mode noise. After amplification by a differential amplifier, the signal-to-noise ratio improves dramatically. The ability to extract a signal from noise is extremely important in instrumentation systems; virtually all low-level transducer signals are first amplified by a differential amplifier. To gain a clearer understanding of this important circuit, we first study it using discrete transistors; the integrated circuit differential amplifier is discussed in Chapter 5.

The basic circuit for a differential amplifier is shown in Figure 4–35. Notice that the emitter-bias is used; generally it is supplied by positive and negative supplies of the same magnitude. The transistors are symmetrical amplifiers sharing a common resistor, called the "tail" resistor. The tail resistor R_T has identical dc currents through it from each transistor. The small dc base current means that the base voltage of each transistor is slightly below ground; as an approximation we will assume it is -0.3 V. Dropping another 0.7 V for the base-emitter diode implies that point A is about -1.0 V. This approximation simplifies the analysis of the dc portion of a differential amplifier. The tail current can be found by applying Ohm's law to the tail resistor:

$$I_T = \frac{V_A - V_{EE}}{R_T} \cong \frac{(-1\text{ V}) - (V_{EE})}{R_T}$$

The tail current is then split between the transistors. The emitter and collector current are approximately one-half of the tail current. The voltage drop across the collector resistor can be found directly from Ohm's law, and the dc analysis is essentially complete.

Next, we consider the ac analysis. There are two inputs, one for each base. In the circuit shown, the output is taken from only one collector with respect to

FIGURE 4–35
Basic differential amplifier.

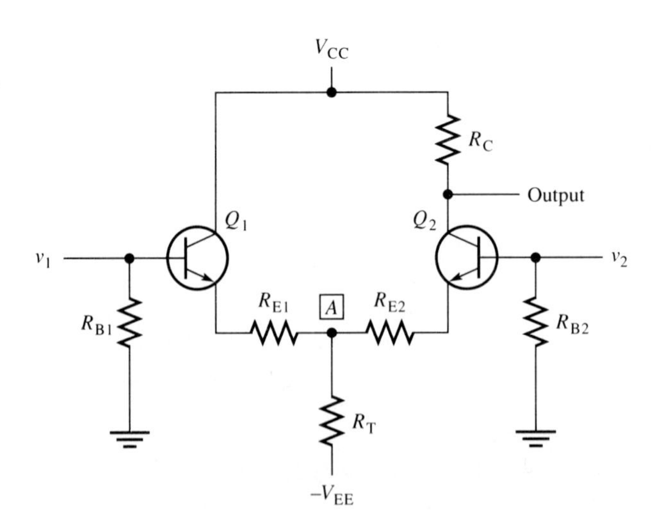

FIGURE 4–36
AC equivalent input circuits.

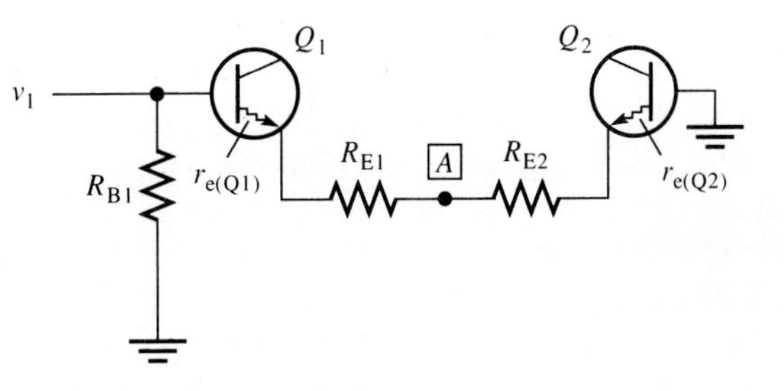

(a) ac equivalent input circuit for v_1.

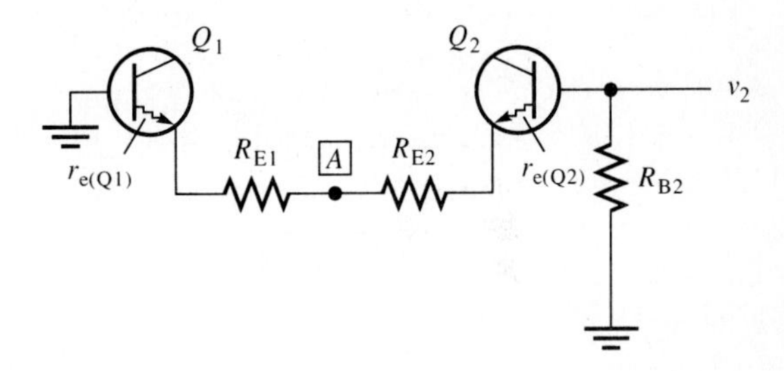

(b) ac equivalent input circuit for v_2.

ground, a configuration called *single-ended output*. Imagine that an input signal slightly raises the voltage on one base while simultaneously lowering the voltage an equal amount on the other base. This is a normal-mode input signal, and it causes the current through one transistor to be increased while the current in the other transistor is decreased by an equal amount. The net result is that the voltage at point A stays the same! Thus the portion of the input signal applied to the Q_2 base is amplified. Since the input was actually *twice* the voltage applied to Q_2, the overall differential gain is a factor of 2 less than a normal CE amplifier:

EQUATION 4–15

$$A_{v(d)} = \frac{1}{2}\left(\frac{R_C}{r_e + R_E}\right)$$

Now imagine identical signals applied to both transistors. The superposition theorem allows us to consider each signal independently of the other. Ignoring the much-larger-valued tail resistor, the input signal at the base of Q_1 "sees" the ac equivalent circuit shown in Figure 4–36(a). The bias resistor for Q_2 is removed because the superposition theorem requires us to short all other sources—in this case the input to Q_2. The signal at point A is one-half the input at Q_1 because of the symmetrical voltage divider consisting of $r_{e(Q1)}$, R_{E1}, R_{E2}, $r_{e(Q2)}$. Likewise, the ac equivalent circuit for Q_2 consists of the same resistors looking from the base of Q_2, as shown in Figure 4–36(b). The input signal at the base of Q_2 is also divided in two at point A. When these two signals are superimposed (added), the original input waveform appears at point A. Consider what happens to the current in Q_2 when its base and point A change together. The current in the transistor is unchanged, meaning the output is unchanged. The common-mode signal has been rejected.

In the preceding analysis, we ignored the large tail resistor. If we don't ignore the tail resistor, a more exact value for the common-mode gain is obtained:

EQUATION 4–16

$$A_{v(cm)} = \frac{R_C}{2R_T + R_E + r_e}$$

Note that a larger tail resistor causes the common-mode gain to be lower. We can make the tail resistor appear to be large by substituting a current source for it (recall that a current source ideally has infinite resistance). In integrated circuits, the tail resistor is actually a current source—forcing the common-mode gain to be smaller.

A differential amplifier's ability to reject the unwanted common-mode signals while favoring the desired normal-mode signal is called the amplifier's **common-mode rejection ratio** (CMRR). The CMRR is often expressed in decibels (dB) and is defined as

EQUATION 4–17

$$\text{CMRR} = 20 \log \frac{A_{v(d)}}{A_{v(cm)}}$$

In this equation, $A_{v(d)}$ represents the differential voltage gain and $A_{v(cm)}$ represents the common-mode voltage gain. The CMRR is a dimensionless number.

EXAMPLE 4–11 Compute the differential gain, the common-mode gain, and the CMRR for the differential amplifier shown in Figure 4–37.

FIGURE 4–37

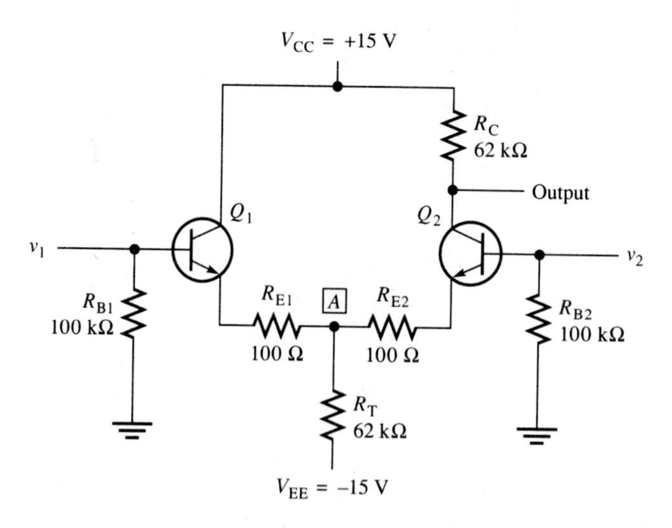

SOLUTION It is necessary to know the emitter current in order to find r_e. To find the emitter current, we can use the approximation that point A is at -1 V because of emitter bias (see Section

4–7). Then applying Ohm's law to the tail resistor

$$I_T = \frac{V_A - V_{EE}}{R_T} = \frac{-1\ \text{V} - (-15\ \text{V})}{62\ \text{k}\Omega} = 226\ \mu\text{A}$$

The tail current is divided between the transistors. Therefore, I_E is

$$I_E = \frac{I_T}{2} = \frac{226\ \mu\text{A}}{2} = 113\ \mu\text{A}$$

The ac emitter resistance r_e is found using Equation 4–8:

$$r_e = \frac{25\ \text{mV}}{113\ \mu\text{A}} = 221\ \Omega$$

The differential gain is

$$A_{v(d)} = \frac{1}{2}\left(\frac{R_C}{r_e + R_E}\right) = \frac{1}{2}\left(\frac{62\ \text{k}\Omega}{221\ \Omega + 100\ \Omega}\right) = 96.6$$

The common-mode gain is

$$A_{v(cm)} = \frac{R_C}{2R_T + R_E + r_e} = \frac{62\ \text{k}\Omega}{2(62\ \text{k}\Omega) + 100\ \Omega + 221\ \Omega} = 0.5$$

The CMRR can be found from Equation 4–17:

$$\text{CMRR} = 20 \log \frac{A_{v(d)}}{A_{v(cm)}} = 20 \log \frac{96.6}{0.5} = 45.7\ \text{dB}$$

4–10 PHOTOTRANSISTORS

A **phototransistor** is a light-sensitive bipolar transistor that is used as a light detector in low-light-level applications. It is essentially a photodiode-amplifier combination that operates much like an ordinary bipolar transistor. Base current is generated when light photons pass through a transparent window to a large reverse-biased base-collector junction, causing valence electrons to move into the conduction band. The resulting base current is amplified by the current gain of the transistor. Some phototransistors are constructed with no external base connection; others have a base lead that can be used for external bias. If the base is connected to an external bias supply, the phototransistor can be operated as a normal transistor, with or without light. To operate as a light detector, the base lead is left open and the collector is positively biased, as shown in Figure 4–38.

The sensitivity of phototransistors is measured in collector current per optical power per unit area. The intensity of optical power per unit area is called **irradiance,** abbreviated H and typically measured in units of milliwatts per square centimeter. Under normal operating conditions, a typical phototransistor has a sensitivity of about 1 mA/mW/cm^2. There is no simple relationship between the power supplied to a source and light power; the power supplied to a source

FIGURE 4–38
A simple phototransistor circuit. The base lead is left open.

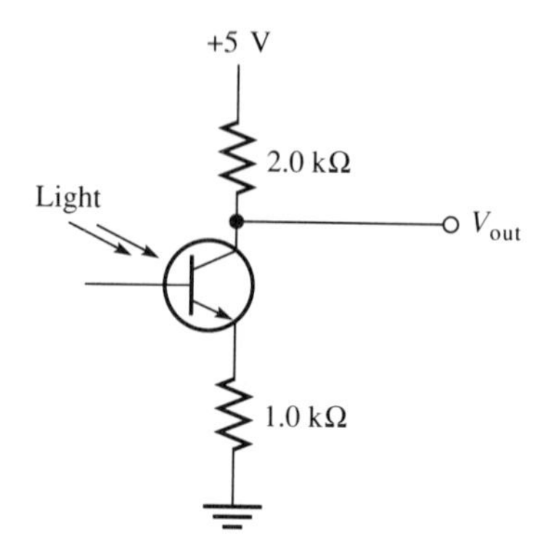

depends on the efficiency of the light source, the spectrum and spectral response, spatial distribution of the light, and losses in the light path. The following example shows how these effects can be estimated:

EXAMPLE 4–12 Estimate the collector current from a silicon phototransistor that has a sensitivity of 1 mA/mW/cm² and is located 30.5 cm (1 ft) from a 100 W tungsten light source. Assume the light is distributed spherically and there are no losses in the light path.

SOLUTION We will assume a tungsten filament bulb is 85% efficient when converting the input power into light. Imagine that the bulb is located at the center of a sphere of radius r. The radiant flux from the bulb is distributed over a spherical area of $A = 4\pi r^2$. The irradiance at a distance r is found by multiplying the input power by the efficiency and dividing by the surface area of the sphere. That is

$$H \cong \frac{P_{in}\epsilon}{4\pi r^2} = \frac{100{,}000 \text{ mW} \times 0.85}{12.56 \times (30.5 \text{ cm})^2} = \frac{7.3 \text{ mW}}{\text{cm}^2}$$

where P_{in} = power supplied to bulb, mW
ϵ = conversion efficiency of bulb
r = distance between source and phototransistor

The spectral response of the phototransistor causes the output to be reduced. This factor depends on the temperature of the source and is approximately 25% at a source temperature of 2700°C. Using this factor and the given sensitivity, we find that the output current is

$$I_C = 0.25 \times \frac{7.3 \text{ mW}}{\text{cm}^2} \times \frac{1.0 \text{ mA}}{\text{mW/cm}^2} = 1.83 \text{ mA}$$

Darlington phototransistors are more sensitive but have a much longer response time. Compared to the photodiode, phototransistors are relatively slow (typically 10 μs rise time) but more sensitive to light. Various light sensors are compared further in Section 13–9.

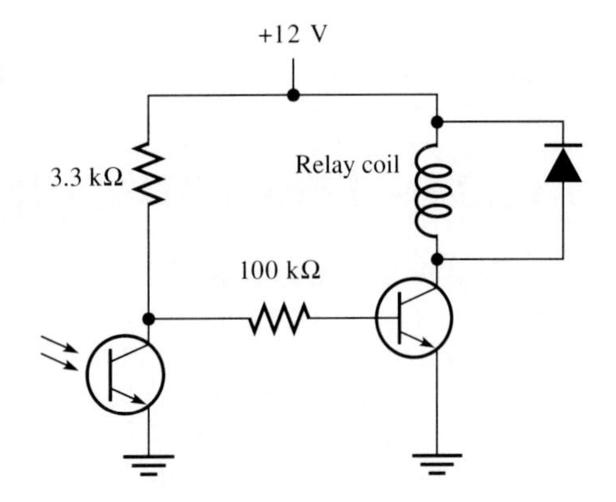

FIGURE 4–39
A phototransistor circuit that de-energizes a relay. The diode is used to prevent damage to the transistor when it turns off due to the collapsing field of the coil. The relay is off when light is on.

Phototransistors and other light sensors have many applications to instrumentation systems; typical applications include a light source and phototransistor to sense an object by blocking a light beam. Practical examples include detecting the end of a roll of paper or other material, detecting the breaking of a thread, checking the level of a hopper, detecting tipped cartons, and counting objects. The signal from the phototransistor frequently needs to turn power either on or off. Figure 4–39 shows a circuit that uses a phototransistor to de-energize a relay that controls power.

OPTO-ISOLATORS

An opto-isolator (also called an optocoupler) consists of a light source (usually an infrared-emitting diode, or IRED), a transparent dielectric, and a phototransistor or other light sensor in the same package. The IRED and phototransistor are coupled optically but electrically isolated from each other. In instrumentation applications, opto-isolators are used to break ground loops from transducers and to isolate low-power, sensitive circuits from high-power loads (see Section 12–5). For example, an opto-isolator can prevent interference from affecting a logic circuit when the logic circuit is used to control a motor or other noisy power device. The opto-isolator can also prevent high voltages from damaging the logic circuit in case of catastrophic failure in the power circuit.

Important specifications for opto-isolators include isolation resistance (typically about 10^{12} Ω), isolation capacitance, and isolation voltage. Isolation capacitance is small (1 to 2.5 pF) but may pass the very fast transients that occur in high power switching circuits. Isolation voltage is the maximum voltage that the dielectric can withstand under specified conditions. Another specification is the transfer characteristic. The transfer characteristic is determined by source and phototransistor responses and the dielectric characteristics. Since these tend to be nonlinear, opto-isolators are more commonly applied to digital than analog circuits.

A variation on the opto-isolator is the optical interrupter, so named because of a slot between the light source and light sensor that allows the beam to be interrupted by a card, tape, or other opaque material. This is a highly reliable, low-cost transducer for high-speed counting or sensing the presence of a thin object.

4–11 FIELD-EFFECT TRANSISTORS

Field-effect transistors (FETs) use an electrostatic field brought about by an applied voltage to control current. Their inputs draw no current (other than leakage), implying extremely high input impedance. There are two basic types of FETs: the JFET (junction field-effect transistor) and the MOSFET (metal-oxide-semiconductor FET). The MOSFET is sometimes referred to as an IGFET (insulated-gate FET). Figure 4–40 illustrates the structure of a JFET. Both the JFET and MOSFET begin with a single piece of doped silicon called the *channel*. On one end of the channel is a terminal called the *source* and on the other end of the channel is a terminal called the *drain*. Connected to the channel is the opposite type of material, called a *gate*. For the JFET, the gate forms a *pn* junction with the channel, whereas the MOSFET contains an insulating silicon dioxide (glass) layer between the gate and the channel. When a voltage is applied between the drain and source terminals, a current flows in the channel. Electrons, or holes, are removed from the drain terminal and replenished at the source terminal (hence the names). This channel current, called I_D, is controlled by the voltage that is applied to the gate terminal. In addition to the source, drain, and gate terminals, many MOSFETs have a base, or substrate, connection. This is not an active terminal; rather it is the crystal foundation of the transistor. The substrate is kept reversed-biased in any MOSFET, usually by connecting it directly to the source. In some MOSFETs, this is done internally; in others, the substrate connection is brought out as a separate lead. Representative symbols for JFETs and MOSFETs are shown in

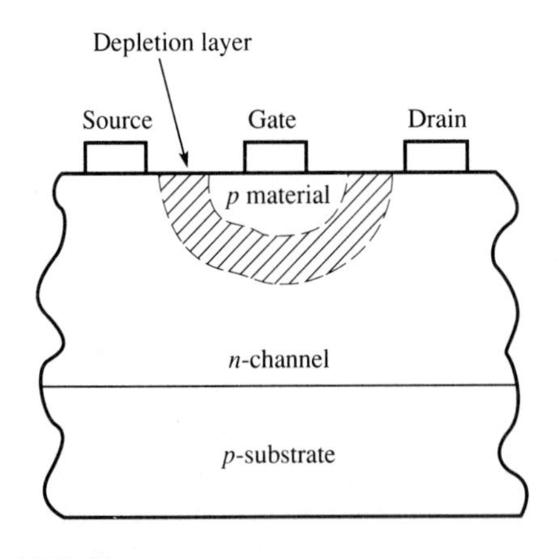

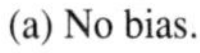
(a) No bias.

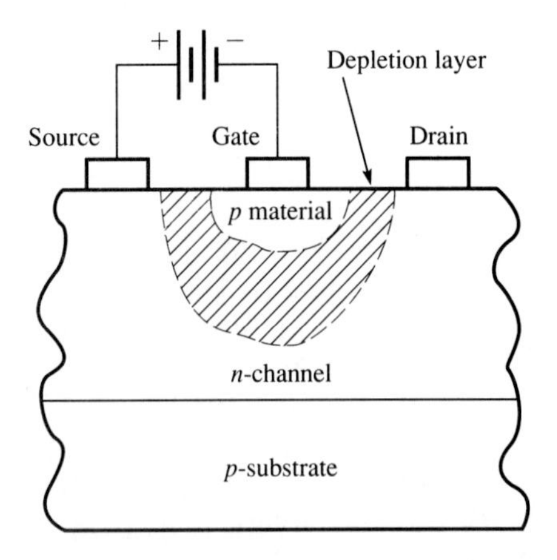

(b) Negative bias applied to the gate. Notice that the channel becomes narrower.

FIGURE 4–40
Structure of an *n*-channel JFET.

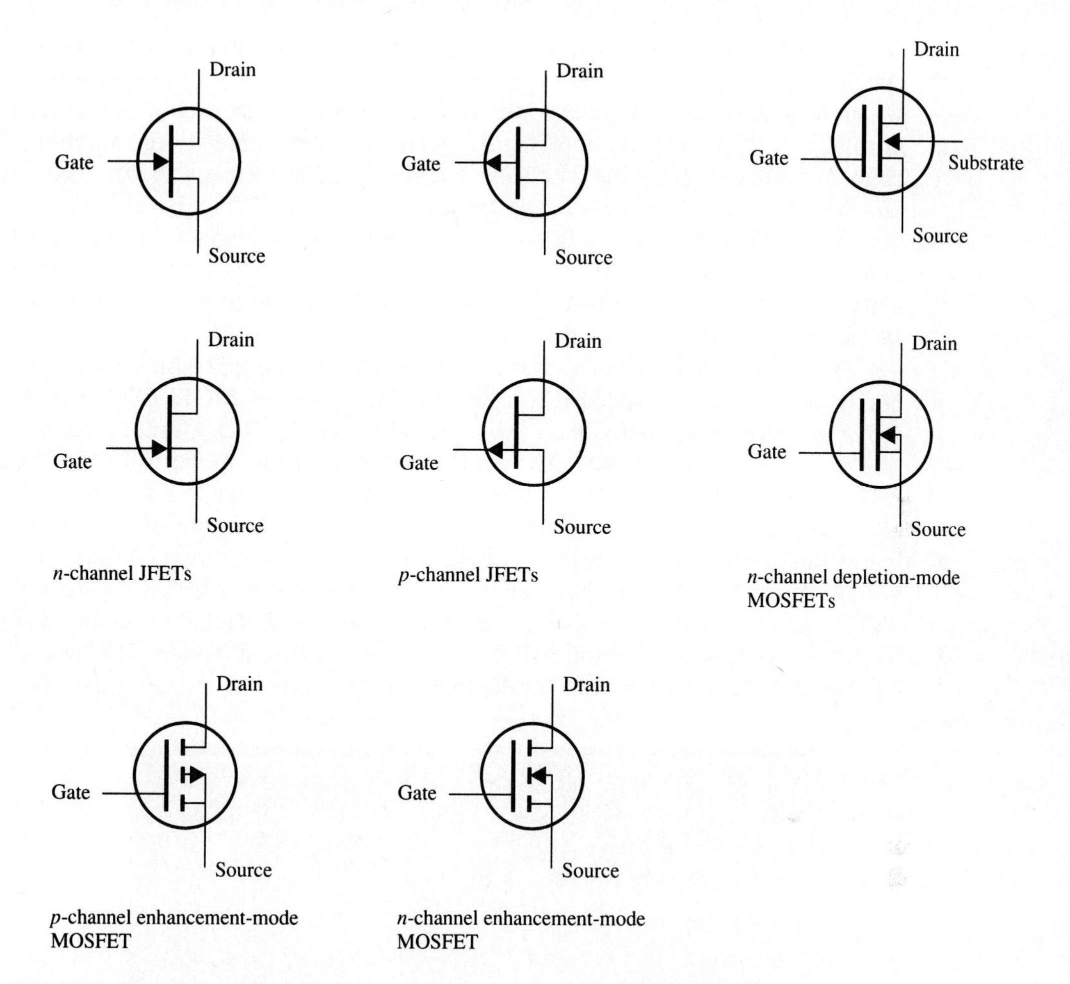

FIGURE 4–41
Representative schematic symbols for FETs.

Figure 4–41. Note that the arrow on the gate points in for an *n*-channel and out for a *p*-channel device. When the arrow is drawn off-center, it is placed closest to the source terminal. Some FETs have an interchangeable source and drain, in which case the gate arrow is drawn between them.

The output, or drain, characteristics for a typical *n*-channel JFET are shown in Figure 4–42. Instead of showing the output characteristic for base current, each curve represents a constant gate-source voltage. Notice that the drain current is not a linear function of the drain-source voltage. A given gate-source voltage causes a nearly constant drain current, except when the drain-source voltage is small. Observe that the curves are not equal distances apart. This is because most small-signal field-effect transistors are nonlinear devices, an advantage in certain applications. Also note that the maximum drain current flows when the gate-source voltage is zero. This current is called I_{DSS}, a current that is listed on specification sheets for JFETs. It is the current that flows from drain to source when the gate is shorted to the source. Unlike the bipolar transistor, the JFET is a normally "on" device. By applying a reverse-bias to the gate-source junction, the channel is closed off by the gate's electric field, and less current flows. The input impedance is always high because the input to the JFET is a reverse-biased *pn* junction. When the gate-source voltage is made large enough, the channel is pinched off and drain current is cut off. This voltage is called $V_{GS(OFF)}$ and is usually specified as the voltage that reduces the drain current to between 0.1% and 1% of I_{DSS}.

MOSFETs have similar characteristics, except the gate-source voltage depends on the type of MOSFET. There are two types of MOSFETs: *depletion-mode* (d-MOSFETs) and *enhancement-mode* (e-MOSFETs). Output characteristics are shown in Figure 4–43. The depletion-mode type is similar to JFETs in that it has a conductive channel with no bias present. The major advantage of the depletion-mode MOSFET is that, because of the insulated gate, it can be biased with either a forward or a reverse gate-source voltage. A reverse gate-source voltage tends to pinch off the channel as in the JFET, but a forward gate-source voltage tends to draw more charge carriers into the channel and increase conductivity. The enhancement mode MOSFET is a normally off device. The gate must be forward-biased to permit conductivity in the channel. By contrast, JFETs

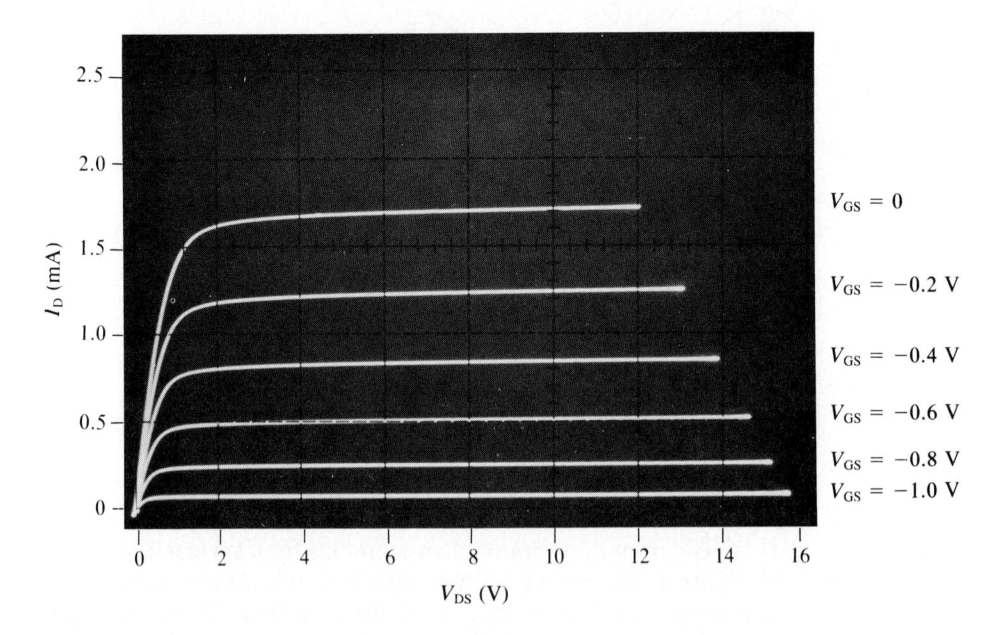

FIGURE 4–42
Drain characteristics for a typical JFET, the 2N5458. Compare this photograph with the bipolar transistor shown in Figure 4–19. Steps are 0.2 V apart. The top line represents a gate source voltage of zero; the bottom line represents cutoff.

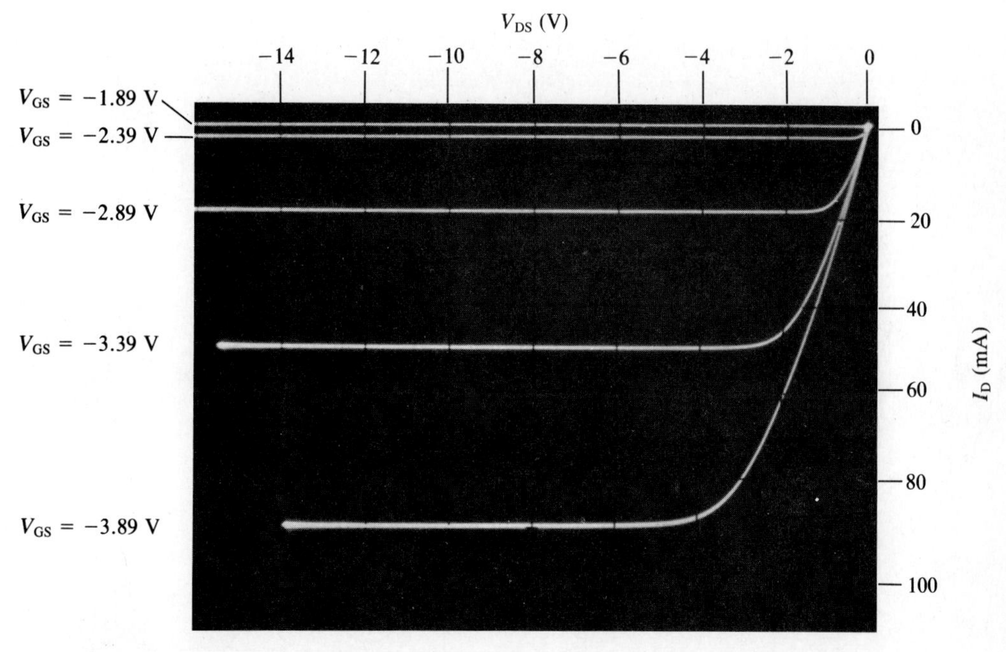

FIGURE 4–43
Drain characteristics for a typical *p*-channel e-MOSFET. This photograph is taken from the screen of a transistor curve tracer and is oriented as it appears on the instrument. Notice that the highest level just turns on the transistor and the channel becomes more conductive as V_{GS} increases. Steps are 0.5 V apart. The top line represents the threshold voltage; for this transistor, the threshold voltage is −1.89 V.

should never be operated in the enhancement mode, as this would cause the gate-source diode to be forward-biased and could destroy the device. Depletion-mode MOSFETs are drawn with a solid line for the channel and enhancement-mode MOSFETs are drawn with a broken line for the channel. (The broken line is a reminder that the channel is not normally conducting.)

4–12 FET BIASING

As you might suspect, the bias circuits depend on the type of FET. With depletion-mode devices, which include JFETs and d-MOSFETs, self-bias is widely used. Self-bias is illustrated in Figure 4–44(a). The drain current, I_D, flows through the source resistor, creating a voltage $V_S = I_D R_S$ at the source terminal. Since the gate is at ground potential (no current flows through R_G), the required reverse bias is on the gate. Self-bias tends to compensate for different device characteristics. For example, if more drain current flows, the voltage across R_S increases. This voltage tends to bias the FET off, compensating for the original increase in drain current.

FIGURE 4–44
Representative FET bias circuits.

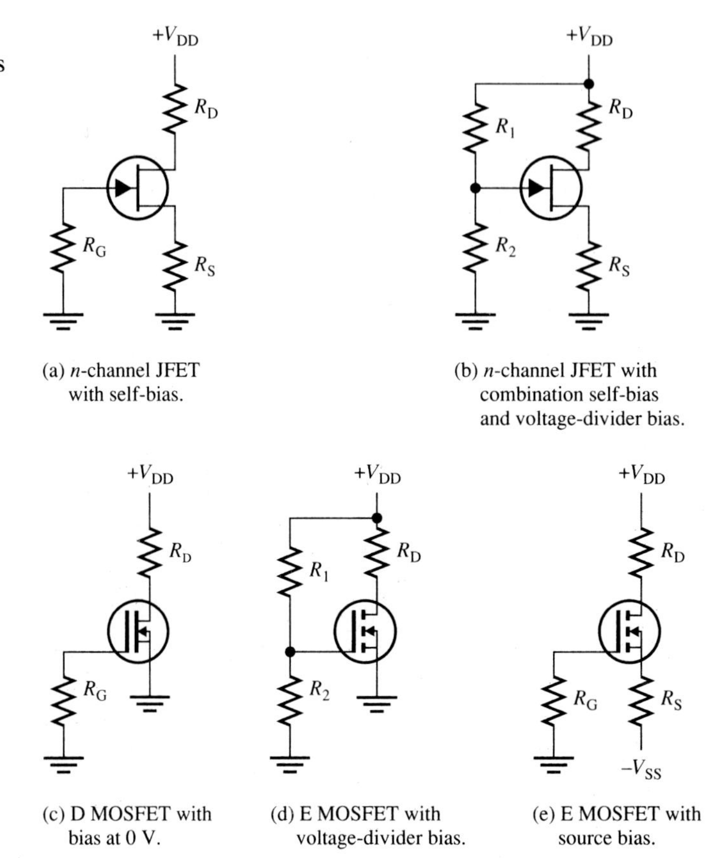

An even more stable form of bias combines self-bias with voltage-divider bias, as illustrated in Figure 4–44(b). The voltage divider connected to the gate biases the gate at some positive voltage. The source voltage must be more positive than the gate in order to have the proper gate-source negative bias. To accomplish this, the source resistor is made larger than in self-bias. The net result of this type of bias is that transistor variations have less affect on the operating point than self-bias.

A simpler form of biasing is available for depletion-mode MOSFETs. In addition to the bias methods described before, the operating point in d-MOSFETs can be set at $V_{GS} = 0$, and the signal can be allowed to vary above and below this point. The circuit requires only a gate resistor to tie the gate to a solid ground, as shown in Figure 4–44(c). This type of bias is valid only for depletion-mode MOSFETs.

Enhancement-mode MOSFETs are normally off devices, similar in this respect to bipolar transistors. Therefore it is not surprising to find that bias circuits that work for bipolar transistors can also be used for e-MOSFETs. The analogous voltage-divider bias and emitter bias circuits shown with bipolar transistors are drawn for enhancement-mode MOSFETs in Figure 4–44(d) and (e).

EXAMPLE 4–13 Assume the JFET circuit shown in Figure 4–45 has a source current of 750 μA. What are the gate-source voltage and the drain-source voltage?

FIGURE 4–45

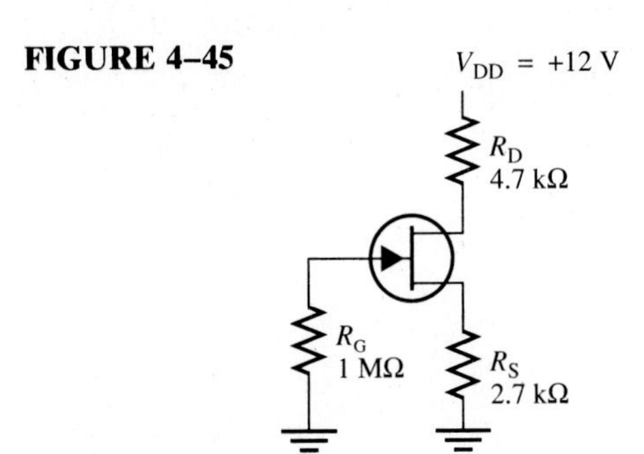

SOLUTION The gate is held at 0 V by R_G and draws no current. The source voltage is found from Ohm's law:

$$V_S = I_S R_S = (750\ \mu A)(2.7\ k\Omega) = 2.02\ V$$

V_{GS} is

$$V_{GS} = V_G - V_S = 0\ V - 2.02\ V = -2.02\ V$$

The drain current is identical to the source current. Applying Ohm's law to the drain circuit gives the drain voltage V_D:

$$V_D = V_{DD} - I_D R_D = 12\ V - (750\ \mu A)(4.7\ k\Omega) = 8.48\ V$$

V_{DS} is

$$V_{DS} = V_D - V_S = 8.48\ V - 2.02\ V = 6.46\ V$$

4–13 FET CIRCUIT APPLICATIONS

FET AMPLIFIERS

FETs are particularly useful as amplifiers when very high input impedance and low noise are necessary, as in multimeters and the front end of receivers. Practical common-source FET amplifiers are shown in Figure 4–46. Compare the common-source amplifier shown in Figure 4–46(a) with the common-emitter amplifier for bipolar transistors in Figure 4–29. The input impedance is much higher for the JFET amplifier because the gate-source junction is reversed-biased rather than forward-biased, as in the bipolar transistor. It turns out that, with practical devices, the bipolar amplifier can have greater voltage gain over a linear region than the JFET; however, the extremely high input impedance in the JFET allows smaller coupling capacitors to be used in order to achieve the same low-frequency response.

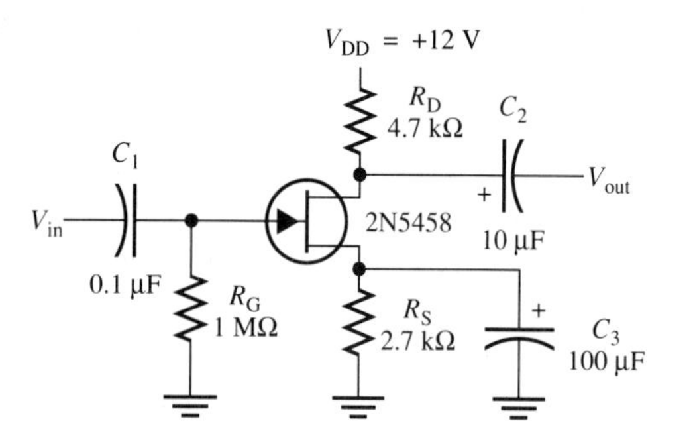

(a) A common-source JFET amplifier.

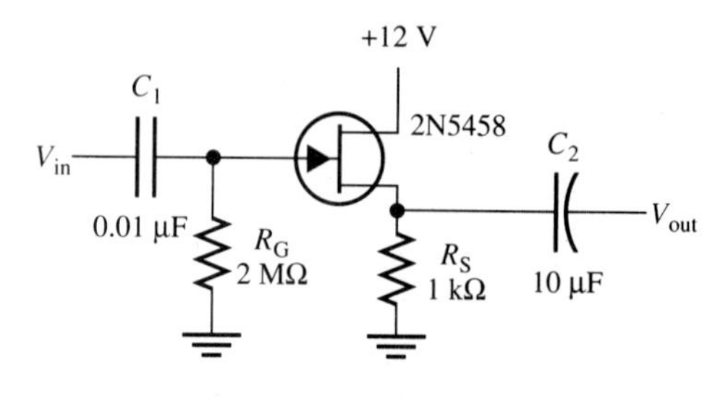

(b) A common-drain (source-follower) JFET amplifier.

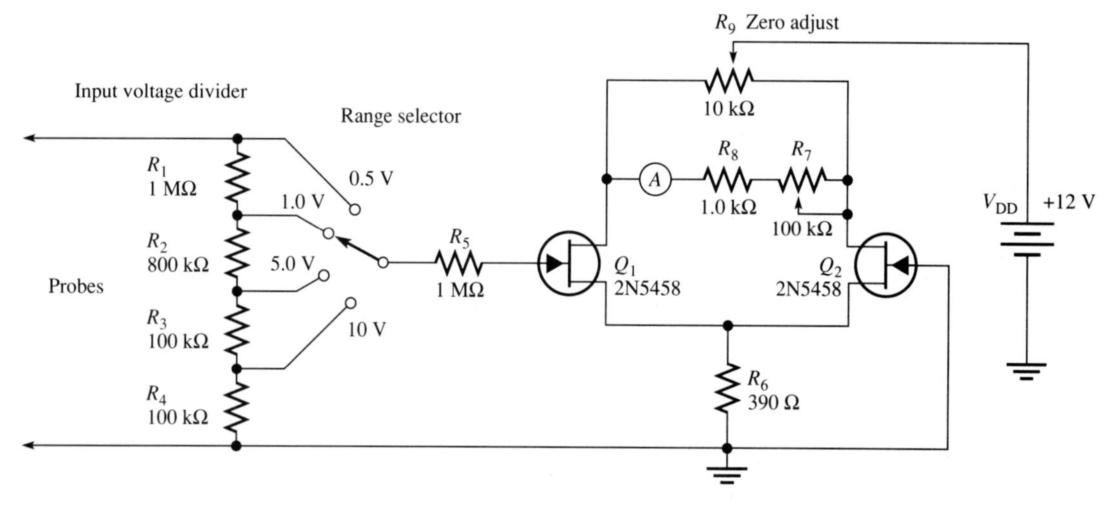

(c) A differential-amplifier with JFETs. This amplifier is connected as a basic voltmeter.

FIGURE 4–46
JFET amplifiers.

A JFET can be configured as a high-impedance differential amplifier, with the advantage of high common-mode noise rejection. Figure 4–46(c) shows a basic four-range electronic voltmeter using a FET differential amplifier. The range selection is accomplished by the input voltage divider. Because of the very high input impedance of the FET, the voltage divider is essentially unloaded. The input impedance is constant for all ranges; in this example it is 2 MΩ. The differential amplifier drives a sensitive ammeter connected between the drains. Electronic voltmeters are discussed further in Section 7–4.

JFET SWITCHES

As previously mentioned, switching circuits are important in many instrumentation systems. There are two basic types of electronic switches—analog switches, which allow current to flow in either direction, and digital switches, which allow

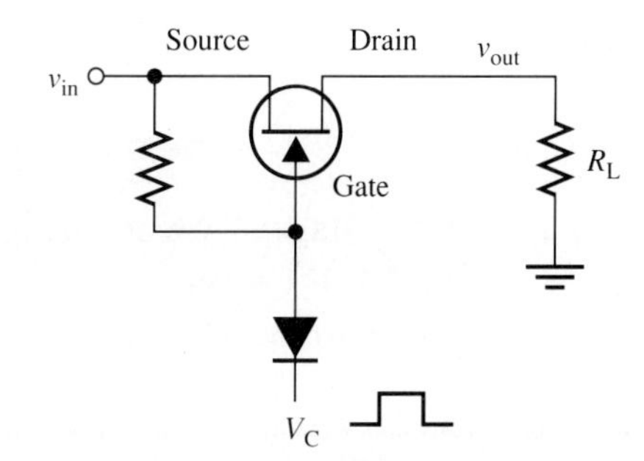

FIGURE 4–47
JFET analog switch.

current to flow in only one direction. Analog switches should pass a signal without changing it and without regard to the waveform polarity. A JFET analog switch (transmission gate) is shown in Figure 4–47. The JFET is switched between the on and off states with a control voltage applied to the gate. To turn the JFET on, V_{GS} is made equal to 0 V—a problem because the input (source) voltage varies. The solution is to connect a resistor between the source and gate terminals and use a diode to prevent V_{GS} from becoming positive. To turn the JFET off, the channel must be pinched off. This is done by making the control voltage more negative than the lowest value of the signal plus the pinch-off voltage, V_p. (The magnitude of V_p is the same as $V_{GS(OFF)}$.)

JFETs are particularly suited as analog switches because the gate draws essentially no current; they have an extremely high off resistance (resulting in a high degree of isolation from the control signal) and a very low on resistance (typically 25 Ω). The ratio of off to on resistance is very high, offering good isolation of signals in the off mode. The JFET appears to be in series with a load resistance, as shown in Figure 4–48. The insertion loss is essentially independent of frequency and is due solely to the internal resistance of the switch, $r_{DS(on)}$, and the load resistance, R_L. Insertion loss in decibels is computed from the equation

EQUATION 4–18

$$I_{dB} = 20 \log \frac{R_L}{R_L + r_{DS(on)}}$$

where I_{dB} = insertion loss in dB for the on state
R_L = load resistance, Ω
$r_{DS(on)}$ = drain-source resistance in the on state, Ω

The on-state channel resistance is a function of $V_{GS(off)}$ and I_{DSS}, parameters that can be found on manufacturers' specification sheets. The equation for $r_{DS(on)}$

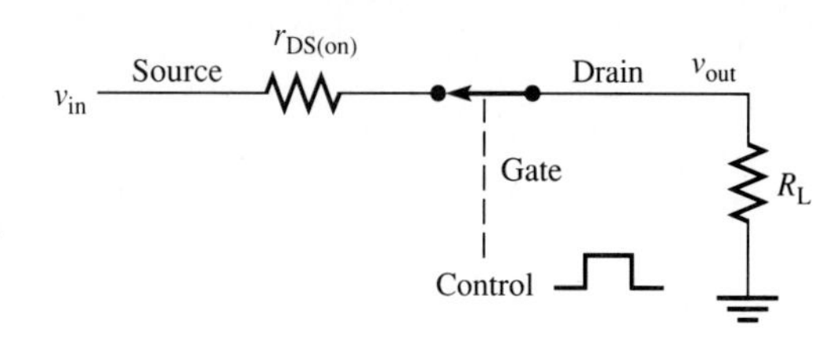

FIGURE 4–48
Simplified equivalent circuit for JFET analog switch. In the on state, a resistance of $r_{DS(on)}$ appears in series with R_L.

is

EQUATION 4–19

$$r_{DS(on)} = -\frac{V_{GS(off)}}{2I_{DSS}}$$

where $V_{GS(off)}$ = gate voltage at which channel conduction approaches zero, V

I_{DSS} = drain current when zero gate voltage exists, A

EXAMPLE 4–14 Compute the insertion loss for a JFET with an $r_{DS(on)}$ of 25 Ω using a 100 Ω load.

SOLUTION

$$I_{dB} = 20 \log \frac{R_L}{R_L + r_{DS(on)}}$$

$$= 20 \log \frac{100\ \Omega}{100\ \Omega + 25\ \Omega}$$

$$= -1.9 \text{ dB}$$

JFET switches are fast, and switching is accomplished from control voltages that can be obtained from standard logic families with an interface such as that shown in Figure 4–49. A disadvantage of JFET switches is that they are also more prone to switching transients appearing on the signal lines when control signals change states. All in all, the advantages far outweigh the disadvantages, and JFET switches are widely used in switching applications for instrumentation systems.

Specially designed FET analog switches are available in IC form with switching speeds of less than 1 ns and capacitances of under 3 pF. They are useful in circuits such as multiplexers, fast A/D and D/A converters, choppers, and

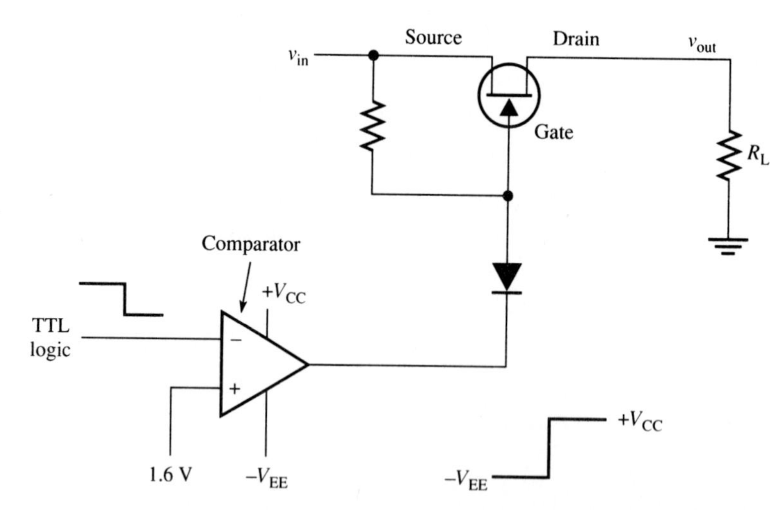

FIGURE 4–49
Digital control of an analog signal. The comparator inverts the logic signal. A logic high causes the JFET to turn off; a logic low causes the JFET to turn on.

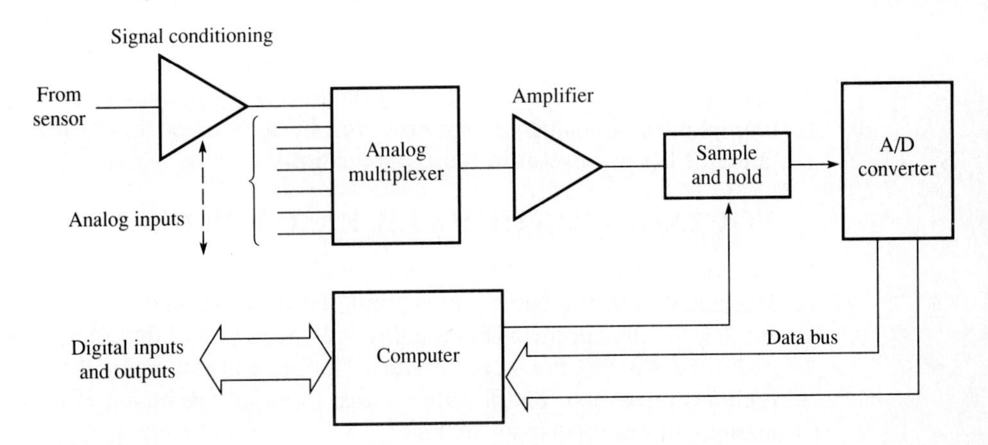

FIGURE 4–50
A data-acquisition system that uses an analog multiplexer. The analog multiplexer is an IC that uses FET switches and on-chip decoding logic to allow digital control of the switches.

sample-and-hold circuits. Figure 4–50 shows how IC analog switches could be used in a data-acquisition system.

MOSFET SWITCHES

MOSFETs are also widely used for analog switching applications. They offer simpler circuits than JFETs and both positive and negative control. A basic MOSFET switch is shown in Figure 4–51. For switching high currents, power MOSFETs are available that can switch as high as 10 A with a control circuit that supplies essentially no current to the gate. A disadvantage to MOSFET switches is that the on resistance tends to be higher than that of JFETs—on the order of 100 Ω, as opposed to 25 Ω for the JFET. This means that the insertion loss will be higher, an effect noticeable with low-impedance circuits.

FIGURE 4–51
A *p*-channel MOSFET analog switch.

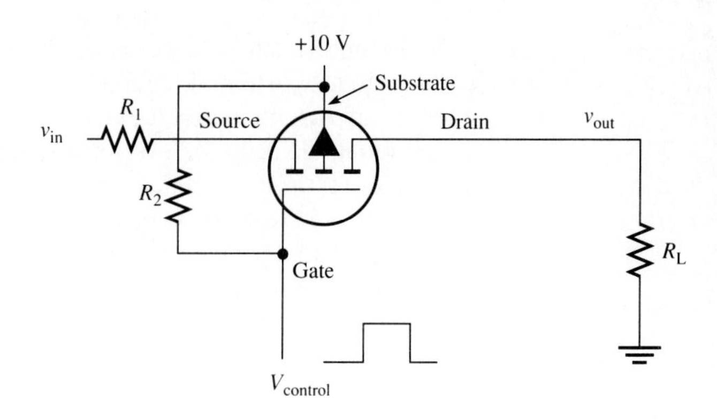

A related analog switch is called a solid-state relay (SSR). SSRs are available as a package containing an optical isolator and power MOSFET output. In the past, mechanical relays have been necessary for applications such as low-level multiplexing of analog signals; however, because of extremely high off-state resistance, SSRs are useful in these applications.

VOLTAGE-CONTROLLED RESISTORS

One of the most useful features of JFETs is the ability to control the channel resistance with the gate voltage, thus becoming a voltage-controlled resistance (VCR). Resitive control is in the low-voltage region of the drain-source characteristic curves called the *ohmic region*. The slope of the characteristic curves in this region is dependent on the gate voltage. Specially designed JFETs are offered by manufacturers for use as VCRs.

The ability to control the resistance of a device with a voltage is important for automatic gain control (AGC) circuits and voltage-controlled attenuators. In instrumentation, automatic gain control is useful to prevent an input signal from saturating an amplifier.

4–14 TRANSISTOR CURVE TRACER

An instrument that is useful for observing the characteristics of semiconductor devices is the **transistor curve tracer** (often referred to as simply the curve tracer). A transistor curve tracer is illustrated in Figure 4–52. In addition to characterizing bipolar and field-effect transistors, the instrument can measure parameters for diodes and thyristors. It is designed to make a variety of measurements on each of these devices, including most of the important parametric tests. Some advanced curve tracers can perform these measurements automatically and include automated setup and sequencing through a variety of measurements, data storage, and direct hard-copy output of measurements for documentation.

The reasons for measuring the characteristics of a device are varied. In engineering work, it is useful to know certain parameters to assure a complete understanding of circuit performance. Component manufacturers measure characteristics in order to develop better products and to characterize production runs. Sometimes a curve tracer is used for an incoming test, quality control, or to sort components. And, of course, there are educational reasons to study the parameters of various active devices.

The block diagram for a transistor curve tracer is shown in Figure 4–53. The device under test (DUT) is placed in one of the test sockets and the controls are set for the specific test. Most tests are three-terminal measurements; others, such as diode tests and certain transistor tests, require only two terminals. For three-terminal measurements, a control voltage or current is provided to one of the terminals of the DUT in equal-value steps by a **step generator.** Normally, the step generator produces a stair-step output with a constant value of voltage or current at each step; for certain measurements, it can be set to produce a steady current or

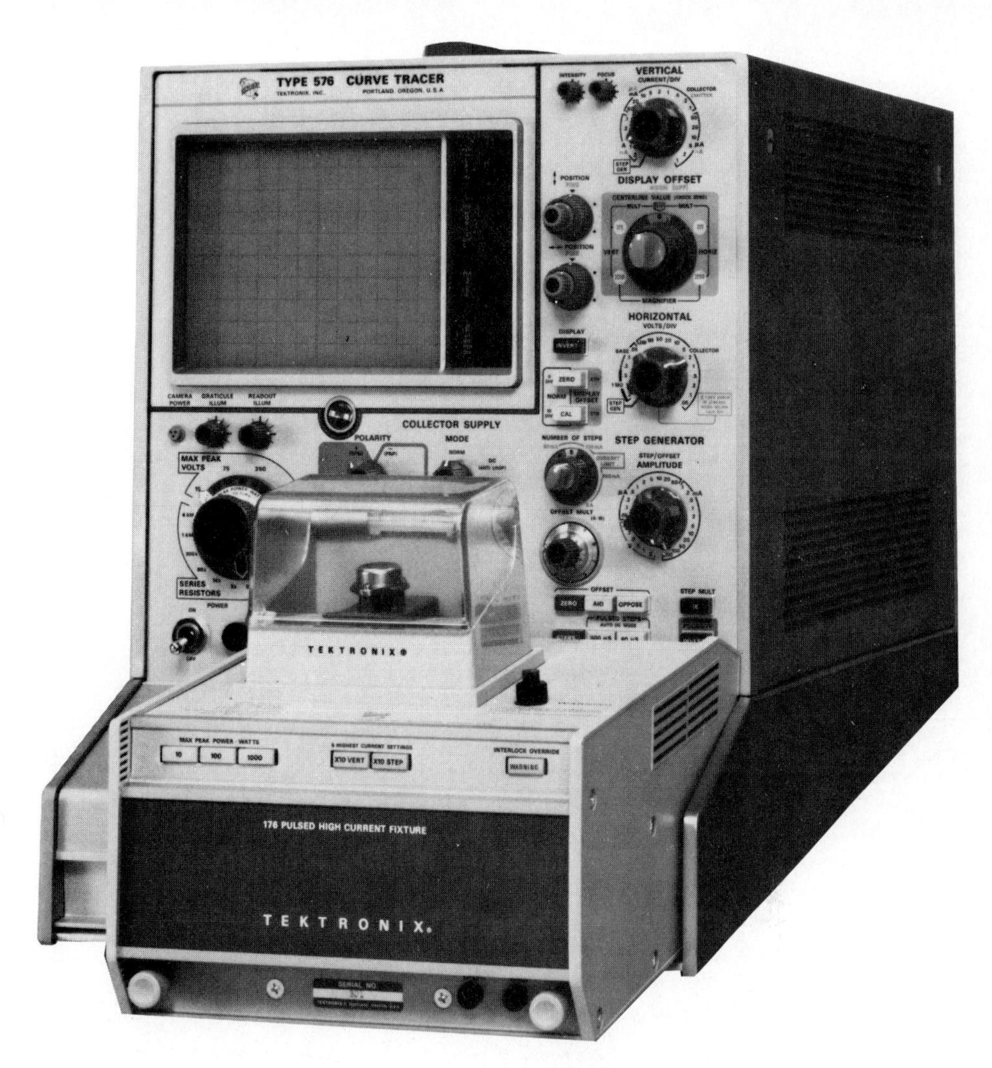

FIGURE 4–52
Transistor curve tracer (copyright by Tektronix, Inc. Used with permission).

voltage. Steps are synchronized with a collector supply voltage that sweeps back and forth—one time for each step. The resulting display shows a family of curves; each curve represents a different step from the step generator.

BIPOLAR TRANSISTOR MEASUREMENTS

For bipolar transistors, the most common measurement is β. To measure dc β, the collector current and voltage conditions are set up based on the circuit or manufacturer's specifications. Polarity of the collector supply and step generator is selected for the particular transistor; *npn* transistors are plotted in the first quad-

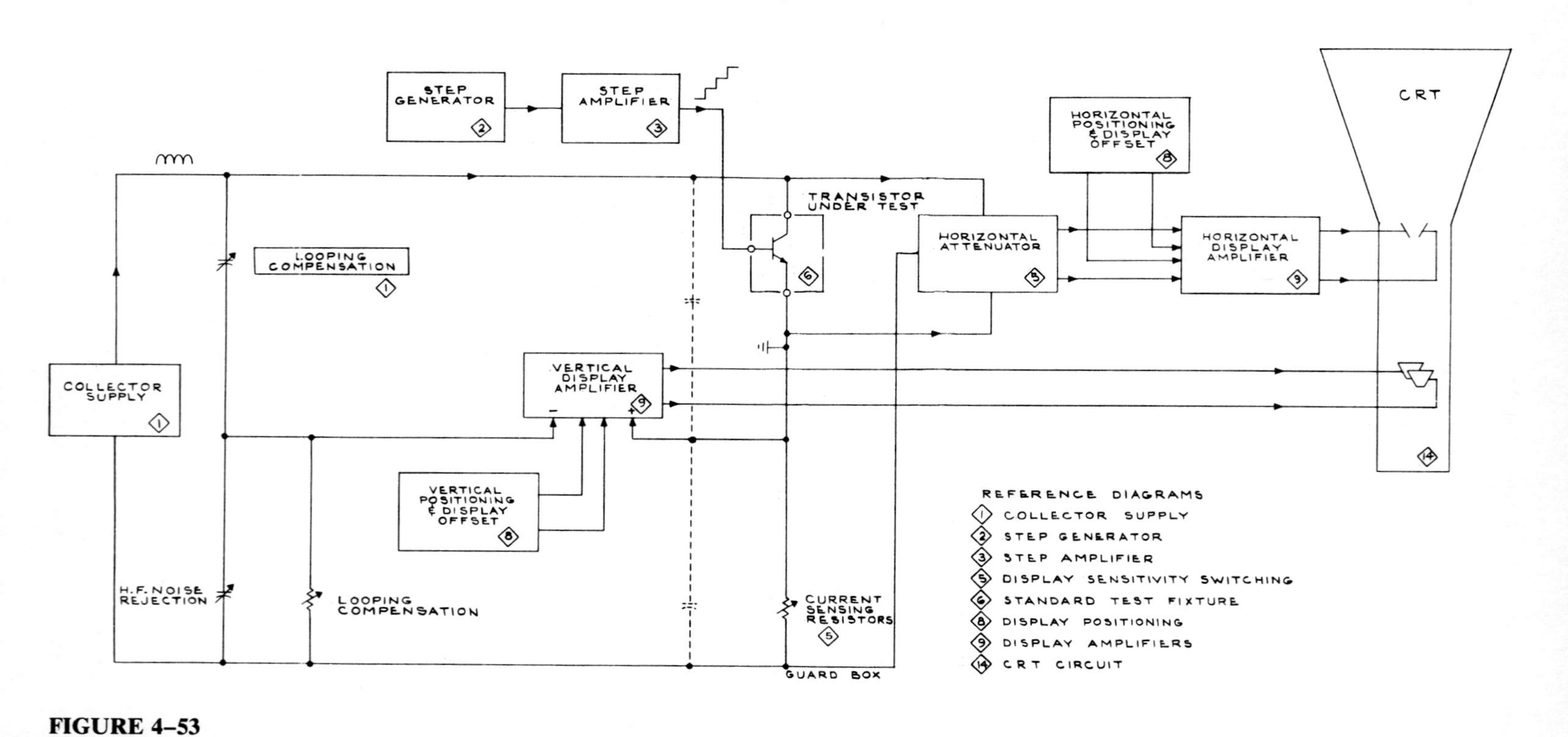

FIGURE 4–53
Block diagram of curve tracer (copyright by Tektronix, Inc. Used with permission).

rant and *pnp* transistors are plotted in the third quadrant. Typically, for small-signal transistors, the collector voltage is set for less than 25 V. The step generator is adjusted for a single step with a base current within specified limits—the base current may need to be computed from the expected β. The collector-emitter voltage is plotted on the horizontal axis and the collector current is plotted on the vertical axis, as was shown in Figure 4–19(b). The β is determined by reading the collector current in the flat region and dividing by the step generator's current per step.

To measure the ac β, the number of steps is adjusted to show a family of curves (such as shown in Figure 4–19(c)). The ac β is the ratio of change in the collector current to change in base current. A vertical line corresponding to the specified operating voltage is selected. The *change* in collector current is determined from two adjacent steps. The corresponding change in the base current is the current per step read on the step-generator control. The curve tracer can measure other parameters for bipolar transistors, including the breakdown voltage between various terminals, collector current cutoff, and the collector-emitter saturation voltage.

FIELD-EFFECT TRANSISTOR MEASUREMENTS

Field-effect transistors require a wide range of voltages and current to test them. FETs are typically tested for drain current with zero gate voltage (I_{DSS}), transconductance (g_m), pinch-off voltage (V_p), and reverse gate leakage (I_{GSS}). JFETs are also tested for gate-source breakdown voltage ($V_{(BR)GSS}$), but MOSFETs may be damaged when breakdown occurs. The setup for viewing the characteristics is similar to the setup for the bipolar transistor, with the leads inserted into the socket that corresponds—gate into base socket, drain into collector socket, and source into emitter socket. The polarity of the step generator and collector supply depends on the device; for an *n*-channel JFET, the drain (collector) supply is set for positive polarity and the gate (base) step generator is set for negative steps.

FETs normally are voltage-controlled devices. To observe the drain characteristics (such as those shown in Figure 4–40), the step generator is adjusted for voltage at the gate instead of current, as in the case of the BJT. The step generator is set to provide a series of voltage steps in even increments from 0 V to $V_{GS(OFF)}$. From the drain characteristic curve, the transconductance can be obtained; as previously defined, it is the *change* in output current divided by the corresponding *change* in input voltage. Transconductance is highest in the region where drain current is highest. It is measured at a point to the right of the knee. Other parameters that can be determined from the drain characteristic curves include the drain current at zero bias (I_{DSS}) and the pinch-off voltage (V_p).

SUMMARY

1. A semiconductor diode allows current to flow when it is forward-biased and blocks current flow when it is reverse-biased.
2. Diode characteristics are described by a graph of their current-voltage relationship.
3. Rectifier diodes are used in power supplies to change ac to dc.
4. Signal diodes are used in general-purpose applications such as clipping and clamping circuits.
5. Diodes can be tested with an ohmmeter or on a curve tracer.
6. In a bipolar junction transistor, a small base current controls a larger collector current. The ratio of these currents is called the β of the transistor.
7. The characteristics of a transistor are described by a family of curves.
8. The load line is a graph that represents all possible operating points for a circuit element.
9. Bipolar transistors are operated with forward bias on the base-emitter junction and reverse bias on the base-collector junction.
10. For a common-emitter amplifier, the input ac signal is applied between the base and emitter. The output signal is taken from between the collector and emitter.
11. The differential amplifier has important applications in instrumentation systems where a weak signal is contaminated with interfering electrical noise.
12. Field-effect transistors (FETs) use a gate voltage to control current in the drain circuit.
13. FETs are useful in applications where extremely high impedance is necessary and in switching applications.
14. Semiconductors can be tested with a transistor curve tracer. Curve tracers are capable of making a variety of measurements of diodes, transistors, and other devices.

QUESTIONS AND PROBLEMS

1. Compare a rectifier diode, a signal diode and a zener diode.
2. A diode has the forward characteristic curve shown in Figure 4–54. Compute the static and dynamic resistance of the diode at a forward current of 5.0 mA.
3. For the circuit in Figure 4–9(b), what effect would an increase in temperature have on the clipping level?
4. Compare half-wave, full-wave, and bridge rectifier circuits. What is the advantage of each?
5. Sketch the output waveform for the circuit shown in Figure 4–55.
6. Explain the difference between a clipping circuit and a clamping circuit.
7. In the circuit of Figure 4–9(d), what effect would reversing the diode have on the output waveform?
8. Explain the difference between a coupling and a bypass capacitor.
9. The transistor test circuit shown in Figure 4–17(b) has V_{BB} = 5.0 V and V_{CC} = 12 V. A voltmeter measures 2.2 V across R_2. (a) What is the collector current? (b) What is the base current? (c) What is V_{CE}? (d) What is the β_{dc} for the transistor?
10. The CE amplifier circuit in Figure 4–56 uses voltage-divider bias and a *pnp* transistor. Compute the dc base, emitter, and collector voltages. You can ignore the capacitors since they are open to dc.

FIGURE 4–54

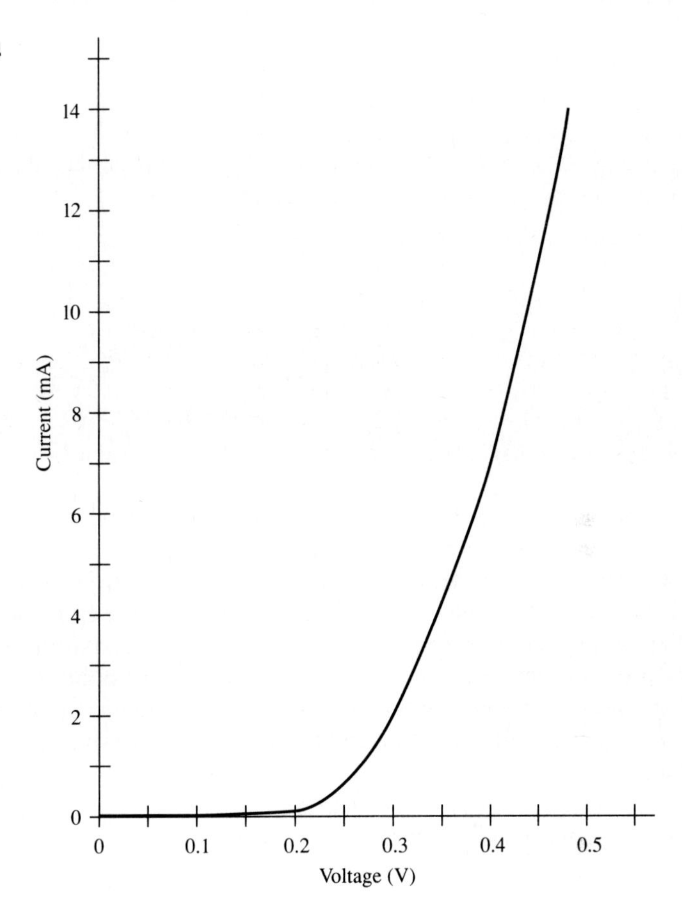

FIGURE 4–55

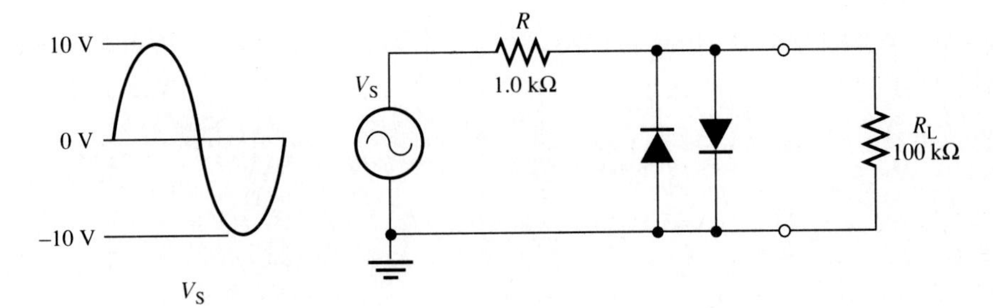

11. Find the voltage gain, input impedance, and output impedance for the unloaded CE amplifier shown in Figure 4–56. The ac resistance of the collector circuit is just R_C. Assume $\beta = 100$.

12. Compare a common-emitter amplifier with a common-collector amplifier. Where is the signal applied and removed for each? How do the gains compare?

FIGURE 4–56

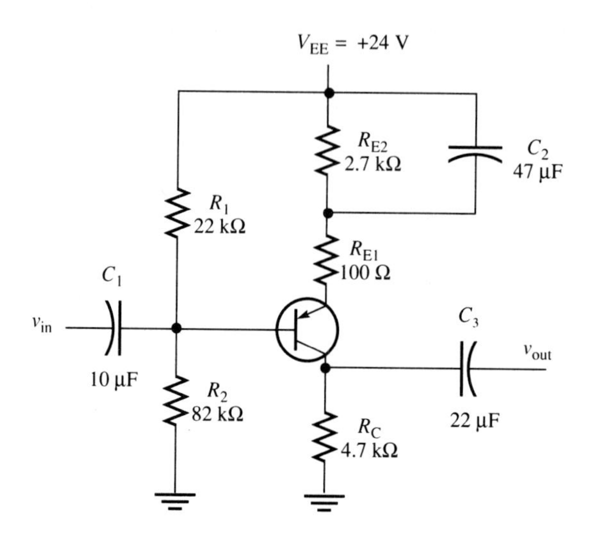

13. Explain the difference between differential gain and common-mode gain for a differential amplifier.
14. If a differential amplifier has a common-mode gain of −0.5 and a differential gain of −120, compute the common-mode rejection ratio.
15. Compute the differential gain, the common-mode gain, and the CMRR of the differential amplifier shown in Figure 4–57. Assume β = 100 for both transistors.
16. What is the difference between a depletion-mode MOSFET and an enhancement-mode MOSFET? Which type can use zero bias?
17. Assume a JFET switch has an insertion loss of 2.5 dB with a 150 Ω load resistor.
    (a) What is $r_{DS(on)}$ for the JFET?
    (b) If $V_{GS(off)} = -4$ V, compute I_{DSS}.

FIGURE 4–57

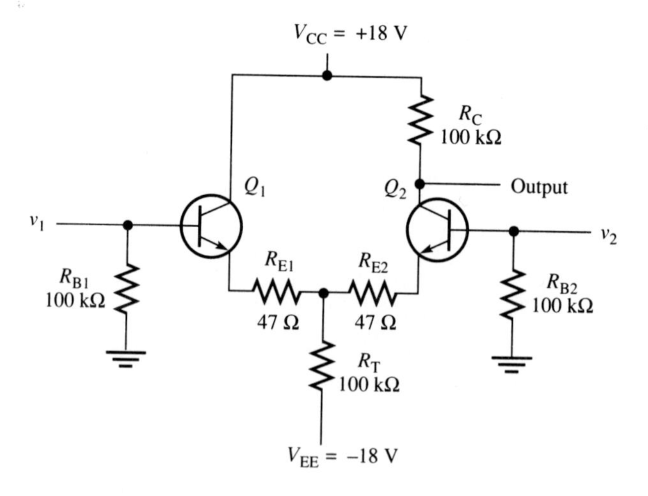

Chapter 5

Operational Amplifiers

OBJECTIVES

The development of integrated circuit operational amplifiers (op-amps) in the early 1960s had a profound influence on instrumentation and electronics. Prior to IC op-amps, transistorized amplifiers were used for various instrumentation requirements. Op-amps simplified instrumentation system design and made available instruments with features that were impractical twenty years ago. Some applications of op-amps in instrumentation systems include linear and nonlinear amplifiers, filters, oscillators, and comparators, to name a few. They provide the user with a highly reliable and very economical package, frequently with performance features that are superior to discrete components. Because of its many applications, the op-amp is the most versatile and widely used linear circuit in electronic equipment.

When you complete this chapter, you should be able to

1. Describe the characteristics of an ideal op-amp.
2. List advantages of negative feedback.
3. Apply ideal op-amp characteristics to linear amplifier circuits to explain their operation.
4. Compute the voltage gain for inverting and noninverting op-amp circuits.
5. Explain terms and specifications used to describe op-amps.
6. Discuss the application of instrumentation amplifiers and isolation amplifiers.
7. Use scaling and summing amplifiers to solve basic algebraic equations.
8. Determine the trigger levels for comparators and Schmitt trigger circuits.
9. Differentiate between regenerative feedback oscillators and relaxation oscillators.
10. Explain the Barkhausen criterion for oscillator circuits.
11. Design an active Butterworth low-pass or high-pass active filter that meets specified roll-off characteristics.

HISTORICAL NOTE

In 1921, Harold S. Black began working on the task of linearizing, stabilizing, and otherwise improving amplifiers for his employer, Western Electric Co., for the purpose of developing multichannel amplifiers that could carry long-distance conversations. The problem facing Black was the requirement for an amplifier that could reduce distortion by a vast amount, a task that seemed insurmountable. In one interesting development, he invented an amplifier called the feed-forward amplifier, which isolated the distortion products in a separate amplifier; he then used the results to cancel the distortion in the original amplifier. Unfortunately, the feed-forward amplifier was too complicated and temperamental to be practical. In 1927, while traveling to work on the old Lackawanna Ferry, he suddenly realized that if he fed some of the output back to the input in opposite phase, he had a means of canceling the distortion. This idea, sketched on a copy of the *New York Times,* led to the development of the negative feedback amplifier, one of the most important and widely applied ideas in electronics.

5–1 INTEGRATED CIRCUITS

Integrated circuits (ICs) are combinations of circuit elements fabricated in a single package to perform some function. Compared to discrete circuits, the advantages of integrated technology are low cost, high reliability, low power consumption, improved performance, and small-size circuits. Most ICs fabricated on a single substrate (''chip'') are called **monolithic** (''one stone'') and may contain thousands of transistors, diodes, resistors, and capacitors and their interconnections. Some circuit elements, such as large capacitors, inductors, and transformers, cannot be fabricated in a monolithic IC. **Hybrid** ICs can be designed to include various discrete components along with monolithic ICs, all mounted and interconnected on an insulating ceramic base.

ICs can also be classified according to their function—the principle division is between linear, digital, and microwave types. Linear ICs contain transistors designed to operate in the active region and are used for amplifying and a wide variety of signal-conditioning applications. Digital ICs contain transistors that are designed to be either on or off to provide either a high or low output level. Digital ICs are virtually all monolithic devices. Microwave ICs are used at frequencies up to 15 GHz; they depend to a great extent on hybrid technology.

5–2 OP-AMP PRINCIPLES

An **operational amplifier** (op-amp) is a linear IC with a high-gain dc amplifier designed to be used with external resistors and capacitors to determine its operation and gain. Op-amps were originally designed using discrete components before integrated circuits were available. They were used in analog computers for the purpose of providing a versatile circuit that could perform mathematical operations. Today, it is possible to manufacture IC transistors with specifications as good as the best low-noise discrete transistors. There are literally hundreds of types of op-amps available in various IC packages. They are designed as both general-purpose and as special-purpose op-amps for a wide variety of applications, including linear and nonlinear signal-processing functions. Closely related to op-amps are instrumentation amplifiers and comparators, with many specifications common to both.

The input stage of nearly all op-amps is a high-gain, dual-input differential amplifier, which is typically followed by an intermediate stage, a level-shifting stage, and a push-pull output stage that provides low impedance drive for the load. Since the input stage is a differential amplifier, op-amps amplify the *difference* voltage that is applied between the inputs. Common-mode signals (see Section 4–10) are rejected. The output signal is single-ended (referenced to ground), enabling the signal to swing either in the positive or negative direction. For this reason, both positive and negative power supply voltages are generally required, although some specialized op-amps have been designed to operate with one supply. The two supplies are of opposite polarity but usually have the same magnitude, commonly ± 12 to 15 V. The signal cannot reach the extreme limits of the supply voltage but generally can reach within a volt or two of the supply before limiting occurs. The ground pin is usually not connected to the op-amp but is the center point between the power supplies and forms the reference for the output. For simplifying schematics, the power supply connections are often not shown but are necessary for proper operation.

The schematic symbol for the entire op-amp is a triangular shape, as shown in Figure 5–1. Notice that the inputs are labeled with a (+) and a (−) sign and the signals are correspondingly labeled v_+ and v_-. The inputs should be read as the noninverting (+) and the inverting (−) inputs and are so labeled because of the phase of the output signal. The output, which is shown at the apex of the triangle, responds in the same direction to the signal applied to the noninverting input (a positive input produces a positive output). On the other hand, the output is the opposite of the input when a signal is applied to the inverting input (a positive input produces a negative output). Many op-amps also have frequency compensation connections.

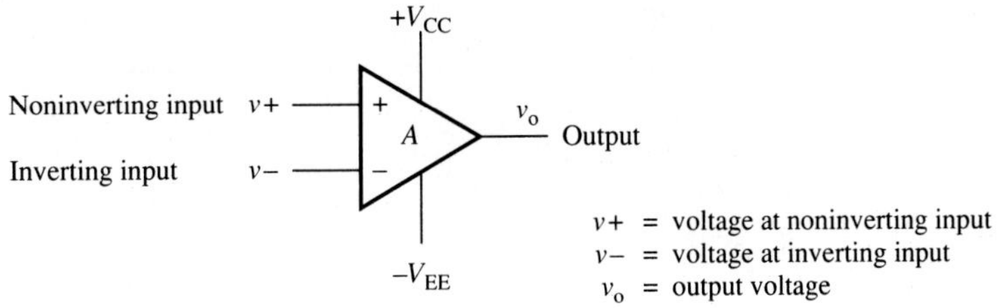

FIGURE 5–1
Schematic symbol for an op-amp.

$+V_{CC}$
Noninverting input $v+$
Inverting input $v-$
A
v_o Output
$-V_{EE}$

$v+$ = voltage at noninverting input
$v-$ = voltage at inverting input
v_o = output voltage

5–3 NEGATIVE FEEDBACK

In order to clarify op-amp circuits, it is necessary to understand the concept of feedback in amplifiers. Feedback occurs when a portion of the output signal is mixed with the input signal. If the feedback signal is returned 180° out of phase with the input signal, it is termed **negative feedback.** The general idea is illustrated in Figure 5–2. Negative feedback decreases the gain of an amplifier but produces a number of desirable effects, including increased stability, reduced distortion, and greater bandwidth. In addition, the input and output impedance of the amplifier is changed by feedback. An example of feedback was seen in Figure 4–30 when an unbypassed emitter resistor was present in a common-emitter amplifier. The result was decreased gain and increased stability, causing the amplifier to be less sensitive to manufacturing variations between transistors.

Feedback requires that a sample of the output signal be taken and returned to the input. The primary requirement of the sampling network is to return a feedback signal that is proportional to the output signal. The sampled signal can be either a voltage sample or a current sample, which will depend on how the feedback network is connected. Figure 5–3 shows these two possibilities. In Figure 5–3(a), the sample network is connected in parallel with the load; hence it is a sample of the output voltage. In Figure 5–3(b), the sample network is connected in series with the load; hence it is a sample of the output current. The sampling networks shown are purely resistive; however, other components are sometimes used, depending on the application.

The sampled signal is returned to the input through a summing network. The purpose of the summing network is to combine the original signal with the inverted feedback signal. This network can be connected in a manner that causes

FIGURE 5–2
Block diagram of a feedback model.

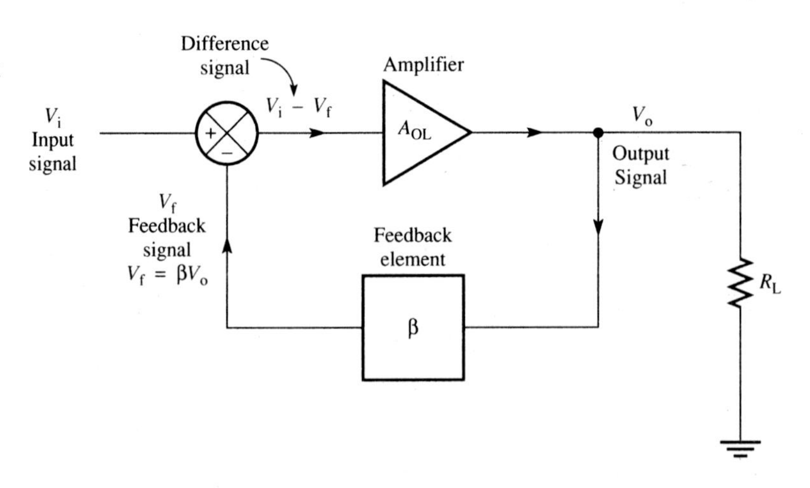

FIGURE 5–3
Feedback can be proportional to output voltage or current.

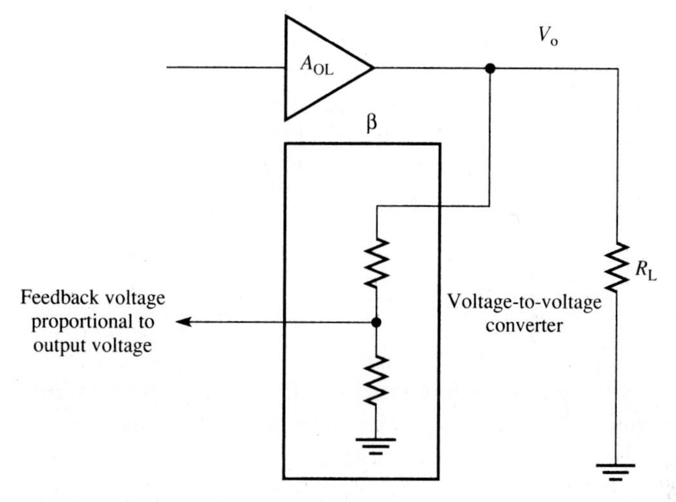

(a) Feedback element generates a voltage proportional to V_o.

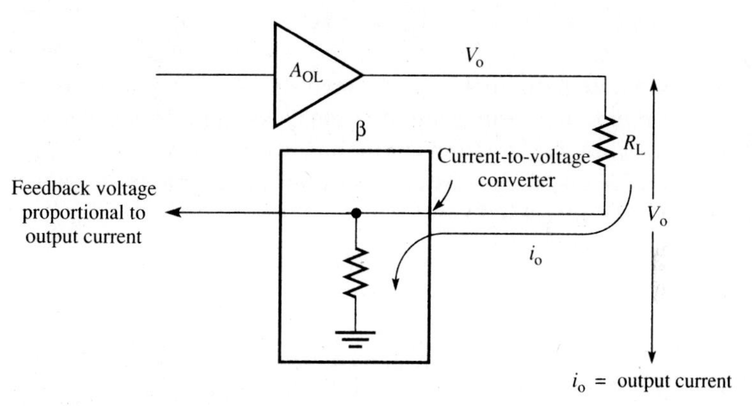

(b) Feedback element generates a voltage proportional to i_o.

the feedback to subtract a sample of either the output voltage or the output current from the input signal.

When the gain of an amplifier is measured without feedback, it is called the **open-loop gain** (abbreviated A_{OL}), which is illustrated in Figure 5–4. The open-loop gain is extremely high in op-amps. The 741C, for example, has an open-loop gain of 100,000. With negative feedback, the gain is reduced significantly but is very stable. The overall gain with feedback is called the **closed-loop gain** (A_{CL}). The output voltage is given by the equation

EQUATION 5–1

$$V_{out} = A_{OL}(v_+ - v_-)$$

where V_{out} = output voltage, V

A_{OL} = open-loop gain

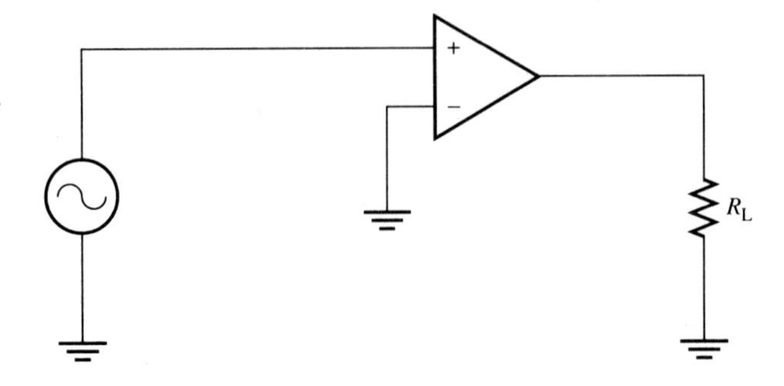

FIGURE 5–4
Op-amp operating in open-loop configuration.

Except for a few specialized circuits, op-amps are normally operated with negative feedback. The closed-loop gain is controlled by the feedback network, which is external to the op-amp. As a result, the user is in control of the gain by the design of the feedback network.

5–4 THE IDEAL OP-AMP

To obtain an approximate analysis of op-amp circuits, we begin with a model of op-amp behavior called the **ideal op-amp.** This model makes several simplifying assumptions for understanding basic op-amp circuits *providing the circuits are stable and linear*. The most important characteristics of an ideal operational amplifier are as follows:

1. The open-loop voltage gain is infinite.
2. The impedance measured between the input terminals is infinite.
3. The output impedance is zero.
4. The dc output voltage is zero when the dc input voltage is zero.
5. The bandwidth is infinite.
6. The common-mode gain is zero.

The first three characteristics lead directly to three important deductions. For the open-loop voltage gain to be infinite implies that an infinitely small input signal can control the output over its full range. If the input signal is infinitely small, then *the voltage difference between the input terminals must be zero*. The assumption that the input impedance is infinite leads immediately to the implication that *the inputs must draw no current*. The assumption that the output impedance is zero is equivalent to saying that the Thevenin impedance is zero. This implies that the *output voltage cannot be changed by variations in the load*.

For analysis of various op-amp circuits, these deductions can be summarized as three rules that are basic to an intuitive understanding of *linear* op-amp circuits that have negative feedback:

1. The voltage difference across the inputs is zero.
2. The inputs draw no current.
3. The output impedance is zero.

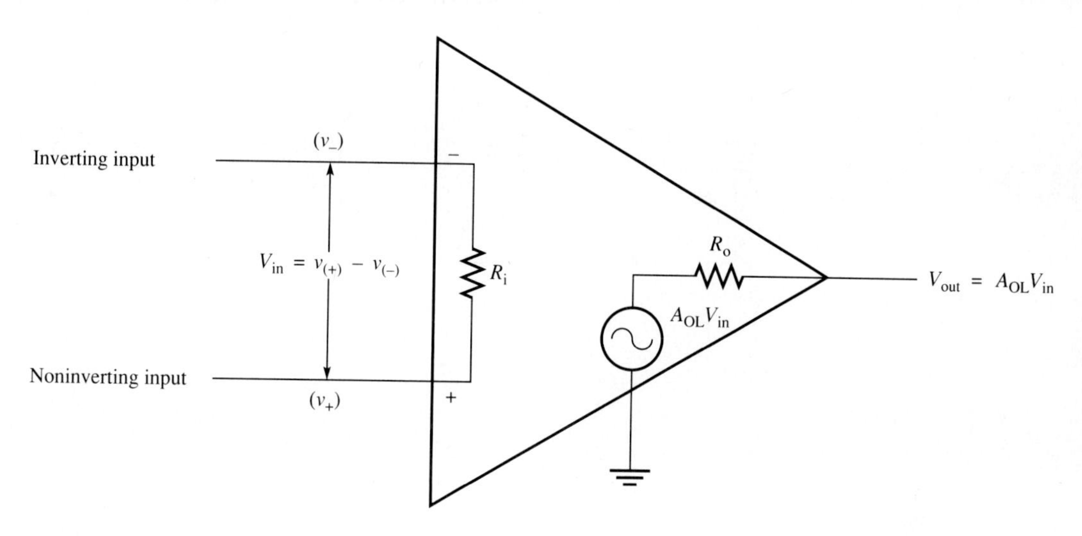

FIGURE 5–5
Equivalent circuit of an op-amp.

These rules, along with laws from basic electronics, can be applied to a variety of op-amp circuits *provided they are using negative feedback and are stable and linear*. The rules do not apply to multivibrators, Schmitt triggers, and oscillators (covered in Section 5–9), which use positive feedback or nonlinear circuits such as comparators. Keep in mind that the rules are for the *ideal* case—practical op-amps do have a tiny voltage difference between the inputs, which is essential to their operation. In addition, no real amplifier can have an infinite input impedance or a zero output impedance. The equivalent circuit of an op-amp, shown in Figure 5–5, includes the input and output impedances (shown as R_i and R_o) as well as the open-loop gain.

In the next section, we compare real op-amp specifications to these ideal characteristics. Generally, op-amps can be categorized into the following four areas for consideration of the most important specifications:

1. General-purpose op-amps, which feature low-cost and moderate specifications
2. High-accuracy op-amps with low-drift and low-noise specifications
3. High-speed op-amps with high slew rate and bandwidth and fast settling time
4. Low-power op-amps with low supply currents and wide-ranging supply voltages, including single-ended operation

5–5 OP-AMP SPECIFICATIONS

Manufacturers of op-amps list the operating specifications on a data sheet such as the one illustrated in Figure 5–6. The data sheet usually includes detailed graphs of various performance characteristics under various conditions such as temperature variations. In addition, a schematic diagram, a connection diagram, applica-

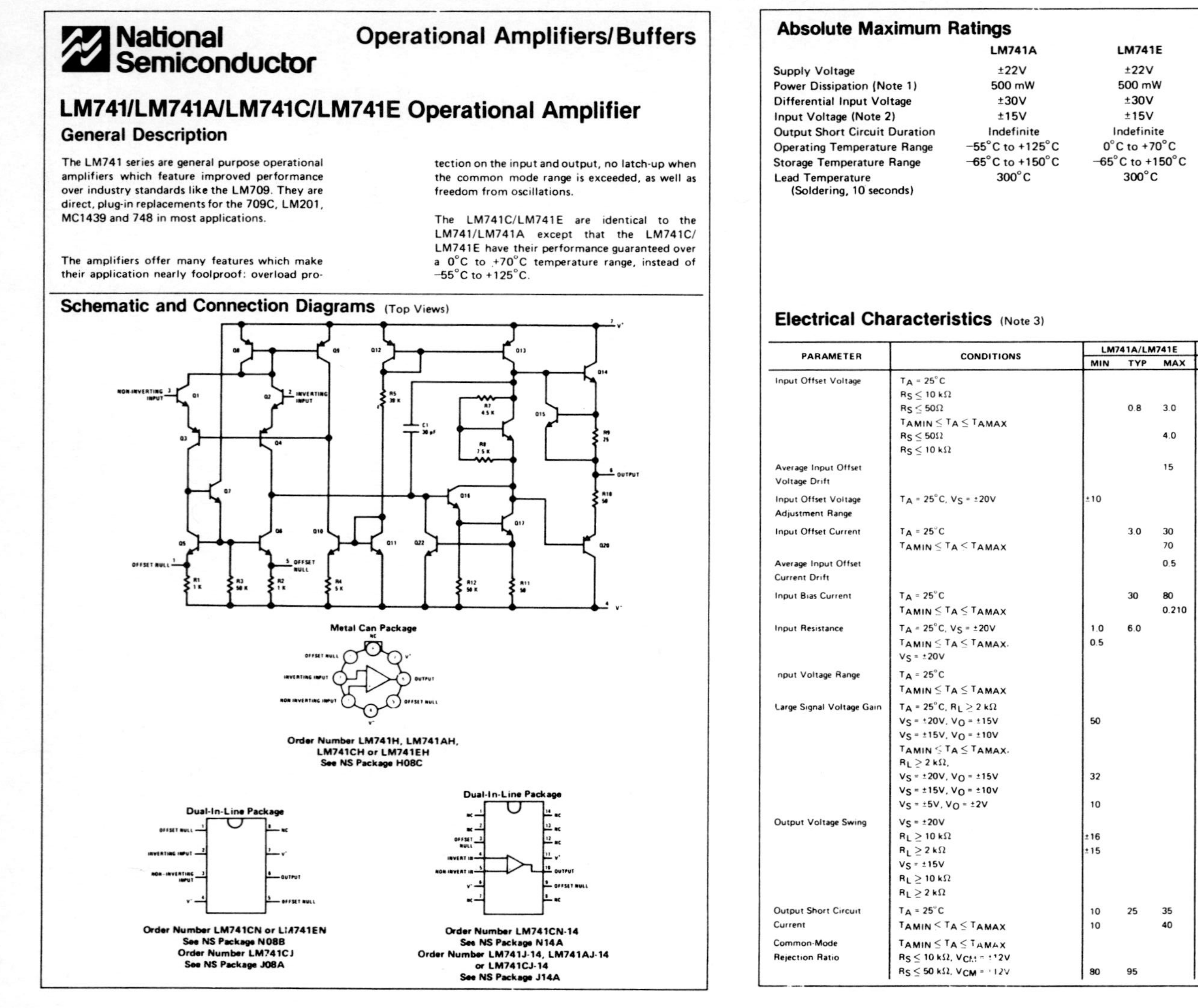

National Semiconductor — Operational Amplifiers/Buffers

LM741/LM741A/LM741C/LM741E Operational Amplifier

General Description

The LM741 series are general purpose operational amplifiers which feature improved performance over industry standards like the LM709. They are direct, plug-in replacements for the 709C, LM201, MC1439 and 748 in most applications.

The amplifiers offer many features which make their application nearly foolproof: overload protection on the input and output, no latch-up when the common mode range is exceeded, as well as freedom from oscillations.

The LM741C/LM741E are identical to the LM741/LM741A except that the LM741C/LM741E have their performance guaranteed over a 0°C to +70°C temperature range, instead of −55°C to +125°C.

Schematic and Connection Diagrams (Top Views)

Metal Can Package

Order Number LM741H, LM741AH, LM741CH or LM741EH
See NS Package H08C

Dual-In-Line Package

Order Number LM741CN or LM741EN
See NS Package N08B
Order Number LM741CJ
See NS Package J08A

Dual-In-Line Package

Order Number LM741CN-14
See NS Package N14A
Order Number LM741J-14, LM741AJ-14 or LM741CJ-14
See NS Package J14A

Absolute Maximum Ratings

	LM741A	LM741E	LM741	LM741C
Supply Voltage	±22V	±22V	±22V	±18V
Power Dissipation (Note 1)	500 mW	500 mW	500 mW	500 mW
Differential Input Voltage	±30V	±30V	±30V	±30V
Input Voltage (Note 2)	±15V	±15V	±15V	±15V
Output Short Circuit Duration	Indefinite	Indefinite	Indefinite	Indefinite
Operating Temperature Range	−55°C to +125°C	0°C to +70°C	−55°C to +125°C	0°C to +70°C
Storage Temperature Range	−65°C to +150°C	−65°C to +150°C	−65°C to +150°C	−65°C to +150°C
Lead Temperature (Soldering, 10 seconds)	300°C	300°C	300°C	300°C

Electrical Characteristics (Note 3)

PARAMETER	CONDITIONS	LM741A/LM741E MIN	LM741A/LM741E TYP	LM741A/LM741E MAX	LM741 MIN	LM741 TYP	LM741 MAX	LM741C MIN	LM741C TYP	LM741C MAX	UNITS
Input Offset Voltage	$T_A = 25°C$										
	$R_S \leq 10\ k\Omega$					1.0	5.0		2.0	6.0	mV
	$R_S \leq 50\Omega$		0.8	3.0							mV
	$T_{AMIN} \leq T_A \leq T_{AMAX}$										
	$R_S \leq 50\Omega$			4.0							mV
	$R_S \leq 10\ k\Omega$						6.0			7.5	mV
Average Input Offset Voltage Drift				15							μV/°C
Input Offset Voltage Adjustment Range	$T_A = 25°C$, $V_S = \pm 20V$	±10				±15			±15		mV
Input Offset Current	$T_A = 25°C$		3.0	30		20	200		20	200	nA
	$T_{AMIN} \leq T_A < T_{AMAX}$			70		85	500			300	nA
Average Input Offset Current Drift				0.5							nA/°C
Input Bias Current	$T_A = 25°C$		30	80		80	500		80	500	nA
	$T_{AMIN} \leq T_A \leq T_{AMAX}$			0.210			1.5			0.8	μA
Input Resistance	$T_A = 25°C$, $V_S = \pm 20V$	1.0	6.0		0.3	2.0		0.3	2.0		MΩ
	$T_{AMIN} \leq T_A \leq T_{AMAX}$, $V_S = \pm 20V$	0.5									MΩ
nput Voltage Range	$T_A = 25°C$							±12	±13		V
	$T_{AMIN} \leq T_A \leq T_{AMAX}$				±12	±13					V
Large Signal Voltage Gain	$T_A = 25°C$, $R_L \geq 2\ k\Omega$										
	$V_S = \pm 20V$, $V_O = \pm 15V$	50									V/mV
	$V_S = \pm 15V$, $V_O = \pm 10V$				50	200		20	200		V/mV
	$T_{AMIN} \leq T_A \leq T_{AMAX}$, $R_L \geq 2\ k\Omega$,										
	$V_S = \pm 20V$, $V_O = \pm 15V$	32									V/mV
	$V_S = \pm 15V$, $V_O = \pm 10V$				25			15			V/mV
	$V_S = \pm 5V$, $V_O = \pm 2V$	10									V/mV
Output Voltage Swing	$V_S = \pm 20V$										
	$R_L \geq 10\ k\Omega$	±16									V
	$R_L \geq 2\ k\Omega$	±15									V
	$V_S = \pm 15V$										
	$R_L \geq 10\ k\Omega$				±12	±14		±12	±14		V
	$R_L \geq 2\ k\Omega$				±10	±13		±10	±13		V
Output Short Circuit Current	$T_A = 25°C$	10	25	35		25			25		mA
	$T_{AMIN} < T_A \leq T_{AMAX}$	10		40							mA
Common-Mode Rejection Ratio	$T_{AMIN} \leq T_A \leq T_{AMAX}$										
	$R_S \leq 10\ k\Omega$, $V_{CM} = \pm 12V$				70	90		70	90		dB
	$R_S \leq 50\ k\Omega$, $V_{CM} = \pm 12V$	80	95								dB

FIGURE 5–6
Data sheets for the 741 op-amp (courtesy of National Semiconductor Corporation).

tion hints, and typical circuits using the particular op-amp are included. There are many different op-amps available because no one circuit can optimize all the parameters listed. A data sheet for the ubiquitous 741 op-amp, a very popular general-purpose op-amp, is shown in Figure 5–6. This op-amp can give satisfactory results for many noncritical applications and is very inexpensive. At the other end of the performance spectrum are precision op-amps that have optimized certain characteristics. One example is the OP-177, a precision op-amp that is useful in critical applications such as thermocouple amplifiers. The specification sheet for the OP-177 is shown in Figure 5–7 for comparison. Some of the important specifications from the 741 data sheet are described. To give you an idea of how the specifications for a general-purpose op-amp such as the 741C compare to specialized op-amps, Table 5–1 lists some specifications for selected op-amps.

The **input offset voltage** is the dc voltage that must be applied between the input terminals to obtain zero *output* voltage. Ideally the output voltage should be zero with no input; however, in practical amplifiers a small dc offset voltage is often present at the output terminal with no input voltage. This is due to inputs that are not perfectly matched. For the 741C, the input offset voltage is typically 2.0 mV with a maximum of 6 mV. By comparison, the OP-07 op-amp has a maximum input offset voltage of 25 μV, and the OP-177 has an ultra-low maximum offset voltage of just 10 μV. Most op-amps have provision for adjusting the output voltage to zero using an external circuit connected as specified by the manufacturer.

The **input offset current** is the difference between two bias currents. Although the ideal model indicates the inputs draw no current, all transistors require a small but finite current for proper operation. The differential input amplifier requires bias current in each input. The input offset current is the difference between these two bias currents. The input offset current for the 741C is a maximum of 200 nA. For FET op-amps, the input bias currents are much smaller than bipolar op-amps, and the input offset current can be less than 1 nA. If you really need low offset current, the OP-80E, a CMOS op-amp, has an input offset current of 50 fA.

The **input bias current** is the *average* of the two bias currents required by the inputs. Note that op-amps must be supplied with this current through the external circuit. That is, the user must provide a dc path to ground for the inputs to allow proper bias. The 741C has a maximum input bias current of 500 nA, whereas an AD611 BiFET op-amp has a bias current of 25 pA, and the OP-80E checks in at a maximum of 0.25 pA. The extremely low bias current and offset of the OP-80E is important in applications such as pH probe buffer amplifiers because of the extremely high Thevenin source resistance of a pH probe (see Section 17–10 for more on pH probes).

The **input resistance** (also called the differential input resistance) is measured between the input terminals and is defined as the ratio of the input voltage to the input current under stated conditions. The ideal amplifier has no input current; therefore, the input resistance is infinite. For the 741C, the input resistance is 2 MΩ, but for FET input op-amps, the input resistance can be well over 100 MΩ.

The **input voltage range** (also called the input common-mode voltage range) is the range of common-mode signals that can be applied to the input terminals for

PMI

OP-177

ULTRA-PRECISION
OPERATIONAL AMPLIFIER

Precision Monolithics Inc.

FEATURES

- **Ultra-Low Offset Voltage**
  - **$T_A = 25°C$ .... 10μV Max**
  - **$-55°C \le T_A \le 125°C$ .... 20μV Max**
- **Outstanding Offset Voltage Drift .... 0.1μV/°C Max**
- **Excellent Open-Loop Gain and Gain Linearity .... 12V/μV Typ**
- **CMRR .... 130dB Min**
- **PSRR .... 120dB Min**
- **Low Supply Current .... 2.0 mA Max**
- **Fits Industry Standard Precision Op Amp Sockets (OP07/OP77)**

ORDERING INFORMATION [†]

PACKAGE				OPERATING
CERDIP 8-PIN	PLASTIC 8-PIN	LCC 20-PIN	SO 8-PIN	TEMPERATURE RANGE
OP177AZ*	–	–	–	MIL
OP177BZ*	–	OP177BRC/883	–	MIL
OP177EZ	–	–	–	XIND
OP177FZ	OP177FP	–	–	XIND
OP177GZ	OP177GP	–	OP177GS	XIND

MIL = –55°C to +125°C XIND = –40°C to +85°C

* For devices processed in total compliance to MIL-STD-883, add /883 after part number. Consult factory for 883 data sheet.

† Burn-in is available on commercial and industrial temperature range parts in cerDIP, plastic DIP, and TO-can packages. For ordering information, see PMI's Data Book, Section 2.

GENERAL DESCRIPTION

The OP-177 features the highest precision performance of any op amp currently available. Offset voltage of the OP-177 is only 10μV MAX at room temperature and 20μV MAX over the full military temperature range of –55°C to +125°C. The ultra-low V_{OS} of the OP-177, combines with its exceptional offset voltage drift (TCV_{OS}) of 0.1μV/°C MAX, to eliminate the need for external V_{OS} adjustment and increases system accuracy over temperature.

The OP-177's open-loop gain of 12V/μV is maintained over the full ±10V output range. CMRR of 130dB MIN, PSRR of 120dB MIN, and maximum supply current of 2mA are just a few examples of the excellent performance of this operational amplifier. The OP-177's combination of outstanding specifications insure accurate performance in high closed-loop gain applications.

This low noise bipolar input op amp is also a cost effective alternative to chopper-stabilized amplifiers. The OP-177 provides chopper-type performance without the usual problems of high noise, low frequency chopper spikes, large physical size, limited common-mode input voltage range, and bulky external storage capacitors.

Continued

PIN CONNECTIONS

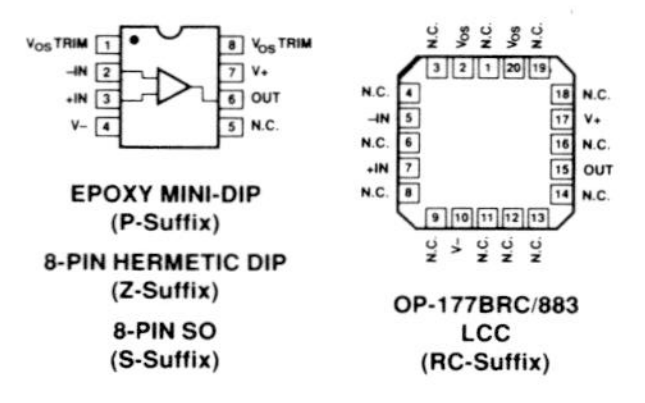

EPOXY MINI-DIP (P-Suffix)
8-PIN HERMETIC DIP (Z-Suffix)
8-PIN SO (S-Suffix)

OP-177BRC/883 LCC (RC-Suffix)

SIMPLIFIED SCHEMATIC

GENERAL DESCRIPTION *Continued*

The OP-177 is offered in both the –55°C to +125°C military, and the –40°C to +85°C extended industrial temperature ranges. This product is available in 8-pin ceramic and epoxy DIPs, as well as the space saving 8-pin Small-Outline (SO) and the Leadless Chip Carrier (LCC) packages.

ABSOLUTE MAXIMUM RATINGS

Supply Voltage .... ±22V
Differential Input Voltage .... ±30V
Input Voltage (Note 1) .... ±22V
Output Short-Circuit Duration .... Indefinite
Storage Temperature Range
  Z and RC Packages .... –65°C to +150°C
  P Package .... –65°C to +125°C
Operating Temperature Range
  OP-177A, OP-177B .... –55°C to +125°C
  OP-177E, OP-177F, OP-177G .... –40°C to +85°C
Lead Temperature Range (Soldering, 60 sec) .... 300°C
Junction Temperature (T_j) .... –65°C to +150°C

PACKAGE TYPE	θ_{jA} (NOTE 2)	θ_{jC}	UNITS
8-Pin Hermetic DIP (Z)	148	16	°C/W
8-Pin Plastic DIP (P)	103	43	°C/W
20-Contact LCC (RC)	98	38	°C/W
8-Pin SO (S)	158	43	°C/W

NOTES:
1. For supply voltages less than ±22V, the absolute maximum input voltage is equal to the supply voltage.
2. θ_{jA} is specified for worst case mounting conditions, i.e., θ_{jA} is specified for device in socket for CerDIP, P-DIP, and LCC packages; θ_{jA} is specified for device soldered to printed circuit board for SO package.

ELECTRICAL CHARACTERISTICS at $V_S = \pm 15V$, $T_A = +25°C$, unless otherwise noted.

PARAMETER	SYMBOL	CONDITIONS	OP-177A MIN	OP-177A TYP	OP-177A MAX	OP-177B MIN	OP-177B TYP	OP-177B MAX	UNITS
Input Offset Voltage	V_{OS}		–	4	10	–	10	25	μV
Long-Term Input Offset Voltage Stability	$\Delta V_{OS}/Time$	(Note 1)	–	0.2	–	–	0.2	–	μV/Mo
Input Offset Current	I_{OS}		–	0.3	1.0	–	0.3	1.5	nA
Input Bias Current	I_B		–0.2	–	1.5	–0.2	–	2.0	nA
Input Noise Voltage	e_n	f_o = 1Hz to 100Hz (Note 2)	–	118	150	–	118	150	nV_{RMS}
Input Noise Current	i_n	f_o = 1Hz to 100Hz (Note 2)	–	3	8	–	3	8	pA_{RMS}
Input Resistance Differential-Mode	R_{IN}	(Note 3)	26	45	–	26	45	–	MΩ
Input Resistance Common-Mode	R_{INCM}		–	200	–	–	200	–	GΩ
Input Voltage Range	IVR	(Note 4)	±13	±14	–	±13	±14	–	V
Common-Mode Rejection Ratio	CMRR	V_{CM} = ±13V	130	140	–	130	140	–	dB
Power Supply Rejection Ratio	PSRR	V_S = ±3V to ±18V	120	125	–	115	125	–	dB
Large-Signal Voltage Gain	A_{VO}	$R_L \ge 2kΩ$, V_O = ±10V (Note 5)	5000	12000	–	5000	12000	–	V/mV
Output Voltage Swing	V_O	$R_L \ge 10kΩ$ $R_L \ge 2kΩ$ $R_L \ge 1kΩ$	±13.5 ±12.5 ±12.0	±14.0 ±13.0 ±12.5	– – –	±13.5 ±12.5 ±12.0	±14.0 ±13.0 ±12.5	– – –	V
Slew Rate	SR	$R_L \ge 2kΩ$ (Note 2)	0.1	0.3	–	0.1	0.3	–	V/μs
Closed-Loop Bandwidth	BW	A_{VCL} = +1 (Note 2)	0.4	0.6	–	0.4	0.6	–	MHz
Open-Loop Output Resistance	R_O		–	60	–	–	60	–	Ω
Power Consumption	Pd	V_S = ±15V, No Load V_S = ±3V, No Load	– –	50 3.5	60 4.5	– –	50 3.5	60 4.5	mW
Supply Current	I_{SY}	V_S = ±15V, No Load	–	1.6	2.0	–	1.6	2.0	mA
Offset Adjustment Range		Rp = 20kΩ	–	±3	–	–	±3	–	mV

NOTES:
1. Long-Term Input Offset Voltage Stability refers to the averaged trend line of V_{OS} vs. Time over extended periods after the first 30 days of operation. Excluding the initial hour of operation, changes in V_{OS} during the first 30 operating days are typically less than 2.0μV.
2. Sample tested.
3. Guaranteed by design.
4. Guaranteed by CMRR test condition.
5. To insure high open-loop gain throughout the ±10V output range, A_{VO} is tested at $-10V \le V_O \le 0V$, $0V \le V_O \le +10V$, and $-10V \le V_O \le +10V$.

FIGURE 5–7
A precision op-amp, the OP-177 (courtesy of Precision Monolithics, Inc.).

TABLE 5–1
Representative specifications for selected op-amps.

TYPE	CLASS	INPUT OFFSET VOLTAGE MAX (mV)	INPUT OFFSET CURRENT MAX (nA)	INPUT BIAS CURRENT MAX (nA)	GAIN MIN (×1000)	CMRR MIN (dB)	SLEW RATE TYPICAL (V/μs)	COMMENT
ALD-1706	FET	4.5	0.025	0.03	100	—	0.17	Very low supply current (20 μA)
AMP02E	BiFET	0.1	10	10	10	115[a]	4	Low input offset voltage
INA-110AG	FET	0.5	0.05	0.10	—[b]	100[c]	4[c]	High-speed instrumentation amplifier
LF351	BiFET	10	0.1	0.20	25	70	13	Low cost, high impedance with fast slew rate
LF411A	BiFET	0.5	0.20	0.10	50	80	10	General-purpose, high impedance
LH0132G	BiFET	5	0.15	0.6	1	50	500	High-speed, high-input Z for sample and hold
LM311	BJT	7.5	250	250	40	—	—	Comparator, can operate from +5 V for logic
LM363	BJT	0.4[d]	5	20	1,000	104[d]	1[d]	Precision instrumentation amplifier
LM607A	BJT	0.8	4	4	5,000	124	0.7	Precision, low drift, low supply current
LM741C	BJT	6.	200	500	20	70	0.5	General purpose, was industry standard
OPA-111	FET	0.25	0.0015	0.001	1,000	100	1	High-performance replacement for 741
OP-07A	BJT	0.06	4	4	200	106	0.1	Very low offset voltage, very stable
OP-27A	BJT	0.025	35	40	1,000	114	1.7	High output drive, low noise replacement for 741
OP-42E	BiFET	0.75	0.040	0.2	500[c]	88	50	Excellent slew rate and fast settling time
OP-80E	FET	1.5	0.000080	0.001	100	50	0.4	Ultralow-bias current
OP-177A	BJT	0.01	1.0	1.5	5,000	130	0.1	Very high CMRR, very low input offset voltage
OP-260A	BJT	3.5	12,000[e]	12,000[e]	—	52	1,000	High speed, uses current feedback
OPA-627M	FET	0.1	0.020	0.020	560	106	45	Very low noise
SE-5539	BJT	5.	2,000	20,000	0.22	70	600	High frequency amplifier
XR4741M	BJT	3.0	30.	200	25	74	1.6	General purpose quad op-amp. (Compare to 741C)

[a] At G = 100 to 1000
[b] Set with internal resistors
[c] Depending on gain
[d] At G = 1000
[e] Inverting input

which the amplifier is operational. The 741C has a minimum input voltage range of 12 V.

The **large-signal voltage gain** is the ratio of the output voltage when it swings over a specified range with a specified load to the differential input voltage. This gain is equivalent to the open-loop gain. The large-signal gain of the 741C is 200,000. Special-purpose op-amps can have gains of over 3,000,000.

The **output voltage swing** is the peak output voltage swing that can be obtained before clipping. It is specified with a given load resistance and referenced to ground. The maximum output voltage swing is dependent on the power supply voltage and is generally 1 to 2 V less than the supply voltage.

The **common-mode rejection ratio** (CMRR) is the ratio of the differential gain, $A_{v(d)}$, to the common-mode gain, $A_{v(cm)}$, as discussed in Section 4–9. Ideally, an op-amp should respond only to differential signals, not to common-mode signals. The CMRR is frequently expressed in decibels (dB) because it is a very large number. The common-mode rejection ratio is a measure of how well the op-amp can reject noise or other signals that drive both inputs together. Because it is dependent on the differential gain, CMRR is often shown on performance characteristic graphs supplied with the manufacturer's specification sheet. For the 741C, the CMRR is specified as a minimum of 70 dB. By comparison, the OP-177 has a minimum CMRR of 130 dB. High CMRR is important when a low-level signal is contaminated with common-mode noise.

The **slew rate** is the maximum rate of change of the output voltage when a large-amplitude step function is applied to the inputs. It is a large-signal parameter that is measured in volts per microsecond. The slew rate is related to the internal currents and capacitance of the op-amp. For the 741C, the slew rate is 0.5 V/μs, whereas fast op-amps have rates many times faster than this. For example, the OPA603 (a current-feedback op-amp) has a minimum slew rate of 1000 V/μs. High slew rate is important in applications such as video circuits, fast data acquisition, and automatic testing. These topics are described in Section 4.

5–6 AMPLIFIER CIRCUITS

INVERTING AMPLIFIER

Analysis of the inverting amplifier can be made by applying the first two ideal op-amp rules listed in Section 5–4. Figure 5–8 illustrates an inverting amplifier. Notice that the noninverting input is grounded. Rule 1 implies that the inverting input must also be at ground potential, since there is no voltage difference across the inputs. Since the inverting input is near 0 V, it can be seen that the input voltage is across R_1 and the output voltage is across R_2. Thus, the closed-loop gain of the inverting amplifier is

EQUATION 5–2

$$A_{CL} = -\frac{V_{out}}{V_{in}} = -\frac{R_2}{R_1}$$

where A_{CL} = closed-loop gain

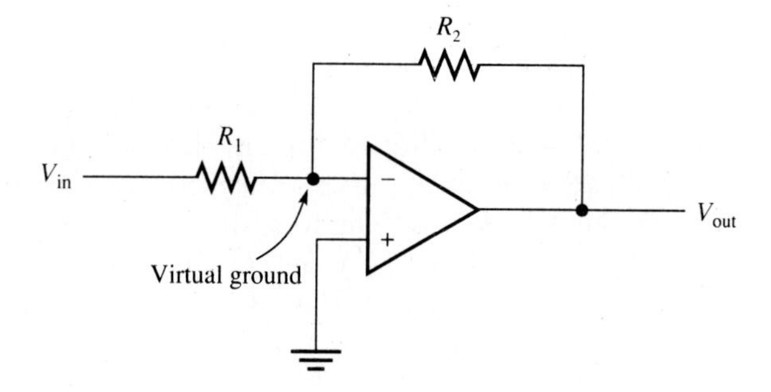

FIGURE 5–8
Inverting amplifier.

The negative sign indicates that the output voltage is out of phase with the input voltage.

The input impedance of the amplifier is easy to find. Notice that the virtual ground means that the ac input signal is applied directly across R_1. Therefore, R_1 is the input impedance of the amplifier.

EXAMPLE 5–1 An inverting amplifier is needed that has an input impedance of 10 kΩ and a voltage gain of −12. Draw the circuit.

SOLUTION Because of the virtual ground, the input resistor R_1 is equal to the required input impedance. Solve Equation 5–2 for R_2:

$$A_{CL} = -\frac{R_2}{R_1}$$

$$R_2 = -R_1 A_{CL}$$

$$= -(10 \text{ k}\Omega)(-12)$$

$$= 120 \text{ k}\Omega$$

To balance the dc bias current, a compensating resistor is added to the noninverting side. This resistor is approximately equal to the parallel combination of R_1 and R_2. The completed circuit is shown in Figure 5–9.

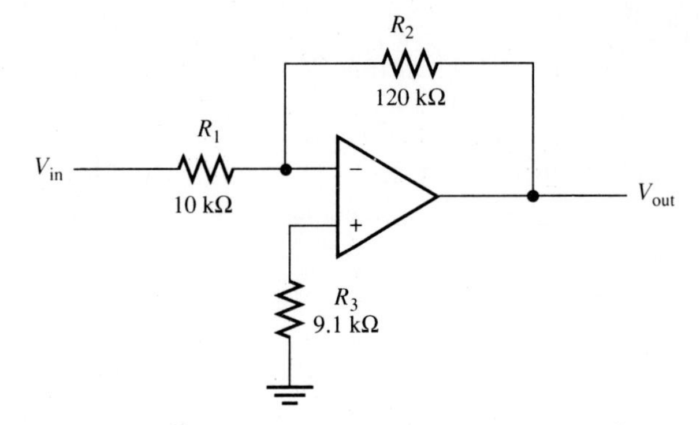

FIGURE 5–9
Circuit for Example 5–1.

For the inverting amplifier, the junction at the inverting input is called a **virtual ground** because the voltage at the inverting input is near ground potential due to negative feedback. The concept of virtual ground may seem strange at first. Rule 1 was developed for an ideal amplifier; an actual amplifier *does* have a voltage difference across the inputs. Without this tiny voltage, the op-amp wouldn't have anything to amplify! To obtain an idea of the size of this voltage, assume a 741 has a 10 V output and an open loop gain of 200,000. The differential voltage at the input of the amplifier is

$$v_d = \frac{10 \text{ V}}{200{,}000} = 50 \ \mu\text{V}$$

Because of the very large open-loop gain of the op-amp, only 50 μV of difference signal is across the inputs to drive the output to 10 V. The assumption that the inputs have *no* voltage difference is seen to be in error by only 50 μV due to the very high gain of the op-amp. The concept of virtual ground is sometimes applied to circuits that have a dc voltage at the virtual ground point. In these cases, the concept of virtual ground means an ac ground, implying that ac signals see zero volts.

THE NONINVERTING AMPLIFIER

The noninverting amplifier is shown in Figure 5–10. A portion of the output voltage is returned to the inverting input via the R_1-R_2 voltage divider. Using ideal Rule 2 (the inputs draw no current), we see that the amount of feedback voltage can be computed directly from the voltage divider. That is

EQUATION 5–3

$$V_f = V_{out}\left(\frac{R_1}{R_1 + R_2}\right)$$

By Rule 1, the feedback voltage is equal to the input voltage. By substitution

$$V_{in} = V_{out}\left(\frac{R_1}{R_1 + R_2}\right)$$

Solving for the gain gives

EQUATION 5–4

$$A_{CL} = \frac{V_{out}}{V_{in}} = \left(\frac{R_1 + R_2}{R_1}\right) = 1 + \frac{R_2}{R_1}$$

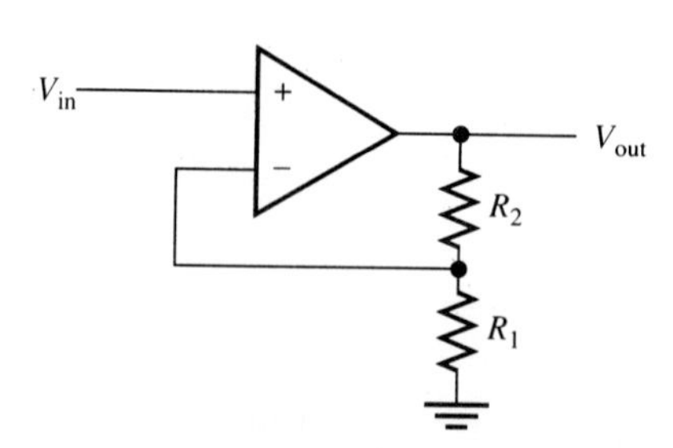

FIGURE 5–10
Noninverting amplifier.

Let's see how the negative feedback keeps the gain constant. Assume for a moment that something occurred to reduce the gain of the op-amp. As a result, the output voltage is lower, causing the feedback voltage to drop. The input *difference* voltage to the op-amp is now larger, causing the output voltage to increase. This increase in the output voltage almost exactly compensates for the original drop due to the change in the op-amp's gain. This correction occurs almost instantaneously, and the overall gain remains constant, as set by the feedback circuit.

The noninverting configuration has much higher input impedance than the inverting configuration. For an ideal op-amp used as a noninverting amplifier, the input impedance is infinite (the input draws no current). This approximation is adequate for most circuit analysis.

VOLTAGE FOLLOWER

The voltage follower, shown in Figure 5–11, is a special case of a noninverting amplifier. The feedback resistor, R_2 is reduced to zero and the resistance to ground, R_1, is increased to infinity. This configuration is called a voltage follower because the output is a replica of the input. Although the voltage gain is unity, the power gain is not unity. Consider the equations for the input and output power for the noninverting amplifier:

$$P_{out} = \frac{V_{out}^2}{R_2}$$

$$P_{in} = \frac{V_{in}^2}{R_1}$$

Taking the ratio of P_{out} to P_{in} leads to

$$\frac{P_{out}}{P_{in}} = \frac{V_{out}^2}{V_{in}^2}\frac{R_1}{R_2}$$

The voltage gain of a voltage follower is unity, leading to

$$\frac{P_{out}}{P_{in}} = \frac{R_1}{R_2}$$

For the voltage follower, we have assumed that R_1 is infinite and R_2 is zero; the implication is that the power gain is infinite. Of course, this cannot happen in a real op-amp; in practice, the input impedance for a general-purpose 741 op-amp is on the order of $2 \times 10^6\ \Omega$ and the output impedance can be around $20\ \Omega$, leading to a power gain of about 10^6. With specialized op-amps, the power gain can be much higher.

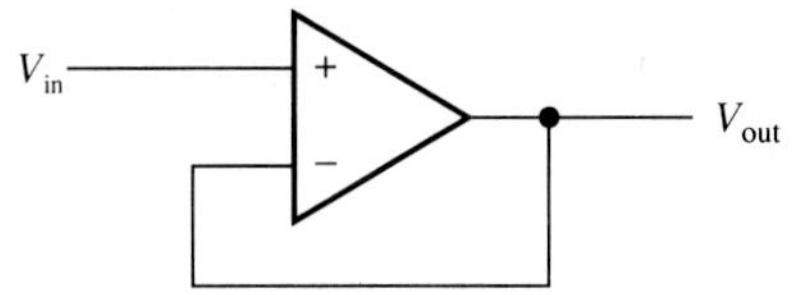

FIGURE 5–11
Voltage follower.

V_{in} + − V_{out}

National Semiconductor

BI-FET II™ Technology

LF411A/LF411 Low Offset, Low Drift JFET Input Operational Amplifier

General Description

These devices are low cost, high speed, JFET input operational amplifiers with very low input offset voltage and guaranteed input offset voltage drift. They require low supply current yet maintain a large gain bandwidth product and fast slew rate. In addition, well matched high voltage JFET input devices provide very low input bias and offset currents. The LF411 is pin compatible with the standard LM741 allowing designers to immediately upgrade the overall performance of existing designs.

These amplifiers may be used in applications such as high speed integrators, fast D/A converters, sample and hold circuits and many other circuits requiring low input offset voltage and drift, low input bias current, high input impedance, high slew rate and wide bandwidth.

Features

- Internally trimmed offset voltage — 0.5 mV(max)
- Input offset voltage drift — 10 μV/°C(max)
- Low input bias current — 50 pA
- Low input noise current — 0.01 pA/$\sqrt{Hz}$
- Wide gain bandwidth — 3 MHz(min)
- High slew rate — 10V/μs(min)
- Low supply current — 1.8 mA
- High input impedance — $10^{12}\Omega$
- Low total harmonic distortion $A_V = 10$, $R_L = 10k$, $V_O = 20$ Vp-p, BW = 20 Hz – 20 kHz — <0.02%
- Low 1/f noise corner — 50 Hz
- Fast settling time to 0.01% — 2 μs

Typical Connection

TL/H/5655-1

Ordering Information

LF411XYZ

- **X** indicates electrical grade
- **Y** indicates temperature range "M" for military, "C" for commercial
- **Z** indicates package type "H" or "N"

Connection Diagrams

Metal Can Package

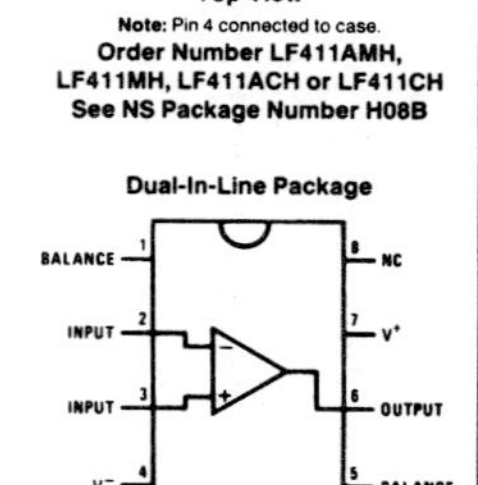

TL/H/5655-5

Top View

Note: Pin 4 connected to case.

Order Number LF411AMH, LF411MH, LF411ACH or LF411CH
See NS Package Number H08B

Dual-In-Line Package

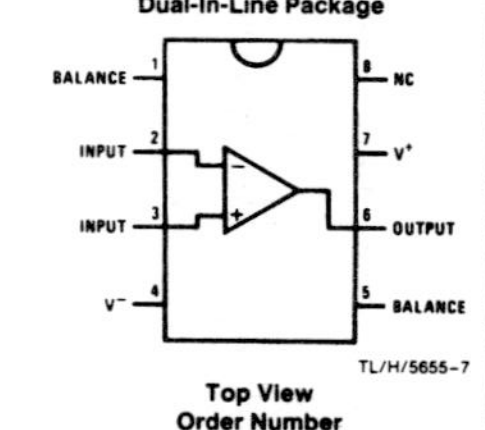

TL/H/5655-7

Top View

Order Number LF411ACN or LF411CN
See NS Package Number N08E

Simplified Schematic

TL/H/5655-6

3-76

Absolute Maximum Ratings

If Military/Aerospace specified devices are required, please contact the National Semiconductor Sales Office/Distributors for availability and specifications. (Note 8)

	LF411A	LF411
Supply Voltage	±22V	±18V
Differential Input Voltage	±38V	±30V
Input Voltage Range (Note 1)	±19V	±15V
Output Short Circuit Duration	Continuous	Continuous

	H Package	N Package
Power Dissipation (Notes 2 and 9)	670 mW	670 mW
$T_j$max	150°C	115°C
θ_jA	162°C/W (Still Air) 65°C/W (400 LF/min Air Flow)	120°C/W
θ_jC	20°C/W	
Operating Temp. Range	(Note 3)	(Note 3)
Storage Temp. Range	$-65°C \le T_A \le 150°C$	$-65°C \le T_A \le 150°C$
Lead Temp. (Soldering, 10 sec.)	260°C	260°C

ESD rating to be determined.

DC Electrical Characteristics (Note 4)

Symbol	Parameter	Conditions		LF411A Min	LF411A Typ	LF411A Max	LF411 Min	LF411 Typ	LF411 Max	Units
V_{OS}	Input Offset Voltage	$R_S = 10\ k\Omega$, $T_A = 25°C$			0.3	0.5		0.8	2.0	mV
$\Delta V_{OS}/\Delta T$	Average TC of Input Offset Voltage	$R_S = 10\ k\Omega$ (Note 5)			7	10		7	20 (Note 5)	μV/°C
I_{OS}	Input Offset Current	$V_S = \pm 15V$ (Notes 4, 6)	$T_j = 25°C$		25	100		25	100	pA
			$T_j = 70°C$			2			2	nA
			$T_j = 125°C$			25			25	nA
I_B	Input Bias Current	$V_S = \pm 15V$ (Notes 4, 6)	$T_j = 25°C$		50	200		50	200	pA
			$T_j = 70°C$			4			4	nA
			$T_j = 125°C$			50			50	nA
R_{IN}	Input Resistance	$T_j = 25°C$			10^{12}			10^{12}		Ω
A_{VOL}	Large Signal Voltage Gain	$V_S = \pm 15V$, $V_O = \pm 10V$, $R_L = 2k$, $T_A = 25°C$		50	200		25	200		V/mV
		Over Temperature		25	200		15	200		V/mV
V_O	Output Voltage Swing	$V_S = \pm 15V$, $R_L = 10k$		±12	±13.5		±12	±13.5		V
V_{CM}	Input Common-Mode Voltage Range			±16	+19.5		±11	+14.5		V
					−16.5			−11.5		V
CMRR	Common-Mode Rejection Ratio	$R_S \le 10k$		80	100		70	100		dB
PSRR	Supply Voltage Rejection Ratio	(Note 7)		80	100		70	100		dB
I_S	Supply Current				1.8	2.8		1.8	3.4	mA

AC Electrical Characteristics (Note 4)

Symbol	Parameter	Conditions	LF411A Min	LF411A Typ	LF411A Max	LF411 Min	LF411 Typ	LF411 Max	Units
SR	Slew Rate	$V_S = \pm 15V$, $T_A = 25°C$	10	15		8	15		V/μs
GBW	Gain-Bandwidth Product	$V_S = \pm 15V$, $T_A = 25°C$	3	4		2.7	4		MHz
e_n	Equivalent Input Noise Voltage	$T_A = 25°C$, $R_S = 100\Omega$, f = 1 kHz		25			25		nV/$\sqrt{Hz}$
i_n	Equivalent Input Noise Current	$T_A = 25°C$, f = 1 kHz		0.01			0.01		pA/$\sqrt{Hz}$

3-77

FIGURE 5–12
Data sheet for LF411 (courtesy of National Semiconductor).

Voltage followers are useful as buffers because they can have the highest input impedance of any op-amp circuit and have very low output impedances. They are used whenever it is necessary to isolate the source and load but maintain the exact signal level. A common application of voltage followers is in video-processing circuits, where the video level must be maintained at a specific level.

In the past, some op-amps (the LM110 series for example) were specially designed as voltage followers; today, there are new general-purpose op-amps that are more useful as they can be used at any impedance level, and have good performance for fast or slow signals. An excellent replacement for either the LM110 or the LM741 is the LF411, a general purpose op-amp with increased speed and bandwidth over its predecessors. The LF411 uses matched JFETs at the inputs, resulting in very high input impedance and low bias current. It is pin for pin compatible with the LM741, allowing older designs to be upgraded easily. Figure 5–12 shows the data sheet for the LF411.

BALANCED DIFFERENTIAL AMPLIFIER

The balanced differential amplifier circuit is a combination of a noninverting and inverting amplifier. It amplifies the *difference* between the two voltages applied to the inputs. The circuit, shown in Figure 5–13, uses matched resistors for the two resistors labeled R_1 and for the two resistors labeled R_2 (to achieve a high CMRR). The voltage gain is simply R_2/R_1. To understand its operation, we will apply the superposition theorem, which is summarized as follows:

Step 1: Reduce V_2 to zero, and compute the output voltage with V_1 operating alone. It can be seen in Figure 5–14(a) that the amplifier is reduced to an inverting amplifier.

Step 2: Reduce V_1 to zero, and compute the output voltage with V_2 operating alone. The amplifier is a noninverting amplifier, as shown in Figure 5–14(b), but with the addition of an input resistive voltage divider that attenuates the input voltage.

Step 3: The output voltage is the sum of the voltages due to the inputs acting as if they were alone. The result (Figure 5–14(c)) shows that the output voltage is the differential input voltage multiplied by the gain, given by R_2/R_1.

FIGURE 5–13
Differential amplifier.

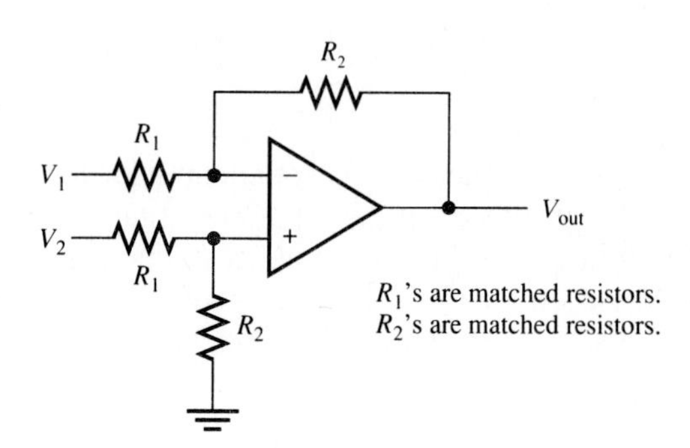

FIGURE 5–14
Applying the superposition theorem to a differential amplifier.

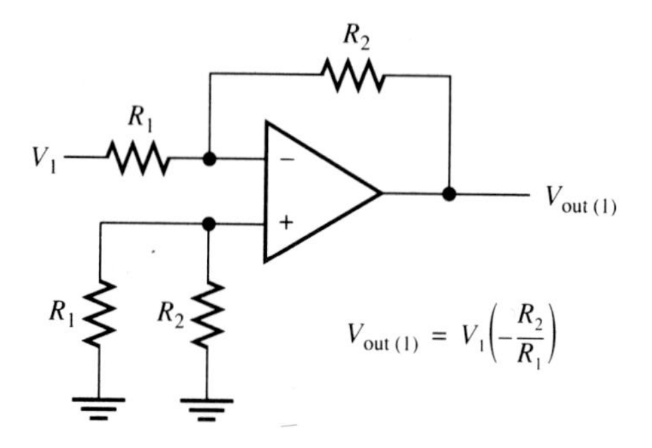

(a) Step 1. Reduce V_2 to zero. Amplifier is an inverting amplifier with a gain of $\left(-\frac{R_2}{R_1}\right)$.

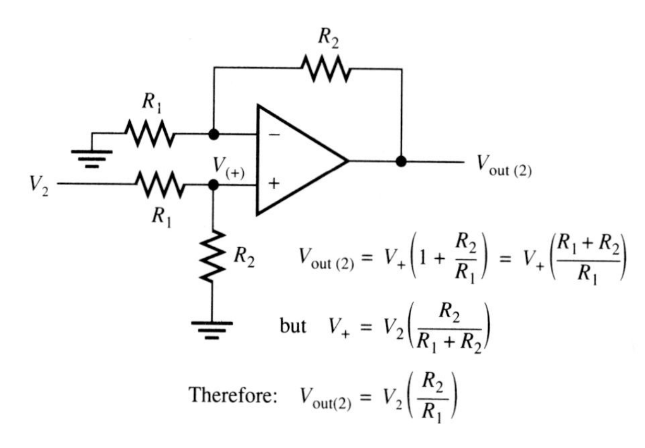

(b) Step 2. Reduce V_1 to zero. Amplifier is a noninverting amplifier with an input attenuator.

$$V_{out} = V_{out(1)} + V_{out(2)}$$

$$= V_1\left(-\frac{R_2}{R_1}\right) + V_2\left(\frac{R_2}{R_1}\right)$$

$$= (V_2 - V_1)\left(\frac{R_2}{R_1}\right)$$

(c) Step 3. Algebraically add results of Steps 1 and 2.

The op-amp differential amplifier is similar to the transistor differential amplifier discussed in Section 4–9. Again, the differential amplifier operates on the *difference* of the input voltage. The signal is connected between the two inputs. Noise voltage is a common-mode signal and is applied equally to both inputs. Thus the differential amplifier selectively amplifies the signal while tending to reject common-mode noise. Differential and common-mode signals are illustrated in Figure 5–15.

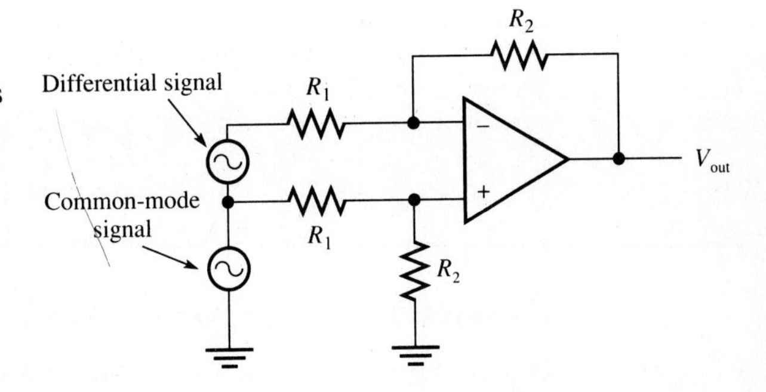

FIGURE 5–15 Comparison of differential with common-mode signals for a balanced differential amplifier.

EXAMPLE 5–2 Assume the circuit shown in Figure 5–16 has a differential input voltage of 30 mV and a 60 Hz interference induced in a common mode of 100 mV.

(a) Compute the output signal and noise, assuming the op-amp is a 741C.

(b) Compare the result if the 741C is replaced with an OP-177A.

FIGURE 5–16

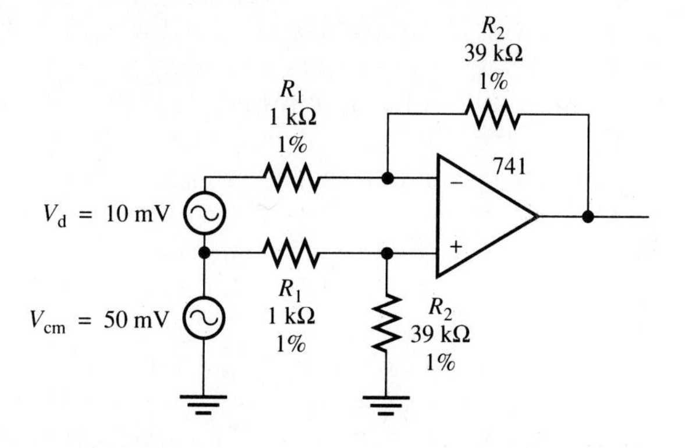

SOLUTION (a) The differential gain of the amplifier is $-\dfrac{R_2}{R_1}$.

$$A_d = -\frac{39\ \text{k}\Omega}{1\ \text{k}\Omega} = -39 \qquad v_{out(d)} = v_{in(d)}A_d = (30\ \text{mV})(-39)$$

$$= -1.17\ \text{V}$$

The 741C has a minimum CMRR of 70 dB.

$$\text{CMRR} = 20 \log \frac{A_d}{A_{cm}}$$

$$A_{cm} = \frac{A_d}{10^{\text{CMRR}/20}} = \frac{-39}{10^{70/20}} = 0.0123$$

$$v_{out(cm)} = v_{in(cm)}A_{cm} = (100\ \text{mV})(0.0123)$$

$$= 1.23\ \text{mV}$$

(b) The A_d is determined by the external resistors. The signal output is therefore the same for both op-amps. The CMRR for the OP-177A is a minimum of 130 dB.

$$A_{cm} = \frac{A_d}{10^{CMRR/20}} = \frac{-39}{10^{130/20}} = 1.23 \times 10^{-5}$$

$$v_{out(cm)} = v_{in(cm)} A_{cm} = (100 \text{ mV})(1.23 \times 10^{-5}) = 1.23\ \mu\text{V}$$

INSTRUMENTATION AMPLIFIER

One of the most common problems in instrumentation systems is the combination of the input signal with noise, causing an adverse effect on the measurement. The signal from many transducers (such as strain gages) is very small and can be subjected to this unwanted noise between the transducer and the amplifier. Furthermore, many transducers have a very high output impedance and can easily be loaded down when connected to an amplifier. A common-mode signal at the input can show up at the output; the amount of common-mode signal in the output can be estimated by dividing the common-mode input signal by the CMRR and multiplying the result by the gain of the amplifier. Although the differential amplifier favors differential signals, a common-mode rejection of about 80 dB just isn't high enough for some applications, and the relatively low input impedance causes loading of some transducers and sources.

The solution to these problems is the **instrumentation amplifier** (IA), a specially designed differential amplifier with ultra-high input impedance and extremely good common-mode rejection (up to 130 dB); it is also able to achieve high, stable gains. Instrumentation amplifiers can faithfully amplify low-level signals in the presence of high common-mode noise. They are used in a variety of signal-processing applications where accuracy is important and where low drift, low bias currents, precise gain, and very high CMRR are required. An application is to strain-gage measurements in which the desired signal is a very small difference voltage in the presence of a relatively large dc bias voltage. Variations in the supply voltage are seen by the amplifier as common-mode signals and are rejected, whereas the signal voltage is amplified. Another widely seen source of common-mode signals is the ac power line. Low-level transducers can pick up 60 Hz noise on both leads, causing a significant interference problem with single-ended (nondifferential) amplifiers.

The input impedance of an ordinary differential amplifier can easily be increased by simply isolating the inputs with voltage followers. This form of instrumentation amplifier is shown in Figure 5–17(a). Although this circuit has high input impedance, it requires extremely high precision matching of all four external resistors to achieve a high CMRR. Further, it still has two resistors that must be changed if variable gain is required, and they must track each other with high precision over the operating temperature range. Although this amplifier is useful for some general-purpose instrumentation requirements, it is not the optimum solution.

A clever alternate configuration that solves the difficulties of the preceding circuit and provides higher gain is the three-op-amp IA shown in Figure 5–17(b).

FIGURE 5–17
Basic instrumentation amplifiers.

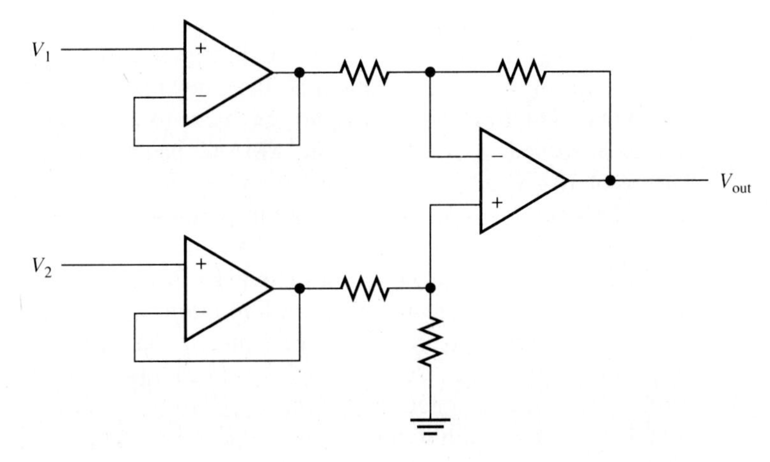

(a) Instrumentation amplifier using voltage followers and differential amplifier.

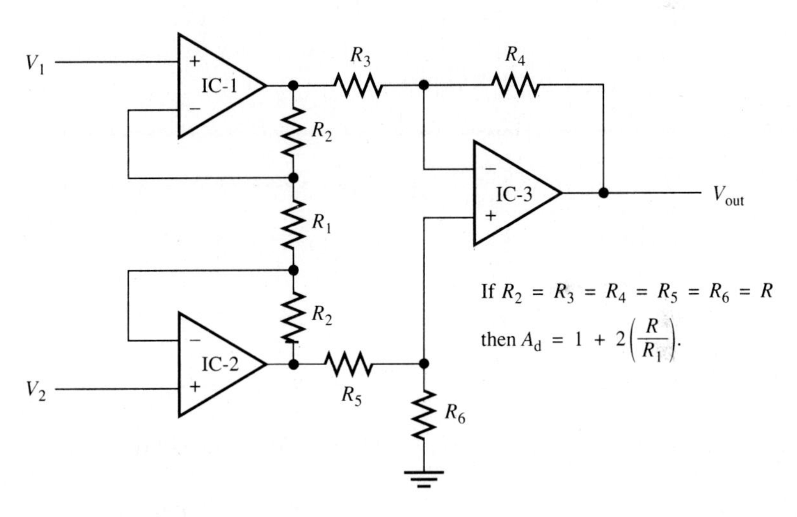

(b) Improved instrumentation amplifier.

The inputs are buffered by IC-1 and IC-2, providing a very high input impedance. The differential gain is controlled by just one resistor, R_1. In this design, the common-mode response depends on very precisely matching resistors. The common-mode response is 0 if R_4/R_3 is exactly equal to R_6/R_5 and is -86 dB if they are within 0.01% of each other. These resistors can be critically matched during manufacture and are generally set for a gain in the differential amplifier of 1.0. Many variations of this circuit are available in one IC package, including precision components such as laser-trimmed resistors fabricated by the manufacturer. Because these precision components can be fabricated very accurately within the IC, very high quality amplifiers can be produced. It is not generally possible to match the performance of a high-quality IA with op-amps.

An example of a high-quality instrumentation amplifier using the three-op-amp IA circuit is the AMP-02, shown in Figure 5–18. The gain is controlled by a single external resistor (supplied by the user) and can be set for any value from 1 to 10,000. The internal resistors (equivalent to R_2 through R_6 of Figure 5–17(b)) are all matched to be 25 kΩ. The AMP-02 has a typical differential input impedance of 10 GΩ and 115 dB of common-mode rejection at a gain of 100 to 1000.

Another type of instrumentation amplifier, called a *differential transconductance amplifier,* uses current feedback instead of the voltage feedback shown in the previous circuits. Current feedback allows changing the gain of the circuit without affecting the bandwidth—a limitation of IAs using voltage feedback.

Instrumentation amplifiers are available that have been optimized for certain application problems, such as high-speed or low-temperature-coefficient operation, or that are designed for specific types of transducers. Some IAs (such as the LH0036) have an additional output, called a *guard drive output,* that is designed to drive the shield of input cables. This feature reduces the input capacitance and noise pickup in extremely stringent applications such as electrometer circuits. Others have digitally programmed gain and multiplexed inputs. These special-purpose IAs usually make it unnecessary for the user to do a detailed design for a new application problem.

EXAMPLE 5–3 (a) Determine the differential- and common-mode gain for an AMP-02 instrumentation amplifier if R_1, is 330 Ω.
(b) Determine the output voltage if $v_1 = 10$ mV and $v_2 = -10$ mV.

SOLUTION (a) The differential gain is

$$A_d = 1 + 2\left(\frac{R}{R_1}\right)$$

The internal resistors of the AMP-02 are 25 kΩ.

$$A_d = 1 + 2\left(\frac{25\text{ k}\Omega}{0.330\text{ k}\Omega}\right) = 152.5$$

$$A_{cm} = \frac{A_d}{10^{\text{CMRR}/20}} = \frac{152.5}{10^{115/20}} = 2.7 \times 10^{-4}$$

(b) $v_{out} = (A_d)(v_1 - v_2) = (152.5)(10\text{ mV} - (-10\text{ mV}))$

$= 3.05$ V

ISOLATION AMPLIFIER

Sometimes it is necessary to isolate the output and input of an amplifier to prevent a common connection from the power supplies or instrumentation system to the transducer. Applications include biomedical instrumentation when patient safety is a consideration. This includes both signal-acquisition applications, such as

PMI

AMP-02

HIGH ACCURACY 8-PIN INSTRUMENTATION AMPLIFIER

Precision Monolithics Inc.

FEATURES

- Low Offset Voltage .... 100µV Max
- Low Drift .... 2µV/°C Max
- Wide Gain Range .... 1 to 10,000
- High Common-Mode Rejection .... 115dB Min
- High Bandwidth (G = 1000) .... 200kHz Typ
- Gain Equation Accuracy .... 0.5% Max
- Single Resistor Gain Set
- Input Overvoltage Protection
- Low Cost
- Available in Die Form

APPLICATIONS

- Differential Amplifier
- Strain Gauge Amplifier
- Thermocouple Amplifier
- RTD Amplifier
- Programmable Gain Instrumentation Amplifier
- Medical Instrumentation
- Data Acquisition Systems

ORDERING INFORMATION [†]

$T_A = +25°C$			
V_{IOS} MAX (µV)	V_{OOS} MAX (mV)	PLASTIC 8-PIN	OPERATING TEMPERATURE RANGE
100	4	AMP02EP	XIND
200	8	AMP02FP	XIND

† Burn-in is available on commercial and industrial temperature range parts in CerDIP, plastic DIP, and TO-can packages. For ordering information, see PMI's Data Book, Section 2.

GENERAL DESCRIPTION

The AMP-02 is the first precision instrumentation amplifier available in an 8-pin package. Gain of the AMP-02 is set by a single external resistor, and can range from 1 to 10,000. No gain set resistor is required for unity gain. The AMP-02 includes an input protection network that allows the inputs to be taken 60V beyond either supply rail without damaging the device.

Laser trimming reduces the input offset voltage to under 100µV. Output offset voltage is below 4mV and gain accuracy is better than 0.5% for gain of 1000. PMI's proprietary thin-film resistor process keeps the gain temperature coefficient under 50 ppm/°C.

Due to the AMP-02's design, its bandwidth remains very high over a wide range of gain. Slew rate is over 4V/µs making the AMP-02 ideal for fast data acquisition systems.

A reference pin is provided to allow the output to be referenced to an external DC level. This pin may be used for offset correction or level shifting as required. In the 8-pin package, sense is internally connected to the output.

For an instrumentation amplifier with the highest precision, consult the AMP-01 data sheet. For the highest input impedance and speed, consult the AMP-05 data sheet.

PIN CONNECTIONS

RG1 1, −IN 2, +IN 3, V− 4, REFERENCE 5, OUT 6, V+ 7, RG2 8

EPOXY MINI-DIP (P-Suffix)

N.C. 1, RG1 2, N.C. 3, −IN 4, +IN 5, N.C. 6, V− 7, N.C. 8, N.C. 9, REFERENCE 10, OUT 11, SENSE 12, V+ 13, N.C. 14, RG2 15, N.C. 16

16-PIN SOL (S-Suffix)

BASIC CIRCUIT CONNECTIONS

+IN, −IN, R_G, R_{G1}, R_{G2}, V+, V−, OUT, REFERENCE

$$G = \frac{V_{OUT}}{(+IN) - (-IN)} = \left(\frac{50k\Omega}{R_G}\right) + 1$$

FOR SOL CONNECT SENSE TO OUTPUT

PMI

ABSOLUTE MAXIMUM RATINGS

Supply Voltage .... ±18V
Common-Mode Input Voltage .... [(V−) − 60V] to [(V+) + 60V]
Differential Input Voltage .... [(V−) − 60V] to [(V+) + 60V]
Output Short-Circuit Duration .... Continuous
Operating Temperature Range .... −40°C to +85°C
Storage Temperature Range .... −65°C to +150°C
Junction Temperature Range .... −65°C to +150°C
Lead Temperature (Soldering, 10 sec) .... +300°C

PACKAGE TYPE	Θ_{JA} (Note 2)	Θ_{JC}	UNITS
8-Pin Plastic DIP (P)	96	37	°C/W
16-Pin SOL (S)	92	27	°C/W

NOTE:
1. Absolute maximum ratings apply to both DICE and packaged parts, unless otherwise noted.
2. Θ_{JA} is specified for worst case mounting conditions, i.e., Θ_{JA} is specified for device in socket for P-DIP package; Θ_{JA} is specified for device soldered to printed circuit board for SOL package.

ELECTRICAL CHARACTERISTICS at $V_S = \pm 15V$, $V_{CM} = 0V$, $T_A = +25°C$, unless otherwise noted.

PARAMETER	SYMBOL	CONDITIONS	AMP-02E MIN	AMP-02E TYP	AMP-02E MAX	AMP-02F MIN	AMP-02F TYP	AMP-02F MAX	UNITS
OFFSET VOLTAGE									
Input Offset Voltage	V_{IOS}	$T_A = +25°C$	–	20	100	–	40	200	µV
		$-40°C \le T_A \le +85°C$	–	50	200	–	100	350	
Input Offset Voltage Drift	TCV_{IOS}	$-40°C \le T_A \le +85°C$	–	0.5	2	–	1	4	µV/°C
Output Offset Voltage	V_{OOS}	$T_A = +25°C$	–	1	4	–	2	8	mV
		$-40°C \le T_A \le +85°C$	–	4	10	–	9	20	
Output Offset Voltage Drift	TCV_{OOS}	$-40°C \le T_A \le +85°C$	–	50	100	–	100	200	µV/°C
Power Supply Rejection	PSR	$V_S = \pm 4.8V$ to $\pm 18V$, G = 1000	115	128	–	110	115	–	dB
		G = 100	115	125	–	110	115	–	
		G = 10	100	110	–	95	100	–	
		G = 1	80	90	–	75	80	–	
		$V_S = \pm 4.8V$ to $\pm 18V$, $-40°C \le T_A \le +85°C$, G = 1000	110	120	–	105	110	–	dB
		G = 100	110	120	–	105	110	–	
		G = 10	95	110	–	90	95	–	
		G = 1	75	90	–	70	75	–	
INPUT CURRENT									
Input Bias Current	I_B	$T_A = +25°C$	–	2	10	–	4	20	nA
		$-40°C \le T_A \le +85°C$	–	12	30	–	20	40	
Input Bias Current Drift	TCI_B	$-40°C \le T_A$ +85°C	–	150	–	–	250	–	pA/°C
Input Offset Current	I_{OS}	$T_A = +25°C$	–	1.2	5	–	2	10	nA
		$-40°C \le T_A \le +85°C$	–	1.8	15	–	3	20	
Input Offset Current Drift	TCI_{OS}	$-40°C \le T_A \le +85°C$	–	9	–	–	15	–	pA/°C
INPUT									
Input Resistance	R_{IN}	Differential, G ≤ 1000	–	10	–	–	10	–	GΩ
		Common-Mode, G = 1000	–	16.5	–	–	16.5	–	
Input Voltage Range	IVR	$T_A = +25°C$ (Note 3)	±11	–	–	±11	–	–	V
		$-40°C \le T_A \le +85°C$	±11	–	–	±11	–	–	
Common-Mode Rejection	CMR	$V_{CM} = \pm 11V$, G = 1000	115	120	–	110	115	–	dB
		G = 100	115	120	–	110	115	–	
		G = 10	100	115	–	95	110	–	
		G = 1	80	95	–	75	90	–	
		$V_{CM} = \pm 11V$, $-40°C \le T_A \le +85°C$, G = 1000	110	120	–	105	115	–	dB
		G = 100	110	120	–	105	115	–	
		G = 10	95	110	–	90	105	–	
		G = 1	75	90	–	70	85	–	

FIGURE 5–18
High-quality instrumentation amplifier, the AMP-02 (courtesy of Precision Monolithics, Inc.).

PMI AMP-02 HIGH ACCURACY 8-PIN INSTRUMENTATION AMPLIFIER

ELECTRICAL CHARACTERISTICS at $V_S = \pm15V$, $V_{CM} = 0V$, $T_A = +25°C$, unless otherwise noted. *Continued*

PARAMETER	SYMBOL	CONDITIONS	AMP-02E MIN	AMP-02E TYP	AMP-02E MAX	AMP-02F MIN	AMP-02F TYP	AMP-02F MAX	UNITS
GAIN									
Gain Equation Accuracy	$G = \frac{50k\Omega}{R_G} + 1$	G = 1000	–	–	0.50	–	–	0.70	%
		G = 100	–	–	0.30	–	–	0.50	
		G = 10	–	–	0.25	–	–	0.40	
		G = 1	–	–	0.02	–	–	0.05	
Gain Range	G		1	–	10k	1	–	10k	V/V
Nonlinearity		G = 1 to 1000	–	0.006	–	–	0.006	–	%
Temperature Coefficient	G_{TC}	$1 \le G \le 1000$ (Notes 1, 2)	–	20	50	–	20	50	ppm/°C
OUTPUT RATING									
Output Voltage Swing	V_{OUT}	$T_A = +25°C$, $R_L = 1k\Omega$	±12	±13	–	±12	±13	–	V
		$R_L = 1k\Omega$, $-40°C \le T_A \le +85°C$	±11	±12	–	±11	±12	–	
Positive Current Limit		Output to-Ground Short	–	22	–	–	22	–	mA
Negative Current Limit		Output-to-Ground Short	–	32	–	–	32	–	mA
NOISE									
Voltage Density, RTI	e_n	$f_O = 1kHz$							$nV/\sqrt{Hz}$
		G = 1000	–	9	–	–	9	–	
		G = 100	–	10	–	–	10	–	
		G = 10	–	18	–	–	18	–	
		G = 1	–	120	–	–	120	–	
Noise Current Density, RTI	i_n	$f_O = 1kHz$, G = 1000	–	0.4	–	–	0.4	–	$pA/\sqrt{Hz}$
Input Noise Voltage	$e_{n\,p\text{-}p}$	0.1Hz to 10Hz							$\mu V_{p\text{-}p}$
		G = 1000	–	0.4	–	–	0.4	–	
		G = 100	–	0.5	–	–	0.5	–	
		G = 10	–	1.2	–	–	1.2	–	
		G = 1	–	10	–	–	10	–	
DYNAMIC RESPONSE									
Small-Signal Bandwidth (–3dB)	BW	G = 1	–	1200	–	–	1200	–	kHz
		G = 10	–	300	–	–	300	–	
		G = 100	–	200	–	–	200	–	
		G = 1000	–	200	–	–	200	–	
Slew Rate	SR	G = 10, $R_L = 1k\Omega$	4	6	–	4	6	–	V/μs
Settling Time	t_S	To 0.01% ±10V Step G = 1 to 1000	–	10	–	–	10	–	μs

NOTES:
1. Guaranteed by design.
2. Gain tempco does not include the effects of external component drift.
3. Input voltage range guaranteed by common-mode rejection test.

3 8/89, Rev. C

FIGURE 5–18
(continued)

electrocardiogram (ECG or EKG[1]) monitoring, and catheterization instruments that require electrodes placed in the body (see Section 17–13). Isolation amplifiers are also useful in impedance buffering and preamplification of low-level signals that may have several hundred volts of common-mode interference present.

An isolation amplifier operates on the same basic principles as other op-amps except it includes a high-accuracy isolation stage. This requires additional inputs for isolated power, feedback connection, and I/O connections. The basic symbol, drawn as a broken triangle to represent the isolation between the input and output terminals, is shown in Figure 5–19(a). Signal isolation is achieved either optically or with transformer coupling. Power for the amplifier must also be isolated by an internal dc-to-dc converter or by a separate "floating" battery supply connected to a second set of dc power supply terminals. Figure 5–19(b) shows an isolation amplifier connected as a noninverting amplifier. Note that

[1] ECG and EKG are both accepted abbreviations.

FIGURE 5–19
Isolation amplifier.

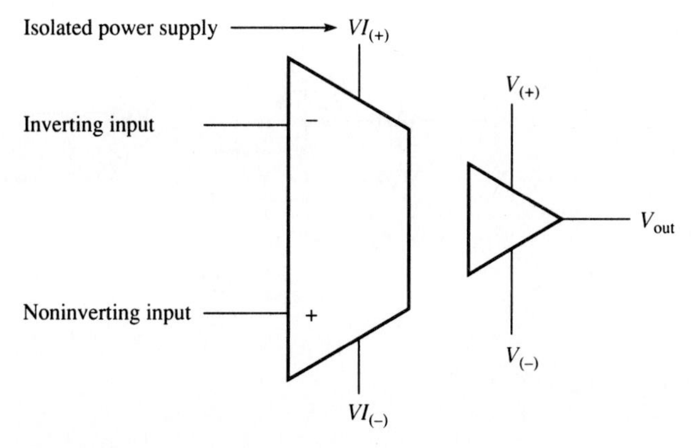

(a) Symbol.

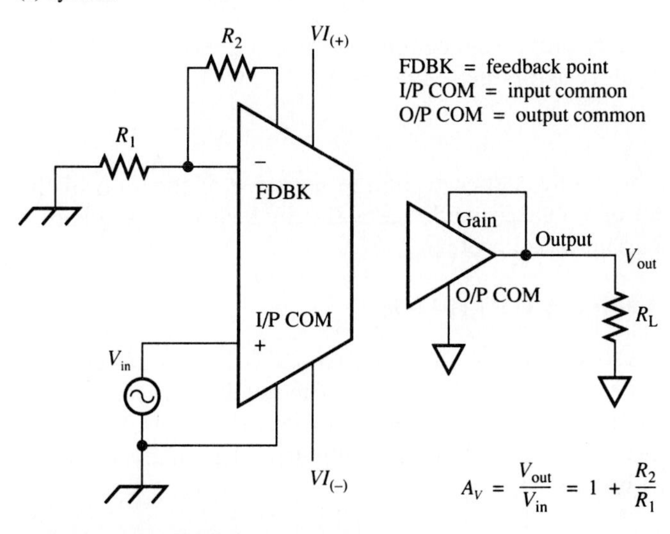

(b) Noninverting amplifier.

grounds between the input and output are shown with different symbols to emphasize that they are isolated.

SUMMING AMPLIFIER

A variation of the bàsic inverting amplifier is the summing amplifier shown in Figure 5–20. If all the resistors are equal, the output voltage is equal to the negative sum of the input voltages (provided the output is not in saturation). This can be seen by noting that the voltage at the inverting input is at virtual ground (from Rule 1). According to Kirchhoff's current law, the current into this node is equal to the current leaving the node. Since no current enters or leaves the op-amp input (Rule 2), all the input current flows through the feedback resistor R_F. Thus we can write

$$i_1 + i_2 + i_3 + i_4 = i_F$$

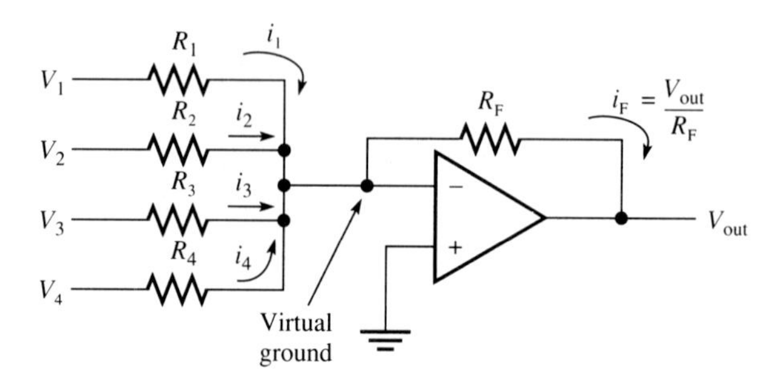

FIGURE 5–20
Summing amplifier.

By Ohm's law

$$\frac{V_1}{R_1} + \frac{V_2}{R_2} + \frac{V_3}{R_3} + \frac{V_4}{R_4} = -\frac{V_{out}}{R_F}$$

The negative sign occurs because $V_{out} = -i_F R_F$
If all resistors are equal, then

EQUATION 5–5

$$V_1 + V_2 + V_3 + V_4 = -V_{out}$$

The result is seen to be the negative of the sum of the input signals. The concept of virtual ground makes the analysis easy and shows that the inputs are independent of each other.

SCALING AMPLIFIER

If the resistors in the summing amplifier have different values, then the input voltages are weighted differently. The feedback resistor controls the overall gain of the amplifier, whereas the input resistors control the current for each input. This circuit is called a scaling amplifier. The analysis proceeds along the same path as before.

$$i_1 + i_2 + i_3 + i_4 = i_F$$

By Ohm's law

$$\frac{V_1}{R_1} + \frac{V_2}{R_2} + \frac{V_3}{R_3} + \frac{V_4}{R_4} = -\frac{V_{out}}{R_F}$$

Solving for V_{out} yields

EQUATION 5–6

$$V_{out} = -\left(\frac{R_F}{R_1} V_1 + \frac{R_F}{R_2} V_2 + \frac{R_F}{R_3} V_3 + \frac{R_F}{R_4} V_4\right)$$

EXAMPLE 5–4 A scaling amplifier can be used to convert a binary number into a voltage proportional to the number. This is a form of digital-to-analog (D/A) conversion that is discussed further in Section 6–6. Assume the input voltage is 5.0 V for each 1 in the binary number and 0 V for each 0 in the binary number. Compute the output voltage for the circuit shown in Figure 5–21 if the input binary number is 1101.

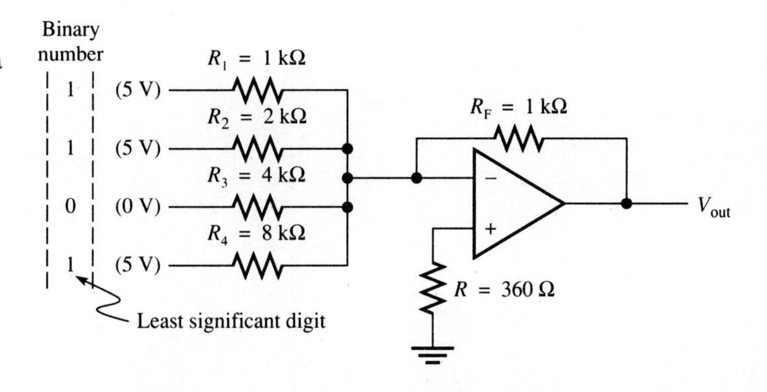

FIGURE 5–21
Scaling amplifier to convert a binary number to an analog voltage.

SOLUTION Substituting into Equation 5–6 gives

$$\begin{aligned} V_{out} &= -\left(\frac{R_F}{R_1} V_1 + \frac{R_F}{R_2} V_2 + \frac{R_F}{R_3} V_3 + \frac{R_F}{R_4} V_4\right) \\ &= -\left(\frac{1\text{ k}\Omega}{1\text{ k}\Omega} 5\text{ V} + \frac{1\text{ k}\Omega}{2\text{ k}\Omega} 5\text{ V} + \frac{1\text{ k}\Omega}{4\text{ k}\Omega} 0\text{ V} + \frac{1\text{ k}\Omega}{8\text{ k}\Omega} 5\text{ V}\right) \\ &= -8.125\text{ V} \end{aligned}$$

Notice that a 360 Ω resistor has been added to the noninverting input to balance the bias paths for the input differential amplifier. This is a common technique to help give 0 V out when there is no input voltage.

SOLVING EQUATIONS WITH A SCALING AMPLIFIER

The scaling amplifier can be combined with other circuits to produce the solution to algebraic equations as is sometimes required in certain signal-processing applications. The inverting scaling amplifier can be used to combine variables with different weighting. A linear equation of the form $y = ax_1 + bx_2 + cx_3$ can be set up using a scaling amplifier. If all the coefficients for the equation are negative, then a single scaling amplifier can be used to provide the solution.

EXAMPLE 5–5 Draw a circuit that will provide an output proportional to the following function:

$$V_{out} = 3.0V_1 + 3.7V_2 - 7.5V_3$$

Assume that the input voltages will be between −1 V and +1 V.

SOLUTION There are several possible ways to combine the circuits needed for the solution. If all the coefficients have negative signs, then a single inverting scaling amplifier can be used. Since two of the coefficients are positive, let us first combine these two terms in a single scaling amplifier, followed by a second amplifier to combine the negative term. The positive terms will have undergone two sign reversals, leaving the net result correct.

For the first term, $R_F/R_1 = 3.0$ and for the second term $R_F/R_2 = 3.7$. The remaining solution is shown in Figure 5–22. Again, resistors on the noninverting inputs are used to balance the bias paths for the differential amplifiers.

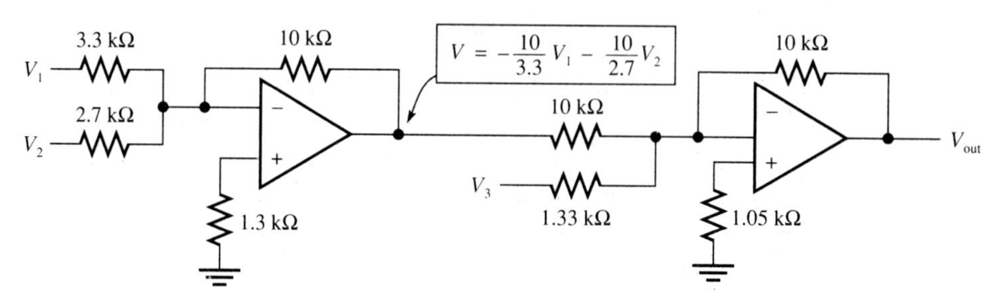

FIGURE 5–22
A circuit that produces an output given by $V_{out} = 3.0V_1 + 3.7V_2 - 7.5V_3$.

5–7 COMPARATORS AND SCHMITT TRIGGER CIRCUITS

Although operational amplifiers normally are used with negative feedback, there are certain configurations that use no feedback or positive feedback. Because of the extremely high open-loop gain, these applications cause the output to go into saturation—either positive or negative. Although the basic comparator shown will saturate near the power supply rails, it is possible to clamp the output voltage at less than the power supply rails and speed up the switching time.

THE BASIC COMPARATOR

A basic comparator is a circuit that allows us to determine which of two voltages is larger. One of the inputs is typically a reference voltage, V_{ref}, and the other input may be a time-varying signal. The output takes on one of two states—generally either saturated high or saturated low. A basic comparator circuit is shown in Figure 5–23(a). Notice that there is no feedback. If the input signal V_{in} is greater than V_{ref} by less than a millivolt, the gain of the op-amp causes the output to go into saturation near the positive supply voltage. Likewise, if V_{in} is below V_{ref}, the output goes into saturation near the negative supply voltage. The transfer curve for the basic comparator is shown in Figure 5–23(b).

Although general-purpose op-amps can be used for comparators (and frequently are), special integrated circuits have been designed by manufacturers for use specifically as comparators. These circuits are listed in specification books as comparators and, because frequency-compensation components have been omitted, can respond much faster than general-purpose op-amps. A comparator such as the Fairchild μA760 can change states in 16 ns, whereas a general-purpose op-amp can take 1000 times longer! Comparators can be optimized for speed, input

FIGURE 5–23
Basic comparator.

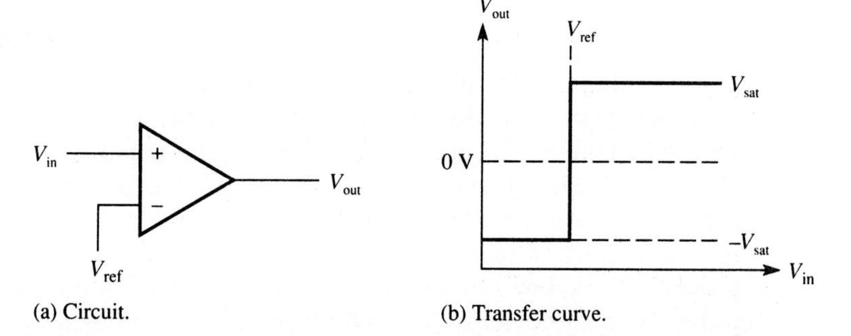

(a) Circuit. (b) Transfer curve.

accuracy, extremely low signal levels, fanout, power supply range, or other parameters. The LM111 series, for instance, is designed to operate over a wide range of supply voltages, including single-ended power supplies, and is designed to drive high current loads. This series can directly interface with transistor-transistor-logic (TTL), which operates from +5.0 V.

EXAMPLE 5–6 Design a comparator circuit that causes one LED to turn on if the input voltage is less than 0.8 V and another LED to turn on if the input voltage is greater than 2.0 V. The power supply is a single-ended +5.0 V supply.

SOLUTION An LM111 comparator is selected because it is designed to operate from a single-ended power supply. The circuit is illustrated in Figure 5–24. The voltage divider is used to set the comparison voltages (standard 10% resistor values were selected). The 330 Ω resistors are used for current limiting. If the input voltage is greater than the upper threshold (2.0 V), the upper comparator saturates high and causes LED_1 to turn on. If the input falls below the lower threshold (0.83 V), the lower comparator saturates high, causing LED_2 to turn on.

FIGURE 5–24

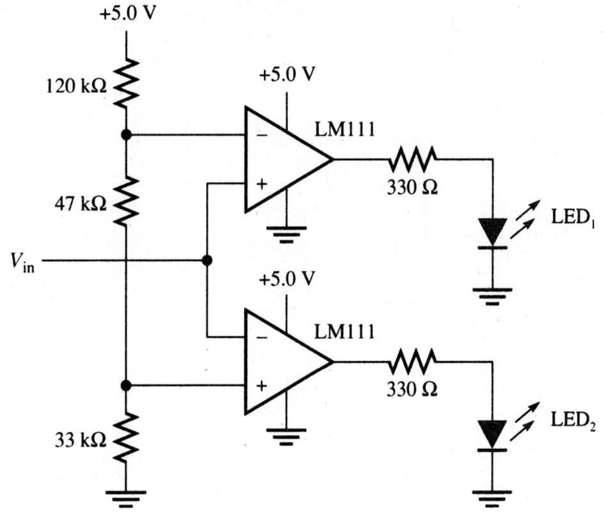

WINDOW COMPARATOR

A window comparator is a circuit designed to test if a voltage is between two reference voltages—or, in other words, within a "window." Figure 5–25(a) shows an example that uses the LM139. The LM139 series uses open-collector outputs, which allow the outputs to be connected together through a single-load resistor. Open-collector means that the output transistor in the comparator has no load resistor, so it is supplied by the user externally to the IC. This configuration causes the output voltage to be pulled to the lower saturation voltage if *either* comparator has a low output. (Connecting the outputs directly together is not allowable on most devices.) The reference voltages are established in this circuit with a voltage divider and provide a lower-level and an upper-level comparison. Whenever the signal is between the reference voltages (in this case +5 V and +10 V), the output transistors in both comparators are turned off, and the output is pulled high through the 10 kΩ load resistor. The transfer curve is shown in Figure 5–25(b).

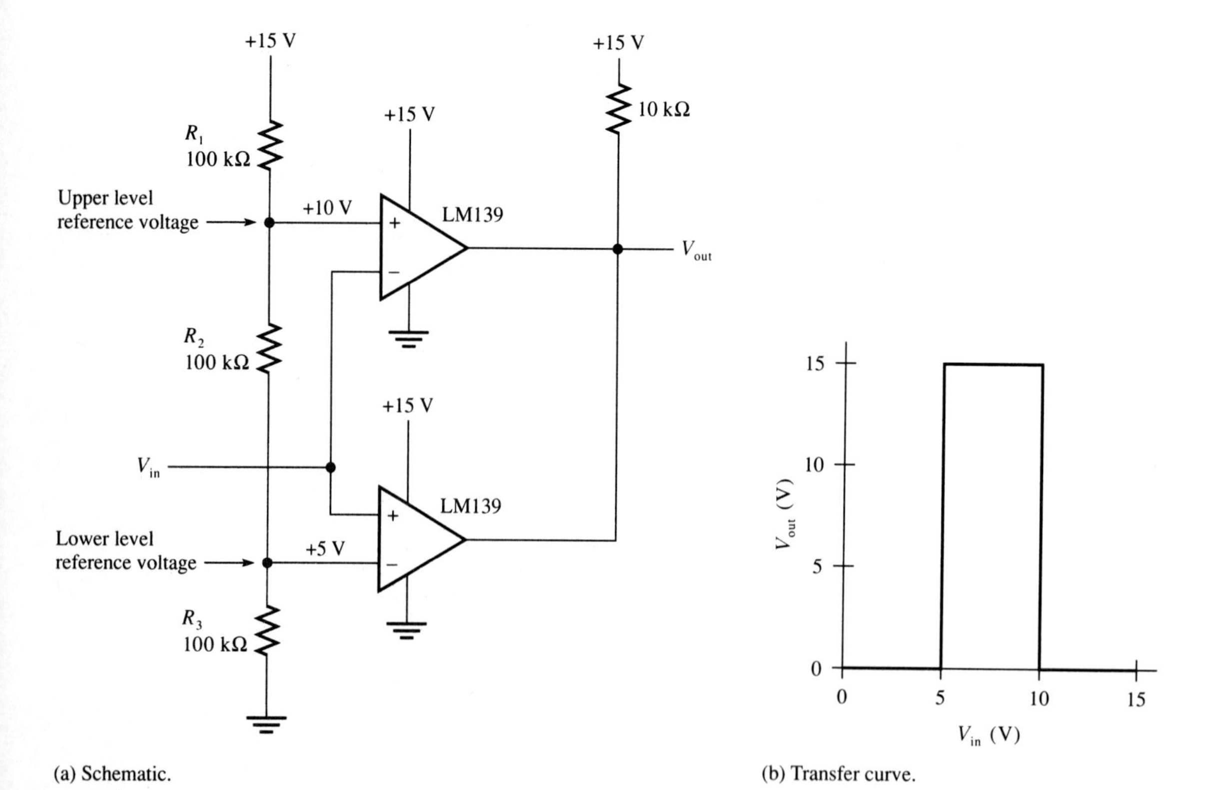

(a) Schematic.

(b) Transfer curve.

FIGURE 5–25
Window comparator.

SCHMITT TRIGGERS

The basic comparator switches at a given level as the input crosses the threshold. For a slowly changing input, especially if the input is noisy, the output can make several transitions while the input crosses the threshold. This sort of problem

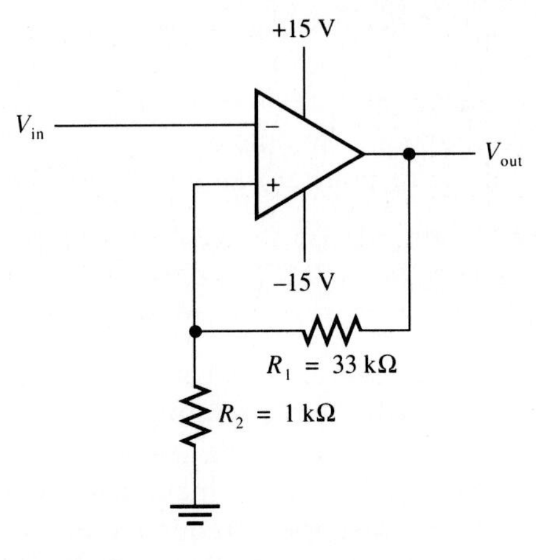

$$\text{Upper threshold level} \cong +V_{sat}\left(\frac{R_2}{R_1 + R_2}\right)$$

$$\cong 14\text{ V}\left(\frac{1\text{ k}\Omega}{1\text{ k}\Omega + 33\text{ k}\Omega}\right)$$

$$\cong 0.41\text{ V}$$

$$\text{Lower threshold level} \cong -V_{sat}\left(\frac{R_2}{R_1 + R_2}\right)$$

$$\cong (-14\text{ V})\left(\frac{1\text{ k}\Omega}{1\text{ k}\Omega + 33\text{ k}\Omega}\right)$$

$$\cong -0.41\text{ V}$$

(a) Circuit.

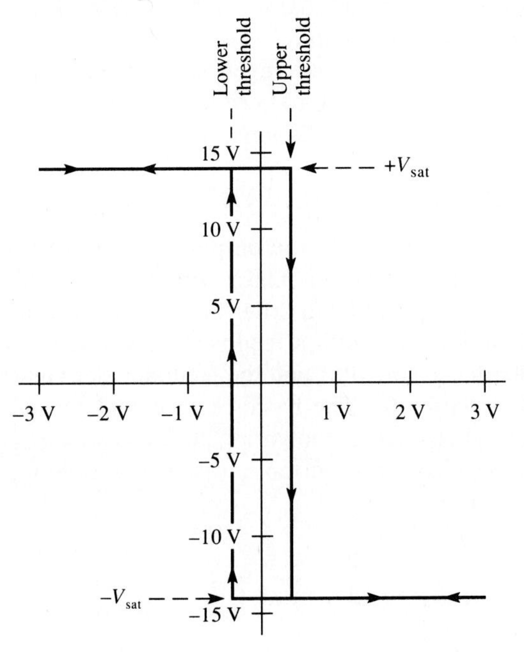

(b) Transfer curve.

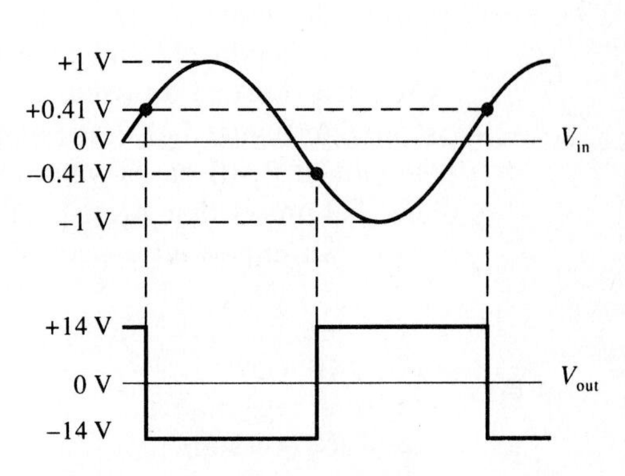

(c) Response to a sinusoidal waveform. Note that the changes on a different threshold depend on whether the input is increasing or decreasing.

FIGURE 5–26
Schmitt trigger.

occurs in devices such as level detectors. A solution to this problem is to introduce positive feedback to a comparator in a circuit called a **Schmitt trigger.** A Schmitt trigger has two comparison thresholds—one that is active only as the signal increases through the threshold and a second, lower threshold that is active only as the signal decreases through it. When a system response shows a repeatable dependence on how the input changes, it is called **hysteresis.** Hysteresis is seen in the transfer characteristic for a system because the output can take on different states, depending on the immediate past history of the input. In addition to minimizing switching transients due to noise, hysteresis is applied to certain waveform-generating circuits. Figure 5–26 shows an example of a Schmitt trigger comparator circuit, the transfer curve, and the response to an input waveform.

5–8 OSCILLATOR CIRCUITS

Electronic instruments require different waveforms for a variety of signal-processing applications as well as for testing and troubleshooting. A common application of oscillators is to generate periodic signals such as sine waves, square waves, and pulses, which are needed in many instruments, including function generators (of course!), digital multimeters, oscilloscopes, frequency counters, and so forth. An **oscillator** is any circuit that produces a controlled periodic signal without an input signal. The output wave of an oscillator can be modified by other circuits to produce periodic waves of various shapes.

Oscillators can be classified into two types: the regenerative feedback form of oscillator and the relaxation oscillator. Common to both types is positive feedback, which provides a signal to maintain the oscillations.

REGENERATIVE FEEDBACK OSCILLATORS

Regenerative feedback oscillators use a sample of the output waveform to reinforce the input signal. This feedback must be in phase with the input signal (positive feedback) and have sufficient magnitude in order to sustain oscillations. We can think of an oscillator as an amplifier with a feedback network, as illustrated in Figure 5–27. The amplifier has a gain of A and the feedback network has a gain of β. (β has a value of less than 1.0.) The basic requirement to sustain oscillations is that $A\beta = 1$. This requirement is known as the *Barkhausen criterion*. An important aspect of the Barkhausen criterion is that it is a function of

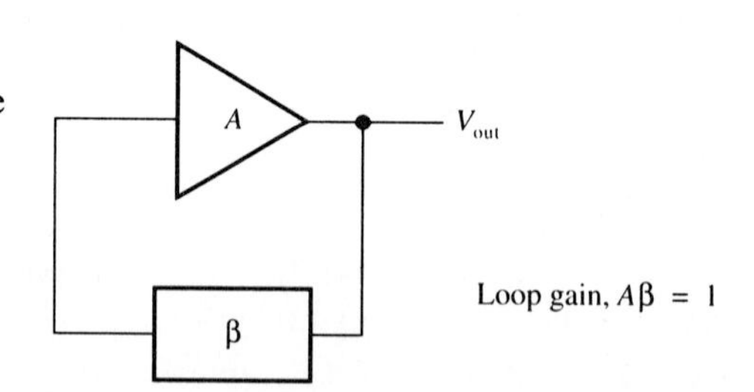

FIGURE 5–27
Model of regenerative feedback oscillator.

FIGURE 5–28
Phase-shift oscillator.

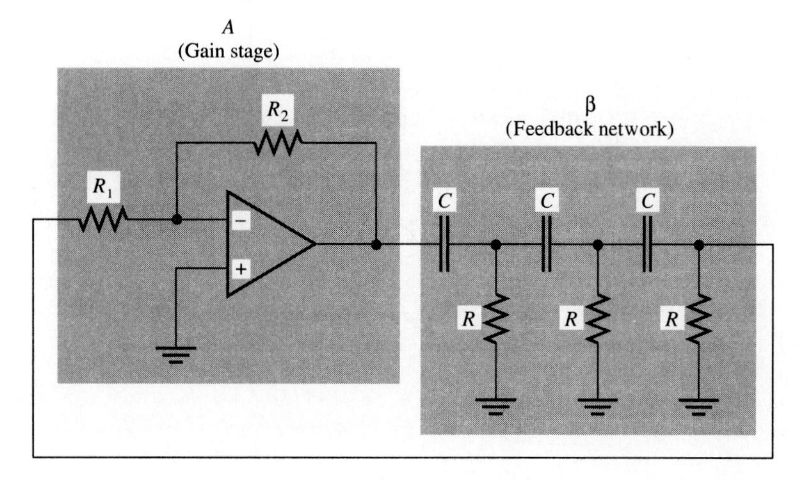

frequency. For most oscillators, there is only one frequency where both the magnitude and phase obey the Barkhausen criterion. The frequency at which the Barkhausen criterion is satisfied is the oscillation frequency.

PHASE-SHIFT OSCILLATOR

There are many circuits that have been designed as regenerative feedback oscillators. One of the simplest to understand is the *phase-shift oscillator* shown in Figure 5–28. The phase-shift oscillator uses an inverting amplifier to shift the phase 180° and a phase-shift feedback network to shift the phase of the feedback signal another 180°. At first glance, it would appear that each RC section must shift the phase by 60°; however, this ignores the loading effect of the latter sections. The oscillation frequency for a phase shift oscillator is found from the equation

EQUATION 5–7

$$f = \frac{1}{\sqrt{6}RC}$$

where f = oscillation frequency, Hz
R = resistance, Ω
C = capacitance, F

The derivation of this equation is given in Appendix A. The derivation reveals that the feedback gain is 1/29. This means that the magnitude of the gain of the amplifier, set by the ratio of R_2/R_1, must be 29 in order to satisfy the Barkhausen criterion.

WIEN-BRIDGE OSCILLATOR

Surprisingly, a low-distortion sine wave is difficult to generate. One circuit widely used in laboratory oscillators for generating good-quality sinusoidal waves at frequencies below 1 MHz is the Wien-bridge oscillator shown in Figure 5–29(a).

FIGURE 5–29
Wien-bridge oscillator.

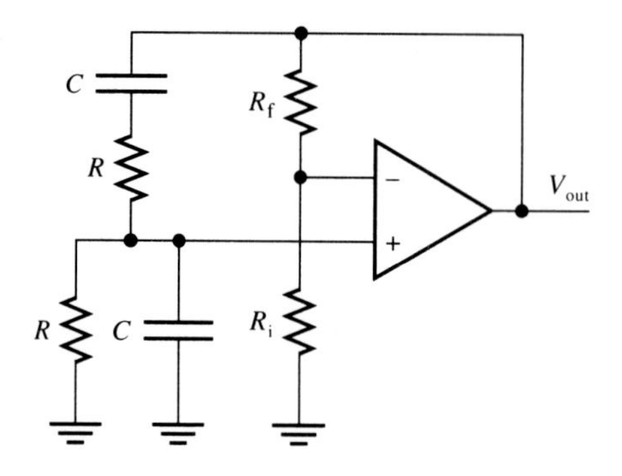

(a) Basic circuit.

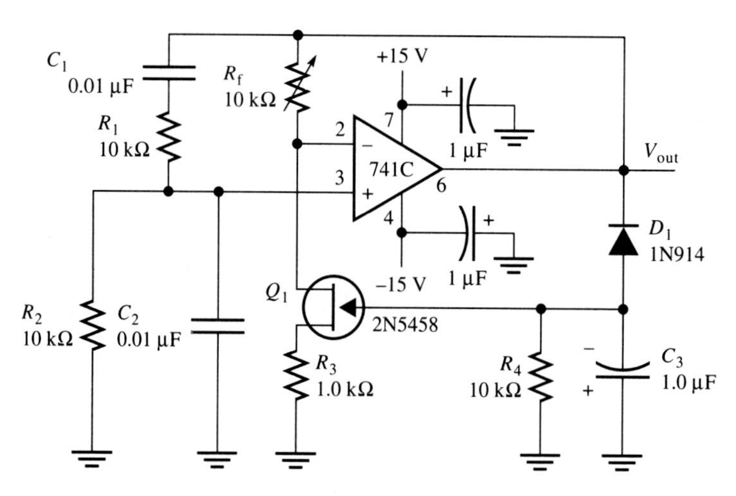

(b) Practical circuit with stabilized gain.

The bridge is a frequency-selective network that returns a specific frequency to the noninverting input and attenuates all other frequencies. The gain of the feedback network is maximum when $X_C = R$. This frequency is the oscillation frequency and can be found by the equation

EQUATION 5–8

$$f = \frac{1}{2\pi RC}$$

where f = oscillation frequency, Hz
R = resistance, Ω
C = capacitance, F

At the oscillation frequency, the gain of the feedback network is 1/3. To meet the Barkhausen criterion, the gain of the amplifier must be 3.00. To maintain an undistorted sinusoidal wave, the gain must be stable and stay at exactly 3.00. With the basic network, maintaining this gain is nearly impossible. If the gain is slightly high, the output goes into saturation; if it is slightly low, oscillations cease. In order to overcome this difficulty, various feedback networks have been designed to regulate the gain automatically. Frequently, gain is maintained by a FET by taking advantage of the voltage-controlled resistance characteristic. An example of an FET-stabilized network is shown in Figure 5–29(b). This network allows the startup gain to exceed 3.00 until the amplitude of the oscillations is sufficient to begin to bias the JFET off through the diode negative peak detector. At this point the gain decreases to exactly 3.00, the point where oscillations are just maintained. Notice that the time constant for the R_4C_3 combination must be long compared to the oscillation frequency to avoid gain change during the oscillation cycle.

RELAXATION OSCILLATORS

Relaxation oscillators, also called **astable multivibrators,** are free-running oscillators used to generate different waveforms, including square waves. Relaxation oscillators depend on hysteresis to cause them to oscillate. Figure 5–30 shows a classic example of a relaxation oscillator that—although not widely used—has a neon lamp as the hysteresis element. A constant-current source is used to charge a capacitor. (In noncritical applications, the constant-current source can be replaced by a resistor.) The lamp is initially off (open circuit). As the voltage across the capacitor rises, it reaches the firing voltage of the neon lamp; the lamp fires (gas ionizes) and discharges the capacitor. The process then repeats. The output voltage is a sawtooth waveform with a rate of rise given by

EQUATION 5–9

$$\frac{dV}{dt} = \frac{i}{C}$$

where $\frac{dV}{dt}$ = rate of change of voltage, V/s

i = current, A

C = capacitance, F

FIGURE 5–30
Classic sawtooth relaxation oscillator.

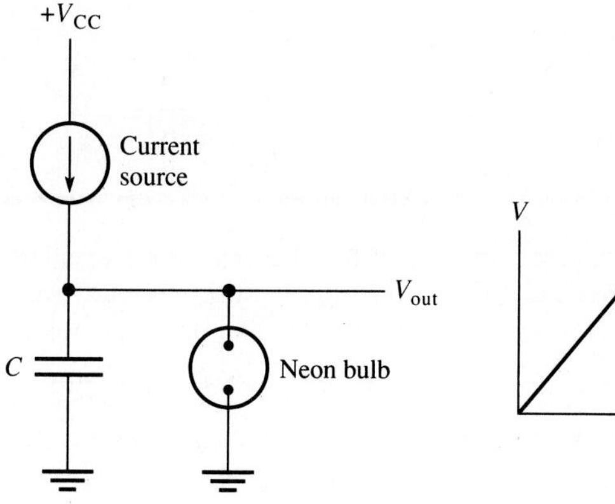

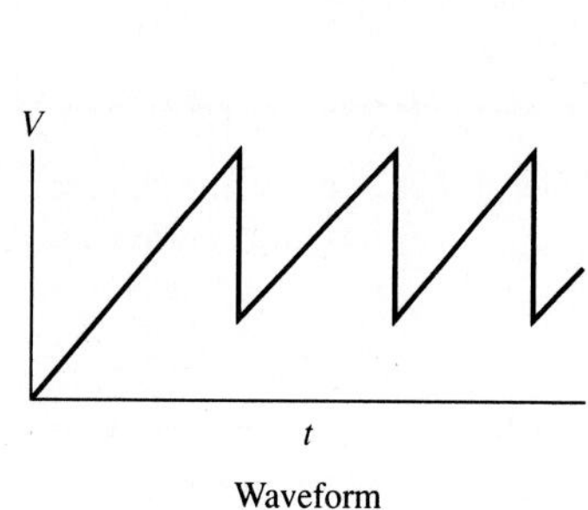

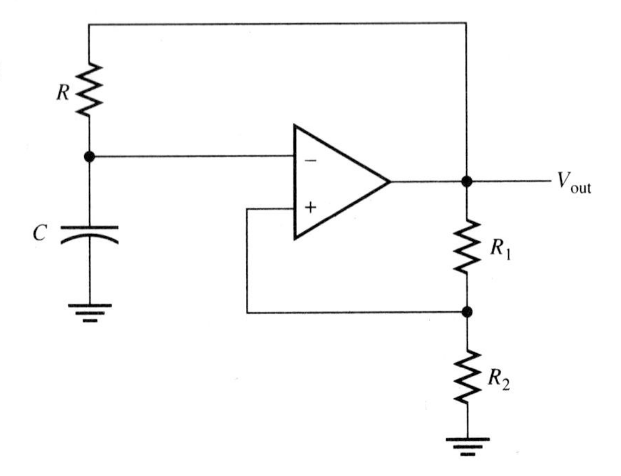

FIGURE 5–31
Square-wave generator using a Schmitt trigger.

The neon lamp fires at a voltage of approximately 70 V and presents a relatively low resistance discharge path for the capacitor. When the capacitor has discharged to some level, the lamp is extinguished, and the process repeats.

A similar idea can be used to produce a square wave. The circuit is another form of relaxation oscillator, illustrated in Figure 5–31. It uses a series *RC* combination to establish the time constant and a Schmitt trigger as the hysteresis element. When the output of the Schmitt trigger is in positive saturation, the capacitor begins to charge through *R* until the voltage reaches the upper threshold for the Schmitt trigger. The Schmitt trigger immediately changes state and goes into negative saturation. The capacitor begins to discharge. When the capacitor voltage reaches the lower threshold, the Schmitt trigger again changes state and the process repeats. The frequency is determined by the threshold levels and the *RC* time constant and can be computed from the equation

EQUATION 5–10

$$f = \frac{1}{2RC \ln(2R_2/R_1 + 1)}$$

where f = oscillation frequency, Hz

If R_2/R_1 is selected to be equal to 0.859, the equation for the frequency simplifies to

$$f = \frac{1}{2RC}$$

EXAMPLE 5–7 Specify the component values for the relaxation oscillator shown in Figure 5–31. The design requirement is for it to oscillate at a frequency of 10 kHz. The trip point voltages should be set for −0.5 V and +0.5 V. The power supplies are ±15 V.

SOLUTION Assume the saturation voltage is 1 V less than the power supply voltages. Then the positive saturation is $V_{sat} = +14$ V. R_1 and R_2 determine the trip points, as shown in Figure 5–26(a).

The upper trip point is

$$\text{UTP} = V_{\text{sat}}\left(\frac{R_2}{R_1 + R_2}\right)$$

By substitution

$$0.5 \text{ V} = (14 \text{ V})\left(\frac{R_2}{R_1 + R_2}\right)$$

The equation for the lower trip point is equivalent, so it is unnecessary to set up an equation for it. Solving for the ratio of R_2/R_1, we find

$$\frac{R_2}{R_1} = 0.0370$$

Look for resistor values that are standard values with the preceding ratio. By choosing R_1 to be 270 kΩ, R_2 is then 10 kΩ. Since these are standard values of resistors and are appropriate for avoiding loading effects, select these values.

By trial and error, the same technique can be used to find appropriate values for R and C. It is preferable to avoid capacitance values greater than 1 μF (although this may not be possible at very low frequencies). Choosing C to be 0.068 μF and solving for R gives standard values for both:

$$f = \frac{1}{2RC \ln(2R_2/R_1 + 1)}$$

$$R = \frac{1}{2fC \ln(2R_2/R_1 + 1)}$$

$$= \frac{1}{2(10 \text{ kHz})(0.068 \text{ μF}) \ln(2 \cdot 0.0370 + 1)}$$

$$= 10 \text{ kΩ}$$

At the 10 kHz frequency, the LM741C will exhibit slew-rate limiting; because of this, it will not produce a square-wave output. An op-amp that is 100 times faster than the LM741C is the LM318C, with a minimum slew rate of 50 V/μs. Figure 5–32 shows the final design using this op-amp.

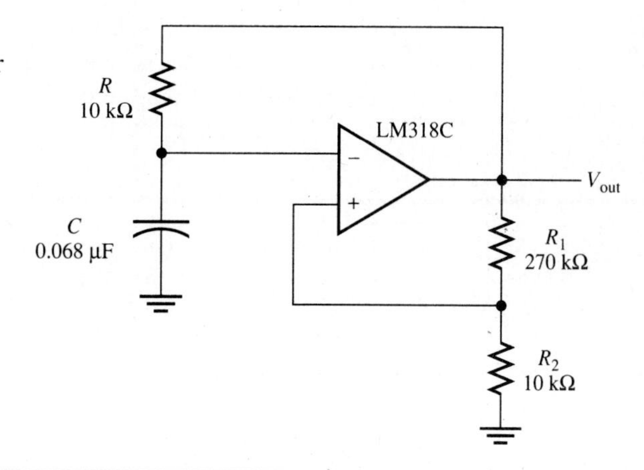

FIGURE 5–32
Square-wave generator for 10 kHz.

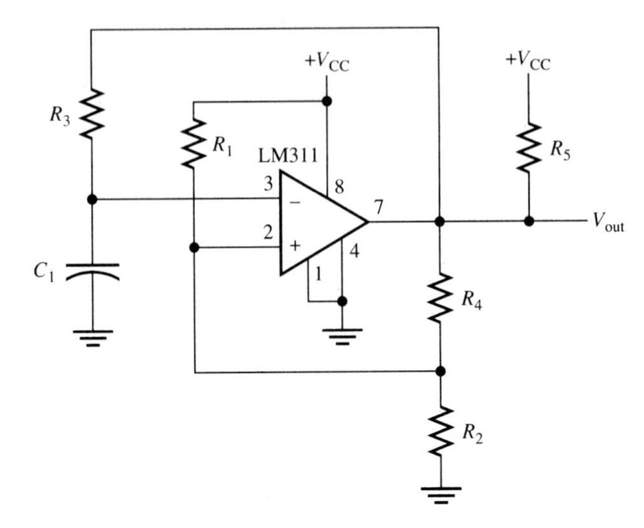

FIGURE 5–33
A relaxation oscillator using the LM311 comparator.

An interesting variation of the relaxation oscillator described in Example 5–7 can be made with a comparator such as the popular LM311. The LM311 has a number of features that make it different than an op-amp. Although designed to operate from both positive and negative supplies, it can be operated from a single +5 V supply, enabling it to interface directly with TTL logic circuits. Both the inputs and outputs can be isolated from system ground, and the output can drive loads referenced to ground, the positive supply, or the negative supply. One terminal (pin 1) is designated as "comparator ground" and can be connected to either circuit ground or to some other voltage level. If the noninverting input (pin 3) is higher than the inverting input, the output is internally connected to the comparator ground; otherwise it assumes an open-collector state. The LM311 also contains a terminal labeled strobe that allows the comparator function to be turned on and off with a separate signal. The strobe is inactive if it is left open or it is connected to a positive voltage.

Figure 5–33 shows a relaxation oscillator using the LM311 comparator. Notice the pull-up resistors on the output and on the noninverting terminal. This arrangement is different than the basic op-amp relaxation oscillator, and Equation 5–10 is *not* valid for this circuit.

EXAMPLE 5–8 Analyze the relaxation oscillator shown in Figure 5–34. Determine the trip points, the period of the waveform, and the duty cycle.

SOLUTION Begin by finding the trip points. Notice that the voltage at the noninverting input (pin 2) is due to two sources—the +15 V from the power supply and the output voltage (pin 7). When the output is in positive saturation, it is pulled up to +5 V through R_5. Solving each

FIGURE 5–34
Circuit for Example 5–8.

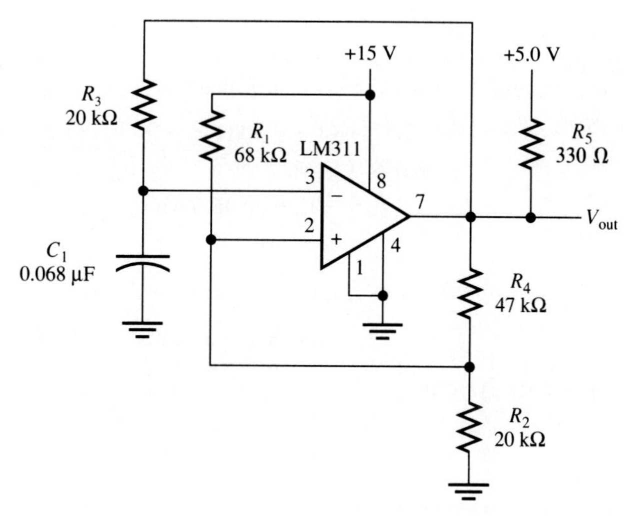

source with the voltage divider rule and applying the superposition theorem to both, we obtain

$$\text{UTP} = (+15\text{ V})\frac{20\text{ k}\Omega \| 47\text{ k}\Omega}{68\text{ k}\Omega + 20\text{ k}\Omega \| 47\text{ k}\Omega} + (+5\text{ V})\frac{20\text{ k}\Omega \| 68\text{ k}\Omega}{47\text{ k}\Omega + 20\text{ k}\Omega \| 68\text{ k}\Omega}$$

$$= 2.57\text{ V} + 1.23\text{ V} = 3.80\text{ V}$$

Since pin 1 is on ground, the lower output voltage is ground; this means that the lower trip point is determined by the +15 V supply acting alone. Thus

$$\text{LTP} = 2.57\text{ V}$$

To compute the frequency and duty cycle, we must find the time required to charge and discharge C_1. From basic electronics, the equation for the instantaneous voltage on a capacitor in a series *RC* circuit is

$$v(t) = (V_f - V_i)(1 - e^{-t/RC}) + V_i$$

where $v(t)$ = instantaneous voltage, V

V_f = final voltage a capacitor would reach if it were allowed to charge indefinitely, V

V_i = initial voltage on the capacitor, V

We substitute into this equation to find the charge time, t_{ch}, when the capacitor initially starts at the lower trip point and charges toward 5.00 V. We are interested in finding the time required for it to reach the upper trip point. Therefore

$$3.80 = (5.00 - 2.57)(1 - e^{-t_{ch}/1.36\text{ ms}}) + 2.57$$

$$0.494 = e^{-t_{ch}/1.36\text{ ms}}$$

$$t_{ch} = -1.36\text{ ms} \ln 0.494$$

$$= 0.96\text{ ms}$$

The same equation is used to find the discharge time, t_{dis}, as the capacitor discharges beginning from the upper trip point and discharging toward ground. This time we look for the time required to reach the lower trip point:

$$2.57 = (0 - 3.80)(1 - e^{-t_{dis}/1.36\text{ ms}}) + 3.80$$
$$0.676 = e^{-t_{dis}/1.36\text{ ms}}$$
$$t_{dis} = -1.36\text{ ms} \ln 0.676$$
$$= 0.53\text{ ms}$$

The period is the sum of the charge and discharge time. Thus

$$T = 0.96\text{ ms} + 0.53\text{ ms} = 1.49\text{ ms}$$

The duty cycle is

$$DC = \frac{0.96\text{ ms}}{1.49\text{ ms}} \times 100\% = 64\%$$

Because the Schmitt trigger relaxation oscillator free-runs back and forth between states with no input signal, it is an astable multivibrator. An astable multivibrator has no stable states—it is a free-running circuit that does not require triggering.

In addition to the astable multivibrator, there are two other types of multivibrators that do require a trigger signal. These are the *monostable* and *bistable multivibrators*. The monostable multivibrator has one stable state and a transitional, or quasi-stable, state. It requires a trigger signal to move it from the stable state into the quasi-stable state, where it remains for a predetermined length of time before returning to the stable state. The bistable multivibrator has two states, which are stable. When it is triggered, the bistable changes to the other stable state and remains there until another trigger moves it back into the original state.

Although circuits for all three multivibrators can be constructed using op-amps, there are a number of specialized ICs called *timers* that are popular for constructing various multivibrators. For digital circuits, flip-flops can be used for bistable multivibrators. Timers and flip-flops are covered in Chapter 6.

5–9 SIGNAL-PROCESSING CIRCUITS

Operational amplifiers are particularly suited for analog signal-processing applications necessary in instrumentation systems. These and related circuits can also be used for changing the shape of a waveform by integrating or differentiating or limiting rectification, and so forth. Some popular circuits for signal-processing applications are illustrated in this section.

INTEGRATING AND DIFFERENTIATING

The word *integrate* means to sum. If we kept a running sum, over time, of the area under a horizontal straight line, the area would increase linearly. As an example,

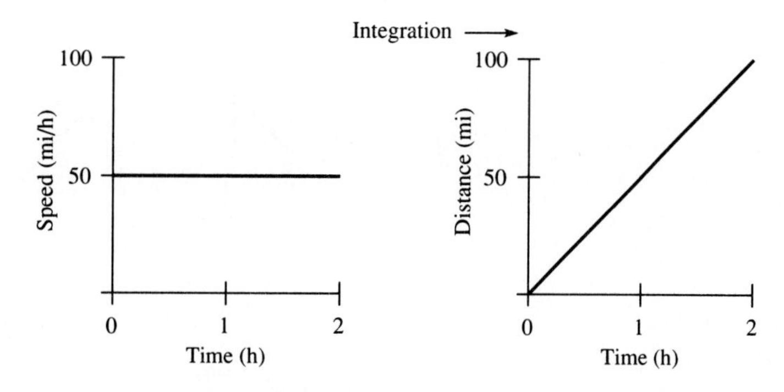

(a) Process of integration.

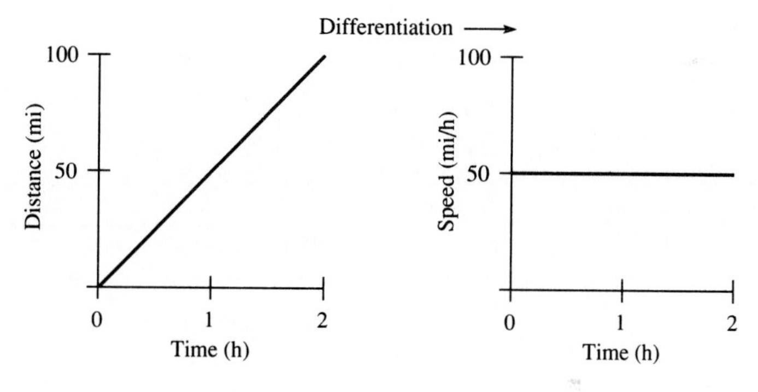

(b) Process of differentiation.

FIGURE 5–35
Comparison of integration and differentiation.

consider the distance traveled if the speed of a vehicle is maintained at a constant rate. If we plot the speed of the vehicle and the distance traveled on the same plot, we would see the relationship shown in the pair of graphs in Figure 5–35(a). The distance traveled is seen to increase linearly as time increases. The plot of distance as a function of time represents the sum (or integral) of the area under the graph of speed.

The opposite of integration is differentiation. The word *differentiate* means to find the instantaneous slope of a mathematical function. When time is the independent variable, the derivative is the rate of change of the function. If the distance covered changes at a linear rate, then rate of change of distance is a constant speed, as shown in Figure 5–35(b).

Differentiation and integration can be shown by the waveforms in Figure 5–36. Consider the rate of change of the triangular waveform shown in Figure 5–36(a). From 0 to 1 μs, the slope is constant; then it abruptly changes to a negative slope from 1 μs to 3 μs. A plot of the derivative is shown in Figure 5–36(b). If we integrate this signal, we obtain the original waveform.

FIGURE 5–36
Example of a signal and its derivative.

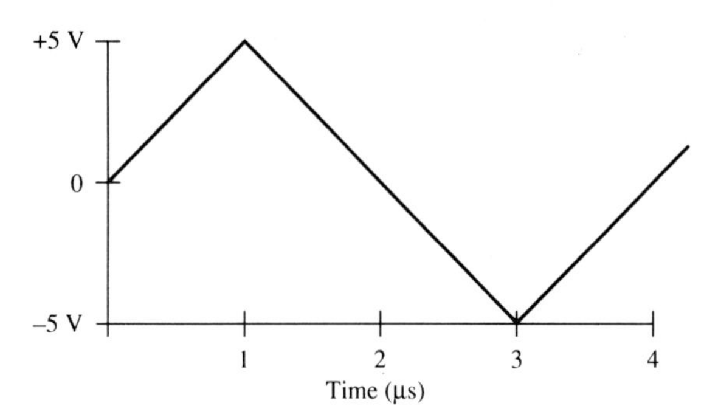

(a) Signal.

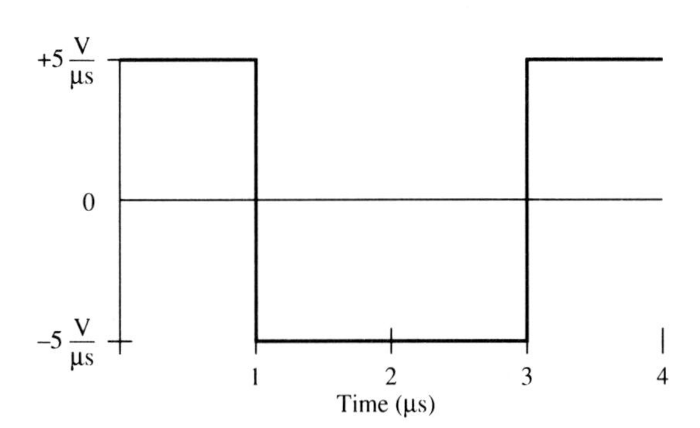

(b) Derivatative of signal shown in (a).

THE DIFFERENTIATOR

The electronic component that permits us to do integration and differentiation is the capacitor. From basic electronics, the current in a capacitor is given by the capacitance times the rate of change of voltage across the capacitor. That is

EQUATION 5–11

$$I = C\frac{dv}{dt}$$

where I = current, A

C = capacitance, F

$\frac{dv}{dt}$ = rate of change of voltage, V/s

Recall that the inverting amplifier circuit has a virtual ground at the inverting terminal, as shown in Figure 5–37(a). The input voltage is converted to a current by R_1 in accordance with Ohm's law, and the output is taken across R_2. If we replace R_1 with a capacitor, as shown in Figure 5–37(b), the current in the capacitor is given by Equation 5–11. This current does not enter the op-amp (Rule 2) but passes through R, generating the output voltage, V_{out}. By Ohm's law

$$V_{out} = -IR$$

The negative sign reflects the inversion process. We can write the expression for the output voltage by substitution:

EQUATION 5–12

$$V_{\text{out}} = -RC\frac{dv}{dt}$$

This expression shows that the output voltage is proportional to the negative rate of change of the input voltage times a constant. Notice that if the input voltage does not *change,* the output voltage will be zero. The faster the input voltage changes, the larger the output voltage will be. The circuit has performed differentiation.

The differentiating circuit described can give problems due to its tendency to oscillate and has problems with noise. Oscillations are caused by the high gain due to low capacitive reactance at high frequencies. To avoid oscillations, practical circuits typically include a small resistor in series with the input capacitor. This resistor lowers the gain at high frequencies. Noise problems occur because the capacitive reactance at high frequencies is small, which means that the gain at these frequencies is high. This problem can be corrected with the addition of a small capacitor in parallel with the feedback resistor. A practical op-amp differentiator is shown in Figure 5–37(c).

THE INTEGRATOR

By rearranging Equation 5–11, we can write

$$dv = \frac{I}{C}\,dt$$

Start again with the inverting amplifier, as in Figure 5–38(a). By replacing R_2 with a capacitor, C, the circuit is changed to the integrating circuit shown in Figure 5–38(b). The current that charges the capacitor is given by

$$I = -\frac{V_{\text{in}}}{R}$$

The minus sign reflects the inversion process. By substitution

$$dv = -\frac{V_{\text{in}}}{RC}\,dt$$

Integrating both sides of this equation gives

$$V_{\text{out}} = -\frac{1}{RC}\int_0^t V_{\text{in}}\,dt$$

This equation shows that the output voltage of the integrator is the negative integral (sum) over time of the input voltage multiplied by a constant.

The ideal integrator described poses a problem for practical circuits. This is because the capacitor is open to dc, causing the gain at dc to be equal to the open-loop gain of the op-amp. A small dc offset voltage will cause the operating point to

FIGURE 5–37
Differentiator circuit.

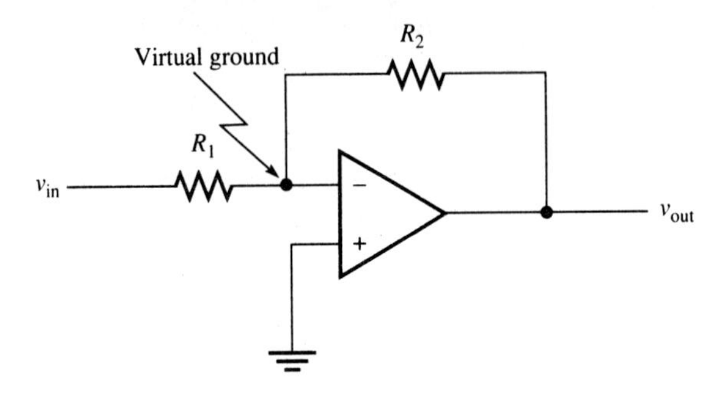

(a) Inverting amplifier.

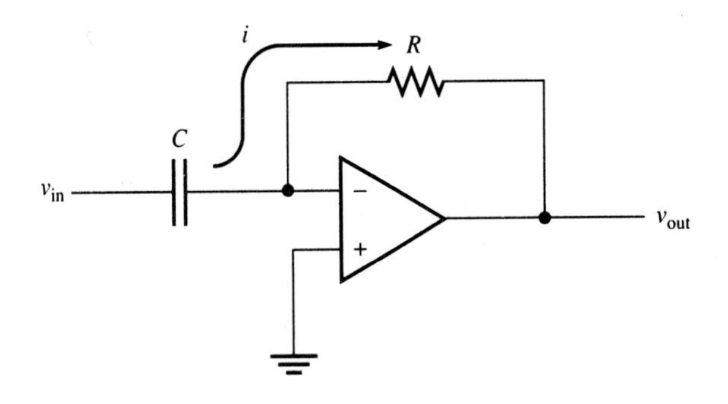

(b) Replacing R_1 with a capacitor. Current in the capacitor and resistor is identical.

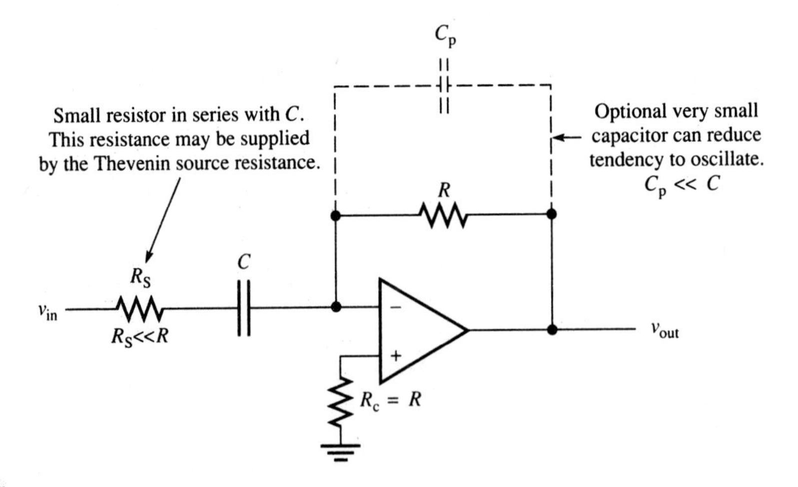

(c) Practical differentiating circuit that reduces tendency to oscillate. R_c is a compensating resistor used to zero the offset voltage.

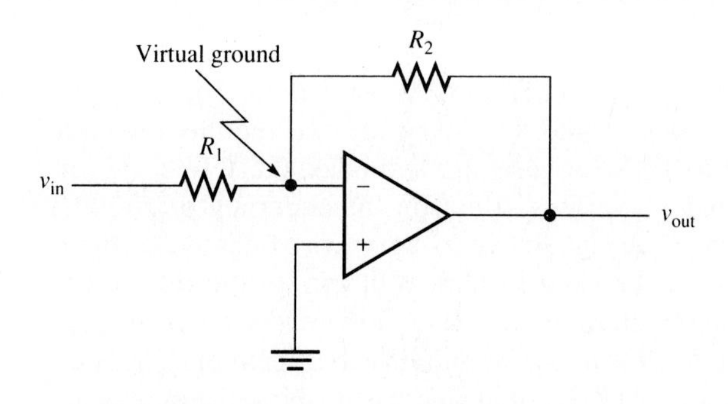

(a) Inverting amplifier.

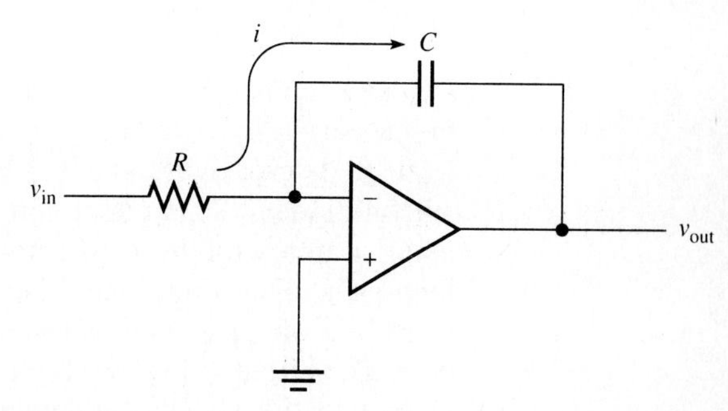

(b) Replacing R_2 with a capacitor.

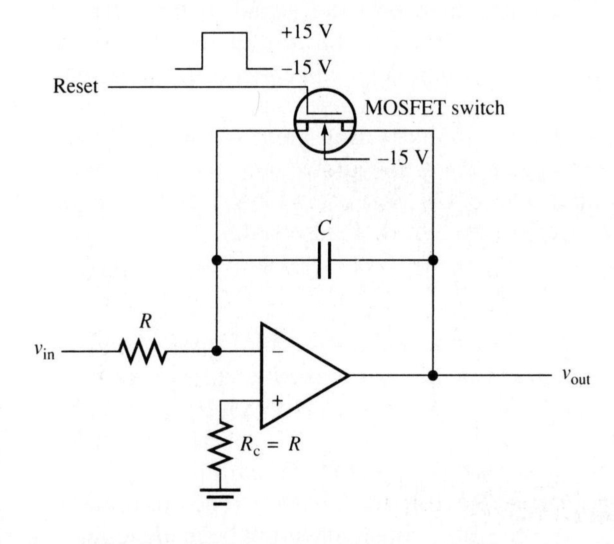

(c) Practical integrator that uses a MOSFET switch as a reset.

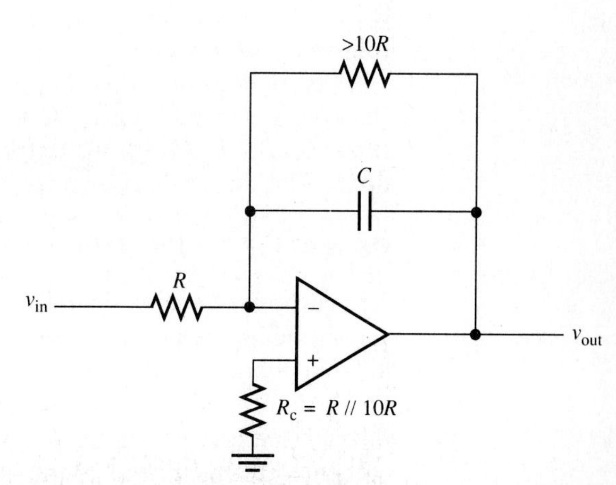

(d) Practical integrator with large resistor used to prevent integrator from drifting into saturation.

FIGURE 5–38
Integrator circuit.

drift toward saturation over a period of time. To avoid this problem, an electronic reset of the circuit can be accomplished by restarting the capacitor with 0 V at the beginning of each integration period. This can be done with a MOSFET switch, which discharges the capacitor whenever a reset signal is applied, as shown in Figure 5–38(c). Another common way of fixing the circuit is to connect a relatively large resistor in parallel with the capacitor, as shown in Figure 5–38(d). This resistor should be at least ten times larger than the input resistor. This reduces the integrator gain but prevents the integrator from drifting into saturation.

RECTIFICATION

In many ac instruments, it is necessary to rectify an ac waveform in order for it to be measured by a device that responds to dc. Ordinary passive rectifiers have a forward drop to overcome (0.2 V to 0.3 V for germanium diodes; 0.6 V to 0.7 V for silicon diodes). This offset can have a serious effect on the accuracy of circuits using ordinary diode rectifiers when the signal level is very low. Because of their high gain, operational amplifiers can be used to almost eliminate the diode forward-bias offset voltage, allowing effective rectification of even low-level signals.

The circuit shown in Figure 5–39 is a precision half-wave rectifier. The way to analyze the circuit is to determine when the diode is forward-biased and when it is reverse-biased. When the diode is forward-biased, it connects negative feedback to the input, whereas when it is reverse-biased, it appears open. Begin the analysis by assuming the noninverting input is slightly higher than the inverting input. The tremendous gain of the op-amp increases the signal at the output, causing the diode to be forward-biased. The result is that the diode provides a feedback path to the inverting input and the amplifier appears to be a voltage follower with a gain of 1.

Now consider what happens when the noninverting input is slightly lower than the inverting input. The op-amp again has tremendous gain, but the output tends to go into negative saturation. The diode is reverse-biased and appears open. The output is connected to ground through the load resistor.

PEAK DETECTION

The purpose of a peak detector is to store the maximum value of a varying signal. The circuit is shown in Figure 5–40. A precision half-wave rectifier appears to be a low-impedance charging path that allows the positive peaks to charge the capacitor. When the diode is off, it appears open and the capacitor's discharge path is through the load resistor. The time constant for this discharge path is made long compared to the period of the input signal, so that the capacitor tends to remain charged to the peak value of the input. If this time constant cannot be made long, a buffer can be added to the output to prevent the capacitor from discharging too fast. A negative peak detector can be made by simply reversing the diode in the circuit shown.

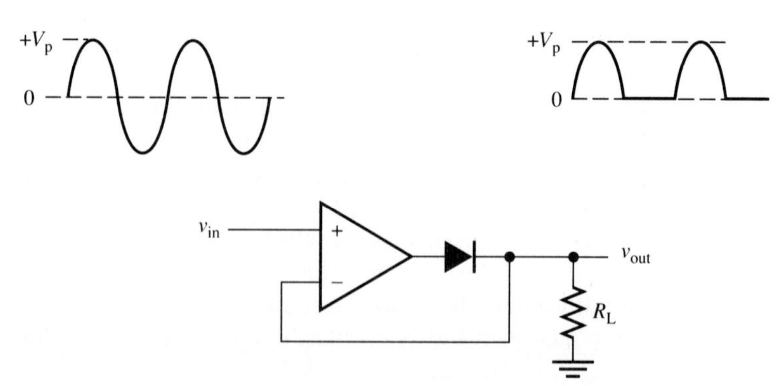

FIGURE 5–39
Precision half-wave rectifier.

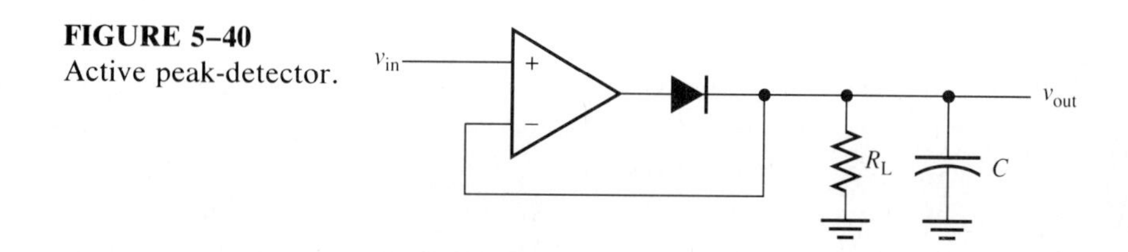

FIGURE 5–40
Active peak-detector.

CLIPPING AND LIMITING

Diode-clipping circuits were described in Section 4–3. Diode clippers are used to remove undesired extremes from waveforms by biasing a diode in a manner that allows it to conduct only when the signal goes beyond the desired level. Diode-clipping circuits work because of the low impedance of a forward-biased diode and the high impedance of a reverse-biased diode.

Diode-clipping circuits are not precise due to the nonideal characteristics and temperature dependence of real diodes. An active clipping circuit, constructed with an op-amp, can almost completely eliminate these problems and give precise clipping action. An active clipping circuit is shown in Figure 5–41(a). The clipping level is set to V_{ref} by the voltage from the variable resistor, which is connected as a voltage divider to the noninverting input of the op-amp.

When clipping is accomplished on both the positive and negative peaks, it is referred to as double-ended clipping, or limiting. Applications for limiters include digital logic circuits; they require signals to be within a certain range. A limiter circuit can be constructed using a zener diode in place of the feedback resistor of an inverting amplifier, as shown in Figure 5–41(b). When the input voltage is positive, the output swings negative and forward-biases the zener diode. Since one end of the zener is at virtual ground, this action puts the output voltage one diode drop (0.7 V) above ground. When the input is negative, the output moves in the positive direction, causing the zener to conduct at the zener breakdown voltage. This time the output is limited to the positive value of the zener breakdown voltage. In between these two voltages, the zener has high impedance; the amplifier therefore has high gain, causing the output to move quickly to the limits set by the zener.

SAMPLE-AND-HOLD CIRCUIT

A very useful circuit for many instrumentation applications is the **sample-and-hold circuit.** The input signal is sampled for a very short time period (typically from 1 to 10 μs), and the sampled value is retained for some amount of time until another sample is acquired. Figure 5–42 shows a basic sample-and-hold circuit that uses a JFET switch to charge (or discharge) a holding capacitor to the value of the analog signal at the sample time. The JFET switch is controlled by a voltage applied to the gate. When the JFET conducts, the holding capacitor assumes the same voltage as the input signal. When the control voltage is negative, the switch is open and the voltage is retained by the capacitor. The output op-amp isolates the capacitor, preventing it from discharging through the load resistor. The input op-

FIGURE 5–41
Clipping and limiting circuits.

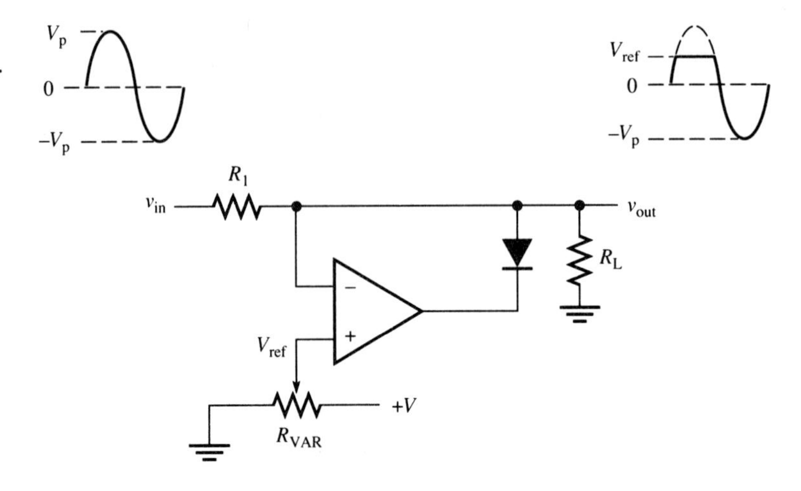

(a) Basic precision-clipper.

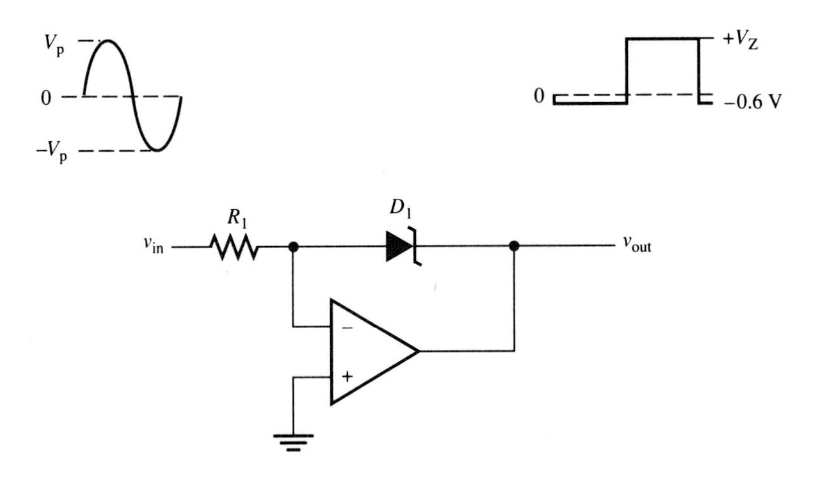

(b) Limiter using a zener diode.

amp buffers the input and provides a low-impedance charging path for the capacitor when the switch is closed.

Specialized op-amp circuits have been designed that incorporate a sophisticated sample-and-hold circuit in an IC except for the holding capacitor. For example, the LF198 series is a high-accuracy, wide-bandwidth sample-and-hold circuit that can be directly connected with common logic families for control.

LOGARITHMIC AMPLIFIERS

The mathematical operation of obtaining a logarithm (log, for short) or its inverse, an antilogarithm (antilog), is useful for signal processing and for applications

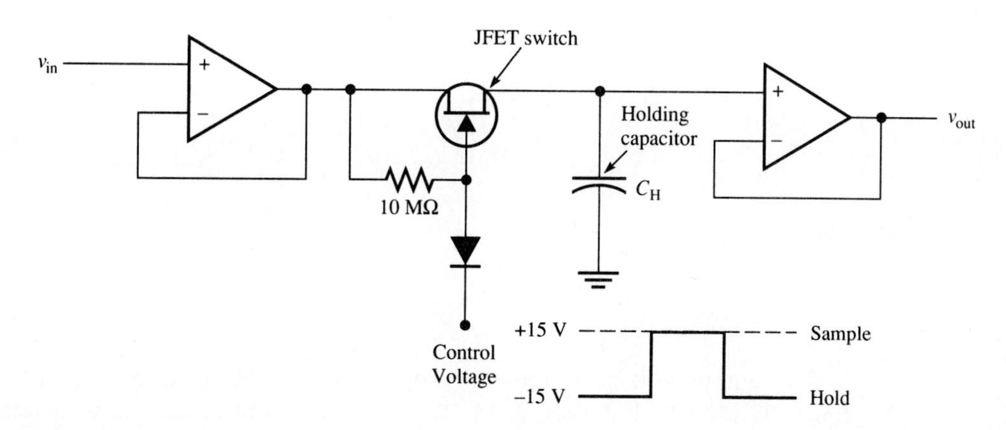

FIGURE 5–42
Basic sample-and-hold circuit.

requiring multiplication, division, exponentiation, and extracting of roots. In data-acquisition applications, it is useful for compressing the dynamic range of a signal. If the input signal extends over many decades, the effects of converting the signal to a logarithmic function can mean that an expensive 20 bit A/D converter could be replaced by an 8 bit A/D converter.

The log of a signal can be generated by using the basic logarithmic amplifier (log amp) shown in Figure 5–43. Recall that a diode has a logarithmic characteristic as given by the diode equation and shown in Figure 4–2. The log amp exploits the logarithmic characteristic of a diode by placing it in the feedback loop of an inverting amplifier. To understand the circuit, observe that the inverting input is a virtual ground point. The input resistor, R_i, converts the signal voltage into a current given by Ohm's law. This current is passed on to the diode, which converts the current to the output voltage. The current in the diode is given by the diode equation repeated here for reference:

$$I = I_S(e^{V/\eta V_T} - 1)$$

At room temperature, the diode equation reduces to

EQUATION 5–13

$$I = I_S e^{39V}$$

where I = diode current, A
I_S = reverse saturation current, A
V = voltage across the diode, V

Taking the natural logarithm of both sides and substituting for the diode current yields

EQUATION 5–14

$$V = -26\text{ mV}\left(\ln\frac{V_i}{R_i} - \ln I_S\right)$$

FIGURE 5–43
Basic logarithmic amplifier.

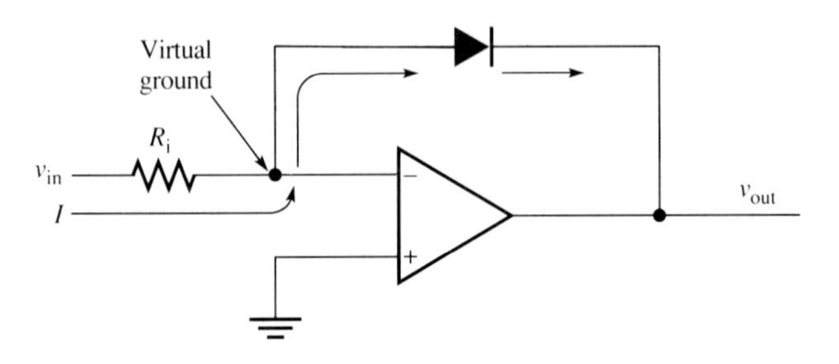

The output voltage is directly proportional to the logarithm of the input voltage, as seen in Equation 5–14. Unfortunately, the basic circuit given has certain undesirable characteristics—namely, it is temperature-sensitive, it tends to have error at very low diode currents, and the output level is not a convenient value. These problems have been addressed by manufacturers, who have designed precision IC logarithmic and log ratio amplifiers with temperature compensation and high accuracy. The Burr-Brown LOG100 is an example. It has a maximum accuracy specification of 0.37% at full-scale output (FSO) and deviation from linearity of 0.1% maximum over 5 decades. It can be connected as either a log, a log ratio, or an antilog amplifier.

5–10 ACTIVE FILTERS

A **filter** is a frequency-selective network that passes certain frequencies within a region called the **passband** while rejecting others in a region called the **stopband.** Filters can be designed to pass either low or high frequencies; they can also be designed to pass or reject a limited band of frequencies. The responses of the four ideal types of filters are illustrated in Figure 5–44(a). The ideal filter passes all signals within its passband without attenuation or phase shift and produces zero output for all other frequencies.

Actual filters can only approximate the ideal response, as shown in Figure 5–44(b). An actual filter cannot produce an abrupt transition between the passband and stopband; nor does it have perfectly flat response within the passband. In addition, actual filters may cause phase shifts that are not the same at all frequencies. Depending on the application, any of these characteristics can be optimized, usually at the expense of another characteristic.

Passive filters are designed with only passive elements such as resistors, capacitors, and inductors, whereas **active filters** contain transistors, op-amps, or other active elements in addition to passive elements. Both types have advantages; passive filters can be used at higher frequencies than active filters, but active filters can achieve frequency-response characteristics that are nearly ideal, they are easier to adjust, and they avoid the attenuation and loading problems associated with passive filters. Active filters are useful at frequencies up to a few megahertz; above this they are limited by bandwidth.

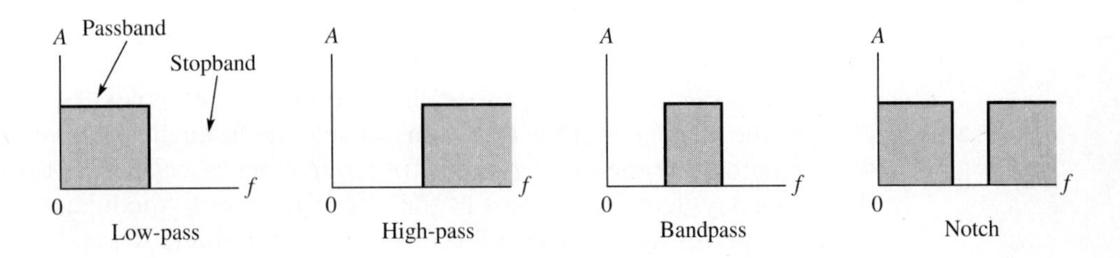

(a) Ideal filter responses.

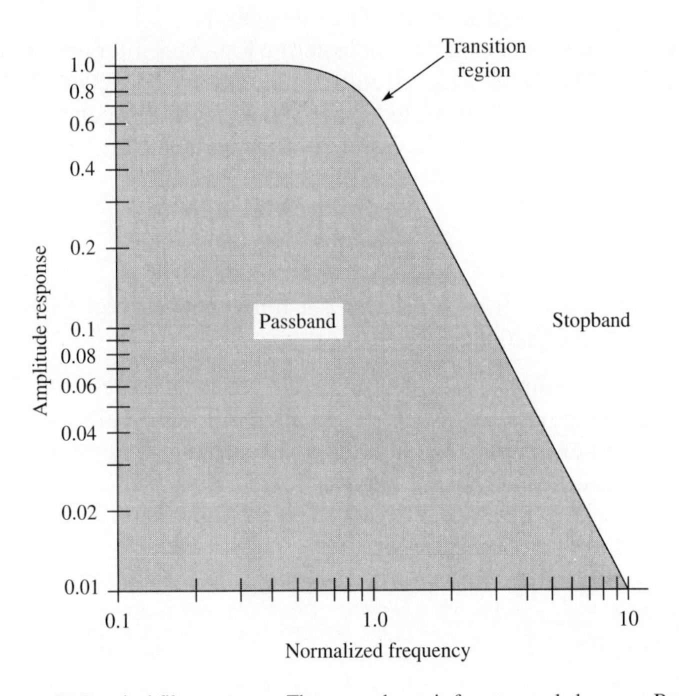

(b) Practical filter response. The curve shown is for a two-pole low-pass Butterworth filter.

FIGURE 5–44
Filter responses.

Active filters can be designed to optimize flatness of the response in the passband, steepness of the transition region, or minimum phase shift by choice of filter type. There are numerous specific design types of active filters—each emphasizing certain desirable characteristics. The Butterworth form of filter has the flattest passband characteristic, but it is not as steep as other filters and has poor phase response. By comparison, the Chebyshev filter has a steep transition from passband to stopband but at the expense of less than ideal phase characteristics and ripple in the passband. Another popular active filter is the Bessel filter, which has excellent phase response and step response but less than ideal transition from stopband to passband.

The **order** of a filter, also called the number of poles, is an integer that governs the steepness of the transition outside the frequencies of interest. A quick way to determine the order is to count the number of capacitors that are used in the frequency-determining part of the filter. In general, the higher the order, the better its performance. The roll-off rate for active filters depends on the type of filter but is approximately −20 dB/decade for each pole. (A **decade** is a factor-of-10 change in frequency.) A four-pole filter, for example, has a roll-off of approximately −80 dB/decade and uses four capacitors in the filter sections. (Note that inductors are rarely used in active filter designs.)

Figure 5–44 illustrates a two-pole active low-pass filter and a two-pole active high-pass filter. Each of these circuits is a section. To make a filter with more poles, the sections can be cascaded, but the gains of each section must be adjusted according to the values listed in Table 5–2. The cutoff frequency is given by the equation

EQUATION 5–15

$$f = \frac{1}{2\pi RC}$$

You can design your own Butterworth low-pass or high-pass active filter by using the following guidelines:

1. Determine the number of poles necessary based on the required roll-off rate. Choose an even number, as an odd number will require the same number of op-amps as the next-highest even number. For example, if the required roll-off is −40 dB/decade, specify a two-pole filter.
2. Choose R and C values for the desired cutoff frequency. These components are labeled R and C on Figure 5–45. The resistors should be between 1 kΩ and 100 kΩ. The value chosen should satisfy the cutoff frequency, as given by the equation

$$f = \frac{1}{2\pi RC}$$

3. Choose resistors R_1 and R_2 that produce the gain for each section according to the values listed in Table 5–2. Gain is controlled only by R_1 and R_2. Solving for the closed-loop gain of a noninverting amplifier gives the equation for R_2 in terms of R_1. That is

$$R_2 = (A_v - 1)R_1$$

TABLE 5–2
Gain required for Butterworth high- and low-pass filters.

	GAIN REQUIRED		
POLES	SECTION 1	SECTION 2	SECTION 3
2	1.586		
4	1.152	2.235	
6	1.068	1.586	2.483

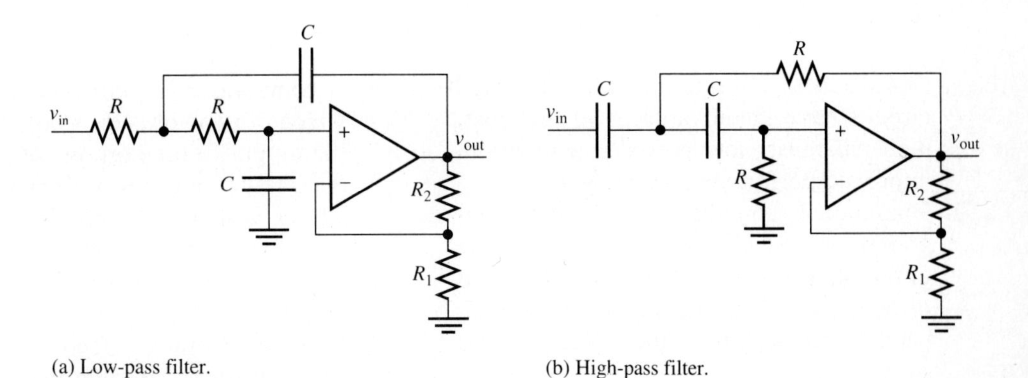

(a) Low-pass filter.

(b) High-pass filter.

FIGURE 5–45
Two-pole active filter.

EXAMPLE 5–9 A low-pass Butterworth filter with a roll-off of approximately −80 dB/decade and a cutoff frequency of 2.0 kHz is required. Specify the components.

SOLUTION **Step 1:** Determine the number of poles required. Since the design requirement is for approximately −80 dB/decade, a four-pole (two-section) filter is required.

Step 2: Choose R and C. Try C as 0.01 μF and compute R. Computed $R = 7.96$ kΩ. Since the nearest standard value is 8.2 kΩ, choose $C = 0.01$ μF and $R = 8.2$ kΩ.

Step 3: Determine the gain required for each section and specify R_1 and R_2. From Table 5–2, the gain of Section 1 is required to be 1.152 and the gain of Section 2 is required to be 2.235.

Choose resistors that will give these gains for a noninverting amplifier. The choices are determined by considering standard values and are shown on the completed schematic in Figure 5–46.

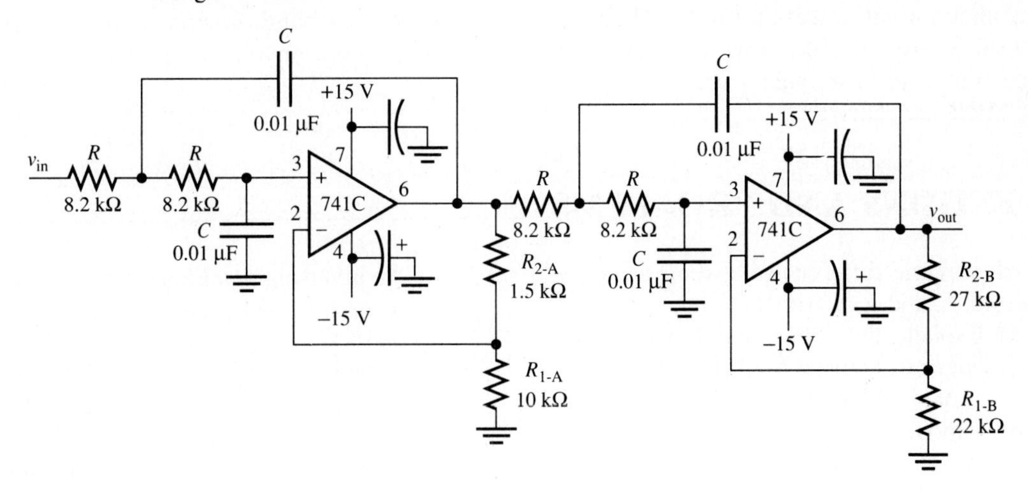

FIGURE 5–46

SUMMARY

1. Compared to discrete circuits, integrated circuits have the advantages of low cost, high reliability, low power consumption, improved performance, and small size.
2. Operational amplifiers are linear integrated circuits containing a high-gain dc differential amplifier. They are designed to be used with external resistors and capacitors to determine the operation and gain.
3. Negative feedback results in increased stability, reduced distortion, and greater bandwidth. It also affects the input and output impedance of an amplifier.
4. The ideal amplifier has infinite open-loop voltage gain, infinite input impedance, zero output impedance, zero dc offset, infinite bandwidth, and a common-mode gain of zero.
5. An ideal amplifier using negative feedback has zero voltage difference between the inputs and the inputs draw no current.
6. Many basic operational amplifier circuits are based on the inverting and noninverting amplifier configurations.
7. Instrumentation amplifiers are specially designed differential amplifiers with high input impedance and extremely good common-mode rejection ratios. They are used where accuracy, low drift, low-bias currents, precise gain, and very high CMRR are required.
8. A comparator is a circuit that determines which of two voltages is larger and causes the output to take on one of two states.
9. A Schmitt trigger is a special comparator circuit with two comparison thresholds. The thresholds change depending on the immediate past history of the state of the output.
10. Both regenerative feedback oscillators and relaxation oscillators use positive feedback to maintain oscillations. Regenerative feedback oscillators use a sample of the output waveform to reinforce the input signal, whereas relaxation oscillators depend on hysteresis to cause them to oscillate.
11. There are three types of multivibrators—the astable, the monostable, and the bistable multivibrator. The classification is based on the number of stable states that each type has.
12. Operational amplifiers can also be used for changing the shape of a waveform by integrating, differentiating, limiting, or rectification.
13. A filter passes certain frequencies while rejecting others. Active filters can be designed to optimize flatness of the response in the passband, steepness of the transition region, or minimum phase shift.

QUESTIONS AND PROBLEMS

1. What is the difference between a monolithic IC and a hybrid IC?
2. (a) Explain the difference between the open- and closed-loop gain of an operational amplifier.
   (b) Why isn't it reasonable to operate linear op-amp circuits with open-loop gain?
3. Compute the voltage gain for each of the following.
   (a) An inverting amplifier with a feedback resistor of 47 kΩ and an input resistor of 10 kΩ.
   (b) A noninverting amplifier with a feedback resistor of 100 kΩ and an input resistor of 27 kΩ.
4. An inverting summing amplifier is shown in Figure 5–47. Compute the output voltage if the inputs are as shown.

FIGURE 5–47

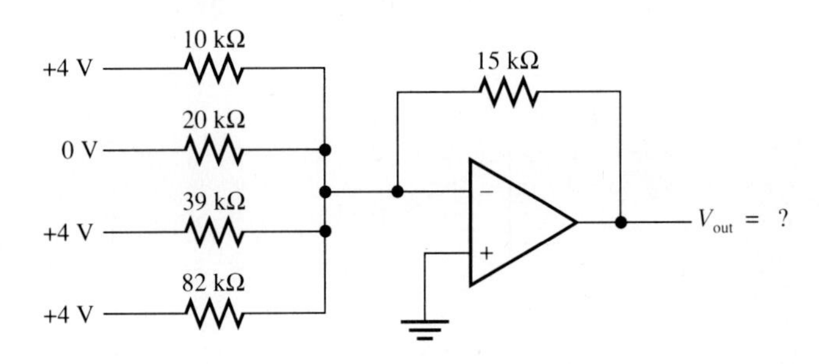

5. Assume the circuit shown in Figure 5–48 uses an ideal op-amp.
   (a) What is the voltage at the noninverting input?
   (b) What is the voltage at the inverting input?
   (c) What is the output voltage?

FIGURE 5–48

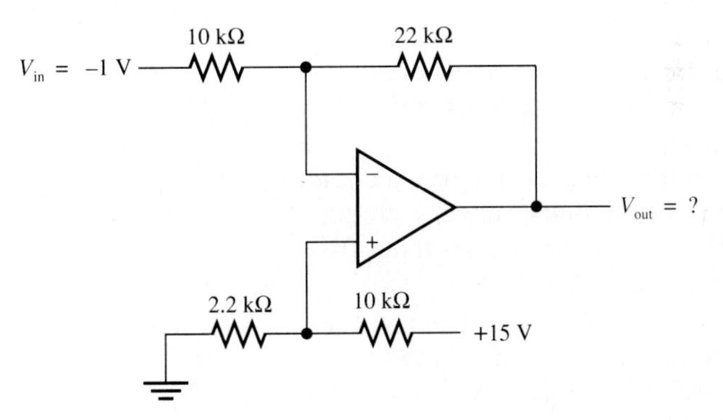

6. The circuit shown in Figure 5–49 is desired to have a current of 100 μA in R_2 when the input voltage is 10 V. Using the ideal op-amp rules, determine the required resistance of R_1 and V_{out}.

FIGURE 5–49

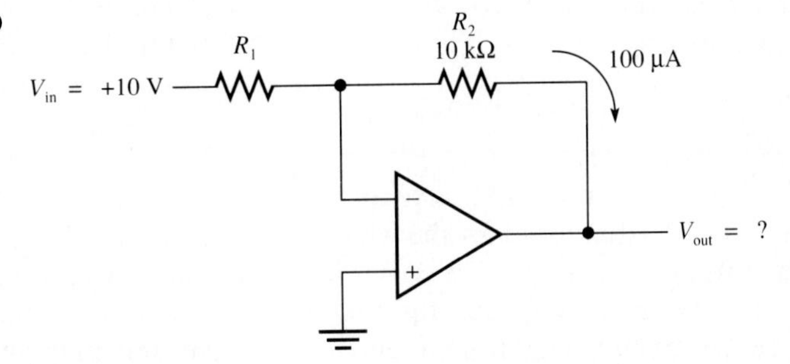

7. The slew rate of a particular op-amp is given as 2.5 V/μs. If the output voltage starts at 0 and changes at the maximum rate, what time is required for the voltage to reach 10 V?
8. Compare the definition of input bias current with input offset current.
9. A certain op-amp has a CMRR of 90 dB. Assume the differential gain at some low frequency is 100. Compute the common-mode gain at the same frequency.
10. A transducer is connected in the circuit in Figure 5–50(a) and a second identical transducer is connected in the circuit of

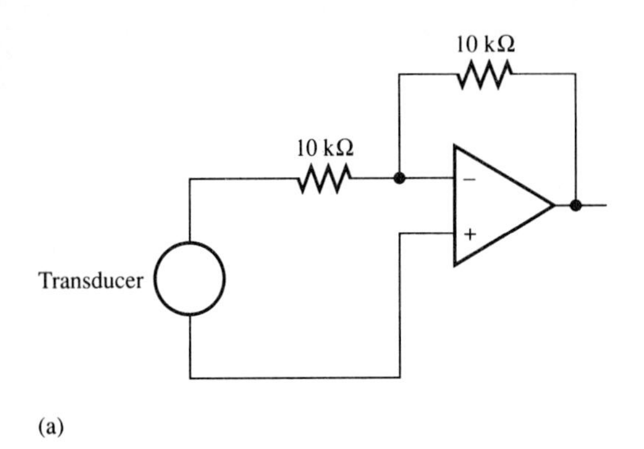

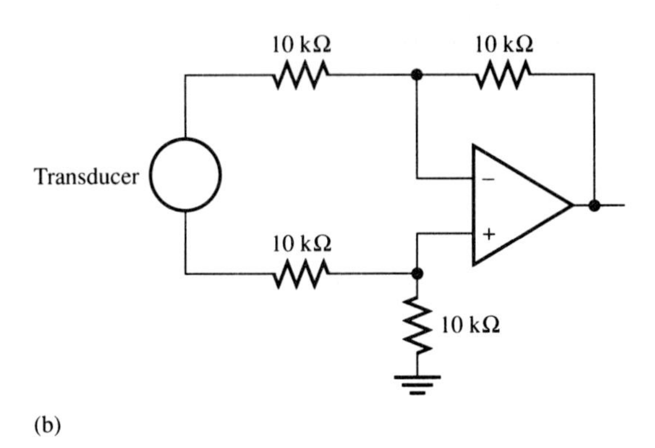

FIGURE 5–50

Figure 5–50(b). Which connection will have the lower common-mode noise? Explain your answer.

11. What advantage does an integrated circuit instrumentation amplifier have over an instrumentation amplifier constructed from operational amplifiers?
12. Assume a signal of 10 mV from a transducer is amplified by a differential amplifier with a CMRR of 80 dB. What is the largest common-mode signal that can be tolerated if the desired output signal can have no more than 1% noise?
13. Draw an operational amplifier circuit that combines the voltages given by the following equation:

$$v_{out} = 10v_1 - 22v_2 + 15v_3$$

14. Assume that R_2 in the circuit of Figure 5–25 is changed to a 39 kΩ resistor. Sketch the new transfer curve for the window comparator.
15. Compute the threshold voltages for the Schmitt trigger circuit shown in Figure 5–51.
16. Explain why the ideal op-amp rules cannot be applied to a Schmitt trigger circuit.
17. What is the difference between a regenerative feedback oscillator and a relaxation oscillator?
18. A phase-shift oscillator, such as the one

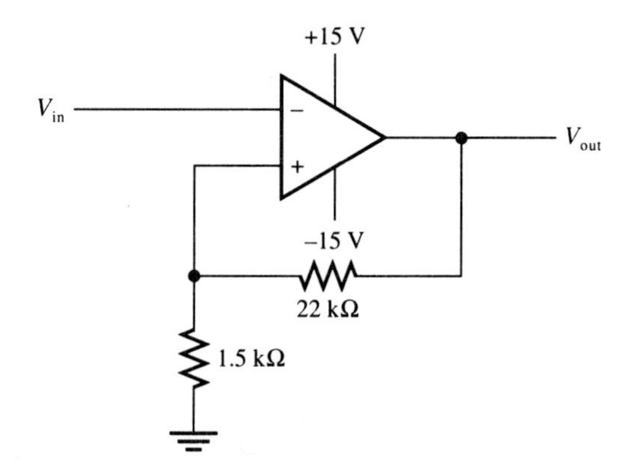

FIGURE 5–51

shown in Figure 5–28, has a feedback resistor R_2 of 100 kΩ. What is the largest value of R_1 that can be used to maintain oscillations?
19. How could the active clipping circuit shown in Figure 5–41(a) be changed to clip only the negative peaks?
20. For the relaxation oscillator in Example 5–8, sketch the waveform at the noninverting input and at the output. Label time and voltage on your sketch.
21. What is meant by the Barkhausen criterion for oscillator circuits?
22. Design a four-pole low-pass Butterworth filter with a cutoff frequency of 3.0 kHz. Draw the circuit and show the values of all components on your drawing.

Chapter 6

Digital Circuits

OBJECTIVES

Instrumentation systems require data acquisition, processing, and display. Although data is frequently acquired in analog form, it is usually converted to digital form for processing, storage, and display. Most processes in nature occur in a continuous manner; however, the sampling process requires that discrete steps must be specified. Any quantity for which a numerical value is assigned is called a digital quantity.

The advances in microprocessors and computers have had a dramatic effect on instrumentation systems. Instruments are available with increased performance, including built-in microprocessors for simplifying controls, self-testing, data processing, and more. Digital processing of signals includes filtering, spectral analysis, smoothing, and other operations. Many instruments have digital readouts, which avoid the potential of human error in reading scales and allow faster reading speed. This chapter covers the digital circuit fundamentals with emphasis on circuits used in instrumentation.

When you complete this chapter, you should be able to

1. Convert base ten numbers to binary and vice versa.
2. Draw the truth tables for basic logic gates.
3. Simplify logic expressions using Boolean algebra and the Karnaugh map method.
4. For a combinational logic problem, show the equivalence of a truth table, a Boolean algebra expression, and a circuit schematic diagram.
5. Compare the *D* latch with the *D* flip-flop.
6. Describe the *JK* flip-flop. Compare edge triggering with pulse triggering and discuss the differences between synchronous and asynchronous inputs.
7. Explain how to make an asynchronous counter using either *D* or *JK* flip-flops.
8. Analyze a synchronous counter to determine the count sequence.
9. Explain fundamentals of a Moore state machine.
10. Compare various A/D and D/A converter circuits and explain operating principles.
11. Give the meaning of key manufacturers' specifications for digital logic circuits.

HISTORICAL NOTE

The first person to apply electronic processing to digital computing was Dr. John V. Atanasoff. Dr. Atanasoff considered the problem of electronic computing for several years prior to constructing his first prototype computer in 1939. He developed an innovative capacitor bank memory unit that underwent continuing refresh, which he called "jogging." Later, with the help of graduate student Clifford E. Berry, he constructed the first computer that used electronics to perform switching and arithmetic functions. Their computer, called the ABC (for Atanasoff-Berry Computer) was also the first electronic computer to separate the processing of data from memory and use memory refresh, a capacitor drum memory, clocked control, and vector processing, among other innovations. It was completed in 1942, but further work was abandoned because of World War II. In November 1990, Dr. Atanasoff received the National Medal of Technology from President Bush in recognition of his achievements. (A fascinating account of this computer is found in *Scientific American,* August 1988, p. 90–96.)

6–1 BASIC PRINCIPLES

BINARY NUMBERS

Digital electronics is based on switching theory, in which two separate states are allowed. The states are typically designated by the terms *true* or *false, high* or *low,* 1 or 0, or other assignments that can be used to describe two states or conditions. Transistor circuits, using either bipolar or FET technology, can be implemented as open or closed electronic switches in integrated circuit form. These circuits are fast, reliable, and relatively inexpensive. Because logic is implemented as a two-state system, both logic and arithmetic functions are implemented using a two-state, or **binary,** counting system.

The number of symbols used by a counting system to represent numbers is called the **base,** or **radix,** of the system. The binary system has a radix of two and uses only the numbers 0 and 1. Binary values are referred to as bits—a contraction of the words binary digit.

Binary is a **weighted** (or positional) number system. Weighted number systems are counting systems in which the position (column) of a digit determines its value. The column value to the immediate left of the radix point has a value of 1 in all weighted number systems. The value, or weight, of each column increases from right to left by an amount equal to the radix. The familiar base ten uses 10 symbols (0–9), and each column has a value 10 times as great as the column to the immediate left. Thus column values are 1, 10, 100, 1000, and so forth. In binary, the column values to the left of the radix point increase by a factor of 2: 1, 2, 4, 8, 16, and so forth.

The value of a number in a weighted number system is found by multiplying the digit value times the column value for all digits and summing the results. The conversion of a binary number to its decimal equivalent is illustrated in Example 6–1.

EXAMPLE 6–1 Convert each binary number to its decimal equivalent.
(a) 110011 (b) 100111

SOLUTION

(a) $110010 = 1 \times 2^5 + 1 \times 2^4 + 0 \times 2^3 + 0 \times 2^2 + 1 \times 2^1 + 0 \times 2^0 = 32 + 16 + 0 + 0 + 2 + 0 = 50$

(b) $100111 = 1 \times 2^5 + 0 \times 2^4 + 0 \times 2^3 + 1 \times 2^2 + 1 \times 2^1 + 1 \times 2^0 = 32 + 0 + 0 + 4 + 2 + 1 = 39$

A decimal number may be converted to a binary number by a process called *reverse division*. Reverse division is used in order to leave the answer in normal form with the most significant bit on the left and the least significant bit on the right. The base ten number is written inside the reverse division sign, as illustrated in Figure 6–1(a). The divisor is the radix of the number system desired for the answer—in this case we are converting to binary, so the divisor is 2. After each division is performed, the whole-number quotient is written directly to the left of the previous number and the remainder is written directly above the number. The process is continued until the number has been reduced to 0. Figure 6–1(b) shows a completed example. The result of reverse division leaves the most significant bit (MSB) to the left and the least significant bit (LSB) to the right, as in standard representation of numbers.

Reverse division is a useful method for converting a base 10 number to any other number system. Simply divide the base ten number by the radix of the system to which you want to convert the given number in the manner shown in Figure 6–1.

EXAMPLE 6–2 (a) Convert the base ten number 217 to binary by reverse division.
(b) Convert the base ten number 57 to binary by reverse division.

SOLUTION

(a)

	1	1	0	1	1	0	0	1	
0	1	3	6	13	27	54	108	217	2

(b)

	1	1	1	0	0	1	
0	1	3	7	14	28	57	2

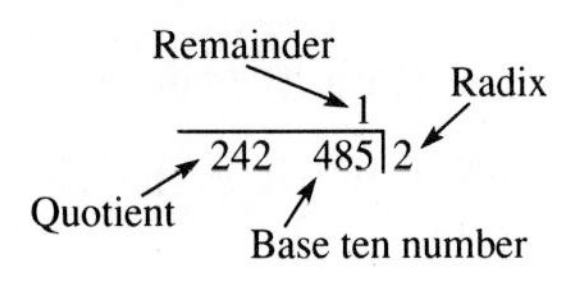

(a) Setup.

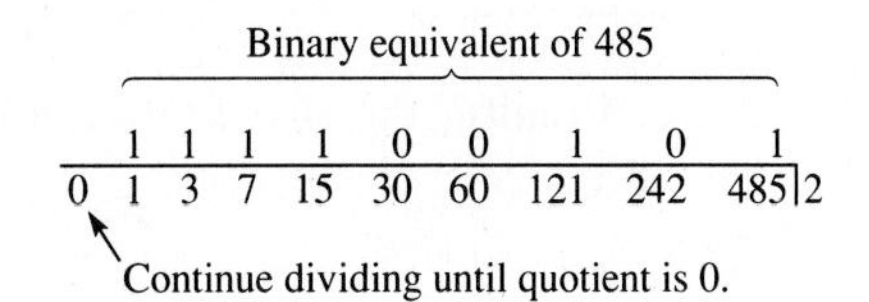

(b) Completed example.

FIGURE 6–1
Conversion of a base 10 number to binary using reverse division.

BOOLEAN ALGEBRA

Boolean algebra is the algebra of logic. It is similar to standard algebra except Boolean variables and constants can take on one of only two states—either 0 or 1. The rules for Boolean algebra can be expressed in terms of three logic operations: AND, OR, and NOT. These operators are basic to Boolean algebra. The AND operator is equivalent to the multiply operator of ordinary algebra and the OR operator is equivalent to the sum operator of ordinary algebra. As a result of the two-state nature of Boolean algebra, there are a few differences in Boolean rules from ordinary algebra.

The basic operators, AND, OR, and NOT, use the mathematical symbols $\cdot$, $+$, and $\overline{\ }$. The dot symbol for AND is sometimes not shown between variables but is implied when variables are written next to each other. The not, or complement, symbol is drawn as an overbar (properly termed a vinculum) above the variable or variables to be notted.

A **truth table** is a tabular listing of all possible inputs for a logic circuit along with the respective outputs. The name truth table comes from a time when logical arguments were validated with a truth table. Its use today is somewhat misleading because it should simply be viewed as a tabular listing of the input/output relationship of a combinational logic circuit. The truth tables for a two-input AND gate and a two-input OR gate are shown in Table 6–1(a) and (b). (The term **gate** is analogous to a gate that can be opened or closed. For digital applications, a gate is a logic element with at least one input and a single output.) The truth table for an inverter (NOT function) is shown in Table 6–1(c).

Like ordinary algebra, Boolean algebra obeys the commutative law, associative law, and distributive law. There are two forms of each law that can be written by exchanging the AND and OR operators. The commutative law states that the order of operation for either the AND operation or the OR operation doesn't matter. In other words

$$A \cdot B = B \cdot A$$

and

$$A + B = B + A$$

The associative law states that AND operations can be grouped in any order; likewise OR operations can be grouped in any order. That is

$$A \cdot (B \cdot C) = (A \cdot B) \cdot C$$

and

$$A + (B + C) = (A + B) + C$$

Finally, the distributive law is written as follows:

$$A \cdot (B + C) = A \cdot B + A \cdot C$$

and

$$A + (B \cdot C) = (A + B) \cdot (A + C)$$

AND

INPUTS		OUTPUT
A	B	X
0	0	0
0	1	0
1	0	0
1	1	1

(a)

OR

INPUTS		OUTPUT
A	B	X
0	0	0
0	1	1
1	0	1
1	1	1

(b)

INVERTER

INPUT	OUTPUT
A	X
0	1
1	0

(c)

TABLE 6–1
(a) Truth table for two-input AND gate. (b) Truth table for two-input OR gate. (c) Truth table for inverter.

Notice that there is not an equivalent law for the second form of the distributive law in ordinary algebra.

In addition to the laws cited, there are rules that can be used to simplify Boolean expressions. The rules, which can be applied to any number of variables, are listed in Table 6–2 and are summarized as follows:

Rule 1: $A + 0 = A$, states that the value of a variable is unchanged if it is ORed with 0.

Rule 2: $A + 1 = 1$, illustrates that when a variable is ORed with 1, the result is 1.

Rule 3: $A \cdot 0 = 0$, indicates that when a variable is ANDed with 0, the result is 0. This is equivalent to disabling the output.

Rule 4: $A \cdot 1 = A$, states that when a variable is ANDed with 1, the result is equal to the value of the variable. This is equivalent to enabling the output.

Rule 5: $A + A = A$, states that when a variable A is ORed with itself, the output is unchanged.

Rule 6: $A + \overline{A} = 1$, states that when a variable is ORed with its complement, the result is always 1. This is true because one of the inputs must always be 1.

Rule 7: $A \cdot A = A$, states that when a variable is ANDed with itself, the result is equal to the variable. Thus the logic does not change; in digital circuits this is equivalent to a buffer.

Rule 8: $A \cdot \overline{A} = 0$, indicates that when a variable is ANDed with its complement, the result is 0. This is true because one of the inputs is always 0.

Rule 9: $\overline{\overline{A}} = A$, indicates that a variable complemented twice is equal to itself.

Rules 10, 11, and 12 are useful reductions for simplifying logic equations. These rules can be proved by application of the other rules. Notice that Rule 12 is the second form of the distributive law.

TABLE 6–2
Basic rules of Boolean algebra.

Rule 1: $A + 0 = A$
Rule 2: $A + 1 = 1$
Rule 3: $A \cdot 0 = 0$
Rule 4: $A \cdot 1 = A$
Rule 5: $A + A = A$
Rule 6: $A + \overline{A} = 1$
Rule 7: $A \cdot A = A$
Rule 8: $A \cdot \overline{A} = 0$
Rule 9: $\overline{\overline{A}} = A$
Rule 10: $A + AB = A$
Rule 11: $A + \overline{A}B = A + B$
Rule 12: $(A + B)(A + C) = A + BC$

Note: A, B, or C can represent a single variable or a combination of variables.

DEMORGAN'S THEOREM

One of the most useful theorems for simplifying logic expressions was developed by Augustus DeMorgan, an English mathematician and compatriot of Boole. DeMorgan's theorem is applied to situations where it is useful to add or remove an overbar that bridges more than one variable or term. DeMorgan's theorem is written in two parts:

EQUATION 6–1

$$\overline{A + B} = \overline{A} \cdot \overline{B}$$

and

EQUATION 6–2

$$\overline{A \cdot B} = \overline{A} + \overline{B}$$

DeMorgan's theorem can be broken into three steps:

1. Change the sign separating the variables.
2. Complement the variable on each side of the sign change.
3. Complement the entire expression.

Remove any double bars as a result of applying the last two steps.

EXAMPLE 6–3 Apply DeMorgan's theorem to the logic expression $\overline{(\overline{A} + B) \cdot \overline{C}}$ to remove overbars covering more than one variable.

SOLUTION Given $\overline{(\overline{A} + B) \cdot \overline{C}}$. Apply DeMorgan's theorem to the quantities on each side of the AND sign:

$$\overline{\overline{\overline{(\overline{A} + B)} + \overline{\overline{C}}}}$$

Remove double bars:

$$\overline{\overline{A} + B} + C$$

Apply DeMorgan's Theorem again:

$$\overline{\overline{A \cdot \overline{B}}} + C$$

Remove double bars:

$$A \cdot \overline{B} + C$$

LOGIC GATES AND SYMBOLS

The fundamental element of logic circuits is the logic gate. Combinational logic circuits are the physical implementation of Boolean expressions using two different voltage levels to represent the two possible logic states. The AND, OR, and NOT (or inverter) functions have been described using truth tables and Boolean expressions. The implementation of these functions is accomplished using gates that respond to HIGH or LOW input voltages in a manner that produces a HIGH or LOW output voltage corresponding to the truth-table representation. The terms HIGH and LOW refer to voltage levels, whereas 1 and 0 refer to the logic itself. In positive logic, HIGH = 1 and LOW = 0.

Because of the confusion that can be invoked with negative logic, many people prefer to use the concept of assertion-level logic to indicate *action*. In this context, **asserted level** means the appropriate logic level that causes the action indicated by a variable name associated with the logic. Variable names are generally written with an overbar to indicate that a LOW level causes the action and no overbar to indicate that a HIGH level causes the action. For example, the word SET can be written SET to indicate a HIGH-level signal causes the action of setting or it can be written $\overline{\text{SET}}$ to indicate that a LOW-level signal causes the action of setting. In the case of SHIFT/$\overline{\text{LOAD}}$, the overbar over the word $\overline{\text{LOAD}}$ indicates that if the line is held LOW, the load operation will occur; the absence of the overbar on the word SHIFT indicates that if the line is held HIGH, the shift operation will be performed.

The symbols used to represent the basic logic gates are shown in Figure 6–2. Two symbols are shown for each gate—the old, distinctive-shaped symbol and the more recent ANSI-IEEE-488 rectangular outline symbol. Although the basic AND, OR, and INVERTER functions can be used to implement any Boolean expression, other types of gates are more versatile for simplifying logic circuits.

FIGURE 6–2
Logic gate symbols.

Distinctive shape symbols

Rectanglular outline symbols

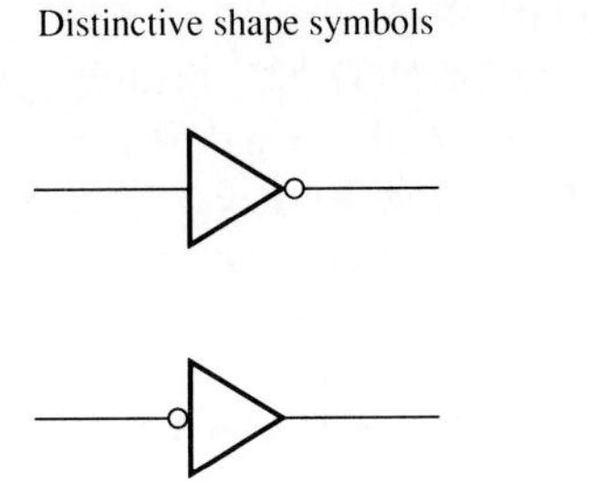

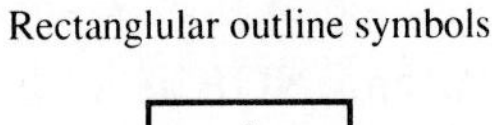

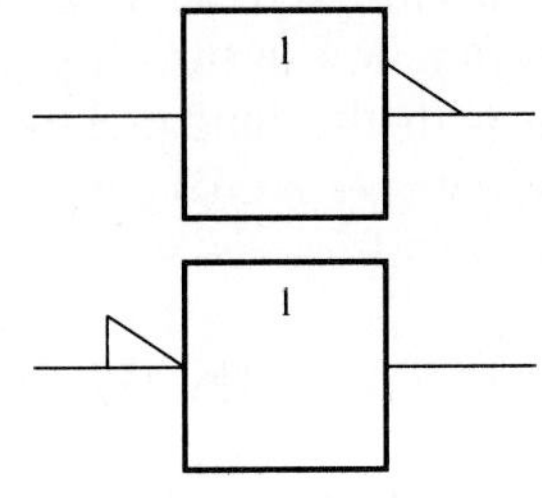

(a) Inverter.

FIGURE 6–2 (*continued*)

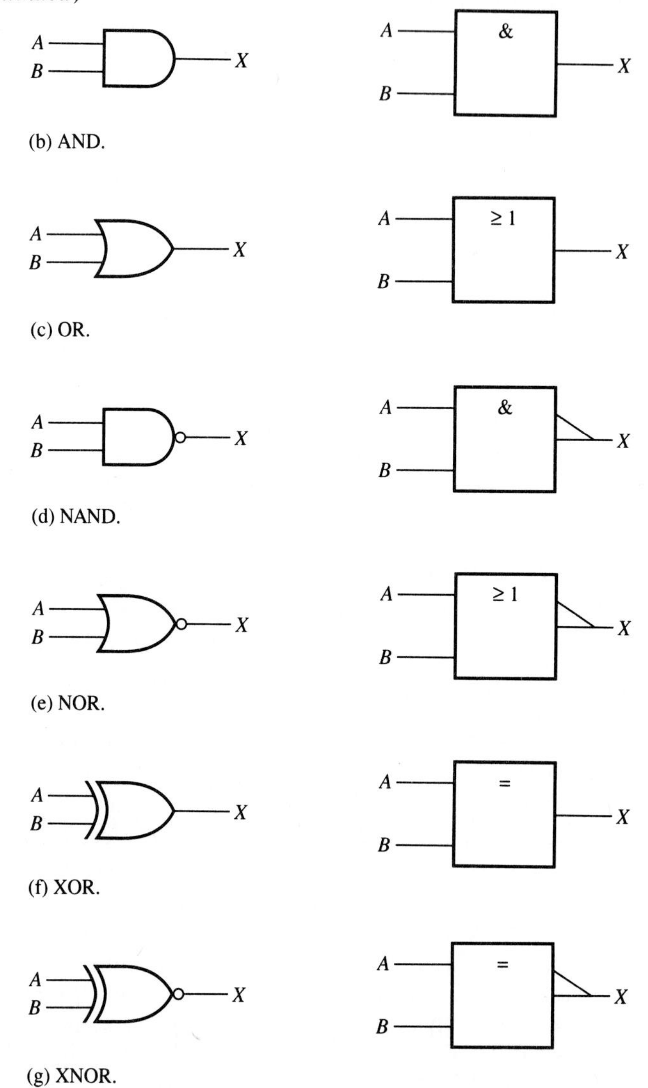

(b) AND.

(c) OR.

(d) NAND.

(e) NOR.

(f) XOR.

(g) XNOR.

These gates include the NAND and NOR functions. NAND is a contraction of NOT AND and NOR is a contraction of NOT OR. These gates simply add an inversion function to the output of AND and OR gates. Another gate frequently classified with the fundamental gates is the XOR (exclusive-OR) gate and the XNOR (exclusive-NOR) gate. The operation of these gates is summarized in the truth tables in Table 6–3(a) and (b).

SIMPLIFICATION OF BOOLEAN EQUATIONS

The operation of a combinational logic circuit can be summarized by either a truth table, a Boolean equation, or a logic drawing. In the past, when logic gates were

TABLE 6–3
(a) Truth table for XOR gate.
(b) Truth table for XNOR gate.

XOR

INPUTS		OUTPUT
A	B	X
0	0	0
0	1	1
1	0	1
1	1	0

(a)

XNOR

INPUTS		OUTPUT
A	B	X
0	0	1
0	1	0
1	0	0
1	1	1

(b)

expensive and required considerable space and power, it was considered advantageous to find the simplest representation of a circuit in order to reduce the number or types of gates. Although the need for minimization has lessened somewhat in recent years due to array logic and other techniques, it is still often useful to find the simplest circuit. Boolean equations can be simplified by applying the rules for Boolean algebra and DeMorgan's theorem. This is illustrated in the following example.

EXAMPLE 6–4 Simplify each of the following expressions using Boolean algebra:
(a) $X = \overline{A}(\overline{A} + B) + BC$ (b) $X = AC + \overline{A}\overline{B}C + \overline{A}BC$
(c) $X = D(E + F + D)$

SOLUTION

(a)
$$
\begin{aligned}
X &= \overline{A}(\overline{A} + B) + BC \\
&= \overline{A}\overline{A} + \overline{A}B + BC && \text{(distributive law)} \\
&= \overline{A} + \overline{A}B + BC && \text{(Rule 7)} \\
&= \overline{A} + BC && \text{(Rule 10)}
\end{aligned}
$$

(b)
$$
\begin{aligned}
X &= AC + \overline{A}\overline{B}C + \overline{A}BC \\
&= AC + \overline{A}C(\overline{B} + B) && \text{(distributive law)} \\
&= AC + \overline{A}C && \text{(Rule 6 and Rule 4)} \\
&= C(A + \overline{A}) && \text{(distributive law)} \\
&= C && \text{(Rule 6 and Rule 4)}
\end{aligned}
$$

(c)
$$
\begin{aligned}
X &= D(E + F + D) \\
&= DE + DF + DD && \text{(distributive law)} \\
&= DE + DF + D && \text{(Rule 7)} \\
&= D(E + F + 1) && \text{(distributive law)} \\
&= D(1) && \text{(Rule 2)} \\
&= D && \text{(Rule 4)}
\end{aligned}
$$

(Note that Rule 10 could be applied in the third line to obtain the answer directly.)

KARNAUGH MAPS

There are a number of techniques that have evolved for the simplification of logic circuits, including various computer methods. One technique for simplifying combinational logic circuits that works well for circuits with four or fewer inputs is the Karnaugh map method, described by M. Karnaugh in a paper published in 1953.

The Karnaugh map is a form of truth table drawn so that adjacent boxes (or cells) differ from each other in only one variable. Table 6–4 shows examples of Karnaugh maps for different numbers of variables. Each cell on the map represents one row, or **minterm,** on the truth table. Adjacent cells share a common border. The common border wraps around the outside edges of the map—both from left to right and from top to bottom. Across any border, the cells differ from each other by only one term.

The map is used for simplification by entering the output logic (0 to 1) from the truth table into the appropriate cells corresponding to the truth table. Two adjacent cells that each contain a 1 on the map represent two output terms. Notice that the map is constructed such that two adjacent cells contain one variable differing by the complement, whereas all other variables are common between the two cells. The variable that differs in the two cells is superfluous to the Boolean equation and can be eliminated.

TABLE 6–4
Examples of Karnaugh maps with (a) two, (b) three, and (c) four variables.

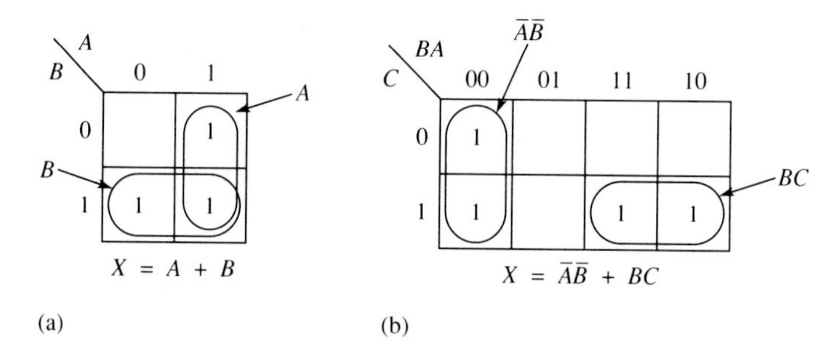

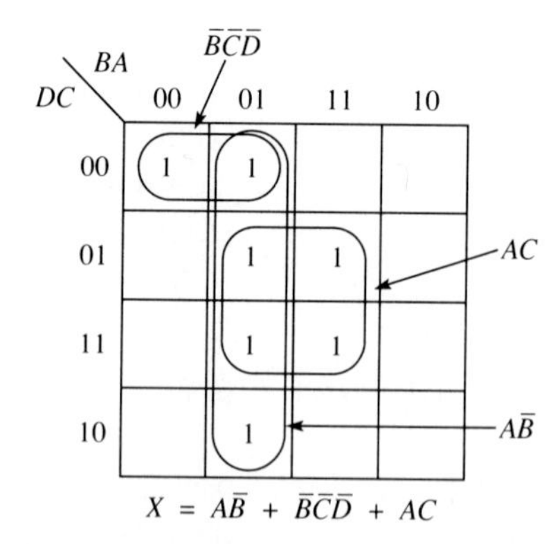

(c)

Reading a Karnaugh map is done by grouping cells by drawing a loop around adjacent 1s and observing which variable changes between the cells. The variable that changes is eliminated from the output expression. Cells must always be grouped by integer powers of 2: 1, 2, 4, 8, and so forth. Groups must be adjacent (square or rectangular in shape). The larger the group, the simpler the logic expression that will result. All 1s on the map must be taken at least once but may be taken more than once if it helps obtain a larger group elsewhere.

The output expression is read as a sum of the products (SOP) formed by reading the terms from each group. The combinational logic circuit can be then constructed by implementing the reduced Boolean expression as read from the Karnaugh map. An example is the best way to illustrate this.

EXAMPLE 6–5 Using the Karnaugh map method, find the Boolean expression for the truth table shown in Table 6–5. Draw the circuit.

TABLE 6–5

INPUTS			OUTPUT
A	*B*	*C*	*X*
0	0	0	1
0	0	1	0
0	1	0	1
0	1	1	0
1	0	0	1
1	0	1	0
1	1	0	1
1	1	1	0

SOLUTION Transfer the truth table to a Karnaugh map:

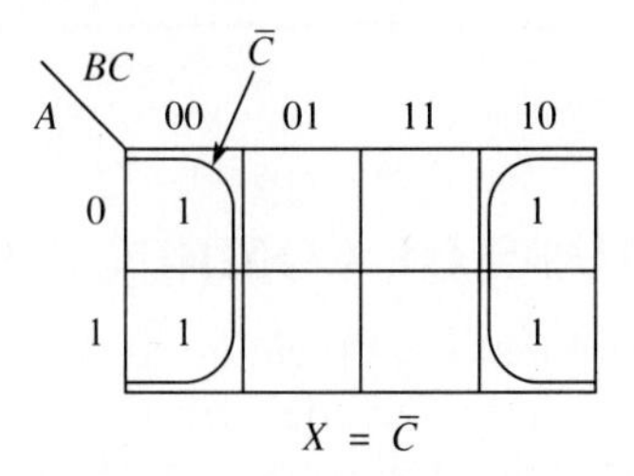

Take the largest group of adjacent 1s; read the map. $X = \overline{C}$.

EXAMPLE 6–6 Draw a truth table, Karnaugh map, and circuit to determine if a 4 bit binary number (0–15) is a multiple of 4. Write the simplified Boolean expression.

SOLUTION Begin with a truth table. Multiples of 4 are entered as 1s in Table 6–6. Then map the truth table onto a Karnaugh map and read the map.

TABLE 6–6

D	*C*	*B*	*A*	*X*
0	0	0	0	0
0	0	0	1	0
0	0	1	0	0
0	0	1	1	0
0	1	0	0	1
0	1	0	1	0
0	1	1	0	0
0	1	1	1	0
1	0	0	0	1
1	0	0	1	0
1	0	1	0	0
1	0	1	1	0
1	1	0	0	1
1	1	0	1	0
1	1	1	0	0
1	1	1	1	0

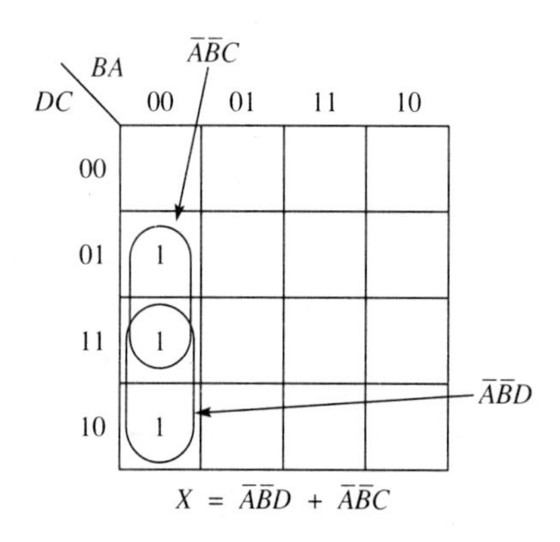

FUNCTIONS OF COMBINATIONAL LOGIC

Certain combinational logic circuits are so widely used that they are available as single ICs. These include arithmetic functions, magnitude comparators, multiplexers and demultiplexers, encoders and decoders, and other functions. Arithmetic functions and magnitude comparators are organized to operate on a binary **word**—a group of bits considered as an entity.

Arithmetic functions include adders, parity generators and checkers, and arithmetic logic units. **Adders** are combinational logic circuits that perform the addition of two binary words and include carry-in and carry-out capabilities. **Parity generators** are used to generate a bit that is "tagged" onto a binary word to allow testing that it has been received correctly in a digital data transmission system. Parity can be selected as *odd* or *even*. Odd parity means that the total number of ones in a word, including the parity bit, is odd. If the receiver finds that the number of 1s set is not in accordance with the prescribed parity, a **parity checker** can detect the error and set a bit. **Arithmetic logic units** are complicated combinational logic circuits used to perform a variety of arithmetic and logic functions on two binary words. The particular operation is selected by a select line; operations include add, subtract, shift, compare, complement, and other functions.

Magnitude comparators are used to compare the size of two binary words and assert one of three outputs, indicating $A > B$, $A = B$, or $A < B$. One application of magnitude comparators in instrumentation and control systems is to enable some action to be taken if a quantity reaches some predetermined limit. Magnitude comparators, like many of the arithmetic functions, can be cascaded (expanded) to provide a greater number of bits than is available in a single IC.

A **multiplexer** (abbreviated MUX) or data selector is a combinational circuit with a number of data inputs and one data output. Select (address) lines are used to indicate which of the input data lines is routed to the output. Multiplexers are available in IC packages with 2, 4, 8, or 16 inputs. For example, an eight-input MUX requires three select lines ($2^3 = 8$) to choose a particular input. The application of multiplexers to implementing an arbitrary combinational logic truth table is discussed in the next section.

A **demultiplexer** (abbreviated DMUX) has a single data input and several output lines. The input line is routed to a particular output determined by the binary code on the select inputs. The concepts of multiplexing and demultiplexing are shown in Figure 6–3. As you can see, the DMUX performs the opposite function of a MUX.

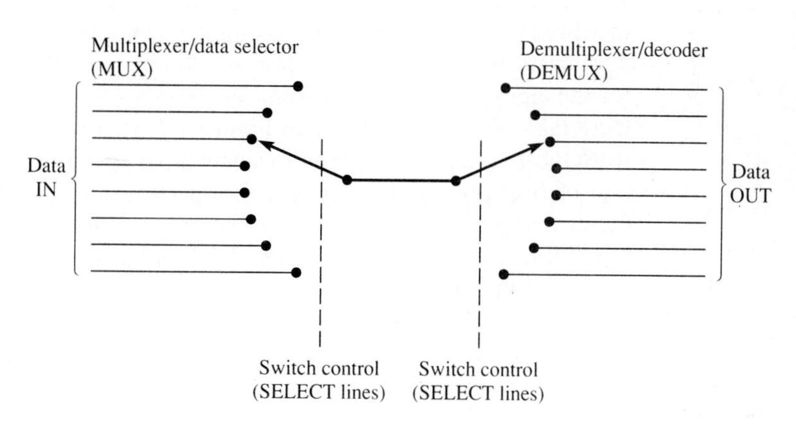

FIGURE 6–3
Concept of multiplexing/demultiplexing using mechanical switches. The concept shown is implemented with electronic switches.

In CMOS logic, certain ICs can serve as either a MUX or a DMUX on the same IC. An example is the 14051B, an IC that can function as a set of digitally controlled bidirectional switches. A common out/in line allows the signal to be routed in either direction, which determines if the IC is a MUX or a DMUX.

An **encoder** is a combinational circuit that converts a single input signal into some coded information at the output. For example, a decimal-to-BCD encoder has 10 inputs to represent the numbers 0 through 9 and 4 outputs to represent the BCD-coded value. Only one input can be active at one time. The output represents the BCD code of the active input. A difficulty arises if more than one input is active at the same time as an incorrect output can be produced. The remedy for this is the **priority encoder,** which assigns each input a priority and converts the highest-priority active input into an output code. An example of a priority encoder is the 74LS148, which accepts 8 inputs and converts the highest priority input into a 3 bit binary code at the output.

A **decoder** serves the reverse function of an encoder. It asserts only one of its several outputs, depending on the code that appears at the input. For example, a BCD-to-decimal decoder has 4 data inputs for the BCD code and 10 outputs to represent the decimal numbers 0 through 9. The asserted output line indicates the input code. An example of a decoder is the 74LS42 BCD-to-decimal decoder.

6–2 IMPLEMENTING COMBINATIONAL LOGIC WITH MULTIPLEXERS AND ROMs

MULTIPLEXERS

A multiplexer can be used to implement directly the truth table of a combinational logic problem. As an example, consider an arbitrary truth table using inputs D, E, and F. To implement this truth table with a multiplexer, the input variables are connected directly to the select inputs, as illustrated conceptually in Figure 6–4(a). Each input of the MUX is connected to the logic level as determined by the truth table for that set of inputs. Figure 6–4(b) shows an example schematic using the 74151A eight-input MUX.

Actually, an eight-input MUX is not required to implement a circuit containing three input variables. Any N-input MUX can generate the output function for $2N$ inputs. To illustrate, we choose a four-input MUX to implement a truth table for the eight possible inputs from the previous example. Begin by reorganizing the truth table in pairs, as shown in Figure 6–5(a). The two most significant bits on the truth table (D and E) are used to select a particular input. Connected to that line is one of four possibilities—either a logic 0, a logic 1, or the remaining input, F, either complemented or not complemented. The selection of which of these four possibilities to use is determined by looking at the outputs two at a time. For example, in Figure 6–4(a), when $D = 0$ and $E = 0$, the first data line (D_0) is selected, representing the first *two* lines on the truth table. The desired outputs are identical to the F input; therefore this input is connected to the D_0 input. If both outputs are opposite to the F input, then $\overline{F}$ is selected, as was done on the last data

FIGURE 6–4
Implementing a three-input function with an eight-input MUX.

Truth table for an arbitrary function

Inputs			Output
D	E	F	X
0	0	0	0
0	0	1	1
0	1	0	0
0	1	1	0
1	0	0	1
1	0	1	1
1	1	0	1
1	1	1	0

MUX 8:1

D_0 D_1 D_2 D_3 D_4 D_5 D_6 D_7

X

Data select

C B A

(a) Conceptual drawing.

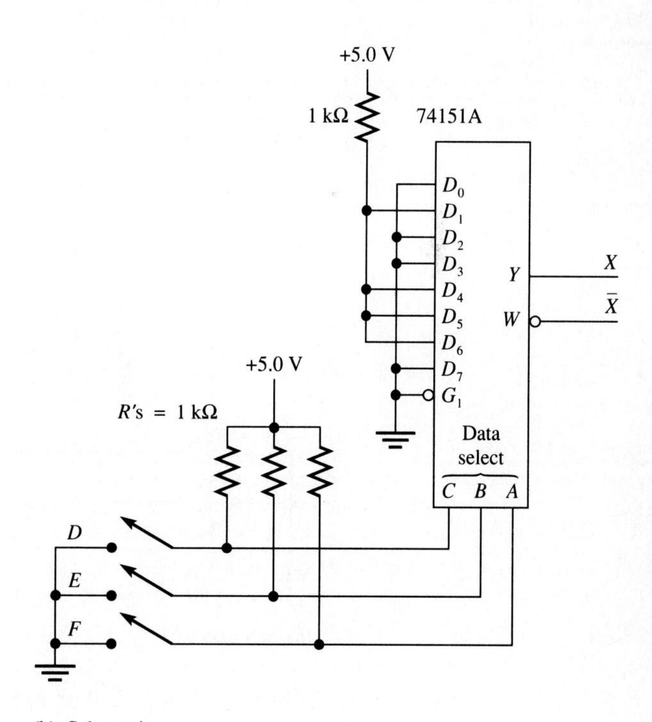

(b) Schematic.

FIGURE 6–5
Implementing a three-input function with a four-input MUX.

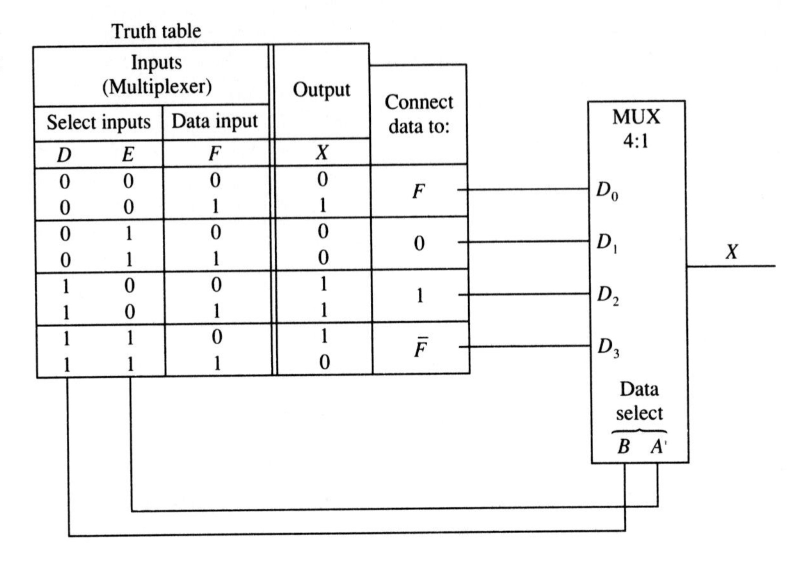

Truth table

Inputs (Multiplexer): Select inputs		Data input	Output	Connect data to:
D	E	F	X	
0	0	0	0	F
0	0	1	1	
0	1	0	0	0
0	1	1	0	
1	0	0	1	1
1	0	1	1	
1	1	0	1	$\overline{F}$
1	1	1	0	

(a) Conceptual drawing.

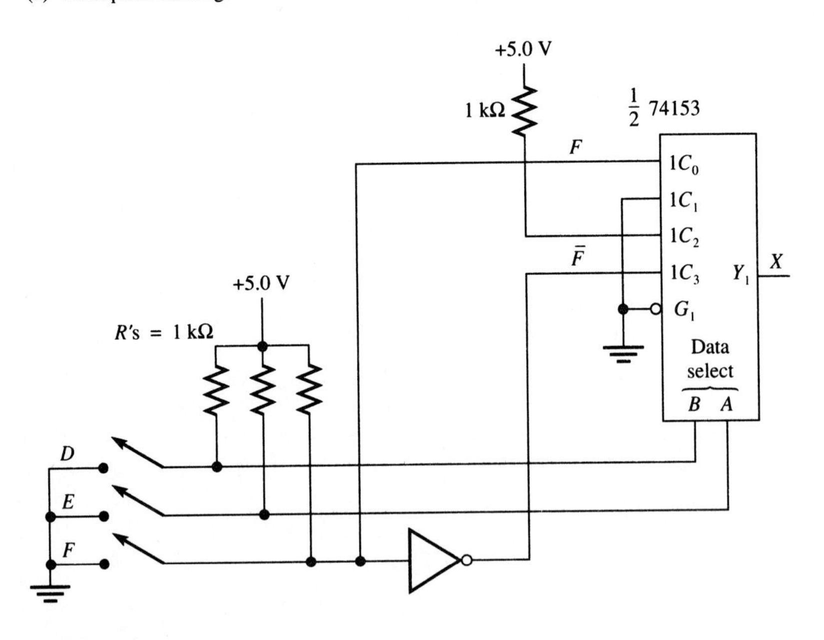

(b) Schematic.

input (D_3). If both outputs are 0 or 1, a LOW or HIGH is selected, as shown on D_1 and D_2.

ROMs

Implementing logic using gates and multiplexers is efficient only for relatively small combinational circuits. Circuits with many inputs or containing multiple outputs, such as lookup tables and code converters, require a large number of gates to implement, making it impractical to implement complex combinational logic with gates. A more efficient way to implement large combinational logic circuits is to program the truth table directly into a read-only memory (ROM), a permanent form of memory. In addition to ROMs permanently programmed at the time of manufacture, others are available that can be programmed by the user—a permanent type called programmable read-only memory (PROM) and a semipermanent type called erasable programmable read-only memory (EPROM). Although, technically, memory devices are typically considered part of sequential logic, ROMs can be used to implement a truth table directly and so are considered in this section.

A ROM is a memory device that can store information even when power is removed. It can store various quantities of information ranging from 64 bits to millions of bits. All digital memories consist of a series of locations or cells that can hold a single binary bit—either a 1 or a 0. Cells can be thought of as being arranged in an array with rows and columns, as illustrated in the simplified drawing shown in Figure 6–6. Any one row of the array can be selected by applying a binary code to the address input lines. The data stored at the selected row is connected to the associated column lines and appears on the output data lines when enable line is asserted.

To implement a truth table in a ROM, each line of the truth table is stored in the memory. The address lines represent the inputs on the truth table, and the desired output is stored for each combination of inputs. For example, a common need in an instrumentation system is to linearize the output of a transducer. The nonlinear data from the transducer is digitized and applied to the address lines of the ROM. The corrected data is taken directly from the ROM.

PLDs

Programmable logic devices (PLDs) offer a convenient way to implement logic in hardware by allowing the user to program a specific application into a general-purpose IC. A wide variety of PLDs are available to meet various applications, including combinational and sequential logic circuits of varying complexity. One type of PLD is called programmable array logic (PAL). The devices are programmed using Boolean expressions written as an SOP or using a truth table directly. One method uses product terms that are written into a programmable AND array with the output terms fed into a fixed or programmable OR array. Thus AND/OR logic can be implemented for any arbitrary application, such as code converters, random logic, adders, decoders, and other combinational logic.

Figure 6–7 illustrates a circuit implementation with a programmable PAL, an integrated circuit that can be programmed by the user for the required application. PALs are programmed using a specific application program language that generates a binary file called a JEDEC[1] file. This file is loaded into the memory of a PAL programmer. The programmer configures the PAL's AND/OR array logic to im-

[1] JEDEC stands for Joint Electron Device Engineering Council.

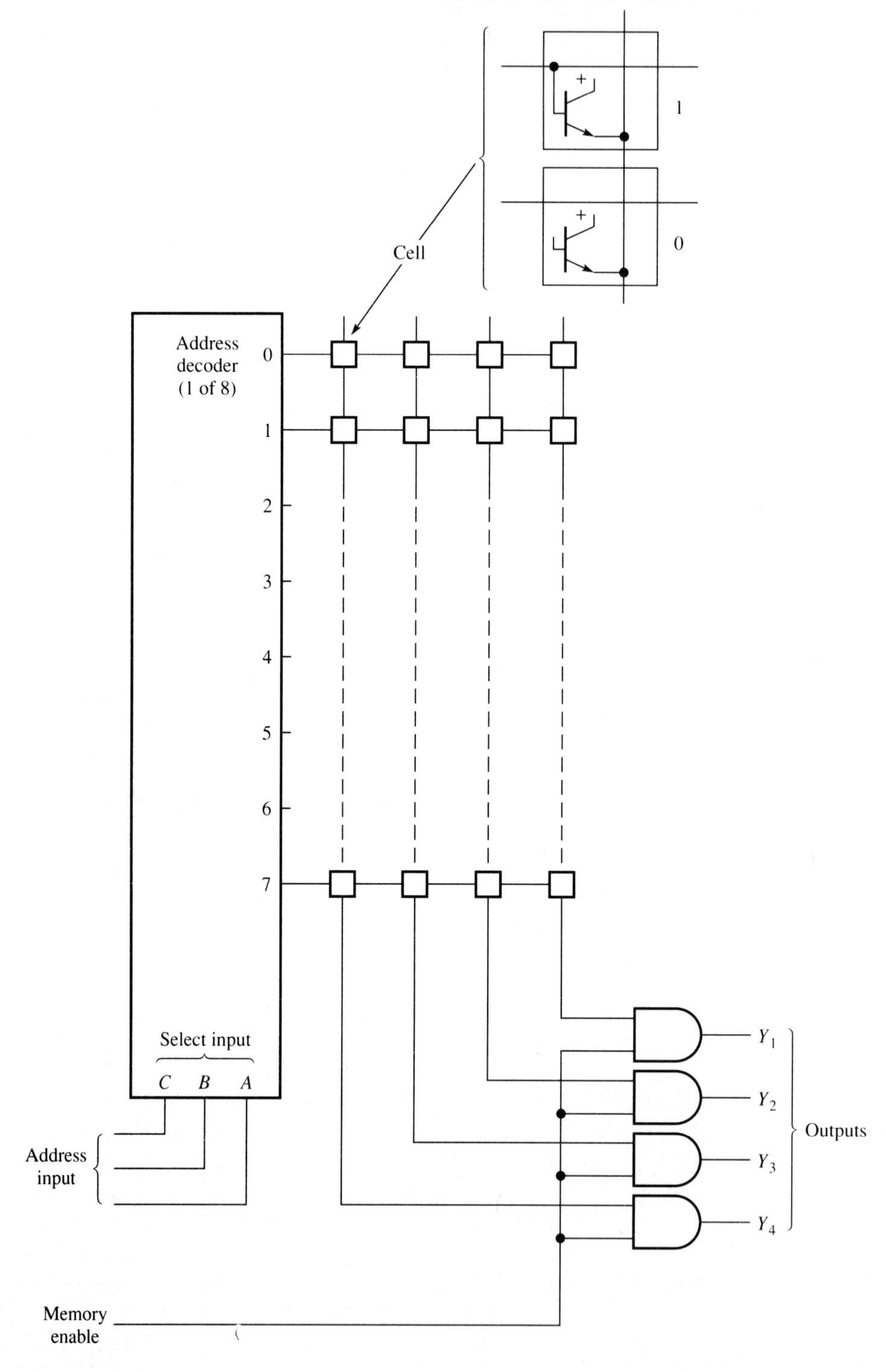

FIGURE 6–6
A simplified 8 × 4 ROM array.

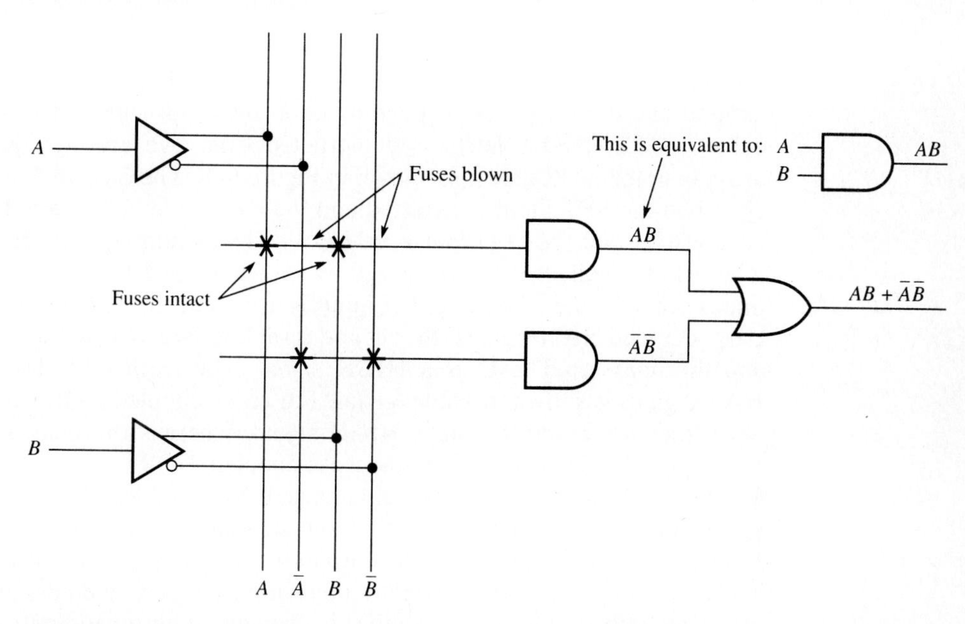

FIGURE 6–7
A programmable PAL. The example shown has a programmable AND array and fixed OR array. Product terms are realized by retaining the fuses at the intersections for the desired function; the other fuses are blown by "super voltages" applied by the programming device. Once the device is programmed, changes are limited (you can't mend a blown fuse). Newer technology uses CMOS cells instead of fuses. These devices, E^2PAL (electronically erasable PAL), can be reprogrammed.

plement the truth table. Many variations of PALs are available to implement complex combinational functions. The example shown is simple and is not an efficient use of a PAL, but it illustrates the method.

6–3 SEQUENTIAL LOGIC CIRCUITS

Up to this point, we have examined how certain combinations of inputs produce a desired output. With **combinational logic,** the output is based solely on the present inputs; there is no memory. Sequential logic is much more interesting. With **sequential logic,** the output depends on the present inputs and on what has occurred earlier. Sequential logic circuits have memory, allowing greater versatility in the types of circuits that can be constructed. Examples of sequential circuits include counters, shift registers, state machines, and computers.

The most basic form of memory is the cell. A **cell** is a small sequential circuit capable of storing a single binary bit. There are many forms of memory, including various semiconductor types as well as magnetic memories such as tape, disk, and magnetic bubbles. The common element in all memories is the ability to store information and recall it on command.

LATCHES

The most elementary form of solid-state memory is a **latch,** a circuit that can assume one of two states, depending upon the input signal. The simplest latch, called a **SET-RESET latch,** is constructed from two cross-coupled NAND or cross-coupled NOR gates, as shown in Figure 6–8. The outputs are labeled Q and $\overline{Q}$. When the SET input is asserted and the RESET is not asserted, the Q output goes HIGH and the $\overline{Q}$ output goes LOW. This action is equivalent to storing a logic 1 in the cell. A 0 can be stored by asserting the RESET line while the SET line is not asserted. When neither input is asserted, the cell stores the value. For cross-coupled NAND gates, the set and reset lines are asserted with a LOW signal and the unasserted level is a HIGH signal. The truth table for cross-coupled NAND gates is shown in Table 6–7(a). For cross-coupled NOR gates, the set and reset lines are asserted with a HIGH signal, whereas the unasserted level is a LOW signal. The truth table for cross-coupled NOR gates is shown in Table 6–7(b). Note that for both types of latches, if SET and RESET are asserted at the same time, both Q and $\overline{Q}$ will take on the same logic level. This violates the Boolean logic of requiring these outputs to be complements of each other and leads to an uncertain output if both inputs are removed together. For this reason, asserting both set and reset together is considered an invalid entry.

A modification to the basic cross-coupled latches is the addition of steering gates and an inverter, as illustrated in Figure 6–9(a). This circuit is called a D (for delay) latch. An enable input allows the data present on the D input to be transfer-

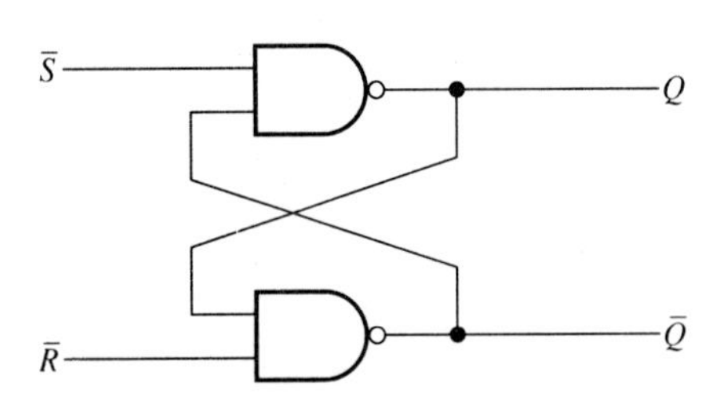

Equation for top NAND gate:

$$Q = \overline{\overline{S} \cdot \overline{Q}}$$

Applying DeMorgan's theorem:

$$Q = S + Q$$

Thus, Q appears on both sides of the equation.
If $\overline{S} = 1$, then $S = 0$ and $Q = 0 + Q$
output is latched.

(a) Cross-coupled NAND gates (active LOW inputs).

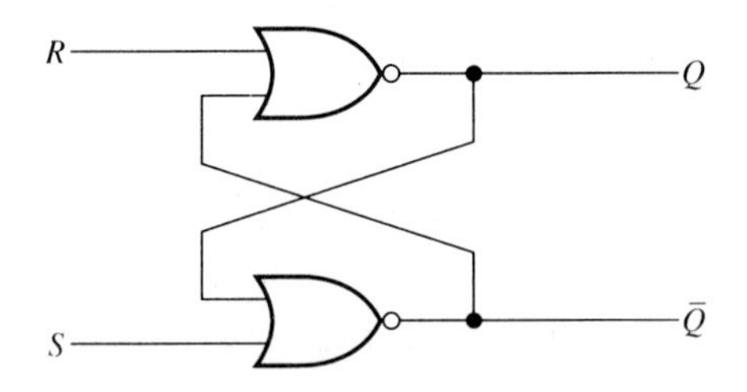

Equation for top NOR gate:

$$Q = \overline{R + \overline{Q}}$$

Applying DeMorgan's theorem:

$$Q = \overline{R} \cdot Q$$

Q appears on both sides of the equation.
If $R = 0$ (not asserted), then $\overline{R} = 1$ and $Q = 1 \cdot Q$
output is latched.

(b) Cross-coupled NOR gates (active HIGH inputs).

FIGURE 6–8

TABLE 6–7
Truth tables for latches. (a) Truth table for active-LOW $\overline{S}\overline{R}$ latch. (b) Truth table for active-HIGH SR latch.

$\overline{S}$	$\overline{R}$	Q	$\overline{Q}$
0	0	Invalid	
0	1	1	0
1	0	0	1
1	1	Latched	

(a)

S	R	Q	$\overline{Q}$
0	0	Latched	
0	1	0	1
1	0	1	0
1	1	Invalid	

(b)

red to the output when the enable line is asserted HIGH. When the enable line is not asserted, the output is latched. An easy rule to remember for a D latch is Q follows D whenever enable is asserted. This latch is sometimes referred to as a transparent D latch. D latches are available in an IC package such as the 7475A quad D latch.

FIGURE 6–9

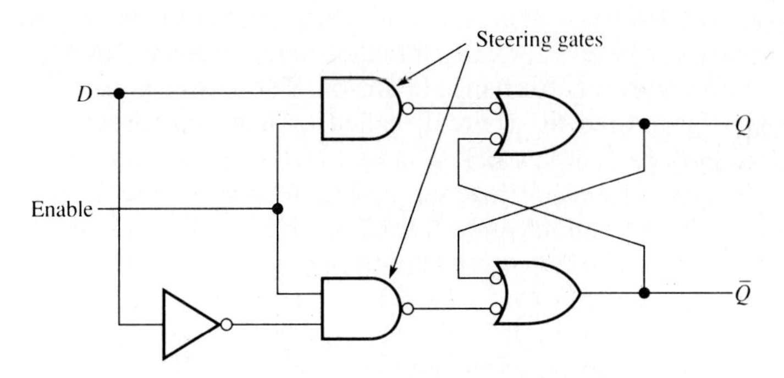

(a) D latch.

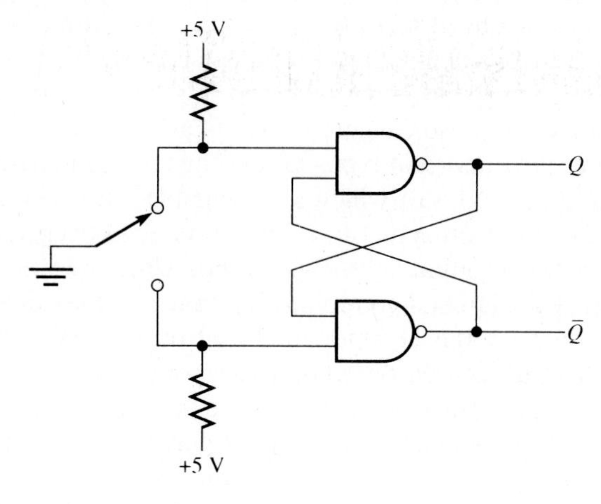

(b) SPDT switch debounce circuit using cross-coupled NAND gates.

In many instrumentation problems, it is necessary to use a mechanical switch closure as an input signal. Because of physically bouncing against the contact, a single-switch closure may appear to a high-speed electronic circuit as numerous switch closures. A useful application of latches is a circuit to prevent circuits from responding to multiple-switch bouncing. A switch-debounce circuit for a single-pole, double-throw (SPDT) switch is shown in Figure 6–9(b). When a switch moves from one position to the other, the latch stores the value on the very first contact. Bouncing leaves the latch in the "no change" mode, causing the output to remain steady. (Note that switches do not bounce all the way back to the other contact.) The 7544/8544 is an example of an IC containing cross-coupled NOR gates plus NOR steering gates especially designed to debounce switches. It includes a strobe input (the same as an enable) that permits sampling the switch information at a predetermined time or can be used to disable the switch, preventing unauthorized use.

FLIP-FLOPS

Flip-flops are *clocked* devices that can store a binary bit. They are sometimes confused with latches, but most manufacturers index them separately and we treat them separately here. Clocking simplifies many problems in digital systems as it allows different events to change in unison. The source used to synchronize these changes is generated by a circuit called a **clock** that produces periodic timing pulses. Flip-flops can be clocked in one of two basic ways—edge-triggered and pulse-triggered. An edge-triggered flip-flop responds to its inputs only at a clock transition. The transition can be a leading edge (LOW to HIGH) or a trailing edge (HIGH to LOW). Pulse-triggered flip-flops respond to short pulses that are considered to have two edges—one edge is a leading edge, the other, a trailing edge. For the purpose of classification, these two transitions are considered to be one event with one response as far as the output is concerned.

The question might arise as to how a pulse-triggered flip-flop differs from simply applying a pulse to the enable line of a latch. Basically, the latch is enabled by a signal that remains at some level for a period of time. The enable signal can remain at the active level indefinitely and can change at any time. Pulse triggering differs from this because it is periodic and involves two transitions rather than a level (although some pulse triggered circuits are sensitive to levels).

There are two important types of flip-flops used in digital circuits. These are the *D* and *JK* types, both of which are available with a variety of options in IC form. The most common type of *D* flip-flop is edge-triggered. Recall that the *D* latch transfers the *D* input to the *Q* output whenever the enable is asserted; otherwise it is in the latched condition. By contrast, the edge-triggered *D* flip-flop transfers the data on the *D* input to the *Q* output only at the clock edge, or *transition*. Flip-flops can be designed to respond to either a leading or a trailing clock edge. The clock edge synchronizes the transfer of the data, so the *D* input is referred to as a **synchronous** input. Edge triggering is indicated with an arrow on the schematic symbol. Symbols are illustrated in Figure 6–10(a).

FIGURE 6–10
Symbols and function table for a *D* flip-flop.

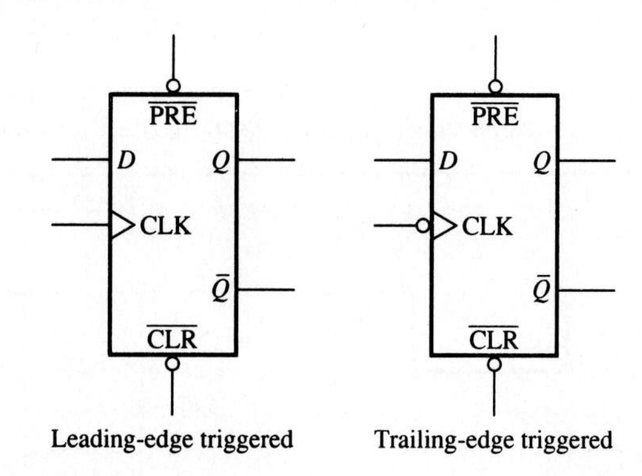

(a) Symbols for *D* flip-flops.

Inputs				Outputs	
$\overline{PRE}$	$\overline{CLR}$	CLK	D	Q	$\overline{Q}$
L	H	X	X	H	L
H	L	X	X	L	H
L	L	X	X	H*	H*
H	H	↑	H	H	L
H	H	↑	L	L	H
H	H	L	X	Q_O	$\overline{Q}_O$

H = high logic level
X = either low or high logic level
L = low logic level
↑ = Positive-going transition of clock.
* = This configuration is nonstable; that, is it will not persist when either the preset and/or clear inputs return to their inactive (high) level.
Q_O = The output logic level of Q before the indicated input conditions were established.

(b) Function table for the leading-edge triggered *D* flip-flop.

D flip-flops are available with two asynchronous inputs. An **asynchronous** input affects the output immediately, regardless of any clock transition. The two asynchronous inputs that are commonly designed into *D* flip-flops are labeled preset ($\overline{PRE}$) and clear ($\overline{CLR}$). The $\overline{PRE}$ input causes the Q output to become a logic 1 (or *set*) when it is asserted LOW, and the $\overline{CLR}$ input causes the Q output to become a logic 0 (or *clear*) when it is asserted LOW. The truth table for a *D* flip-flop is shown in Figure 6–10(b). Examples of *D* flip-flops include the TTL version 7474 dual *D* flip-flop and the CMOS version, the 4013B dual *D* flip-flop.

EXAMPLE 6–7 Given the input waveforms shown, compare the Q output waveform for the D latch and the leading-edge-triggered D flip-flop. Assume the $\overline{PRE}$ and $\overline{CLR}$ inputs are not asserted.

(a) Inputs for *D* latch :

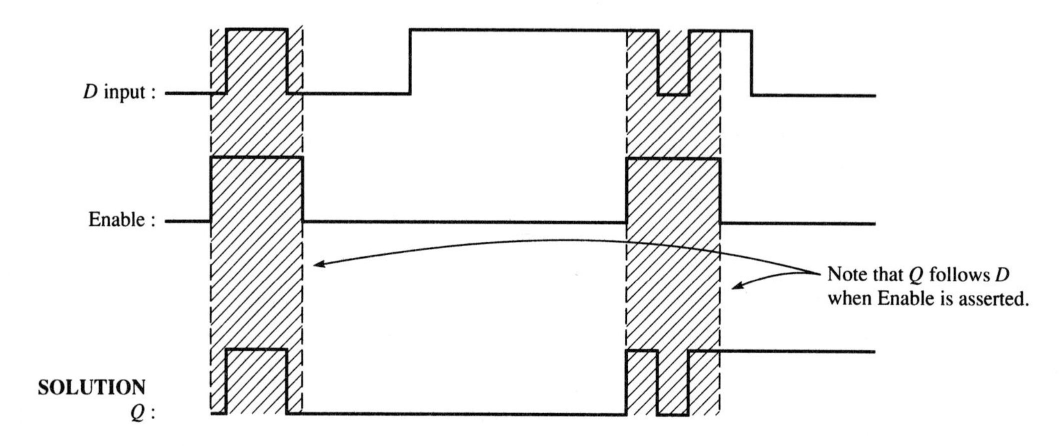

(b) Inputs for *D* flip-flop (leading-edge triggered) :

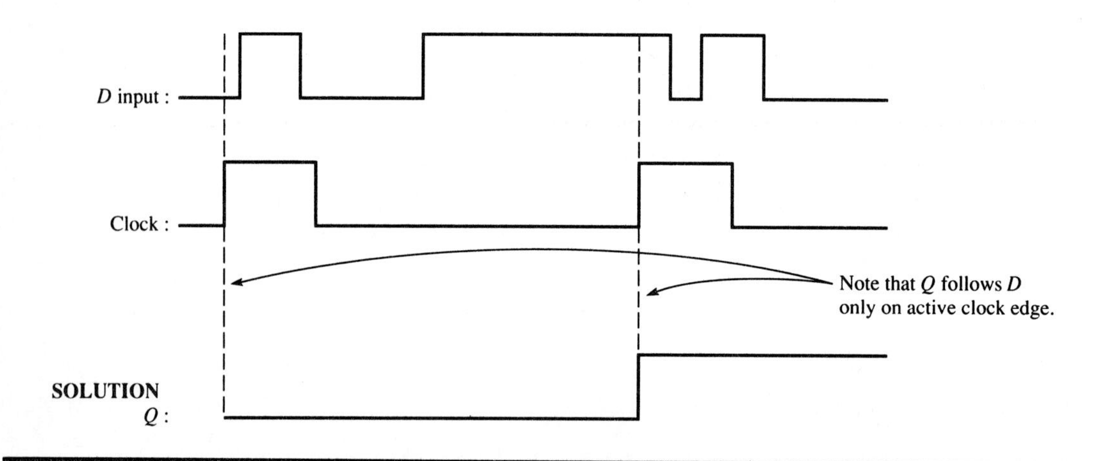

The *JK* flip-flop has greater versatility than the *D* flip-flop due to two synchronous inputs labeled *J* and *K*. The additional input allows the *JK* flip-flop to have two added functions. One of the new functions occurs when both *J* and *K* are LOW; the output is said to be latched—that is, clock pulses are ignored and the flip-flop remains in the last state it assumed. The other function occurs when both *J* and *K* are HIGH; the output changes whenever the clock is asserted. This is called the *toggle mode;* it is a bit like someone throwing a switch. The main idea with the toggle mode is that the output will change every time, no matter what

FIGURE 6–11
Schematic symbol for pulse-triggered *JK* flip-flop.

state it is in. The remaining lines on the truth table follow this rule: If *J* and *K* are different, *Q* follows *J* whenever the clock is asserted.

Like the *D* flip-flop, the *JK* flip-flop can be leading-edge-triggered or trailing-edge-triggered. In addition, the *JK* flip-flop is available as a pulse-triggered flip-flop (the 7476 or 4027B, for example). Pulse triggering, also called master-slave triggering, is shown on the schematic symbol using no edge on the clock and a delay indicator on the output, as shown in Figure 6–11. Data is clocked into the master section of the flip-flop on the first edge of the clock (usually the leading edge). The delay indicator shows that the data transfers on the second edge of the pulse (usually the trailing edge). The inputs must not change during the time the clock pulse is at the active level. For this reason, the clock pulses should be kept narrow on pulse-triggered flip-flops. One solution to this problem is provided by a special triggering feature called *data lockout*. This effectively causes changes in the inputs to be ignored except at the leading and trailing edges of the clock. Examples of a *JK* flip-flop with data lockout are the 74110 and 74111 *JK* flip-flops.

EXAMPLE 6–8 Given the input waveforms shown, determine the *Q* output waveform for the pulse-triggered *JK* flip-flop. Assume the $\overline{\text{PRE}}$ and $\overline{\text{CLR}}$ inputs are not asserted.

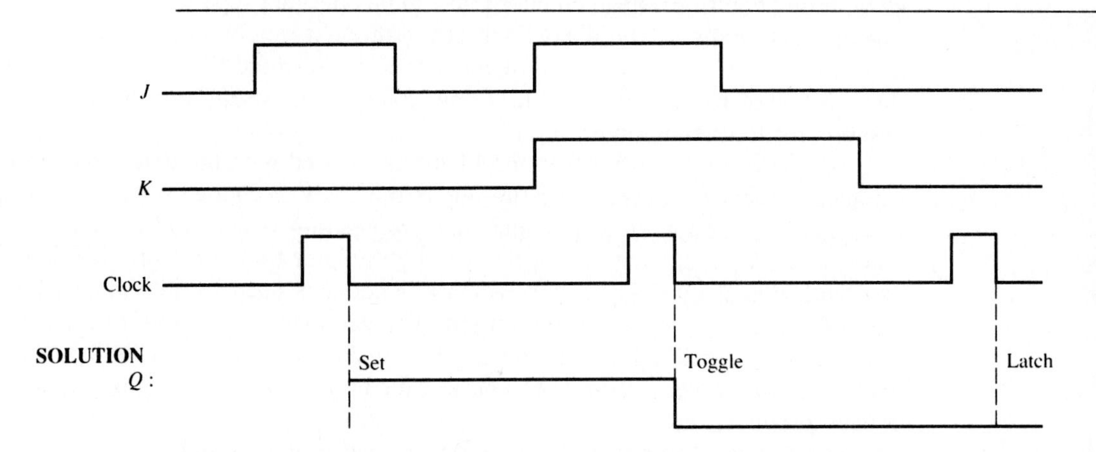

Note that the output changes on the trailing edge. Data is clocked into the master on the leading edge and to the output on the trailing edge.

ASYNCHRONOUS COUNTERS

A **counter** is a circuit that follows a predetermined sequence of states, changing from one state to the next state upon receiving a clock signal. Counters are composed of a series of flip-flops that hold the current output state until they are clocked. They can be divided into two types—synchronous and asynchronous—depending on how they are clocked.

An asynchronous counter is designed with flip-flops placed in the toggle mode. Flip-flops designed to toggle are sometimes called toggle, or *T*, flip-flops. Either *JK* or *D* flip-flops can be made to toggle. Recall that a *JK* flip-flop can be placed in the toggle mode by connecting both the *J* and the *K* inputs HIGH. Likewise, a *D* flip-flop can be made to toggle by connecting the $\overline{Q}$ output back to the *D* input. This causes, at each clock transition, the opposite state to be clocked into the flip-flop. (Notice that a *D* latch would not work properly trying to toggle it in this manner. Why?)

An asynchronous counter works by cascading a series of *T* flip-flops together. The output of each flip-flop is used to clock the succeeding stage. The effect is that the various flip-flops change asynchronously—much like the popular "wave" caused by spectators in a stadium. Asynchronous counters are more commonly referred to as *ripple counters*. Because of the manner in which they are connected, each flip-flop divides the frequency of the previous flip-flop by 2. This is shown in the example of a ripple counter given in Figure 6–12(a). The timing diagram is shown in Figure 6–12(b), with the arrows on the timing diagram indicating cause and effect. Notice how the waveforms can represent a binary count sequence if the slowest waveform is considered to be the MSB and the fastest waveform is considered to be the LSB.

Ripple counters are available in IC packages. The 7490A, 7492A, and 7493A are widely used ripple counters—each contains four master-slave flip-flops and gated zero reset. The flip-flops are connected internally as one single-stage counter and one three-stage counter that can be operated separately or connected together externally by the user. Each of the three-stage counters is wired differently to allow a variety of count sequences to be selected by the user. Because they are inexpensive and versatile, ripple counters are useful in instrumentation systems for frequency division.

Ripple counters suffer from problems associated with the delay that occurs because of the time it takes each flip-flop in the chain to change states. The delay associated with each state is called the **propagation delay,** and although very short, it causes each stage to toggle at a slightly later time. One problem associated with this slight delay occurs when a decoder is used to determine when a particular count is present at the output. The decoder may recognize a false, or error, state because all the flip-flops did not change together. This error state exists for only a very short time and is called a glitch. Ripple counters almost always produce glitches in the decoded outputs, which can be troublesome in certain systems. Another problem associated with the time delay is the speed at which ripple counters can be operated. At high frequencies, the output waveforms

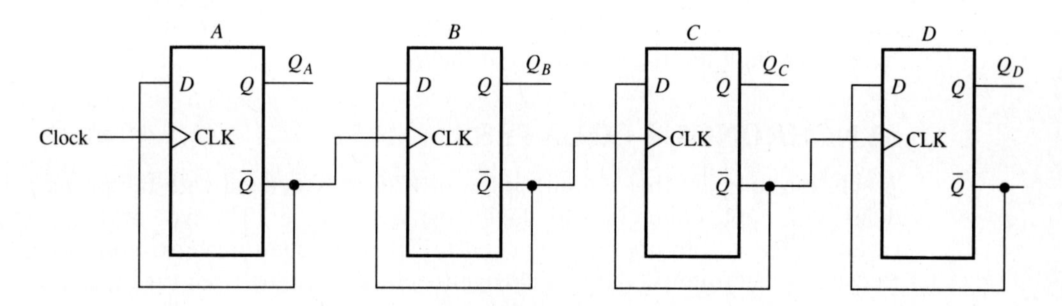

D flip-flops in toggle mode.

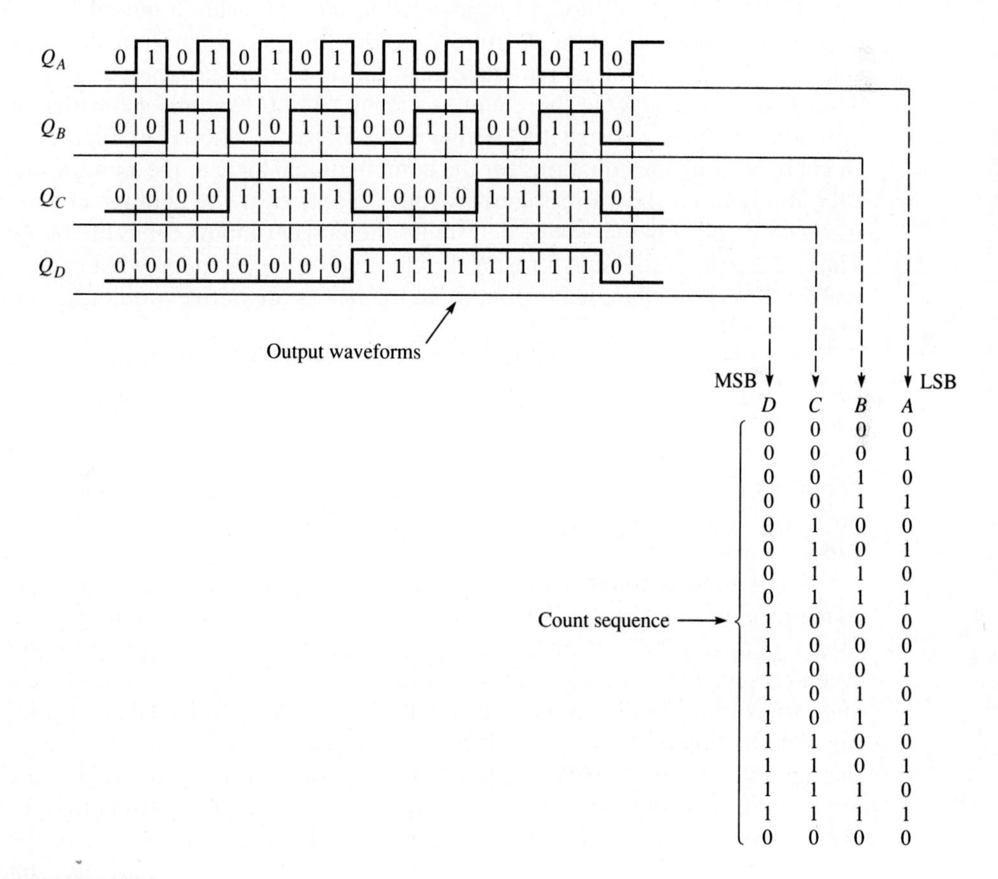

FIGURE 6–12
Four-stage binary ripple counter.

become confused because of the false states, which can occur during a significant fraction of the time the counter is in that state.

SYNCHRONOUS COUNTERS

A solution to many (but not all) of the problems associated with the propagation delay with ripple counters is to clock all stages together. The type of counter that results when all flip-flops are clocked together is called a **synchronous counter.** Because synchronous counters change together, the time from the clock pulse to the next count transition is much faster than in ripple counters. This greater speed reduces (but doesn't always eliminate) the problem of glitches in the decoded outputs. A common technique for eliminating glitches is to use a gray-code sequence, in which only one flip-flop changes for each clock transition.

Counters can be thought of as cycling through a finite number of possible outputs, or **states.** A **state diagram** is a pictorial representation describing the sequence of the counter. The state diagram for a synchronous counter can be determined by analyzing the count sequence. As an example, consider the synchronous counter constructed with *JK* flip-flops, as shown in Figure 6–13(a). Begin by writing the equations for the combinational logic at the *J* and *K* inputs for each flip-flop, as shown in Figure 6–13(b). Next, start the counter in some arbitrary state and, for that state, determine the current inputs for each *JK* flip-flop. Then, using the truth table for a *JK* flip-flop, you can determine what each flip-flop will do. This procedure leads directly to the next state of the counter, as shown in Figure 6–13(c).

A variety of synchronous counters are available in IC form, with options such as up/down (reversible) counting, asynchronous clear, parallel load inputs, carry out and borrow inputs, and two-clock operation. The 74ALS192 and 74ALS193 are examples of 4 bit counters with all these features. The dual-clock operation allows one clock to cause an up transition and the other clock to cause a down transition. The carry out and borrow inputs allow the counters to be cascaded to any desired size.

Sometimes a count sequence that is needed is not available in an IC package. Synchronous counters can be designed to count any arbitrary sequence. In order to design a counter, it is necessary to know how to force a flip-flop to go from a given present state to a required new state. The inputs required to produce a given state transition are shown on a table called a **transition table.** The transition table for a *JK* flip flop is shown in Table 6–8. Note that the output changes are listed on the left side of the transition table and the inputs required to produce the change are listed on the right side of the transition table. One of the strengths of the *JK* flip-flop for design of synchronous counters is the fact that one-half of the input entries on the transition table are *X*'s (don't care). This frequently enables a simpler circuit to be designed using *JK* flip-flops for small synchronous counters.

Assume you need to design a synchronous counter using *JK* flip-flops. The procedure is illustrated in Figure 6–14 for a simple counter. From the required sequence, it is often useful to draw a state diagram. All states that are in the main sequence of the counter should be shown; states that are not in the main sequence

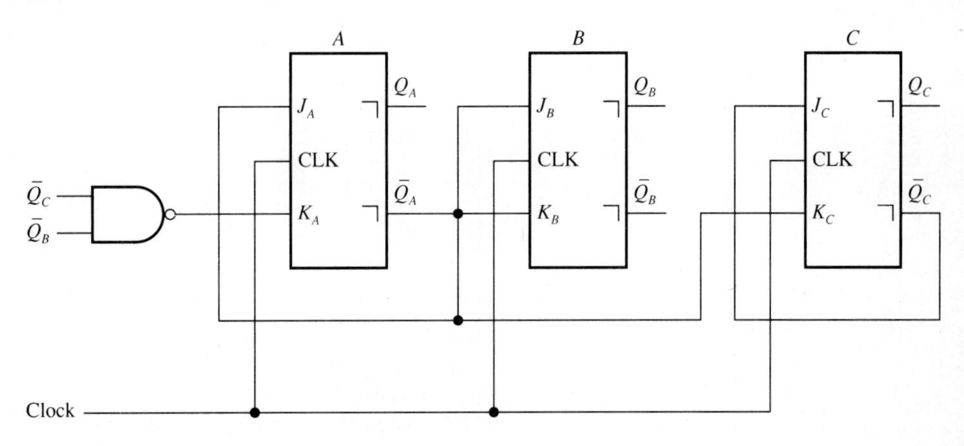

(a) Synchronous counter constructed with *JK* flip-flops.

Outputs			Inputs					
Q_C	Q_B	Q_A	$J_C = \bar{Q}_C$	$K_C = \bar{Q}_A$	$J_B = \bar{Q}_A$	$K_B = \bar{Q}_A$	$J_A = \bar{Q}_A$	$K_A = Q_C + Q_B$
0	0	0	1	1	1	1	1	0
1	1	1	0	0	0	0	0	1
1	1	0	0	1	1	1	1	1
0	0	1	1	0	0	0	0	0
1	0	1	0	0	0	0	0	1
1	0	0	0	1	1	1	1	1
0	1	1	1	0	0	0	0	1
1	1	0	Previously tested state					
0	1	0	1	1	1	1	1	0
1	0	1						

Main sequence

Steps in analysis:

1. Write equations for each input.
2. Start counter in arbitrary state.
3. Determine current inputs for the state.
4. From *JK* truth table, determine next state. Draw state diagram.

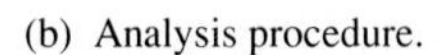

(b) Analysis procedure.

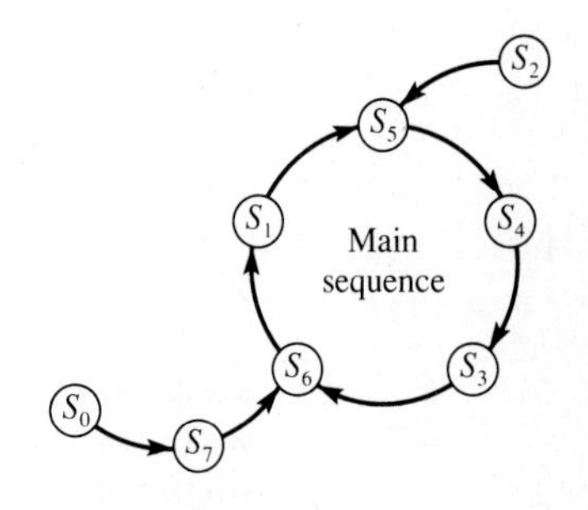

(c) State diagram.

FIGURE 6–13
Analysis of a *JK* synchronous counter.

should be shown only if the design requires these unused states to return to the main sequence in a specified way. If the sequence can be obtained from an already existing IC, this is almost always more economical and simpler than designing a special sequence. If an existing sequence is not available, proceed with the design.

TABLE 6–8
Transition table for *JK* flip-flop.

Transition table for *JK* flip-flop.

Output transitions			Inputs	
Q_N		Q_{N+1}	J_N	K_N
0	→	0	0	X
0	→	1	1	X
1	→	0	X	1
1	→	1	X	0

Q_N = output before clock
Q_{N+1} = output after clock
J_N, K_N = inputs required to cause transition
X = don't care

From the required sequence, a next-state table is constructed. The next-state table is just another way of showing the information contained in the state diagram. Because the largest state number is 7 (binary 111), three flip-flops are required. The next-state table clearly shows the change (transition) required by each flip-flop in order to go from one state to the next state. For convenience, the flip-flops are labeled *A, B,* and *C*. Flip-flop *C* represents the MSB in the count sequence.

For small counters, we can use Karnaugh maps to determine the logic required for our counter. In Step 3, six maps are completed, one for each input of the three flip-flops. Each square of a Karnaugh map represents a state of the counter. In a given state, the input that will force the correct sequence is read from the right side of the transition table and entered onto the square representing that state on the Karnaugh map. These entries represent the desired input to the flip-flop that will cause it to execute the required output transition. In effect, the counter sequence is just moving from square to square on the Karnaugh map at each clock pulse.

Although the Karnaugh map takes on a different meaning than it did with combinational logic, it is read the same way. When the maps are completed, the logic can be read from the map. As before, all 1s must be taken and no 0s can be in a group; however, squares with an *X* can be taken if it helps obtain a larger group with the 1s. This logic is then used to set up the circuit, as shown in Step 4 of Figure 6–14. This procedure can be extended to more complicated designs.

SHIFT REGISTERS

A **shift register** is a series of flip-flops connected so that the output of one flip-flop forms the input of the next flip-flop. The register holds a binary number that can be shifted from one flip-flop to the next on each successive clock pulse. One application of a shift register is to delay a digital signal. As an example, consider the serial-in, serial-out (SISO) register constructed with *D* flip-flops shown in Figure

Step 1 : Design a synchronous counter that follows the sequence 0-1-3-2-6-4-0. Unused states (5, 7) should return to main sequence. Use *JK* flip-flops. State diagram for main sequence:

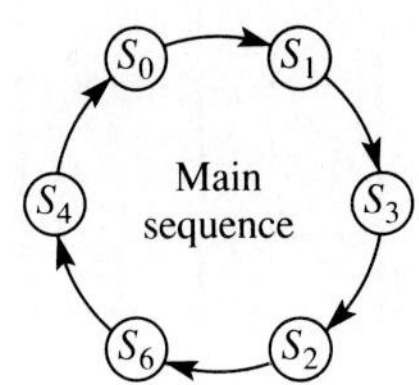

Step 2 : Next-state table.

Present State			Next State		
Q_C	Q_B	Q_A	Q_C	Q_B	Q_A
0	0	0	0	0	1
0	0	1	0	1	1
0	1	1	0	1	0
0	1	0	1	1	0
1	1	0	1	0	0
1	0	0	0	0	0

Unused states are not shown

← On this line (S_0), Q_C transition is 0→0, Q_B transition is 0→0, and Q_A transition is 0→1. Use transition table to find values for S_0 square on Karnaugh maps. S_0 square is upper left corner on maps.

Step 3 : Karnaugh maps (don't care (X) is put in unused states).

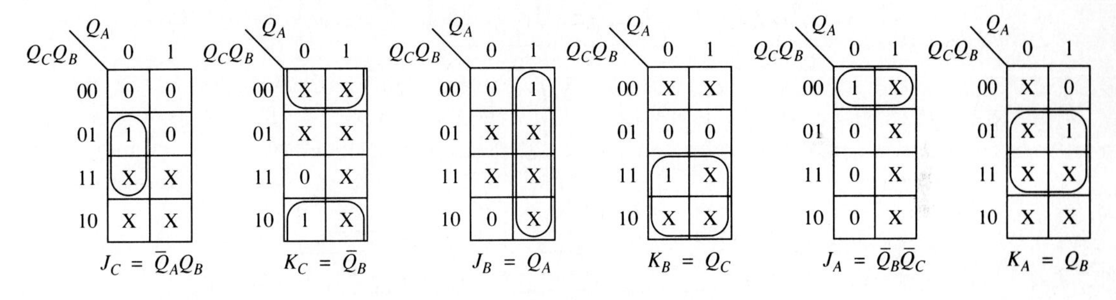

Step 4 : Draw circuit and check. Unused states are found to return to main sequence, so the design is satisfactory.

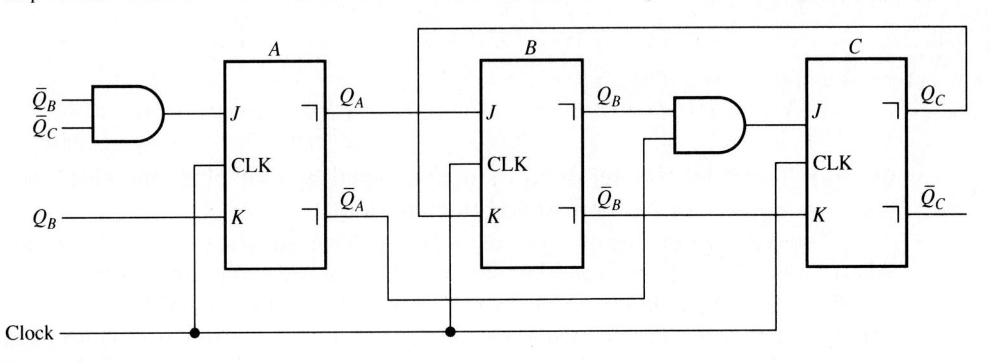

FIGURE 6–14
Design of a synchronous counter.

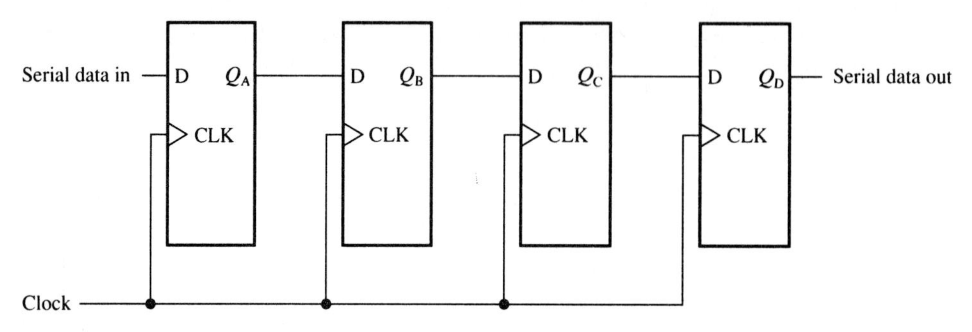

(a) A serial-in, serial-out (SISO) shift register.

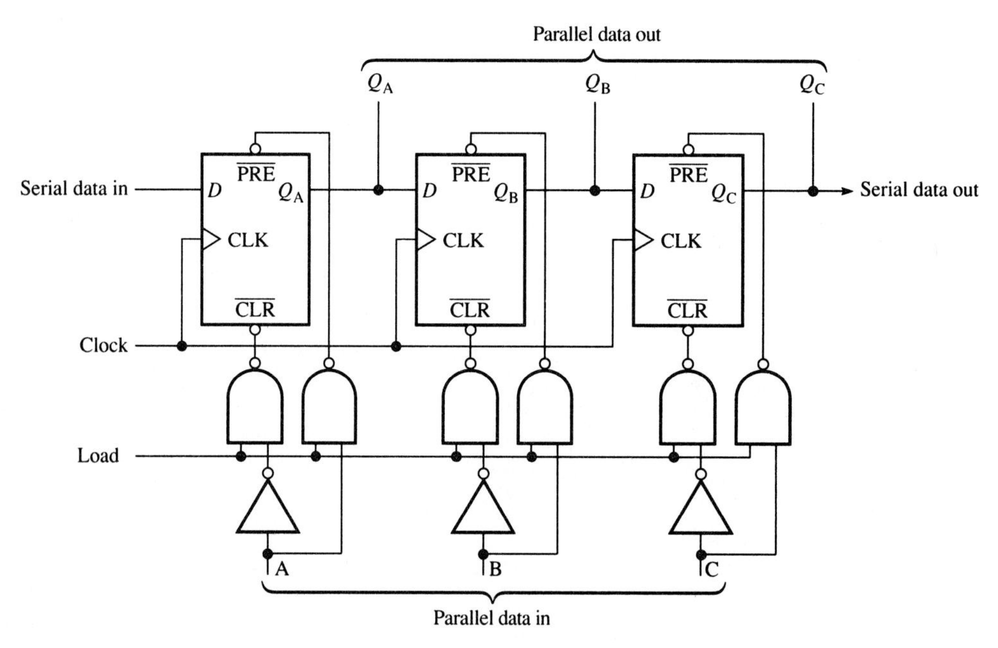

(b) A parallel or serial-in /parallel or serial-out shift register.

FIGURE 6–15
Examples of shift registers using *D* flip-flops.

6–15(a). This 4 bit register delays a digital signal by four times the clock period (delay = number of flip-flops times clock period).

There are a number of types of shift registers. Some can be loaded in parallel—that is, all flip-flops are either set or cleared at the same time (sometimes this is called a *broadside* load). If the output of each flip-flop is available, then the data can be removed in parallel also. An example of a shift register with both parallel and serial inputs and parallel and serial outputs is shown in Figure 6–15(b). Shift registers can also be made to shift the data either to the left, right, or both (bidirectional) using a control signal.

Shift registers are available in IC form; they are available with all the options discussed previously, including serial and parallel inputs and outputs and shift directions. Another important application for shift registers is in data conversion—from serial to parallel form or parallel to serial form. In order to convert parallel data to serial form, the register is loaded with the data in parallel form. Each succeeding clock pulse then moves the bit pattern through the register in serial fashion, where it is removed from the last flip-flop. A similar procedure can be used to convert serial data to parallel data. In this case, the serial data is applied to the input of the shift register one bit at a time and is shifted through the register by applying a series of clock pulses. When the register is filled, the data is removed simultaneously from all the flip-flops.

Shift registers are also used as counters by feeding the output of an SISO register back to the input. One important type of shift register counter is called a *ring counter,* so named because it can be drawn as a simple ring, as illustrated in Figure 6–16(a). The ring counter recirculates the same data over and over around

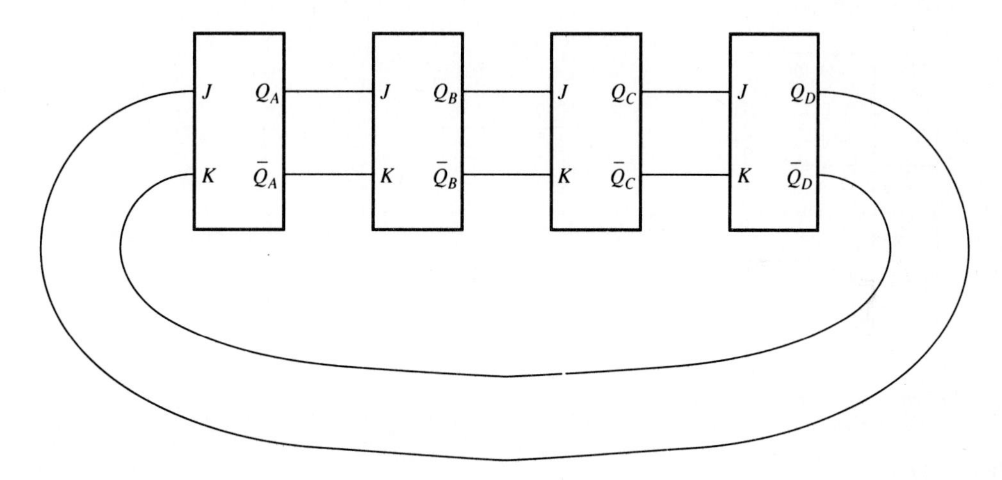

(a) Ring counter drawn to emphasize name. Clock, preset and clear inputs are not shown.

Clock pulse	Output Q_A	Q_B	Q_C	Q_D
1	0	1	1	1
2	1	0	1	1
3	1	1	0	1
4	1	1	1	0
5	0	1	1	1

(b) Sample output sequence for a pattern initially loaded with one zero.

FIGURE 6–16
Ring counter.

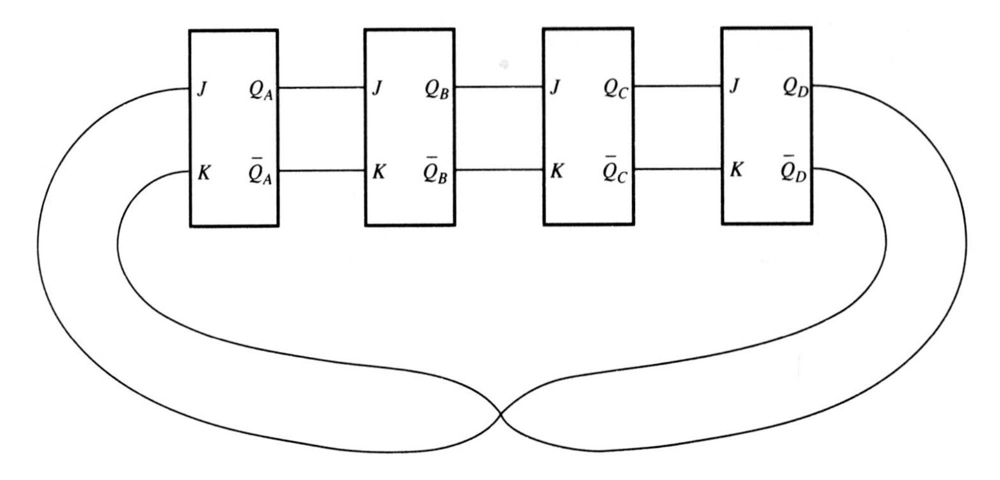

(a) Twisted-ring counter (Johnson counter) drawn to emphasize name. Clock, preset, and clear inputs are not shown.

Clock pulse	Output Q_A	Q_B	Q_C	Q_D
1	0	0	0	0
2	1	0	0	0
3	1	1	0	0
4	1	1	1	0
5	1	1	1	1
6	0	1	1	1
7	0	0	1	1
8	0	0	0	1
9	0	0	0	0

(b) Sample output sequence starting with counter initially cleared.

FIGURE 6–17
Twisted-ring counter.

the ring. It must initially be loaded with the pattern to be shifted. A typical sequence of a ring counter that was loaded with one zero is shown in Figure 6–16(b).

A second type of shift register counter is called a *twisted-ring,* or *Johnson, counter*. The twisted-ring counter is constructed with *JK* flip-flops, with a ''twist'' introduced in the ring at some point, as illustrated in Figure 6–17(a). This causes the flip-flop immediately after the twist to invert the pattern as it circulates. A typical count sequence for a twisted-ring counter that started initially cleared is shown in Figure 6–17(b).

6–4 STATE MACHINE FUNDAMENTALS

A sequential logic circuit presents a pattern of bits at its output that is determined by the initial conditions of the system and the input signals that were applied to the circuit. The **state** (or condition) of a sequential logic circuit is related to two components—the inputs to the circuit and information stored in memory. A sequential logic circuit in which the output is related to the inputs and the information is stored in the memory of the circuit is called a **state machine.** Figure 6–18(a) illustrates the relationship between the input and output variables, as well as the

FIGURE 6–18

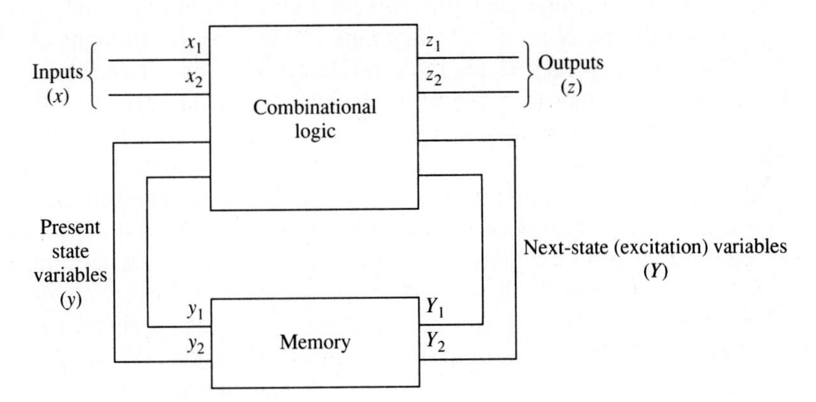

(a) General model of a state machine.

State table

Present state	Next state Y_2Y_1				Output z_2z_1			
y_2y_1	x_2x_1 0 0	x_2x_1 0 1	x_2x_1 1 1	x_2x_1 1 0	x_2x_1 0 0	x_2x_1 0 1	x_2x_1 1 1	x_2x_1 1 0
0 0	0 0	0 1	0 1	1 1	0 0	0 1	0 1	1 1
0 1	0 1	1 0	1 1	0 0	0 1	1 0	1 1	0 0
1 1	1 1	0 0	1 0	1 0	1 1	0 0	1 0	1 0
1 0	1 0	1 1	0 0	0 1	1 0	1 1	0 0	0 1

Next-state table for a 2-bit controlled counter. The sequence of the counter is controlled by the inputs x_2x_1. For this example, the output is the same as the next state. Count functions shown are:

x_2x_1	Function
0 0	Latch; count does not advance.
0 1	Binary sequence — count up.
1 1	Gray code sequence.
1 0	Binary sequence — count down.

(b) Example of a state table. Note that inputs (x_2x_1) and present state (y_2y_1) are entered in gray code to simplify mapping.

present state and next state (or excitation) variables. The next state and output variables must be defined for every combination of input and present state variable. A tabular summary of the relationship between the present state, next state, inputs, and outputs of a sequential logic circuit is called a **state table.** An example of a state table is shown in Figure 6–18(b).

One type of state machine, called the Moore state machine, has outputs that are completely determined by the present state variables—that is, the memory elements determine the present output. The next state of a Moore circuit is determined by *both* the inputs and the present state. Another type of state machine is called the Mealy state machine. Its outputs are determined by both the present state of the machine *and* the current input conditions. For simplification, we discuss only the Moore state machine, which can be thought of as a controlled counter that assumes states (counts) determined by the inputs.

The best way to explain the design of a simple Moore state machine is by example. The example selected is a common problem for instrumentation systems. Assume that you need to control a device (called the listener) by sending information from a computer (called the talker) as part of an instrumentation and control system. You need to design a handshake circuit—a circuit that allows the information to be exchanged in an orderly manner by generating control signals that indicate when the listener is ready to receive information and when the data has been received. The computer, in turn, provides a signal to the listener that indicates data from the computer is valid. The handshake selected for this example is a three-wire handshake originally designed by Hewlett-Packard and is a portion of the bus used in instrument control known as the general-purpose interface bus (GPIB), the Hewlett-Packard Interface Bus (HPIB), or the IEEE-488 bus. This bus is described in more detail in Section 14.9.

Begin by drawing the handshake signals needed in the proper time relationship. Figure 6–19(a) shows the input and output signals referenced to the listener; arrows show cause and effect. Note that control signals are usually asserted low in interface applications—that is, the word associated with the variable is true when the line is low. In this example, the two output variables, labeled $\overline{\text{NRFD}}$ and $\overline{\text{NDAC}}$, are identical to the present state variables. The present state variables follow a sequence that can be represented as a series of binary numbers by assigning one of them as the MSB and the other as the LSB. The present states, representing the output sequence, are shown under the timing drawing in Figure 6–19(a).

The input variable labeled $\overline{\text{DAV}}$ is asserted by the computer whenever the data lines have valid information and is asserted in response to the listener indicating that it is ready for data. The other input, labeled $\overline{\text{BUSY}}$, is not actually a GPIB signal but is used as an internal indicator that the listener is processing the information. It is controlled by a timer circuit internal to the digital circuit. The outputs (state variables) change in response to these two inputs and the present state.

All this information can be summarized in the state diagram drawn in Figure 6–19(b), which represents the desired operation of the circuit. The states are entered onto the circles and represent the outputs of the circuit. All possible combinations of inputs are shown beside arrows indicating the desired action that

FIGURE 6–19
Design of a handshake circuit.

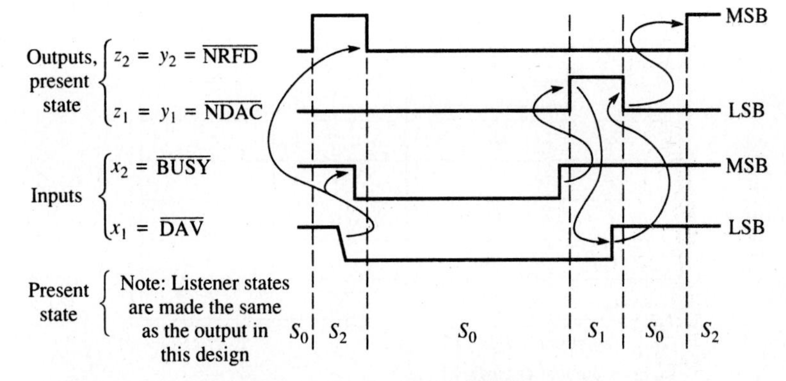

(a) Signals for a handshake circuit listener. The $\overline{BUSY}$ signal is internally generated in the listener and is not a handshake signal.

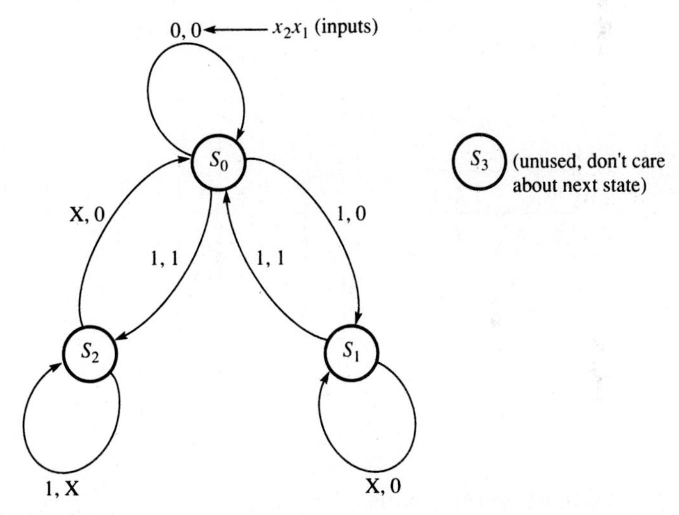

(b) State diagram. Inputs are indicated by numbers written next to arrows. Outputs are the same as the states. Input 0, 1 should never occur and is not indicated.

State table

Present state y_2y_1	Next state Y_2Y_1, x_2x_1 = 0 0	x_2x_1 = 0 1	x_2x_1 = 1 1	x_2x_1 = 1 0	Output z_2z_1
0 0	0 0	X X	1 0	0 1	0 0
0 1	0 1	X X	0 0	0 1	0 1
1 1	X X	X X	X X	X X	1 1
1 0	0 0	X X	1 0	0 0	1 0

← This row (1 1) represents S_3 and is unused; therefore use don't care (X). Check final design for lockup.

↑ These inputs (0 1 column) do not occur; therefore use don't care (X).

(c) State table. Present states and inputs are written in gray code to avoid errors in mapping.

Transition table for *JK* flip-flop.

Output transitions		Inputs	
Q_N	Q_{N+1}	J_N	K_N
0	→ 0	0	X
0	→ 1	1	X
1	→ 0	X	1
1	→ 1	X	0

Q_N = output before clock
Q_{N+1} = output after clock
J_N, K_N = inputs required to cause transition
X = don't care

(d) Transition table for *JK* flip-flops (Table 6–8 repeated for reference).

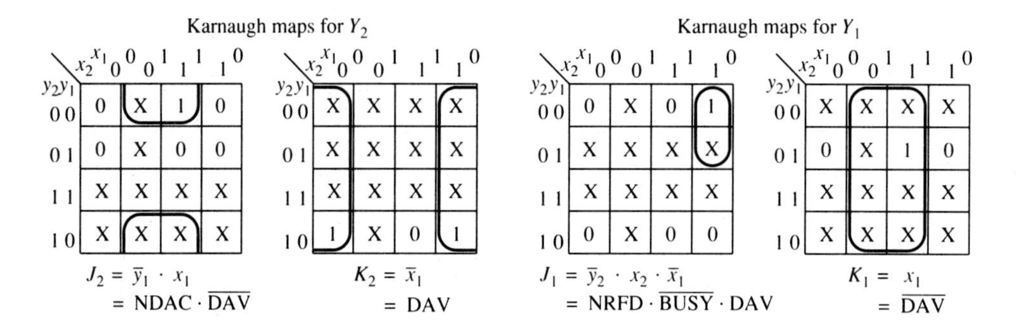

(e) Karnaugh maps for both inputs of *JK* flip-flops.

FIGURE 6–19 *(continued)*

should be taken. For instance, if the circuit is in state 1 (bottom right-hand corner) and the inputs are both 1s ($\overline{\text{BUSY}}$ and $\overline{\text{DAV}}$ both HIGH), the diagram indicates that the circuit should go to state 0. Notice the X's representing don't-care situations, which are combinations that are not possible. For instance, the desired timing diagram does not allow the condition where both outputs (states) are 1s at the same time. Therefore, there is no state 3, so the exit from state 3 doesn't matter.

The state diagram is a useful pictorial summary of the operation of the circuit. For designing the circuit, it is easier to represent the information in a state table. The state table shown in Figure 6–19(c) contains the same information as the state diagram. Note that the output is not shown separately, as it will be the same as the state in this case. In order to avoid errors in transferring the information to a Karnaugh map, the possible combinations of inputs can be written in gray code on the next-state table, as was done in the stable table shown. This allows every position of the next-state listing to have a corresponding square on the Karnaugh map.

As in the case of small-counter design, the information for a small-state machine can be extracted from the state table and mapped onto Karnaugh maps (more properly termed Karnaugh state maps). Because of their versatility, *JK* flip-flops are selected for the design of this small-state machine. A separate map is required for each input of the *JK* flip-flops. The values entered onto the squares of

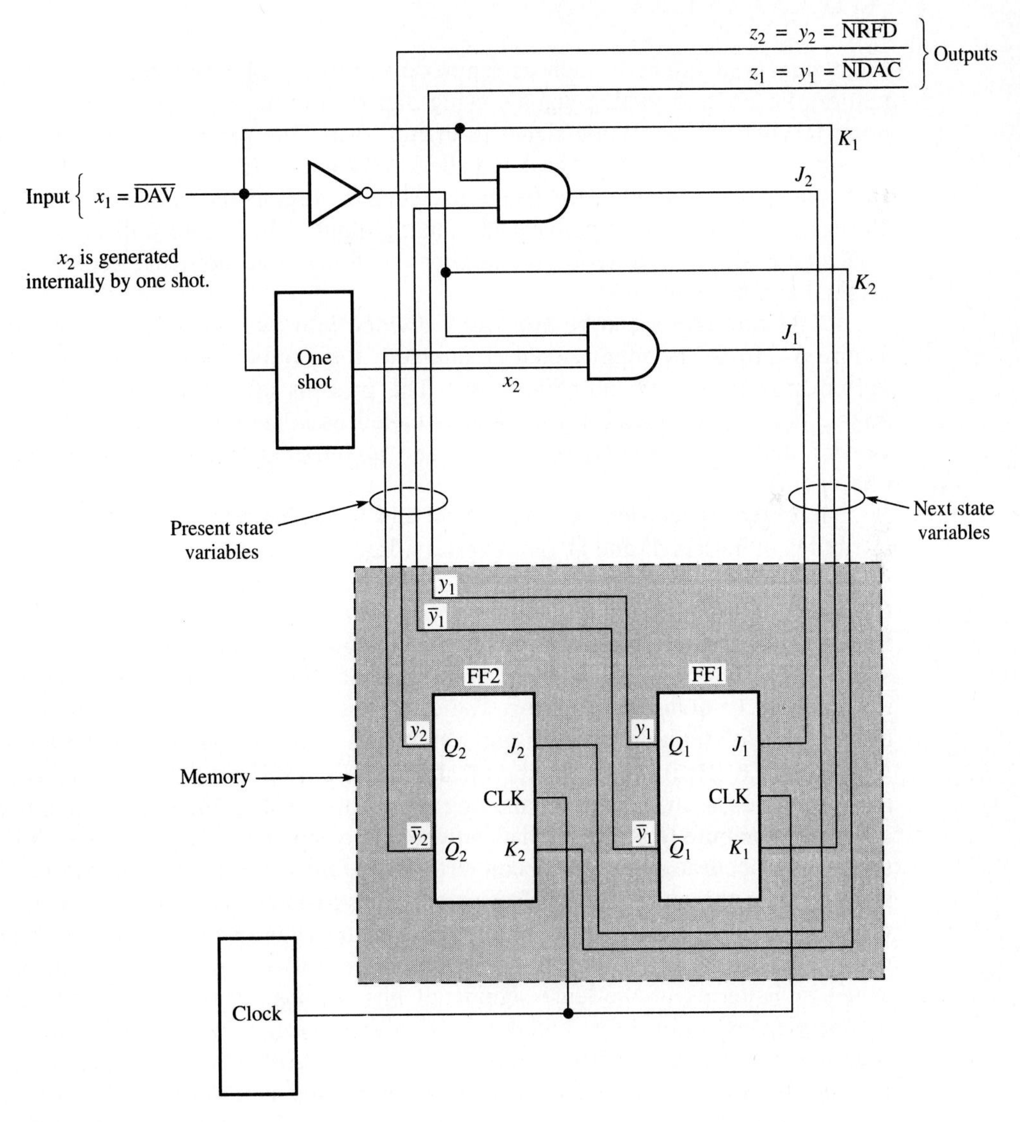

FIGURE 6–20
Circuit for listener. Compare this drawing with the general model shown in Figure 6–25.

the Karnaugh maps represent the inputs to the flip-flops that will cause the circuit to go to the next state. To complete the maps, the transition from the present state to the next state is observed for each flip-flop. The inputs that force the required transition are obtained from the transition table. For reference, the *JK* transition table is shown in Figure 6–19(d), and the completed maps are shown in Figure 6–19(e). The reduced expressions for the inputs are obtained from the maps, and the circuit can easily be drawn. The completed circuit is shown in Figure 6–20.

6–5 A/D CONVERTERS

Most signals of interest, such as temperature, force, or pressure, are analog in nature. To process analog signals using digital circuits, it is necessary at some point to convert the analog signal to digital form. The process of converting an analog signal to a digital signal is called **A/D conversion.** A/D converters are essential in many modern instruments, such as digital multimeters and digital storage oscilloscopes, to convert the analog input voltage into a discrete value. The digital signal is a series of numbers of finite precision that represent the original analog signal.

A/D convertion can be arbitrarily divided into two processes, as shown in Figure 6–21(a). The input signal is sampled; the sampled quantity is then converted into a discrete, numeric value. The process of converting the sampled analog quantity into a digital number is called **quantization.** In actual A/D converters, the two processes—sampling and quantizing—are often performed as one operation.

Accuracy, precision, and resolution (discussed in Section 2–2) are important attributes of both A/D and D/A converters. For an A/D converter, accuracy is the measured difference between the digital output value and the true analog input. Because the numeric values assigned to an analog quantity are chosen from a finite set of numbers, a digital number is an approximation of the original analog signal. The difference between the original analog signal and the quantized value is known as the **quantization error.** The quantization process has an inherent error associated with it and appears in the system as a contribution to the system noise.

Figure 6–21(b) illustrates the ideal transfer curve for a 3 bit A/D converter that produces a maximum quantizing error of ±1/2 LSB. Although various sources contribute to error (such as power supply ripple or temperature stability), the overall accuracy of an A/D converter is specified by three basic errors. The total error is the sum of the offset error, gain error, and linearity error. These errors are illustrated in Figure 6–22. Overall accuracy is specified in terms of these errors and can be listed by the manufacturer in parts per million, percent of full scale, or in terms of the least significant bits. Precision, on the other hand, is independent of these errors and depends only on the number of bits in the digital output. For example, an 8 bit converter has 256 possible states; its resolution is 1 part in 256, which can be expressed as a percentage as 0.4%.

Incidentally, an interesting point can be made about the difference between precision and accuracy by considering the noise inherent in all systems. All analog signals contain some level of noise. Increasing the number of bits that are digitized could give a false sense of accuracy due to increased resolution. A high-resolution A/D converter that is converting a noisy analog signal merely uses more bits to represent the noise already present in the analog signal. Noise is noise whether it is digitized or not. However, at the price of longer conversion time, predictable noise (such as power line interference) can be reduced by integrating A/D converters, such as the dual-slope converter.

For most applications of A/D converters, the two most important characteristics are accuracy and speed. The speed of conversion is inversely proportional to the conversion time. Depending on the type of converter, the conversion time may be constant or may depend on the difference between the current amplitude and the final amplitude. Conversion time is related to aperture time and may be equal to it. **Aperture time** is the time the converter takes to sample the input signal

FIGURE 6–21

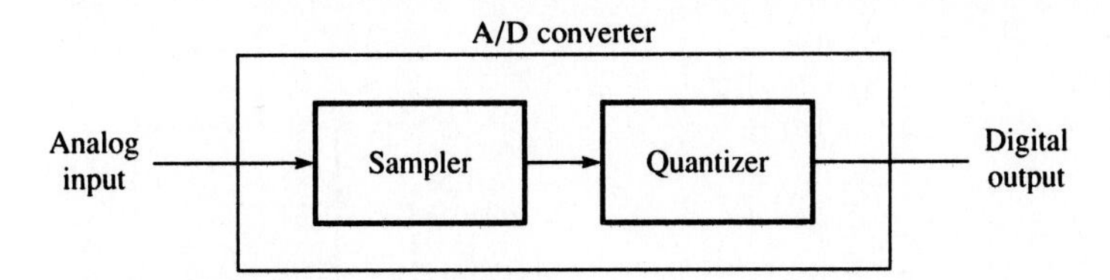

(a) Analog-to-digital conversion process.

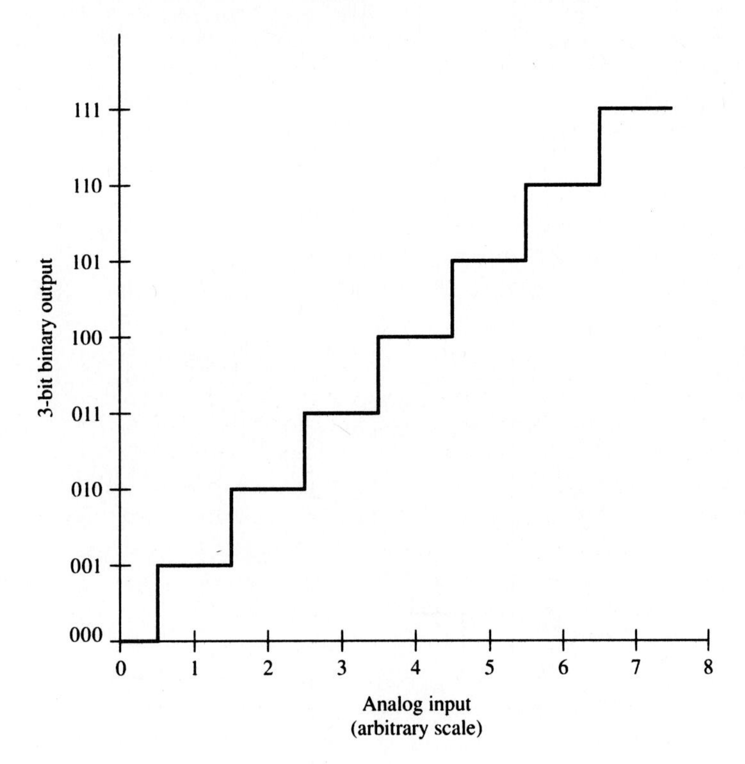

(b) Ideal transfer curve for a 3-bit A/D converter.

FIGURE 6–22
A/D errors.

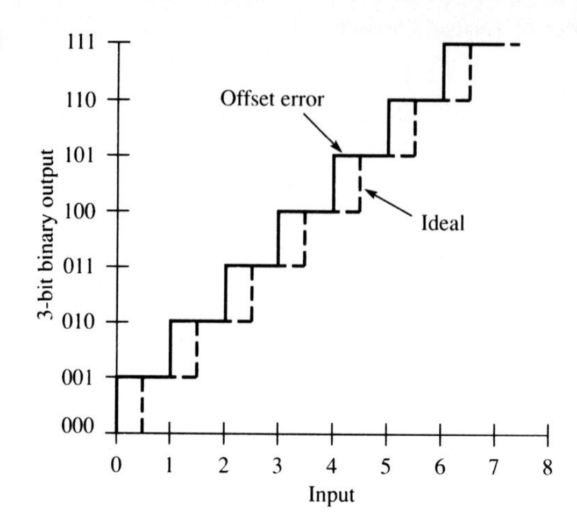

(a) A/D converter with offset error.

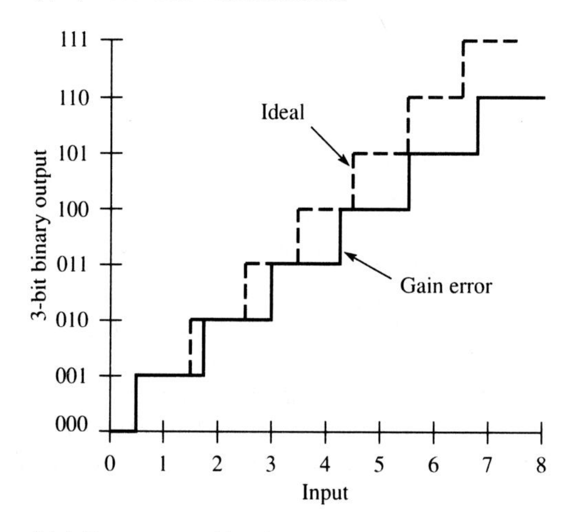

(b) A/D converter with gain error.

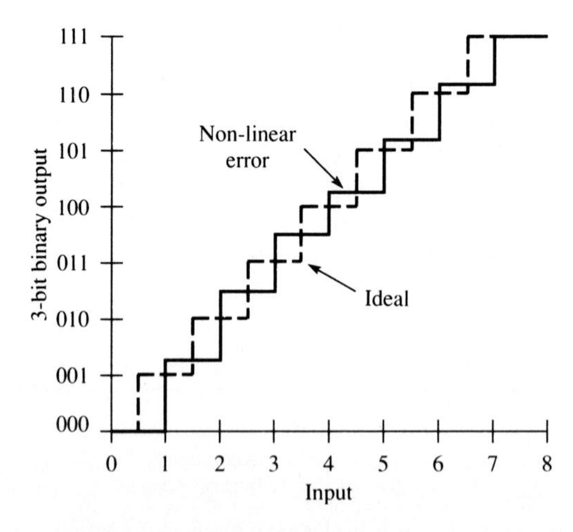

(c) A/D converter with linearity error.

and may or may not be the same as the conversion time. It is usually important that the analog input signal not change more than ±1/2 LSB during the aperture time. If a sample-and-hold circuit is used, the aperture time may be shorter than the conversion time, since the analog signal is held constant while the conversion process is completed.

The increasing densities of ICs and the increase in speeds has made available A/D and D/A converters unheard of a few years ago. Recent advances in developing large-scale GaAs (gallium arsenide) integrated circuits have lead to 14 bit converters operating at 1 GHz conversion speed. The applications for digital signal processing at high speeds are innumerable, and the impact on instrumentation will be to increase the application of digital instruments in areas previously reserved for analog instruments.

TRACKING A/D CONVERTERS

The **tracking A/D converter** is based on a closed-loop technique, as illustrated in Figure 6–23(a). It features an up/down digital counter, a D/A converter, and a comparator in the feedback loop. At the start of the conversion process, the digital counter begins counting. The output of the counter is digitized and compared with the analog input signal. If the output of the D/A converter exceeds the analog input signal, the comparator senses that the counter value is higher than the input and sends a signal that causes the counter to count down. If the output of the D/A converter is less than the analog input signal, the comparator directs the counter to count up. As a result, the counter is constantly trying to track the input signal. As the input signal rises, the counter increments the count. Likewise, when the input signal falls, the counter decrements the count. The output is taken from the counter and tends to oscillate about the true value, as illustrated in Figure 6–23(b). This is a disadvantage of this type of A/D converter. Another disadvantage of the converter is that it is slower than other techniques and the conversion time is not a constant for every input signal. Also, aperature time is longer than for other converters.

SUCCESSIVE-APPROXIMATION CONVERTERS

Successive-approximation converters are also based on a closed loop, similar to the tracking A/D converter. The successive approximation method is illustrated in Figure 6–24. Within the loop is a successive-approximation register (SAR) that produces a digital output for a D/A converter. As with the previous converter, the D/A converter sends an analog signal to a comparator representing the current output of the successive-approximation register. The comparator detects the difference between the analog input signal and the output of the D/A converter, producing a greater-than or less-than indication for the control and clock logic. The control logic uses this indication to increment or decrement the count in the successive-approximation register.

The process begins with an external start signal sent to the control logic. The control logic initially turns off all bits except the MSB. The MSB is tested by converting it to an analog value and comparing it with the input analog signal,

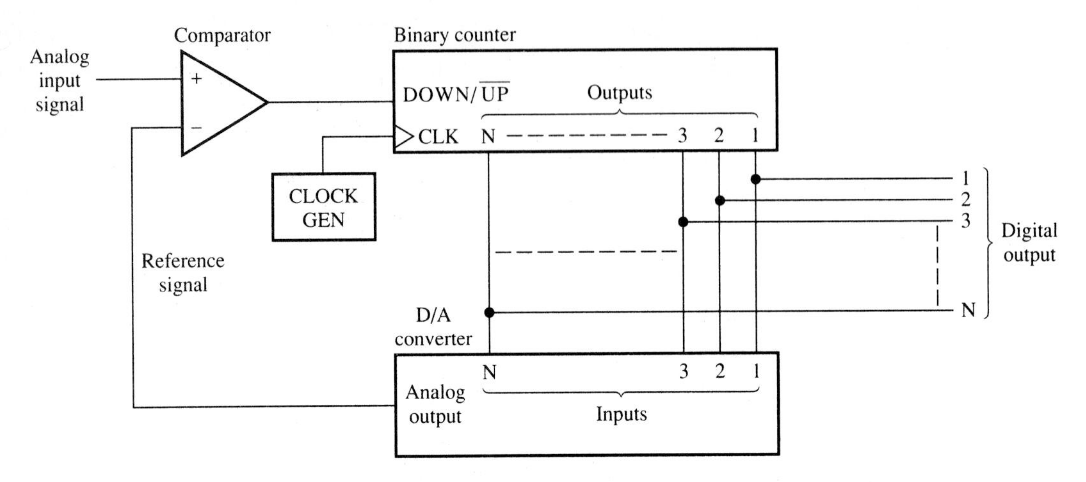

(a) Block diagram.

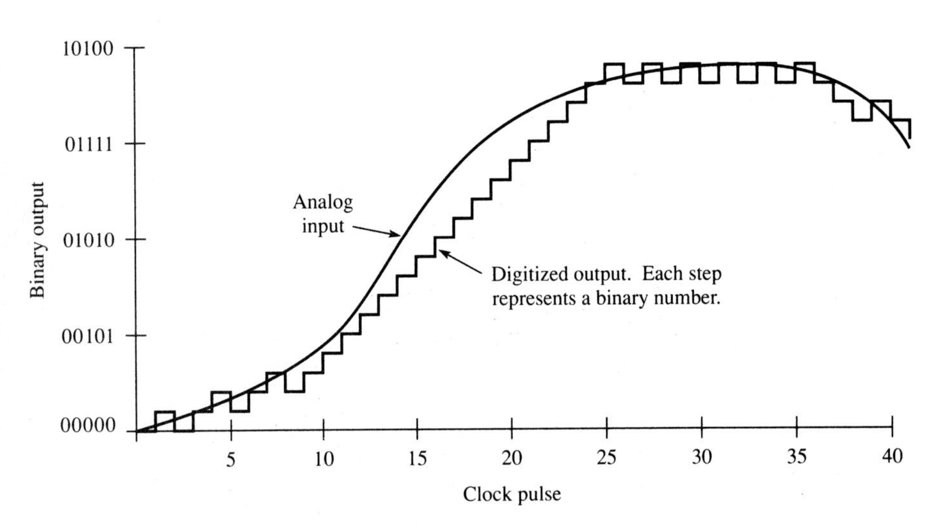

(b) Representative input and output signal. Note the oscillating behavior of the output and the inability of the output to follow a rapidly changing input signal.

FIGURE 6–23
Tracking A/D converter.

which is held constant during the conversion process. This is like opening a phone book in the middle and checking which half to investigate further. If the input signal is larger than the MSB, it is left on; otherwise it is turned off. Each bit in the SAR is turned on and tested, in turn, dividing the error by half. When the LSB has been tested, the conversion is complete, causing a conversion complete signal to be generated. The digital output signal is contained in the SAR. The output number represents the analog input as a fraction of the internal reference voltage.

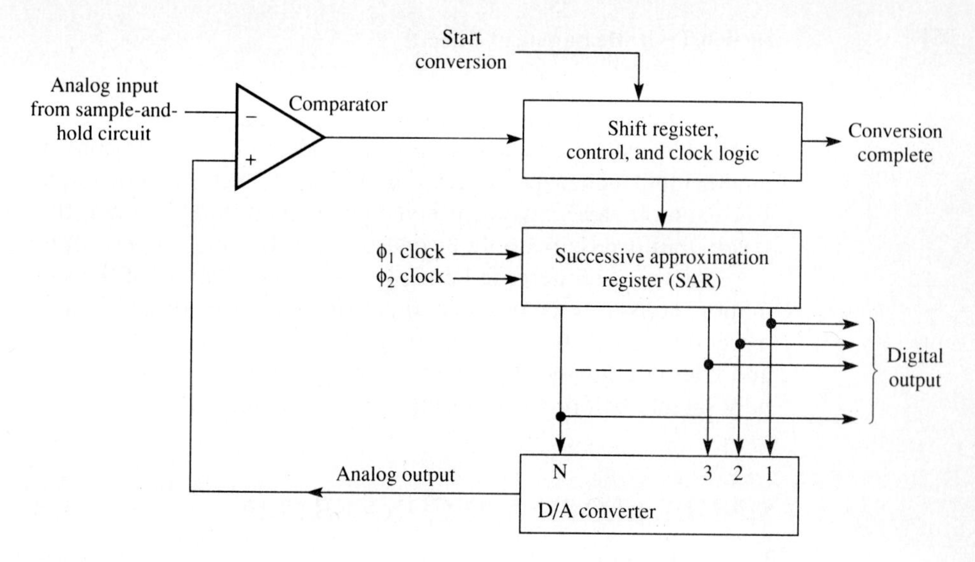

(a) Block diagram.

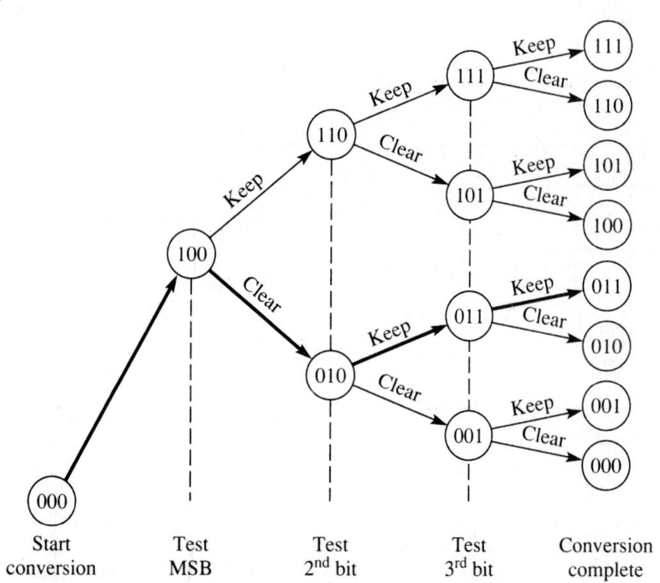

(b) State diagram for a 3-bit SAR showing the conversion path for an output of 011.

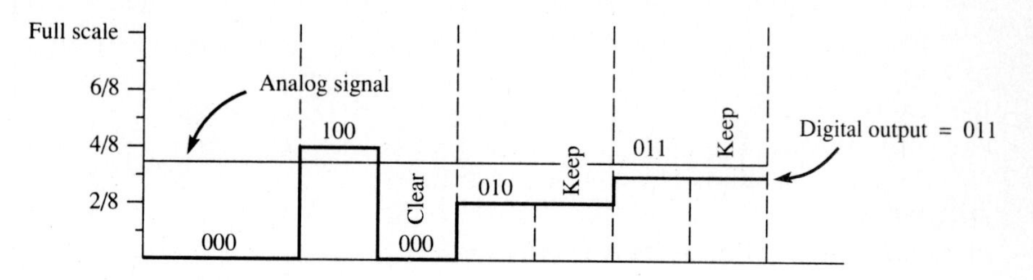

(c) Timing diagram for converting 3.4 V to digital. Full scale represents 8.0 V.

FIGURE 6–24
Successive approximation A/D converter.

Although it requires more control logic, the successive-approximation method is faster than the tracking method and has a fixed conversion time for any analog input signal. Its speed is limited by the settling time required for each test. For example, an 8 bit successive-approximation A/D converter has 256 possible levels, but it requires only 8 (phase 1) clock cycles to test all bits.

At a cost of increased complexity, a modification of the successive-approximation converter is possible that will speed up the conversion process. The method is called the two-step method and basically splits the process into coarse and fine bit conversions. The two-step method allows the converter to acquire a new sample before the old sample has completed the process.

SINGLE-SLOPE A/D CONVERTER

The **single-slope A/D converter** does away with the D/A converter used in the tracking and successive-approximation methods. Instead, it uses a precision ramp

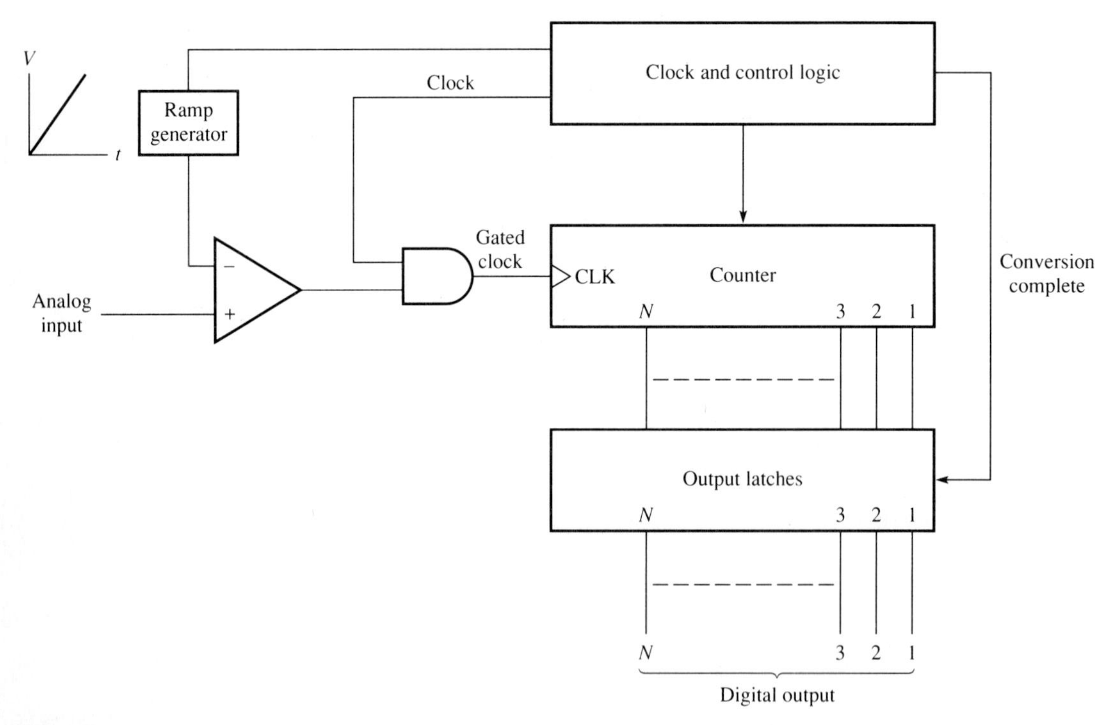

FIGURE 6–25
Block diagram of a single-slope A/D converter.

generator to produce a reference voltage that increases linearly with time for the comparator. Figure 6–25 shows the idea. At the start of a conversion cycle, the ramp generator is set to 0 V and the counter is reset. A start pulse enables the clock for the counter and simultaneously starts the ramp generator. When the ramp reaches the input analog voltage, the comparator disables the clock to the counter. At this time, the counter output is then proportional to the input voltage. The count is then latched for the display.

DUAL-SLOPE INTEGRATION A/D CONVERTER

The **dual-slope A/D converter** has a run-up and a run-down cycle, each of which is generated by an integrator. The integrator generates a constant-slope ramp during run-up and a variable-slope ramp during run-down. The block diagram is shown in Figure 6–26 for reference. At the start of a conversion, S_1 is connected to the analog input and charges the capacitor with a constant current for a fixed amount of time. The run-up time is a fixed interval controlled by an accurate clock and counter. When the time has elapsed, the capacitor holds a voltage proportional to the input voltage. Switch S_1 is now electronically connected to a reference voltage of opposite polarity, and the counter is started again from zero. The capacitor

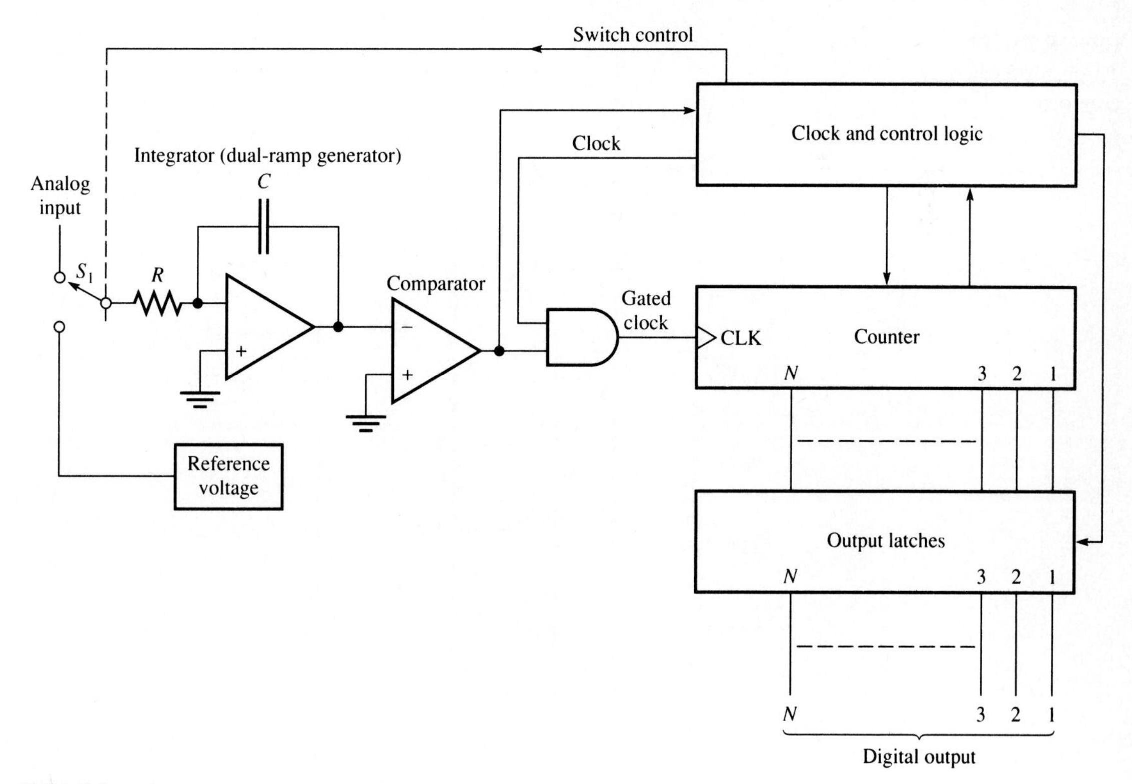

FIGURE 6–26
Block diagram of a dual-slope A/D converter.

discharges with constant current during run-down. The run-down time is dependent on the amplitude of the voltage stored on the capacitor. During run-down, the counter is clocked until the zero crossing is detected, causing the counter to be stopped. When this occurs, the counter has a number proportional to the input voltage. The advantage of the dual-slope method is high noise immunity because of the integration of the input signal. In addition, errors inherent in the single-slope method are canceled in the dual-slope method. The major drawback is that the conversion time is relatively long.

FLASH CONVERTERS

The fastest type of A/D converters are called simultaneous, parallel, or **flash converters.** Flash converters rely on a string of comparators designed to test the input voltage directly with a series of reference voltages set up in a resistive voltage divider. Figure 6–27 shows a flash converter for 3 bits that requires seven comparators and a priority encoder. All comparators simultaneously test the analog input against a fraction of the reference voltage and indicate if the input signal is greater than a given reference level. In turn, the priority encoder converts the highest responding input into a digital output. For instance, if the analog input is

FIGURE 6–27 Three bit flash A/D converter.

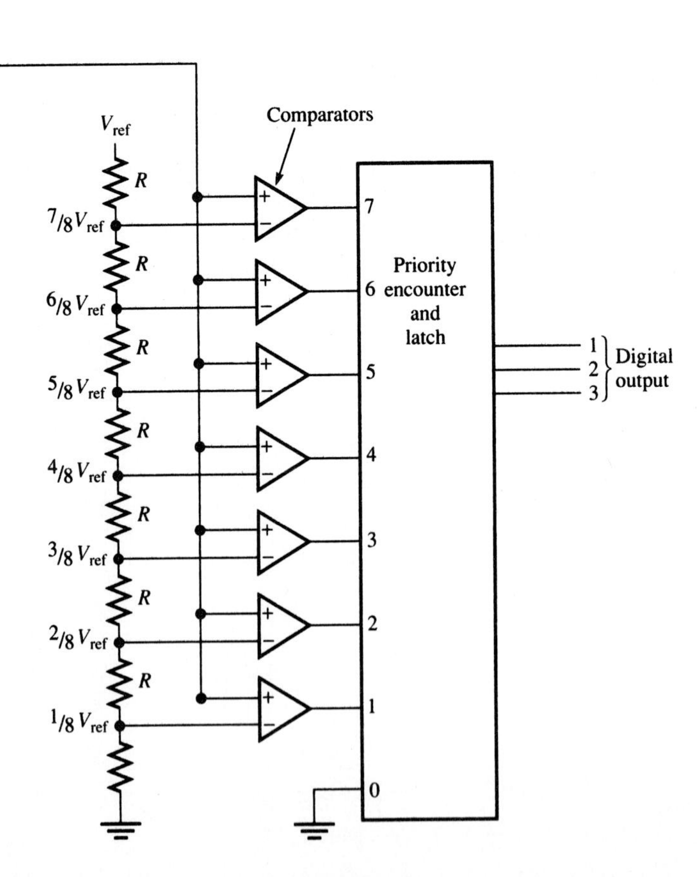

51% of the reference voltage, then the lower four comparators respond and the upper three comparators do not. The priority encoder translates this particular response into a binary 100.

The advantage of flash converters is that the conversion is very fast; however for high resolution, the overall complexity and cost is high (n bits require $2^n - 1$ comparators). For an 8 bit flash converter, 255 comparators are required ($2^8 - 1 = 255$).

6–6 D/A CONVERTERS

The process of converting a digital signal to an analog signal is called **D/A conversion.** The analog output is usually a voltage or current but may be some other physical quantity. In instrumentation, D/A conversion has taken on new importance in recent years because of the increasing use of arbitrary waveform generators in automatic testing (see Section 9–2). Another instrumentation application is changing the output of a computer or controller to a voltage suitable to drive a chart recorder or other device. Usually the input digital quantity is a binary number, but D/A converters can be designed for binary-coded decimal (BCD) or other codes.

The accuracy of a D/A converter is measured by comparing the actual analog output for each digital code with the theoretical analog output. The transfer characteristic for the ideal 3 bit D/A converter is shown in Figure 6–28. The output analog values correspond to the input digital number plus or minus 1/2 LSB. As in the case of A/D converters, the total error for a D/A converter is the sum of the offset error, gain error, and linearity error. Overall accuracy is specified by the manufacturer in terms of these three errors.

The conversion of a digital signal into an analog quantity is generally accomplished using electronic switches and a resistor network, as shown in the basic

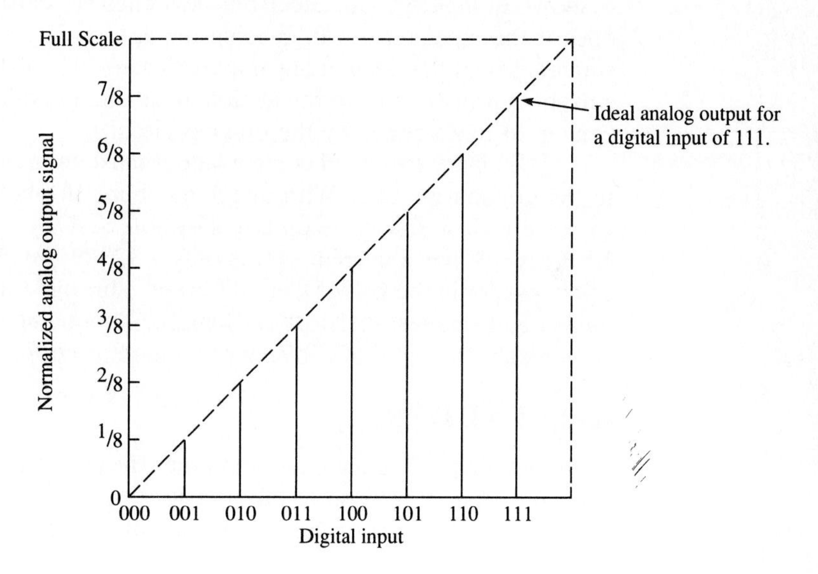

FIGURE 6–28
Transfer characteristic for an ideal 3-bit D/A converter.

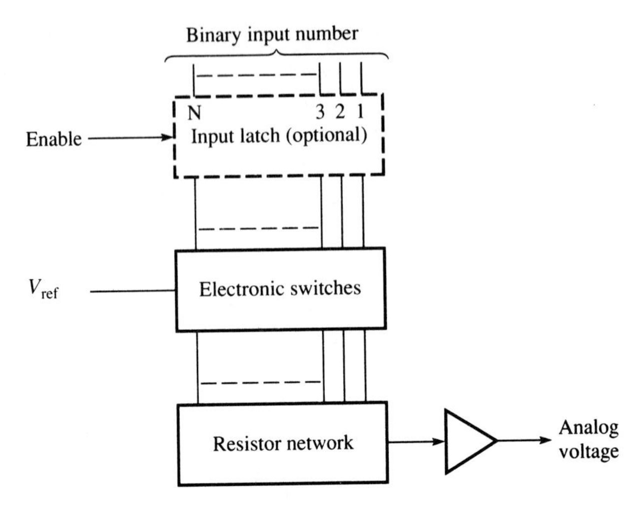

FIGURE 6–29
Basic D/A diagram.

D/A diagram in Figure 6–29. An input latch may be used to store the binary word to be converted. Electronic switches connect either a precision voltage reference or ground to a resistance network, depending on whether the input bit is a 1 or a 0. Each bit is weighted in proportion to the column value of the bit by the resistive network and produces a current that is proportional to the binary column value. An operational amplifier can then sum the individual currents and transform them into an output voltage. There are variations of this idea, but essentially it is used by virtually all D/A converters.

BINARY-WEIGHTED CURRENT LADDER

The **binary-weighted current ladder** uses precision resistor values to set up a current that is proportional to the column weight of the binary number. The circuit is shown in Figure 6–30. Electronic switches are closed if the input bit is a 1 and open if the input bit is a 0, causing the appropriate currents to be added by the summing amplifier. (Summing amplifiers were described in Section 5–6.) The total current is proportional to the analog value and is available as an output or can be converted to a voltage by the op-amp circuit.

The binary-weighted current ladder must have currents that are proportional to the column weights. With larger numbers of bits, this requirement places an extremely close tolerance on the allowable resistors. For an 8 bit converter, the LSB resistor has a current that is only 1/128 of the current in the MSB resistor. Each resistor in the network is a different value and the resistor ratios between the largest and smallest resistors are high, compounding the tolerance requirements. As a result, this type of converter is seldom used for more than 4 bit converters.

R-2*R* LADDER

An interesting technique that overcomes the problems associated with large resistor ratios and various sizes of precision resistors is the ***R*-2*R* ladder,** which requires only two resistor values for the entire ladder. The absolute values of the

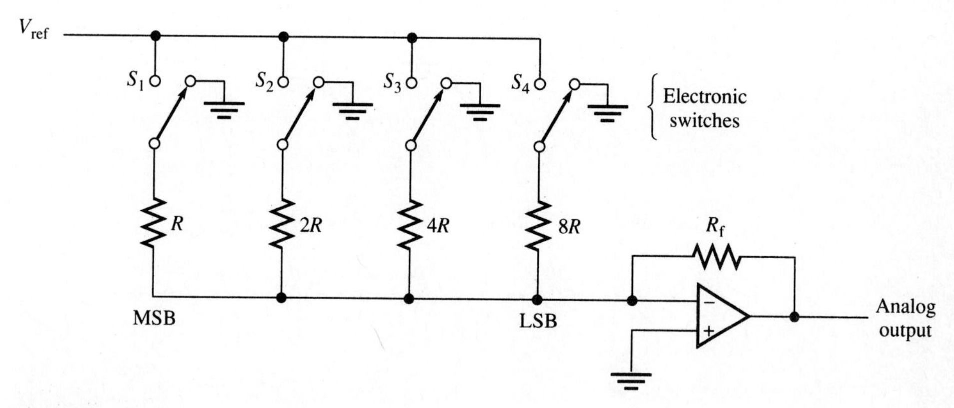

FIGURE 6–30
Binary-weighted current ladder.

resistors are not as critical as their ratio, which should be 1 : 2. A 4 bit R-$2R$ ladder circuit is shown in Figure 6–31 with nodes labeled N_0, N_1, N_2, and N_3. Looking from each node in the direction of a switch, the resistance is $2R$. This is true regardless of the switch position (recall that the Thevenin resistance of an ideal voltage source is 0). In addition, the resistance looking to the left of N_1, N_2, and N_3 is also $2R$, regardless of the switch position.

Consider the 4 bit R-$2R$ ladder shown in Figure 6–32(a). Assume the binary input number is 1000_2—the MSB is a 1 and the other bits are 0s. Replace the circuit to the left of N_3 with the Thevenin equivalent, as shown in Figure 6–32(b). Since the voltage source is 0, we can replace the circuit to the left of N_3 with a resistance of $2R$. Applying Thevenin's theorem again gives the circuit shown in

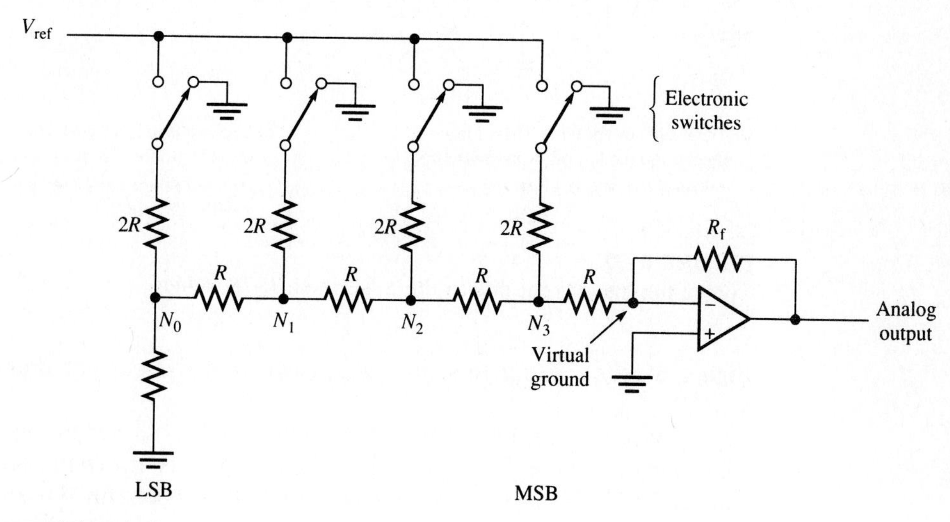

FIGURE 6–31
R-$2R$ ladder D/A converter.

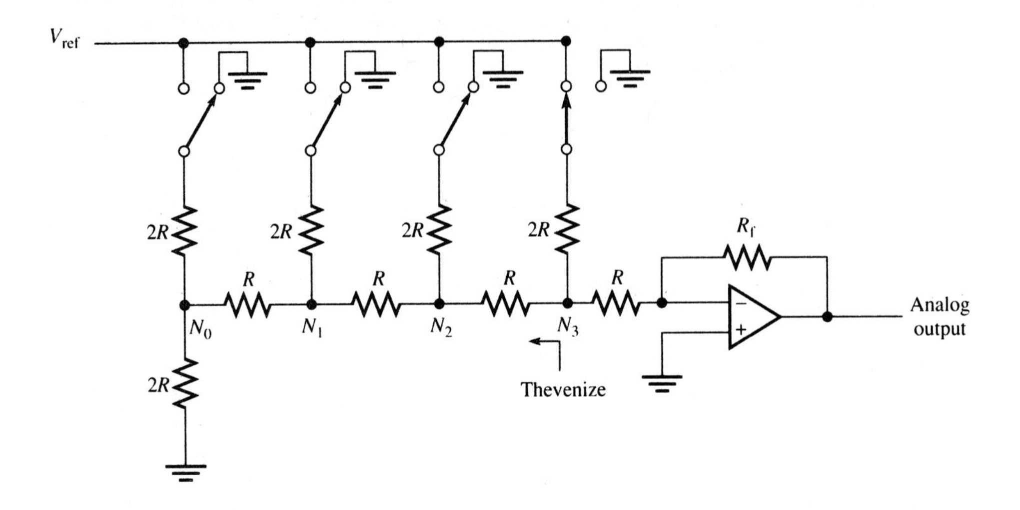

(a) R-$2R$ ladder with a binary input 1000. To solve the circuit, we will apply Thevenin's theorem twice.

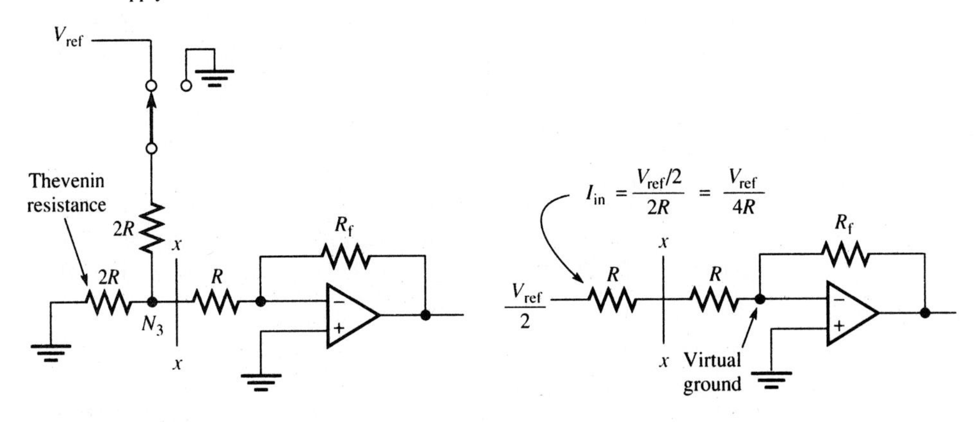

(b) The results of applying Thevenin's theorem to the left of N_3 in Figure (a). The result is simply $2R$ to ground.

(c) The results of applying Thevenin's theorem to the left of x-x in Figure (b). The current into the virtual ground is seen to be $V_{ref}/4R$.

FIGURE 6–32
Computing the current due to the MSB in an R-$2R$ ladder.

Figure 6–32(c). From this, it can be seen that the current due to the MSB is $V_{ref}/4R$.

Now consider the effect with the input 0100_2 as shown in Figure 6–33(a). The Thevenin equivalent for the circuit to the left of N_2 is drawn in Figure 6–33(b). We can apply Thevenin's theorem a second time and obtain the equivalent circuit shown in Figure 6–33(c). The calculation of the current into the reference ground is shown and results in a current of $V_{ref}/8$. The second MSB is seen to cause exactly one-half as much current to flow as the MSB, in keeping with the column values of the binary number system. Continuing in a similar manner will show each input to the left on the ladder to have one-half the weight of the previous input. The output of the ladder for an arbitrary input can be found by the superposition theorem, resulting in an analog output proportional to the digital input.

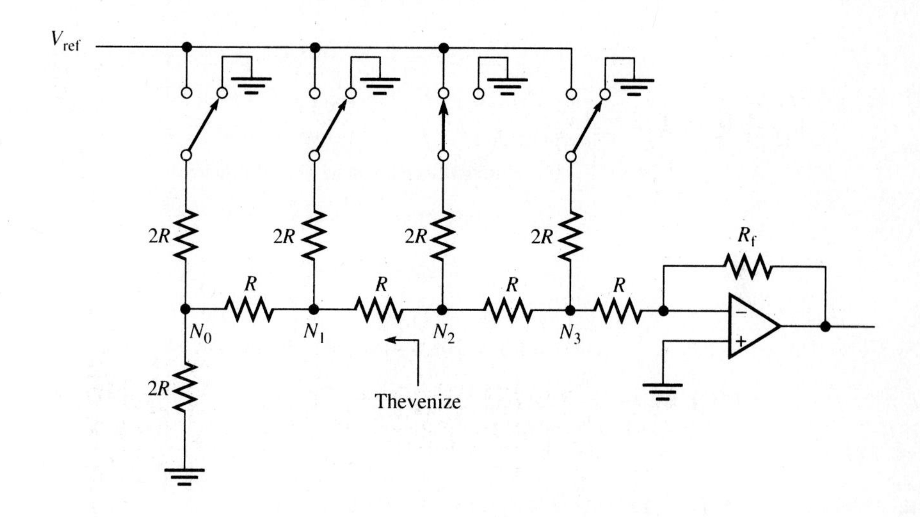

(a) R-$2R$ ladder with a binary input number of 0100. To solve the circuit, we apply Thevenin's theorem twice.

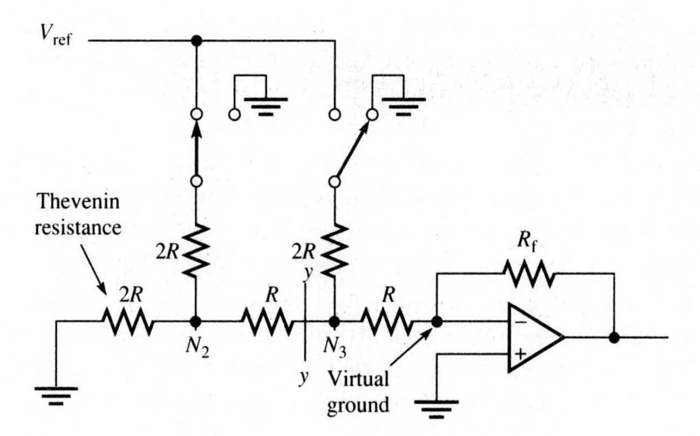

(b) The result of applying Thevenin's theorem to the circuit to the left of N_2 in Figure (a).

FIGURE 6–33
Computing the current due to the second bit in an R-$2R$ ladder. Notice the result is exactly one-half the current due to the MSB.

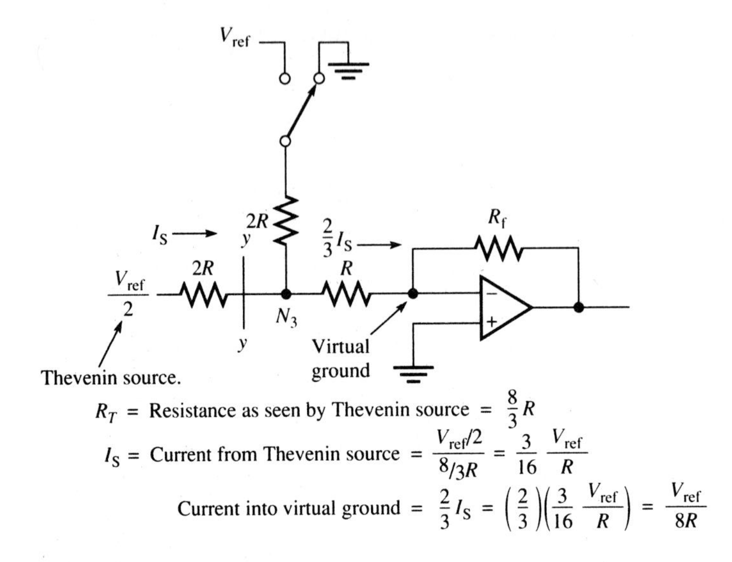

(c) The result of applying Thevenin's theorem to the left of *y-y* in Figure (b).

FIGURE 6–33 (*continued*)

6–7 LOGIC SPECIFICATIONS

The practical implementation of digital logic requires consideration of the voltage and current relationships. As you have seen, there are two major categories of transistors—bipolar junction transistors and field-effect transistors. Within each of these categories, several families of digital integrated circuits have evolved. For bipolar technology, TTL and ECL logic are the most popular, whereas CMOS is the most prevalent family using field-effect transistors. Certain specifications are important for any family and will be reviewed here.

Some of the most important performance specifications for any logic family include the input and output voltage and current specifications for both the HIGH and LOW logic levels, the power dissipation, and speed and loading effects. Input and output voltage and current specifications are important to understanding interfacing between logic families and determining the drive requirements from other circuits. Frequently in instrumentation problems, it is necessary to use digital logic to control an output or to connect a switch or transducer to a digital logic circuit. The specifications of the logic family must be considered by the designer.

VOLTAGE AND CURRENT SPECIFICATIONS

Several input and output voltage levels are defined by manufacturers. The output voltage level of a logic gate is specified to be either above V_{OH} or below V_{OL}, where V_{OH} is the minimum output high-level voltage and V_{OL} is the maximum

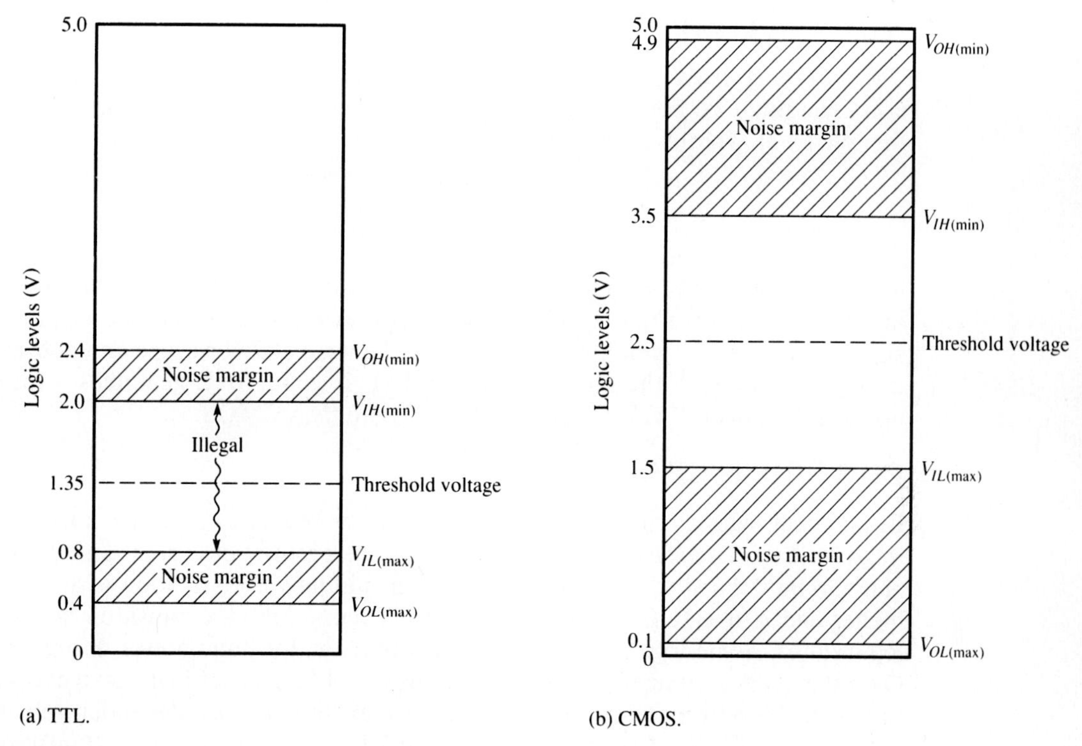

FIGURE 6–34
Logic voltage levels for TTL and CMOS. TTL is always operated from +5.0 V. Logic levels for HCMOS (high-speed) operating from +5.0 V are shown. CMOS can be operated from other voltages.

output low-level voltage. The input levels are given as V_{IH} and V_{IL}, where V_{IH} is the minimum input high-level voltage and V_{IL} is the minimum input low-level voltage. The specifications depend on the technology. Figure 6–34 illustrates the voltage definitions for TTL and CMOS logic families.

In addition to the voltage levels described, manufacturers define current specifications for the input and the output. The direction of current flow is taken with respect to the gate or IC under consideration. Assuming conventional flow of current (plus to minus), current that is entering the IC is shown with a negative sign and current leaving the IC is shown with a positive sign. (The acronym PINNOUT for positive in, negative out may help you remember this.) The input and output current specifications are shown on the manufacturer's data sheet. When one type of logic is driven from another type or is interfaced to discrete components, it is important to keep in mind the current as well as the voltage requirements. For example, standard TTL has a specified maximum input current low (I_{IL}) of −1.6 mA. Although in practice this current is usually less, the designer should assume this current flows from the input in order to pull it low. A current path *must* be provided by the driving circuit or the input will not be in the desired low state.

NOISE MARGIN

The difference between the input and output voltage specifications is called the **noise margin.** Two noise margins are defined—one for the high level and one for the low level. That is

$$V_{NH} = V_{OH(min)} - V_{IH(min)}$$
$$V_{NL} = V_{IL(max)} - V_{OL(max)}$$

The overall noise margin is the smaller of the two noise margins. For TTL logic, the high- and low-level noise margins are both 0.4 V, and so the overall noise margin is also 0.4 V. The noise margin is significant because unwanted signals less than this amplitude cannot alter the logic state or cause an undesired transition. Although the noise margin for TTL is good, TTL generates a lot of noise and its overall noise performance is worse than CMOS.

POWER DISSIPATION

In practical logic circuits the voltage and current are not 0 in either logic state (high or low). The static power dissipation is the product of current and voltage and can be defined for both states. The dynamic power dissipation is generally higher and is due to power dissipation that occurs during a transition between logic states. Some logic, such as TTL, is dominated by static power dissipation. Over the years a number of new generations or subfamilies of logic have evolved in which the power dissipation and other characteristics are dependent on the particular subfamily. For example, standard TTL typically dissipates 10 mW per gate, whereas low-power Schottky (LS) TTL dissipates 2 mW per gate. CMOS, on the other hand, dissipates the majority of its power during transitions. Consequently, the power dissipation exhibits a frequency dependency. The static dissipation can be as low as 1 μW per gate for static dissipation, but the dynamic dissipation can be higher than TTL.

SPEED

The quest for increasing the quantity and speed of computations has led to the search for faster logic circuits. The time for the output to respond to a change in the input signal is called the **propagation delay.** Because the time to turn off a transistor is not the same as the time required to turn it on, the time required for a high-to-low transition is generally not the same as the time required for a low-to-high transition. Therefore, the propagation delay is generally specified for both transitions.

The speed of TTL logic is dependent on the particular subfamily. The advanced low-power Schottky (ALS) TTL subfamily has a typical power dissipation of 1 mW per gate and a typical propagation delay of 4 ns into a 50 pF/2 kΩ load. Advanced Schottky (AS) is even faster, with typical propagation delay times of 1.5 ns when driving the same load. Although CMOS logic has typically been slower than TTL, high-speed CMOS (HCMOS) approaches TTL speeds. HCMOS has a typical propagation delay of 8 ns.

Currently, the fastest logic family is emitter-coupled logic (ECL), with several types available. The fastest has propagation delays of less than 1 ns.

SPEED/POWER PRODUCT

Generally, if we increase the resistance in a circuit, we decrease the power dissipation and slow the switching speed. Thus a compromise between power dissipation and switching speed is necessary. The **speed/power product** is a performance measurement that is the product of the power dissipation times the propagation delay. It has units of joules, a measure of energy. The smaller the speed/power product, the better. For example the speed power/product of an ALS TTL gate with a power dissipation of 1 mW and a propagation delay of 4 ns is 4 pJ.

SUMMARY

1. Boolean algebra is applied to logic and uses three basic logic operations: AND, OR, and NOT. Boolean algebra obeys the commutative law, associative law, and distributive law.
2. DeMorgan's theorem states

$$\overline{A + B} = \overline{A} \cdot \overline{B}$$

and

$$\overline{A \cdot B} = \overline{A} + \overline{B}$$

3. Assertion-level logic is used to indicate the appropriate logic level that causes a logic action. Variable names are generally written with an overbar to indicate that a LOW level causes the action and no overbar to indicate that a HIGH level causes the action.
4. The Karnaugh map is a means of simplifying logic expressions. It is a form of truth table drawn so that adjacent boxes (or cells) differ from each other in only one variable.
5. A multiplexer can be used to implement a truth table of a combinational logic problem by connecting the input variables to the select inputs. Each input of the MUX is connected to the logic level, as determined by the truth table for that set of inputs.
6. Programmable logic devices (PLDs) offer a convenient way to implement logic by allowing the user to program a specific application into a general-purpose IC.
7. With sequential logic circuits, the output depends on the present inputs and on the contents of memory. Examples of sequential circuits include counters, shift registers, state machines, and computers.
8. Two important types of flip-flops are the D and JK types. The edge-triggered D flip-flop transfers the data on the D input to the Q output only at the clock edge, or *transition*. The JK flip-flop has two additional modes—latch and toggle mode—making it more versatile than the D flip-flop.
9. Counters are composed of a series of flip-flops that hold the current output state until they are clocked. Asynchronous counters consist of a series of flip-flops placed in the toggle mode and clocked one after the other. Synchronous counters change together and can be made faster than asynchronous counters.
10. Shift registers consist of a series of flip-flops connected in a chain. They are widely used when data must be converted from serial to parallel form or parallel to serial form.
11. A state machine is a sequential logic circuit in which the output is related to the inputs and the information stored in the memory.
12. A/D conversion is the process of converting an analog signal to a digital signal. The input signal is sampled; the sampled quantity is then converted (quantized) into a discrete, numeric value.

13. D/A conversion is the process of converting a digital signal to an analog signal. D/A conversion is useful for control applications.
14. The output voltage levels of logic are specified as V_{OH} and V_{OL}. The input levels are given as V_{IH} and V_{IL}. Noise margin is specified for both the high and low levels and represents the difference between the input and output voltage specifications.

QUESTIONS AND PROBLEMS

1. Convert the following binary numbers into decimal numbers.
(a) 0010_2 (b) 0110_2 (c) 1100_2
(d) 1110_2 (e) 11100011_2
(f) 10000100_2
2. Convert the following decimal numbers into binary numbers using reverse division.
(a) 15 (b) 33 (c) 75
(d) 109 (e) 200 (e) 256
3. Using Boolean algebra, simplify each of the following expressions.
(a) $A(AB + C)$ (b) $A(\overline{A} + B)$
(c) $(A + B)(A + C)$
(d) $ABC(A\overline{B} + BC)$
(e) $AB + \overline{A}C + A\overline{B}C$
4. Apply DeMorgan's theorem to each of the following expressions.
(a) $\overline{(A + \overline{B} + C)}$ (b) $\overline{A + \overline{BC}}$
(c) $\overline{(A + B) \cdot (\overline{C} + D)}$
(d) $\overline{A\overline{C}\,\overline{(A + B)} + D}$
(e) $\overline{\overline{(ABC)} + \overline{D} + EFG}$
5. Read the minimum sum of products for each of the Karnaugh maps shown.

(a)

AB \ CD	00	01	11	10
00			1	1
01	1			
11	1			
10		1	1	1

(b)

AB \ CD	00	01	11	10
00	1			1
01				
11	1	1	1	1
10		1		

6. Using the Karnaugh map method, find the Boolean expression for the truth table.

A	B	C	X
0	0	0	0
0	0	1	0
0	1	0	1
0	1	1	0
1	0	0	0
1	0	1	1
1	1	0	1
1	1	1	1

7. Write the reduced Boolean expression for each circuit in Figure 6–35.

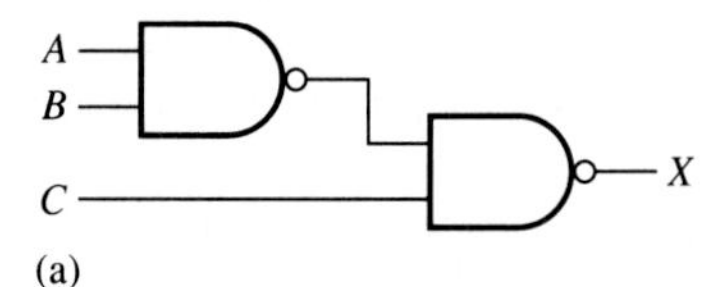

(a)

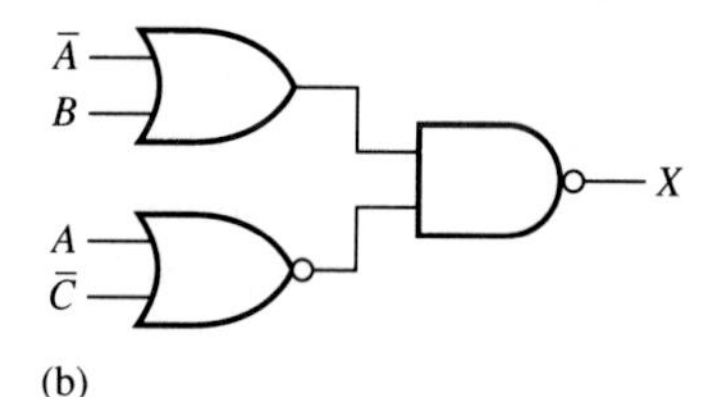

(b)

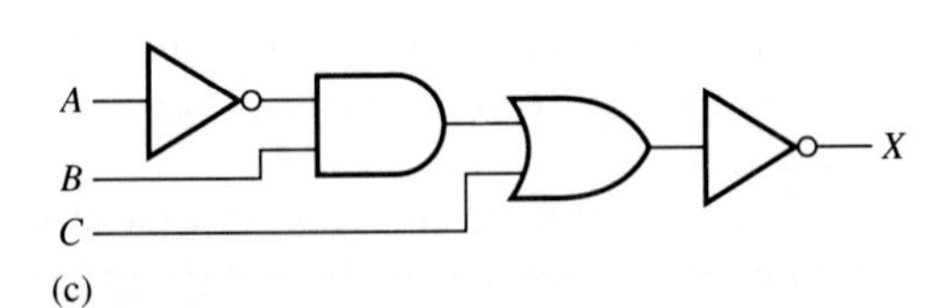

(c)

FIGURE 6–35

8. Show how to connect the MUX to implement the truth table shown in Figure 6–36.
9. Explain the difference between a latch and a flip-flop.

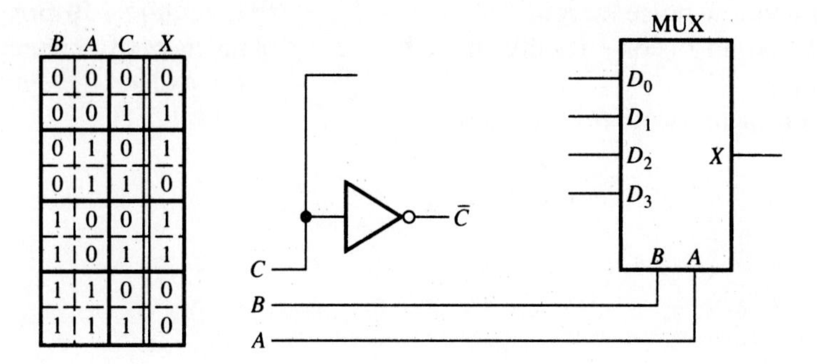

B	A	C	X
0	0	0	0
0	0	1	1
0	1	0	1
0	1	1	0
1	0	0	1
1	0	1	1
1	1	0	0
1	1	1	0

FIGURE 6–36

10. What are the advantages of a *JK* flip-flop over a *D* flip-flop?
11. What is the difference between a synchronous input and an asynchronous input on a *JK* flip-flop?
12. What is meant by a transition table?
13. Determine the count sequence for the counter shown in Figure 6–37 and draw a state diagram. Include the unused states in the diagram. (Q_C = MSB)

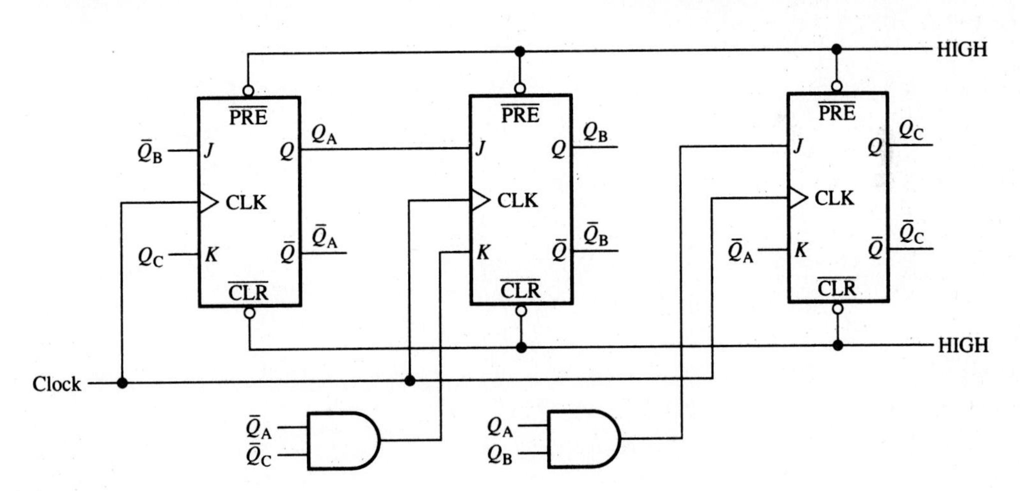

FIGURE 6–37

14. Complete the state diagram shown in Figure 6–14, Step 4, by showing the unused states.
15. Draw the state diagram for a 5 bit twisted-ring counter that starts in state 0.
16. What is the percent resolution of a 12 bit A/D converter?
17. Explain why a tracking A/D converter exhibits oscillatory behavior for the LSB.
18. What is the advantage of a successive-approximation A/D converter over a tracking A/D converter?

19. A hypothetical logic family has the following specifications: $V_{OH} = 3.0$ V, $V_{OL} = 0.5$ V, $V_{IH} = 2.7$ V, $V_{IL} = 1.0$ V. Compute the high and low noise margin.
20. Standard TTL has an I_{OL} of -16 mA and an I_{OH} of 400 μA.
   (a) What is the meaning of the negative sign with respect to the I_{OL} specification?
   (b) Assume you want to drive an LED that requires 10 mA of current. Explain how to connect the LED so as not to exceed the output specifications for the IC.

Section Three

BASIC INSTRUMENTS

Chapter 7

Meters and Bridges

OBJECTIVES

The volt-ohm-milliammeter (VOM) was based on an ammeter designed over 100 years ago by Jacques D'Arsonval, a French physicist. For many years, the portable VOM was the basic instrument for measuring voltage, current, or resistance. There are applications where the analog VOM is superior to newer digital meters, and analog meters are still used as an integral part of many instruments; however, because of many advances in digital electronics, portable digital meters have become more popular for general use than VOMs. Digital multimeters (DMMs) can be made that have greater accuracy, lower cost, and more options than are available on VOMs. In addition, digital meters are not subject to the subjective reading errors common to all analog meters. This chapter covers both analog and digital meters, beginning with a discussion of analog meters. The chapter concludes with impedance measurements, including dc and ac bridge measurements.

When you complete this chapter, you should be able to

1. Explain the construction and operation of a D'Arsonval meter movement.
2. Describe how to measure the internal resistance of a permanent-magnet moving-coil (PMMC) meter using the full-scale deflection method.
3. Compute the shunt resistance needed to make a PMMC meter measure larger currents than its basic sensitivity and the series resistance needed to make a PMMC meter function as a voltmeter.
4. Explain how a PMMC meter can be used as a series ohmmeter.
5. List normal operating steps for using an ammeter, voltmeter, or ohmmeter, including precautions to guard against damage to the meter.
6. Explain how to use a VOM for basic measurements (voltage, current, resistance) as well as decibel measurements and capacitor checks.
7. List features of a DMM and describe how to make basic measurements with DMMs (voltage, current, resistance) as well as diode, bipolar transistor, and SCR checks.
8. Describe key specifications for DMMs, including resolution, accuracy, response time, and common-mode rejection.
9. Compare the advantages and disadvantages of a VOM with a DMM.
10. Describe the key specifications and measurement capabilities of the electrometer, picoammeter, and nanovoltmeter.
11. Discuss power measurement for both low and high frequencies.
12. Describe instruments used for impedance measurements, including the vector impedance meter, component analyzer, and both dc and ac bridges.

HISTORICAL NOTE

Digital meters first appeared on the market in 1958. The first one was a digital voltmeter (DVM) with a three-digit display (999 full scale). Accuracy was quoted as 0.25% of full scale with little knowledge of long-term stability. Today, DVMs are available that have 8½ digits in the display and accuracy specifications of a few parts per million.

In the same year, a new type of current meter was announced by Hewlett-Packard. It was a clip-on milliammeter that sensed the strength of the magnetic field produced by the current being measured instead of using standard moving-coil methods. It avoided the need to break the circuit to measure current and introduced extremely small loading into the circuit. Clip-on current meters still provide an effective way to measure current when it is not practical to break the circuit.

7–1 PMMC METERS

THE D'ARSONVAL METER MOVEMENT

Analog meters indicate the quantity to be measured by a pointer and scale that is interpreted by the user. The **D'Arsonval ammeter** uses a simple electromechanical movement to indicate the current. The most popular type of analog meter is a permanent-magnet moving-coil (PMMC) type that responds to direct current and moves a pointer against a calibrated scale by an amount proportional to the current in the meter. PMMC meters respond directly to current but can easily be designed to read voltage, resistance, or some other quantity. The common-panel meter shown partially disassembled in Figure 7–1(a) is a PMMC meter.

The D'Arsonval-type meter is a PMMC meter that consists of a moving coil suspended between the poles of a permanent magnet, as shown in Figure 7–1(b). Current in the coil induces a magnetic field that interacts with the magnetic field of the permanent magnet. The interaction of the fields produces a force proportional to the current, which rotates the coil. In turn, this force is opposed by spiral springs mounted on each end of the coil. A pointer is attached to the same axis that holds the coil. The position of the pointer along the calibrated scale is interpreted by the user to measure the current.

The sensitivity of a PMMC meter is measured by the full-scale deflection current I_{fs}. Laboratory meters can be made highly sensitive (as little as 1 μA full-scale deflection) but are not generally portable. They usually have a mirrored scale to avoid **parallax,** a reading error that occurs when the observer's eye is not exactly in line with the pointer and scale (see Figure 7–2). Industrial meters, on the other hand, are highly ruggedized and are designed to be portable. In general, the sensitivity of D'Arsonval meters is limited by the size of the coil and its resistance. A meter with a sensitivity of 10 μA is quite expensive; sensitivities of 50 μA, 100 μA, 200 μA, and 1 mA are common sizes of basic movements.

There are other types of PMMC meters available. For example, the taut-band meter is a popular variation of the D'Arsonval movement in which a tightly stretched band supports the coil and eliminates the need for pivots, bearings, and

FIGURE 7–1

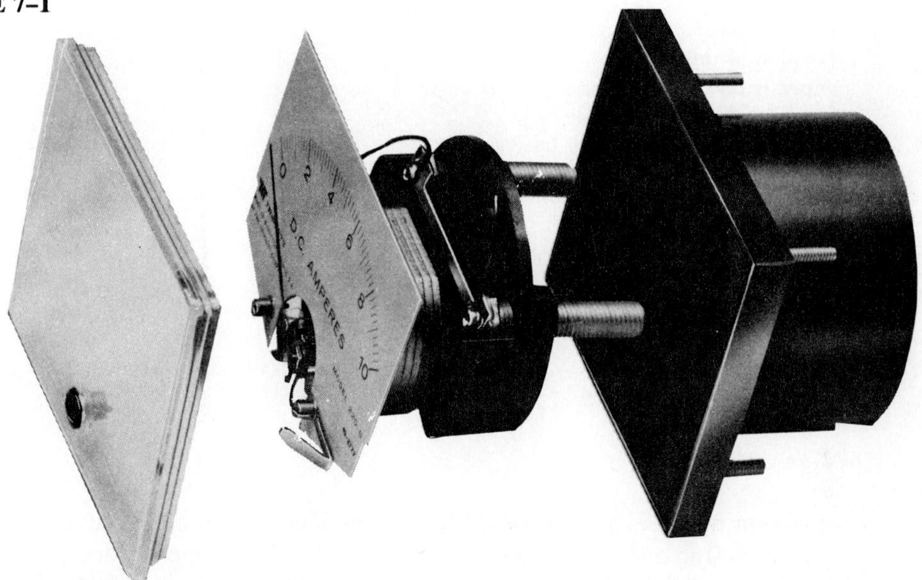

(a) Disassembled panel meter (courtesy of Triplett Corporation)

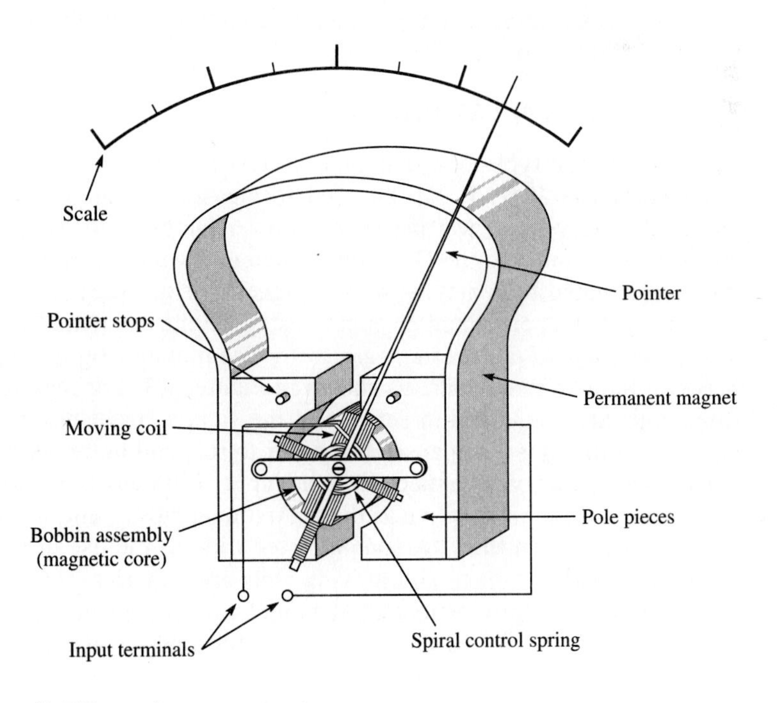

(b) D'Arsonval meter movement.

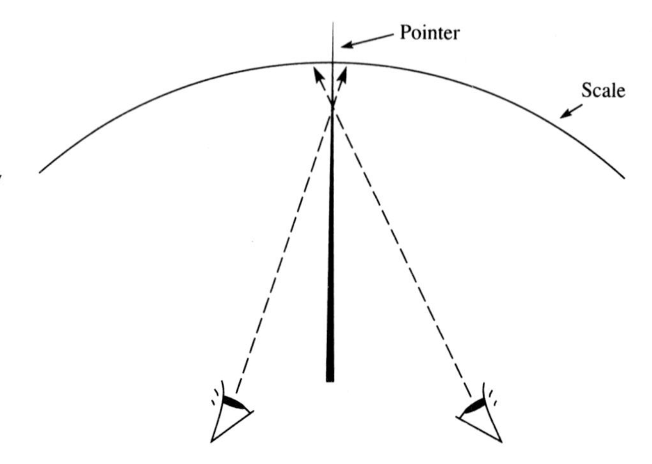

FIGURE 7–2
Parallax error. The apparent position of the pointer depends on the observer's location. A mirrored scale can prevent parallax error by allowing the observer to align the pointer and its reflection.

spiral springs. It is popular because the simplified construction reduces the cost of the meter and allows it to be more rugged. Taut-band meters are available with sensitivities of 2 μA, a significant improvement over normal D'Arsonval meter movements.

All moving-coil instruments are sensitive to temperature variations. An increase in temperature causes the strength of the magnetic field to decrease and the coil resistance to increase, with a net effect of causing the meter to read low. Various compensating techniques can be used to improve the temperature performance of PMMC meters.

ELECTRODYNAMOMETER

One of the oldest types of moving-coil meters still in use is the **electrodynamometer,** invented in 1842. It can be used to measure dc or ac current and can be modified to serve as a wattmeter or a power factor meter (see Section 7–9). Its movement is similar to the D'Arsonval meter movement, using springs to balance the force created by a moving coil's magnetic field, except that it uses two fixed coils instead of a permanent magnet to establish the magnetic field. The fixed coils are called the *field coils,* or *current coils*. The moving coil is placed in the region between the two fixed coils, as shown in Figure 7–3. For measuring current, the three coils are connected in series and the same current flows through all three coils. Note that this also causes the meter to respond in the upscale direction, no matter what polarity is connected to the input. Although the dynamometer movement is not nearly as sensitive as the D'Arsonval movement, it can measure much larger currents than the D'Arsonval meter without the use of a shunt.

Since both the fixed and moving coils are in series, the magnetic flux produced from each is proportional to the input current, causing the movement of the pointer to respond to I^2. The meter's inertia tends to dampen the effect of ac variations, causing it to respond to the average of the current squared. By making an appropriate scale, the meter can be calibrated to read the rms value of a current

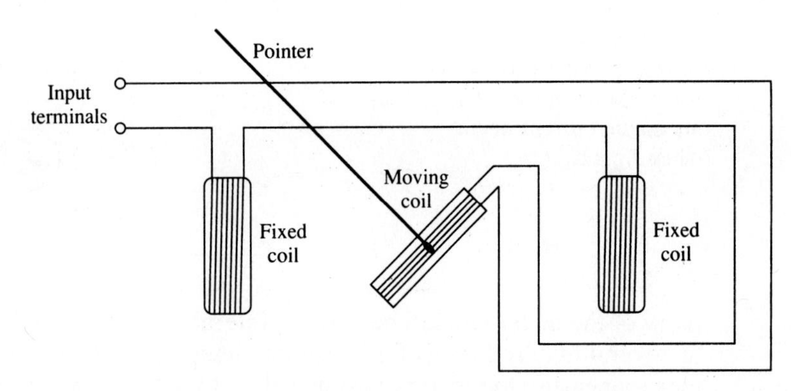

FIGURE 7–3 Electrodynamometer movement.

directly. This is another important advantage of the meter, as it can be used in either dc or low-frequency ac circuits without modification.

AMMETERS

The D'Arsonval meter is sensitive to a relatively small current—usually ranging from 50 μA to 1 mA—as are other PMMC meters. It is a relatively simple matter to modify the basic meter to measure larger currents. This is accomplished by providing a parallel path through a shunt resistance, R_{sh}, allowing a fraction of the current to bypass the meter, as shown in Figure 7–4. The meter has an internal resistance of R_m and a meter current of I_m. The total current is split into a fraction that goes through the internal resistance of the meter and another fraction that goes through the shunt resistance. At maximum current, we can apply Kirchhoff's current law to the parallel junction of R_{sh} and R_m

$$I_T = I_{fs} + I_{sh}$$

and

EQUATION 7–1

$$I_{sh} = I_T - I_{fs}$$

where I_T = total current entering ammeter, A
I_{fs} = full-scale meter current, A
I_{sh} = shunt current, A

To determine the shunt resistance required, note that the voltage across the shunt is equal to the voltage across the meter. That is

$$I_{sh}R_{sh} = I_{fs}R_m$$

Substituting this information into Equation 7–1 and rearranging gives

EQUATION 7–2

$$R_{sh} = \frac{I_{fs}R_m}{I_T - I_{fs}}$$

Equation 7–2 gives the value needed for a shunt resistance that will decrease the basic sensitivity of an ammeter. The meter resistance and basic current sensi-

FIGURE 7–4
Adding a shunt resistor to a basic PMMC meter. The maximum current occurs when $I_m = I_{fs}$.

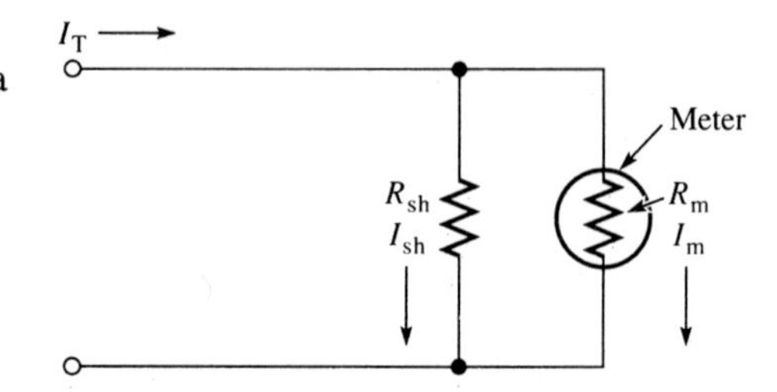

tivity of the meter need to be known. Internal resistances range from a few ohms to several hundred ohms for sensitive meters. For such meters, it is not a good idea to measure the meter resistance directly with an ohmmeter as the ohmmeter can put out enough current to damage or uncalibrate the meter under test. A better way to determine the meter resistance is to use a meter calibrator, an instrument that places a controlled and calibrated current into the meter under test. When the current is adjusted for full-scale deflection with a known current, the voltage across the meter terminals is measured, and the resistance is determined by applying Ohm's law.

If a meter calibrator is not available, a method known as the full-scale deflection method can be used to determine the meter resistance. The setup is illustrated in Figure 7–5(a). Resistor R_A is a large-series resistor that limits the current to the meter. (The value is not critical but depends on the sensitivity of the meter. For a 100 μA meter, a value of 47 kΩ or so is fine.) The power supply is increased slowly until the meter is reading full scale. The voltage setting of the supply is noted. Resistor R_B, a potentiometer, is then placed across the meter, as shown in Figure 7–5(b), and adjusted until the current in the meter is less than one-half of full-scale deflection. Now the voltage on the power supply is exactly doubled. This causes the current in the meter to go nearly to full-scale deflection. R_B is then adjusted until the meter reads exactly full scale. Now the resistance of the meter and the resistance of R_B are the same. The power supply is turned off, and R_B is measured to determine the internal resistance of the meter.

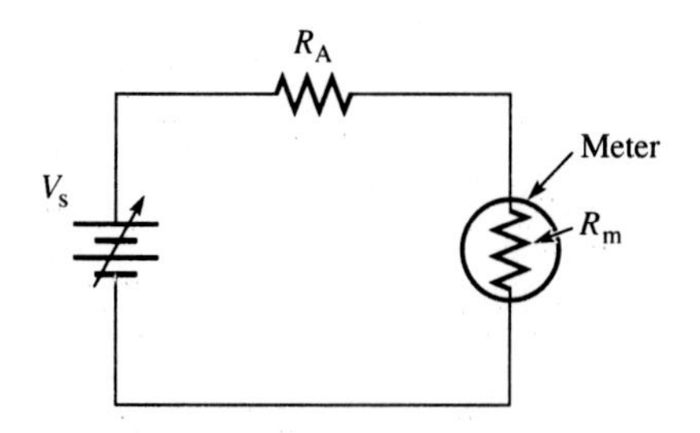

(a) V_s is raised until the meter reads full-scale current.

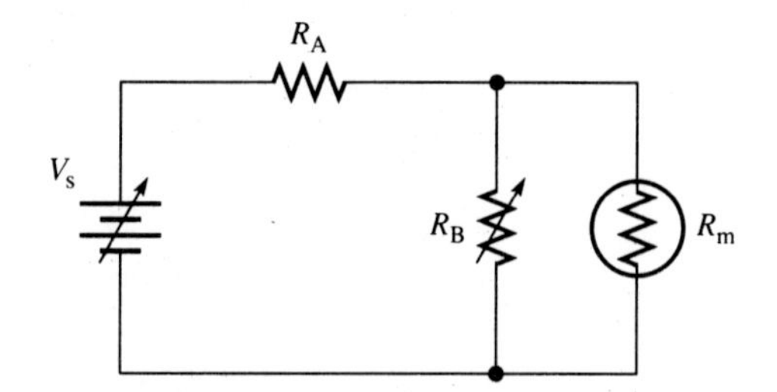

(b) R_B is inserted and adjusted for less than half-scale current. V_s is then doubled. R_B is readjusted for full-scale current. At this point, $R_B = R_m$.

FIGURE 7–5
Determining the internal meter resistance by the full-scale deflection method.

EXAMPLE 7–1 Assume a PMMC meter has a full-scale deflection current of 100 μA and an internal resistance of 200 Ω.

(a) What shunt resistance is needed to make the meter read 1.0 mA at full scale?

(b) What shunt resistance is needed to make the meter read 100 mA at full scale?

SOLUTION (a)

$$R_{sh} = \frac{I_{fs}R_m}{I_T - I_{fs}}$$

$$= \frac{(100\ \mu A)(200\ \Omega)}{1000\ \mu A - 100\ \mu A} = 22.2\ \Omega$$

(b)

$$R_{sh} = \frac{(100\ \mu A)(200\ \Omega)}{100{,}000\ \mu A - 100\ \mu A} = 0.200\ \Omega$$

Notice that the shunt resistance can be a very small resistor for large current ranges. Frequently, a resistance such as this is made from a length of wire cut to the exact length required.

For a multiple-range ammeter, the switching arrangement shown in Figure 7–6(a) can be used. Although this arrangement places the proper shunt resistor in the circuit for each range, there is a problem. The problem occurs when the rotary switch is between ranges. It is possible to have an open circuit around the meter as the switch is moved unless an expensive make-before-break switch is used. An open circuit would cause all current to go through the meter, possibly destroying it.

A method that avoids the problem of the make-before-break switch is the Ayrton shunt arrangement illustrated in Figure 7–6(b). With the Ayrton shunt, there is always a resistance path around the meter, allowing a less expensive rotary switch to be used. This is the switching arrangement used in most portable meters.

USING AMMETERS

Ammeters are always connected in series with the circuit under test, never in parallel. This means that the circuit path must be broken in order to insert an ammeter. The polarity of the meter must be taken into account to avoid connecting it backward. If the needle is deflected backward, it can bend on the stop, causing the meter to read incorrectly. With multiple-range meters, select the highest range to begin with and decrease the range as needed to obtain a reasonable deflection. The greatest accuracy occurs when the meter is reading close to full-scale deflection.

Any ammeter should have much less resistance than the circuit under test, or else the circuit will be affected by the insertion of the meter. The insertion of an ammeter adds resistance to the circuit under test and causes less current to flow than would have had the ammeter not been added. As long as the resistance of the circuit is much higher than the ammeter resistance, loading should not create a

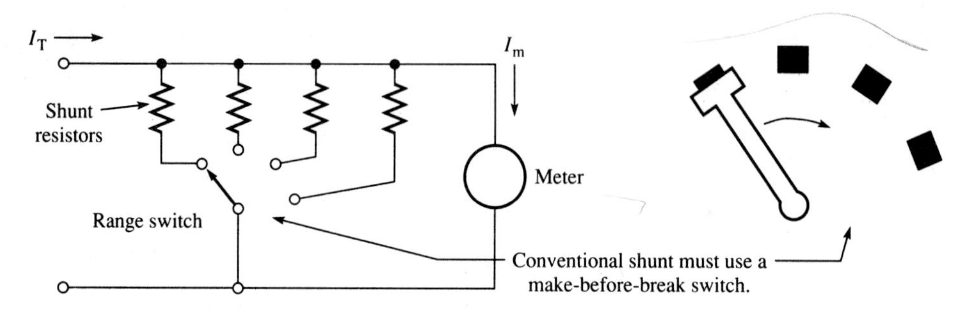

(a) Conventional shunt for an ammeter.

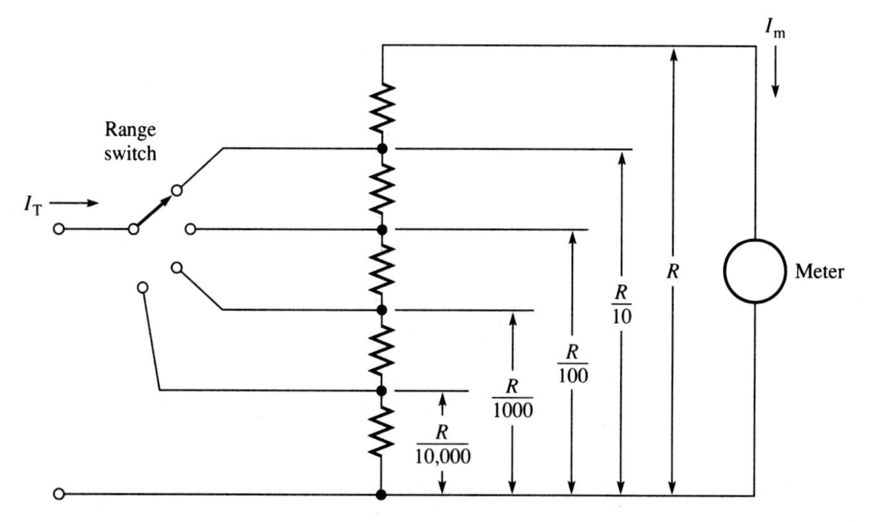

(b) Ayrton (universal) shunt for an ammeter. The Ayrton shunt prevents the meter from being connected without a shunt resistor.

FIGURE 7–6
Conventional and Ayrton shunts for ammeters.

serious problem, but you should always be aware of the loading effect of any instrument on the circuit under test. For example, if the meter resistance is 1% of the circuit resistance, a 1% reduction in current will occur. This loading error is in addition to the error caused by reading and calibration of the meter itself.

7–2 ANALOG VOLTMETERS

dc VOLTMETERS

A basic PMMC meter can be converted to a dc voltmeter by the addition of a series multiplier resistor, R_s, as illustrated in Figure 7–7. The size of the resis-

FIGURE 7–7
Converting a basic PMMC meter into a voltmeter.

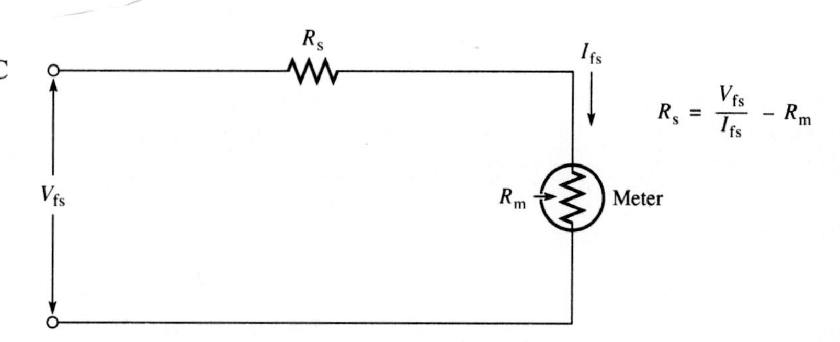

tance needed can easily be computed from Ohm's law and Kirchhoff's voltage law. The maximum voltage to be measured is assumed to be at the full-scale current of the meter. From Kirchhoff's voltage law

$$V_{fs} = I_{fs}(R_m + R_s)$$

Solving for R_s yields

EQUATION 7–3

$$R_s = \frac{V_{fs}}{I_{fs}} - R_m$$

where V_{fs} = voltage reading at full scale, V
R_s = series resistance required to convert the meter into a voltmeter, Ω

EXAMPLE 7–2 Compute the series resistance required to convert an ammeter with a full-scale current of 250 μA and an internal resistance of 250 Ω into a voltmeter that reads 10 V full scale.

SOLUTION

$$R_s = \frac{V_{fs}}{I_{fs}} - R_m$$

$$= \frac{10\text{ V}}{250\ \mu\text{A}} - 250\ \Omega = 39{,}750\ \Omega$$

For multiple-range voltmeters, the series resistors can be selected by the arrangement illustrated in Figure 7–8(a). In this arrangement, the total resistance is the sum of the selected resistor and the meter resistance. The value of each multiplier resistor is independent of the other resistors and depends only on the voltage ranges designed for the meter. The resistors can easily be specified by independently applying Equation 7–3 for each resistor.

A more practical arrangement of the multiplier resistors is the modified multiplier shown in Figure 7–8(b). Resistor R_D is always in the circuit, and subse-

FIGURE 7–8
Two methods for connecting multiplier resistors on a multirange voltmeter.

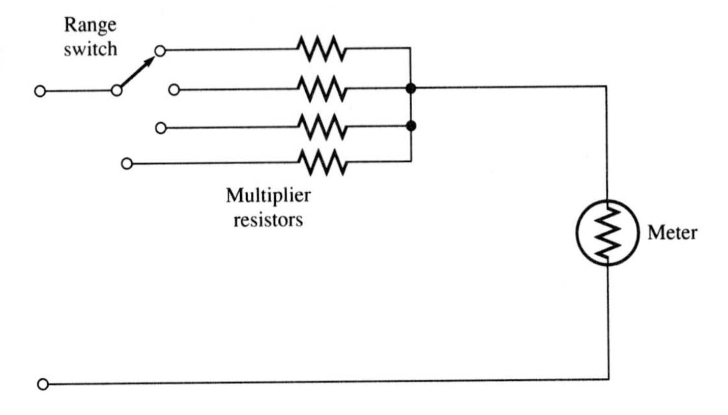

(a) Conventional multiplier arrangement.

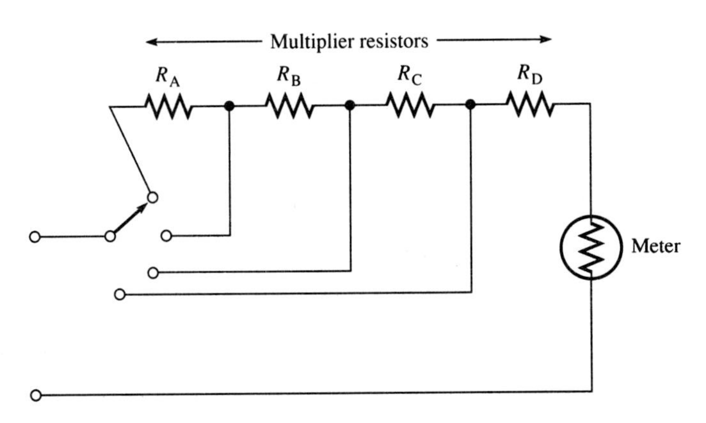

(b) Modified multiplier arrangement.

quent resistors are added by the switching arrangement. R_D is specially selected for the particular meter resistance and desired range, but all other resistors can be standard values.

USING VOLTMETERS

Voltmeters are always connected in parallel with the circuit under test. A series connection will not harm the meter, but for measuring voltage such a connection is not meaningful. As in the case of ammeters, the polarity of the meter must be taken into account to avoid connecting it backward. Start by selecting a range that you are sure is larger than the possible voltage to be measured *before* connecting the meter; decrease the range as needed to obtain a reasonable deflection. The greatest accuracy occurs when the meter is reading close to full-scale deflection.

A voltmeter can load a circuit under test. It should have much greater resistance than the circuit under test; otherwise a parallel path for current is introduced through the meter that can lower the measured voltage. This is a problem when measuring in high-impedance circuits. One common measure of a voltmeter's loading effect is given as the ohms-per-volt rating of the meter, also called the *sensitivity* of the voltmeter. The ohms-per-volt rating is simply the reciprocal of the full-scale current. That is

EQUATION 7–4

$$S = \frac{1}{I_{fs}}$$

where S = sensitivity, Ω/V
I_{fs} = full-scale meter current, A

The advantage of listing the sensitivity of a voltmeter as an ohms-per-volt rating is that the user can quickly determine the meter's equivalent resistance across a circuit simply by multiplying the sensitivity by the selected full-scale voltage range of the meter. The easiest way to find the loading error is to compute the equivalent Thevenin resistance of the circuit under test and solve for the voltage of the Thevenin circuit with the meter in place. This approach is illustrated in Figure 7–9. Note that it is the Thevenin voltage that the user is attempting to measure at the output terminals of the Thevenin circuit. Ideally, this is done with a meter that has infinite internal resistance and does not load the Thevenin circuit. In practice, any meter places some load on the equivalent Thevenin circuit; using the voltage-divider rule, the loading error is easy to calculate. If the internal meter resistance is very large compared to the Thevenin resistance, the loading error is small and is usually ignored.

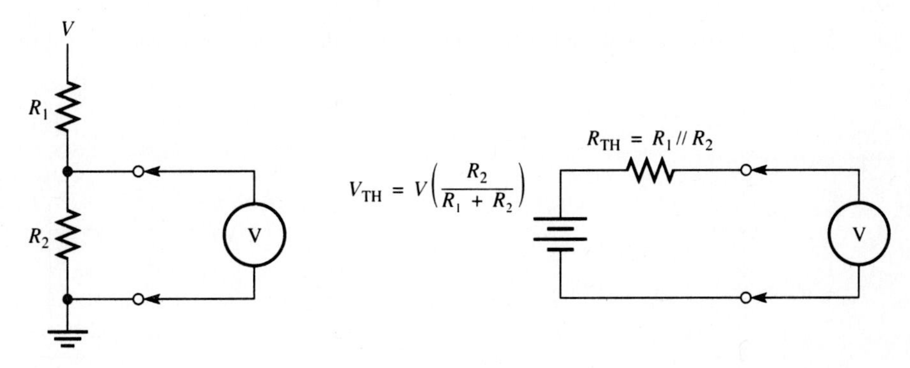

(a) Test circuit.

(b) Thevenin equivalent circuit. The total resistance is the sum of the Thevenin resistance and the internal meter resistance.

FIGURE 7–9
Computing loading effect by applying Thevenin's theorem.

EXAMPLE 7–3 (a) Compute the sensitivity of the meter in Example 7–2.
(b) Compute the loading error when using this meter to measure V_{AB} in Figure 7–10.

FIGURE 7–10

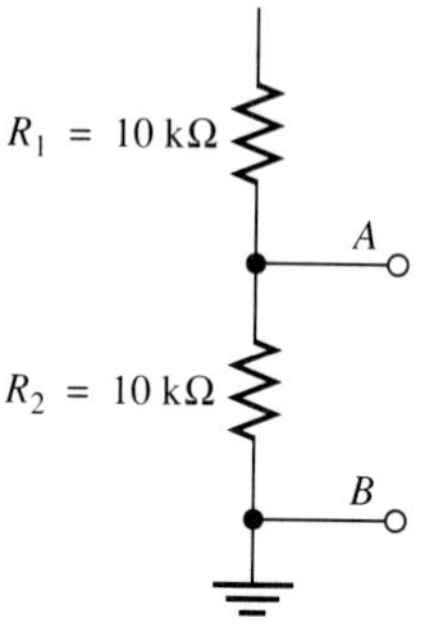

+20 V
$R_1 = 10\ \text{k}\Omega$
A
$R_2 = 10\ \text{k}\Omega$
B

SOLUTION (a) The full-scale current was given as 250 μA. Therefore, the sensitivity is

$$S = \frac{1}{I_{fs}} = \frac{1}{250\ \mu\text{A}} = 4\ \text{k}\Omega/\text{V}$$

The same result can be obtained by dividing the total resistance of the meter (40 kΩ) by the full-scale voltage (10 V).
(b) This is a voltage-divider circuit with a computed voltage V_{AB} equal to the Thevenin voltage of 10 V, as shown in Figure 7–11(a). The meter appears as a 40 kΩ resistance in series with the Thevenin circuit, as illustrated in Figure 7–11(b). The measured voltage can be found by applying the voltage-divider rule:

$$V_{meas} = 10\ \text{V}\left(\frac{40\ \text{k}\Omega}{45\ \text{k}\Omega}\right) = 8.89\ \text{V}$$

The meter has significantly loaded the circuit. The error from the computed value is 1.11 V, resulting in a loading error of 11.1%. Remember that this error is only the loading error and does not include either the calibration or reading error of the meter.

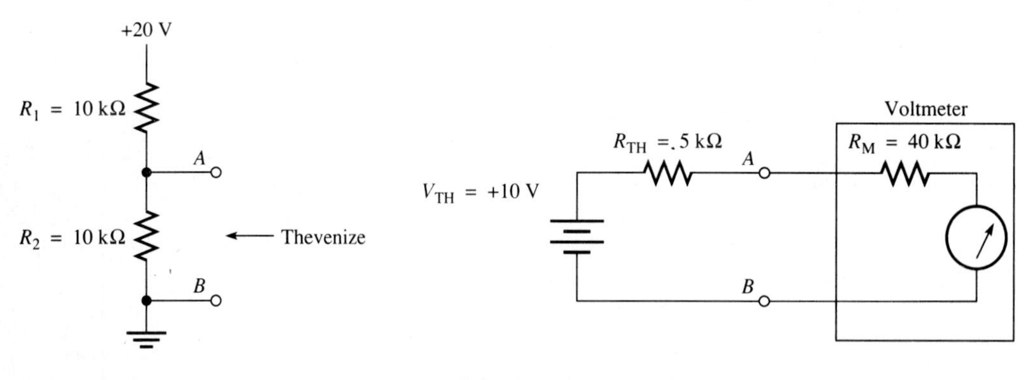

(a) Test circuit.

(b) Thevenin equivalent circuit.

FIGURE 7–11

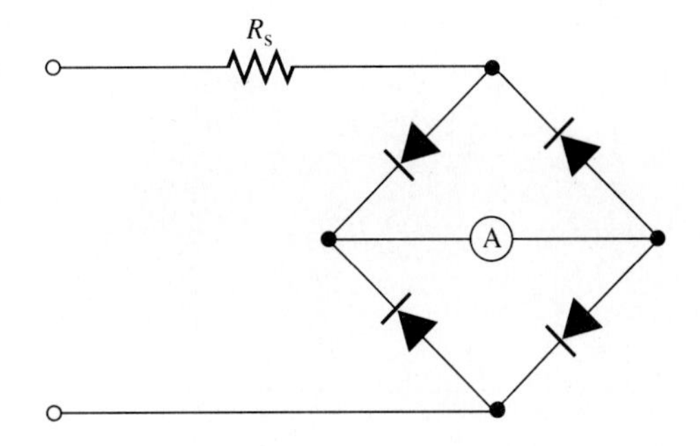

FIGURE 7–12
Full-wave bridge circuit for detecting ac voltage.

ac VOLTMETERS

The D'Arsonval meter responds to the average current passing through the coil. Recall that the average of an ac waveform is 0 V. In order for a D'Arsonval meter to be used as an ac meter, the waveform must first be rectified. Rectification is typically done with a full-wave bridge, as shown in Figure 7–12. The meter will then respond to the average value of the rectified waveform.

Although most meters respond to the average of a waveform, the scale is usually calibrated to read the rms value. This means that the meter will read correctly only for sinusoidal waves. Some meters are available that do respond to the true rms value of a waveform without the necessity of modifying the scale. These meters are specified by the manufacturer as *true rms–reading meters*. They can read the rms value correctly for any waveform within the frequency, voltage range, and crest factor requirements. (**Crest factor** is the ratio of the peak value of a waveform to its rms value.)

As with dc voltmeters, ac voltmeters load the circuit under test. The loading effects for a given meter are more severe than with dc meters, so you need to be aware of the meter sensitivity and the resistance of the circuit under test. Another factor to consider is the frequency response of the meter. Most ac voltmeters do not have a particularly high frequency response.

7–3 ANALOG OHMMETERS

Analog meters can be designed to measure resistance by allowing the unknown resistance to complete a series circuit and calibrating the ammeter scale to read the resistance. A simple series-type ohmmeter circuit is illustrated in Figure 7–13(a). For the series ohmmeter, the current in the meter is inversely proportional to the total resistance of the circuit, including the series-limiting resistance, the meter resistance, and the resistance being measured. An improved series-type ohmmeter is shown in Figure 7–13(b). The switch and parallel shunt resistors extend the usable range of the meter.

The zero-adjust control is necessary to compensate for the changing voltage from the internal battery as it ages. It is adjusted to read 0 Ω when the ohmmeter leads are shorted together. This corresponds to full-scale current in the meter; the

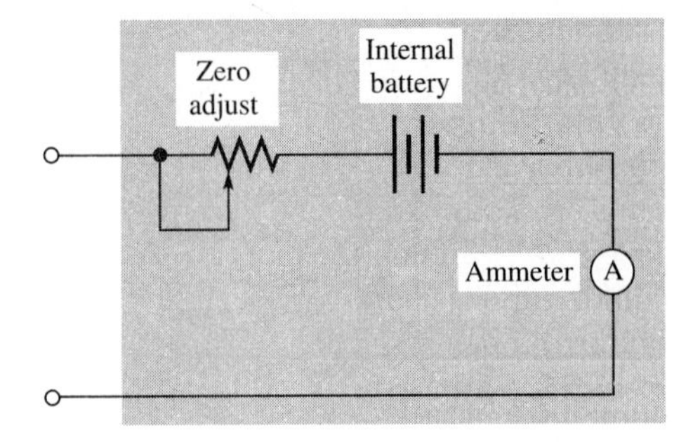

(a) Basic series ohmmeter circuit.

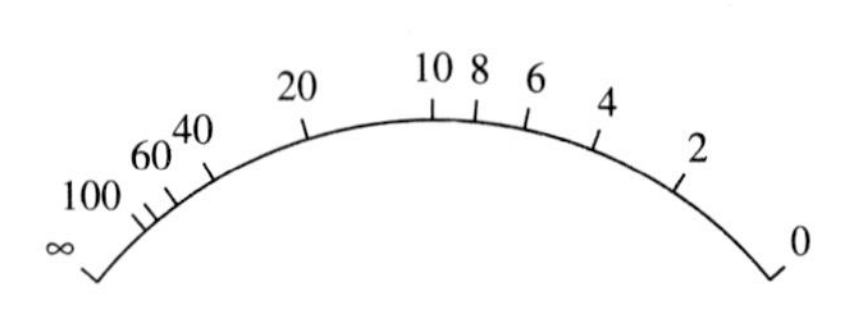

(c) Series ohmmeter scale.

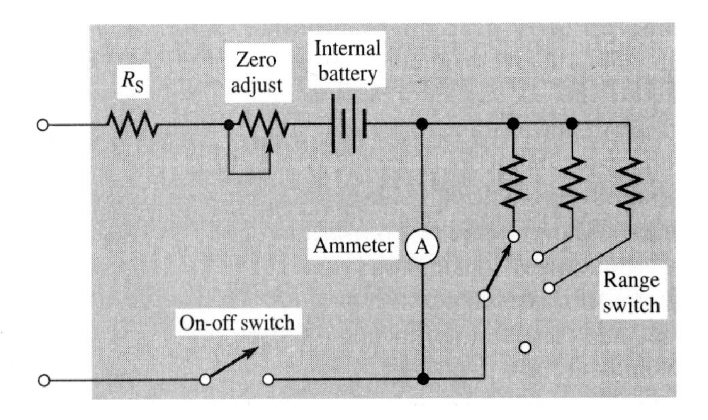

(b) Improved series ohmmeter circuit.

FIGURE 7–13
Ohmmeter circuits.

0 Ω position is therefore marked on the right hand-side of the scale. The far left side of the scale represents an infinite resistance, since no current flows with an open. The half-scale current occurs when the unknown resistance is equal to the total internal resistance of the meter. The deflection of the meter can be expressed as the ratio of the internal meter resistance to total resistance of unknown and internal resistance. That is

EQUATION 7–5

$$D = \frac{R_o}{R_o + R_{uk}}$$

where D = fraction of full scale meter deflection
R_o = total internal resistance of ohmmeter, Ω
R_{uk} = unknown resistance, Ω

The scale for a series-type of ohmmeter is a nonlinear, reverse scale (larger resistances to the left side), as shown in Figure 7–13(c). This scale was drawn assuming that R_0 is 10 kΩ. Notice how the larger values are crowded at the left side of the scale. To make an ohmmeter meter sensitive for reading smaller value resistors, a small-value parallel shunt resistor can be switched around the meter to provide more current to the resistor being measured. This reduces R_0 and increases the deflection for a given unknown resistance. For higher-resistance measurements, a larger battery voltage can be selected.

The series-type ohmmeter is by far the most popular type of analog ohmmeter. Another type of ohmmeter is the shunt type shown in Figure 7–14. The unknown resistance is connected in parallel with the meter. Parallel shunt resistors are selected with the range switch to control the current through the meter. When the resistance being measured is 0, all current flows through the resistor and none flows through the meter. When no resistance is connected to the meter (leads open), all the current flows through the meter and shunt resistors. The scale is nonlinear, but, in contrast to the series meter, the resistance increases on the right side. Shunt meters must be turned off when not in use to prevent the battery from discharging.

USING OHMMETERS

Because an ohmmeter supplies a current to the resistance to be tested, it must be operated in an unenergized circuit; otherwise the meter calibration is invalid. The unknown resistance must be isolated from any other components that could provide an alternate current path; if you are not sure, disconnect one end of the resistance to be measured. You should not hold both probes while making a resistance measurement as the resistance of your body will change the measured value (it is acceptable to hold one end). A range should be selected that will provide readings toward the right side of the scale for better accuracy. Before making a measurement, short the probes and check the zero-adjust; it should be rechecked if the range switch is moved.

THE MEGOHMMETER

A variation on the ohmmeter is the megohmmeter, or megger, a portable instrument designed to measure very high resistances such as insulation. The megger is

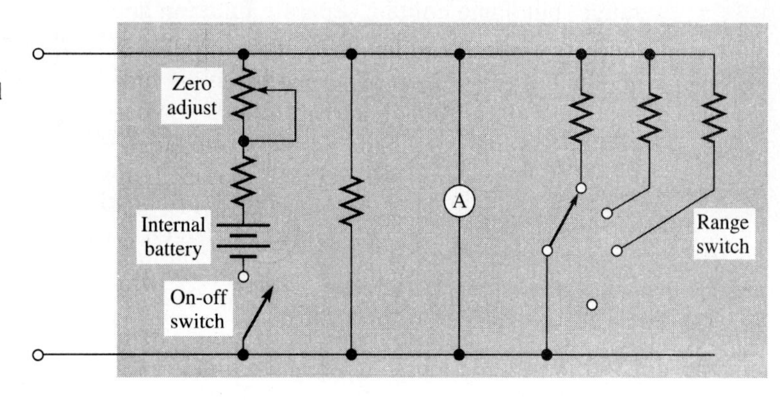

FIGURE 7–14
Shunt-type ohmmeter. Shunt-type meters are used primarily for low-value resistors.

equipped with a handcrank that is used to turn a generator to deliver a high-enough voltage to force current in the resistance to be measured. Two coils and a pointer are mounted on a movable element that is free to turn. One coil tends to move the pointer clockwise; the other, counterclockwise. One of the coils and the series-limiting resistor are connected in series with the unknown resistance. As the generator is cranked, current in the first coil creates a magnetic field that tends to move the pointer to the right. If the generator is cranked more vigorously, more current flows in the first coil, creating a larger force. The second coil is designed to oppose this force, making the meter reading essentially independent of the generator voltage.

Meggers are useful as a quick test that insulation breakdown has not occurred; this is difficult to spot visually or measure with standard meters. For production areas, specialized meters that operate from the ac line are designed to test insulation. In addition to resistance, insulation is tested by high-voltage testers known as HI-POT voltage testers.

7–4 THE VOM

The ammeter, voltmeter, and ohmmeter functions can be combined into a single multipurpose meter called a VOM (for voltmeter-ohmmeter-milliammeter). VOMs are portable analog instruments that use battery power as necessary. Although VOMs are less popular than in the past, there are still many in use. They are preferred in certain applications, including monitoring trends when the readings tend to fluctuate or oscillate slowly. VOMs are also superior to electronic meters in radio-frequency and strong electromagnetic fields such as those that occur in radio transmitting facilities and some industrial environments. Because of the various functions and ranges available, VOMs have several multiple scales. Scales depend on the function selected as well as the range. Often the scales are color-coded in some way to make it easier to identify the proper scale to read. Figure 7–15 shows a typical VOM (the Triplett Model 60) with key specifications.

VOMs usually have a single multipurpose switch to select the function and range, but some contain separate function and range switches. Usually the functions include AC and DC VOLTAGE and CURRENT as well as OHMS. To measure ac voltage or current, a separate switch is used. Note that certain functions may require moving the probes to a separate set of jacks. In the AC position, a rectifier is placed into the circuit; typically, both ac and dc voltage will cause the meter to respond. If you want to isolate the ac only, you need to have a series-blocking capacitor in the path. This is provided by many instrument manufacturers on a separate jack in the front labeled OUTPUT, a somewhat misleading name as it is an input to the VOM. To measure the ac component of a signal, the test lead is moved from the + jack to the OUTPUT jack.

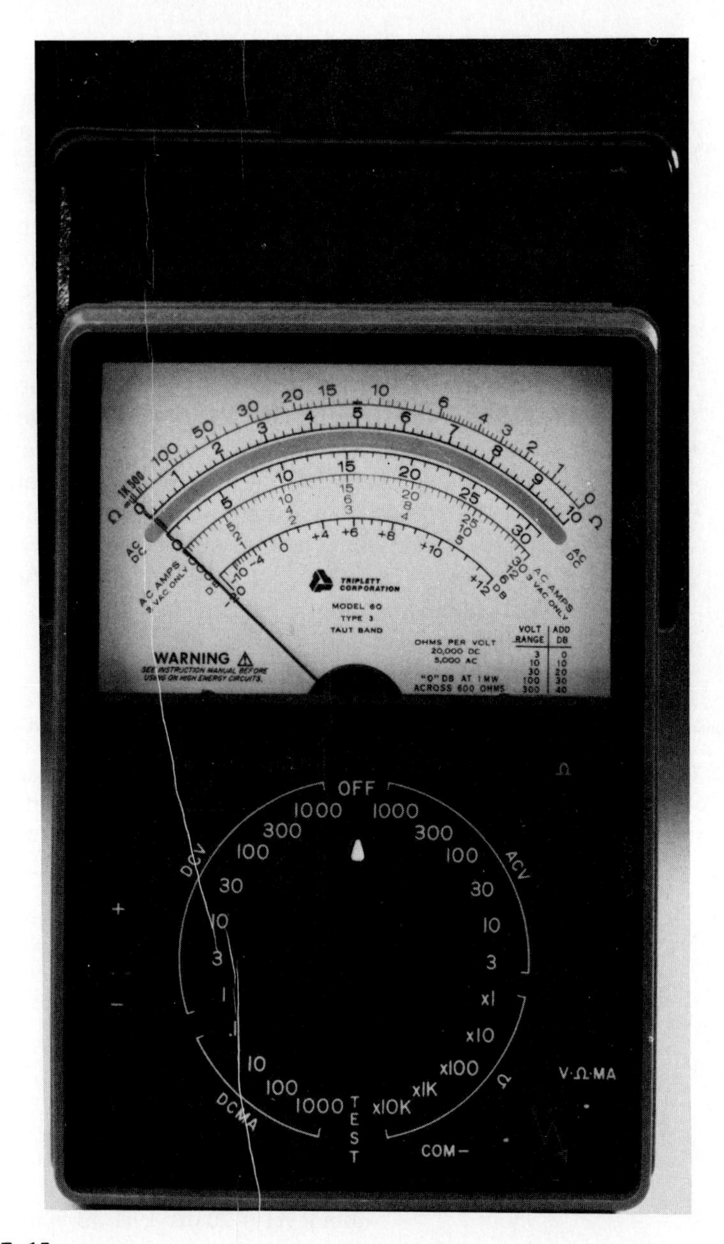

FIGURE 7–15
Triplett Model 60 VOM. The sensitivity is 20,000 Ω/V for dc ranges and 5000 Ω/V on ac ranges. Accuracy on dc voltage scales is 1½% of full-scale value (courtesy of Triplett Corporation).

THE DECIBEL SCALE

In addition to the voltage, current, and resistance scales, many VOMs have a decibel (dB) scale for making direct measurements of audio power with sinusoidal waveforms. The dB scale is designed for use in 600 Ω audio circuits with a reference power level of 1 mW. This corresponds to a voltage of 0.775 V in 600 Ω or 1 dBm (see Section 2–4). This level is marked on the meter scale as 0 dB. The dB scale can be used for absolute power measurements provided it is connected to a 600 Ω load. Measurements across other loads are relative—they are valid only if you want to monitor the change in power in the load.

To use the dB scale, select the AC VOLTAGE function and connect the meter as you would for measuring ac voltage starting with a high range (if a dc component is present, use the OUTPUT jack to isolate the ac component or use a low-reactance external capacitor at very low frequencies). Read the dB scale instead of the voltage scale. You will need to add the number of decibels indicated on the chart on the meter face corresponding to the selected ac range; the addition of the dB value is equivalent to multiplication of power. For the Triplett model 60, the 0 dB reference is on the 3 V ac scale; this means the pointer extends over 0 dB when it is indicating 0.775 V ac on the 3 V ac scale.

EXAMPLE 7–4 An audio amplifier with a 600 Ω load is tested with a sinusoidal wave at 1 kHz. Compute the dBm power level at the output if the reading is 11 dB when the range selected is the 30 V ac scale. Assume you are using a VOM with the 0 dB reference on the 3 V ac scale.

SOLUTION The chart on the meter face indicates 20 dB should be added to the reading of the 30 V ac scale. The chart value can be computed based on 0 dB on the 3 V range as follows:

$$\text{dB} = 20 \log \left(\frac{V_2}{V_1}\right)$$

$$\text{dB} = 20 \log \left(\frac{30\text{ V}}{3\text{ V}}\right) = 20\text{ dB}$$

The total dBm is the sum of the reading and the quantity to be added:

$$\text{dBm level} = 20\text{ dB} + 11\text{ dB} = 31\text{ dB}$$

EXAMPLE 7–5 The output of an audio amplifier with a 16 Ω load is measured at 100 Hz and indicates 19 dB. At 10 kHz, the level shown is 16 dB. What is the ratio of the power at 100 Hz to the power at 10 kHz?

SOLUTION The power level difference is 3 dB.

$$\text{dB} = 10 \log \frac{P_2}{P_1}$$

$$3 \text{ dB} = 10 \log \frac{P_2}{P_1}$$

$$\frac{P_2}{P_1} = 2.0$$

Three decibels represent a power ratio of 2 : 1. Since this measurement was not taken in a 600 Ω load, the absolute values cannot be used directly; however the *difference* in the two readings is valid because they are measured across the same load.

USING VOMs

The precautions listed for ammeters, voltmeters, and ohmmeters are applicable to VOMs. When measuring resistance, the zero-adjust should be checked each time the range is switched. For voltage and current readings, it is important to set the range switch higher than the expected reading. If the approximate voltage or current to be measured is not known, start with the highest possible range and switch to a lower range if necessary to obtain reasonable deflection. Meter movements can be damaged if the meter is connected when the range switch is set too low. To prevent arcing, the range switch should not be changed while the instrument is under load. As with any deflection instrument, accuracy is highest if the selected range provides the greatest on-scale reading. Before connecting the instruments to the circuit, make a quick check that the pointer is exactly at zero. Some instruments, such as the Triplett Model 60 (illustrated in Figure 7–15), have a "confidence-test" feature that allows a periodic check of the meter and its fuses.

A VOM is handy for making a quick functional check of a capacitor. Two tests are described: The first is useful for capacitors larger than about 0.01 μF; the second is useful for smaller capacitors. For these tests, a VOM or analog ohmmeter is preferred rather than a digital meter. The tests will sometimes indicate a faulty capacitor is good; however, you can be sure that if a capacitor fails the test, it is bad. A capacitor that passes the tests may fail when working voltage is applied.

For the first test, the capacitor is removed from the circuit and discharged by placing a short across the terminals. The ohmmeter is placed on a high-resistance scale connected across the capacitor, as illustrated in Figure 7–16(a). For electrolytic capacitors, the ohmmeter must be connected with the proper polarity. Do not assume the common lead from the ohmmeter is the negative side. The meter should start by indicating a low resistance and then gradually show an increase in the resistance as the capacitor charges. (The rate of change depends on the size of the capacitor.) If you put the meter in a higher range, the ohmmeter charges the capacitor more slowly and the capacitance "kick" will be emphasized. For small

FIGURE 7–16
Two basic capacitor tests. The ohmmeter test is useful for larger-value capacitors.

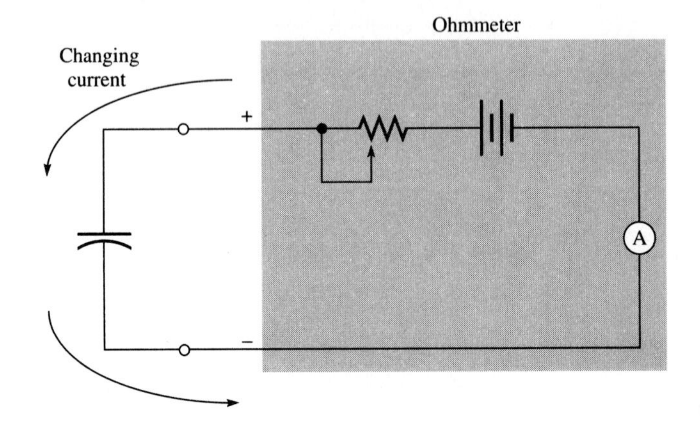

(a) Ohmmeter test of a capacitor.

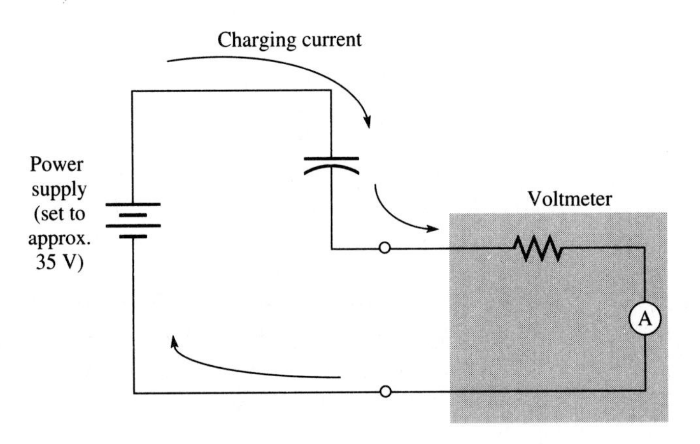

(b) Voltmeter test of a capacitor.

capacitors (under 0.01 μF), this change may not be seen. Large electrolytic capacitors require more time to charge, so you can use a lower range. Capacitors should never remain near zero resistance, as this indicates a short. An immediate high-resistance reading indicates an open for larger capacitors.

The second test increases the charging time by taking advantage of the very high internal resistance of a voltmeter. The VOM is changed to the voltage function. The VOM is connected in *series* with the capacitor, as indicated in Figure 7–16(b). When voltage is first applied, it is divided between the capacitor and the voltmeter according to Kirchhoff's voltage law. As the capacitor charges, the voltage appears across it, and the meter indicates a very small voltage. Large electrolytic capacitors may have enough leakage current that they can appear bad (no voltage), especially with a high-impedance meter. In this case, use the test as a relative test, comparing the reading with a similar capacitor that you know is good. A complete test can be performed with a component analyzer, discussed in Section 7–10.

To extend the usefulness and range of VOMs, a number of accessories are available. These include external current shunts, clamp-on current-measuring accessories, high-voltage probes, and high-frequency probes. The current shunts extend the current ranges up to approximately 100 A. The clamp-on ac-current accessory enables the meter to measure ac current by sensing the magnetic field surrounding a current-carrying conductor. This is handy in applications where it is inconvenient to break the circuit in order to insert an ammeter. A similar accessory is available for digital meters (see Section 7–6).

FET VOMs AND ELECTRONIC METERS

Although the VOM is an excellent general-purpose instrument, it has excessive loading errors when measuring high impedance circuits. By placing a high-impedance FET amplifier at the input of a VOM, the sensitivity can be greatly increased. The sensitivity can be increased from about 20,000 Ω/V for standard VOMs to a constant 10 MΩ for a VOM with an FET input. This means that loading is not a factor in circuits with resistive loads less than about 100 kΩ; however, capacitive loading still needs to be considered for higher frequencies. Also note that FET VOMs are more sensitive to noise than passive VOMs. The circuit for a basic electronic voltmeter using FETs in a differential amplifier was shown in Figure 4–34(c).

Various electronic meters can also be constructed using a basic meter and operational amplifiers. For example, a combination voltmeter and ammeter can be constructed from an operational amplifier configured as a high-impedance, noninverting amplifier as illustrated in Figure 7–17. When current is selected, a small voltage is developed across the 0.1 Ω resistor. For both current and voltage, the range is selected from an input voltage divider constructed from precision resistors. When voltage is selected, the voltage divider determines the input impedance of the electronic meter. The operational amplifier is selected to have very

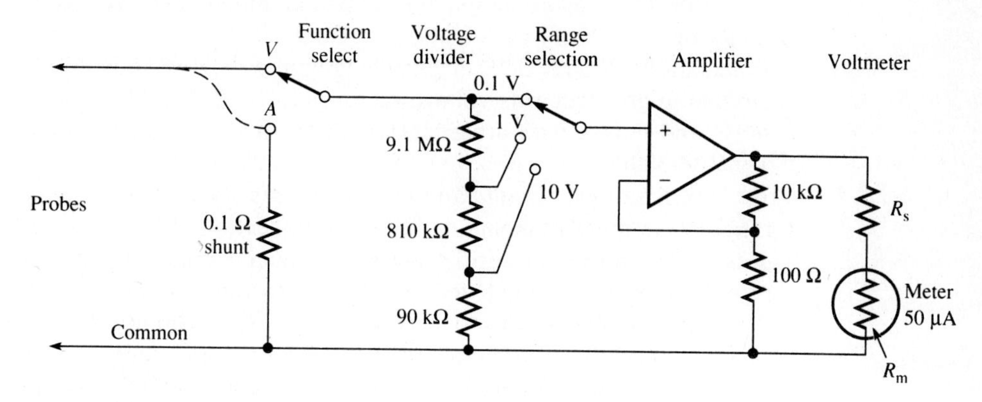

FIGURE 7–17
An electronic meter that can measure voltage or current.

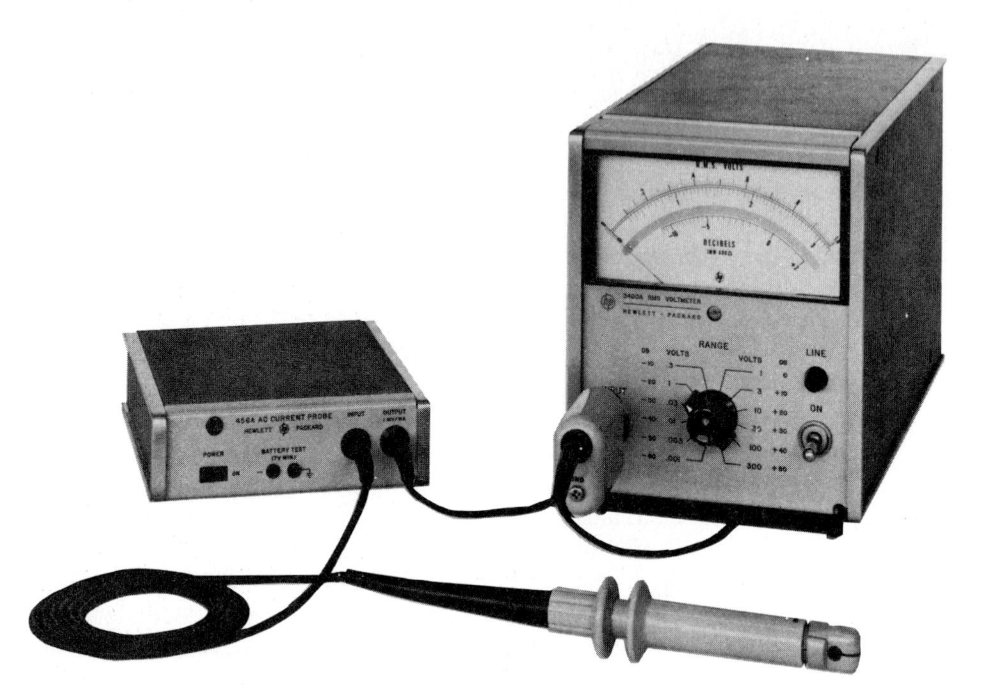

FIGURE 7–18
Electronic voltmeter (courtesy of Hewlett-Packard).

low bias current, so it presents almost no load on the divider. The output of the amplifier is applied to a meter calibrated to read voltage or current.

By incorporating a shunt resistor across the terminals of the electronic voltmeter, it can be converted into a sensitive ammeter. A small current causes a small voltage drop to appear across the terminals of the shunt resistor. This voltage is amplified and used to drive the meter. For larger currents, the op-amp is unnecessary and can be switched out by the appropriate switching arrangement.

For ac measurements, the nonlinear diode response of a rectifier circuit limits the sensitivity of VOMs. Electronic voltmeters overcome this limitation. For example, the HP 3400 electronic voltmeter, shown in Figure 7–18, uses active circuits to give true rms voltage readings to 10 MHz and can measure broadband noise and pulses with crest factors up to 10 : 1. It has 12 voltage ranges, from 1 mV to 300 V.

Active diode circuits and peak detectors are also used to extend the usable range of ac measurements to full-scale deflections of 100 μV or so. The peak-detector circuit is placed in a probe to reduce capacitive and inductance effects prior to the detector and extend the frequency range to microwave frequencies. At the same time, this serves to increase the usable sensitivity. Peak-detector circuits convert the ac waveform into a dc level representative of the maximum excursion of the waveform. Any basic dc meter can use a peak detector to measure radio-frequency ac, but the calibration is valid only for sinusoidal waveforms. Some meters are calibrated to indicate the peak or peak-to-peak value.

7–5 DIGITAL METERS

Digital voltmeters (DVMs) and digital multimeters (DMMs) have improved the accuracy, stability, and versatility of older analog meters many times over. Digital multimeters are used in calibration laboratories as transfer standards where long-term stability, accuracy, and precision need to be the highest attainable. For these meters, stability and drift are specified as a few parts per million, and accuracies are on the order of 0.001%. On the other hand, low-cost portable instruments can be as small as a pocket calculator and are even easier to use than their analog counterparts because of features such as autoranging, automatic polarity selection, and the elimination of scales to interpret. Figure 7–19 illustrates a portable, battery-powered DMM, and Figure 7–20 shows a highly accurate bench-type DMM that can be used as a transfer standard.

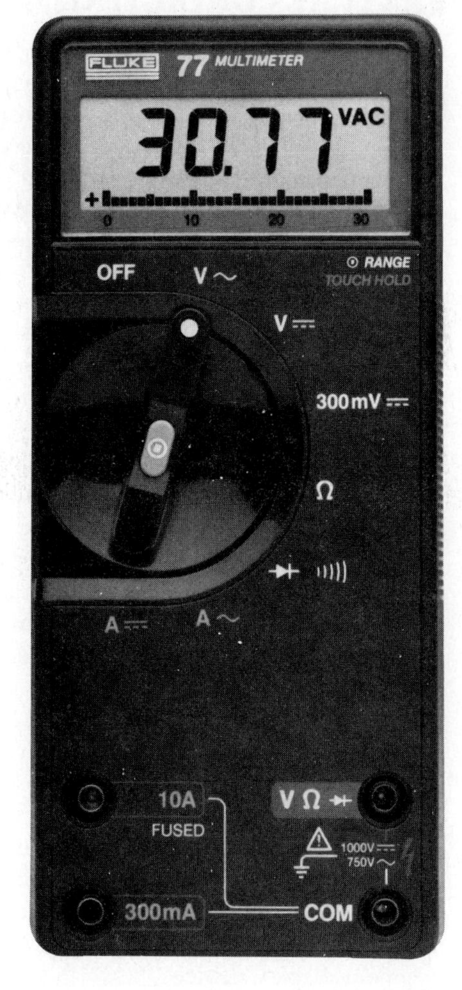

FIGURE 7–19
Fluke 77 handheld DMM. The Fluke 77 is an autoranging 3½-digit meter with a basic accuracy of 0.3%. It features "touch-hold" to capture readings, continuity beeper, and range hold (reproduced with permission from the John Fluke Mfg. Co., Inc.; © John Fluke Mfg. Co.).

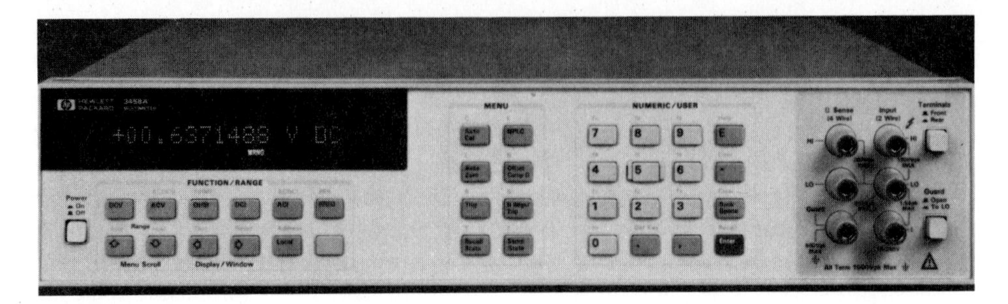

FIGURE 7–20
HP 3458A, an 8½-digit DMM suitable as a transfer standard. This DMM can take 100,000 samples per second. Accuracy of the voltmeter is 8 ppm for 1 y from calibration (courtesy of Hewlett-Packard).

FEATURES OF DIGITAL METERS

Autoranging is a feature in which the instrument selects the correct range for displaying the measurand with the maximum number of significant figures. Automatic polarity selection enables the instrument to read either polarity of voltage or current and display the polarity. These features speed the measurement process and reduce the possibility of reading errors. In addition, DMMs offer other features, such as overload protection, comparison measurements, peak hold, logic sensing, and various display enhancements (such as a bar graph or dB scales), that enable the user to spot signal trends and null conditions. Various optional probes are available to extend the measuring capability further. For industrial applications, ruggedized DMMs are available, with some manufacturers claiming their meter can be dropped from 10 ft to a concrete floor without damage. Ruggedized meters are also designed for protection against sprayed water, heat, or accidental overload.

Some advanced digital multimeters offer IEEE-488 interface connectors, which can allow automated measurements in test systems (see Section 14–9). With automated control, it is possible for a DMM to make more than 100 measurements per second and send the results to a computer for analysis. Lower-priced meters are useful in field servicing or as panel meters for monitoring applications. Often the readout for a digital panel meter (DPM) is converted to another physical parameter such as temperature or pressure, even though the basic meter responds to voltage.

Figure 7–21 shows a simplified block diagram of a DMM. At the heart of all digital multimeters is an A/D converter for changing the analog input value to a discrete number. The input signal is first processed, depending on the selected function, and then sent to the A/D converter. The most popular converter is a dual-slope converter such as the one described in Section 6–5; however, advanced instruments use more sophisticated techniques such as multislope converters and pulse-width conversion to reduce errors. The output of the A/D converter goes to a readout driver. The readout is generally some type of seven-segment display device.

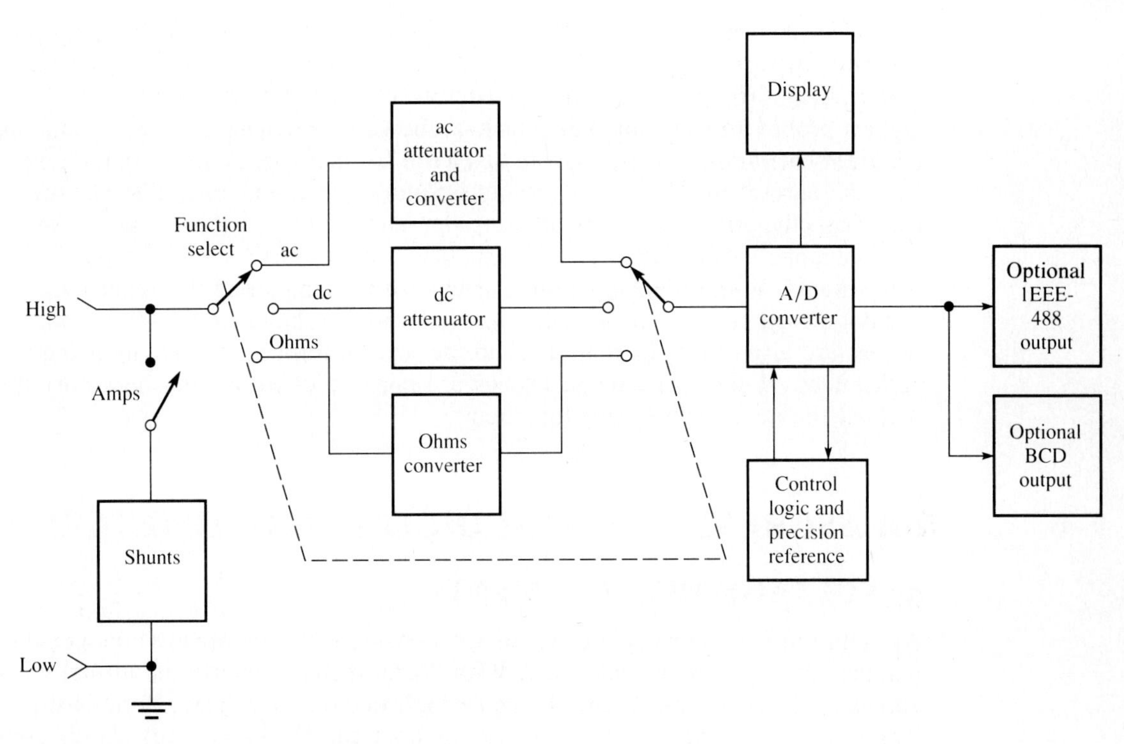

FIGURE 7–21
Simplified block diagram of a DMM.

SPECIAL MEASUREMENT PROBES

Options for extending the measurement capability of DMMs include high-voltage probes, high-frequency probes, temperature probes, and clip-on current probes. High-voltage probes are designed to reduce very high voltages (in the 30 kV range) to a potential safe for the meter. The probe uses a large-ratio voltage divider (1000 : 1 is typical) mounted in a large plastic insulating handle. The handle has protruding fins to provide increased surface area, providing more protection for the user. The probe must be kept clean and dry for the insulation to protect the user effectively from shock, and it should not be used as extremely high altitudes because of the lower ionization potential of the air. Probes are designed for dc voltages or ac and dc voltages with the maximum voltage and altitude specified by the manufacturer.

High-frequency probes are designed to convert radio frequencies to a dc level. The probes use a passive peak detector to provide a dc voltage that is proportional to the rms value of the ac signal. As an example, the Philips PM9221 high-frequency probe can respond to frequencies from 100 kHz to 1 GHz with an amplitude of 30 V.

Temperature probes are offered as an accessory for certain models of DMMs. They are used to convert temperatures in the range from about −60°C to 200°C into a voltage that can be directly read by the DMM.

Clip-on current probes and current transformers are designed to allow measurement of current without the necessary of breaking the circuit and inserting a meter. A split ring is opened and connected around one conductor. Transformer current probes are used for ac and pick up the varying magnetic field surrounding a current-carrying conductor. (Note that a transformer current meter cannot measure the current in a line cord without isolating one conductor.) The conductor represents the primary winding of a step-up transformer, and the secondary winding is wrapped around the ring. The induced voltage is measured by the DMM using the AC VOLTAGE function. Current probes can extend the upper range of current measurements for the meter to ac currents in the range of 200 A. Current probes are also made that can measure dc, ac, or composite currents using the Hall effect. Hall-effect current probes are capable of measuring high currents without the need for breaking the circuit.

7–6 MEASUREMENTS WITH DIGITAL MULTIMETERS

dc VOLTAGE MEASUREMENTS

As in the case of an analog meter, voltage measurements are always connected in parallel with the circuit under test. With digital meters, polarity is normally set automatically by the meter and the sign is indicated on the display. If the common (black) lead is connected to the lower potential and the other (red) lead is connected to the higher potential, the sign will be positive; otherwise it will be negative. Many DMMs have autoranging, a feature in which the DMM selects a range to exhibit the optimum number of significant figures.

Lower-priced meters have manual ranging, requiring the user to select the appropriate range for the measurement. The number of ranges is limited because ranges for digital meters are designed for decade steps rather than the 1-2-5 sequence common in analog instruments. For manual ranging instruments, the correct range will show a reading with the optimum number of significant figures. A 3½-digit meter can show numbers up to 1999 and the decimal. *Ignoring the decimal,* a reading that is a factor of 10 less than the largest number that can be displayed—in this case any number less than 200—is not using the optimum range. For example, assume a 3½-digit meter indicates 5.0 V. The selected range is not optimum because a more sensitive range can show an additional significant figure; for instance, the next-lower range could show 5.04 V. On the other hand, if the selected range is too sensitive, the meter will indicate this with a space before the decimal or with some other indication of overload.

Digital meters can load a circuit under test, but because of the extremely high input impedance, this is not as serious as for VOMs. If the meter is used to measure voltage in a high-impedance circuit, the meter-loading effect should be taken into account. A typical input impedance for a hand-held DMM is 10 MΩ, although the input impedance of high-quality laboratory DMMs can be as high as 10 GΩ. The loading error caused by using a meter in a high-impedance circuit can be easily computed by calculating the effect of the meter's input impedance in

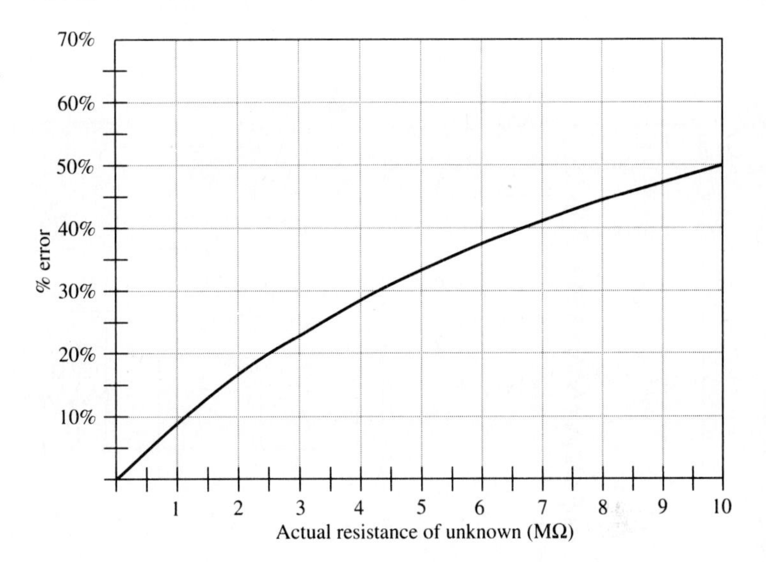

FIGURE 7–22
Percentage error for a 10 MΩ DMM measuring a voltage across a high resistance.

parallel with the resistor to be measured. Figure 7–22 illustrates the effect of loading error using a 10 MΩ meter to measure in high-impedance circuits. A good rule of thumb to determine if loading is a problem is to have a meter impedance that is a factor of 10 times greater than the source impedance for each significant figure you wish to maintain in the result. For example, if you are measuring across a 10 kΩ source and want to have two significant figures in the result, the meter impedance should be 100 times larger than the source or 1 MΩ. For this example, to maintain 3-digit accuracy, the meter should be 10 MΩ.

MEASURING ac VOLTAGE

Most DMMs rectify ac and measure the average of the rectified ac waveform. The value is converted to an rms value for display by scaling the result by a factor of 1.11 (the ratio of the rms to average for a sinusoidal waveform). This does not create any problem as long as the waveform being measured is an undistorted sinusoidal waveform. As long as you are sure this is true, readings can be made accurately. If you attempt to measure a nonsinusoidal waveform, such as a sine wave that has been chopped by a triac or SCR, the results are inaccurate and may be entirely misleading.

A feature offered on some new DVMs and DMMs is true rms–reading capability. Usually this is shown on the front of the meter with a label that indicates a ''true rms'' meter. It is possible to convert an average responding meter to a true rms meter with an external circuit—one example is Analog Devices' AD736, which converts true rms to dc. The device has limited dynamic range, so inputs greater than 1 V rms have to be attenuated by a known amount to be measured accurately. The complete circuit for doing this is shown in Figure 7–23. Even true rms–reading meters can be misleading if the input waveform is not within a

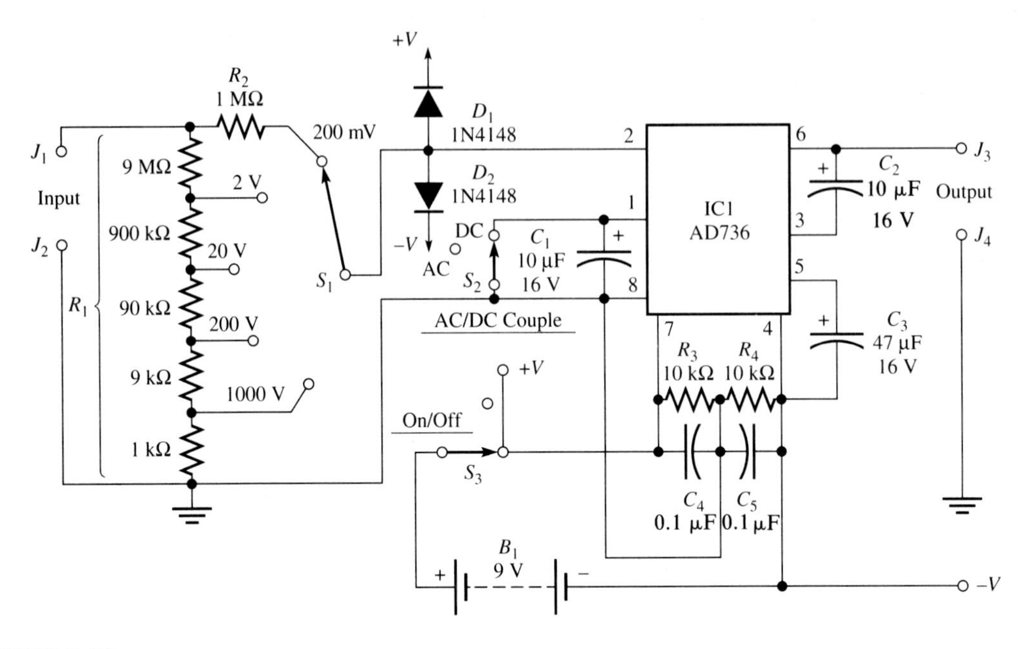

FIGURE 7–23
Circuit to convert an average reading meter to a true rms–reading meter (reprinted with permission of Radio-Electronics).

specified crest factor. The crest factor for short pulses can be very large; a pulse with a duty cycle of 1% can have a crest factor of greater than 10.

Another factor to keep in mind for ac voltage measurements is the bandwidth of the meter. Meters have both a lower and an upper frequency limit that cannot be exceeded and maintain the accuracy of the meter. Typically the lower frequency is 45 Hz; frequencies below this will cause the meter to exhibit erratic and constantly changing values. The upper-frequency limit depends on the meter design. It can be less than 1 kHz to over 1 MHz, so it is important to check the specification before attempting ac measurements. Some meters are specified for a different frequency response between the ac voltage function and the ac current function. Note that meter bandwidth is not the same as amplifier bandwidth, which is taken at the 3 dB points. The upper limit for a meter is normally specified along with an accuracy value. Readings above the upper-frequency limit will be lower than the actual voltage.

As with dc measurements, loading must be considered when measuring high-impedance circuits. With ac circuits, loading can take the form of either resistive or capacitive loading (capacitive loading is discussed in Section 8–5). Typical hand-held DMMs have an input impedance on ac scales of 10 MΩ shunted by 100 pF. Although these numbers are good for controlling loading effects, you should be careful to study the ac voltage specifications carefully to determine the frequency specifications, as discussed previously.

COMPOSITE VOLTAGE MEASUREMENTS

Frequently a sinusoidal waveform is superimposed on a dc level, as illustrated in Figure 7–24. The sinusoidal waveform alternates between a positive and negative peak. The dc level is represented by a horizontal line drawn midway between the peaks. For a composite waveform consisting of an ac and a dc component, the meter will read the dc portion only when the DC VOLTAGE function is selected and the ac portion of the waveform when the AC VOLTAGE function is selected.

CURRENT MEASUREMENTS

For DMMs, either the AC CURRENT or DC CURRENT function is selected, and it is necessary to move the probes to separate current jacks. As in any current measurements, the meter is connected in series with the circuit under test. If the meter is inadvertently connected in parallel, very high current will flow through the meter, causing either a fuse to blow or damage to the meter. The precautions listed with analog meters also apply to digital meters.

As in the case of ac voltage, ac current is shown on the display as the rms value. Meters that are not true rms meters convert the average to the equivalent rms value based on an assumed sinusoidal input. Only true rms reading meters can measure nonsinusoidal waveforms accurately.

RESISTANCE MEASUREMENTS

DMMs measure resistance slightly differently than analog ohmmeters. Instead of a voltage source and nonlinear scale, they typically use a current source to provide an accurate, fixed current through the unknown resistance. The voltage developed across the unknown is converted to a resistance value (by Ohm's law) and displayed.

DMMs frequently have continuity testing as a part of the ohmmeter function. An audible beep indicates a conducting path between the probes, making the test a quick check of wiring without having to look at the display. The continuity tester is a handy feature for tracing wiring and testing for open or shorted paths on

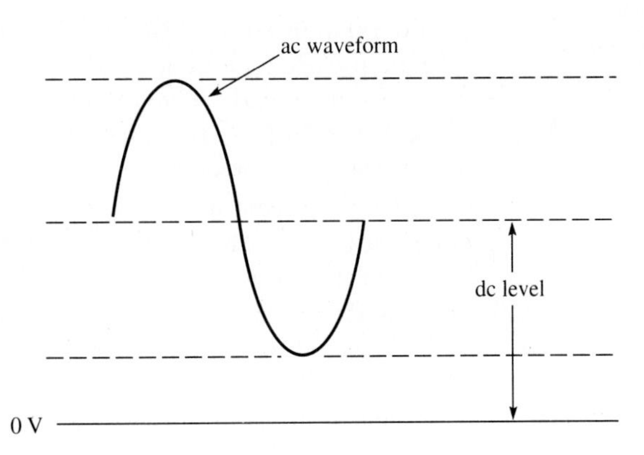

FIGURE 7–24
A sinusoidal waveform superimposed on a dc level. The dc level is midway between the maximum excursions of the ac waveform.

circuit boards. Some meters use two tones to differentiate between a reading that is less than 1% of the selected range and a reading that is less than 10% of the selected range.

FOUR-WIRE RESISTANCE MEASUREMENTS

Lead resistance from a meter can lead to a high percentage error when measuring very small resistance, such as wire resistance. In order to avoid measuring lead resistance, four-wire measurements are available in high-quality bench DMMs. In a four-wire measurement, two leads are connected across the resistance being measured and provide an accurate current through the unknown. Two more "sense" leads are connected at the same point to measure the voltage developed across the unknown as a result of the current source. The voltage-sensing leads have almost no current in them, which keeps the *IR* drop in the leads from contributing to the error in the measurement. Four-wire measurements can have resolutions of 100 μΩ.

DIODE, BIPOLAR TRANSISTOR, AND SCR TESTS

Diodes can be checked with a special function available on many DMMs. One common diode test is to measure the forward and reverse resistance with an ohmmeter. The forward resistance depends on the diode and the internal current source in the meter but can range from a few thousand ohms upward. The reverse resistance is usually too large to measure and shows as an open. Unless you know the specific forward-bias readings for the diode being tested, the ohmmeter test is only a gross test of a diode. A better test is to measure the voltage drop across a forward-biased diode. Many DMMs provide this feature with a special diode function that biases the diode and displays the voltage drop on the meter. When connected across a good diode biased in the forward direction, the meter will show a drop of between 0.6 and 0.8 V for a silicon diode and will show a drop of between 0.2 and 0.4 V for a germanium diode. When the diode is reverse-biased, it will show an open. Unlike the resistance test, this method for testing diodes has the advantage of being able to test diodes that are connected in a circuit without removing them.

Bipolar transistors can also be tested using the diode test to assure that they are not open or shorted. Recall that the base-emitter junction and the base-collector junction are *pn* junctions, the same as diodes. You can also use this type of test to determine quickly if an unknown transistor is an *npn* or a *pnp* and to identify the leads. Look for a lead that has a forward diode drop to two other leads. This is the base lead. If the most positive probe is on the base when the transistor is forward-biased, the transistor is an *npn*; otherwise it is a *pnp*. To identify the emitter and collector leads, note that the forward drop on the base-collector diode is slightly less than the forward drop on the base-emitter diode.

SCRs can also be checked with the diode test. The test is performed with the SCR removed from the circuit. For the test, the meter is connected between the anode and cathode. Measure the resistance with the gate open; then reverse the leads and repeat the measurement. Both readings should be very high (almost infinite) because the gate is open and no conduction should occur. A clip lead is then connected between the anode and gate of the SCR. The meter is connected between the anode and cathode, with the most positive lead connected to the anode. The clip lead provides a forward-biased path (from the meter's internal battery), and the meter should indicate a low reading. The clip lead is then removed without changing the meter. The reading should remain a low value because the gate loses control once conduction begins. If the meter is disconnected and reconnected, the reading should be back to a very high value.

7–7 DIGITAL MULTIMETER SPECIFICATIONS

RESOLUTION

The number of digits shown for a digital multimeter is generally expressed as a mixed number such as 3½ or 5½. The whole number indicates the number of nines the display is able to show. The fraction indicates that the display is unable to show all possible decimal numbers in the most significant position. The fraction 1/2 indicates that the most significant digit (MSD) can be either a 0 (normally blanked off) or a 1; a 3/4 (unusual) indicates that the MSD can be either a 0, 1, 2, or 3. A 3½ digit display can shown numbers from 0000 to 1999; this represents a resolution of 1 part in 2000, or 0.05%. A 4¾ display can show the numbers from 00000 to 39999 for a resolution of 1 part in 40,000, or 0.0025%. Notice that the resolution specification is *not* the same as accuracy. It is a mistake to purchase a DMM on the basis of resolution and ignore the accuracy specification.

Although the resolution is expressed by the number of digits that can be displayed, it can also be expressed as the smallest difference in voltage, current, or resistance that the meter can display. For example, a 3½ digit meter on the 20 V range has 10 mV resolution. On the same meter, the 2 V range gives a resolution of 1 mV. Manufacturer's specifications often list the available ranges and resolution on this basis.

ACCURACY

The accuracy of digital meters is specified in various ways. Typically it is specified as a percentage of the reading plus or minus a number of counts, to reflect the behavior of the circuitry. Occasionally, manufacturers specify accuracy as a percentage of the reading plus a percentage of the full-scale reading. Others express it in parts per million (ppm). However it is stated, accuracy is valid only if the instrument is given the specified warm-up time and is operated within the stated temperature range and if the time since calibration of the instrument is valid. The accuracy of a given measurement can be in error for other subtle effects of which the user needs to be aware, such as the presence of common-mode signals or the measurement of a waveform beyond the bandwidth of the meter or for which the meter is not calibrated. For example, some meters read true rms values only for a sinusoidal waveform. If you attempt to measure another waveform with the meter, the error can be substantial.

Assuming you are measuring a sinusoidal waveform on the ac voltage scale and it is within the frequency limits of the meter, you may still be making an inaccurate measurement. The accuracy depends on the actual reading on the scale, as can be seen in Figure 7–25. At readings in the lower ranges, the ±1 digit dominates the accuracy specification, as shown. Accuracy also depends on the operating temperature and the time since calibration. Manufacturers may specify an operating temperature or range for which accuracy specifications are valid (normally 25°C). A derating factor for other temperatures (either above or below this temperature) may be given in the specification. The time specification can be given as both a 90 d and a 1 y interval for valid calibration. Advanced DMMs can be calibrated with no internal adjustments or removing of covers. A calibration standard is connected to the unit, and calibration constants are stored in a nonvolatile memory, making the calibration process simple and fast.

EXAMPLE 7–6 Specifications for two meters read as follows:

Meter 1: Accuracy: ±0.05% rdg (reading) + 1 digit at 25°C.
Meter 2: Accuracy: ±0.05% rdg ± 0.03% fs (full scale) at 25°C.

Compute the possible error for the given specifications for a 3½-digit meter operating at 25°C with an input of 100.0 V.

SOLUTION Both meters can display the input number with four digits because the MSD is a 1. The optimum reading will therefore be 100.0 V. The first meter has an accuracy of ±0.05% rdg + 1 digit, which translates into an error of 0.05 V + 0.1 V for a total error of 0.15 V.

The second meter is specified for both percentage of reading and percentage of full scale. Since the full-scale voltage on this range is 199.9 V, the error is 0.05 V + 0.06 V = 0.11 V. For this measurement, the second meter is more accurate.

SENSITIVITY

Sensitivity is the ratio of output response to the input cause under static conditions. For meters, the sensitivity is related to the lowest range available for a given function. It can be expressed as divisions per volt or millimeters per volt, for example.

RESPONSE TIME

The **response time** of a meter is the time required for the meter to respond to an input signal change. It is affected by the meter's A/D conversion time and settling time. For automatic measurement, the response time is an indication of how many readings can be made per second.

COMMON-MODE REJECTION

The leads from meters can be subject to common-mode noise pickup that can be interpreted by the meter as part of the measurand. Common-mode interference

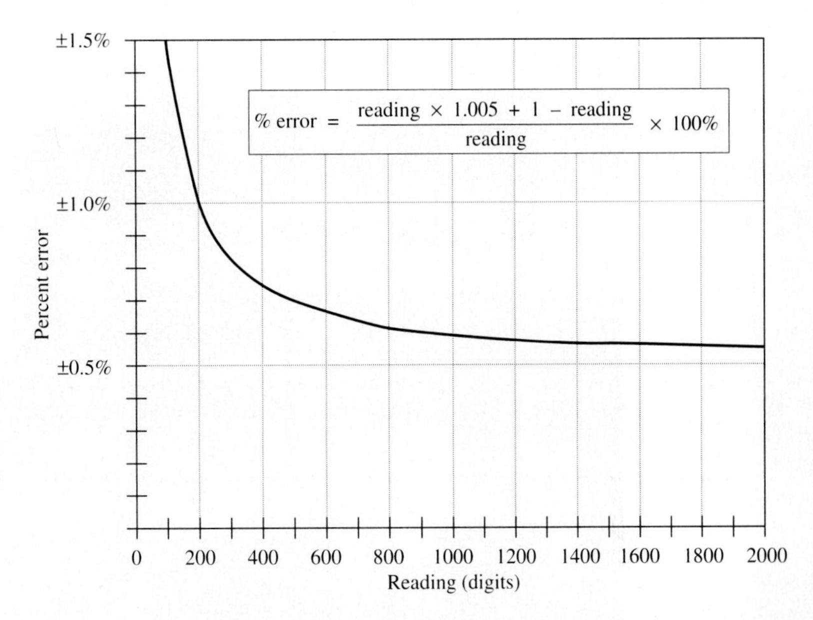

FIGURE 7–25
Total error of a meter specified is ±0.5% ± 1 digit.

occurs when both the high and low input lines receive the same interfering signal. Common-mode noise (discussed in Section 4–10) can be rejected by meters by using a differential amplifier in the input stage of an amplifier. It is usually specified from dc to 60 Hz and can be over 120 dB of rejection.

7–8 ELECTROMETERS, PICOAMMETERS, AND NANOVOLTMETERS

For certain measurements, particularly low-level measurements with very high impedance sources, DMMs are inadequate to make accurate measurements. The theoretical limits of measurement depend on the noise generated by the resistances in the circuit (discussed in Section 12–1) and the source resistance. In general, when it is necessary to make measurements near the theoretical limits, the use of an electrometer, picoammeter, or nanovoltmeter is called for. Electrometers are multifunctional instruments, whereas the picoammeter and nanovoltmeter are single-function instruments that are designed to optimize current and voltage measurements, respectively.

THE ELECTROMETER

The **electrometer** is a specialized multimeter with the capability of measuring voltage, current, resistance, or charge. It has extremely high input impedance, typically above 100 TΩ (10^{14} Ω), making it ideal for measuring high-impedance circuits or extremely small currents (in the nanoamp range). In addition to extremely high input impedance, it is designed with low input offset current, low drift, and low current noise. The readout can be either digital or analog. Figure 7–26 shows a digital electrometer.

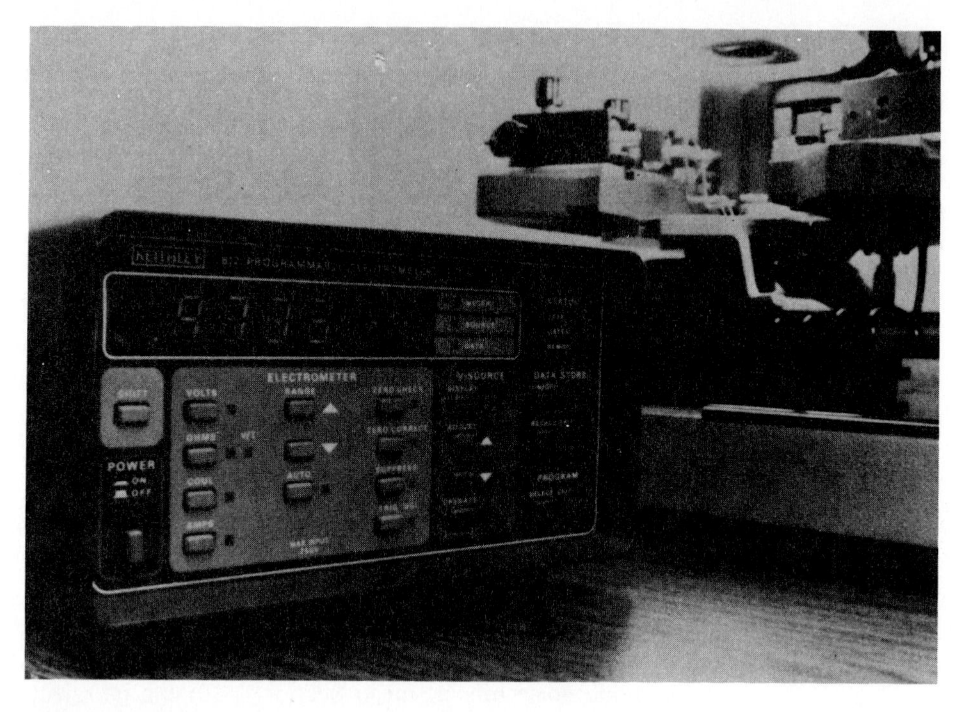

FIGURE 7–26
Keithley Model 617-HIQ programmable electrometer. The 617-HIQ is a fully autoranging digital electrometer with seven functions. The most sensitive of the 11 current scales has a full-scale range of 2 pA with a resolution of 100 aA (0.1×10^{-15} A) (courtesy of Keithley Instruments, Inc.).

In the voltmeter and ohmmeter functions, the most important specifications are the input resistance and the input offset current. The input offset current, the current drawn by the input with no signal current source, is typically less than 50 fA (5×10^{-14} A). For measuring voltages, the electrometer's extremely high input resistance and very small offset current mean that circuit-loading effects are minimized. An electrometer, for instance, can measure the voltage across a 500 pF capacitor without significantly discharging it. In the ohmmeter function, the high input resistance enables the electrometer to measure resistances up to 100 TΩ.

In the ammeter function, the electrometer features a very small voltage burden. The voltage burden is the voltage drop across the meter terminals when connected in series with a current source. The voltage burden and offset current specifications are the important specifications for measuring very small currents. Currents are small as 1 fA (10^{-15} A) can be measured with an electrometer. Currents in this range are found in the process of calibrating certain radiation detectors. The electrometer can also measure charge by current integration, a function not available on DMMs.

Measurements of very high impedance sources or of low-level signals require special techniques not available in ordinary DMMs. One of these techniques is called **guarding,** a clever method of reducing leakage current and input capacitance effects. For high-impedance voltage measurements, the unknown voltage is connected to the high-impedance terminal on the electrometer through a short triax cable to avoid the leakage-current paths that occur with normal test leads. The electrometer provides a ×1 analog output, which is essentially a low-impedance voltage that is identical to the input voltage. By connecting the guard lead to the inner shield of the triax, leakage currents are almost eliminated because the shield and the voltage to be measured are at the same potential. The outer shield is connected to ground. Leakage current between the shields is supplied by the low impedance source rather then the high impedance source being measured. Figure 7–27 illustrates guarding.

Electrometers also provide means of adjusting for the zero offset and drift that occur with very high impedance instruments. Zero offset is the instrument reading with no input signal. It can be generated with the instrument or in the source and can be connected to the instrument by a variety of means. Zero drift is a gradual change in the zero offset with no input signal. It may be specified as a

FIGURE 7–27
Comparison of unguarded and guarded sources.

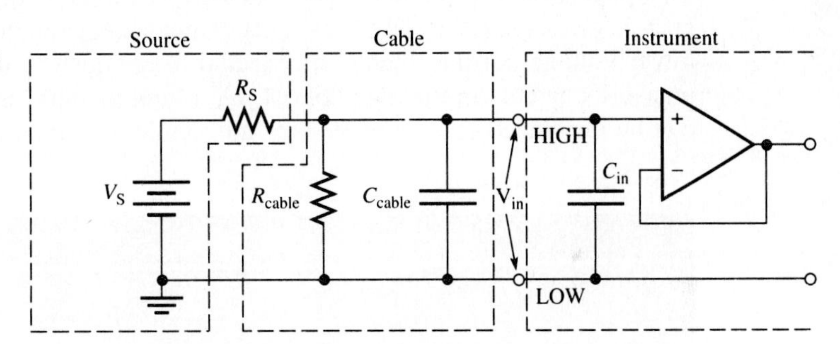

(a) Unguarded circuit. R_S represents a high source impedance typical of many transducers. The cable resistance and capacitance represent a load on the source that affects V_{in} and the time constant for the circuit.

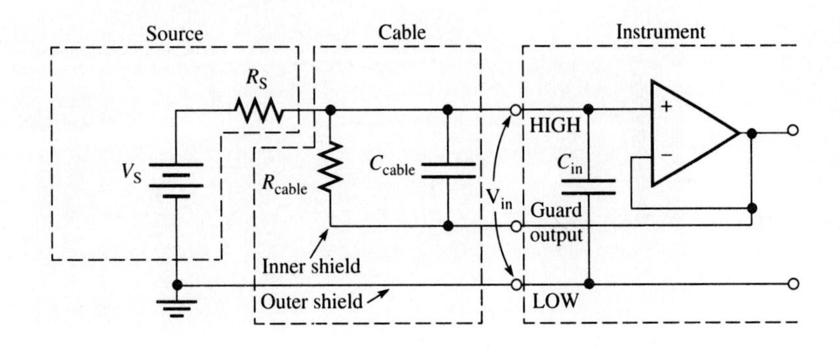

(b) Guarded circuit. A triax cable is used. The inner shield is driven by a low-impedance source that is at the same potential as the signal. This reduces leakage current to near zero. The time constant is also significantly lower because guarding reduces the effective capacitance.

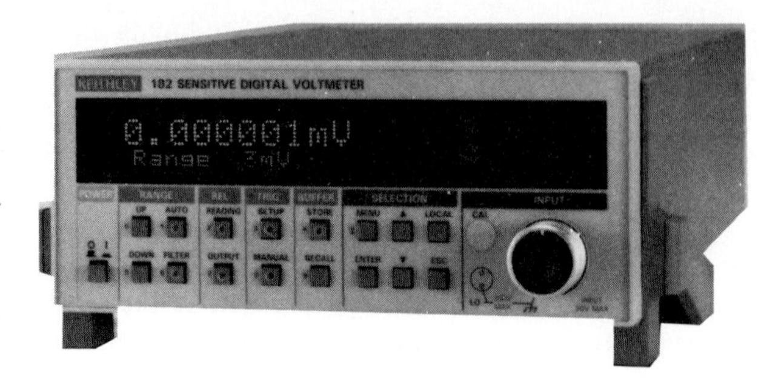

FIGURE 7–28
Keithley Model 182 nanovoltmeter. The 182 can be controlled by a computer over the IEEE-488 bus; it has 1.0 nV usable sensitivity and up to 6½ digits of resolution with over 10 GΩ of input resistance (courtesy of Keithley Instruments, Inc.).

function of both time and temperature by the instrument manufacturer. The input signal is disconnected with the zero-check switch, and a front-panel zeroing control is adjusted to minimize the offset.

THE PICOAMMETER

The **picoammeter** is a sensitive ammeter optimized for measuring very small currents. It usually has less sensitivity than an electrometer but is designed with a lower voltage burden, higher speed, and lower cost. A digital picoammeter can measure currents in the range of 0.1 pA. Many models have a digital readout and can be interfaced to a computer for automatic data collection.

FIGURE 7–29
Keithley Model 237 source-measurement unit. The source-measurement unit integrates an electrometer, DMM, current source, and voltage source in one unit (courtesy of Keithley Instruments, Inc.).

THE NANOVOLTMETER

The **nanovoltmeter** is a very sensitive voltmeter optimized to measure very small voltages from relatively low-source impedances, as opposed to high-impedance sources for the electrometer. It can be used to measure voltages as small as 1 nV. The input impedance is about the same as a good DMM but is lower than an electrometer. It has lower voltage noise (but higher current noise) than an electrometer, and the cost is less because of lower complexity. Figure 7–28 illustrates a nanovoltmeter.

SOURCE-MEASUREMENT UNITS

Source-measurement units are recently introduced instruments capable of simultanously sourcing voltage and measuring current or sourcing current and measuring voltage. A source-measurement unit consists of four instruments in one: an electrometer, a DMM, a current source, and a voltage source. The integration of these instruments eliminates the problems associated with setting up and cabling these instruments, a common problem for sensitive measuring instruments. Source-measurement units find application in automated testing and in parametric testing of semiconductor devices. Figure 7–29 illustrates a source-measurement unit.

7–9 POWER MEASUREMENTS

LOW-FREQUENCY POWER MEASUREMENTS

In many electronic systems, the rate that energy is transferred is an important consideration. Power is a measure of the rate that energy is transferred or transformed into a different form and is measured in watts (1 W = 1 J/s). In a reactive circuit, there are three types of power, as illustrated in the phasor drawing in Figure 7–30. These are true power, reactive power, and apparent power. True power is that quantity taken from the source and converted to heat in a nonreversible process. The unit for true power is the watt. Reactive power is stored temporarily by the reactance, a reversible process, with no net power taken from the source. It can be either purely inductive or purely capacitive (no resistive component to either). In a pure reactance, the reactive power is simply the product of voltage and current associated with a reactance and is identified with the unit volt-ampere-reactive (VAR). Apparent power is the product of current and voltage as seen by the source. If it is known to contain a reactive component, then it is given the unit volt-ampere (VA). The defining equation for determining the true power dissipated in a reactive circuit is

EQUATION 7–6

$$P_t = IV \cos \phi$$

where P_t = true power dissipated in a resistive circuit, W

I = current in the circuit, A

V = voltage across the circuit, V

ϕ = phase angle between voltage and current

FIGURE 7–30
Phasor diagrams for a series circuit. The power phasors are found by multiplying the voltage across each component by the current in the circuit.

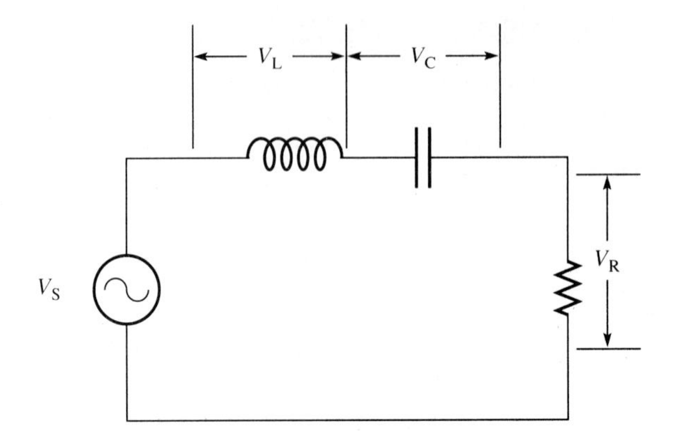

(a) Circuit.

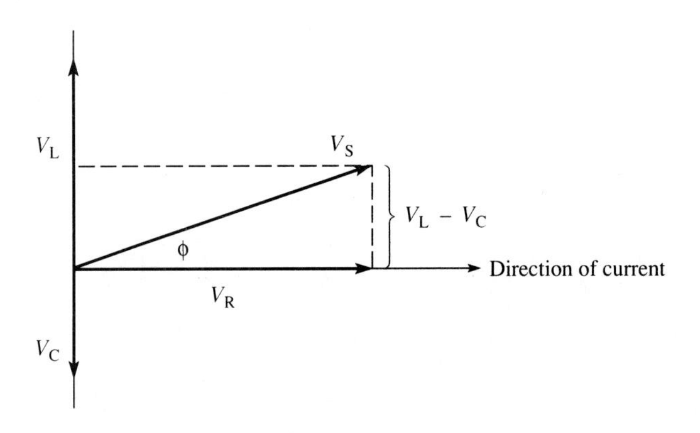

(b) Voltage phasor diagram.

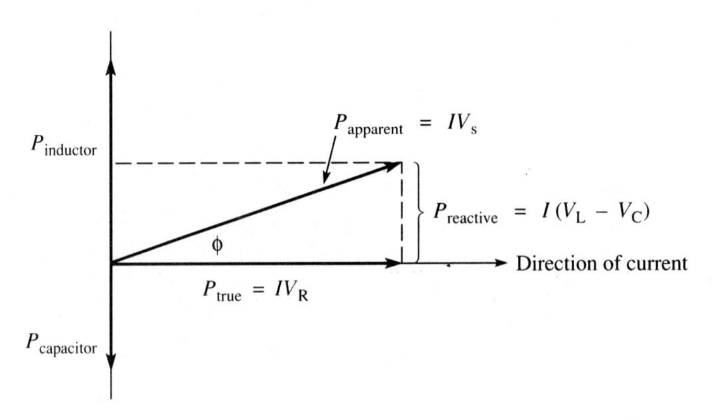

(c) Power phasor diagram.

Here cos ϕ is defined as the power factor, a number that varies from 0 (purely reactive circuits) to 1 (purely resistive circuits).

In single-phase, low-frequency circuits, power is usually determined by measuring the voltage and the current. An oscilloscope can determine both the voltage and current. The current can be found with a special current probe or with a small-value sense resistor inserted in series with the load. The phase angle between the voltage and current is determined by one of the methods described in Section 8–7.

A meter that is used to measure true power by sensing both current and voltage simultaneously is the *dynamometer wattmeter,* shown schematically in Figure 7–31. It uses the dynamometer meter movement described earlier (in Section 7–1) but is connected differently. When designed as a wattmeter, the moving coil is connected to sense the voltage instead of the current by adding a series multiplier resistor, as shown. The two fixed-current coils are connected in series and set up a magnetic field that is proportional to the current. The meter deflection is proportional to the product of the currents in the current coils and the voltage coil; hence, the meter can be calibrated to read the power directly in watts. Dynamometer wattmeters are not affected by the waveform in the circuit—they respond to the average true power and can be used to determine power from dc to the lower audio frequencies.

A standard dynamometer wattmeter has four connections to be made to measure power—two designated as the current terminals and two designated as the voltage terminals. Each set of terminals has one terminal marked with a $\pm$. To connect the wattmeter properly in a circuit, the terminals marked $\pm$ should be connected to the same side of the input line and load, either before or after the current coils. The best connection is determined by whether the measurement is for a high-current, low-voltage case or a low-current, high-voltage case. With the arrangement of the coils shown in Figure 7–31(a), the current that passes through the current coils consists of two components—the load current and the current in the voltage coil. This arrangement causes the meter to read high, an error that is emphasized when the load current is low. The other possibility is the arrangement shown in Figure 7–31(b). This connection avoids the problem of the extra current in the current coils, but now the voltage coil is connected across *both* the load and the current coils, causing it to sense more voltage than is across the load. Once again, the meter will read high, particularly with a large load current. To minimize these errors, the arrangement shown in Figure 7–31(a) should be used when the current is high and the voltage is low and the arrangement shown in Figure 7–31(b) should be used when the current is low and the voltage is high.

A variation of the standard dynamometer that avoids the problem of high readings is the compensated dynamometer wattmeter. It uses two additional coils connected in series with the voltage coil; thus they have the same current in them as the voltage coil. These extra coils are wound in a manner to oppose the magnetic field produced by the current coils, causing the effects of the extra current in the voltage coil to be canceled. The compensated wattmeter has only three connections instead of four, since one side of the current winding is connected internally to one side of the voltage winding.

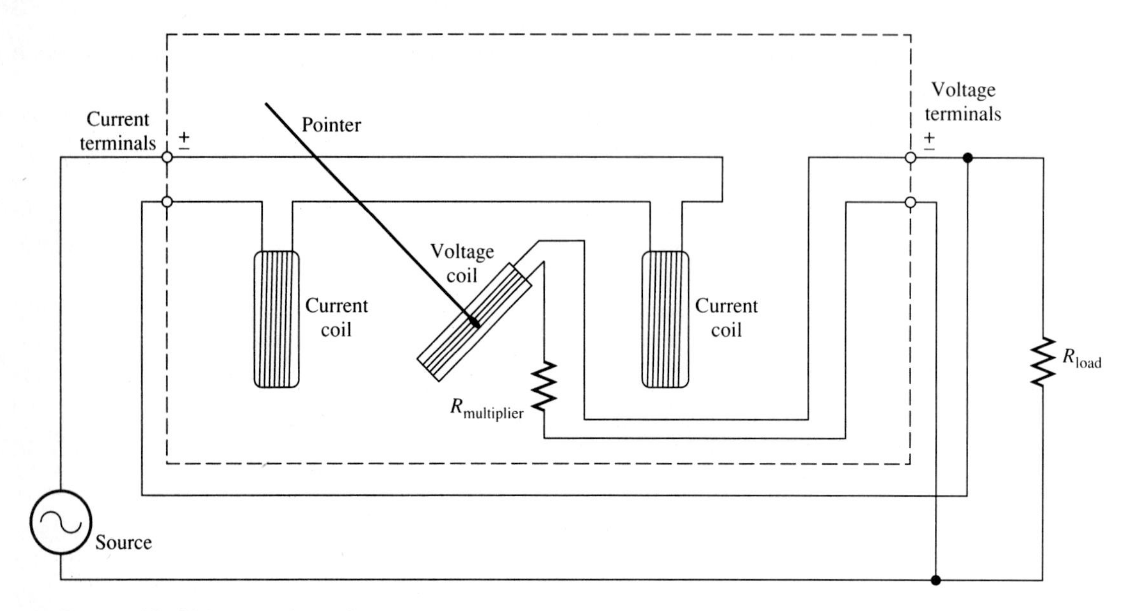

(a) Connected for high-current, low-voltage-power measurement.

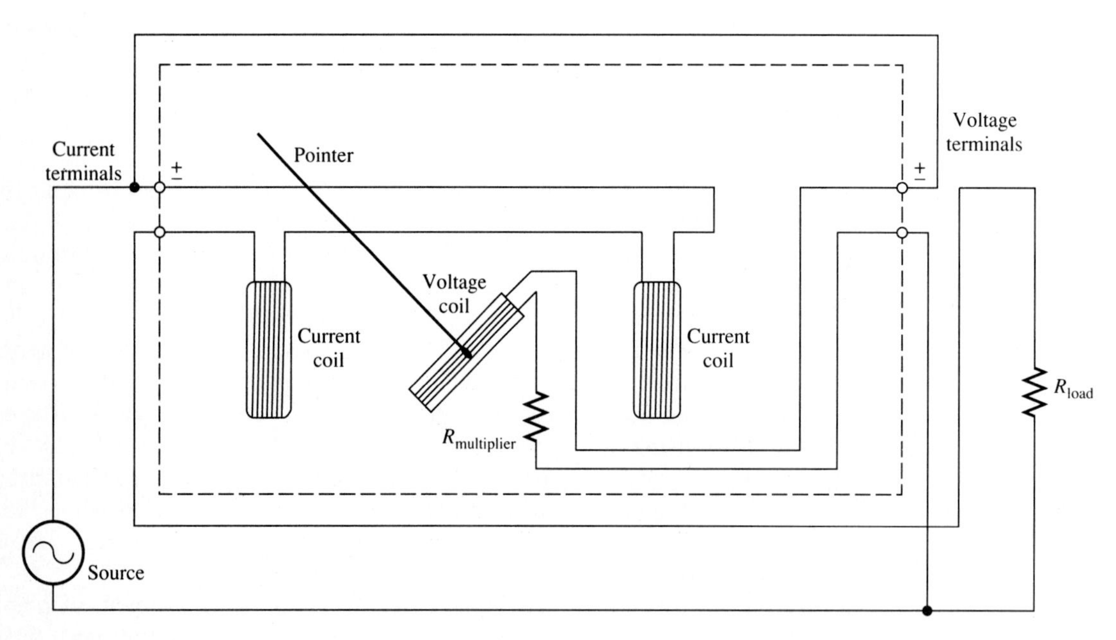

(b) Connected for low-current, high-voltage-power measurement.

FIGURE 7–31
Electrodynamometer wattmeter.

By adding a second moving coil, the dynamometer wattmeter can be modified to measure the power factor directly. The two moving coils are permanently connected together at right angles to each other. One of the coils is connected in series with an inductor and responds to current that is out of phase with the line current; the other coil is connected in series with a resistor and responds to current that is in phase with the line current. The coils tend to rotate to a position that is determined by the fraction of current in each—the meter face is calibrated directly in terms of the power factor.

HIGH-FREQUENCY POWER MEASUREMENTS

At radio frequencies, the measurement of power is usually done by means of connecting a calibrated radio-frequency (rf) voltmeter across a known resistive load and computing the power dissipated using $P = V^2/R_L$. If the load is not purely resistive, the power found will be the apparent power rather than the true power. For high-power measurements, such as radio transmitters, this often involves the use of a special very high wattage resistor called a dummy load. The dummy load dissipates the rf power in place of the antenna.

For low-level rf power measurements, a high-frequency voltmeter and known resistive load can be combined into a single instrument in which the meter and load are within the instrument. The resistor is designed to maintain a constant resistance over the frequency range of the instrument. This type of instrument absorbs the power from the source and displays the absorbed power directly on the scale, which is calibrated in watts.

Another technique for measuring power in rf circuits involves converting the power to be measured into heat in a resistance and observing the temperature rise. This technique is called **calorimetry.** The resistor is well insulated to prevent heat loss, and the resulting temperature increase is measured and converted into an equivalent input power. A similar but more sensitive method is the bolometer method. A bolometer contains a special temperature-sensitive resistor that changes resistance as it is heated. The change in resistance is detected by a bridge circuit, and the result is converted into a power reading.

The bolometer method is common for microwave power measurements. At microwave frequencies, power is the best measure of the signal amplitude. In addition, the measurement of the forward and reflected power in a system is an indicator of system performance. For these reasons, power measurement is a vital measurement in any microwave system. Although the bolometer is limited as to the total power it can measure directly, it can be used to determine larger power than it can measure directly by inserting a directional coupler in the rf path. The directional coupler samples the incident or reflected wave and passes a small fraction of the total power onto the bolometer for measurement. Directional couplers respond only to the wave that is traveling in a particular direction. In high-power microwave systems, such as a microwave radar transmitter, the directional coupler is often permanently installed in the waveguide. Separate directional couplers can be used for sampling both the forward power and the reflected power, enabling the continuous measurement of the transmitter power and reflection coefficient.

7–10 IMPEDANCE MEASUREMENTS

Impedance measurements are important characterizations for components, devices, networks, and systems. Impedance measurements are made as performance verification to assure that the component or device falls within certain specifications and to characterize a device under certain operating conditions. An example of a performance-verification test is when a manufacturer checks a component for quality control and either accepts or rejects the device (this is called go/no-go testing). Quantitative characterization may be used as part of a design effort, as in the case of a wideband attenuation network, or to evaluate the high-frequency performance of certain components.

The impedance of a simple two-terminal device is the complex ratio of voltage to current. The definition of impedance can be extended to linear networks that have more than two terminals; typically we mean a set of input terminals and a separate set of output terminals. A network such as this is said to have two **ports**—meaning that it has two pairs of terminals. **Transfer impedance** is defined as the ratio of any voltage to any current of the same frequency when the voltage and current are measured at two different ports of a common system. If current is used as the stimulus at the input port and a voltage appears in response at the open-circuited output port, the ratio of the output voltage to input current is called the **open-circuit transfer impedance.** This concept is applied to the analysis of many networks that have separate input and output terminals. These networks form basic building blocks in communication systems, automatic control systems, and other electronic systems. A network analyzer (described in Section 10–7) can be used to measure the transfer impedance of a multiple-port network.

In general, the impedance for a device or component is a complex value composed of a real part and an imaginary part that depend on frequency. The real part represents the energy dissipative, or resistive, portion of impedance and the imaginary part represents the energy storage, or reactive, portion of the impedance. Expressed in rectangular form, impedance is usually shown as an equivalent circuit composed of a resistance in series with a reactance because these quantities can be expressed as a sum:

EQUATION 7–7

$$\mathbf{Z} = R \pm jX$$

where $\mathbf{Z}$ = complex impedance, Ω

R = resistive component of impedance, Ω

X = reactive component of impedance, Ω

In polar form, impedance is written as $Z \angle\theta$, where Z represents the absolute magnitude of the impedance vector and θ represents the angle. The important points are that, in general, impedance measurements must include *both* a resistive and a reactive quantity, and impedance is dependent on the frequency. The impedance characterization of any device is not complete unless the frequency at which the measurement was made is stated.

By definition, ideal passive components (resistors, capacitors, and inductors) contain only pure resistive or reactive elements. Unfortunately, there are no ideal passive components; all components contain resistive and reactive combinations. An inductor contains resistance and capacitance, resistors and capacitors contain inductance in the leads, and so forth. These unwanted effects depend on the many factors. Typically, the impedance of a real component is given in terms of ideal resistive and reactive elements. At *one* frequency, the equivalent circuit can be drawn either as a series resistor and a reactance or a parallel resistance and reactance.[1] This means that impedance-measuring instruments must be able to indicate two parameters for impedance at the test frequency. To be useful over a wide frequency range, the equivalent circuit for a component or device will need to be more complicated and will include more than two ideal components.

Because real components are not ideal, the impedance indicated by a test instrument may not be the same as the effective impedance of a component in a particular application. The indicated impedance can be affected by factors such as the test frequency, temperature, the test-signal level, and environmental factors. The test-signal level can affect the results because of distortion that can occur at high levels. Even the test fixture can affect the indicated impedance; for example, a small-value inductor can have its inductance changed by the presence of nearby ferrous materials. Connecting cables can also affect the measurement, due to capacitance effects.

A **vector impedance meter** is an instrument that indicates the impedance at a stated frequency or frequency range from a measurement of voltage and current in the unknown. The current vector is determined by measuring the voltage across a small test resistor. The vectors are separated into their x and y components by a phase detector and digitized. The unknown impedance is calculated from the results.

COMPONENT ANALYZER

A **component analyzer** is used primarily to determine if a capacitor or inductor (and sometimes other components) has failed. A dynamic component analyzer, called a Z meter by its manufacturer, is shown in Figure 7–32. It is particularly useful in servicing applications, as it can check for standard faults as well as perform dynamic tests. It can determine the magnitude of the reactive or resistive portion of impedance from a special test signal applied to the component and uses this information to indicate the value of the component. Value change is a common fault in capacitors and can occur from thermal stress and other factors. In addition, component analyzers check for other failure modes; for capacitors, these include tests for leakage current and the dielectric absorption. **Leakage current** flows when a dc voltage is applied to a capacitor. A small leakage current

[1] The parallel equivalent circuit is frequently shown in terms of admittance vectors rather than impedance vectors.

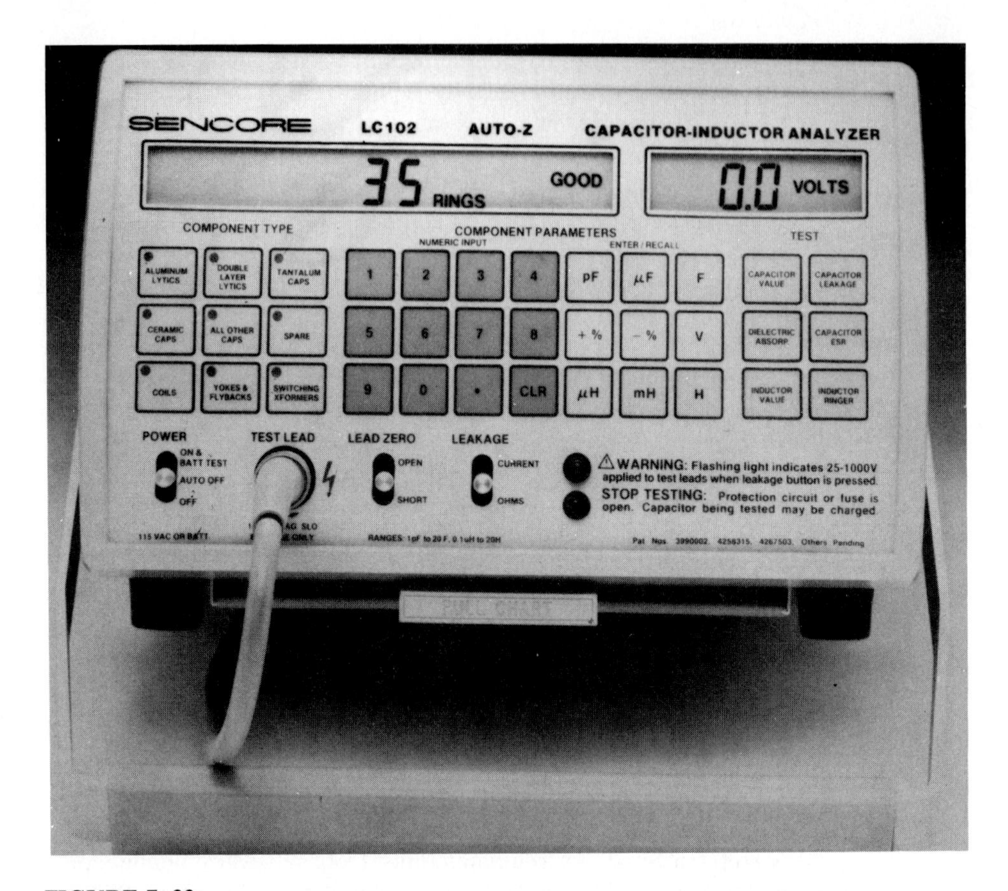

FIGURE 7–32
The Sencore LC102 dynamic component analyzer. Testing for the value of a capacitor or inductor is done by pushing a button and reading the display (courtesy of Sencore).

is normal for electrolytic capacitors but can become significant as the capacitor ages. **Dielectric absorption** is the result of internal dipoles remaining in a polarized state after the capacitor discharges, preventing the capacitor from releasing all its stored energy, even when a direct short is placed across the terminals.

Although value change is less common with inductors than capacitors, it can occur in inductors due to shorted windings or from a physical crack in the core material. An open winding is another failure; it is normally readily found with an ohmmeter, although occasionally it is an intermittant problem that can be missed. Frequently, an open is caused by excessive current in the inductor, a problem that may point to another circuit malfunction. If an inductor indicates the correct value, it is most likely okay. One way to measure inductance, devised by Sencore for its *Z* meter, is to observe the back emf across an inductor when a current ramp is applied. This test can directly indicate the inductance and can also indicate a winding breakdown that might not show up with a simple static test.

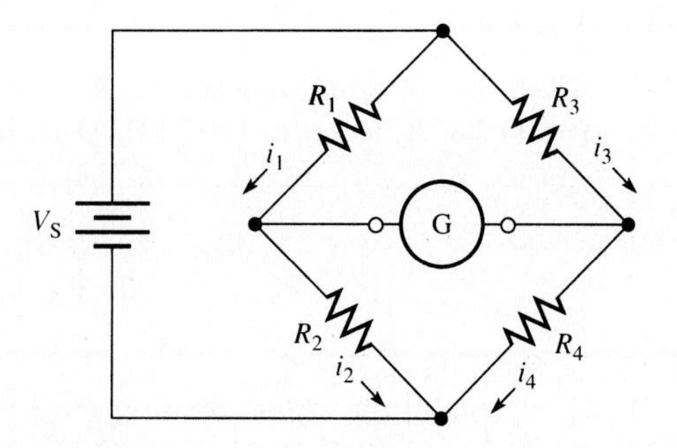

FIGURE 7–33
A basic Wheatstone bridge. Detection is shown using a galvonometer, a sensitive meter that detects current in either direction.

THE WHEATSTONE BRIDGE

The traditional instrument used for making accurate measurements of the impedance (resistance, reactance, or both) of a component is the bridge circuit, which is still used but is frequently automated. The **Wheatstone bridge** is the most basic of bridge circuits and can be used to measure an unknown resistance. It consists of four resistance arms with a source of energy (called the **excitation**) and a detector, indicated in Figure 7–33 by a sensitive meter. An important application of the Wheatstone bridge is for the sensitive measurement of resistive transducers (discussed in Section 13–4).

When any bridge is balanced, no current flows through the meter, and we can deduce that the voltage drops in corresponding arms of the bridge must be equal. For the Wheatstone bridge, we can therefore write $i_1R_1 = i_3R_3$ and $i_2R_2 = i_4R_4$. We also note that at balance $i_1 = i_2$ and $i_3 = i_4$. By algebra, we find that $R_1/R_2 = R_3/R_4$ at balance. This can be summarized by stating that *at balance, the ratio of the resistors in any two adjacent arms of the Wheatstone bridge must equal the ratio of the resistors in the remaining arms taken in the same sense.*[2] Another equivalent observation can be made: *at balance, the products of the diagonal resistances are equal.* That is

$$R_1R_4 = R_2R_3$$

The magnitude of the excitation voltage does not alter this conclusion, but it will affect the sensitivity for detecting any imbalance condition.

To find the value of an unknown resistance, it is placed in one of the arms of the bridge. The remaining arms consist of two standard resistors and a calibrated variable resistor that can be used to balance the bridge. The unknown resistor is connected by terminals located on the front of the bridge. Let us assume that R_{uk} is an unknown resistor and that it replaces R_4 in the bridge. The variable resistor is adjusted until balance is achieved. The value of $R_{unknown}$ can then be found from $R_{unknown} = R_3(R_2/R_1)$. In standard laboratory Wheatstone bridges, the ratio of R_2/R_1 can often be selected by fixed amounts in order to extend the range of the bridge. This ratio is referred to as a *multiplier*.

[2] "Taken in the same sense" means that if the first resistor ratio is formed from top to bottom, the second ratio must also be formed from top to bottom.

EXAMPLE 7–7 A Wheatstone bridge is balanced when the multiplier (R_2/R_1) is set to 10.00 and the variable resistance, R_3, is set to 1.817 kΩ. What is the value of the unknown resistance?

SOLUTION

$$R_{unknown} = R_3(R_2/R_1)$$
$$= 1.817\ \text{k}\Omega \times 10.00 = 18.17\ \text{k}\Omega$$

One of the chief advantages of the Wheatstone bridge is its great sensitivity to change. A high-precision bridge can measure resistance values with six digits of resolution and with an accuracy of 0.01%, although typical manual bridges have accuracies on the order of 0.1%. The chief disadvantage of bridge measurements is that, to obtain accuracy, the bridge must be balanced. The balancing procedure can be time consuming for manual bridges; however, automatic-balancing bridges can complete the process in hundreds of milliseconds even when range changing is required.

THE KELVIN BRIDGE

When making measurements of very small resistances (less than 1 Ω), lead resistance can be a problem that results in significant error in the measurement. A Wheatstone bridge is capable of detecting these very small resistances, but it cannot discern the difference between lead resistance and the unknown resistance. The situation is illustrated in Figure 7–34. Lead resistance may appear to be part of the unknown R_{uk} or part of R_3 and clearly affects the balance point. The **Kelvin bridge** (sometimes called the Kelvin double bridge) solves this problem with a second set of ratio arms that cancel the effect of lead and contact resistance. The Kelvin bridge can measure resistances as small as 1/10,000 Ω with an error of less than 0.1%. Applications include measuring switch contact resistance, motor winding resistance, or carbon brushes and quality control in wire manufacture.

The circuit for a Kelvin bridge is shown in Figure 7–35. For very small resistances, it is necessary to use relatively large currents in the bridge in order to

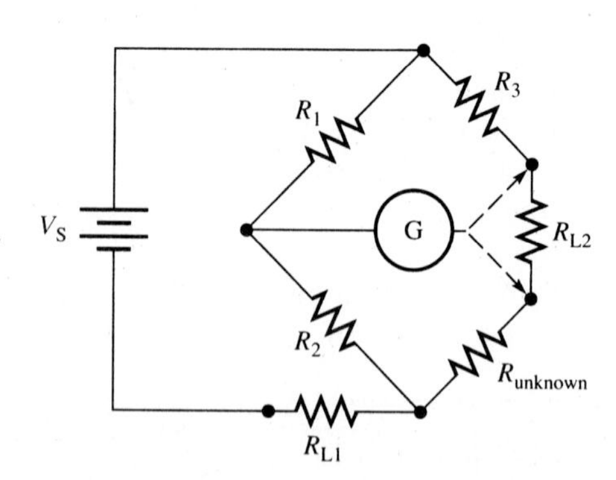

FIGURE 7–34
A Wheatstone bridge showing lead and contact resistances, R_{L1} and R_{L2}, from connecting resistor R_{uk}. R_{L2} can make R_{uk} look larger or smaller depending on the meter connection.

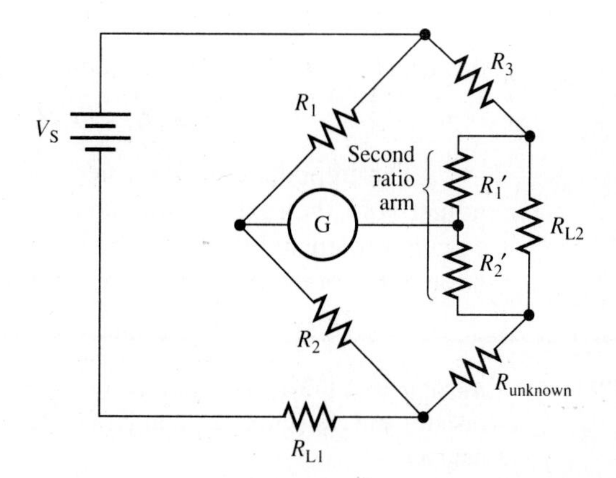

FIGURE 7–35
A Kelvin bridge. The second ratio arm cancels the effect of lead and contact resistance represented by R_{L2}.

maintain the desired sensitivity. Resistances R_{L1} and R_{L2} represent lead resistance due to connecting the unknown, R_{uk}, to the bridge. R_{L1} is outside the bridge circuit and has no effect on the measurement. However, R_{L2} develops a voltage due to the high current in the bridge arms that could affect the balance point if the meter were connected on one side or the other. It is this lead resistance with which the second ratio arm, containing R_1' and R_2', is in parallel. If the second ratio arm in the Kelvin bridge is set to the same ratio as the other side of the bridge, such that $R_1/R_2 = R_1'/R_2'$, then error due to any voltage drop in R_{L2} is not seen by the meter and is therefore eliminated.

ac BRIDGE CIRCUITS

An **ac bridge circuit** is a basic bridge circuit that uses a combination of capacitors, inductors, and resistors in the arms. It is excited by a sine-wave generator and detection is typically done with an ac meter or an oscilloscope. There are many variations of ac bridge circuits with various complex impedances in the arms, but nearly all ac bridges can be represented by the general impedance model shown in Figure 7–36. These bridges share a common relationship—namely, that *at balance, the ratio of the impedance in any two adjacent arms of an ac bridge must equal the ratio of the impedance in the remaining arms taken in the same sense.*

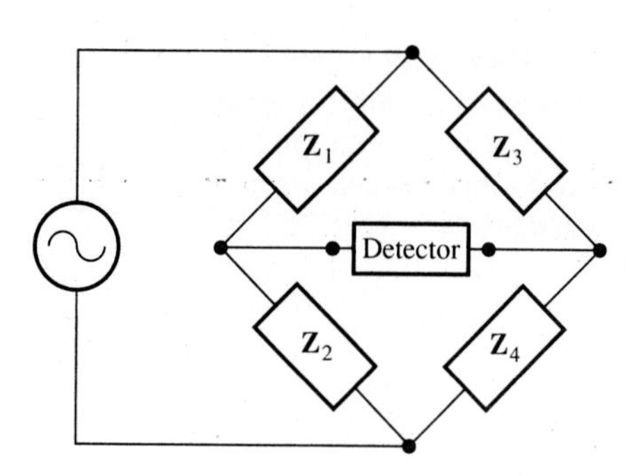

FIGURE 7–36
A general model of an ac bridge.

As in the case of the Wheatstone bridge, we can deduce that *at balance, the products of the diagonal impedances are equal.* That is

$$\mathbf{Z}_1\mathbf{Z}_4 = \mathbf{Z}_2\mathbf{Z}_3 \qquad \text{(balanced bridge)}$$

The impedances given here represent complex numbers. It is necessary to adjust *both* the real and imaginary components of impedance to achieve balance. In general, an ac bridge can be balanced only if a similar type of reactance is in an adjacent arm or an unlike reactance is in a diagonal arm.

EXAMPLE 7–8 Assume that the ac bridge shown in Figure 7–36 has $\mathbf{Z}_1$ and $\mathbf{Z}_2$ such that each has a series resistor and capacitor and that $\mathbf{Z}_3$ and $\mathbf{Z}_4$ are pure resistors. What are the two conditions for balance?

SOLUTION We can write the impedance of each leg in rectangular form as

$$\mathbf{Z}_1 = R_1 - jX_1 \qquad \mathbf{Z}_2 = R_2 - jX_2$$
$$\mathbf{Z}_3 = R_3 \qquad \mathbf{Z}_4 = R_4$$

The products of the diagonals are equal at balance. Therefore

$$(R_1 - jX_1)R_4 = (R_2 - jX_2)R_3$$

Expanding the expressions, we can write

$$R_1R_4 - jX_1R_4 = R_2R_3 - jX_2R_3$$

Equating the real and the imaginary parts, we find that the conditions for balance are

$$R_1R_4 = R_2R_3$$

and

$$X_1R_4 = X_2R_3$$

Many different ac bridges have been designed to optimize certain measurements. In general, ac bridges use capacitors for the reactive standards, even to test inductors. For example, the Maxwell bridge, shown in Figure 7–37, is useful

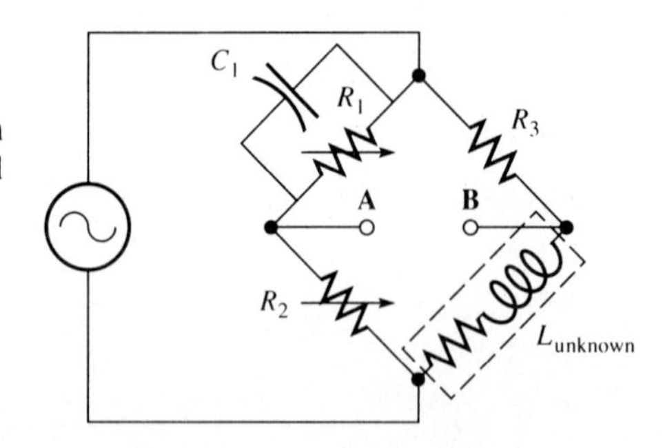

FIGURE 7–37
A Maxwell bridge, a circuit used to compare an unknown inductance against a standard capacitor.

FIGURE 7–38
The GenRad 1689M Digibridge™, an automated bridge. The 1689M is a programmable microprocessor testor, capable of making 50 *RLC* measurements per second (courtesy of GenRad).

for determining inductance by comparing it with a standard capacitor and resistors because ordinary capacitors come closer to ideal components than inductors and cost less. The Maxwell bridge can be used to find the inductive reactance and the resistive components of impedance for an inductor by varying only two resistors and keeping the capacitor fixed. The Maxwell bridge is best suited to measuring inductors with relatively low Q at the frequency of the bridge. The adjustments of the resistors tend to interact, as they will with any bridge for measuring a low Q or high D impedance, making balancing a manual bridge a slow process. When balance is achieved, the reactance can be converted directly to inductance and displayed.

Modern instrumentation has blurred the distinction between meters and bridges. In the past, bridges were used for accurate measurements, whereas meters were used when speed was required. Modern bridge measurements are made by automated instruments that achieve balance by feedback techniques. Some automated bridges, using digital readouts and microprocessor control, can make as many as 50 measurements per second. An example of an automated bridge is shown in Figure 7–38.

SUMMARY

1. PMMC meters are basically ammeters but can be designed to measure voltage, resistance, or some other quantity.
2. The sensitivity of a PMMC meter is measured by the full-scale deflection current, I_{fs}. The meter resistance is the internal resistance of the coil.
3. To modify a PMMC meter to measure

larger currents than its basic sensitivity, a parallel shunt divides the input current into a fraction that goes through the internal resistance of the meter and another fraction that goes through the shunt.

4. To modify a PMMC meter to measure voltage, a resistor is inserted in series that limits the current in the meter to the full-scale current when the full scale voltage is applied to the meter.
5. Ammeters and voltmeters can have loading error. An ammeter should have much lower resistance than the test circuit, whereas a voltmeter should have much greater resistance than the test circuit to avoid loading error.
6. Analog ohmmeters use an unknown resistance to complete a circuit. The scale is calibrated to read the resistance. The most popular type is the series ohmmeter, which uses a reverse, nonlinear scale.
7. VOMs are portable analog instruments that measure voltage, resistance, and current. VOMs are superior to electronic meters in certain applications, such as when rf interference is present.
8. Digital multimeters (DMMs) range from highly accurate 8½-digit meters used in calibration laboratories to low-cost portable instruments. DMMs are more accurate, are more stable, and include more features than analog meters.
9. The measurement capability of DMMs can be extended through the use of optional features and probes.
10. The accuracy of digital meters is typically specified as a percentage of the reading plus or minus a number of counts.
11. Electrometers are very sensitive instruments optimized to measure voltage, resistance, current and charge. Picoammeters and nanovoltmeters are very sensitive single-function instruments designed to measure current and voltage, respectively.
12. The Wheatstone bridge consists of four resistance arms with a source of excitation and a detector; it can be used for the measurement of an unknown resistance.
13. The Kelvin bridge is a modification of the Wheatstone bridge in which a second ratio arm is used to cancel the effects of lead and contact resistance. The Kelvin bridge is useful for measuring very small resistances.
14. AC bridges can be used to determine resistance, inductance, or capacitance. To balance the bridge, both the real and imaginary components of impedance must be balanced.

QUESTIONS AND PROBLEMS

1. What is parallax error? How can you avoid parallax error?
2. A PMMC meter has a full-scale deflection current of 1.0 mA and an internal resistance of 50 Ω. Compute the shunt resistor required to increase the current rating to 100 mA full scale.
3. Assume the meter in Problem 2 is to be used as a voltmeter with a full-scale reading of 10 V.
   (a) What multiplier resistor is required?
   (b) What is the ohms-per-volt rating of the voltmeter?
4. How can an ammeter loading affect a circuit under test?
5. A basic 100 μA meter has an internal resistance of 250 Ω. Compute individual multiplier resistors needed to make the meter read 1, 2, 5, and 10 V full scale.
6. A series ohmmeter uses the same meter

movement specified in Problem 5 along with an internal series 29,750 Ω resistance and 3 V battery. There is no shunt resistance. Draw a scale for the meter. Mark the 15 kΩ, 30 kΩ, and 60 kΩ positions on the scale as well as 0 Ω and infinite ohms locations.

7. A PMMC meter has an internal resistance of 100 Ω and a full-scale deflection current of 50 μA.
   (a) What is the lowest full-scale voltage that could be displayed?
   (b) What is the Ω/V rating at the meter?
8. Explain the advantage of the Ayrton shunt for a multiple-range ammeter.
9. What precautions should be taken when using a VOM to measure dc current?
10. Explain what is meant by the term loading error for a dc voltmeter.
11. What error will result from measuring a square wave with an average reading meter?
12. A VOM has 20,000 Ω/V rating and is used on the 30 V scale.
    (a) What is voltmeter resistance?
    (b) Assume the meter is used to measure a Thevenin source with a 10 kΩ Thevenin resistance. What is the reading on the meter if the actual Thevenin voltage is 15 V?
13. Assume a reverse-biased diode is in series with a 1.0 MΩ resistor and a 5.0 V source. A student decides to determine the current in the circuit with a VOM. The student finds no voltage across the resistor but 0.8 V across the *reverse*-biased diode. Suggest the most likely cause of the problem.
14. A full-wave bridge ac voltmeter uses a PMMC meter with a sensitivity of 50 μA and an internal resistance of 250 Ω.
    (a) What series resistance must be used to produce a full scale reading of 10 V rms?
    (b) The meter is used to measure the voltage across the resistors shown in Figure 7–39. What voltage does it measure across each resistor?
    (c) What is the percent error due to meter loading?

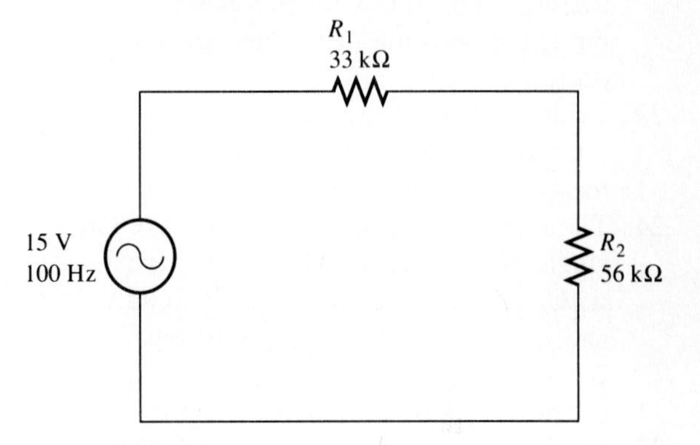

FIGURE 7–39

15. Assume you want to measure the amplitude of a sinusoidal voltage waveform riding on a dc level. How would you use a VOM to make this measurement?
16. The dB scale on a VOM is used to compare two power levels in the same resistance. The first reading is 22 dB and the second reading is 32 dB. Compare the ratio of the two powers.
17. To use the dB scale of a VOM to measure actual power in a circuit, it is necessary to make measurements across a 600 Ω load. Explain.
18. Assume the circuit in Figure 7–17 uses a PMMC meter with a full-scale current of 50 μA and an internal resistance of 1 kΩ. The closed-loop gain of the amplifier is 101. Compute the required series resistance R_s to complete the circuit.
19. Explain what is meant by 1/2 digit when expressing the resolution of a DMM.
20. What is the resolution of a 4½-digit meter on the 20 V range? Express your answer as the smallest difference in voltage that the meter can display.
21. Compare the advantages and disadvantages of using a VOM with a DMM for voltage measurements.

22. A meter has a specified accuracy of: ±0.1% rdg + 2 digits at 25°C. The meter is a 4½-digit meter with an input voltage of 5 V and is on the 20 V scale. Compute the error possible with this meter assuming it is operating at the design temperature.
23. Under what circumstance is a nanovoltmeter more suitable than an electrometer for a low-level measurement?
24. The voltage and current are measured in a single-phase circuit with a reactive load with a power factor of 0.27. The load voltage is 220 V rms and the load current is 2.8 A. Determine the true power, apparent power, and reactive power in the load.
25. Explain how a bolometer can measure the power from a high-power microwave transmitter.
26. What is the stimulus and what is the response for measuring the open-circuit transfer impedance of a two-port network?
27. What is the condition for balance for a Wheatstone bridge?
28. Sketch a Kelvin bridge and explain why the lead and contact resistances are canceled.
29. Assume that $\mathbf{Z}_1$ and $\mathbf{Z}_2$ of the ac bridge in Figure 7–36 consist of a parallel resistor and capacitor and that $\mathbf{Z}_3$ and $\mathbf{Z}_4$ are pure resistors. What are the two conditions for balance?

Chapter 8

Oscilloscopes

OBJECTIVES

The word oscilloscope comes from two root words—"oscillo," short for oscillations, and "scope," derived from a Greek word meaning "an instrument for viewing." Thus the oscilloscope is literally an instrument for viewing oscillations by forming a graph of voltage versus time. The oscilloscope has become far more than this definition implies. It can be used for measuring details about complex waveforms from dc to microwave frequencies and can measure nonelectrical phenomenon by using an appropriate transducer. Oscilloscopes are among the pioneering electronic instruments; yet today they are still the principle instrument for observing and measuring waveforms in both linear and digital systems. In recent years, the oscilloscope has become even more versatile with the addition of automated measurement, comparison, calculation, and control features. High-performance digital scopes contain new features that are unavailable in analog scopes. In this chapter, both analog and digital scopes are covered. When you complete this chapter, you should be able to

1. Given a basic block diagram of a general-purpose oscilloscope, explain the function of each block.
2. Describe the elements of an electrostatic CRT, including the heater, cathode, electron gun, deflection system, aquadag, phosphors, and graticule.
3. Specify the major controls for a general-purpose oscilloscope and the purpose of each control.
4. Explain how to use a delayed-sweep oscilloscope to make a differential delayed-sweep measurement.
5. Compare the common types of oscilloscope probes, including performance trade-offs for various measurements.
6. Given the equation for a sinusoidal waveform, find the peak amplitude, average, and rms values of the wave and the phase angle.
7. Explain how to make timing, amplitude, and phase angle measurements as well as pulse measurements with a general-purpose oscilloscope.
8. Compare digital storage oscilloscopes with general-purpose analog oscilloscopes.

HISTORICAL NOTE

The first true digital oscilloscope was inspired by a high school science project in 1970. The student, Jole Shackleford, corroborated with his stepfather, Robert Schumann, founder of Nicolet Instruments, to work on the project. Schumann, intrigued with the possibilities of a digital scope, set to work on the project at his company. After months of development, the first commercial digital oscilloscope, the 1090A Explorer, was marketed by Nicolet in 1972. Nicknamed "Schumann's folly" by company executives, one executive bet Schumann that the company would not sell 100 digital scopes. The loser was to roll a peanut with his nose for 1 mile down State Street in Madison, Wisconsin. Although Schumann won the bet, it was never paid off.

8–1 OSCILLOSCOPE BASICS

INTRODUCTION

An oscilloscope (scope, for short) uses a cathode-ray tube (CRT) to display a faithful reproduction of a voltage-versus-time waveform. It does this by causing an electron beam to draw a graph on the inside of a phosphor-coated screen. The waveform "stands still" because the same information is written repeatedly and because of the persistence of vision of the human eye combined with the persistence of the phosphors that coat the CRT. The vertical axis of the graph is calibrated in volts and the horizontal axis is calibrated in units of time, although these can be converted to other units for certain applications.

Oscilloscopes are used to examine both analog and digital signals, allowing the user to obtain both time and amplitude information. Oscilloscopes can display a wide range of repetitive and nonrepetitive signals—from low frequency to high frequency and from low amplitude to high amplitude. In order to accomplish this, the trace must be initiated (**triggered**) at the proper time, whether it is a single event, a very slowly repetitive waveform, a complicated digital pattern, or a line from a television signal. Nearly all general-purpose oscilloscopes have two or more channels in order to compare waveforms. They also have a number of other features that will be explained in this chapter. For example, most oscilloscopes can show the sum or difference between two signals directly on the screen. Because of its great versatility, the oscilloscope is one of the most important electronic test instruments.

BLOCK DIAGRAM

There are four major subsystems in an oscilloscope, as shown in the block diagram in Figure 8–1. The input signal is connected to the *vertical* section, which can be set to attenuate or amplify the input signal to provide the proper voltage level to the vertical deflection plates of the CRT. The *trigger* section samples the input waveform and sends a synchronizing trigger signal at the proper time to the horizontal section. The trigger occurs at the same relative time to superimpose each succeeding trace on the previous trace. This action causes the signal to

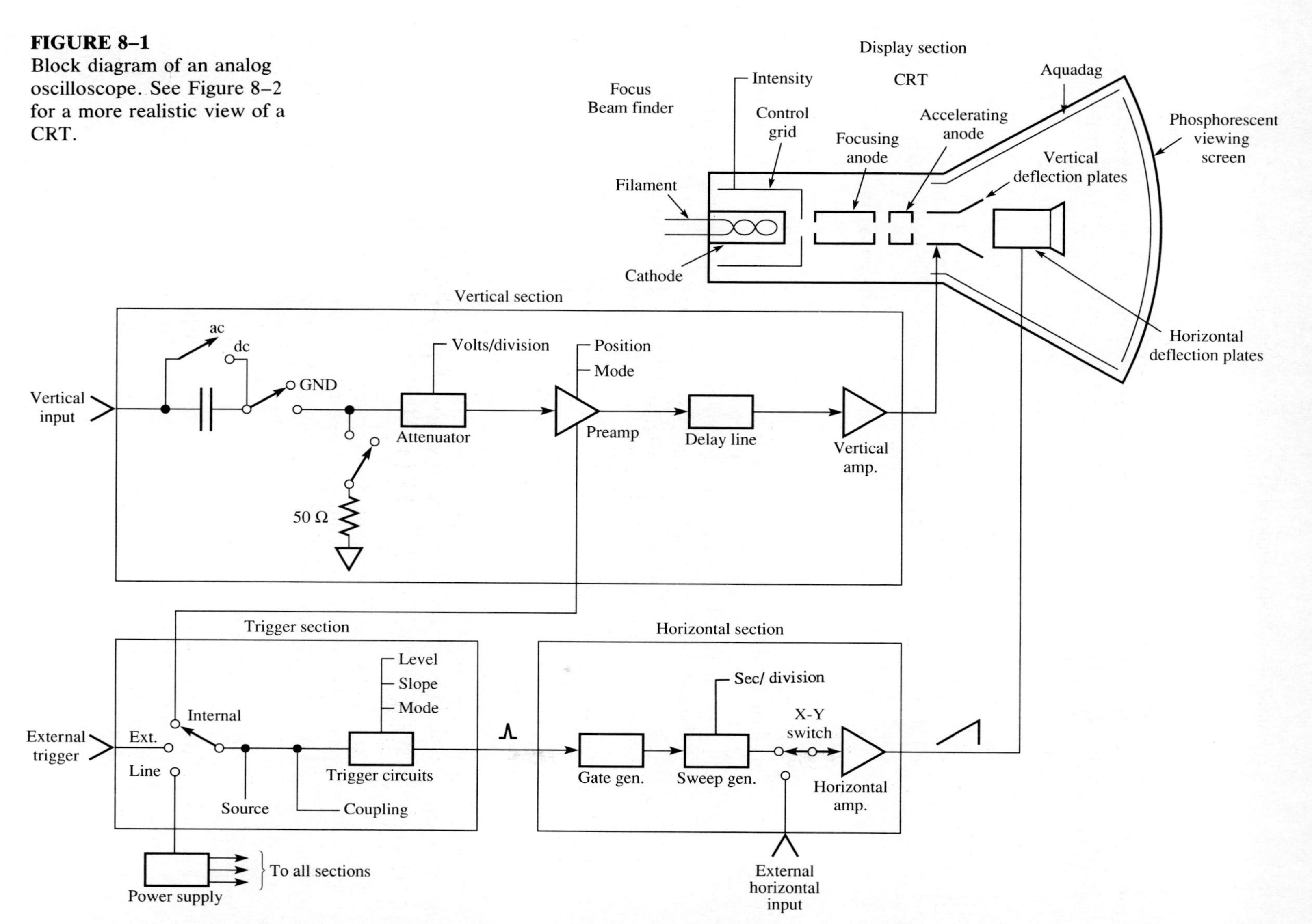

FIGURE 8–1
Block diagram of an analog oscilloscope. See Figure 8–2 for a more realistic view of a CRT.

appear to stop, allowing the viewer to examine the signal. The *horizontal* section contains the time-base (or *sweep*) generator, which produces a linear ramp, or "sawtooth," waveform that controls the rate the beam moves across the screen. The horizontal position of the beam is proportional to the time that elapsed from the start of the sweep, allowing the horizontal axis to be calibrated in units of time. For this reason, the horizontal section is often called the time base. The output of the horizontal section is applied to the horizontal deflection plates of the CRT. Finally, the *display* section contains the CRT and beam controls. It enables the user to obtain a sharp presentation with the proper intensity. The display section frequently contains other features. Sometimes, the user can lose the displayed waveform by accidently positioning it offscreen. One common feature in the display section is a beam-finder button that enables the user quickly to locate the position of the trace.

The major controls for each section are included on the block diagram in Figure 8–1. Controls for a particular section are usually grouped together on the front panel and frequently have color-coding around related controls. Variable controls that can be used to uncalibrate the oscilloscope are often color-coded in red, and some oscilloscopes provide a visual warning light on the front panel when controls are out of the calibrate position.

Signals should always be coupled into the oscilloscope through a probe. The probe reduces the loading effect on circuits. Probes also have a short ground lead that should be connected to a nearby circuit ground point to avoid oscillation and power line interference. Probes are covered in Section 8–5.

POWER SUPPLY

Supporting each of the preceding sections is a power supply section containing both low- and high-voltage supplies. The power supply converts ac line voltage into the regulated dc voltage required by the various other sections. Low voltages provide power for the preamplifiers, trigger section, and logic circuits. The vertical and horizontal power amplifiers require higher voltages—typically 100 V—to provide sufficient drive for the deflection plates of the CRT. Very high voltages are required by the CRT and are generated by a separate high-voltage power supply. The cathode is typically operated at 1 to 2 kV below ground, whereas the post-deflection acceleration electrode may be operated at 10 kV or more. The CRT grid voltages have very low power requirements, so their voltages can be obtained from the other supplies using a voltage divider.

SPECIAL FEATURES

Many specialized features have been developed for oscilloscopes. One special feature is the inclusion of a horizontal magnifier that allows the user to expand a section of the display in order better to observe details. The magnification amount can be calibrated to enable the sweep time to remain calibrated. Some oscilloscopes can show both the magnified and unmagnified trace at the same time. Another feature is to add a second time base, which allows the user to examine a particular portion of a waveform that occurs well after the trigger.

With the increasing complexity of electronic systems, manufacturers are paying more attention to ease of use. Some newer oscilloscopes display numeric parameters on the screen in addition to the waveforms. These on-screen readouts include control settings such as the time per division or volts per division and may include measurement parameters about the waveform under test, such as period, frequency, duty cycle and the like. Some oscilloscopes contain built-in counters and DVMs. Another feature is pairs of lighted markers (called **cursors**) that can be positioned on the display by the user to indicate amplitude and time values automatically. For time measurements, the first cursor, called the *reference cursor,* is positioned on the left side of the trace. The second cursor, called the *delta cursor,* is positioned at the end of the time interval. The time between the cursors can be displayed on the scope face. Features such as automated waveform measurements and cursors are designed to simplify the making of measurements and repetitive operations. For the novice user, the additional controls may give a feeling that the oscilloscope is very complex, but by concentrating on the basic blocks and with practice, the operation of any oscilloscope can be mastered.

8–2 THE CATHODE-RAY TUBE

BASIC ELEMENTS

A **cathode-ray tube** (CRT) is a vacuum tube in which a trace is made visible on a phosphorescent viewing screen that glows when a narrow beam of electrons strikes the phosphors. There are three basic elements to a CRT. These are (1) an electron gun, which produces and focuses the beam of electrons (called the **primary electrons**), (2) a means of deflecting the beam according to the signals that are present, and (3) the screen, which converts the electron beam into visible light. The inside of the glass tube is coated with a conductive graphite material called *Aquadag,* which is connected to a positive voltage and collects electrons (called **secondary electrons**) emitted from the phosphorous screen and returns them to the high-voltage supply. A typical CRT is shown in the diagram in Figure 8–2.

Ideally, the viewer would like the CRT to be large, precise, bright, and uniform. In addition, it should have high sensitivity, low capacitance for higher speeds, and short overall length. The manufacturer also wants it also to be reliable and low in cost. These various parameters interact with each other, so CRT designs are a compromise between the various attributes.

THE ELECTRON GUN

The electron gun consists of a heater and cathode, a control grid, and a series of electrodes that accelerate and define the electron beam. The type, spacing, and voltage requirements of the electrodes depend on the type of gun assembly, but typically they require increasing voltages further from the cathode. The purpose of the heater is to warm the cathode to a sufficiently high temperature to cause electrons to escape from the surface. This process is called **thermionic emission** of

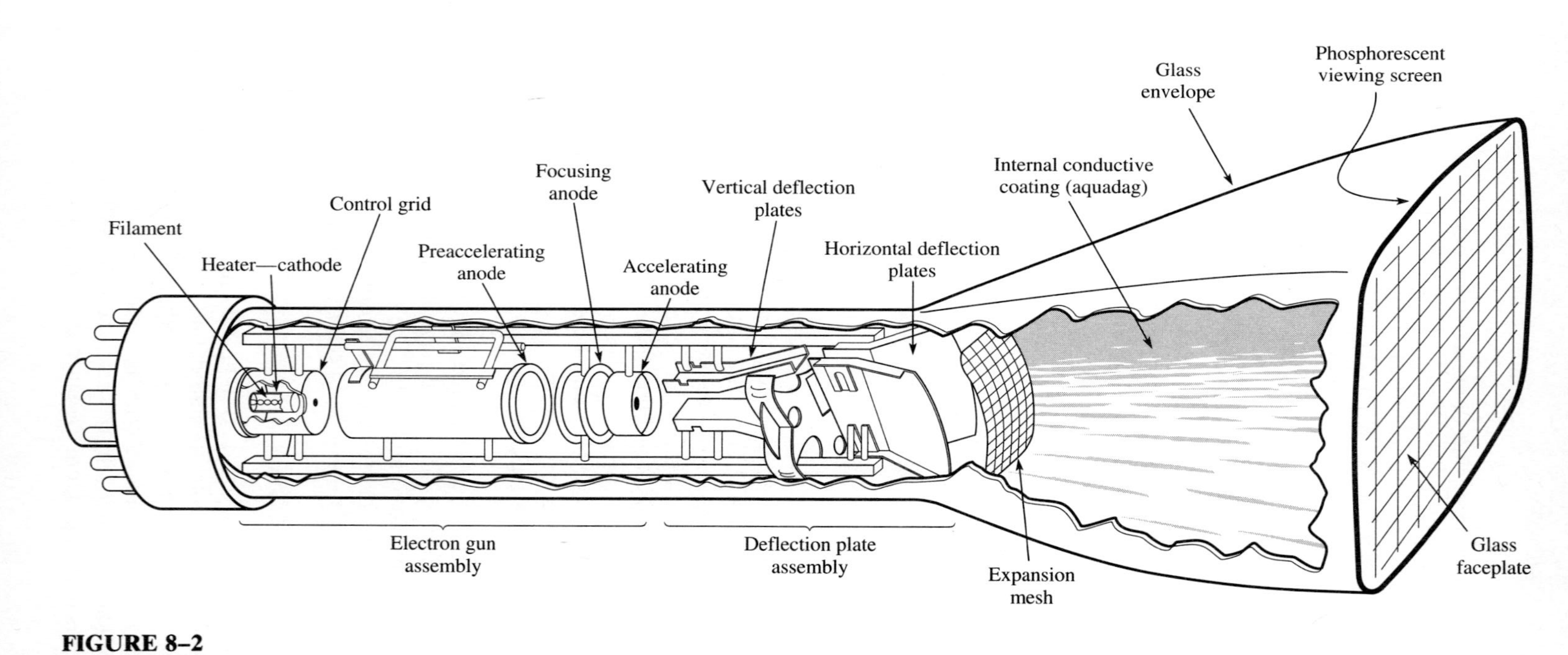

FIGURE 8–2
Cutaway view of a cathode-ray tube.

electrons and is similar to evaporation of molecules from the surface of a liquid. As the temperature rises, the number of electrons that can escape from the cathode increases. The heater usually requires a 6.3 V ac source supplying several hundred milliamperes of current. The electrons are drawn from the cathode by a positive electrostatic field passing through a control grid. The control grid, unlike vacuum-tube grids, is a metal cup with a hole in the cap through which the electrons are allowed to escape. The bias on the control grid is adjusted by the intensity knob, which, in turn, determines the beam current (or brightness of the spot). If the control-grid voltage is made sufficiently negative, the beam can be completely cut off. The control-grid voltage can also be varied by an input to the z-axis, not available on all oscilloscopes. Following the control grid is an accelerating electrode and, typically, two anodes that form a lens to focus the beam. The first anode is called the *focusing anode*. The focus knob controls the voltage to this anode. The second anode is an *accelerating anode*. The combination of geometry and electric fields from these electrodes acts to concentrate the beam into a fine spot in a way similar to the control of light by optical lenses. Some CRTs use external magnetic fields to focus the beam, but magnetic focus is primarily used in TV tubes.

DEFLECTION SYSTEMS

An electron beam can be deflected either electrostatically or magnetically. Magnetic deflection systems permit higher beam currents and are useful for the large deflection angles common in television; however, electrostatic deflection is almost always used for general-purpose oscilloscopes as it can be used at much higher frequencies than magnetic deflection. Inside the CRT are two pairs of metallic deflection plates—one pair for each axis. The electron beam is moved (deflected) in the x-axis direction by the voltage on the horizontal deflection plates and in the y-axis direction by the voltage on the vertical deflection plates. The horizontal plates are connected to the horizontal section, which provides the sweep voltage. The vertical plates are connected to the vertical section, which is indirectly connected to the signal voltage. For the past several years, Tektronix has used a quadrapole deflection system that simplifies the manufacturing process and avoids the need for precision stamped deflection plates.

Some tubes have a screen grid located after the deflection plates. This grid provides additional acceleration of the electrons after deflection and allows the tube to have a high writing rate. **Writing rate** is an indication of the ability of a CRT to give off light; a high writing rate is necessary to show rapidly changing waveforms. After the screen grid, most modern electrostatic tubes use an expansion mesh to increase the deflection factor without increasing the deflection voltage or lengthening the CRT funnel. This allows the CRT to be up to 50% shorter.

In an electrostatic deflection system, the amount the beam is moved from the center of the screen is proportional to the deflection voltage and is inversely proportional to the accelerating voltage as well as geometric factors for the tube determined from the equation

EQUATION 8–1

$$D = \frac{Ll}{2d}\frac{V_{\mathrm{d}}}{V_{\mathrm{a}}}$$

where D = the deflection, cm
V_d = the deflection plate voltage, V
L = the length from the center of the deflecting field to the face of the CRT, cm
l = the length of the deflecting plates, cm
V_a = the accelerating voltage between the cathode and the final anode, V
d = the separation distance between the deflection plates, cm

The geometric factors are illustrated in Figure 8–3. Rearranging Equation 8–1 gives the deflection sensitivity. The deflection sensitivity is the ratio of the deflection plate voltage to the deflection distance measured on the screen:

EQUATION 8–2

$$\frac{V_d}{D} = \frac{2dV_a}{Ll}$$

The units for deflection sensitivity are volts/centimeter. Typical CRTs have a deflection sensitivity of about 10 V/cm. This sensitivity implies that a 10 V difference between the vertical plates would deflect the beam on the face of the CRT by 1 cm. This basic sensitivity of the CRT is greatly increased at the signal input by the oscilloscope's vertical amplifier section. A typical oscilloscope may have a vertical sensitivity of as little as 1 mV/cm.

Most electrostatic CRTs use postdeflection acceleration, which means that the electron beam is accelerated further after passing the deflection plates. A common method of applying postdeflection acceleration voltage is through a resistive helix that surrounds the CRT. Another method uses conducting mesh electrodes located after the deflecting plate and is designed not to alter the path of the electrons. The postdeflection acceleration voltage is a potential several times higher than the final electrode. Since postdeflection acceleration is done after the beam has passed the deflection plates, it offers the advantage of allowing good

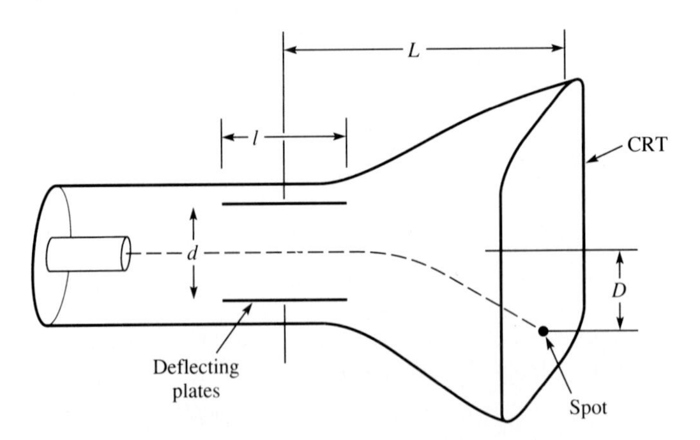

FIGURE 8–3
Geometric factors for electrostatic deflection.

deflection sensitivity while providing a brighter image. The postacceleration potential is not part of Equation 8–1.

Another new development in CRTs, called a **microchannel plate,** has enabled a major improvement in the writing rate on higher-priced oscilloscopes by amplifying the beam current just before it strikes the phosphor. Microchannel plate (MCP) is a 0.050 in.–thick glass plate with millions of tiny, closely spaced holes (microchannels). The holes are offset slightly from the axis of the tube and are internally treated to promote the production of secondary electrons. The electron beam enters the holes and produces a great increase in beam current due to the secondary electron emission. The amplified beam then strikes the phosphor and produces a trace. This technology enables the user to view very fast signals that would otherwise be invisible.

EXAMPLE 8–1 A CRT with 3.0 cm plates separated by a distance of 1.0 cm is operated with an accelerating voltage of 1.0 kV. The length of the CRT from the deflecting field to the face of the CRT is 22 cm. What voltage is required between the deflection plates to deflect the beam by 4 cm on the screen?

SOLUTION Solving Equation 8–2 for V_d gives

$$V_d = \frac{2dV_aD}{Ll}$$

$$= \frac{2(1.0\text{ cm})(1000\text{ V})(4.0\text{ cm})}{(22\text{ cm})(3.0\text{ cm})} = 121\text{ V}$$

PHOSPHORS

A **phosphor** is a crystalline compound coated on the inside of the faceplate of a CRT that emits visible light when it is bombarded by the energetic electrons from the beam. The color of the light emitted is determined by the type of phosphor. The property of a phosphor to emit light when struck by electrons is called **fluorescence.** Phosphors have another property, called **phosphorescence,** which is the property of a material to emit light after the source of excitation (the electron beam) has been removed. The light from a phosphor does not reach its maximum the instant it is struck by the electron beam; neither does it disappear immediately. **Persistence** refers to the length of time that a phosphor continues to glow after excitation by the electron beam. Figure 8–4 is a persistence graph for a typical oscilloscope tube phosphor showing that light continues to be emitted after the source of excitation (cathode current) has been removed. Phosphors are specified using a two-letter symbol indicating the main characteristics of phosphor screens. This system, called the Worldwide Phosphor Type Designation System (WTDS) is replacing the older P-number designation system that had been used in the United States since 1945. The first letter in the WTDS indicates the fluorescent color of the phosphor based on a chromaticity chart, whereas the second letter

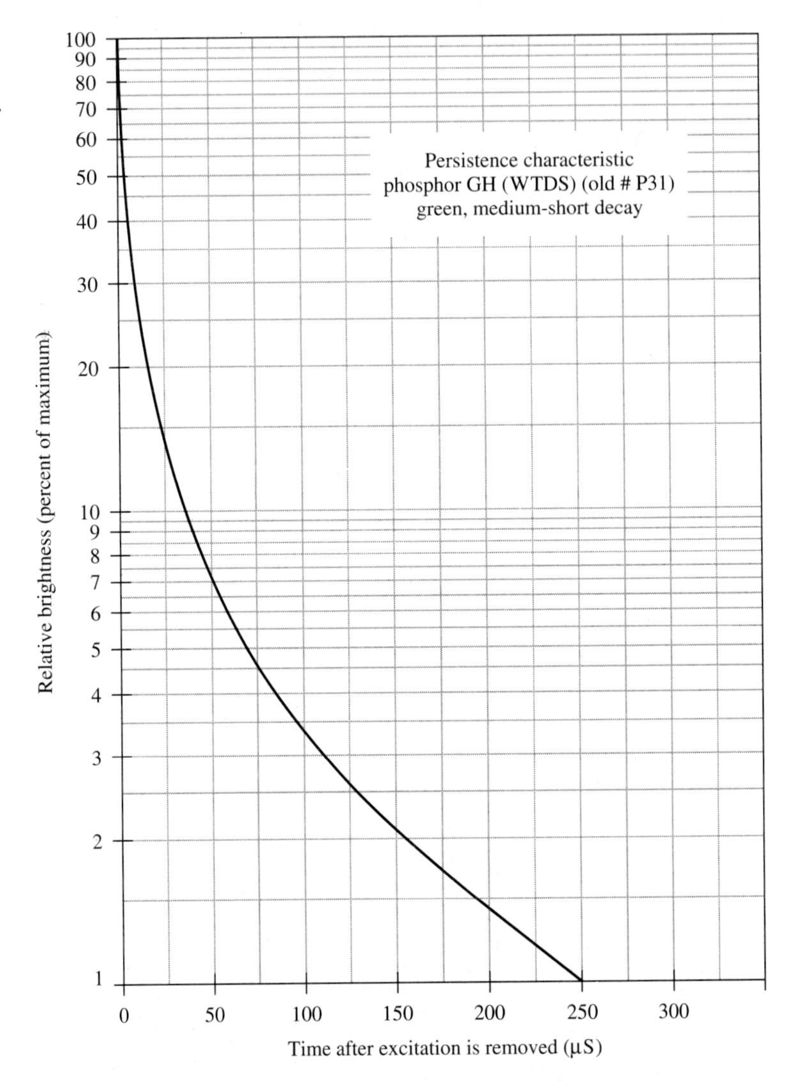

FIGURE 8–4
Persistence of typical oscilloscope phosphor.

denotes other differences in screen characteristics. The efficiency of a phosphor is the ratio of the light energy emitted to the excitation energy of the electron beam. The type of phosphor determines the efficiency, relative luminance, photographic response, and persistence as well as other factors. Various phosphors are used in CRTs, depending on the application.

Some CRTs have a thin, reflective coating of aluminum on the rear of the screen. The aluminum is almost transparent to high-energy electrons but serves to direct more of the light toward the viewer. In addition the coating helps remove primary electrons from the beam.

GRATICULES

The **graticule** is a calibrated screen with ruled divisions, similar to graph paper. On older scopes, the graticule was a separate screen, but on most newer scopes the graticule is ruled directly on the inner surface of the CRT to avoid parallax. This implies that any misalignment of the graticule with the horizontal beam must be corrected electronically. Typically, the graticule is a screen 10 divisions long and 8 divisions high, with major divisions at each centimeter. Minor divisions are scribed along the *x*- and *y*-axes. In addition, to facilitate making rise-time measurements, graticules are frequently marked with a vertical percentage scale. Some oscilloscopes are supplied with separate controls for illuminating the graticule, a feature that is particularly useful for photographic applications.

PROBLEMS IN THE DEFLECTION ASSEMBLY

It is not possible to build a "perfect" CRT. Problems with CRTs include internal noise, which can affect trace focus, and deviations of the beam from the graticule

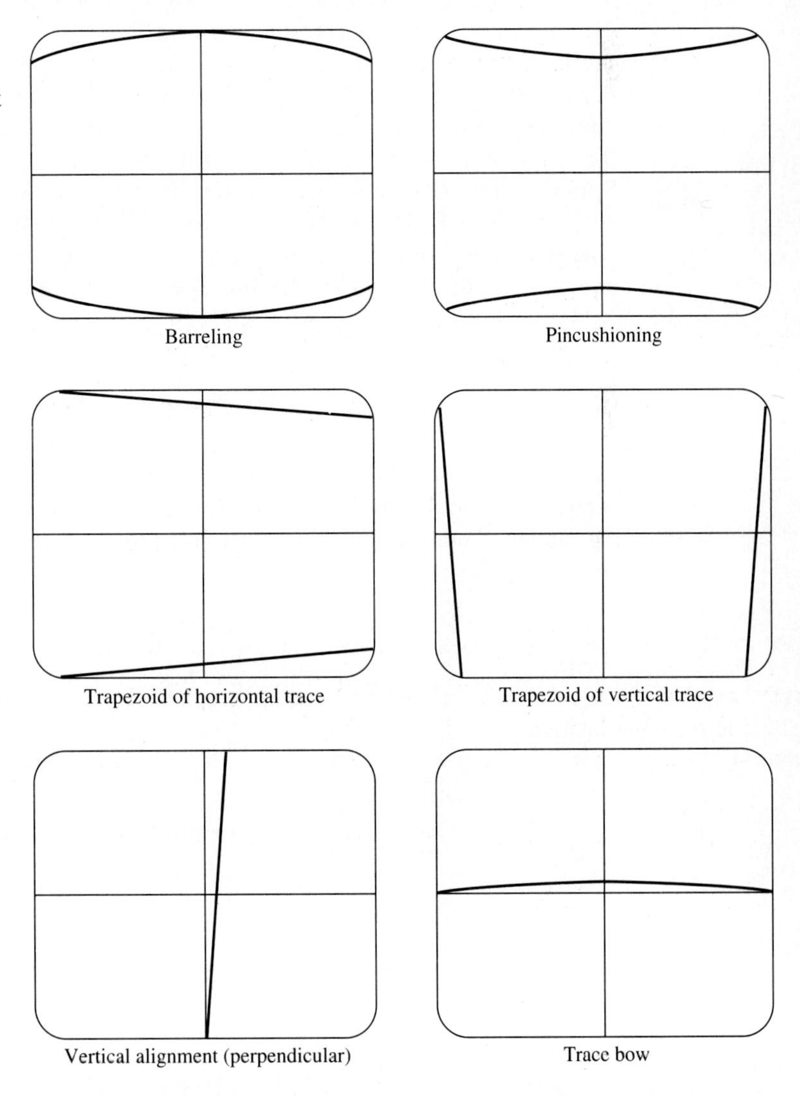

FIGURE 8–5
CRT pattern distortions. Most pattern distortion is not as severe as the examples shown.

lines. The deviations, called *pattern distortions,* are most frequently noticed at the extreme edges of the CRT and are shown in exaggerated form in Figure 8–5. Because of these distortions, it is best to make measurements within the center region of the oscilloscope. Horizontal nonlinearities can be observed by connecting a crystal marker to the oscilloscope and looking for irregular spacing of the markers.

8–3 OSCILLOSCOPE CONTROLS

DISPLAY CONTROLS

The display system on an oscilloscope contains controls for adjusting the electron beam, including FOCUS and INTENSITY controls. FOCUS and INTENSITY are adjusted for a comfortable viewing level with a sharp focus. The display section may also contain the BEAM FINDER, a control used in combination with the horizontal and vertical POSITION controls to bring the trace on the screen. Another control over the beam intensity is the z-axis input. A control voltage on the z-axis input can be used to turn the beam on or off or adjust its brightness. Some oscilloscopes also include the TRACE ROTATION control in the display section. TRACE ROTATION is used to align the sweep with a horizontal graticule line. This control is usually adjusted with a screwdriver to avoid accidental changes. Figure 8–6 illustrates the effect of trace rotation.

VERTICAL CONTROLS

Vertical controls include the VOLTS/DIV (vertical sensitivity) control and its vernier, the input COUPLING switch, and the vertical POSITION control. Dual-trace and multiple-channel oscilloscopes have a duplicate set of these controls for each channel and various switches for selecting channels or other vertical operating modes. The vertical input is connected through a selectable attenuator to a

FIGURE 8–6
Effect of the trace rotation control.

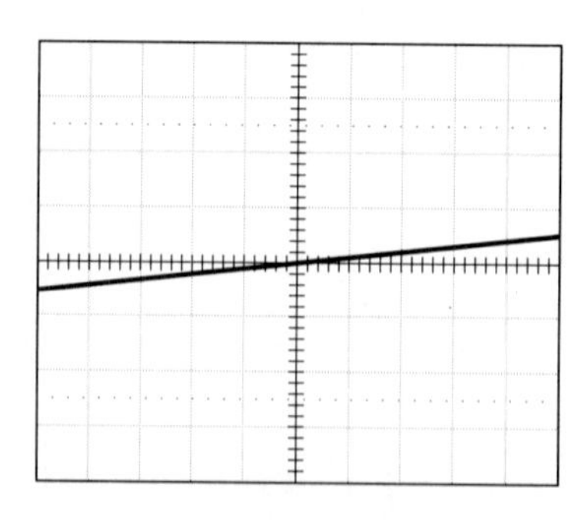

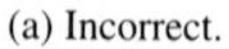

(a) Incorrect.

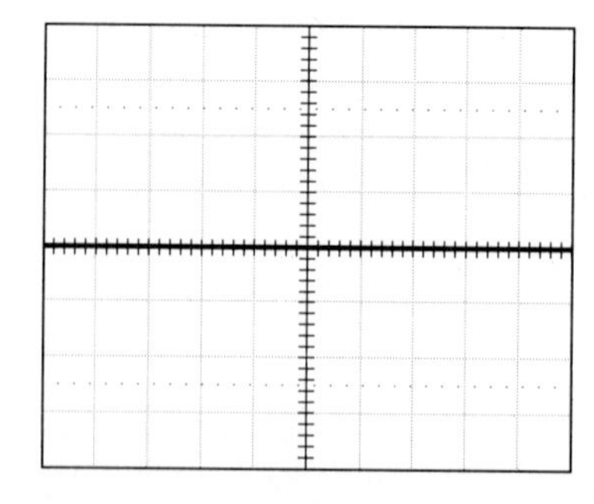

(b) Correct.

high-input impedance dc amplifier. The VOLTS/DIV control selects a combination of attenuation and gain to determine the vertical sensitivity of the scope. For example, a low-level signal will need more gain and less attenuation than a higher-level signal. The vertical sensitivity is adjusted in fixed VOLTS/DIV increments to allow the user to make calibrated voltage measurements. In addition, a concentric vernier control is usually provided to allow a continuous range of sensitivity. This knob must be in the detent (calibrated) position to make voltage measurements. The detent position can be felt by the user as the knob is turned because the knob tends to "lock" in the detent position. Some oscilloscopes have a warning light or message to indicate when the vernier is not in its detent position.

The input-coupling switch is a multiple-position switch that can be set for AC, GND, or DC and sometimes includes a 50 Ω position. The GND position of the switch internally disconnects the signal from the scope and grounds the input amplifier. This position is useful if you want to set a ground-reference level on the screen for measuring the dc component of a waveform. The AC and DC positions are high-impedance inputs—typically 1 MΩ shunted by 15 pF of capacitance. High-impedance inputs are useful for general probing at frequencies below about 1 MHz. At higher frequencies, the shunt capacitance can load the signal source excessively, causing measurement error. Attenuating divider probes are good for high-frequency probing because they have very high impedance (typically 10 MΩ) with very low shunt capacitance (as low as 2.5 pF). More information on this is given in Section 8–5.

The AC position of the coupling switch inserts a series capacitor before the input attenuator, causing dc components of the signal to be blocked. This position is useful if you want to measure a small ac signal riding on top of a large dc signal—as in power supply ripple, for example. The DC position is used when you want to view *both* the ac and dc components of a signal. This position is best when viewing digital signals because in the AC position, the input *RC* circuit forms a differentiating network. The AC position can distort the digital waveform because of this differentiating circuit. The 50 Ω position places an accurate 50 Ω load to ground. This position provides the proper termination for probing in 50 Ω systems and reduces the effect of the variable load that can occur in high-impedance termination. The effect of source loading *must* be taken into account when using a 50 Ω input. It is important not to overload the 50 Ω input because the resistor is normally rated for only 2 W—implying a maximum of 10 V rms of signal can be applied to the input.

The vertical POSITION control varies the dc voltage on the vertical deflection plates, allowing the trace to be positioned anywhere on the screen. Each channel has its own vertical POSITION control, enabling the two channels to be separated on the screen. You can use vertical POSITION when the coupling switch is in the GND position to set an arbitrary level on the screen as ground reference.

Some oscilloscopes also have a vertical magnifier. The magnifier increases the scope sensitivity by providing more gain to the input signal at the expense of the overall bandwidth. The vertical magnifier is useful for measuring very low level signals that are not bandwidth-limited.

EXAMPLE 8–2 An oscilloscope has a dc voltage connected to its input and shows the horizontal sweep on the bottom graticule line when the input coupling switch is in the GND position. When the input coupling is changed to the DC position, the horizontal sweep moves up 5.2 divisions. If the vertical sensitivity is set to 50 mV/div, what dc voltage is being measured?

SOLUTION The vertical deflection is 5.2 div. Multiplying this by the sensitivity gives the dc voltage:

$$(5.2\ \cancel{\text{div}})(50\ \text{mV}/\cancel{\text{div}}) = 260\ \text{mV}$$

HORIZONTAL CONTROLS

The horizontal controls include the SEC/DIV control and its vernier, the horizontal magnifier, and the horizontal POSITION control. In addition, the horizontal section may include delayed sweep controls (described in Section 8–4). The SEC/DIV control sets the sweep speed, which controls how fast the electron beam is moved across the screen. The control has a number of calibrated positions divided into steps of 1-2-5 multiples, which allow you to set the exact time interval at which you view the input signal. For example, if the graticule has 10 horizontal divisions and the SEC/DIV control is set to 1.0 ms/div, then the screen will show a total time of 10 ms. The SEC/DIV control usually has a concentric vernier control that allows you to adjust the sweep speed continuously between the calibrated steps. This control must be in the detent position in order to make calibrated time measurements. Many scopes are also equipped with a horizontal magnifier that affects the time base. The magnifier increases the sweep time by the magnification factor, giving you increased resolution of signal details. Any portion of the original sweep can be viewed using the horizontal POSITION control in conjunction with the magnifier. This control actually speeds the sweep time by the magnification factor and therefore affects the calibration of the time base on the SEC/DIV control. For example, if you are using a 10× magnifier, the SEC/DIV dial setting must be *divided* by 10.

EXAMPLE 8–3 Assume the SEC/DIV control is set to 50 μs/div and the 10× magnifier is *on*. What is the time required for the sweep to go 10 divisions?

SOLUTION The SEC/DIV control setting is divided by the magnification factor to find the displayed time per division. The sweep time is the number of divisions multiplied by the time per division:

$$\left(\frac{50\ \mu\text{s}/\cancel{\text{div}}}{10}\right)(10\ \cancel{\text{div}}) = 50\ \mu\text{s}$$

After the sweep has moved across the screen, it must return rapidly to its original position on the left side. During the return time, the beam is turned off by applying a negative pulse called the **blanking** pulse to the control grid (or a positive voltage to the cathode). In this way the retrace is not observed on the screen. When the next sweep begins, the CRT is unblanked.

TRIGGER CONTROLS

The trigger section is the source of most problems when operating an oscilloscope. Trigger controls determine the proper time for the sweep to begin in order to produce a stable display. These controls include the MODE switch, SOURCE switch, trigger LEVEL, and SLOPE, COUPLING, and variable HOLDOFF controls. In addition, the trigger section includes a connector for applying an EXTernal trigger to start the sweep. Trigger controls may include HIGH or LOW FREQUENCY REJECT switches and BANDWIDTH LIMITING.

The MODE switch is a multiple position switch that selects either AUTO or NORMAL (sometimes called TRIGGERED) and may have other positions such as TV or SINGLE sweep. In the AUTO position, the trigger generator selects an internal oscillator that will trigger the sweep generator as long as no other trigger is available. This mode ensures that a sweep will occur even in the absence of a signal, since the trigger circuits free-run in this mode. This allows you to obtain a baseline for adjusting ground reference level or for adjusting the display controls. In the NORMAL, or TRIGGERED, mode, a trigger is generated from one of three sources selected by the SOURCE switch—the INTERNAL signal, an EXTERNAL trigger source, or the ac LINE. If you are using the internal signal to obtain a trigger, the normal mode will provide a trigger only if a signal is present and other trigger conditions (level, slope) are met. This mode is more versatile than AUTO because it can provide stable triggering for very low to very high frequency signals. The TV position is used for synchronizing either TV fields or lines, and SINGLE is used primarily for photographing the display.

The trigger LEVEL and SLOPE controls are used to select a specific point on either the rising or falling edge of the input signal for generating a trigger, as illustrated in Figure 8–7. The trigger SLOPE control determines which edge will generate a trigger, whereas the LEVEL control allows the user to determine the voltage level on the input signal that will start the sweep circuits.

The SOURCE switch selects the trigger source—either the INTERNAL signal, an EXTERNAL trigger source, or the ac LINE. In the INTERNAL position, a sample of the signal being viewed is used to start the sweep. Some multiple-channel scopes allow you to choose the triggering channel or use some combina-

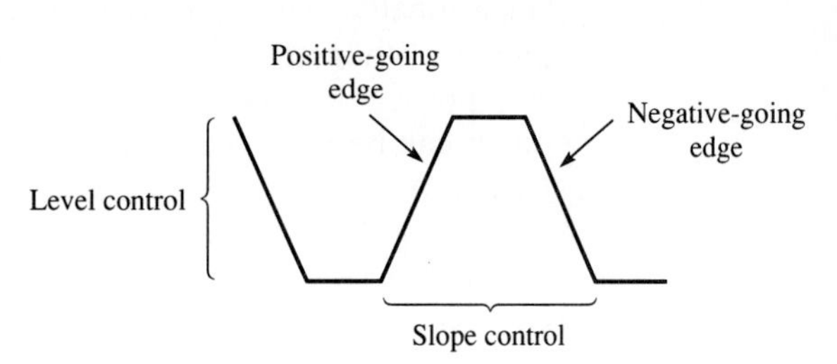

FIGURE 8–7
Slope and level controls (copyright Tektronix, Inc. Used with permission).

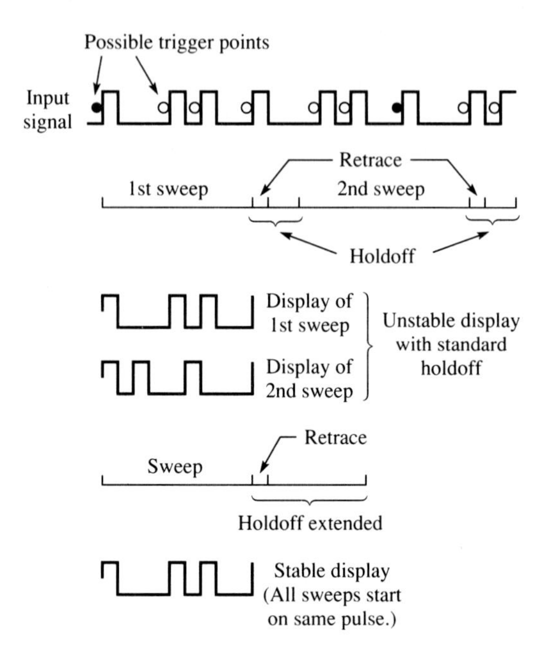

FIGURE 8–8
Variable holdoff control (copyright Tektronix, Inc. Used with permission).

tion of channels for triggering. In the EXTERNAL position, a time-related external signal is used for triggering. Trigger attenuation (divide-by-10) is provided on many oscilloscopes. Large external triggers can be attenuated in order to extend the range of the trigger-level control. The external trigger can be coupled with either ac or dc coupling. Couple the trigger signal with ac coupling if the trigger signal is riding on a dc voltage. Use dc coupling if the triggers occur at a frequency of less than about 20 Hz. The LINE position causes the trigger to be derived from the ac power source. This synchronizes the sweep with signals that are related to the power line frequency.

The variable HOLDOFF control allows you to exclude otherwise valid triggers until the holdoff time has elapsed. For some signals, particularly complex waveforms or digital pulse trains, obtaining a stable trigger can be a problem. This problem can arise when one or more valid trigger points occur before the signal-repetition time, as illustrated in Figure 8–8. If every event that the trigger circuits qualified as a trigger were allowed to start a sweep, the display could appear to be unsynchronized. By adjusting the variable HOLDOFF control, the trigger point can be made to coincide with the signal-repetition point.

Occasionally, a trigger signal contains interfering high- or low-frequency components. When this happens, the trace may flutter back and forth on the screen or produce the appearance of unsynchronized sweeps. A high-frequency component can be removed using HIGH FREQUENCY REJECT. Sometimes it is necessary to observe a high-frequency signal in the presence of a low-frequency signal, such as might occur in measuring a low-level transducer signal. In this case, LOW FREQUENCY REJECT can remove the low-level interference from the trigger signal.

z-AXIS

Many oscilloscopes are equipped with a separate input for intensity modulation, which is labeled as *z*-axis. Intensity modulation is useful if you would like to emphasize a portion of the display, such as a particular reference time. Time markers can be used to brighten the display at a particular point in time for proper synchronization. Intensity modulation is also advantageous in swept-frequency measurements for displaying frequency markers (see Section 8–7). In this application, a particular frequency can be highlighted on the display, which is helpful in measuring filters and in IF alignment.

DUAL-CHANNEL OSCILLOSCOPES

Most oscilloscopes have two (or in some cases more) separate channels and controls to view more than one signal at a time. There are two types of dual-channel oscilloscopes—dual-beam and dual-trace. A **dual-beam** oscilloscope has two independent beams in the CRT and independent vertical deflection systems, allowing both signals to be viewed at the same time. A **dual-trace** oscilloscope has only one beam and one deflection system; it uses electronic switching to show the two signals. Dual-beam oscilloscopes are generally restricted to high-performance research instruments and are more expensive than dual-trace oscilloscopes. An example of a standard-performance dual-trace oscilloscope is the Tektronix 2225. It is shown, with key specifications, in Figure 8–9(a). For contrast, an advanced-performance oscilloscope—the Tektronix 2467B—is shown, with key specifications, in Figure 8–9(b). The 2467B has an extremely fast writing speed, allowing it to show very fast signals even when they occur infrequently.

A dual-trace oscilloscope has controls labeled CHOP or ALTERNATE to switch the beam between the channels so that the signals appear to occur simultaneously. The CHOP mode rapidly switches the beam between the two channels at a fixed, high-speed rate so that the two channels appear to be displayed at the same time. The ALTERNATE mode first completes the sweep for one of the channels and then displays the other channel on the next (or *alternate*) sweep. When viewing slow signals, the CHOP mode is best, since it reduces the flicker that would otherwise be observed. High-speed signals can usually be best observed in ALTERNATE mode in order to avoid seeing the chopping frequency.

Another feature on most dual-trace oscilloscopes is the ability to show the algebraic sum and difference of the two channels. For most measurements, you should have the vertical sensitivity (VOLTS/DIV) on the same setting for both channels. You can use the algebraic sum if you want to compare the balance on push-pull amplifiers, for example. Each amplifier should have identical signals except they are out of phase with each other. When they are added, the resulting display should be a straight line, indicating balance. You can use the algebraic difference when you want to measure the waveform across an ungrounded component. The probes are connected across the ungrounded component with probe grounds connected to circuit ground. Again, the vertical sensitivity (VOLTS/DIV) setting should be the same for each channel. The display will show the algebraic difference in the two signals. The algebraic difference mode also allows

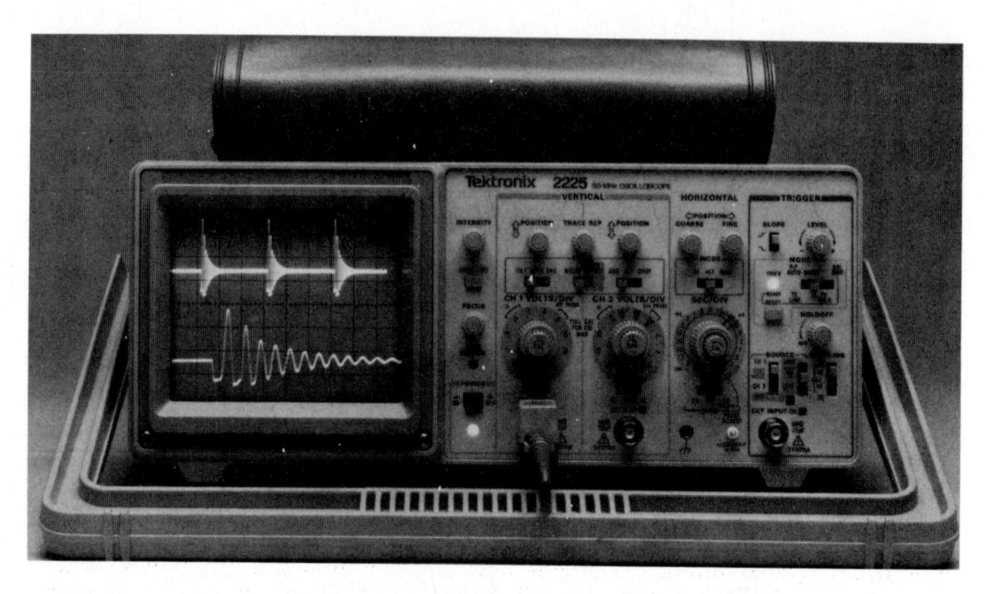

The Tektronix 2225 is a representative general-purpose portable oscilloscope. It is a dual-trace oscilloscope with 50 MHz bandwidth for each channel. The four basic control groups are located together. The 2225 has an alternate magnification feature that allows the user to view the magnified and unmagnified sweeps at the same time. Horizontal magnification can be selected from three levels: 5×, 10×, and 50×. Another feature is vertical magnification of a factor of 10× by pulling the central knob on the VOLTS/DIV control. Key specifications are:

Vertical System	
Bandwidth:	50 MHz
Vertical sensitivity:	500 μV/div (with 10× vertical magnification)
Deflection factor accuracy:	±3%. Add ±2% with 10× vertical magnification.
Horizontal System	
Time base range:	0.5 s/div to 0.05 μs per division; horizontal magnifier extends time base to 5 ns per division.
Triggering System	
Modes:	Peak-to-peak auto, normal, TV field, TV line, single sweep
Level:	Can be set to any point on the trace that is displayed
Variable holdoff range:	Increases sweep holdoff time at least by a factor of 1 to 8
Trigger couplings:	AC, DC, HF reject, LF reject

(a) Tektronix 2225 oscilloscope and key specifications (copyright Tektronix, Inc. Used with permission).

FIGURE 8–9

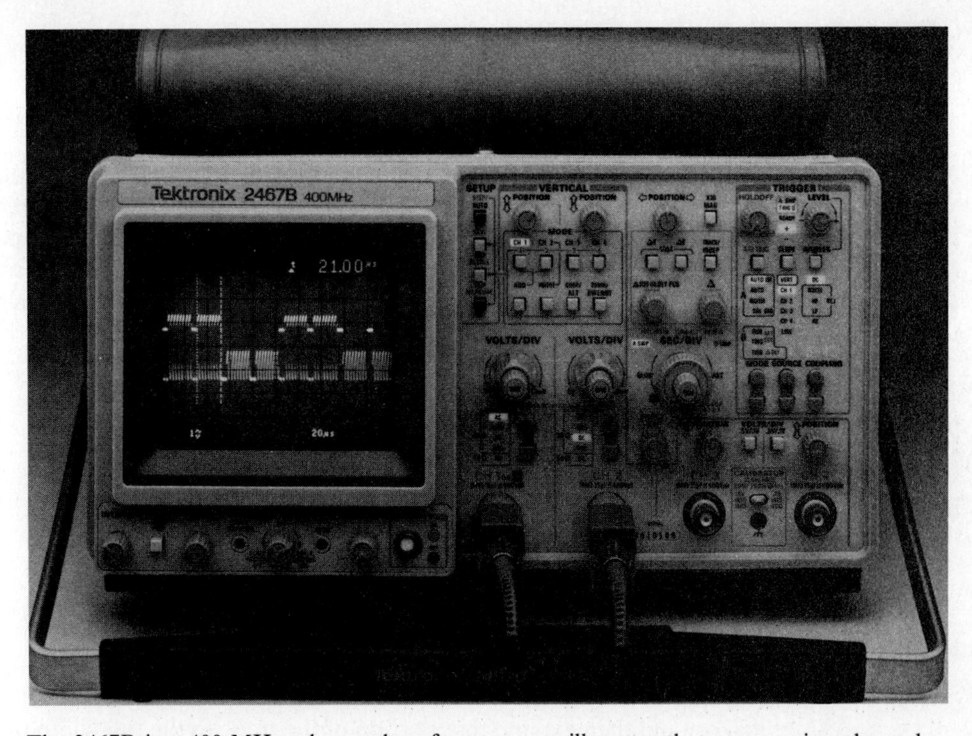

The 2467B is a 400 MHz advanced-performance oscilloscope that uses a microchannel plate (MCP) CRT to produce an extremely high writing rate. Using the 10× magnifier, the fastest sweep rate is 500 ps/div. It has four independent channels, on-screen readout of front-panel settings, and automated setups and measurements and can be interfaced to a computer using the IEEE-488 bus. Automated measurements are simplified by the use of cursors, a technique of adding reference lines to the display. Automated measurements provide a CRT readout of various waveform parameters. In addition, complete setup of all front panel controls can be stored in 30 nonvolatile memories. Even though there are many more controls, notice how the grouping of controls is similar to the 2225. Key specifications are:

Vertical System	
Bandwidth:	400 MHz
Vertical sensitivity:	2 mV/div
Deflection factor accuracy:	±2%
Horizontal System	
Time base range:	0.5 s/div to 5.0 ns/div; horizontal magnifier extends time base to 0.5 ns/div.
Trigger System	
Modes:	Auto, normal, single sweep, TV field, TV line
Level:	±18 times the setting of the vertical range
Variable holdoff range:	Extends slowest sweep to 1.5 s/div
Trigger coupling:	AC, DC, HF reject, LF reject, noise reject

(b) Tektronix 2467B oscilloscope and key specifications (copyright Tektronix, Inc. Used with permission).

FIGURE 8–9 *(continued)*

you to cancel any unwanted signal that is equal in amplitude and phase and is common to both channels.

Dual-trace oscilloscopes usually have an X-Y mode, which causes one of the channels to be graphed on the *x*-axis and the other channel to be graphed on the *y*-axis. This is necessary if you want to change the oscilloscope base line to represent a quantity other than time. Applications include viewing a transfer characteristic (output voltage as a function of input voltage), swept-frequency measurements (see Section 8–7), and showing Lissajous figures for phase measurements. Lissajous figures are patterns formed when sinusoidal waves drive both channels. They are discussed in Section 8–8.

8–4 DELAYED-SWEEP OSCILLOSCOPES

A AND B SWEEPS

Many oscilloscopes have an additional horizontal-mode feature called **delayed sweep.** Delayed sweep is basically an additional calibrated time base for the oscilloscope. It allows the user to pick out a portion of the sweep with an intensified marker and then expand that portion of the trace to the full width of the screen for detailed analysis. The main time base, called the **A sweep,** is the normal time base for the oscilloscope. The second time base is called the **B sweep** (although various manufacturers have other dial labels). When the horizontal MODE switch is in the A position (also called NO DLY), the A sweep generator provides the time base for the scope. When the horizontal MODE switch is in the A INTEN BY B position (also called ALT), a brighter region is visible on a portion of the trace. The length of this intensified region is controlled by the B sweep time, often a concentric control on the SEC/DIV knob. The location of the intensified region is controlled by a separate DELAY control. When the horizontal MODE switch is in the B position (also called DLY'D), the intensified region is expanded to the full width of the screen. The scope is still triggered by the A sweep trigger, but it is not used to display the signal. Instead, the B sweep is triggered after the delay period set on the DELAY control.

Delayed sweep is useful when you need to examine a portion of a waveform that occurs well after the trigger time. In radar, for example, there is a relatively long time interval following a pulse. By using delayed sweep, the trigger can be obtained from a convenient trigger source, and a detailed view of a delayed pulse, including the leading edge, can be displayed on the screen. You can also examine the leading edge of the trigger pulse with the B sweep by expanding the next trigger pulse.

DIFFERENTIAL DELAYED-SWEEP MEASUREMENTS

Differential delayed sweep can make accurate time measurements and remove CRT nonlinearity as a source of error. This usually results in more accurate time-interval measurements. To make a differential delay-time measurement, you need an oscilloscope with delayed sweep and a calibrated DELAY dial. (On many newer oscilloscopes, the DELAY dial reading is presented as video information on the face of the scope.) If the oscilloscope is equipped with a separate B trigger circuit, it should be set for B-runs-after-delay mode rather than set to use the B trigger circuits. The horizontal MODE switch is placed in the A INTEN BY B position and the intensified line is adjusted to the desired starting point for the time measurement. The intensified zone should be a relatively small portion of the A sweep. Switch the horizontal MODE to B and position the timing point exactly over the center vertical graticule mark using the DELAY dial. Note the dial reading. Then turn the DELAY control until the desired ending timing point appears in exactly the same point on the screen and take a second reading of the dial. The time between the first and second dial readings multiplied by the A sweep time is the time difference between the selected starting and ending points. For more information on this technique, you should consult the operator's manual that comes with your oscilloscope.

EXAMPLE 8–4 Assume you are making a differential delayed-sweep measurement to determine the time between two points on a waveform. The A sweep SEC/DIV control is set to 20 μs/div. The horizontal MODE control is set to the B mode and the starting time is moved with the DELAY control to center screen. The DELAY control reads 1.55. Then the DELAY control is moved until the stopping time is center screen. The new reading of the delay dial is 4.90. What is the time between the starting and stopping points?

SOLUTION The time is the difference between the delay dial readings multiplied by the A SEC/DIV setting:

$$\Delta t = (4.90 \text{ div} - 1.55 \text{ div})(20\ \mu\text{s/div}) = 67\ \mu\text{s}$$

8–5 OSCILLOSCOPE PROBES

A **probe** is used to pick off a signal and couple it to the oscilloscope. It makes a mechanical connection to the test circuit and passes the signal through a flexible, shielded cable to the oscilloscope. The shielding helps protect the signal from external noise pickup. The probe is also provided with a short ground lead, which should be connected to a convenient ground on the circuit under test. The proper probe can extend the oscilloscope's measuring ability by reducing the effects of circuit loading and extending the amplitude capability of the scope. Oscilloscope probes are provided with the instrument as part of the system. *The wrong probe or a probe that is not adjusted properly can affect the measurement to the point of making it worthless or misleading.*

EFFECT OF LOADING

In order to have a faithful reproduction of the test signal, the measurement system should have no effect on the circuit being tested and should introduce no noise or distortion to the signal. Unfortunately, this ideal system does not exist. Any measurement system must absorb some finite amount of energy from the circuit under test. When a probe is connected to a test circuit, the circuit sees an equivalent load consisting of the input impedance of the probe. Oscilloscope probes have both resistance and capacitance to ground. As a result, they cause some attenuation due to *resistive* loading at all frequencies and *capacitive* loading at higher frequencies, an effect that reduces the probe impedance at higher frequencies. Capacitive loading can become the dominant factor for measurements at high frequencies and also affects the rise time of pulses. In serious cases, the circuit under test could stop working or break into oscillation when the probe is connected.

One way of alleviating the effects of capacitive loading is to use an attenuator probe that contains a passive divider network, as shown with the representative probe circuit in Figure 8–10. When the probe is connected to the oscilloscope input and properly compensated, the input capacitance is effectively reduced by approximately the attenuation ratio of the probe. This serves to increase the frequency response over that of a nonattenuating probe. The displayed signal is a more accurate representation of the circuit test signal because of the reduced loading effects. For this reason, attenuating probes are normally supplied with an oscilloscope and represent a good choice for general circuit probing. Probes are normally marked with the effective input resistance, capacitance, and attenuation ratio. A typical general-purpose attenuator probe has an input resistance of 10 MΩ, an input capacitance of about 14 pF, and an attenuation ratio of 10 : 1.

An oscilloscope's ability to pass high frequencies is specified by the bandwidth. **Bandwidth** is the frequency range between a low and high frequency, where the attenuation of a sinusoidal waveform is attenuated by 3 dB from the average response. Since oscilloscopes can pass frequencies from dc to some upper limit, only the upper frequency is reported for the bandwidth. The probe affects the bandwidth of the oscilloscope's measuring system by adding capacitance. Probes, like oscilloscopes, are rated for bandwidth. Manufacturers frequently specify the bandwidth of an oscilloscope when it is used with a probe rated at a bandwidth equal or greater than the scope's bandwidth. In other words,

FIGURE 8–10
Circuit for a typical 10 : 1 attenuated probe. The *RC* time constant of $R_{in}C_{in}$ is equal to R_sC_p, when the probe is adjusted properly.

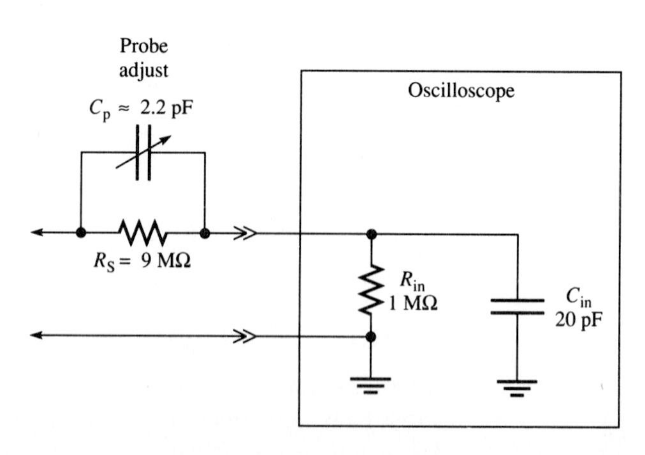

the specified bandwidth of the oscilloscope system is measured at the probe tip provided that the probe's bandwidth is equal to or greater than the oscilloscope's bandwidth. A higher-bandwidth probe can be used with an oscilloscope as long as it is otherwise compatible with the oscilloscope.

The input resistance of a probe appears to be in parallel with the input capacitance of that probe. The total input impedance of the probe can be found from the product-over-sum rule:

EQUATION 8–3

$$Z_{in} = \frac{(R_{in})(-jX_{C(in)})}{(R_{in}) + (-jX_{C(in)})} = \frac{(R_{in})(-jX_{C(in)})}{\sqrt{R_{in}^2 + X_{C(in)}^2}}$$

The capacitive reactance is found from

EQUATION 8–4

$$X_{C(in)} = \frac{1}{2\pi f C_{in}}$$

These equations are worst-case equations because they ignore inductive elements designed to offset capacitive loading to some degree.

EXAMPLE 8–5 A 10× probe has 20 pF of capacitance to ground and an input resistance of 10 MΩ. What is the input impedance of the probe at a frequency of 100 kHz?

SOLUTION First find the capacitance reactance at the frequency of interest:

$$X_{C(in)} = \frac{1}{2\pi f C_{in}} = \frac{1}{2\pi(100\text{ kHz})(20\text{ pF})} = 79.6\text{ k}\Omega$$

Equation 8–3 is the product-over-sum rule (using phasors) for a parallel resistance and capacitance. By substitution, we obtain

$$Z_{in} = \frac{(R_{in})(-jX_{C(in)})}{\sqrt{R_{in}^2 + X_{C(in)}^2}} = \frac{(10\text{ M}\Omega)(-j79.6\text{ k}\Omega)}{\sqrt{10\text{ M}\Omega^2 + 79.6\text{ k}\Omega^2}} = 79.6\text{ k}\Omega$$

This example illustrates that above the cutoff frequency, the input impedance of a high-impedance probe is determined primarily by the capacitive reactance.

To understand the effect of probe loading, consider the problem of measuring the amplitude of a 10 MHz sinusoidal waveform from a known 1 kΩ source. If you make this measurement with a typical attenuating probe that has an input impedance of 10 MΩ and a capacitance of 15 pF, the probe will attenuate the signal by nearly 50% over and above its attenuation ratio, resulting in a 50% measurement error. Some improvement can be made using a higher-impedance 10 MΩ, 6 pF probe. The frequency response of these two probes is shown in Figure 8–11 for a 1 kΩ source impedance. Notice that the effective impedance of the probe is dropping significantly (−20 dB/decade) at the frequency where the capacitive reactance is equal to impedance of the circuit test point (26 MHz for the 10 MΩ, 6 pF probe).

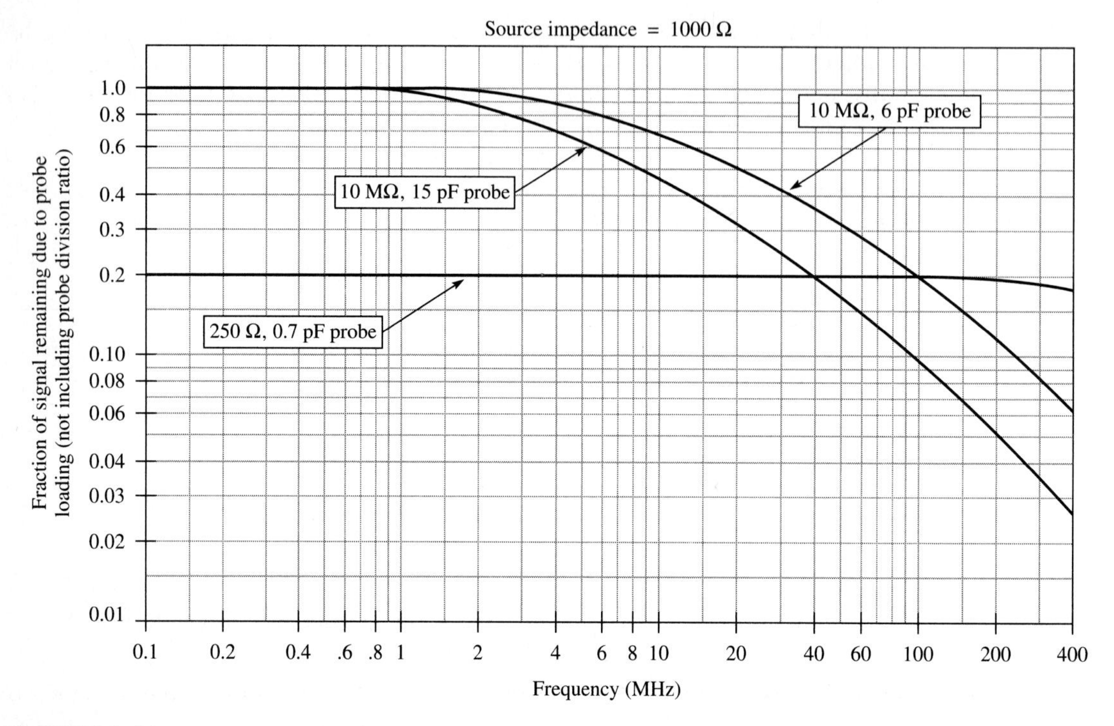

FIGURE 8–11
Probe attenuation due to frequency response and source loading. (Division ratio is not included.)

A way to alleviate the effects of capacitive loading is to increase the resistive loading by using a low-input-resistance probe. For the 1 kΩ source impedance discussed before, a 250 Ω probe with 0.7 pF of capacitance raises the frequency response significantly. The use of a 250 Ω probe loads the source to 20% of the input signal but provides a known, constant attenuation to well beyond 100 MHz. The response for this probe is also shown in Figure 8–11. This attenuation is due to loading and must be taken into account when determining the actual input signal. Figure 8–12 illustrates how this loading occurs. Low-impedance probes are useful for amplitude measurements, providing that you are probing a known low-source impedance. The resistive loading effect is computed by the voltage-divider equation:

EQUATION 8–5

$$v_{obs} = v_{test}\left(\frac{Z_{in}}{Z_{in} + Z_s}\right)$$

where v_{obs} = the signal observed on the oscilloscope, V
v_{test} = the unloaded test voltage of the point being probed, V
Z_s = the source impedance, Ω
Z_{in} = the probe input impedance, Ω

Clearly, if Z_s is large compared to Z_{in}, a large attenuation of the test signal will occur. For this reason, when you are probing an unknown impedance, a high input impedance is necessary to avoid attenuating the signal under test by an unknown amount. However, if the source impedance is known, a low-impedance probe virtually eliminates capacitive loading effects.

EXAMPLE 8–6 Compute the cutoff frequency for a 1 kΩ, 0.7 pF probe.

SOLUTION The cutoff frequency, f_{CO}, is the frequency at which the capacitive reactance is equal to the resistance. Setting $X_C = R$ and solving for f_{CO} gives

$$X_C = R = \frac{1}{2\pi f_{CO} C}$$

$$f_{CO} = \frac{1}{2\pi RC}$$

Substituting the given values yields

$$f_{CO} = \frac{1}{2\pi(1\text{ k}\Omega)(0.7\text{ pF})} = 227\text{ MHz}$$

TYPES OF PROBES

A number of types of probes are available, each having advantages for particular applications. These include passive attenuating and nonattenuating probes, active probes, current-sensing probes, high-voltage probes, and special-purpose probes. Passive attenuating probes are available with 10×/100×/1000× attenuation ratios with input resistances ranging from 50 Ω to 10 MΩ. Higher voltage ratings and bandwidths are associated with higher attenuation ratios. Most oscilloscopes come from the manufacturer with 10× high-impedance attenuating probes. Nonattenuating probes (1×) are useful if you need maximum voltage sensitivity of the scope; however, the bandwidth of these probes is typically from 4 to 34 MHz,

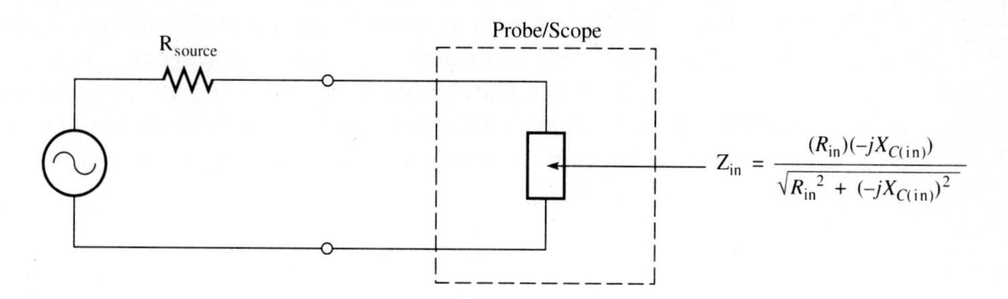

FIGURE 8–12
Effect of loading on a source.

meaning that high-frequency or fast-rising pulses will be affected by loading. Some probes can be switched between 10× and 1×, giving the user a choice between higher bandwidth or greater sensitivity. A ground reference position of the probe switch is used to indicate circuit ground for positioning the trace on a convenient graticule line. Usually the 10 MΩ input impedance of the standard 10× probe is satisfactory for general probing, but it is less sensitive. The sensitivity of the 1× probe is better, but these probes may cause loading problems in high impedance circuits.

Active probes contain a very high impedance, low capacitance amplifier (typically a FET type). The bandwidth can be over 1 GHz. Power for the amplifier is supplied by the oscilloscope or a separate probe power supply. Active probes have the highest input resistance and lowest capacitance of all types of probes and therefore have the lowest loading effect on the test signal. They are available as differential probes with dual probe tips and built-in differential signal processing for high CMRRs in the probe itself. This minimizes measurement errors of differential amplifiers due to differences in probes, cables, and input attenuators. Disadvantages are limited dynamic range (the maximum linear operating range) and higher cost.

Current waveforms sometimes contain more valuable information than voltage waveforms. Current-sensing probes eliminate the need for using a sense resistor to determine the current in a conductor and have very low loading of the circuit under test. Accuracy of amplitude and rise-time measurements is improved when the loading of the circuit under test is minimized. In addition, current waveforms can be very different from the voltage waveform in a circuit. AC current probes are operated by connecting the jaws of the probe around the conductor. They use a transformer to convert the coupled flux from an alternating current into an alternating voltage. The frequency response can be from a few hundred hertz to 1000 MHz. DC current probes use a combination of a Hall-effect device with a transformer to respond to both dc and ac. The Hall-effect device is an active element and requires a power source. They operate from dc to 50 MHz. Current probes are available for measuring currents in ranges from under 1 mA to thousands of amperes. They can be used with current amplifiers to extend their sensitivity.

Specialty probes are available for a variety of applications. These include high-voltage probes capable of measuring up to 40 kV, environmental probes that can operate in temperature extremes, logic probes for digital circuits, and temperature probes designed to measure the temperature at the probe tip. In addition, special isolated probes allow you to make floating voltage measurements in the presence of high common-mode voltages rather than referencing ground, as in the usual case. This type of probe is useful when you have to measure across the floating inputs of a circuit.

MEASUREMENTS WITH PROBES

When making any kind of measurement with a probe, a ground path is required to complete the measurement circuit. To avoid noise pickup and undesired inductance in the ground circuit, the ground path should be as short as possible, consistent with the need to move the probe around the circuit. For very fast signals, the ground lead should be connected to a test-circuit ground located the shortest possible distance from the test point—especially when making low-level measurements. A long ground lead can have enough inductance to create a series resonant circuit that can cause ringing in circuits with fast signals, distorting the observed waveform. The ideal grounding situation occurs when a coaxial test point is provided, allowing the ground to surround the test point.

Two of the most common measurements made with oscilloscopes are voltage amplitude and pulse rise time. Pulse-amplitude measurements are less prone to error than CW (continuous-wave) measurements because the input capacitance of the probe does not affect the pulse amplitude unless the pulse is very narrow. On the other hand, CW amplitude measurements can have error due to bandwidth limiting when the frequency being measured is attenuated due to the measuring system bandwidth. Recall that the definition of bandwidth is the frequency range between a low and high frequency where the attenuation of a sinusoidal waveform is 3 dB from the average response. A 3 dB attenuation implies 70% of the signal is passed. In other words, a voltage measurement of a sinusoidal waveform with the same frequency as the measuring system's bandwidth will have a *30% error* due to bandwidth limiting.

In addition to bandwidth limiting, probe loading can also affect amplitude measurements when the probe impedance is near the circuit impedance. To keep the ratio between the probe and circuit impedance high, select a low-impedance point to probe if you have a choice. As a general rule, for low frequencies, the probe's input-resistance component should be at least two orders of magnitude larger than the circuit-source impedance. At high frequencies, the probe's capacitance component is the primary consideration. The best probe for a measurement is a trade-off between the resistive and capacitive components of loading. In cases where the loading needs to be determined, Equation 8–3 will enable you to compute the input impedance of the probe and Equation 8–5 can be solved for the actual voltage of the test point, v_{test}.

Most amplifiers can be represented by a dominant pole step-response equivalent to the response of a single-pole, low-pass filter. (See the discussion of filters in Section 5–10.) If a theoretically perfect pulse (zero rise time) were applied to an amplifier, the observed output rise time would represent the step-response of the amplifier. Rise time is defined as the time required for the leading edge of a pulse to change from 10% to 90% of the peak value,[1] as illustrated in Figure 8–13. An important relationship exists between the rise time of an amplifier and its bandwidth. The rise time is related to the bandwidth by the following equation:

EQUATION 8–6

$$t_r = \frac{0.35}{B}$$

where t_r represents the rise time, μs

B is the bandwidth, MHz

[1] For some digital circuits, the 20% and 80% levels are selected.

FIGURE 8–13
Definition of rise time, t_r (see IEEE Standard 194-1977, "Standard Pulse Terms and Definitions").

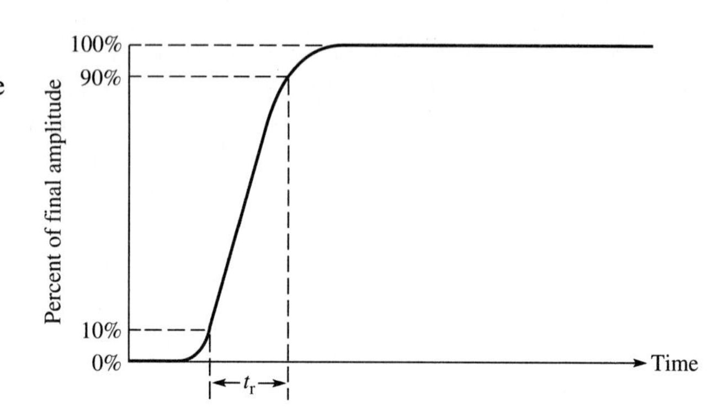

This equation is derived in Appendix A and is valid for amplifiers with single-pole response.

The step-response, or rise, time of an oscilloscope is related to its bandwidth by Equation 8–6. If you could apply a "perfect," or zero, rise-time pulse to an oscilloscope, the observed rise time on the face of the oscilloscope would represent the response of the measurement system. If you know the oscilloscope's bandwidth, you can predict the rise time you would observe for the perfect pulse wave using Equation 8–6.

EXAMPLE 8–7 What is the equivalent rise time of a scope with a 60 MHz bandwidth?

SOLUTION Substituting into Equation 8–6 gives

$$t_r = \frac{0.35}{B}$$

$$= \frac{0.35}{60 \text{ MHz}} = 5.8 \text{ ns}$$

When an oscilloscope is used to measure a fast-rising signal, the rise time of the oscilloscope affects the observed signal. The displayed rise time is not the true rise time of the signal but rather is changed by the oscilloscope (and probe) in accordance with the following equation:

EQUATION 8–7

$$t_{r(\text{displayed})} = \sqrt{t_{r(\text{true})}^2 + t_{r(\text{scope})}^2}$$

where $t_{r(\text{displayed})}$ = the observed rise time
$t_{r(\text{true})}$ = the actual rise time of the pulse
$t_{r(\text{scope})}$ = the rise time of the oscilloscope and probe

Equation 8–7 illustrates that the displayed rise time is seriously affected by the instrument when the rise time of the test signal is near that of the oscilloscope. For example, if you were using an oscilloscope with a 3 ns rise time to measure a pulse with a 4 ns rise time, the observed rise time would be 5 ns. This represents a 25% error. The error associated with other rise-time ratios is illustrated graphically in Figure 8–14. Rise-time measurements require proper termination to avoid reflections and, as you can see in Figure 8–14, a measuring-system bandwidth at least four times larger than the rise time to be measured is needed to reduce errors to 3%. For systems with less bandwidth than this, a calculation of the true rise time can be done using Equation 8–7; however, the error associated with this procedure can be significant when the true rise time and the scope and probe rise times are close.

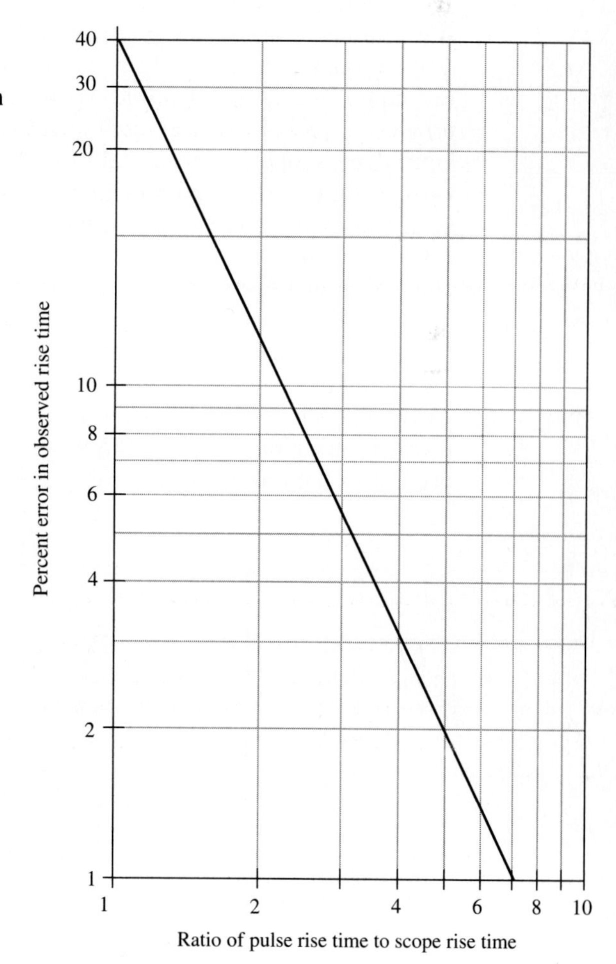

FIGURE 8–14
Errors associated with measuring rise time with an oscilloscope.

PROBE COMPENSATION AND CALIBRATION CHECK

One of the most common operator errors is using an incorrectly compensated probe. Probe compensation should be checked at the start of every session with an oscilloscope or any time a probe is changed. This is done by varying a capacitor in the probe while viewing a test signal. Most oscilloscopes have a PROBE COMP or calibrator point on the front panel that provides a 1 kHz square wave as a test signal. The signal serves two main functions—as a check on the amplitude calibration of the oscilloscope and to assure that the frequency response of the scope is correct. Some calibrators also provide a frequency and/or current reference.

To compensate a probe, connect the probe to the calibrator and set the scope controls to observe the test square wave on the channel you intend to use. The AC-GND-DC coupling switch should be set to DC, the SEC/DIV control set to 1 ms/div, and the VOLTS/DIV set so that the test square wave will be deflected over about four divisions. Some probes have a small access hole in which a nonmetallic screwdriver can be inserted; others are adjusted by unlocking a sleeve and rotating a part of the probe. The probe capacitor is adjusted for a good flat-topped square wave without overshoot or rounding, as illustrated in Figure 8–15. Basically, this adjustment allows you to adjust the parallel *RC* circuit inside the probe so it has a time response equal to that of the parallel *RC* circuit of the scope input. The scope's vertical calibration should be checked next. Verify that the vernier VOLTS/DIV control is in the CAL position, and measure the peak-to-peak amplitude of the square wave. It should agree with the value marked on the calibrator output; otherwise, the scope should be recalibrated according to the manufacturer's specification.

Before probing an unknown circuit, make sure that the circuit's ground is the same as the scope's ground. In some circuits, particularly inexpensive consumer TVs and the like, the chassis can easily be at a dangerous voltage with respect to earth. The scope will normally be at earth ground as long as a three-wire plug is properly grounded through the power line ground. Before connecting the probe ground to the circuit, check by connecting the probe's ground wire to a known

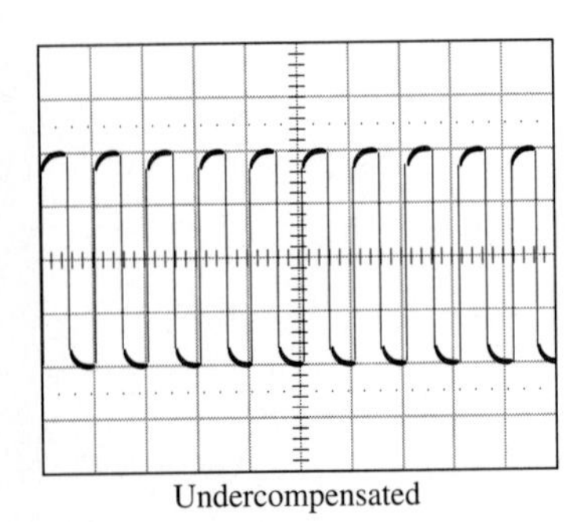
Undercompensated

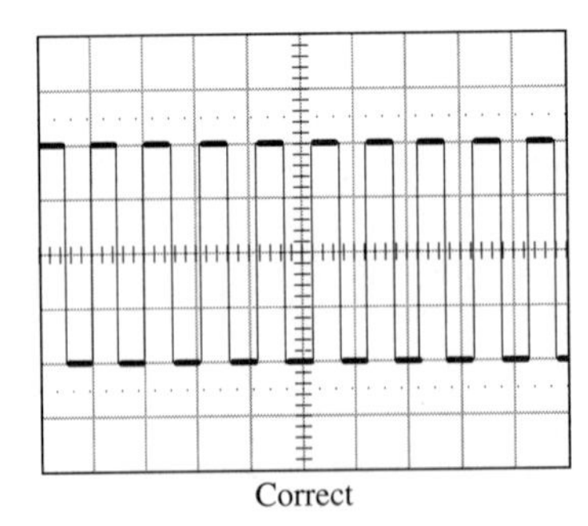
Correct

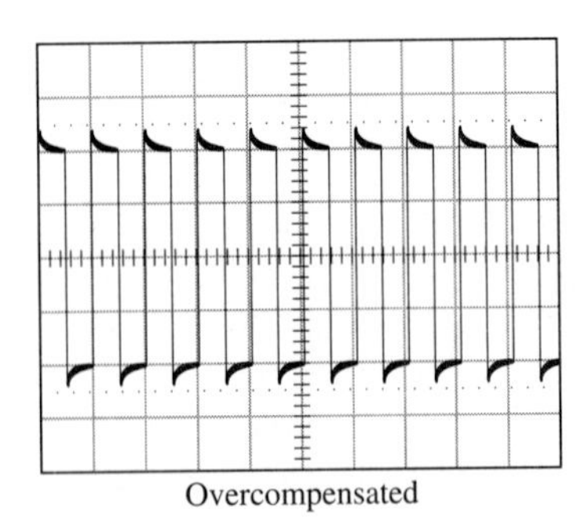
Overcompensated

FIGURE 8–15
Probe low-frequency compensation.

ground point and measuring the test circuit ground with the probe tip. If the circuit is at the same ground, you will see no voltage on the scope.

8–6 OSCILLOSCOPE CALIBRATION

PERFORMANCE CHECKS

Laboratory instruments should be checked periodically against standards to ensure that measurements are within the accepted tolerance. In addition, whenever an instrument has been repaired, it should be given a performance check before returning it to service. Oscilloscope performance checks can usually be done without removing the covers. The detailed procedure for a performance check is given by the manufacturer in the instrument's manual. For oscilloscopes, it will generally include a check of power supply voltages and ripple, the deflection accuracy of the vertical channels, trace rotation, a measurement of the bandwidth of each channel, a test of the sweep-rate accuracy, and other tests for triggering, jitter, and so forth. Failure of any performance test means the instrument needs to be serviced or recalibrated.

When an oscilloscope is given a periodic performance check, certain preventive maintenance procedures should be done to prevent future failure. Air filters should be cleaned, the interior should be cleaned, fan motors should be checked, and a general inspection should be done to look for heat-damaged components, improperly seated circuit boards or connectors, and so forth. Low-pressure air hosing of the chassis to remove dust and dirt will help reveal potential problems and any physical damage. Electrolytic capacitors should be inspected for any sign of leakage. Dust and film around high-voltage anodes should be removed. Switches and other controls should be tested to ensure they are functional. The CRT should be inspected around the neck pins for any sign of spider cracking. Mounting nuts should be in place and tight. Line cords should be inspected for cracking. This type of preventive maintenance can correct many of the problems that eventually lead to failure.

CALIBRATION

As discussed in Section 3–8, calibrating an instrument means comparing it to a known precise standard. The test instruments used for performance checks must be precision instruments with specified accuracies exceeding the oscilloscope under test. The test instruments required for calibration are specified by the instrument manufacturer. They generally include a digital voltmeter to check power supply voltages, an amplitude calibrator to set the vertical gain, a time-mark generator for setting the horizontal sweep timing and y-axis alignment, and a square-wave generator for making high-frequency compensation and vertical adjustments. It is important to use the proper cables, terminators, and couplers as specified by the manufacturer to avoid calibration error.

8–7 OSCILLOSCOPE MEASUREMENTS

Begin any session with the oscilloscope by checking the probe compensation on each channel, as described in Section 8–5. Adjust the probe for a flat-topped square wave while observing the scope's calibrator output. This is a good signal to use for checking the focus and intensity and verifying trace alignment. Check the front panel controls for the type of measurement you are going to make. Normally, the variable controls (VOLTS/DIV and SEC/DIV) should be in the calibrated (detent) position. The vertical coupling switch is usually placed in the dc position unless the waveform in which you are interested has a large dc offset. Trigger holdoff should be in the minimum position unless it is necessary to delay the trigger to obtain a stable sweep. It is a good idea to review the position of each of the front panel controls before starting with a measurement.

WAVEFORM CHARACTERIZATION

Periodic waves were described in Section 1–2 and are characterized by the period *T*. They are used extensively in the measurement field because many practical electronic circuits generate periodic waves, including square, rectangular, triangular, and sawtooth waves. Square waves, for example, are useful for testing amplifiers and compensation networks. The sawtooth wave, as you have seen, is used to move a beam across an oscilloscope with the displacement proportional to time and is useful for swept-frequency measurements. The oscilloscope performs a vital function when it allows you to see and analyze the shape of periodic waves.

There are many other nonperiodic signals that need to be analyzed. Whenever meaningful information is carried by electrical waves, the waveform must change in some unpredictable way or else it contains no information. A television set would not be very interesting if the picture never changed! It is more difficult to obtain information on a scope about nonperiodic waves because of the need to synchronize the display. Oscilloscopes can display television signals, for example, by triggering from certain periodic synchronization pulses contained in the TV signal. A common method for observing circuit action that processes signals is to inject a specific periodic signal for testing purposes. In the case of a digital computer, for example, it can be forced into a repetitive loop to produce predictable behavior. Other methods of observing nonperiodic waves include photographing the display or using a storage oscilloscope as discussed in Sections 8–8 and 8–9.

TIMING MEASUREMENTS

Many measurements, such as period, pulse width, and rise time, require time measurements using one channel of the oscilloscope. Other timing measurements, such as phase angle or the time between cause and effect in digital circuits, require critical time comparison of two signals and need a dual-channel oscilloscope. To make a normal one-channel time measurement, the signal should be positioned using the vertical POSITION control on the center horizontal graticule. Time measurements are more accurate if the time to be measured is spread out on the screen. The horizontal POSITION control is used to move the starting time point to a convenient vertical graticule mark near the left side of the screen. Count the number of divisions between the starting and stopping points, including minor divisions, and multiply the number of divisions by the setting of the SEC/DIV control. If the horizontal MAGNIFIER is on, the result is divided by the magnification factor.

EXAMPLE 8–8 An oscilloscope screen shows one complete cycle of a waveform in 6.8 divisions. The SEC/DIV control is set to 50 μs/div and the 10× magnifier is *on*. What is the frequency of the wave?

SOLUTION The period is found by multiplying the setting of the SEC/DIV control by the number of divisions measured on the display and dividing by the magnification factor:

$$T = \frac{(50\ \mu\text{s/div})(6.8\ \text{div})}{10} = 34.0\ \mu\text{s}$$

The frequency is the reciprocal of the period:

$$f = \frac{1}{T} = \frac{1}{34.0\ \mu\text{s}} = 29.4\ \text{kHz}$$

An oscilloscope can be used to determine the frequency of a signal by measuring the period and computing the frequency. Figure 8–16 shows how to make a normal frequency measurement with a standard oscilloscope. The basic idea is to measure the period and compute the frequency.

Timing between two separate signals is similar to timing with one channel. For critical timing, the time delay in each measurement path should be identical. For normal in-circuit testing, standard 10× probes provide equal paths and prevent circuit loading that could affect timing. It is best to measure the time difference across the center horizontal graticule with the signals spread out on the x-axis of the display. With dual-trace oscilloscopes, it is important to keep in mind how the scope is triggered. When measuring the time difference between signals shown on two different channels, the scope should be triggered from only *one* of the channels, not both. Composite triggering should *not* be selected, because it uses both channels for triggering and destroys the time relation between the signals. Figure 8–17 illustrates a time measurement between two digital signals.

The measurement of time by counting divisions across the oscilloscope and multiplying by the SEC/DIV setting has inherent errors due to the sweep generator, the power amplifier, and the CRT. Some oscilloscopes combine counter and oscilloscope functions in one instrument to increase the resolution of time measurements. Counters typically measure time by starting and stopping a high-speed square-wave oscillator (called a **clock**) with trigger pulses and counting the number of clock pulses between the start and stop triggers. The trigger point and level can affect the time, and the operator can be misled by a poor-quality trigger pulse

FIGURE 8–16
Measuring frequency with an oscilloscope.

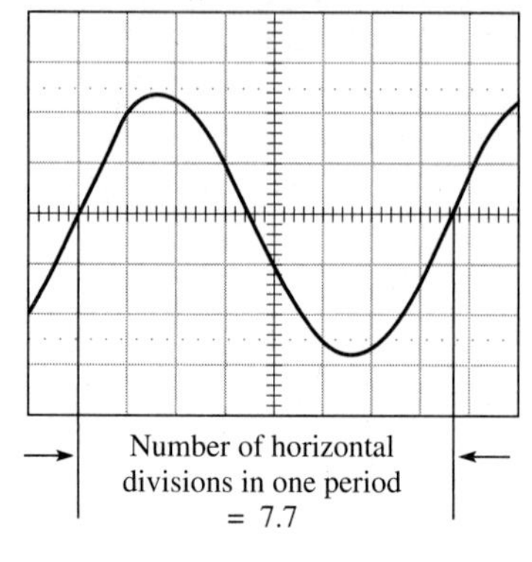

Steps:

1. Using the SEC/DIV control, spread the waveform across the screen so that one complete cycle is visible.
2. Count the horizontal divisions from any point on the waveform until one repetition of the same point. (See figure.)
3. Multiply the number of horizontal divisions by the SEC/DIV control and divide by any magnification factor. This is the period of the signal, T.
4. Compute f from $f = \dfrac{1}{T}$

or by not knowing the precise trigger points. By combining the functions and using a very accurate crystal-controlled time base for the oscilloscope, the timing resolution can be as high as 10 ps. The trigger points can be displayed with an intensified marker to ensure that the time is measured between specific points of the displayed waveform.

FIGURE 8–17
Time measurement between two signals.

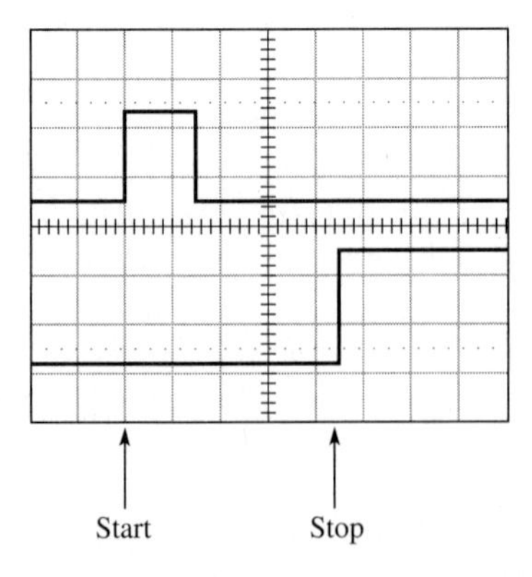

Number of divisions between start and stop = 4.4. Multiply the number of divisions by the SEC/DIV setting and divide by the magnifier setting to determine the time. The oscilloscope should be triggered by only one channel. Do not use composite triggering.

AMPLITUDE MEASUREMENTS

Amplitude measurements are most accurate if they occur in the middle portion of the oscilloscope screen and a vertical sensitivity is selected that deflects the signal to be measured over most of the screen (4 to 6 divisions is best). The probe should be compensated. Adjust the trigger controls for a stable display and check that VARIABLE controls are in their calibrated (detent) positions. Use the vertical POSITION control to set the lower part of the signal to be measured on a convenient horizontal graticule line near the bottom of the display. (It is best to avoid using the very lowest and highest divisions as they are more likely to be in the nonlinear region of the CRT.) To measure the peak-to-peak voltage, count the number of major and minor vertical divisions from the bottom to the top of the waveform and multiply by the VOLTS/DIV setting. For a sinusoidal wave, you will need to convert the measured peak-to-peak voltage to an rms voltage for comparison with most meters.

PHASE ANGLE MEASUREMENTS

Phase angle measurements always involve the comparison of two signals of the same frequency. It is meaningless to refer to the phase of a signal without having a reference to another signal. As in two-channel time measurements, phase measurements must have equal delays in each signal path. Standard 10× probes have equal delays and contribute very little loading effect that could disturb the measurement. A typical phase measurement is done by overlapping the two signals so that they appear to have the same amplitude. This can be done by taking one (or both) of the vertical channels out of calibration and using the vertical-position control to place both waveforms on center screen.

The phase shift between two sinusoidal waveforms can be computed from

EQUATION 8–8

$$\theta = \left(\frac{\Delta t}{T}\right) \times 360^\circ$$

where θ = the phase shift, degrees

Δt = the time difference between corresponding points on the two waveforms, s

T = the period of the waveforms, s

The period is measured by counting the number of divisions for one complete cycle and multiplying by the SEC/DIV dial setting. To measure Δt, both signals are viewed at the same time. Use the vertical-amplitude vernier and vertical-position controls to make both waveforms appear to have the same amplitude and position on the scope face. The waveforms are spread out horizontally to increase the resolution using the SEC/DIV control, as shown in Figure 8–18. The time difference between the two waveforms is again measured by counting the number of divisions between corresponding points on the waves and multiplying by the SEC/DIV dial setting.

Another method involves calibrating the horizontal axis in degrees rather than time. In this method, one cycle is spread across the scope face by adjusting

FIGURE 8–18
Measurement of Δt.

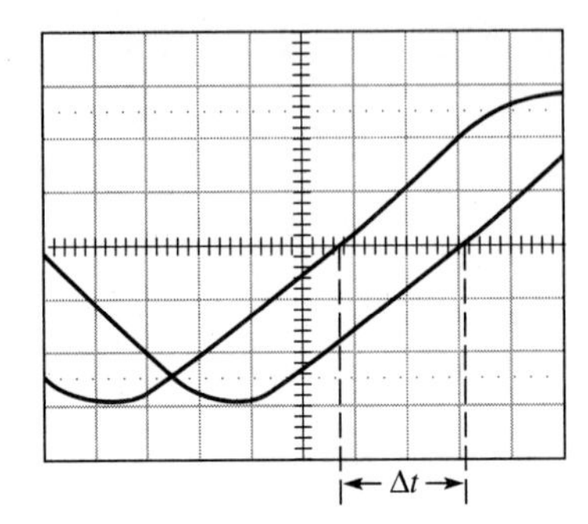

the horizontal position and SEC/DIV vernier. Since there are 360° in a cycle, each division represents 36° (assuming the graticule has 10 divisions). As before, both waveforms are adjusted to have the same vertical amplitude and position. The phase shift is determined by counting the number of divisions between corresponding points on the waveforms and multiplying by 36°/div. You can increase the resolution of this method by carefully setting up the original waveform for one-half cycle across the screen. In this way, the value of each division is 18°. The 10× magnifier is another convenient way to increase the resolution of the phase measurement. The measurement points are first put on each side of the center graticule line and then the magnifier is activated.

EXAMPLE 8–9 An oscilloscope shows one cycle of a sinusoidal waveform in 6.7 div. The SEC/DIV control is set to 2.0 μs/div and the horizontal magnifier is off.

The oscilloscope is readjusted to simultaneously show a second sinusoidal waveform of the same frequency. The second waveform is separated by 1.3 div when the SEC/DIV control is set for 0.5 μs/div. Compute the phase shift between the signals.

SOLUTION Begin by finding the period of the waveforms:

$$T = (2.0\ \mu\text{s/div})(6.7\ \text{div}) = 13.4\ \mu\text{s}$$

Next, find the time between the two waveforms:

$$\Delta t = (0.5\ \mu\text{s/div})(1.3\ \text{div}) = 0.65\ \mu\text{s}$$

From Equation 8–8, the phase angle is

$$\theta = \left(\frac{\Delta t}{T}\right) \times 360^\circ$$

$$= \left(\frac{0.65\ \mu\text{s}}{13.4\ \mu\text{s}}\right) \times 360^\circ = 17.5^\circ$$

An older method for measuring phase angles involved interpreting Lissajous figures. A **Lissajous figure** is the pattern formed by the application of a sinusoidal waveform to both the x- and y-axis of an oscilloscope. The scope's sweep genera-

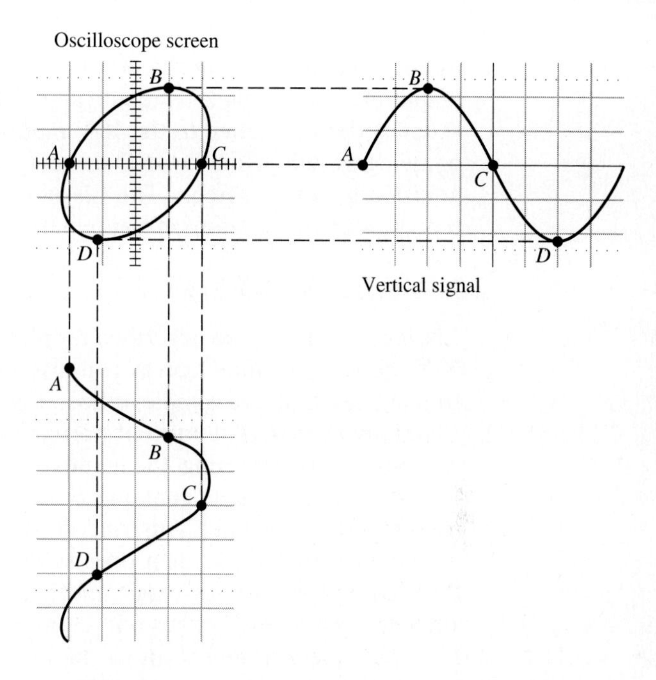

FIGURE 8–19
Formation of a Lissajous figure from two sinusoidal waveforms ($\theta = 67.5°$).

tor is not used; the beam is moved only by the signals on the two channels. Two signals at the same frequency having equal amplitudes will produce a 45° line on the scope face if they are exactly in phase. If the same signals are exactly 90° apart, the waveform will appear as a circle. The way Lissajous figures are formed is shown graphically in Figure 8–19. Other phase angles can be determined by applying the formula

EQUATION 8–9

$$\theta = \arcsin \frac{OA}{OB}$$

The definitions of *OA* and *OB* are shown in Figure 8–20.

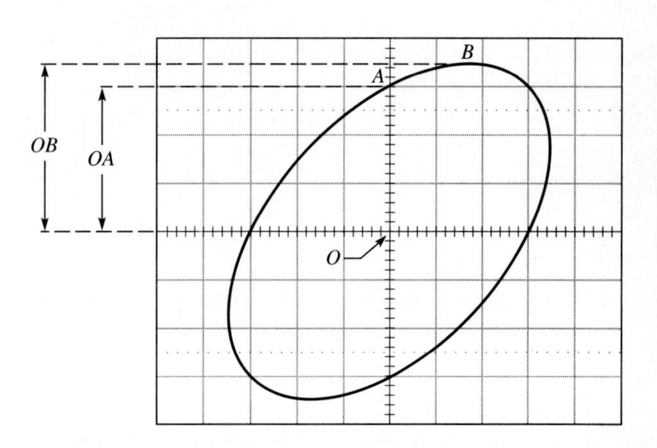

FIGURE 8–20
Definition of *OA* and *OB*.

To make a phase measurement by this method, both waves are centered on the oscilloscope face in the normal mode (time domain) and adjusted so that the amplitude of each signal is identical (using the variable VOLTS/DIV control). The scope is then switched to the X-Y mode to view the Lissajous figure. If the input coupling switch is set to GND on one channel at a time, the input signals can be adjusted to center screen. The Lissajous figure is then referenced to the center of the graticule for making measurements.

X-Y MEASUREMENTS

Lissajous figures were described for phase measurements as one application of the X-Y mode. Lissajous figures can also be used as a sensitive comparison of two frequencies that are simple ratios. Figure 8–21 illustrates some representative Lissajous figures. The ratio of horizontal frequency to vertical frequency can be determined by counting the number of times the side of the Lissajous figure touches a horizontal tangent line divided by the number of times the figure touches a vertical tangent line. This rule does not hold when the return trace coincides with the forward trace, as it does in the cases of 0° and 180° phase differences. If the frequencies are not exact multiples, the Lissajous figure will not be stable, since the phase differences will cause the figure constantly to change.

X-Y measurements allow the user to graph the signal in one channel as a function of the signal in the other channel. The horizontal signal, instead of representing time, can represent any variable for which a voltage proportional to the desired quantity can be generated. An application in which the X input repre-

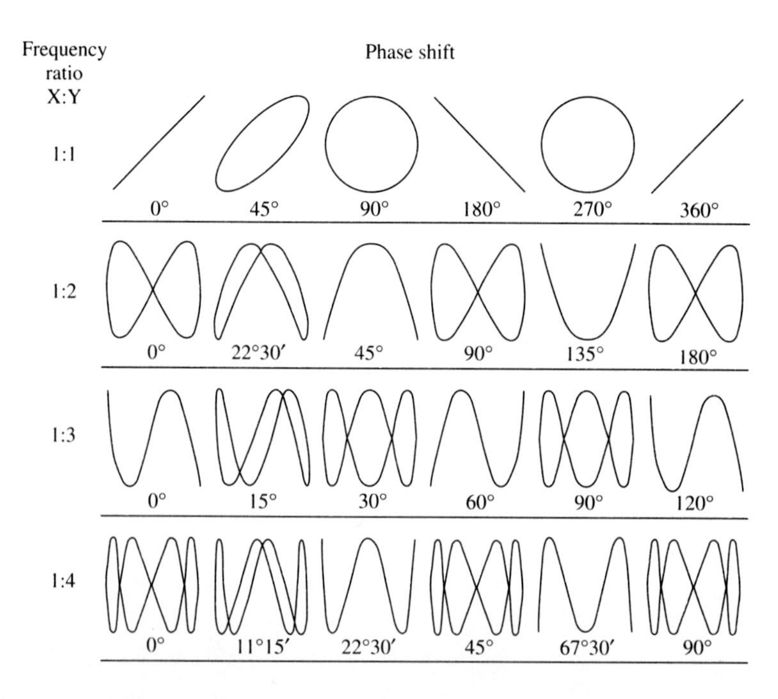

FIGURE 8–21
Frequency and phase measurements with Lissajous figures (copyright Tektronix, Inc. Used with permission).

sents frequency instead of time is swept-frequency measurements. The horizontal beam on the oscilloscope is moved by a ramp signal that also controls the frequency of a voltage controlled oscillator. This causes the oscilloscope's horizontal axis to represent frequency. Swept-frequency measurements are widely employed for making frequency response measurements on amplifiers, filters, and tuned circuits. They are discussed further in Section 9–6, and a diagram of a swept-frequency measurement is shown in Figure 9–11.

PULSE MEASUREMENTS

A pulse is a signal that rises from one level to another, remains at the second level for some time, and then returns to the original level. The time required for a pulse to rise from 10% to 90% of its maximum level is called the rise time (review Figure 8–13), and the time to return from 90% to 10% of the maximum level is called the fall time. Pulse width, abbreviated PW, is measured at the 50% level.

Measurement of pulses is normally done using dc coupling, which directly couples the signal to the oscilloscope and avoids problems caused by inserting a capacitor in series with the input signal. As always, probe compensation should be checked before making pulse measurements. For time and voltage measurements, be sure that the appropriate VARIABLE controls (usually red) are in their calibrated (detent) position.

To measure the rise time of a pulse, it is important that pulse distortions be minimized by properly terminating the input signal. (Pulse characteristics are described in Section 9–6; pulse distortions are illustrated in Figure 9–14.) The pulse to be measured is first set up between the 0% and 100% markers on the graticule using the variable VOLTS/DIV and the vertical POSITION controls. (If the graticule does not have vertical percentage markers, use five divisions, with each division representing 20%.) The leading edge of the pulse is expanded to cover most of the screen. Measure the number of divisions from the time the rising edge crosses the 10% line until it crosses the 90% line, as illustrated in Figure 8–22. Multiply the number of divisions by the SEC/DIV setting, and if you are using the magnifier, divide the result by the magnification factor.

FIGURE 8–22
Measurement of pulse rise time.

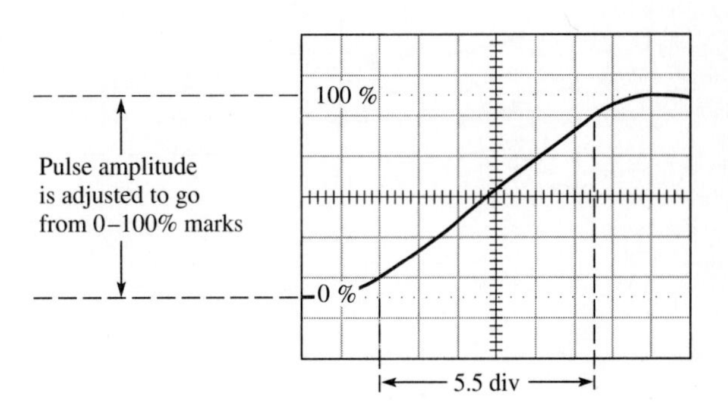
100 %
Pulse amplitude
is adjusted to go
from 0–100% marks
0 %
5.5 div
Multiply the number of divisions by the setting of the SEC/DIV
control and divide by the magnification factor.

As you know, all measurements involve some error due to the limitations of the measurement instrument. Figure 8–14 indicated the error that occurs when the oscilloscope's rise time is near that of the pulse to be measured. An oscilloscope with a 60 MHz bandwidth has a rise time of approximately 6 ns, as shown in Example 8–7. Measurements of pulses with rise times faster than about 24 ns on this particular oscilloscope will have measurable error. A correction formula, useful for estimating the true rise time, can be applied to the measured value. This is done by rearranging Equation 8–7 to compute the true rise time of a pulse:

EQUATION 8–10

$$t_{r(true)} = \sqrt{t_{r(displayed)}^2 - t_{r(scope)}^2}$$

MEASUREMENTS IN DIGITAL SYSTEMS

Digital systems frequently have signals that are nonrepetitive and are continuously changing states. Often, analyzing a digital system requires an ability to ascertain when a particular event occurred. A logic analyzer (see Section 11–3) can help, but viewing the details of a waveform frequently requires an oscilloscope. Some new products combine the function of an oscilloscope with a logic analyzer, including logic triggering, multiple-channel inputs, and processor emulation. With many events occurring in time, it is useful to be able to isolate a particular event with a trigger probe. A **trigger probe** generates a pulse when a particular set of predetermined input conditions are met. The conditions are selected by the user with switches on the probe that can select either a logic HIGH, LOW, or DON'T CARE (X) at any of several points in a circuit. The trigger conditions for the oscilloscope can be set up with a logic analyzer, but if one is not available, the trigger conditions can be set up with a trigger probe or a comparator can be connected to look for a particular condition, as illustrated in Figure 8–23.

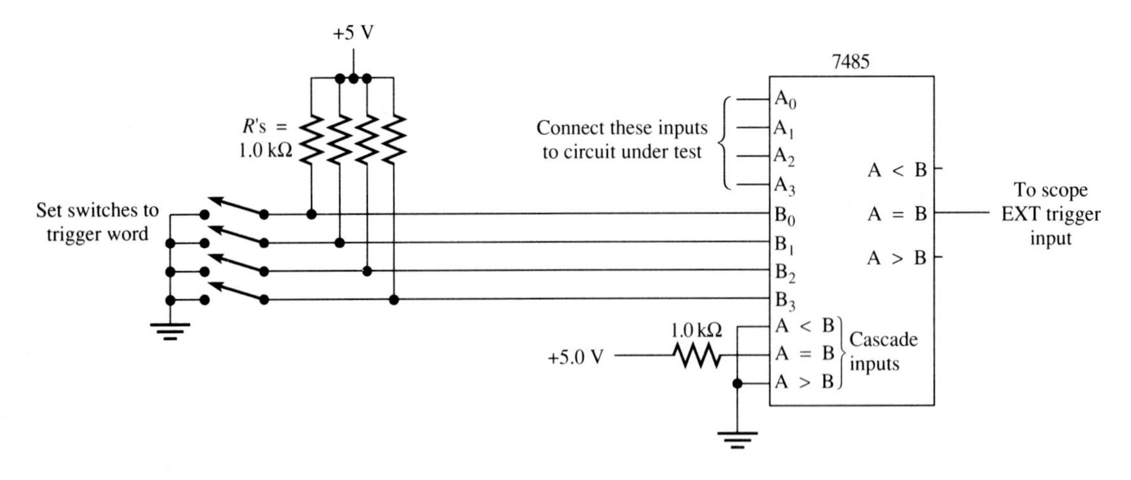

FIGURE 8–23
A comparator can be used as a trigger probe. When the circuit logic is the same as the trigger word, the oscilloscope will be triggered.

EXAMPLE 8–10 Assume you want to measure the time required from the read ($\overline{RD}$) for the disable interrupt instruction (DI) in an 8080 microprocessor until the interrupt enable line (INTE) goes low as a result. Show the connection of a trigger probe and oscilloscope to make the measurement.

SOLUTION A short test program in assembly language causes the computer to cycle through the desired instruction, as illustrated in Figure 8–24(a). The program shown is a simple loop program that executes just three instructions repetitively. The $\overline{RD}$ line goes low each time an instruction is fetched from memory. To interpret which instruction is being fetched, the trigger probe is connected to three address bits, which are unique to the DI instruction, as shown in Figure 8–24(b). The oscilloscope sweep is triggered only when the DI instruction is being fetched from memory. The time measurement can now be accomplished by observing the $\overline{RD}$ on channel 1 and the INTE response on channel 2.

```
0000 FB         START: EI             ;INTE GOES HIGH
0001 F3                DI             ;INTE GOES LOW
0002 C30000            JMP START      ;LOOP
```

(a) Test program.

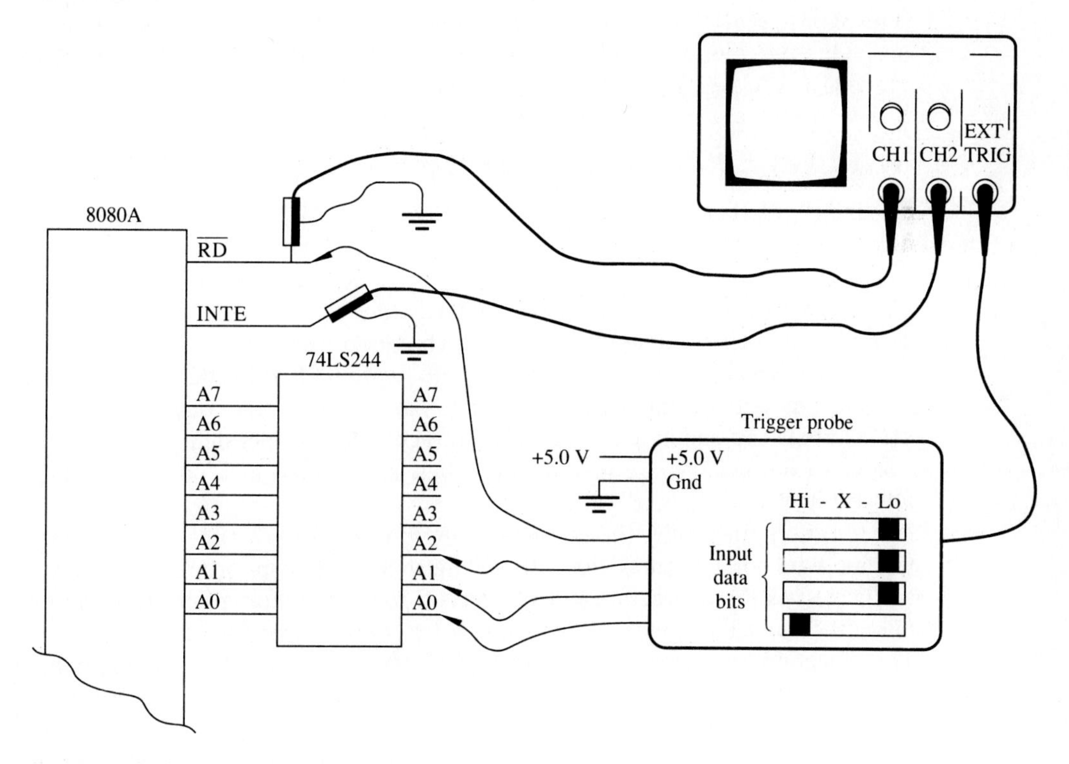

(b) Using a trigger probe to qualify a trigger.

FIGURE 8–24

8–8 ANALOG STORAGE OSCILLOSCOPES

Storage oscilloscopes are capable of holding an image for a period of time; they are used to capture and display very fast, infrequent, or single-event signals and find application in the analysis of intermittent failures. There are two basic types of storage oscilloscopes: digital and analog. The fundamental difference between them is that a digital storage oscilloscope quantizes the input waveform and stores it in a digital memory, whereas the analog storage oscilloscope retains an image of the input waveform as an electrostatic charge inside a special CRT for a limited period of time. (The later is frequently referred to as CRT storage.) There are several methods for capturing and storing a waveform in a CRT, including variable persistence, bistable storage, and fast-transfer storage; the most common are variable-persistence CRTs and bistable storage CRTs. All CRT storage uses secondary emission of electrons to build up and retain an electrostatic charge on an insulated target and read and write the image with two separate electron guns: a **write gun,** which is equivalent to the gun on a standard CRT and is used to send a highly collimated beam of electrons to the phosphorous screen, and a **flood gun,** which produces an uncollimated cloud of low-energy electrons that are not energetic enough to write the image on the target but can be made to cause the phosphor to glow. The flood-gun electrons are directed in a wide parallel beam by a conductive aquadag collimator coating on the inside of the CRT. A cutaway view of a storage CRT is illustrated in Figure 8–25.

VARIABLE-PERSISTENCE CRTs

The most widely used analog storage oscilloscope is the variable-persistence type. The basic features of a variable-persistence storage tube are diagrammed in Figure 8–26. The write-gun assembly and phosphor screen are similar to a conventional CRT. The write gun is operated at a high negative voltage—typically −5 kV—and produces a well-collimated beam of high-energy electrons. The phosphorous screen has a high positive voltage associated with it. Just inside the phosphorous screen is mounted a special storage mesh that has dielectric coating on the inside surface that can retain a charge. The storage mesh is operated at about −10 V. Closer to the neck is a collector mesh that is operated at a positive voltage of about 100 V.

In the write cycle, the energetic electrons from the write gun pass through the fine-wire collector mesh and storage mesh to the phosphorous screen, forming an image. As the energetic electrons pass through the storage mesh to the phosphor, they cause secondary electrons to be emitted from the dielectric coating. These secondary electrons are collected by the collector mesh, leaving behind a positively charged electrical "image" on the storage mesh. This electrical image replicates the visual image formed on the phosphorous screen. The write gun is then turned off, but the stored positive charge is retained on the storage mesh wherever the original high-energy electrons passed.

Now the flood guns, mounted along the inside funnel of the CRT, are turned on. They produce a uniform cloud of low-energy electrons, which are directed in a

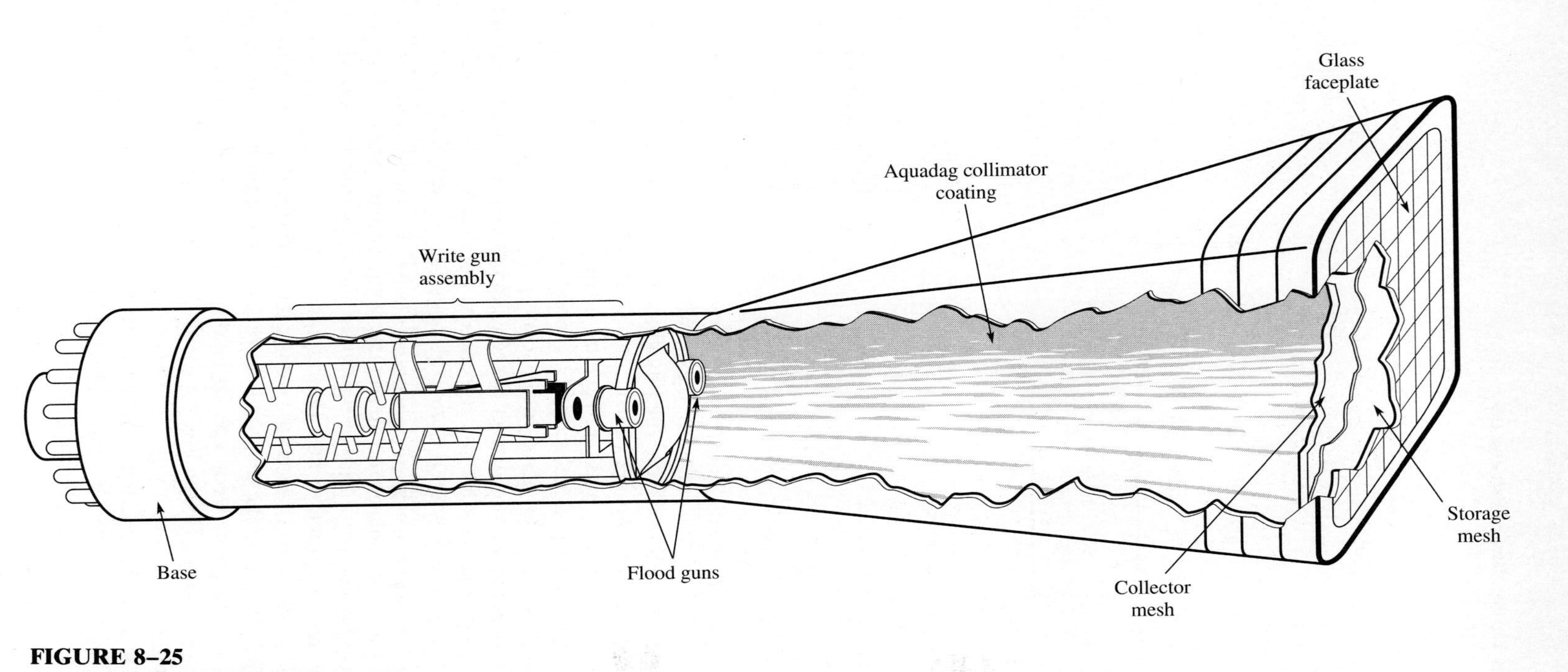

FIGURE 8–25
Cutaway view of a CRT storage tube.

FIGURE 8–26
Variable-persistence CRT.

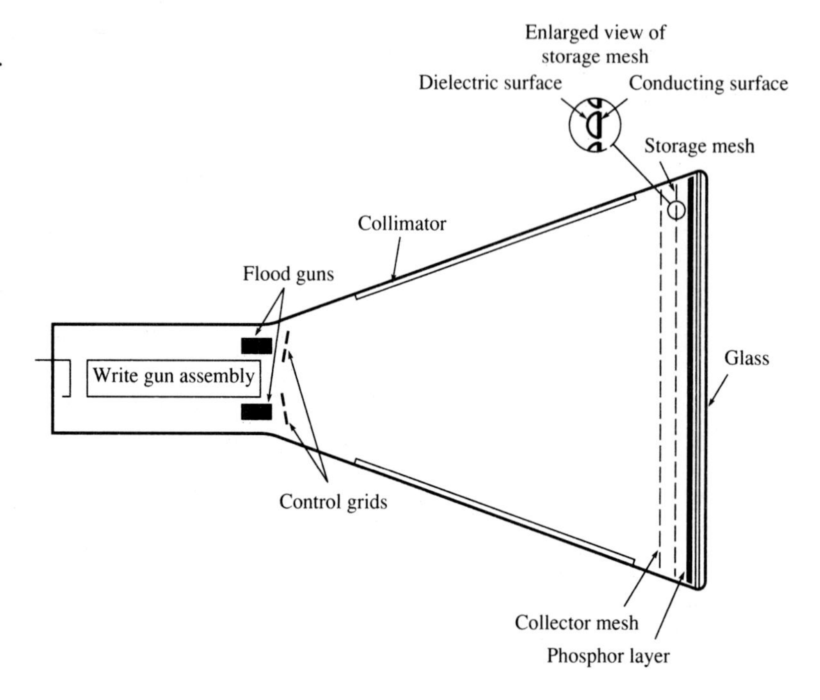

parallel path toward the screen by the collimator. Electrons from the flood guns are attracted to the collector mesh. Most of the low-energy electrons from the flood gun are blocked from the phosphorous screen by virtue of the small negative potential on the storage mesh; however, the low-energy electrons can pass through the storage mesh in regions made positive by the secondary emission. These electrons are accelerated by a positive voltage on the phosphor, causing an image of the original signal to be seen.

The image on the storage screen can be observed with reasonable brightness for several minutes after it is stored. The persistence can be varied by discharging the storage mesh with a series of negative pulses that discharge the positive regions. The pulse width is controlled by a front-panel PERSISTENCE control—longer pulses discharge the mesh faster than shorter pulses. The entire pattern can be erased by momentarily bringing the storage screen to a positive voltage, causing the entire screen to become bright.

Variable-persistence storage is useful for stopping very slow periodic signals to avoid the image flicker that occurs with conventional oscilloscopes. Measurements that require sweep speeds below about 10 ms/div can benefit from storage of the waveform. Depending on the CRT and the application, persistence can be adjusted over a large range. The persistence can be varied to fade out just before the next sweep is written or adjusted to show several previous traces for comparison. The CRTs are also useful in applications such as spectrum analyzers and medical diagnostics, where slow sweep speeds are required.

BISTABLE STORAGE

The bistable storage CRT does not use a storage mesh to hold the image. Instead, it has a special phosphor coating that has two stable states and is capable of holding a charge for several hours. The basic features of the bistable storage CRT are diagrammed in Figure 8–27. A transparent backplate is sandwiched between the phosphor and the glass face of the tube. It is operated with a small positive voltage with respect to ground. In the writing phase, the write gun aims a collimated beam of electrons at the phosphorous screen, forming a direct image of the signal applied to the oscilloscope's input. These energetic electrons cause secondary emission to occur in the phosphor, leaving a positive charge in the grains of phosphor.

Now the write gun is turned off and the flood guns are turned on. The flood guns produce a cloud of low-energy electrons that cover the viewing screen and ensure that the image remains visible. The low-energy electrons are attracted to the positively charged regions of the phosphor. These low-energy electrons pass right through the phosphor and are collected by the transparent metal-film backplate. As they pass through the phosphor, it glows, revealing the original image stored during the writing phase. Some oscilloscopes have the storage area split in two, enabling signals to be stored, viewed, or erased independently.

Bipolar storage oscilloscopes are used in applications requiring very fast single-shot or low-repetition-rate signals, such as in laser and nuclear research, high-speed transient analysis, and cases where a one-time event needs to be

FIGURE 8–27
Bistable storage CRT.

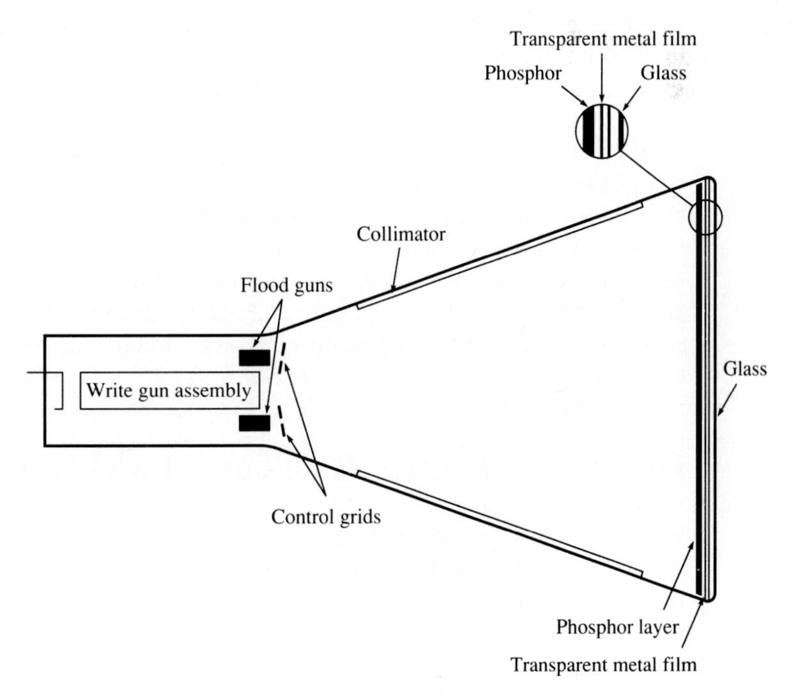

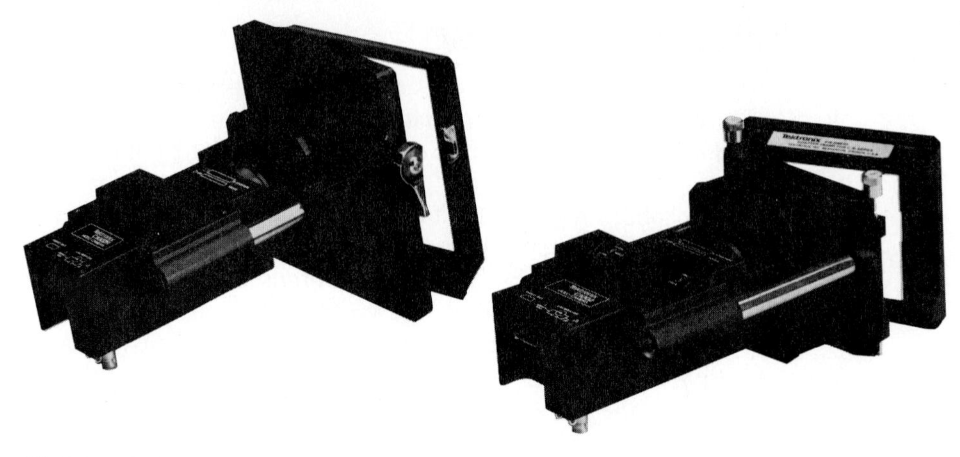

FIGURE 8–28
Digitizing camera system (copyright Tektronix, Inc. Used with permission).

examined (for instance, failure analysis). Bipolar storage oscilloscopes are capable of much higher bandwidths for single-shot events than digital storage oscilloscopes; however, the cost of bipolar storage oscilloscopes is high and the CRTs are fragile. Storage times depend on the CRT but can be as long as 24 h.

DIGITIZING CAMERA SYSTEM

A practical method for combining the benefits of high analog scope bandwidths with computer processing is to use a digitizing camera system. Figure 8–28 illustrates a digitizing camera system. The camera is attached to the analog oscilloscope and responds to light from the trace. It converts the light image into a video signal, which is cabled to a frame store interface. The raw video can be stored on a videocassette recorder or digitized for processing by a computer. The bandwidth of the oscilloscope is limited only by the photographic writing rate, typically 60% of the overall scope bandwidth. This limitation can be overcome on scopes with microchannel plate (MCP) CRTs, a form of light multiplier built into the CRT screen,[2] as described in Section 8–1. Applications include very high speed single-shot events for which computer processing is needed or where a record of a series of images is necessary.

8–9 DIGITAL STORAGE OSCILLOSCOPES

Recent advances in analog-to-digital conversion technology and high-speed memories have given digital storage oscilloscopes (DSOs) very impressive specifications and features not possible a few years ago. These advances have made the

[2] The Tektronix 2467B, illustrated in Figure 8–9(b), has an MCP CRT.

price of DSOs very competitive with analog oscilloscopes, have produced new features that are generally unavailable on analog oscilloscopes, and have simplified digital scope operation (for instance computer-aided measurements and automatic recall of settings). A digitizing oscilloscope is shown in Figure 8–29; the front cover also shows a digitizing oscilloscope.

A DSO converts a waveform into a series of numbers that are stored in a memory. A single digitized waveform is called a **memory record.** Typically, several different records can be stored by a DSO. Within the record, each byte represents one data point on the digitized waveform. Its location in memory corresponds to the position on the screen at which the data point will be shown. In a digital scope, a record can be longer than the information that is displayed on the screen. To view the record, it is converted back into analog form for display on a conventional CRT.

Applications for digital storage oscilloscopes include automated test measurements, biomedical research, and mechanical shock and fracture studies. DSOs sample an input signal, convert the samples to digital words, and display the signal. However, for many applications, the digitizing oscilloscope has important advantages over its analog cousin: It can make measurements on transient signals that are otherwise very difficult to measure; it can do computations with wave-

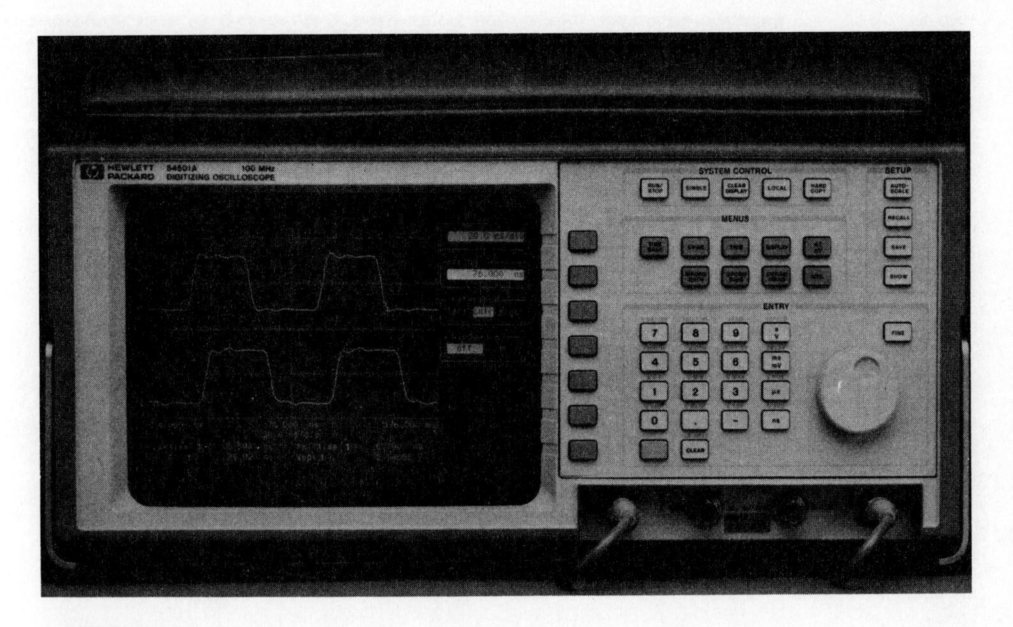

FIGURE 8–29
HP 54501A digitizing oscilloscope has four channels and 100 MHz bandwidth. The new instrument has the same high-performance features as the HP 54100 and 54200 series of products, such as full IEEE-488 programmability, fast hard-copy output to HP graphics printers, automatic measurements, and advanced-logic triggering (courtesy of Hewlett-Packard).

forms with its built in processing and analysis capability; and it can make comparisons of a live waveform with a waveform stored in its internal memory. In addition, front-panel setups can be saved and automatic measurements can be made on either stored or live waveforms; additionally, many digitizing oscilloscopes have provisions for a paper plot of the output (''hard copy''), including a listing of front-panel setups and specific measurement results. One manufacturer has integrated high-speed thermal array recorders directly into their digital storage oscilloscopes, as shown in Figure 8–30. The recorder can be used for continuous recording of data on-line or can plot data previously stored in memory.

BLOCK DIAGRAM

A DSO uses an A/D converter to change the input waveform into a series of discrete time-ordered samples, which are stored in a memory. Figure 8–31 shows a block diagram of a DSO. The input waveform is first conditioned for the A/D converter with a buffer amplifier. A portion of the analog signal is picked off to generate a trigger for the time base. The time base generates a series of pulses that control a sampling gate and A/D converter. The A/D converter then quantizes (converts to a binary number) a sample of the input waveform at a rate determined by the time-base SEC/DIV control. Samples are then stored in a memory that

FIGURE 8–30
Gould digitizing oscilloscopes with built-in thermal array recorders (courtesy of Gould Electronics).

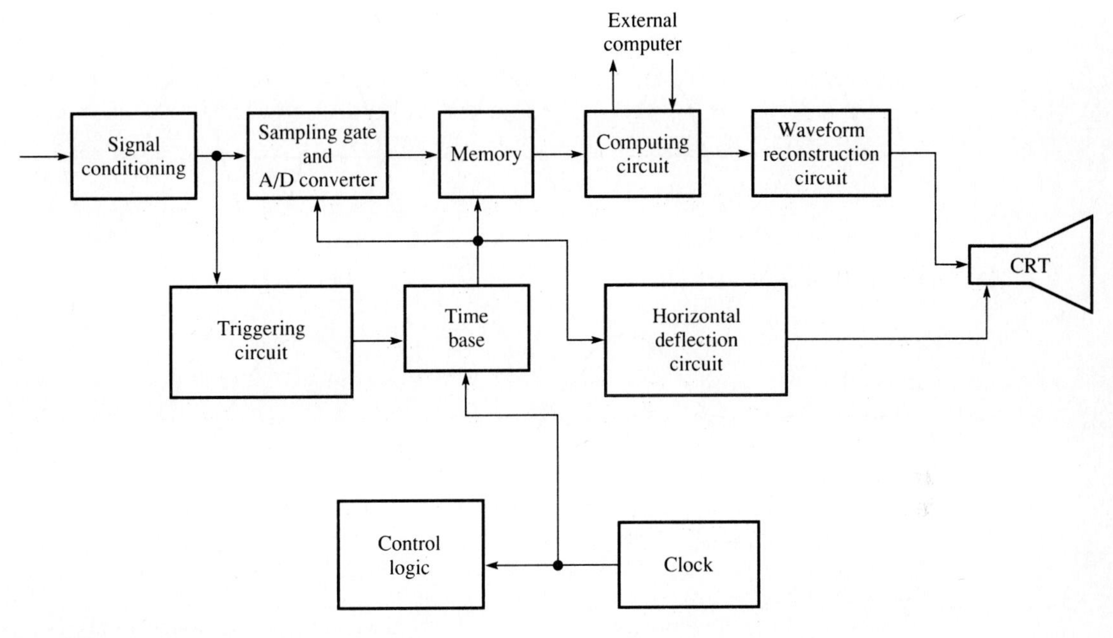

FIGURE 8–31
Block diagram of digital storage oscilloscope.

operates at the same rate as the A/D converter; the memory must be capable of storing data very rapidly. Once stored, the data can be retained indefinitely or processed with reconstruction algorithms to be displayed on a CRT. Once digitized, the data can be sent at a slower rate to the CRT, allowing a less expensive CRT than in an analog scope, since the CRT does not require as high a bandwidth.

RESOLUTION

The resolution of a digital oscilloscope can be divided into two parts—amplitude resolution and time resolution. The amplitude resolution is a function of the number of bits per sample. For a typical 8-bit digitizer, the resolution is 1 part in 256 ($2^8 = 256$). For a 1 V peak signal, this represents a 4 mV resolution. Although additional bits mean greater vertical resolution, there are higher costs associated with more bits. The time resolution and the equivalent bandwidth for real time sampling is dependent on how fast the A/D converter can acquire samples. The Nyquist theorem is a fundamental theorem for information sampling. It states that the sampling frequency must be greater than twice the highest frequency to be reproduced in order to reconstruct the unknown waveform. In other words, the Nyquist criterion sets the bandwidth of sampled data at one-half the sampling rate. A signal digitized less frequently than the Nyquist criterion can have **aliasing** error, which causes the reconstructed waveform to have the wrong frequency, as illustrated in Figure 8–32. The Nyquist theorem does not imply that a waveform

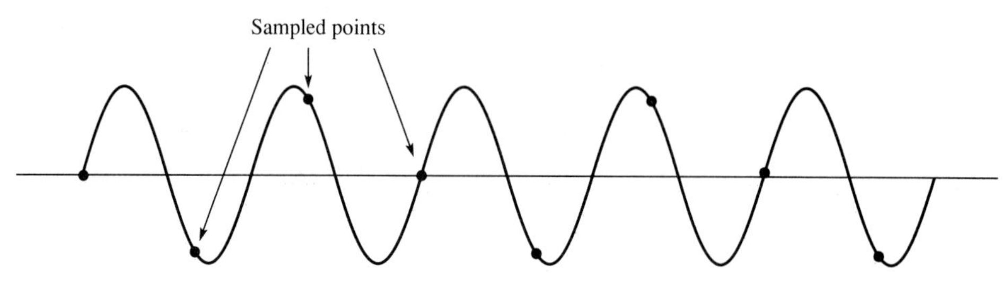

(a) Original waveform.

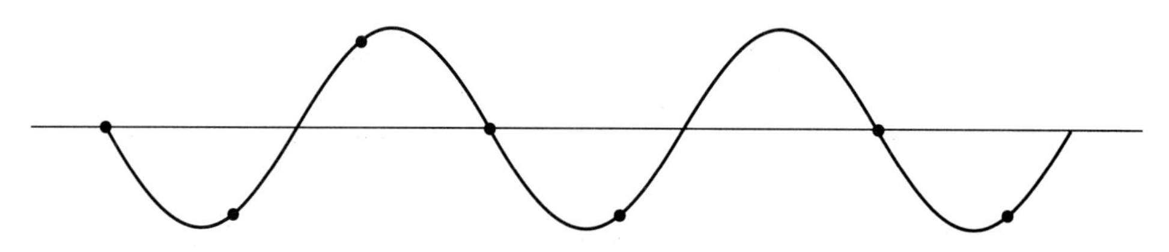

(b) Reconstructed waveform.

FIGURE 8–32
Aliasing due to sample rate being too low. The original waveform is sampled only 1½ times per cycle instead of the required minimum of more than 2 times per cycle, causing the reconstructed waveform to be at the wrong frequency.

can be completely reconstructed if you sample at greater than twice the signal frequency. If you try to reconstruct a waveform from data taken near the Nyquist limit, serious errors in amplitude, phase, or frequency may result. Four points per period allow the phase amplitude and frequency of an unknown waveform to be determined but leave the shape of the waveform undefined. A reasonable picture of the waveshape needs 20 or more samples per period. From this, it can be seen that the digitizing rate cannot be directly compared to bandwidth. For example a digitizing rate of 20 MSa/s (megasamples per second) will produce 20 data points per period on a 1 MHz signal, sufficient to determine the amplitude, frequency, phase, and shape of the wave.

SAMPLING METHODS

The **sampling rate** is defined as the number of samples taken each second and is specified in units of megasamples per second. Digital oscilloscopes can sample the input waveform in either real time or in equivalent time. Most digitizing oscilloscopes can acquire data in either mode. The two methods are compared in Figure 8–33. **Real-time sampling** means that samples are all collected sequentially in a single acquisition as the waveform is received. For single-shot events, real-time sampling is necessary, and the sample rate determines whether the oscilloscope can capture the event. **Equivalent-time sampling** means that the waveform is reconstructed from samples acquired over a number of cycles of the waveform.

As such, equivalent-time sampling can be accomplished only on periodic waveforms. The samples are interspersed on each acquisition to build a very high resolution composite picture of the input waveform, resulting in a very high equivalent bandwidth of the oscilloscope. Equivalent-time sampling can be done repetitively without waiting for a trigger or it can be done after some predetermined time from a trigger event.

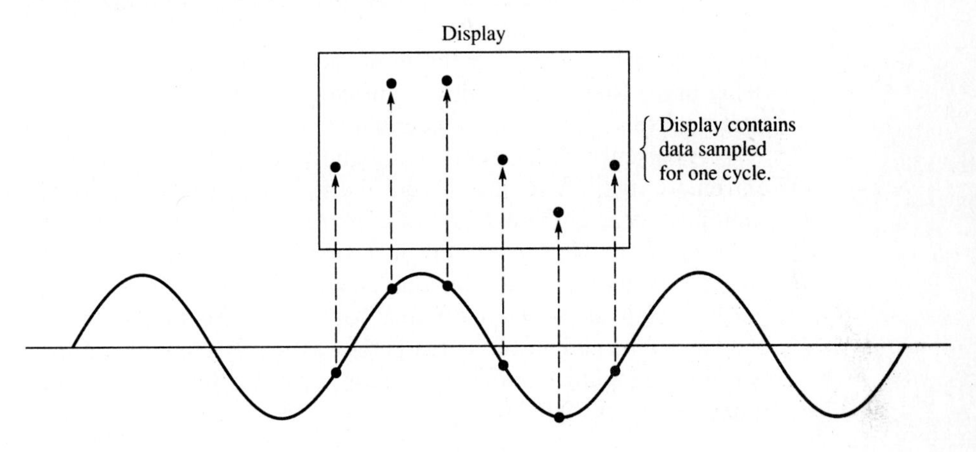

(a) Real-time sampling.

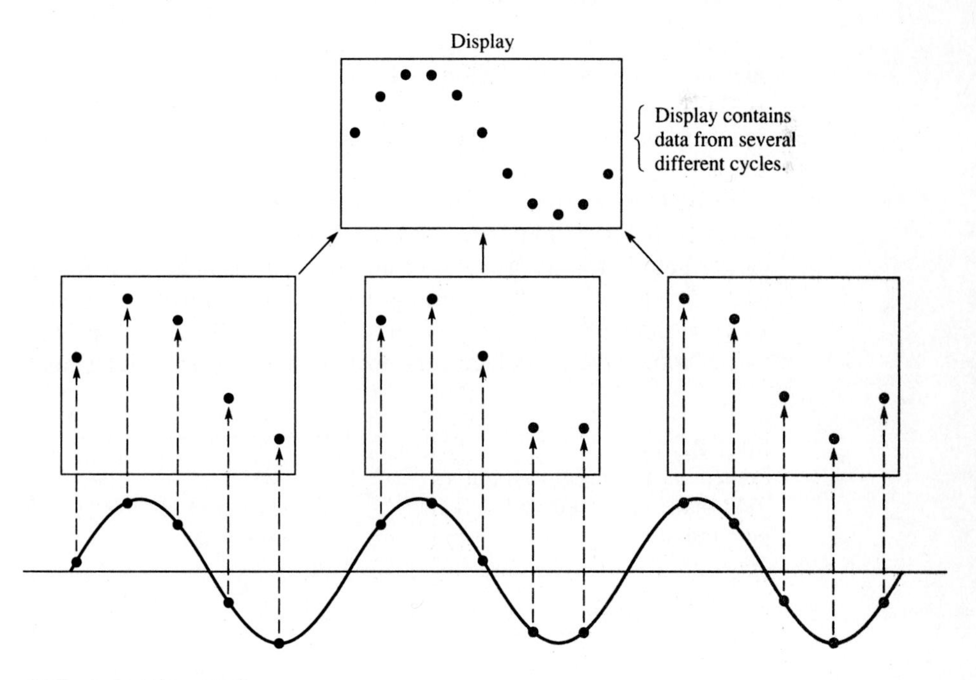

(b) Equivalent-time sampling.

FIGURE 8–33
Comparison of real-time and equivalent-time sampling.

Equivalent-time sampling can be accomplished either sequentially or as random repetitive sampling. In sequential sampling, one point is taken on the waveform for each trigger event. The sampled point is moved further from the trigger each time and "builds" the waveform one point at a time. With random repetitive sampling, the signal is sampled at a rate determined by the internal oscilloscope clock. Samples are stored in memory in locations that are in the proper relationship to the trigger to give the appearance of a single continuous record. Sequential equivalent-time sampling can attain extremely high bandwidths (20 GHz) but loses the pretrigger viewing capability at these high bandwidths due to limitations of delay lines. Random repetitive sampling retains the pretrigger advantage but has greater display jitter than sequential sampling.

As discussed, the equivalent bandwidth for equivalent-time sampling can be extremely high. Also, the Nyquist criterion does not apply to equivalent-time sampling because the data is not captured in real time. Equivalent-time sampling is useful in reading extremely fast signals, such as ultrahigh-frequency (UHF) measurements and time-domain reflectometry (TDR). In TDR measurements a pulse is sent along a conducting path and measured as it reflects back. Any changes in impedance along the path are directly readable on the display. Equivalent-time sampling can resolve events separated by as little as 0.25 ps with amplitudes as low as 500 μV.

PERCEPTUAL ALIASING

A factor to consider with sampled data is called perceptual aliasing. **Perceptual aliasing** causes sampled data to *appear* to have changed frequency. This problem is illustrated in Figure 8–34. Perceptual aliasing occurs when sampled data is viewed as a set of dots. The human eye tends to connect the closest dots in space even if they are not the closest in time. This is most noticeable when the frequency of the data is near the sampling frequency. Drawing straight lines between sequential data points is better but still misleading, especially for sinusoidal waves. A better solution is to provide an interpolation scheme that fits the data to a sine wave or a pulse waveform. **Interpolation** is the addition of data points between the original samples. The interpolator's algorithm needs to be optimized for the stored waveform. Most manufacturers provide some form of interpolation to aid in re-

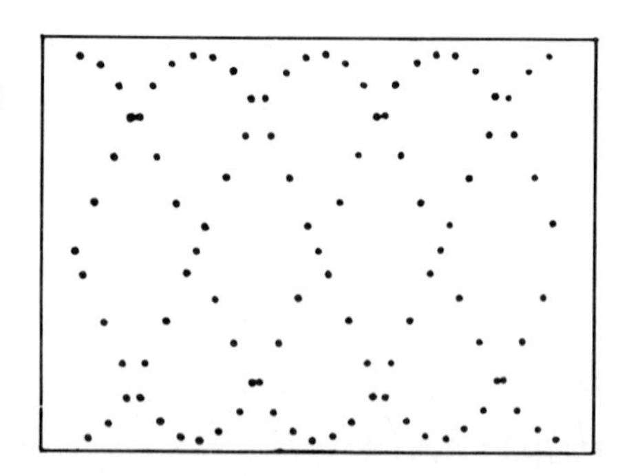

FIGURE 8–34
Perceptual aliasing (copyright Tektronix, Inc. Used with permission).

construction of the waveforms. With the proper interpolator, reconstruction of the waveform can provide a truer picture of the original data.

USEFUL STORAGE BANDWIDTH

For years, bandwidth has been one of the principle specifications of oscilloscopes. The bandwidth should be higher than that of the signal of interest to assure that the signal is not measured inaccurately because of the limitations of the measuring system. With digitizing oscilloscopes, the meaning of bandwidth is not the same as for analog oscilloscopes, and specifications can be misleading if you try to compare the bandwidth between digital and analog scopes.

Once a signal is digitized, the ability to reconstruct the original waveform depends on the number of samples within one period of the waveform and the type of waveform. The useful storage bandwidth (USB) is defined as the frequency limit at which, for most measurements, minimal error is encountered. For a sinusoidal waveform, this frequency limit depends on the interpolation scheme. If a sine interpolator is used, a reconstructed sinusoidal waveform can be perceived and measured with as little as 2.5 samples per period (close to the Nyquist criterion). This implies that the maximum usable stored bandwidth is

$$\text{USB (MHz)} = \frac{\text{maximum sampling rate (MSa/s)}}{2.5} \qquad \text{(with sine interpolator)}$$

With a linear interpolator, it requires at least 10 data points per cycle to reconstruct the trace adequately, and if no interpolation is used, about 25 samples are needed. Thus, with no interpolator, the USB is reduced by a factor of 10 from the optimum:

$$\text{USB (MHz)} = \frac{\text{maximum sampling rate (MSa/s)}}{25} \qquad \text{(with no interpolator)}$$

The effects of interpolation on a sinusoidal waveform can be seen from the photographs in Figure 8–35. Without interpolation, the digitized waveform is almost unrecognizable with 2.5 samples per cycle. A clear improvement occurs with linear interpolation, but the optimum sine interpolator produces a usable trace at just 2.5 samples per cycle. While the sine interpolator does a good job of reconstructing the signal, the reconstruction algorithm has had to make certain assumptions about the signal shape. If the sampled data is from a triangle or other waveform, applying a sine interpolator can produce misleading results when the sampling rate is so close to the Nyquist limit.

For pulse applications, a similar problem exists. If the rising edge of a pulse is observed, the rise time and location of the pulse edge can be interpreted incorrectly if the sample interval is too short. This problem is illustrated in Figure 8–36. The sample times can make a significant difference in interpreting the rise time or shape of the pulse. The rise time can appear to be as little as 0.8 sample intervals to as much as 1.6 sample intervals, depending on where sampling takes place. Pulse interpolators have no way of correcting for an inadequate sample rate.

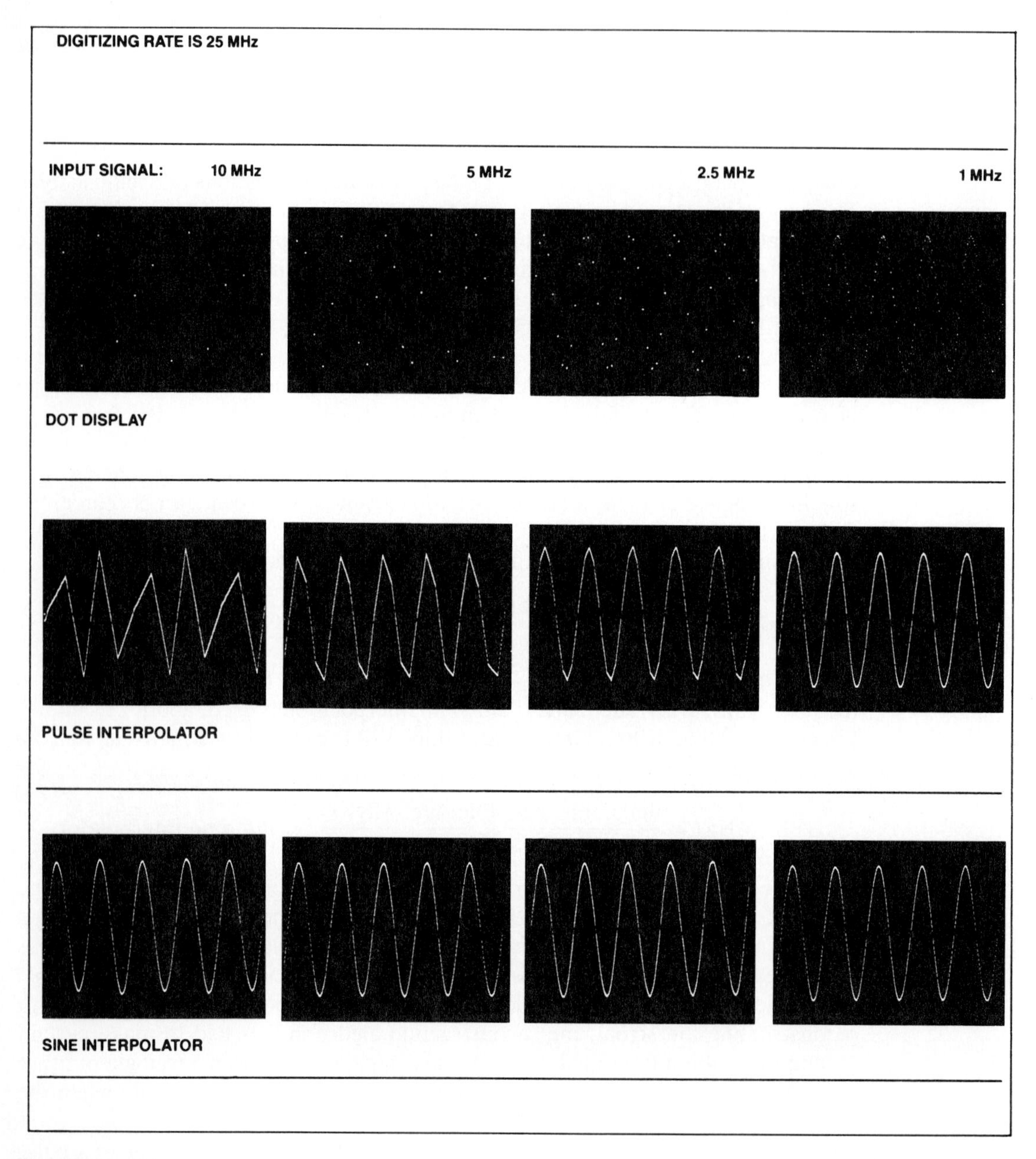

FIGURE 8–35
Effects of interpolation and sample rate on sinusoidal inputs. The photographs in the left column represent 2.5 samples per cycle; those in the right column represent 25 samples per cycle. Notice that the sine interpolator gives usable results with 2.5 samples per cycle; it requires about 10 samples per cycle to achieve the same results with the pulse interpolator (copyright Tektronix, Inc. Used with permission).

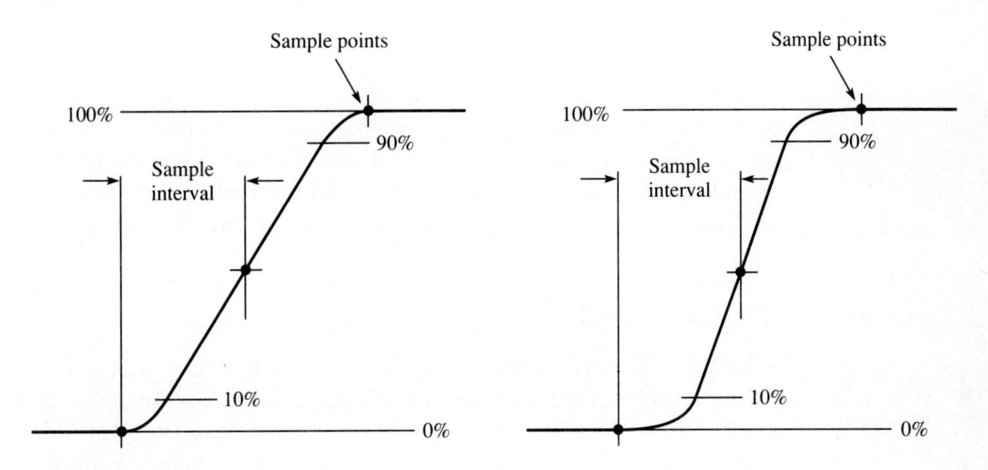

FIGURE 8–36
Rise-time error due to sample interval that is too long. The sample interval and data are *identical* for both waveforms but the rise time is not.

In Section 8–5, the rise time and bandwidth were shown to be related by the equation

$$t_r = \frac{0.35}{B}$$

This equation is derived from the charging and discharging equations for a capacitor in an *RC* circuit. It indicates the equivalent rise time of a pulse for an *RC* circuit in which the cutoff frequency is known (the frequency at which $X_C = R$). Although the equation is useful for specifying the bandwidth limitation of the input buffer amplifier (before A/D conversion), it cannot be used to determine if a digitizing oscilloscope can make an accurate rise-time measurement of a particular pulse.

Computer studies of rise-time measurements for digitizing oscilloscopes have shown that in order to represent the usable rise time (Ut_r) with about the same accuracy as an analog oscilloscope, the minimum sample interval must be multiplied by 1.6. That is

EQUATION 8–11

$$Ut_r = \text{minimum sample interval} \times 1.6$$

where Ut_r = usable rise time of a digitizing oscilloscope

The minimum sample interval is the reciprocal of the maximum sample rate. By substitution, it can be shown that for pulse rise-time measurements, the usable storage bandwidth is

$$\text{USB (MHz)} = \frac{\text{maximum sampling rate (MSa/s)}}{4.6} \qquad \text{(pulse rise time)}$$

From the preceding discussion, it can be seen that the usable bandwidth depends on several factors. These include the sampling rate, the interpolation scheme employed for reconstructing the waveform, the shape of the waveform, and the type of measurement. Although the bandwidth specification is a guide to its performance, for many applications it is easier to analyze the effect of the sampling rate directly rather than to attempt to relate it to bandwidth. Sample rate is usually a more applicable specification for determining if a digitizing oscilloscope can capture a particular waveform.

TRIGGERING

One useful feature of digital storage oscilloscopes is their ability to capture waveforms either before or after the trigger event. Any segment of the waveform, either before or after the trigger event, can be captured for analysis. **Pretrigger capture** refers to acquisition of data that occurs *before* a trigger event. This is possible because the data is digitized continuously, and a trigger event can be selected to stop the data collection at some point in the sample window. With pretrigger capture, the scope can be triggered on the fault condition, and the signals that precede the fault condition can be observed. For example, troubleshooting an occasional glitch in a system is one of the most difficult troubleshooting jobs; by employing pretrigger capture, trouble leading to the fault can be analyzed. A similar application of pretrigger capture is in material-failure studies, where the events leading to failure are most interesting, but the failure itself causes the scope triggering.

Beside pretrigger capture, posttriggering can also be set to capture data that occurs some time after a trigger event. The record that is acquired can begin after the trigger event by some amount of time or by a specific number of events, as determined by a counter. A low-level response to a strong stimulus signal is an example of a time when posttriggering is useful.

ENVELOPE MODE AND GLITCH CAPTURE

Digitizing oscilloscopes offer advantages in monitoring occasional problems such as changes in a signal, glitches, or dropped bits on a data line. The **envelope mode** is a recording mode that stores two samples of the data for each clock pulse. One of the samples represents the largest value of the data since the previous clock pulse, and the second sample represents the smallest value of the data since the previous clock pulse. In this way, the data is bracketed by the envelope, allowing you to see the minimum and maximum values of the data over multiple sweeps. If, during any sweep, the waveform changes from previous sweeps, the envelope mode will clearly show that change, as illustrated in Figure 8–37. An extension of the envelope mode is peak detection. In this mode a narrow spike that happens between sample intervals is caught and displayed. A DSO that does not have envelope mode takes one sample for each data point on the screen. If the glitch occurs between sample times, it will be missed.

Another application for the envelope mode is in establishing allowable limits for a measured waveform in a testing situation. If the waveform exceeds the

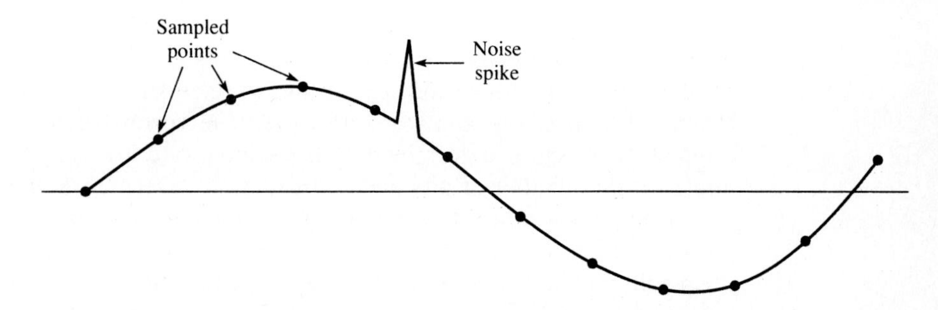

(a) Normal-mode sampling. The noise spike is missed because it falls between sampling times.

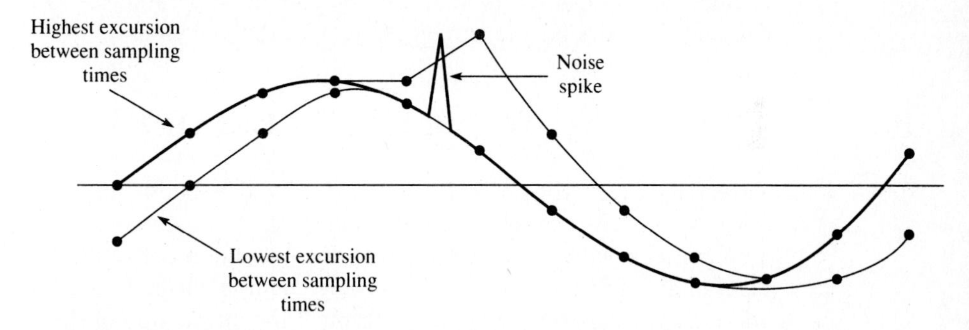

(b) Envelope-mode sampling. The envelope indicates both the lowest and the highest excursions of the data between sample times. The noise spike is evident in the data.

FIGURE 8–37
Comparison between normal-mode and envelope-mode sampling.

established limits, the oscilloscope can store the out-of-limit waveform and alert the operator (Tektronix calls this *save on delta*). This feature can be employed in pass-fail testing or in troubleshooting an infrequent event, such as a noise spike. The user sets up a reference waveform from the front-panel controls with an upper and a lower comparison point by adjusting the vertical position and delay-time controls while in the ENVELOPE CONTinuous mode. This waveform is then stored in a reference memory and used for comparison of the incoming signal.

DATA PROCESSING

The principle advantages of digitizing oscilloscopes are their ability to be used as data-collection devices for automated measurements and their computational capabilities. All modern digital oscilloscopes contain an internal microprocessor. The digital data can be transformed by the microprocessor or sent to an external computer for processing. The computer can permanently store the data, process it further, or document it. Some examples of signal-processing requirements include adding and subtracting channels, computing standard deviations, integration, differentiation, and signal averaging. Signal averaging the data from several different sweeps helps to eliminate random noise from the data. The processed data can be viewed as a series of disconnected dots or dots connected by straight lines, or it can be enhanced with various algorithms to reconstruct the original waveform or to provide waveform analysis. How well the reconstructed data matches the original waveform is determined by the sample rate (the reciprocal of the sample interval), the number of bits per sample, and the reconstruction algorithm.

Computed values from pulse measurements, such as frequency, amplitude, pulse width, rise time, overshoot, and other parameters, can also be automated; the results can be stored or used in automatic testing. The stored data can actually have higher resolution than data that a CRT can display. For many applications, this means that automated measurements can be of higher quality than an analog oscilloscope can show. In addition, the computed parameters can be presented as decimal numbers on the screen along with the waveform.

8–10 TRAVELING-WAVE OSCILLOSCOPES

Traveling-wave tubes (TWTs) are specialized vacuum tubes primarily used as microwave power amplifiers and oscillators. A traveling-wave tube is an evacuated cylinder with an electron gun at one end and a collector at the other end, as diagrammed in Figure 8–38. This arrangement is normally used to amplify microwave energy supplied to the helix through the interaction of the electrons and the radio field.

Although primarily associated with microwave power amplifiers, the concept of the traveling-wave tube has been applied to oscilloscopes for specialized applications. A phosphorescent screen replaces the collector and can include microchannel plate intensification. Deflection of the beam is done by applying the signal directly to the helix. As the electron beam travels down the tube, it is continuously deflected by the signal. An example of a TWT for oscilloscope use is

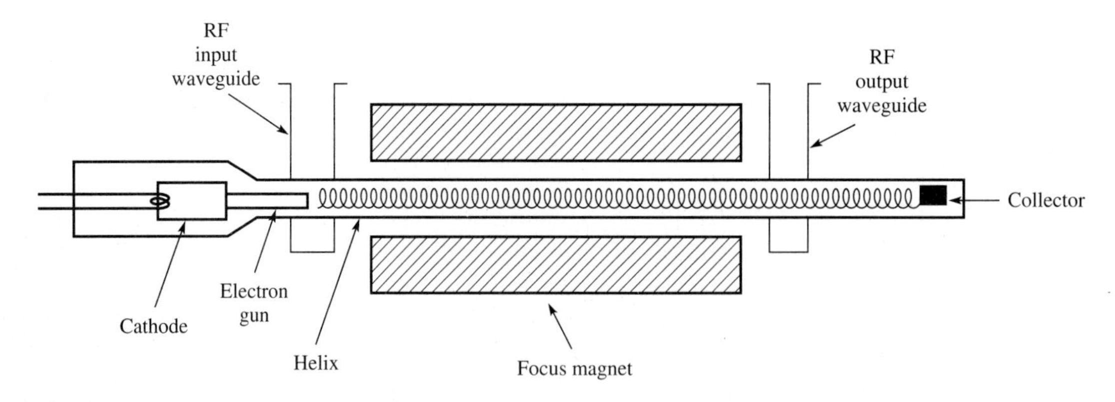

FIGURE 8–38
The traveling-wave tube (TWT). Normally the TWT is used as a microwave amplifier; however, it can be adopted for a CRT.

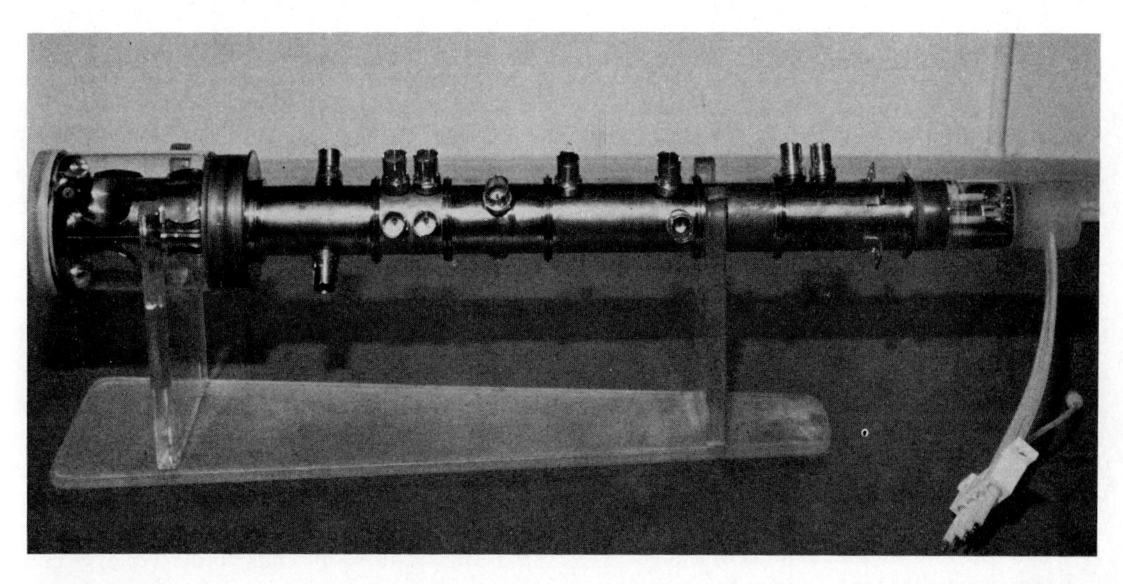

FIGURE 8–39
Traveling-wave tube CRT (courtesy of EG&G Energy Measurements).

shown in Figure 8–39. This tube is from an EG&G Model OS-40 oscilloscope with a bandwidth of 2.5 GHz and sweep lengths as short as 50 ns. Timing accuracy is enhanced by adding a known high-frequency sine wave (as high as 2 GHz) to the horizontal sweep. The added oscillator signal enables relative time measurements that are independent of any sweep nonlinearities. The display is normally recorded on film and then digitized, and the oscillator signal is removed by computers. Applications for TWT oscilloscopes include measurements of very high speed transient phenomena such as explosive tests.

SUMMARY

1. General-purpose oscilloscopes have four basic subsystems—the vertical, trigger, horizontal, and display sections. Each section has its own set of controls.
2. A cathode-ray tube is an electron-beam tube that converts an electrical signal into a visible trace on a phosphor-coated screen.
3. Delayed-sweep oscilloscopes have a second time base that offers greater flexibility and (generally) increased accuracy for making timing measurements.
4. An oscilloscope probe picks off the signal and applies it to the oscilloscope. Accurate measurements require the user to be aware of the limitations of the probe/scope combination. Probe compensation should be checked at the start of every session with the oscilloscope or any time a probe is changed.
5. Common oscilloscope measurements include amplitude, phase shift, period, frequency, rise time, pulse width, and time differences between signals.
6. A digital storage oscilloscope stores a waveform as a series of binary numbers in a memory. This offers the advantage of

capturing infrequent signals, making computations on the stored waveform, allowing unattended recording of signals, making a permanent record of the waveform, and recording data outside preset limits, among others.

7. CRT storage oscilloscopes retain the image in the CRT as an electrostatic charge on a special insulated mesh. Variable-persistence scopes allow you to control the length of time for which the waveform is stored and viewed.
8. Digital sampling oscilloscopes take a set of digital data over several cycles of a periodic waveform. Only a portion of the waveform is captured each cycle. The display is a reconstructed replica of the waveform formed by combining the samples in "equivalent" time. Digitizing oscilloscopes are particularly useful for automated data-acquisition systems.
9. Traveling-wave oscilloscopes are extremely high bandwidth oscilloscopes used for recording very fast, high-level signals.

QUESTIONS AND PROBLEMS

1. Name the four major sections of a general-purpose oscilloscope and list the common controls for each section.
2. What element in the CRT controls the intensity of the beam?
3. (a) Compute the deflection sensitivity for an electrostatic CRT in which the accelerating voltage is $V_a = 1000$ V, $L = 24$ cm, $d = 1.2$ cm, and the length of the plates is 4.0 cm.
   (b) What deflection voltage is required to deflect the beam 3.0 cm from the center of the CRT?
4. An oscilloscope has a dc voltage connected to its input and the horizontal sweep is observed to move up 4.6 divisions when the input coupling switch is moved from its GND position to the DC position. If the vertical sensitivity is 0.5 V/DIV, what dc voltage is being measured?
5. An oscilloscope displays the waveform shown in Figure 8–40. The SEC/DIV control is set to 20 μs/div and the horizontal 10× magnifier is on. Compute the period and frequency of the waveform.

FIGURE 8–40

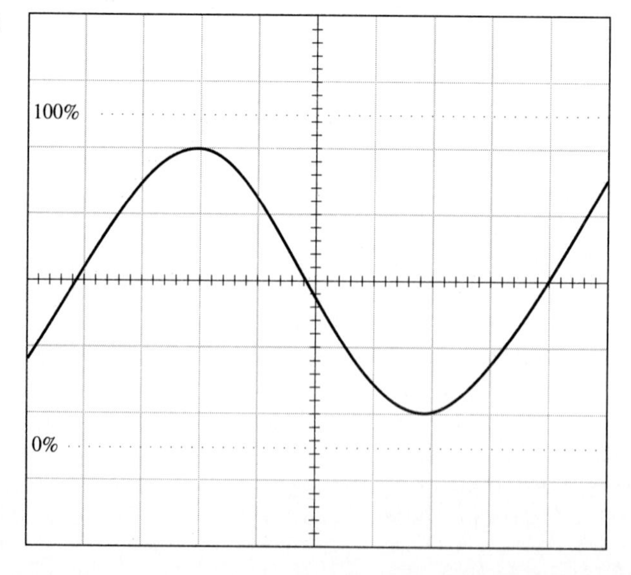

6. Compute the time difference between two signals if a differential delayed-sweep measurement is made. The A sweep SEC/DIV control is set to 10 μs/div. The horizontal MODE control is set to the B mode, and the starting time is moved with the DELAY control to center screen. The DELAY control reads 2.81. Then the DELAY control is moved until the stopping time is center screen. The new reading of the delay dial is 6.55.
7. Explain the difference between the trigger LEVEL control and the trigger SLOPE control.
8. Explain when the trigger HOLDOFF control can be used to stabilize a display.
9. Compare the CHOP and ALTERNATE modes. When should you use one or the other?
10. Compute the input impedance of a 10 MΩ, 10 pF probe at a frequency of 1 MHz.
11. Explain why a low-input-resistance probe can have a higher cutoff frequency than a similar high-input-resistance probe.
12. Why is it necessary to compensate a probe?
13. What is the difference between an active and a passive probe?
14. (a) Compute the equivalent rise time of a 100 MHz scope.
    (b) What is the fastest rise-time signal that the oscilloscope from part (a) could be used to measure if a 3% error due to bandwidth limiting is allowed?
15. Assume you measure a 500 Hz, 5.6 V_{pp} sinusoidal waveform on an oscilloscope. What reading do you expect if you measure this same waveform on a DMM that is accurate at this frequency?
16. (a) Why isn't it necessary to have the VOLTS/DIV control calibrated when making a rise-time measurement?
    (b) What graticule marks are useful for measuring the rise time of a pulse?
17. The household voltage in the United States is 120 V rms at 60 Hz.
    (a) Compute the peak-to-peak value of this voltage.
    (b) If this waveform were applied to an oscilloscope in which the vertical sensitivity was set to 50 V/div, how many divisions of vertical deflection would you expect from peak to peak?
    (c) Compute the period of the waveform.
    (d) If this waveform were applied to an oscilloscope in which the horizontal SEC/DIV control were set to 10 ms/div, how many cycles would you expect to see in 10 divisions?
18. Assume you are troubleshooting a circuit in which you suspect power line interference. You observe an interfering 60 Hz signal. How could the LINE position for the trigger source be used to confirm your suspicion?
19. What is the *z*-axis of an oscilloscope?
20. For the Lissajous figure shown in Figure 8–41, compute the phase difference between the inputs.
21. For the pulse waveform shown in Figure 8–42, the SEC/DIV control is set to 0.5 μs/div and the 10× horizontal magnifier is on.
    (a) Compute the rise time of the pulse.
    (b) If the oscilloscope/probe has a 100 MHz bandwidth, is the measurement bandwidth limited? Explain your answer.
22. Explain when you would need to invert channel 2 for a measurement.
23. What type of CRT storage tube is represented by the cutaway view shown in Figure 8–25?
24. What is the function of the flood guns in a CRT storage tube?
25. Compare a digital storage oscilloscope with a CRT storage oscilloscope. Give at least two advantages of the digital storage oscilloscope.
26. How does equivalent-time sampling increase the effective bandwidth of a digital storage oscilloscope?

FIGURE 8–41

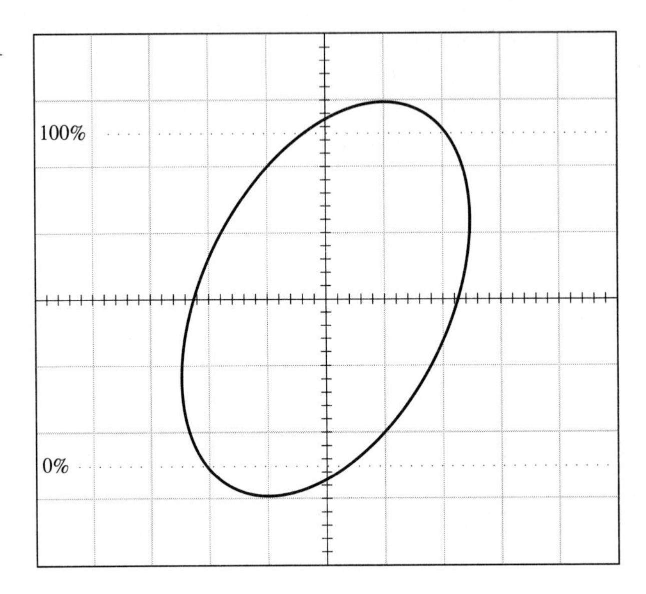

FIGURE 8–42

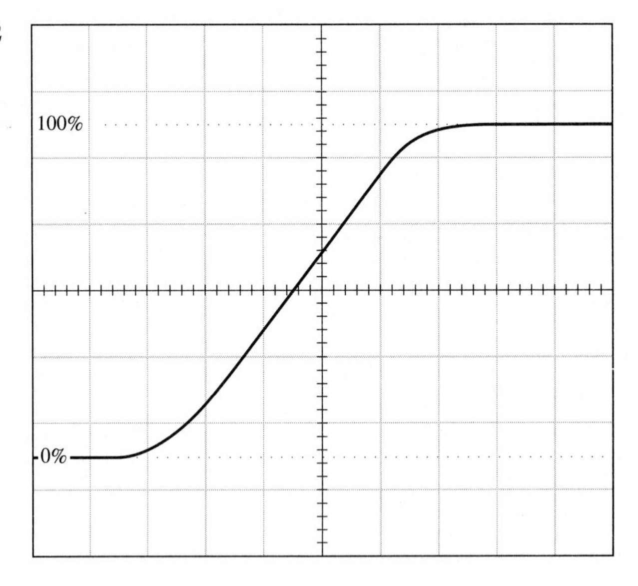

27. (a) What is meant by pretrigger capture?
    (b) How can pretrigger capture help locate an occasional fault?
28. Explain how the envelope mode can be used to capture an occasional glitch in a system.
29. Assume you digitize a periodic waveform that has a frequency of 10 MHz at 80 MSa/s. How many times would you sample each cycle?
30. What is the purpose of adding a high-frequency sinusoidal waveform to the display of a traveling-wave oscilloscope?

Chapter 9

Signal Sources

OBJECTIVES

Many circuit-performance tests require periodic waveforms (notably sine, square, and triangle) to test various functions. Tests frequently involve stimulus-response testing; in this type of test, the circuit is stimulated with one instrument and the response of the circuit is observed with another instrument. The classical stimulus instrument was a low-power laboratory oscillator that produced only sine waves. Today a variety of sources are available; sources are usually classified according to their frequency range and the types of waveforms they produce. Sources include function generators, arbitrary waveform generators, signal generators, swept-frequency oscillators, pulse generators, and digital-pattern generators. In addition to stimulus-response testing, signal sources find application in simulation and as a stable source for test, alignment, and calibration work.

When you complete this chapter, you should be able to

1. Explain the difference between various types of sources, including function generators, arbitrary waveform generators, signal generators, swept-frequency oscillators, and pulse generators.
2. Discuss applications of function generators, arbitrary waveform generators, signal generators, swept-frequency oscillators, and pulse generators.
3. Describe how to set up a sweep oscillator.
4. Compute the output amplitude of various generators due to loading effects.
5. Define terms related to waveform generation, including continuous, triggered, gated, burst, and swept.
6. Discuss the importance of specifications for signal sources, including frequency, spectral purity, output amplitude, modulation, operating characteristics, and output impedance.
7. Compute the maximum power and voltage delivered to a specified load for a given output in dBm.
8. Explain pulse parameters, including baseline, amplitude, rise and fall times, pulse width, preshoot, overshoot, undershoot, ringing, settling time, offset, and duty cycle.
9. Describe how a vector engine can compress data for a digital-pattern generator.

HISTORICAL NOTE

The first product from the Hewlett-Packard Co. was a two-stage *RC* oscillator using triode vacuum tubes that could operate over a five-decade frequency range. Introduced in 1939, it featured both positive and negative feedback loops, which resulted in a low-distortion output, primarily due to the negative feedback. Shortly after World War II, the oscillator was improved to extend the range to six decades (from 10 Hz to 10 MHz) using six decade ranges. The *RC* oscillator was described in the first volume of the *Hewlett-Packard Journal,* issues 3 and 4.

9–1 TYPES OF SIGNAL SOURCES

Signal sources include a variety of low-power, electronic instruments that produce a selection of waveforms. Generally, the output frequency and amplitude are calibrated and can be varied over a specified range. Signal sources include instruments from strictly sine-wave oscillators to those that can synthesize waveforms for any mathematical function. Specialized sources include function and waveform synthesizers, arbitrary function generators, signal generators, swept-frequency oscillators, pulse generators, and digital-pattern generators.

Signal sources are generally categorized as either low-frequency, radio-frequency, or microwave. Traditionally, these regions were divided so that low-frequency sources fell in the range from dc to 1 MHz, radio-frequency sources fell in the region from about 100 kHz to 1.0 GHz, and microwave sources ranged from about 1 GHz on up. New sources have changed the traditional boundaries between regions. For example, Hewlett-Packard offers a signal generator that offers continuous coverage from 0.1 Hz to 4.2 GHz. Some sources are called audio-frequency generators or oscillators because they produce frequencies within the region of human hearing. Strictly speaking, audio waves are acoustical in nature, whereas the waves produced by audio generators are electrical. The frequency range for audio generators is normally 20 Hz to 20 kHz, although it may be extended to 100 kHz.

General categories of signal generators are listed next. As in many classifications, the boundaries tend to be fuzzy, since some new products encompass more than one role. Many function generators are actually synthesizers, and some high-end generators can supply virtually any of the waveforms that formerly required a specialized instrument.

Function generators use a free-running oscillator to provide a selection of sine, square, and triangle waveforms in a single instrument. Better-quality function generators may also produce pulses, ramps, and other waveforms. A **synthesized function generator** is a waveform generator that has one or more internal crystal reference oscillators that operate at fixed frequencies. The frequency is set digitally and synthesized from the reference oscillator to give the frequency precision normally associated only with fixed oscillators.

Arbitrary function generators (also called AFGs, arbitrary waveform generators, or ARBs) are digital synthesizers that allow the user to create custom waveforms as well as select from a collection of stock waveforms. AFGs create waveforms from stored digital data using a controller and D/A converter. High-end AFGs can generate virtually all of the waveforms described for the other specialized signal generators.

Signal generators (sometimes referred to as rf generators) produce high-frequency sine waves and modulated sine waves. There are a large variety of signal generators available with frequencies that go from dc to microwave. Newer signal generators include frequency synthesizers and programmable signal generators. A **frequency synthesizer** provides very high frequency stability by generating the frequency digitally from a fixed-reference oscillator. **Programmable signal generators** allow the user to store a series of front-panel setups in memory to be recalled as needed.

Swept-frequency oscillators (or sweepers) are instruments that generate a sine-wave output that varies at a cyclic rate between two selected frequencies. They allow the user to view signals directly in the frequency domain. Some models produce a logarithmic or other nonlinear sweep output; another option is a stairstep sweep signal. Many signal generators also have swept-frequency capability.

Pulse generators are specialized instruments for producing pulses with fast rise times over a wide frequency range and with varying duty cycles. They frequently have outputs for specific logic levels. Most pulse generators allow the user to control a variety of pulse parameters, such as amplitude, rise time, offsets, triggering, and polarity. In addition, some units produce pulse pairs and trains. Generally, pulse generators produce faster and more controllable pulses than function generators.

Digital-pattern generators (DPGs) produce digital words in a sequence for testing digital equipment. Some DPGs are specifically designed for testing telecommunications equipment and can generate a series of standard test patterns for performance checks. Others are used to generate test patterns for stimulus-response testing.

9–2 SPECIFICATIONS FOR SIGNAL SOURCES

The principle specifications for signal generators deal with frequency, spectral purity, output amplitude, modulation, operating characteristics, and output impedance. These parameters are discussed in this section.

FREQUENCY

The **frequency range** is the span of frequencies over which instrument performance is specified. Accuracy and resolution limits are also given with the frequency specifications. The accuracy specification depends on the internal oscillator's accuracy as well as the type of dial controls (mechanical or digital). A precise digital control is of little value if the internal oscillator is not accurate, nor is it meaningful to have an accurate oscillator if the control is an imprecise mechanical dial.

SPECTRAL PURITY

In Section 1–2, it was shown that a pure sine wave consists of a single fundamental line in the frequency domain. **Phase noise** occurs when the short-term stability of a sine wave causes small frequency shifts. Phase noise tends to broaden the fundamental line. **Harmonic distortion** occurs when other lines are present in the output at multiples of the fundamental. In addition to phase noise and harmonics, the spectral purity is affected by subharmonics and other noise sources. These noise sources are illustrated in Figure 9–1. Related to the frequency specification is the spectral purity specification for sine waves. **Spectral purity** is a measure of the short-term stability of a signal and is generally specified in terms of the single-sideband phase noise in a 1 Hz bandwidth as a function of the offset from the carrier frequency. Spectral purity is specified in dBc, meaning the number of decibels of noise with respect to the carrier (fundamental) level.

To illustrate the importance of spectral purity from a generator, consider the problem of measuring a radio receiver's rejection from an adjacent channel using a

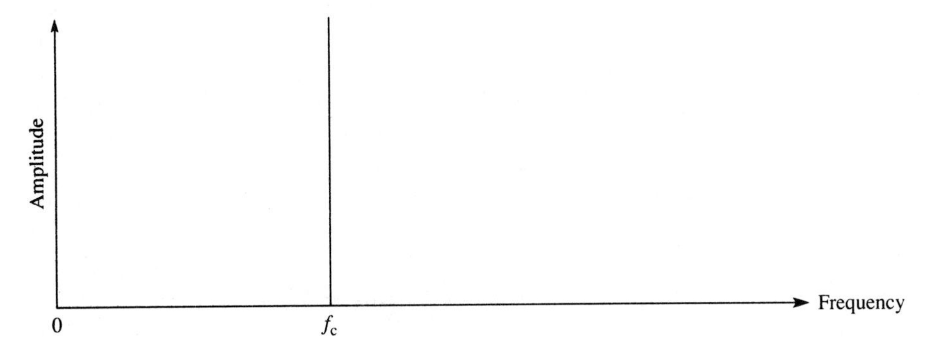

(a) The spectrum of a pure sine wave is a single line.

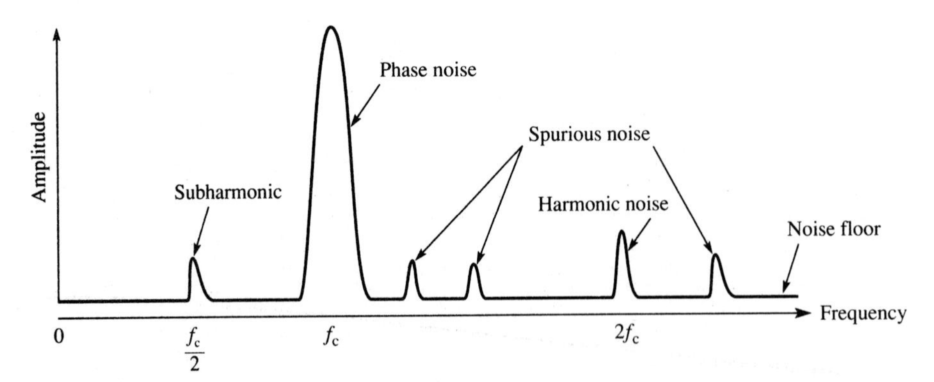

(b) Actual sources have various sources of noise, including phase noise, spurious noise, and harmonic and subharmonic noise.

FIGURE 9–1
Spectrum of ideal and actual sinusoidal sources.

FIGURE 9–2
Noise from a signal generator can seriously degrade certain measurements and tests. Both examples make the receiver look worse than it is (courtesy of Hewlett-Packard).

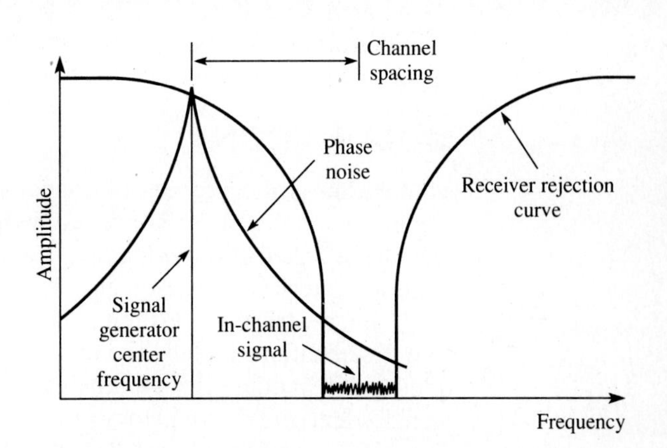

(a) Phase noise from a signal generator can affect the channel rejection test of a radio receiver, as shown.

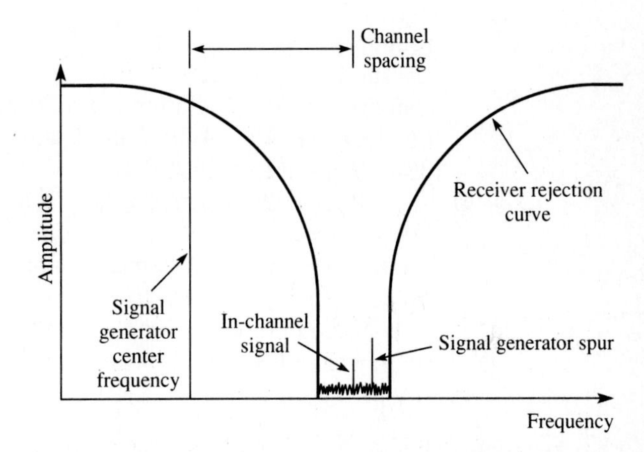

(b) Spurious noise from a signal generator can also affect a receiver's apparent ability to reject an adjacent channel.

signal generator. The signal generator is tuned to the adjacent channel's frequency and the receiver's signal is observed. Figure 9–2(a) illustrates what can happen if the signal generator has phase noise—the spectrum from the generator spills over into the receiver's channel and the receiver looks worse than it really is. Likewise, spurious noise from a signal generator can create an equivalent problem. This is illustrated in Figure 9–2(b).

OUTPUT AMPLITUDE

Output specifications include the voltage amplitude range or the maximum and minimum power delivered to a specified load. Power specifications are generally given in dBm. Typical signal generators produce relatively low outputs; for applications requiring high voltages or power, an amplifier is used. In addition to the ac voltage amplitude range, output specifications include the dc offset range as well as the accuracy, resolution, and flatness of the output across the frequency range.

MODULATION

Modulation is the process of changing some characteristic of a high-frequency waveform (called a *carrier* in communications systems) by a lower-frequency waveform called the *modulating signal.* The modulation signal can be supplied from an external source or may be generated within the generator; typically, a sine, square, triangle, or ramp can be selected as the modulation signal. Modulation specifications include the modulation mode (AM, FM, pulse, etc.) and details about each mode. The parameter being modulated can be the high-frequency signal's amplitude, frequency, or phase. Modulation methods are compared in Figure 9–3. Amplitude modulation (AM) causes the output amplitude to change at a rate determined by the modulating signal. Frequency modulation (FM) and phase modulation (PM) are forms of angle modulation in which the frequency or phase is varied at a rate determined by the modulating signal. Note that whenever frequency is varied, the phase is also varied and vice versa. The various modulation methods are useful for testing the performance of different communication systems. Phase modulation is also useful for evaluating the performance of phase lock loops (PLLs). Amplitude, frequency, and phase modulation are discussed in more detail in Section 12–3.

Other common modulation schemes are available for specific applications. An example of a modulation available on many generators is pulse-width modulation (PWM). Pulse-width modulation uses an amplitude-varying signal to vary the duty cycle of a constant-frequency square wave, a technique used in audio broadcast equipment. Most complex modulation schemes can be developed by the user by programming specific waveforms on specialized arbitrary function generators.

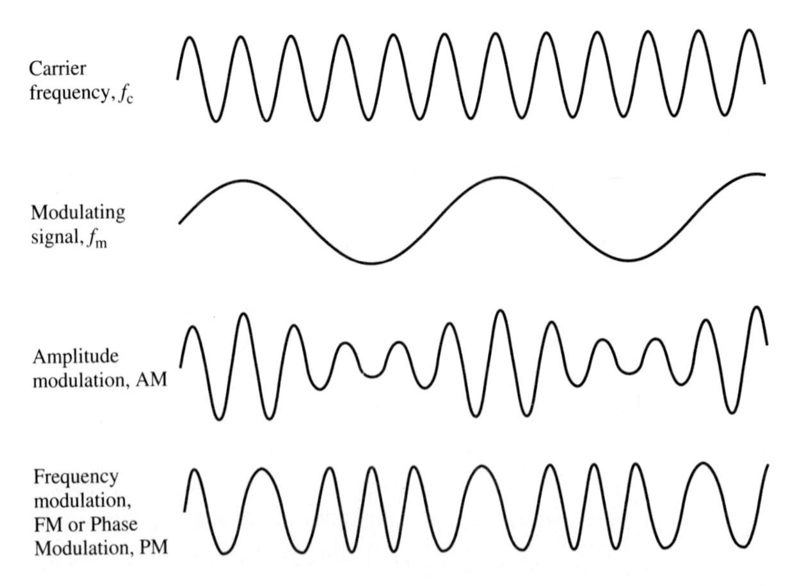

FIGURE 9–3
Comparison of AM and FM modulation. Phase modulation is similar in appearance to FM.

OUTPUT IMPEDANCE

Output impedance specifications include the impedance magnitude and accuracy. No matter how complicated the internal circuitry, signal generators can be modeled as a Thevenin circuit consisting of a voltage source in series with a resistance. The series resistance is the equivalent output impedance of the source. Generators are normally marked with the Thevenin resistance on the output terminals. Typical values are 50, 75, or 600 Ω.

It is important to keep in mind the fact that connecting a load to a source that has a finite output resistance will change the output voltage amplitude of the generator. If overlooked, the change in amplitude can affect the desired measurement to the point of making it meaningless. For example, to measure the frequency response of a circuit without having the generator affect the results, the amplitude must be maintained at a constant value from the generator. The output of the circuit under test is monitored while the frequency is changed. Each change in frequency changes the input impedance of the amplifier and causes a different loading effect to occur. To prevent input loading from affecting the output, the amplitude of the generator should be readjusted to the same value each time a new frequency is tested.

EXAMPLE 9–1 A function generator with an output impedance of 600 Ω is set to 5.0 V_{pp} and then connected to a 1.0 kΩ resistor. What is the voltage across the resistor after the generator is connected?

SOLUTION The function generator looks like a voltage source in series with a resistance, as shown in Figure 9–4. The loaded voltage can be calculated from the voltage-divider theorem. That is

$$V_L = V_{TH}\left(\frac{R_L}{R_{TH} + R_L}\right)$$

$$= 5.0 \text{ V}\left(\frac{1000}{600 + 1000}\right)$$

$$= 3.13 \text{ V}$$

FIGURE 9–4

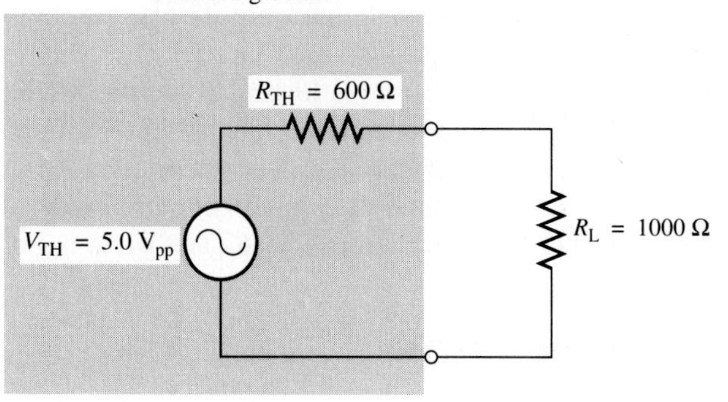

Operating characteristics include the controls, switching speed, temperature range, interfaces available, and the like. One feature that has become more important in recent years is automated control. The front-panel setups and/or sequences are stored in digital memory. Switching speed is a measure of the time required for the generator to respond to commands and change function or range when under automated control.

WAVEFORM MODES

Depending on the complexity, waveform generators can have from one to several waveform modes from which to choose. Common waveform modes are as follows.

Continuous: The output is steady at a specific frequency, amplitude, and offset. This is the basic output of signal generators.

Triggered: The output is initiated by an internal or external (user-supplied) trigger. The trigger can be generated by external equipment or a manual pushbutton or sent over a control bus.

Burst: This is the same as the triggered mode, except the output is programmed for a specific number of cycles. The burst can be triggered manually or with a command in programmable instruments.

Gated: The output is enabled for the duration of an external gate signal.

Swept: The frequency of the output changes in some predetermined way. Sweeps can be triggered or programmed (start-and-stop frequency), linear or logarithmic. Sweep-frequency oscillators are discussed in Section 9–6.

Modulation: The output is modulated by an internal or external waveform. Either AM, FM, or some other modulation scheme may be available.

9–3 FUNCTION GENERATORS

THE BASIC FUNCTION GENERATOR

Function generators are characterized by the various waveforms they produce; they all produce sine, square, and triangle waves and may also include pulses and ramp (sawtooth) waveforms. (Several of these waveforms were illustrated in Figure 1–12.) Pulse and ramp outputs can be obtained by varying the symmetry of square and triangle waves, respectively. Sine and square waves are commonly used for general-purpose testing of circuits such as amplifiers. Triangle waves are useful for testing circuits such as comparators to determine the threshold. Pulses that have a duty cycle that is not near 50% are useful for digital testing because it is easy to tell if the pulses have been inverted—they don't have the same appearance. On the other hand, if a square wave is inverted, it still looks like a square wave.

APPLICATIONS

The basic waveforms (sine, square, and triangle) are used in many tests of electronic circuits and equipment. A common application of a function generator is to inject a sine wave into a circuit to check the circuit's response. The signal is capacitively coupled to the circuit to avoid upsetting the bias network; the response is observed on an oscilloscope. With a sine wave, it is easy to ascertain if the circuit is operating properly by checking the amplitude and shape of the sine wave at various points or to look for possible troubles, such as high-frequency oscillation.

A common test for wideband amplifiers is to inject a square wave into a circuit to test the frequency response. Recall that a square wave consists of the fundamental frequency and an infinite number of odd harmonics (see Section 1–2). The square wave is applied to the input of the test circuit and the output is monitored. The shape of the output square wave indicates if specific frequencies are selectively attenuated. Figure 9–5 illustrates square-wave distortions due to selective attenuation of low or high frequencies. In addition to a basic functional test of the circuit, the rise time of the square wave is an indirect measurement of the bandwidth, as described in Section 8–5.

FIGURE 9–5
Square-wave response of circuits. The amount of distortion is indicative of the frequencies that are attenuated.

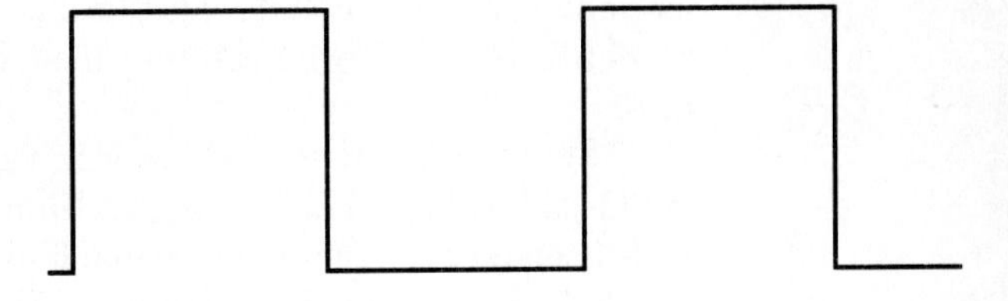

(a) Input.

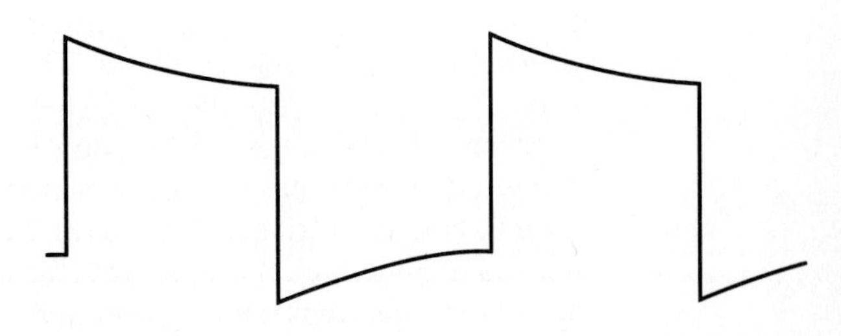

(b) Low-frequency attenuation.

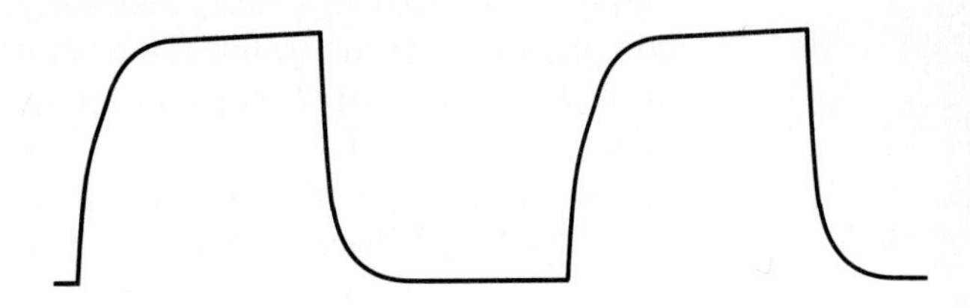

(c) High-frequency attenuation.

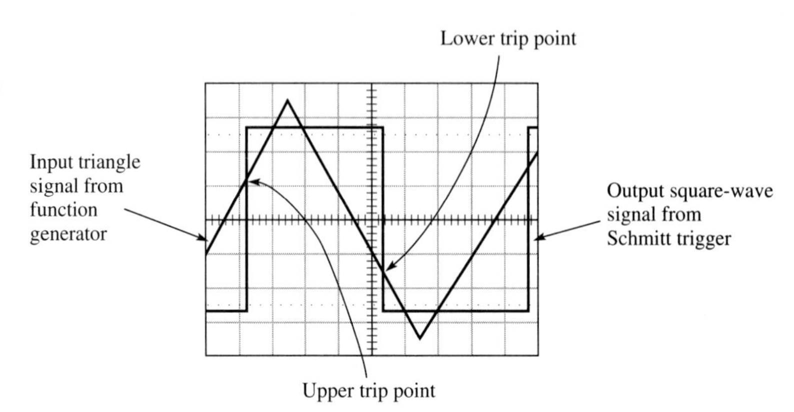

FIGURE 9–6
Testing a Schmitt trigger circuit for the trip points by using a triangle wave. Both signals are centered on the oscilloscope; the trip points can be observed where the signals cross.

The triangle waveform is useful for testing circuits such as Schmitt triggers that have built-in hysteresis. By observing both the input and output waveforms simultaneously from a Schmitt trigger, the trip points can readily be ascertained, as shown in Figure 9–6. The triangle waveform can also be used to test the response of other voltage-controlled circuits, such as voltage-controlled oscillators.

Some function generators are available with frequencies as low as 0.000001 Hz (1 μHz). The upper frequency limit varies widely with the particular model; 50 MHz is not uncommon and much higher frequencies are available. In addition to producing the basic waveforms, many function generators have specialized capabilities such as modulation, sweep, and pulse generation.

The operation of function generators is illustrated in the diagrams shown in Figure 9–7. You should recognize the essential circuits shown in the shaded blocks. Two approaches are illustrated. In each method, a basic waveform is generated from a stable oscillator. The basic waveform can be a square or triangle wave. In Figure 9–7(a), the basic square wave is generated by a relaxation oscillator such as that discussed in Section 5–8. Integration of the square wave produces a triangle waveform, and diode waveshaping (described in Section 4–4) can be used to convert a triangle waveform into a sine wave. Another method frequently used is to generate a triangle waveform as the basic wave (see Figure 9–7(b)) by alternately charging and discharging a capacitor from a positive and negative current source. The triangle wave is then converted to a square wave with a comparator and to a sine wave with a diode array or other sine converter.

Many function generators include the capability for modulation. This requires a separate internal generator for the modulation signal. The additional complexity costs more, but the advantage is that the generator can be used as a more versatile input source for various systems, particularly communication systems. The AM signal, for example, can be used to drive the radio frequency and intermediate stages of a radio receiver or test automatic gain and amplitude control circuits. FM is used for a variety of tests, including frequency shift keying (FSK) circuits used by modems. The modulation index can usually be controlled between 0 and 100%.

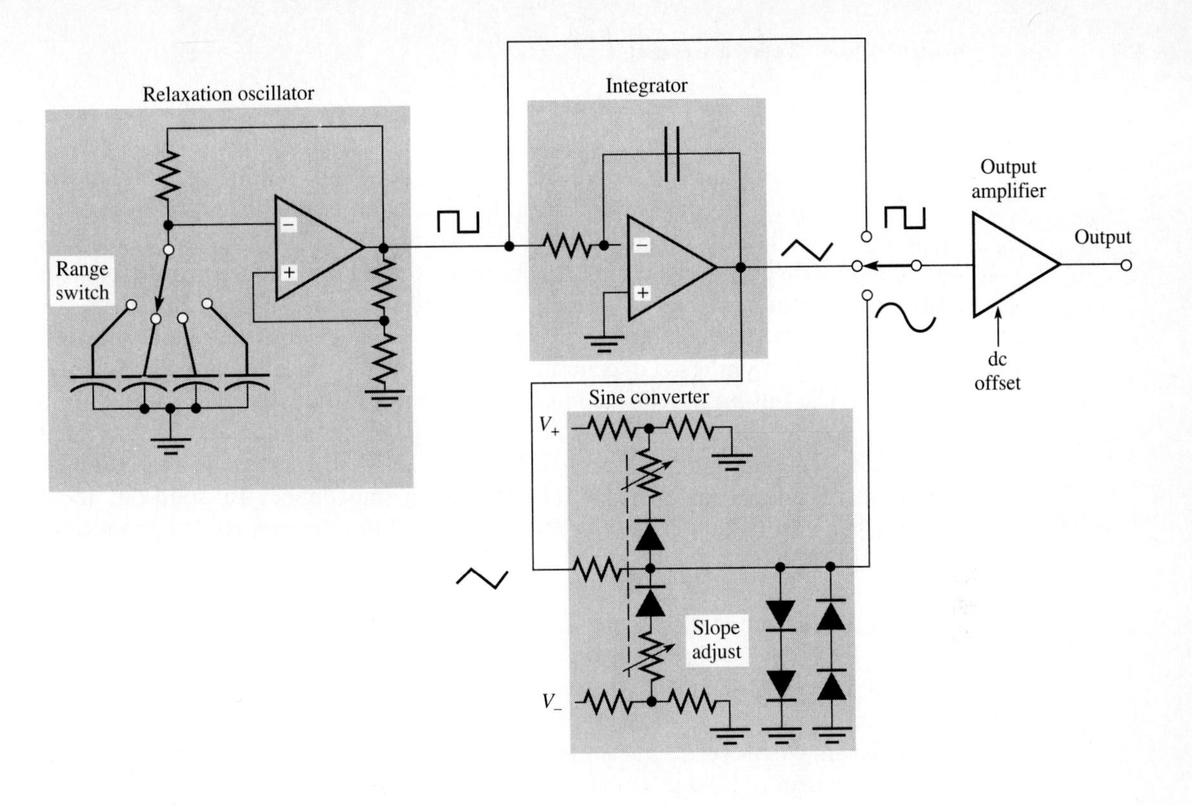

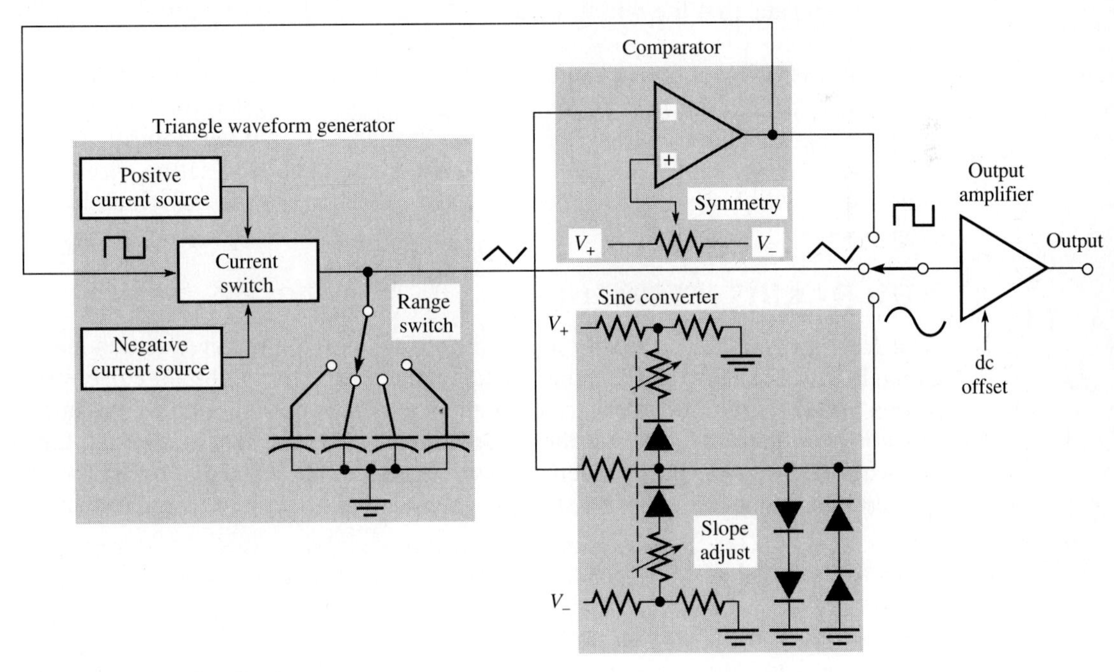

(b) Basic waveform is a triangle wave.

FIGURE 9–7
Function generator block diagrams.

CONTROLS

The basic function generator controls are for frequency, output amplitude, and function selection. Frequency selection may be done by switches for frequency range and a calibrated vernier for fine-frequency adjust. The vernier reading is multiplied by the range selection in order to determine the frequency. Mechanical dials have limited accuracy—typically 3%. Better accuracy—typically 0.1%—is obtained with newer digital inputs. Even higher accuracy for frequency selection is possible with synthesized generators that can be phase-locked to a crystal oscillator. The output amplitude is usually calibrated and may be indicated on a meter. Output attenuators are found on some units to attenuate the output in fixed increments. To maintain calibration, attenuators must be operated in a circuit with the same impedance as the generator's driving impedance. In addition, most function generators have a dc offset control that allows the user to add or subtract a dc level to the output.

To make adjustments, many newer function generators use a digital data-entry method to enter the precise frequency, output amplitude, offset, and other data, or they can be programmed via a computer to control functions. Some function generators have a mode control, allowing the user to select between free-run, gated, triggered, or burst operation. When the gated mode is selected, the output is on whenever a particular voltage level is applied to the gate input, allowing the user to start or stop the generator using an electrical signal. Triggered operation means that the output is turned on by a trigger signal. Burst causes the output to occur for a specified number of cycles.

Another control found on better-quality generators is external synchronization. This enables the user to synchronize the function generator output to an external signal. Phase control may also be available. Computer control, via the IEEE-488 interface bus, is another option that is useful for setups needed in automatic testing and other applications. The IEEE-488 bus is described in Section 14–9.

THE HARRIS ICL8038 IC

The Harris ICL8038 is a function generator on a single integrated circuit that is capable of producing accurate sine, square, triangular, sawtooth and pulse waveforms with a minimum of external components. Several suggestions for implementing circuits are given by the manufacturer. A basic function generator is shown in Figure 9–8. The frequency for the generator shown is determined by the RC time constants. If $R_A = R_B$, the time constant is given by the equation

EQUATION 9–1

$$f = \frac{0.3}{RC}$$

where R = resistance between V_+ and either pin 4 or pin 5

By adjusting two external resistors, the triangle can be modified into an asymmetrical sawtooth and the square wave can be adjusted to produce pulses with a duty cycle from less than 1% to more than 99%. Improved performance can be obtained with the addition of op-amp buffers to the output.

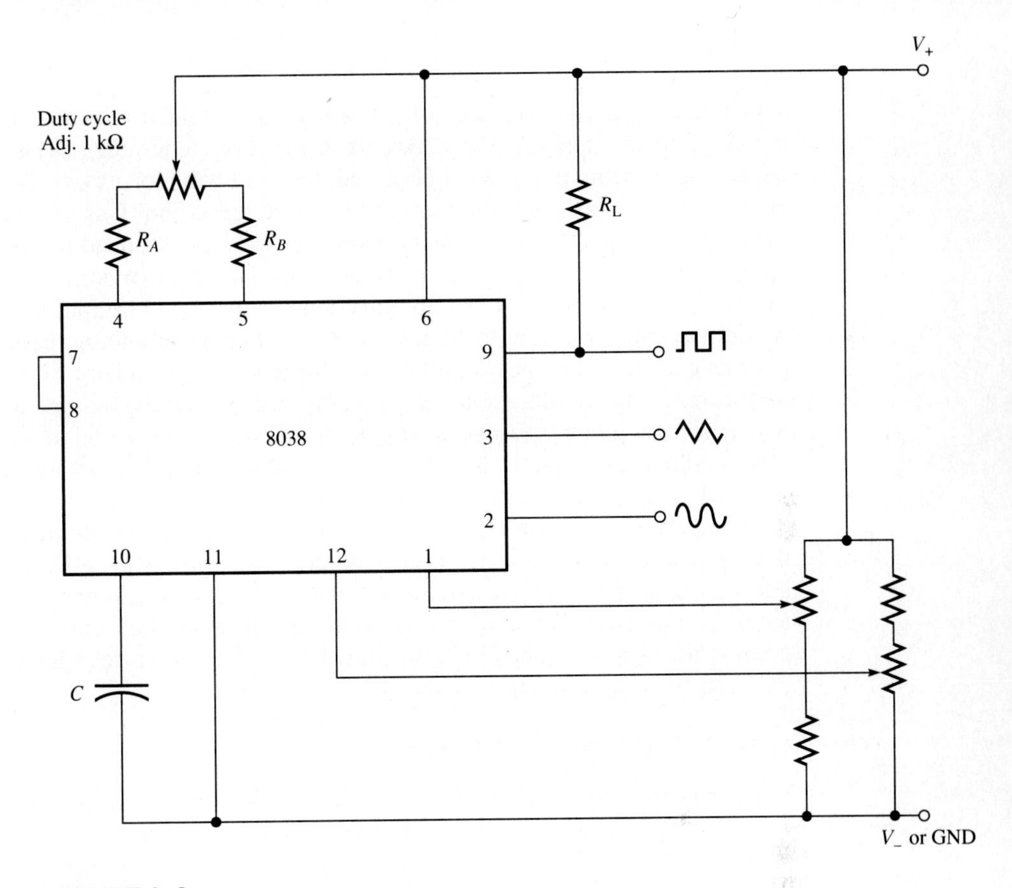

FIGURE 9–8
Harris ICL8038 precision waveform generator. See the manufacturer's specification sheet for values. Capacitor C and resistors R_A and R_B (and the 1 kΩ duty-cycle adjust) determine the frequency. A range switch can be used to change capacitor or resistor values (courtesy of Harris Semiconductor).

EXAMPLE 9–2 Assume that the circuit shown in Figure 9–8 has $R_A = R_B = 22$ kΩ. Compute the value of C that will produce a frequency of 400 Hz.

SOLUTION We will assume the 1 kΩ duty cycle potentiometer is centered. The resistance to each pin is then 22.5 kΩ. Rewriting Equation 9–1 in terms of the capacitance yields

$$C = \frac{0.3}{Rf}$$

$$= \frac{0.3}{(22.5\ \text{k}\Omega)(400\ \text{Hz})}$$

$$= 0.033\ \mu\text{F}$$

9–4 ARBITRARY FUNCTION GENERATORS

Basic function generators are limited to a specific number of repetitive waveforms provided by the manufacturer. Arbitrary function generators (AFGs) enable the user to select from a larger set of stored digitized waveforms or produce complex custom waveforms for testing or circuit characterization. Custom waveforms are ideally suited to automatic test systems and can also be used to test systems by simulating signals from transducers such as vibration or stress transducers or biomedical transducers. They are particularly useful for simulating infrequent or random events for use in testing. Custom waveforms in an automatic test system can save generating the required waveform with specialized circuits. Another application is to simulate nonideal waveforms or simulated failures to determine a test system's tolerance. For example, the response of a digital system can be observed using a pulse train with an extra pulse or glitch generated by the AFG.

AFGs essentially store the information about a custom waveform in a digital memory. The waveform is generated, under microprocessor control, by converting the stored digital numbers into an analog waveform using a D/A converter. In addition to controlling the generation of the waveform, the microprocessor performs a variety of other tasks, such as coordinating data entry and necessary communications over the IEEE-488 bus. Figure 9–9 illustrates a high-quality AFG that can be programmed over the IEEE-488 bus.

CREATING WAVEFORMS

The methods used by arbitrary waveform generators to create a waveform depend on the cost and complexity of the instrument. They vary in complexity from plug-in cards for a personal computer to complex systems that can generate waveforms up to 50 MHz. Some are designed for direct entry from the front panel; others depend on computer entry. Some methods for generating waveforms include the following:

1. Using internally stored waveforms defined by the manufacturer that can be digitally calculated in real time and converted to an analog output by a special waveform synthesis IC. These waveforms can include the usual sine, square, and triangle waveforms as well as a host of other standard test signals. For example, the AFG can generate a standard NTSC video signal directly from a lookup table.
2. Generating the data points from an equation. Some AFGs have the ability to generate a waveform directly from the algebraic equation. This is useful for signals such as Loran system signals that fit a specific mathematical function.
3. Data acquisition from a digital storage oscilloscope (DSO) or waveform digitizer. The DSO or waveform digitizer can communicate directly with the AFG if both are equipped with a standard interface. Single-shot events, such as failure analysis, can be captured by a DSO and played back repeatedly by the AFG.
4. Direct entry by keyboard or digitizing tablet.

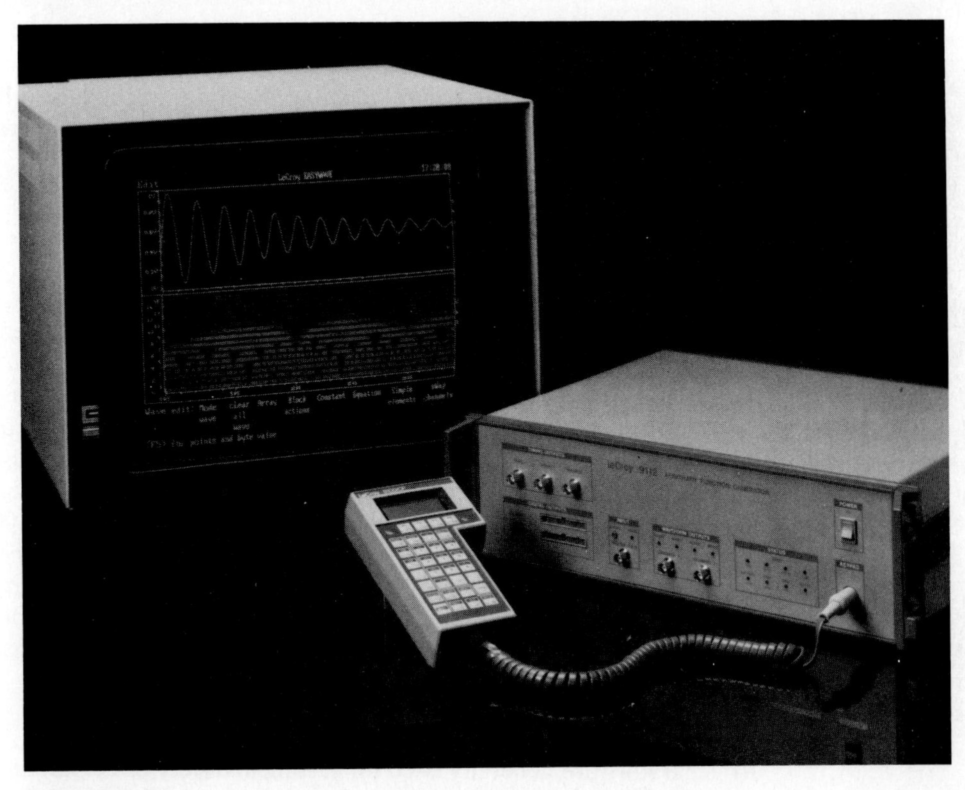

FIGURE 9–9
A high-resolution, dual-channel arbitrary function generator, the LeCroy 9112. This generator has 12 bit amplitude resolution and can generate waveforms using specially developed computer software (courtesy of LeCroy).

5. Special waveform-creating software. Many AFGs are supported by special computer software that enables the user to "cut and paste" waveforms, define scaling parameters, add dc offsets, and perform other algebraic operations on waveforms. Software is available that enables the user to create waveforms by computer techniques such as looping (repeating an operation a given number of times) and linking (connecting strings of waveform segments together).

MULTIPLE CHANNELS AND CHANNEL SUMMING

In certain test situations, it is necessary to have a second channel of information that is precisely timed to the first channel. Some higher-quality AFGs have the ability to output two waveforms that are synchronized to each other by operating both channels from a common time base. This is useful for simulating two dependent inputs to a system in which precise phase control is necessary. An example is the I and Q (in-phase and quadrature) signals generated in digital radio and other communication systems.

Channel summing enables the user to synthesize a new waveform by forming a composite waveform from two channels. One channel can have a low-level, high-resolution signal that can be added to the signal on the other channel, thereby increasing the dynamic range and resolution of the composite waveform. This feature is useful in certain failure-analysis problems.

9–5 SIGNAL GENERATORS

Signal (rf) generators are accurate sine-wave oscillators designed to cover frequencies from about 100 kHz to microwave frequencies. The output amplitude and frequency is accurately known and can be adjusted continuously over a large range of frequencies and amplitudes. In addition, rf generators offer calibrated modulation. Different generators vary widely in the frequency coverage, but a typical rf generator might cover the range of 100 kHz to 1 GHz. High-frequency rf and microwave generators are available that can cover to over 100 GHz.

In order to cover the range of frequencies required, rf generators are designed with multiple tuning bands. Within each band, the output frequency is continuously variable. The output amplitude is adjustable with a precision attenuator, allowing the user to obtain accurate amplitudes to better than −140 dBm.

APPLICATIONS

Rf generators are generally used for testing a variety of communication systems, including AM and FM receivers, digital radios, and telemetry receivers. They can also be used to test transmission lines and antennas. Since rf generators are widely used in testing communication systems, they are usually designed to permit output modulation. The generator provides the user with the choice of a continuous-wave (cw) or a modulated output. Typically, AM and FM modulation is provided; specialized generators add additional modulation schemes such as those used in digital radio or other communications systems. Specifications indicate the modulation percentage, frequencies available, resolution, and more.

OUTPUT SPECIFICATIONS

All rf generators provide the user with a calibrated output amplitude at the generator, *provided the output is properly terminated in the source impedance*. Note that the output amplitude is attenuated if long cables are connected to the generator. The output amplitude is typically specified in dBm and may be set by either a digital input or some combination of a calibrated dial and precision attenuator. The attenuator is an accurate network designed to maintain constant-source impedance at all frequencies. It is housed in a shielded compartment to minimize radio-frequency noise and is connected to the output with coaxial cable.

The attenuator specification is particularly important when the generator is used for making performance tests on the front end of sensitive receivers. Radio receivers are designed to pick up signals that are extremely weak. For sensitivity measurements, it is necessary to have an attenuation range that is large enough to provide accurate, very low amplitude signals with a known source impedance. The attenuator must be accurate throughout the range. A related concern is that radiated emissions (leakage) from the generator do not interfere with the sensitivity measurements.

EXAMPLE 9–3 The specification sheet for a signal generator indicates it has an output amplitude of +20 dBm and an output impedance of 50 Ω. Compute the maximum power and voltage delivered to a 50 Ω load.

SOLUTION

$$20 \text{ dBm} = 10 \log \frac{P_{out}}{1 \text{ mW}}$$

$$P_{out} = 10^2 \text{ mW} = 0.1 \text{ W}$$

$$P_{out} = \frac{V_{out}^2}{R_L}$$

$$V_{out} = \sqrt{P_{out}R_L} = \sqrt{(0.1 \text{ W})(50\ \Omega)} = 2.24 \text{ V}$$

EXAMPLE 9–4 Assume the generator in Example 9–3 can be set to a minimum output of −120 dBm. Compute the voltage amplitude in a 50 Ω load that is represented by this.

SOLUTION

$$-120 \text{ dBm} = 10 \log \frac{P_{out}}{1 \text{ mW}}$$

$$P_{out} = 10^{-12} \text{ mW} = 10^{-15} \text{ W}$$

$$P_{out} = \frac{V_{out}^2}{R_L}$$

$$V_{out} = \sqrt{P_{out}R_L} = \sqrt{(10^{-15} \text{ W})(50\ \Omega)} = 0.22\ \mu\text{V}$$

EXAMPLE 9–5 A generator has a specified spurious noise that is −90 dBc. Compute the spurious noise from the generator into a 50 Ω load when the carrier amplitude is 0 dBm.

SOLUTION The carrier amplitude of 0 dBm represents 1 mW into 50 Ω (223 mV).

$$-90 \text{ dBc} = 10 \log \frac{P_{noise}}{1 \text{ mW}}$$

$$P_{noise} = 10^{-9} \text{ mW} = 10^{-12} \text{ W}$$

$$P_{noise} = \frac{V_{out}^2}{R_L}$$

$$V_{out} = \sqrt{P_{noise}R_L} = \sqrt{(10^{-12} \text{ W})(50\ \Omega)} = 7\ \mu\text{V}$$

9–6 SWEEP OSCILLATORS

A **sweep oscillator** (also called a swept-frequency generator, or sweeper) provides the means to plot the frequency response of a circuit automatically on an oscilloscope display. Sweep oscillators are instruments that generate a sine-wave output that varies at a cyclic rate between two selected frequencies. The output starts at the first selected frequency and changes in either a linear or logarithmic rate until it reaches a second selected frequency; this process is called a **sweep.**

For many measurement problems, the parameter of interest is the frequency response of a circuit. Examples include the response of active filters and equalizers as well as many communication circuits such as video and intermediate frequency amplifiers. Swept frequency measurements are also important in microwave applications such as radar and satellite communications. Low-power sweep oscillators can be used to check the impedance of waveguides and high-frequency transmission lines or tune circuits. Signals can be viewed directly in the frequency domain, and the effect of adjustments can be observed as they are made ("real time").

Basically, a sweep oscillator produces a wideband FM output of constant amplitude. Figure 9–10 shows a simplified block diagram of a sweep oscillator. A relatively low frequency ramp generator (100 Hz or so) produces a sawtooth waveform that drives a voltage-controlled oscillator (VCO) that is essentially a voltage-to-frequency converter. The ramp can be started by an external trigger signal or allowed to free-run. The frequency of the ramp generator is controlled by the operator and determines the sweep rate of the generator. Likewise, the ampli-

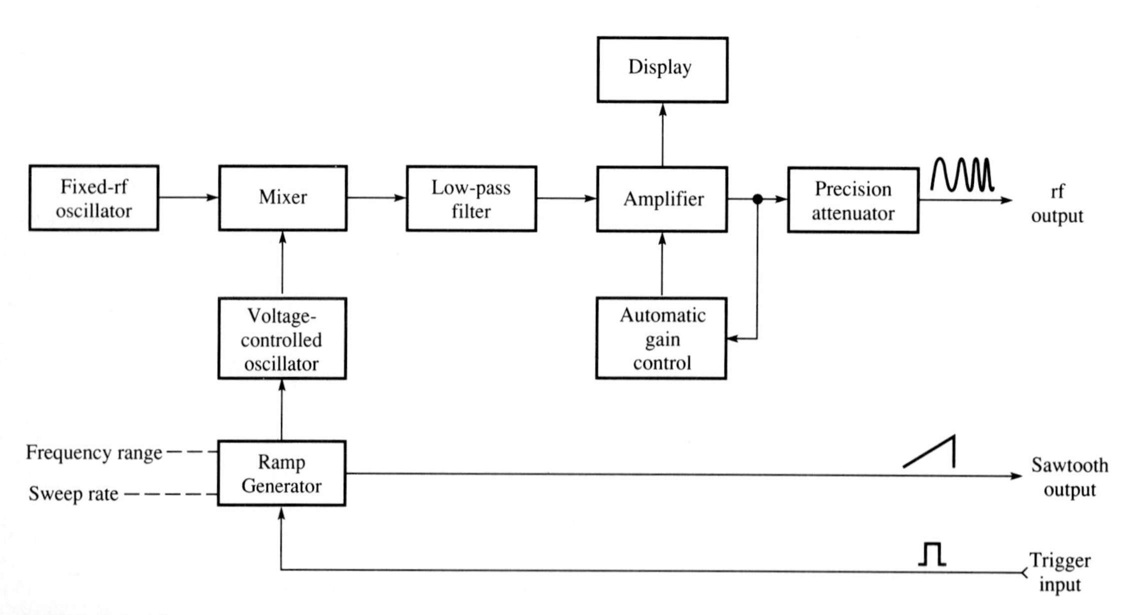

FIGURE 9–10
Simplified block diagram of a sweep-frequency oscillator.

tude of the ramp is controlled by the operator and determines the range of frequencies that are swept. Both linear and nonlinear outputs are available, depending on the generator.

It is difficult to design an oscillator than can be tuned over a large frequency ratio, so mixing is used to extend the range of the output. The idea is to sweep the VCO over much *higher* frequencies than required and then mix the VCO with a fixed-rf oscillator to produce a lower-frequency difference signal for the output. The difference frequency changes by the same amount as the VCO, but since it is a lower frequency, the ratio of high to low frequency is much larger. For example, if the VCO changes from 10 MHz to 15 MHz, the ratio of frequencies is 1.5 : 1. If the VCO is mixed with a 9 MHz fixed oscillator, the *difference* frequency changes from 1 MHz to 6 MHz, a ratio of 6 : 1. The mixing process also produces the sum of the two input frequencies as well as the original inputs; the mixer output is passed through a low-pass filter to remove these unwanted frequencies. The difference frequency is then amplified to a constant amplitude, making the output power constant as a function of frequency. The amplified signal is sent to a calibrated attenuator to produce the output rf signal.

A diagram of a typical sweep setup is shown in Figure 9–11. The leveled rf output of the sweep oscillator is usually sent directly to the device under test (DUT). The output of the DUT is the swept-frequency signal modified by the frequency-response characteristic of the DUT. For example, a high-pass filter will attenuate the low frequencies from the sweep oscillator but pass the high frequencies. The rf signal is detected in order to obtain the envelope of the response. Note that the oscilloscope does not have to have a particularly large bandwidth to be useful in swept-frequency work. The signal sent to the scope is the *detected* signal, not the rf. Detectors need to respond to a wide range of frequencies and relative power levels. The most common detectors are diode detectors or special temperature-compensated thermal detectors.

To observe the frequency response on an oscilloscope, it is necessary to change the time base of the oscilloscope to a frequency base. The oscilloscope is operated in the XY mode, in which the horizontal position of the beam is controlled directly by the X input and the vertical position is controlled by the Y input from the detector. The sweep oscillator's sawtooth waveform is sent to the oscilloscope so that it *simultaneously* determines the output frequency of the sweep

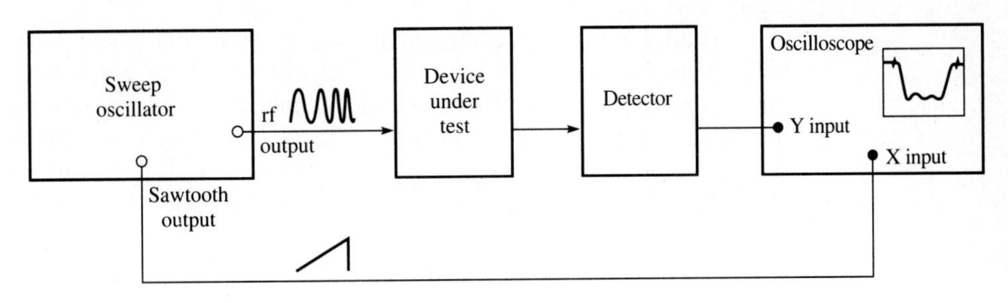

FIGURE 9–11
Swept-frequency measurement setup.

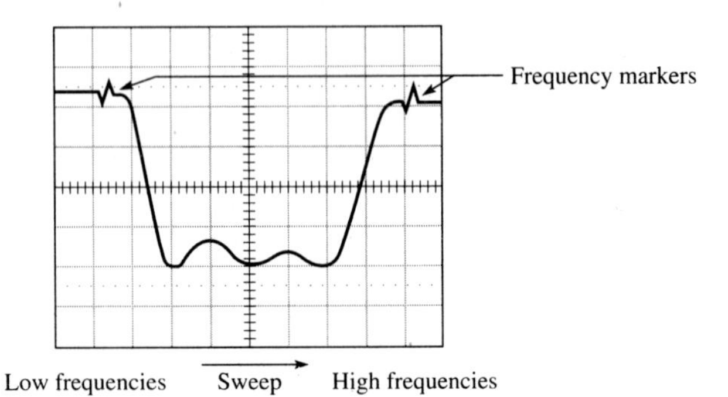

FIGURE 9–12
Oscilloscope presentation showing the passband of a stagger-tuned microwave amplifier. The response is inverted because of the detector. The effect of tuning controls can be immediately observed on the presentation.

oscillator and controls the horizontal position of the oscilloscope beam. The horizontal beam thus moves along the x-axis in proportion to the frequency from the sweep oscillator.

Calibration of the frequency base is normally done with frequency markers. A frequency marker is generated internally or by an external crystal-controlled rf marker generator and may be displayed as a short "burst" on the display or as an intensified region by using the z-axis of the oscilloscope. Many sweep oscillators have built-in markers that can be independently set. Figure 9–12 illustrates a typical oscilloscope display of the frequency response of a staggered-tuned circuit. The rf signal has been detected, so only the envelope of the response is shown.

9–7 PULSE GENERATORS

Many digital systems require test signals with fast rise times, high frequencies, and variable pulse parameters. In digital systems, pulse generators can be used to stimulate circuits and are useful for testing the switching speed of circuits. They must also provide pulses that are compatible with the various logic families (in particular TTL, CMOS, and ECL). For example, CMOS logic can use larger-voltage amplitudes, whereas ECL requires very fast (subnanosecond) edges (rise and fall times). Pulse generators are specialized instruments that meet these various requirements. A simplified block diagram for a pulse generator is illustrated in Figure 9–13.

In addition to supplying pulses for testing digital systems, pulse generators are also applied to simulation and testing of other types of systems. For example, pulse generators are useful for testing amplifiers and components for frequency response and response to transients such as recovery time. Another application for pulse generators is to simulate video returns in radar systems for range calibration, testing the sensitivity of video circuits, and other tests and measurements.

PULSE CHARACTERISTICS

A **pulse** is a voltage (or current) that begins at some level, changes rapidly to a new level, remains at the new level for a limited duration of time, and returns to the

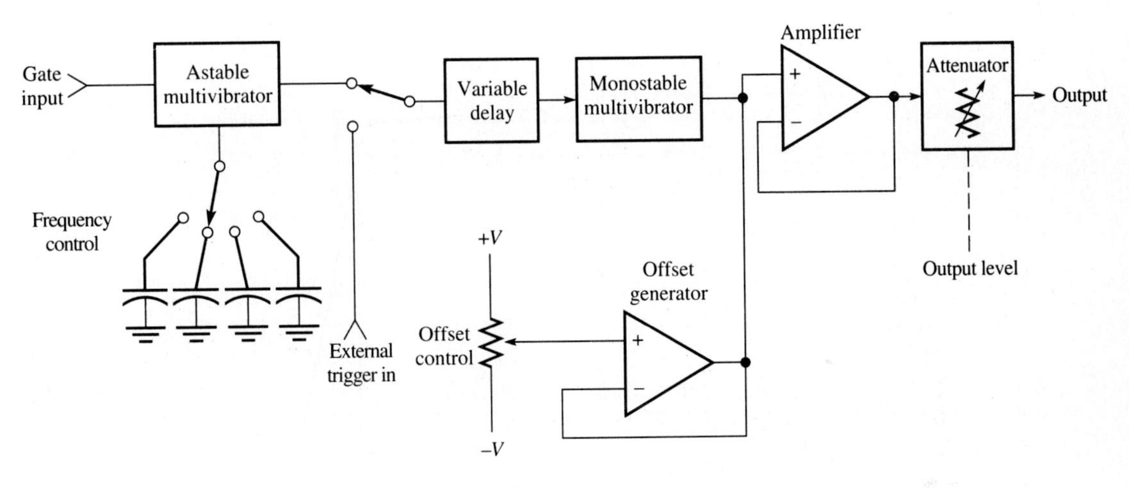

FIGURE 9–13
Simplified block diagram of a pulse generator.

original level. Important pulse parameters are illustrated in Figure 9–14(a). Important pulse definitions are as follows:

Baseline: The amplitude level from which the pulse appears to originate.

Offset: The algebraic difference between the amplitude of the baseline and the amplitude of a reference level (usually ground).

Pulse amplitude: The algebraic difference between the maximum excursion and the baseline of a pulse.

Rise time, t_r: Usually defined as the time required to go from 10% to 90% of the amplitude, although occasionally the 20% and 80% levels are used.

Fall time, t_f: Defined similarly to rise time but from the higher level to the lower level.

Pulse width: The time from the 50% amplitude level on the rising edge to the 50% amplitude level on the falling edge of the pulse.

Jitter: Variation in the starting time, duration, or amplitude of a pulse between successive cycles.

Preshoot: A baseline distortion that occurs immediately before a major transition, expressed as a percentage of the pulse amplitude.

Overshoot: A peak distortion that occurs immediately after a major transition, expressed as a percentage of the pulse amplitude.

Undershoot: A baseline distortion that occurs immediately after a major transition, expressed as a percentage of the pulse amplitude.

Ringing: A damped oscillation that occurs immediately following a major transition.

Settling time: The time from the 90% point on the rising edge until the level remains within some specified range of the amplitude (typically the range is 1 to 5%).

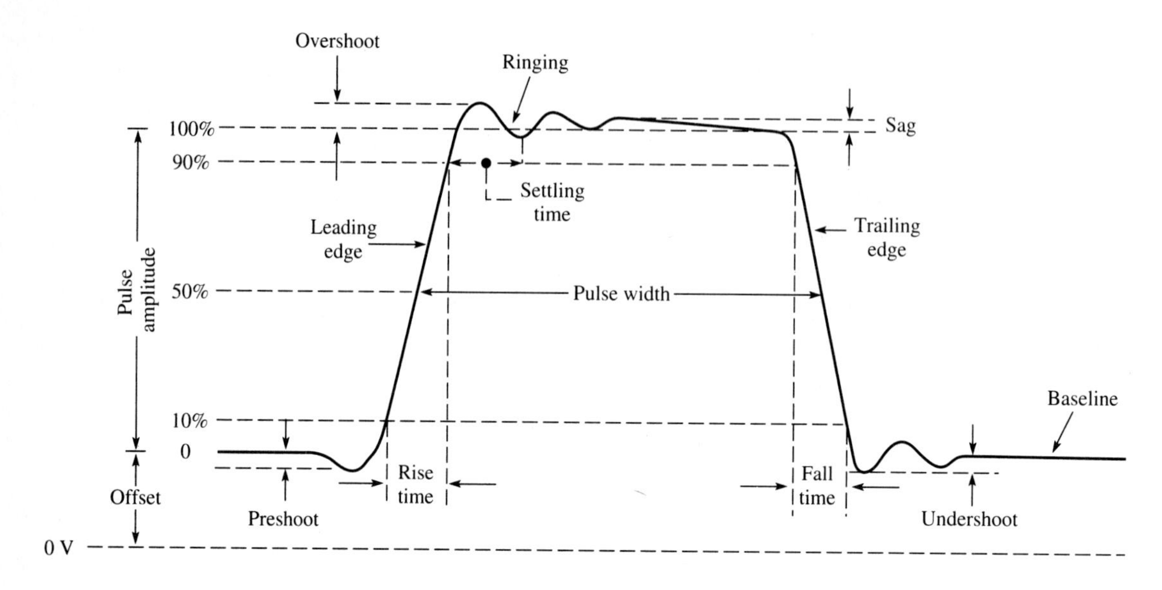

(a) Pulse parameters.

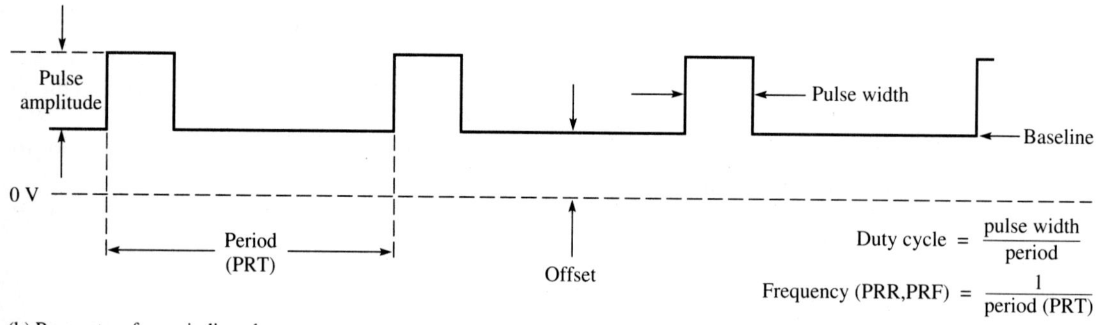

(b) Parameters for periodic pulses.

FIGURE 9–14
Characteristics of pulses and periodic pulses.

Periodic pulses repeat at a regular interval. Definitions for periodic pulses are illustrated in Figure 9–14(b). The definition for *period* is the minimum interval between points on a waveform that reoccur. For pulse trains, period is sometimes referred to as pulse repetition time, or PRT. Similarly, frequency (reciprocal of period) is often called pulse repetition rate (PRR) or pulse repetition frequency (PRF). Another characteristic of periodic pulses is the **duty cycle.** Duty cycle is the pulse width divided by the period; it represents the fraction of time the pulse is high. Duty cycle can be expressed either as a fraction or as a percentage. A duty cycle of exactly 50% defines the special case of a square wave (equal on and off times). Square-wave testing is valuable for testing amplifiers when the low-frequency response is important, as shown earlier in Figure 9–5. A practical example of square-wave testing is the internally generated 1 kHz square wave in an oscilloscope for testing the probe and input amplifier, as discussed in Section 8–5. Pulses with a low duty cycle are preferred for testing the high-frequency dynamic behavior of certain physical systems and for testing cables for reflections due to impedance mismatch.

EXAMPLE 9–6 For the oscilloscope presentation shown in Figure 9–15, determine the PRT, PRR, and the duty cycle of the pulse train.

FIGURE 9–15

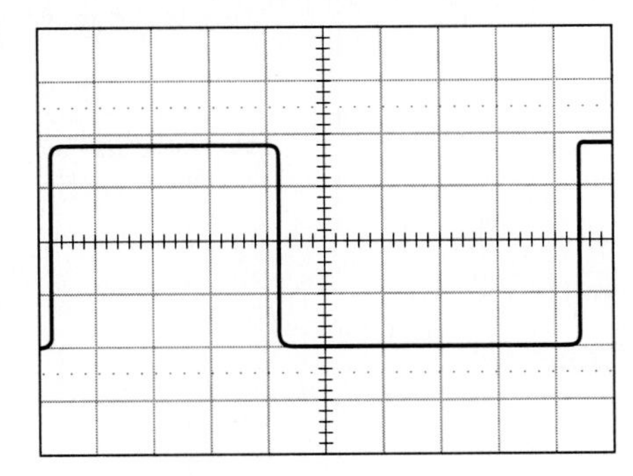

Oscilloscope settings:
SEC/DIV = 20 μs/div
V/DIV = 0.5 V/div
Horizontal 10 × magnifier is ON.

SOLUTION Begin by finding the period. The PRT is the same as the period and is found by multiplying the SEC/DIV setting by the number of divisions between reoccurring points on the waveform. The number of divisions is seen to be 9.2. Since the 10× horizontal multiplier is *on*, the result is *divided* by 10 (as discussed in Section 8–3). That is

$$T = \frac{(20\ \mu\text{s/div})(9.2\ \text{div})}{10} = 18.4\ \mu\text{s}$$

The PRR is the frequency. It is found by computing the reciprocal of the period:

$$f = \frac{1}{T} = \frac{1}{18.4\ \mu\text{s}} = 54.3\ \text{kHz}$$

In order to find the duty cycle, the pulse width must be measured. The pulse width is found by multiplying the SEC/DIV setting by the number of divisions between the 50% rising and falling edges on the pulse. The number of divisions is 4.0. Therefore

$$\text{PW} = \frac{(20\ \mu\text{s/div})(4.0\ \text{div})}{10} = 8.0\ \mu\text{s}$$

The duty cycle is found by dividing the pulse width by the period:

$$\text{DC} = \frac{\text{PW}}{T} = \frac{8.0\ \mu\text{s}}{18.4\ \mu\text{s}} = 0.43 = 43\%$$

EXAMPLE 9–7 For the oscilloscope presentation shown in Figure 9–16(a), determine the settling time (ST) for a 5% final level.

SOLUTION The settling time is measured from the 90% level to the point where the waveform stays within 5% of the amplitude. These points are marked on Figure 9–16(b). The number of divisions between these points is counted and multiplied by the SEC/DIV setting. Notice that the VOLTS/DIV control is shown as uncalibrated. For this measurement, it is easier to read the percentage by using the variable VOLTS/DIV control to set the baseline at 0% and the top line at the 100% graticule line on the oscilloscope:

$$ST = (0.1\ \mu s/div)(2.3\ div) = 0.23\ \mu s$$

FIGURE 9–16

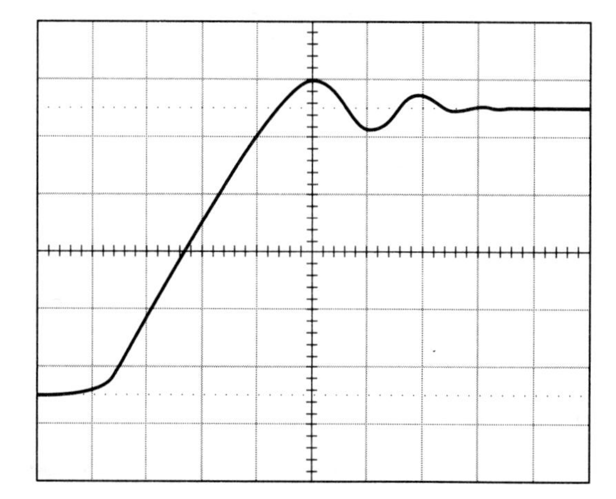

(a) Oscilloscope presentation for Example 9–7.

Oscilloscope settings:

SEC/DIV = 0.1 μs/div
Volts/DIV = uncalibrated
Horizontal 10 × magnifier is OFF.

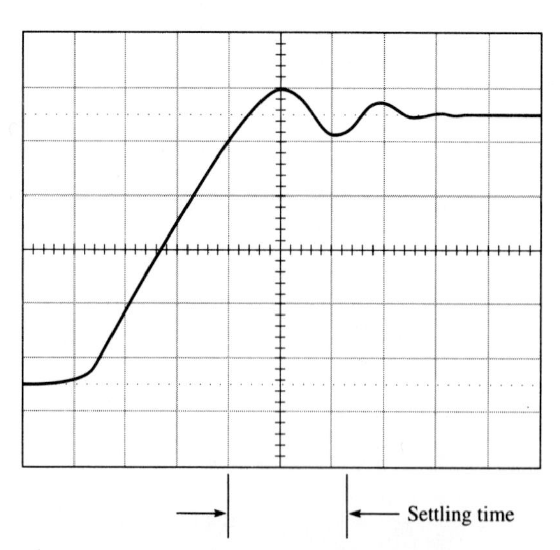

(b) Oscilloscope presentation for solution to Example 9–7.

PULSE-GENERATOR SPECIFICATIONS

The most important specifications for pulse generators include rise and fall time, the maximum operating frequency, output level (amplitude and offset), output impedance, and operating modes. Among other specifications of interest are the duty cycle, output formats, modulation modes, and programmable control. Programmable generators offer many additional features, including waveform storage and computation and complex three- and four-level waveforms to simulate certain pulses used in radar and communication systems.

The fastest pulse generators have rise times of 150 ps and clock speeds of 1 GHz. For high-speed measurements, very fast rise times and speeds are necessary. For example, if you are measuring the slew rate of a high-speed comparator, the rise time of the generator must be much faster than the slew rate or else you are measuring the generator's rise time rather than the comparator's slew rate.

When rise and fall times are very fast, the problems from cable mismatches and ground bounce can create unacceptable error in a measurement. Reflections from even a short piece of cable create ringing, obscuring the desired signal. Recall that there is a relationship between rise time and bandwidth given by the equation

$$t_r = \frac{0.35}{B}$$

where t_r = rise time, s
B = bandwidth, Hz

A pulse generator with a 500 ps rise time has an equivalent sinusoidal frequency of 700 MHz. At frequencies this high, the pulse generator should be terminated in its characteristic impedance to prevent reflections.[1] Some pulse generators offer controlled transitions in order to avoid problems for applications where very high speeds are not necessary. Slowing the rise and fall times reduces the reflections from mismatches.

The output-level specification is given in terms of amplitude and offset. Current specifications may also be given, and the manufacturer may specify which logic families are compatible with a given generator. For pulse generators, the offset specification is referenced to the baseline, as was indicated in Figure 9–14(b). (Note that function generators have a symmetrical output, so the median is the reference for offset.) Typically, the output impedance is 50 Ω.

Programmability enables the user to control pulse parameters from a computer or controller. Virtually all parameters, including waveform modes such as continuous, triggered, burst, or gated modes, can be programmed over the IEEE-488 bus. In addition, programmable instruments frequently have the capability of storing various setups or digitized waveforms. Figure 9–17 illustrates a typical programmable pulse generator with a digital readout.

[1] Notice that this is well within the decision threshold for controlled impedance interfacing given in Figure 1–15.

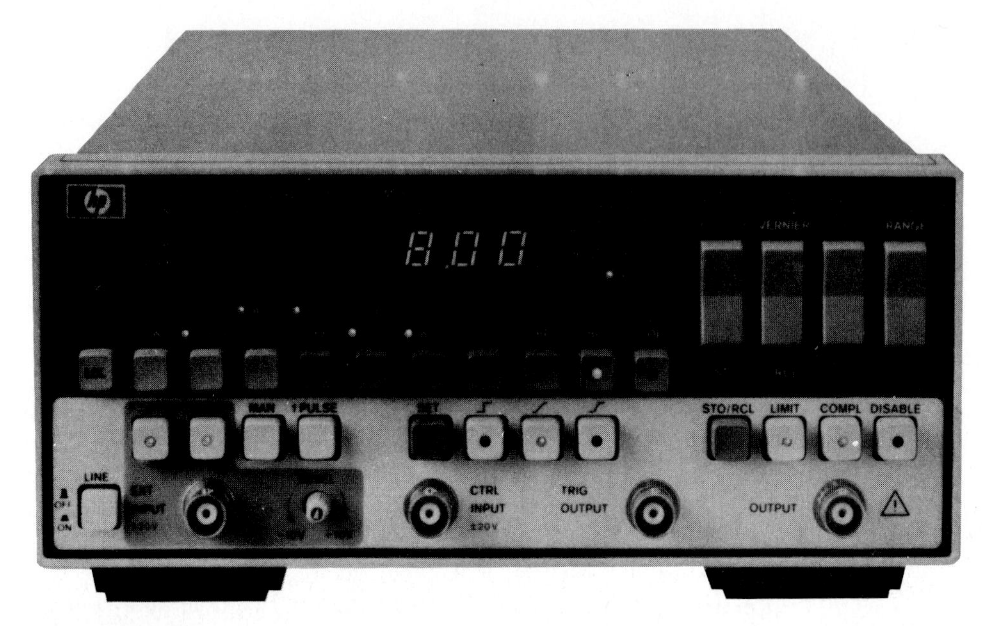

FIGURE 9–17
A programmable pulse generator, the HP-8112A. It is a programmable 50 MHz generator with 5 ns transitions, comprehensive trigger modes, modulation, three-level signals, and other features (courtesy of Hewlett-Packard).

PULSE-GENERATOR CONTROLS

Controls for a typical pulse generator enable the user to adjust nearly all pulse parameters. Once the frequency range is selected, the exact frequency can be adjusted with a continuously variable vernier. Generally, the user can control pulse amplitude, offset, rise and fall time, and pulse width or duty cycle. Depending on the generator, a menu of pulse modes may be selected, including continuous, triggered, gated, or burst modes that are similar to the waveform modes described in Section 9–2. In the triggered mode, all other front-panel controls remain operational, but the generator waits for a trigger. This enables the generator to be synchronized to other instruments, a circuit signal, or a manual pushbutton. Gated, or burst, mode is useful when it is necessary to generate a specific number of pulses, as when testing a shift register. Gated mode can be synchronous or asynchronous. In the synchronous mode, the pulse is *on* for a fixed time after the gate signal and the last pulse are completed, whereas in the asynchronous mode the gate merely enables free-running pulses and may send a partial first or last pulse. Another waveform mode that is frequently offered is double pulses. The delay and spacing between pulses may be adjusted by the user. Outputs include the normal or complemented pulses (inverted) as well as trigger and/or sync outputs. The trigger outputs enable the pulse generator to synchronize other circuits, including another pulse generator.

9–8 DIGITAL-PATTERN GENERATORS

Digital-pattern generators are used to generate complex digital test signals for stimulus of digital systems, including telecommunications equipment, computers, and micrprocessor-based systems. Telecommunication testing is done by specialized instruments that frequently combine pattern generation with error detection and other functions. Digital patterns for testing computer and microprocessor systems are often special sequences that are designed to work with other test equipment such as logic analyzers. In high-quality analyzers, a pattern generator may be included as part of the digital analysis system. The pattern generator forms the stimulus instrument of stimulus-response testing. The required pattern for testing can be generated by an algorithm or from stored data. An example of a digital pattern generator, capable of generating complicated patterns, is shown in Figure 9–18.

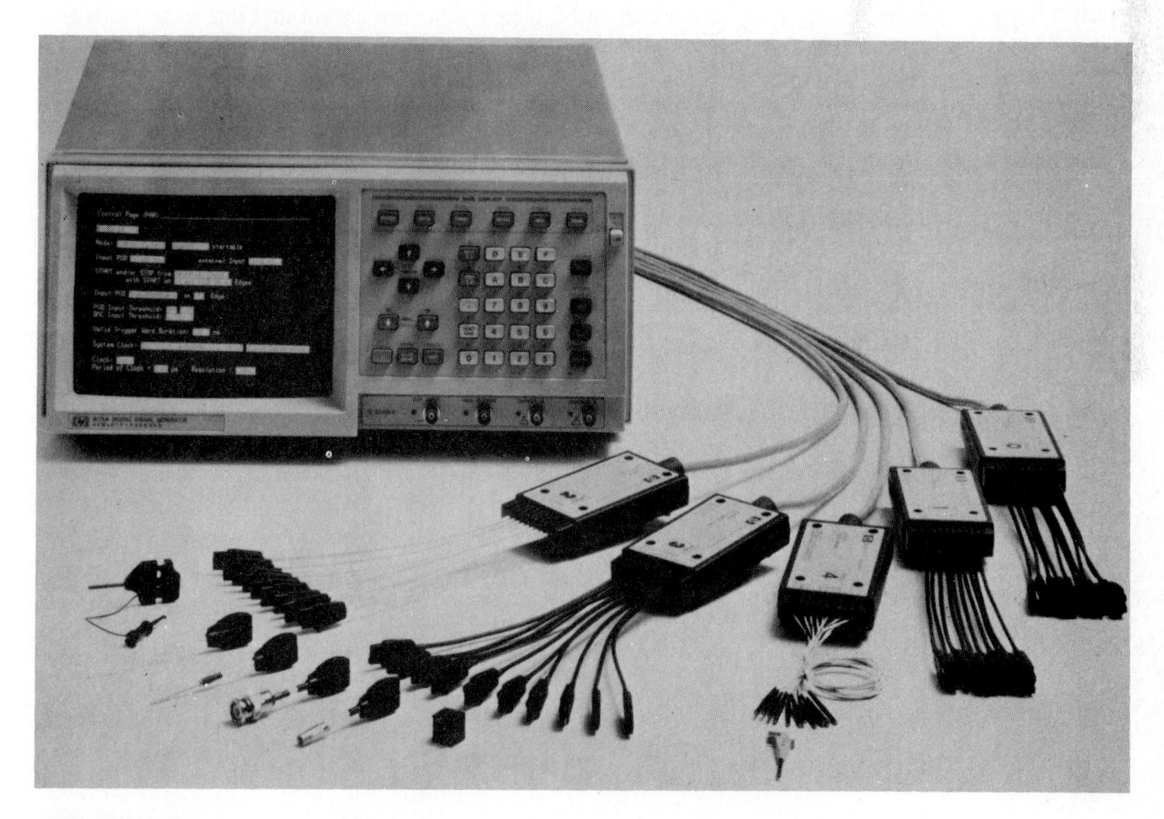

FIGURE 9–18
Digital-pattern generator. The HP 8175A digital pattern generator is pictured here, along with output pods and flexible interface adaptors. The output pods and interface adaptors transmit signals directly to the device under test. This ensures the specified signal quality at the probe tip, a precondition for reliable results (courtesy of Hewlett-Packard).

In order to generate long digital sequences, data compression is frequently used to save storage. One method, called a vector engine, uses a vector table and sequencer to generate the required patterns. The vector table can be thought of as a list of instructions; the instructions tell whether to keep or toggle a particular bit in a pattern. To keep a bit means that the bit is unchanged; to toggle a bit means it is changed. All that is stored is the vector table and a list of pointers to the vector table contained in a sequencer. To start the pattern, the vector engine is first "seeded" with a starting value; the sequencer then points to locations in the vector table. The value obtained from the vector table is used to determine the next output. A simplified example should make the method clear.

EXAMPLE 9–8 Assume you need to generate the 4 bit binary sequence shown in Table 9–1. Show the vector table and sequence that can be used to generate the pattern.

SOLUTION The desired output from Table 9–1 is first listed in terms of a *keep and toggle* table, as shown as Table 9–2. Assuming the seed pattern is 0000, this table is a list of instructions for creating the pattern. Observing Table 9–2, we find only four unique instructions; all of the remaining instructions are redundant. The unique instructions are listed in Table 9–3 as the *vector* table. A list of pointers can now regenerate the *keep and toggle* table by simply listing the instructions taken from the *vector* table in sequence. Table 9–4 is the regenerated *keep and toggle* table with the seed value 0000 shown. Using the seed value, the desired output pattern is generated as the output as shown in Table 9–5.

TABLE 9–1

0	0	0	0
0	0	0	1
0	0	1	0
0	0	1	1
0	1	0	0
0	1	0	1
0	1	1	0
0	1	1	1
1	0	0	0
1	0	0	1
1	0	1	0
1	0	1	1
1	1	0	0
1	1	0	1
1	1	1	0
1	1	1	1

Desired output

TABLE 9–2

K	K	K	T
K	K	T	T
K	K	K	T
K	T	T	T
K	K	K	T
K	K	T	T
K	K	K	T
T	T	T	T
K	K	K	T
K	K	T	T
K	K	K	T
K	T	T	T
K	K	K	T
K	K	T	T
K	K	K	T

Keep and toggle table

TABLE 9–3

					Pointers
K	K	K	T	1	1
K	K	T	T	2	2
K	T	T	T	3	1
T	T	T	T	4	3
					1
Vector table					2
					1
					4
					1
					2
					1
					3
					1
					2
					1

Stored data

TABLE 9–4

0	0	0	0
K	K	K	T
K	K	T	T
K	K	K	T
K	T	T	T
K	K	K	T
K	K	T	T
K	K	K	T
T	T	T	T
K	K	K	T
K	K	T	T
K	K	K	T
K	T	T	T
K	K	K	T
K	K	T	T
K	K	K	T

TABLE 9–5

0	0	0	0
0	0	0	1
0	0	1	0
0	0	1	1
0	1	0	0
0	1	0	1
0	1	1	0
0	1	1	1
1	0	0	0
1	0	0	1
1	0	1	0
1	0	1	1
1	1	0	0
1	1	0	1
1	1	1	0
1	1	1	1

Output

SUMMARY

1. The principle specifications for most signal sources deal with the frequency, spectral purity, output amplitude, modulation, operating characteristics, and output impedance.
2. Modulation is the process of changing some characteristic of a high-frequency waveform by a lower-frequency waveform called the modulating signal.
3. Function generators produce sine, square, and triangle waveforms over a range of frequencies. Basic controls are function selection and calibrated controls for frequency and amplitude selection. The controls may include other capabilities, such as generating other waveforms or pulses or the ability to modulate the output.
4. Arbitrary function generators are synthesizers that allow the user to create custom waveforms or choose from a selection of waveforms stored in memory. Custom waveforms are useful for automatic test systems and other systems that require specialized and simulated signals.
5. Signal generators produce high-frequency sine waves and modulate sine waves. There are a large variety of signal generators available with frequencies that go from dc to microwave. Signal generators are widely used in radio frequency testing. Many signal generators also have swept-frequency capability.
6. Swept-frequency oscillators (or sweepers) generate a sine-wave output that allows the user to view signals directly in the frequency domain. Swept-frequency oscillators enable the user to observe the effects of adjustments to a circuit as they happen.
7. Pulse generators are specialized instruments for producing pulses with fast rise times and varying duty cycles. Some generators allow the user to control rise time, triggering, and other parameters.

QUESTIONS AND PROBLEMS

1. Assume a student is making a rise-time measurement in an *RC* circuit. The student sets a function generator to indicate a 0 to 100% pulse on an oscilloscope and then connects the *RC* circuit to the generator and measures the 10% to 90% rise time. What error is the student making? What should have been done?
2. Assume a signal generator can be set to a minimum output of −127 dBm in a 50 Ω load. Compute the voltage amplitude that is represented by this.
3. A signal generator with a 50 Ω output resistance is connected to a matching load. The output is set for +20 dBm.
   (a) Compute the power dissipated in the load.
   (b) Compute the voltage across the load.
4. Compare amplitude, frequency, and phase modulation.
5. What type of circuit can be used to convert a triangle waveform into each of the following?
   (a) Square wave
   (b) Sine wave
6. Assume that the circuit shown in Figure 9–8 has $R_A = R_B = 10$ kΩ and the 1 kΩ potentiometer is centered. Compute the value of C that will produce a frequency of 1.0 kHz.
7. Assume a function generator is set to a frequency of 1 μHz. How many days elapse for one complete cycle?
8. What are advantages of arbitrary waveform generators over nonsynthesized function generators?
9. A signal generator has a specified spurious noise that is −80 dBc. Compute the spurious noise from the generator into a 50 Ω load when the carrier amplitude is 10 dBm.
10. How does the dc offset control of a function generator differ from the dc offset of a pulse generator?
11. Explain how square-wave testing can be used to indicate the frequency characteristics of an amplifier.
12. What problems occur if a high-frequency signal generator is not operated in the correct load?

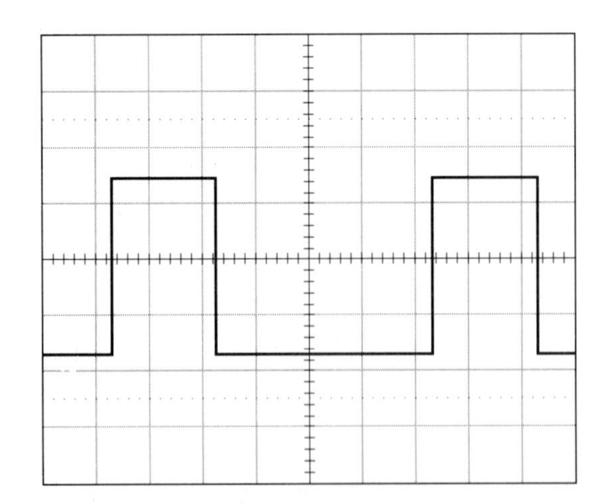

Oscilloscope settings:

SEC/DIV = 0.2 μs/div
Volts/DIV = 2.0 V/div
Horizontal 10 × magnifier is OFF.

FIGURE 9–19

13. Assume you need to choose a signal generator for making sensitivity tests for radio receivers.
    (a) Explain the importance of the attenuator specification in your selection.
    (b) What other specifications are important in choosing a generator?
14. Explain the difference between phase noise and harmonic distortion.
15. Draw the diagram that explains how to set up a sweep oscillator, including the sweep generator, the device under test, the detector and the oscilloscope.
16. How is the horizontal axis of an oscilloscope converted to represent frequency with swept-frequency measurements?
17. For the oscilloscope presentation shown in Figure 9–19, determine the PRT, PRR, and the duty cycle of the pulse train.
18. Assume the oscilloscope controls in Figure 9–19 were uncalibrated. Can you still determine the duty cycle of the pulse train? Explain.
19. Explain how to measure the settling time of a pulse.
20. When might you want to be able to slow the rise time of a pulse generator?
21. For the problem in Example 9–8, determine the output pattern if the seed value were 0011.

Chapter 10

Signal-Analysis Instruments

OBJECTIVES

Signal analyzers consist of instruments that indicate the frequency, time, amplitude, or logic properties of a signal. We have already discussed the oscilloscope, the principal instrument for looking at the amplitude-versus-time characteristics of a signal. The spectrum analyzer is the principle instrument for displaying the amplitude-versus-frequency characteristics of a signal. In this chapter, we discuss the spectrum analyzer, the waveform recorder, and distortion analyzer. We also discuss the network analyzer. The network analyzer uses a stimulus instrument and a signal analyzer to characterize *circuits* rather than *signals*. In the next chapter we introduce an instrument for observing the logic properties of a signal, namely, the logic analyzer.

When you complete this chapter, you should be able to

1. Compare the important features of spectrum analyzers and distortion analyzers.
2. Describe the frequency domain representation of an amplitude or frequency-modulated carrier with a sine wave–modulating signal, including the frequencies and amplitude of the sideband signals.
3. Compute the modulation index of an amplitude-modulated signal given the amplitude of the carrier and modulating signals.
4. Cite the advantages and disadvantages of a swept-tuned spectrum analyzer as compared to a real-time spectrum analyzer.
5. Given a simplified block diagram for a swept-tuned spectrum analyzer, explain the function of each block.
6. Describe the operating controls for a swept-tuned spectrum analyzer.
7. Define harmonic distortion, total harmonic distortion, intermodulation distortion, and transient intermodulation distortion.
8. Given the fundamental and harmonic values of a signal, compute the harmonic distortion.
9. Explain how a distortion analyzer can be used to measure the total distortion or signal-to-noise ratio in an amplifier.
10. Discuss how a network analyzer can be used to characterize the response of a circuit.

HISTORICAL NOTE

The spectrum analyzer was developed during World War II by scientists at the Massachusetts Institute of Technology Radiation Laboratory to aid in developing radar. Radar signals are short pulses with fast rise and fall times. One application of the new spectrum analyzer was to optimize the bandwidth, gain, and fidelity of radar receivers to reproduce returning echoes with the highest possible resolution and accuracy. The first spectrum analyzers were uncalibrated indicators that lacked the broad coverage of modern spectrum analyzers but were adequate for the job at the time.

10–1 INTRODUCTION

SIGNAL ANALYZER INSTRUMENTS

Signal analyzers are instruments that indicate the frequency, time, amplitude (voltage or power), or logic properties of a signal. By this definition, the oscilloscope and logic analyzer are signal-analyzer instruments. In this chapter we look at instruments that analyze signals in the frequency domain. Analyzers that indicate a frequency property include the spectrum analyzer, the wave analyzer, and the distortion analyzer. The spectrum analyzer shows the signal directly in the frequency domain—the screen shows a plot of the amplitude verses frequency. The wave analyzer is essentially a frequency-selective voltmeter with a narrow passband that is tuned to a single-frequency component of a signal while rejecting other components. The distortion analyzer is the opposite of a wave analyzer, indicating the energy in a range of frequencies present outside the specific frequency band occupied by the signal.

Although it is not specifically a *signal*-analysis instrument, a network analyzer includes a signal analyzer in order to measure the frequency properties of a circuit. It does this by stimulus-response testing. Network analyzers consist of a signal generator, a signal separator, and the signal-analyzer instrument. Scaler analyzers measure the magnitude of signals as a function of frequency, whereas vector analyzers measure both magnitude and phase.

The oscilloscope, which was introduced in Chapter 8, is primarily an instrument used to view signals in the time domain. For digital signals, a logic analyzer is the key instrument. Logic analyzers, which are discussed in Section 11–3, are instruments that record and display the digital information from a number of channels simultaneously.

THE FREQUENCY DOMAIN

The concept of the frequency domain was introduced in Section 1–2. Although we generally view signals in the time domain with an oscilloscope, there are many times that the time-domain view does not give a complete picture. The time domain is particularly limited for showing low-level harmonics or the presence of other frequency-dependent distortion. The frequency domain can easily show such distortion. It dissects the signal into sinusoidal components of different

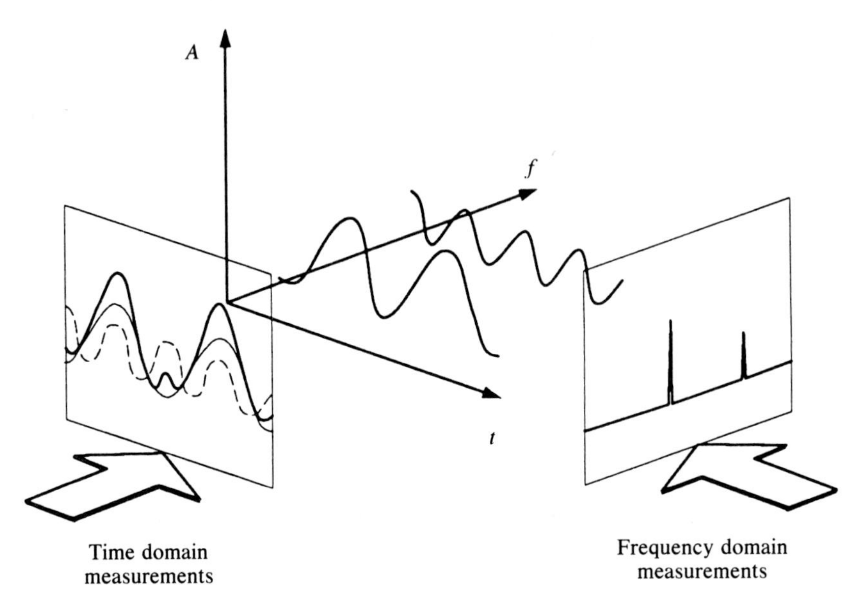

FIGURE 10–1
Frequency- and time-domain comparison (courtesy of Hewlett-Packard).

frequencies; this is particularly useful for observing the output signals from mixers, modulators, filters, and other frequency-dependent circuits. The process of converting a time-domain signal into its frequency distribution is called **spectral analysis.** Figure 10–1 illustrates the time and frequency domains in a three-dimensional view, showing that the frequency domain represents another view of the same signal.

Fourier analysis includes the Fourier series and the Fourier transform. Recall that the Fourier series enables us to express any periodic, time-dependent signal as a series of harmonically related sine waves of the appropriate amplitude, frequency, and phase. Nonperiodic signals cannot be represented by the Fourier series. The **Fourier transform** is an algorithm (or mathematical method) that breaks a continuous, periodic, time-domain signal into its sinusoidal components or vice versa. It is based on a signal that stretches from $-\infty$ to $+\infty$. In practice, the behavior over a limited time period (several seconds or minutes) is sufficient to establish the characteristics of the signal.

Although the original Fourier transform was based on a continuous signal, mathematicians developed a version of the Fourier transform for discrete (sampled) signals that gives the frequency representation as a continuum of weighted, complex exponentials. The method, called the discrete Fourier transform, required a considerable amount of calculation. In the 1960s, a version of the discrete Fourier transform called the **fast Fourier transform** (FFT) considerably reduced the number of calculations required. Mathematicians have continued to find methods for speeding up the FFT, and it is now a common procedure to do spectrum analysis with a computer by transforming data taken in the time domain. This can be done with sampled data from an oscilloscope and a digital computer. Fourier transform analyzers essentially perform the Fourier transform on sampled time-domain data and display the signal in the frequency domain. LeCroy markets a series of digital oscilloscopes that can perform the FFT, effectively giving the user two instruments in one. The result is displayed on a split screen, so that the user can view the time-domain data (as an oscilloscope) and the frequency domain (as a spectrum analyzer) at the same time. This dual capability is particularly useful for transient phenomena, a limitation for swept-tuned analyzers (discussed in Section 10–2). Figure 10–2 shows an example of a split screen display showing both time- and frequency-domain views of the same information.

Although the FFT allows the user measurements that are otherwise impossible in the time domain, there are a few limitations. The frequency range of a Fourier transform analyzer can extend from microhertz to 100 MHz, whereas the frequency range of a spectrum analyzer can extend to several hundred gigahertz, the region of the electromagnetic spectrum for numerous applications such as microwave communications. Generally, Fourier analyzers have less sensitivity and dynamic range than spectrum analyzers. Although they cannot show the data as it occurs in real time, this is not a serious limitation because of the speed of modern processors (a 1000-point FFT can be done in less than 1 s). In the next

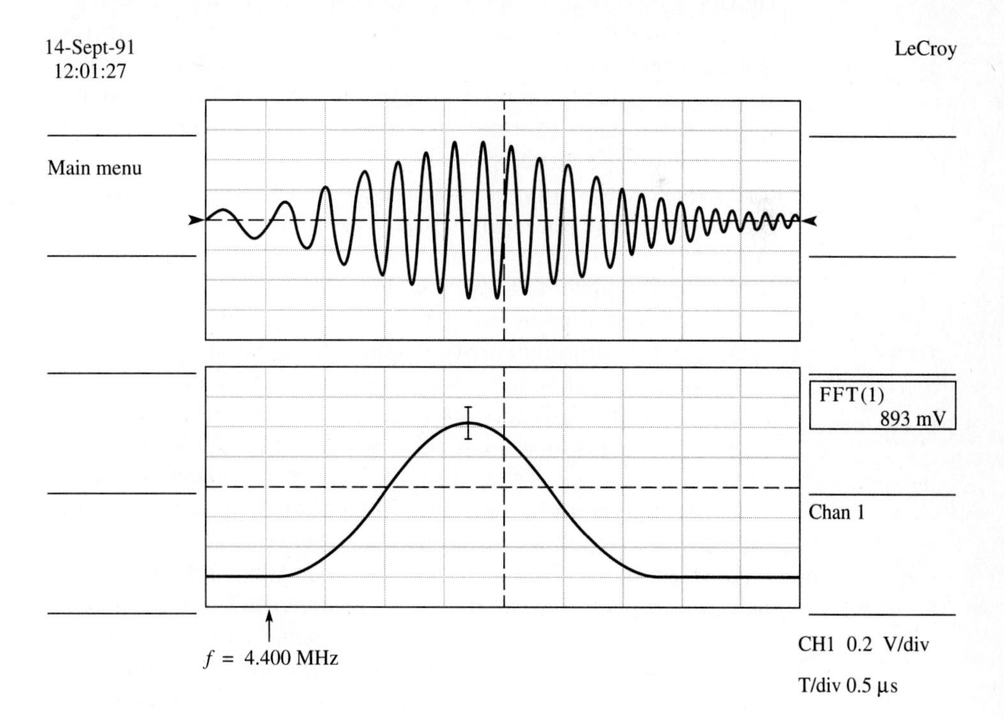

FIGURE 10–2
Plots from the screen of a LeCroy 9450 digital storage oscilloscope. The top waveform shows the time-domain view; the bottom trace represents the frequency domain (courtesy of LeCroy).

section, we consider spectrum analyzers that operate directly in the frequency domain.

10–2 THE SPECTRUM ANALYZER

The spectrum analyzer is an instrument that indicates the frequency characteristics of a signal. By using an analyzer in conjunction with an appropriate stimulus instrument, spectrum analyzers can be used to determine the frequency response of various circuits and networks, including filters, stereo amplifiers, broadcast transmitters, radar, and the like. For TV broadcast stations, the transmitter is stimulated with a video signal from a special tracking generator. The generator is locked onto the sweep of the spectrum analyzer, which is tuned to the particular channel being tested. The analyzer indicates the frequency response of the system. Other applications to communication systems are discussed in Section 10–4.

In addition to electronic systems, many engineering fields and sciences use spectrum analyzers to obtain further insight into measurement problems. These applications include low-frequency work in geology and earth science, where vibration of the earth is stimulated with a shaker and the spectrum of returning signals is recorded. Other sciences that use spectrum analyzers include biomedicine, oceanography, and structural and mechanical engineering. For mechanical testing, spectrum analyzers are used for vibration studies on various structures. A transducer is used to convert the mechanical motion into an electrical signal. The frequency-domain information enables mechanical engineers to determine specific properties of the structure.

There are two types of spectrum analyzers: **real-time** and **swept-tuned** analyzers. Real-time analyzers use a large number of bandpass filters tuned to different frequencies to separate the components of a signal simultaneously into each of the ''bins'' represented by each filter. Although they cost more, real-time analyzers are the only type of analyzer that can be used to observe transient phenomena because they simultaneously display the components of the spectrum; by contrast, swept-tuned analyzers display only one small portion of the spectrum at a time. Real-time analyzers are consequently considerably faster and more expensive than swept-tuned analyzers. Swept-tuned analyzers are more common and offer the advantage of flexibility. A swept-tuned spectrum analyzer is essentially a superheterodyne radio receiver with a CRT display to indicate the power level of the received signal as a function of frequency. It looks at the signal spectrum through a narrow window that is moved in time. To understand its operation, it is helpful to review the operation of the superheterodyne radio.

The superheterodyne radio has been with us for many years. It greatly simplified the tuning of a radio. Prior to the superheterodyne radio, if you wanted to change stations on your AM radio, you had to tune each stage individually, a procedure that was tedious, to say the least. The superheterodyne receiver, shown in the block diagram of Figure 10–3(a), changed all that by first tuning an rf amplifier stage to the desired rf signal and then mixing it with a sinusoidal waveform from a local oscillator in a separate stage. The mixing process produces two

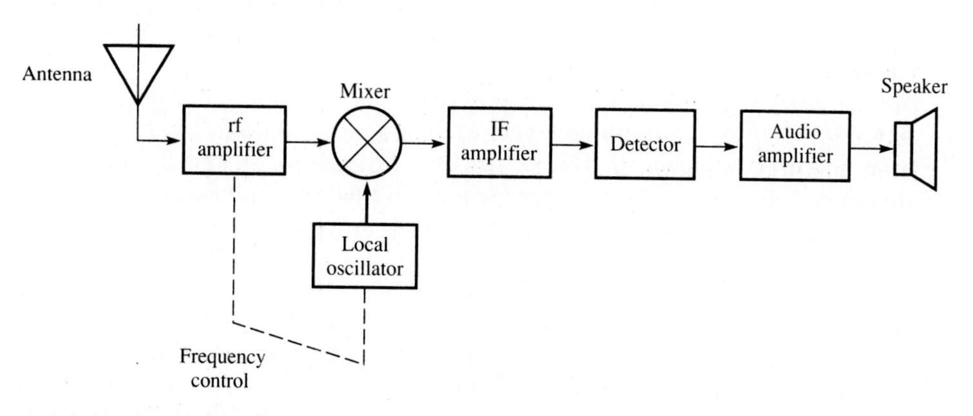

(a) Superheterodyne radio.

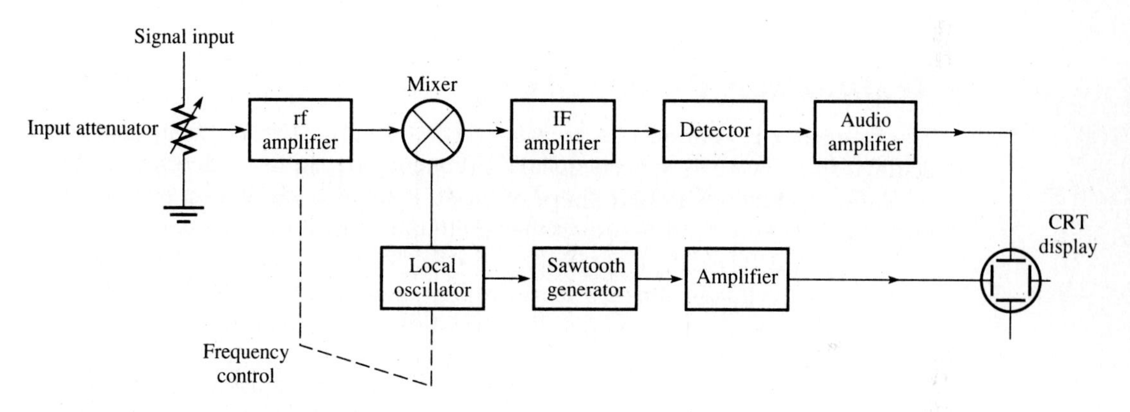

(b) Swept-tuned spectrum analyzer.

FIGURE 10–3
Comparison of superheterodyne radio with swept-tuned spectrum analyzer.

new frequencies—the sum of the signal frequency and the local oscillator frequency, and the difference between the signal and the local oscillator frequencies.

The difference frequency is very important and is given the name **intermediate frequency,** or IF. The mixer has, in effect, converted the incoming frequency to the IF frequency in a process called heterodyning. For any given radio receiver, this frequency is constant (the standard for AM broadcast receivers is 455 kHz); it always represents the difference between the signal and the local oscillator. To tune a radio, the frequency of the local oscillator is changed. Simultaneously, the rf amplifier is tuned because of a common ganged capacitor. As a result, the radio tunes to a different incoming signal but maintains the same IF frequency. Thus the IF stages do not have to be adjusted every time a new station is tuned and they can be optimized for the desired response. Following IF amplifiers, the signal is detected (carrier frequency is removed) and amplified before being sent to the speaker.

The swept-tuned spectrum analyzer is, of course, far more sophisticated than a superheterodyne radio. In addition to more circuits, features, and controls, the spectrum analyzer requires a higher degree of resolution, accuracy, stability, and sensitivity than a radio receiver. The block diagram of a swept-tuned spectrum analyzer is shown in Figure 10–3(b). Compare it to the diagram for the superheterodyne radio. Notice that the local oscillator is voltage-controlled and is electronically tuned by a scan generator. The scan generator produces a sawtooth waveform that both controls the frequency of the local oscillator and moves the beam in the horizontal direction across a CRT. This means that the horizontal position of the beam corresponds to the frequency to which the spectrum analyzer is tuned. Meanwhile, the output of the IF stage is detected, amplified, and sent to the vertical plates of the CRT, producing a plot of amplitude verses frequency for the input signal. The vertical scale can be calibrated as either a linear scale or as a logarithmic scale, whereas the horizontal scale is calibrated linearly in frequency, increasing from left to right.

RESOLUTION BANDWIDTH

The **frequency resolution** is a measure of the ability of a spectrum analyzer to distinguish two closely spaced signals. Frequency resolution is determined by the resolution bandwidth and the shape of the IF response. **Resolution bandwidth** is a term used to describe the width of the IF filter at some decibel level relative to the peak response (typically −3 dB or −6 dB). The *shape* of the IF response is specified by the **bandwidth selectivity,** which is the ratio of the 60 dB bandwidth to the 3 dB bandwidth for a given IF filter. For equal-amplitude signals, the frequency resolution is specified solely by the resolution bandwidth. If the signals have unequal amplitudes, the frequency resolution is determined by both the resolution bandwidth and the bandwidth selectivity. It might appear that an analyzer should be operated with as narrow a resolution bandwidth as possible; however, there is a trade-off as the bandwidth is decreased. A narrow bandwidth requires a longer time constant and slower sweep time. The sweep time is directly related to how long it takes to complete a measurement.

AMPLITUDE MEASUREMENTS

A basic function of an analyzer is to provide the user with the ability to make accurate, calibrated amplitude measurements for the various frequency components of a signal. Amplitudes may be measured on either a logarithmic or linear scale; however, most measurements are done with signals that span a very large dynamic range.[1] For this reason, the vertical scale is usually selected to be a logarithmic scale, calibrated in decibels per division. The vertical-scale factor can be set for various levels, depending on the dynamic range of the input signals.

[1] For spectrum analyzers, the dynamic range is expressed as the decibel ratio of the largest to the smallest signal that can be simultaneously displayed and that allows measurement of the smaller signal within a stated degree of accuracy.

Each division on a logarithmic scale represents a *fixed ratio* between signal levels. (By contrast, each division on a linear scale represents a *fixed difference* between signal levels.) The reference level for decibel measurements is an internally generated calibration signal that is used to set a fixed reference level on the graticule scale. The **reference level** is the signal strength needed to deflect the beam to the top line of the display. The input signal level can then be found by observing the amplitude on the display (in decibels per division from the reference) and subtracting the number of dB from the reference level.

Amplitudes can also be measured relative to a particular signal. For example, in harmonic distortion measurements, the absolute values of the harmonics are not as important as the relative values of the harmonics compared to the fundamental. In this case, the reference is the fundamental, and the number of decibels the harmonics are attenuated is the measurement of concern.

EXAMPLE 10–1 Two signals are observed on a spectrum analyzer. One is six divisions high and the other is two divisions high. The vertical scale is set to 10 dB/div.

(a) What is the ratio of their voltages?

(b) If the larger signal is 0.6 V, what is the voltage of the smaller signal?

SOLUTION (a) The ratio of the voltages is 10 dB/div × 4 div = 40 dB. Expressed as a nonlogarithmic ratio, it is

$$40 \text{ dB} = 20 \log \frac{V_2}{V_1}$$

$$\frac{V_2}{V_1} = 10^2 = 100 : 1$$

(b) Since the smaller signal is 1/100 of the larger signal, its amplitude is 0.006 V = 6 mV.

From this example, you can see the value of the logarithmic scale for viewing signals that have large differences in amplitude. The dynamic range of a typical linear scale is only 40 : 1. If the same two signals were viewed simultaneously on a linear scale, the smaller signal would be lost in the baseline.

10–3 SPECTRUM-ANALYZER CONTROLS

INPUT SIGNAL REQUIREMENTS

The input signal must be within certain limits in both amplitude and frequency. If the amplitude is too high, the sensitive input circuits *can be damaged* or they can create distortion and spurious signals. A dc component to the input signal can also damage the input circuits. The dc component can be eliminated by an external blocking capacitor. It is important to verify the maximum input level from manufacturer's specifications and assure that the signal you are testing is within those specifications before connecting it to the analyzer. Any signal that is larger than the specified limit should be attenuated by a directional coupler or other attenuator before it is connected to the spectrum analyzer. On the other hand, if an input signal is too small, internal noise that is generated by the analyzer can reduce the signal-to-noise ratio. The signal-to-noise ratio can be improved if the bandwidth of the analyzer is kept to the minimum required. Spectrum analyzers also contain limits in regard to the maximum dynamic range of the input signal that can be displayed without distortion.

FRONT-PANEL CONTROLS

Figure 10–4 shows an example of a modern spectrum analyzer, the Tektronix 2712. The primary functions are located at the top of the instrument. These functions include the controls for the reference level, frequency and marker controls, and span per division. Many modern analyzers, such as the 2712, have replaced many of the physical controls on the front panel with menu controls that enable the user to select from a list of options. The use of internal microprocessors enables the manufacturer to provide certain limits to those controls that might otherwise be set incorrectly and can also provide a wider range of options, including automated setups and measurements.

The input signal is connected to a spectrum analyzer through an input attenuator, which provides calibrated attenuation steps for selecting the proper input amplitude. As previously discussed, the front end of a spectrum analyzer can easily be damaged if too large a signal is applied; the maximum is in the range of 1 mW at the mixer with a maximum input signal level to the attenuator in the range of 1 watt. The reference level is determined by the input attenuation as well as the gain of the IF amplifiers. Most modern analyzers control the attenuation and gain

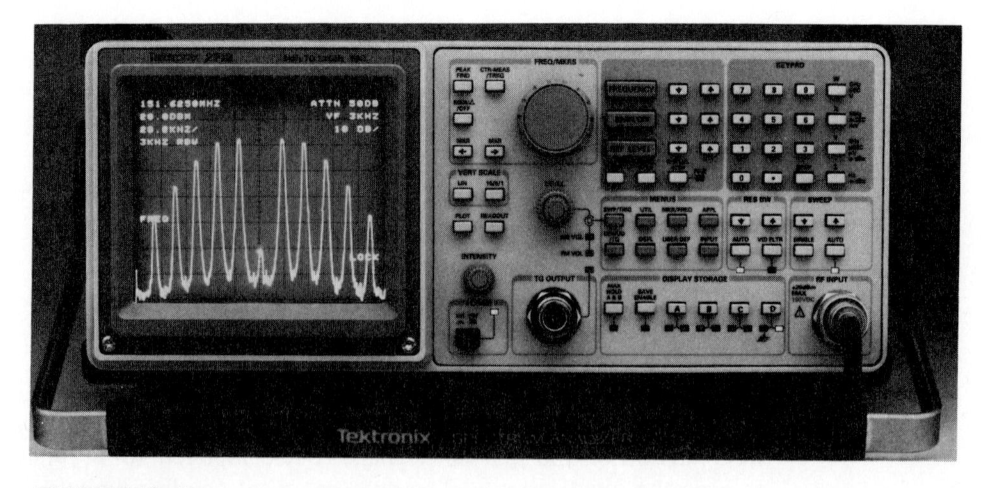

FIGURE 10–4
The Tektronix 2712 spectrum analyzer. The 2712 is a microprocessor-controlled, portable instrument that has a frequency range of 10 kHz to 1.8 GHz (copyright Tektronix, Inc. Used with permission).

with a single control that automatically selects the proper attenuation and gain to prevent overloading the input mixer and optimize the measurement.

The frequency range is selected and the CENTER FREQUENCY is set from the front panel or from a menu. (Some analyzers reference the starting frequency rather than the center frequency). After selecting the frequency, the SPAN, sweep RATE, and RESOLUTION BANDWIDTH controls are adjusted. The SPAN (also known as the DISPERSION control) determines the width of the spectrum. It is calibrated in frequency/division on the horizontal axis. The SPAN control can also be set for maximum span, displaying the maximum frequency range, or zero span, causing the analyzer to stay at a fixed frequency. The sweep RATE control determines the rate that the input spectrum is scanned. It is calibrated in time/division. The RESOLUTION BANDWIDTH control selects different IF filters, causing the effective width of the IF filter to change. For maximum resolution, the resolution bandwidth should be as narrow as possible, but, as previously mentioned, this may require a very long sweep time. The SPAN, sweep RATE, and RESOLUTION BANDWIDTH controls interact with each other. If the sweep rate is too fast, the signal is attenuated because of scan loss, resulting in inaccurate measurements. At the highest resolution, the sweep RATE must be set slow enough for the filter to respond to the system under test. The accuracy of the amplitude measurement can be seriously degraded if the analyzer sweep rate is too fast. As a general rule, the RATE and SPAN controls should be reduced until the vertical signal no longer increases in amplitude. Many new spectrum analyzers can automatically choose the optimum resolution bandwidth, depending on the span and sweep rate selected, or offer the user the option of choosing the resolution bandwidth.

FREQUENCY MARKERS

An important feature of spectrum analyzers is the ability to add frequency markers to the display for direct frequency calibration. Some markers are tunable so that they can be positioned over the signal of interest; the frequency can be read out directly on the display from a built-in frequency counter. Frequency markers are shown on the display as small intensified spots. This enables the frequency of a small signal to be measured accurately in the vicinity of a larger signal. Frequency accuracy depends on the frequency span and the resolution bandwidth.

10–4 COMMUNICATION SYSTEM APPLICATIONS OF SPECTRUM ANALYZERS

INTRODUCTION

Spectrum analyzers are widely applied to communication system measurements; in fact, a spectrum analyzer is the single most useful instrument in this field. It can be used for measurements of modulation levels and modulation index, zero-modulation carrier power and frequency, frequency response, distortion, signal-to-noise levels, and more. Basic concepts of modulation are explained in this section

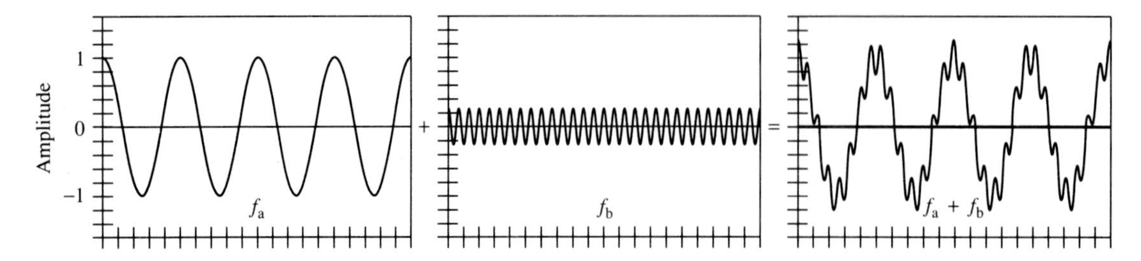

FIGURE 10–5
The addition of two sine waves of different frequency and amplitude.

to familiarize you with some of the most important signals that are observed on spectrum analyzers and the terms used to describe modulation. The emphasis is on the signals rather than modulation methods; refer to a good communication book for more details.

As discussed in Section 9–2, modulation results when one waveform affects certain properties of another waveform. For communication systems, modulation is used to encode information to be transmitted on a carrier frequency. The **carrier** is a high-frequency electromagnetic signal that is the basic transmitting frequency of a broadcast station. When a radio receiver is tuned to the carrier frequency, the station is said to be *tuned,* but no sound or other information is received from the carrier alone.[2] Recall that a continuous-wave (cw) sinusoidal signal can be represented by a single line in the frequency domain. The carrier frequency of a radio station, with no modulation, has the characteristics of a pure sine wave—that is, a single-line spectrum at the assigned frequency of the station. When the signal is modulated, the spectrum changes in accordance with the modulation scheme and the modulating signal. The modulating signal causes the carrier to vary in some predetermined way; three major methods were introduced in Section 9–2.

AMPLITUDE MODULATION

When two or more sinusoidal waveforms are added together, and the resultant waveform is viewed in the time domain, a complex waveform is formed. Figure 10–5 shows two sinusoidal waveforms, added together, which differ in frequency and amplitude; the two waveforms are clearly distinguishable in the resultant waveform. Figure 10–6 shows two waveforms that are similar to each other—equal in amplitude but slightly different in frequency; the two waveforms that make up the resultant waveform are not perceptible. What stands out in Figure 10–6 is a beat frequency. The beat is the difference of the two frequencies, $|f_a - f_b|$. An example of an audio beat frequency occurs when two aircraft propellers are not turning at exactly the same speed—the sound seems to drift in and out, at a rate equal to the difference of their speeds, causing a perceived beat between the two.

If we view the resultant signal of Figure 10–5 in the frequency domain (with a spectrum analyzer), there are only the two frequencies present, as shown in

---

[2] An exception to this is when the carrier is pulsed; information can be received in coded form.

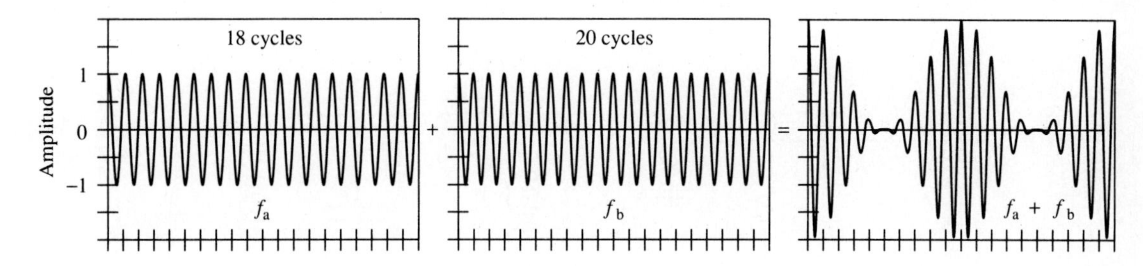

FIGURE 10–6
Linear addition of two signals.

Figure 10–7. However, if either or both of the waveforms are not sinusoidal, then additional frequencies are generated. The additional frequencies are the sum and difference frequencies ($f_a + f_b$ and $f_a - f_b$). The additional frequencies are referred to as **sidebands.** Figure 10–8 shows a circuit for an AM modulator used in a transmitter. The higher frequency (carrier) is amplified by a class C amplifier that greatly distorts the waveform. The output of this stage (called a final) is fed into a resonant tank circuit. The modulating signal is usually much lower in frequency than the carrier frequency. It is mixed with the carrier, at this point, creating the sideband frequencies. The sideband frequencies are given by the equations

EQUATION 10–1

$$f_{LSB} = f_c - f_m$$

EQUATION 10–2

$$f_{USB} = f_c + f_m$$

where f_c = carrier frequency, Hz
f_m = modulating frequency, Hz
f_{LSB} = lower sideband frequency, Hz
f_{USB} = upper sideband frequency, Hz

FIGURE 10–7
Spectral content of $f_a + f_b$ mixed linearly.

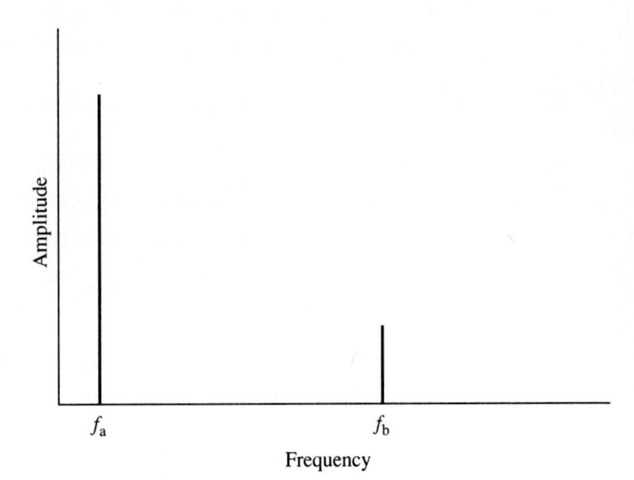

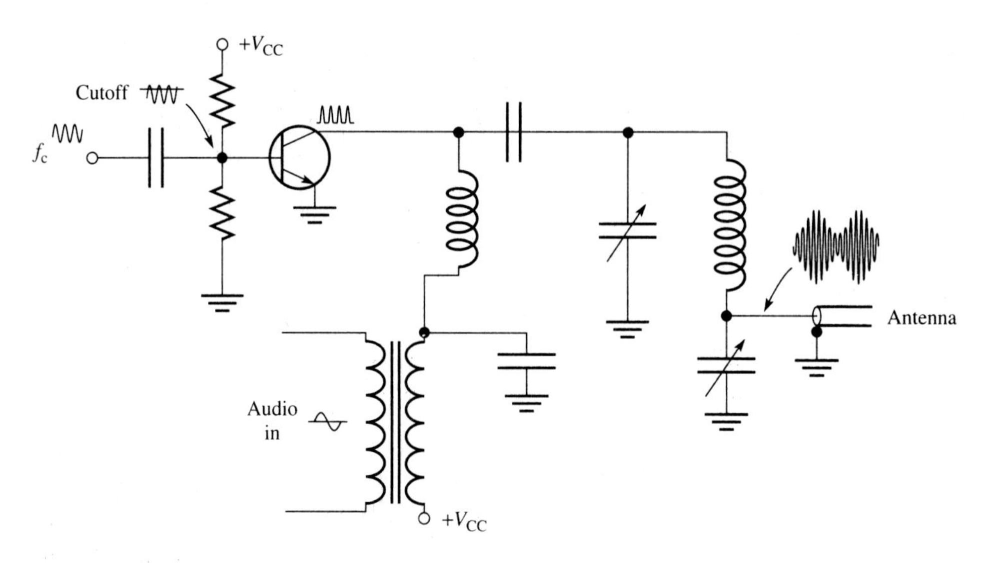

FIGURE 10–8
Class-C final stage in a simple AM transmitter.

Consider, as an example, that an AM radio station is broadcasting at 1 MHz frequency (f_c) and that an audio tone (f_m) of 1 kHz is used to modulate the signal. What is generated in the final amplifier is four frequencies: f_c, f_m, f_{LSB}, and f_{USB}. Only three of the signals are broadcasted; f_m is filtered out. Figure 10–9(a) shows the carrier and both sidebands overlaid on a composite plot. Note that when both sidebands are in phase with each other, they are either in phase with the carrier or 180° out of phase with the carrier. The result of adding the instantaneous values of the sidebands and the carrier together is shown in Figure 10–9(b). The back cover of the text shows a color drawing of this waveform in the time domain; the carrier is shown in red, the lower sideband is green, the upper sideband is blue, and the resultant is black. Figure 10–10 is a composite drawing showing the resultant in both the frequency domain and the time domain. The modulated output voltage can be represented by the equation

EQUATION 10–3

$$v_t = A_c(1 + m_t) \cos \omega_c t$$

where v_t = instantaneous output voltage, V
A_c = amplitude of the carrier, V
m_t = instantaneous amplitude of the modulating signal, V
ω_c = carrier frequency, rad/s
t = time, s

A different way to view AM modulation is by phasor diagrams. Figure 10–11 shows the carrier and sidebands drawn as vectors. Assume that the carrier is rotating counterclockwise ω_c rad/s, and that the reference plane (the observer's

FIGURE 10–9

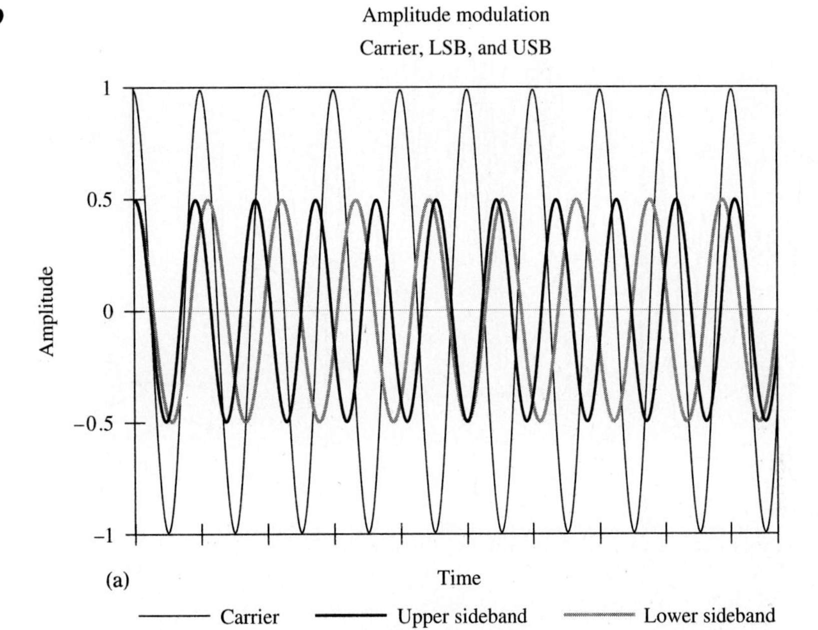
Amplitude modulation
Carrier, LSB, and USB
1
0.5
0
−0.5
−1
Amplitude
(a)
Time
Carrier
Upper sideband
Lower sideband

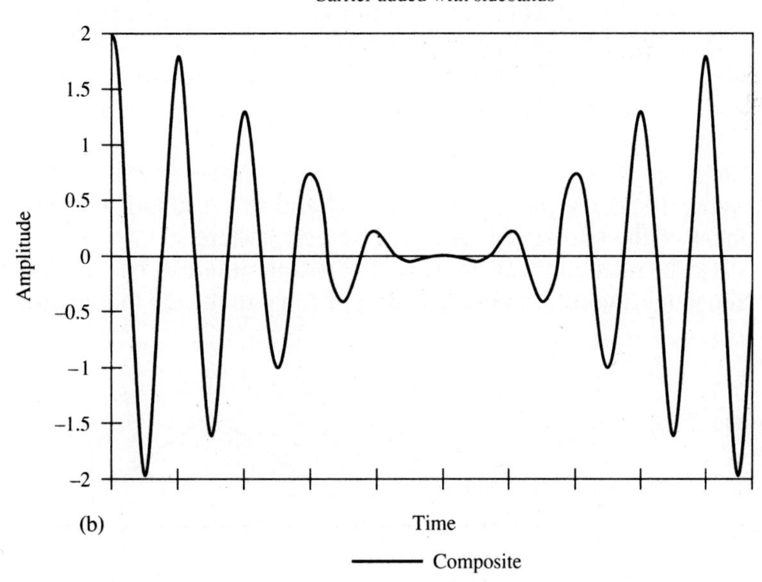
Amplitude modulation
Carrier added with sidebands
2
1.5
1
0.5
0
−0.5
−1
−1.5
−2
Amplitude
(b)
Time
Composite

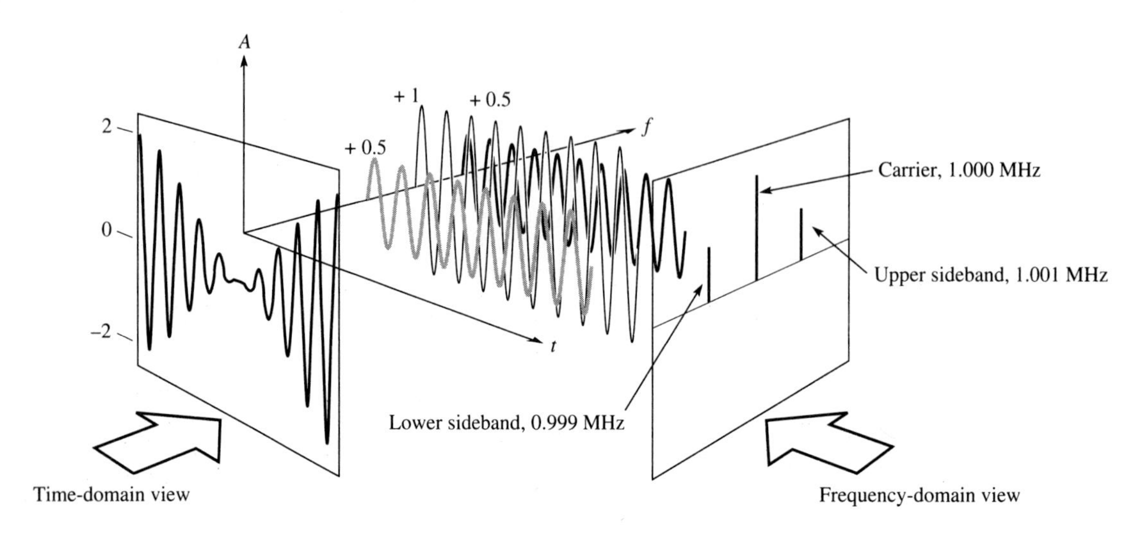

FIGURE 10–10
Time and frequency domain views of amplitude modulation.

view) is rotating at the same rate, causing the carrier vector to appear stationary. The sideband vectors will then appear to rotate with respect to the stationary tip of the carrier vector. The upper sideband is a higher frequency than the carrier; as a consequence, it will rotate in a counterclockwise direction. The lower sideband, with a lower frequency than the carrier frequency, will rotate in a clockwise direction. When the two sidebands move to a position such that they add in a straight line with each other, they will be either in phase or 180° out of phase with the carrier. The resultant phase (representing the output) will change length at a rate given by f_m, but it maintains the same phase as the carrier; this is shown in Figure 10–12 by plotting the carrier, and the resultant on the same graph. The phase of the two signals is the same everywhere.

The resultant waveform is of particular interest, because the beat is the same frequency and the amplitude is proportional to the the modulating signal, f_m. Even

FIGURE 10–11
Vectors for amplitude modulation.

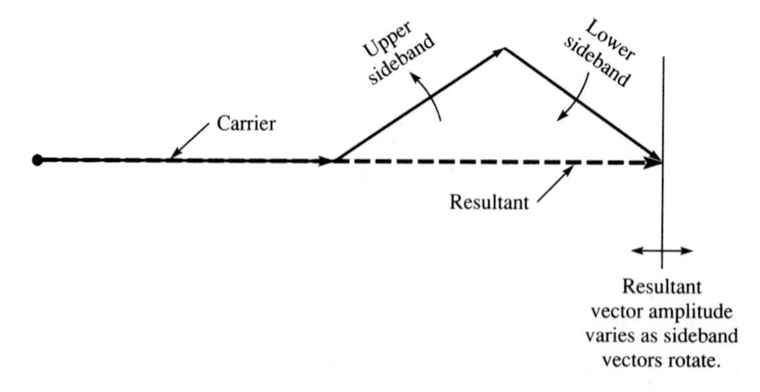

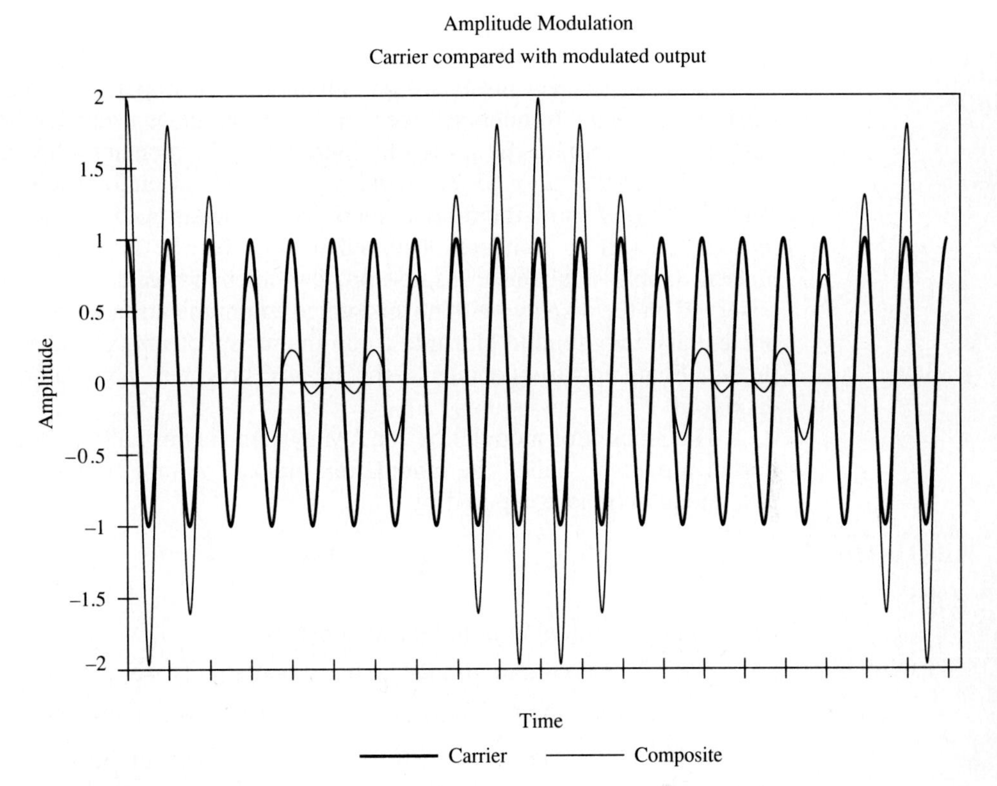

FIGURE 10–12
Amplitude modulation showing that the carrier is always *in phase* with the modulated composite waveform.

though the modulating signal, f_m, was not sent, it can be extracted from the beat as shown in Figure 10–13. The signal is passed through a diode rectifier that removes one half of the waveform. The original modulating waveform can be recovered from the envelope using a low-pass filter.

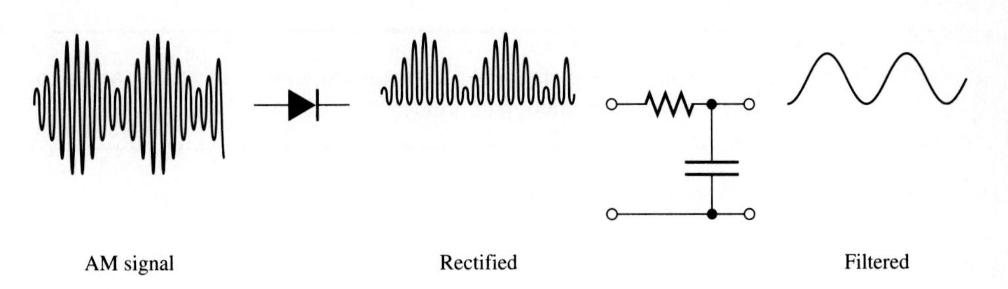

FIGURE 10–13
Recovering the modulation frequency f_m from an AM signal.

If the modulating signal is composed of several sinusoidal waveforms, the spectrum will have additional sidebands. Each frequency that is present in the modulating signal will produce two sidebands—an upper sideband and a lower sideband. If eight frequencies were used to modulate a carrier frequency, the output would contain the 8 original modulating frequencies, the carrier, and 16 sidebands, for a total of 25 frequencies. For received radio signals, the spectrum analyzer *cannot* show the original modulating signals because they are not transmitted; the signals that are observed depend on the type of transmitter. For normal double sideband transmission, the sidebands and carrier signals can be observed on the display; with normal voice communication, many frequencies are present and they tend to blur into a continuous spectrum. The analyzer will show the amplitude and frequencies of the signals; however, phase information is not displayed.

The amount of modulation for AM systems is normally expressed as a fraction or a percent, called the **modulation index.** The modulation index for amplitude modulation is computed by

EQUATION 10–4

$$M_{AM} = \frac{V_{max} - V_{min}}{V_{max} + V_{min}}$$

where M_{AM} = modulation index, dimensionless
V_{max} = maximum voltage of the modulated rf waveform, V
V_{min} = minimum voltage of the modulated rf waveform, V

V_{max} and V_{min} may be measured as either peak or peak-to-peak voltages, provided they are measured in a consistent manner.

The modulation index can also be computed directly from the frequency domain. In the frequency domain, the amplitudes of the upper and lower sidebands are added together and divided by the amplitude of the carrier. That is

EQUATION 10–5

$$M_{AM} = \frac{V_{USB} + V_{LSB}}{V_C}$$

where V_{USB} = upper sideband voltage, V
V_{LSB} = lower sideband voltage, V
V_C = carrier voltage, V

EXAMPLE 10–2 An rf-modulated waveform has a maximum voltage of 15 V_{pp} and a minimum voltage of 3 V_{pp}. Compute the modulation index.

SOLUTION Substituting into Equation 10–4

$$M_{AM} = \frac{V_{max} - V_{min}}{V_{max} + V_{min}}$$

$$= \frac{15 - 3}{15 + 3} = 0.667$$

Notice that this answer can be expressed as a percentage; that is, the percent modulation is 66.7%.

EXAMPLE 10–3 Compute the modulation index from the frequency domain display shown in Figure 10–14.

SOLUTION From Figure 10–14, the amplitude of each sideband is seen to be 4.0 V and the carrier is 10 V. Substituting into Equation 10–5 yields

$$M_{AM} = \frac{V_{USB} + V_{LSB}}{V_C}$$

$$= \frac{4\text{ V} + 4\text{ V}}{10\text{ V}} = 0.8 = 80\%$$

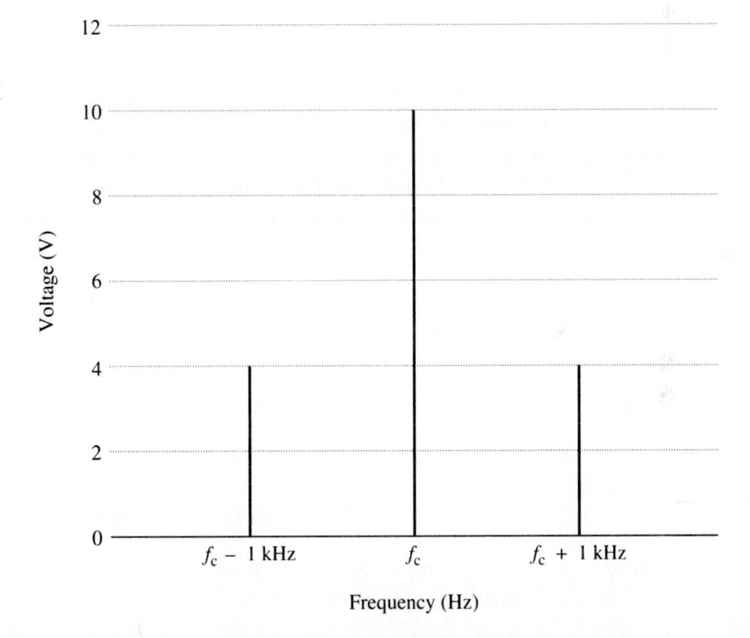

FIGURE 10–14 Frequency-domain representation of an AM signal.

The actual spectrum of a carrier modulated with a sine wave is more complicated than the example discussed because it will contain distortion products created by nonlinearities in the circuit and it will contain noise. Harmonic distortion occurs in any nonlinear circuit and creates multiples of the fundamental frequency; intermodulation distortion occurs in the mixing process when two or more signals are combined in a nonlinear mixer. Noise is present in all electronic systems. These distortion products and the noise add to the spectrum.

The examples discussed showed the spectrum for double-sideband (DSB) modulation. In another type of communication system, the carrier and one of the sidebands are removed completely, saving bandwidth. This is called single-sideband suppressed-carrier transmission.[3] As you have seen, the information content of the signal is completely contained within the sidebands, not the carrier. You can still recover the basic intelligence from the remaining sideband by mixing the received sideband frequency with an oscillator (appropriately called a *beat frequency oscillator*) to form a difference frequency. This process is common in shortwave radios specially designed for single-sideband reception. In order to recover the original modulating frequency, the beat frequency oscillator needs to be set to the proper frequency; otherwise the recovered signal can have too high or too low a pitch.[4]

FREQUENCY AND PHASE MODULATION

Frequency modulation transmits intelligence by swinging the output frequency back and forth at a rate determined by the modulating frequency. For example, if a carrier is modulated with a 1 kHz waveform, the output will swing back and forth 1000 times per second. Frequency modulation is a form of phase modulation; with phase modulation, the output phase deviates in both directions from the carrier by a certain amount. With either frequency or phase modulation, the output amplitude remains constant, an important advantage, since most interference causes the amplitude of the signal to change rather than the frequency. Unlike AM, the total output power remains constant, with or without modulation.

The equation used to describe PM is

EQUATION 10–6

$$v(t) = A \cos(\omega_c t + \phi)$$

where $v(t)$ = instantaneous voltage, V

A = amplitude, V

ϕ = added phase angle, which varies with f_m, rad

If the modulation frequency is sinusoidal, Equation 10–6 becomes

EQUATION 10–7

$$v(t) = A \cos(\omega_c t + \beta \sin \omega_m t)$$

where β = maximum phase deviation, usually referred to as the modulation index

$\omega_m = 2\pi f_m$, rad/s

MODULATION INDEX

Like amplitude modulation, phase modulation can be illustrated with a phasor diagram, as exemplified by Figure 10–15. If the phase deviation is small, then only two sidebands appear. When the two sidebands are in phase with each other, they are 90° from the carrier phase. The resultant phase-modulated output is the sum of

---

[3] In color television broadcast systems, the black-and-white portion of the signal is transmitted using AM; color information is transmitted in the sidebands only, and the carrier is suppressed.

[4] Because of this problem, single-sideband transmission is satisfactory for voice communication but not for music.

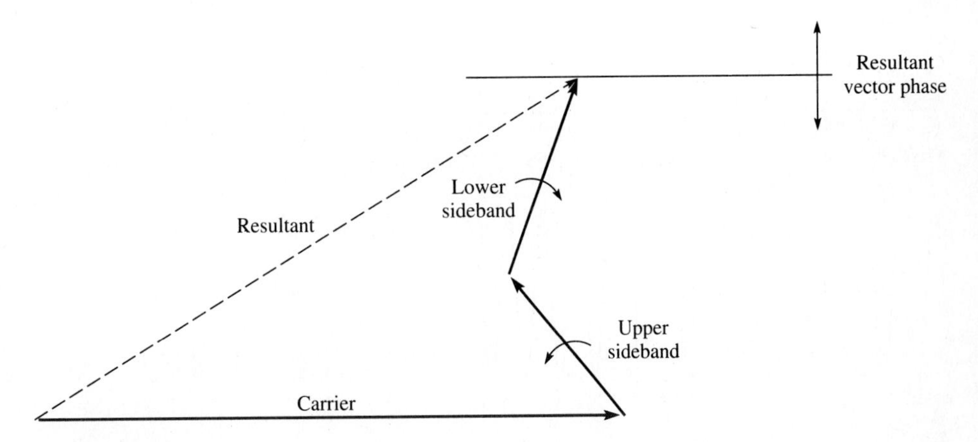

FIGURE 10–15
Vectors for phase modulation.

both sidebands and the carrier. As you can see, the resultant does not change amplitude; it changes only in phase because of the quadrature phasing of the sidebands. Figure 10–16 shows the result in the time domain; note the changing phase on the output signal.

Frequency modulation is the time derivative of a changing phase. Imagine a sinusoidal waveform that you would like to cause the phase to change. In order to change the phase, you could change the frequency for a short time and then return it to its original value. The original sinusoidal waveform now has the same frequency but a different phase. If the phase is to advance, the frequency must advance for a short time in order for the phase to catch up. As an analogy, phase is to velocity as frequency is to acceleration. If the phase is caused to deviate continuously because of sinusoidal modulation, the frequency will deviate in a like manner. There are other forms of frequency modulation that are not a derivative

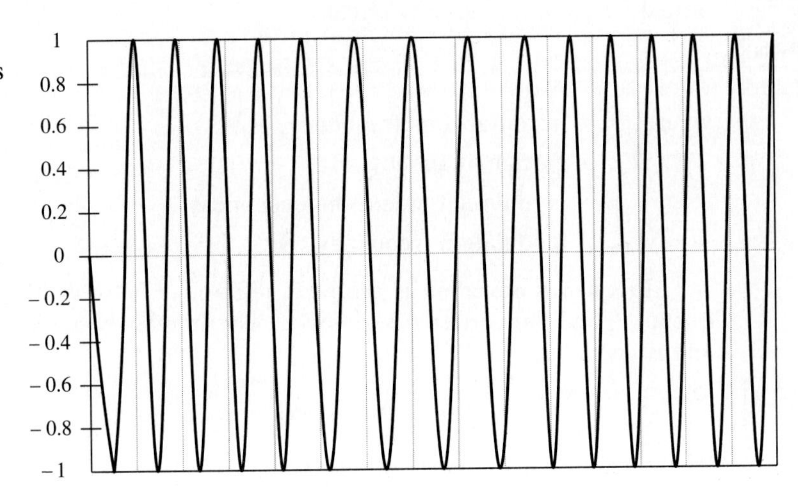

FIGURE 10–16
Phase change causes frequency change.

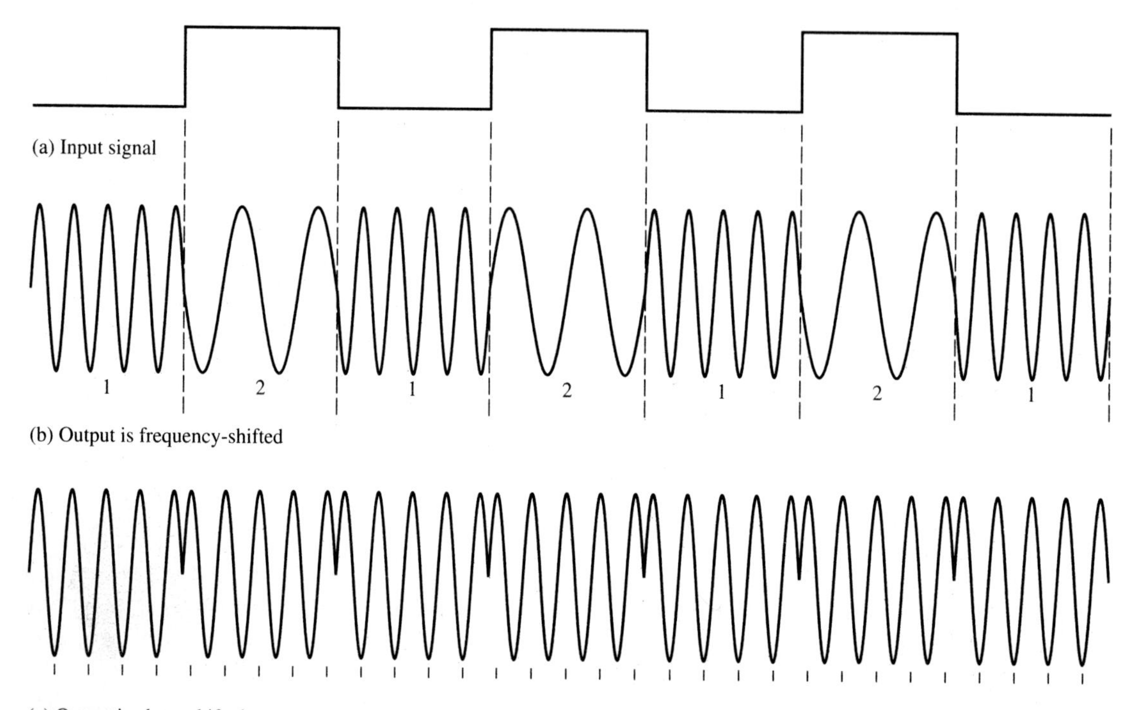

FIGURE 10–17
Example of frequency-shift keying (b) and phase-shift keying (c).

of the phase. Figure 10–17 shows frequency-shift and phase-shift keying methods where steady-state dc values of the modulating signal cause a change in the frequency (Figure 10–17(b)) or phase (Figure 10–17(c)).

Given a sinusoidal modulation signal, the equation for the instantaneous output frequency of an FM signal is

EQUATION 10–8

$$f_t = f_c + \beta f_m \cos \omega_m t$$

where f_t = instantaneous frequency, Hz
f_c = carrier frequency, Hz
β = modulation index, dimensionless
f_m = modulation frequency, Hz

Frequency deviation is the peak difference between the instantaneous frequency f_t and the carrier frequency f_c. The frequency deviation is defined as Δf and is given by

EQUATION 10–9

$$\Delta f = \beta f_m$$

where Δf = frequency deviation, Hz

Rearranging, the modulation index β is given by

EQUATION 10–10

$$\beta = \frac{\Delta f}{f_m}$$

The spectrum of an FM signal is much more complicated than the spectrum from AM modulation. An interesting aspect of FM is that the number of sidebands is not only a function of the number of frequencies used to modulate the carrier (as in AM) but is also a function of the frequency deviation. When a modulating signal causes the output frequency to deviate, sideband pairs (above and below the carrier) are created from the various phase combinations that result from the mixing process. In theory, an infinite number of sideband pairs are created; in practice, only a limited number have significant amplitudes. The number of significant sideband pairs is determined by a mathematical means known as Bessel functions. The significant sideband pairs are the ones necessary to ensure that the information is transmitted with good fidelity—a rule of thumb is to include those sidebands that add up to 98% of the power as significant. If $\beta << 1$, for a given modulation frequency only a single pair of sidebands is significant (contains at least 98% of the power). As β increases, the number of significant sideband pairs also increases. A good approximation is

EQUATION 10–11

$$n = \beta + 1$$

where n = number of significant sideband pairs

The sideband frequencies are symmetrical above and below the carrier frequency and depend on the amplitude of the modulating waveform. As the modulation depth increases (increased amplitude of the modulating wave), so does the number of significant sidebands. At full modulation, the sidebands are found at multiples of the modulation frequency. Thus if the modulation frequency is a constant 15 kHz and the carrier frequency is 100 MHz, sidebands are found at 100.015 MHz, 100.030 MHz, 100.045 MHz, and so forth. Likewise, sidebands are found below the carrier at 99.985 MHz, 99.970 MHz, 99.955 MHz, and so forth.

The bandwidth for an FM signal depends on the modulation index and the modulation frequency. The relationship can be written in equation form as

EQUATION 10–12

$$B = 2(\beta + 1)f_m$$

EQUATION 10–13

$$= 2(\Delta f + f_m)$$

where B = bandwidth, Hz

For commercial broadcasting, the Federal Communications Commission (FCC) sets the maximum frequency deviation, $\Delta f_{(max)}$, at 75 kHz. If we assume the maximum modulation frequency that is transmitted is limited to 15 kHz, the bandwidth can be computed from Equation 10–13 as

$$
\begin{aligned}
B &= 2(\Delta f_{(\max)} + f_{\mathrm{m}}) \\
&= 2(75 \text{ kHz} + 15 \text{ kHz}) = 180 \text{ kHz}
\end{aligned}
$$

The amplitude of the FM output signal is constant regardless of the amplitude of the modulation (phase or frequency). Unlike AM, in FM the amplitude of the output carrier *does* change in amplitude with β. In fact, the carrier completely vanishes for certain values of β (namely, 2.4048, 5.5201, 8.6654, 11.7915, and others). At these values, all the power is contained in the sidebands.

EXAMPLE 10–4 What is the modulation index for a commercial broadcasting FM station at the maximum deviation when the modulating frequency is (a) 1 kHz? (b) 50 Hz?

SOLUTION (a) The maximum deviation for commercial broadcasting is 75 kHz. From Equation 10–10

$$
\begin{aligned}
\beta &= \frac{\Delta f}{f_{\mathrm{m}}} \\
&= \frac{75 \text{ kHz}}{1.0 \text{ kHz}} = 75
\end{aligned}
$$

(b)

$$
\beta = \frac{75 \text{ kHz}}{50 \text{ Hz}} = 1500
$$

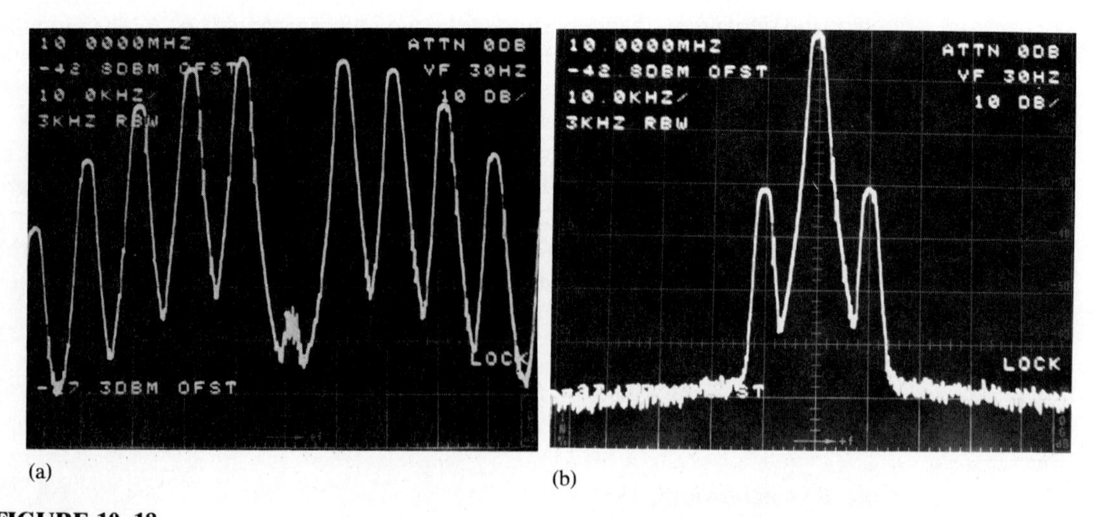

(a) (b)

FIGURE 10–18
Comparison of FM and AM with the same carrier frequency (10 MHz) and the same modulation frequency (10 kHz). Note that the FM signal has a much larger bandwidth.

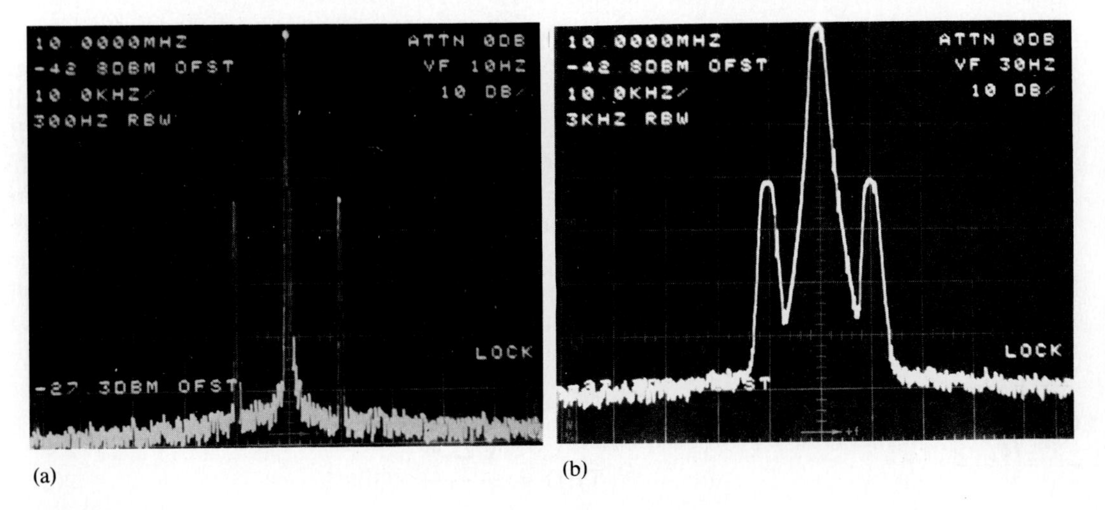

(a) (b)

FIGURE 10–19
Comparison of the effect of the resolution bandwidth (RBW) on the spectrum. Although the larger RBW produces a faster sweep, the sidebands are broader and the noise floor is higher. This shows that it is more difficult to detect harmonics or spurious frequencies if the RBW is larger.

At certain values of the modulation index, sidebands have 180° phase reversal and are given mathematically as a negative quantity. Since the spectrum analyzer does not indicate phase relationships, the sidebands appear as positive values on the analyzer. An FM spectrum photographed from the display of a spectrum analyzer is shown in Figure 10–18(a). Compare this spectrum to the AM spectrum shown in Figure 10–18(b), which has the same carrier and modulation frequency. Figure 10–19 shows the effect of the resolution bandwidth control on the analyzer for the AM spectrum. Notice that the sidebands are poorly resolved and the noise floor is higher when the resolution bandwidth is made larger.

DISTORTION MEASUREMENTS

There are several kinds of distortion that are easily indicated by a spectrum analyzer but are nearly impossible to see on an oscilloscope (except in very severe cases). These include harmonic distortion, intermodulation distortion, and transient intermodulation distortion.

Harmonic distortion is characterized by the appearance of lines in the output spectrum at multiples of the fundamental frequency when the input is a pure sinusoidal waveform. It occurs when a transducer, amplifier, or other portion of the signal path has a nonlinear response. Unless harmonic distortion is very large, the oscilloscope does not readily show it; however, a spectrum analyzer can easily show the harmonic components. Figure 10–20 illustrates harmonic distortion. By directly measuring the amplitude of the harmonics, the amount of harmonic distortion can be determined. The total harmonic distortion is a specification frequently quoted for amplifiers and other equipment such as tape recorders. **Total harmonic distortion** (THD) is the ratio of the rms value of all harmonics to the rms value of the total signal. Since the harmonic portion of the total signal is normally very small, THD can be expressed as a ratio of the rms value of the harmonics to the rms value of the fundamental. It is frequently expressed as a percentage:

EQUATION 10–14

$$\text{THD} = \frac{\sqrt{\sum_{i=1}^{n} v_i^2}}{v_f} \times 100\%$$

where THD = total harmonic distortion, %
v_i = rms value of ith harmonic, V
v_f = rms value of fundamental, V

For tape recorders, the distortion-free operating range is considered to be the region for which the output signal level is less than 3% THD. When the recorder is operated at higher levels, the distortion rises dramatically because of tape saturation.

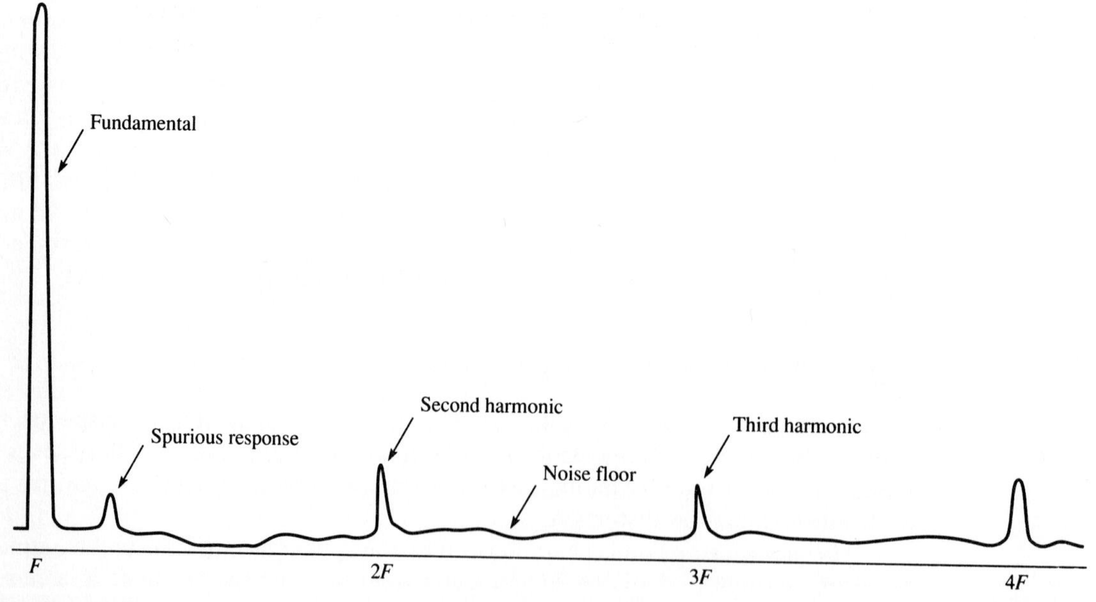

FIGURE 10–20
Harmonic distortion.

EXAMPLE 10–5 Compute the THD of a signal in which the fundamental has an rms value of 20 V and the first four harmonics have the following rms voltages: 70 mV, 45 mV, 25 mV and 10 mV. Assume the remaining harmonics are negligible.

SOLUTION From Equation 10–14

$$\text{THD} = \frac{\sqrt{v_1^2 + v_2^2 + v_3^2 + v_4^2}}{v_f} \times 100\%$$

$$= \frac{\sqrt{70\text{ mV}^2 + 45\text{ mV}^2 + 25\text{ mV}^2 + 10\text{ mV}^2}}{20\text{ V}} \times 100\%$$

$$= 0.44\%$$

Intermodulation distortion is a nonlinear distortion characterized by the appearance of sum and difference frequencies developed from the input frequency components. It can be observed when two pure sinusoidal waveforms are input to an amplifier and one modulates the other. In a perfect linear amplifier, the two signals do not affect each other. In actual amplifiers, nonlinearities cause cross modulation to occur—for a quality linear amplifier it can be less than 0.1%. Intermodulation distortion can be measured by a spectrum analyzer by connecting the analyzer to the output of the unit to be tested while feeding two pure tones to the input. The analyzer will show the pure tones as two lines; the intermodulation distortion will show up as small lines at the sum and difference frequencies and may also be indicated at harmonics of the sum and difference frequencies, although harmonic components are usually not considered to be part of intermodulation distortion.

Intermodulation distortion in TV transmitters produces loss of picture fidelity at the receiver. A test of intermodulation distortion is to drive the transmitter with two pure sinusoidal test signals connected through a resistive network and look at the spectrum for sum and difference components of intermodulation distortion. The spectrum can be picked up by the analyzer using an antenna on the analyzer; otherwise, the analyzer is connected directly through a coupler to the transmitting antenna's transmission line. The intermodulation components should be −50 dBc. (The dBc is the number of decibels below the carrier level.)

Transient intermodulation distortion (TIM) is distortion that occurs in negative feedback amplifiers during input transients such as those occurring in musical passages. TIM is due to the inability of the feedback circuit to respond instantaneously to a transient waveform applied to the input, creating momentary intermodulation distortion. During transients, the negative feedback correction signal is slightly delayed because of the transit time through the amplifier, as illustrated in Figure 10–21. For a very short period of time, the input is out of sync with the feedback signal, and the amplifier responds to the feedback signal that existed just prior to the transient. If a large dynamic change occurs in the input, the output changes rapidly before the correction signal arrives, giving rise to distortion. TIM

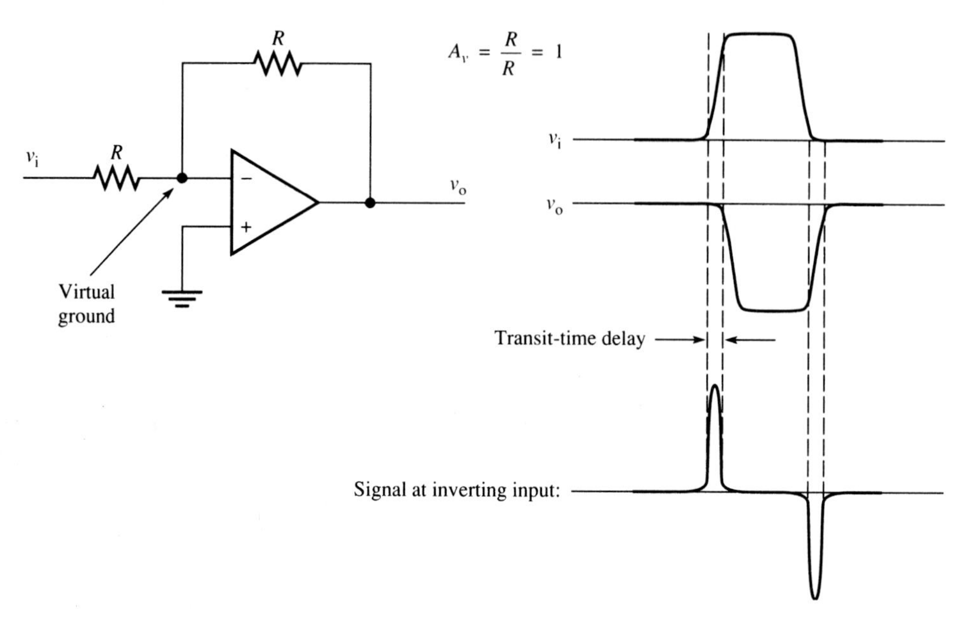

FIGURE 10–21
Transient intermodulation distortion. The signal at the inverting input is normally very tiny due to the virtual ground. During transients, this signal can be larger due to propagation delay through the amplifier. The large transient signal is amplified by the open-loop gain of the amplifier.

can be observed on a spectrum analyzer when a sine-wave signal is mixed with a lower-frequency square-wave signal.

10–5 THE WAVEFORM RECORDER

The waveform recorder stores a digital record of a waveform in memory for later manipulation or analysis. It is essentially the front end of a digital oscilloscope, containing a sample-and-hold circuit followed by an A/D converter and memory. Data stored in memory can be sent slowly to a chart recorder or converted back to analog data and viewed repetitively on a conventional analog oscilloscope. The waveform recorder is capable of capturing several types of difficult-to-measure waveforms, including nonrepetitive waveforms, fast-rising but low-frequency waveforms, and waveforms that are fast and repeat at random intervals. These types of waveforms are difficult to record using conventional analog oscilloscopes and chart recorders. The analog oscilloscope is easiest to use with relatively fast repeating waveforms. Digital-storage oscilloscopes are also capable of capturing these types of waveforms as discussed in Chapter 8. As in the case of the digital oscilloscope, the waveform recorder has various triggering modes that allow it to store waveforms before the trigger point. It does this by continuously recording the input signal, discarding the oldest data, until the trigger is detected. At the trigger time, only part of the oldest data is discarded, leaving in the record the data that occurred just prior to the trigger point as well as data after the trigger.

10–6 THE DISTORTION ANALYZER

The distortion analyzer is an instrument that consists of a high-Q (narrow-bandwidth) notch filter followed by a broadband amplifier and detector, as shown in Figure 10–22. A distorted sinusoidal waveform consists of noise and/or harmonics. Harmonics are present in all nonsinusoidal waves; a perfect sinusoidal waveform has no harmonics. If a "perfect" sinusoidal wave is passed through a nonlinear amplifier, the output will contain harmonics. The harmonics and circuit noise are a measure of the distortion added by the amplifier.

There are two common methods of measuring distortion using a distortion analyzer. In both methods, a sinusoidal waveform is fed into the amplifier under test. The first method measures the total bandwidth voltage (rms), including the fundamental, noise, and any harmonic terms that exist in the output; then the fundamental is filtered out and the voltage is measured again. The ratio of the two measurements is computed, converted to a percent, and displayed as the distortion. The second method uses the notch filter to exclude the fundamental from the remaining output signal. The rms voltage of the signal, exclusive of the fundamental, is measured. Then the notch filter is converted into a narrow-bandpass filter and is used to tune only the fundamental. The rms voltage of the fundamental is

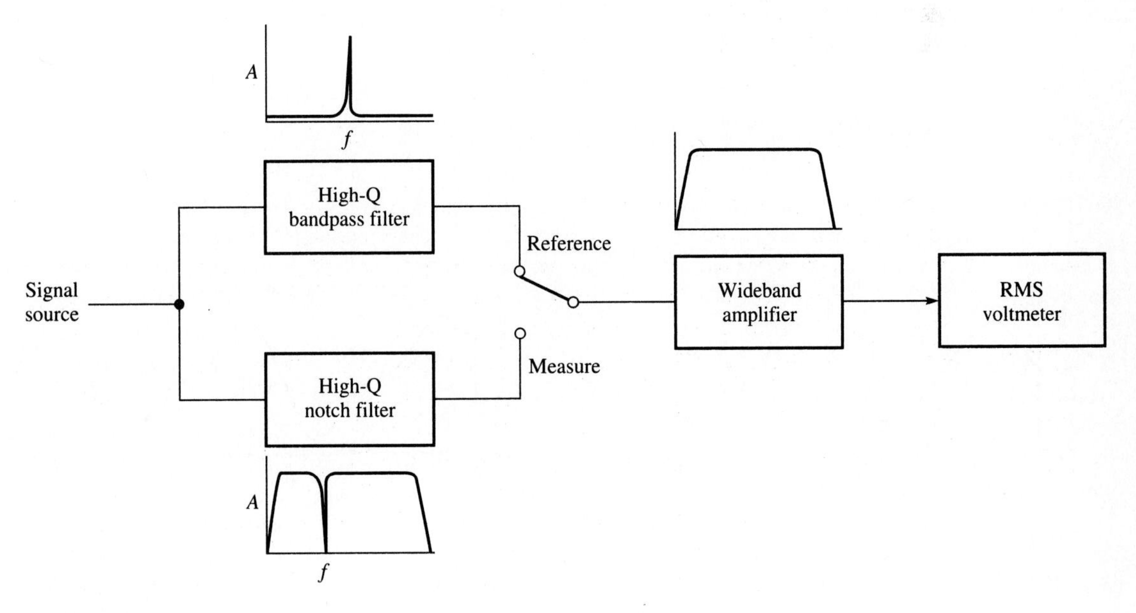

FIGURE 10–22
Block diagram of a distortion analyzer.

measured. The distortion is computed from the ratio of the output without the fundamental to the fundamental and converted to a percent.

You might wonder what a signal containing a small amount of harmonic distortion looks like on an oscilloscope. Unfortunately, unless a significant amount of distortion is present in a signal, it is difficult to detect on an oscilloscope. Consider the harmonic distortions illustrated in Figure 10–23. In Figure 10–23(a), the second harmonic is present at the 5% level. The distortion is nearly impossible to see in the time domain. Even at the 10% level (shown in Figure 10–23(b)), it is difficult to see and would normally escape notice. At the 20% and 40% levels, the distortion is apparent but cannot be quantitatively measured on an oscilloscope. The frequency domain view of these signals will readily show the distortion.

As another example, consider the badly distorted sinusoidal waveform shown in Figure 10–24. The distortion shown consists of harmonics and noise; however, the nature of the distortion is difficult to discern in the time domain. The spectrum analyzer reveals the spectral characteristics as shown in Figure 10–25(a). Figure 10–25(b) shows the response of the notch filter. If the signal is passed through the notch filter, the spectrum is the same as before except the fundamental has been removed. The broadband voltage measurement in Figure 10–25(c) is compared to the total broadband voltage in Figure 10–25(a) to deter-

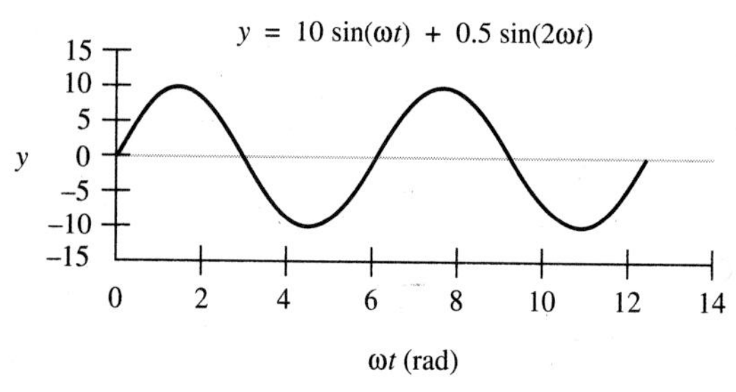

(a) 5% distortion.

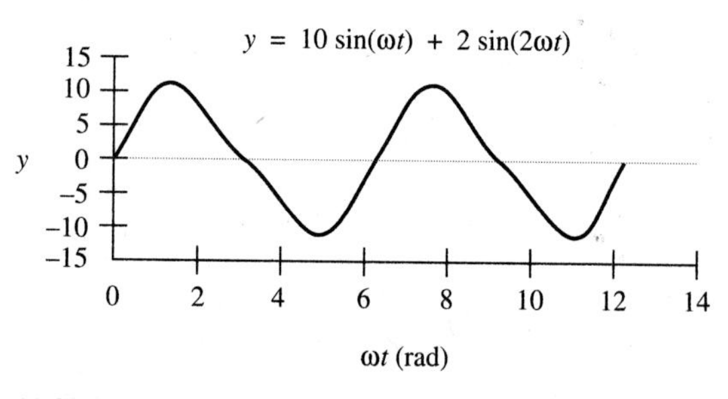

(c) 20% distortion.

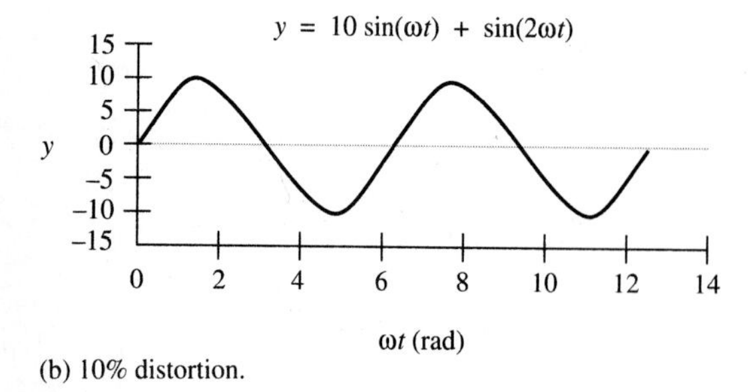

(b) 10% distortion.

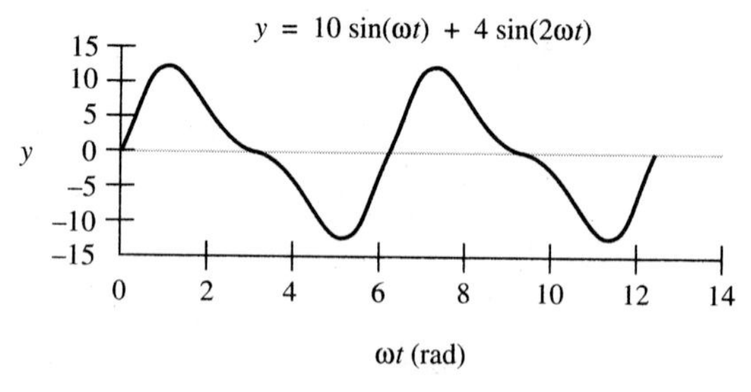

(d) 40% distortion.

FIGURE 10–23
Sine waves with harmonic distortion.

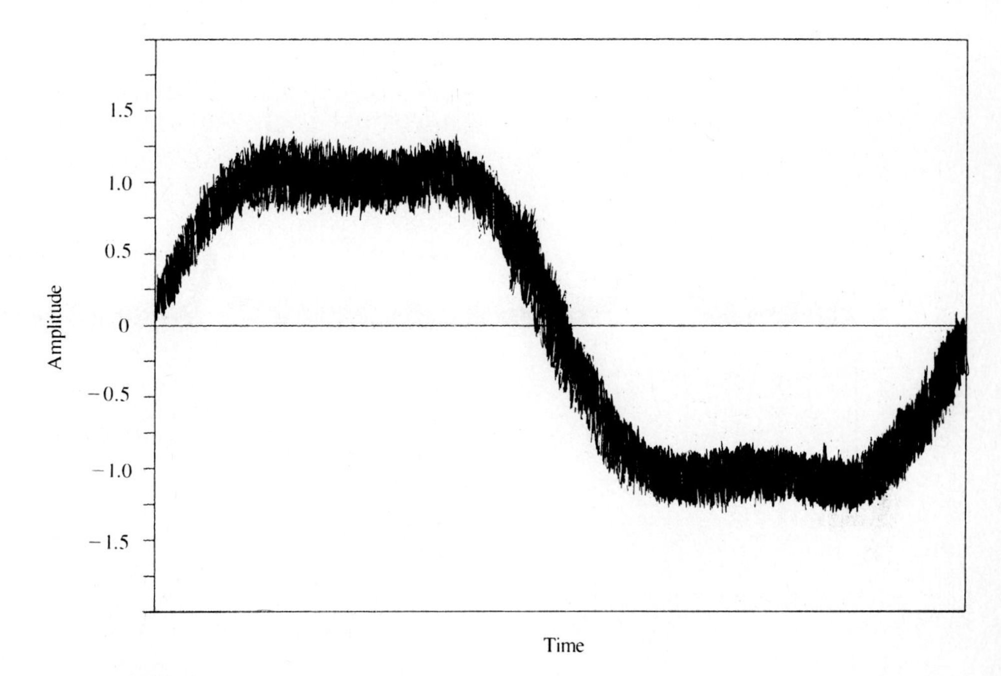

FIGURE 10–24
A noisy signal harmonic distortion.

mine the fraction of the sinusoidal signal that represents distortion. The result is normally expressed as a percentage.

The trick to making accurate measurements with a distortion analyzer is to keep the fundamental frequency and the notch filter frequency the same. To achieve accuracy, many distortion analyzers contain the signal source and also perform the analyzer function. By turning off the signal source, the analyzer can measure the noise content.[5] The noise content is frequently expressed in terms of the signal-to-noise ratio in decibels:

EQUATION 10–15

$$S/N = 20 \log \frac{V_f}{V_n}$$

where S/N = signal-to-noise ratio, dB
V_f = rms voltage of the fundamental, V
V_n = rms voltage of the noise, V

[5] Harmonics of the signal will disappear if the signal source is shut off, leaving just the noise.

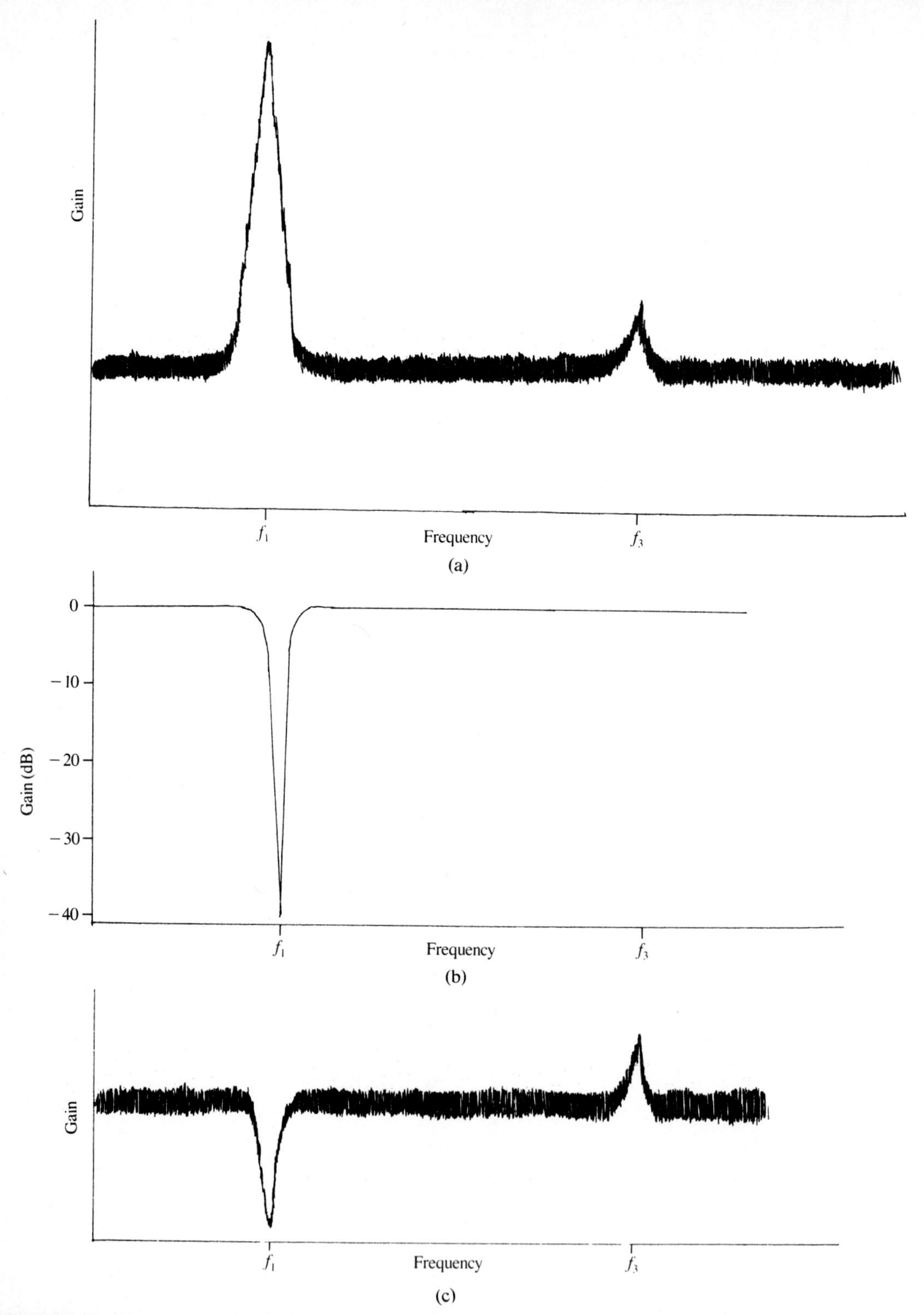

FIGURE 10–25
Frequency response of the notch filter.

EXAMPLE 10–6 A distortion analyzer is set up to make a measurement of the signal-to-noise ratio in a power amplifier. The output fundamental voltage is measured and found to be 450 V rms. The signal source is turned off and the output voltage is measured and found to be 70 mV rms. Compute the signal-to-noise ratio.

SOLUTION From Equation 10–15

$$\begin{aligned} S/N &= 20 \log \frac{V_f}{V_n} \\ &= 20 \log \frac{450\ \text{V}}{0.070\ \text{V}} \\ &= 76\ \text{dB} \end{aligned}$$

10–7 THE NETWORK ANALYZER

Network analyzers use stimulus-response testing to measure various characteristics of active and passive linear networks. Figure 10–26 shows a simplified block diagram of a network analyzer. It consists of a stimulus instrument (usually a sweep oscillator), a signal splitter, and a measurement instrument (generally a spectrum analyzer). The output is a frequency-dependent characteristic of the network, such as the network's transfer function. In addition to frequency domain presentations, some network analyzers permit the data to be transformed into the time domain for analysis. Figure 10–27 shows a network analyzer.

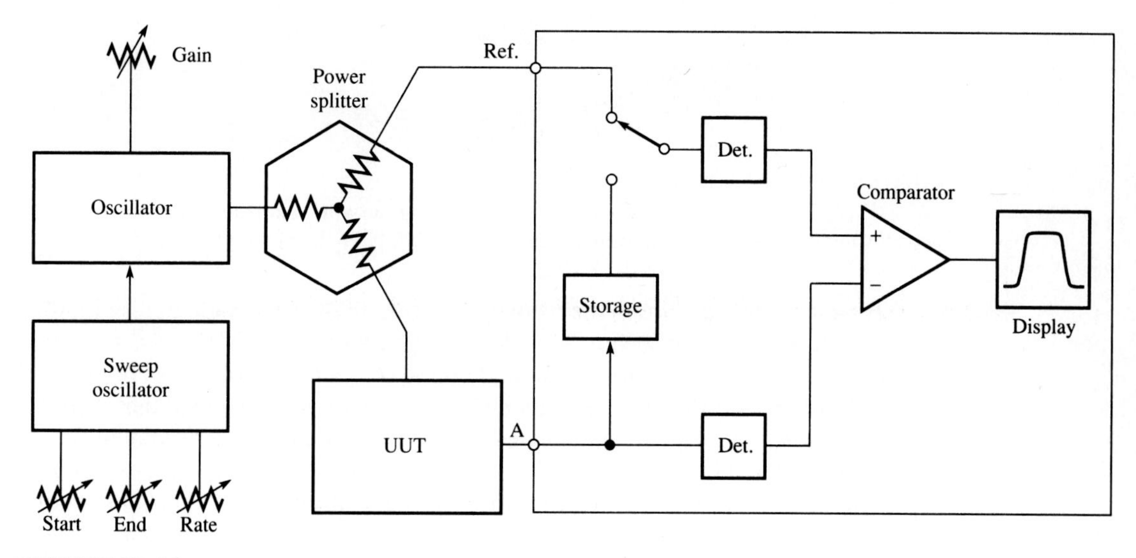

FIGURE 10–26
Block diagram of network analyzer.

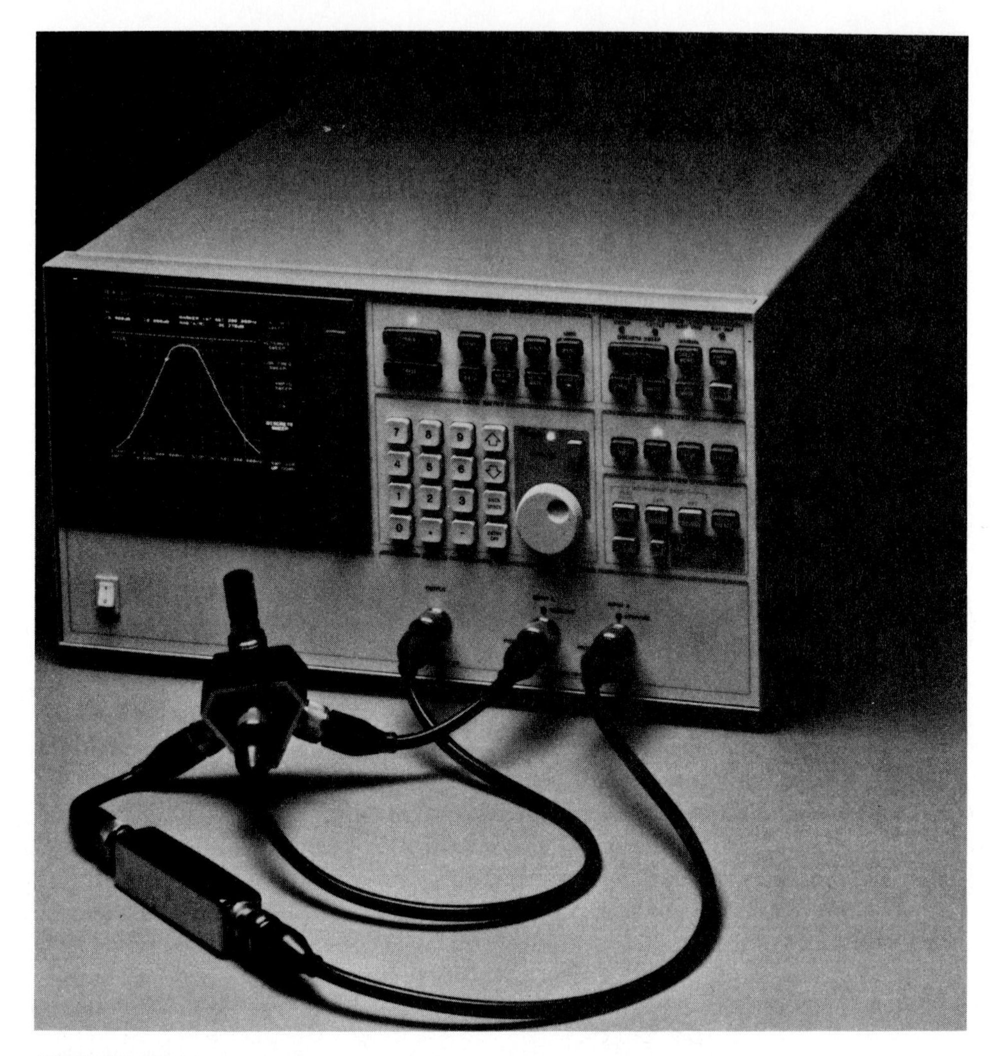

FIGURE 10–27
The HP 3577B network analyzer combines precision network analysis with high-throughput device evaluation (courtesy of Hewlett-Packard Co.).

Examples of measurements using a network analyzer include determination of parameters such as insertion loss, standing wave ratio (SWR), transfer function, complex impedance characteristic, and the frequency or phase response of a linear circuit. Most often, the signal source is a sinusoidal waveform that is made to sweep through a range of frequencies; however, in some cases, it may be set to one particular frequency. Measurement of the transfer or impedance characteristics is important for radio frequency and microwave circuits. For this application, the analyzer samples the input excitation voltage and the input current using a current transformer. The impedance is found by dividing the input voltage by the input current at each frequency.

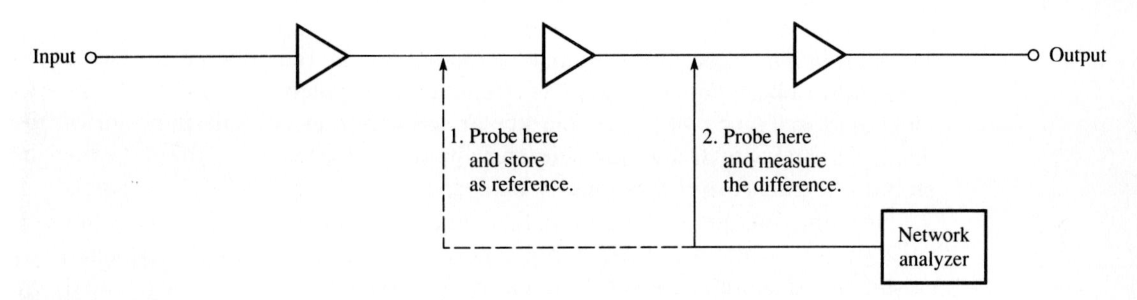

FIGURE 10–28
Using a network analyzer to measure in-circuit gain and phase.

Frequency or phase response measurements can be done in a similar manner. The measurement instrument has two dual-purpose detectors (gain and phase) and a differential amplifier. One detector is used to measure a signal path, and the other is used to measure a reference path. The reference path does not have to be an actual physical path but can be simulated by stored data taken from an earlier reading. The output levels between the paths are compared to determine the differences in gain or phase. One application for the use of stored data for the reference data is in comparing two circuits with each other or comparing the output to the input of a cascaded amplifier system, as shown in Figure 10–28. The

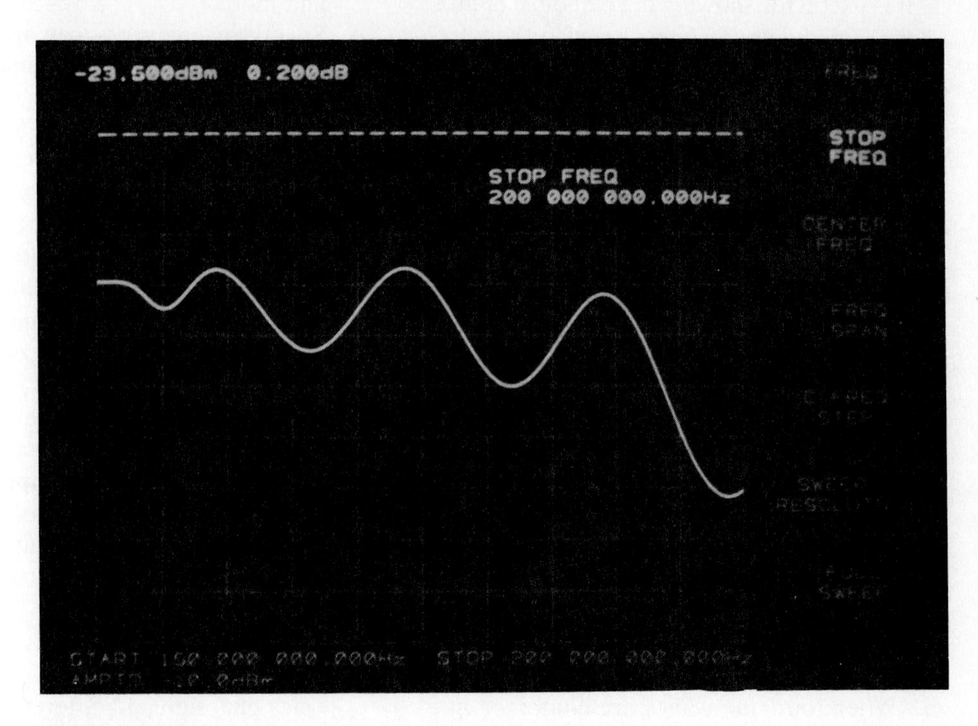

FIGURE 10–29
Display on network analyzer.

output level is held to a constant amplitude as the generator frequency is changed. The start and stop frequencies can be adjusted, as can the sweep rate. The amplitude and/or phase data is plotted as a function of frequency on a CRT display, as illustrated in Figure 10–29. A marker or markers can be superimposed on the display to indicate frequency reference points. Alternatively, the data may be stored in digital form for computer analysis.

Another method of generating the stimulus for a network analyzer is to use a noise source. White noise (equal noise power per hertz) causes all frequencies to be stimulated simultaneously. Using a white-noise source, the spectrum analyzer is made to sweep over a range of frequencies of interest several times and the results are averaged. The frequency response of the network can be observed as that portion of the spectrum that passes the noise. A periodic random noise (PRN) source is an artificially generated source that repeats a sequence exactly. It can be tailored to produce noise in the spectral region of interest, resulting in a faster measurement.

SUMMARY

1. The spectrum analyzer shows the frequency domain representation of a signal; it is useful for showing the response of mixers, modulators, filters, and other frequency-dependent circuits. The input frequency is plotted on the horizontal axis of a CRT, whereas the vertical axis represents the amplitude of the input signal.
2. The swept-tuned spectrum analyzer uses a voltage-controlled local oscillator to effectively scan the input signal in the frequency domain. The real-time analyzer uses multiple filters to observe all input frequencies simultaneously.
3. For unequal signals, frequency resolution is specified by the resolution bandwidth and the bandwidth selectivity.
4. In communication systems, spectrum analyzers can be used to measure modulation levels, modulation index, zero-modulation carrier power and frequency, frequency response, distortion, signal-to-noise levels, and more.
5. Spectrum analyzers can measure various types of distortion, including harmonic distortion, intermodulation distortion, and transient intermodulation distortion.
6. The wave analyzer can tune a narrow passband, which enables it to be tuned to a single frequency component of a signal while rejecting other components.
7. The distortion analyzer indicates the energy in a range of frequencies outside the specific frequency band occupied by the signal.
8. Network analyzers can be used to measure insertion loss, standing wave ratio (SWR), transfer function, complex impedance characteristic, and the frequency or phase response of a linear circuit.

QUESTIONS AND PROBLEMS

1. When four sine waves are combined in a nonlinear mixer, the spectrum will contain 16 principle components (original frequencies plus sum and difference frequencies). List the 16 components.
2. Assume that a spectrum analyzer can

show signals over an 80 dB voltage range by selecting a logarithmic scale factor of 10 dB/div over 8 div. Compare this dynamic range with that of a linear scale that also has 8 div. Assume the minimum signal that can be measured is 0.1 div.

3. Assume a 2.00 MHz carrier is amplitude-modulated with a 50 kHz modulation frequency.
   (a) What are the frequencies of the sidebands?
   (b) What happens to the sidebands if the amplitude of the 50 kHz signal is increased?
4. A spectrum analyzer is used to observe an amplitude modulated waveform. The signal is a carrier that is modulated by a steady sinusoidal waveform. The analyzer indicates that the carrier is 12 dB above the level of the sidebands.
   (a) Compute the voltage ratio of the carrier to the sideband level.
   (b) What is the modulation index?
5. Explain what is meant by the term modulation index for frequency modulation.
6. Assume an FM station with a 2.00 MHz carrier frequency is testing with a 3 kHz modulating signal. The deviation is 15 kHz.
   (a) What is the modulation index?
   (b) How many significant sidebands are there?
   (c) What are their frequencies?
7. (a) Describe the spectrum analyzer display of a 5 kHz sine wave used to amplitude-modulate a 1.0 MHz carrier.
   (b) Describe the spectrum analyzer display of a 5 kHz sine wave used to frequency-modulate a 1.0 MHz carrier if the deviation is 75 kHz.
8. For the signal shown in Figure 10–30, determine the modulation index.
9. Sketch the frequency domain view of each signal shown in Figure 10–23.
10. Assume two sidebands have 180° phase reversal and are given mathematically as a negative quantity. What polarity will the sidebands have on a spectrum analyzer? Why?

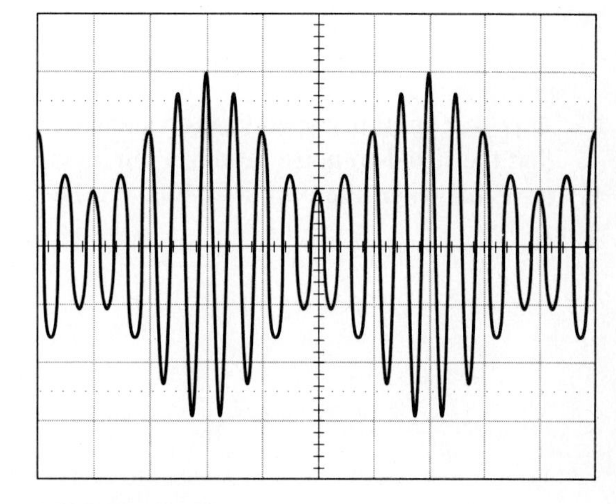

FIGURE 10–30

11. Explain the difference between the SPAN, sweep RATE, and RESOLUTION BANDWIDTH controls on a spectrum analyzer.
12. Explain how you would use a spectrum analyzer to measure the total harmonic distortion of an amplifier.
13. Compute the THD of a signal in which the fundamental has an rms value of 10 V and the first five harmonics have the following voltages: 100 mV, 15 mV, 65 mV, 10 mV, and 35 mV. Assume the remaining harmonics are negligible.
14. What is intermodulation distortion? What would you expect to see on the display of a spectrum analyzer if intermodulation distortion were present in a test circuit when you connected two pure sine waves to the input of the test circuit?
15. Assume a sinusoidal waveform was sent on a single-sideband suppressed-carrier transmitter.
    (a) Describe the signal you would expect to see on a spectrum analyzer.
    (b) If the frequency of the sine wave was slowly varied, what would you expect to see?

16. Explain how a distortion analyzer can measure the signal-to-noise ratio of an amplifier.
17. A power amplifier specification indicates that the signal-to-noise ratio is a minimum of 68 dB when the output is 1000 V. The gain is adjusted for this output and then the signal source is turned off. What is the maximum allowable output voltage when the source is turned off?
18. How does a network analyzer measure the frequency response of a circuit?

Chapter 11

Digital Analysis Instruments

OBJECTIVES

Digital analysis instruments can be divided into two categories: (1) those that convert an analog quantity into a digital quantity in order to process and display it as a discrete value and (2) specialized instruments that are designed specifically for analysis of digital circuits. Examples of digital instruments that can be used to convert an analog quantity into a digital quantity are the DMM and the DSO. An instrument that is designed for analysis of digital circuits is the logic analyzer.

As microcomputers and other digital circuits have gained in popularity and importance, the logic analyzer is the digital instrument that has become as important to monitoring digital systems as the oscilloscope is to analog circuits. It is capable of showing the simultaneous time relationships between multiple digital channels, an important advantage because digital systems contain multiple data, address, and control lines. The grouping of these lines into digital words and bytes simplifies the task of assimilating the information and helps pinpoint any problem with an individual line.

In this chapter, you will study the universal electronic counter, the logic analyzer, and hand-held digital instruments—the logic pulser, current tracer, and logic probe. When you complete this chapter, you should be able to

1. Explain the block diagram of a universal electronic counter and discuss its principle features and operation.
2. Describe common operational modes of the universal electronic counter, including the frequency, period, frequency ratio, time-interval, and event-counting modes.
3. Discuss methods used in microwave counters for converting very high frequencies to lower frequencies for digital circuits.
4. Discuss the input and operational specifications for counters.
5. Compare logic analyzers and oscilloscopes; cite advantages of each for digital circuit analysis.
6. Explain the block diagram of a logic analyzer and discuss its principle features and operation.
7. Explain how a logic analyzer can be used as a timing analyzer and as a state analyzer.
8. Compare asynchronous and synchronous sampling techniques for logic analyzers.

HISTORICAL NOTE

In the late 1960s, digital circuits of increasing complexity were difficult to troubleshoot with the existing instrumentation. The oscilloscope, which had been the principle instrument for analog design, had certain limitations for digital circuits; for example, it was difficult to capture a particular sequence of logic signals from a long sequence. In 1973, Hewlett-Packard announced the Model 5000A logic analyzer, the first digital instrument designed entirely for displaying logic signals. The 5000A analyzer had two input channels and could display 32 bits of digital data on each of two rows of red LEDs. Special trigger conditioning enabled the user to select a 3-bit combination of HIGHs or LOWs as the trigger word. The analyzer was also capable of serving as a triggering source for an oscilloscope to enable the scope to display the desired digital sequence.

11–1 ELECTRONIC COUNTER MEASUREMENTS

INTRODUCTION

Electronic counters can measure frequency, period, frequency ratios, and time intervals and do event counting. Counters designed for these various functions are called **universal electronic counters.** Essentially, electronic counters operate by counting the number of events that occur during a precisely controlled time interval. A high-performance universal counter is shown in Figure 11–1. The desired function is selected by the user from a front-panel control. The counter shown has other capabilities not found in lower-quality units, such as the capability to mea-

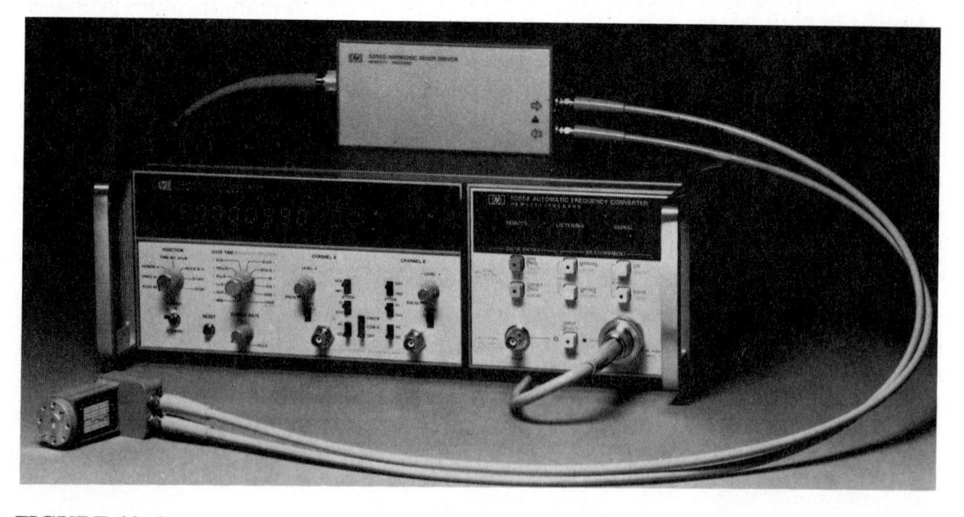

FIGURE 11–1
A high-performance universal electronic counter, the HP-5345A. The counter is shown with an optional frequency converter that can extend the frequency range of CW and pulse measurements to 100 GHz (courtesy of Hewlett-Packard).

sure rise and fall times, pulse width, and duty cycle and to perform statistical and other data-reduction calculations.

Universal counters cannot operate at frequencies above about 1 GHz because of the speed limitation of digital circuits. At very high frequencies, special conversion techniques are necessary to bring down the input frequency to one suitable for the digital circuits in the counter. Instruments that include the down converter within the basic instrument are primarily designed for microwave measurements; accordingly, they are called **microwave-frequency counters.**

A feature found on high-end frequency counters is an arithmetic processor that enables post-measurement data manipulation. The processor can do mathematical operations such as finding the standard deviation of a series of measurements or computing the average. If the input is from a transducer and represents a physical quantity such as speed, temperature, or flow rate, the arithmetic processor can provide the appropriate unit conversions, offset, and normalizing factors to give a direct indication of the physical quantity. As in many modern instruments, counters are available that can be controlled from a computer or controller using the IEEE-488 bus. With the bus, the measurement can be programmed and remotely controlled from the computer; the data is then processed by the counter and sent to the computer.

BLOCK DIAGRAM

The block diagram of a universal electronic counter operating in the frequency mode is shown in Figure 11–2. The signal to be measured is applied to input signal–conditioning circuits, which converts the input into a digital signal that is compatible with the rest of the internal counter circuits. The purpose of the input signal–conditioning circuits is to "clean up" the input signal and provide pulses of the correct amplitude and shape for the remaining digital circuits in the counter. The input signal–conditioning circuits contain a high impedance amplifier to prevent loading of the circuit under test. It also contains an attenuator and a Schmitt trigger circuit (explained in Section 5–7) that prevents counting errors when the input signal is noisy. Sometimes the Schmitt trigger thresholds can be controlled by the user. The input attenuator is normally set to attenuate the signal to the lowest level that will satisfactorily trigger the counter.

The trigger level and slope are among the controls in the input group. In addition, the signal can be coupled with ac or dc coupling, as in the case of an oscilloscope. The proper setting of the trigger-level control is important. If it is set too high, no pulses are counted or variations at the top of the input signal can produce extra pulses sent on to the rest of the counter. If the trigger level is set too low, noise can cause extra pulses. Some electronic counters have special circuits that can automatically set the trigger level at the 50% level of the input signal.

Depending on the measurement function selected, the control circuit turns on the main gate to send pulses to the decimal counting unit. The decimal-counting unit is essentially a series of base 10 counters with the appropriate logic to place the decimal place. The output of the decimal counting unit is connected to

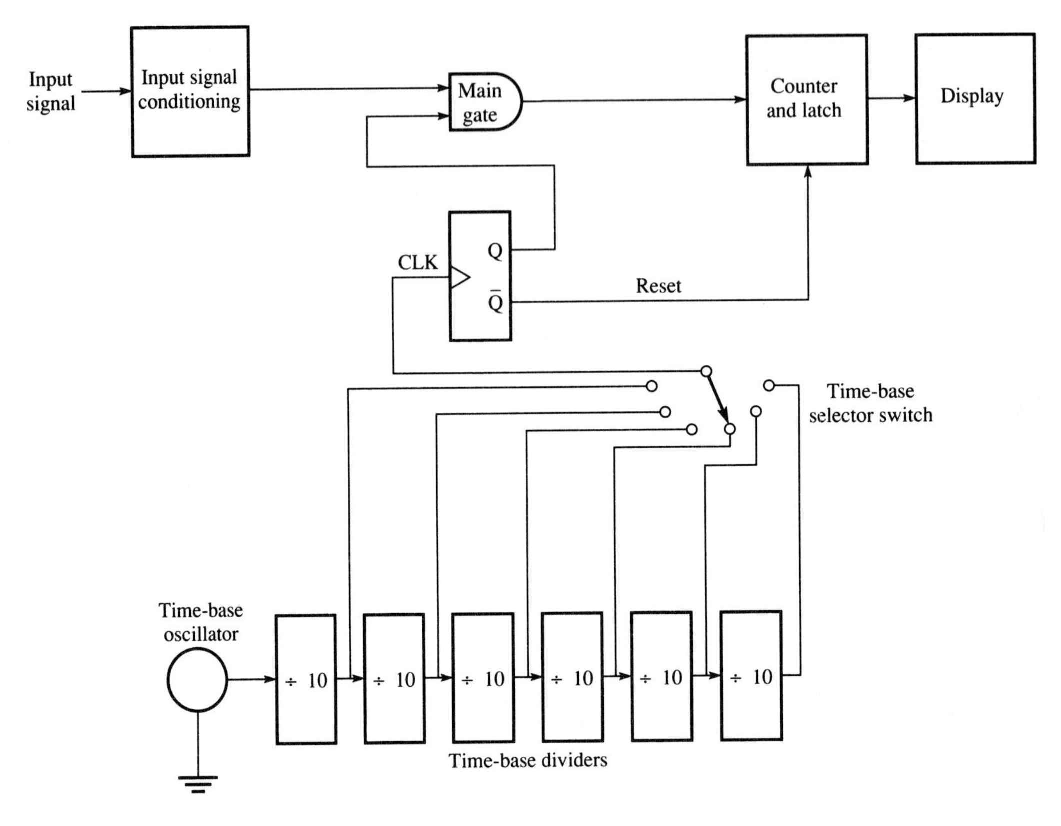

FIGURE 11–2
Block diagram of a universal electronic counter in the frequency-counting mode.

display drivers and to a seven-segment display. The number of digits in the display depends on the counter; typical counters have 6 to 10 digits. Some may indicate large or small numbers with scientific notation.

FREQUENCY MEASUREMENT

The most common function for electronic counters is the measurement of frequency. Frequency counters can count frequencies ranging from a few hertz to over 100 GHz (using conversion techniques) with exceptional resolution and accuracy. For frequency measurements, the main gate is opened for a precise interval of time, determined by an internal oscillator and decade-divider network. The input signal is counted for the time the gate is open. If the gate is opened for exactly 1 s, the count is a direct reading of the input frequency in hertz. To change ranges the gate time can be changed by factors of 10. If the gate is opened for 1 ms instead of 1 s, the count represents the frequency in kilohertz. The gate time is accurately controlled by the time-base oscillator and the time-base-divider network.

A stable crystal-controlled oscillator generates a precise frequency used for the time base, such as 10 MHz. The oscillator drives a pulse-forming circuit that generates precisely timed pulses at the period of the oscillator. For example, a 10 MHz oscillator controls pulses that are 0.1 μs apart. The pulses are then divided by a series of divide-by-10 dividers. Pulses of any specified period, from 0.1 μs to 1 s, can be obtained from the divider network and used to gate the unknown frequency. The decimal counter counts the input frequency, keeps track of the decimal place, and stores the count. After each count sequence is completed, the displayed frequency is updated.

A method for increasing the accuracy of periodic signals is to take several readings and compute the average. Although it requires more time to obtain a value, the method tends to cancel errors that are random, such as the error caused by the least significant digit. Another option available on some counters is normalized, or scaled, counting, in which the measured frequency is multiplied by some factor supplied by the user. This is useful if the count is proportional to some value. For example, the rpm of a gear might be displayed on the counter by counting the number of gear teeth that pass a point each second and multiplying by a constant.

HIGH-FREQUENCY MEASUREMENTS

For high frequencies, the input signal can be divided using a prescaler. A **prescaler** is a digital frequency divider that reduces the input signal frequency by some known factor, typically from 2 to 16. To determine the unknown frequency, the factor used for prescaling is multiplied by the reading. For example, if a prescaler with a division ratio of 10 is used in front of a 100 MHz universal counter, the counter range will be extended to 1 GHz. The displayed reading is multiplied by 10 to obtain the actual frequency. Notice that prescaling reduces the resolution of the measurement, but for that matter, so does reducing the gate time. The resolution of a frequency measurement with any counter is represented by the least significant digit in the decimal-counting unit. With the prescaler, each digit on the display, including the least significant, must be multiplied by 10 to obtain the original frequency. The reduced resolution can be restored by increasing the gate time.

Typically, prescalers are limited to about 2 GHz, which is in the low-microwave region. Above this are microwave frequencies where it is necessary to use other techniques to extend the range of counters. One method is to use a **heterodyne converter,** which is basically an analog mixer that combines a high frequency with the input signal to lower the frequency to a range that the counter's logic circuits can handle. The high frequency for mixing is generated by a harmonic generator that is an integer multiple of the internal time-base oscillator of the counter. The harmonic generator provides a series of harmonics, one of which is selected by a tunable filter for mixing with the input frequency. Control of the filter is done by a processor within the counter. The heterodyne frequency converter is located in the front end of a microwave-frequency counter and can handle input frequencies to about 20 GHz. The remainder of the microwave-frequency counter is the same as that of the universal-frequency counter except it is necessary to add a processor to control the conversion process and display the correct result.

Another high-frequency technique is to use a transfer oscillator. A **transfer oscillator** uses a phase-locked loop to lock a low-frequency oscillator onto the microwave input. The frequency of the oscillator is an integer number of cycles of the input frequency. The oscillator is used as the input signal to a conventional frequency counter. With the known harmonic relationship between the input signal and the oscillator, the unknown frequency can be displayed directly.

Another type of converter, called a **harmonic heterodyne converter,** combines the heterodyne converter and transfer oscillator techniques. The input frequency is mixed with a frequency from a harmonic of a synthesized oscillator and the difference frequency is counted. The harmonic number is determined by a phase-lock-loop technique. The displayed frequency is found by multiplying the harmonic number by the synthesized frequency and adding it to the measured frequency. It is necessary to know which harmonic is selected in order to calibrate the counter; modern microwave frequency counters can select the harmonic and provide the correct calibration automatically. Microwave frequency counters use this technique to operate to 100 GHz and measure from cw to the short bursts of radio frequencies found in a radar burst. Pulsed signals are measured by setting the gate signal to the pulse width and arming the counter with the start of the waveform. Alternately, an arming signal can be provided to enable the counter at the right time for measuring the unknown frequency.

EXAMPLE 11–1 An electronic counter is connected to an unknown frequency.

(a) If 174 pulses are counted in a gate interval of 100 μs, what is the frequency?
(b) Determine the resolution of the measurement.

SOLUTION The frequency is found by dividing the count by the time interval. That is

$$f = \frac{174}{100\ \mu\text{s}} = 1.74\ \text{MHz}$$

The resolution is represented by the least significant figure in the decimal counting unit. The number 4 is the least significant figure. It represents the hundredths column of megahertz, or ±10 kHz.

EXAMPLE 11–2 Assume a seven-digit frequency counter counts 900,155 counts in a gate time of 10 s.

(a) What is the frequency?
(b) What is the resolution of the measurement?

SOLUTION (a) The unknown frequency is the number of counts divided by the gate time. A gate time of 10 s implies that the frequency is 90,015.5 Hz.
(b) The resolution is 0.1 Hz.

ACCURACY OF FREQUENCY MEASUREMENTS

The key to the accuracy of an electronic counter is the accuracy of its internal time base oscillator. It is frequently housed in a temperature-stable environment to keep it at its specified frequency. The internal oscillator is normally a high-frequency crystal-controlled type—most counters have provisions either for using the internal oscillator as a reference for other instruments or for connecting an external oscillator in place of the internal oscillator to form the time base. The internal oscillator is usually accessible on a connector for calibrating instruments and other applications where a stable reference is needed. The accuracy of the internal oscillator is affected by aging; the accuracy can be improved by frequent calibration. The accuracy specification is normally written as ± 1 LSD $\pm$ the time-base error. For example, a counter might have an accuracy specification of ± 1 count $\pm$ 10^{-6} times the frequency. At low frequencies, the ± 1 count affects the percentage error the most, whereas at high frequencies the time-base is more important.

The resolution of microwave frequency counters is determined in the same manner as for universal counters—by the value of the LSD on the counter. Accuracy is not determined by the type of conversion but rather by the LSD and the time-base error.

EXAMPLE 11–3 The accuracy specification for a very accurate frequency counter that has been calibrated within 3 mo is given as ± 1 count $\pm$ 10^{-7} times the frequency.

(a) What is the maximum percentage error for a frequency of 10 Hz and a gate time of 10 s?

(b) What is the maximum percentage error for a frequency of 100 MHz and a gate time of 10 ms?

SOLUTION (a) At 10 Hz for 10 s, there are 100 counts. The maximum count error is ± 1 count $\pm$ $(10^{-7} \times 10 \text{ Hz}) = 1.000001$ counts. The percentage error is

$$\% \text{ error} = \frac{1.000001 \text{ counts}}{100 \text{ counts}} \times 100\% = 1.00\%$$

Notice that this error is due to the resolution of the measurement, not the accuracy of the internal clock.

(b) At 100 MHz for 10 ms, there are 1.00×10^6 counts. The maximum count error is ± 1 count $\pm$ $(10^{-7} \times 100 \text{ MHz}) = 11$ counts.

$$\% \text{ error} = \frac{11 \text{ counts}}{1.00 \times 10^6 \text{ counts}} \times 100\% = 0.0011\%$$

Even though the absolute error (maximum count error) is higher, the percentage error is much smaller and is due primarily to the time-base error, not the resolution.

PERIOD MEASUREMENT

Recall that the period of a waveform is defined as the time between any two corresponding points on successive cycles of a waveform. Two corresponding points, such as the positive zero-crossing points, are used to gate the internal high-frequency oscillator. To measure the period of an unknown waveform, the main gate is controlled by the input signal instead of the clock. Figure 11–3 shows a universal counter operating in the period-measurement mode. During the time the input signal holds the main gate open, internal oscillator pulses are counted by the decimal-counting unit. The count is directly proportional to the period; the decimal-counting unit needs only to place the decimal in the proper place. If the internal oscillator is counted for exactly one cycle of an unknown, the period can be found by multiplying the count accumulated during one cycle by the period of the internal oscillator.

If the frequency or period of a waveform is known, the other can be found because they are reciprocals of each other. You may wonder why the period measurement is needed if it can be computed directly from the frequency measurement. The answer is that for low frequencies, the period measurement is more accurate. As you saw in Example 11–3, large errors can occur in directly measuring a low frequency because the measurement is limited by the uncertainty of one count. For this same problem, a period measurement is much more accurate because the high-frequency clock is counted many times during one cycle, giving much higher resolution than the direct frequency measurement. The reverse is true at very high frequencies; period counting can have high errors due to the resolution. For example, if a 100 Hz signal is measured by the frequency counter using a 1 s gate, the ±1 count produces a 1% error. If, instead, the period of the

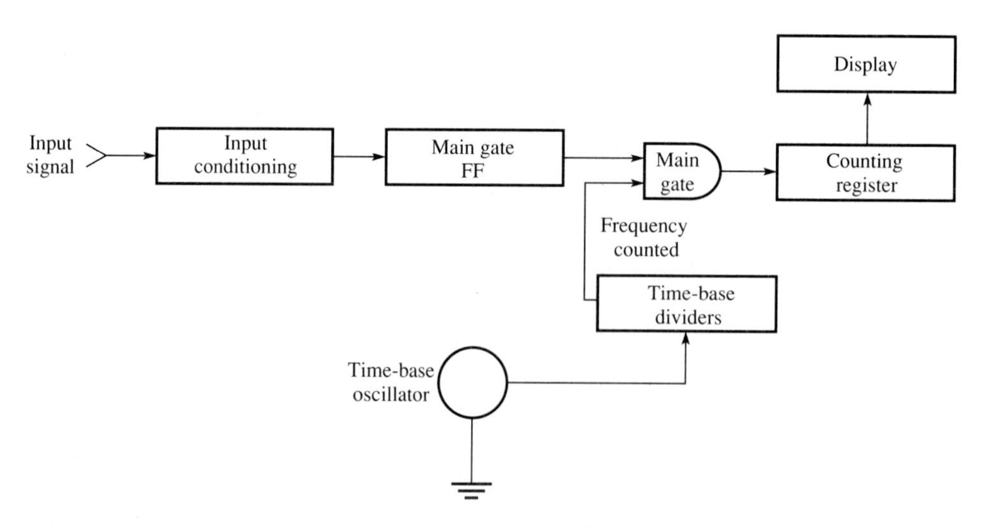

FIGURE 11–3
Basic block diagram of a universal electronic counter in its period-measurement mode (courtesy of Hewlett-Packard).

100 Hz signal were measured by counting a 10 MHz clock, 100,000 clock cycles would occur in one period, reducing the ±1 count error to only 0.001%.

One common procedure for obtaining higher-resolution measurements is to use averaging. Averaging also enables making period measurements of higher-frequency signals. The idea is to divide the input frequency by some factor and use the lower frequency to control the gate time. With the longer gate time, more clock pulses are counted. For example, if the input frequency is divided by 1000, the gate time is 1000 times longer and 1000 times as many clock pulses will be counted. The decimal point needs to be moved three positions to account for the extra gate time, but the resolution is increased as a result.

FREQUENCY RATIO MEASUREMENTS

Frequency ratio measurements display the ratio between two signal frequencies. The lower frequency of the two is used to generate a gate pulse. The higher frequency is counted during the gate time, and the result is displayed as the ratio. To increase the resolution, the lower frequency can be divided by the decade-divider network. (Conversely, the higher frequency can be multiplied.) The number of digits of resolution is determined by the ratio of the input frequencies and the amount by which the lower frequency was divided. The ratio mode is illustrated in Figure 11–4.

TIME-INTERVAL MEASUREMENTS

Time-interval measurements are made by universal counters using external start and stop pulses, much like an electronic stopwatch. The start and stop pulses open the main gate, sending timing pulses to be counted during the time the gate is open. The number of pulses counted is proportional to the time the main gate is open. The time-interval mode is shown in Figure 11–5.

There are two basic time-interval measurements—single shot (nonrepetitive) and periodic (repetitive). To obtain high resolution with nonrepetitive timing, the clock frequency should be high. A 10 MHz clock has pulses separated by 100 ns, limiting the resolution to this value. To obtain higher resolution, higher-

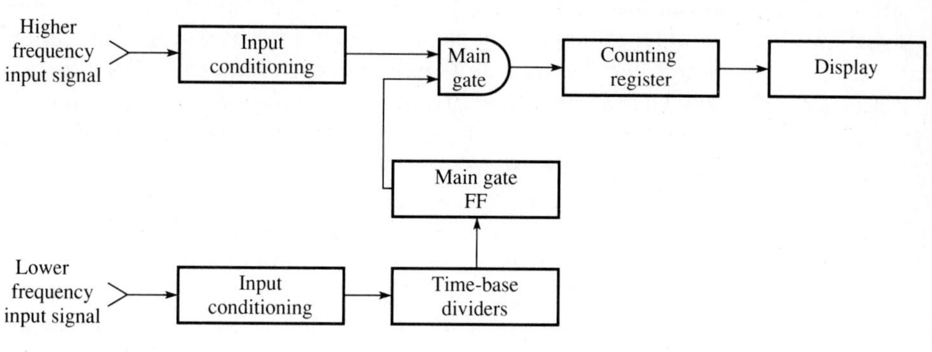

FIGURE 11–4
Ratio measurement mode (courtesy of Hewlett-Packard).

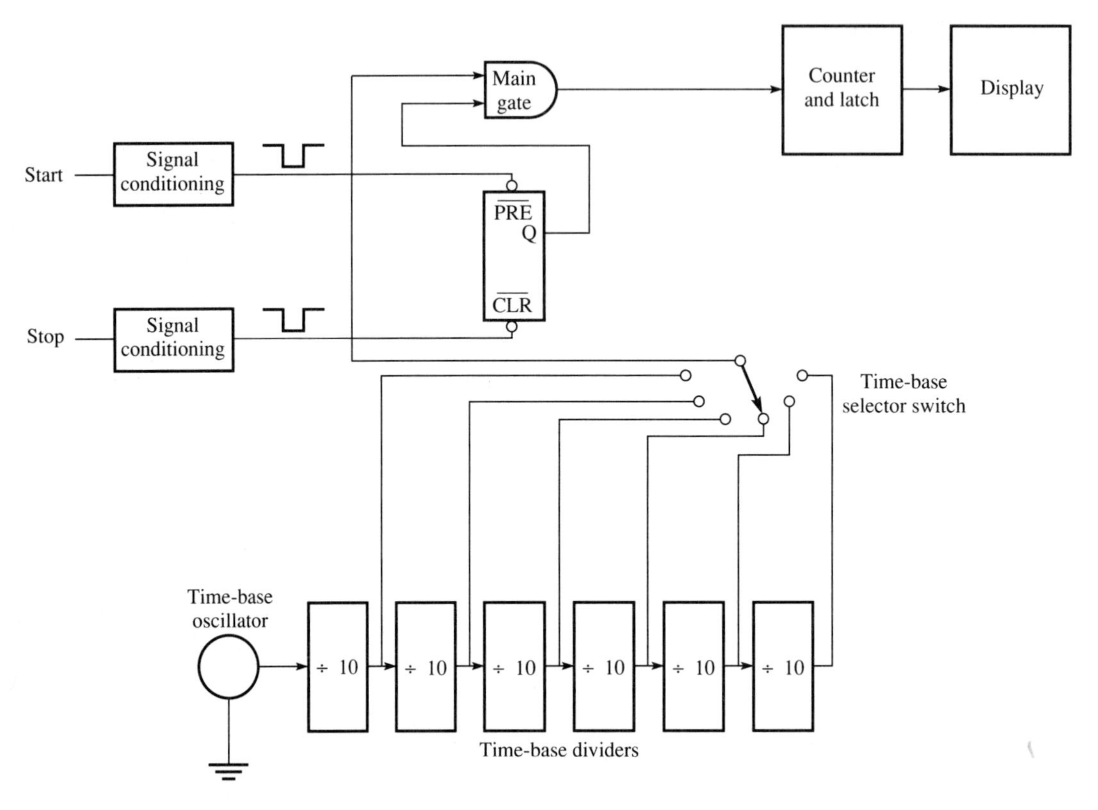

FIGURE 11–5
Time-interval mode.

frequency clocks are required, but the increase in resolution is warranted only if the accuracy is equivalent. For example, a 500 MHz clock requires an accuracy on the order of 10^{-9} to provide meaningful resolution.

For repetitive time-interval measurements, averaging can be done. Averaging is meaningful only if the counter counts only the leading edge of pulses rather than allowing partial pulses to be counted. Counters that are designed to perform time-interval averaging can do this correctly, but a manual computation from a counter that simply counts pulses can lead to error. The situation leading to error is shown in Figure 11–6. In addition to this potential error, it is also necessary that the clock period is not an integer multiple or submultiple of the input-signal period that is being measured.

Time-interval measurements can be made by starting the counter with a pulse on one channel and stopping the counter with a pulse from a second channel. If TIME A to B is selected, the time interval is shown from the trigger on channel A to the trigger on channel B. The trigger level will affect the measurement, and in precise work, the cable delays in each channel will need to be accounted for. (Some counters allow you to insert certain delays automatically to account for differences in the two channels.) Figure 11–7 illustrates a two-channel

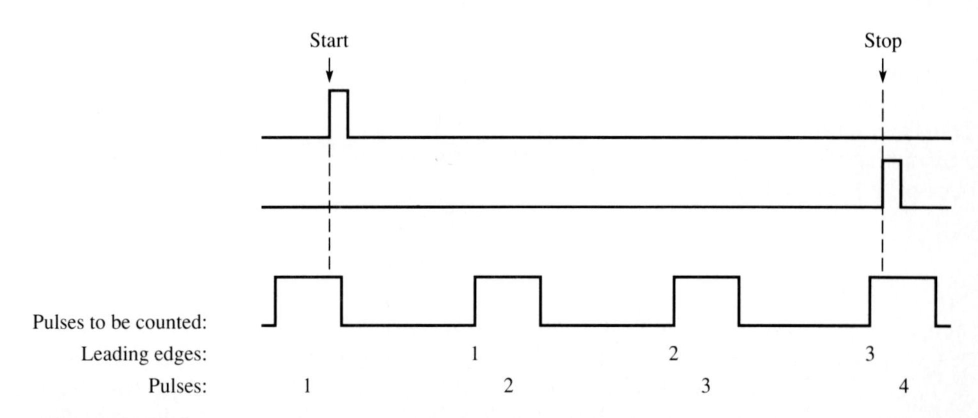

FIGURE 11–6
A one-count error can occur if the counter simply counts pulses rather than a leading or trailing edge. Notice that the time shown is actually less than three periods, but the number of pulses that a pulse counter sees is four.

time-interval measurement in which the positive slope of start and stop pulses is selected.

Time intervals on one channel can also be measured in the time-interval mode. The start and stop pulses can be derived from a single signal by selecting rising and falling edges for starting and stopping the clock. If the start pulse is generated by a positive slope and the stop pulse generated by a negative slope, the counter will measure the pulse width directly. Another technique for measuring time intervals on one channel requires an external timing generator. The start pulse is enabled by an arming signal. A lockout time can be set up by the external timing generator to prevent the wrong pulse from stopping the counter. When the lockout time expires, the stop time is enabled. With this technique, accurate time measurements between arbitrary pulses in a chain can be measured.

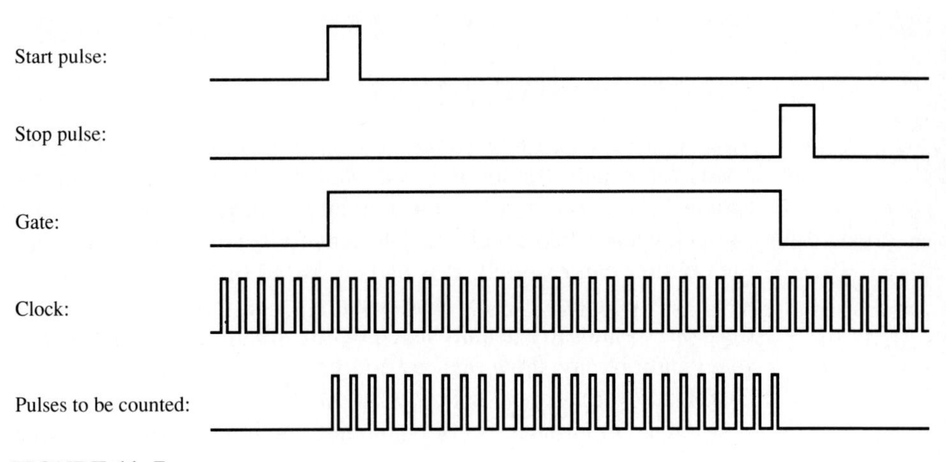

FIGURE 11–7
A two-channel time-interval measurement.

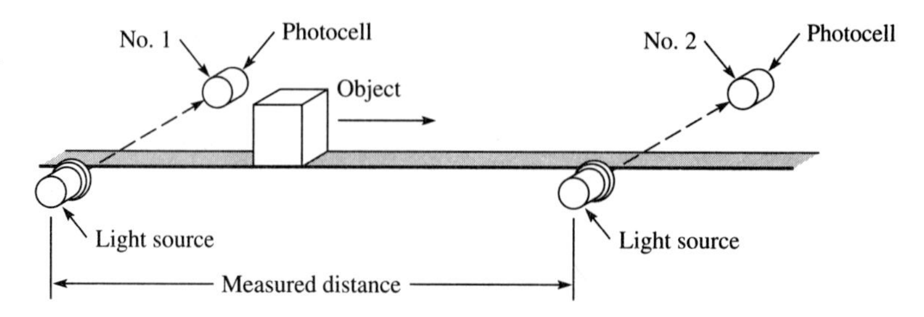

FIGURE 11–8 Measurement of the velocity of an object. The object breaks the light beams, generating a start and a stop pulse for the counter.

Some counters, using the arithmetic processor, can show the reciprocal of the time interval, displaying the result as units per second. By using the proper unit conversion, the counter can indicate velocity on the display. Figure 11–8 illustrates the idea. Two light beams are positioned a known distance apart in the path of a moving object. As the object crosses the first light beam, it generates a start pulse; a little later, it crosses the second beam and generates the stop pulse.

EVENT COUNTING

A running total of pulses can be displayed in the counting mode. The measurement is similar to frequency counting, except that the gate is controlled by a start-stop command. Start and stop commands can be issued by the user from a front-panel control. The count continues from the previous point each time it is started unless it is reset by a separate reset command. In the event-counting mode, the input signal may occur at random times and could even have variations in the input level. Either the rising or falling edge of the input can be used as a trigger. Some counters allow the start and stop functions to be done with a gate signal rather than manual control. The number of pulses in a burst, or number of events within a time window, can be counted by using this method. The count can be reset automatically between each measurement.

11–2 ELECTRONIC COUNTER SPECIFICATIONS

INPUT SPECIFICATIONS

The input specifications for electronic counters include the frequency and signal levels for which the counter is designed. The frequency range of the counter is specified by the minimum and maximum frequencies. The input impedance and signal level are also specified. The sensitivity is the minimum signal level to which the counter can respond. It is usually based on the minimum rms sine wave for a specific frequency range. The sensitivity is frequency dependent, and several frequency bands are usually listed by the manufacturer. For example, a sensitivity specification may show that the counter can respond to a 25 mV rms sine wave to 50 MHz, 75 mV for 50 MHz to 200 MHz, and 120 mV to 400 MHz.

Other input specifications include the input impedance, dynamic range, and triggering level. At low frequencies, the input impedance is normally high to prevent loading. At high frequencies, above about 10 MHz, capacitive loading becomes important and 50 Ω input impedance is preferred because of its low shunt capacity. The dynamic range is the input amplifier's linear operating range. Although linear operation is not essential, as in the case of an oscilloscope, exceeding the upper end of the dynamic range may cause the input amplifier to saturate and give false counts or cause damage to the amplifier. The trigger-level control allows the operator to shift the hysteresis level of the Schmitt trigger above or below ground to ensure that any signal within the input amplifier's dynamic range may be counted. The trigger level is usually the value of the center of the hysteresis band. It can be adjusted so that pulse trains that contain either a positive, negative, or zero dc offset can be counted.

OPERATIONAL SPECIFICATIONS

The basic modes for the universal counter have already been described. For each of these modes, operational specifications are given by the manufacturer. These include range, resolution, accuracy, and other specifications pertinent to the particular measurement. The range specification indicates the minimum to maximum value of the measurand. For example, the specification for the period mode could be written 2 ns to 20,000 s, indicating that the counter could measure periods within those limits.

As given in Section 1–1, the resolution is the minimum discernible change in the measurand that can be detected. For example, for the time-interval measurement, resolution is a function of the frequency of the time-base oscillator and the noise on the input signal. A resolution specification could be written 2 ns ± noise trigger error (obtained from a graph).

The accuracy specification for frequency was discussed previously. In general, universal electronic counters have inaccuracy due to resolution, time-base, trigger, and systematic errors. Not all these errors affect all measurement modes, but all modes are affected by the resolution (least significant digit) and the accuracy of the time-base oscillator (except frequency ratios and event counting). The time-base error is caused by the error present at the initial calibration and by changes in the oscillator frequency due to short- and long-term effects. Short-term changes tend to be random and can be minimized by averaging. Long-term effects are caused by aging of the crystal and can be minimized by frequent calibrations. Other accuracy factors are listed, depending on the particular mode. An example of an accuracy specification that includes all four types of error is for the time-interval mode; a specification could be written as ± resolution ± time base error ± trigger level error ± 700 ps. The maximum values of the first three errors listed are dependent on other conditions and are given graphically. The ±700 ps error represents channel delay error and other systematic errors.

11–3 LOGIC ANALYZER FUNDAMENTALS

DIGITAL SYSTEM PROBLEMS

Digital systems have special measurement problems unlike those of analog systems. In analog systems, information represents the magnitude of a quantity—usually a voltage or a current. If we digitize the same information into a binary word, a number of lines are required for the binary bits. Most digital systems have a large number of data, address, and control lines that change constantly. The time relationships between these signals is critical to the proper performance of the system. The signals sent over these lines are sent as a group—for example, 16 or more lines are used to determine the address in a typical small microprocessor; another group of lines is needed for the control and data buses. The grouping of these lines into digital words and bytes simplifies the task of assimilating the information and helps pinpoint any problem with an individual line.

An instrument designed for showing the pattern of 1s and 0s in digital circuits is the logic analyzer, such as the modular analyzer illustrated in Figure 11–9. Essentially a **logic analyzer** is a multichannel digital storage instrument with one bit of vertical resolution and with flexible triggering and display modes. It takes digital samples on as many as 96 channels at intervals determined by a clock signal and stores the samples in a memory for subsequent analysis. By contrast, oscilloscopes are generally limited to showing from two channels to four channels of information but with hundreds of levels of vertical resolution. Like logic analyzers, digital oscilloscopes can store the sample data in memory (but ordinary analog oscilloscopes cannot). Both logic analyzers and oscilloscopes have application to specific digital problems, but a logic analyzer is likely to be used first to find a

FIGURE 11–9
The Hewlett-Packard 16500A logic analyzer. The 16500A is a modular system that can be expanded according to need. It features powerful triggering modes and can support most microprocessors using preprocessor modes. It can also be used as a digital pattern generator for testing digital circuits (courtesy of Hewlett-Packard).

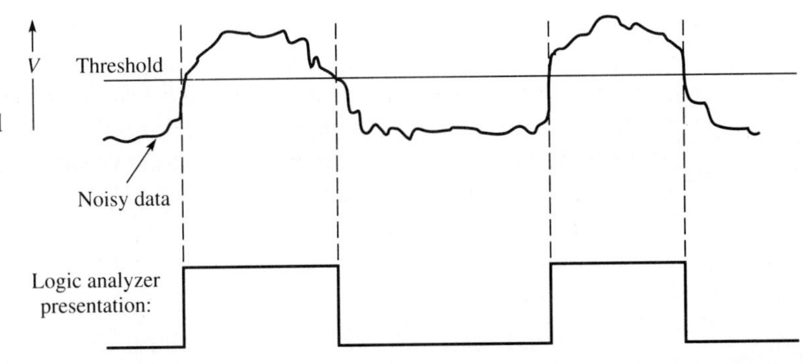

FIGURE 11–10
The logic analyzer makes noisy data look "clean" because of limited vertical resolution.

digital system problem. The limited vertical resolution of a logic analyzer causes the display to be idealized—it will not show the details of a signal that can be observed on an oscilloscope (ringing and the like)—but the logic analyzer does offer certain advantages for digital systems. Analyzers can show the data in the timing mode as digital waveforms and in the state mode in any of several forms: binary, octal, hexadecimal, ASCII, or even as disassembled computer instructions.

An analyzer determines if the data present at each probe tip is a logic 1 or 0 by comparing the voltage to a predetermined threshold voltage. The threshold voltage can be adjusted by the user. TTL, for example, typically has a threshold set to 1.4 V, halfway between the low and high thresholds. The analyzer will tend to make even slowly rising or noisy data look "clean" because of the limited resolution, as shown in Figure 11–10. Some advanced analyzers offer two thresholds and will store data as a low, intermediate, or high value. For TTL, the thresholds would typically be set to 0.8 V and 2.0 V. The analyzer can indicate if the data is not valid; this is the case when the line is released in three-state logic.

LOGIC ANALYZER APPLICATIONS

The most important application of logic analyzers is diagnosing microprocessor and digital systems, including new systems under development and systems requiring repair. Problems in microprocessor systems can generally be categorized as either software, hardware, interface, or I/O problems. The category of a problem is not always clear when troubleshooting begins. Software problems are usually due to programming errors and can be looked for by using state analysis to observe the code that is actually being executed by the processor. Hardware, interface, and I/O problems are generally observed in the timing analyzer mode. Hardware problems can be caused by a great variety of reasons, including bad ICs, open or shorted connections, bent pins, incorrect parts, and so forth. Interface problems are due to incompatibility between software and hardware—signals could be at the wrong time or of the wrong logic level. I/O problems occur between the computer and the outside world. They can be caused by incorrect handshake timing, glitches,[1] incorrect wiring, and the like.

[1] A glitch is an unwanted, short perturbation in a pulse waveform.

Frequently, there are special triggering requirements to make possible capturing the specific data from a digital sequence. The sequence must be stored, since it may not repeat. Logic analyzers are designed to handle these problems by providing many channels of digital storage and flexible triggering capabilities. In addition, they frequently have display and data-search features such as cursors, glitch highlighting, trigger point identification, pattern searches, channel groupings, and statistical analysis.

BLOCK DIAGRAM

To understand the logic analyzer, it is helpful to begin with the block diagram. Figure 11–11 shows a simplified block diagram of a logic analyzer. The connections to the system under test are made through an active pod containing buffers and level-detection circuits. The probe buffers convert the voltage on each probe to either a logic 0 or a logic 1 based on the setting of the logic threshold. The sampled data is then sent to the log-in register of the analyzer, where it is compared with a previously set trigger word. The comparison causes an action to be taken by the trigger control to either start, continue, or stop recording the data. Assuming the action is to start or continue recording, the sampled data is then saved by the storage memory in the next available location. This sequence continues until either the storage memory is filled or the trigger control detects a stop condition. After storage is complete, the sampled data can be displayed in two

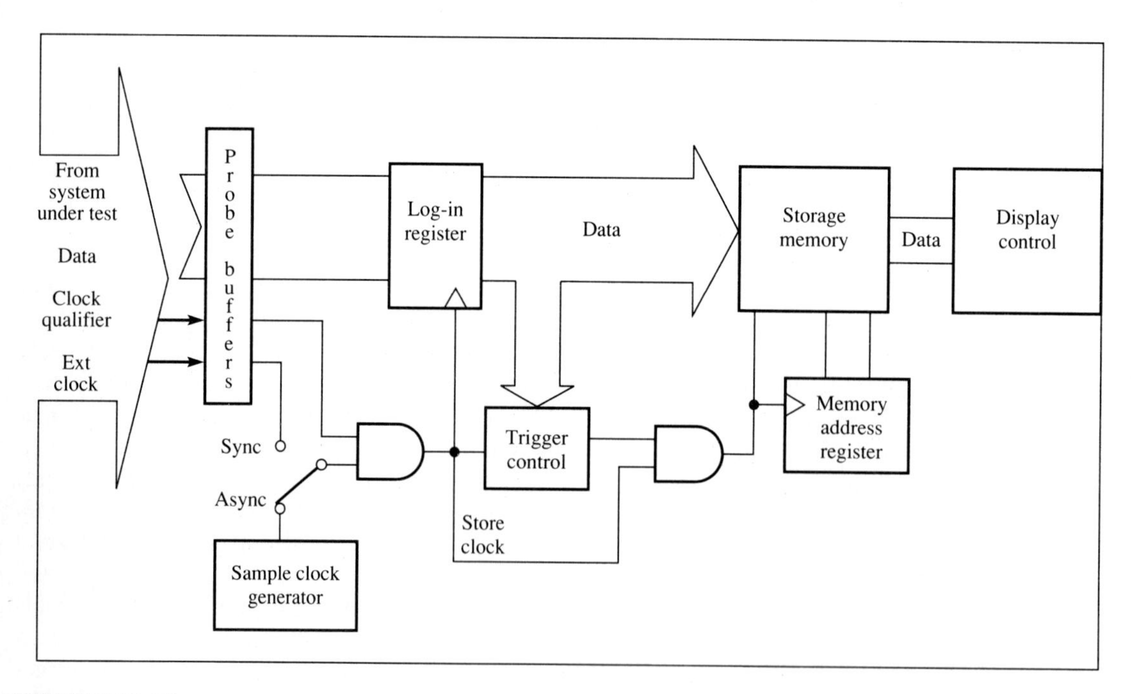

FIGURE 11–11
Logic analyzer block diagram (copyright Tektronix, Inc. Used with permission).

basic formats—either as a timing diagram or as a state listing. High-end analyzers can also present data in a performance-analysis format to indicate statistical information such as the number of accesses to a subroutine. These modes are described in the next section.

11–4 LOGIC ANALYZER OPERATION

CONTROLS

A modern analyzer uses menus and user prompts to aid in configuring the analyzer for a particular measurement. Because of active protection circuits in the analyzer's pods, the analyzer is first turned on and is then connected to the system under test. Each pod is coded by color or some other scheme to help keep track of the various channels. Individual wires on the pods are also frequently color-coded. After the analyzer is first turned on, it will go through a self-test and then indicate that it is ready for data entry. The user first goes to a setup or configuration menu. This menu includes the clock interval, trigger conditions, and other parameters that must be defined before a measurement can take place. Another menu allows the user to select how the data will be displayed. This can be a timing diagram, state listing, disassembled code, or other option. Depending on the analyzer, other menus may be called for automated setups or other available options.

TIMING ANALYZER MODE

As a timing analyzer, the logic analyzer displays the time relationship between multiple input channels on a CRT. The timing mode is useful for analyzing the hardware operation of digital circuits by indicating the relative time between channels or showing the presence of glitches. Proper timing can be verified, including setup and hold times or propagation delay timing. An example of the timing mode is shown in Figure 11–12. The display shows a series of input channels and indicates the logic present on each channel as a function of time and simultaneously shows two analog waveforms. In the timing analyzer mode, data is sampled asynchronously and stored as binary samples. **Asynchronous sampling** means that the sample clock is not related to the clock used by the system under test. In addition to the timing mode, data can be displayed in a state table as shown on the split-screen display in Figure 11–13. Depending on the analyzer, other display modes may be available. In the accumulative mode, the waveform display is not erased between successive acquisitions, making changes easy to spot. In the overlay mode, more than one channel may be plotted on one display line.

The sampling interval for the timing mode is set by the user. The sample times are controlled by the internal clock. Top-quality, high-speed analyzers can check setup or hold times and do other parametric testing; however, analyzers are generally used for functioning testing rather than parametric testing. For functional testing, it is typical to choose a sampling interval that guarantees a minimum

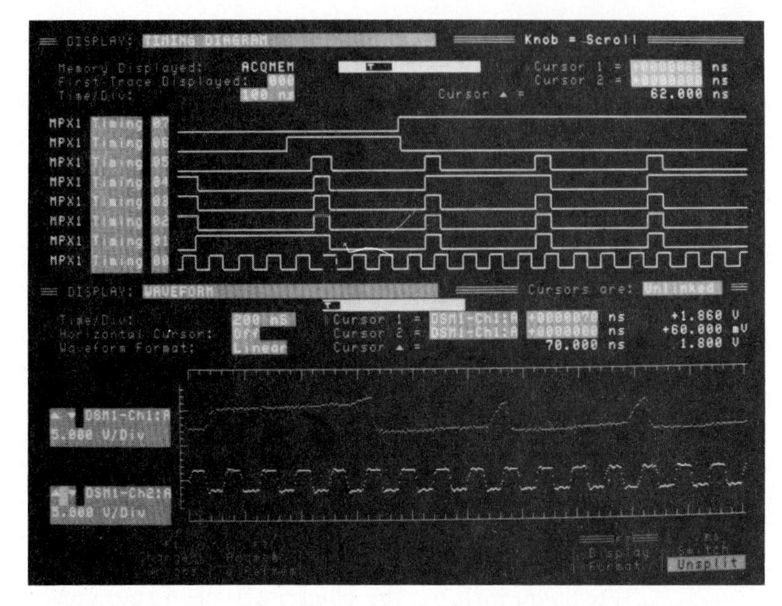

FIGURE 11–12 Logic analyzer display in timing and waveform mode (copyright Tektronix, Inc. Used with permission).

of two or three samples for both the high and low logic on the fastest signal. Usually this is 5 to 10 times faster than the period of the fastest waveform to be sampled. For example, a 100 kHz signal with a 20% duty cycle has a period of 10 μs. The pulse is in the HIGH state for 2 μs. A sample interval of 1 μs or less will guarantee that all pulses will be sampled at both the high and low points. A shorter sampling interval can be selected; however, there is a trade-off, as illustrated in Figure 11–14. The shorter sampling interval will produce higher resolution, but it produces a shorter record due to the finite length of the storage memory.

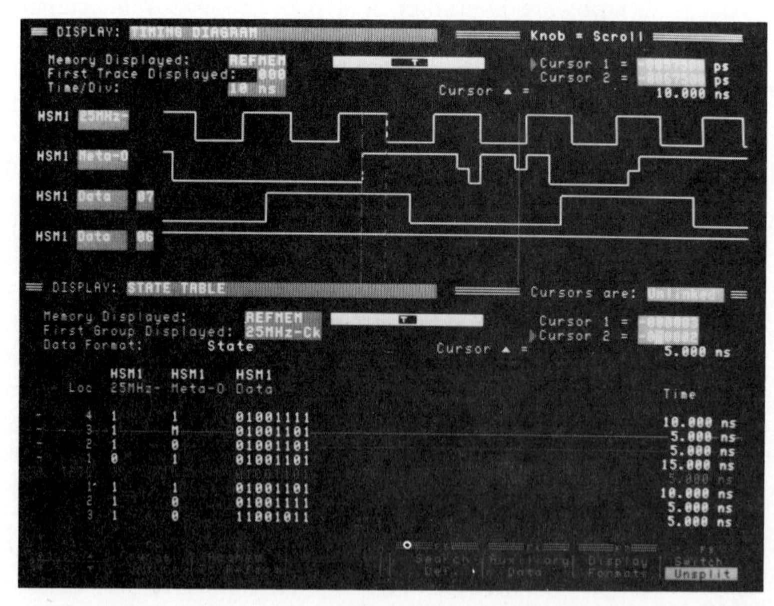

FIGURE 11–13 Logic analyzer display in timing and in state table mode (copyright Tektronix, Inc. Used with permission).

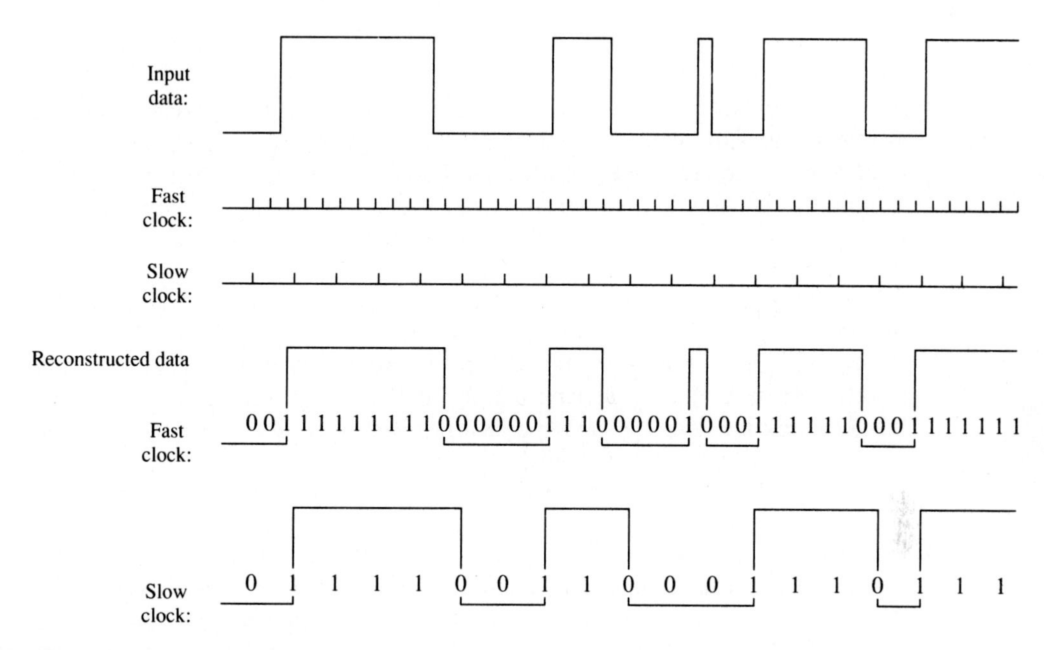

FIGURE 11–14
Comparison of fast and slow sampling rates. The short signal near the center was missed by the slower rate.

A more costly sampling technique, requiring sophisticated clocking and timing correlation, is transitional sampling, found on high-end logic analyzers. With **transitional sampling,** individual clock pulses are assigned a reference time; when an input line changes (a transition), a sample of all input lines is taken along with the clock time. Data that is stored is "time stamped" for reconstruction. In contrast to asynchronous sampling, redundant data is not stored. This technique is particularly powerful if data occurs in bursts, a mode that frequently occurs in interrupt-driven control systems. The analyzer will look only at the data, not at the pauses between data. This enables the reconstruction of a high-resolution record in a finite memory length.

In digital systems, it is frequently necessary to zero in on a specific word or memory location from which data is to be observed. The trigger section of the analyzer allows the operator to specify trigger conditions for storing the data. Flexibility in specifying trigger conditions is one of the advantages of logic analyzers—the true power of a logic analyzer lies in the number and quality of the triggering features. The trigger conditions can be specified from either the data itself or from a special set of qualifiers. A **qualifier** is a digital input that is not stored but is used for setting up the required conditions for triggering. One useful triggering feature is the ability to use Boolean logic to qualify the conditions for triggering. Triggering can be initiated on the specific pattern of high, low, and don't-care lines programmed as a Boolean expression of AND/OR logic, or the operator can choose an edge following a specific pattern. For example, a set of qualifier lines can be connected to the address bus of a computer and programmed for the specific address that is desired. When the programmed address occurs, the trigger is generated and data is stored from the data bus. Because the data is continuously stored but retained only when triggering occurs, the data can be observed *before* the trigger point. This is an important advantage when the cause of a failure is being looked for; the trigger is set up for the fail condition and the conditions just prior to the failure can be observed. For some analyzers, triggering can also be initiated by a glitch. Glitches in digital systems can be extremely frustrating to troubleshoot because they often occur at random times and from an uncertain origin. Glitch triggering enables you to see the activity that occurred at the time of the glitch as an aid in tracking down the cause. Tektronix offers a slightly different fault-triggering method on their high-end logic analyzers—the trigger conditions can be specified as a hardware problem, such as setup time or insufficient pulse width on a given line. The point of these advanced triggering methods is to offer the user a way to troubleshoot a difficult problem.

The timing analyzer mode is valuable for observing timing problems in general, including stuck bits (HIGH or LOW), decoding errors, intermittent problems, glitches, and the like. For example, a digital system usually uses data buses for more than one purpose, so they contain valid information only at specific times. To indicate that valid information is present, a strobe is sent on a separate line. If the strobe occurs too early, the data may not be valid. The analyzer can aid in diagnosing this problem by showing the time relationship between the strobe and the data lines.

STATE ANALYZER MODE

In the state analyzer mode, the trigger is generated from the unit under test rather than from the analyzer's clock. This type of triggering is called synchronous triggering. There are many examples of signals in a digital system that do not occur periodically. For example, the fetching and execution of instructions in a computer proceeds at a rate determined by the time required to execute each instruction. Assume you want to observe the sequence of instructions rather than the time relationship between the signals. The analyzer is to acquire a sample only when a new instruction is placed on the data bus. The trigger can be qualified by a Boolean expression programmed by the operator in the trace specification. For example, instructions for the 8085 microprocessor could be observed by qualifying the $IO/\overline{M}$ line LOW and the S_0 and S_1 lines HIGH. This ensures that only the desired instructions will be seen on the display rather than a mix of instructions and status bits, which are multiplexed on the lines. In the state analyzer mode, the data is sequential but does not necessarily occur at equal intervals of time. This is an example of an instrument that is operating in the data domain, as described in Section 1–2.

One application of the state analyzer mode is in viewing a computer program as it is executed. Certain computer programming errors (*bugs*) cannot be observed until a program operates in real time. To avoid the problem of storing multiple-loop instructions, some analyzers allow qualifiers for the data to be stored, allowing the elimination of redundant states. Another feature that is offered is the ability to count states between the stored states or to measure time between stored states (*time tagging*). State analysis is often used in conjunction with a disassembler. A disassembler converts the binary instruction code into an assembly language mnemonic that can be understood by a programmer. The instructions that a computer executes are not always the ones intended by the programmer. By observing the dissassembled code, the programmer can locate programming errors and take corrective action. Figure 11–15 shows a portion of a listing taken from the display of a logic analyzer that shows disassembled code for a 68000 microprocessor. Because of the capability of showing microprocessor mnemonics, some companies have integrated logic analyzers in their microprocessor-development stations as an aid in software development. As new microprocessors are brought out, support for the instruction set and mechanical connection to the system is offered by various companies to keep older analyzers up to date.

As in the case of the timing analyzer, triggering for the state analyzer can be qualified by a specific combination of inputs on a special trigger probe. Some analyzers allow sophisticated sequential logic operations to qualify a trigger. For example, assume you wanted to observe the data written to a specific location in a computer memory by a certain subroutine (a subroutine is a portion of a computer program). Assume that other subroutines can also write to that same location. How do you capture only the data written by the subroutine of interest? This is a

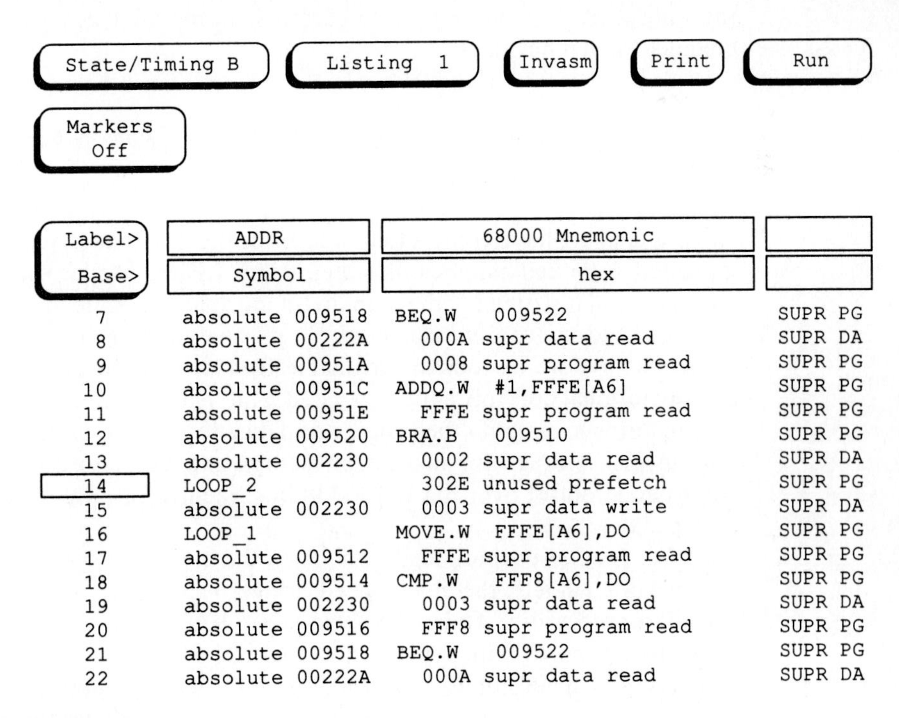

FIGURE 11–15
Disassembled 68000 code shown on the display of a logic analyzer.

problem that can be solved by two-level qualification. First the address of the subroutine of interest is used as a trigger qualifier. When the subroutine is called, the second level of logic waits until the address of the data is placed on the address bus; then it triggers the analyzer. There are cases where even more complicated triggering may be required than in this example. Some analyzers allow many more decision levels of triggering to qualify a trigger.

Many analyzers can analyze code at full clock speed without affecting the system operation. If a system fault (*crash*) occurs, the data immediately preceding the crash can be stored to do fault analysis. Frequently, when a problem exists in a computer, it is difficult to tell if the problem is a result of a software bug or a hardware timing problem. The listing of the instructions that were actually executed compared to the listing of the code can reveal software problems. The timing analysis mode can then be used to check the timing specifications of the system.

For very fast systems, the clocking rate may exceed the analyzer's ability to store data. Near the analyzer's performance limits, the analyzer's setup and hold time are important—especially when you want to use an external clock with a high-speed system. If the circuit that is being tested changes states faster than the analyzer's required setup and hold time, incorrect information may be stored. There may also be channel-to-channel skew in the analyzer that can affect the data if one channel is not digitized at exactly the same time as another channel. Channel-to-channel skew error is the difference in delay time from the probe tip to the output of the analyzer's comparators. Some analyzers can be adjusted for channel-to-channel skew.

11–5 SPECIAL FEATURES AND SPECIFICATIONS OF LOGIC ANALYZERS

Logic analyzers have come a long way since the first two-channel analyzer, which indicated a limited number of states using LEDs. Modern analyzers are very sophisticated instruments that can do timing analysis, state analysis, performance analysis, and microprocessor control. As analyzers have continued to grow in complexity, more options are available from which the user can choose. Setup and operational controls have expanded, and manufacturers have responded with new features to simplify the process of data entry. Some of these features include built-in disk storage for instrument setups and data storage, the ability to do setups and control over the IEEE-488 bus, menus for ease of entry, and built-in help packages.

For general-purpose logic analyzers, the important specifications include the number of channels, the clock speed, triggering options, and specialized support for microprocessors. Today, analyzers are available with 96 channels (in some cases many more) with typical clock speeds of 100 MHz for state and timing analysis (and speeds up to 2 GHz available for special applications). High-quality analyzers offer even more triggering options than are found on other analyzers. Trigger qualification can include a number of conditions strung together in a

Boolean statement. For example, a trigger qualifier might require a specific sequence of conditions, including data words, elapsed time, or a number of events. Triggering can also be initiated if a specified time (such as setup or hold time) is greater or less than some amount.

Although the specifications just listed are important in general, the application will dictate the specifications needed for a particular system. Logic analyzers have found diverse applications in development and analysis of complex digital systems—from hardware timing and state analysis to software disassembly and performance analysis. For example, a hardware-debugging problem in which very short glitches are suspected requires an ultrafast sampling rate. On the other hand, a problem with a two-processor system requires time correlation between the processors, cross-triggering capability, and disassemblers for the particular processors in the system. System integrators need to be able to control microprocessors; software people want to be able to observe disassembled code. As a result of these and other diverse requirements for analyzers, manufacturers have produced analyzers that have modular capability to allow the user to optimize the analyzer for the particular application. Another example of a modular logic analyzer that is designed for a wide variety of digital system applications is the Tektronix Prism 3000 Series, illustrated in Figure 11–16.

One of the tools for complex troubleshooting in complex instruction set computers (CISC) that have 16- and 32-bit microprocessor systems is a **preprocessor,** also known as a disassembler pod. The preprocessor is designed to disassemble (take apart) the instruction codes that are being executed by the system and convert them into assembly language mnemonics. In some cases, the preprocessor is plugged into the system's microprocessor slot and the microprocessor plugs into the pod, simplifying connecting the analyzer to the target system. Any given preprocessor is designed to support only one microprocessor, but a given logic analyzer may have a number of preprocessors available. This avoids causing the analyzer from going out of date whenever a new microprocessor is introduced; all that is necessary is to acquire a new preprocessor pod. The direct connection of the pod also offers the advantage of reduced capacitive loading effects on high-speed processors, a problem with longer probe lengths.

A powerful feature offered on high-end logic analyzers is a built-in digitizing oscilloscope, giving the effect of two instruments in one. For added versatility, the analyzer function can be used to trigger the oscilloscope function, enabling time correlation between the analyzer and oscilloscope when they are displayed together on the same screen. The user can then examine a digital line in detail and determine pulse parameters such as rise time, preshoot, overshoot, pulse width, and other parameters that cannot be determined strictly with the logic analyzer function. Since the built-in oscilloscope is digital, it can make pulse parameter measurements automatically.

Other optional features are available to examine and store waveforms. Displays include split screens for viewing state and timing simultaneously or two processors at the same time. Most analyzers offer pattern-recognition, search, and cursor features. With cursors, the time between events can be measured, for example. Statistical information can be tabulated by some analyzers. If a system is

FIGURE 11–16
Tektronix Prism 3000 Digital Instrumentation System. The Prism 3000 has integrated high-speed data acquisition, microprocessor analysis, microprocessor control, and real-time performance analysis in one module. The high-speed data-acquisition module can acquire data at rates up to 2 GHz (copyright Tektronix, Inc. Used with permission).

placed in a repeating loop, statistics can be taken to reveal variations in timing for repeated measurements. Waveform storage, hard-copy printouts, and the ability to send data from one analyzer to another are other options that are available.

PC-BASED ANALYZERS

The analyzers discussed so far are basically stand-alone bench-type instruments. A variety of logic analyzers are available as PC- (personal computer) based plug-in cards that convert a PC into an analyzer. With the appropriate software, the PC's large memory can be an advantage for storing setups and various waveforms for reference. For example, a technician can call up waveforms obtained from a known good unit for comparison. Generally, PC-based analyzers cost less but are less convenient to use than dedicated instruments and have some performance limitations. One disadvantage of PC-based systems is the lack of portability. To use the instrument, the entire computer—including keyboard, monitor, and chassis—must be transported to the location where it is required. Another drawback to PC-based instruments is that the computer is tied up for the time the analyzer is needed. On the plus side, if a PC is already available, the incremental cost of the analyzer is low. PC-based analyzers can be configured with as many as 80 channels and can perform timing and state analysis. For state analysis, disassembly pods are available. If the computer is observing the same processor used as its own resident processor, the analyzer function can be set up to collect the data and the computer can exercise the system using the collected data. PC-based analyzers can also be used in conjunction with digital pattern generators.

11–6 LOGIC PULSER, CURRENT TRACER, AND LOGIC PROBE

The logic pulser and current tracer are hand-held instruments illustrated in Figure 11–17. They are useful for finding certain difficult faults, such as a short between a data line and ground. A problem like this can be very difficult to find because there are many possible paths for current. The current tracer responds to pulsating current by detecting the changes in the magnetic field. A hand-held logic pulser can provide nondestructive pulses of very short duration into the short circuit on a board in which the power has been removed. The current tracer, used in conjunction with the pulser, allows the user to follow the path of current directly to the short. This method of troubleshooting is also useful for "stuck" nodes in a circuit. A *stuck node* is a junction in a circuit that stays high or low despite the logic. The node has more than one path for current, so following the current path to its source will normally reveal the problem. The sensitivity of the current tracer can be varied to allow the user to trace different types of faults, including shorts in all types of electronic circuits.

The logic probe is another hand-held instrument that is useful for quickly checking the logic levels in digital circuits. It can be set for different logic families and provides the operator with an indication if the logic is HIGH or LOW or if pulses are present. Better probes can indicate the presence of very short pulses that are difficult to observe on an oscilloscope. It does this by "stretching" a very short pulse, allowing it to be easily observed. Since only one point can be observed at a time, it is most useful for troubleshooting when the circuit is latched in one state. In these cases, it provides a quick way to move around a circuit and test the logic, although it is limited to fairly simple problems.

Some newer probes can do more than provide just a logic indication. One combination probe is shown in Figure 11–18. It combines the function of a logic probe with an autoranging DMM to give the user an indication of the exact voltage level. It can also measure resistance and do diode and continuity tests.

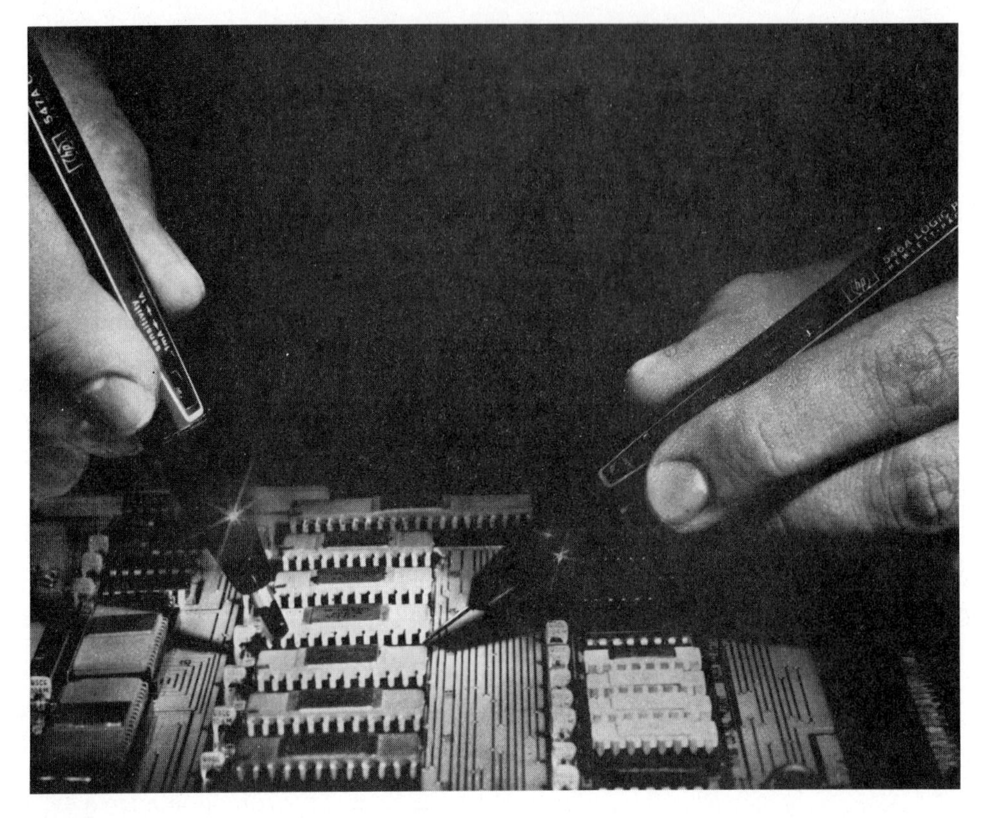

FIGURE 11–17
Hewlett-Packard 546A logic pulser and 547A current tracer. The pulser can act as a voltage or current source in digital troubleshooting applications. The current tracer can follow the path of current to a stuck node or other fault (courtesy of Hewlett-Packard).

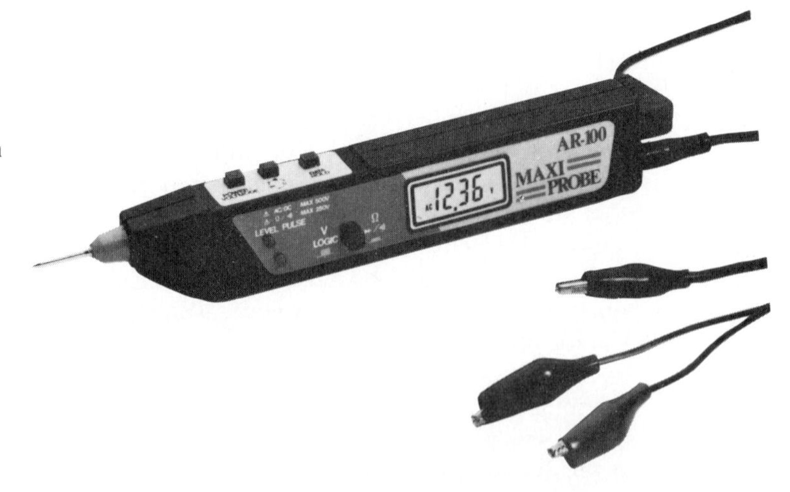

FIGURE 11–18
The American Reliance AR-100 is a hand-held combination logic probe and DMM (courtesy of American Reliance).

SUMMARY

1. Universal electronic counters can measure frequency, period, frequency ratios, time intervals, and do event counting. High-end counters can measure rise and fall time, pulse width, and duty cycle and can make statistical and data-reduction calculations.
2. An electronic counter contains input-conditioning circuits, control circuits, a high-frequency oscillator, a main gate, decimal-counting circuits, and a display. The input-conditioning circuits provide pulses of the correct amplitude and shape. Control circuits determine when the main gate sends pulses to the decimal-counting unit. The decimal-counting unit is connected to the display drivers and display. A stable oscillator generates a precise frequency for the time base.
3. Two techniques for reducing a high frequency for the electronic counter include using a prescaler or a heterodyne frequency converter.
4. The accuracy specification for frequency measurements is based on the accuracy of the oscillator and the resolution. It is normally written as ± 1 LSD (least significant digit) $\pm$ the time-base error.
5. Time-interval measurements can be made by starting the counter with a pulse on one channel and stopping the counter with a pulse from a second channel. Single-channel measurements can be made by obtaining the start and stop pulses from one channel.
6. A logic analyzer is a multichannel digital storage oscilloscope with one bit of vertical resolution and with flexible triggering and display modes. The most important application of logic analyzers is troubleshooting microprocessor and digital systems.
7. Two modes for using logic analyzers are as timing analyzers or as state analyzers.
8. Trigger conditions for a logic analyzer can be specified from either the data itself or from a special set of qualifiers.
9. The logic pulser and current tracer are useful for finding certain difficult faults, such as a short between V_{CC} and ground.
10. The logic probe can be set for different logic families and provides the operator with a quick indication if the logic on a single line is high or low or if pulses are present.

QUESTIONS AND PROBLEMS

1. A frequency counter is gated on for 10 ms and counts 540 pulses from a periodic input signal.
   (a) What is the input frequency?
   (b) If the gate time is changed to 100 ms, approximately how many counts would you expect from the same source during the gate time?
   (c) In what way does the change in the gate time affect the resolution?
2. An electronic counter has a 6-digit display and counts 500,397 pulses from a periodic input signal in a gate time of 0.1 s.
   (a) What is the frequency?
   (b) How can the display show this result?
   (c) What is the resolution of the measurement?
3. Assume the electronic counter in Question 2 has a time base accuracy of 3 ppm and ± 1 count. What is the accuracy of the measurement?

4. Compute the percent error for Example 11–3(a) for a time base of 1 s.
5. Explain why period counting is more accurate for measuring low frequencies than direct counting.
6. Describe the four types of errors that affect the universal electronic counter.
7. Compare a prescaler with heterodyne frequency mixing.
8. What is a glitch? How could you use a logic analyzer to find a glitch in a digital system?
9. Explain how you would search for the cause of a stuck node in a circuit.
10. Compare a timing analyzer with a state analyzer. Which is most useful for debugging a computer programming error?
11. Why is transitional sampling preferable for acquiring the timing diagram of an interrupt-driven system?
12. Compare the advantages and disadvantages of PC-based logic analyzers.
13. Explain how a logic pulser and current tracer can be used to find a short circuit.

Section Four

MEASUREMENT SYSTEMS

Chapter 12

Noise and Noise-Reduction Techniques

OBJECTIVES

Noise is defined as any undesired voltage or current. In communication systems, noise is often associated with static causing the desired signal to be less intelligible. Noise sources are both natural and artificial. Natural noise sources are highly random and include thermal noise, atmospheric noise, and semiconductor noise, whereas artificial noise includes sources such as motors, fluorescent light, power lines, and transmitters. Interference is when noise causes the unsatisfactory operation of a circuit.

There are two basic methods for reducing the effects of noise in electronic measuring systems. The first method is to reduce the noise at either the source or receiver through shielding, filtering, or proper grounding. The second method is to use signal-recovery methods to extract the desired signal from the noise. Signal-recovery methods can be very sophisticated and are beyond the scope of this book. When you complete this chapter, you should be able to

1. Describe sources of noise, related terminology, and types of noise sources.
2. Compare various intrinsic noise sources. Describe thermal, shot, and contact noise.
3. Given a specified bandwidth and temperature, compute the thermal noise for a resistor.
4. Give examples of sources of interference and compare the effects on circuits of electrostatic and electromagnetic sources. Explain how to reduce the effects of interference.
5. Given the input and output S/N ratio for an amplifier, compute the noise figure.
6. Describe how to measure noise with a true rms reading meter, an oscilloscope, or a spectrum analyzer.
7. List steps that can be taken to reduce noise in electronic sytems.
8. Discuss the advantages and disadvantages of various cabling systems (coax, twisted pair, ribbon) and their effect on noise.
9. Explain how to properly ground equipment racks for safety as well as for minimizing noise in sensitive equipment.

HISTORICAL NOTE

In 1965, Amo A. Penzias and Robert W. Wilson of the Bell Telephone Laboratories were making a careful study of radio noise capable of interfering with satellite communication systems. They were investigating microwave frequencies and discovered that wherever they pointed their precisely calibrated horn antenna, a background radiation was present corresponding to that of a perfect "blackbody" radiation source at a temperature of 3 K, which permeates the universe. (At one point, they had to evict pigeons from the antenna, but the noise remained.) Earlier, George Gamow and others had predicted a thermal background as fossil radiation evidence of a cosmic "big bang" theory of the formation of the universe; Penzias and Wilson's experiments led theoreticians to develop models of the very early history of the universe—within seconds of its formation. Penzias and Wilson were awarded a Nobel Prize in 1978 for their momentous discovery.

12–1 SOURCES OF NOISE

Noise can be characterized as any disturbance that tends to obscure a desired signal. Noise can be generated within a circuit or picked up from external natural or artificial sources. When noise is generated within a circuit, it is called **intrinsic noise.** When noise is picked up from an external source, it is called **extrinsic noise.**

Interference is noise that tends to obscure the useful signal. It is usually caused by electrical sources but can be induced from other physical sources such as mechanical vibration, acoustical feedback, or electrochemical sources. In addition to characterizing noise by its source, it is useful to distinguish noise by its frequency spectrum and amplitude distribution. When noise power has a flat frequency distribution, it is called **white noise** (from the analogy of white light containing all frequencies). If you double the bandwidth of a system, you double the white-noise power. Noise that is inversely proportional to frequency is called **$1/f$**, or **pink, noise.** Pink noise is present in many physical systems but is particularly important in low-frequency systems. In electrical systems, pink noise is generated when a current flows through a nonhomogeneous material, such as in a carbon-composition resistor (a mixture of carbon and other semiconductive materials). Pink noise is also generated in switches and other contact surfaces that are composed of dissimilar metals; it is then referred to as *contact noise.*

12–2 INTRINSIC NOISE

Intrinsic noise sets a lower limit on measurements and it is present in all electronic measuring systems. There are three important sources of intrinsic noise: (1) thermal noise generated by random motion of electrons in any resistance, (2) contact noise caused by the flow of current across the imperfect boundary formed between two materials, and (3) shot noise caused by the flow of current across a potential barrier, such as a *pn* junction.

THERMAL NOISE

Thermal noise is produced by the motion of free electrons in a resistance due to temperature. It is generated even when the resistance is not connected to a circuit but is due to the random fluctuations in charge at either end of the resistance. Thermal noise was originally studied by J. B. Johnson in 1928 and is often called Johnson noise. The noise power in thermal noise is constant per unit of bandwidth across the usable electronic spectrum. Because of this, it is a form of white noise. The maximum noise power available from a thermal noise source is given by the equation

EQUATION 12–1

$$P_n = kTB$$

where P_n = noise power, W
k = Boltzmann's constant, 1.38×10^{-23} J/K
T = absolute temperature, K
B = bandwidth of interest, Hz

Boltzmann's constant is the average energy per particle per kelvin. The energy of particles is directly proportional to their absolute temperature; hence the absolute temperature appears in the equation. Since the energy is distributed across the frequency spectrum, limiting the bandwidth of interest reduces the thermal noise power.

Thermal noise can be expressed in terms of a voltage for a resistor, as in Figure 12–1. Since noise is random and unpredictable, the noise voltage is given as the rms value. In order to compute the thermal noise that is delivered from a resistance, it is convenient to draw a Thevenin circuit composed of a noiseless resistor in series with a fictitious noise voltage source. A load resistor with no noise of its own is assumed. The maximum power than can be transferred occurs when the Thevenin source resistance R is equal to the load resistance R_L, as shown in Figure 12–1. The noise voltage is divided between R and R_L. Therefore, the noise power delivered to the load resistor is

EQUATION 12–2

$$P_n = \frac{(V_n/2)^2}{R}$$

Substituting Equation 12–2 into Equation 12–1 and solving for V_n results in

EQUATION 12–3

$$V_n = \sqrt{4kTBR}$$

FIGURE 12–1
Thermal noise in a resistor.

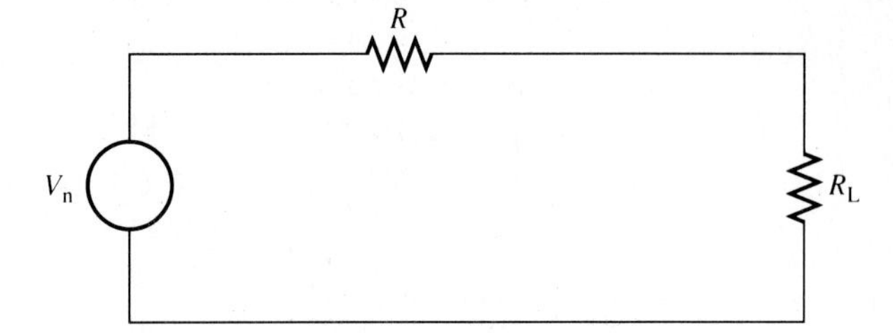

FIGURE 12–2
Noise current.

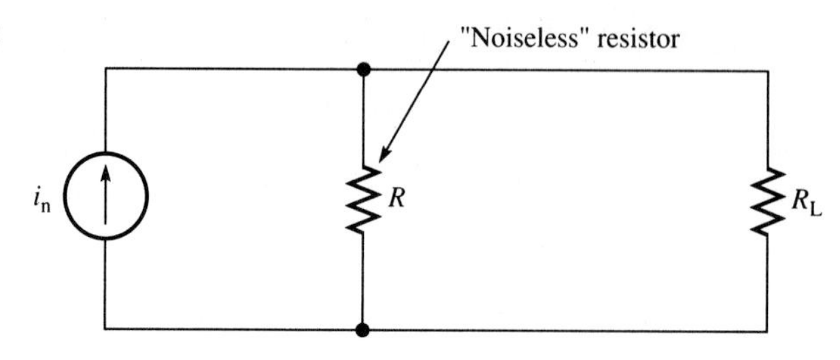

where V_n = equivalent rms noise voltage of a resistance, V
R = resistance, Ω

This equation has allowed us to find the equivalent Thevenin circuit of a thermal noise source of resistance R. This is the noise voltage found at the terminals of a resistor; it sets a lower boundary on the noise voltage from any resistive source. Notice that a perfect conductor has no thermal noise. Thermal noise in a practical system is limited by the bandwidth of the system but is small. For example, a room-temperature 10 kΩ resistor operating in a system with a 1 MHz bandwidth generates about 10 μV of noise. Note that the *thermal* noise has nothing to do with the physical size or composition of a resistor—it is the same for the most expensive low-noise resistor as for an ordinary carbon resistor, provided both resistors are the same value and are measured at the same temperature and bandwidth.

Thermal noise can also be modeled as a Norton current source. In this case, the thermal noise is considered to be a current source with a noiseless resistor in parallel with the current source, as shown in Figure 12–2. The magnitude of the current source is given by

EQUATION 12–4

$$i_n = \sqrt{\frac{4kTB}{R}}$$

where i_n = rms noise current of a resistance, A

EXAMPLE 12–1 Compute the rms thermal noise voltage for a 1 MΩ resistor. Assume a bandwidth of 100 kHz and a temperature of 100°F.

SOLUTION

$$100°\text{F} = 38°\text{C} = 311\ \text{K}$$

Substituting into Equation 12–3

$$v_n = \sqrt{4 \times 1.38 \times 10^{-23}\ \text{J/K} \times 311\ \text{K} \times 10^5\ \text{Hz} \times 10^6\ \Omega}$$
$$= 41\ \mu\text{V}$$

CONTACT NOISE

All resistors have noise voltages in excess of the thermal noise due to other noise-generation mechanisms. This additional noise is called **contact noise;** it is dependent on the quantity of current and the type of resistor. Contact noise is also called excess noise, flicker noise, or pink noise. Pink noise can be formed by passing white noise through a low-pass filter with roll-off of -3 dB per octave. Causes of contact noise are not well understood; however, it has been observed in many experiments. Types of resistors are (1) carbon-composition, (2) carbon film, (3) metal film, and (4) wirewound resistors. The noisiest of these is the carbon-composition resistor, and the quietest are the metal film and the wirewound types. In addition, variable resistors generate noise due to the wiper junction.

Contact, or flicker, noise is found in both field-effect transistors (FETs) and bipolar junction transistors (BJTs). Causes of such noise vary with the type of device. In bipolar transistors, it is a function of base current and leakage currents and increases as these currents rise.

Another type of frequency-dependent noise occurs at higher frequencies and is related to the transient time of charge carriers in the transistor. These effects are important when the period of the signal is comparable to the transit time of the charge carriers in the device. This occurs at very high frequencies—typically more than 500 MHz. Above these frequencies, FETs have an advantage over bipolar transistors in terms of noise because of faster transit times.

SHOT NOISE

The flow of current is not continuous in a circuit but rather is associated with random variations in the number of charge carriers passing some voltage boundary. Charge is limited by the smallest unit of charge available—that of the charge on an electron. Shot noise, like thermal noise, has the same power per unit of bandwidth; hence it is a type of white noise. When amplified, it sounds something like lead shot raining on a metal roof—hence the term **shot noise.** Shot noise is given in terms of a current and is found from the equation

EQUATION 12–5

$$i_{sh} = \sqrt{2eI_{dc}B}$$

where i_{sh} = rms noise current for shot noise, A

e = electronic change, 1.60×10^{-19} C

I_{dc} = dc current for which the shot noise is computed, A

B = bandwidth of interest, Hz

Shot noise occurs in virtually all active devices. The shot noise depends on a number of variables, so it is convenient to represent noise sources by assuming a noise-free device with external noise sources connected to it. One way of specifying the random noise contribution in an *active* device is to assign an "effective noise temperature" to the input. This temperature, labeled T_e, is added to the effective input noise temperature T_{in} to obtain an equivalent operating temperature of an active device. The noise temperature of a device does not mean that the device is actually operating at that temperature; rather, it gives an equivalent temperature of a thermal source with the same noise power. The output noise power of a transistor can then be written as

EQUATION 12–6

$$P_n = Gk(T_e + T_{in})B$$

where G = gain of the transistor
T_e = effective noise temperature of active device, K
T_{in} = effective input noise temperature, K
k = Boltzmann's constant, 1.38×10^{-23} J/K

COMBINING NOISE SOURCES

When two or more noise sources are connected, the superposition theorem can be applied to find the total noise. Since the rms voltages are represented by random variations in sources, the result is found by taking the square root of the sum of the squares. For series noise sources, the following equation can be used directly:

EQUATION 12–7

$$V_t = \sqrt{V_1^2 + V_2^2 + V_3^2 + \cdots + V_n^2}$$

where V_t = total rms noise, V
V_1 = voltage of source 1, V
V_2 = voltage of source 2, V
V_3 = voltage of source 3, V
.
.
.
V_n = voltage of source n, V

You can use this equation to combine as many noise sources as necessary to find the noise in a system.

EXAMPLE 12–2 The circuit shown in Figure 12–3 generates noise in the resistor from two sources: thermal noise in the resistor and shot noise in the diode. Compute the total noise voltage in the resistor at a bandwidth of 1 MHz from these sources. Assume the temperature of the resistor is 27°C.

FIGURE 12–3

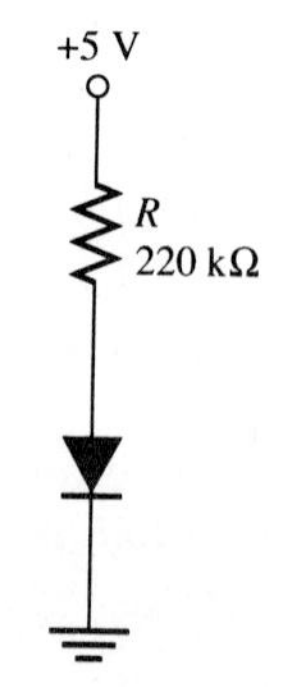

SOLUTION First compute the thermal noise voltage:

$$V_n = \sqrt{4kTBR}$$
$$= \sqrt{4 \times 1.38 \times 10^{-23}\ \text{J/K} \times 300\ \text{K} \times 10^6\ \text{Hz} \times 220 \times 10^3\ \Omega}$$
$$= 60.3\ \mu\text{V}$$

Next, compute the direct current in the diode.

$$I_{dc} = \frac{5.0\ \text{V} - 0.7\ \text{V}}{220\ \text{k}\Omega}$$
$$= 19.5\ \mu\text{A}$$

This direct current contributes to the shot noise, which can now be computed using Equation 12–5:

$$i_{sh} = \sqrt{2eI_{dc}B}$$
$$= \sqrt{2 \times 1.6 \times 10^{-19}\ \text{C} \times 19.5 \times 10^{-6}\ \text{A} \times 10^6\ \text{Hz}}$$
$$= 2.5\ \text{nA}$$

The voltage in R due to shot noise can be found using Ohm's law:

$$V_n = i_{sh}R$$
$$= 2.5\ \text{nA} \times 220\ \text{k}\Omega$$
$$= 550\ \mu\text{V}$$

The noise from both sources is

$$V_t = \sqrt{V_1^2 + V_2^2}$$
$$= \sqrt{(60.3\ \mu\text{V})^2 + (550\ \mu\text{V})^2}$$
$$= 553\ \mu\text{V}$$

12–3 EXTRINSIC NOISE

Extrinsic noise is induced from an external source and can cause unsatisfactory operation of a circuit (interference). The source of noise may be from another circuit on the same circuit board (often referred to as cross talk), or it may be external to the equipment. For interference to occur, there needs to be a source of noise and a means of coupling it into the circuit. The external source may come from conduction, capacitive coupling, magnetic coupling, or radiation. To reduce the effects of interference, the interference can be suppressed at the source, the source can be isolated by shielding or filtering, the coupling path can be reduced, or the receiving circuit can be made less sensitive to noise.

CONDUCTIVE COUPLING

Conductive coupling of noise requires at least two or more conductive paths to the noise source. There cannot be a complete circuit for the noise if the conductive

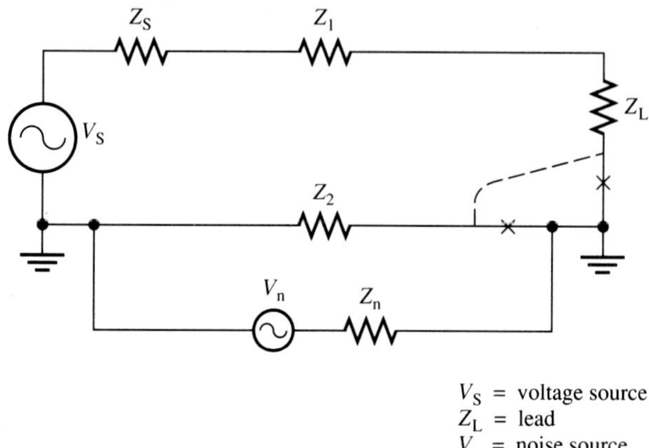

FIGURE 12–4
Hypothetical noise-source model. Noise is caused by a difference in reference (ground) potentials due to conductive coupling. The problem can be corrected by breaking the circuit at the x's and reconnecting as shown with the dashed line.

paths are reduced to only one conductor. Figure 12–4 shows a hypothetical noise model. The noise is induced into the circuit when the ground reference points are at different potentials—a very common problem in systems, especially when components are separated by long distances.

EXAMPLE 12–3 A computer in a CAD (computer-aided design) room is used to develop programs for a milling machine (see Figure 12–5). This machine operates by computer control and has its own memory for storing the tooling information, but the data file is sent to it via a 20 m cable from the CAD room. Both ends of the data cable have been connected to their respective chassis (the chassis is used as circuit ground reference). For safety purposes, the chassis has also been connected to the building's ground at each end. There is enough

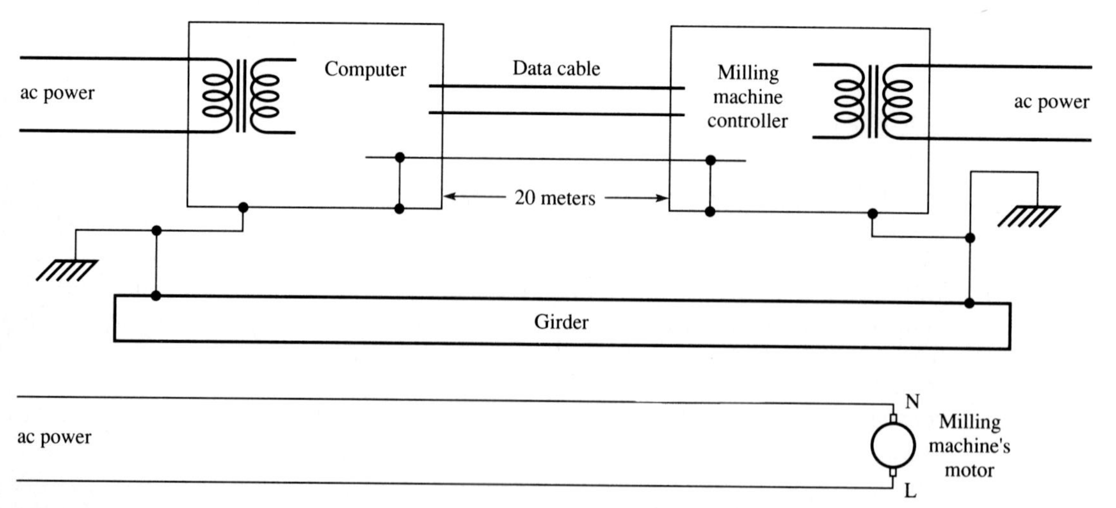

FIGURE 12–5
Example of ground potential differences causing conductive noise.

interference from a large motor starting to cause serious digital interface problems. The motor draws a few hundred amperes when it is started. This causes magnetically induced currents in the building's support girders, which produce small voltage potentials between the ground points within the structure. The chassis reference (also the signal reference) of the milling machine is forced to a different potential than the computer's chassis reference, causing erroneous data. What would be the easiest way to fix this problem?

SOLUTION Connect the power for the computer and the milling machine's controller to the same power source. (The motor should be connected to a different source.) The safety ground to the milling machine should be returned to the same point as the computer. This keeps the reference potentials equal.

One way to determine if a noise source is conductive is to view the noise on the reference with an oscilloscope. If the average voltage of the interfering signal is not at 0 V, then a conductive path is suspected (see Figure 12–6). Care must be taken to not "short out" the noise when connecting the oscilloscope's ground reference to the circuit. The oscilloscope must be referenced to the *signal* source's reference while measuring the load end's reference.

Some suggestions to reduce conductive noise follow:

1. Be sure the signal ground return and the power ground return are on different wires.
2. Be sure equipment that can be a source of high interference (motors, arc welders, etc.) is on different power-distribution circuits from sensitive measuring equipment.

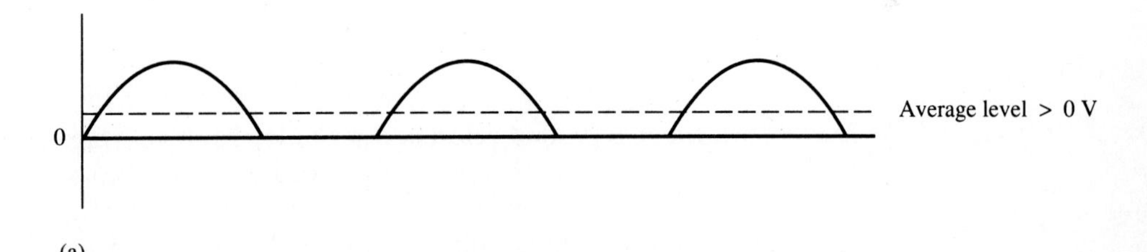

(a)

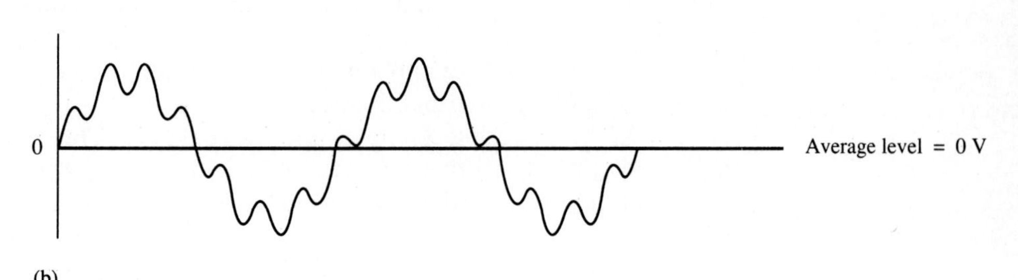

(b)

FIGURE 12–6
The signal in (a) is caused by conductive currents, whereas the signal in (b) is not as likely to be caused by such currents.

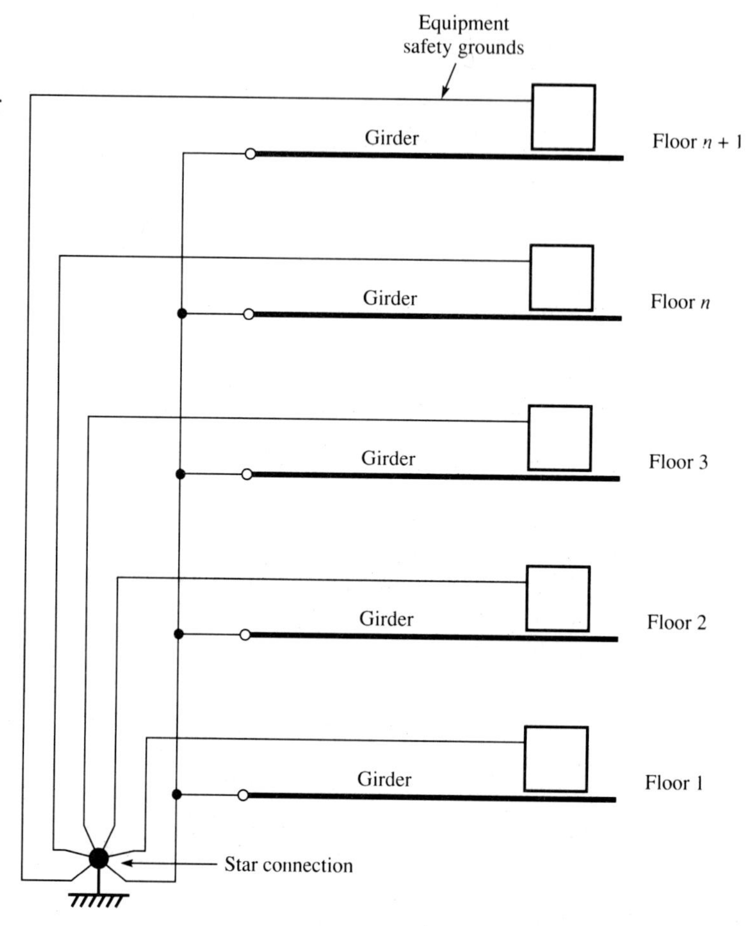

FIGURE 12–7
Grounding technique for a building. The electrical safety ground for each floor should return to a common "star" point.

3. Establish the signal ground reference at the lowest-level stage and connect all the succeeding stages to this ground by separate conductors. The ground return wire should be in close proximity to the supply wire.
4. Evaluate the earth grounds for separate buildings (or, in a tall building, the separate floors) for a common ground point (star connection), as shown in Figure 12–7.
5. Break the electrical path so only one conduction path exists. Never do this by breaking the safety ground path, however.
6. Short out V_n by lowering Z_2 (refer to Figure 12–4). This can be done by strapping equipment together with heavy bus wiring (less than 10 gage).
7. Block high-frequency conductive noise through filtering, as shown in Figure 12–8.
8. Bypass high-frequency conductive noise with bypass capacitors.

ELECTRIC AND MAGNETIC FIELDS

When a current flows in a conductor, electric and magnetic fields are present. We can create a picture of a field by drawing vectors in space that represent the

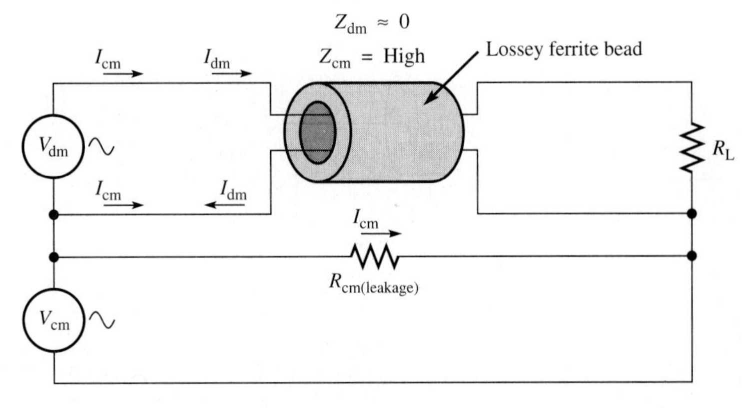

FIGURE 12–8
High-loss ferrite bead blocks I_{cm} noise.

position and direction of the fields, as shown in Figure 12–9(a). The current produces a magnetic field surrounding a long conductor, which decreases linearly with distance away from the conductor. An electric field is present whenever there is a potential difference between two points and the electric field lines join the points. In a current-carrying conductor, electric field lines leave the conductor in a direction perpendicular to the conductor. If a grounded, nonmagnetic shield is placed around the conductor, as in Figure 12–9(b), the electric field lines terminate on the shield; however there is no effect on the magnetic field lines.

The characteristics of the electric and magnetic fields are determined primarily by the source and the distance from the source. Close to a source (less than $\lambda/2\pi$), the field can be either predominately electrical or predominately magnetic.

EQUATION 12–8

$$\lambda = \frac{3 \times 10^8}{f}$$

where f = frequency, Hz
λ = wavelength, m

This region is known as the **near field.** High-current, low-voltage sources cause the field impedance to be less than 377 Ω, and so the near field is primarily magnetic. Conversely, low-current, high-voltage sources cause the field impedance to be greater than 377 Ω, a near field that is primarily electric. Equation 12–9 shows the concept.

EQUATION 12–9

$$\frac{\frac{dV}{dt}}{\frac{dI}{dt}} = Z \cong \frac{\Delta V}{\Delta I}$$

At distances near $\lambda/2\pi$ and greater, the field impedance becomes a constant 377 Ω, independent of air or free space. Beyond this region, known as the **far field,** the field travels as a plane wave, with neither magnetic or electric field dominant. The field is said to be an electromagnetic field, and interference from it is given the

FIGURE 12–9

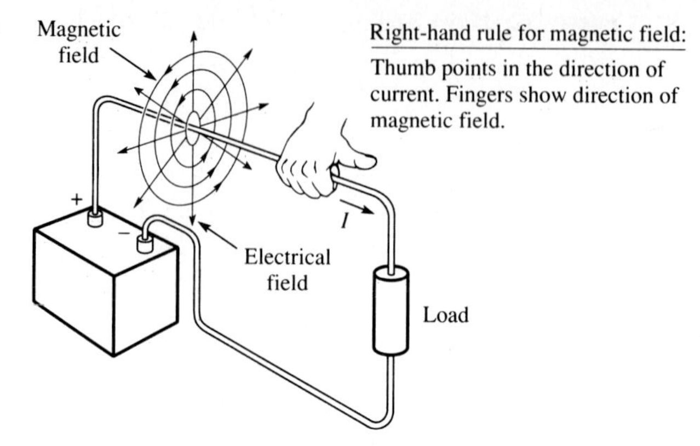

(a) Magnetic and electrical field around a long unshielded conductor. The magnetic field lines encircle the conductor in the direction given by the right-hand rule.

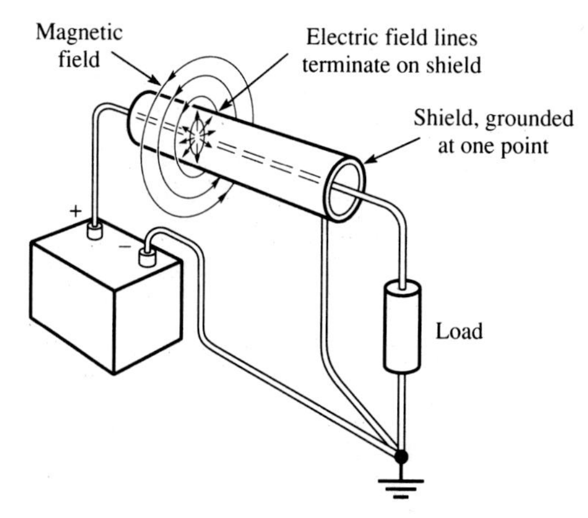

(b) Magnetic and electrical field lines around a shielded conductor. The shield carries no current and is grounded at one point.

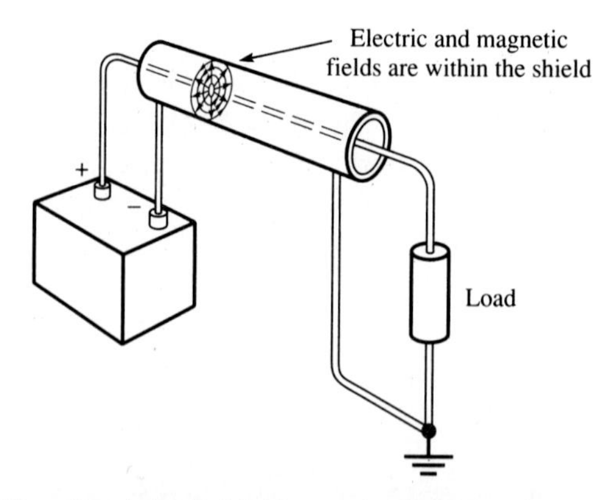

(c) Magnetic and electrical fields around a shielded conductor. The shield carries a current equal to the current in the conductor in the opposite direction.

name electromagnetic interference, or EMI (the older term RFI, for radio-frequency interference, is still in common use).

Noise can be induced in a conductor by either capacitive coupling, from the electric field, inductive coupling, from the magnetic field, or electromagnetic coupling, from a combination of these fields. EMI can originate from a number of sources, including radio transmitters, microwave ovens, motors, and natural sources. Shielding for EMI can be done at either the source or the receiver.

ELECTRIC FIELD INTERFERENCE

Electric field interference is also called **capacitively coupled interference.** All configurations of conductors have capacitance between them, allowing a coupling path for noise. Parallel conductors abound in electronic circuits and include ac wiring, power supply distribution lines, signal lines, and multiple conductors on printed circuit (PC) boards. Figure 12–10 shows two metallic conductive surfaces. The capacitance between the surfaces is

EQUATION 12–10

$$C = \epsilon \frac{A}{d}$$

where C = capacitance, F

ϵ = permittivity of the material between the surfaces, F/m

A = surface area, m^2

d = distance between the surfaces, m

This equation is for the special case of a parallel plate capacitor with two plates. The general equation is given in Equation 13–1.

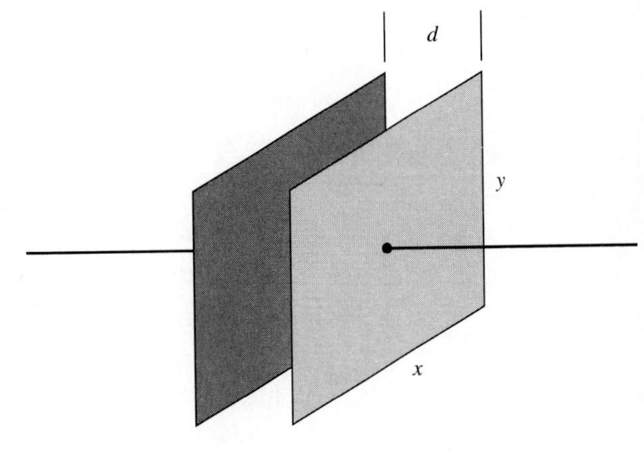

FIGURE 12–10
Parallel conductive surfaces form a capacitor.

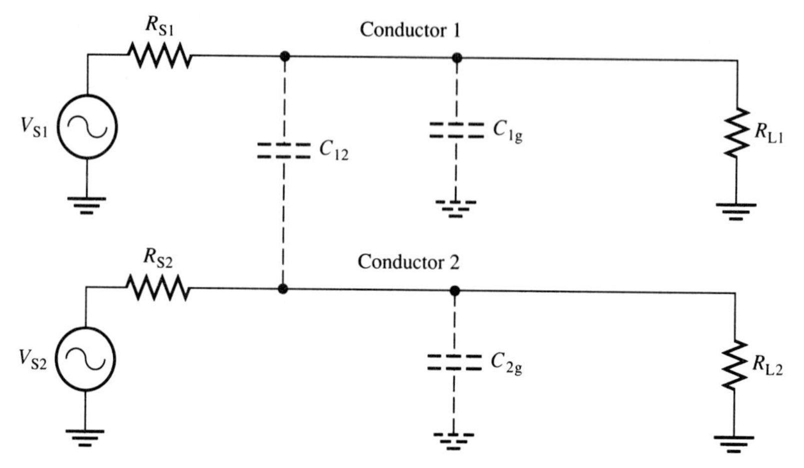

(a) Capacitive coupling of voltage from V_{S1} to R_{L2}.

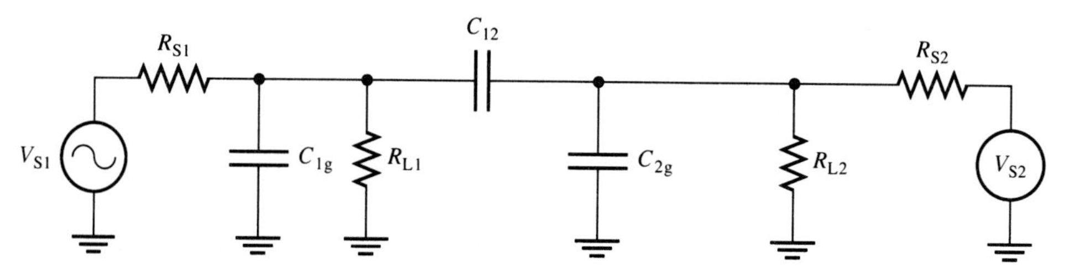

(b) Equivalent circuit.

FIGURE 12–11
Parallel conductors and the stray capacitance.

Recall that the impedance of the capacitor is

EQUATION 12–11

$$X_C = \frac{1}{2\pi fC}$$

where X_C = capacitive reactance, Ω
f = frequency, Hz
C = capacitance, F

The capacitance between two parallel conductors and the stray capacitance to ground is illustrated in Figure 12–11(a). If an alternating current is present in conductor 1, conductor 2 is affected by the capacitive coupling between the circuits. Electric signals from the first circuit induced into the second are directly proportional to the capacitance between the conductors, as illustrated in the equivalent circuit shown in Figure 12–11(b). The equivalent resistance R is $R_{L2} \| R_{S2}$. If the resistance of circuit 2 is much smaller than its stray capacitive reactance to ground, the induced noise voltage can be computed by applying the voltage divider to R and C_{12}. The induced voltage V_{RL2} due to V_{S1} is given by

EQUATION 12–12

$$V_{RL2} = V_{S1} \frac{R}{(R - jX_{C12})}$$
$$= V_{S1} \frac{R}{(R - j/2\pi f C_{12})}$$

Additionally, if R is small compared to X_{C12}, this equation reduces to

EQUATION 12–13

$$V_{RL2} = j2\pi f R C_{12} V_{S1}$$

Equation 12–13 shows that the induced voltage V_{RL2} between conductors can be reduced by making the capacitance between the conductors smaller, reducing the impedance of the affected circuit, decreasing the source voltage, or lowering the frequency of the noise source. The capacitance between conductors can be made smaller by separating the conductors (larger d) or by decreasing the total area (smaller A). An example of decreasing the area is shown in Figure 12–12(a), where two traces run parallel with each other, separated only by the thickness of the circuit board. If the traces run perpendicular, as shown in Figure 12–12(b), then the cross-sectional area is less, reducing capacitive coupling. Lowering the impedance of the circuit can be done at either the source end of the line (R_{S2}) or the load end (R_{L2}), as shown in Figure 12–11.

Lowering the operating frequency or source voltage is usually not practical; however if the frequency of the desired signal is vastly different (by at least a factor of 10) from that of the interference, then filtering techniques can be used at the end of the line, preferably just before R_L. The filter should bypass the interference signal to ground and at the same time block it from reaching R_L. Figure 12–13 is an example of a simple filter. More-complex filters will be needed if line impedance matching, frequency response, and/or phase-shift characteristics are of importance.

Another method of reducing capacitive coupling is through the use of shielding. A shield is a conductive barrier that is placed between the source of the interference and its potential receptor. The shield must be grounded in such a way as to *return* the electric field of the noise *back to its source*. Ground connections need to have very low contact resistance (less than or equal to 4 mΩ) and be able

FIGURE 12–12
Capacitance in (a) is greater than in (b).

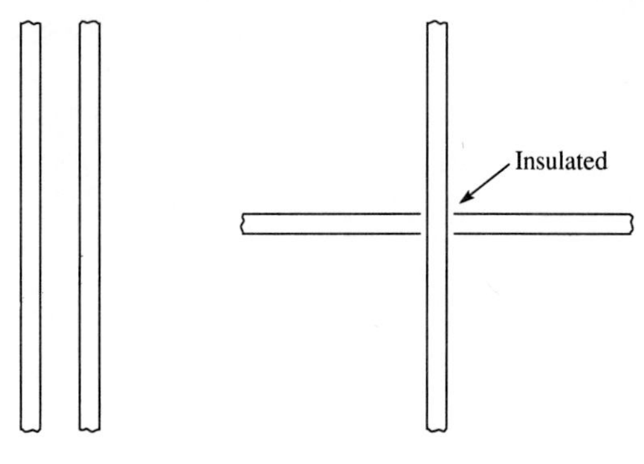

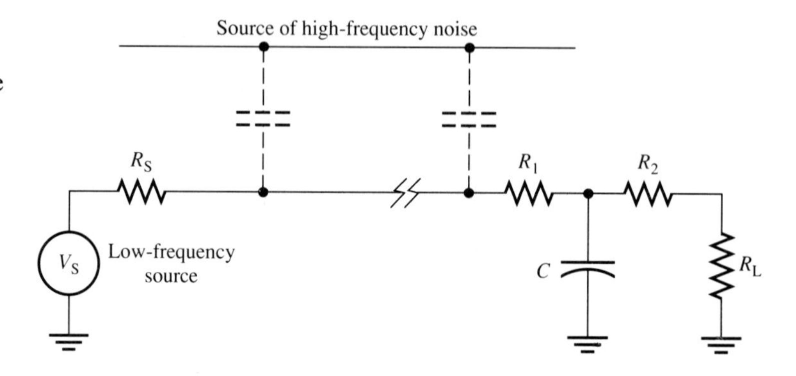

FIGURE 12–13
Capacitor C bypasses high-frequency noise and the resistors block the noise.

to maintain this resistance over time, withstanding the effects of possible corrosion or abrasion. Figure 12–14 illustrates the concept of shielding; the electric field is contained between the shield and the noise source, as shown. The following rules must be used when considering shielding:

1. The shield must be grounded and not left electrically floating. A floating shield is worse than no shield at all because it can increase the coupling capacitance as well as ϵ. The unused wires in a cable should be grounded at one end (the signal-source end) and not left disconnected.

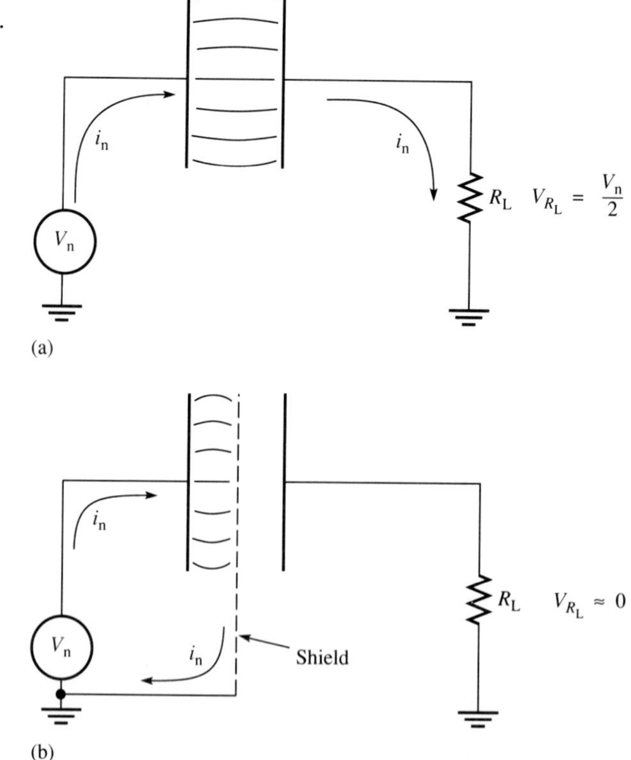

FIGURE 12–14
The effects of shielding.

2. The shield must return the noise back to the noise source; otherwise it may actually make the electric field interference worse.
3. If both ends of a shield on a cable are grounded, then a conduction path can be established, causing conduction problems (as described earlier); therefore, consider potential problems that may arise (often referred to as *ground loops*).
4. Unless the shield has magnetic shielding properties, it will not have any effect on magnetically induced interference.

Figure 12–15 illustrates a coax cable used to carry a signal to R_L. Coax, as discussed in Section 1–4, is shielded cable in which a conductor is surrounded by a metal sheath or braid separated by a dielectric. The cable follows along a steel girder and picks up electric interference from the girder. As shown in Figure 12–15(a), connecting the shield to the wrong ground applies the noise source to the shield, reducing the distance (d) between the plates of the noise-coupling capacitance. Connecting the shield as shown in Figure 12–15(b) causes the noise to be returned to the noise source, completing its electrical path.

FIGURE 12–15
An improperly connected shield can make electrostatic interference worse.

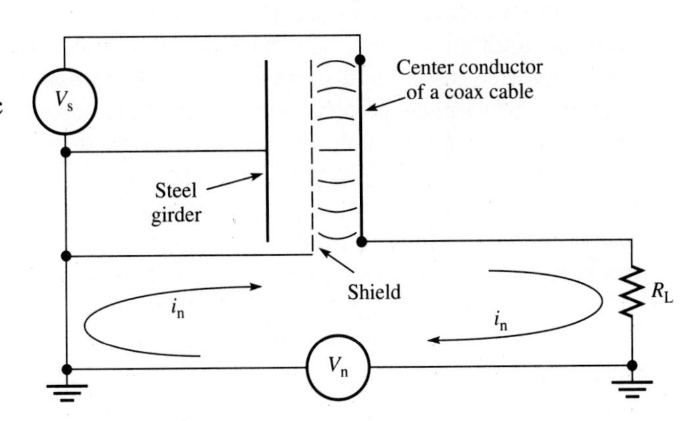

(a) Shield connected to the wrong ground.

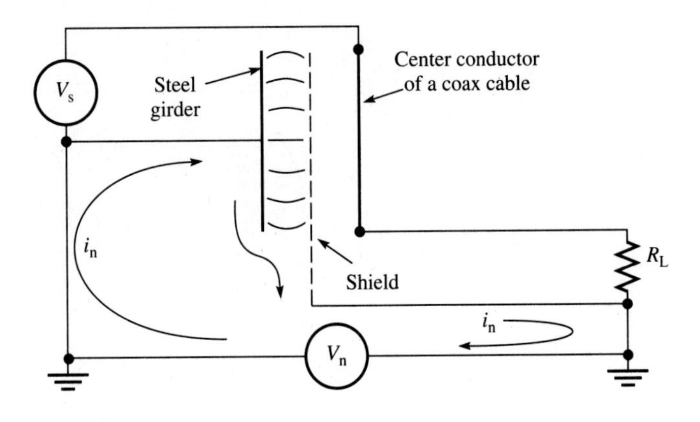

(b) Shield connected to the correct ground.

MAGNETIC FIELD INTERFERENCE

One early observation of noise occurred shortly before the Civil War when the transatlantic cable was completed. The cable and the earth formed the largest closed loop that had been constructed to that time. Signals could be heard on the cable even when no one was sending a message! The noise was caused by fluctuations in the magnetic field of the earth, which, in turn, induced currents in the cable that were mistaken for telegraph messages.

Faraday's law states that a changing magnetic field in a loop of wire will induce a current in the loop. Conversely, a magnetic field is produced whenever charge is moved. Magnetic field interference occurs when a current in one circuit can produce a current in another circuit due to the mutual inductance between the circuits. The induced current is related to the loop area of the receiving circuit, the rate of change of current in the source, and the distance separating the source and receiver. It follows that the induced voltage is increased with higher frequencies and larger currents. Magnetic field interference can come from inductors, transformers, conductors, or any low-impedance source in a circuit.

Mutual inductance is present when magnetic flux lines from one circuit are linked to another circuit. The mutual inductance is dependent on the physical distance, orientation, and dimensions of the inductive elements as well as the presence of magnetic materials in the vicinity, including shielding materials. The strength of a magnetic field is inversely proportional to the cube of the distance from the source. Magnetic field lines tend to concentrate in magnetic materials, such as iron, causing the field to be minimized outside the magnetic material. For example, air-core inductors produce more noise than equivalent iron-core inductors because they generate magnetic fields that are not contained within the core.

If a conductor is coiled as shown in Figure 12–16, then its self-inductance (L) is given by Equation 12–14. Notice that L is proportional to the area of the loop.

EQUATION 12–14

$$L = \frac{\mu A N^2}{l}$$

FIGURE 12–16
Self-inductance is proportional to the cross-sectional area.

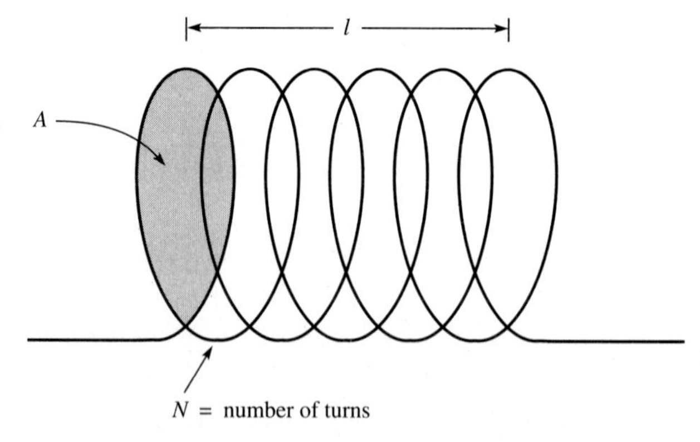

where L = inductance, H
μ = permeability of the core, H/m
A = cross-sectional area, m^2
l = length of the coil, m
N = number of turns, dimensionless

If the length is reduced to some unit value, then Equation 12–14 reduces to

EQUATION 12–15

$$L' = \mu A N^2$$

where L' = inductance per unit length, H/m

Figure 12–17 illustrates that a wire forming a single coil will have its greatest inductance when the coil is opened to its fullest area. A larger inductance implies a larger impedance, as given by Equation 12–16.

EQUATION 12–16

$$X'_L = 2\pi f L'$$

where X'_L = inductive reactance per unit length, Ω/m
f = frequency, Hz
L' = inductance per unit length, H/m

Substituting Equation 12–15 into 12–16 and letting $N = 1$ yields

EQUATION 12–17

$$X'_L = 2\pi f(\mu A)$$

Equation 12–16 indicates that the path of least impedance to a high-frequency signal is the path with the least cross-sectional area (also seen as the least inductance).

An interesting demonstration of how loops affect high frequencies can be constructed with a loop of coaxial cable as shown in Figure 12–18. The shields are soldered together at the top of the loop. Both rf and dc currents are fed into the center conductor of the cable, and at the other end, they flow through the terminating resistor and onto the shield. The dc current will take the path of least resistance and will return to the source via the solder bridge, whereas the rf signal will take the path of least impedance and return on the shield itself. The shield and center conductor form a flattened loop with a smaller cross-sectional area than if the rf signal returned across the solder bridge.

FIGURE 12–17
Loop (a) has greater inductance (L) than loop (b), even though the circumferences are the same.

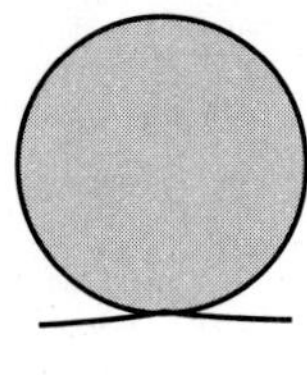

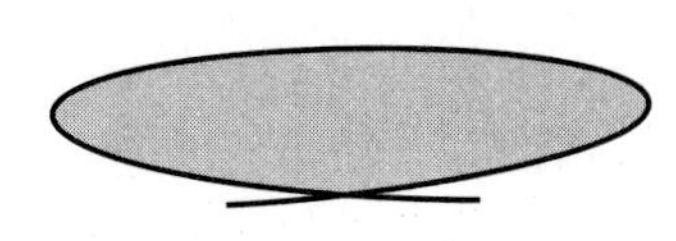

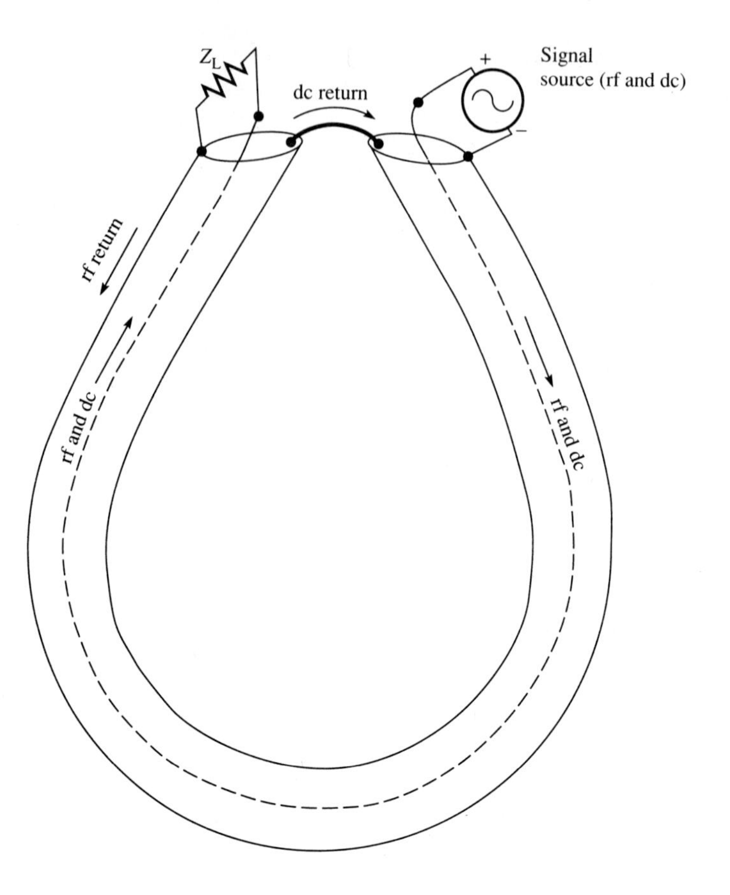

FIGURE 12–18
A loop of coax cable has its shield soldered together at both ends. The dc signal returns to the source across the solder bridge, whereas the high-frequency rf returns to the source via the shield.

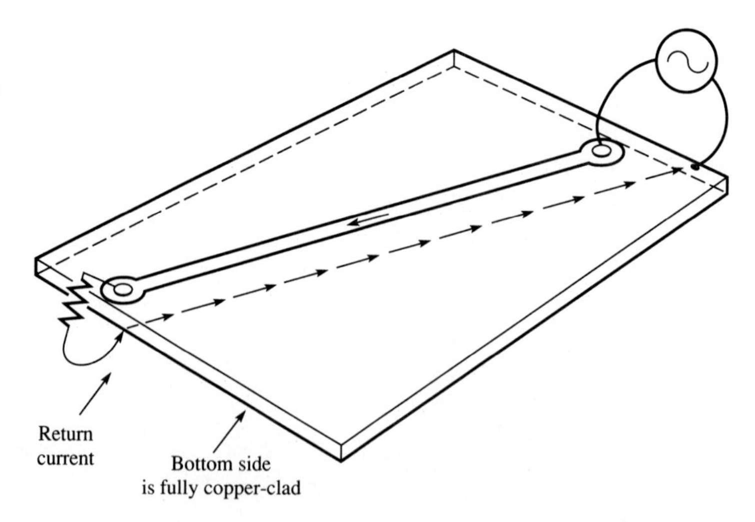

FIGURE 12–19
Return current is directly under the supply path, where the loop area is the smallest.

The current-loop effect is noticed on circuit boards as well. Figure 12–19 shows a circuit board with a solid copper clad (**ground plane**) on the bottom side and a single trace on the top side. An rf signal is applied to the trace with a return path on the ground plane. The return current on the ground plane is directly below the supply trace where the path loop area is the smallest.

Sometimes solutions to problems can cause other problems. Consider a circuit board that is designed with split ground planes, in order to isolate ground loops, as shown in Figure 12–20. One ground plane is connected to a microprocessor and contains digital circuits associated with it; it is known as the *digital* section. Sensitive measurement circuits are connected to the other ground plane; it is known as the *analog* section. The reason for the split ground plane is to keep ground noises from the digital section from being injected into the ground plane used by the analog section. The two ground planes are connected to assure they are at equal voltage potential. To avoid the heavy digital currents on trace B from coupling to the sensitive circuits on trace A, the two traces were placed on opposite sides of the circuit board. Trace B crosses a gap on the plane; the return currents have to travel around the gap back to the source. This forms a loop, which magnetically couples current into trace A (notice that trace A is not far away from the heavy ground return currents). A solution to this problem would be to continue the ground plane under trace B.

Effective shielding for magnetic field interference requires the use of materials that provide a good magnetic path for magnetic flux lines. Shielding can be done at the source or receiver.

Figure 12–21 shows two separate circuits, which have some mutual inductance between them. Either circuit can induce current into the other, but for simplicity, we will only consider the current I_{12} induced into circuit 2 by the

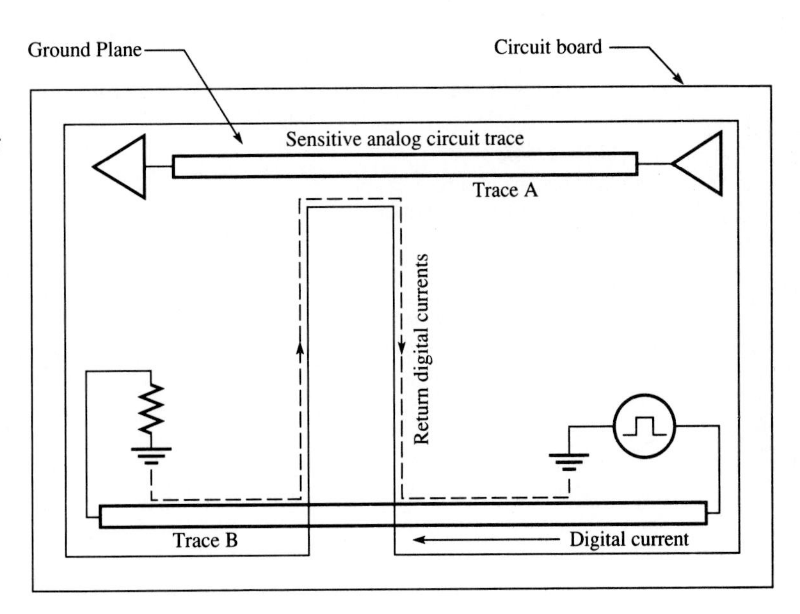

FIGURE 12–20
The split ground plane causes the return current from trace B to pass near trace A.

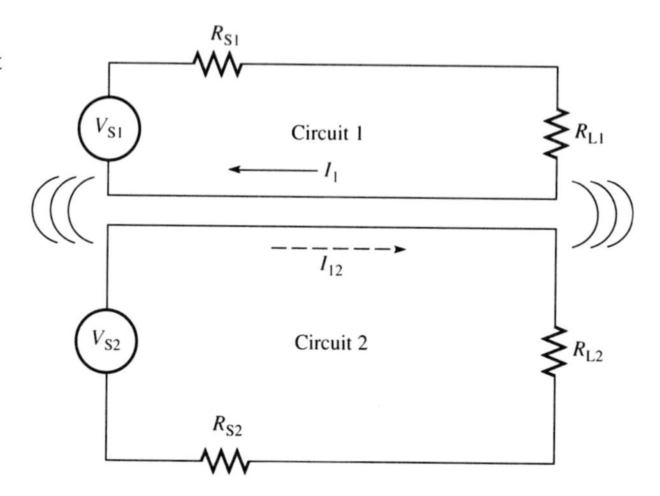

FIGURE 12–21
ac current I_1 induces current I_{12} in the adjacent circuit through magnetic induction. Lowering R_{S2}, R_{L2}, or both will increase I_{12}.

changing current in circuit 1. Recall that for situations where interference is capacitively coupled, decreasing R_2 ($R_{L2} \| R_{S2}$) reduces the effects of the interference; but if the interference is magnetically induced, then reducing R_2 *increases* the induced current in circuit 2.

For Figure 12–22, the voltage induced from circuit 1 to circuit 2 is given in Equation 12–18.

EQUATION 12–18

$$V_1 = V_S + L_1 \frac{di_1}{dt} + M \frac{di_2}{dt}$$

where V_1 = voltage in circuit 1 across R_1, V
L_1 = self-inductance, H
M = mutual inductance (the inductive effect of circuit 2), H
$\frac{di}{dt}$ = a rate change in current over a time interval

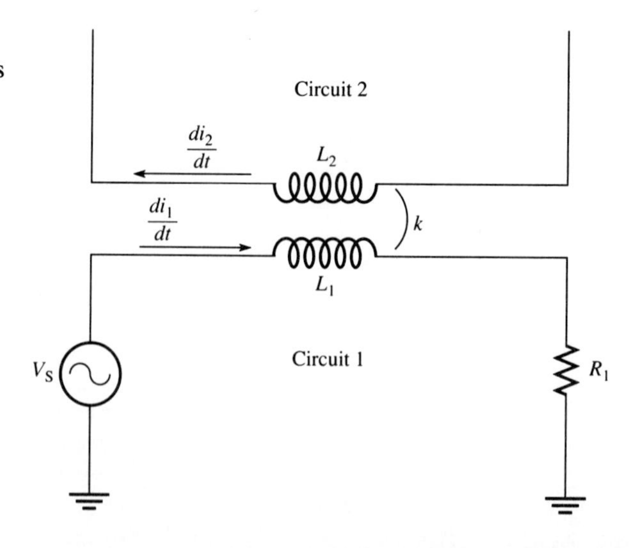

FIGURE 12–22
The voltage drop across R_1 is proportional to the induced current in L_1.

Also

$$M = k\sqrt{L_1 L_2}$$

where k = coefficient of coupling ($k \leq 1$)

Some of the considerations of magnetically coupled interference are as follows:

1. Lowering the loop resistance increases the interference.
2. Your body movements about the circuit do not affect the coupling.
3. Nonmagnetic materials acting as shields have no effect on the interference.

Switching currents in integrated circuits (ICs), particularly in CMOS devices, tend to affect other circuits through the supply and ground paths; Figure 12–23 illustrates this concept. IC_1 produces high-frequency current spikes (a source of noise) as it switches logic states. The current will return to its source (IC_1) through the shortest loop, in this case, through IC_2. The resistance of IC_2 will cause a voltage drop between ground and V_{CC}, which can be seen with an oscilloscope. IC_1 may also interfere with IC_2 through direct conduction currents. In addition to direct conduction interference, a loop can be formed from IC_1 that will cause electromagnetic interference to other adjacent circuits. By placing bypass capacitors across each IC (power to ground), the noise current is routed back to

FIGURE 12–23

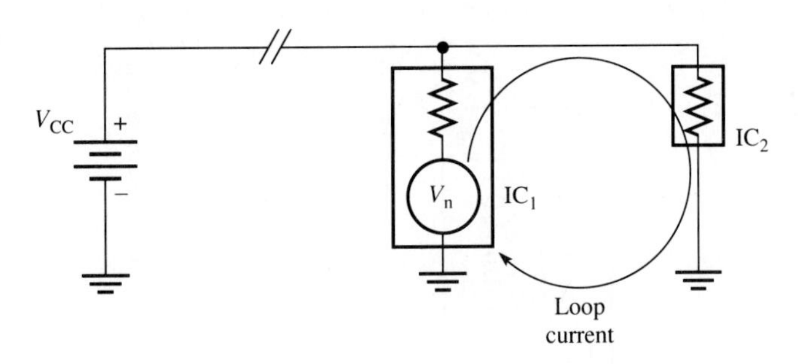

(a) Loop currents caused by switching.

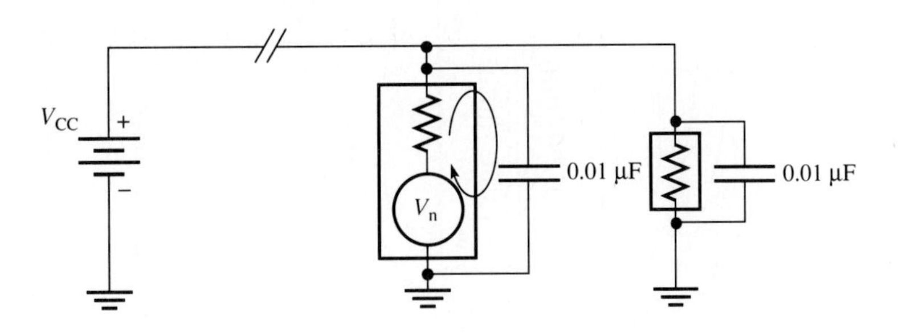

(b) Loops are reduced in size by bypass capacitors.

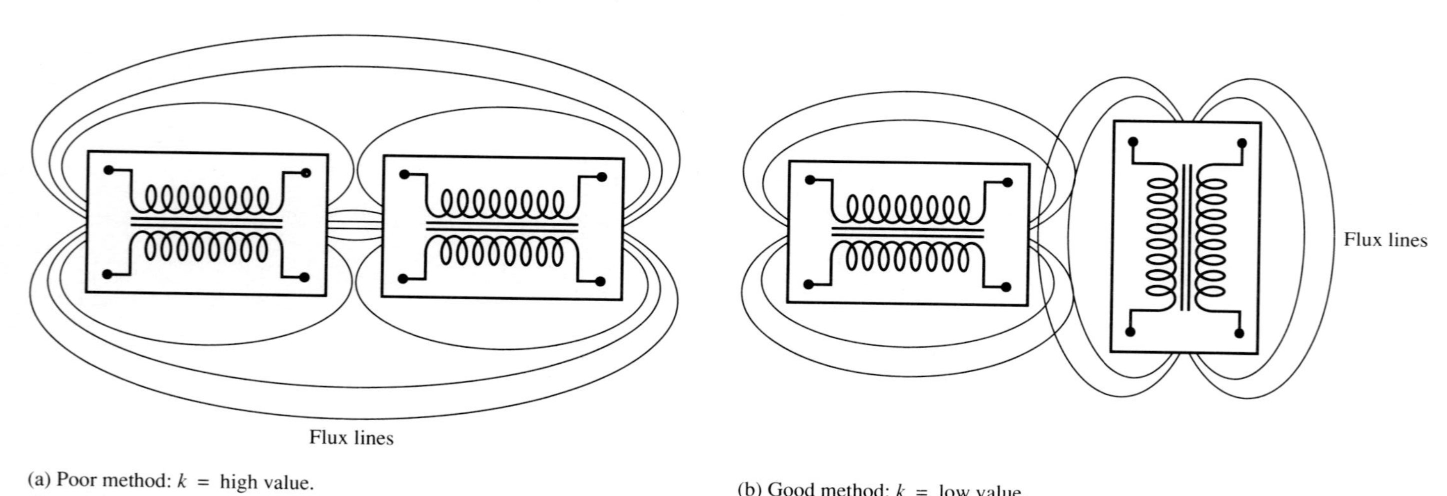

(a) Poor method: k = high value.

(b) Good method: k = low value.

FIGURE 12–24
Magnetic interference can be improved by component layout.

the source via the smallest loop possible. In addition, the bypass capacitors tend to shunt extrinsic noise coupled to the power supply rails to ground.

Recapping, some methods used to reduce magnetic interference are as follows:

1. Reduce loop areas.
2. Reduce the noise current source.
3. Reduce noise frequency, if possible.
4. Reduce the coefficient of coupling (k) by separating circuits or by changing the orientation of the components, as shown in Figure 12–24.
5. Add filtering and bypassing capacitors.

EMI

Thus far, we have discussed fields of interference that are predominantly magnetic or electric; both of these are near-field effects. But as you move further away from the source of interference, to distances greater than $\lambda/2\pi$, the electrostatic and magnetic fields tend to become equal to each other. This is referred to as far-field radiation, and the natural impedance is about 377 Ω. Examples of far-field radiation are noise pulses from electric welders, lightning, radio transmitters, or radiation from a circuit that has high-frequency parasitic oscillations.

The receiver must have an effective antenna large enough to capture the energy from the radiating source. The antenna surface needs to have either a linear size greater than $\lambda/20$ m or a total area of $\lambda^2/100$ m^2.

Any material will either absorb, reflect, or allow an EMI field to pass through. An X-ray picture of a tooth is possible because a high-frequency wave passes through the tissues of the head but is absorbed to varying extents by the teeth, fillings, and bones of the face. Radar is an example of reflected radiation in which the returned wave is an echo of the incident wave.

To understand absorption, you need to understand skin effect. **Skin effect** refers to a nonuniform layer of current on the surface of a conductor caused by time variation in the current. Current density is greatest on the surface of a conductor when in the presence of an electromagnetic field, as shown in Figure 12–25. This is because the conductor tends to short out the electric field as the signal penetrates deeper. The currents in the conductor produce a magnetic field, in opposition to the incident field; the magnetic field is then attenuated at the same rate as the electric field is attenuated. One unit of skin depth is given by

EQUATION 12–19

$$T_s = \frac{1}{2\pi\sqrt{\mu\sigma f}}$$

where T_s = unit skin depth, m

μ = magnetic permeability, H/m

σ = resistivity, S/m

f = frequency, Hz

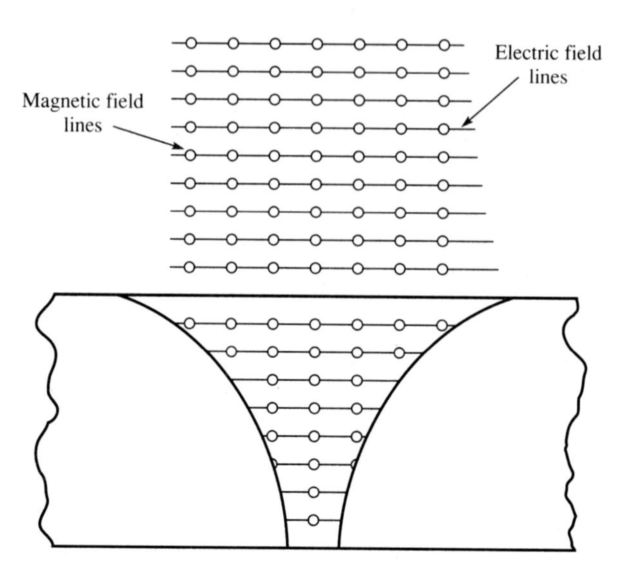

FIGURE 12–25
Attenuation of EMI through skin effect.

Each unit skin depth, by definition, has an attenuation of −8.68 dB (called 1 neper). The number of skin depths in a conducting material is simply the thickness of the material divided by one unit skin depth.

EXAMPLE 12–4 A 0.25 mm–thick copper plate is used as an EMI shield. Copper has a resistivity of 5.8×10^7 S/m and a magnetic permeability of 10^{-7} H/m. A frequency of 10 MHz is the radiation source. What is the attenuation of the signal through the copper shield?

SOLUTION Given $\mu = 10^{-7}$ H/m, $\sigma = 5.8 \times 10^7$ S/m, and $f = 10$ MHz

$$T_s = \frac{1}{2\pi\sqrt{\mu\sigma f}} = 21 \times 10^{-6} \text{ m} = 0.021 \text{ mm}$$

Finding the number of skin depths:

$$\frac{0.25 \text{ mm}}{0.021 \text{ mm}} = 11.9 \text{ skin depths}$$

$$\text{Attenuation} = -8.68 \text{ dB/skin depth} \times 11.9 \text{ skin depths} = -103.3 \text{ dB}$$

Everything will absorb electromagnetic radiation to some extent; the difference is in the thickness of the unit skin depth. For example, paper can attenuate the EMI by the same amount as the copper shield in Example 12–5, but it will need to be several feet in thickness to achieve the same result.

If a conductor is to be effective as a shield against EMI, it must be rf tight. For example, wires in a cable can radiate and receive interference. Not only does the entire length of the cable need shielding, but the wires at the connectors must be shielded, as shown in Figure 12–26.

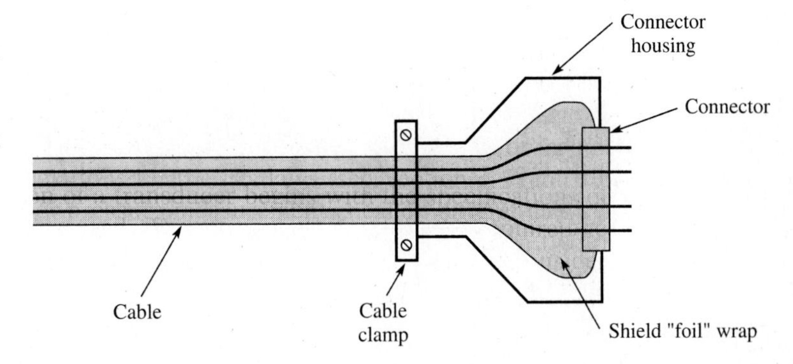

FIGURE 12–26
The shield in the cable and the shield in the connector must join to form a continuous, rf-tight barrier.

As stated earlier, high-frequency EMI causes the current to flow on the surface of a conductor. Current induced on the surface of a shield must be returned to its source and remain on the noise side of the shield. Figure 12–27(a) shows a shield used to enclose radiation from a noise source. For convenience of assembly, the shield was soldered on the outside to chassis ground. As shown, the skin effect causes the current to flow through the cooling hole and continue outside of the shield to the ground point. This causes secondary radiation effects. A better method is to ground the inside of the shield through a flange and mounting hardware, as shown in Figure 12–27(b).

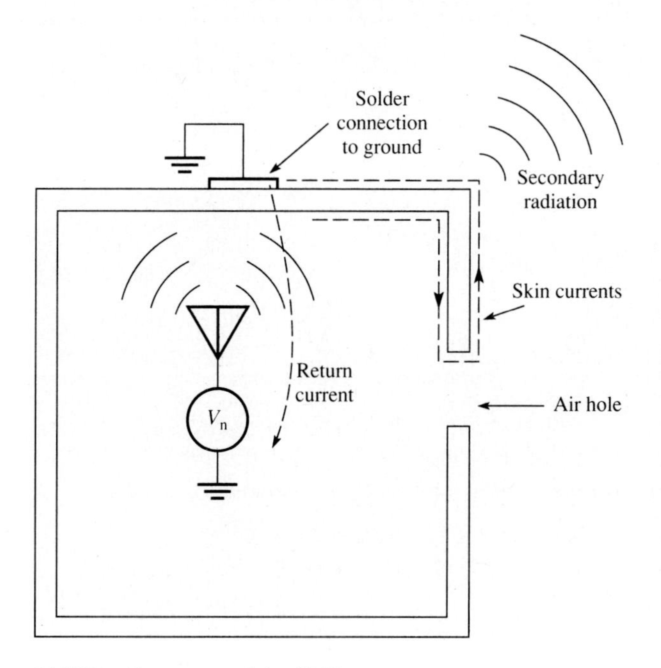

(a) Wrong place to ground the shield.

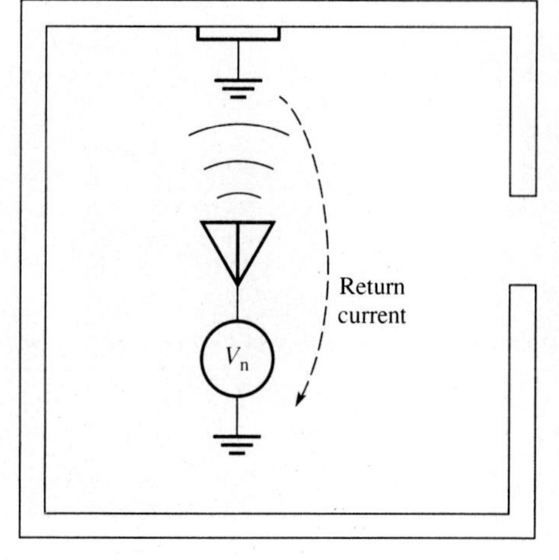

(b) Correct place to ground the shield.

FIGURE 12–27
The skin effect will cause the surface currents to escape the enclosure, as shown in (a).

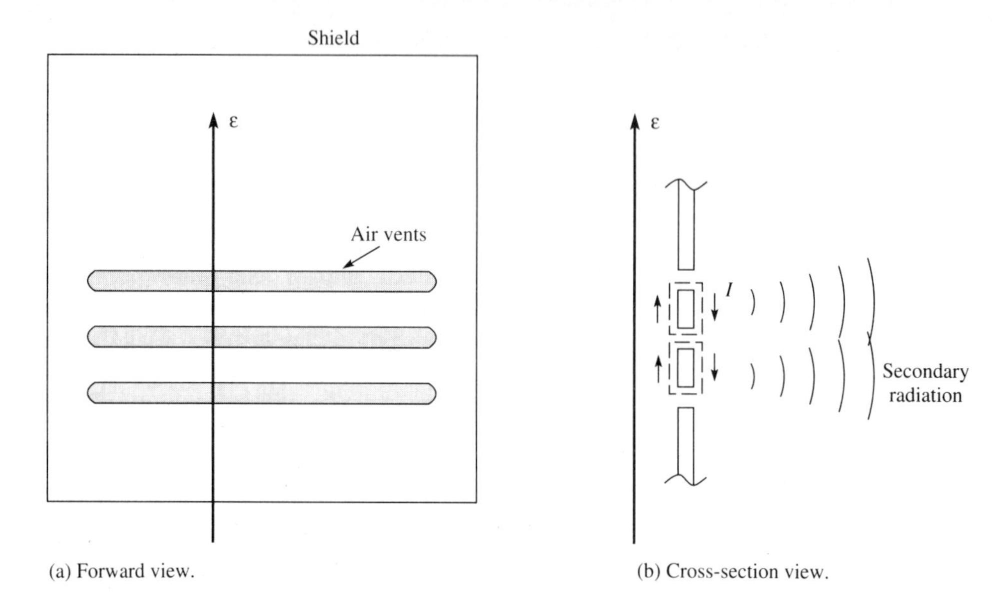

(a) Forward view.

(b) Cross-section view.

FIGURE 12–28
Slots will produce secondary radiation due to induced currents on the skin of the shield.

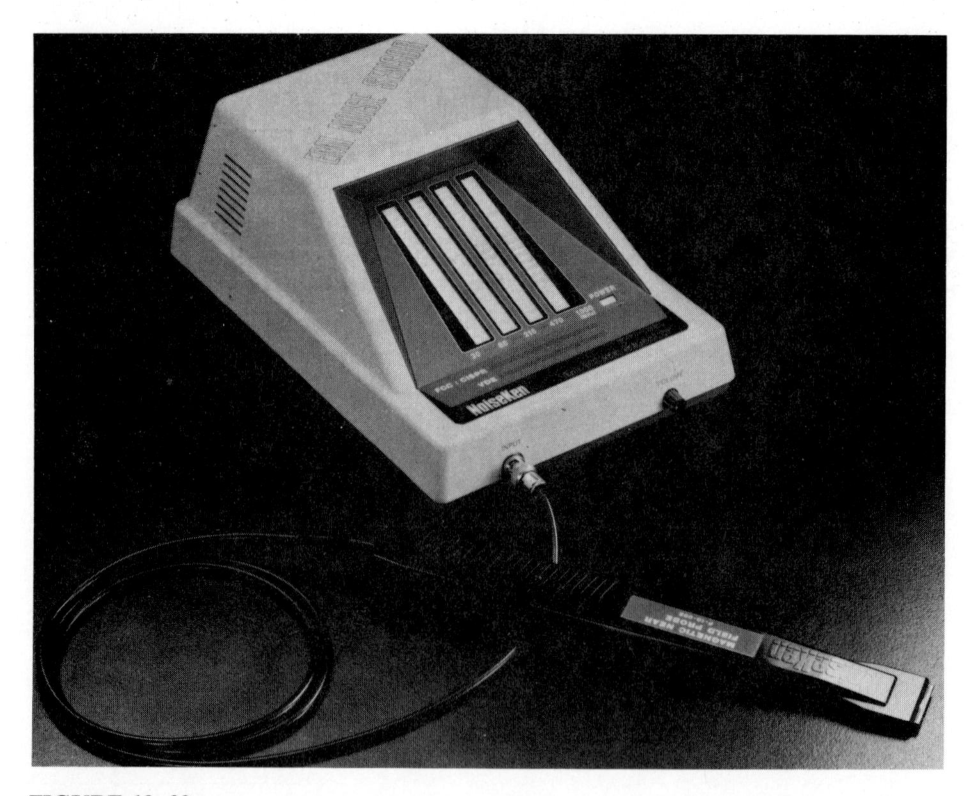

FIGURE 12–29
Noise sensor. The noise sensor is supplied with active electric- and magnetic-field probes and indicates the field strength as a bar graph of light-emitting diodes. Each bar corresponds to a frequency band (courtesy of Noise Laboratory Co., Ltd.).

Many enclosures have slots to permit ventilation of the electrical equipment inside. These slots will reradiate EMI, as shown in Figure 12–28. If the electric component of the EMI field is perpendicular to the slot, as shown in Figure 12–28(a), skin currents are formed around the material separating the slots, as shown in Figure 12–28(b). If the slots are closer together, the loop area will be smaller; in effect, the antenna is smaller. If the EMI shield is not continuous (rf tight), then there will be some leakage, which may be unavoidable due to cooling needs and other enclosure requirements. There are acceptable levels of EMI that have been established by the governing agencies around the world. EMI radiation is measured in a quiet room[1] using a special antenna called a biconical, or log periodic, antenna and a spectrum analyzer. Measurements are made in dBm. Another method for detecting and measuring noise is with a noise sensor as shown in Figure 12–29. The noise sensor can quickly locate noise sources and allows the user to evaluate the effectiveness of noise suppression efforts. Noise measurements are discussed further in Section 12–4.

POWER LINE INTERFERENCE

Frequently, the source of electromagnetic interference is 60 Hz power line pickup. Power line interference, sometimes called *hum* in audio systems, can be caused by an external source that induces unwanted voltage in the circuit or it can be internally generated from a power supply. The frequency of power line interferences does not change—it is either the power line frequency or a multiple of the power line frequency. Although 60 Hz interference may have several sources, 120 Hz interference is almost always due to power supply problems. Other harmonics of the 60 Hz can be a source of noise.

In audio systems, a poorly grounded system is susceptible to hum pickup in addition to having a potential shock hazard. Power line interference is generally avoided by using shielded cable and by careful attention to grounding (discussed in Section 12–7). A broken shield can be responsible for introducing power line interferences into a system; this typically occurs at a connector. Another common source of power line interference is due to ground loops. A ground loop is formed when multiple ground paths exist, providing more than one path for ground current. The finite impedance of the ground path is responsible for generating a potential difference between different ground points in the system.

Noise may also be conducted along power lines, which can interfere with instrumentation operation. The power lines can act as an antenna for electromagnetic interference. Power line interference can generally be filtered out by installing rf filters and transient suppressors on the incoming ac power lines. If power line interference is induced into signal lines in common mode, it can be reduced dramatically with a differential amplifier on the input; otherwise 60 Hz filtering can be used to remove power line interference selectively.

[1] This room is a copper-screened room that shields outside EMI sources.

12–4 MEASURING NOISE

Noise is unavoidable. Proper shielding and design can obtain the lowest noise level, but there will always be intrinsic noise in electronic systems, even if the power source is turned off. In interpreting and measuring noise, it is necessary to take into account the bandwidths involved and the type of noise. White noise, in theory, is infinite in bandwidth, having an even distribution from dc to infinity. This infinite bandwidth implies infinite noise power when applied to a resistor. In practice, electronic systems are bandwidth-limited, so the noise is *not* infinite in energy. An important aspect of measuring noise is that the noise energy will increase as the bandwidth of the measurement system increases. For example, two meters may obtain different results when measuring noise in V_{rms} on the same circuit because the bandwidth of the two meters may differ; the measurement equipment must have a bandwidth larger than the system being measured. Often filters are used to limit the bandwidth of the noise. For example, noise above 20 kHz is inaudible and does not diminish the signal quality in an audio system; therefore the measurement system is filtered above 20 kHz.

The type of noise can affect the way noise is measured and calculated. For example, white noise, in the time domain, has a Gaussian voltage or power distribution about a dc level (the dc level is the signal level at any given period of time) and is best measured in terms of rms values of volts or power. Rms measurements are useful because they disregard the polarity of the noise. However, if the noise is an impulse (spike), then the peak or peak-to-peak value may have more meaning. For example, ignition noise may add very little to the rms value, but the effects can be very noticeable over a radio receiver. A look at the noise, using an oscilloscope and/or a spectrum analyzer, may be needed to determine the significant *spectrum* and the *type* of noise so the proper method of measuring the noise may be used.

SIGNAL-TO-NOISE RATIO

The effects of noise on a signal are best analyzed as a ratio of the signal compared to the noise. This ratio is called the **signal-to-noise ratio (S/N)** and is often expressed in decibels. The noise is often referred to as the **noise floor** because its level stays fixed as the signal is varied in amplitude above it. There are three basic ways to improve the S/N ratio: (1) increase the signal strength, (2) decrease the noise level, and (3) limit the noise bandwidth. The larger S/N, the less the noise will affect the signal. If reducing the noise is not possible or practical or if increasing the signal will cause distortions elsewhere, then reducing the noise bandwidth may be the only approach to improving S/N.

Although signal-to-noise ratio is normally expressed as a power ratio, it can be expressed as a voltage ratio. If voltage ratios are used, then care must be taken to retain a constant impedance at the point of measurement. Measuring the noise with one instrument and measuring the signal with another instrument may cause errors if the loading effects on the circuit by the instruments are not the same. If the measurements are in volts, then S/N is calculated by Equation 12–20.

EQUATION 12–20

$$S/N = 20 \log \frac{V_s}{V_n}$$

where S/N = signal-to-noise ratio, dB
V_s = voltage of the signal, V
V_n = voltage of the noise, V

If the noise is expressed as a ratio of power, then it is given by Equation 12–21.

EQUATION 12–21

$$S/N = 10 \log \frac{P_s}{P_n}$$

where S/N = signal-to-noise ratio, dB
P_s = power of the signal, W
P_n = power of the noise, W

EXAMPLE 12–5 An amplifier has a signal output measured with a true rms voltmeter to be 0.707 V rms. The signal was removed and the noise was measured to be 70.7 μV rms. Find the S/N ratio.

SOLUTION

$$S/N = 20 \log \frac{V_s}{V_n}$$

$$= 20 \log \frac{0.707}{70.7 \times 10^{-6}}$$

$$= 80 \text{ dB}$$

SENSITIVITY

Radio-receiver performance is often expressed as **sensitivity.** In the absence of any signal, noise is present at the audio output of the receiver. As the radio-carrier signal is increased, the noise is decreased. Sensitivity is determined by first measuring the noise at the output of the audio amplifier using a true rms voltmeter, without any radio-carrier signal present. The carrier is supplied and increased until the noise is down by −20 dB; at that point, the signal strength of the carrier is measured. This level is called the sensitivity. The sensitivity is expressed in microvolts at −20 dB quieting.

NOISE FACTOR

The **noise factor** is a means of specifying the added contribution of an amplifier to the signal-to-noise ratio due to noise generated within the amplifier. As such, it is a measure of the quality of the amplifier and includes the overall effect of all noise sources within the amplifier. Like S/N, noise factor is a dimensionless number that can be expressed as a power ratio or as decibel equivalent. Noise factor can be defined as

EQUATION 12–22

$$F = \frac{S_i/N_i}{S_o/N_o}$$

where F = noise factor, dimensionless

S_i/N_i = power ratio of signal to noise at the input, dimensionless

S_o/N_o = power ratio of the signal to noise at the output, dimensionless

S_i = signal at the input, W

S_o = signal at the output, W

N_i = noise at the input, W

N_o = noise at the output, W

For Equation 12–22 to be valid, the amplifier must be linear and the bandwidth must be the same for both the input and output signal. Rearranging Equation 12–22 gives

$$F = \frac{S_i}{S_o} \times \frac{N_o}{N_i}$$

S_i/S_o is the reciprocal of the power gain of the amplifier. Therefore, the noise factor can be written

EQUATION 12–23

$$F = \frac{N_o}{A_p N_i}$$

where N_i = input noise power delivered to the amplifier, W

N_o = output noise power delivered by the amplifier into an output termination, W

A_p = power gain of the amplifier, dimensionless

Noise factor is frequently expressed as a decibel ratio. As a decibel ratio, noise factor is generally called noise figure.[2] The expression for noise figure written as a decibel ratio is

EQUATION 12–24

$$F_{dB} = 10 \log F$$

In an ideal system, the input noise is increased by the gain of the amplifier and shows up as the output noise power. An ideal amplifier, with no internally generated noise, has a noise factor of unity. In an actual amplifier, the total noise increases through each stage; however, noise factor is most dependent on the first stage in a system. This is because the noise introduced in stages after the first undergoes less amplification and therefore makes a smaller contribution to the output noise.

---

[2] The terms *noise figure* and *noise factor* are frequently used interchangeably in the literature. To avoid confusion between the terms, we will use the term noise factor for the nonlogarithmic ratio and noise figure for the logarithmic ratio.

All real amplifiers contribute their own internally generated noise from several causes. The output noise from such an amplifier can be expressed as the sum of the amplified input noise and the internal noise generated in the amplifier. That is

EQUATION 12–25

$$N_o = A_p N_i + N_a$$

where N_a = internal noise power generated in the amplifier, W

By substitution, the noise factor can be written

EQUATION 12–26

$$F = \frac{A_p N_i + N_a}{A_p N_i}$$

$$= 1 + \frac{N_a}{A_p N_i}$$

The second term in Equation 12–26 is the amplifier's contribution to the noise factor. This equation shows that an ideal amplifier, with no noise (N_a) contributed by the amplifier, has a noise factor of unity.

The amount of noise in a system is a function of the bandwidth. To measure noise factor, the noise can be measured in a very narrow band and compared with the power gain of the system at that frequency. This measurement of noise factor is called the **single-frequency noise factor.** Noise factor can also be measured over a large bandwidth compared to the system under test. The result is called the **integrated noise factor.**

EQUIVALENT NOISE TEMPERATURE

Sometimes it is convenient to specify the noise performance of an amplifier using an equivalent temperature instead of the noise factor. The observed noise at the input of an amplifier can be expressed as if it were generated by thermal noise in the source resistance at some temperature. As you have seen, the thermal noise voltage associated with a resistor is given by the equation

$$V_n = \sqrt{4kTBR}$$

There is an equivalent temperature for a source resistance that delivers the same input noise to an amplifier as the observed input noise. The amplifier then adds a noise contribution to this input noise. This added amplifier noise can be thought of as being due to an increase in the reference temperature of the source resistor by an amount that produces the observed output noise. This temperature is defined as the equivalent noise temperature of the amplifier. The standard reference temperature of the source resistance is 290 K. The increase in this temperature due to noise generated in the amplifier defines an equivalent input-noise temperature. The equivalent input-noise temperature of the amplifier can be found from the equation

EQUATION 12–27

$$T_{eq} = 290(F - 1), \text{ K}$$

where T_{eq} = equivalent input-noise temperature of the amplifier, K

Notice that for the ideal amplifier, with a noise factor of 1.0, the equivalent noise temperature is zero.

NOISE MEASUREMENT WITH A TRUE rms METER

Noise measurements can be made using a very sensitive true rms meter. In order to obtain the needed accuracy with a meter, the following should be observed:

1. The meter needs to be a *true* rms meter. Some meters have an ac scale indicating rms, but this may be an averaging meter and not a true rms meter. If in doubt, refer to the operator's manual.
2. The meter needs to be sensitive enough to measure the expected levels of noise. For example, you will not be able to measure a few microvolts of noise with a millivolt meter.
3. The passband of the meter needs to be wide enough to measure the spectrum of the noise. Remember, noise energy is proportional to bandwidth.
4. The normal two-wire lead set that comes with most common meters may not provide enough shielding against interference, causing errors in the readings. You may need to provide other means of connecting to the meter, such as coax cable.
5. The loading effects of the meter should be taken into consideration. For example, probing a high-impedance circuit with a meter could cause the noise to be shunted to ground through the lower impedance of the meter. The measured amplitude of the noise would then be less than actual.

The technique for measuring noise is straightforward. Disconnect any signal sources from the circuit so the output is "quiet." Connect the true rms meter from a reference point to the point to be measured. Most often the reference point will be chassis common (ground), but this is not always the case; if you were measuring a stereo system for example, the meter should be across the speaker. The rms value of the noise can then be measured directly with the meter in volts rms. The desired level of signal is then applied to the circuit and, using the same connections, the rms value of the signal is read in volts rms. Signal-to-noise ratios can then be calculated as in Equation 12–20.

NOISE MEASUREMENT WITH AN OSCILLOSCOPE

Noise measurement with an oscilloscope can be done accurately if the noise has a Gaussian voltage distribution around the signal and is nonperiodic. That is, for any given signal level, the noise must be distributed equally above and below the mean value of the signal, as seen in Figure 12–30.

Before discussing how to measure noise using an oscilloscope, it would be beneficial to mention how *not* to measure noise with an oscilloscope. You cannot measure noise by reading the peak-to-peak levels of the noise, because the observed peak-to-peak reading is far too ambiguous. For example, if the intensity of

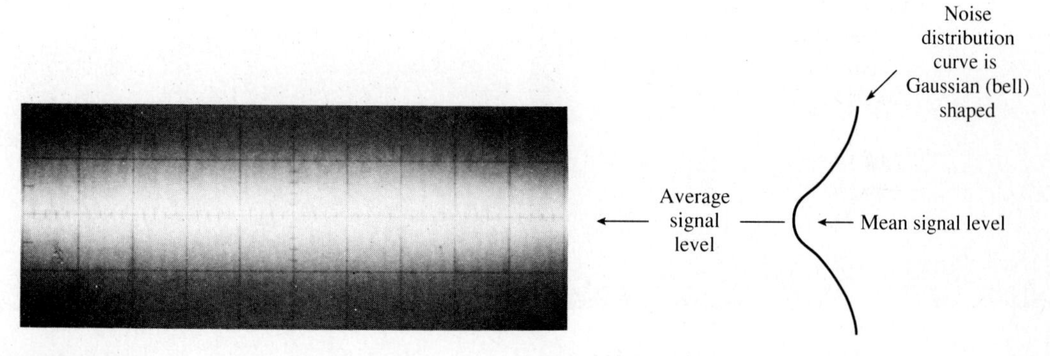

FIGURE 12–30
Noise that is Gaussian distributed.

the oscilloscope is turned down, the peak-to-peak level of the noise appears much less than if the intensity is turned up.

The following method provides a high degree of accuracy in measuring noise that is nonperiodic and Gaussian. It requires a low-noise square-wave oscillator and a dual-channel oscilloscope.

1. Connect the oscilloscope as shown in Figure 12–31.
2. Select channel 1 subtract channel 2. The vertical sensitivity (VOLTS/DIV) of each channel must match throughout this test. They will be changed as needed to view the noise, but channel 2's attenuation must match channel 1's attenuation.
3. Set the function generator for a square-wave signal at about 1 kHz.
4. The oscilloscope should not be locked to any signal. Adjust the frequency of the square wave so it traverses quickly across the display, giving the effect of two parallel lines (traces) as shown in Figure 12–32(d).
5. Increase the sensitivity of both channel 1 and channel 2 (by the same amount) to enlarge the noise display; at the same time, reduce the amplitude of the function generator so both traces remain on the screen. Increase the sensitivity as far as you can on both channels until the scope has no further gain or the noise patterns are filling the screen.

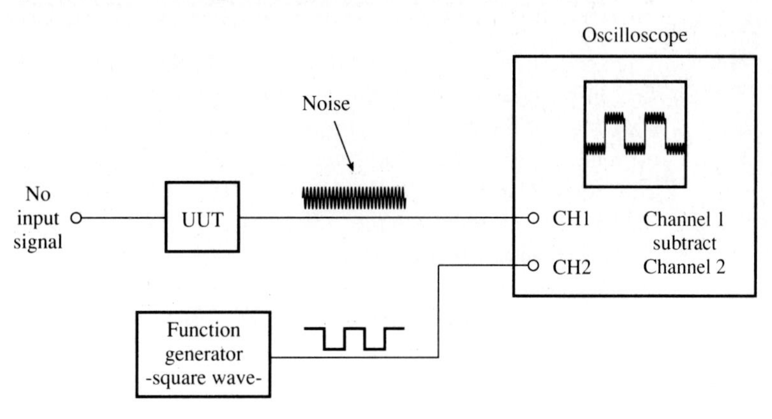

FIGURE 12–31
Connections to an oscilloscope to measure noise.

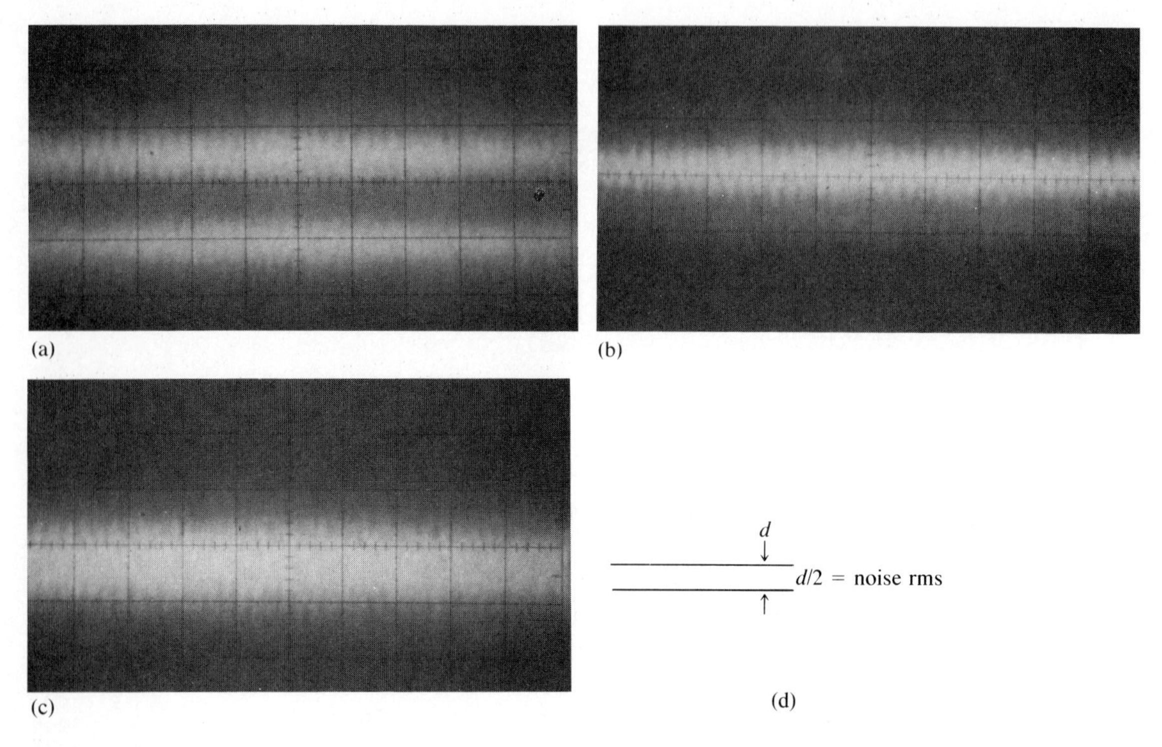

FIGURE 12–32
A square wave and noise source are mixed in the CH1-CH2 mode of an oscilloscope. The square-wave signal free-runs at a rapid rate, so the appearance is two bands, as in (d).

6. Reduce the output of the function generator until the two bands of noise just blend together, as in Figure 12–32(c); then stop.
7. Remove the source of the noise and measure the peak-to-peak level of the square wave, as shown in Figure 12–32(d). Divide this value by 2; the result is the rms value of the noise.

NOISE MEASUREMENT WITH A SPECTRUM ANALYZER

The spectrum analyzer was introduced in Section 10–2. It can be used to measure the amplitude of the noise as well as display the spectrum of the noise. The spectrum of the noise is an important factor as it gives information about the types of noise sources. For example, if the noise rolls off in amplitude at higher frequencies, then the noise might be pink noise or the amplifier might be bandwidth-limited. If the noise is pink noise, then it can possibly be improved by changing certain resistors from carbon to film types. If the noise has a roll-off due to the circuit's bandwidth, then the noise level should be measured at specific frequencies within the bandwidth. If the noise is higher in amplitude at a specific frequency, then the "noise" may actually be interference from an adjacent circuit, such as from a local oscillator used elsewhere to decode the signal.

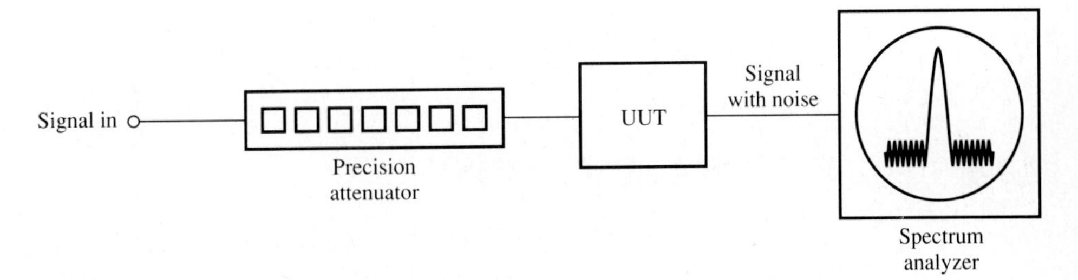

FIGURE 12–33
Connection to a spectrum analyzer to measure S/N ratio.

S/N measurement can easily be accomplished using a spectrum analyzer. The gain of the spectrum analyzer is adjusted so the level of the signal of interest is set to 0 dB (at the top of the CRT display); the noise level is viewed on the graticule in terms of decibels down from the reference signal. The accuracy of this measurement can be increased by connecting a precision step attenuator to the circuit, as shown in Figure 12–33. Figure 12–34(a) shows what a spectrum may look like with one frequency used as a reference and the associated noise; the precision attenuator is set to 0 dB. The reference signal is then attenuated by the attenuator *just* until it is indistinguishable from the noise, as shown in Figure 12–34. The S/N is read directly from the attenuator settings. This method has the

FIGURE 12–34
Use an attenuator to reduce the signal in (a) to look like that in (b). Read the S/N ratio in decibels from the attenuator.

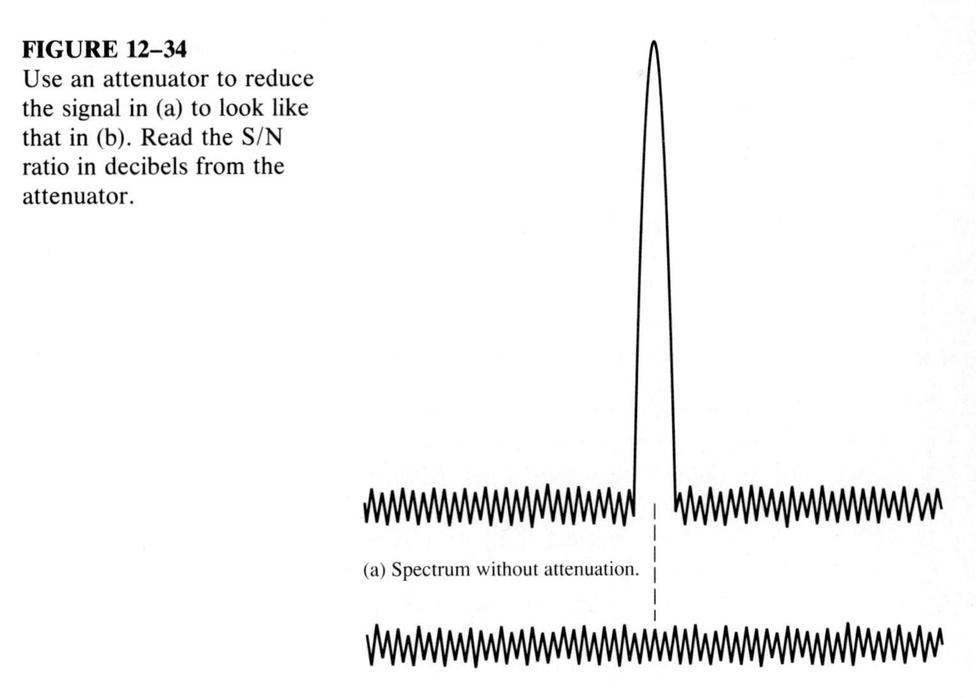

benefit of attenuating the effects of the noise of the test signal; it does not lessen the added noise from the spectrum analyzer's circuits.

12–5 ELIMINATING NOISE

Noise can never be completely eliminated in a circuit, but its effects can be mitigated. The first step in reducing a noise problem is to understand the source of noise. You may not find just one cause for the noise; there are many sources that may contribute a millivolt here and another millivolt there, and their accumulative effects may be a problem. Noises from one source may mask the effects created by another source, so you will need to correct noise when it is found before continuing with your investigation. Many suggestions have already been made for eliminating noise, but there is an orderly method that can be helpful. First, look for direct connections where currents can be conducted. Next look for signal-return loops; then make the loops as small as possible. Third, look for capacitive coupling. Rerouting traces and/or wires will usually cure the problem. If it does not, then properly grounded shielding can be employed. Fourth, look for far-field effects (EMI) and use shielding, filtering, bypassing, and the other techniques described earlier.

Intrinsic noise, which is generated internally, is reduced by different means than extrinsic noise. Intrinsic noise was discussed in Section 12–2. Typically, it is reduced by limiting the bandwidth of the system to the minimum necessary, raising the input signal level, changing the source resistance, and selecting low-noise-input amplifiers. Specially designed op-amps are available (the Linear Technology LT1028, for example) that introduce less noise than a 50 Ω resistor. The optimum op-amp for a particular application depends on the source resistance and the bandwidth requirements of the system.

SENSOR NOISE

Many sensors inherently produce low signal levels, ranging from microvolts to tens of millivolts. Clearly, the addition of even a small amount of noise can seriously degrade the desired signal from a sensitive sensor because of the need to measure quantities at the lowest possible threshold. The sensor itself may respond to an interfering signal or stray magnetic field. Noise may also be picked up as interference along the transmission path, particularly when the signal must be sent through long wiring runs or past electrically noisy motors, relays, and other equipment.

The practical limit on the sensitivity of detectors is set by thermal noise generated in the sensor. Equation 12–1 reveals that there are only two options available to reduce thermal noise power from a sensor: (1) reduce the temperature of the sensor or (2) limit the bandwidth of the signal. For applications in which a voltage sensor is used, choose one that has low source resistance, if possible. When the sensor is a current source, choose one that has a high source resistance because it will avoid coupling thermal noise currents from the source.

Noise can also be generated in detectors from other causes. Passive transducers modify a reference voltage or current in response to a physical stimulus. The reference voltage source can be a source of noise, including power line interference or noise conducted from other sources. For example, if the power supply is also connected to another circuit, noise generated in that circuit may be coupled through the power supply to the sensor. This noise can be particularly troublesome for sensitive sensors. The sensor may be sensitive to electrical interference, including step ring and commutator noise, or electromagnetic interference. In addition, mechanical vibration can induce noise in sensors such as strain gages. The physical parameter being measured may also contribute to unwanted background signals. A microphone, for example, may receive extraneous acoustic signals from sources other than the source being measured. When background noise is set by the physical parameter to be measured, there is nothing to be gained by reducing the noise within the electronic measuring system.

COMMON-MODE INTERFERENCE

Common-mode interference is a form of noise that occurs when noise voltages are induced in both signal conductors of a transmission line by the same amount with respect to a common ground. The source of common-mode interference includes electrostatic and electromagnetic induction at frequencies ranging from power line to radio frequencies. This type of noise can be canceled by using a differential amplifier such as the one described in Section 4–10. Figure 12–35 illustrates common-mode noise.

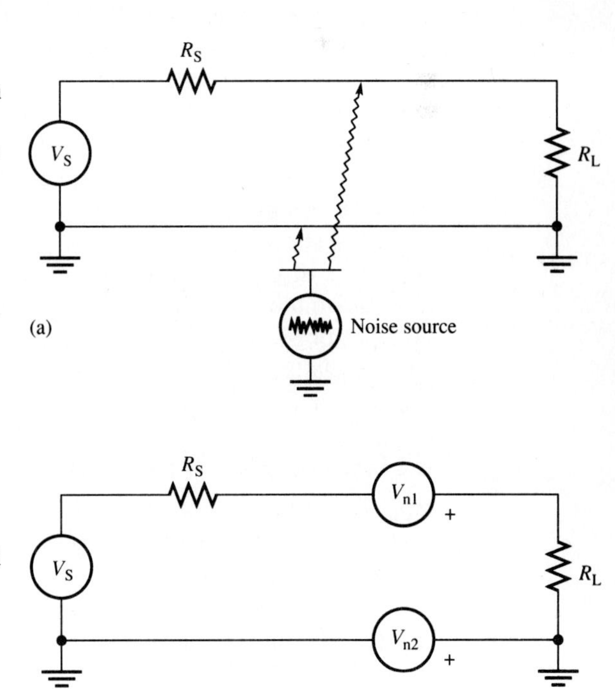

FIGURE 12–35
Noise is introduced into both signal wires as in (a) and is referred to as common-mode noise. Figure (b) is the noise model. If $V_{n1} = V_{n2}$ (and they are in phase), then the noise has no effect on producing a noise voltage across R_L. Also, note that if V_{n2} is shorted out by the common grounds, then the common mode noise is converted to differential mode noise, which will produce a noise voltage across R_L. This can be avoided by lifting the ground connection from R_L.

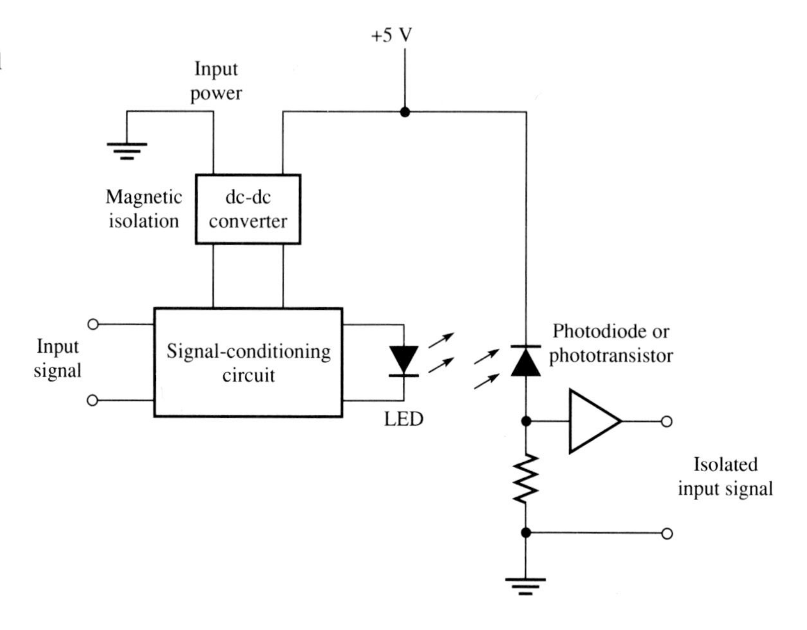

FIGURE 12–36
Isolating signals using optical isolators.

To prevent common-mode interference, it is helpful to know the source and type of noise. The source of noise can be tracked down with the aid of a noise sensor, an instrument described in Figure 12–29.

The type of shielding necessary to shield a noise source depends on the type of radiation. A computer, for example, can radiate noise from the power supply, the CPU clock, or various buses and interconnecting cables. Control measures include installation of filters on noisy lines and adding shielding.

Isolation is another method used to control pickup from most common-mode sources. It is typically accomplished by using signal conditioning and optical isolators. An optical isolator that uses an LED and photocell to isolate the noisy circuit is shown in Figure 12–36. Power for the signal conditioning and optical isolator is also isolated using a dc-dc converter with its own magnetic isolation. Optical isolators can be used for simple switch-closure sensors or for ac and dc voltages or ac current. Circuit boards with optical isolators for PCs are available commercially.

OTHER CONSIDERATIONS

Figure 12–37 shows two other points to consider. First, some bus driver devices, such as those found in emitter coupled logic (ECL), contain either pull-down source resistors or pull-up source resistors. These resistors tend to carry heavy currents and can cause magnetic fields that can couple into other circuits. These components should be placed as close as possible to the source, and the trace lengths should be kept short.

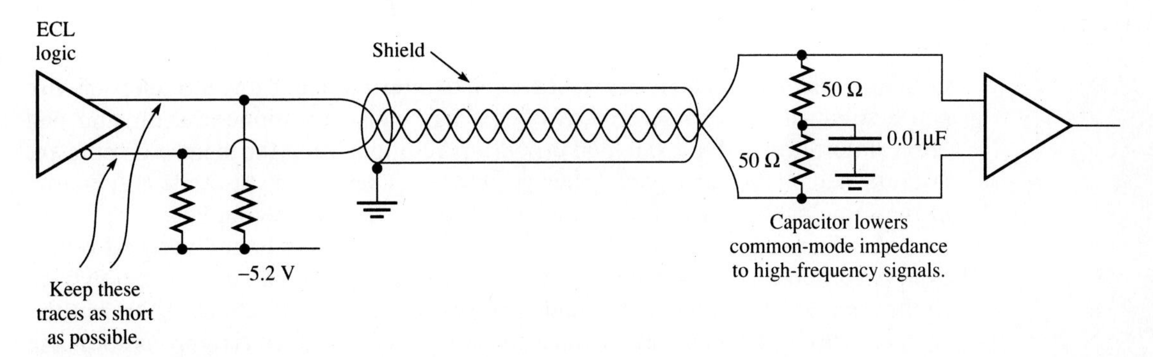

FIGURE 12–37
Other considerations.

As mentioned earlier, one method of reducing the effects of noise is to make the receiving amplifier differential. However, the level of the common-mode signal that can be rejected is limited on many devices. One way to reduce the common-mode signal is shown in Figure 12–37 (at the terminating end). Most twisted-pair cables have the characteristic impedance of about 100 Ω. The technique is to replace the termination resistor with two 50 Ω resistors in series and a 0.01 μF capacitor to ground at their junction. This reduces the common-mode impedance of the circuit significantly.

As previously mentioned, it is best to ground coax cables only at the source end. The other end is connected to a differential circuit. This, in practice, causes problems when dealing with frequencies above 1 MHz. The skin effect of a shield causes the terminating end not to be at ground. The voltage is then proportional to the frequency applied. If a network analyzer or spectrum analyzer were used to sweep the cable, you would observe the cable response increase in proportion to the frequency. This is just opposite of what you might expect from the normal capacitive loading effects of coax cable. Figure 12–38 shows an *RC* termination that tends to bleed off the charges accumulated on the otherwise floating cable end.

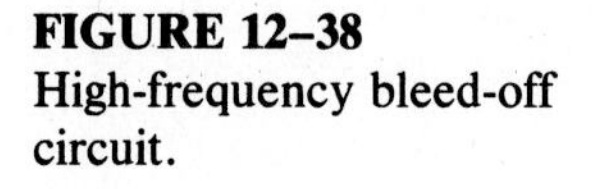
FIGURE 12–38
High-frequency bleed-off circuit.

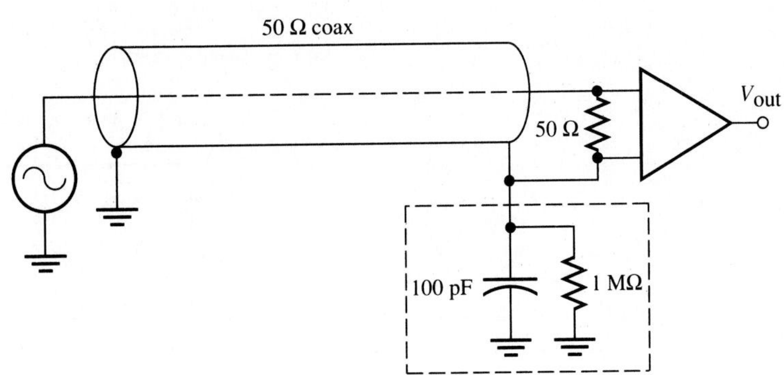

12–6 CABLING

Cabling problems can be the largest source of interference. Long stretches of wire can exhibit all the characteristics of extrinsic noise mentioned so far and can create interference if not shielded properly. Additionally, if the signal currents are not returned on the same cable, then ground loops can occur. Multiple conductors in the same cable can exhibit cross talk through capacitive coupling.

Another source of interference that has not been mentioned is **signal reflection.** An unterminated signal line will cause the incident signal to be reflected back to the source. This can cause **standing waves,** which can effectively short out the signal at various frequencies. It can also cause an effect called **ringing,** as shown in Figure 12–39.

COAX CABLE

Coax cable is pictured in Figure 12–40. It contains an inner conductor surrounded by an outer conductor. The outer conductor is often referred to as the **shield.** To avoid large loops, the shield must be the return conductor for the signal on the center conductor. The coax cable, due to its concentric design, contains the electric field within its dielectric, as shown in Figure 12–41. If the return path is in

FIGURE 12–39
Logic errors caused by cable ringing.

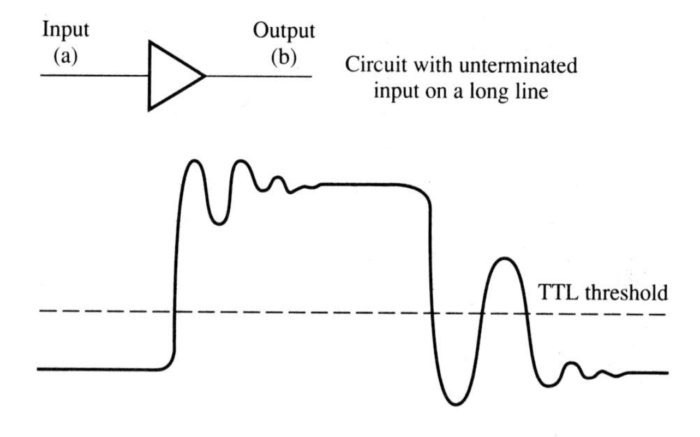

(a) Input signal at (a).

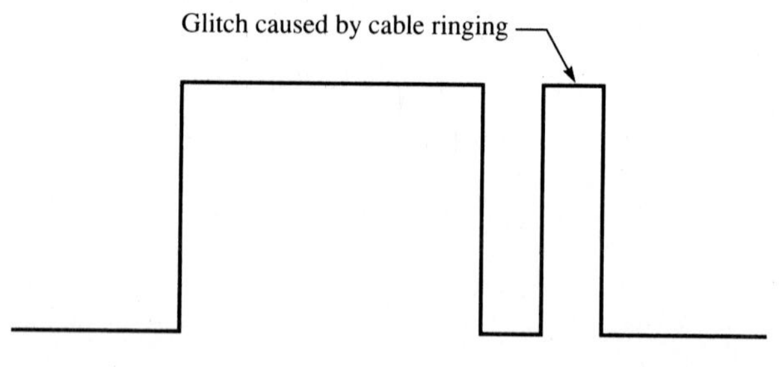

(b) Output signal at (b).

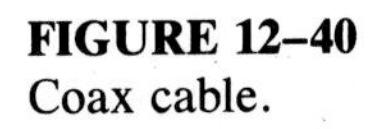

FIGURE 12–40
Coax cable.

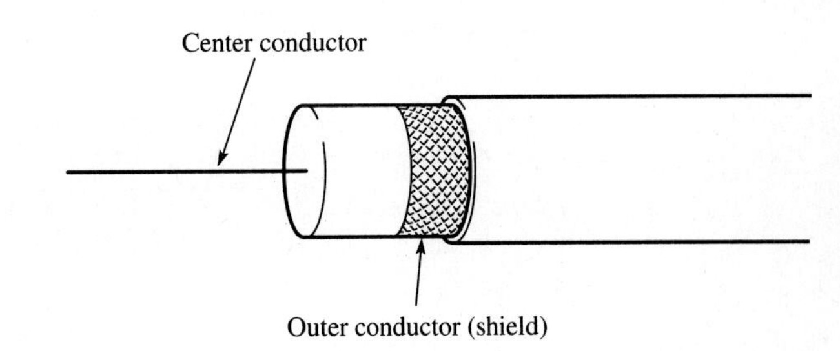

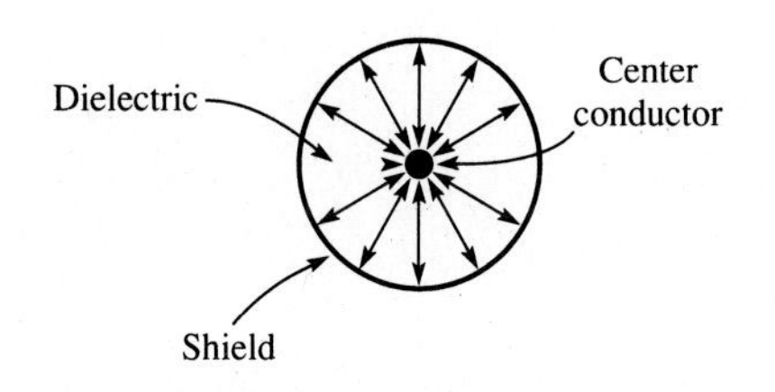

(a) The electric field is contained within the dielectric of a coax cable.

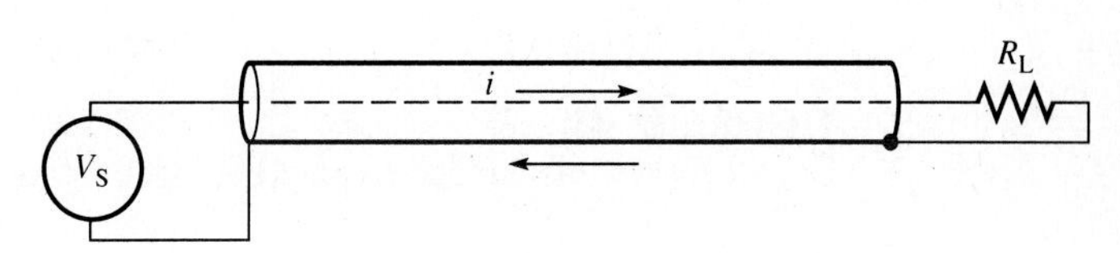

(b) The magnetic field of the center conductor opposes the magnetic field of the outer conductor (shield), leaving a net result of zero flux.

FIGURE 12–41
The coax cable reduces electromagnetic effects.

FIGURE 12–42
Placing lossey ferrite beads on the coax cables can reduce recirculating "ground-loop" currents.

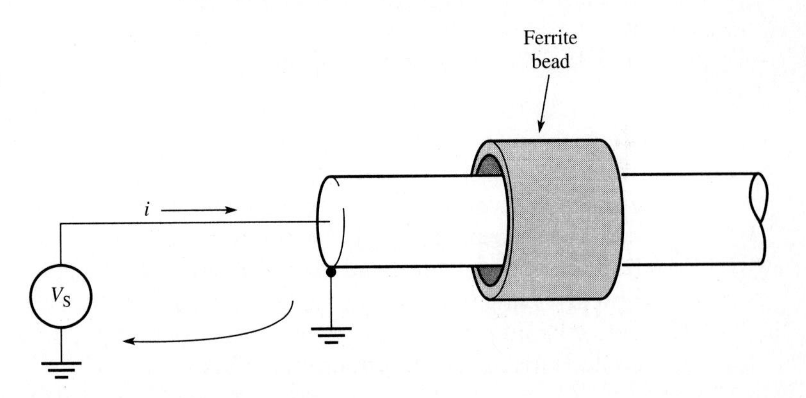

the shield, then the net magnetic field is zero because of the opposing currents. Often coax cables are connected to a common ground at one end (preferably the source end).

A **lossey ferrite bead** can be placed on the coax to assure that all the current going down the center cable comes back on the shield but other currents are blocked as shown in Figure 12–42. A lossey ferrite bead acts like a transformer coupled to a low-impedance load. The bead has no effect on differential currents, but it will exhibit a few ohms of impedance to high-frequency common mode currents.

Coax cable works very well in keeping signal *sources* from becoming sources of interference to surrounding circuits, but it does not do as well in

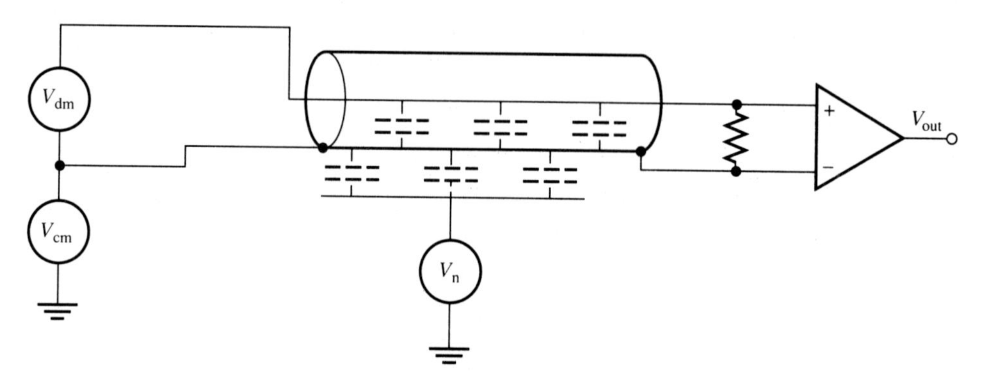

FIGURE 12–43
Coax cable is unbalanced to external electrostatic interference.

shielding from interference. As we have seen, the coax has no effect in shielding magnetic interference from external sources. In addition, some noise can be coupled into the center conductor as well as the shield, but *not equally,* causing differential problems. Figure 12–43 shows a circuit with a balanced input. The balanced input is to cancel the effects of V_{cm}, letting V_{dm} dominate. The noise source has greater coupling to the shield than to the center conductor, thereby causing V_{out} to contain some components of V_n.

TWISTED-PAIR CABLE

Twisted-pair cable is shown in Figure 12–44. It is recommended that there be about three complete turns per inch. Twisted-pair cable keeps the signal loop area small and is very effective in canceling the effects of magnetic interference. The current induced in the wire is opposed by the field in the next half-turn, and this effect is distributed along the length of the wire. Because twisted-pair cable is a balanced line, electric and magnetic fields affect each wire equally. Most application involving twisted-pair cable will terminate at a differential amplifier. Electric field containment is poor, so the cable can radiate; however, shielded twisted-pair cable is available to correct this effect.

Figure 12–45 shows a shielded twisted-pair cable that tends to meet all the requirements for noise protection. Twisted pairs are wrapped in foil with a drain wire connected to the ground at the source end only. The drain wire is a low-impedance path that connects to the foil continuously over the length of the cable, keeping the voltage on the shield consistent and homogeneous.

FIGURE 12–44
Twisted-pair cable.

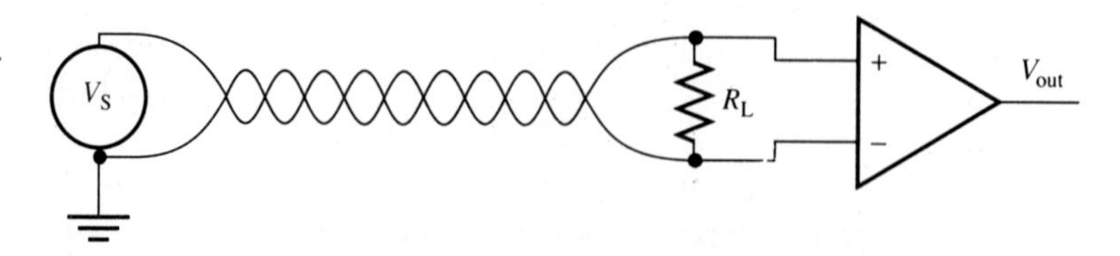

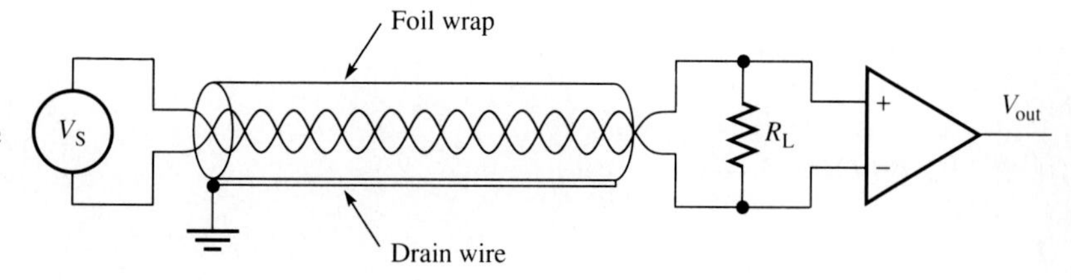

FIGURE 12–45
Combination of twisted-pair cable with an electrostatic shield grounded at the source end.

RIBBON CABLE

Ribbon cable is pictured in Figure 12–46. It is a flat package of wires placed side by side. It is generally used in low-impedance digital circuit applications, where the effects of electric and magnetic interferences are negligible. Its primary advantage is its consistency as to where the wires lay with respect to each other; for example, the signal wire containing data bit D3 can be made to lay between data bit D2 and D4, whereas in a round cable, it may weave anywhere among all the active data lines. In addition, assembly of the connectors on the ends of the cable is reliable and easy to do; it takes only seconds to fasten a cable connector, as compared to several minutes with standard round-cable connectors (see Figure 12–47). To avoid cross talk from adjacent wires, every other wire can be connected to ground. This makes an electrostatic shield between the wires, as shown in Figure 12–48. The function of the shield is to return the interference back to the source, so the shield lines are to be grounded at the source end of the cable. If the signal is bi-directional on the cable (either end can act as the source), then the cable should be grounded at both ends; other grounding methods will be needed to assure that the ground potentials at each end of the cable are the same.

12–7 GROUNDING

Grounding has three basic purposes: (1) to protect people from unsafe voltages on electrical equipment, (2) to protect equipment from damage or malfunction due to

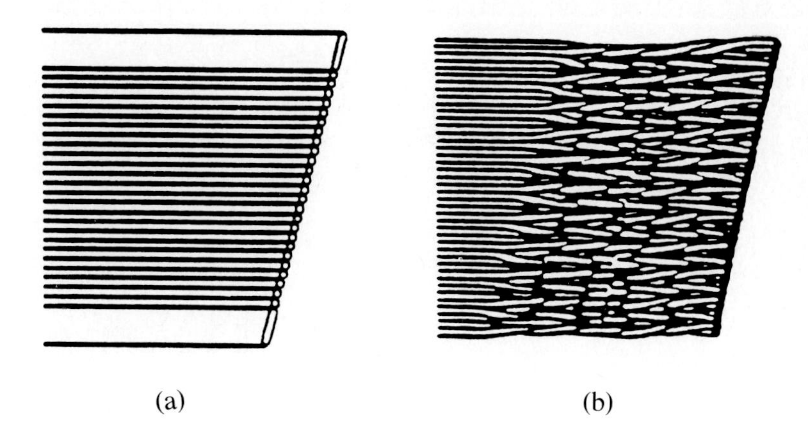

FIGURE 12–46
Ribbon cable (courtesy of Beldon Wire and Cable).

FIGURE 12–47
Connector for ribbon cable requires little cable "prepping" (courtesy of Panduit Electronics Group).

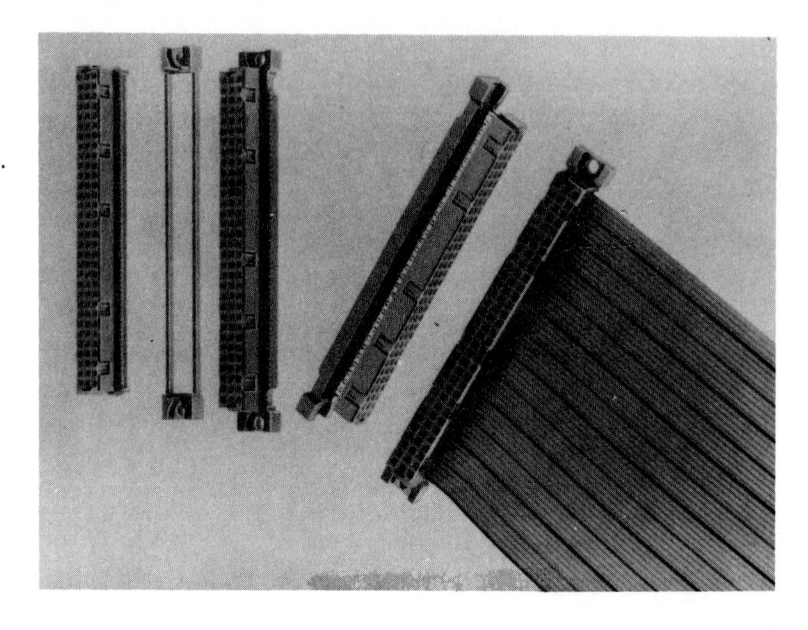

FIGURE 12–48
Every other wire is used as a shield.

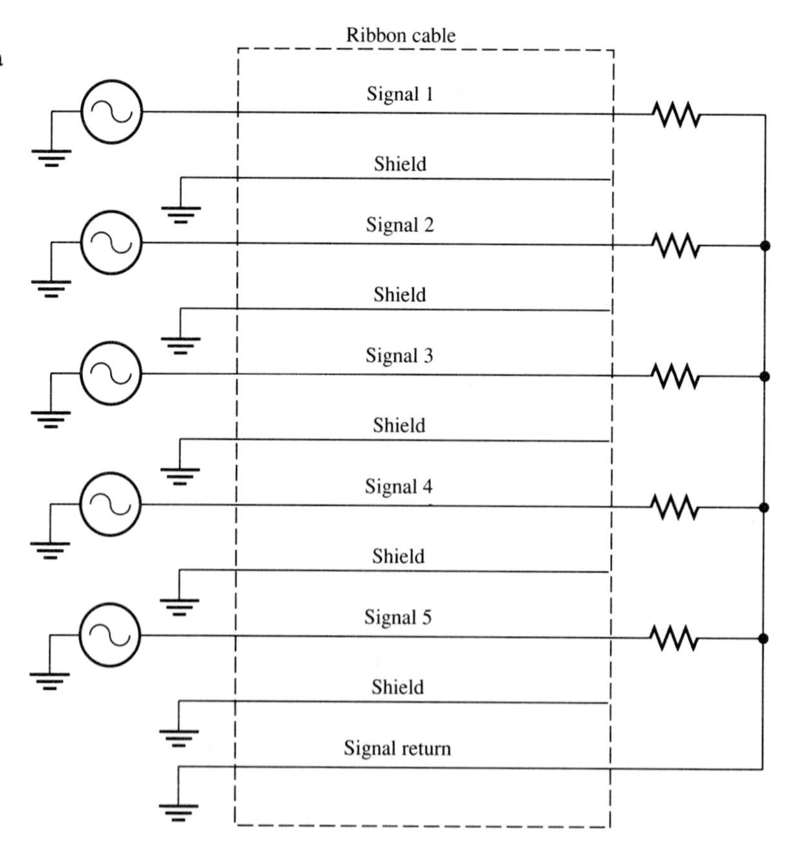

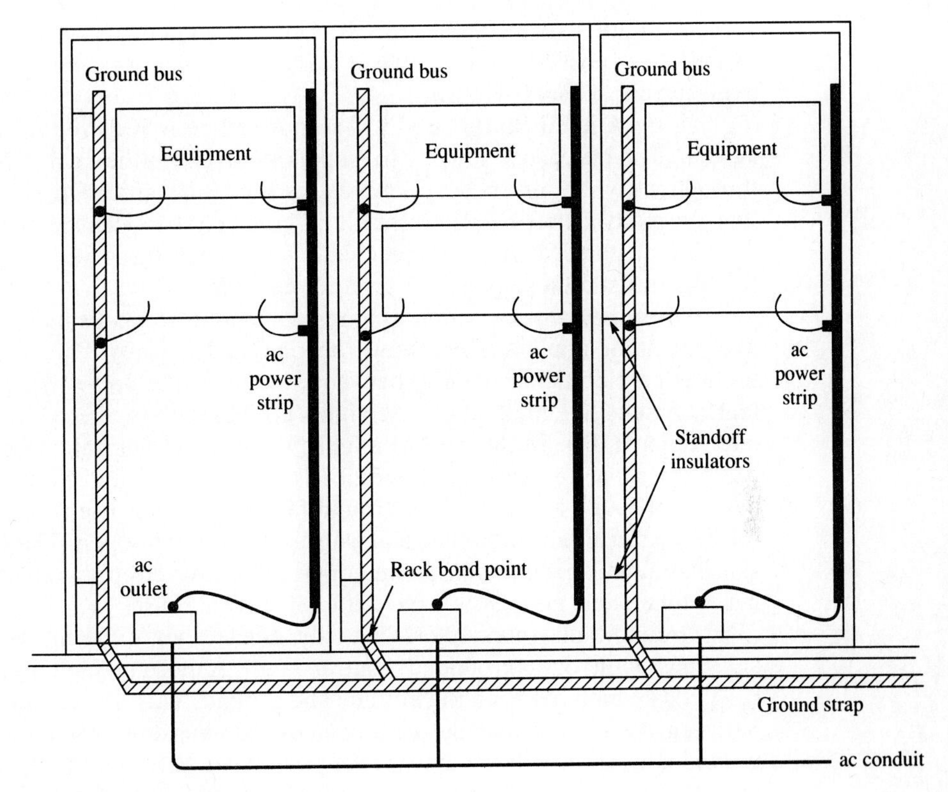

FIGURE 12–49
Grounding an equipment rack.

transients, and (3) to reduce noise coupled into electronic systems. Improper grounding can reduce the safety margin of equipment and create a potentially hazardous condition. In addition, improper grounding can couple significant noise into a system.

An ac power line is common to virtually all electrical equipment in a facility, including lights, appliances, motors, and electronic instrumentation systems. Sensitive systems should be operated from separate supply lines. These lines should preferably be encased in shielded conduit. In three-phase systems, all instrumentation should be operated from the same phase to minimize ground-leakage current flowing between equipment. Power lines should not be within 2 ft of any signal lines.

Recommended grounding of equipment racks is shown in Figure 12–49. A vertical ground bus is placed in each rack that is constructed from a copper bus bar. The main ground is connected to each rack.

SAFETY GROUNDING

Electronic chassis—as well as conduits, metal cableways, and frames of equipment—should be connected to grounds so that they cannot become shock hazards in the event of a wiring error or fault. The ground connection should be as low a resistance as possible and capable of carrying a ground-fault current to ensure that over-current sensors will trip.

In the United States, a 115 V ac power line is required to be a three-wire system. The black, or "hot," wire supplies current to the load, which is returned through the white, or "neutral," wire. The remaining green wire is a safety ground that normally carries no current. It is connected to neutral only at the main distribution panel. It is also connected to metal enclosures, racks, cable trays, and the like to keep these points at ground potential.

The most common failure that results in an electrical shock occurs when an instrument or appliance enclosure becomes electrically "hot"—that is, when there is an electrical potential between the enclosure and earth ground **(ground fault).** A sensitive method of detecting a ground fault is offered by a ground-fault interrupter (GFI). In the case of a fault, the hot and neutral wires will no longer carry the same current; the difference is carried to ground through other paths, perhaps through a person. The ground-fault interrupter uses a differential transformer to sense this difference and causes the circuit to open. The GFI circuit can sense an unbalanced condition and open the power circuit in as little as 10 to 30 ms, a time short enough to save a life or prevent a fire.

Ground-fault interrupters actually sense two different types of faults: a normal fault or a grounded neutral fault. A normal fault occurs when the fault adds a path between the hot line and ground. This divides the return current between the ground and the neutral lines, tripping the ground-fault breaker. A grounded neutral fault occurs when a path exists between the neutral return line and the ground line. A grounded neutral fault of 2 Ω or less can cause the GFI breaker to trip.

Test instruments are connected to the protective ground through the third pin when the instrument is plugged into an ac outlet. The protective ground is generally connected to the chassis, keeping it at ground potential. The instrument signal ground is frequently connected to this point, forcing the instrument to make measurements with respect to chassis ground. Defeating the instrument ground, by clipping the third prong, connecting the instrument to an isolation transformer, or using a "cheater connector," is a dangerous practice that should never be done. In addition to being an unsafe practice, defeating the safety ground can be hazardous to equipment by allowing fault current to flow through printed circuit boards instead of the chassis. Floating measurements can be made using an isolation amplifier or differential inputs.

Double insulation is a technique used by some manufacturers to afford protection from an instrument or appliance becoming hot. The insulation is typically plastic, which insulates the circuits from the body of the instrument. Double-insulated equipment is not required to have a third ground pin but must be labeled *double insulated.* Double-insulated equipment can present a hazard if allowed to become conductive by submerging it or spilling a conductive fluid on it. The subject of safety grounding is of particular concern for biomedical instruments, especially those instruments that are connected to humans for monitoring the body. The topic of safety grounding for biomedical instruments is covered in Section 17–14.

SUMMARY

1. Noise sources are either intrinsic or extrinsic. Intrinsic noise is caused by the circuit itself, whereas extrinsic noise is caused from external interference. Thermal, contact, and shot noise are examples of intrinsic noise.
2. Thermal noise is produced by the motion of free electrons in a resistance. It is proportional to temperature and is independent of the material. It is a form of white noise.
3. Contact noise is dependent on the material but not on the temperature. In most cases, this noise is larger than thermal noise. The amplitude of the noise varies inversely with frequency.
4. Shot noise is a form of white noise that is generated by moving charge carriers through some voltage boundary, such as a transistor junction. The noise is proportional to the current.
5. Noises are combined by the square root of the sum of the noise sources squared.
6. Interference is introduced into a circuit by conduction, capacitive coupling, magnetic coupling, and radiation.
7. Conductive interference is common when pieces of equipment are connected together. It is most effectively eliminated by opening the return path(s) to the source (ground loops). Other methods used to reduce conductive interference include filtering, bypassing, and grounding with large conductors.
8. Electric field interference, also called capacitively coupled interference, is proportional to the area and inversely proportional to the distance between the conductors. Separating objects or minimizing the exposed surface areas is most effective in eliminating the interference. Shielding can be used, but care must be taken to return the noise back to the source. Filtering the noise is another way to reduce its effect. This type of interference is reduced as the impedance of the affected circuit is lowered. Body movements around the circuit can have an effect on the interference.
9. Magnetic field interference is proportional to the area of the loop. Using bypass capacitors, the loop is made smaller, an effective way to reduce this type of interference. Using twisted wires in cables is also an effective method of reducing interference. This type of interference is reduced as the impedance of the affected circuit is increased. People moving around circuits with magnetic interference have no effect on the interference.
10. Electromagnetic field interference (EMI) is a far-field effect. The electric and magnetic fields are equal and support each other in a radiated wave. The most effective way to reduce this type of interference is through shielding. The shield will absorb the wave, producing currents on the skin of the shield (skin effect). Skin currents can reradiate rf energy; therefore, care must be taken to drain the skin currents from the shield on the same side as the noise source.
11. The signal-to-noise ratio is the ratio of the signal (rms) to the noise (rms).
12. Sensitivity is the signal strength (in radio) required to quiet the noise by a specified amount, usually by −20 dB.
13. Noise factor is expressed as the ratio of the S/N at the input of an amplifier to the S/N at the output of the amplifier. It is a means of specifying the amplifier's contribution to the noise.
14. Equivalent noise temperature is a way of specifying the noise performance of an amplifier. The amplifier is described as an equivalent resistor producing thermal noise.
15. Cables are the greatest concern for noise; they tend to radiate and are susceptible to all types of extrinsic noise. Proper termi-

nation is essential to avoid reflections and standing waves. Coax cables do well in containing electrostatic fields between the conductors, but they are not satisfactory in maintaining a balance to common mode sources of interference. Shielded twisted-pair cable is best for magnetic and electrostatic shielding.

16. Grounding provides protection from electric shock, reduces extrinsic noise, and helps avoid damage to equipment.

QUESTIONS AND PROBLEMS

1. What is intrinsic noise? What three types were discussed in the text? Give examples of each.
2. What is extrinsic noise? Name four methods by which extrinsic noise can be coupled into a circuit.
3. When does noise become interference? Give at least three examples of interference.
4. Explain why white-noise power increases as the bandwidth of the measurement (or circuit) is increased.
5. A resistor does not require any current from an external source to produce noise. What type of noise is this, and what primary factor affects this noise? Is the noise affected by the type of material of the resistor—i.e., carbon versus wire?
6. Current in a resistor adds additional noise. What type of noise is this? Does the composition of the resistor affect this type of noise?
7. Compute the thermal noise voltage developed in a 100 kΩ resistor, assuming 1.0 MHz bandwidth and 115°F temperature.
8. Two resistors, which are in parallel, are connected to a diode, as shown in Figure 12–50. The bandwidth of the measurement is 5.0 MHz. Assume the temperature is 60°C. Compute the thermal noise voltage of each of the resistors and the shot noise of the diode.
9. When combining the noise of the two resistors in Figure 12–50, should the noise of the resistors be calculated independently and then combined using Equation 12–7, or should the resistors be combined ($R_1 \| R_2$) and then the noise calculated? Use the superposition theorem to prove your answer.
10. For Figure 12–50, combine the noise of the resistors and the diode at V_{out}.
11. Figure 12–51 shows a voltage source driving a long set of signal wires. R_S is a 150 Ω source resistor and R_L is the 150 Ω load resistor. R_1 and R_2 represent the loss in the wires. V_S and R_L share a common ground. The ground potential at R_L differs from the ground at V_S by 20 mV rms.
    (a) Compute the ground loop current—i.e., the current through R_2 created by V_n.
    (b) What is the level of the noise across R_L?

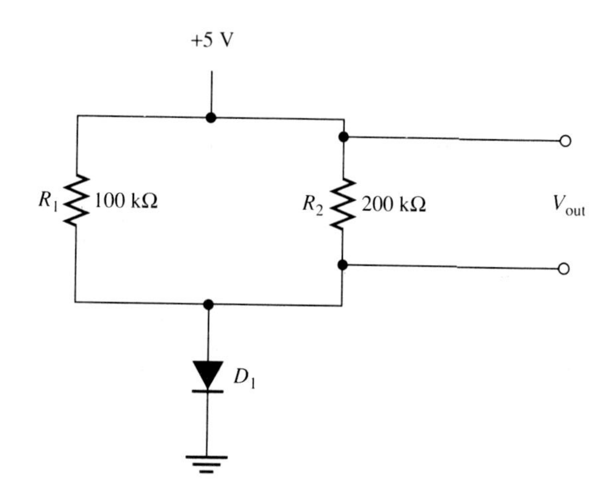

FIGURE 12–50

FIGURE 12–51

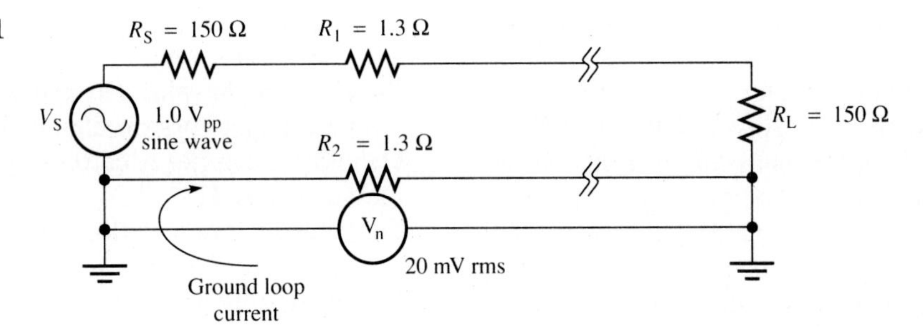

12. (a) In Figure 12–51, what is the most practical way of eliminating the ground loop?
    (b) Assume the ground loop can't be removed, list two other ways that the interference from the ground potential difference can be reduced.
13. (a) What is a ferrite bead?
    (b) How does it block common-mode noise while not interfering with differential signals?
14. The permittivity of a vacuum is $\epsilon_0 = 8.85 \times 10^{-12}$ F/m. The permittivity of any insulating medium divided by the permittivity of vacuum yields relative permittivity:

$$\epsilon_r = \frac{\epsilon}{\epsilon_0}$$

Therefore

$$\epsilon = \epsilon_0 \epsilon_r = (8.85 \times 10^{-12})\epsilon_r$$

Substituting into Equation 12–10 gives

$$C = \frac{(8.85 \times 10^{-12})\epsilon_r A}{d}$$

What is the capacitance between two square plates (0.2 m by 0.2 m) that are parallel with each other and separated by 0.5 mm? (The relative permittivity of air (ϵ_r) is 1.00.)
15. List two examples of what can be done to reduce capacitive coupling effects.
16. Explain how a shield can be worse than no shield at all if it is connected incorrectly.
17. The circuit board shown in Figure 12–52 is a two-sided circuit board. The top side contains a single trace as shown; the bottom side is a ground plane. The two sides are insulated from each other. A voltage source is connected to one end of the circuit trace and to the ground plane as

FIGURE 12–52

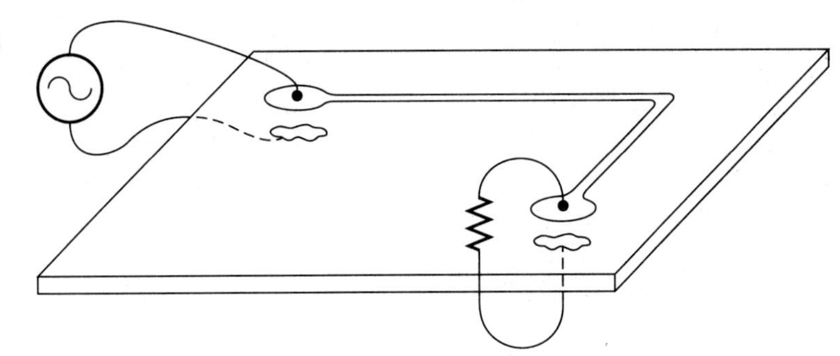

shown; a load resistor is connected to the other end of the trace and to the ground plane. Redraw Figure 12–52. If the voltage source is dc, draw a dotted line showing the return path of current on the ground plane. If the voltage source is very high frequency, use a dashed line to show the return path of the current on the ground plane. Are the paths the same? Explain.

18. What is the inductance of a coil with 300 turns wound on a wooden cylinder that is 6 cm long and has a diameter of 2 cm? The permeability of the wooden cylinder is $4\pi \times 10^{-7}$ H/m.
19. What is the impedance of the coil in Problem 18 at 2 MHz? What can you do to the geometry of the coil to reduce the impedance? Why is keeping the loop size small important in preventing magnetically coupled interference?
20. Explain the purpose and function of bypass capacitors on digital circuits.
21. Why is copper a better shield than paper against EMI?
22. Explain how air slots can cause EMI to leak into an enclosure.
23. What is the signal-to-noise ratio (decibels) of a 1 V rms signal if the noise is 2 mV rms?
24. An amplifier has a power gain of 100, but the noise seems to be amplified 102 times. What is the noise factor for this amplifier? What contributed to the increase in noise?
25. What should you do with the extra wires that aren't used in a cable?
26. Why should you *never* cut the safety ground from electrical equipment?

Chapter 13

Transducers

OBJECTIVES

The detection and measurement of a physical quantity by an electronic measuring system requires an interface between the physical quantity and the measurement system. That interface is the transducer. All measurements require a source of energy and all measurements in some way modify the quantity to be measured. The transducer is one of the key elements of a measuring system.

In this chapter, we first take up some transduction principles, then look at the selection criteria for transducers, and finally consider specific physical quantities. When you complete this chapter, you should be able to

1. Discuss the criteria for selecting a sensor for a particular measurement as they relate to measurement specifications and operational and environmental considerations.
2. List common transduction principles for conversion of a physical parameter into an electrical quantity and give examples of transducers.
3. Define temperature, heat, stress, strain, pressure, displacement, velocity, and acceleration; give the measurement unit for each.
4. Explain how to make sensitive measurements with a Wheatstone bridge, including avoiding noise from common-mode signals, and explain how to make guarded measurements.
5. Transform readings on the fahrenheit, celsius, and Kelvin temperature scales.
6. Compare advantages and disadvantages of various types of transducers, including those for measurement of temperature, strain, motion, pressure, and light.
7. Explain the difference between gage pressure, absolute pressure, and differential pressure.
8. Describe common displacement and motion transducers, including potentiometric displacement transducers, linear variable differential transformers, fiber-optic transducers, and capacitive transducers.
9. Describe the electromagnetic spectrum in terms of the frequency and type of radiant energy. Compare photometric and radiometric quantities, including luminance intensity with radiant intensity and illuminance with irradiance.
10. Give the principle differences between photovoltaic sensors, photoconductive sensors, and photoemissive sensors.

HISTORICAL NOTE

The first application of a solar cell in space was done by American scientists using the second satellite launched by the United States—the *Vanguard I,* on March 14, 1958. The U.S. physicist John A. O'Keefe used the satellite to detect slight variations in gravitational intensity over the earth's surface; he deduced that the earth deviated from the oblate spheroid theory predicted. The satellite's source of power for radio transmission was solar cells—they converted the sunlight that struck them into electrical energy sufficient to power a radio transmitter.

13–1 TRANSDUCER CHARACTERISTICS

Electronic measuring systems are often used to measure nonelectrical quantities such as the temperature of a liquid or the pressure in a container. A *transducer* was defined in Section 1–3 as a device that receives energy from a measurand and responds to that energy by converting it into some usable form for the measuring system. This definition can be broadly applied to include devices such as the familiar mercury thermometer, which converts temperature into the height of a mercury column. For electronic measuring systems, our primary interest is restricted to devices that convert physical quantities into electrical form or vice versa. Although electrical transducers can be found at the input or the output[1] of a system, our primary interest is in the input transducer, which converts the quantity to be measured into an electrical output that is a function of the input. In general, there is a mathematical relationship between the input and the output of the transducer.

ACTIVE AND PASSIVE TRANSDUCERS

Most transducers are considered to be **active**—meaning they require an external source of energy. An example is a simple potentiometer. The potentiometer's resistance is changed by the position of the knob. In order to sense the resistance, a small current must be sent through the potentiometer from an outside source. The external source of voltage or current is given the name *excitation.* Certain other transducers can convert the measurand directly to an electrical signal. These transducers are considered to be **self-generating,** or **passive,** as no outside power source is required for their operation. Such transducers can convert a physical quantity directly into an electrical quantity. An example is the thermocouple, a simple device that converts a temperature difference between two wire junctions into a current. For electronic measuring systems, the fact that a given transducer is active or passive usually isn't the most important consideration; rather, the output must be a well-understood function of the input and it must be in electrical form. It can be a voltage, current, frequency, capacitance, resistance,

[1] A microphone can be considered to be an electrical input transducer, whereas a speaker can be considered to be an electrical output transducer.

TABLE 13–1
Representative input sensors that convert input energy into an electrical parameter.

ENERGY SOURCE	REPRESENTATIVE SENSOR	COMMENT
Mechanical	Rotating vane flowmeter	Pulse-repetition frequency of output is proportional to flow rate.
Thermal	Thermocouple	Output proportional to temperature difference of two junctions.
Nuclear radiation	Ionization chamber	Current between electrodes is proportional to ionizing radiation.
Electromagnetic	Antenna	Can change electric energy to electromagnetic and vice versa.
Magnetic	Hall-effect sensor	Voltage produced in a current-carrying conductor in a magnetic field.
Chemical	pH sensor	Measure of the hydrogen ion concentration in a solution.

pulse width, or other electrical variable that is related to the measurand. As discussed in Section 1–1, the ratio of the output signal to the input signal is the transfer function—for input transducers used in electronic measuring systems, this is the ratio of an electrical quantity to a physical quantity.

MEASURANDS

There are six forms of energy that can be converted into an electrical signal with a transducer. These forms include mechanical, thermal, nuclear radiation, electromagnetic radiation, magnetic, and chemical. Each of these sources of energy can be converted into an electrical stimulus with a transducer. For example, nuclear radiation can be converted directly into an electrical signal with a Geiger counter, or the radiation can first be converted to light using a fluorescent substance which is then changed to an electrical signal with a photomultiplier or other light sensor. Table 13–1 lists representative sensors that convert each form of energy into an electrical signal.

There are an immense variety of transducers that use many different principles of operation; they can be classified by either their operating principle or by the quantity that is to be measured. We begin by examining some important transduction principles, then discuss common attributes and selection criteria of transducers in general, and finally consider specific transducers for particular physical quantities.

13–2 TRANSDUCTION PRINCIPLES

The conversion of a physical variable into an electronic signal can be done with a number of transduction principles. Active transducers usually contain a passive circuit element such as a resistor, capacitor, or inductor that changes value as a physical variable changes. The passive element is used to modify a dc or ac excitation voltage in response to the physical variable and may be contained in a bridge circuit or other circuitry to produce a voltage or current output.

RESISTIVE TRANSDUCTION

The variation of resistance is one of the most common transduction principles. The resistance of a resistive element can be varied by several methods, including sliding a wiper along a rheostat, applying mechanical stress, varying light intensity to a photosensitive material, and changing the temperature. Resistive transduction is used by certain displacement transducers, strain gages, resistive temperature devices (RTDs), thermistors, and photoconductive devices. Some examples of resistive transduction principles are shown in Figure 13–1.

A **strain gage** is essentially a thin metallic conductor that is firmly attached to a solid object to detect strain in the object. It changes its resistance as it is stretched or compressed. Stretching increases the length of the wire and decreases its cross-sectional area, resulting in greater resistance; compression does the opposite. The change in resistance is measured in a bridge circuit. Strain gages are used in many types of transducers, most noticeably in **load cells,** a force-measuring transducer found in scales ranging from postal scales to those for weighing trucks. Load cells measure force indirectly by the deformation produced in a beam or other structure. The deformation is converted to a resistance change by strain gages that are bonded to the surfaces of the beam. Strain gages are discussed in Section 13–6.

Resistive transduction can be used to measure temperature and light. Temperature transducers include RTDs and thermistors. RTDs are generally constructed from platinum, and their resistance increases as temperature increases. Thermistors are constructed from semiconductors that exhibit a strong negative-

FIGURE 13–1
Resistive transduction.

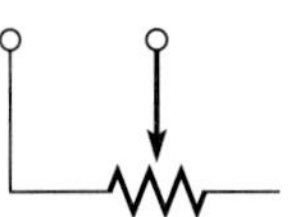
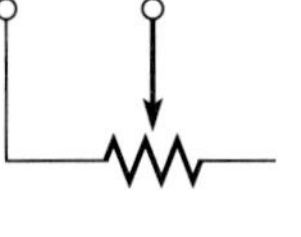

(a) Sliding contact.

Stress

(b) Strain gage.

Heat

(c) RTDs and thermistors.

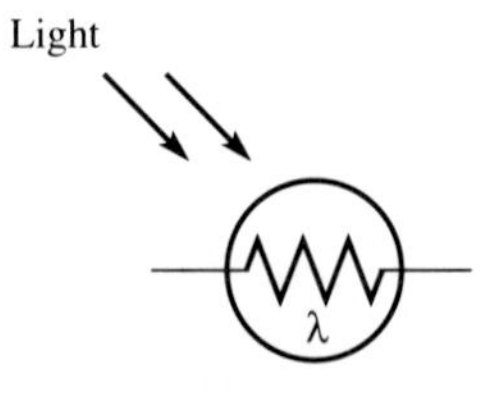

(d) Photoconductive device.

resistance temperature dependence. Photoconductive devices work because absorption of light can increase the charge-carrier concentration in the conduction band of semiconductors, which causes their resistance to decrease.

CAPACITIVE TRANSDUCTION

The capacitance of a parallel-plate capacitor is given by the equation

EQUATION 13–1

$$C = \epsilon_0 \epsilon_r \frac{(n - 1)A}{d}$$

where C = capacitance, F
ϵ_0 = permittivity of free space, 8.85×10^{-11}, F/m
ϵ_r = relative permittivity, dimensionless
n = number of plates
A = plate area, m^2
d = distance between the surfaces, m

Note that permittivity, ϵ, is equal to $\epsilon_0 \epsilon_r$.

Capacitive transduction occurs when the measurand changes the capacitance, as illustrated in Figure 13–2. Transducers can take advantage of any of the quantities on the right side of the equation to vary the capacitance. For example, a capacitive microphone uses acoustical pressure to vary the spacing between the plates, turning an audio signal into a variation of capacitance. A capacitive level indicator can change the relative permittivity between the plates because a nonconducting liquid is introduced in the region between the plates—a method applied to aircraft fuel measurements. In addition, the effective plate area can be varied by displacement of one of the plates, as in a capacitive displacement transducer.

To give a usable electrical signal from a capacitive transducer, the change in capacitance can vary an ac signal or be part of a bridge circuit. The change in capacitance can also be made to change the frequency of a resonant circuit.

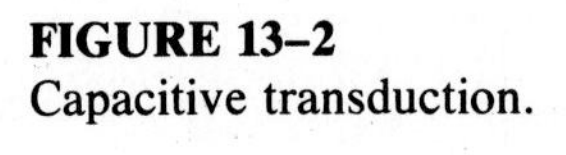

FIGURE 13–2
Capacitive transduction.

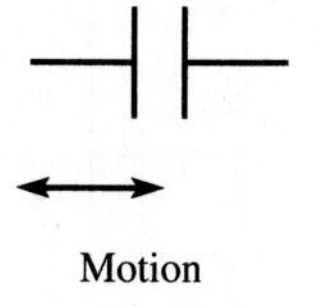

(a) Distance between plates.

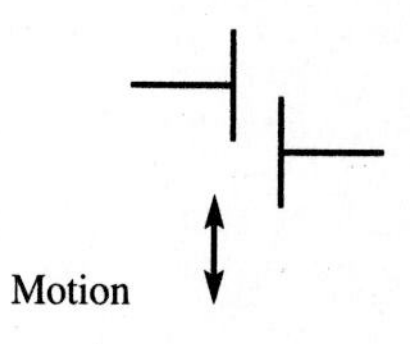

(b) Effective plate area.

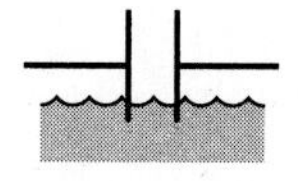

(c) Dielectric.

INDUCTIVE TRANSDUCTION

Inductive transduction occurs when the measurand changes the self-inductance or mutual inductance of a coil. One common method for changing the inductance is to move a magnetic core, linked to a sensing element, in response to a change in displacement. An advantage of inductive displacement and velocity transducers over potentiometric transducers is that there is no wear due to sliding contacts in the inductive transducer. The linear variable differential transformer (LVDT) described in Section 13–7 is an example of a transducer that uses inductive transduction.

MAGNETIC TRANSDUCTION

Magnetic transduction occurs in several materials. The **Hall effect** occurs when a current-carrying conductor is placed in a magnetic field such that the magnetic and electric fields are at right angles to each other. A voltage appears across the conductor in a direction perpendicular to both the current and the magnetic field, as illustrated in Figure 13–3. A change in the magnetic field produces a proportional change in the Hall voltage. The Hall voltage can be used to determine the magnetic field strength or alternatively can be used to determine the current in the conductor.

SELF-GENERATING TRANSDUCTION

Self-generating transducers convert a physical quantity directly into an electrical quantity. Examples of self-generating transducers include sensors that can produce a voltage from the thermoelectric effect, magnetic induction, the piezoelectric effect, or the photoelectric effect. In some way, each of these conversions absorbs some energy from the measurand and produces an electrical output as a

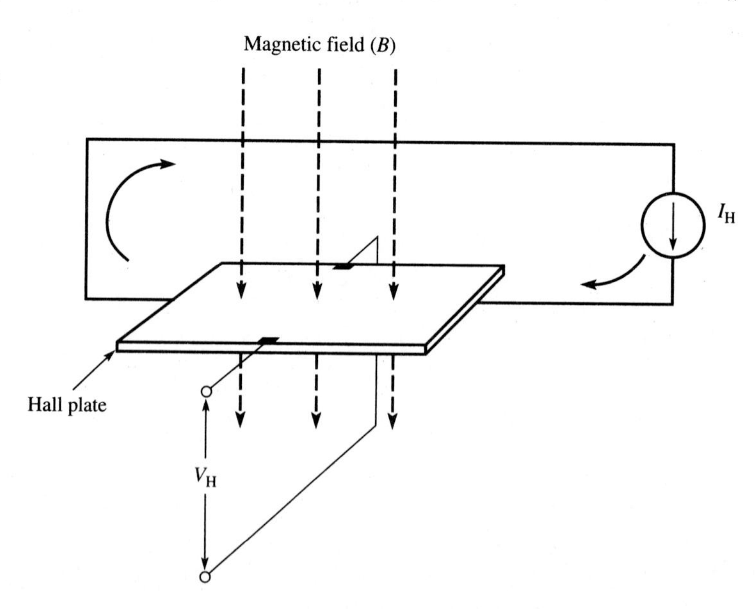

FIGURE 13–3
Magnetic transduction. The principle illustrated is the Hall effect.

result. A familiar example is the solar cell—an array of solar cells is capable of producing enough electricity to power a small electric car.

13–3 TRANSDUCER SELECTION

The selection of a transducer begins with the specifications of the physical quantity to be measured. The user must ascertain the required accuracy of the measurement, the duration of the test, and may need to consider cyclic behavior or other factors. In addition, consideration needs to be given to the environment in which the transducer is placed. Finally, the calibration procedure should be considered. Each of these categories is discussed as part of the selection criteria for transducers. The general selection criteria listed in this section should be considered representative; additional factors may be necessary for specific transducers.

MEASUREMENT REQUIREMENTS

The measurement requirements for a transducer are as follows.

Range

The **range** is the set of values a transducer is designed to measure. The minimum and maximum values of the transducer's range are called the **endpoints.** Some transducers can be adjusted to cover a different range by attenuating the measurand—for example, a sensitive radiation transducer can be used if the measured quantity is attenuated using radiation-absorbing filters. It is not always possible to find a single transducer to cover the entire range of measurand values; in these cases transducers with overlapping ranges must be selected.

Input Threshold

The **input threshold** is the smallest detectable value of the measured quantity starting near the zero value of the variable. For an input to be discerned, it must be possible to assign a unique number to the input. The selection of a transducer requires that it respond in a discernible manner to the threshold.

Dynamic Behavior

The **dynamic behavior** of a transducer specifies how the transducer can respond to a changing input. No transducer could follow an instantaneous change. (For that matter, no measurand can change instantaneously!) The transducer's dynamic performance is usually specified as a frequency response or response time, depending on the type of transducer. The response time is the time required to reach a specified percentage (typically 90% to 99%) of the final value for a given change of the input. The response time is measured in much the same way as the time constant for an *RC* or *RL* circuit. (Recall that the time constant is the time required for the output to reach 63% of its final value.)

Accuracy and Resolution

Accuracy, defined in Section 2–2, is the difference between the measured and accepted value. The accuracy requirements for a particular measurement can greatly affect the total cost of the measurement system. In addition, certain transducers, such as strain gages and pressure transducers, have a fatigue life that can change the accuracy, depending on the duration and cyclic behavior of the measurand. In some cases, the accuracy isn't as important as the ability to detect a small change (resolution), as when quantities are being compared. For example, in underground-tank testing, the interface between the liquid and air can be located by observing the small temperature difference between the air and liquid. In other cases, consistency is the most important criterion. Other accuracy errors include nonlinearities due to a zero shift or drift due to aging, which can affect the long-term repeatability of measurements.

Repeatability and Hysteresis Error

Repeatability is the maximum difference between consecutive measurements of the same quantity when the measured point is approached each time from the *same* direction for full-range traverses. It is usually expressed as a percentage difference of the full-scale output. **Hysteresis error** is the maximum difference between consecutive measurements for the same quantity when the measured point is approached each time from a *different* direction for full-range traverses. An example is when backlash in gearing causes the readings of a dial to be different, depending on whether the gearing was turned in one direction or the other.

OPERATIONAL AND ENVIRONMENTAL CONSIDERATIONS

The operational and environmental considerations for a transducer are as follows.

Natural Hazards

The transducer, wiring, and connectors must all be able to withstand the effects of exposure to the required environment. Natural hazards include the effects of dust and dirt, high or low temperatures, water (including salt water), and humid conditions. Wire insulation is available that is resistant to these effects as well as solvents, acids, bases, and so forth. In contrast, the transducer should not present a hazard to the environment in which it is placed—including causing electrical problems such as explosion hazard or shock hazard.

Human-caused Hazards

Human-caused hazards include high-radiation environments, corrosive or dangerous chemicals, immersion, abrasion, vibration, and explosive environments, to name a few. The electrical signal from the transducer may be interfered with if the signal cables are routed in an electrically noisy environment, another possible hazard for proper operation, particularly with transducers that have a low output signal.

Power Requirements

Power requirements depend on the type of transducer. Passive transducers, such as photocells, convert some of the incident source energy into electrical energy.

Others require a source of excitation, which can be a dc or ac source. If the transducer is being operated in a remote or noisy environment, the power leads become a source of potential problems.

Signal-conditioning Requirements

If the transducer produces a very small signal or it is located at a remote location or in a noisy environment, amplification or other signal conditioning may be required at the transducer. The transducer output may need to be converted into a compatible format for the remainder of the instrumentation system.

Physical Requirements

In some installations, the space available may be limited or the measurand may be over a limited region. If the quantity to be measured is concentrated, such as in the case of a collimated light source, the physical size of the transducer can affect the output.

Loading Effects

Loading effects cause the measurand to be disturbed in some manner by the presence of the transducer. All measurements in some way modify the quantity to be measured. For example, a rotating-vane flowmeter extracts a small amount of the energy from the fluid to turn the vane and thus changes the flow from its undisturbed value. Loading can also occur in the electrical-measuring circuit. Electrical loading can occur when a transducer with an equivalent high Thevenin resistance is connected to an amplifier with a finite input impedance. In this case the transducer signal is reduced by the connection to the amplifier. Many transducers have a very high equivalent Thevenin resistance (pH electrodes, for example), so it is important to be aware of the need for compatible amplifiers.

Human Factors

Human factors that need to be considered in the selection of a transducer are the operating skill of the persons installing and using the transducer, the ease of installation, the cost of the transducer, and the required maintenance.

CALIBRATION REQUIREMENTS

Our final consideration for selecting a transducer is the calibration requirements. The calibration interval and type of calibration necessary need to be considered. The *calibration interval* is determined by the operating life of the transducer and other factors such as long-term sensitivity shift, zero shift, and the accuracy requirements of the application. The calibration interval, complexity, and need to refer to calibration data can affect the total cost of the transducer.

The ideal calibration will precisely predict the response of the transducer in the application setting. This prediction may be difficult, particularly when there are large differences between the calibration process and the application such as different loading, dynamic response, or environmental conditions. Calibration methods vary widely, depending on the transducer, but should be made in a manner that is traceable to NIST. The errors associated with the standard should be much smaller than the transducer that is being calibrated. The calibration can consist of comparing the transducer to a known reference instrument or a physical standard (such as a known mass for a load cell) or using a physical reference (such as the triple point of water[2] for a temperature transducer).

Frequently, the physical parameter is varied and the response of the transducer is observed and compared to a known reference. Generally, the specific calibration points should extend over the full range of the measurand to avoid the need for extrapolation. For example, a pressure transducer should be subjected to the same range of pressures in the calibration process that it will be subjected to in its intended application to reveal any nonlinearities or other problems. The data should be taken in both an increasing and decreasing direction to reveal any hysteresis present. Data taken during calibration is called a **calibration record.** A line connecting the data points is called a **calibration curve** for the particular transducer. If the calibration of the transducer is not done under the same conditions as the operating conditions, systematic error can result whenever the transducer is used. For example, a radiation transducer that is calibrated with a radioactive source that has a different spectrum than the spectrum that is to be monitored will lead to flawed data if the difference is not accounted for.

A calibration performed in a manner that gives the transducer time for the output to settle to a fixed value is called a **static calibration.** Transducers used in dynamic measurements can be tested for their response with a **dynamic calibration.** A dynamic calibration is often a comparison of the transducer that is being calibrated with a known reference transducer that is faster. Another dynamic test is called a **step-function response test.** In it, the transducer is subjected to a rapid change in the input measurand, typically from 10% to 90% of the transducer's range. For example, a temperature transducer is very quickly moved to a much warmer or colder environment and its response is observed. The time required for the output to settle to the new temperature is a measure of the transducer's response time.

13–4 BRIDGE MEASUREMENTS

As you have seen, many transducers—such as strain gages, resistance temperature devices, and certain displacement transducers—use resistive transduction principles. Typically, these transducers produce a very small resistance change in response to the measurand. A circuit that can detect these very small changes is the Wheatstone bridge, described in Section 7–10. Because of their great sensitivity and other advantages, bridge circuits are widely used for transducer measurements. A bridge circuit that consists of four resistance arms with a source of energy (called the **excitation**) and a detector is shown in Figure 13–4. This circuit is shown with *two* pairs of leads to the power supply, a common method of ensuring that the voltage at the bridge is stable. Current to the bridge flows on the

---

[2] The triple point of water is the temperature at which the solid, liquid, and gas phases of water are in equilibrium.

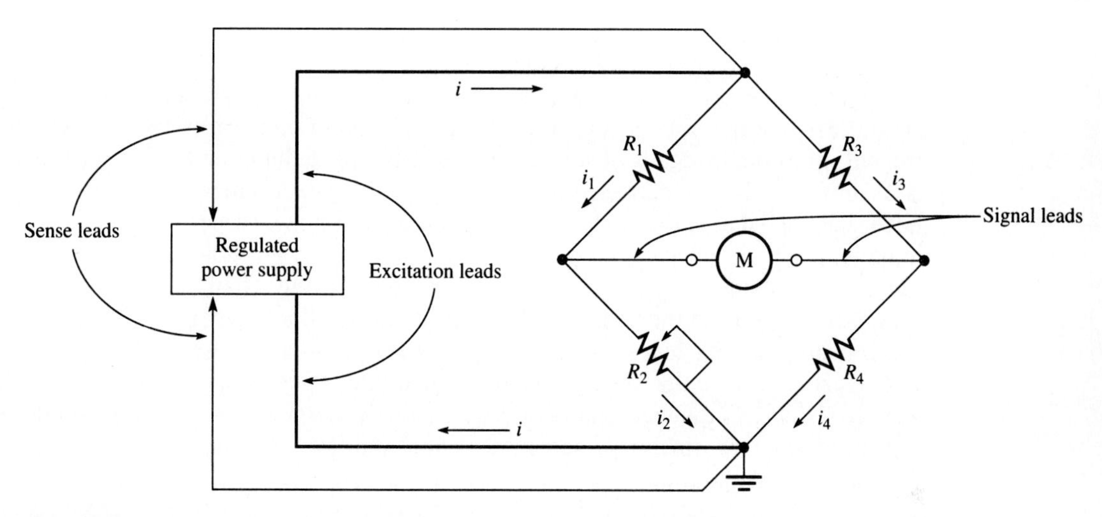

FIGURE 13–4
A basic Wheatstone bridge. A sensitive meter, typically a galvanometer or voltmeter, is used to detect an imbalance. Sense and excitation leads are shown to the power supply, a common technique for maintaining regulated voltage at the bridge.

excitation leads. Any variation in the voltage due to IR drop in the excitation leads is detected by the **sense leads** and is used to regulate the supply. Notice that the sense leads carry almost no current, so they have almost no *IR* drop.

The balance condition can be detected with great sensitivity using an instrumentation amplifier (IA) connected across the differential output, as shown in Figure 13–5. For general applications, the circuit shown is a good choice (CMRR

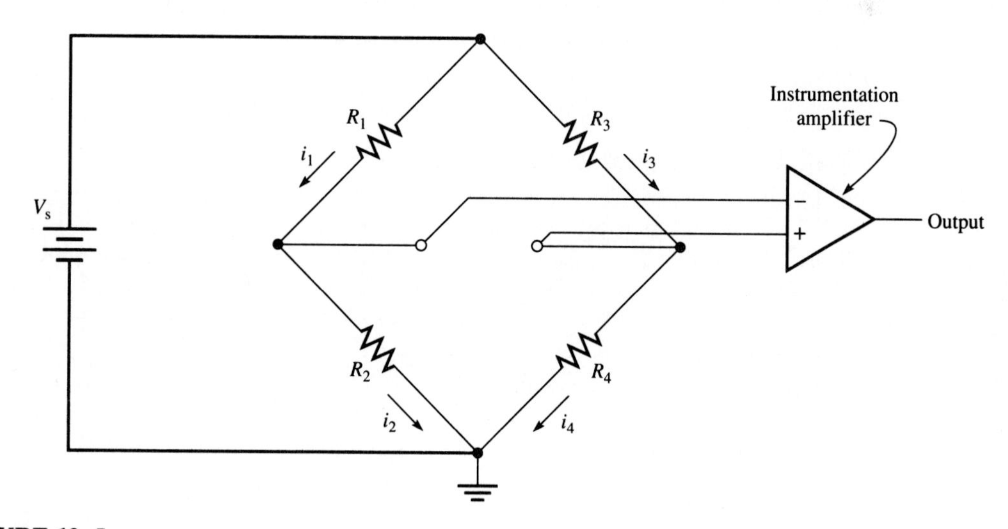

FIGURE 13–5
Detecting imbalance with an instrumentation amplifier. The IA uses differential inputs and adds almost no loading effect.

is typically greater than or equal to 110 dB). The IA is selected for low-drift, high-common-mode rejection and gain stability.

When a Wheatstone bridge is used to detect the change in resistance of a single resistive transducer, the transducer is placed in one arm of the bridge, and the output of the bridge is observed. Frequently, more than one active transducer is used in a bridge to increase the sensitivity of the measurement. It is not always necessary to balance the bridge to determine the unknown resistance; instead, the magnitude of the off-balance condition can be used as an indicator of the resistance or to detect a change in resistance. The output change is not a linear function of the resistance change (it is approximately linear when $R_1 \gg R_{unknown}$), but this nonlinearity has some advantages in certain measurements. It is possible to construct bridges that tend to linearize the output of nonlinear transducers such as thermistors. Transducer manufacturers have developed a number of techniques for increasing the linearity, sensitivity, and stability of bridge circuits. These include specialized amplifiers, changing the excitation source, and matching the thermal or other characteristic of the transducer with the bridge resistors. For example, strain gages respond to more than just strain; they also respond to temperature changes. To cancel the temperature effects, a "dummy" gage can be placed in the same environment but not subjected to the strain of the measuring strain gage. The dummy gage is placed on the same side of the bridge as the measuring gage. Temperature effects change the resistance of both gages in a like manner, causing no change in the output.

Many resistive transducers produce only a tiny change in resistance for a given input change. For sensitive measurements, the detecting instrument must

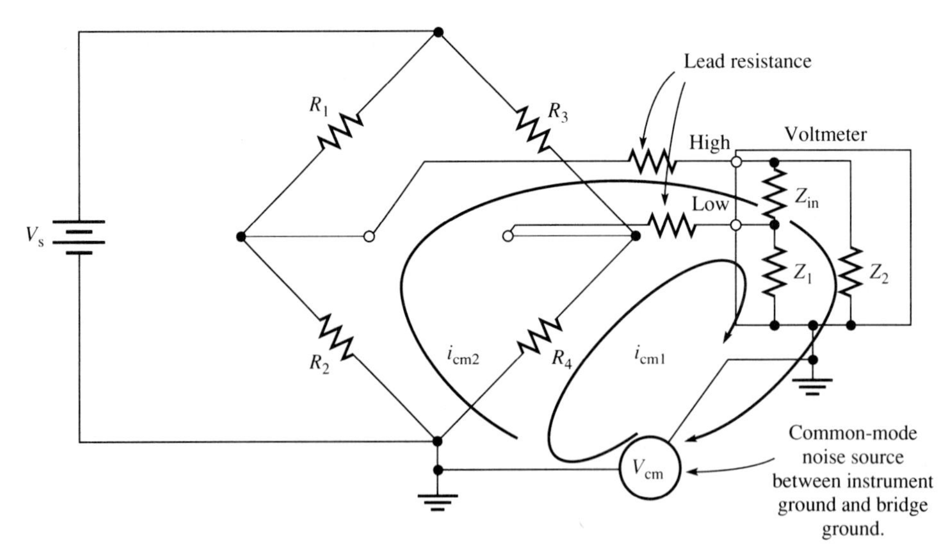

(a) A common-mode source, created by ground currents, can be converted into a differential source by different impedances in the paths shown.

FIGURE 13–6
Common-mode noise problem and solutions for Wheatstone bridge circuits.

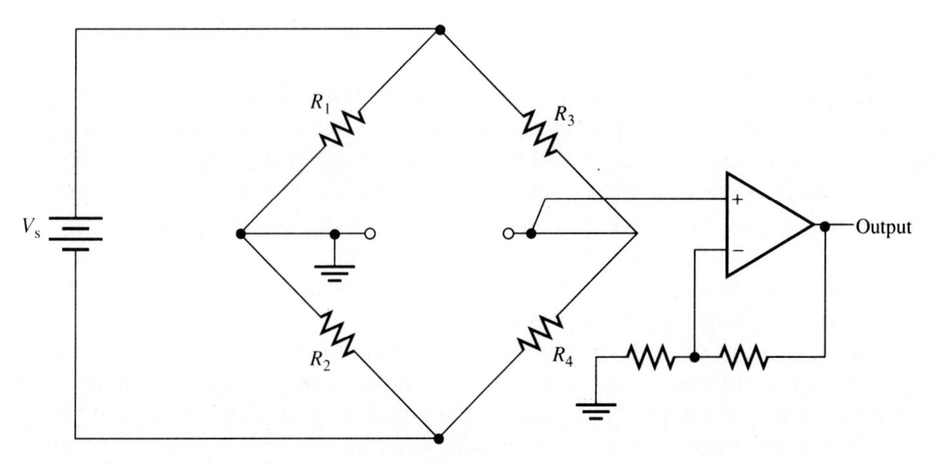

(b) An arrangement that provides a very high CMRR. The method requires a floating excitation supply.

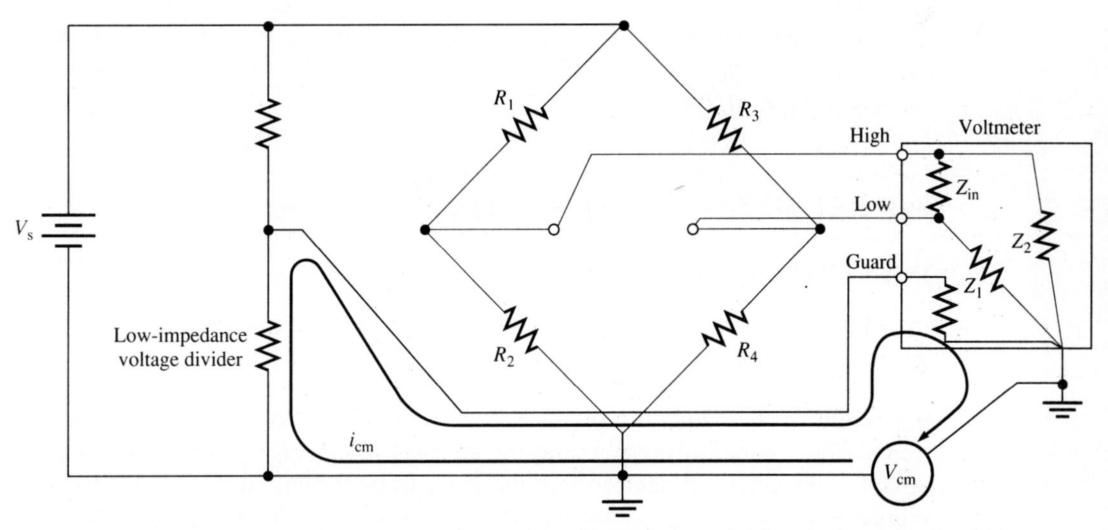

(c) An arrangement that shunts common-mode current around the signal path. A guarded detecting instrument is required.

FIGURE 13–6 *(continued)*

have good common-mode rejection (CMR) because each side of the bridge includes a common-mode signal from the excitation supply. When the detecting instrument is located some distance from the transducer, an even more serious common-mode problem can occur when ground current causes a common-mode source to appear between the bridge ground and the instrument ground. Notice that the output (signal leads) of the bridge in Figures 13–4 and 13–5 are not connected to circuit ground, a condition referred to as a **floating** output. When a voltmeter is connected as a detecting instrument to the bridge outputs, the voltmeter signal connections should be isolated from circuit ground; however, there is always some high impedance to ground. This situation was discussed in Section 12–5 and is diagrammed in Figure 13–6(a). Assume that the impedance to ground is different between the high and low inputs (the usual case). Current from the common-mode ground source finds a path through the voltmeter's leads and generates a differential voltage due to the impedance difference in the return paths. This means that a common-mode signal has been converted into a differential-mode signal, and the meter will respond to this potential. One possible solution is shown in Figure 13–6(b). The excitation supply is floating—in other words, its reference is isolated from the circuit ground by using an isolation transformer or a battery. This method can provide CMRRs greater than 160 dB. The left side of the bridge is connected to circuit ground and the right side of the bridge is connected to a single-ended, noninverting op-amp.

Another solution is to use a guarded voltmeter. The guard connection is made so that it shunts common-mode current away from the meter's inputs. One of the best ways to do this is to connect the guard lead to a low-impedance point that is at the same potential as the low side of the meter. This is done by adding a low-impedance divider to the bridge, as illustrated in Figure 13–6(c). Most of the common-mode current will flow through the low-impedance path provided by the voltage divider. There are other variations of sensing circuits that are designed to optimize dc offset, voltage or temperature drift, nonlinearity, noise performance, or other characteristic.

13–5 MEASUREMENT OF TEMPERATURE

INTRODUCTION

In industrial-process control, temperature is the most frequently controlled and measured variable. It is necessary to monitor temperature in a wide variety of industries because the proper operation of most industrial plants—and frequently the safety of their operation—is related to the temperature of a process. The transducers for measuring temperature fall into two categories. If the transducer is directly connected or inserted into the body to be measured, the transducer is a **thermometer.** If the temperature is measured by observing the body to be measured rather than by direct contact, the transducer is a **pyrometer.** Pyrometers indirectly determine temperature by measuring the radiated heat or sensing the optical properties of the body.

DEFINITION OF TEMPERATURE

The molecules of all substances are in constant motion due to thermal energy. The temperature where molecular motion totally ceases and there is no thermal energy is called absolute zero,[3] a point that cannot be reached in practice. In a solid, molecules are constrained to a particular relationship to other molecules and molecular motion is vibrational energy. In liquids, the molecules have sufficient

[3] According to quantum theory, however, molecules still have a small kinetic energy even at absolute zero.

energy to move from their fixed locations and move around each other. As more heat energy is added to a substance, the velocity of molecules increases to the point that the molecules overcome the attractive forces between them and are free to roam independently of each other, forming a gas. Theoretically, there is no upper limit to temperature—as temperature is increased, molecules break apart into atoms and atoms lose electrons, forming a plasma. This is the condition that occurs in the stars.

Temperature (from the Greek words for "heat measure") is related to the *average* translational kinetic energy of molecules due to heat. Notice that this is entirely different than the concept of heat. Heat is a measure of the total internal energy of a substance, measured in joules or calories. Thus a substance with a higher temperature may contain *less* heat if it has a smaller total internal energy. For example, a hot cup of coffee has a higher temperature than an iceberg because of the average velocity of its molecules, but the iceberg contains more thermal energy because of its mass. In the macroscopic view, temperature can be defined as a condition of a body that determines the transfer of heat to or from other bodies.

TEMPERATURE SCALES

Although he did not invent the thermometer, Gabriel Fahrenheit, a Dutch instrument maker, was recognized for producing the first mercury thermometers; they were the first ones accurate enough for scientific work. His scale was calibrated on the basis of the lowest temperature he could obtain (a mixture of ice water and ammonium chloride at 0°F) and the temperature of the human body as 96° (although 98.6° was later found to be more accurate). The scale, used primarily in the United States, indicates the freezing point of water as 32° and the boiling point of water as 212° at standard pressure.[4] Absolute zero is at a temperature of −459.6° on the fahrenheit scale.

The celsius scale is more widely used worldwide than the fahrenheit scale. The celsius scale defines the freezing point of water as 0° and the boiling point as 100° at standard pressure. This scale has absolute zero at −273.15°.

The Kelvin scale is an absolute scale in which all temperatures are positive; it is the temperature scale used in most scientific work. Thus absolute zero is defined as 0 K. Degree markings on the Kelvin scale are defined as 1/273.16 of the triple point of water. The number of degrees between the freezing point of water and the boiling point of water is the same on both the celsius and the Kelvin scales; thus the magnitude of degrees on the two scales is the same. Conversion of temperatures between the scales can be done with the following equations:

$$F = \tfrac{9}{5}C + 32$$

$$C = \tfrac{5}{9}(F - 32)$$

$$K = C + 273$$

[4] Standard pressure is 760 mm of Hg, or 14.7 lb/in^2.

where F = temperature, °F
C = temperature, °C
K = temperature, K

EXAMPLE 13–1 A temperature of 65°F is what temperature in celsius?

SOLUTION

$$C = \tfrac{5}{9}(F - 32)$$
$$C = \tfrac{5}{9}(65 - 32)$$
$$= 18.3°C$$

THE THERMOCOUPLE

A thermocouple junction is created when two dissimilar metal wires are joined at one end. When the junction is heated, a small thermionic voltage that is directly proportional to the temperature appears between the wires. This effect was discovered by Thomas Seebeck in 1821 and is named the **Seebeck effect.** The emf is produced by contact of the two dissimilar materials and is proportional to the junction temperature.

If a circuit is completed by joining both ends of the wires and one junction is at a different temperature than the other, a current will flow in the circuit, as illustrated in Figure 13–7. The amount of current is a function of the temperature *difference* between the two junctions and the type of metals used in the wires. To be useful as a temperature measurement, one junction is the sensing, or "hot," junction, whereas the other junction is the reference, or "cold," junction. If the cold junction is at a known temperature, such as that of melting ice, the current in the circuit can be calibrated in terms of the temperature of the sensing junction.

If you break the circuit and try to measure the thermionic voltage created at a junction with a voltmeter, you encounter a problem. This is because when you connect the leads of a voltmeter to the dissimilar metals of the junction, you create two new junctions (called parasitic junctions) that are themselves thermocouples, as shown in Figure 13–8. Even if both meter leads are at the same temperature, the voltmeter responds only to the *difference* between the temperatures of the

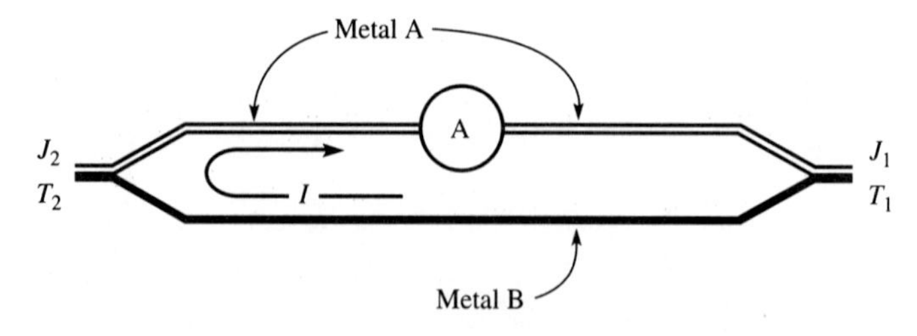

FIGURE 13–7 A basic two-wire thermocouple. Current in the circuit depends on the type of wires and the temperature difference between the junctions.

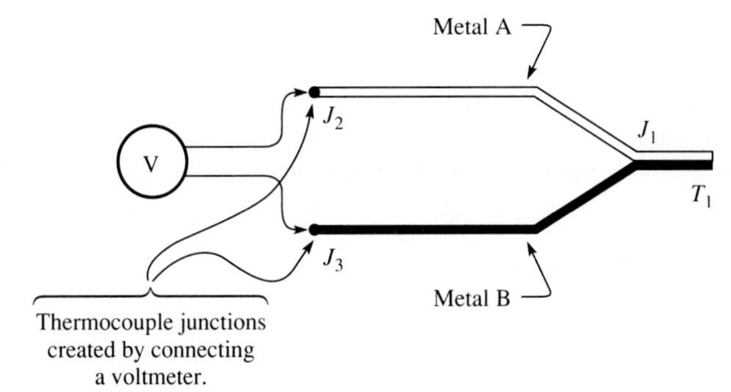

FIGURE 13–8
Connection of a voltmeter creates two new junctions. The voltmeter reads the algebraic sum of the junction voltages.

meter leads and the original junction that you are attempting to measure. The voltage read by the meter is given by the approximate equation

EQUATION 13–2

$$V \cong \alpha(T_1 - T_2)$$

where V = thermoelectric (Seebeck) voltage, V
α = Seebeck coefficient, V/°C
$T_1 - T_2$ = temperature difference between junctions, °C

It would appear that it is necessary to know the temperature of the voltmeter leads in order to use a thermocouple to measure an unknown temperature. The solution to the dilemma is to move the junctions from the voltmeter onto an isothermal block and place the block at a known reference temperature. The voltage from the unknown junction will now be proportional to the type of materials and the temperature difference between the unknown and the reference block. This idea is illustrated in Figure 13–9. The temperature can be a precisely controlled reference, such as the temperature of melting ice. Although melting ice is a suitable reference, it is inconvenient and not necessary for most measurments. For high-temperature measurements, the accuracy of the reference may be sufficient if the block is at room temperature. The isothermal block is usually made from a thermally conductive ceramic material. Its temperature can be monitored

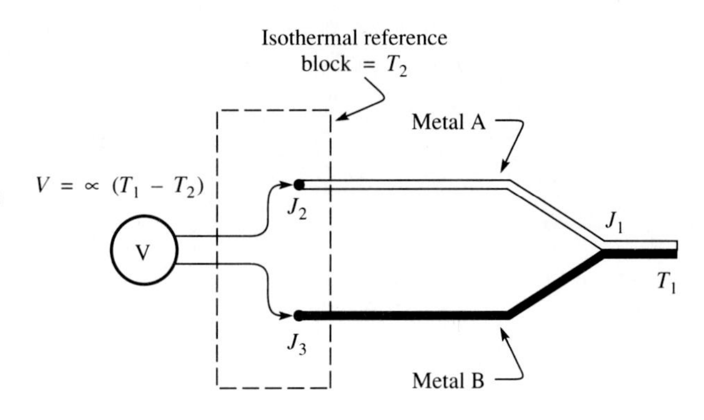

FIGURE 13–9
Connection of the voltmeter leads to an isothermal block. The block is at temperature T_2 and is equivalent to the basic thermocouple in Figure 13–4(a).

with an IC temperature sensor or thermistor (discussed later), and compensation for the block temperature can be done by a microprocessor. It is not necessary to keep the cold junction at a constant temperature, as long as the reference is known. You may wonder why bother with the thermocouple at all if sensors exist that can measure the temperature directly. The reason is that the range of the sensor for measuring the reference temperature is limited, but the thermocouple can operate over a much larger range and at much higher temperatures.

Special electronic compensation circuits are available that both track the cold-junction temperature and scale the output voltages to common types of thermocouples. These circuits provide automatic compensation and offer the same accuracy as an ice bath but are much simpler to implement (±0.5% accuracy). The circuits consist of two parts—a cold-junction compensating integrated circuit and an op-amp amplifier to provide signal conditioning and amplification. In applications where the Seebeck coefficient is small (as in type-S thermocouples), chopper-stabilized amplifiers are used because of their stability.

Although Equation 13–2 indicates a linear relationship, actual thermocouples deviate from this ideal. Figure 13–10 shows the relationship between the

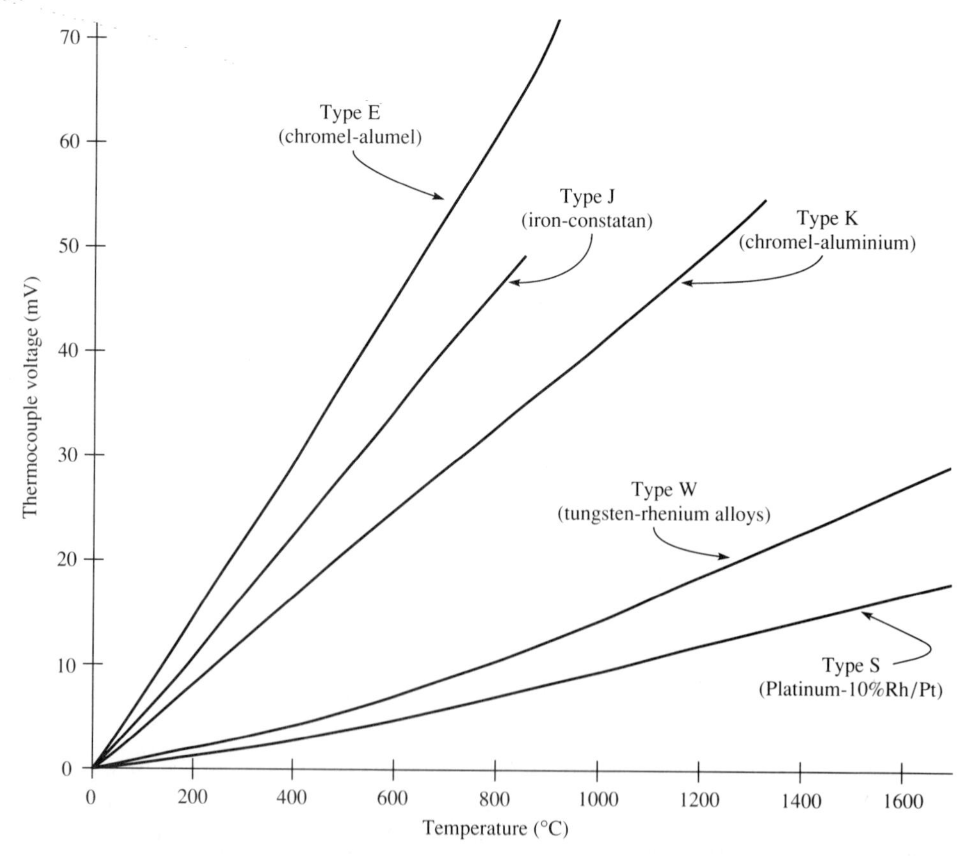

FIGURE 13–10
Output voltage of some common thermocouples with 0°C as the reference temperature.

TABLE 13–2
Comparison of standard thermocouples.

TYPE	MATERIAL	TEMPERATURE RANGE	EMF at 100°C (ref = 0°C)	COMMENTS
J	Iron vs. copper-nickel	0°C to 750°C	5.268 mV	Not recommended for low temperatures
K	Nickel-chromium vs. nickel-aluminum	−200°C to 1250°C	4.095 mV	Most popular; wide temperature range
E	Nickel-chromium vs. copper-nickel	−200°C to 900°C	6.317 mV	Highest Seebeck coefficient
T	Copper vs. copper-nickel	−200°C to 350°C	4.277 mV	Low temperature and damp environments
S	Platinum–10% rhodium vs. platinum	0°C to 1450°C	0.645 mV	High-temperature applications
R	Platinum–13% rhodium vs. platinum	0°C to 1450°C	0.647 mV	High-temperature applications
B	Platinum–30% rhodium vs. platinum–6% rhodium	0°C to 1700°C	0.033 mV	High-temperature applications

temperature and the thermoelectric voltage for several types of common thermocouples. The output voltage is shown for a reference temperature (T_1) of 0°C. The slope of the line represents the **Seebeck coefficient** in Equation 13–2, but the slope is not constant for the full range of temperature for any given thermocouple. One of the more linear types is the K-type, with a Seebeck coefficient specified as 39.4 μV/°C. The K-type has a linear coefficient over the range of 0°C to 1000°C and is widely used for this reason. If greater accuracy is required, refer to thermocouple reference tables published by the NIST.

Standard thermocouples cover different ranges of temperature and have different sensitivity, linearity, stability, and cost. For reference, standard thermocouples are compared in Table 13–2. For instance, type-J thermocouples containing iron are relatively inexpensive but limited in range. Type-R and type-S thermocouples (platinum-rhodium) are particularly stable, the type-E thermocouple has advantages for measuring low temperatures but has higher nonlinearity than others, and type-W (tungsten-rhenium) thermocouples are suited for very high temperatures. Exposed-junction thermocouples are fragile and corrode easily; to prevent this, sheathed probes are made in a metal or ceramic insulated tube. Additional protection is given to the wires by overbraiding.

EXAMPLE 13–2 Assume a thermocouple with a Seebeck coefficient of 58.5 μV/°C has an output of 24 mV when the reference junction is at room temperature (21°C). What is the temperature of the measuring junction?

SOLUTION Rearranging Equation 13–2

$$T_2 = \frac{V}{\alpha} + T_1$$

$$= \frac{24 \text{ mV}}{58.5\ \mu\text{V}/°\text{C}} + 21°\text{C} = 431°\text{C}$$

MEASUREMENTS WITH THERMOCOUPLES

In order to provide circuits optimized for temperature measurements, manufacturers have devised special thermocouple digital thermometers. Essentially, the thermometer is a digital voltmeter that uses a computer memory to store the Seebeck coefficients and a microprocessor to recall the appropriate coefficient and convert the measured emf to a temperature. The output of a thermocouple is fairly linear for small regions but not over its entire range. Sophisticated thermometers use polynomial curve fitting to reduce the error from a straight-line fit. Accuracy can be improved if the thermometer, along with the thermocouple to be used, is calibrated against a standard temperature near the one to be measured.

Because thermocouples produce a relatively small output voltage, typically a few millivolts, special precautions must be observed to prevent noise from affecting the data. Thermocouples are susceptible to interference because the wire can act as an antenna for interference pickup. To avoid interference, keep the thermocouple wire as short as possible; twisting the lead wires and shielding may be necessary. Connect the shield to the guard terminal of the thermometer. When longer runs are required, the thermocouple wires should be extended with commercially manufactured extension wire designed to match the Seebeck coefficients for the particular thermocouple.

Other problems associated with thermocouple measurements may be due to the environment in which they are in contact. In harsh chemical environments, thermocouples may deteriorate. Water can cause problems because of dissolved electrolytes and impurities. The thermocouple wires should be protected from harsh environments and liquids by special shielding. At extreme temperatures, the metal of the thermocouple can boil off, changing the alloy and the Seebeck coefficient. These types of deteriorations require the thermocouple to be replaced periodically. As a check on deterioration, the thermocouple's electrical resistance can be logged. To measure the resistance, the ohmmeter should be used on the same range every time—readings are taken with the leads on one set of contacts and then reversed. The average of the readings is used. This procedure cancels out the effect of the thermocouple's own emf.

It is worth noting that there are operational amplifiers on the market specifically designed to work with thermocouples. Analog Devices markets two operational amplifiers, the AD-594 and AD-595, that linearize and amplify the signal from type-J and type-K thermocouples, respectively. The integrated circuit also contains a built-in ice-point compensation and has an output of 10 mV/°C. It uses a differential amplifier on the input with a single-ended output.

THE RTD

A resistance-temperature detector (RTD) exploits the fact that the resistivity of metals is a positive function of temperature. The change in resistivity causes a change in the resistance of a conductor. Nearly all RTDs are constructed from fine wire of platinum, although wires made from nickel, germanium, and carbon-glass are occasionally used for specialized applications. The highest-quality RTDs are made from platinum wire that is mounted in a manner to avoid strain-induced change in the resistance. Platinum RTDs (PRTDs) are the most accurate thermometers made for temperatures between the boiling point of oxygen at −182.96°C to the melting point of antimony at 630.74°C, and special laboratory PRTDs are used as interpolation standards between these temperatures. The useful temperature range extends to as low as −240°C and as high as +750°C.

The nominal resistance of standard RTDs is designed to be 100 Ω at 0°C, although special RTDs with resistances from 50 Ω to 2000 Ω are also available. The resistance at any temperature can be determined by the alpha, a number that specifies the resistance change per ohm of nominal resistance per degree change in temperature. The resistance can be found at any temperature, t, from the linear equation

EQUATION 13–3

$$R_t = R_n(1 + \alpha t)$$

where R_t = resistance of RTD at some temperature, Ω

R_n = nominal resistance at 0°C, Ω

α = resistance coefficient, Ω/Ω/°C

t = temperature for which resistance is computed, °C

For standard American gages, the resistance coefficient is 0.00392 Ω/Ω/°C. The European standard is 0.00385 Ω/Ω/°C.

EXAMPLE 13–3 A platinum RTD with an α of 0.00392 Ω/Ω/°C has a nominal resistance of 100 Ω at 0°C.
(a) Compute the resistance at 450°C.
(b) At what temperature is the resistance 142.6 Ω?

SOLUTION (a)

$$\begin{aligned} R_t &= R_n(1 + \alpha t) \\ &= 100.0[1 + (0.00392)(450)] \\ &= 276.4\ \Omega \end{aligned}$$

(b) Rearranging Equation 13–3 yields

$$\begin{aligned} t &= \frac{1}{\alpha}\left(\frac{R_t}{R_n} - 1\right) \\ &= \frac{1}{0.00392}\left(\frac{142.6}{100.0} - 1\right) \\ &= 108.7^\circ\text{C} \end{aligned}$$

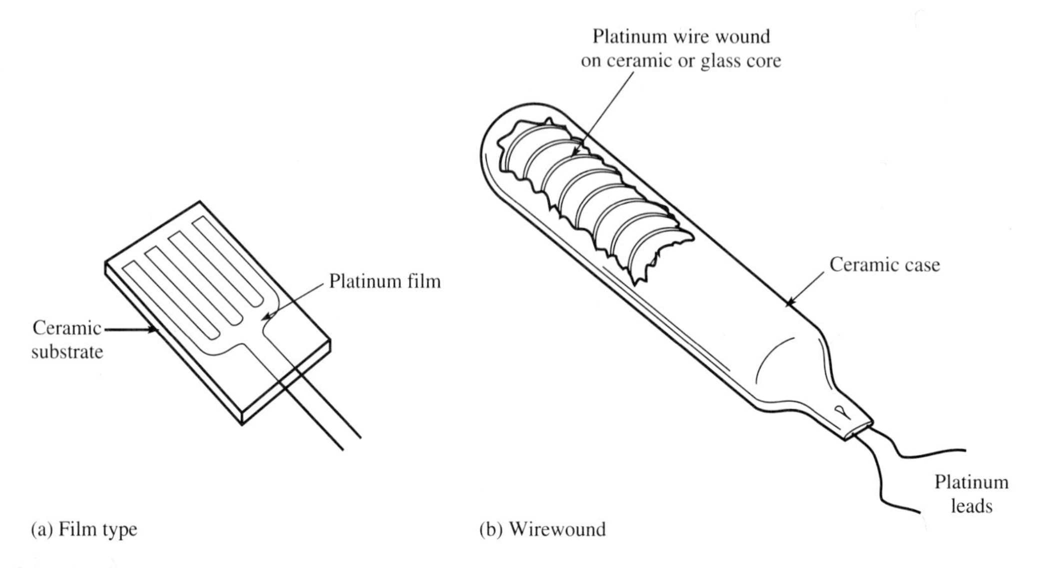

FIGURE 13–11
Platinum resistance thermometers. These are available in a wide variety of shapes and sizes.

RTDs are made from either wirewound or film elements mounted on a core and sealed within a capsule. Examples of wirewound and film elements are shown in Figure 13–11. Wirewound assemblies are typically supported on a ceramic or glass core with a special winding technique to avoid inducing strain in the wire that could change the resistance. Film elements are made from a platinum sensing layer on a ceramic substrate. New miniature thin-film RTDs can be made smaller than a match head with accuracy equal to that of wirewound units and better response time at a lower cost.

The resistance measurement of RTDs is generally done by a Wheatstone bridge or a four-wire ohms measurement. In the bridge method, the RTD is placed in one leg of the bridge and the output voltage is sensed. The output voltage is a nonlinear function of the resistance, so a correction must be applied to determine the resistance of the RTD and the associated temperature. Since the resistance of RTDs is typically 100 Ω or so, care must be taken to avoid problems with lead resistance. The four-wire resistance method uses two leads to source a constant current into the RTD and two leads to sense the voltage developed across the RTD. This method avoids the nonlinearity problems associated with resistance measurements in the Wheatstone bridge. Because the sense wires carry very little current, the wire-resistance problem is alleviated.

THERMISTORS

Like RTDs, thermistors (thermal resistors) change resistance as a function of temperature; however, instead of a positive temperature coefficient, they have a negative coefficient (resistance decreases with increasing temperature). Thermistors are available in a variety of packages—including glass beads, probes, discs, washers, rods, and so forth. Typically they are manufactured as small, encapsulated beads made of a sintered mixture of transition metal oxides (nickel, manganese, iron, cobalt, and so forth). These oxides exhibit a very large resistance change for a change in temperature. At room temperature, the resistance is about 2000 Ω. In addition to the advantage of great sensitivity (400 times greater than an IC thermocouple), thermistors are chemically stable, have fast response times, and are physically small. The small physical size and fast response of thermistors makes them ideal for monitoring temperatures in a limited space, such as the case temperature of a power transistor or the internal body temperature of an animal. The negative temperature coefficient, opposite to that of most semiconductors, makes them ideal for applications where temperature compensation of semiconductor devices is required. The major limiting factors for thermistors are the limited range—from −50°C to a maximum of 300°C—fragility, decalibration if exposed to high temperatures, and a nonlinear response. Figure 13–12 illustrates a typical thermistor response.

Although a thermistor is nonlinear, the response can be represented quite accurately with the Steinhart-Hart equation:

EQUATION 13–4

$$\frac{1}{T} = A + B(\ln R) + C(\ln R)^3$$

where T = temperature, K
R = resistance of thermistor, Ω
A, B, C = curve-fitting constants

FIGURE 13–12
Typical thermistor response.

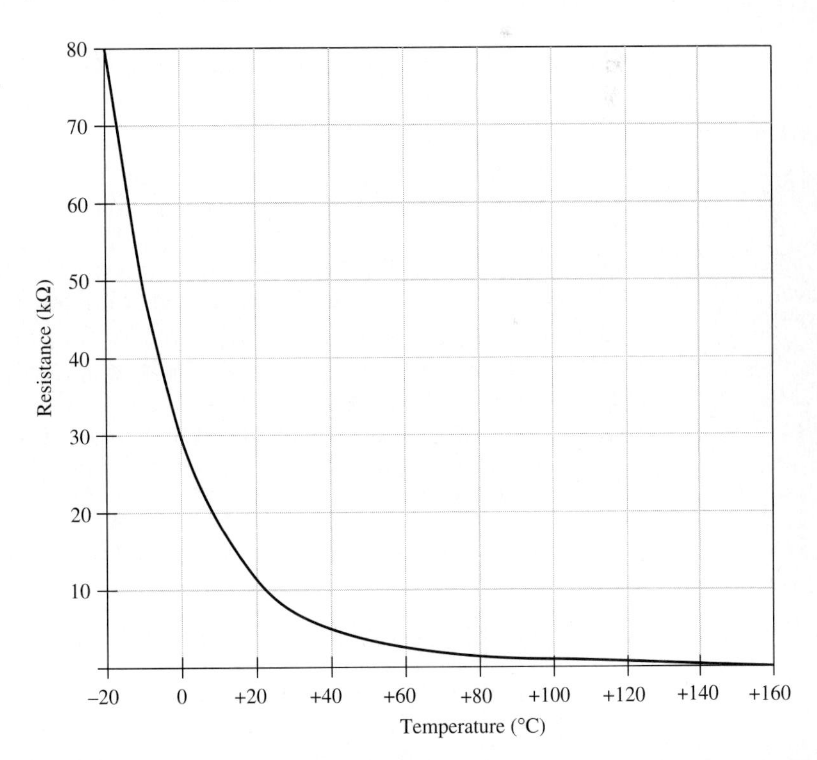

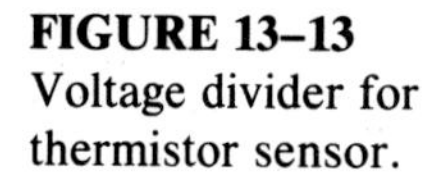
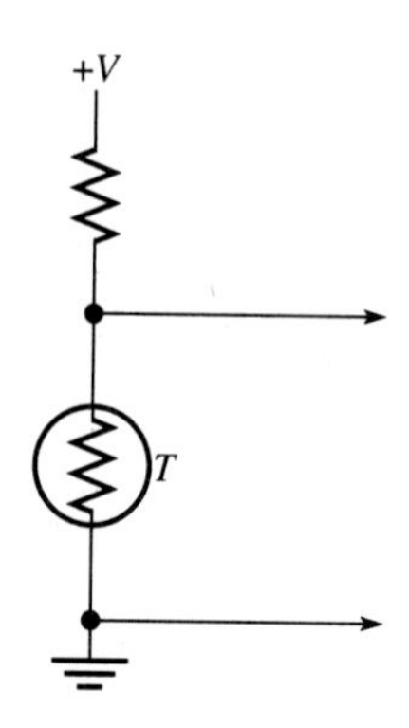

FIGURE 13–13
Voltage divider for thermistor sensor.

The constants A, B, and C can be found for a given thermistor by writing three equations for the thermistor at three different temperatures and solving the simultaneous equations for the constants. With curve-fitting, thermistors can be useful for measuring temperature to ±0.1°C accuracy. In addition, special linearizing networks are available for applications where a linear response is needed.

Thermistors can be placed in one leg of a Wheatstone bridge to provide precise temperature information; the Wheatstone bridge is particularly sensitive near balance, and a temperature change of as little as 0.01°C can be detected. For less-demanding operations, a thermistor can be placed in the voltage-divider arrangement shown in Figure 13–13. Because of the high sensitivity, thermistors can be measured directly with an ohmmeter or with special meters calibrated for thermistors.

SEMICONDUCTOR TEMPERATURE SENSORS

The junction voltage of a forward-biased diode with a constant current is a function of temperature as discussed in Section 4–1. For silicon diodes, the voltage across the diode decreases by approximately 2.2 mV for each degree celsius rise in temperature. By detecting this change in voltage, transistors and diodes can be used as temperature sensors. The exact sensitivity depends on certain parameters such as the size of the junction, doping level, and current density and varies between devices. Diode temperature sensors are inexpensive and can provide an indication of temperature; however, normally their range is limited to −40°C to +150°C. Some new diode sensors have been developed that can measure temperatures in the cryogenic regions with a range of 1.4 K to 475 K. These sensors must operate with extremely low current to minimize self-heating effects.

Certain DMMs are set up to measure temperature directly by using an *npn* transistor as a sensor. For example, the Tektronix DM 501 has a TEMP PROBE connector. An *npn* transistor, such as a 2N2484, can be used in place of the probe to get ±5°C accuracy with no calibration and better accuracy with a simple calibration.

INTEGRATED CIRCUIT TEMPERATURE SENSORS

To improve the accuracy, linearity, and sensitivity of simple diode sensors, manufacturers have developed IC temperature sensors. IC sensors are not as accurate as resistance thermometers or thermocouples, but they are convenient and low in cost. IC sensors are available in conventional transistor and IC packages with either a voltage or a current output that is proportional to temperature. An example of a sensor with a voltage output is National Semiconductor's LM135. The circuit operates as a two-terminal zener diode with a breakdown voltage that is proportional to the absolute temperature: +10 mV/K. The LM135 operates over the range from −55°C to 150°C. A device with a current output is Analog Devices AD590. This two-terminal device is connected in series with a low-voltage power supply, and the series current is equal to 1 μA/K.

IC sensors are an improvement over diode sensors but still have the problem of limited range and fragility. The advantages are easy calibration, low cost, and an output that can be read directly in degrees when connected to a DMM.

RADIATION PYROMETERS

The radiation pyrometer is a noncontacting temperature sensor that can detect infared radiation from a source, thus making it possible to measure temperatures from a remote location. It is generally used to observe high temperatures such as hot ovens, but with recent developments, it is capable of measuring temperatures to as low as −50°C. It operates by filtering all but the infared radiation from the field of view and focuses the radiation onto a temperature-sensing element. The temperature sensor converts the absorbed radiation into a voltage or current. This is converted to a reading that indicates the temperature of the source. Pyrometers are primarily used for measuring high temperatures in inaccessible locations or in environments in which a thermocouple cannot operate.

13–6 MEASUREMENT OF STRAIN

INTRODUCTION

If you apply a force to an elastic material, it will deform to some extent. **Elasticity** is the ability of a material to recover its original size and shape after a deforming force has been removed. If a relatively small force is applied to the length of a block, the block will change length by an amount that is proportional to the applied force. The applied force can be either a positive tensile force or a negative compressive force, as shown in Figure 13–14. As long as the material remains elastic, the change in length is proportional to the applied force. This relationship is known as Hooke's law:

EQUATION 13–5

$$F \propto \Delta l$$

where F = applied force, N

Δl = change in length due to the applied force, m

The tendency of a body to return to its original shape is limited. As more and more force is applied to a body, it reaches its **elastic limit**—a point where permanent deformation results. For materials such as steel, the elastic limit is reached if

FIGURE 13–14
The change in length, Δl, is proportional to the applied force, F.

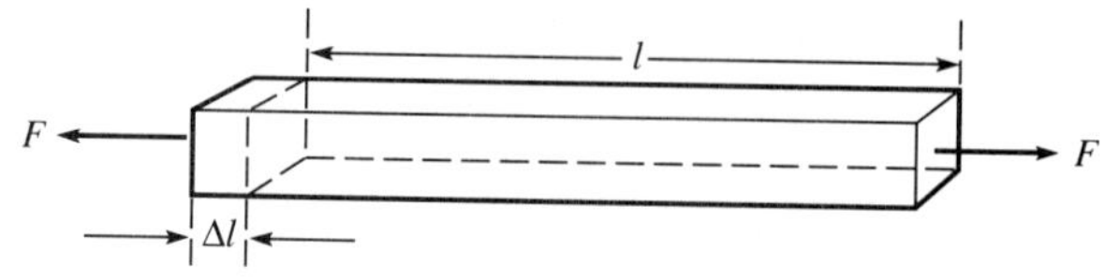

(a) Tensile forces (positive).

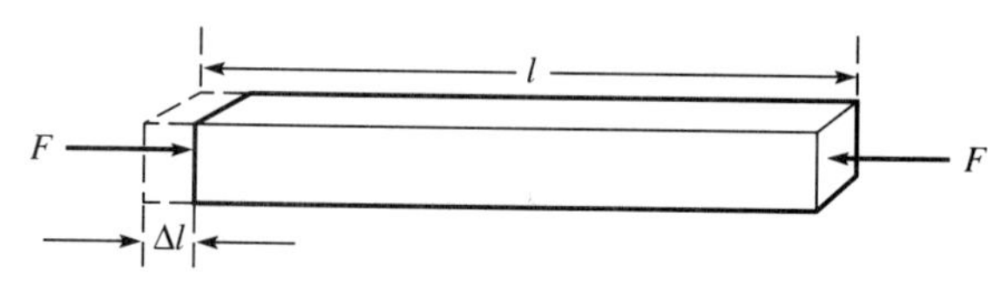

(b) Compressive forces (negative).

the change in length is more than a small percentage of its initial length. Once this point is reached, additional force will cause deformation and fracture. Some materials, such as modeling clay, have no elasticity and will not return to their original dimensions regardless of how small a force is applied. Clay is said to be a *plastic material.*

The deformation of a material depends on its length, cross-sectional area, and composition. Let's examine the effect of length first. If two homogeneous blocks of the same diameter and material are compressed by the same force, the longer block will be compressed by an amount that is proportionally greater. However, if we divide the change in length by the original length of the block, we obtain a quantity that is independent of the block's length:

$$F \propto \frac{\Delta l}{l}$$

The quantity $\Delta l/l$ is called **strain** and is written in equation form as

EQUATION 13–6

$$\epsilon = \frac{\Delta l}{l}$$

where ϵ = strain, a dimensionless number (often expressed as in./in.)

In practice, the magnitude of strain is a very small number; hence it is common practice to express strain in units of microstrain. Microstrain is $\epsilon \times 10^{-6}$ and is written as $\mu\epsilon$.

The second factor that affects the block's change in length is the area of the block. Imagine a block that is supporting a load. The load causes the block to be compressed by some amount. If a second identical block is added to share the load, the force is distributed equally between the two blocks; the change in length is half as much as before. Evidently, the effect of the force on the strain is reduced by the area of the block; that is

$$\frac{F}{A} \propto \frac{\Delta l}{l}$$

where A = area, m^2

In order to change the proportional relationship in an equation, we need to introduce a constant that is related to the composition of the block. This constant is called *Young's modulus, E,* and is a property only of the material. Young's modulus is a measure of the stiffness of a material and, for a given cross-sectional area, of the material to resist a change in length when loaded. Hooke's law can then be modified to

EQUATION 13–7

$$\frac{F}{A} = E\frac{\Delta l}{l}$$

where E = Young's modulus, N/m^2

The quantity F/A is called **stress** and refers to the force per unit area *on a given plane within a body.* Stress is written mathematically as

EQUATION 13–8

$$\sigma = \frac{F}{A}$$

where σ = stress, N/m^2

Equation 13–7 indicates that the stress is equal to Young's modulus times the strain. Notice that stress has the same units as pressure, namely, force per area. The stress-strain relationship shown in Equation 13–7 is usually written

EQUATION 13–9

$$\sigma = E\epsilon$$

The relationship between stress and strain depends on the material, including any heat treatment it may have had. A stress-strain diagram is shown in Figure 13–15 for low-carbon steel. Notice that it begins with a proportional region where Hooke's law applies. If the stress is removed along this region, the steel will return to its original shape. At the elastic limit, increases in strain are no longer proportional to increases in stress. With continued increases in stress, a point is reached called the **yield point,** where additional strain occurs without a corresponding increase in stress. After this point, permanent deformation occurs. This region is called the **plastic range.** Continued stressing of the material leads to a point called the *ultimate,* or *tensile, strength* of the material.

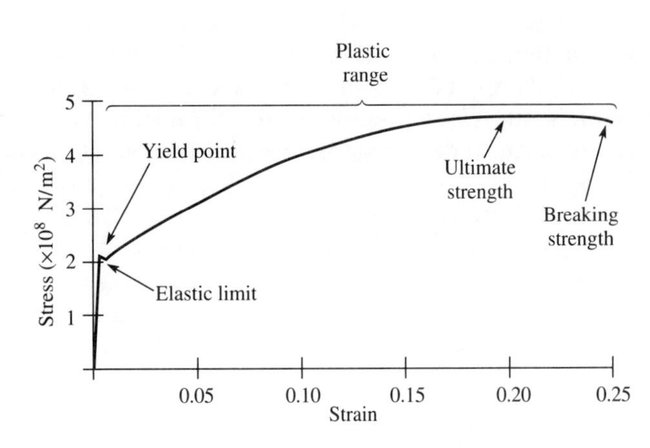

FIGURE 13–15 Stress-strain diagram for low-carbon steel. By conventional engineering practice, strain is plotted as the independent variable.

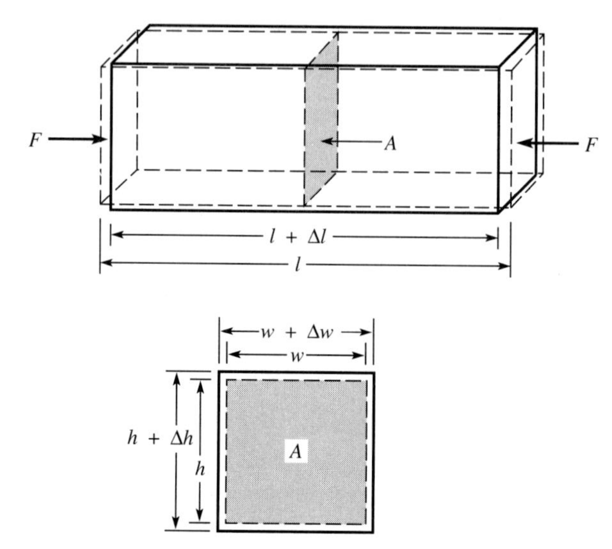

FIGURE 13–16
Normal, or axial, strain. The length, width, and height all change.

You have seen how a longitudinal force that compresses a block produces a strain in the block and causes it to shorten by some small amount, Δl. The force that produced the strain was applied perpendicular to the plane of the material and is a measure of the deformation per unit length; hence it is called **normal, or axial, strain.** Normal strain is illustrated in Figure 13–16. Another aspect to this compression is that the block also expands at right angles to the axis of the block, as shown. Dividing each change in dimension by the unstressed dimension gives a pure number that is related to the strain and a constant. The constant is called **Poisson's ratio,** abbreviated ν. Poisson's ratio is a dimensionless number that is characteristic of the material. The relationship between these quantities is given by the equation

EQUATION 13–10

$$\frac{\Delta w}{w} = \frac{\Delta h}{h} = -\nu \frac{\Delta l}{l}$$

where ν = constant of proportionality; Poisson's ratio

A second type of strain is caused by bending forces and is called **bending strain.** Bending strain is illustrated in Figure 13–17. With the force applied as shown, the top of the beam will have a tensile strain, whereas the bottom of the beam is under compressive strain. Strain gages used to measure bending strain can be used to determine the vertical forces, a technique employed in load cells for scales.

A third type of strain is due to forces that cause an angular distortion to a material. These forces are parallel to the plane, as illustrated in Figure 13–18, and cause **shear strain.** Shear strain, γ, can be defined as the angular change (measured in radians) by which the block is twisted. Assume an arbitrary plane is drawn at some angle through the block, as shown in Figure 13–19. The force that is applied to the block can be divided into two components, which depend on the orientation

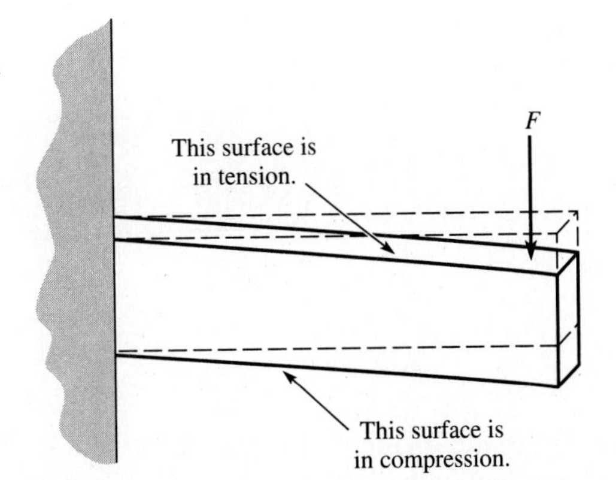

FIGURE 13–17
Bending strain on a cantilever beam.

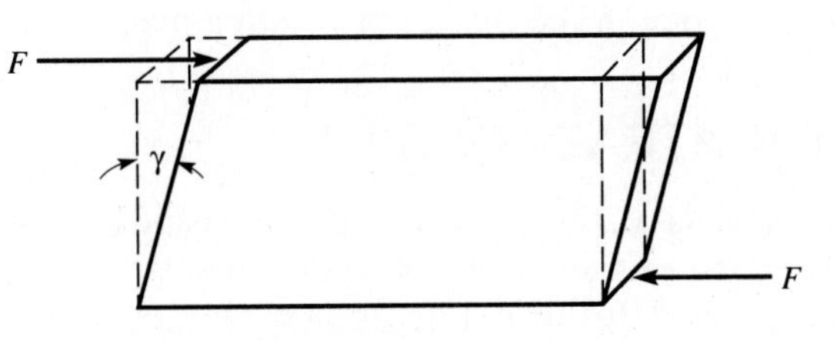

γ = angular change, rad = shear strain

FIGURE 13–18
Shear strain.

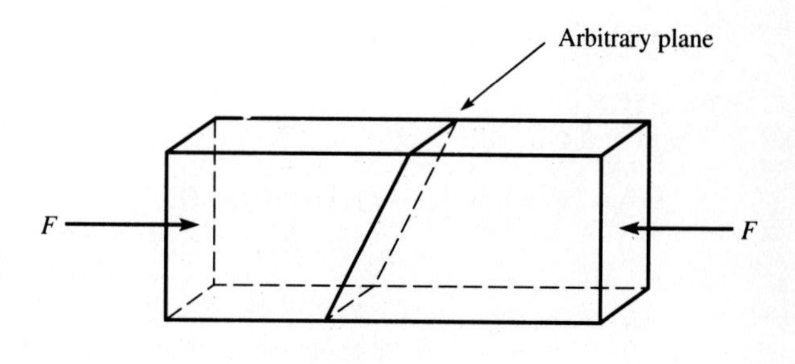

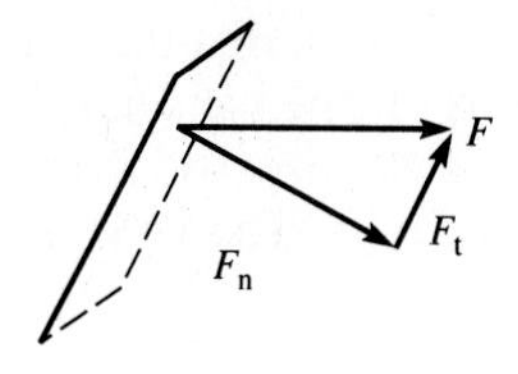

$$\boldsymbol{F} = \boldsymbol{F}_n + \boldsymbol{F}_t$$

FIGURE 13–19
The force on an arbitrary plane can be divided into normal and tangential components. The normal component produces normal stress, whereas the tangential component produces shear stress in the plane.

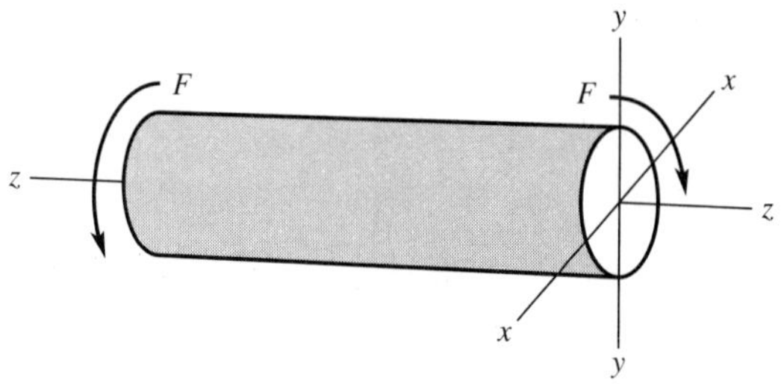

FIGURE 13–20
Torsional strain.

of a designated plane within the block. One component is normal to the plane and the other is parallel to the plane. Like strain, shear stress, τ, is developed along an arbitrary plane, whereas normal stress, σ, is developed perpendicular to the plane.

The fourth type of strain is due to twisting forces and is called **torsional strain.** Torsional strain is illustrated in Figure 13–20. It is a strain that is of special importance in the design of rotating machinery.

THE STRAIN GAGE

As defined earlier, a resistive strain gage is a thin metallic conductor that is firmly attached to a solid object to detect strain in the object. When a force causes the test object to be deformed, the strain gage undergoes the same deformation, causing the resistance of the gage to change. Strain is a directly measurable quantity, but stress, usually the measurand of interest, is not. The measurement of stress requires knowledge of material constants such as Young's modulus and Poisson's ratio.

Consider a thin metallic conductive bar that has a resistance R. Recall from basic electronics that the resistance of wire (or a metallic bar) can be found from the equation

EQUATION 13–11

$$R = \frac{\rho l}{A}$$

where R = resistance, Ω
ρ = resistivity, Ω-m
l = length of bar, m
A = cross-sectional area of bar, m^2

Note that the resistivity, ρ, is a constant for the material at a given temperature. If the bar undergoes a compression force, the length of the bar is decreased and the area of the bar is increased. This causes the resistance of the bar to decrease. The new resistance of the bar is given by

EQUATION 13–12

$$R - \Delta R = \frac{\rho(l - \Delta l)}{A + \Delta A}$$

The volume does not change for a metallic conductor as it is compressed. With this premise and Equation 13–12, it can be shown that the change in resistance is equal to

EQUATION 13–13

$$\Delta R = \mathrm{GF}\, R\, \frac{\Delta l}{l}$$

where GF = a constant called the gage factor, dimensionless

For materials that exhibit a change in resistance strictly due to dimensional change, the gage factor is approximately 2.0. This approximation is good for metals. The gage factor may change in response to effects such as temperature change, the composition of foil material, and any impurities in the foil material. For certain special gages made from semiconductor crystals, the gage factor can be much larger than that of standard foil gages (from 20 to 200). The gage factor is a measure of the sensitivity of the strain gage, so a larger gage factor implies a larger change in resistance for a given strain—an advantage because a larger change in resistance is easier to measure accurately. When semiconductor gages, with large gage factors, were first introduced, the high output was considered an important advantage; however, these gages are sensitive to temperature variations and tend to be nonlinear. With inexpensive, high-quality amplifiers available, conventional metallic gages are more widely used because of their inherent accuracy.

Metallic strain gages are made from very thin conductive foils with a conductor folded back and forth to allow a long path for the conductive elements while maintaining a short gage length. A typical strain gauge and strain gage nomenclature are illustrated in Figure 13–21. Gage lengths vary from as small as 0.2 mm to over 10 cm. The back-and-forth pattern is designed to make the gage sensitive to strain only in the direction parallel to the wire and insensitive in the direction perpendicular to the wire. Most gages exhibit some transverse sensitivity, but the effect is generally small, usually less than a small percentage of the longitudinal sensitivity. Typically, the foils are made from special alloys such as constantan, a combination of 60% copper and 40% nickel. Standard foil resistances are 120 Ω and 350 Ω; some gages are available with resistances of up to 5000 Ω.

There are two basic types of strain gages, bonded and unbonded; bonded strain gages are more widely used for measuring strain. **Bonded strain gages** are made with a carrier adhesive on an electrically insulating backing that serves as electrical isolation and a means of transmitting the strain in the specimen to the gage. The gage is attached to the specimen with a special adhesive. The adhesive is a critical part of the transducer because it must transmit the strain to the gage without "creep" and must maintain a void-free line throughout the operating life and temperature range of the gage. Special adhesives have been developed by manufacturers for installation of strain gages.

There are several configurations for **unbonded strain gages,** but generally they consist of pretensioned wires assembled in some type of fixture. One type uses four resistance wires stretched in a frame containing a movable armature.

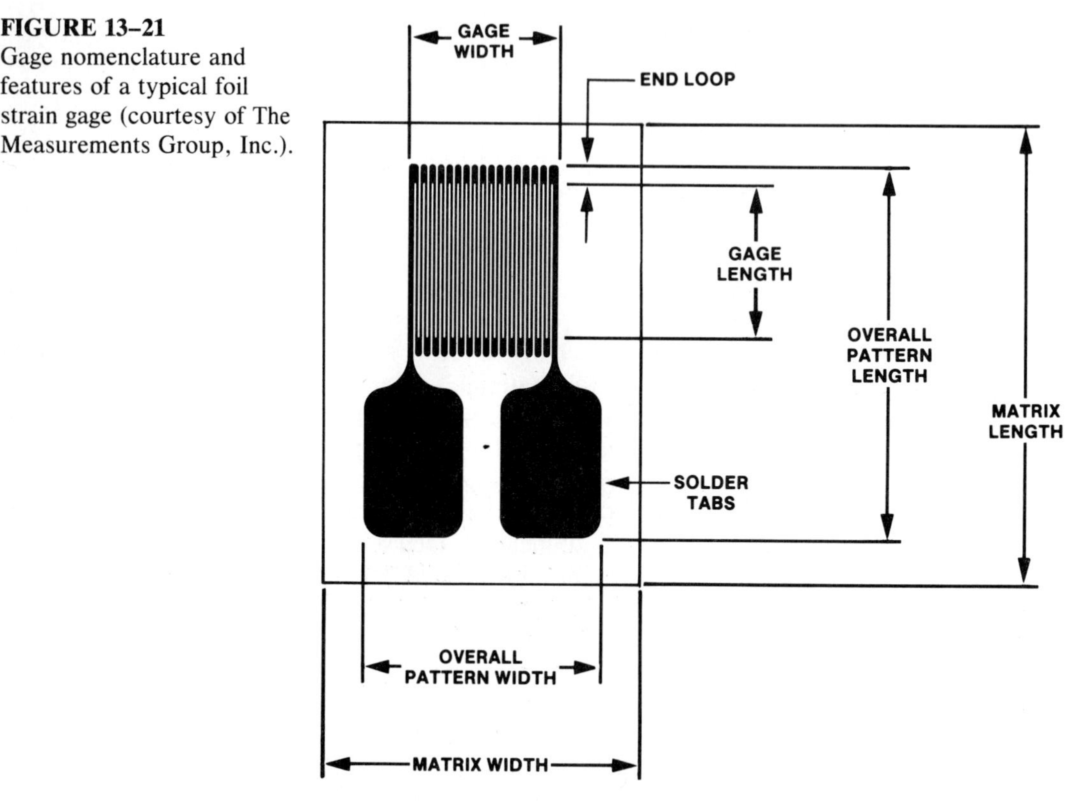

FIGURE 13–21
Gage nomenclature and features of a typical foil strain gage (courtesy of The Measurements Group, Inc.).

The wires are connected such that when the armature moves, two of the wires are placed in tension and two are put in compression. The change in resistance can be measured on a Wheatstone bridge. Unbonded strain gages can be used to measure displacement directly but have a number of disadvantages for strain measurements, including weight, fragility, sensitivity to vibration, and attachment difficulties.

MEASUREMENTS USING STRAIN GAGES

In a single bar of conductive material, the change in resistance within the elastic region is an extremely small quantity and requires sensitive instrumentation. The change in resistance is too small for ordinary ohmmeter measurements. To measure the tiny resistance changes, a Wheatstone bridge circuit is normally employed. It is powered by a stable dc source, normally set for 15 V or less to avoid self-heating of the gages. A Wheatstone bridge is capable of detecting resistance changes by employing the strain gages as an element in one or more arms of the bridge. The preferred configuration for most strain gage measurements is with active strain gages in all four arms of the bridge; however, we examine other arrangements as well.

In addition to the requirement for high sensitivity of the measurement instruments, strain gage measurements are complicated by temperature effects that must be accounted for in order to determine the signal due to the actual strain in the specimen. There are two temperature effects of importance—gage-resistance changes due to heating and specimen expansion and contraction. Temperature effects that change the resistance of the gage produce apparent strain; the gage itself cannot discriminate between mechanically and thermally induced temperature effects. Temperature compensation is often employed in the bridge circuit to simplify data analysis.

A Wheatstone bridge with a strain gage in one arm is called a **quarter-bridge** configuration; a quarter bridge is illustrated in Figure 13–22. The bridge shown includes a dummy gage that does not experience the strain; it is included for temperature compensation. To determine the strain, the output voltage with no strain is observed; the bridge can be balanced (*zeroed*) at this point by adjusting one of the bridge resistors. The gage is then subject to strain, and the output voltage is observed again. The strain is found by the equation

EQUATION 13–14

$$\epsilon = \frac{4V_r}{GF(1 + 2V_r)}$$

V_r is a dimensionless number that represents the difference in the ratios of the output to input voltage between the strained and unstrained condition:

EQUATION 13–15

$$V_r = \left(\frac{V_{out}}{V_{in}}\right)_{strained} - \left(\frac{V_{out}}{V_{in}}\right)_{unstrained}$$

where V_{out} = output voltage from the unloaded bridge, V
V_{in} = excitation voltage, V

Equation 13–14 takes into account the nonlinearity of the bridge. This error is very minor at the low levels normally encountered in experimental stress analysis work. If the bridge is initially balanced (V_{out}, unstrained = 0 V) and we ignore the very minor nonlinearity of the bridge near balance, Equation 13–14 reduces to

EQUATION 13–16

$$\epsilon = \frac{4V_{out}}{GF\ V_{in}}$$

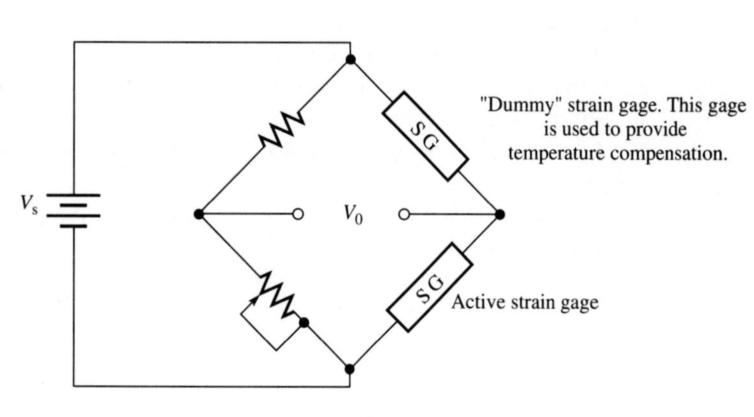

FIGURE 13–22
Quarter-bridge arrangement. The bridge contains one active strain gage.

EXAMPLE 13–4 A strain gage with a nominal resistance of 350 Ω and a GF of 2.00 is connected into the quarter-bridge circuit shown in Figure 13–22. The excitation voltage for the bridge is set to 15 V. The other bridge resistors, including the dummy gage, are 350 Ω. Assume the bridge is initially balanced with no strain and has an output of 45 μV under strain.

(a) Compute the exact applied strain.

(b) Compute the applied strain assuming the bridge is linear.

SOLUTION (a) From Equation 13–15, we can find the ratio of the output voltage to the dc excitation voltage for the strained and unstrained condition:

$$V_r = \left(\frac{V_{out}}{V_{in}}\right)_{strained} - \left(\frac{V_{out}}{V_{in}}\right)_{unstrained}$$

$$= \left(\frac{45\ \mu V}{15\ V}\right)_{strained} - \left(\frac{0\ V}{15\ V}\right)_{unstrained}$$

$$= 3 \times 10^{-6}$$

$$\epsilon = \frac{4V_r}{GF(1 + 2V_r)}$$

$$= \frac{4 \times 3 \times 10^{-6}}{2.0(1 + 2 \times 3 \times 10^{-6})} = 5.99996\ \mu\epsilon$$

We retain more significant figures in the answer than are justified to show the effect of the assumption that the bridge is linear when the output change is small.

(b) Using the linear approximation given in Equation 13–16, we find that

$$\epsilon = \frac{4V_{out}}{GF\ V_{in}}$$

$$= \frac{4 \times 45\ \mu V}{2.0 \times 15\ V} = 6.00\ \mu\epsilon$$

The result justifies the assumption of linearity for small changes in the bridge. The error is approximately 0.1% for each 1000 μϵ.

When two active gages are used, the configuration is called a **half-bridge.** A half-bridge configuration produces increased sensitivity over a quarter-bridge arrangement and is best applied to measuring bending beams. For bending measurements, the two gages are installed so that one is in tension and the other is in compression by installing one directly on top of the other and on opposite sides of the beam. They are connected in adjacent arms of the Wheatstone bridge, as illustrated in Figure 13–23. This arrangement causes temperature effects, which change the resistance of both gages in the same way, to be canceled because they are in adjacent arms of the bridge.

For axial strain measurements, the connection of the active strain gages is to the opposite diagonals, causing bending effects to be canceled and axial strain to be additive. In this arrangement, temperature effects that change the resistance are additive, since the gages are installed in opposite diagonals. A common proce-

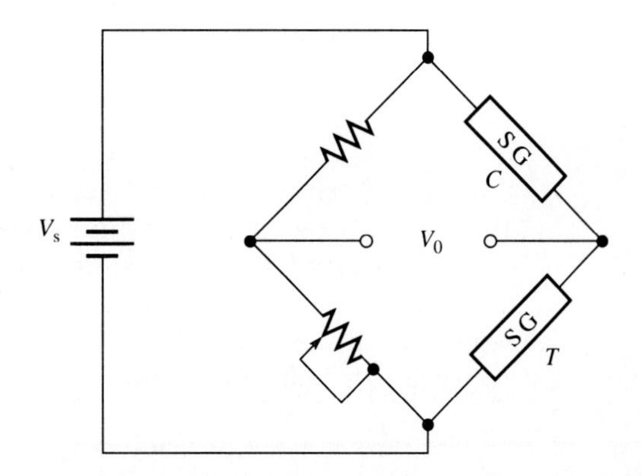

FIGURE 13–23
Half-bridge arrangement. The bridge shown contains two active gages, one in tension and the other in compression. If both gages are exposed to a change in temperature, compensation is automatic.

dure is to include the use of compensating (dummy) gages for nullifying temperature effects in the remaining arms. Compensating gages can be mounted on a small unstrained block of the same material as the specimen or they can be mounted in such a manner that they do not experience strain.

The chief advantage of a half-bridge over a quarter-bridge is an increase in sensitivity by a factor of 2. If we ignore the minor nonlinearity error for the bridge, the strain for a half-bridge arrangement can be found from the equation

EQUATION 13–17

$$\epsilon = \frac{2V_{out}}{\text{GF } V_{in}}$$

A **full-bridge** configuration uses active gages in all four arms. Two strain gages in opposite diagonals are in tension and the other two are in compression. Temperature compensation can still be included by placing compensating elements in the excitation lines and at the output junctions, as shown in Figure 13–24. Again ignoring the minor nonlinearity error, the strain for a full-bridge

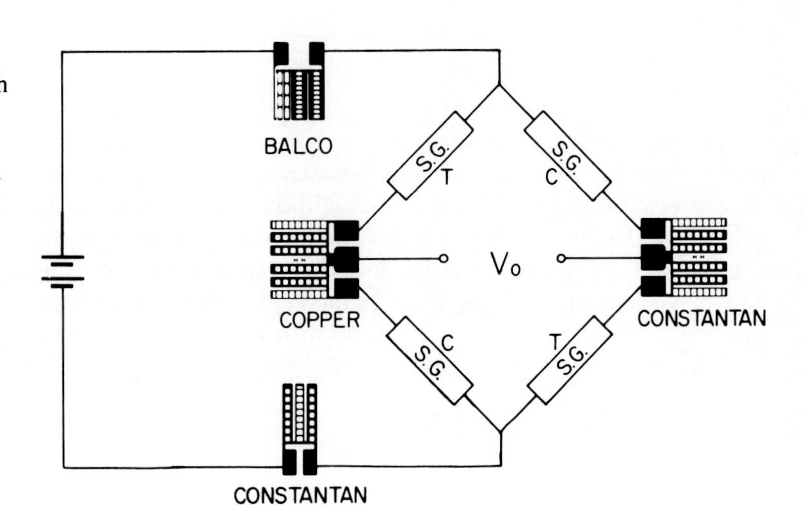

FIGURE 13–24
Full-bridge arrangement with temperature compensation (courtesy of The Measurements Group, Inc.).

arrangement with four equal strain gages can be found from the equation

EQUATION 13–18

$$\epsilon = \frac{V_{out}}{GF\ V_{in}}$$

Load cells are generally constructed with 350 Ω strain gages mounted in a full-bridge arrangement. Notice that the resistance measured from excitation leads to output leads will also measure a nominal 350 Ω in the unstrained condition. When the strain gages are mounted in a load cell in the full-bridge configuration, the full-scale output of the load cell is normally designated to be 1, 2, or 3 mV of signal per volt of excitation.

EXAMPLE 13–5 A 2 mV/V load cell is excited with a 10 V source and monitored with a digital voltmeter. The load cell is used in a scale designed for a full-scale output with a 100 lb load. If the required resolution of the measurement is 0.1 lb, what is the required resolution of the voltmeter, expressed as a voltage?

SOLUTION The required resolution of the measurement, expressed as a percentage, is

$$\text{Resolution} = \frac{0.1\ \text{lb}}{100\ \text{lb}} \times 100\% = 0.1\%$$

The full-scale output from the bridge is

$$\text{FSO} = \frac{2\ \text{mV}}{\text{V}} \times 10\ \text{V} = 20\ \text{mV}$$

The resolution of the voltmeter, expressed as a voltage, is

$$\text{Voltmeter resolution} = 0.1\% \times 20\ \text{mV} = 20\ \mu\text{V}$$

As you can see from this result, the measurement requires a sensitive meter and the absence of noise.

Manufacturers of strain equipment have developed instrumentation that performs the basic functions necessary for making a strain measurement. Basic functions include operational controls that enable the user to set the gage factor directly into the arm resistance, scale the output signal, zero the bridge, and perform calibration. In addition, the instrument supplies the excitation voltage, completes the quarter- and half-bridge configurations, and provides for the readout. Other functions may be employed by sophisticated instrumentation, including multiple-channel acquisition, active filtering, tape playback, and computer analysis of data, to name a few. Figure 13–25 illustrates an automated strain indicator that can perform these functions.

13–7 MEASUREMENT OF PRESSURE

Consider a fluid such as water in a flat-bottomed container. The weight of the water exerts a force on the bottom of the container that is distributed over the

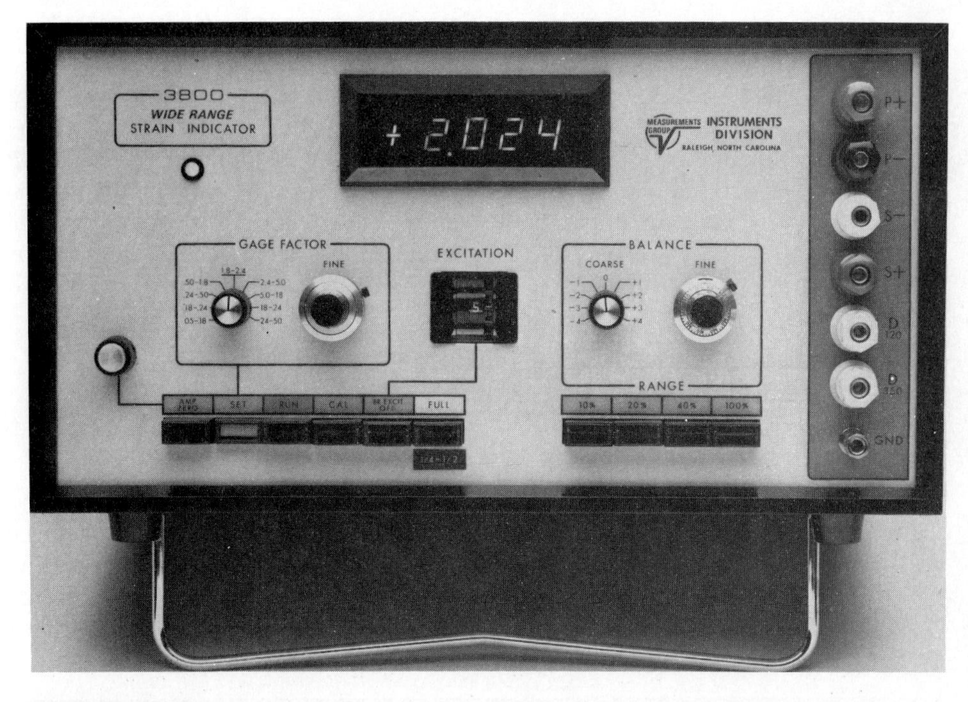

FIGURE 13–25
Automated strain indicator (courtesy of The Measurements Group, Inc.)

entire bottom surface. Each square meter of the bottom area carries the same weight as every other square meter. The force per unit area is defined as **pressure.** That is

EQUATION 13–19

$$P = \frac{F}{A}$$

where P = pressure, N/m^2 (Pa)
F = force, N
A = area, m^2

As indicated, pressure is measured in newtons per square meter. One newton per square meter has been given the special name **pascal** (abbreviated Pa). This unit is small, so it is common to see the unit written with a prefix of kilo- or mega-. In the English system, if force is measured in pounds and the area is measured in square inches, the pressure is in pounds per square inch (psi); 1 psi is approximately 6.895 kPa. Pressure can also be measured in terms of the height of a column of mercury it can support, a common procedure for atmospheric pressure. Atmospheric pressure is the force on a unit area due to the weight of atmosphere; it can be written as either 760 mm of mercury, 29.92 in. of mercury, 14.7 psi, or 101 kPa.

In a static fluid,[5] pressure acts equally in all directions and increases in proportion to the depth. It also depends on the density of the liquid and any additional pressure acting on the surface. If the surface is exposed to the atmosphere, atmospheric pressure acts on the surface. Note that the pressure in a liquid does not depend on the quantity of liquid, only the depth, density, and surface pressure. Ignoring surface pressure, we can find the pressure of a liquid in a tank by multiplying the weight density times the depth of the liquid. (The weight density is equal to the mass density times the acceleration of gravity.) That is

EQUATION 13–20

$$P = \rho g h$$

where P = pressure at the bottom of a liquid, Pa
ρ = mass density, kg/m^3
g = acceleration of gravity, 9.80 m/s^2
h = height of liquid, m

It is common practice to express pressure measurements in terms of the equivalent pressure at the bottom of a column of a liquid of a stated height. Liquids used in pressure measurements are generally either mercury (because of its very high density) or water. Thus 29.9 in. of mercury is the pressure at the bottom of a mercury column 29.9 in. high.

EXAMPLE 13–6 Mercury has a density of 13.6 g/cm^3. Convert 700 mm of mercury to pascals.

SOLUTION

$$P = \rho g h$$

$$= \left(13.6 \frac{\cancel{g}}{\cancel{cm^3}}\right)\left(10^6 \frac{\cancel{cm^3}}{m^3}\right)\left(10^{-3} \frac{kg}{\cancel{g}}\right)\left(9.80 \frac{\cancel{m}}{s^2}\right)(700\ \cancel{mm})\left(10^{-3} \frac{\cancel{m}}{\cancel{mm}}\right)$$

$$= 93.3\ kPa$$

With gases, pressure is also exerted equally in all directions. If a gas is inside a closed container, the pressure of the gas is the force per unit area that the gas exerts on the walls of the container. In the atmosphere, as we move above sea level, the pressure decreases because of the decreased weight of the air that is supported. At the top of Mt. Everest, air pressure is only one-third that of sea level.

GAGE PRESSURE AND ABSOLUTE PRESSURE

If you experience a flat tire and check the tire pressure, the gage reads zero. Although we might say there is no air in the tire, the fact is that the tire has air in it that is at the same pressure as the atmosphere. The gage is simply reading the *difference* between the atmospheric pressure and the pressure inside the tire. This

[5] Pressure is different in a moving fluid. We concern ourselves only with static pressure.

difference is known as **gage pressure** (shown in the English system as psig). Pressure readings that include the atmospheric contribution are called **absolute pressures** (shown in the English system as psia). Absolute pressure is referenced to a vacuum. Most pressure gages are designed to read gage pressure; it is important to keep in mind which pressure you are using. For instance, pressure values used in calculations for the gas laws (see Section 3–7) must be in absolute pressure. Another pressure measurement is called **differential pressure.** As the name implies, differential pressure is the difference between two pressures (shown in the English system as psid).

PRESSURE TRANSDUCERS

The variations in pressure transducer design are almost endless, but virtually all pressure transducers operate on the principle of balancing an unknown pressure against a known load. A common technique is to use a diaphragm to balance the unknown pressure against the mechanical restraining force keeping the diaphragm in place. A **diaphragm** is a flexible disk that is fastened on its periphery and changes shape under pressure. A spring may be used to push against the diaphragm and provide a load. The amount of movement of the diaphragm is proportional to the pressure. Diaphragms can be used on a wide range of pressures, from about 15 to 6000 psi. In simple pressure gages, the displacement of the diaphragm is mechanically linked to an indicator.

Other mechanical methods for converting pressure into displacement include a bellows and a Bourdon tube, both constructed from resilient metals. A **bellows** is a thin-walled corrugated tube that is sealed on one end that expands or contacts under pressure. A **Bourdon tube** is an elliptical or circular metal tube, closed on one end, that is made into a spiral, helix, twisted, or C shape. Pressure inside the tube tends to straighten it, causing the end to deflect. Examples of pressure-sensing elements are shown in Figure 13–26.

In an electronic measuring system, it is necessary to convert the mechanical motion of the pressure-sensing element into an electrical signal. There are many conversion techniques possible; the most common are the potentiometric, reluctive, capacitive, and strain gage methods. The potentiometric method converts the displacement of the sensor into a resistance. The wiper arm of a potentiometer is mechanically linked to the pressure-sensing element, causing the resistance to change as a function of pressure. The potentiometric method is simple and less expensive than other conversion techniques; however, it suffers from the disadvantage of mechanical wear and is electrically noisy.

The reluctive method changes the inductance of one or two coils by moving some part of the magnetic circuit. One technique is to change the position of a magnetic core within a transformer. Another is to move a diaphragm between two coils, increasing the inductance of one while decreasing the inductance of the other. The coils are electrically connected in a bridge circuit.

In capacitive transducers, the motion changes the capacitance of an internal capacitor. This is usually accomplished by moving one of the plates (diaphragm) of a capacitor. The other plate is stationary. An increase in pressure causes the

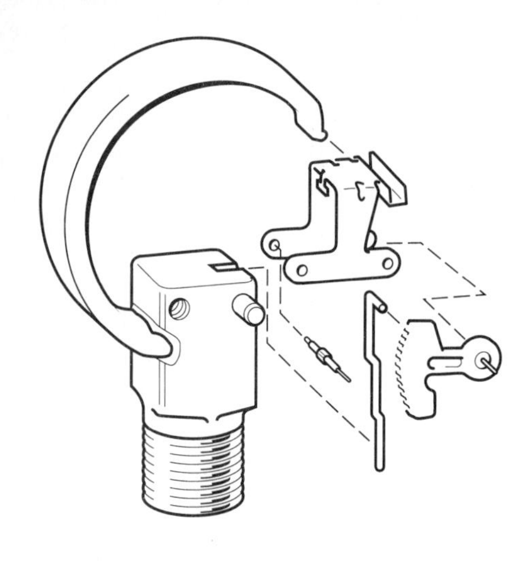

(a) Spring-suspended Bourdon tube movement.

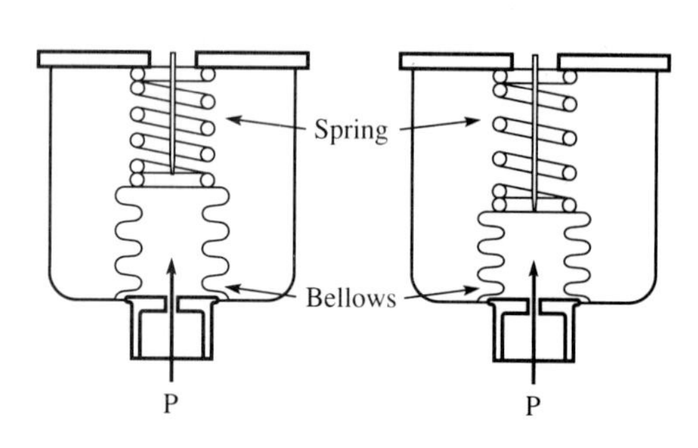

(b) Bellows movement.

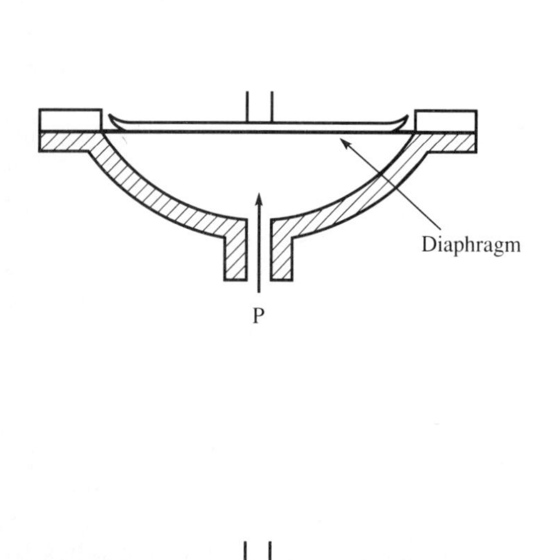

(c) Diaphagm movement.

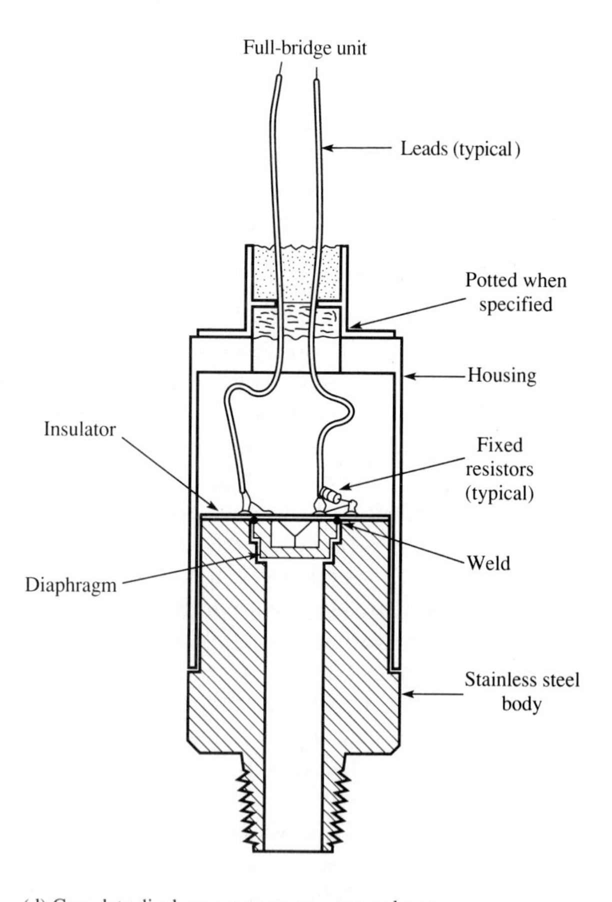

(d) Complete diaphragm-type pressure transducer.

FIGURE 13–26
Examples of pressure-sensing elements (courtesy of Omega Engineering, Inc.).

plates to move together, increasing the capacitance. A related method is to center a moving plate between two fixed plates. As the diaphragm moves, the capacitance of one of the capacitors increases while the other decreases. The capacitors are electrically connected into a bridge circuit. Capacitive pressure transducers have a high-frequency response (due to low mass), implying that they can respond quickly to changes in pressure.

A strain gage can be used as a sensing element by bonding it to the diaphragm. Pressure on the diaphragm introduces strain, which is sensed by the gages and converted to an electrical resistance. Typically, gages are bonded on both sides of the diaphragm and connected in a full- or half-bridge arrangement. In the full-bridge arrangement, two gages are mounted on one side of the diaphragm; these will be in compression while the two mounted on the other side are in tension. In the half-bridge arrangement, one gage is on either side of the diaphragm. The bridge is completed with two fixed resistors. The full-bridge arrangement gives a higher output (up to about 100 mV) for a given pressure. In addition to the active strain gages, the transducer may also contain temperature compensation and zero-balance resistors.

13–8 MEASUREMENT OF MOTION

Motion can be rectilinear—along a straight line—or it can be circular—about an axis. The measurement of motion includes displacement, velocity, and acceleration. *Displacement* is a vector quantity that indicates the change in position of a body or point. *Velocity* is the rate of change of displacement, and *acceleration* is a measure of how fast velocity changes. Angular displacement is measured in degrees or radians.

DISPLACEMENT TRANSDUCERS

Displacement transducers can be either contacting or noncontacting. Contacting transducers typically use a sensing shaft with a coupling device to follow the position of the measurand. The sensing shaft can be connected to the wiper arm of a potentiometer. The electrical output signal can be either a voltage or a current. Potentiometric displacement transducers are simple and can be designed to measure rather large displacements, but they are subject to wear and dirt and are electrically noisy. Displacement can also be converted into an electrical quantity using a variable inductor and monitoring the change in inductance. The inductance can be changed by moving the core material, varying the coil dimensions, or a sliding contact.

A related displacement transducer is the linear variable differential transformer (LVDT). The sensing shaft is connected to a moving magnetic core inside a specially wound transformer. A typical LVDT is shown in Figure 13–27. The primary of the transformer is in line and is located between two identical secondaries. The primary winding is excited with ac (usually in the range of 1 to 5 kHz). When the core is centered, the voltage induced in each secondary is equal. As the

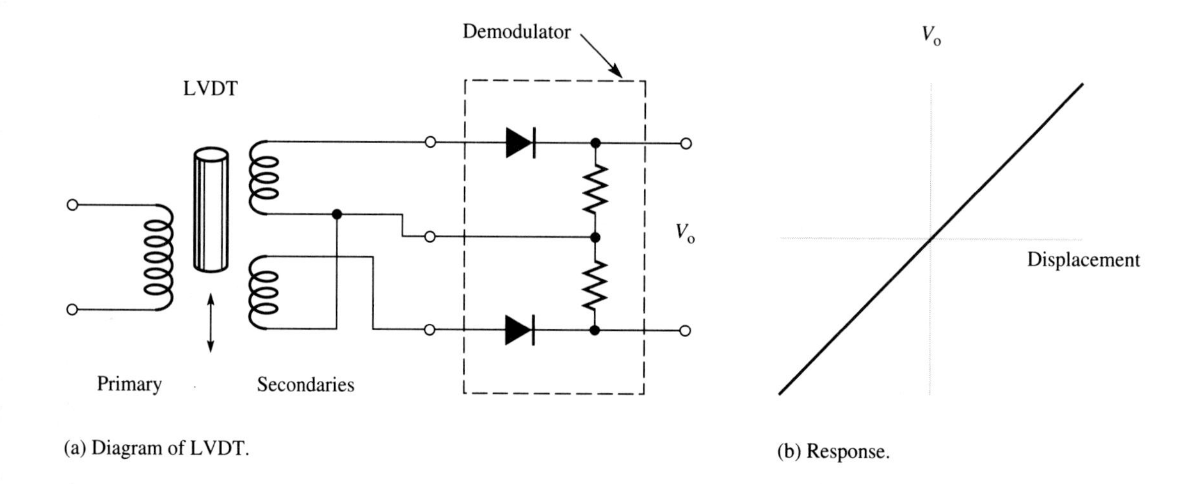

(a) Diagram of LVDT.

(b) Response.

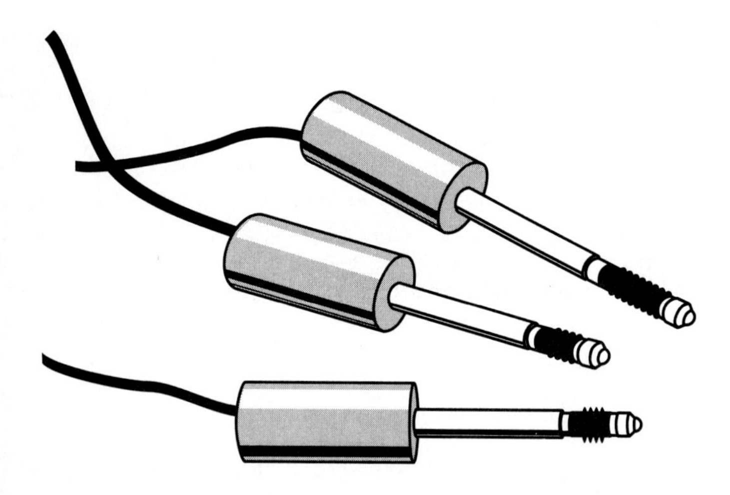

(c) Gauging LVDTs (courtesy of Omega Engineering, Inc.).

FIGURE 13–27
LVDT displacement transducers.

core moves off-center, the voltage in one secondary will be greater than the other. With the demodulator circuit shown, the polarity of the output changes as the core passes the center position. The transducer has excellent sensitivity, linearity, and repeatability.

Noncontacting displacement transducers include optical and capacitive transducers. Photocells can be arranged to detect light through holes in an encod-

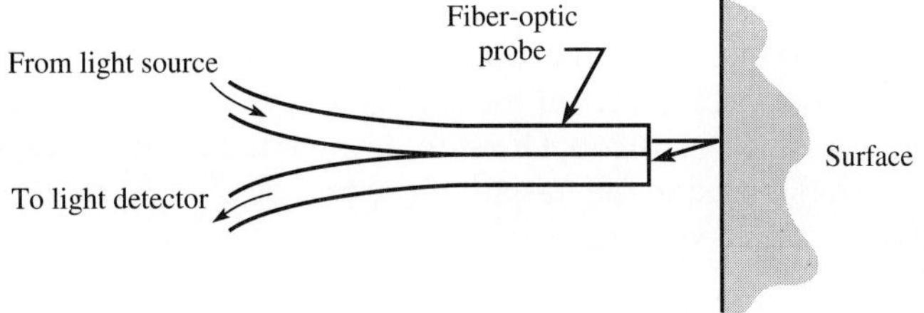

FIGURE 13–28
Fiber-optic proximity detector.

ing disk or to count fringes painted on the surface to be measured. Optical systems are fast, but noise, including that from background light sources, can produce spurious signals in optical sensors. It is useful to build hysteresis into the system using a Schmitt trigger if noise is a problem.

Fiber-optic sensors make excellent proximity detectors for close ranges. Reflective sensors use two fiber bundles, one for transmitting light and the other for receiving light from a reflective surface, as illustrated in Figure 13–28. Light is transmitted in the fiber bundle without any significant attenuation. When it leaves the transmitting fiber bundle, it forms a spot on the target that is proportional to the distance. The receiving bundle is aimed at the spot and collects the reflected light to an optical sensor. The light intensity detected by the receiving bundle depends on the physical size and arrangement of the fibers as well as the distance to the spot and the reflecting surface, but the technique can respond to distances approaching 10^{-6} in. The major disadvantage is limited dynamic range.

Capacitive sensors can be made into very sensitive displacement and proximity transducers. The capacitance is varied by moving one of the plates of a capacitor with respect to the second plate. The moving plate can be any metallic surface, such as the diaphragm of a capacitive microphone or a surface that is being measured. The capacitor can be used to control the frequency of a resonant circuit to convert the capacitive change into a usable electrical output.

VELOCITY TRANSDUCERS

Since velocity is the rate of change of displacement, velocity can be determined by using a displacement sensor and measuring the time between two points. A direct measurement of velocity is possible with certain transducers that have an output proportional to the velocity to be measured. They sense either linear or angular velocity. Linear velocity transducers can be constructed using a permanent magnet inside a concentric coil forming a simple motor by generating an emf proportional to the velocity. Either the coil or the magnet can be fixed and the other moved with respect to the fixed component. The output is taken from the coil.

There is a variety of transducers designed to measure angular velocity. Tachometers are angular velocity transducers that provide a dc or ac voltage output. Dc tachometers are basically small generators with a coil that rotates in a constant magnetic field. A voltage is induced in the coil as it rotates in the magnetic field. The average value of the induced voltage is proportional to the speed of rotation and the polarity is indicative of the direction of rotation, an advantage with dc tachometers. Ac tachometers can be designed as generators that provide an output frequency proportional to the rotational speed.

Another technique for measuring angular velocity is to rotate a shutter over a photosensitive element. The shutter interrupts a light beam focused on a photocell, thereby causing the output of the photocell to fluctuate at a rate proportional to the rotational speed.

ACCELERATION TRANSDUCERS

Acceleration is usually measured by use of a spring-supported seismic mass mounted in a suitable enclosure, as shown in Figure 13–29. Damping is provided by a dashpot. The relative motion between the case and the mass is proportional to the acceleration. A secondary transducer such as a resistive displacement transducer is used to convert the relative motion to an electrical output. In the ideal world, the mass does not move when the case accelerates because of its inertia; in practice it does because of forces applied to it through the spring. The accelerometer has a natural frequency, the period of which should be shorter than the time required for the measured acceleration to change. Accelerometers used to measure vibration should also be used at frequencies less than the natural frequency.

An accelerometer that uses the basic principle of the LVDT can be constructed to measure vibration. The mass is made from a magnet that is surrounded by coils. Voltage induced in the coils is a function of the acceleration.

Another type of accelerometer uses a piezoelectric crystal in contact with the seismic mass. The crystal generates an output voltage in response to forces induced by the acceleration of the mass. Piezoelectric crystals are small in size and have a natural frequency that is very high; they can be used to measure high-frequency vibration. The drawback to piezoelectric crystals is that the output is very low and the impedance of the crystal is high, making it subject to problems from noise.

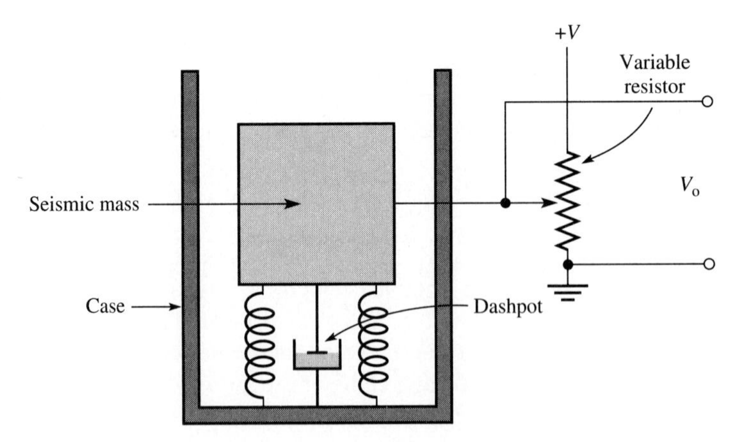

FIGURE 13–29
A basic accelerometer. Motion is converted to a voltage by the variable resistor.

13–9 MEASUREMENT OF LIGHT

THE ELECTROMAGNETIC SPECTRUM

The light we see is only a small portion of the vast electromagnetic spectrum. The theory of electromagnetic radiation was developed in the 1860s when James C. Maxwell combined the known laws of electricity with those of magnetism into a set of laws that govern the behavior of light. Light has characteristics similar to those of radio waves but at a very much shorter wavelength. Electromagnetic waves travel at a velocity of 3.00×10^8 m/s in a vacuum. The frequency is related to the wavelength by the equation

EQUATION 13–21

$$f = \frac{c}{\lambda}$$

where c = velocity of electromagnetic radiation, m/s
f = frequency, Hz
λ = wavelength, m

Exploring the electromagnetic spectrum (see Figure 13–30), we find radio waves that extend from wavelengths hundreds of meters long to wavelengths that are millimeters long (microwaves). At shorter wavelengths is the infrared region that extends to the region of visible light. The wavelength of visible light is a thousand times shorter than the shortest radio waves; the light to which our eyes respond has a wavelength between approximately 390 nm (violet) and 760 nm (red). Immediately above visible light is the ultraviolet region, and beyond this is the X-ray and gamma-ray portion of the electromagnetic spectrum. The visible spectrum is shown in color on the back cover.

The boundaries between these regions are not well defined—they are simply names associated with parts of a continuous spectrum of which visible light is a tiny fraction of the total. Although our eyes do not respond to infrared and ultraviolet radiation, they are often loosely referred to as infrared and ultraviolet "light." These regions are of interest in instrumentation because many detectors are sensitive across the boundaries of the regions. For example, photoelectric cells are sensitive in the ultraviolet and visible portions of the spectrum.

The approximations used for the study of electromagnetic radiation depend on the wavelength of the radiation and the resolution of the sensing equipment. If the wavelength is comparable to the resolution of the measuring equipment, a wave model for the behavior of the radiation is appropriate. This is the physical model for the region of the electromagnetic spectrum from radio waves through the infrared region. On the other hand, if the wavelength is very short compared to the resolution of the measuring equipment, a particle model is more appropriate. The particles of radiant energy are called *photons*. In the ultraviolet region, a photon model is generally used. The energy of a photon is related to its frequency by the equation

EQUATION 13–22

$$E = hf$$

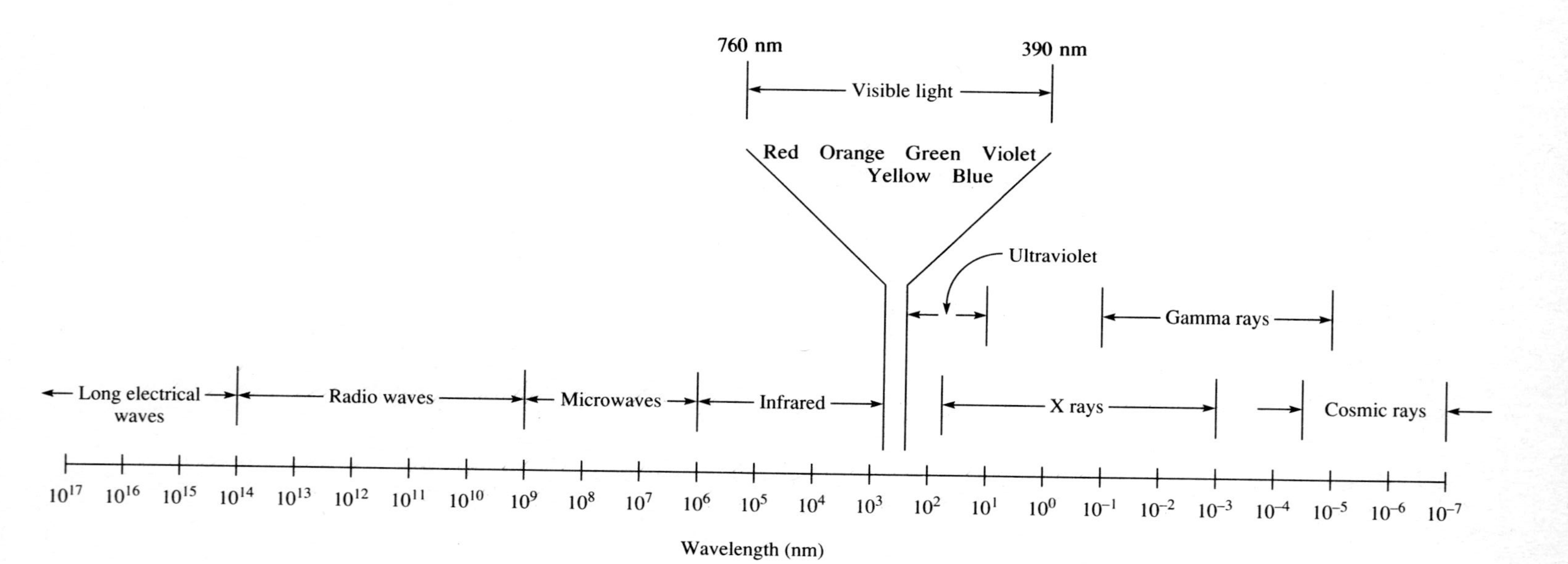

FIGURE 13–30
The electromagnetic spectrum.

where E = energy of a photon, J

h = Planck's constant; 6.626×10^{-34} J/Hz

f = frequency, Hz

This equation indicates that photons in the ultraviolet region of the spectrum are more energetic than photons of visible light. Likewise, photons of blue light are more energetic than photons of red light. Note that an increase in the intensity of light does not increase the energy of the photons—only the frequency affects the energy.

The spectrum of common light sources is distinctly different and dependent on how the light was generated. Incandescent sources (the tungsten-filament light bulb, for example) produce a continuous spectrum that is dependent on the temperature of the filament. At higher temperatures, the spectral content shifts toward the blue end. By varying the operating voltage of a tungsten light, the specific color temperature can be shifted. Sunlight also produces a continuous spectrum that is temperature-dependent, but it includes certain absorption lines that identify the elements in the solar atmosphere.

Fluorescent light sources produce a spectrum with very sharp lines at certain wavelengths superimposed on a generally continuous background. These lines are characteristic of the mercury vapor within the tube and are primarily toward the ultraviolet end of the spectrum. Neon lights are similar—they produce spectra that are characterized primarily by bright lines.

The light from LEDs has a narrow distribution, which depends on the color of the LED. (See Figure 4–4, for example.) They have a light distribution that is somewhat dependent on the geometry, with most light being emitted in the forward direction. LEDs are important sources for optical fiber communications.

LIGHT-MEASUREMENT UNITS

The measurement of light visible to the human eye is called **photometry,** and the measurement of the total optical spectrum is called **radiometry.** Electronic instruments that are designed for the same spectral response as the human eye are said to be photometric instruments, whereas those that measure in the infrared, visible light, and ultraviolet regions are said to be radiometric instruments. The units for measuring light are divided between radiant units and photometric units. There are equivalent units for each, but, in general, they are not interchangeable.

The photometric measurement of light usually is done either as the luminance intensity or the illuminance. **Luminance intensity** is defined as the total light from a source through a given area; the SI unit is the candela (cd). The comparable radiometric measurement is **radiant intensity,** usually measured in watts/steradian. **Illuminance** is the photometric measurement of light received on a surface on which there is a uniformly distributed flux; the metric unit is the lux and the English unit is the foot-candle. The comparable radiometric measurement is *irradiance, H,* previously defined in Section 4–11. Irradiance is typically measured in milliwatts per square centimeter.

OPTICAL TRANSDUCERS

There are a large variety of optical transducers, and the spectral responses of the sensors differ significantly. In addition, there are differences in sensitivity to light, geometric considerations, bandwidth, cost, and ability to operate in different environments. In spite of the great variety, nearly all transducers can be classified into one of three basic forms of photodetection: (1) photovoltaic sensors, which are self-generating semiconducting devices that convert light directly into an emf, (2) photoconductive sensors, which act as light-sensitive resistors, and (3) photoemissive sensors, which contain a light-sensitive cathode that emits electrons when struck by light. The principle of photoemission is employed principally by phototubes and photomultipliers. Phototubes and photomultipliers are discussed in Section 17–3.

Photodiodes are constructed from a *pn* diode junction using nonmetallic or metallic compounds with an overlying light-sensitive layer. When a photon of light passes through the transparent layer, it can be absorbed; the process moves an electron from the valence band into the conduction band, creating an electron-hole pair. If the energy of the photon is high enough, the electron will exceed the bandgap, and a current can flow if the circuit is closed with an external resistor. The *IV* curve for a photodiode is shown in Figure 13–31. Notice that when the incident light is zero, the response curve is that of a normal diode. As light intensity increases, reverse current increases.

To increase the frequency response, the diode can be operated with reverse bias. When a photodiode is operated without bias, it is a photovoltaic sensor; with reverse bias it acts as a photoconductive sensor. When the diode is reverse biased, little current flows until a photon is absorbed with sufficient energy to create

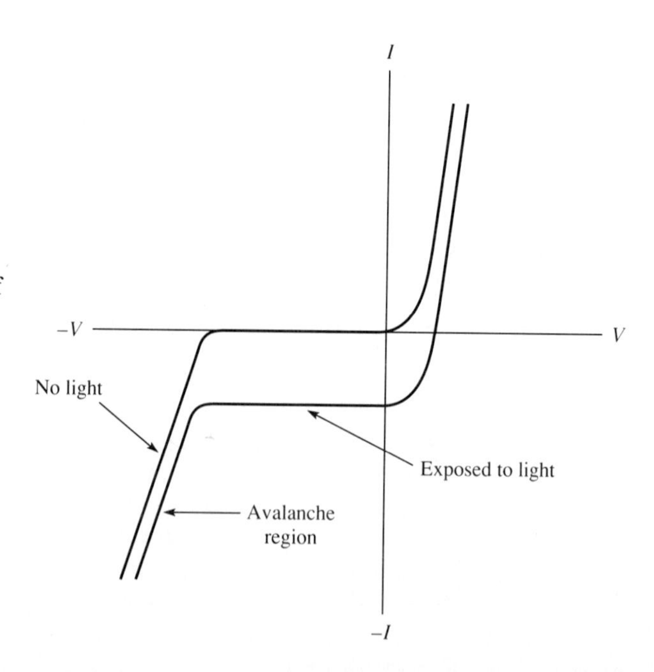

FIGURE 13–31
The *IV* curve for a photodiode. The photodiode acts like a constant current source when it is reverse-biased (up to the avalanche region). Output current is proportional to the light intensity and the area of the photodiode.

the hole-electron pair. If this pair is created in the depletion region, the charges quickly separate, creating current in the external circuit. On the outside of the depletion region, the migration of charge is slower, creating a long tail on the time-response curve.

A variation on the photodiode that produces higher-frequency response is made by separating the *pn* junction at the time of manufacture with a thin intrinsic layer. This forms a three-layer sandwich of positive-intrinsic-negative materials called a PIN diode. PIN diodes have bandwidths in the range of 1 GHz with very low noise.

Phototransistors are more sensitive than photodiodes because they have internal gain, typically from 100 to 1000 times that of a photodiode. Still even more sensitivity is available with photodarlingtons. Phototransistors are similar to ordinary transistors, but base current is provided by photons striking the reverse-biased base-collector junction. Phototransistors are useful where a sensitive sensor with a small active area is needed to detect light, but, because of nonlinear response and poor temperature characteristics, they are not useful for precision light-measurement applications. Phototransistors and optoisolators were discussed in Section 4–11.

Photothyristors are available for controlling larger currents. A light-activated SCR (or LASCR) is turned on by light rather than gate current. The gate lead is frequently brought out of the package to enable the threshold light level to be adjusted.

Photoconductive sensors include cadmium sulfide (CdS), cadmium selenide (CdSe), lead sulfide (PbS), and other cells. CdS and CdSe cells are popular when it is necessary for the cell to respond to visible light. The peak response for CdS cells is around 600 nm and the peak for CdSe cells is round 720 nm; PbS has a peak in the infrared region at 2200 nm. Photoconductive sensors consist of a photosensitive crystalline material sandwiched between two conductive electrodes. Absorption of photons in the crystalline material causes the resistance of the crystalline material to decrease. They are inexpensive and sensitive and can withstand high voltages, making them suited for control applications such as outdoor-lighting control. Their principle disadvantage is that they are slow.

SUMMARY

1. Input transducers for electronic measuring systems convert a physical quantity, property, or condition into a usable electrical signal. The measurand can be in one of six energy forms—mechanical, thermal, electrical, nuclear radiation, electromagnetic radiation, magnetic, or chemical.
2. Common transduction principles include resistive, capacitive, inductive, magnetic, and self-generating transduction.
3. Measurement considerations for transducer selection include range, input threshold, dynamic behavior, accuracy and resolution requirements, and repeatability and hysteresis error.
4. Environmental considerations for transducer selection include natural hazards, human-caused hazards, power requirements, signal-conditioning requirements, physical requirements, loading effects, and human factors.

5. The Wheatstone bridge is frequently used to detect the change in resistance of resistive transducers. More than one active transducer can be employed in a bridge to increase the sensitivity of the measurement.
6. Temperature is related to the average translational kinetic energy of molecules due to heat agitation. Scales for measuring temperature include the fahrenheit, celsius, and Kelvin scales.
7. Temperature transducers include thermocouples, RTDs, thermistors, semiconductor temperature sensors, integrated circuit temperature sensors, and radiation pyrometers.
8. Deformation of material ($\Delta l/l$) due to applied force is called strain; the deforming force per unit area is called stress.
9. The resistance strain gage is a thin metallic conductor that changes resistance as it undergoes a strain. Strain is a directly measurable quantity, but stress is not.
10. Force per unit area is defined as pressure. Gage pressure is the difference between the atmospheric pressure and the measured pressure. Absolute pressure includes the atmospheric contribution. Differential pressure is the difference between two pressures.
11. Virtually all pressure transducers operate on the principle of balancing an unknown pressure against a known load. Common pressure transducers use a restraining force against a diaphragm; movement of the diaphragm is proportional to the pressure.
12. The linear variable differential transformer (LVDT) is a displacement transducer that detects the position of a magnetic core inside a specially wound transformer.
13. The electromagnetic spectrum is divided into a number of regions, depending on wavelength, beginning with radio waves and extending through gamma and cosmic radiation. Visible light is bounded by the infrared region and the ultraviolet region. Many transducers operate in one or more of these regions.
14. Photometry is concerned with the measurement of light visible to the human eye, whereas radiometry is concerned with the measurement of the total optical spectrum.
15. Optical transducers can be grouped into photovoltaic sensors, photoconductive sensors, and photoemissive sensors.

QUESTIONS AND PROBLEMS

1. Explain the criteria you would consider important if you needed to select a pressure transducer to measure the pressure in an underground gasoline tank that has a depth of 10 m. The pressure reading will be converted to a liquid-level indication by a computer.
2. For the transducer in Question 1, what criteria do you consider least important in selecting a transducer?
3. Assume you need to measure the temperature of warm water used in an industrial process. The water needs to be warmed to 60°C but the incoming water can be from 10°C to 20°C.
   (a) What is the measurand?
   (b) What is the range?
   (c) What is the input threshold?
   (d) What other factors would you consider before specifying a particular transducer?
4. What is a calibration curve? What problems can occur during the calibration process?

5. Give an example of different transducers that use a resistive, capacitive, and inductive transduction principle, respectively.
6. Explain the difference between excitation leads and sense leads on a Wheatstone bridge circuit.
7. Explain the difference between a strain gage and a load cell.
8. Under what conditions is a linear assumption (using a constant Seebeck coefficient) warranted for determining the temperature of the measuring junction of a thermocouple?
9. A thermocouple with a Seebeck coefficient of 7.0 μV/°C has an output of 5.5 mV when the reference junction is at the temperature of ice/water (0°C). What is the temperature of the measuring junction?
10. When you connect a voltmeter to a thermocouple junction, the meter does not read the thermionic emf of the junction. Why not?
11. Compare the sensitivity and linearity of a thermistor with that of a thermocouple.
12. A platinum RTD with an α of 0.00392 Ω/Ω/°C has a nominal resistance of 100 Ω at 0°C. The resistance is measured as 88.5 Ω. Compute the temperature of the RTD.
13. Two strain gages with nominal resistances of 350 Ω and each with a GF of 2.00 are connected into a half-bridge circuit. The other bridge resistors are 350 Ω. The excitation voltage for the bridge is set to 10 V. Assume the bridge is initially balanced with no strain and has an output of 1.5 mV under strain. What strain is applied?
14. Why does a half-bridge arrangement tend to cancel temperature effects in bending measurements but not in axial measurements?
15. Assume a full-bridge arrangement is used with equal strain gages to measure a strain of 200 μϵ. The GF is 2.06. What is the output voltage for each volt of excitation voltage?
16. Compare advantages of bonded strain gages with unbonded strain gages.
17. A 2 mV/V load cell is excited with a 15 V source and monitored with a digital voltmeter. The meter can resolve 100 μV. The load cell is used in a scale designed for a full scale output with a 500 lb load. What is the resolution of the scale?
18. The mass density of water at 4°C is 1.000 g/cm^3. The mass density of mercury is 13.6 g/cm^3. Atmospheric pressure at sea level is approximately 760 mm of mercury.
    (a) Express atmospheric pressure at sea level in terms of the height of a column of water.
    (b) Express atmospheric pressure at sea level in pascals.
19. Explain the difference between gage pressure, absolute pressure, and differential pressure.
20. What is an LVDT? What does it measure?
21. Compare the terms *photometry* and *radiometry*. What are the measurement units for each?
22. The wavelength of a certain electromagnetic radiation is 4 cm.
    (a) What is the frequency in free space?
    (b) What type of radiation is this?
23. What is the approximate wavelength emitted by a green LED?

Chapter 14

Data Acquisition, Recording, and Control

OBJECTIVES

The previous chapter introduced a variety of input transducers that can convert a physical parameter into a voltage for use in an electronic system. Most (but not all) transducers convert the physical variable, such as pressure, temperature, or strain, into an analog voltage. Modern data-acquisition systems generally convert the data into digital form for processing. A basic data-acquisition system includes the transducers, interface circuitry (amplifiers, filters, multiplexer, etc.), analog-to-digital (A/D) converters, processors, and data output (human interfaces or machine control). Many of these components of the data-acquisition process have been introduced in previous chapters. In this chapter, data acquisition is developed as a system.

When you complete this chapter, you should be able to

1. Draw a block diagram of a typical data-acquisition system and describe the function of each block.
2. Explain why it is necessary to use a low-pass filter on the input.
3. Explain why the (sin x)/x correction filter produces essentially a flat response for a sampled data system.
4. Give examples of applications of data-acquisition systems, including applications to machine control and recording devices (chart recorders, magnetic storage, etc.), and applications for the analog output.
5. Describe the IEEE-488 bus and its application to data acquisition. Explain the difference between a listener, a talker, and a controller.
6. Describe the VXI bus as it applies to data acquisition and instrumentation systems.

HISTORICAL NOTE

In 1960 a high-performance A/D converter that could sample at 50,000 samples per second cost about $8000, used several hundred watts of power, and occupied the space of an oscilloscope. Today, a similar high-performance A/D converter sells for a few dollars, uses less than 1 W, and is housed in a single integrated circuit. Further, it can be interfaced directly to many microprocessors. Advances like this have affected the way data is acquired and processed and have moved data acquisition away from analog processing to digital processing.

14–1 OVERVIEW

Data-acquisition systems vary widely in scope and depend on both the properties of the data itself (number of channels, dynamic range, frequency content) and the application requirements. The application requirements include resolution and accuracy as well as the environment in which the system resides. At the lower end, a single channel of recording with direct readout may be needed for applications such as a chart recorder or meter. At the upper end, complicated multichannel systems may be needed, in which data will be sampled, processed, and sent to a central computer. In this chapter, we will look at a typical multichannel data-acquisition system and the problems that are associated with such a system.

Figure 14–1 shows a block diagram of a data-acquisition system. The particular system shown has various transducers connected to input conditioning. Depending on the transducer signal, input conditioning can be used for amplification, offset, or filtering. The gain of the input conditioning circuits may be controlled by the central processing unit to set the input to the proper level. The input signals are then routed to an analog-to-digital (A/D) converter through a selector switch (MUX). If the analog data is a sinusoidal waveform, then a low-pass filter may be needed to prevent *aliasing effects* (see Section 6–5). The A/D converter converts the analog signal into a digital signal that can be used by the central processing unit (CPU). The CPU controls, records, analyzes, and reports on the data it acquires. The data can be used in digital form or converted to analog form by a digital-to-analog (D/A) converter. If the analog signal is sinusoidal, then a correction filter may be needed at the output to reshape the frequency response of the system. The analog output may be used for control of the system, or it can be connected to other analog devices such as chart recorders, oscilloscopes, or meters.

Analog Devices manufactures data-acquisition systems in integrated circuit form (the AD363/AD364). Each of these systems contain an analog input stage (AIS) and a separate 12-bit A/D converter. The AIS is shown in the functional block diagram in Figure 14–2. It includes a 16-channel multiplexer, a differential amplifier, and a sample-and-hold amplifier integrated into a hybrid IC. Initially, the channel is selected by an address generated by an external computer or controller; the address is latched in an internal address register in the AIS. The signal from the selected channel is then applied to the sample-and-hold circuit, which is

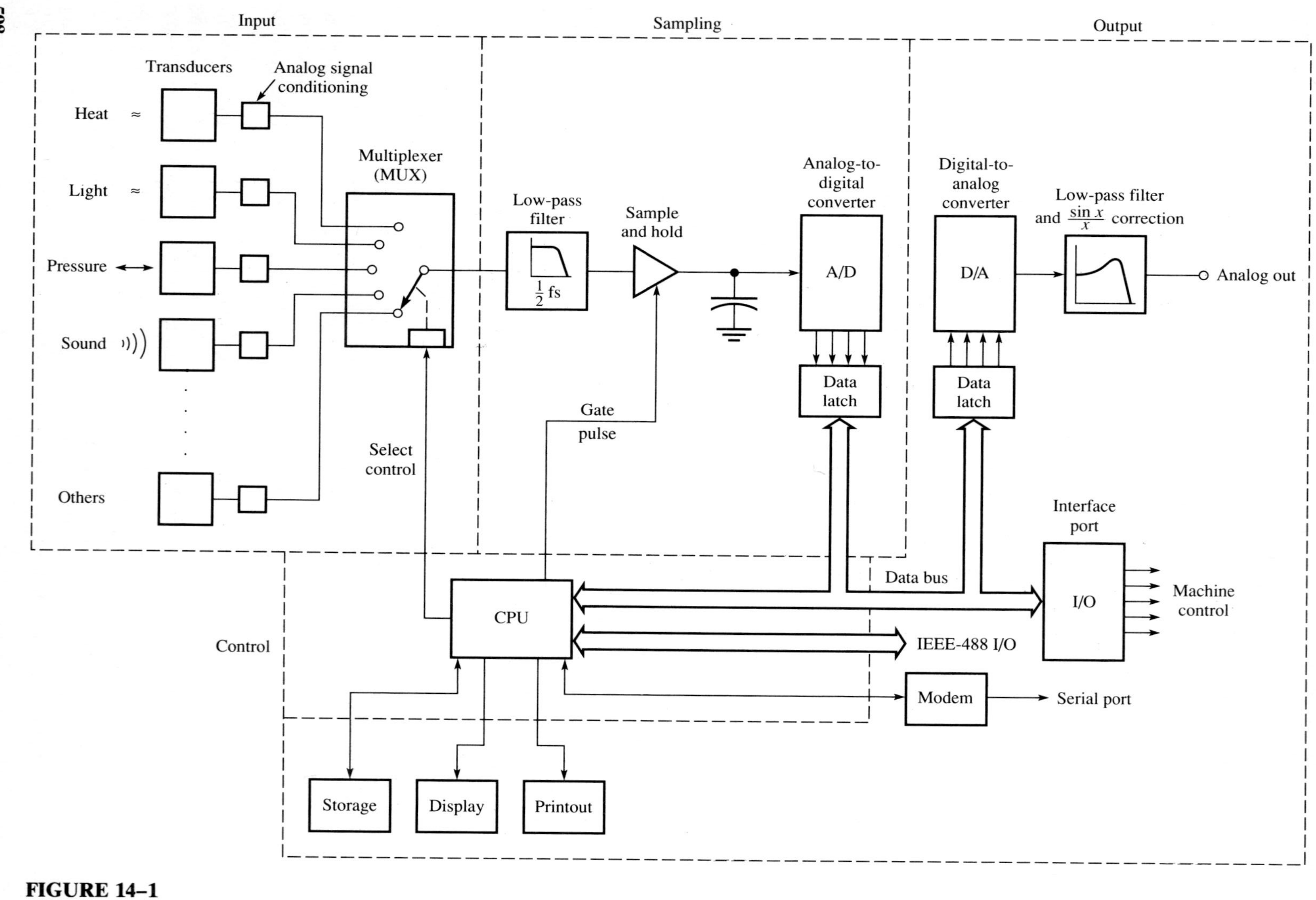

FIGURE 14–1
Data-acquisition system.

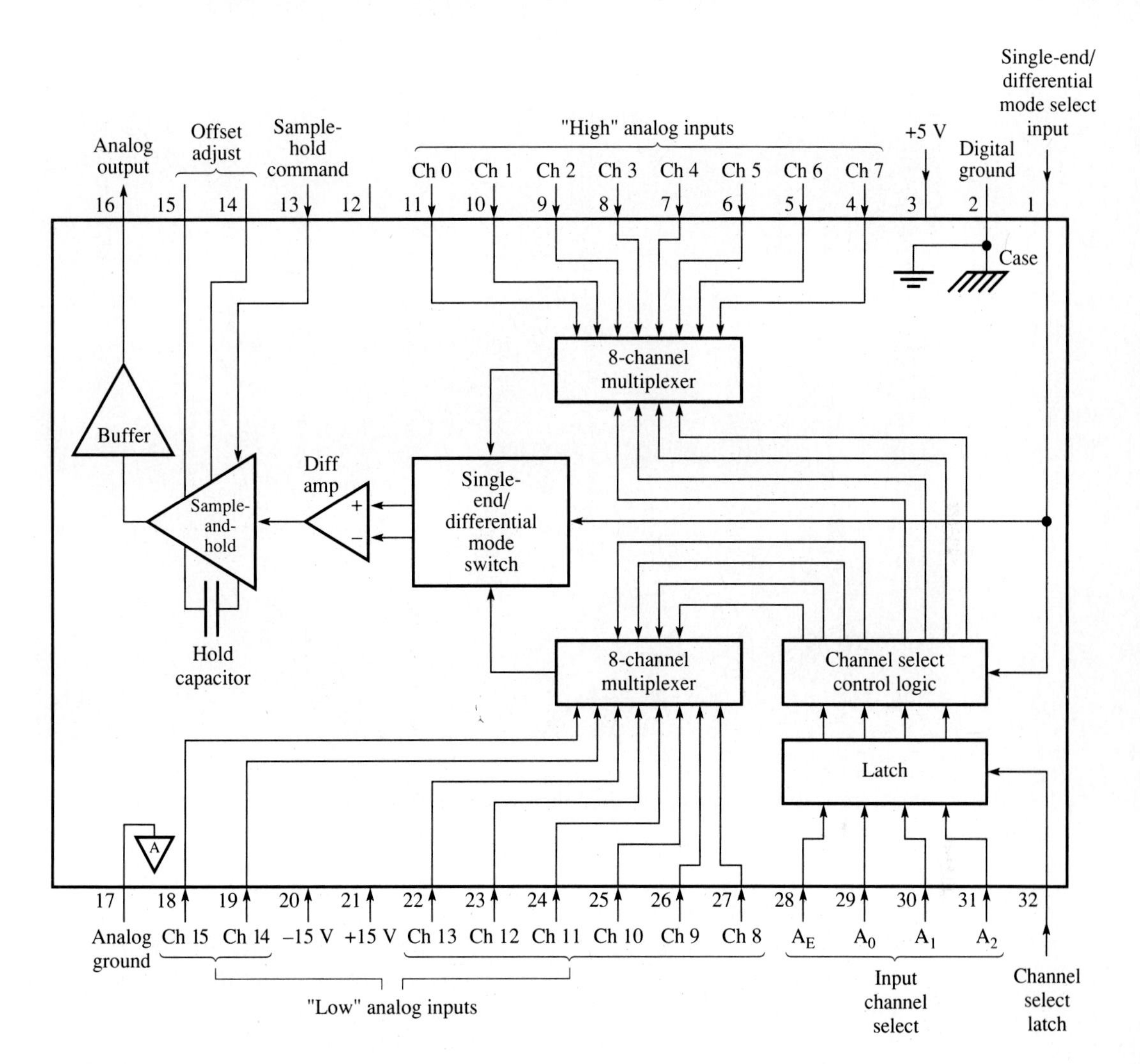

FIGURE 14–2
AIS functional block diagram (courtesy of Analog Devices).

controlled by the A/D converter. The sample-and-hold circuit typically requires 10 μs to acquire an input signal with sufficient accuracy for a 12-bit conversion. The overall system can function at 25 kHz.

There are several companies that make various kinds of data-acquisition boards that simply plug into the option slots of common personal computers (Figure 14–3). These boards contain the differential input buffers, multiplexers, sampling sections, and (optionally) the analog output section. The user of the system supplies the transducers, recording devices, and other peripherals. The connections from the transducer to the board are normally made with twisted-pair (differential) shielded wires, described in Section 12–6. The advantage of personal

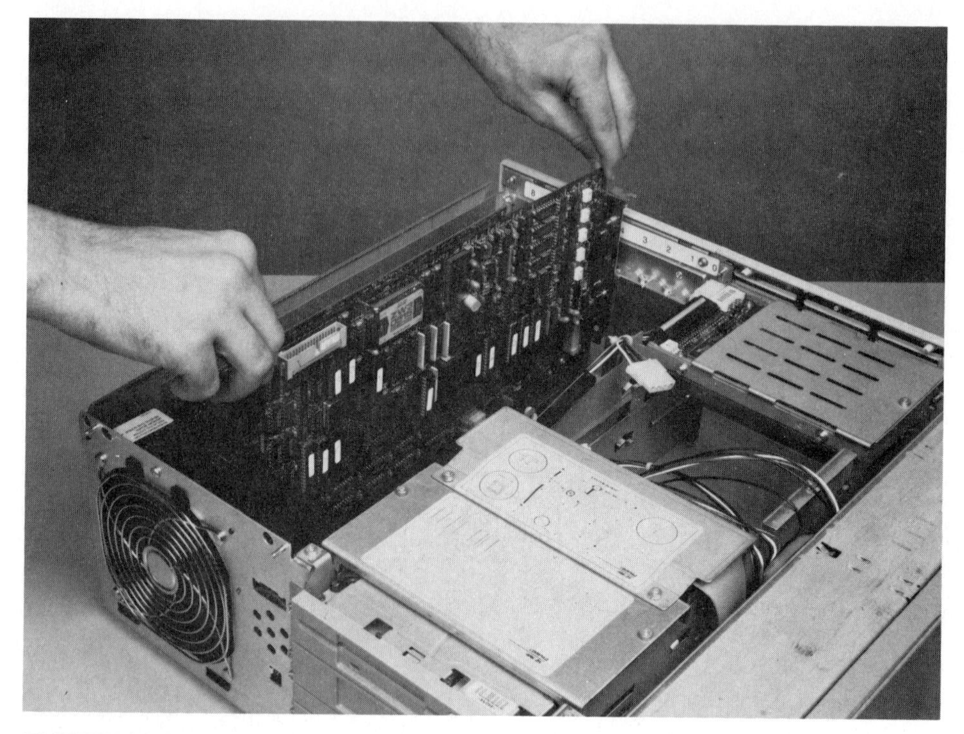

FIGURE 14–3
A data-acquisition board is installed into a personal computer (courtesy of National Instruments).

computer–based systems is that high-quality data-acquisition capability can be added to an existing personal computer at relatively low cost. Storage and processing of data is simplified by the computer. The trade-off is that the computer becomes tied up for the data-acquisition tasks and may not be available for other functions; a better solution may be to have a dedicated data-acquisition system.

General-purpose data-acquisition systems are available from manufacturers as stand-alone systems. An example is the Analog Devices μMAC-1060, a flexible data-acquisition system that includes machine monitoring and control, remote data acquisition, and other features. A photograph of the system is shown in Figure 14–4.

14–2 INPUT

The input section consists of transducers, signal conditioning, and a multiplexer. If there is only one transducer, then the multiplexer is not required, but having a multiplexer permits future expansion of the system. The transducers may be of different types, all the same type, or a mix. Even in a small system, such as one used in a greenhouse, there may be several types of transducers placed at different

FIGURE 14–4
The Analog Devices μMAC 1060™ data-acquisition computer. This single-board computer is optimized for data-acquisition and control applications. It includes analog I/O, digital (ON/OFF) I/O, pulse, event-counting I/O, CPU, serial ports, SBX port expansion, and user-programmable memory. The computer can be programmed in C language (courtesy of Analog Devices).

points in the building. These include temperature transducers, which are used to obtain average temperature data throughout the building as well as determining the areas with maximum and minimum temperature deviation. In addition, there may be humidity transducers and photo transducers. These transducers will provide data that can be used in controlling the greenhouse environment.

The multiplexer unit is controlled by the CPU. It selects the appropriate transducer required by the execution code in the program. The inputs are buffered, as shown in Figure 14–5, to reduce the loading effects the transducers may have on the common bus. The selector switch, shown in Figure 14–1, is depicted as a mechanical switch for clarity, but in most systems the switch comprises analog gates such as those described in Section 4–14. The analog gate, if it is selected, will pass the signal to the common bus; otherwise its output is disconnected. Single-chip analog multiplexers are available that minimize circuit effects on frequency response and settling time.

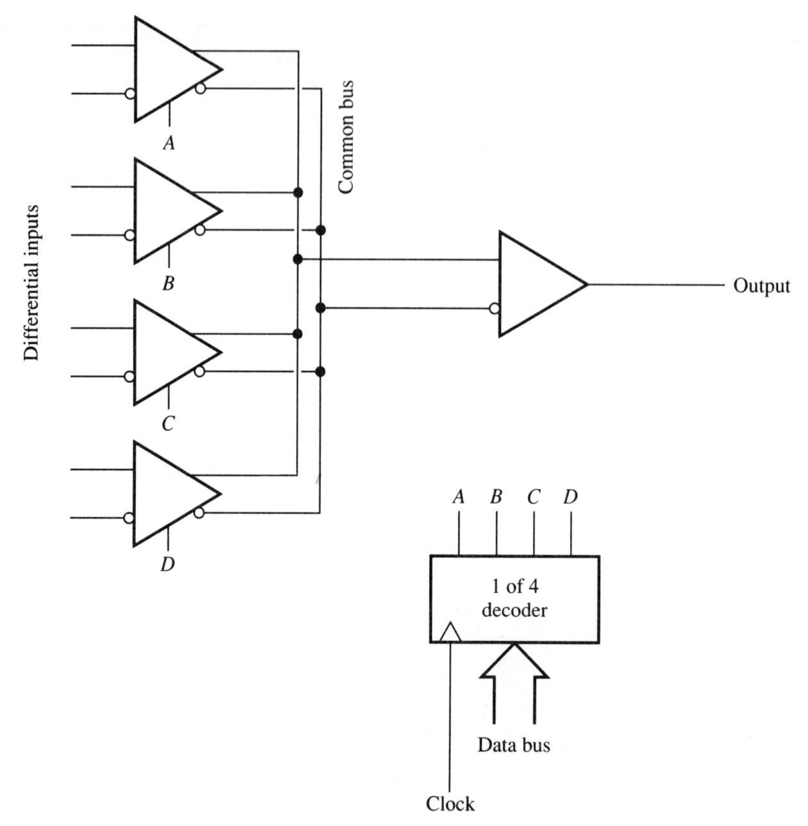

FIGURE 14–5
Input multiplexer.

14–3 SAMPLING AND LOW-PASS FILTERING

Transducers convert the physical parameters to be measured into a voltage, current, or resistance. If the measurand does not change rapidly, the input signal can be considered to be a constant or dc level for practical considerations. Sometimes the measurand varies rapidly, as in vibration measurements, for example. In this case, the signal contains ac components.

As with dc signals, rapidly changing ac signals must be sampled and digitized. If a time-varying signal is digitized as it is received, it may change levels before the conversion is complete, causing substantial error. A common technique for eliminating this error is through the use of a sample-and-hold circuit. The sample-and-hold circuit is capable of rapidly sampling the input signal and freezing its value by temporarily storing its voltage in a *hold* capacitor. The A/D converter then digitizes the stored value, after which another sample may be captured.

An example of a sample-and-hold operation on a sinusoidal waveform is shown in Figure 14–6. Notice that the waveform is sampled and held at the beginning of each sample time. The original sinusoidal signal is now represented by a series of sampled points that are spaced in equal increments of time. In order

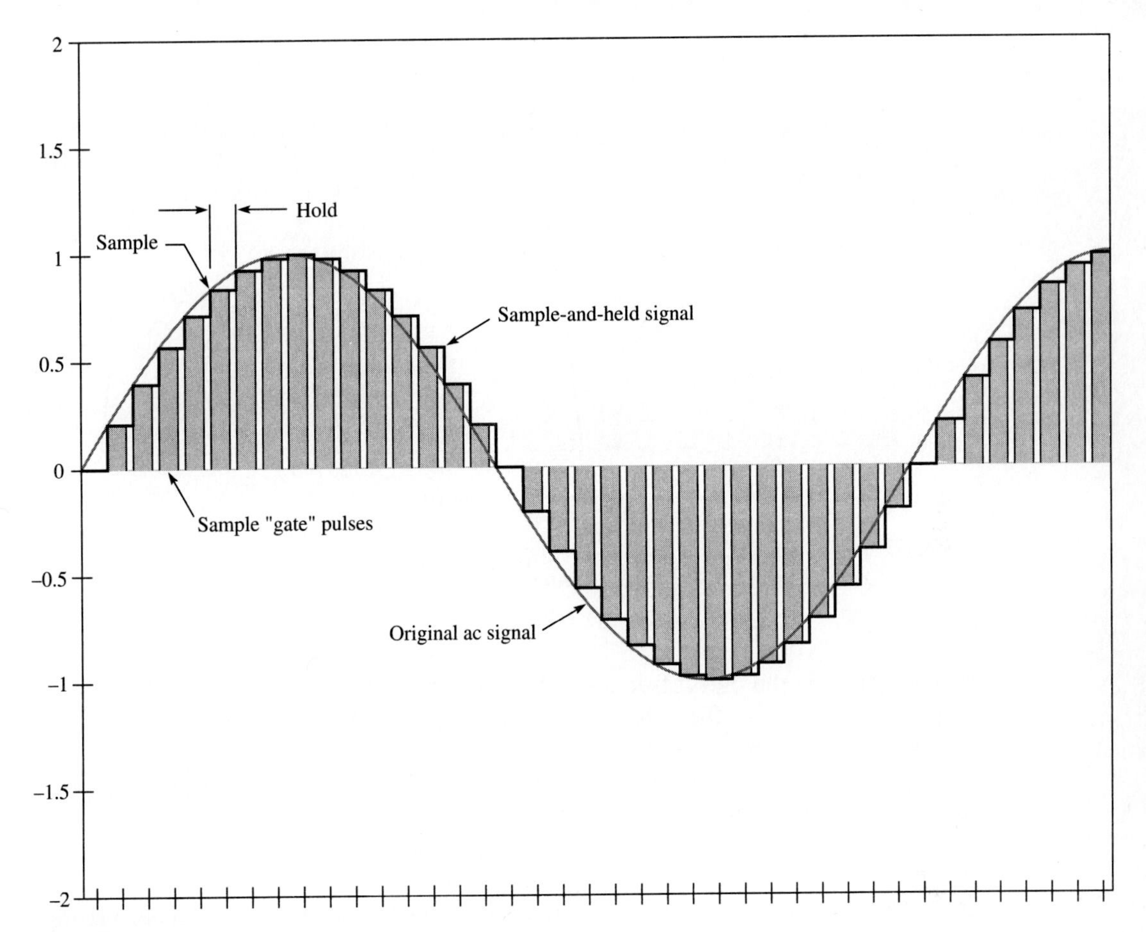

FIGURE 14–6
A sinusoidal signal that is "sampled" and "held" at fixed intervals.

to avoid errors in the sampled data, the Nyquist criteria (see Section 8–8) requires that the sampling must occur at a rate that is a *minimum* of twice the rate of the highest frequency component of the signal. This highest component is the cutoff frequency of the signal; it is *not* the standard −3 dB frequency we associate with filters and the like. This highest-frequency component must approach −40 dB in order to avoid aliasing errors. The Nyquist criteria can be stated in equation form as

EQUATION 14–1

$$f_s \geq 2f_c$$

where f_s = sample rate frequency, Hz
f_c = maximum frequency, Hz

The sampling process produces a harmonic spectrum centered around multiples of the sampling frequency, as illustrated in Figure 14–7. At the lower end of

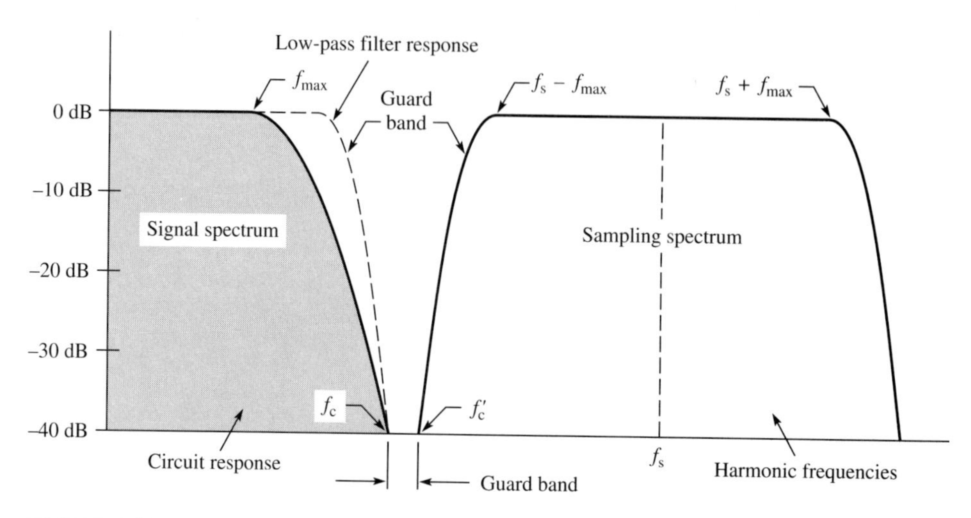

FIGURE 14–7
A filter is needed to limit the input frequencies to $\leq f_s/2$.

this spectrum is the cutoff frequency for the first harmonic of the sampling spectrum, labeled f_c'. Like the cutoff for the signal spectrum, this lowest component of the sampling spectrum should be at -40 dB.

If we increase the maximum frequency at cutoff for the signal (f_c), the lower cutoff frequency (f_c') will decrease by the same amount, causing the guard band to diminish as shown in Figure 14–8. Without the low-pass filter, an increase in maximum frequency above one-half the sample frequency will cause aliasing errors. The need for the low-pass filter is thus established: It prevents aliasing errors by keeping a reasonable guard band between the signal spectrum and the sampling spectrum. The response characteristics of the low-pass filter need to be flat through the passband and have a steep roll-off. The maximum signal frequency, f_{max}, should occur in the flat portion of the bandpass; the rate of roll-off is dependent on f_{max} and the desired width of the guard band.

FIGURE 14–8
Spectrum overlap, which causes aliasing errors.

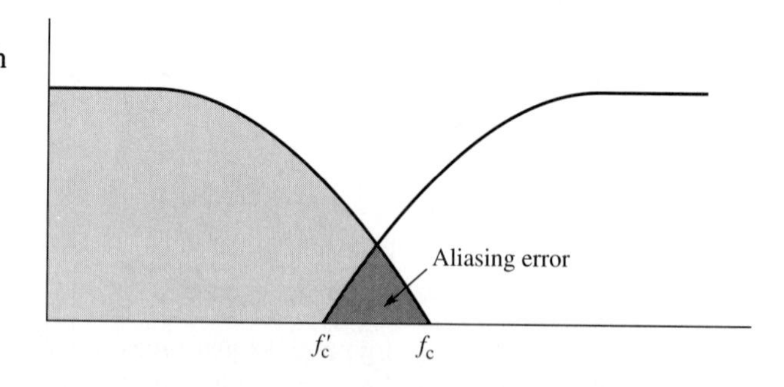

14–4 A/D CONVERSION

As stated earlier, a sample-and-hold circuit freezes the input signal at a constant level so that an A/D converter can accurately convert the signal to digital form. A sample-and-hold circuit is not always needed if the A/D reaction time is fast compared to the rate change of the input signal. The CPU can sample the output of the A/D and hold it in its memory. A sample-and-hold circuit can reduce sampling errors and reduce spectral noise when sampling ac signals.

A/D converters were discussed in Section 6–5, but a few comments should be reiterated: (1) The response time of the converter should be fast enough to convert the input voltage to a digital signal accurately. (2) The resolution of the converter needs to match the precision of the measurement. (3) The maximum and minimum limits of the A/D should be around the center of the input signal's range. (4) The input signal may need to be scaled, using amplifiers, so the full dynamic range of the A/D can be utilized (precision) without overdriving (distortion). Sealing can be done with a programmable gain amplifier, under control of the CPU.

14–5 THE CPU AND I/O

A CPU is the heart of a data-acquisition system. It controls the collection, processing, and storage of data. In addition, a CPU may control peripherals based on the data it collects and processes. Depending on requirements, data may be presented to humans as a series of numbers, graphs, a trace on a chart recorder, or in other formats.

An example of data acquisition and control is found in a city's water-treatment plant. The chlorine content may be monitored by a data-acquisition and control system. A valve that meters the flow of chlorine is adjusted by the CPU to keep the concentration levels of the gas to within specified limits. Each time a measurement is taken, the data is stored for later use. In addition, if the system measures concentrations outside of certain limits, an alarm will sound and the proper officials will be notified.

Another example of data acquisition and control is found in a bottling company. The quality of gaskets in bottle caps is essential in making a good seal, which preserves the product. A laser is used to scan the bottle side of the cap. Light is reflected from the surface of the cap but is absorbed by the gasket. A photo transducer is used to detect the intensity of the reflected light, and the analog waveform is sent to the data-acquisition system, where it is digitized and compared to previously recorded waveforms of good gaskets. If the newly acquired waveform is comparable to the stored "good" waveforms, then the gasket is assumed to be good; otherwise, the cap is rejected. In addition to controlling the selection mechanism in real time, the acquisition system can be used to store historical information on the manufacturing process, such as percentage of rejects to the total produced and the types of defects. The CPU can be programmed to analyze the recorded data and determine certain parameters about the gasket, such as minimum, maximum, and average width of the gasket. By using statistics, manufacturing engineers can maintain control over the manufacturing process to assure the rejection rate of the caps is kept low.

Sometimes it is more efficient to acquire data from a measuring instrument instead of directly from a transducer. Many instruments can process the information within the instrument, sending only the required data to the data-acquisition system. In this case, the communication to the CPU can be through the IEEE-488 bus (see Section 14–9 for details). This general-purpose instrumentation bus is used to convey control and measurement data from instruments containing IEEE-488 ports. A practical example is an electron discharge drill. This drill uses an electric arc to drill complex patterns (such as a square hole) through metals. The data-acquisition system monitors the force of the drill with a pressure transducer and the power supply current through the IEEE-488 port. By maintaining the force and current at predetermined levels, the highest-quality work can be done in the shortest time.

The IEEE-488 bus is limited to relatively short distances without special interfaces. A data-acquisition system may be spread over several hundred miles, as in a power-distribution network grid. Communication between the acquisition subsystems is essential in order to maintain control. This communication is coordinated by the CPU. Methods for communication include modem connections over phone lines, microwave links, buried fiber-optic cables, and satellites. The communication links between the subsystems are often made from the data-acquisition system's serial port to the data communication equipment (DCE). Local data can then be acquired by the data-acquisition system, sent to the DCE via the serial port and transmitted, in processed form, to the central data-processing system. The serial port is shown in Figure 14–1 connected to a modem (modulator-demodulator).

14–6 DATA RECORDING

Sometimes it is necessary to store data from an acquisition system as a permanent record. Generally, this is done using a hard-copy printout or plot directly from the computer. Two other methods are storing the data as a continuous paper record using a graphic recorder and storing data as a magnetic record using magnetic tape or disk storage.

GRAPHIC RECORDERS

Graphic recorders provide a permanent record of data as a function of time. They include various technologies from older pen-and-ink chart recorders to higher-speed units such as optical, electrostatic, and thermal-array recorders. Controls on a typical graphic recorder allow the user to select the sensitivity, set the zero point, and select the chart speed. Standard pen-and-ink chart recorders are limited to signals that change no faster than about 120 Hz at low amplitudes. Optical, electrostatic and thermal-array recorders can respond to frequencies up to several kilohertz (with high-speed electrostatic recorders going to 35 kHz). Depending on the application, graphic recorders are available having multiple channels (typically from 2 to 12). Applications for multiple-channel recorders include monitoring of a number of sensors simultaneously. The drive can be continuous or stepped (moved under control of an external pulse).

The traditional chart recorder provides a continuous record of a signal by drawing a line (or printing coded marks) on a moving paper. The paper can be in the form of a long strip or a circular graph. Circular charts are used in applications that have very low frequency, limited changes; their advantage is that they show the entire chart at once. Circular charts are used primarily to monitor slowly varying signals such as the outside temperature or the line frequency of an electrical utility. Typically, one revolution of the chart corresponds to either 24 h or to 7 d. Strip-chart recorders can use a very long chart (as long as 980 ft) to record data. The chart is moved at a constant rate under a pen (called the **stylus**). The stylus moves back and forth under control of the input signal, causing the record to be written as a continuous line on the chart. The **galvanometer** recorder has a mechanism that is similar to the D'Arsonval meter movement described in Section 7–1. The stylus is connected to the pointer. Another common mechanism for chart recorders is the **self-balancing** recorder. The stylus is connected to a slide-wire potentiometer that is connected as a voltage divider, providing a voltage proportional to the pen's position. A servo motor is driven by a voltage derived from the difference (error) between the stylus position and the input signal. The self-balancing recorder continuously drives the error voltage to zero, moving the stylus to the balance position.

High-speed graphic recorders, such as the thermal and electrostatic recorders, use a nonmoving print head, giving a considerable speed advantage over moving-pen recorders, because they do not have to overcome inertia and other mechanical limitations. At high chart speed, events as close as 2 ms can be distinguished between channels. These recorders are available as "intelligent" instruments that can be controlled by a computer using the IEEE-488 bus. They can be programmed to record only the data of interest, stop and start on command, change the chart speed when higher or lower resolution is required, and perform automatic calibrations and other functions. The print head can be used to mark the paper with text, time signals, or event markers that highlight certain points along the chart.

Another type of graphic recorder is the X-Y recorder. It is useful for plotting data when time is not one of the variables. X-Y recorders enable the user to graph one variable as a function of another. For example, an X-Y recorder could be used to plot the transfer curve (V_{out} versus V_{in}) for an amplifier. Instead of plotting data on a long strip, the recorder makes the plot on a rectangular sheet of paper. The X-Y recorder has two independent balancing servo mechanisms that drive the stylus—one for each axis. The zero point can be located at any point but is typically located on one corner or the center of the plot. Examples of applications where X-Y recorders are useful include the recording of relatively slowly changing data, such as strain in a test material as the stress is increased or graphing the temperature in a well as a function of the depth.

MAGNETIC RECORDERS

Two common forms for magnetic recording are tape recorders and disk recorders. Tape is an inexpensive medium for storing either digital or analog data for later use and is capable of directly reproducing the original electrical signal. Magnetic tape recorders use a recording head that magnetizes a thin ribbon of magnetic tape when it is passed by a gap in the head. The magnetic tape can be moved across the recording head continuously (when monitoring a continuously changing variable), or it can be started when data is available and then stopped. Data can be time-compressed by recording it in real time and then playing back the tape at a higher speed. An example of magnetic tape storage is the black-box flight recorder used to record voice and numerical data from the instruments of an aircraft.

Disk storage is used primarily for digital recording. The data is recorded on a circular magnetic disk that revolves under the recording/playback head. The head can be positioned radially over the disk, whereby the recorded data is placed in adjacent **tracks** arranged in concentric rings like those on a phonograph record.[1] The disk differs from the phonograph record, however, because the tracks on a disk are not continuous but are arranged in **sectors** that are like pie slices. Both sides of the disk can be used to store data. The primary advantage of disk-recording over tape-recording is improved access time to the data during playback; the head can skip across the tracks to the wanted data, whereas the tape recorder needs to fast-forward or rewind to the data. Its greatest disadvantage is its susceptibility to vibration. Mechanical shock or vibration can cause the recording head to be bumped onto an adjacent track, losing (or scrambling) the data. Disk-recording involves more complicated machine control and transport mechanisms than tape-recording; therefore, it is more expensive and prone to more failures. For data-acquisition systems involving personal computers (or large mainframe computers), disk recording is frequently used because the disk system is already a part of the computer system.

Magnetic recording has been applied to data storage since it was first used to keep a record of rudder deflections on a flight test in 1936; later (in 1947) magnetic recording was applied to storage of digital computer information. Disadvantages include susceptibility to stray magnetic fields and sensitivity to high temperatures, dust, and variations in humidity.

14–7 D/A CONVERSION

Some applications involve displays or control functions that require analog voltages or currents, so the digital signal needs to be converted to analog (D/A

[1] The phonograph record is an excellent example of continuous and linear data recording. The recording method is mechanical.

converters were covered in detail in Section 6–6). Television is an example in which an analog output is needed. Digital techniques can be used to process the video image or generate computer graphics, but because the present method for broadcasting the picture is analog, the digital signal will need to be converted to analog. As another example, analog control signals are used to focus a laser beam. A lens is held in equilibrium between the force of an electromagnet and a spring. Adjustment of the lens is done by changing the current through the electromagnet.

14–8 OUTPUT FILTERING AND (sin *x*)/*x* CORRECTION

As mentioned in Section 14–3, the repetitive sampling causes harmonics of the original signal to be centered about multiples of f_s (f_s, $2f_s$, $3f_s$, . . . , nf_s). In addition, there is another adverse effect caused by holding the signal until the next sample, and that is a reduction in the frequency response. While the circuit holds the sampled signal, it is unable to respond to changes in the signal until the next sample (Figure 14–6). This causes the frequency response of the sample-and-hold circuit to roll off at higher input frequencies, as shown in Figure 14–9. At multiples of f_s, the response drops to zero because the samples occur at the same point on the input waveform, giving a dc output. The response of the sample-and-hold circuit is given by

EQUATION 14–2

$$A_{dB} = 20 \log \frac{\sin x}{x}$$

where A_{dB} = response gain, dimensionless

$x = \frac{\pi f}{f_s}$, dimensionless

f = frequency of interest, Hz

f_s = sample frequency, Hz

Figure 4–10 shows the response of the sample-and-hold circuit overlaid on the frequency spectrum generated by the repetitive sampling. The low-pass filter at the output of a D/A converter serves two purposes. One is to filter out all the

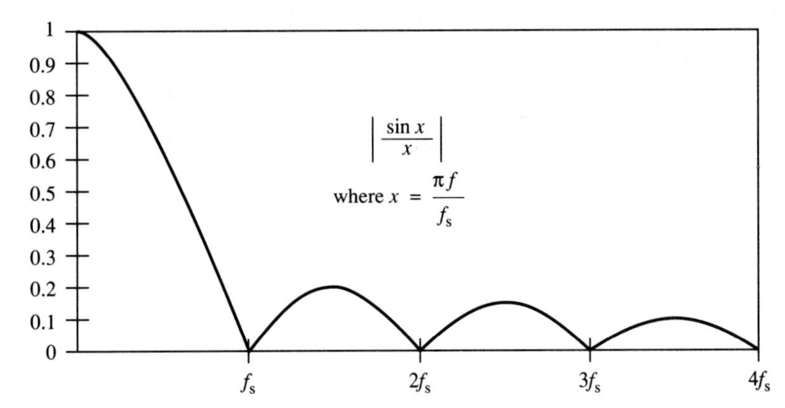

FIGURE 14–9
Frequency response of a sample-and-hold circuit.

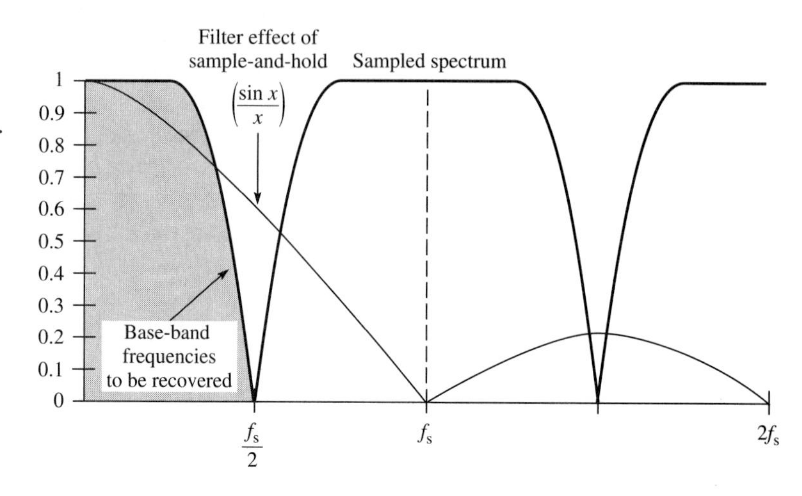

FIGURE 14–10
Sampled spectrum and response characteristics of the sample-and-hold circuit.

harmonics above the reconstructed base-band frequencies (the shaded area in Figure 14–10), and the second is to correct for the $(\sin x)/x$ frequency response distortion. Figure 14–11 illustrates the required shaping of the output filter's frequency response. Its response is the reciprocal of $(\sin x)/x$ combined with a sharp roll-off. Together, the $(\sin x)/x$ filtering due to the sample-and-hold circuit and the compensating $x/(\sin x)$ filter produce a flat response out to f_{max}, and an additional filter stage gives a sharp roll-off from f_{max} to f_c, as illustrated.

14–9 THE IEEE-488 BUS

During the early 1970s, problems in interfacing automatic test equipment became so severe that engineers from various organizations met to recommend a standard

FIGURE 14–11

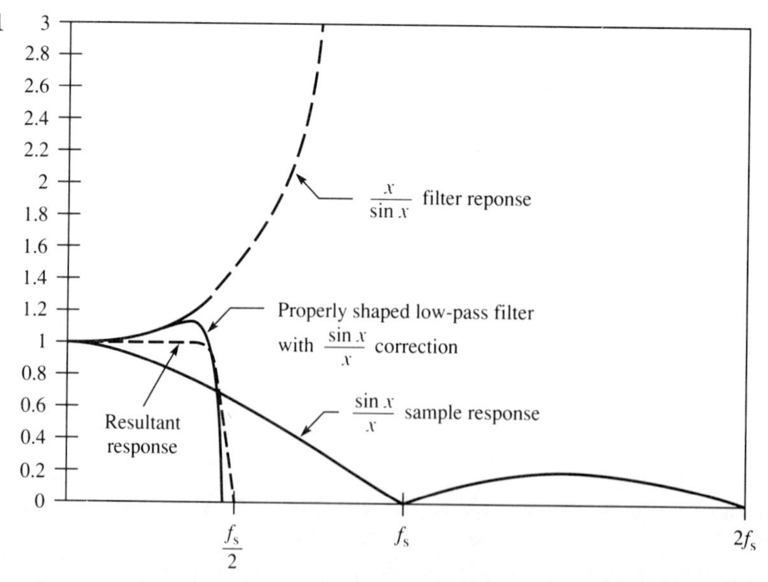

interface that could be used for programmable instruments and controllers. Following international discussions, an advisory committee of the IEEE (Institute of Electrical and Electronic Engineers) met in March 1972 to recommend a plan for a standard digital interface. Following these meetings, in April 1975 the IEEE published an electrical/mechanical interfacing standard (IEEE Std 488-1975)[2] that can be used by a variety of instruments. The standard was based on work done by Gerald Nelson and David Ricci of Hewlett-Packard, which had been published in the *Hewlett-Packard Journal* in October 1972. In 1978, the IEEE published a revised specification known as the ANSI/IEEE Std 488-1978, referred to as the IEEE-488. The International Electrotechnical Commission (IEC) later adopted this specification for international use, except for a change in the definition of the bus connector. The international definition of the standard is called the IEC 625-1. In 1987, the standard was revised again to what is now called the ANSI /IEEE Std 488.1-1987.[3] The 488.1 standard deals primarily with the mechanical interconnection specification and the electrical protocol. The standard enables engineers to interconnect products manufactured by a variety of companies in order to build automatic test systems, particularly the rack-and-stack style of test systems. Also in 1987, the American National Standard Institute adopted the ANSI/IEEE Std 488.2-1987 to standardize the software in terms of codes, formats, protocols, and common commands used by different manufacturers.

The IEEE-488 bus is well suited to flexible measuring systems, including data-acquisition systems, because cabling is easy to change or extend, the bus has good speed performance, it is suitable for simple to complex systems, and a wide variety of instruments and controllers is available from various manufacturers.

THE BUS STRUCTURE

The IEEE-488 bus is an interconnection scheme for interfacing instruments. The objective of the IEEE-488 bus was to enable different instruments using different data-transfer rates, different message lengths, and a wide range of capabilities to communicate with each other on a single interface. It is a party-line communication system using a total of 16 lines, with 8 of the lines assigned to data, 3 to data handshake, and 5 to bus management. Data is transferred by a bit-parallel/byte-serial approach in which the 8 data bits are transmitted in parallel (byte) and multiple bytes are transmitted serially (see Figure 14–12). Parameters are specified for maximum data-transfer rate, transmission distance, and the maximum number of devices that can be connected to the bus; the maximum data-transfer rate is specified as 1 megabyte per second. The transmission path length is not to exceed 20 m total, and it is recommended that the interconnecting cables themselves be limited to 2 m each, but this is not a design requirement. A maximum of 15 instruments may be connected to the bus, although special IEEE-488 extender instruments are available to extend the bus to longer lengths or to add more than 15 instruments. Figure 14–13 shows different configurations for interconnections.

---

[2] IEEE Std 488-1975, IEEE Standard Digital Interface for Programmable Instrumentation. The standards are published by The Institute of Electrical and Electronics Engineers, Inc., 345 East 47th Street, New York, NY 10017.

[3] The ANSI/IEEE Std 488.1-1978 and the ANSI/IEEE Std 488.2-1978 should be used together when developing a hardware and software system.

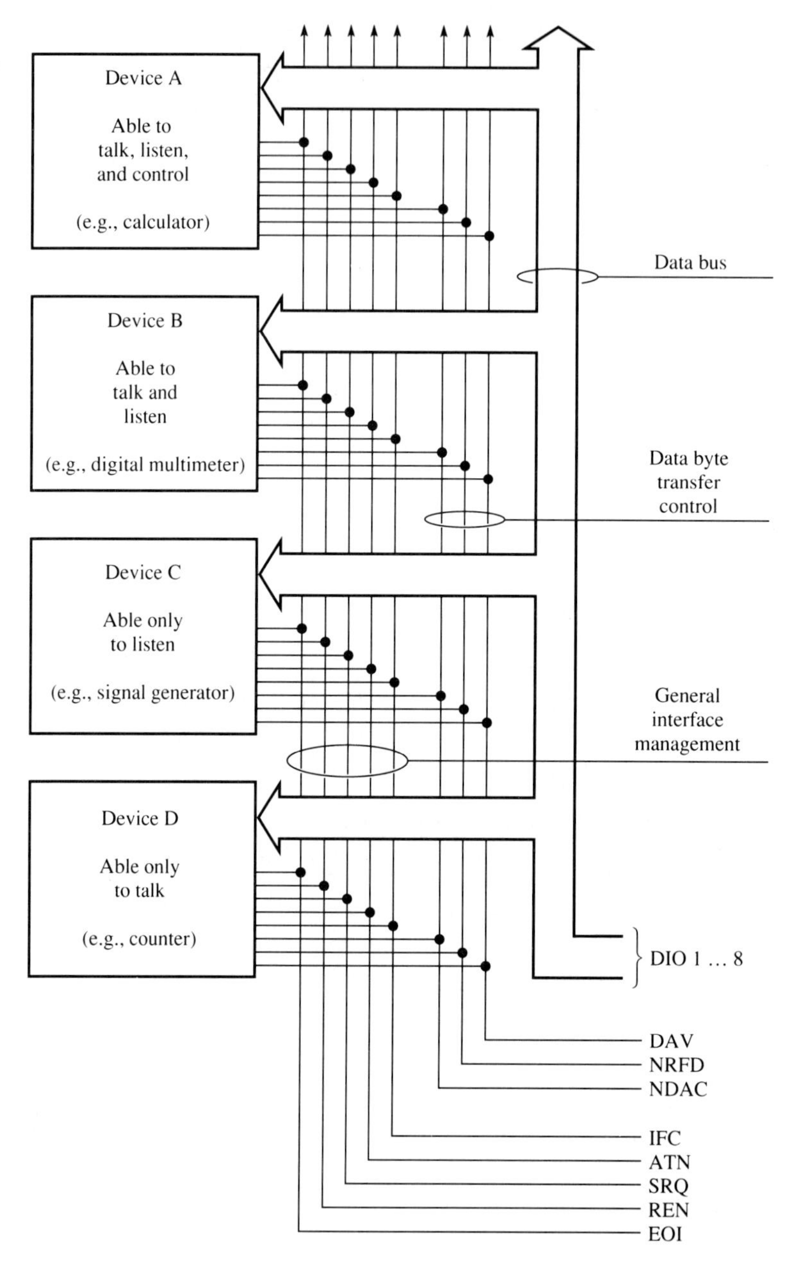

FIGURE 14–12
Interface bus structure (Courtesy of IEEE).

The output of each of the attached extenders is to have no more than a total of 20 m of interconnecting cables and no more than 15 devices connected to the bus. Other special interface/conversion equipment includes fiber-optic cables, serial data buses such as EIA Std RS-232C, or telephone lines.

The IEEE-488 bus is a bidirectional bus in which the data lines are used to establish the role of each device and to transmit data. Devices are not treated

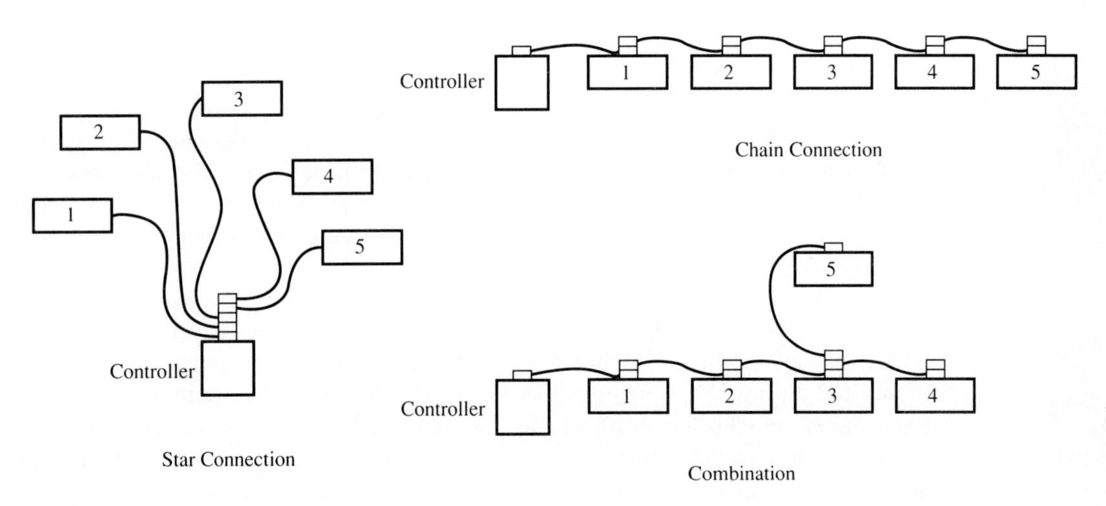

FIGURE 14–13
Connection scheme for IEEE equipment. Total cable length is not to exceed 20 m.

according to their actual end use (voltmeter, signal generator, etc.) but rather in terms of their function with respect to the bus. Only three functions are recognized. Devices are programmed to be a (1) talker, (2) listener, (3) controller, or any combination of these.

Talkers are devices such as multimeters or oscilloscopes, which can send data to other devices or to the controller. To avoid conflicts, only one device is allowed to talk at any one time. **Listeners** are devices such as power supplies and printers, which receive data from other instruments or the controller but do not send data. Devices that both talk and listen can be programmed to do either. For example, a digital multimeter is a talker when it sends voltage readings to a controller and is a listener when it is instructed to change ranges. The **controller** is responsible primarily for information management on the bus. It sets up devices to perform a particular task, monitors progress of measurements, and interprets results. Many controllers have separate access to computer peripherals such as display devices, printers or disk storage. Only one controller may be active at any time, but more than one can be connected to the bus. A controller can pass control to another controller. It can get control back only by requesting control from the current controller and then only if the request is granted. A minimum system could be set up with no controller at all—only a talker and a listener. For example, a digital multimeter can be set as a talker and a printer set as a listener. This is done with switches, setting the digital multimeter as "talk always" and setting the printer as "listen always."

DATA LINES

Line assignments for the IEEE-488 buses are summarized in Figure 14–12. The 16 lines are divided into three sets called the data bus lines, the data byte transfer control lines (or handshake), and the general interface management lines. Eight of the 16 lines are reserved for data input and output. These lines are labeled DIO-1 through DIO-8. The data lines are used for data transfer, addressing, and commands. When used as data transfer, the bus is bidirectional, thereby allowing data flow between the instruments and the controller or directly between the assigned talker and the assigned listeners. When used as an address bus, the controller addresses instruments that are to receive commands. When used as a command bus, the controller issues commands to the instruments previously addressed. Commands establish whether a device is a talker, listener, controller, or some combination of these.

Each instrument has an assignable and unique address on the bus. The address is selected by switches (shown in Figure 14–14). Newer types of instrumentation do not have switches; the addresses are assigned by the user from the front-panel controls via configuration menus. Address can be from 0 through 30 (31 total addresses). Address 31 is often used by some manufacturers for self-diagnostics. Of the 31 addresses, only 15 can be used on the same IEEE-488 bus without the use of a bus extender. The address switches (located on each instrument) set the **primary** address of that instrument. There could be a "host frame" that contains other resident instruments. The host frame would be given the primary address and the instruments contained within would each be assigned a unique **secondary** address. An example is a frame containing several relay boards. The purpose of this frame may be to switch the output of a power supply to different transducers. The frame has an address switch for selecting its primary address. Each of the relay boards in the frame may have its own secondary address. The board in slot 1 may have a secondary address of "1," the board in slot 2 may have a secondary address of "2," and so on. The secondary addresses

FIGURE 14–14

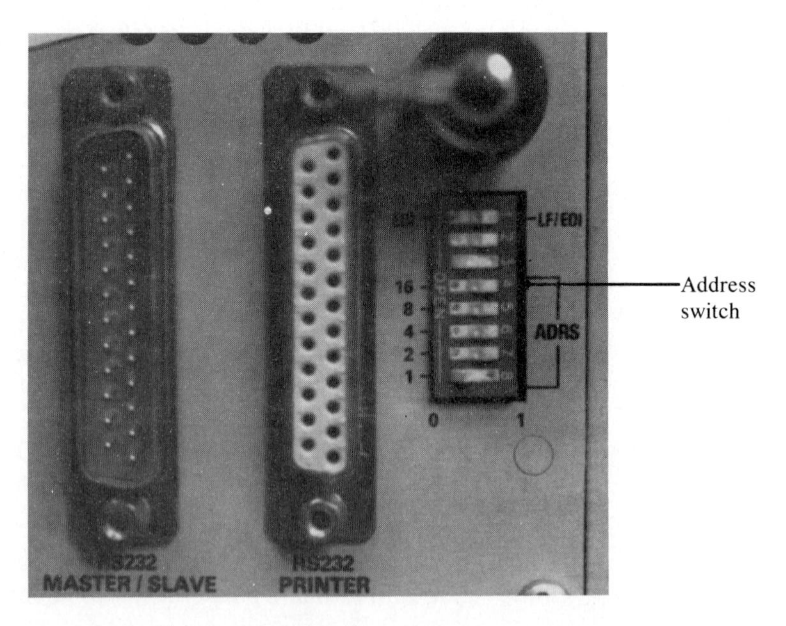

are most often defined by the manufacturer of the equipment and are not changeable by the user. The secondary addresses interact with the software running the IEEE-488 processor of the equipment—in this case, the relay frame.

A lot of timing, control, and data transfer occurs between equipment using the IEEE bus. Fortunately, there are several sources of software programs and drop-in computer boards that make IEEE programming and control simple to use. Usually, single statements are used (with parameters) to configure the equipment, issue commands, or transport readings; the software and hardware take care of the rest.

14–10 VXI BUS

VXI stands for VME bus extension for instrumentation. The VME bus is an accepted standard bus architecture used in modular computer systems. The VME bus was developed so computer components from different vendors could be assembled into an integrated *system*. A CPU from one vendor can interface with parallel input/output (I/O) of another vendor, a RAM card from someone else, and perhaps a machine-control card developed by your own company; all these units interface onto a common bus within a frame. Each of the boards that plugs into the frame has specification requirements as to its physical size, power requirements, connector pin-out, cooling, and communication protocol. A computer can be built by selecting modular option *cards* and assembling these cards into a frame.

The concept of building a modular computer system evolved into a modular instrumentation system by extending the VME bus to include instrumentation buses as well. The VXI bus is a multivendor system that permits instrumentation modules (cards) to be plugged into the same frame. For example, an arbitrary waveform generator from one company can be interfaced to the user's I/O card and to a waveform analyzer from another company.

The concept behind the VXI bus is similar to the IEEE-488 bus in that equipment can interface together to form an instrumentation system. Most of the instruments that are connected via the IEEE-488 bus are complete stand-alone systems; they have their own power supplies, front-panel controls, displays, processor units, and enclosures. This adds size, weight, and a considerable amount of redundancy to an instrumentation *system*. As an example of redundancy: An oscilloscope used in an instrumentation system has its own CRT for displaying the waveform, and the computer acting as the controller and data-acquisition processor also has a CRT capable of displaying the same waveform. Both displays are not needed. Also, the computer can select any number of the oscilloscope's front-panel controls, so the controls on the oscilloscope become redundant when integrated into a system. The instrumentation contained in the VXI frame is essentially the functional elements of the traditional instruments, stripped of the front-panel controls, displays, and the power supplies. Twelve instruments can be packaged into one frame, and several frames can be grouped together to form a larger system.

FIGURE 14–15
VXI bus module size (copyright Tektronix, Inc. Used with permission).

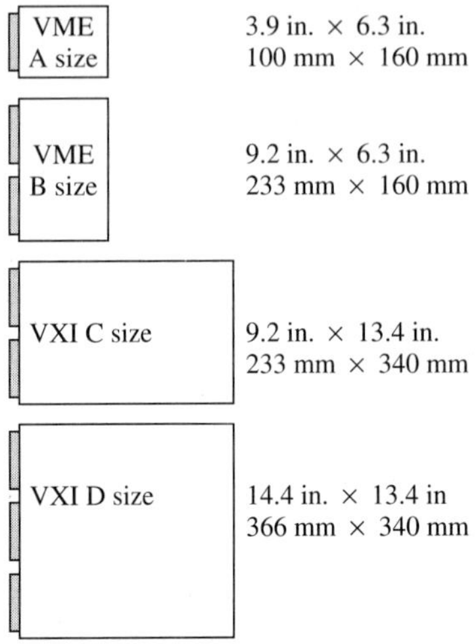

VME A size	3.9 in. × 6.3 in. 100 mm × 160 mm
VME B size	9.2 in. × 6.3 in. 233 mm × 160 mm
VXI C size	9.2 in. × 13.4 in. 233 mm × 340 mm
VXI D size	14.4 in. × 13.4 in 366 mm × 340 mm

In the summer of 1987, Colorado Data Systems, Hewlett-Packard, Racal-Dana Instruments, Tektronix, and Wavetek agreed to support instruments based on a jointly developed VXI bus. Since then, other companies have joined the effort: Bruel & Kjaer, John Fluke, GenRad, Keithley Instruments, and National Instruments. Other companies are providing VXI bus products that include backplanes, A/D modules, D/A modules, IEEE-488 and RS-232 interface modules, and object-oriented software.

VXI BUS MECHANICAL SPECIFICATIONS

There are four standard sizes for VXI modules, as shown in Figure 14–15. Each slot will accommodate a module that is 1.2 in. thick, but if an instrument needs more than 1.2 in., it can take up multiple slots. The modules stand on edge; the cooling holes are on the top and bottom edge of each module. Power is supplied to the modules via the backplane.

VXI BUS STRUCTURE

The VXI bus structure is shown in Table 14–1. The basic system is the VME computer bus. The *extension for instrumentation* buses are the clock and sync

TABLE 14–1

BUS	TYPE
VME computer bus	Global
Clock and sync bus	Unique
Star bus	Unique
Trigger bus	Global
Local bus	Private
Analog sum bus	Global
Module identification bus	Unique
Power distribution bus	Global

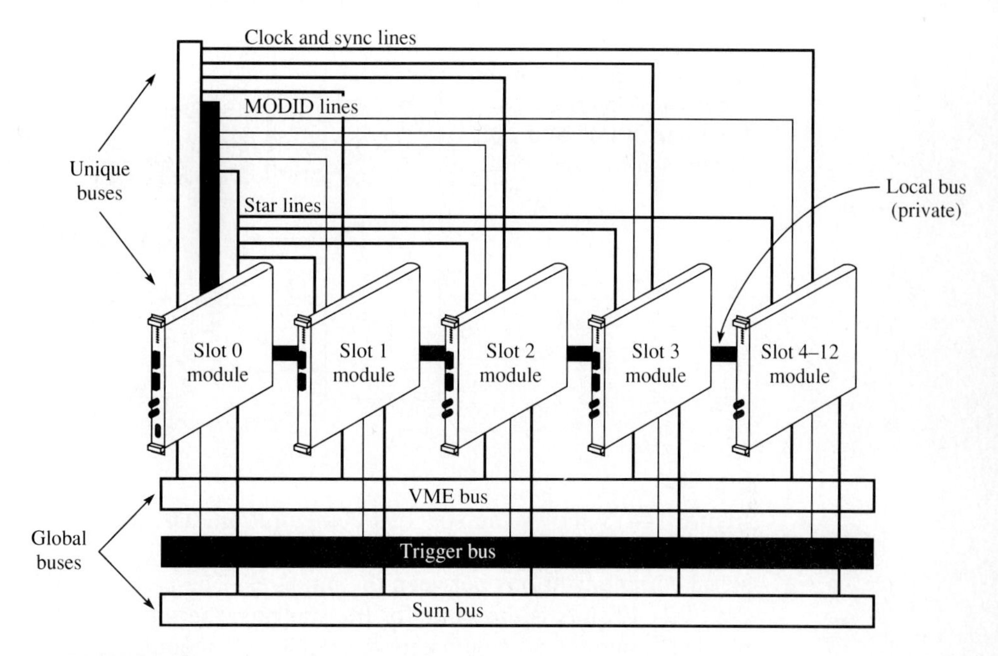

FIGURE 14–16
VXI bus electrical architecture (copyright Tektronix, Inc. Used with permission).

bus, star bus, trigger bus, local bus, analog sum bus, module identification bus, and power distribution bus. Figure 14–16 shows the layout of the buses.

VME Computer Bus

The VME computer bus is based on the IEEE-1014-1987 standard. It contains four buses: data-transfer, arbitration, priority-interrupt, and utilities.

The data-transfer bus is a high-speed asynchronous data bus that can transfer 8, 16, or 32 bits between modules. This allows extended addressing up to 4 gigabytes.

Unlike the IEEE-488 bus, which can have only one controller assigned at a time, the VME bus can have several controllers, such as the CPU, DMA (direct memory access), I/O controllers, and any other device that needs to control the bus. To avoid conflicts, the arbitration bus provides the means to transfer control.

The VME priority-interrupt bus provides a means whereby devices can request service from a VME interrupt handler. This is similar to the IEEE-488 service request (SRQ).

The utilities bus provides power distribution, clocks, initialization, and failure detection.

Clock and Sync Bus

There are two clocks and a clock synchronization signal. One clock is a 10 MHz clock, and the other is a 100 MHz clock accompanied by a sync signal. The clocks are ECL (emitter-coupled logic) and are buffered on the backplane, as shown in Figure 14–17. The reason for buffering the clocks is to provide a high amount of intermodule isolation and to relax the clock-loading rules for the modules.

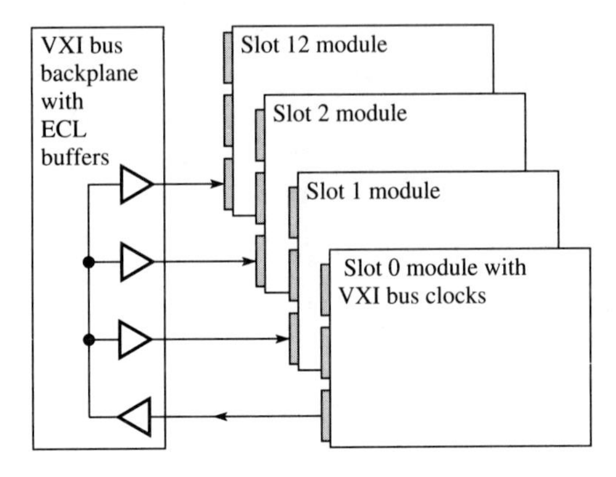

FIGURE 14–17
VXI buffered clock bus on backplane (copyright Tektronix, Inc. Used with permission).

Star Bus

Figure 14–18 shows the star bus configuration. The star bus contains two high-performance ECL lines called STARX and STARY. The bus provides high-speed serial communications between the modules. The path length between slot 0 and the other 12 slots is the same for the star bus; therefore, it keeps the timing skew to a maximum of 5 ns between slot 0 and any module.

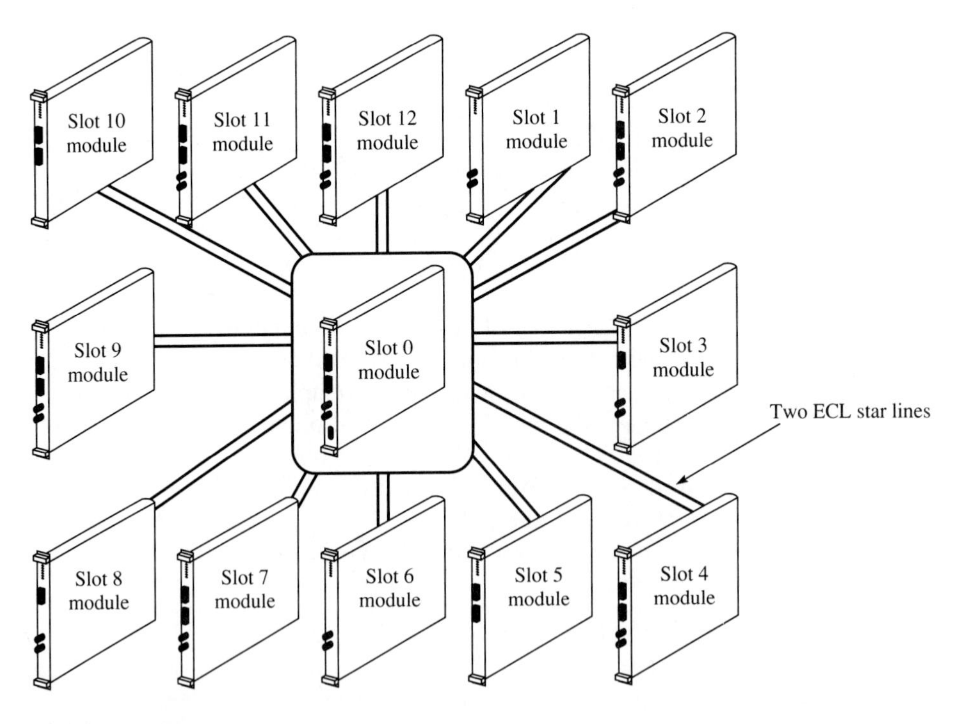

FIGURE 14–18
Star bus configuration (copyright Tektronix, Inc. Used with permission).

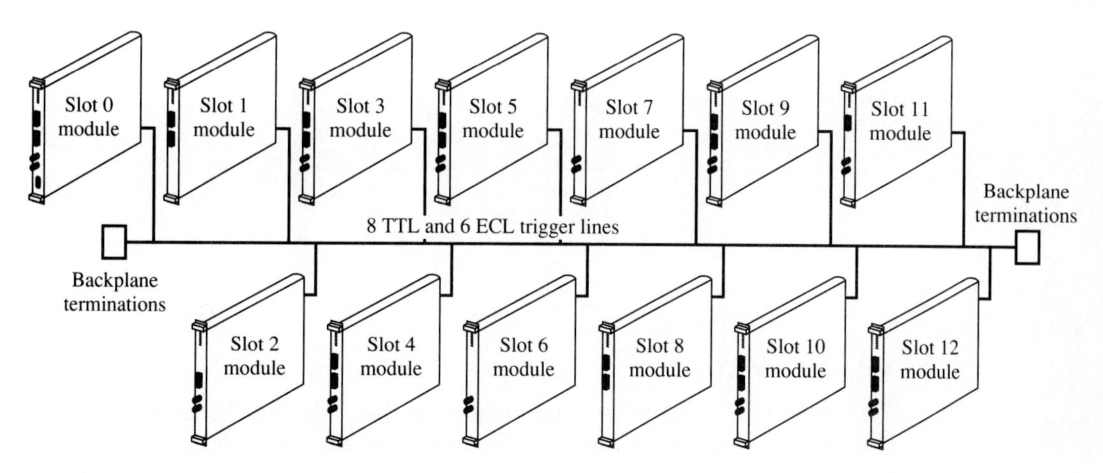

FIGURE 14–19
VXI bus TTL and ECL trigger lines (copyright Tektronix, Inc. Used with permission).

Trigger Bus

The trigger bus is a general-purpose logic bus that may be used for triggering, handshaking, clocking, or data transmission. The bus is subdivided into eight TTL trigger lines and six ECL trigger lines. Figure 14–19 shows the trigger bus configuration.

Local Bus

The local bus consists of 72 lines on each module, which are partitioned into 36 lines on each side of the module, as shown in Figure 14–20. The purpose of the bus is to decrease the need for jumpers between modules and to provide local communications between two or more modules without using a global bus. Figure 14–21 shows the bus configuration.

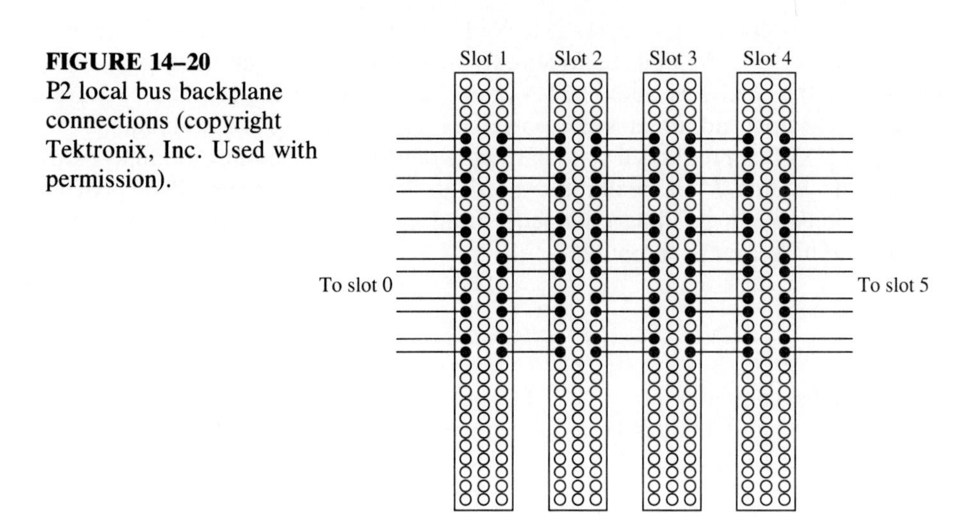

FIGURE 14–20
P2 local bus backplane connections (copyright Tektronix, Inc. Used with permission).

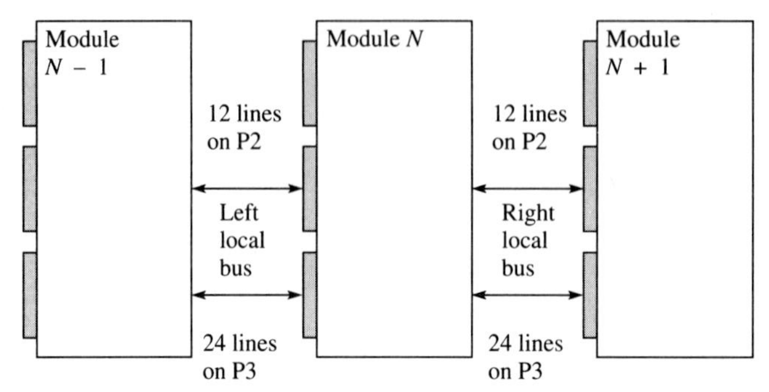

FIGURE 14–21
VXI P2 and P3 local bus (copyright Tektronix, Inc. Used with permission).

Analog SUMBUS

The SUMBUS is an analog summing node that is bused the length of the frame. Each module can drive this line using an analog current source, or each module may receive information from this line through a high-impedance receiver. A complex waveform, for example, can be generated using three arbitrary waveform generators (AWG); their outputs add together (sum). The advantage of this technique is that several low-cost AWGs can be used to create complex waveforms that otherwise might require a high-cost AWG.

Module Identification Bus

There are 12 identification wires connected to slot 0, one to each of the 12 other slots, 1 through 12. If a module is plugged into a slot, that unique identification wire is grounded, indicating that a module is present. This helps to provide quick system configuration at start-up and aids in diagnostics. For example, on power-up, the module in slot 0 scans all the identification bus lines to locate modules. It then attempts to communicate to the modules as part of start-up diagnostics. If it detects the presence of a module via the identification bus but is unable to communicate with it, then a failure is indicated.

Power Bus

The power bus can provide up to 268 W of power to a single module. Voltages of +5 V, +12 V, −12 V, and +5 V standby (battery backup) already exist on the VME bus; the VXI bus adds +24 V and −24 V for analog circuits, −5.2 and −2 for ECL, and additional VME power for greater power requirements.

The VXI bus provides a solid hardware platform for building modular instrumentation systems. More flexibility, speed, and computing power will be provided with this bus when building data-acquisition systems as compared with boards that plug into personal computers.

SUMMARY

1. Data acquisition involves computer processing. Data can be stored and analyzed for information purposes or for control of external equipment.
2. The input section of a data acquisition consists of transducers, analog signal conditioning, and multiplexing (switching).
3. Input transducers convert properties of

the physical world to proportional electrical signals.

4. Input-signal conditioners perform a variety of functions as required to interface the transducer to the data-acquisition system. For example, the conditioner may provide a constant current source to a strain gage so that a voltage, proportional to the resistance in the gage, can be developed and sent to the data-acquisition system.
5. The multiplexer selects one of many transducers as an input.
6. The sampling section consists of a low-pass filter, sample-and-hold circuit, and analog-to-digital (A/D) converter. Depending on the application, the low-pass filter and sample-and-hold circuits may not be needed.
7. Frequencies above one-half the sample frequency are filtered by the low-pass filter to prevent aliasing errors.
8. Aliasing occurs when a changing signal is not sampled frequently enough to reconstruct the original signal from the sampled data. The sample rate is to be at least twice the highest frequency of the input signal.
9. The CPU section of the system (which may be a computer) is used to control the analog signal conditioning and multiplexers, capture data from the A/D converter, process the data, store and report on the data, and control peripheral devices through special I/O ports, serial modem interfaces, and the IEEE-488 bus.
10. Output may involve several optional devices. The most common output devices are the display monitor, printer/plotter, and magnetic storage devices. In addition, analog outputs can be provided for external interfacing with other equipment.
11. Magnetic storage devices include tape recorders and magnetic disks. The tape recorder is preferred in harsh environments involving vibration. Disk storage is often used with computer systems because of availability and because data retrieval is faster.
12. Graphic recorders plot the data on paper. The paper can be a continuous strip, circular, or rectangular. Visual representation of data utilizes the brain's capacity to recognize patterns and trends.
13. Digital-to-analog (D/A) conversion requires a low-pass filter to remove the harmonics of the base-band frequencies caused by the sampling process. In addition, a filter to correct for $(\sin x)/x$ error is required. These errors are caused by holding a sampled value until the next sample.
14. The IEEE-488 is a general instrumentation interface bus. It is used to control instruments remotely as well as exchange data between instruments. The bus is 8-bit-parallel/byte-serial. In addition to the data lines, there are three lines used to synchronize the transfer of data (handshake) and five lines used for control.
15. Devices on the IEEE-488 bus operate in one of three modes. They may be controllers, talkers, or listeners. There can be only one controller active at at time. The primary job of the controller is to establish which instrument is to be a talker and which instrument should listen.
16. Instruments on the IEEE-488 bus are assigned unique primary addresses. Secondary addresses cannot be changed; they are given to devices contained in the same instrument and grouped under the primary address.
17. VXI stands for VME bus extension for instrumentation. Tight time coordination and high-speed data transfer provide advanced ATE system-measurement capabilities in an open, multivendor-supported architecture.

QUESTIONS AND PROBLEMS

1. A data-acquisition and control system is connected to a fish tank. It is to measure and control the temperature (68°F to 74°F), measure and control the light (artificial light is to emulate the rise and fall of the sun), measure the pH of the water and issue a warning if it is not 7.0 ± 1, and dispense food twice per day. In addition to the measurement and control aspects of the system, a strip chart showing temperature, light, and pH is to be made. A hard copy of the data is to be printed on a line printer.
   (a) Draw a block diagram of the system indicating input, CPU, and output sections.
   (b) Is input filtering needed? Why or why not?
   (c) How would you control the electrical heater and how would you isolate the 120 V power line from the computer?
2. When is an input filter needed before the A/D converter?
3. What is aliasing error?
4. A simplified filter is used ahead of a D/A converter, as shown in Figure 14–22. Its roll-off is −20 dB/decade. Its Bode plot shows the −3 dB point to be at 5.0 MHz.
   (a) Using semilog graph paper, plot the Bode curve.
   (b) At what frequency is the gain −40 dB?
   (c) Plot the lower sideband created by the sample frequency so the −40 dB point intercepts with the −40 dB point of (a). From this, determine the minimum sampling frequency f_s.
   (d) If the filter is designed to have a −40 dB/decade roll-off, then what is the minimum sampling frequency (assume the −40 dB points intercept)?
5. An automated lettuce picker uses an analog sensor to measure the X-rays coming from a focused X-ray beam. As the lettuce picker moves down the furrow, the beam is focused through the lettuce head, with the sensor on the other side. As lettuce ripens, the concentration of water in the head increases; this causes the X-ray flux reaching the sensor to decrease. Three actions are taken by the data-acquisition system: (1) If the water content is too low (X-ray flux too high), then no action is taken; the lettuce needs to ripen. (2) If the water content is too high (X-ray flux too low), then the head is over-ripened; it will be picked and placed into a discard bin.

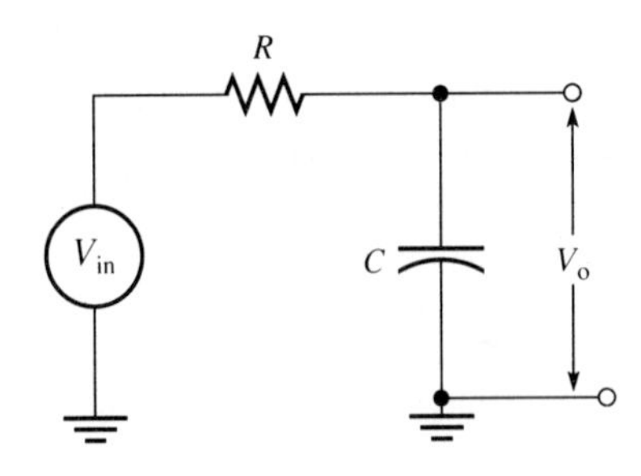

(a) Simplified filter.

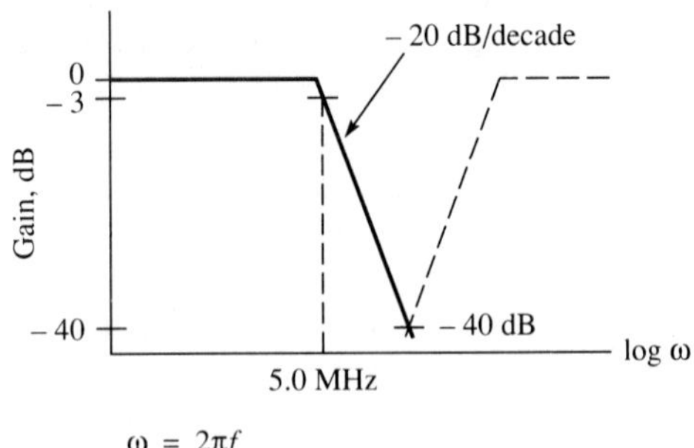

(b) Bode plot.

FIGURE 14–22

(3) If the water content is within an acceptable range, the lettuce will be picked for market.

(a) Draw a simplified block diagram of the system.

(b) Picking is done once each day. What type of information should be gathered that could help the farmer plan next year's crop?

(c) How could you modify the system to give the farmer information as to what areas of the farm are most productive?

6. How many steps of resolution are obtained with a 4 bit A/D converter? If the A/D converter's full scale range is from 1.0 V to 4.0 V, what is the resolution (in volts) of each quantizing step? *Note:* A quantizing step is the vertical displacement, as shown in Figure 14–6.
7. Repeat Problem 6 for an 8 bit A/D converter.
8. Using semilog paper, plot $A_{dB} = 20 \log |(\sin x)/x|$, where $x = \pi f/f_s$. Assume f_s = 100 kHz. Plot gain versus f from 0 Hz to 500 kHz.
9. Compare magnetic tape storage of data with disk storage. What are the advantages of each?
10. How much $(\sin x)/x$ correction (in decibels) is needed if f = 800 kHz and f_s = 1.8 MHz?
11. (a) What is the IEEE-488 bus?
    (b) What is the maximum number of instruments on the bus?
    (c) What is the maximum total length of the bus?
12. For the IEEE-488 bus:
    (a) Define primary address and secondary address.
    (b) How are they selected?
    (c) Define talker, listener, and controller.
13. (a) What are the advantages of the VXI bus over the IEEE-488 bus?
    (b) How does the VXI bus prevent conflicts between multiple controllers?

Chapter 15

Automatic Test Equipment

OBJECTIVES

From the beginning, manufacturers of electronic equipment and components have tested their products to assure quality. Testing is also required for maintenance and fault detection in electronic systems. As electronic systems have become increasingly complex, testing requirements have intensified. The basic goal of automated test systems is to test integrated circuits, discrete components, circuit boards and systems at ever-faster rates with better accuracy and lower cost. The demand for high quality and large quantity has led the testing system to feed back information on the manufacturing processes. Enabling the improvement of the manufacturing process itself is part of the goal of automated testing.

When you complete this chapter, you should be able to

1. Describe the different types of automated test equipment (ATE), applications, benefits, and limitations.
2. Describe test methods used for testing passive and active devices (both analog and digital).
3. Give a basic block diagram of an auto test system and explain the function of each block.
4. Describe two-, three- and six-wire measurement techniques used in in-circuit testers.
5. Describe the design considerations for test fixtures.
6. Explain how surface-mounted devices are tested.
7. Compare conventional automatic test methods with boundary scan technology.

HISTORICAL NOTE

In the late 1960s, several manufacturers were working on automated test systems. By 1972, the Fluke Manufacturing Co. marketed an automatic test system called the Terminal/10. It was a calibration system that used BCD to communicate with test equipment. The new HPIB (later called the IEEE-488 bus) that had recently been developed by Hewlett-Packard was available as an option to communicate with newer instruments that had this capability. The system was controlled by a Digital Equipment Corporation PDP-8 that could be programmed in a high-level language such as BASIC or FORTRAN. The Terminal/10 automated the stimulus to the instrument to be calibrated and verified calibration parameters. Usually, the test instrument had to be read by the operator and the reading was manually input into the computer. The computer could then determine if the instrument was within the required calibration. If it was not, the operator was instructed which adjustment to change to bring the instrument into calibration. Today, a similar calibration can be performed using no human interaction at all (including adjustment); instruments that do not have a communication interface can be read using a computer vision system.

15–1 INTRODUCTION TO ATE SYSTEMS

Automatic test equipment (ATE) has evolved from the requirement of electronic manufacturers to check the quality of the products they produced. In the early days, when products were simpler and production was slower, a technician tested products on a bench by manually connecting the unit under test (UUT) to the test equipment, selecting appropriate settings on the measurement equipment, taking measurements, and interpreting the results. As production increased, it was necessary to automate these steps. Automated test systems have been developed to speed the testing cycle, reduce human error, make the tests more comprehensive, and develop a data base for characterizing devices and circuits. In addition to improving the quantity and quality of production, ATE is used in maintenance and servicing of equipment after it has been placed in operation.

There are many different types of ATE for a wide variety of applications. These systems generally fall into two categories: *manufacturing test systems* and *rack-and-stack*. The most common is the manufacturing tester.

MANUFACTURING TESTERS

Manufacturing testers, illustrated in Figure 15–1, consist of a large mainframe that is complete with its own power supplies, digital processor, test signal sources, and signal acquisition. The digital processor controls the hardware and performs the programmed tests on the UUT.

Different types of UUTs that are to be tested on the manufacturing tester have their own fixtures. Each of the fixtures has a mechanical/electrical interface by which it is clamped to the mainframe of the manufacturing tester. The fixtures contain spring-loaded contact pins that make several connections to the UUT at once in a configuration called a **bed-of-nails.** The pins can either stimulate the

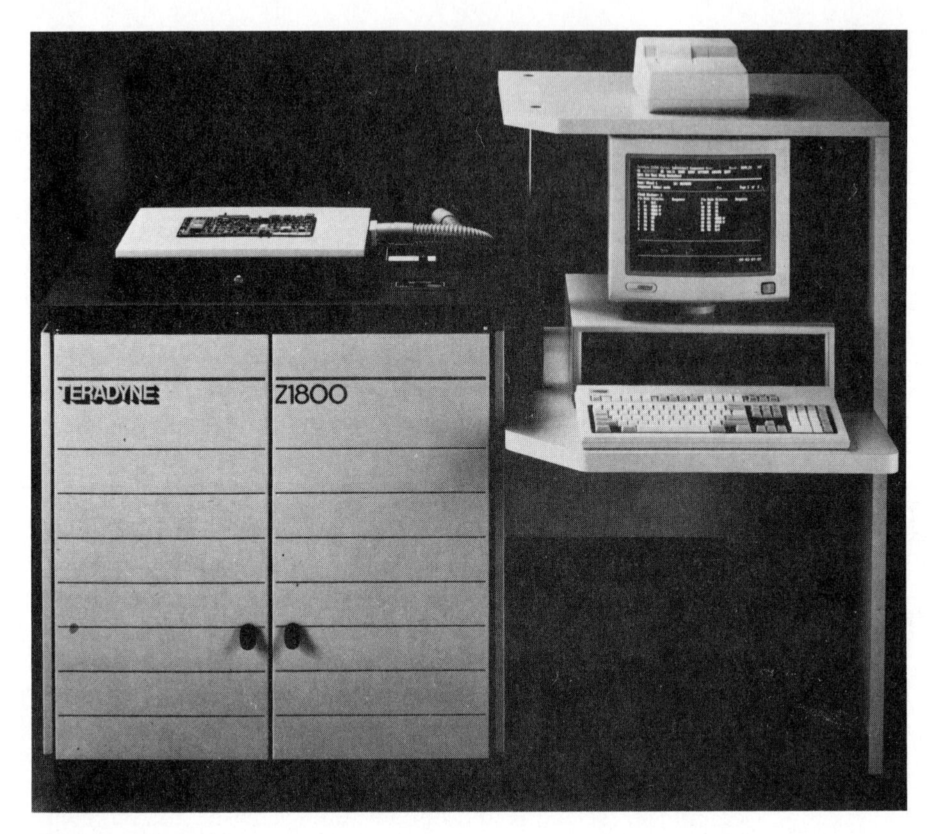

FIGURE 15–1
Z1800 in-circuit tester (courtesy of Teradyne).

UUT with current, voltage, or a digital pattern or they can measure the results as an effect of the stimulus in terms of voltage, current, or a digital pattern. The primary advantage of these types of fixtures is the speed at which they can test for manufacturing errors or failed components. The primary disadvantages are its limited high-frequency response, capacitive loading, and cross talk. These problems are inherent in the bed-of-nails fixture because of the rather long interconnecting wires associated with its construction.

There are several different types of manufacturing testers. They range from rather simple and easy to program units to very complex systems that require highly skilled and talented programmers. Placing these in order from the simplest to the more complex, they are

1. Bare-board testers
2. Manufacturing defects analyzer (MDA)
3. In-circuit component testers
4. Component testers (characterization parameters)
5. Functional testers
6. Combinational testers (doing both in-circuit test and functional testing)

RACK-AND-STACK FUNCTIONAL TESTERS

Large functional testers can cost over $1 million per system. As an alternative to these large and expensive systems, many companies have built their own flexible functional testers using various pieces of equipment stacked in a rack. These are known as **rack-and-stack.** The stack of equipment is generally custom-configured to a specific need, as required for the test. It contains equipment such as oscilloscopes, voltmeters, logic analyzers, function generators, power supplies, and other test and measurement equipment.

Rack-and-stack functional test systems use stimulus instruments to send power and test signals to the UUT and measurement instruments to monitor both analog and digital responses. The instruments are connected to the UUT through a switching system and a test fixture or interface to provide signal paths to the UUT. A typical system is shown in Figure 15–2(a) and (b). Figure 15–2(c) is the block diagram for this system. The control of the test sequence is managed by a computer that also can accept data over the bus for processing. In its controller role, the computer issues commands that include instrument setups and allocation of switching paths and directs the flow of data from measuring instruments. The most commonly used instrument control bus is either the IEEE-488 or the VXI bus, as discussed in Sections 14-9 and 14-10.

The specific test system depends on the nature of the test required, time available for testing, costs, required calibrations, future expandability of the system, skill level of the operators, and so forth. Most rack-and-stack ATE systems for functional testing are configured as a combination of standard instruments using either the IEEE-488 or the VXI bus and a commercial controller. The advent of personal computers has popularized this form of controller for smaller test systems by requiring only the addition of a plug-in interface card to the computer. The fixturing and programming is unique to the specific test that is run. Sophisticated user-friendly software is available to allow the user to set up and control the test sequence and acquire the data with high-level commands. For example, depending on the software set, a DMM might be set up to measure resistance with a command such as

wrt "DMM", "Resistance"

The computer or controller can then read the data from the bus and compare it to acceptable values. The program can continue the test if values are within a specified range or it can branch to an error routine if they are not in range. Some controllers come with sophisticated software packages allowing an operator with a minimum of programming experience to set up a test sequence.

MANUFACTURING TESTERS USED WITH RACK-AND-STACK

An assembly process may have both a manufacturing tester and a rack-and-stack functional tester, as shown in Figure 15–3. The manufacturing tester is used to locate the defective components and assembly errors. The rack-and-stack functional tester is used for alignment of analog circuits at their operating frequency (something the bed-of-nails fixture may not be able to do) and can perform opera-

(a)

Network analyzer

Digital oscilloscope

Tektronix VM700

Analog oscilloscope

Logic analyzer

Video monitor

Computer monitor

NTSC Gen

Pal Gen

Digital Gen

RGB Gen

Programmable power supply

Computer

(b) Equipment identification.

FIGURE 15–2
Rack-and-stack equipment.

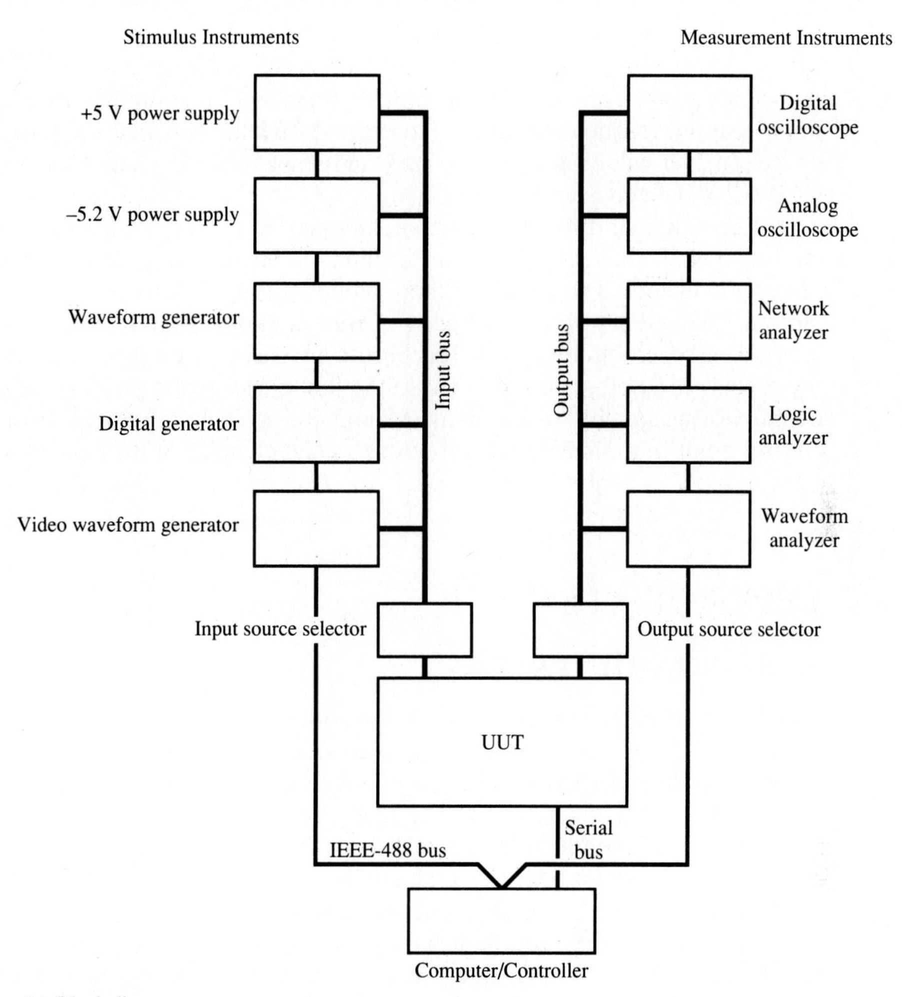

(c) Block diagram.

FIGURE 15–2 (*continued*)

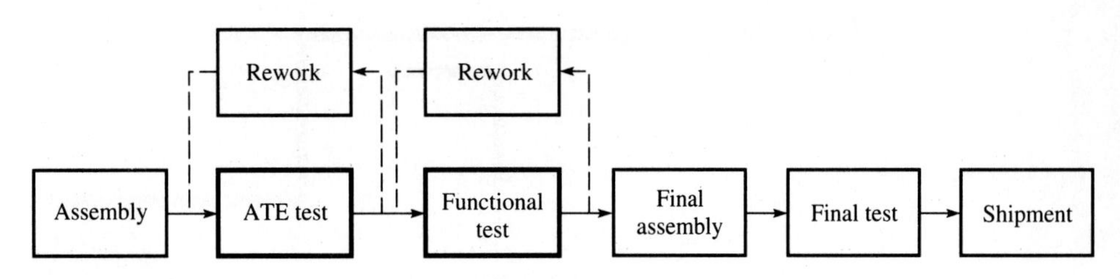

FIGURE 15–3
Manufacturing flow with ATE and ATE functional.

tional and specification checks at these frequencies. For example, in the manufacturing process of testing a radio receiver, the bed-of-nails manufacturing tester will check all the components for proper values and function; however, the maximum stimulus frequency is limited to about 2.0 MHz. The final test and alignment of the receiver will be done on the rack-and-stack tester at its operating frequency (151.955 MHz, for example).

Depending on the control a company may have over its manufacturing processes, manufacturing defects can account for as much as 80% to 90% of the product's defects. The most common manufacturing defects are wrong, missing, and misoriented (in backward) parts. Other common manufacturing defects are shorted and open-circuit traces. It is best to have enough control over manufacturing processes so errors don't occur in the first place, but regardless, some means of monitoring quality is essential. Manufacturing testing should be a process-control point, not a means of correcting the deficiencies of the process—in other words, *build* in quality; don't *test* it in. Automating testing will improve the consistency of the process through fast and accurate process reporting.

15–2 TYPES OF ATE

BARE-BOARD TESTERS

Bare-board testers are used to check for continuity and shorts on circuit boards before they are populated with components. They generally use the bed-of-nails fixture described in Sections 15–7 and 15–8. A test pin is placed at the end of each trace, as shown in Figure 15–4. The test program is developed by "learning" a good board. The production run of boards is tested by comparing the various continuities of the test board to a table of expected results. Unexpected continuities are short circuits, or *shorts,* and the absence of expected continuities results in *open circuits*. This type of test is particularly useful with **multilayer circuit boards.** A multilayer circuit board is one that contains more than two layers of traces. These traces are sandwiched between insulating layers of epoxy, as shown in Figure 15–5. The innermost layers may contain large copper planes, which supply power and ground. Other layers contain the circuit connection (traces). The fabricating process of making the traces (etching) and pressing the layers together into a single board (laminating) can have errors that are not correctable;

FIGURE 15–4
Bare-board tester.

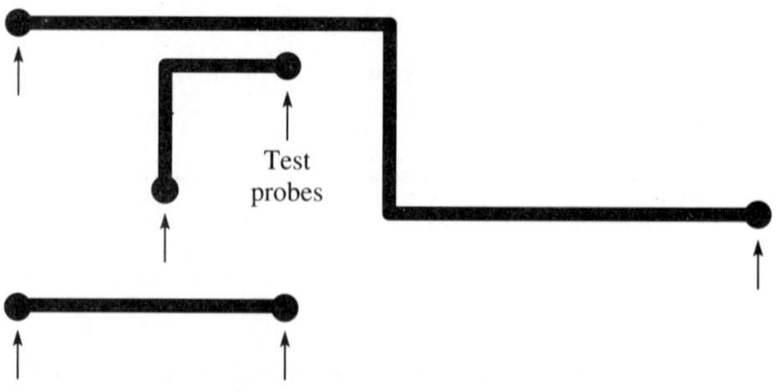

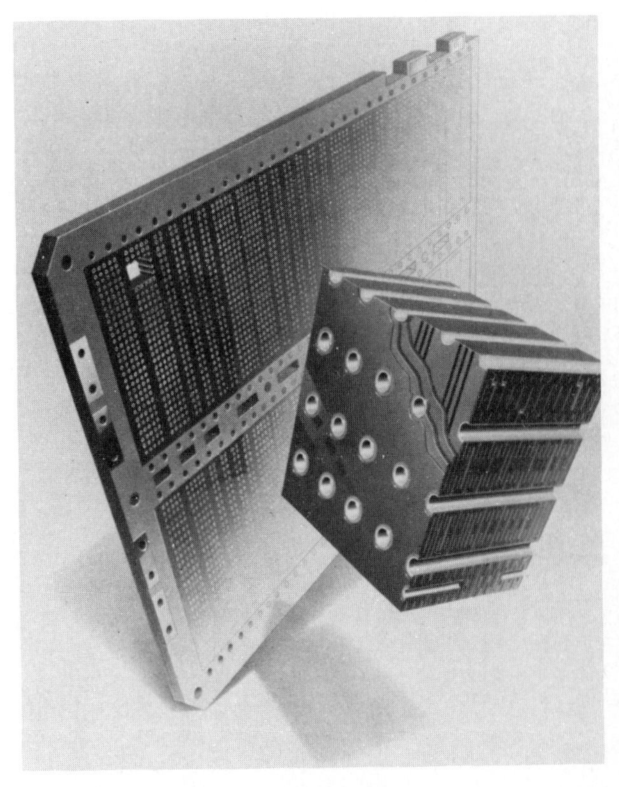

FIGURE 15–5
Multilayer circuit board (courtesy of Teradyne Circuits Operation).

for example, a power plane may shift in position when the circuit board is laminated, causing shorts. Another example is a trace within buried layers that may be open or shorted to other traces. These types of errors may not be correctable, and the failed circuit boards will have to be scrapped.

Bare-board testing is used to find these types of faults *before* the board is populated with parts so that the parts and labor associated with assembling the board are not wasted if an unrepairable fault exists.

CABLE SCANNING

Cable scanning tests continuity and shorts in a manner similar to the bare-board tester, but it is used for testing *cables* and *harnesses*. Cable scan testers are quite inexpensive and have a quick return on investment. A cable is a bundle of interconnecting wires with a connector at each end. A harness is a bundle of interconnecting wires that have more than two connectors (see Figure 15–6(a) and (b)). Both ends of the cable (or all ends of the harness) are connected to the tester, as shown in Figure 15–7. Programming is done by learning a known good assembly. The production run of cables or harnesses is tested by comparing the continuities with the expected continuities. Figure 15–8 shows a type of cable scan tester.

The basic principle behind cable scanners is as follows: A dc signal (used as a stimulus) is applied to pin 1 at one of the connectors. All the rest of the pins at *every connector* are scanned, one at a time, and checked for continuity. Then the

FIGURE 15–6

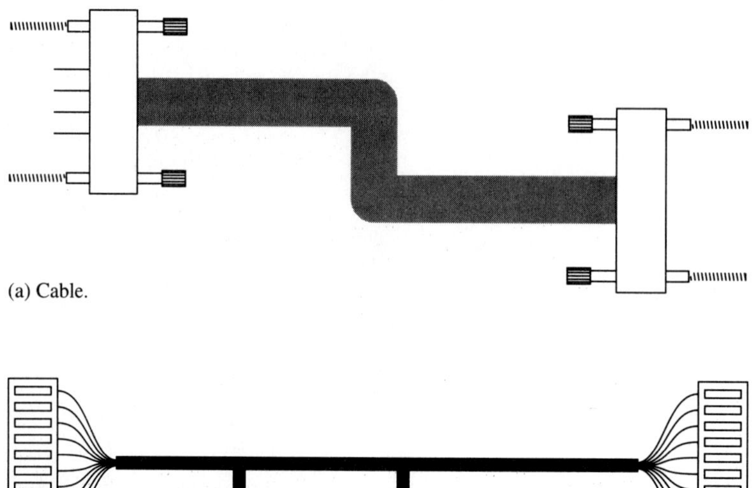

(a) Cable.

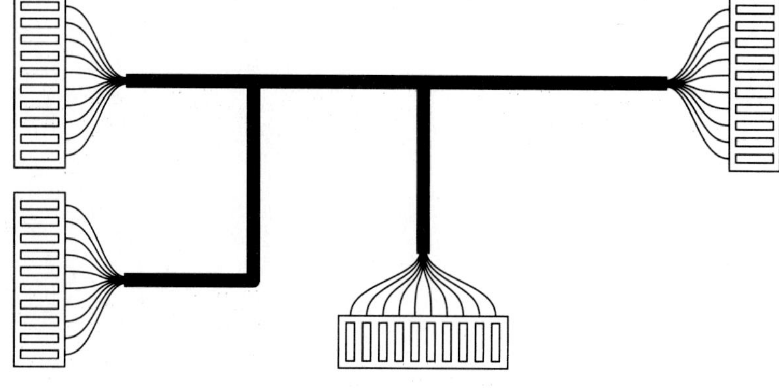

(b) Harness.

FIGURE 15–7
Cable scanning.

Drive here

Scan for continuity all pins on this side and then all pins on the other end of the cable.

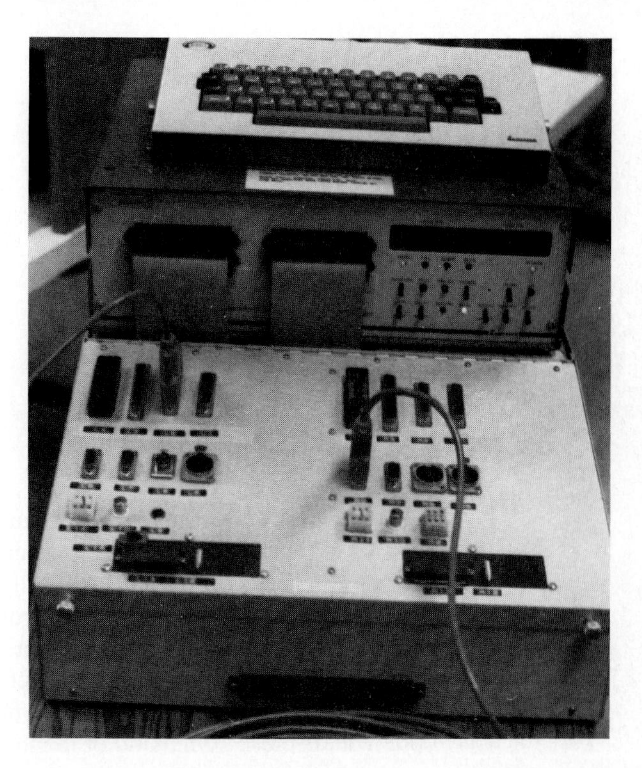

FIGURE 15–8
Equipment for testing cables.

tester moves the dc stimulus to pin 2. All pins (pin 3 and higher) are scanned again. There is no need to check pin 1 again because the continuity there was tested in the previous test.[1] This process is repeated until the entire assembly is checked. An error report is generated so repairs can be made.

MANUFACTURING DEFECTS ANALYZER

A **manufacturing defects analyzer** (MDA) tests assembled circuit-board assemblies.[2] It uses a bed-of-nails fixture, with a pin placed at every wire node. A **node** is a junction where components are connected to a common point (see Figure 15–9). The tester measures impedances at the various nodes and compares the measured impedances to the expected values (learned from a good board). One node at a time is connected (via a switching matrix) to a receiver of ac signals. The pins driving all the other nodes are connected to a constant-level ac signal output. Power to the board is turned *off*. Leakage currents are measured at the receiving-node pin through the components as they are driven from source signals (Figure 15–10).

---

[1] As you move "forward" (moving from pin 1 to pin 2 and so on) in the test, there is no reason to test backward because those tests are redundant. The test cycle becomes quicker as you progress through the test because there are fewer pins left that have not been tested for continuity and shorts.

[2] For the purpose of distinction, a circuit board is a board that is not populated with components, whereas a circuit-board *assembly* is a circuit board that is assembled with components.

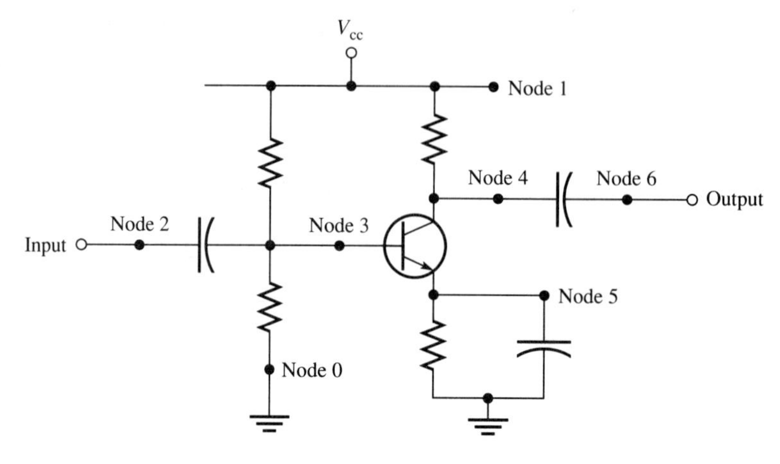

FIGURE 15–9
Locations of nodes.

Compared to other types of automated testers, MDAs are relatively inexpensive ($25,000 to $60,000). Programming is easy, since you learn from a good board. The *actual* defect is not disclosed from this kind of test. Any component around the node that is defective, wrong in value, missing, or oriented incorrectly can affect the impedance of the measured node. It very quickly analyzes which nodes have impedance mismatches compared to the expected values of the good board. Most of the time, a visual inspection of the components around the faulty node will disclose some kind of manufacturing defect. If there are no manufactur-

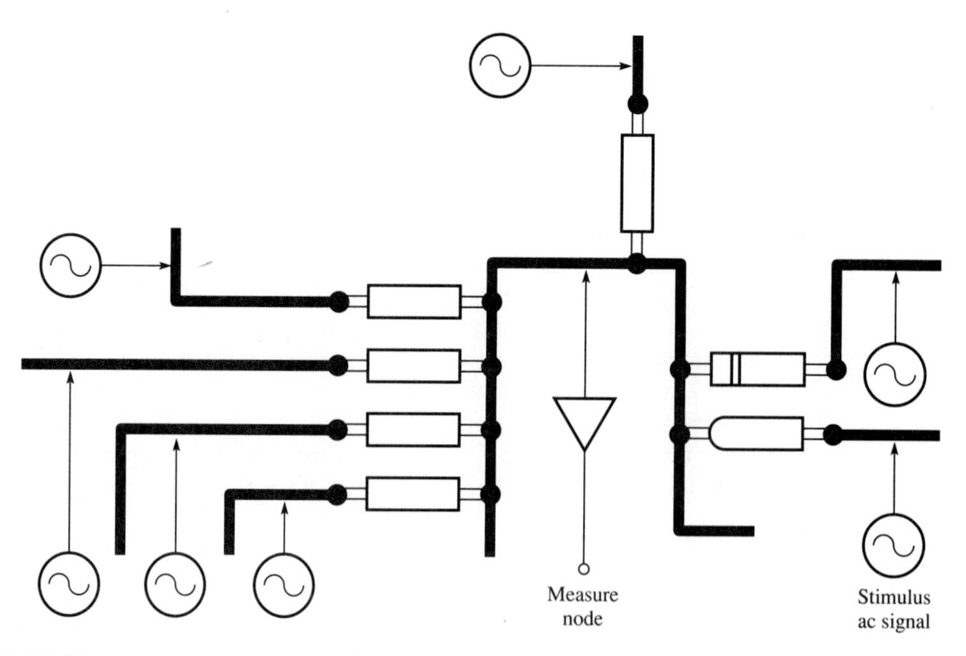

FIGURE 15–10
Manufacturing defects analyzer (MDA).

ing defects in the assembly but a component is defective instead, the indication of which nodes are bad will lead a skilled technician to the suspected part.

The MDA works best with analog and passive devices. Digital logic circuits often have buffered (high-impedance) inputs and outputs that can be switched to high impedance (tri-state). Because of the high-impedance states with digital circuits, errors in assembly are not caught nearly as easily as they are with the analog and passive circuits.

COMPONENT TESTERS

A **component tester** tests individual components for a set of desired parameters. The components are often checked for acceptance to a performance standard (go/no-go testing) or over a range of varying conditions to determine where and how the components may fail to meet some specification **(characterization).** For example, an integrated circuit can be tested with a special load, as shown in Figure 15–11. If the IC is to be accepted, it must pass certain minimum rise times published in the specifications. This is testing for acceptance to a particular performance standard. If, however, the value of the loading capacitor (C_l) were increased in value until the device no longer met the minimum rise times, then you would be characterizing the part to determine the absolute maximum value C_l can have as a capacitive load to the device.

Component testers range in price from less than $100 to more than $1 million. A very simple form of component test is illustrated in Figure 15–12. This component tester is used to select various diodes for V_f (the forward voltage drop). In this example, a relatively constant current source drives current through a diode and the diode voltage drop is measured with a conventional voltmeter. This tester can be built for less than $100. If 30,000 diodes per day were to be selected and categorized into bins of ±5 mV ranges, then some kind of automation would be needed. Figure 15–13 shows an example of a diode sorter. It consists of a computer-controlled material handler with measurement contacts, a computer to control the selection activities, a meter that is capable of reporting measurement results via an IEEE-488 bus, and a set of selection bins. A system like this might cost $10,000. If the diodes were to be selected and tested at the rate of 30,000 per *hour* instead of per *day,* then the cost would increase because the

FIGURE 15–11
Device characterization.

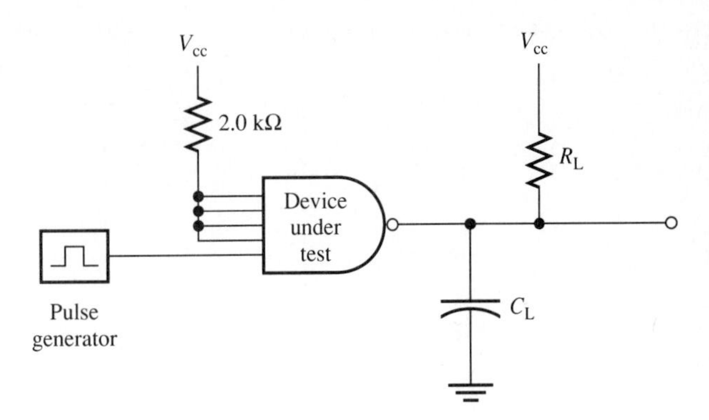

FIGURE 15–12
Method of testing for V_f.

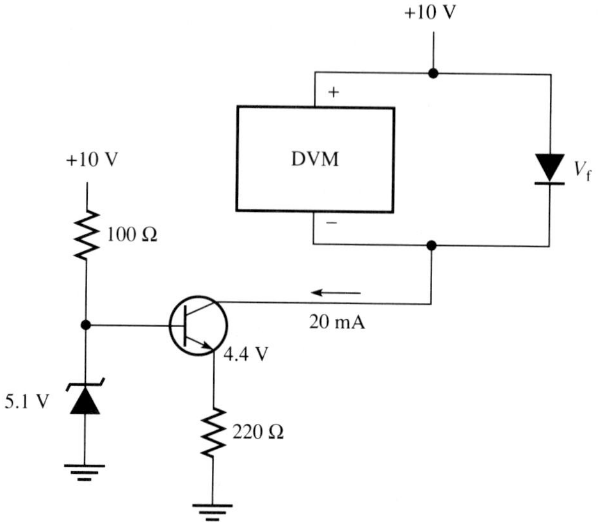

+10 V
+
DVM
−
V_f
+10 V
100 Ω
20 mA
4.4 V
5.1 V
220 Ω

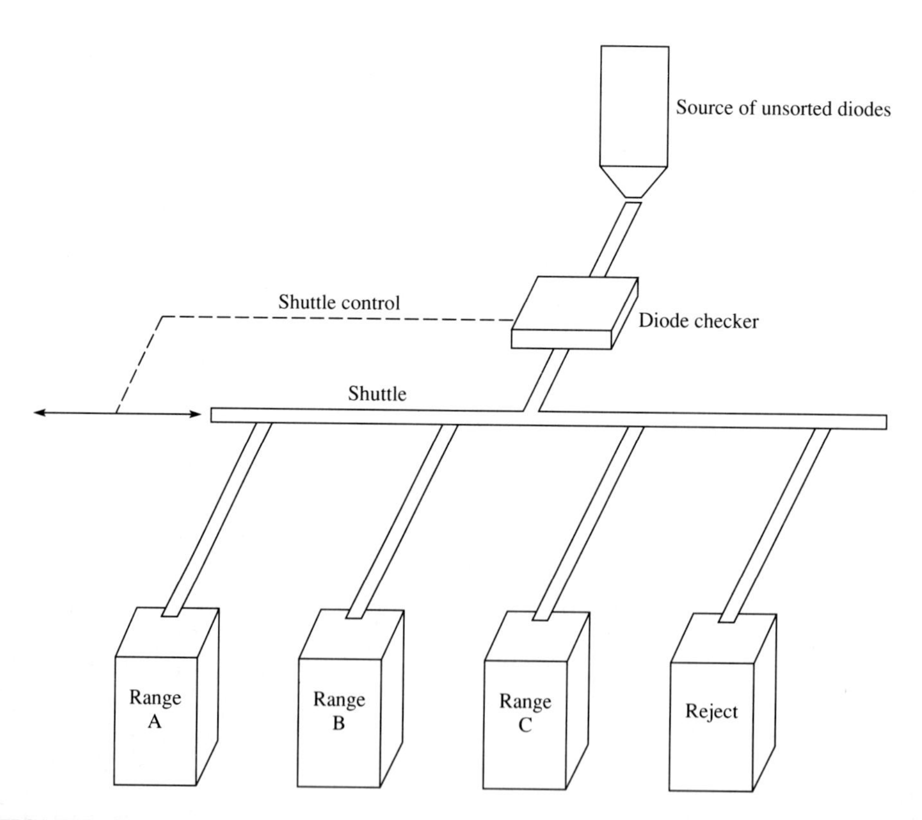

FIGURE 15–13
Diode sorter.

material handler would need to act faster and the settling times of the current source and the meter would need to be better. All these factors add substantially to the cost.

Figure 15–14 shows a Data I/O programmer that is used to program PROMs (programmable read-only memories), PALs (programmable array logic devices), gate arrays, and other electronically programmable hardware/firmware devices. After a device has been programmed, the tester will automatically check the device to see that it was programmed correctly and that it did not fail during the programming cycle (most devices of this type are programmed by supplying a "super voltage" to a programming pin, which can cause a weak device to fail). The programmer can be attached to a material handler, as shown in Figure 15–15.

Figure 15–16 shows a tester that is used as a development tool to completely characterize all the parameters of **LSI** (large-scale integration) chips. LSI chips are often custom chips used to reduce large circuit-board sections down to a single chip. An example of LSI technology is the clock chip used in digital chronographic watches (watches with multiple functions, such as stop and alarm modes). A single chip handles all the functions, the display drivers, switch decoding and debouncing, musical tones, and all the other features of the watch.

After a new IC has been developed, a production run of several pieces is made. The chip's parameters are characterized by gathering and reducing data. Examples are maximum and minimum supply voltage, maximum input voltage that the IC will recognize as a logic low level, the minimum voltage that the IC will recognize as a logic high level, and so on. If done manually, the tests can take a considerable amount of time for each part, with inconsistent results from tired operators. The test system varies the parameters, takes the measurements, and generates the reports, all automatically. This saves time and gives accurate information to the designers and process engineers.

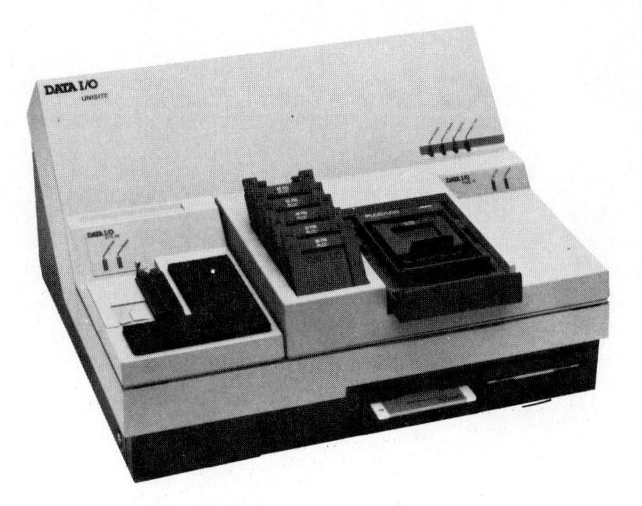

FIGURE 15–14
Data I/O programmer (courtesy of Data I/O Corporation).

FIGURE 15–15
Data I/O Unisite programmer used with the Autolaser 7000 to program, test, mark, and sort programmable ICs (courtesy of Data I/O Corporation and Quality Automation).

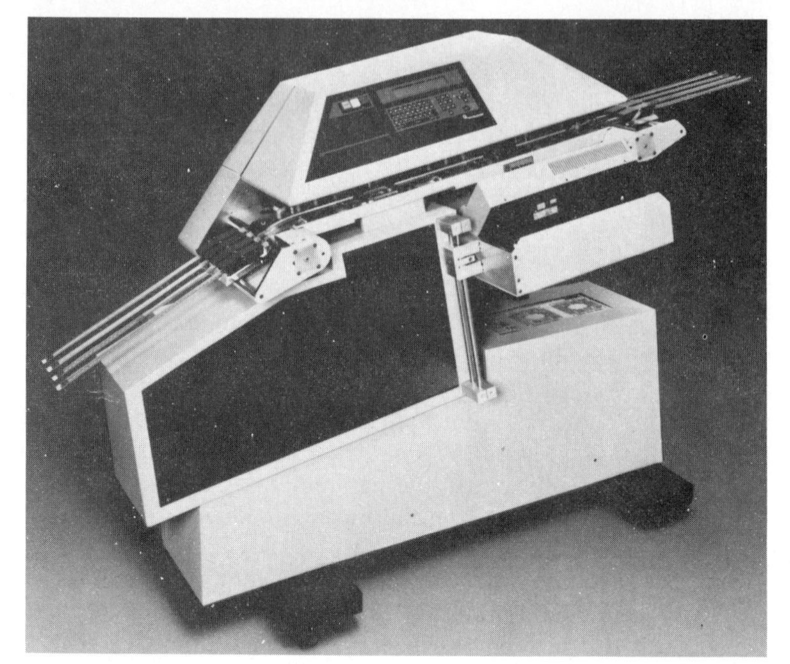

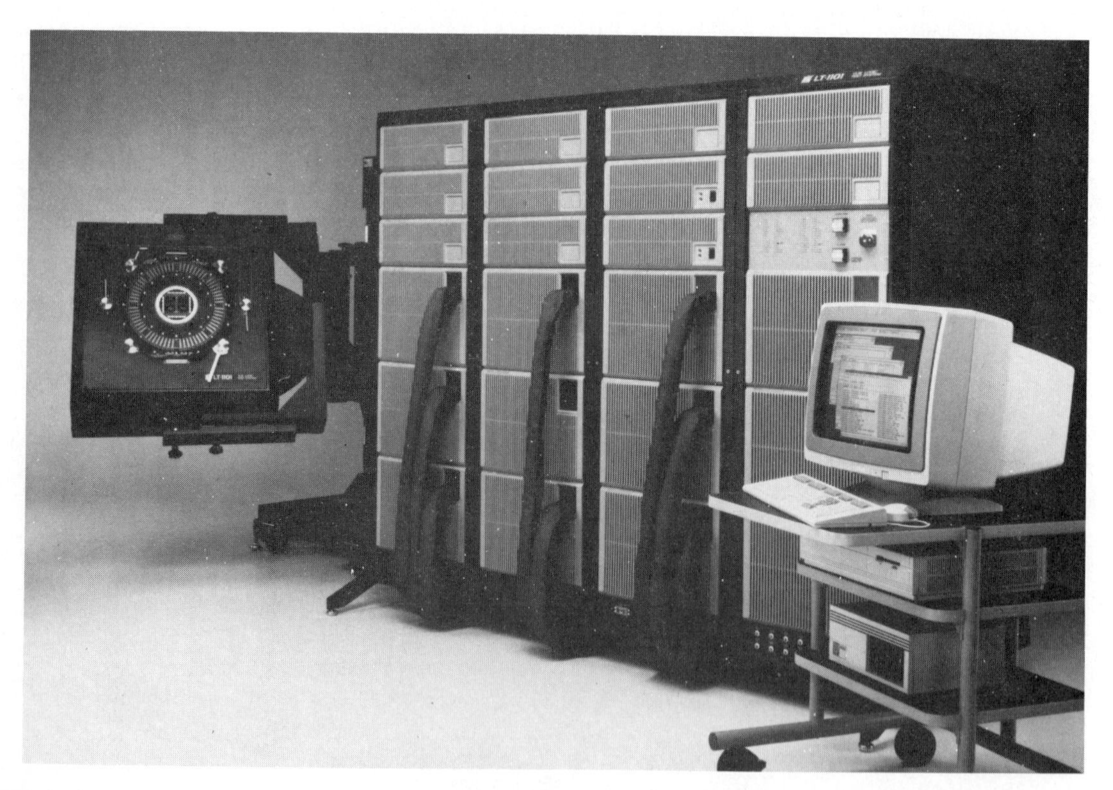

FIGURE 15–16
Model LT-1101 VLSI tester (courtesy of Credence Systems Corporation).

IN-CIRCUIT COMPONENT TESTERS

In-circuit component testers range from inexpensive hand-held testers to large systems costing anywhere from $100,000 to about $800,000. One example of a hand-held in-circuit tester is the Fluke Model 77 DVM (see Figure 7–19), which has a "diode-check" position. With the selector switch in this position, there is enough voltage at the probe tips to overcome the diode junction in most diodes and transistors. If the red lead is connected to the anode and the black lead to the cathode, the diode should conduct. The meter will read the forward voltage drop of the diode. With the leads reversed, the meter should read a substantially higher value and may even read over range. The same meter can be switched to the OHMS (Ω) position. In this position, the voltage at the probe tips is not sufficient enough to bias on transistor junctions or diodes. Resistors can be measured for their values while in the circuit, but the user must be aware of parallel paths that will cause the meter to indicate a lower-than-actual value. But this approach can give some insight to problems, especially if a good circuit-board assembly is compared, part by part, to a defective board.

Another form of hand-held in-circuit component tester is the capacitance meter. These meters are useful in measuring large capacitors ($>0.1\ \mu F$). There are too many sources of stray capacitances to measure with much accuracy below this value.

Generally, hand-held in-circuit component testers have three basic problems: (1) their measurements are not highly accurate, (2) they can't completely test active devices, and (3) they require a skilled technician to interpret the results.

A popular form of ATE system developed by many manufacturers will measure components, in circuit, with good accuracy. The basic philosophy behind these testers is as follows: If you can assure that all the components in an assembly are working, the components are installed where they should be, and there are no opens or shorts to the wiring connecting them, then the assembly should work (assuming that it is designed correctly to begin with). These systems will test all the traces for proper connections and all the components for proper values and functions in a matter of seconds for smaller circuit boards and in less than 5 min for a board containing about 400 ICs. These systems cost over $100,000. The Zehntel Model 1800, shown in Figure 15–1, is an example of one of these systems. It supports a bed-of-nails-type of fixture. Power to a circuit-board assembly is turned off while it tests for shorts, continuity, and correct values of passive components. Power to the circuit-board assembly is turned on while active components are tested (analog and digital). When testing passive devices, a technique called **guarding** is used to greatly improve the accuracy of the measurements. Guarding is a method used by automated test systems that prevents currents from parallel paths from affecting the measurement and is described in Section 15–4.

Active and linear devices are tested with the power applied to the circuit board assembly. The devices are driven with either a voltage source or a current source, and the output is measured as a voltage or current (see Section 15–3). Digital devices are driven with some kind of digital stimulus, and the output is compared to a "table" containing expected results (see Section 15–5).

In-circuit component testers are essential for large manufacturing production runs. A hypothetical example will illustrate the idea. Suppose a company produces 30,000 units of a particular product per week. Of these, about 30% of the circuit boards fail to work the first time. Assume that it takes an average of 20 min to repair each circuit board. This means 9000 boards need to be repaired, and at 20 min each, this equals 3000 hours of rework time per week. If a technician works 40 productive hours per week, then 75 technicians are needed to keep up with the repair load. Assume that 97% of the failures could have been found using ATE automation, leaving 3% which would require the skills of a trained technician to repair. Each one of these boards can be tested in less than 2 min, including the loading and unloading (setup) times of the fixture. Twelve testers, running 16 hours per day, could keep up with the load. The number of defective boards that require technicians to work on them drops from 9000 to 900. This means that instead of hiring 75 technicians, only 8 are required, plus 24 operators and 3 programmers. The savings due to automated testing could be as much as $5 million; Table 15–1 summarizes major cost factors for this example.

FUNCTIONAL TESTERS

A functional tester is a device that will test the various parameters of the UUT (unit under test) to assure conformance to specification and to assure that all the *modes* of operation work as they are designed. A **mode** is the operating condition to which the UUT has been set; for example: one mode for an answering machine may be to pick up a call after one ring, another mode may be to pick up after four rings. The UUT may be anything from a single IC to a complete circuit-board assembly. Functional testers can become quite involved and costly.

The skill level needed to program ATE functional testers is quite high compared with the other forms of ATE. The programmer needs to know all the modes and operational specifications of the UUT.

Some benefits of functional ATE are these:

1. The tests are very quick.
2. The test parameters and methods are consistent.

TABLE 15–1

COSTS WITHOUT THE MACHINES		
75 technicians at $50/h overhead labor rate[a]	=	$7,800,000/y
COSTS WITH THE MACHINES		
12 machines at $30,000/y (depreciated value each)	=	$360,000/y
8 technicians at $50.00/h overhead labor rate[a]	=	$832,000/y
24 operators (12 machines on two work shifts)	=	$998,400/y
3 programmers at $40.00/h	=	$249,600/y
Total		$2,440,000/y

[a] Overhead labor rate is the base pay of the individual plus benefits and other costs associated with the employee (electricity, heating, the building, etc). The actual pay the employee receives may only be a quarter of the overhead labor rate.

3. Full functional testing of all units can be accomplished. (This is necessary to satisfy some military and other contractors.)
4. Parameters can be characterized.

The disadvantages of functional ATEs include the following:

1. Initial cost is high.
2. Much higher level of programming skill is required.
3. Design of the test fixtures can be very complicated compared with other forms of ATE testers. This is primarily due to the fact that functional testing is done at the highest operating speed of the UUT. The capacitive loading of the fixture, as well as cross-talk interference, can greatly influence the performance of the tester and cause errors in the measurements.

COMBINATION TESTERS

A combination tester is capable of doing in-circuit component testing and functional testing on the same machine. This form of tester is the most expensive form of the types discussed; however, such testers provide an excellent comprehensive test of the UUT. Depending on the completeness and accuracy of the tests on the UUT, yields from the tester can approach 100%. The **yield** is the percentage of *good* assemblies out of the *total* number of assemblies tested. As an example, assume 300 circuit board assemblies were tested. The defective assemblies were reworked and tested again. All assemblies tested good, but 2 boards had hidden defects. The *yield* from this batch of circuit board assemblies is then calculated to be 298 ÷ 300, or 99.3%.

15–3 HIERARCHY OF AN IN-CIRCUIT TEST

In-circuit component testers used in manufacturing plants need to test for certain parameters of an assembly in a given sequence, called a hierarchy. A **hierarchy** is the order in which tests are conducted, as shown in Figure 15–17.

A hierarchy in testing is needed for three main reasons: (1) to prevent damage to the tester and/or the UUT, (2) to prevent failures from being "masked" by other errors, and (3) to prevent false errors. An error that is **masked** is one that is not visible to the tester as a result of another failure. A **false error** is one in which a component is reported to be defective, but in fact, something else is defective. For example, suppose that a desk lamp will not turn on. The error could be that the lamp is not plugged in, the light bulb is defective, or both. If the light bulb is defective, the error is *masked* by the lamp not being plugged in; until it is plugged in, the lamp can't be tested. The other error (*false error*) is generated by testing the lamp first and then the power. The light bulb is reported as defective because it won't light, but it is really all right; then the power is reported as defective because the lamp was found to be unplugged. In this example, two problems were reported because the proper hierarchy was not used. The proper hierarchy is to assure that power is connected and then test to see that the lamp lights; in this example, false and masked errors are reduced by half just by using the proper sequence of test.

FIGURE 15–17
Hierarchy of in-circuit testing.

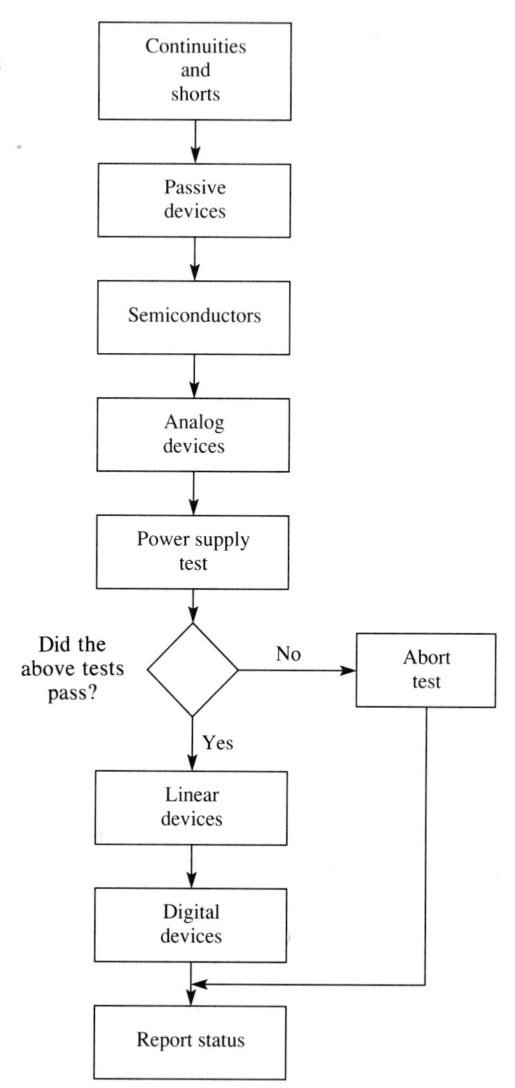

The suggested hierarchy of in-circuit testing is as follows:

1. *Continuities and shorts*. This test assures that all the circuit connections are intact and not shorted together. Power to the assembly will be off while the tests are being conducted. The test should not continue on to the next-higher hierarchy until all errors have been corrected.
2. *Passive components*. This test checks all the passive components to assure they are the correct values. Power to the assembly will be off while the tests are being conducted. The test should not continue to the next-higher hierarchy until all the errors have been corrected.

3. *Semiconductor devices.* This test checks devices such as transistors, diodes, and zener diodes. It may be as simple as testing the diode junction on the parts. Power to the assembly will be off while the tests are being conducted. The test should not continue on to the next-higher hierarchy until all the errors have been corrected.
4. *Analog devices.* This test is similar to the semiconductor tests, but there are more parameters to check. For example, you may choose to test an FET (field-effect transistor) by supplying a fixed voltage to the gate (stimulus) and measuring the current from source to the drain (measurement). This would be a "stimulate by voltage and measure the current" (STEM V/MEAS I) type of test. The four combinations involve either stimulating using voltage or current and measuring the output voltage or current. A BJT (bipolar junction transistor) can have its gain (β) checked by using STEM I/MEAS I. Power will be off to the assembly while voltages or currents are used to test the device. The test should not continue on to the next-higher hierarchy until the errors have been corrected.
5. *Power tests.* This test is intended to protect the UUT. It is a test to assure that any and all adjustable (or programmable) power supplies are set to within specified limits before the supply is connected (via relays) to the UUT. Power to the assembly will be off while the tests are being conducted. The test should not continue on to the next-higher hierarchy until all the errors have been corrected.
6. *Linear devices.* This is the first test in which the power is supplied to the UUT. Linear devices are those devices classified as having an output that is proportional to its input. The output may be either a voltage or current, and the input may be either a voltage or current. In addition to inputs and outputs, there will be connections for power and reference ground. Power will be on to the assembly while the tests are being conducted. The test may continue to the next-higher hierarchy even though errors have been detected.
7. *Digital devices.* Digital devices are those devices classified as having a binary state (on or off) as either an input, an output, or both. For the purpose of automated test, D/A and A/D converters are classified as digital devices. Power to the assembly will be on while the tests are being conducted.

15–4 TWO-, THREE-, AND SIX-WIRE MEASUREMENT METHODS

To assure that the correct components are installed in circuit assemblies, the in-circuit tester will attempt to measure the value of each component. When components are measured with connections made to other components (in-circuit), errors can result due to the existence of parallel paths. In addition, resistive losses in the test fixture can introduce errors. The two-wire method of measurement is easiest to use but is sensitive to these type of errors; three- and six-wire measurements add wires to the test fixture to reduce measurement errors.

THE TWO-WIRE METHOD

Consider the resistance measurement of a single isolated resistor ($R_{unknown}$) as shown in Figure 15–18. This resistor will be tested using a fixed (and known) voltage V_S, a reference resistor (R_{ref}) used to establish a reference current (I_{ref}), and an op-amp. One end of the resistor will be connected to a known voltage source (the stimulus node) and the other end will be connected to the op-amp (measurement node). Assuming the op-amp is ideal

1. The input current into the negative node of the op-amp is negligible; therefore, all the current going through $R_{unknown}$ is equal to the current through R_{ref}.
2. There is no offset voltage between the + terminal and the − terminal; therefore, the inverting terminal is at ground potential (virtual ground).

EXAMPLE 15–1 Using Figure 15–18, find the value of the unknown resistance, $R_{unknown}$, if

$$V_S = 200 \text{ mV} \quad \text{(driving voltage, known)}$$
$$R_{ref} = 10 \text{ k}\Omega \quad \text{(reference resistor, known)}$$
$$V_o = 513 \text{ mV} \quad \text{(resultant voltage, measured)}$$

SOLUTION

$$I_{ref} = \frac{V_o}{R_{ref}} = \frac{513 \text{ mV}}{10 \text{ k}\Omega} = 51.3 \ \mu\text{A}$$

$$R_{unknown} = \frac{V_S}{I_{ref}} = \frac{200 \text{ mV}}{51.3 \ \mu\text{A}} = 3.9 \text{ k}\Omega$$

THE THREE-WIRE METHOD

The two-wire method works only when there are no other parallel components that cause current to be shunted past the device under test (parallel paths). Parallel paths are almost always present on a circuit-board assembly populated with components. Current from these paths will give false readings and must be blocked from entering the measurement node. This is done by adding an additional wire or wires to *guard* the measurement node. Figure 15–19 illustrates the idea.

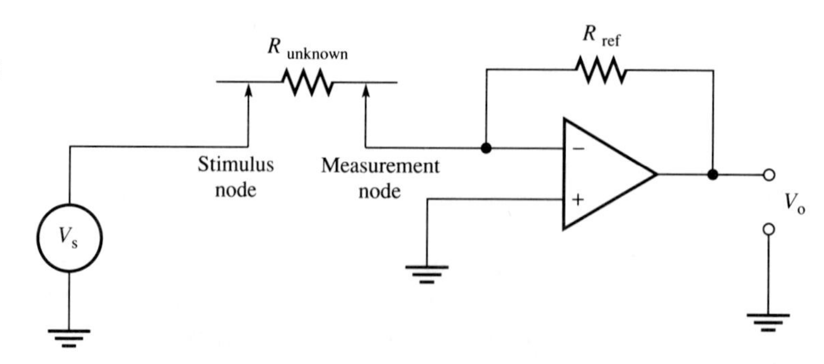

FIGURE 15–18 Two-wire measurement method.

FIGURE 15–19
Three-wire measurement method.

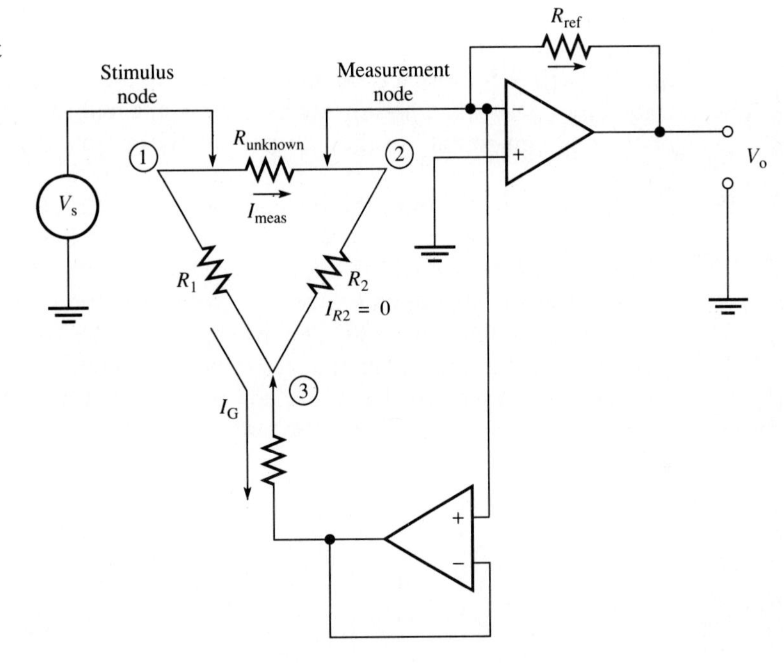

EXAMPLE 15–2 Errors in measuring a resistor can be introduced by parallel paths that add current to the measure node. In Figure 15–19, assume R_1 = 2.2 kΩ, R_2 = 1.0 kΩ and $R_{unknown}$ = 3.9 kΩ. What is the measured resistance of $R_{unknown}$? How will guarding node 2 help?

SOLUTION The measured resistance is

$$\frac{(3.9\text{ k}\Omega)(2.2\text{ k}\Omega + 1\text{ k}\Omega)}{3.9\text{ k}\Omega + 2.2\text{ k}\Omega + 1\text{ k}\Omega} = 1.76\text{ k}\Omega \qquad \text{(instead of 3.9 k}\Omega\text{)}$$

The lower resistance is calculated because total current, entering node 3, equals the current through R_1 and R_2 plus the current through R_3; the higher-than-expected current causes a lower-than-expected resistance measurement. To keep the current from R_2 from entering the measurement node (3), a guard pin is added to node 2. This is how it works: The voltage at node 3 is 0 V because of the virtual ground of the op-amp. If node 2 is forced to 0.0 V (via the guard pin), then the voltage across R_2 is zero; 0 V across a resistor produce zero current; therefore, all the current entering into node 3 comes from R_3, so R_3's value can be accurately calculated.

THE SIX-WIRE METHOD

There are situations where there is enough current flowing in the wires connecting to the stimulus node, measurement node, and/or the guard node that the resistive losses in the interconnecting wires become significant. To overcome this problem, a sense wire is attached to each node. This wire senses the voltage at the node and compensates for the losses encountered in the wiring. There are three sense wires, one for each of the wires used in the three-wire measurement method, making the total wire count six. The sense wires are connected to high-impedance junctions of the test equipment, so they do not supply or sink any appreciable amount of current.

Figure 15–20 shows the concept of the system. The wire marked *stimulus sense* measures the voltage at the sense node. The voltage is fed back to a comparator. The output voltage of the comparator is increased to compensate for the voltage drop in the stimulus wire. The voltage at the guard node is measured by the wire marked *guard sense*. This wire compensates for the voltage losses in the guard wire in the same manner as the stimulus circuits. There are variations to this scheme that simplify the hardware and/or software requirements; they are dependent on the specific design by the manufacturer.

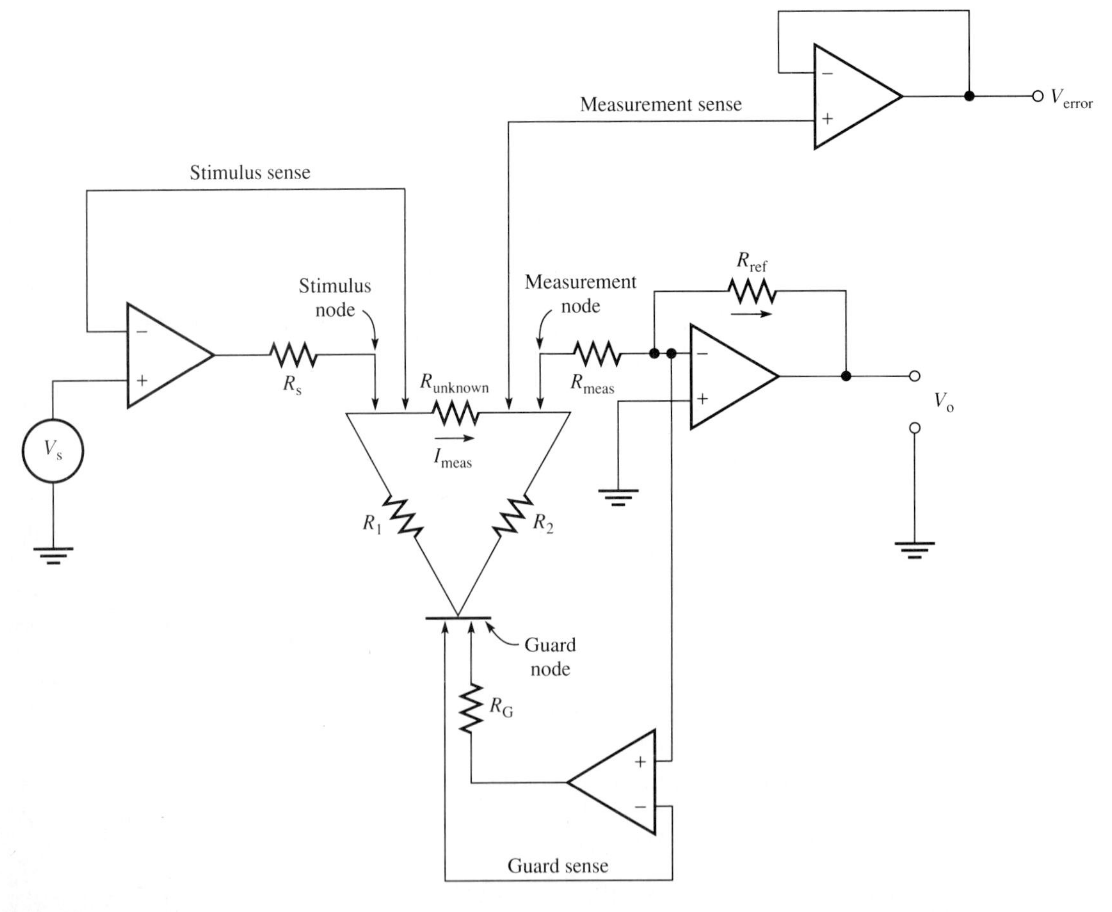

FIGURE 15–20
Six-wire measurement method.

If a low-impedance component is to be measured, then there will be losses in the measure wire as well. Compensating for these losses is not as straightforward as in the stimulus and guard circuits, which are drivers. One method is to measure the voltage at the measure end of $R_{unknown}$ and assume that the input of the op-amp is 0 V (virtual ground). This voltage difference, along with the measured current, will yield a calculated value for R_{meas}, which can then be used in the determination of $R_{unknown}$.

15–5 DIGITAL TEST METHODS

Testing of digital devices requires that power be applied to the UUT. The test should verify that the device functions as intended and that all the connections are made to the device (no *folded* pins). A **folded pin** is a pin that has missed being inserted into a through-hole (see Figure 15–21). If more exhaustive tests of a digital device are to be conducted, the device should be tested before it is inserted into the finished assembly. Many of the digital devices can be tested using predefined test programs that are part of a *library* in the tester. A **library** is a set of files; each file contains the test patterns (stimulus) and expected output (result) for that device.

TESTING LOGIC

Combinatorial logic devices can be specified completely by a truth table (AND gates, OR gates, and other similar devices). One way to test combinatorial logic devices is through the use of *test vectors*. **Test vectors** are the conditions that are placed on the input and control pins in order to stimulate the device under test to known states. Often, test vectors are in tables. The **test vector table** contains all the desired states needed to complete a digital test on the device. An example of test vectors and a test vector table for a three-input NAND gate (shown in Figure 15–22) is given in Tables 15–2 (truth table) and 15–3 (vector table). You should notice the similarities between the truth table and the test vector table. Often the test vector table will contain all the defining states contained in the truth table in order to test all the modes and logic conditions of the device.

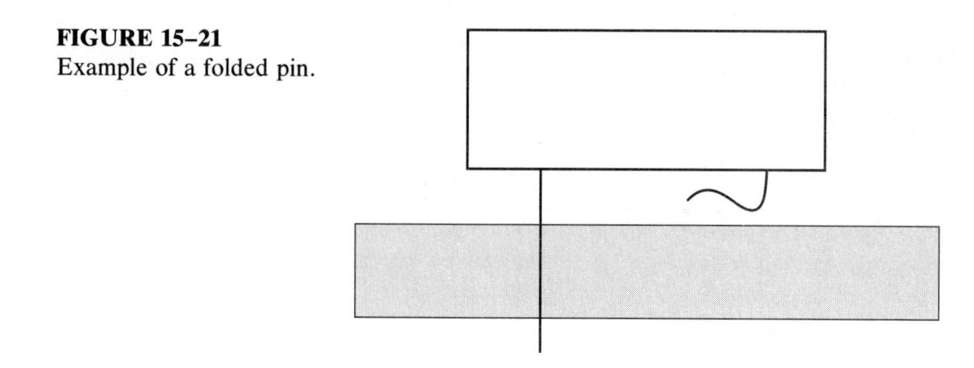

FIGURE 15–21
Example of a folded pin.

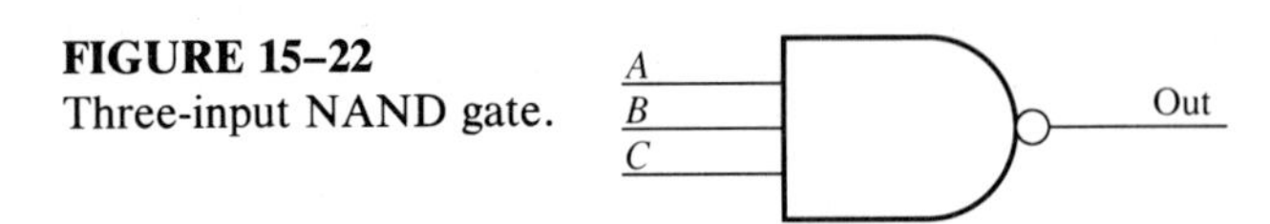

FIGURE 15–22
Three-input NAND gate.

The vectors are executed in order from the top of the list down. Each line in the table is a single test vector. Table 15–4 shows an example of a test vector table that failed at vectors 4 and 6. This is because pins A and B are shorted together producing a logic 1 on the inputs when either input is 1.

Another method of testing logic is to use a **gray-code generator.** Gray code is a binary sequence where each new state causes only 1 bit at a time to change. An example that compares traditional binary counting to gray-code counting is shown in Table 15–5(a) and (b). In Table 15–5(a), each row that has more than one bit changing is marked with an asterisk. You should observe in Table 15–5(b) that only 1 bit changes as you progress from row to row.

The gray code was invented for early mechanical machine-control feedback systems that employed cams and switches to sense the mechanical position of the machines parts or work pieces. If more than one switch is to change state at the same mechanical location, there could be errors. These errors are called glitches. An example of a glitch is as follows. Suppose the logic state 01 was incremented to 10 (traditional binary count from 1 to 2). Two glitch conditions could exist: (1) The lower bit could be cleared before the higher bit was set, giving a sequence of 01, 00, 10. In this case, 00 would be the glitch (as in Table 15–6). (2) The higher bit

TABLE 15–2
Truth table.

A	*B*	*C*	OUTPUT
0	0	0	1
0	0	1	1
0	1	0	1
0	1	1	1
1	0	0	1
1	0	1	1
1	1	0	1
1	1	1	0

TABLE 15–3
Vector test table.

VECTOR TEST	*A*	*B*	*C*	EXPECTED OUTPUT
1	0	0	0	1
2	0	0	1	1
3	0	1	0	1
4	0	1	1	1
5	1	0	0	1
6	1	0	1	1
7	1	1	0	1
8	1	1	1	0

TABLE 15–4
Failed vector table.

VECTOR TEST	*A*	*B*	*C*	ACTUAL OUTPUT
1	0	0	0	1
2	0	0	1	1
3	0	1	0	1
4	0	1	1	0*
5	1	0	0	1
6	1	0	1	0*
7	1	1	0	1
8	1	1	1	0

* = failed vectors.

TABLE 15–5
(a) Binary code. (b) Gray code.

A	*B*	*C*	*D*
0	0	0	0
0	0	0	1
0	0	1	0†
0	0	1	1
0	1	0	0†
0	1	0	1
0	1	1	0†
0	1	1	1
1	0	0	0†
1	0	0	1
1	0	1	0†
1	0	1	1
1	1	0	0†
1	1	0	1
1	1	1	0†
1	1	1	1

† = states where more than one bit changed at a time.

(a)

A	*B*	*C*	*D*
0	0	0	0
0	0	0	1
0	0	1	1
0	0	1	0
0	1	1	0
0	1	1	1
0	1	0	1
0	1	0	0
1	1	0	0
1	1	0	1
1	1	1	1
1	1	1	0
1	0	1	0
1	0	1	1
1	0	0	1
1	0	0	0

(b)

TABLE 15–6
(a) Expected. (b) Glitch 0. (c) Glitch 3.

A	*B*	OUTPUT
0	0	0
0	1	1
1	0	2
1	1	3

* = glitch.

(a)

A	*B*	OUTPUT
0	0	0
0	1	1
0	0	0*
1	0	2
1	1	3

(b)

A	*B*	OUTPUT
0	0	0
0	1	1
1	1	3*
1	0	2
1	1	3

(c)

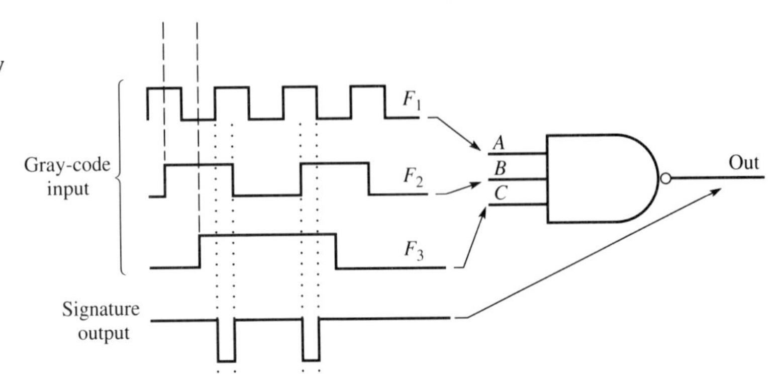

FIGURE 15–23
Signature testing using gray code as an input.

could be set before the lower bit is cleared, giving a sequence of 00, 11, 10. In this case, 11 would be the glitch. Glitches are very short in duration as compared with the normal cycle of the machine code. In modern digital circuits, binary counts may cause glitches in the order of just a few nanoseconds. In the absence of a gray code, glitches are bypassed by reading the results of a test vector after the input states have settled and become valid.

The three-input NAND gate in Figure 15–22 can be tested using a gray-code generator, as shown in Figure 15–23. This is also known as a **frequency test.** Each pin is driven by a frequency, F_1, F_2, or F_3. F_1 is the highest frequency, F_2 is half the frequency of F_1, and F_3 is half of F_2. You should notice that only one input changes state at a time (gray code) in order to avoid glitch errors. A "listen window" is opened just before F_1 goes to a logic high and closes some time after F_3 goes low.

The idea behind the gray-code frequency method is that all states of a digital device will be addressed without having to generate extensive vector tables. The disadvantage of the frequency method is that some devices, acting as state machines, may be put into indeterminate states, giving varying results each time the test is run.

TESTING MEMORY

Testing memory devices (RAMs and ROMs) can be very challenging (especially with memory densities as well as speeds increasing). Testing RAMs requires several passes: loading memory with a pattern, reading back that pattern, loading in a different pattern, reading back that pattern, and so on, with each test checking for a different kind of fault.

If anything can be learned about testing RAM, it is that you can't test everything. For example, a small RAM containing 12 address lines has $2^{12} = 4096$ cells. Each of these cells can either be at a logic state low (0) or a logic state high (1). The total possible combinations for a 4096 RAM is $2^{4096} = 10^{1233}$. If the tester were running at 25 MHz clock speed (assuming each clock cycle would load in a new pattern), then it would take billions of years to test a single chip! This is clearly not possible but there are several tests that can be done that disclose the most common errors. These errors are categorized into the following classes:

1. Failed output drivers (stuck high, stuck low, stuck in high impedance)
2. Storage cell shorts and opens
3. Address decoder failures

Failed Output Drivers

A failed output is rather easy to detect; an output line is either stuck low or high and can't be toggled to the other state, or it remains tri-stated (high impedance) regardless of the state of the enable pin. The enable pin is an input and will cause the output drivers to be connected to the output pins or disconnect them, depending on whether the enable pin is high or low.

Storage Cell Shorts and Opens

One of the easiest test methods for storage cell shorts and opens is to load all the cells with 0s and read these back; then load all cells with 1s and read back again; then load in an alternate pattern, such as 01010101, and test for the same pattern back from memory; and then, lastly, toggle the bits to 10101010, load in, and read back again. These checks test for gross errors of the cells being stuck in a state (high or low) or shorted together. Figure 15–24(a)–(d) shows an example of how these cells are loaded.

Address Decoder Failures

A test called a *walking ones* test is popular for checking to see if the address decoder is capable of storing and retrieving data into the proper locations in memory. A single 1 is marched through all the cell locations and read back from each location. Decoding errors will either misplace the 1 or read back the contents from the wrong cell; in either case, the value returned will be a 0 instead of the expected 1.

The walking ones test is rather slow. If the memory device is bigger than 1 bit wide (8 bits wide, for example), then the process can be speeded up by writing the address location as the stored number into each address. That is, a 1 is written into address location 1, a 2 into address location 2, and so forth, up through the last address.

The three aforementioned tests do well in testing large-scale faults with memory chips, as well as testing for assembly errors (open or shorted data, address, and control lines). Even if these tests pass, other faults may exist, such as the following:

1. *Response-time failures.* The device does not act fast enough for the required demand. An example of response-time failures is when a memory chip fails to give the stored value(s) by the time the processor assumes that the data is "valid" and attempts to use that data.
2. *Disturbance sensitivity.* This is the state of one cell being influenced by another cell. A charge from adjacent cells can be picked up by the cell under test, not in terms of a short but in terms of cross talk. This sensitivity may be great enough that one adjacent cell changing state can cause the test cell to change, or it may happen only if two, three, or four of the adjacent cells all change states at the same time. The disturbance may only be sensitive in one direction—that is, going from low to high but not in the other direction.

00

00000000	00000000	00000000	00000000
00000000	00000000	00000000	00000000
00000000	00000000	00000000	00000000
00000000	00000000	00000000	00000000
00000000	00000000	00000000	00000000
00000000	00000000	00000000	00000000
00000000	00000000	00000000	00000000
00000000	00000000	00000000	00000000
00000000	00000000	00000000	00000000

FF

11111111	11111111	11111111	11111111
11111111	11111111	11111111	11111111
11111111	11111111	11111111	11111111
11111111	11111111	11111111	11111111
11111111	11111111	11111111	11111111
11111111	11111111	11111111	11111111
11111111	11111111	11111111	11111111
11111111	11111111	11111111	11111111
11111111	11111111	11111111	11111111

55

01010101	01010101	01010101	01010101
01010101	01010101	01010101	01010101
01010101	01010101	01010101	01010101
01010101	01010101	01010101	01010101
01010101	01010101	01010101	01010101
01010101	01010101	01010101	01010101
01010101	01010101	01010101	01010101
01010101	01010101	01010101	01010101
01010101	01010101	01010101	01010101

AA

10101010	10101010	10101010	10101010
10101010	10101010	10101010	10101010
10101010	10101010	10101010	10101010
10101010	10101010	10101010	10101010
10101010	10101010	10101010	10101010
10101010	10101010	10101010	10101010
10101010	10101010	10101010	10101010
10101010	10101010	10101010	10101010
10101010	10101010	10101010	10101010

FIGURE 15–24
Testing memory cells for stuck or bad bits.

3. *Refresh problems.* In dynamic RAMs (DRAMs), data is temporarily stored as a charge, which will bleed off in a short time if not refreshed. Cells can lose their charge before the prescribed time, giving an indication of "forgetfulness."

ROMs and PROMs contain permanently stored data. This data is often used by the processor as instructions or as a "look-up table" containing fixed values. There are two common methods of doing quick checks on ROMs. One way is to query each address location for its value and compare that value to those stored in the tester.

A faster method is called the **check-sum** method. In the check-sum method, the data at each address location is added together (accumulated in the tester). The resultant sum is then compared to the expected check sum. If the check sum is incorrect, the location of the error is not pinpointed, but you know that something is wrong. Table 15–7 shows an example of a check-sum calculation.

Adding the values in both parts of the table results in a *sum* used to *check* for bit errors. Note that multiple bit errors may produce a valid check sum. For

TABLE 15–7
Check-sum calculations.

BINARY VALUES IN MEMORY				HEX EQUIVALENT VALUES
A	*B*	*C*	*D*	
0	0	0	0	0
0	0	0	1	1
0	0	1	0	2
0	0	1	1	3
0	1	0	0	4
0	1	0	1	5
0	1	1	0	6
0	1	1	1	7
1	0	0	0	8
1	0	0	1	9
1	0	1	0	A
1	0	1	1	B
1	1	0	0	C
1	1	0	1	D
1	1	1	0	E
1	1	1	1	F
1111000			=	78_{hex}

example, if the data in locations 0 and 1 were swapped, there would be an error (both bit C and D were inverted), but the check sum would still equal 78_{hex}.

CYCLIC REDUNDANCY CHECKS

The **cyclic redundancy check** (CRC) was developed for checking errors in serial data communication systems. The basic idea behind the test is that a block of data is sent serially from the sending device to the receiving device. As the data is sent (bit by bit), it is entered into a special shift register containing exclusive-OR gates. Figure 15–25 shows a CRC standard for arranging the registers and gates. The

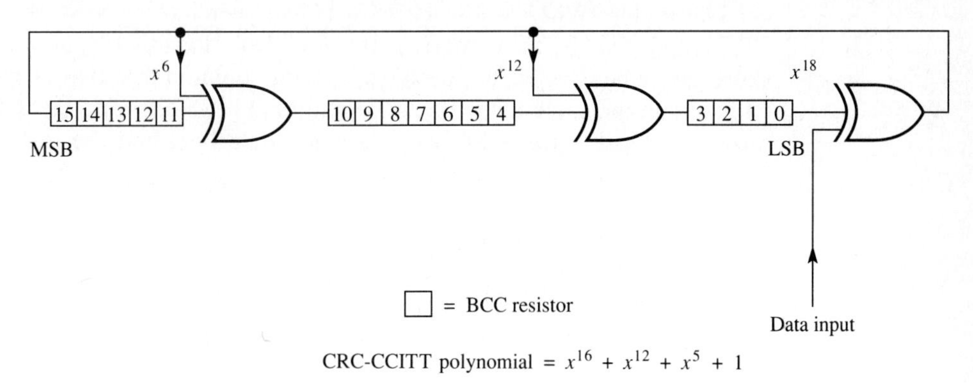

FIGURE 15–25
BCC shift register using CRC-CCITT standard.

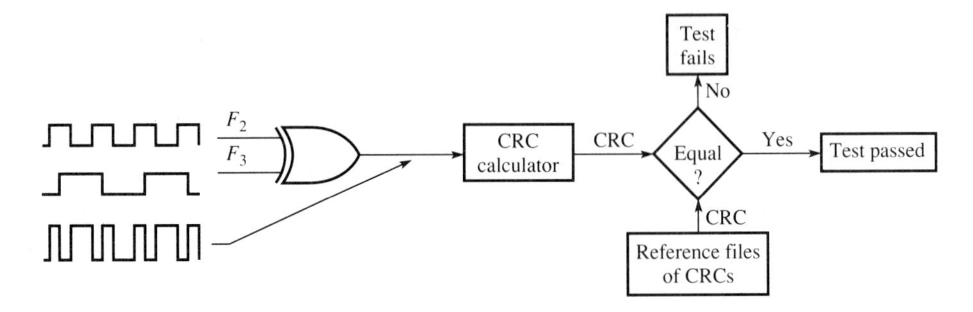

FIGURE 15–26
Gray code and CRC calculator chip used to check an exclusive-OR gate.

registers are called **block check character** (BCC) registers. At the end of each block of data, the contents of the BCC register are first divided by a polynomial; the quotient is ignored but the remainder is used as the CRC value and is sent (serially) to the receiving equipment. It is then cleared, ready to operate on the next block of data. The receiving equipment receives the data (bit by bit) and shifts it into its BCC. After the last block of data is received, the value in the receiver's BBC is divided by the same polynomial used at the sending end; again the quotient is ignored but the remainder is compared to the CRC value sent by the sending end. If the CRCs match, then the data block is accepted and the next block is sent. If they don't match, then the previous block of data is tried again. This sequence repeats until all the desired blocks of data are sent. For more information on CRC, consult one of the many books on data communications.[3]

CRCs can be used by ATE equipment. Figure 15–26 shows an example of an exclusive-OR gate. In this case, the inputs are stimulated by two frequencies: F_2 and F_3. This output is fed into a CRC calculator chip (BBC). The value stored in the CRC chip is read by the ATE computer and compared to a CRC value learned from a "good board." If the CRCs match, then the test passes.

In addition to using check sums to test logic devices, CRCs can be used to check memory devices, such as ROMs. The address lines are incremented through all the addresses (see Figure 15–27). Each data line is connected to a CRC calculator chip. When all the addresses have been interrogated, the value of each CRC chip is read by the computer. It is not likely that any two of the data lines will share the same CRC value, because it is not common that two or more data lines will contain the same data. Each of the outputs exhibits a unique CRC value. Bit error will cause a bad CRC value for that data line, and the test will reject the whole part.

SIGNATURE ANALYSIS

Signature analysis is a method that uses CRCs to check for faults. As in Figure 15–27, each data line is "signing" a unique signature for that particular UUT, but

[3] For example, John E. McNamara, *Technical Aspects of Data Communication,* Bedford, Mass.: Digital Press, 1978.

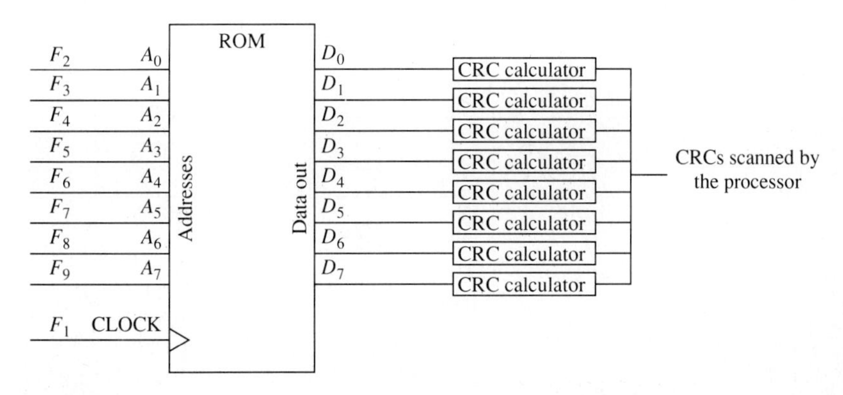

FIGURE 15–27
ROM tested using CRC.

it is easier to represent that signature with a number, which is as unique as the signature itself. This unique number is the CRC value.

In the preceding section, the CRC value was calculated by opening a ''listen window.'' The stimulating test patterns (test vectors) and/or addresses and the listen window are all generated by the same control processor (computer). Therefore, it is rather easy to control when to capture the data and calculate the CRC values. Using an oscilloscope to compare data on a good board to that of a bad board is rather difficult because subtle differences may go unnoticed. Using a signature analyzer can aid in quickly comparing two circuits. The signature analyzer is connected to the clock of the UUT. In addition to the clock, *start-read* and *stop-read* signals must be supplied, which are in synchronous to the clock. The analyzer will capture the signature of the signal and calculate a CRC value. This value can then be compared between working and nonworking boards in an attempt to locate the fault.

BACK DRIVING

The inputs to digital devices must be driven to a high state (1) or low state (0) while the outputs are monitored for response. With a few exceptions, most digital devices have their inputs connected to the outputs of other digital devices. When testing digital devices, bus collision can be a problem. **Bus collision** occurs when two outputs are tied to the same wire (bus), with one output pulling the bus high and the other pulling the bus low. In the case of an ATE tester, this can happen when the tester is driving the input (of a device under test) in opposition to the output from another device, as in Figure 15–28.

The easiest and safest way to avoid bus collision is to place the outputs of the driving devices into a high-impedance state (tri-state), as in Figure 15–29. Tri-state is accomplished with an output-enable pin. Rather than connect this pin directly to ground, pin 1 in Figure 15–29 is tied to a 120 Ω resistor. The ATE tester can, via a test pin, drive the enable input to a logic high, placing U_1 in a high-impedance state so that U_2 can be tested.

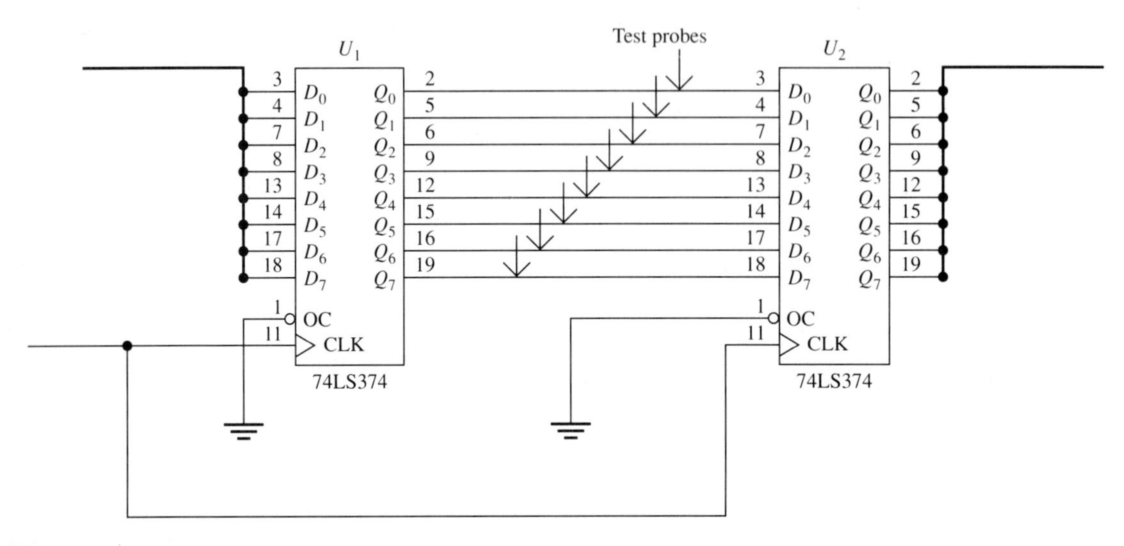

FIGURE 15–28
Back-driving U_1 in order to test U_2.

There are certain devices that don't have an output enable. In this case, the only alternative is to **back-drive** the output device. Back-driving means that there can be a bus collision but that the ATE tester will *force* the bus to the desired logic level regardless of the circuit's logic. Manufacturers of digital devices contend that their devices should survive back-driving conditions, but it may weaken the part, shortening its life. Manufacturers of ATE equipment agree but contend that

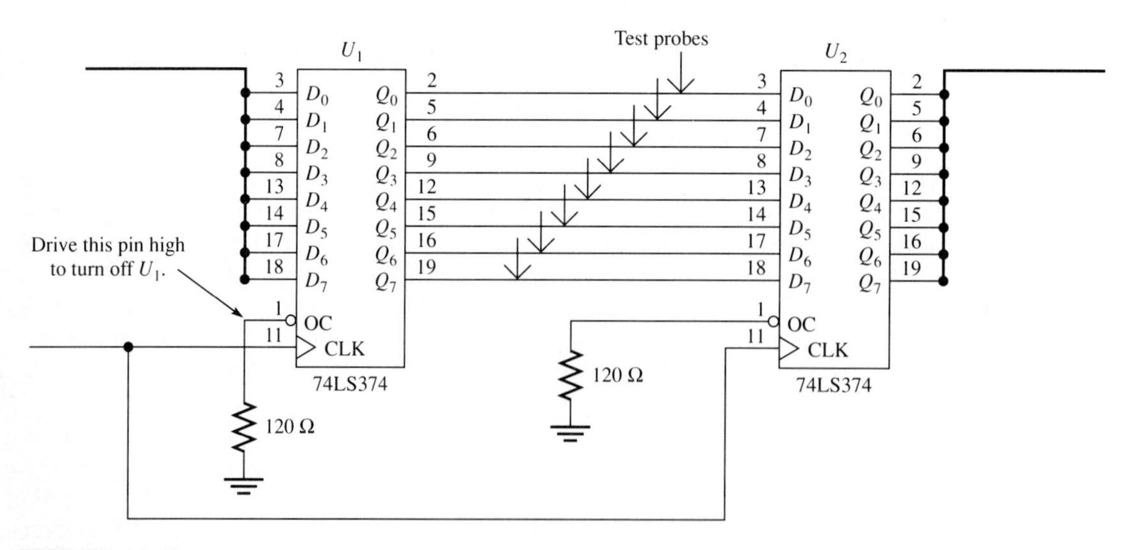

FIGURE 15–29
Disabling U_1's outputs so U_2 can be tested.

if the testing is done correctly, the life expectancy may only be a few hours shorter than a ten year life expectancy.

Correct back-driving conditions focus on two parameters of the device: (1) The test should be performed quickly enough so that the substrate is not heated to temperatures higher than 125°C, and (2) the total current should be kept low enough to avoid melting the bonding wires, particularly the power and ground bonding wires. This is done by limiting the number of output pins that are to be back-driven. Most ATE systems contain test libraries for various types of digital devices. The libraries limit the test times and the number of back-driven pins. Typically, even if the heat-dissipation properties of the package are neglected, the junction of a TTL package will not overheat (125°C) with a 200 mW overload in less than 30 ms. If you keep the back-drive to 200 mW maximum and less than 30 ms in duration, then you should have little trouble.

Figure 15–30 shows a type of TTL totem-pole output. You can see that because of the resistor (R_4), it takes less current to back-drive from a high state to a low state than it takes to back-drive the other way. In other words, it's easier to pull low than to pull high. If you are back-driving a device and you cannot place it into a high-impedance state, then consider (if possible) driving its inputs to a logic state so the outputs are at a high state.

Programmable array logic devices (PALs) can oscillate if back-driven from a low to a high state. PALs should be tri-stated if there is a chance for bus collision. A PAL is a hardware logic device that can be programmed, much in the same manner as ROMs are programmed. PAL programming was described in Section 6–2.

FIGURE 15–30
A type of TTL totem-pole output.

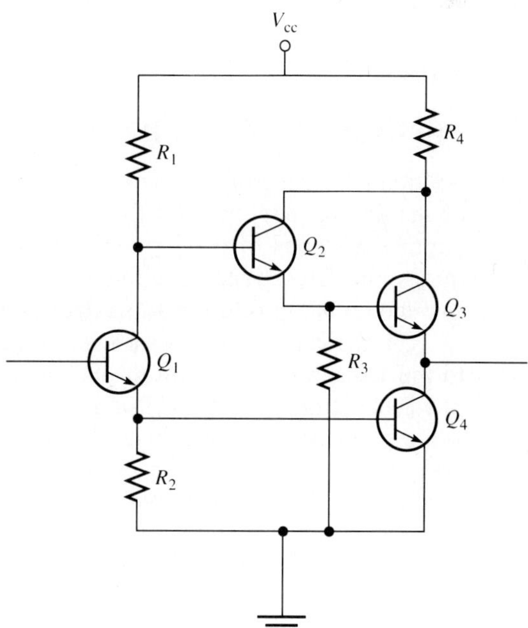

TESTING CMOS DEVICES

CMOS devices can be supplied with voltages greater than 5.0 V (5.0 V is the required voltage for TTL logic devices). But at voltages greater than this voltage, more power is dissipated. It is recommended that CMOS devices be tested at 5.0 V to limit this power dissipation.

Also, certain CMOS devices can go into what is known as **SCR latch-up.** An SCR (silicon-controlled rectifier) is often used as a latching switch; once it is activated, it will stay on until the voltage across it is reduced to near 0 V. Figure 15–31(a) and (b) shows the output structure of many CMOS devices and the diode model. The diode model illustrates that the MOS barriers can become conductive if the output is forced to a voltage that is greater than V_{CC} (supply) or less than ground. Once the barrier is conducting, the device latches with near short-circuit currents and is damaged.

There are three rules to follow in testing CMOS devices:

1. Test CMOS only with +5.0 V. This will ensure that longer-duration input patterns, which are needed for testing the larger integrated devices (LSI and VLSI), will not cause the output channel temperatures to exceed 125°C.
2. Connect any open inputs to either V_{CC} or ground through a 4.7 kΩ resistor. This is to assure that the inputs do not drift outside the supply rails when power is applied to the device under test.
3. Avoid directly driving capacitively shunted logic lines, which could cause *ground bounce*. **Ground bounce** is produced when enough current is flowing out the ground pin to cause a voltage in the ground wire (see Figure 15–32(a) and (b)). This can cause the ground to move higher than that supplied as a logic low, which can cause latch-up.

15–6 PINS

PIN-DRIVER CONCEPTS

Section 15–7 describes the construction of the pins (spring-loaded test probes) and of the fixture that supports these pins and the UUT. This section describes two popular methods of driving and receiving signals from these pins. (To simplify things, the driver/receiver combination will be referred to as just a "driver.") The two methods are referred to as **dedicated drivers** and **multiplex drivers.**

A dedicated driver is shown in Figure 15–33. There is only one pin attached to the driver's input/output through a relay. If the relay is open, the driver has no physical contact to the UUT and therefore minimizes loading effects on the UUT when other drivers are in use. The main advantages of using this kind of driver are: (1) The fixture can be arbitrarily wired and then the nodes "learned" by touching the connections with a test probe and (2) changes involving added or deleted components to the UUT can be implemented by connecting the nearest unused pin driver or removing drivers that are no longer needed. The main disadvantage of using dedicated drivers is cost of the system versus driver capability. Drivers with controlled rise times, adjustable timing (with respect to the other

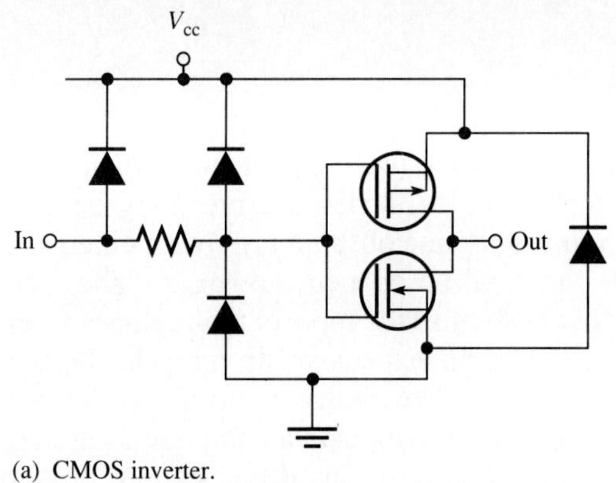

(a) CMOS inverter.

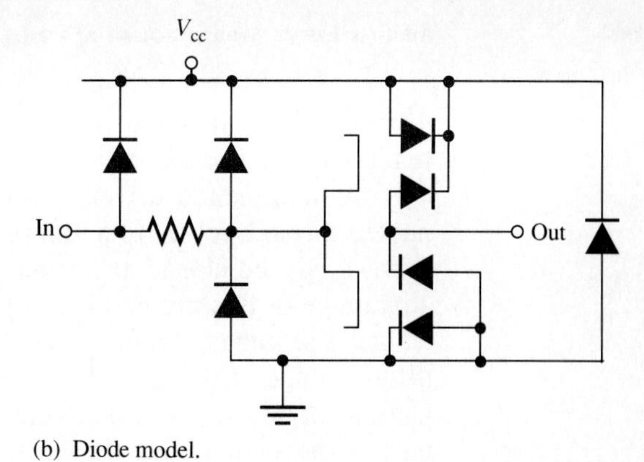

(b) Diode model.

FIGURE 15–31
CMOS outputs and latch-up.

FIGURE 15–32
The effects of capacitive loading on ground potential (ground bounce).

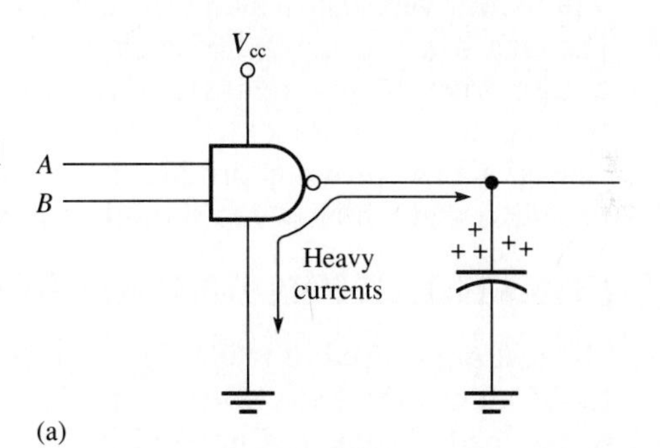

(a)

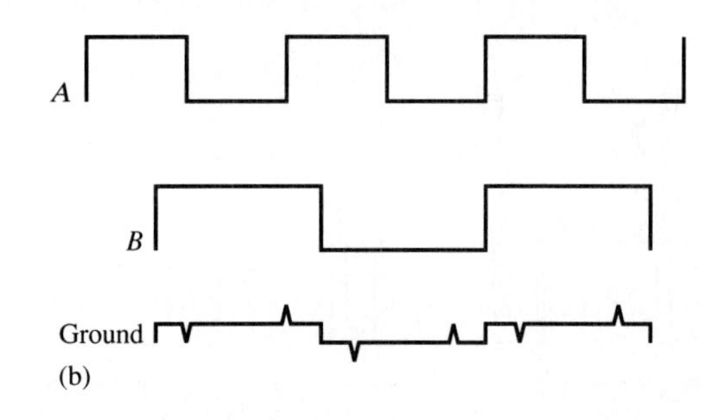

(b)

FIGURE 15–33
Dedicated driver/receiver.

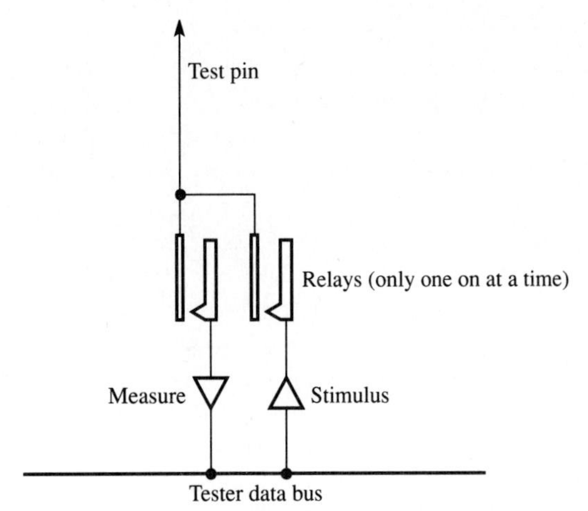

drivers), and adjustable voltage threshold sensitivities are rather expensive. Systems that use dedicated drivers will often sacrifice these advanced capabilities.

A multiplexed driver is shown in Figure 15–34. Each driver may be connected through relays to a number of pins. Only one of the relays (connected to a driver) may be closed at a time. (They may all be opened, however.) The main advantage of this method is a lower-cost system that supports the features (controlled rise times, timing, and thresholds) of an expensive driver. The biggest disadvantage of this type of system is the lack of flexibility in wiring the fixture. Care must be taken to assure that pins, planned to stimulate a component, are not part of the same driver (in the same multiplex group); one driver cannot do two tasks at once. Systems that use multiplex drivers are supported with software that will build a **wire-list** to aid in the construction of the fixture. A wire-list is a listing showing where each end of each wire is to go. One end will be connected to the fixture driver points (relays) and the other end, to the test pins on the board. Each time components are added to or deleted from the UUT, the wire-list program is run again. The program produces a report showing what wires should be removed or added and which wires should be moved.

CONSTRUCTION AND SELECTION

The test pin is used to make electrical contact with the tester to the UUT. Figure 15–35 shows the basic construction of the test pin inserted into a socket. The spring-loaded socket is pressed into a precision hole drilled into the fixture. A square post on the socket is used to connect a wire to the ATE equipment; usually

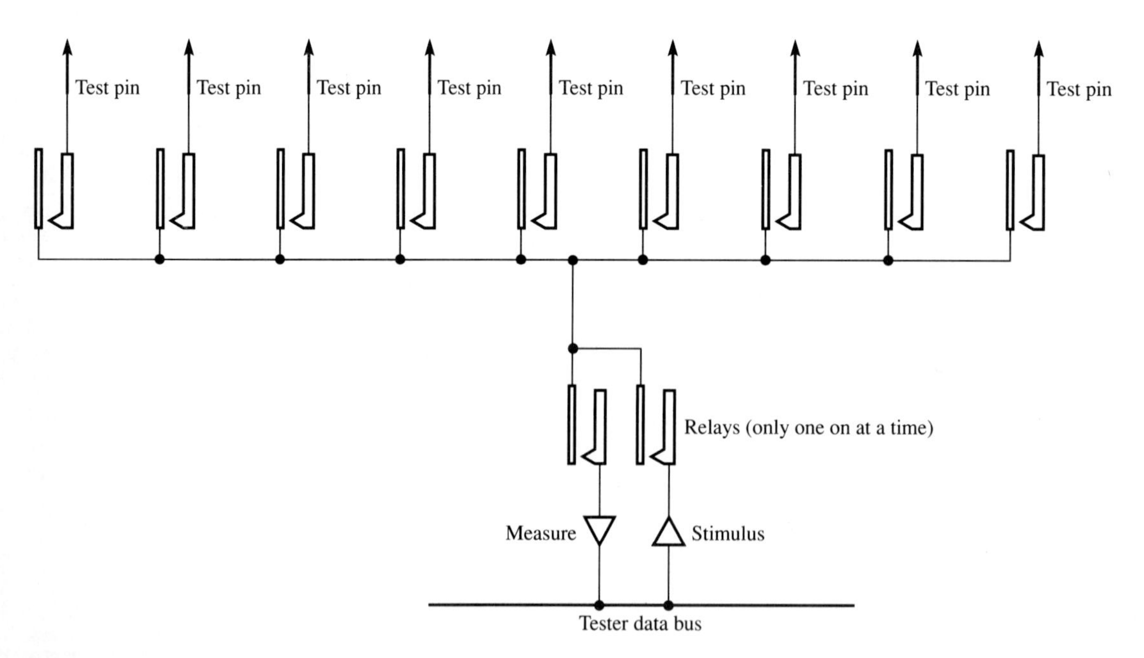

FIGURE 15–34
Multiplexed driver/receiver.

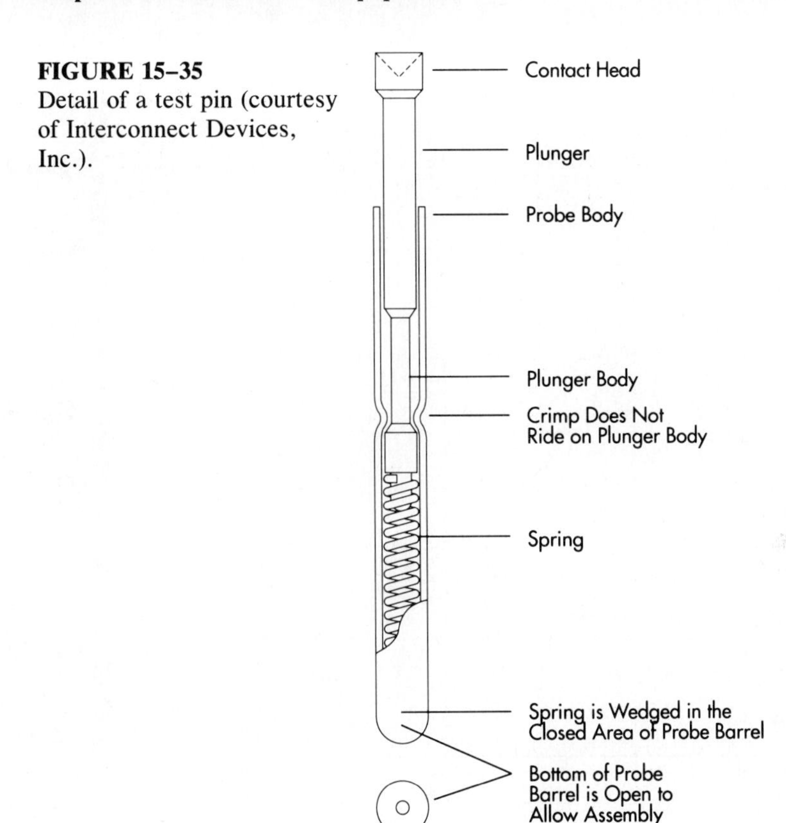

FIGURE 15–35
Detail of a test pin (courtesy of Interconnect Devices, Inc.).

wirewrap wire is used. The socket is designed to accept several types of pins. Figure 15–36 shows these types of pins and their intended application.

15–7 FIXTURE CONSTRUCTION AND DESIGN CONSIDERATIONS

The fixture's primary purpose is to position the UUT in precise mechanical alignment with the test pins. The fixture, unique to a UUT, can quickly be exchanged with other fixtures to test other units. The pins in the fixture are forced to make contact with the circuits in the UUT in one of two ways: (1) through a vacuum, which draws the board down to the pins, or (2) through mechanical means, where a clamp is used to press the board down to the pins. The vacuum is the cheaper and most reliable method of the two.

Figure 15–37 illustrates the vacuum fixture mechanics. The vacuum is applied to the vacuum port through a valve. The valve is controlled by the tester's test program. The board is positioned to precise mechanical alignment in the fixture through the use of **guide pins.** The guide pins slide into **tooling holes,** which are drilled into the board to be tested. The tooling holes are originally drilled into

SELECTION GUIDE

PLUNGER	TIP STYLE	APPLICATION COMMENTS	PLUNGER	TIP STYLE	APPLICATION COMMENTS
A (or) G	Concave	Long Leads, Terminals, and Wire Wrap Posts.	LM	Star	Plated through holes, Lands, Pads—Self-cleaning.
B	Spear Point	Lands, Pads or Plated through holes.	T (or) K	3 or 4 Sided Chisel	Plated through holes—Cuts through contamination.
C (or) F	Flat	Gold Edge Fingers—No marks or indentations.	U	Crown .040	Lands, Pads, Leads, Holes—Self-cleaning.
D (or) J	Spherical Radius	Gold Edge Fingers—No marks or indentations.	V	Crown .060	Lands, Pads, Leads—Self-cleaning.
E	Convex	Plated through holes.	W	Crown .050	Lands, Pads, Leads—Self-cleaning.
FX	Flex Probe	Contaminated Boards or Conformal Coating.	X	Tapered Crown	Lands, Pads, Leads, Holes—Self-cleaning.
H	Serrated	Lands, Pads, Leads, Terminals.	Y	Tulip	Self-cleaning Leads, Wire Wrapped Terminals.

TYPICAL APPLICATIONS

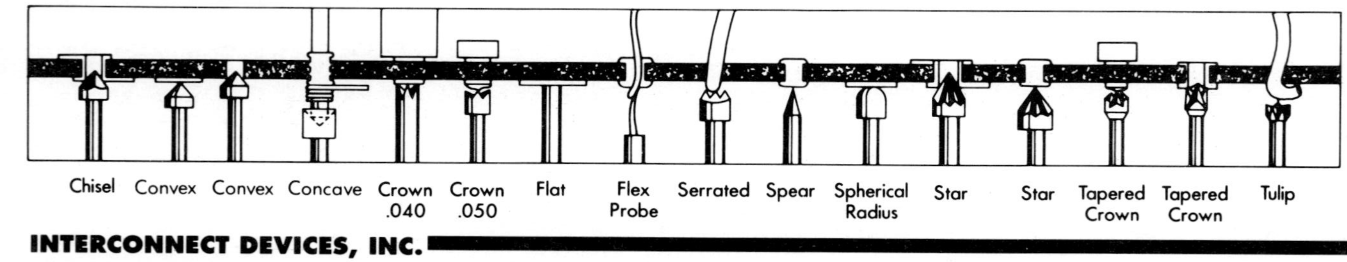

INTERCONNECT DEVICES, INC.

FIGURE 15–36
Types of pins and their applications (courtesy of Interconnect Devices, Inc.).

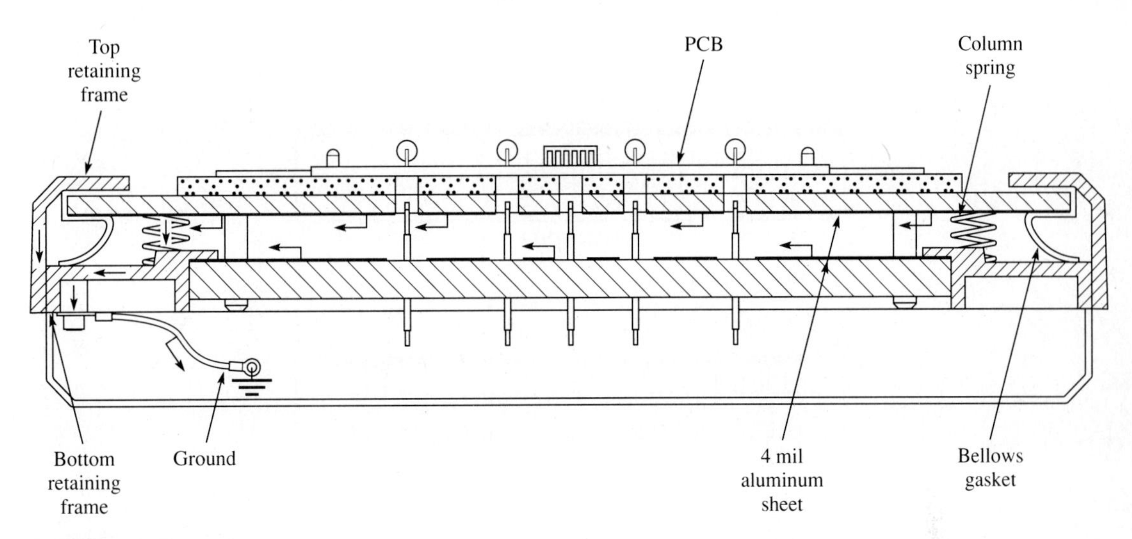

FIGURE 15–37
(Courtesy of Virginia Panel Co.)

the board as part of the board's fabrication. They are used as precision reference points and hold it in place while the other **feed-through holes** and **vias** are drilled. As shown in Figure 15–38, feed-through holes are drilled holes that permit mounting of lead-type components. Vias are electrical paths connecting a circuit trace from one side of a board to a trace on another side of a board (in the case of multilayer boards, a via may connect traces that are internal to the different layers). The vacuum causes the board to be drawn down to the pins, as shown in Figure 15–39(a). The rubber pads, which the board rests upon, prevent the board

FIGURE 15–38
Feed-through holes have component leads involved. Vias connect the top-side traces to the bottom-side traces without a component lead involved.

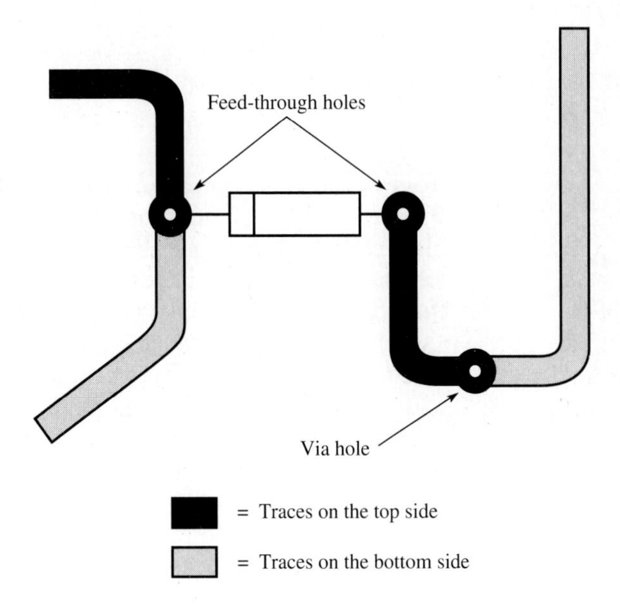

FIGURE 15–39

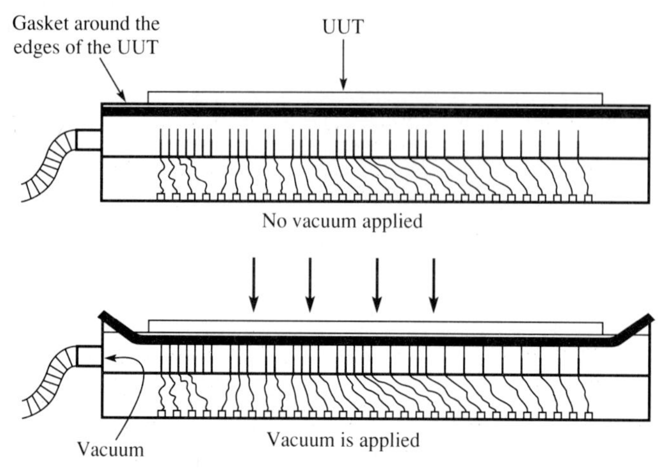

(a) Vacuum actuates the fixture by causing atmospheric pressure to push the UUT down to the test pins.

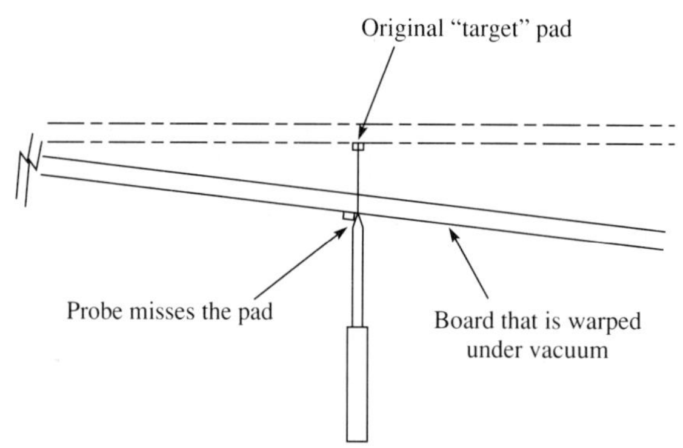

(b) An unsupported board can cause the UUT to warp under vacuum. This can cause the pin to miss the pad.

from warping. Warping can cause the pin to miss its intended target, as shown in Figure 15–39(b). In addition to the rubber pads, many fixtures incorporate a top plate called a **diaphragm plate.** This plate has holes that permit the passage of the test pin and provides support to the board under test. Of course, any open holes in the board under test must be filled or covered in order to stop vacuum leaks.

In most modern facilities, designs are done with the aid of **CAD** (computer-aided design). The schematics are drawn at a workstation by the design engineer, and the schematics are "captured" by the software. The captured schematic permits **netlists** (Figure 15–40), which are lists documenting the interconnections between the various components, to be generated. In addition to schematic capture, the engineer will run a **placement** program, which determines where the components are to go. X-Y coordinates and rotation data are available as output from the placement program (see Figure 15–41). The netlist and the placement

PARTS LIST

Name	Description	Location			
J1\121	CON\J1\121	J1			
J2\100	CON\J2\100	J2			
J3\100	CON\J3\100	J3			
J4\100	CON\J4\100	J4			
J5\100	CON\J5\100	J5			
J6\100	CON\J6\100	J6			
C\470U	CP\900\686X289	C62	C61		
R\110	R\400	R7	R3	R6	R8
R\110	R\400	R4	R2	R1	R5
R\56.2	R\400	R10	R9		
RNS\120\10-9\GND	RN\SIP\10-FL	RN2	RN3	RN6	RN1
RNS\120\10-9\GND	RN\SIP\10-FL	RN4	RN5		
10E111	PLCC\28	U51			
10H125	DIP\16-FL	U22	U19	U25	U28
74F821	DIP\24\300-FL	U17	U31	U32	U18
74F821	DIP\24\300-FL	U33	U34	U49	U50
74F821	DIP\24\300-FL	U20	U21	U36	U35
74F821	DIP\24\300-FL	U37	U52	U53	U38
74F821	DIP\24\300-FL	U23	U39	U24	U40
74F821	DIP\24\300-FL	U41	U42	U54	U55
74F821	DIP\24\300-FL	U26	U43	U44	U27
74F821	DIP\24\300-FL	U45	U56	U57	U46
74F821	DIP\24\300-FL	U29	U47	U30	U48

Node connections

```
NODENAME AUX2V/K0A_EFS     $
    RN9         6  J2        21
NODENAME AUX2V/K1A_EFS     $
    RN10        6  J2        22
NODENAME AUX2V/K2A_EFS     $
    RN11        6  J2        23
NODENAME AUX2V/K3A_EFS     $
    RN12        6  J2        24
NODENAME AUX2V/K4A_EFS     $
    RN13        6  J2        25
NODENAME AUX2V/K5A_EFS     $
    RN14        6  J2        26
NODENAME AUX2V/K6A_EFS     $
    RN15        6  J2        27
NODENAME AUX2V/K7A_EFS     $
    RN16        6  J2        28
NODENAME AUX2V/K8A_EFS     $
    RN17        6  J2        29
NODENAME AUX2V/K9A_EFS     $
    RN18        6  J2        30
NODENAME AUX2V/K0B_EFS     $
    RN19        6  J3        31
NODENAME AUX2V/K1B_EFS     $
```

FIGURE 15–40
Net list.

FIGURE 15–41 X-Y data.

X-Y Data

Name	X	Y	Rotation
J4	17.250	7.300	0
RN28	16.050	7.700	90
RN27	16.050	7.800	90
RN26	16.050	7.900	90
RN25	16.050	8.000	90
RN24	16.050	8.100	90
RN23	16.050	8.200	90
RN22	16.050	8.300	90
RN21	16.050	8.400	90
RN20	16.050	8.500	90
RN19	16.050	8.600	90
RN18	16.050	11.300	90
RN17	16.050	11.400	90
RN16	16.050	11.500	90
RN15	16.050	11.600	90
RN14	16.050	11.700	90
RN13	16.050	11.800	90
RN12	16.050	11.900	90
RN11	16.050	12.000	90
RN10	16.050	12.100	90
RN9	16.050	12.200	90
RN48	16.050	2.800	90

data are both used to generate a program called **routing,** which is used to place the actual routing of the circuit traces on the board. An image drawing generated from the routing software, called a **photo plot,** is often used to produce the film transparencies needed for the actual fabrication of the board's circuit traces. In addition to the photo plot, a file called a **drill file** (or drill tape) is produced. This data file contains all the X-Y locations and drill sizes needed to drill the mounting holes, tooling holes, and vias for each layer of a circuit board.

The test engineer will use several of the files mentioned in the preceding paragraph. One of the most useful will be the drill file. The same machine that drilled the circuit board can be used to drill the holes in the text fixture. Most of the holes will be the same size as the sockets for the test pin. Other hole sizes may be needed; the most common is the tooling hole. The netlist can be used to generate a test program, as illustrated in Figure 15–42. The ATE tester will use the data in Figure 15–42 to perform the actual tests on the components.

Wirewrap techniques are often used in the construction of the fixtures; the most common size of the wire is #30. Figure 15–43(a) and (b) show proper wirewrap connections and poor wirewrap connections. The wires should be long enough to avoid strain on them when the fixture is hinged open and, at the same time, be short enough to avoid the effects of capacitive coupling (between the wires) and large inductance (causing ringing, which can create test errors). In addition to the long wires causing electrical problems, long wires tend to become entangled and can catch in the hinge mechanism of the fixture.

```
"R001 47.00 O, pg 2, SECTION 4C"
'47.00 O
TOL 2;
47.000 O 260-262 WAIT 27 MS;

"R002 3.01KO, PG 2, SECTION 4C"
'3.01KO
3.010 K 262-234/0 WAIT 19 MS;

DIGFIL ON 7;

"R003 100.00mO, PG 2, SECTION 4C"
'100.00mO
TOL +0 -100;
100.000 unknown units
262-288 WAIT 5 MS;
DIGFIL OFF;

"R004 3.01KO, PG 2, SECTION 4C"
'3.01KO
TOL 2;
3.010 K 288-235/0 WAIT 14 MS;

"R005 2.890KO, PG 2, SECTION 4C, R5 AND R6 IN PARALLEL"
'73.20KO
TOL 5;
2.890 K 0-234/262 WAIT 14 MS;

"R007 3.01KO, PG 2, SECTION 4C"
'3.01KO
TOL 2;
3.010 K 0-235/288 WAIT 14 MS;

"R008 5.60KO, PG 2, SECTION 4C"
'5.60KO
5.600 K 288-289 WAIT 5 MS;

"R009 4.30KO, PG 2, SECTION 4C"
'4.30KO
4.300 K 262-292 WAIT 105 MS;

"R010 4.70KO, PG 2, SECTION 4C"
'4.70KO
4.700 K 292-237 WAIT 5 MS;

DIGFIL ON 14;

"R011 7.50KO, PG 2, SECTION 4C"
'7.50KO
TOL 5;
7.500 K 236-3/266, 639 WAIT 52 MS;
DIGFIL OFF;
```

FIGURE 15–42
Test program.

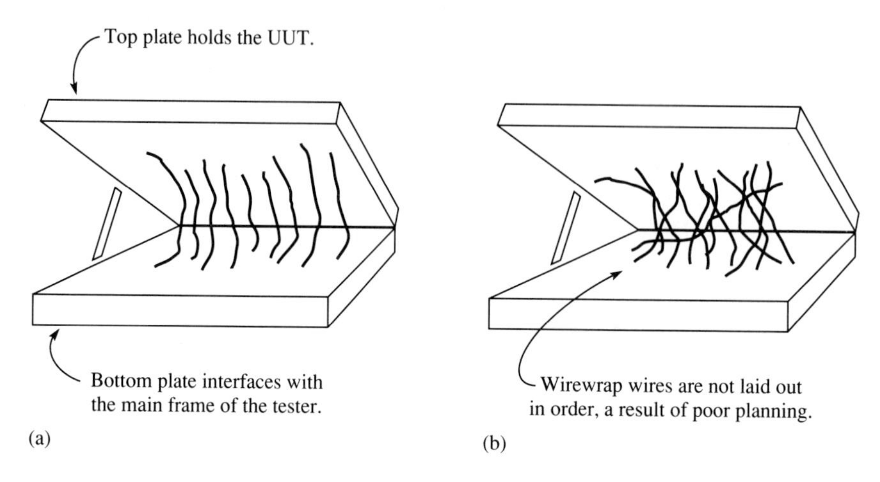

FIGURE 15–43
Proper and improper methods of interconnecting with wirewrap wire.

Some lines, which connect to very sensitive or high-impedance input circuits, will pick up sufficient cross talk from the signals in other wires to cause spurious failures. One method of preventing cross talk is to include a shield wire along with the signal wire, as shown in Figure 15–44(a). Only one end of the wire is connected to signal ground. Also note that some devices have differential

FIGURE 15–44
Use twisted wire (a) to shield or (b) to carry differential signals.

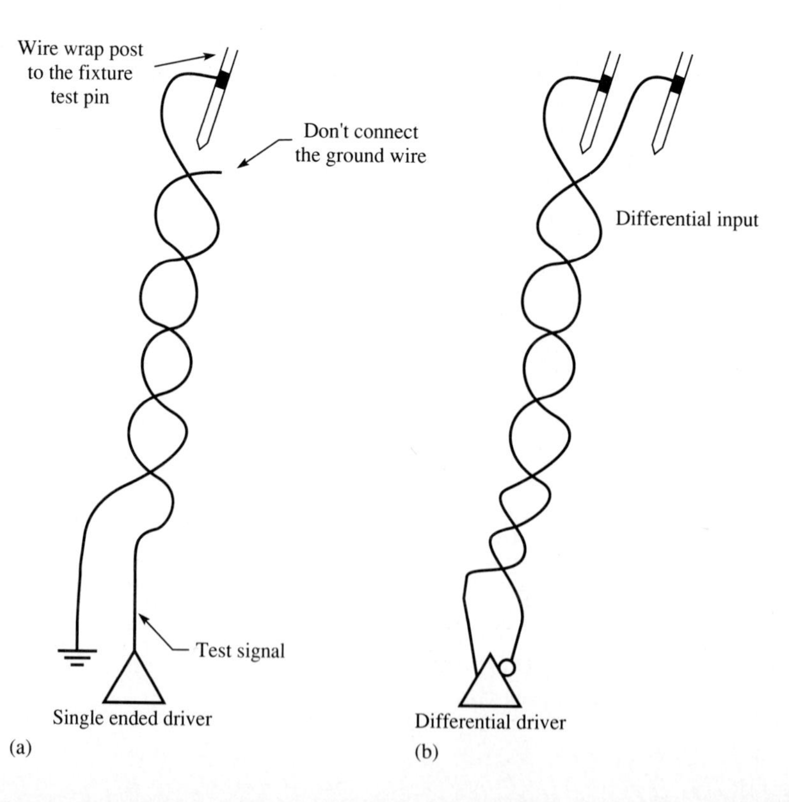

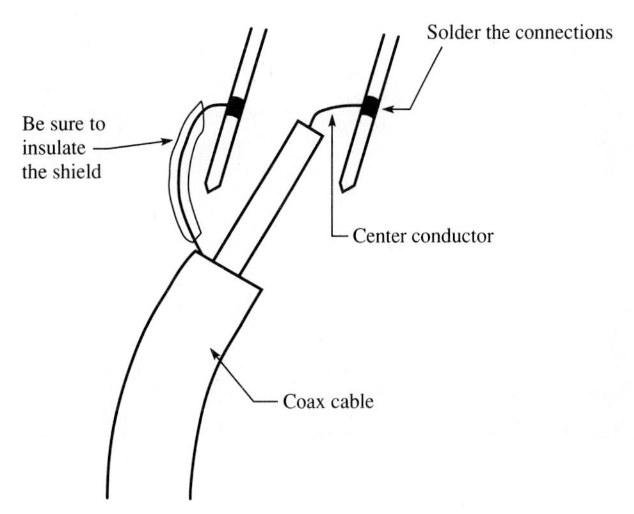

FIGURE 15–45
Connecting coax cable to wirewrap posts.

inputs. Examples of differential input devices are op-amps, high-gain comparators, and ECL logic devices. Some devices also have differential outputs. The two signal wires driving the differential inputs or differential outputs should be twisted together to reduce cross talk (Figure 15–44(b)). Analog signals often require coax wire to be connected to the fixture. The wires should be soldered and the shield should be protected with insulation (to avoid shorting to other pins), as shown in Figure 15–45.

There is no such thing as a fixture that will work right the first time. There will always be something that will need to be rethought and redone or done differently. This process is referred to as **debugging.** Debugging may take as much as 75% of the total test-design time for a new board. Depending on the complexity of the board and development software available, test program development may take only 1 week (per board); actual fixture drilling may take 1 day, wiring of the fixture may take 1 week, and debugging may take 1 month. But the results are rewarding if the work is done correctly and completely.

15–8 TESTING SURFACE-MOUNTED DEVICES

Surface-mounted devices, or SMDs (Figure 15–46(a)), mount to the surface of the circuit board rather than using the conventional through-hole device (Figure 15–46(b)). The three most common methods of mounting components are illustrated in Figure 15–47(a)–(c). Some devices don't have wire leads; they have solder-filled grooves that form a solder bridge from the device to the circuit board. SMD capacitors and resistors don't have wire leads; the ends of the component are solder-plated, as shown in Figure 15–48.

Two important objectives of using surface-mounted technology are (1) space, which can be as little as 30% of the available space on the board using conventional lead-through methods, and (2) better automation of circuit-board assembly.

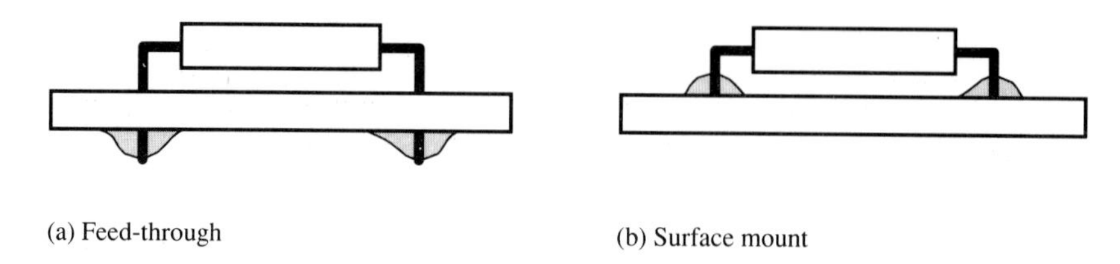

FIGURE 15–46
Through-hole and surface mount.

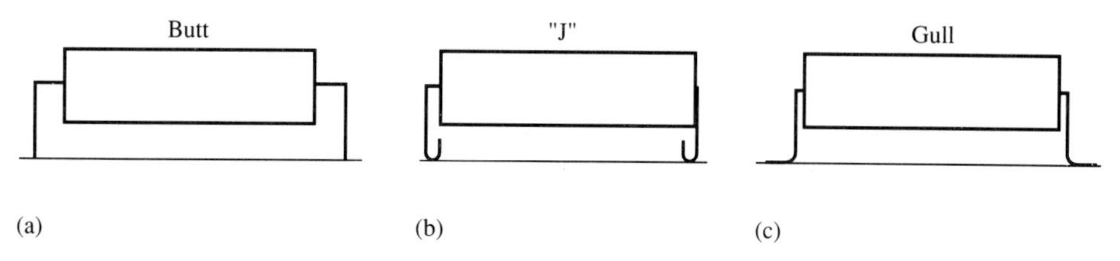

FIGURE 15–47
Common SMD lead-forming techniques.

FIGURE 15–48
Leadless device soldered to a circuit board.

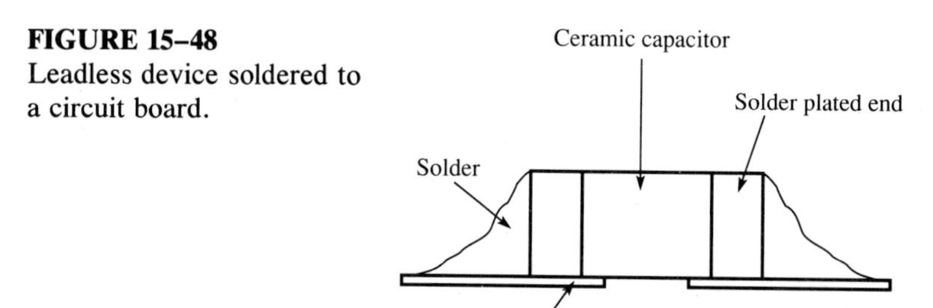

There are added challenges to testing a board containing SMD technology. Unlike lead-through parts, where the leads are accessible on the bottom of the board (solder side), SMD devices do not have every node accessible to the bottom side of the board. Some nodes (common connections to parts) are not accessible on the bottom side. In this case, there are two methods whereby the nodes can be probed: (1) the preferred method is to provide vias to test pads, or (2) the top side of the board can be probed using a **clamshell fixture,** as shown in Figure 15–49. The clamshell fixture is very expensive to build and requires frequent maintenance to keep it reliable. The trade-off for having all the nodes accessible from one side (bottom) of the board means that the board size will increase by 10% to 30% over a board without vias and test pads. Another source of difficulty in testing SMDs is that newer and more dense SMD circuit boards have components on both sides of the board. Probing becomes more difficult because components can get in the way of the probes.

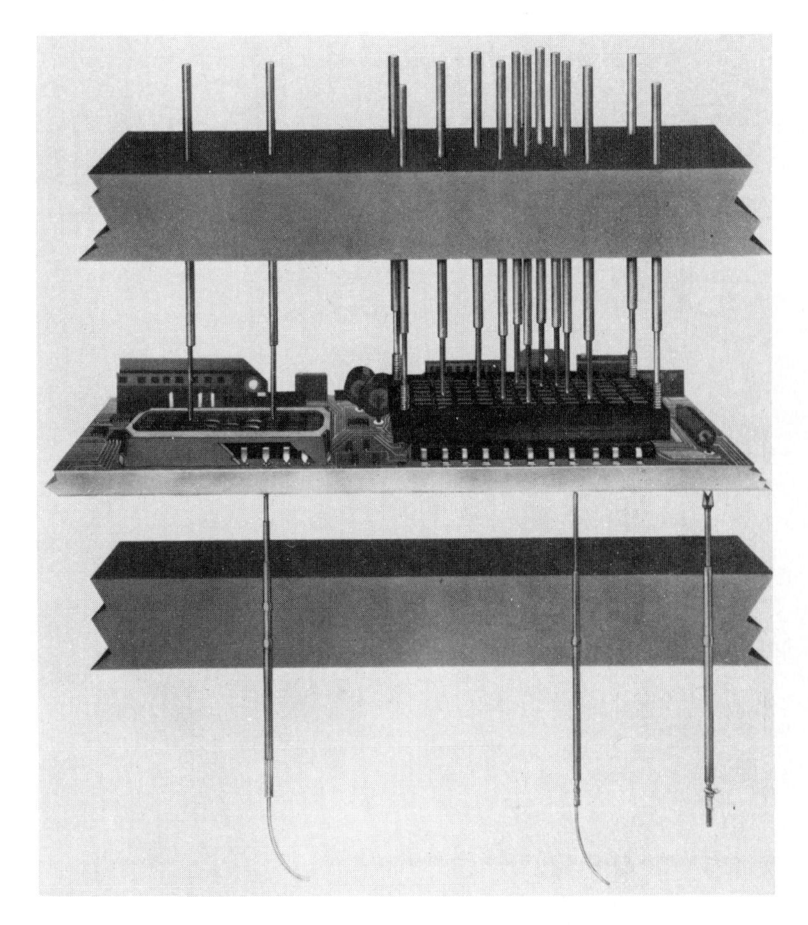

FIGURE 15–49
A clamshell fixture (courtesy of Contact Products, Inc.).

Testability must be planned into the *original* design; attempts to add testability later in a design cycle almost always fail. A few simple design guidelines that help make SMD testing easier are the following:

1. Use test pads. Don't probe the lead of a device because a lead to a broken solder connection may be pushed back down to the contact surface during the test, causing the test to be successful, but the circuit will fail once the pressure of the probe is removed (Figure 15–50). Test pads consist of vias, extended solder pads, expanded traces, and remote test pads (see Figure 15–51).
2. Avoid probing both sides of the board. Have all the nodes accessible from one side of the board.
3. Try to keep 0.10 in. between probes. If the test pads are close together (0.050 in. down to 0.020 in.), try to stagger them, as shown in Figure 15–52. Miniature probes are available for probing 0.050 in., but they should be avoided if at all possible, because they are expensive and fragile.
4. Place the test pads at least 0.2 in. away from tall components (greater than 0.2 in.) to avoid mechanical interference with the probe socket.

FIGURE 15–50
Pressure from the test probe can "mend" a broken connection. Don't probe on component leads.

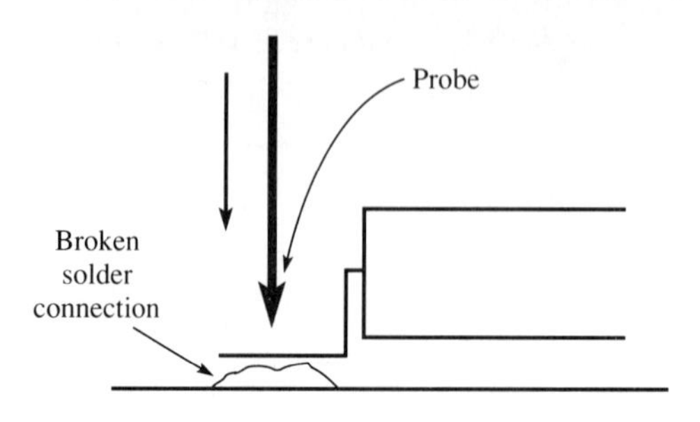

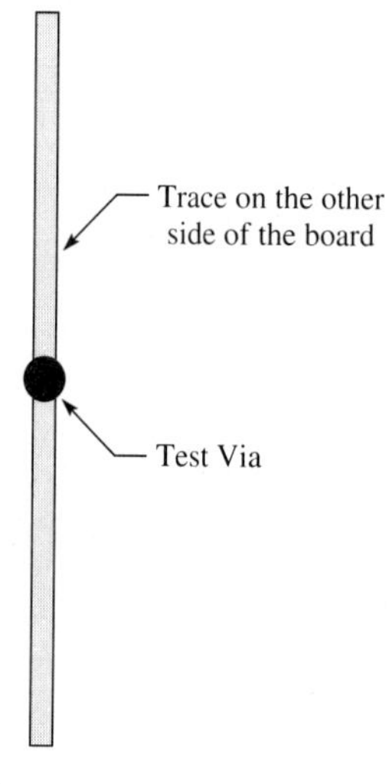

(a) Test Via.

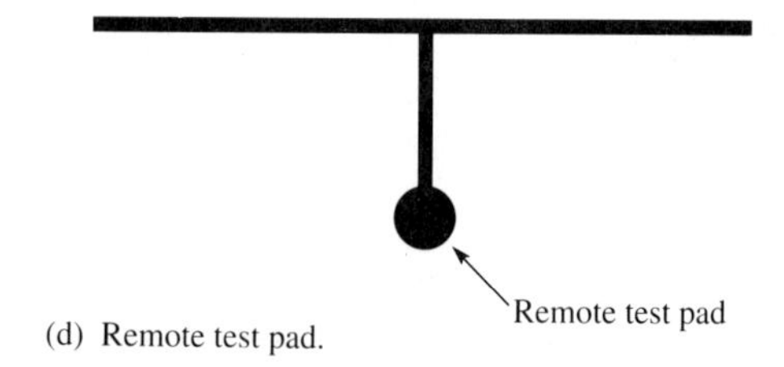

(b) Extended solder pad.

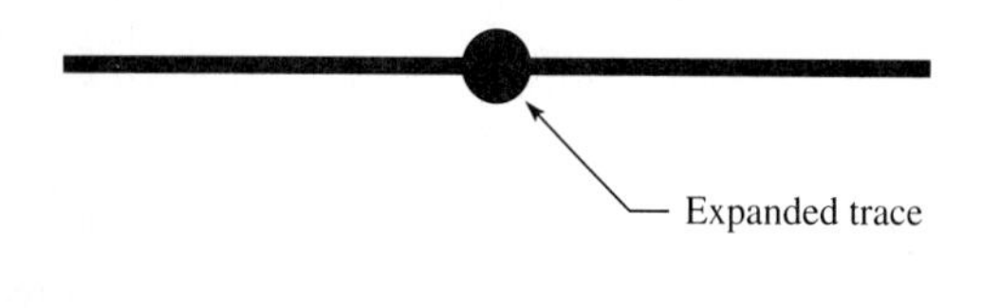

(c) Expanded trace.

(d) Remote test pad.

FIGURE 15–51
Common forms of test pads.

FIGURE 15–52
Stagger pads to yield 0.1 in. separation.

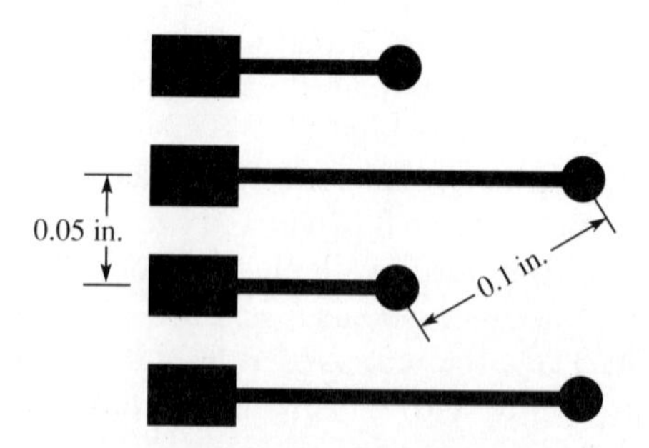

5. Make sure the tooling holes are accurate and large enough (greater than 0.12 in.) to align the board with the fixture. The tooling holes should not be **plated through.** In plated-through holes, copper is plated on the surface of the sides of the hole. It is hard to control the thickness of the plating, so some boards may fit too tight (causing jamming), whereas others may be too loose (causing alignment errors).

15–9 BOUNDARY SCAN

A new test technology called *boundary scan* is gaining popularity with test engineering and IC manufacturers. **Boundary scan** technology places a test circuit between each pin on a chip package and its internal logic, as shown in Figure 15–53. Each of the test circuits, called **boundary scan cells,** are connected to each other via a single-wire serial bus. Data bits are "marched" through the cells as either inputs to or outputs from the cell. In essence, boundary scan is equivalent to a bed-of-nails within the chip.

Boundary scan had its beginning in Europe in 1985 with the Joint European Test Action Group (JETAG). This group was seeking solutions to the problem of testing highly complex digital ICs on dense surface-mounted boards. In 1986, the group expanded to include members from both Europe and North America; they were renamed JTAG. In 1988 the proposal was offered to the IEEE Testability Bus Standards Committee for inclusion in the standard then under development. The IEEE Standard 1149.1-1990[4] was approved in February 1990.

FIGURE 15–53
A boundary-scannable board design (courtesy of IEEE).

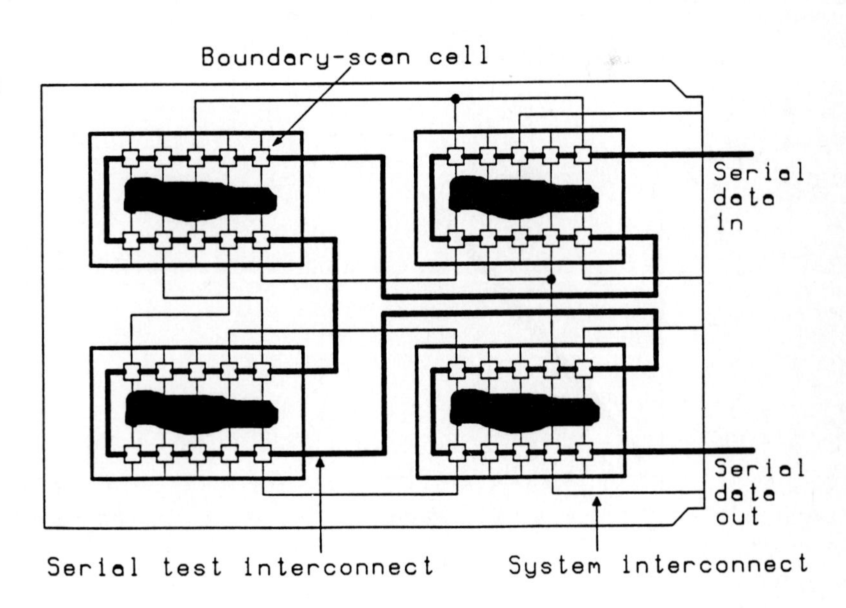

[4] For a copy of the standard, write to The Institute of Electrical and Electronics Engineers, Inc., 345 East 47th Street, New York, NY 10017-2394.

LIMITATIONS OF BED-OF-NAILS TESTING

Figure 15–54 shows the trends in device packaging that concerned the JTAG committee and prompted them to develop the test proposal. Lead counts on very large scale integrated circuits (VLSI) and application-specific integrated circuits (ASIC) are approaching more than 500 pins; as a result, the spacing between the leads is sometimes less than 0.025 in. This makes probing very difficult, particularly since more circuit boards use SMDs rather than through-hole devices, and components are on both sides of the board. These dense SMD boards are often used when conserving space is critical, as in a hand-held video camera. Bed-of-nails testing adds 10% to 30% to the area of the board because of the added test

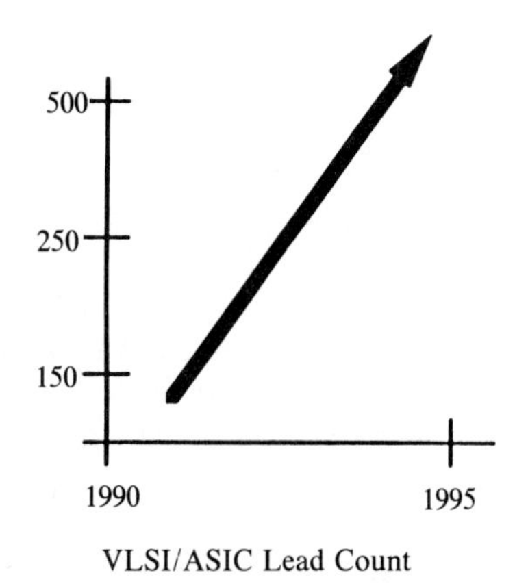

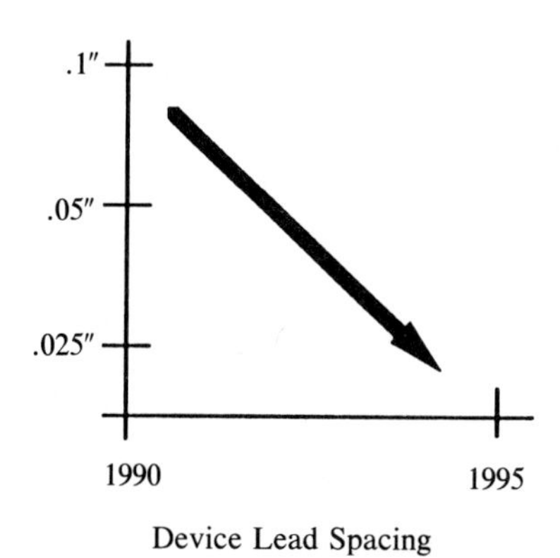

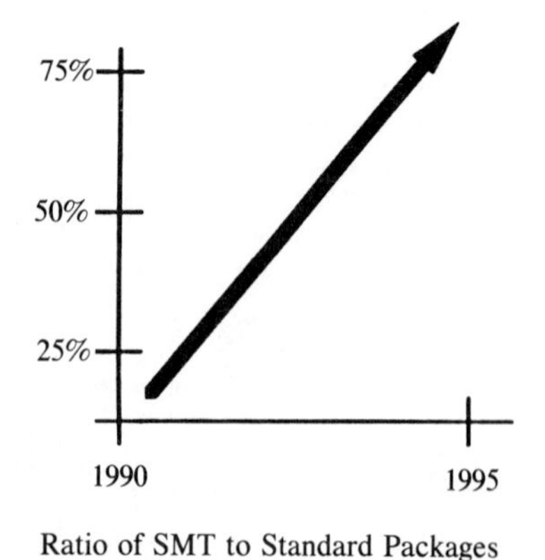

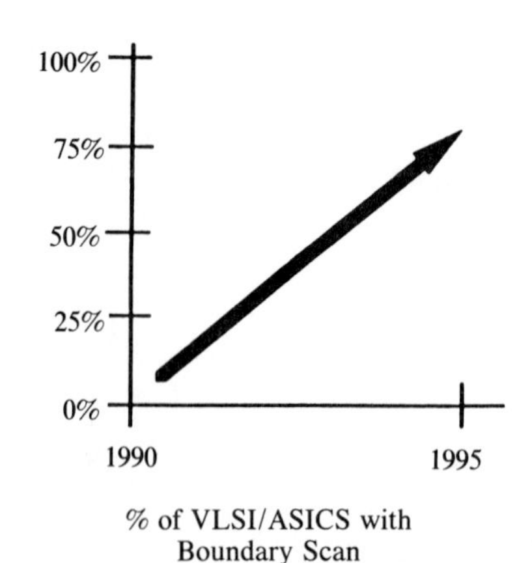

FIGURE 15–54
Device packaging trends (courtesy of Teledyne).

pads and vias. This added space for test is contradictory to the principal goal of SMDs, reducing space.

THE SYSTEM CONCEPT

Boundary scan places a test "cell" between the logic and the connecting pin on the device, as shown in Figure 15–55. Each boundary scan cell has a scan input (SI) and scan output (SO), which link the boundary scan cells together, and each cell has a signal input (PI) and signal output (PO), which is the normal data path to (or from) the chip. The boundary scan cells connected to the input pins of a device are essentially the same as the boundary scan cells connected to the output pins. A test access port (TAP) is included; it is the control interface to the test equipment. The test equipment can be an external computer or it can be a dedicated microprocessor within the system (designed for diagnostics). The TAP has two lines connected to it: one line is the test-mode select (TMS) and the other is the

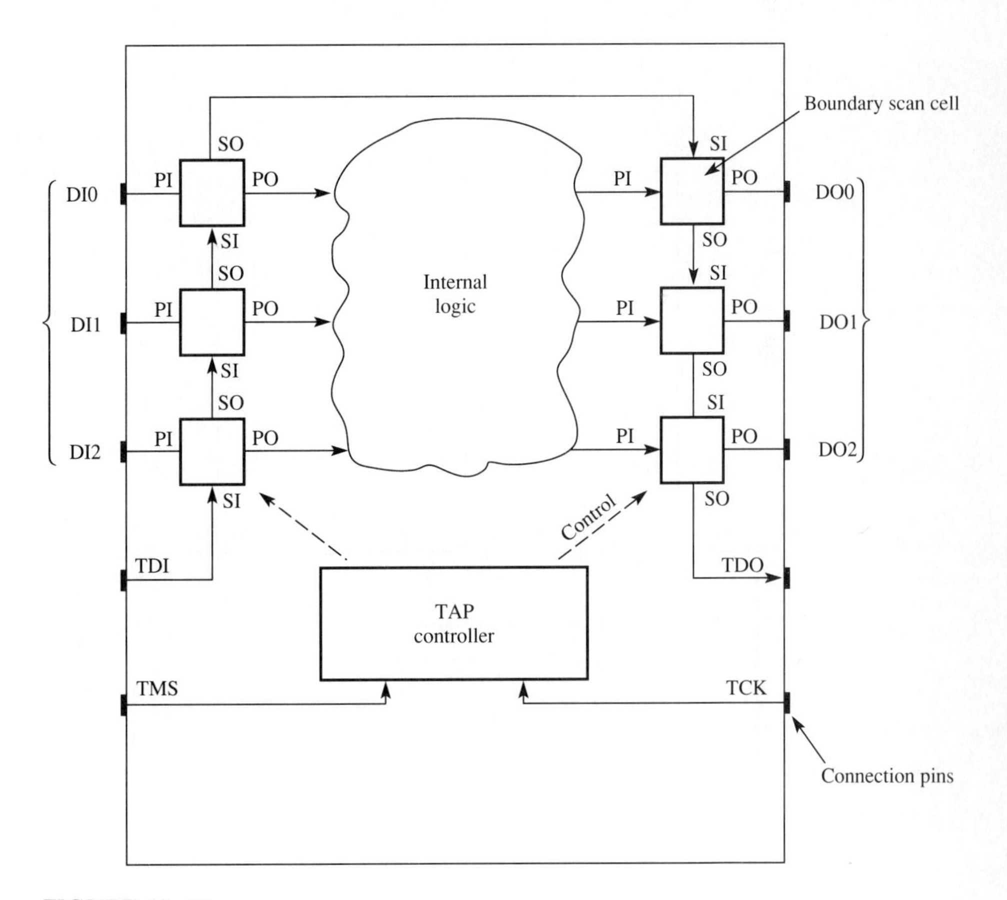

FIGURE 15–55
A block diagram of a chip containing boundary scan. The shaded areas are the boundary scan parts added to a logic device.

test clock (TCK). The TMS is used to configure the operating modes of the boundary scan cells. The TCK is used to clock in the control data to the TMS (serial string of instructions); in addition, it is used to march data from boundary scan cell to boundary scan cell via the SO and SI connections. Test data is placed serially into the DUT via the test data input (TDI), and output data is from the test data output (TDO). Optionally, a test reset (TRST) line may be included that will asynchronously reset the TAP to a known state, as when all the boundary scan cells are cleared of latched data and placed into the bypass mode.

Devices that contain boundary scan cells are connected together as shown in Figure 15–56. An example of the use of boundary scan, which can clarify its intent, is illustrated in Figure 15–57. Two digital logic chips are connected together with four data lines. The added boundary scan cells are shown as shaded areas. Normally, the boundary scan cells are in the bypass mode and have no effect on the operation of the circuits as IC-1 passes data to IC-2. The normal operation is suspended while the boundary scan does its test. The test procedure is explained in the following paragraphs.

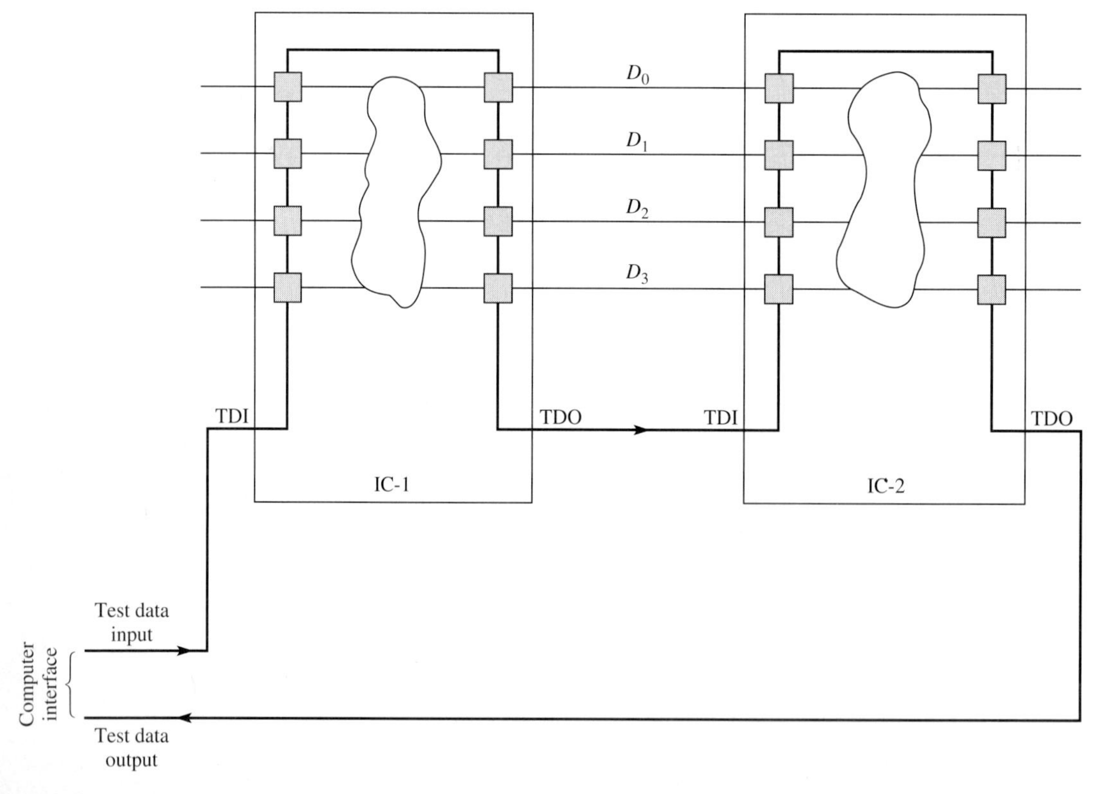

FIGURE 15–56
Example of boundary scan devices connected together.

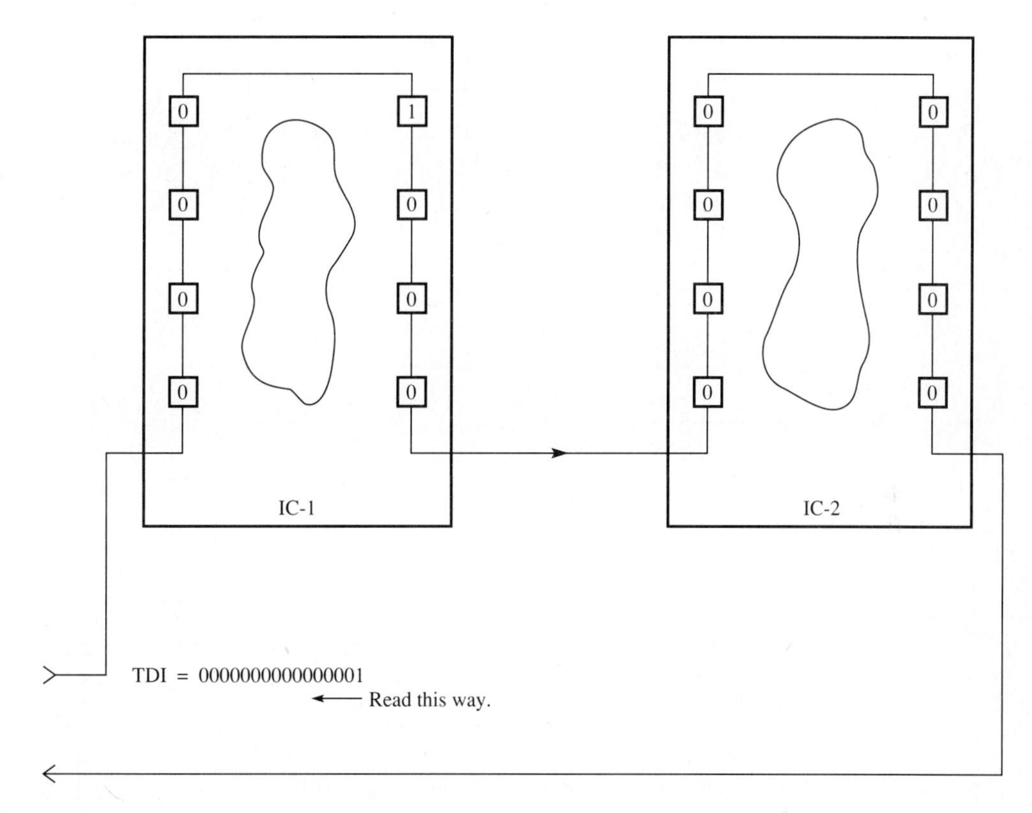

FIGURE 15–57
Testing the loop. It takes 16 TCKs to shift a 1 on TDI before it shows up on TDO. In this illustration, logic 1 has been shifted five times.

TESTING THE LOOP

The continuity of the test loop (TDI to TDO through all the chips) is first verified. This can be done by placing all the boundary scan cells into a shift mode. In the shift mode, any data that is present on the SI input will be latched (during the leading edge of TCK) and will be present on SO until the next clock (TCK). The test loop is tested by placing a single logic one on the TDI input of IC-1 and clocking TCK sixteen times to shift the logic one through all the boundary scan cells and out the TDO as shown in Figure 15–57. A computer is used to count and check the results.

TESTING CONNECTIONS

Connections between devices can be accomplished as shown in Figure 15–58. A test pattern of all 0s, all 1s, or alternating 0s and 1s can be shifted to IC-1's output cells. IC-1 is then placed into an *output load* mode. An output load will place the

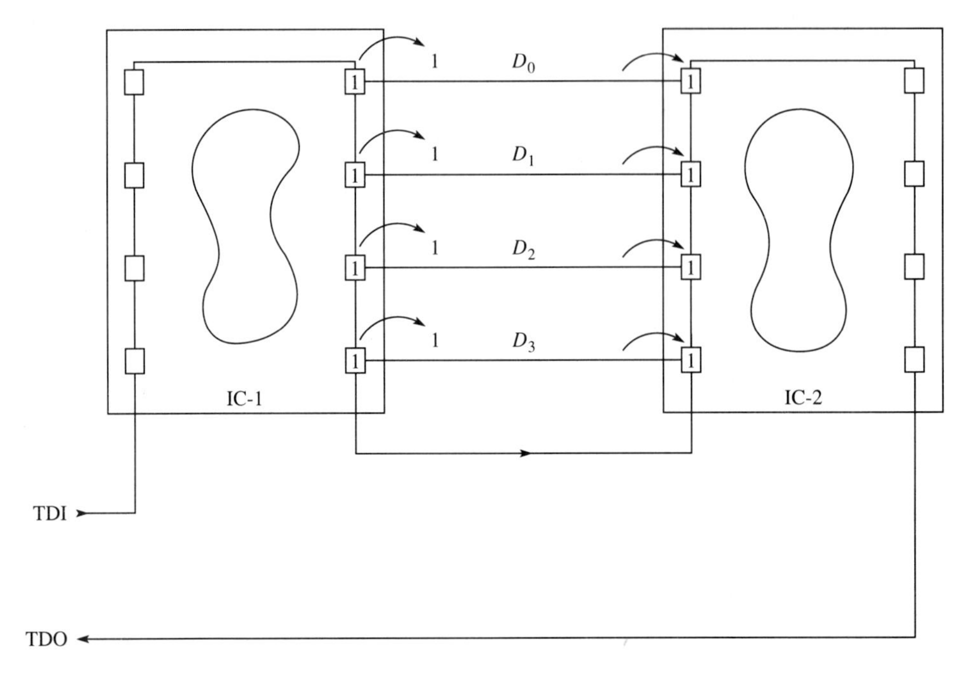

FIGURE 15–58
Testing interconnections. A pattern is shifted to the output cells of IC-1 and then loaded onto the data bus D_0–D_3. The data on the data bus is loaded into the input cells of IC-2. It is then shifted around to TDO and compared to expected values.

contents of the output boundary scan cells onto PO; thus, the contents of all the output cells of IC-1 will be loaded, in parallel, onto the data bus. IC-2 is then placed into an *input load* mode. An input load will place the contents of PI into the boundary scan cell; thus, the data on the data bus will be loaded into all the input boundary scan cells of IC-2. IC-2 is placed into the shift mode and the data, latched in its input scan cells, is marched out TDO and examined by the computer. By loading different test patterns onto the data bus, information can be gained about opens and shorts that exist between the devices.

TESTING LOGIC

Logic can be tested in a similar manner as testing connections. Test data is shifted into the input cells of IC-1 and IC-2. This data is then loaded into the internal logic of the device. Output data from the internal logic is loaded into the output cells of IC-1 and IC-2 and then shifted to TDO and examined by a computer.

THE BOUNDARY SCAN CELL

The boundary scan cell is shown in Figure 15–59. It contains two mode switches and two data latches. It is controlled by the TAP controller. The left-hand switch and latch are used to latch data from either PI (load) or SI (shift); and the other

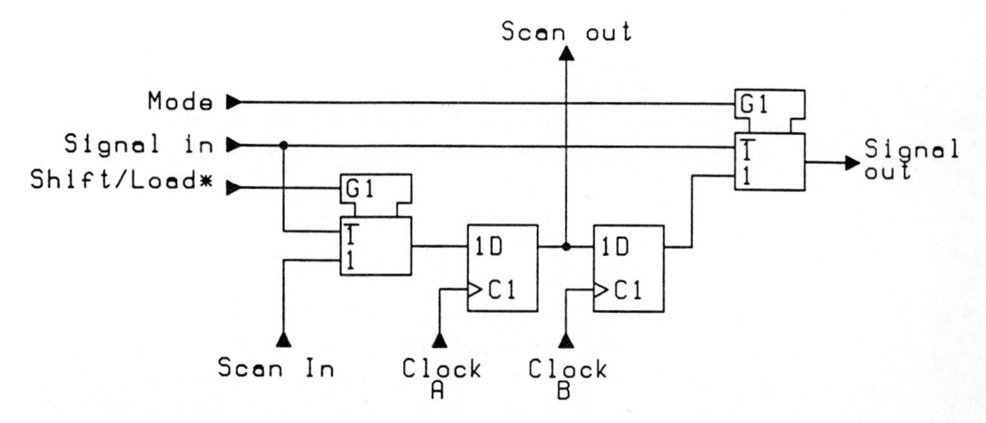

FIGURE 15–59
A boundary scan cell (courtesy of IEEE).

latch and switch are used either to bypass the cell (normal mode) or to load the contents of the cell onto PO (test mode). The architecture of the boundary scan cell for the input side of a chip is essentially the same as for the output side, and the data travels through the cell in the same direction. When the cell is designed onto the silicon of a chip, there are differences on the input-loading parameters and output-drive capabilities between the input cell and the output cell, but the basic concept is the same.

SOFTWARE FOR BOUNDARY SCAN

Currently, software and plug-in boards for personnel computers are available that interface with the protocol specified in the IEEE Standard 1149.1-1990. The software requires skilled personnel to develop a test program, but it guides the programmer through the process of describing the UUT. When the description is complete, preprogrammed routines are used to develop tests on the loop, connections, and the logic. The final program that is used to test a unit should enable a nonskilled operator to locate faults easily and accurately.

ADVANTAGES AND DISADVANTAGES OF BOUNDARY SCAN

The primary advantage of boundary scan is that it can gain access to test nodes that are normally inaccessible using mechanical means (bed-of-nails, for example). This is particularly useful for high-density surface-mounted boards and in cases where parts are on both sides of the board. The primary disadvantages of boundary scan are the following: (1) All the devices on an assembly need to have boundary scan in them in order to locate all possible faults. (2) Board power must be applied for boundary scan to work. A serious short can render a board useless, causing catastrophic damage. (3) Because the test is done by shifting data serially through the ICs, test times involved to perform the test can be considerably longer than with a bed-of-nails approach. (4) Some devices cannot be tested because of the time dependency. The cycle time between "loads" can cause the device to forget the value of the previous load, causing erroneous results.

A combination of bed-of-nails, in-circuit component testing, and boundary scan testing is a compromise. As of 1991, one company had added boundary scan programming to their in-circuit testers to give a mix of the two technologies.

15–10 RELATIVE COSTS OF ATE

There are several parameters associated with the cost of ATE to a manufacturer. There is the cost to initially purchase the equipment (*purchase cost*), the cost to maintain the system (*cost of ownership*), and the cost savings (*return on invested capital*). The cost to purchase is a one-time cost. It includes the base price, cost of options, service kits, and training classes, sales taxes, facilities requirements (power connection and well as air or vacuum), and other costs needed to complete the system. The cost of ownership accrues year after year. Often this cost is at a fixed rate, but it may go up as business increases. The cost of ownership includes such items as test engineering salaries, cost of fixtures, routine maintenance charges, and other day-to-day operating costs. Return on investment is the money that is saved compared to the money that is spent. The earlier in the manufacturing process a problem is found, the cheaper it is to correct. If a problem is delivered to a customer, the cost to correct it is the greatest; the customer may even be lost for repeat business. Figure 15–60 shows representative costs to locate a problem at various stages in the manufacturing cycle.

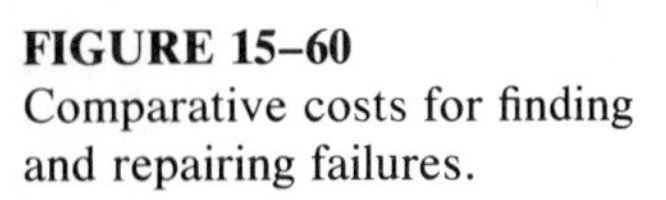
FIGURE 15–60
Comparative costs for finding and repairing failures.

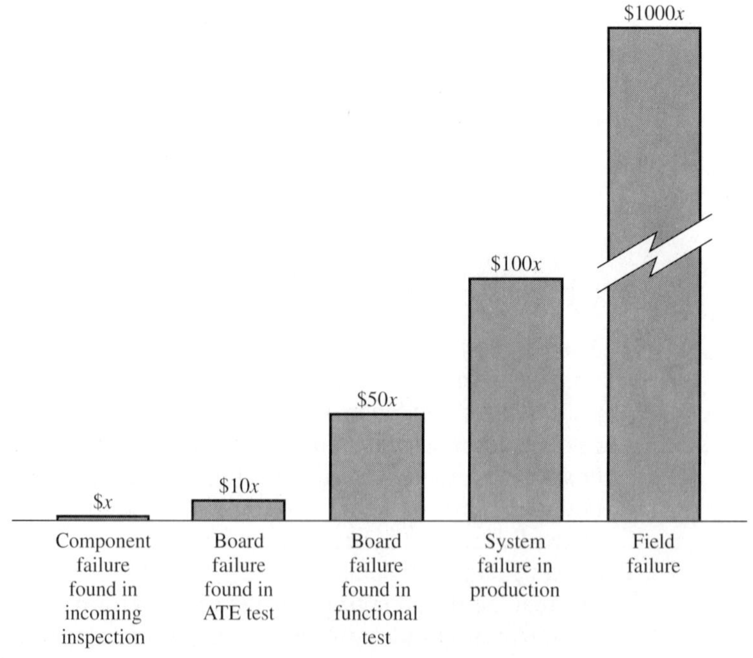

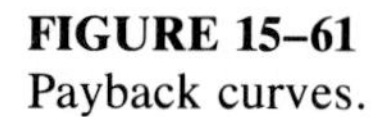

FIGURE 15–61
Payback curves.

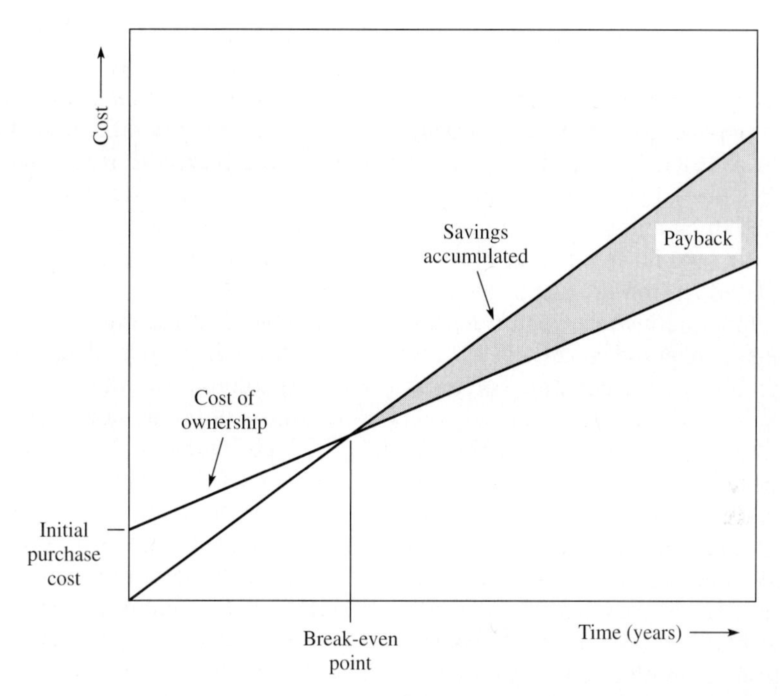

Figure 15–61 shows the initial purchase price of the equipment (the initial step) and the cost of ownership (a fixed-rate climb), which is cumulative over the life of the tester. It also indicates the savings that the automated tester will offer over testing using manual means. The savings accumulate at a greater rate than the cost of ownership. The two lines intersect at the point called the **break-even** point. The payback period is where the savings exceed the cost. The time beyond the payback period is when the tester is earning money. Most companies set the break-even point to be within 3 years, whereas the life of testers is 5 to 10 years, meaning that the tester can generate savings for 2 to 7 years.

SUMMARY

1. ATE stands for automatic test equipment. It refers to systems that can test manufactured electronic assemblies, or it can be used in data-acquisition systems to test physical parameters for recording and/or control.
2. UUT stands for unit under test. It can be a single device, component, or an entire assembly of devices and components.
3. Manufacturing test systems are systems used primarily to test electronic circuit assemblies. They generally consist of a bed-of-nails for probing circuit nodes, power supplies, a computer, and driver/receiver circuits to connect the pins to stimulus and measurement equipment.
4. Rack-and-stack is a term used to describe a test system comprising individual pieces of test and measurement equipment that are connected together and housed in a common equipment rack. The system is a rack containing a stack of equipment and

is typically interconnected with the IEEE-488 bus.

5. Manufacturing test systems are easy to reconfigure for general and specific testing needs but are often limited in frequency response because of stray inductance and capacitance. Rack-and-stack systems are configured to specific needs and can become outdated as requirements change; they do permit the highest-quality equipment to be configured together for the greatest precision and accuracy of measurement.
6. A manufacturing defects analyzer (MDA) is a method of testing a board by measuring impedances. It is very easy to program and use. It will not locate which component is defective; rather it shows problems with nodes. A component that interconnects two "bad" nodes is suspected of being the cause of a problem. MDA is generally effective for circuits consisting of passive components and analog devices. It does not do well with digital circuits.
7. In-circuit testers measure the value of each passive component, diode junctions, analog and linear devices, and digital devices. They are effective for locating which device or trace is defective. A bed-of-nails is used to probe the individual nodes to gain access to the components.
8. Functional testers perform specific parametric checks on the UUT to assure it performs to specifications.
9. Combinational testers are made up of in-circuit testers and functional testers combined into one unit. In-circuit testing is done first, followed by functional testing.
10. Two-, three-, and six-wire measurement methods are common for in-circuit testing. Two-wire measurement consists of a stimulus and a measure wire, three-wire measurement adds a guard node to block parallel currents, and six-wire measurement adds sense wires to compensate for fixture wiring losses.
11. Gray code is a method used to test digital logic and memory. Inputs to a digital UUT are driven with frequencies that will exercise all the logic states of the device. The output is fed to a CRC chip that assigns a unique number as the "signature" of the output.
12. The most accurate test of digital devices is through the use of vector tables, but they require knowledge in advance of the operation of the device in order to write meaningful vectors. Vector tables can run into tens of thousands of test vectors; they are often generated by computer programs using the device type and equations as inputs to the vector processor.
13. Signature analysis is a method of troubleshooting digital circuits. A long string of changing digital information is captured and a CRC is calculated on a "good" board. Test boards are then checked for the same CRC ("signature") as the good board.
14. Forcing the output of a digital device to the opposite state (at the output pin) is called back-driving.
15. Pin drivers or receivers are either multiplexed at their outputs via relays or are dedicated (one driver per pin).
16. SMD stands for surface-mounted device. The devices are soldered to the surface of the circuit board on the same side as the component, whereas through-hole devices have leads that pass through holes on the circuit board and are soldered on the opposite side of the board.
17. Boundary scan is a serial method of scanning the pins at the boundary of the device (where it connects to the pins) through electronic test circuits that are designed onto the silicon of the chip. It is a practical approach to testing when physical probing is not possible.

QUESTIONS AND PROBLEMS

1. Describe three types of ATE equipment. How do they compare with each other? What are their strong points and their weak points?
2. If you were to use rack-and-stack equipment to test and align CB radios, what type of equipment would you have in the rack? Would you include automation; if so, what would be the functions of the automation?
3. When would you consider using a bare-board tester, and when would you think it is not practical? Explain.

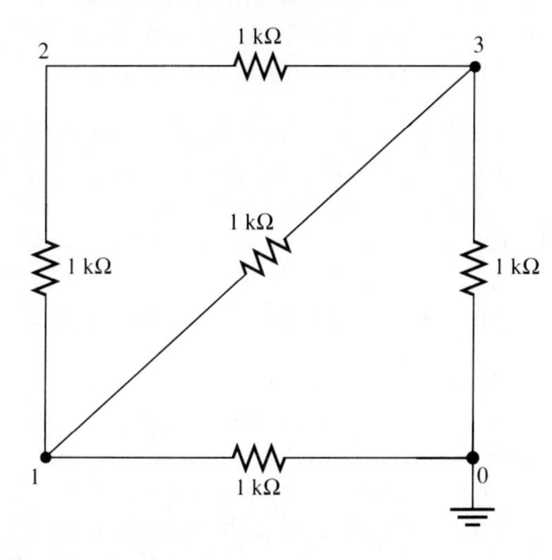

FIGURE 15–62

4. In the circuit shown in Figure 15–62, the reference node is node 0. On a good board, the impedance from the reference to each of the respective nodes is: node 1 = 624 Ω, node 2 = 1 kΩ, and node 3 = 624 Ω. A defective board is placed on the tester. The readings are: node 1 = 604 Ω, node 2 = 684 Ω, and node 3 = 604 Ω. Which component do you suspect to be bad? (*Hint:* Which node differs the most from the good board? Are you able to choose a single resistor or a group of suspect resistors? Explain why.)
5. Figure 15–63 shows a simplified cable scan tester. A voltage is supplied to one of the pins at connector A while a meter is scanning the connections at connector B. Once all the pins on connector B have been scanned and the results recorded, the voltage source at connector A is moved to the next pin and B is scanned again. This process repeats itself until all positions on connector A have been supplied with the voltage. What is wrong with this approach, and what would you do differently?
6. What is the primary difference between the analog devices test and the linear devices test?

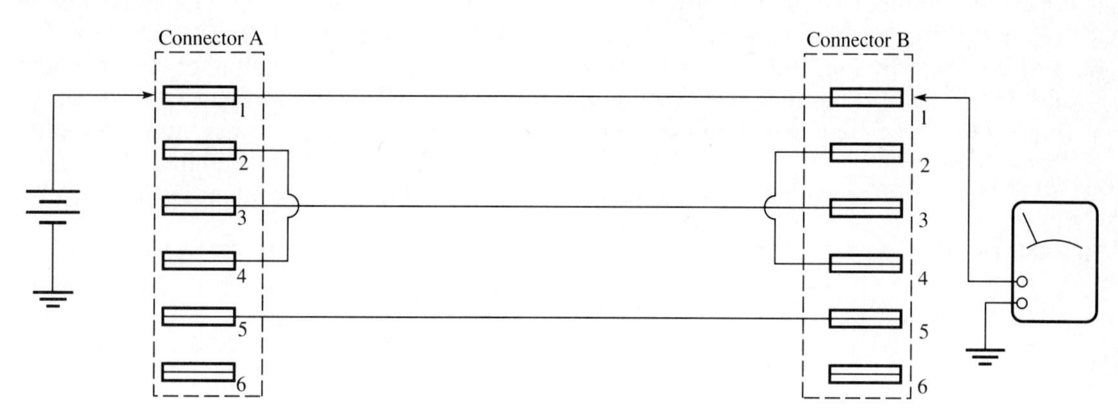

FIGURE 15–63

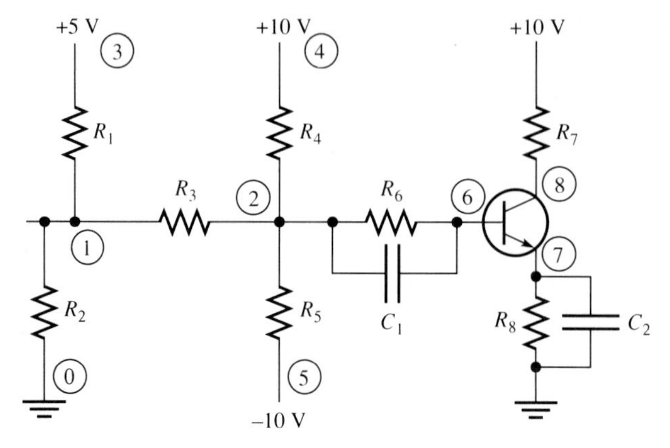

FIGURE 15–64
Circled numbers are node designators.

7. In Figure 15–64, R_3 is to be tested using an in-circuit tester. Will power (+5 V, +10 V, and −10 V) be supplied during the test? What nodes should be guarded? Assume the stimulus node is 1 and the measure node is 2. What is the voltage on the guard points during the test?
8. A 2.7 Ω resistor measures 3.1 Ω with an in-circuit tester. Will guarding correct this problem? What is the probable cause of the high reading and how would you correct it?
9. Explain the principle purpose of Gray code in automatic test equipment. How does it help to avoid glitches?
10. What three tests can be conducted on memory devices that will give reasonable confidence that the memory device is working properly?
11. What is signature analysis and how does it relate to CRC calculations?
12. Explain back-driving, why it is sometimes needed, what precautions are required, and what can be done to a circuit to avoid back-driving.
13. Explain the difference between multiplex drivers and dedicated drivers. What are the advantages of each?
14. What type of test pin would you use to probe a via that is filled with solder, and what type would you use to probe a via that is hollow?
15. What is a clamshell fixture, when is it used, and what design changes can be made to avoid its need?
16. What is the purpose of boundary scan, and under what conditions is it needed? In your estimation, will boundary scan ever replace bed-of-nails testing?
17. Five chips are connected together with the boundary scan serial test data lines, as shown in Figure 15–65. Each chip has eight digital inputs and eight digital outputs, each having a boundary scan cell. How many clock cycles (TCK) are needed to move a logic 1 from TDI on IC-1 before it reappears on TDO of IC-5?
18. Explain the operation of the boundary scan cell shown in Figure 15–59. How is the cell placed in bypass, how is SI shifted to SO, how is PI loaded into the cell, and how is the cell's content loaded onto PO?

FIGURE 15–65

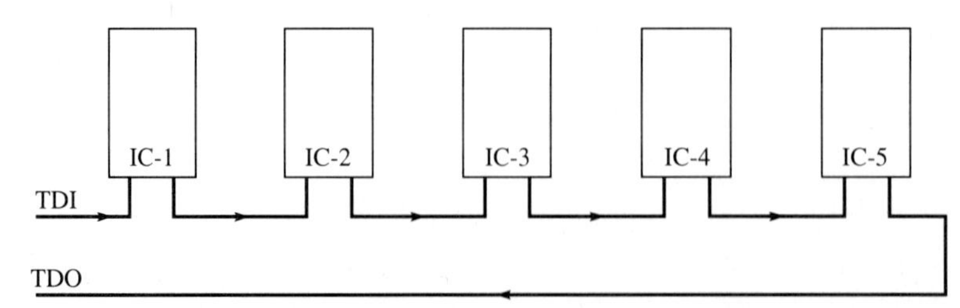

Chapter 16

Video Test Methods

OBJECTIVES

Video was originally developed for information exchange and entertainment; however, it has played an increasingly important role in medical and industrial instrumentation applications. Miniaturization of video optics has enabled the physician to view inside small passageways such as blood vessels. High magnification has permitted long hours of microsurgery without the fatigue of bending over a microscope and has allowed the surgery team to view the details simultaneously on a video monitor. Video enhancing has added perception not seen through ordinary optical methods; high contrast is used so sutures are more visible. Color enhancement can be used in industrial applications to indicate various temperatures, depths, or material density. Television robotics has permitted views otherwise humanly impossible, such as of rocks in a city's storm-drain system or of the water injectors in a coal-steam power plant.

This chapter presents applications of television to instrumentation systems and introduces instruments that are unique to measuring broadcast video signals. Broadcast video is more complex than other forms of video because the waveform contains synchronization information and color in addition to the picture information. Once you understand broadcast video, the other forms are easy to grasp; for this reason, broadcast video is the principal form discussed. In addition, studying broadcast video gives an opportunity to investigate some specific application instrumentation used to measure parameters of a television broadcast video signal. Two instruments in particular will be studied: the waveform monitor (a special type of oscilloscope) and the vectorscope (an oscilloscope that displays color signals as vectors of phase and amplitude). Before these instruments are introduced, video is explained. The U.S. television system, known as the NTSC system (standard RS170A), is discussed.

To understand certain video measurements and instruments, we need to delve into what makes up the television signal and the method by which color is encoded onto the picture signal.

When you complete this chapter, you should be able to

1. Describe the basic concepts of the television sweep system and how the color is superimposed as a phase-modulated signal.
2. Explain how to use a vectorscope.
3. Describe the features of the vectorscope graticule.
4. Define and explain how to measure chroma delay, differential gain, and differential phase and explain their effect on the picture quality.
5. Given a color at maximum saturation, compute the phase and amplitude of the color vector.
6. Explain how to use a waveform monitor.
7. Describe the features of a waveform monitor graticule.
8. Explain how to measure gain and line-time distortions.
9. Explain how line-time distortions affect the quality of the picture.

HISTORICAL NOTE

In the early 1930s, Felix the Cat was the first "star" to appear before RCA-NBC experimental television cameras. Felix whirled around on his phonograph turntable for hours on end while four hot arc lights beat down on him. In those early days the crude TV images of Felix made it look as if he was being viewed through a venetian blind (Figure 16–1). The picture was transmitted from New York City and was received as far away as Kansas. There, and at points in between, it was picked up by video buffs on their primitive 60-line viewers.

16–1 IN THE BEGINNING

The transmission of visual images goes back more than 100 years, to 1875, when George Carey in Boston used the system in Figure 16–2 to transmit simultaneously (in parallel) each separate picture element (**pixel**—the smallest defining element of a picture) by wire, one wire per pixel. A few years after that, a method was developed that reduced the bundle of wires to just one wire pair; this method used the principle of rapidly scanning each picture element in succession and sending the picture elements in series rather than in parallel. In 1880 Maurice Leblanc of France proposed this method; it led to one of the first television patents, which was issued to Paul Nipkow of Germany in 1884. The Nipkow disk, shown in Figure 16–3, is a disk containing a series of tiny holes placed in a spiral. As the disk rotates, the holes scan an object line by line, allowing the light to enter a photo cell. At the receiver end, there is another Nipkow disk spinning in front of a flat neon screen. The current produced by the photocell is used to excite the

FIGURE 16–1
Felix the Cat.

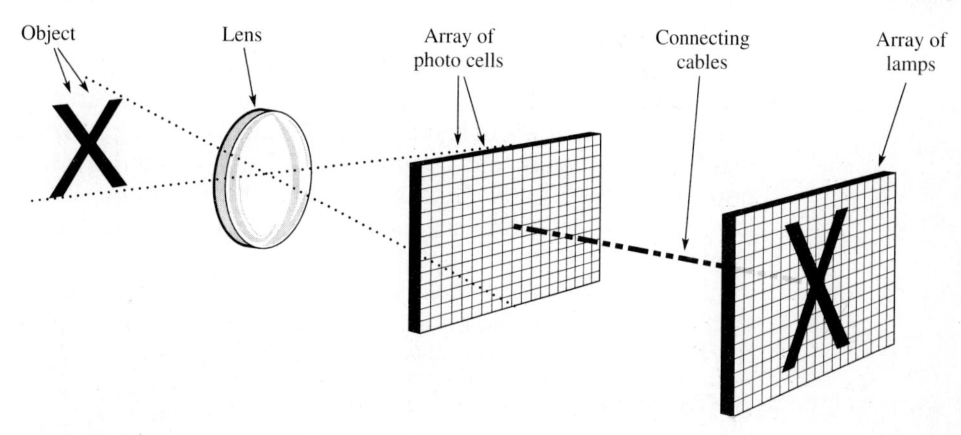

FIGURE 16–2
Early experiments in image transmissions.

neon screen. The two disks are synchronized with each other. The rotating disk at the sending end examines the light intensity of the source object one pixel at a time, line by line, and causes the neon screen to change brightness in proportion to the captured light. The rotating disk at the receiving end masks the neon screen except for light shining through the hole. Because the disks are synchronized, the holes of the receiving disk are in the same positions as the holes in the sending disk as the two disks spin. Therefore, the light showing through the receiving disk at any given position is proportional to the light shining through the sending disk at that same position. Your retina retains the image as the disk rotates past the neon screen, causing you to see the transmitted image.

In the summer of 1928 the Federal Radio Commission issued experimental television licenses to Jenkins Laboratories in Washington (W3XK) and to the General Electric Company in Schenectady, New York (W2XCW). The early systems used the Nipkow disk as a mechanical means for scanning a picture and viewing it again. It used the power company's 60 Hz line frequency to synchro-

FIGURE 16–3
Nipkow scanning disk.

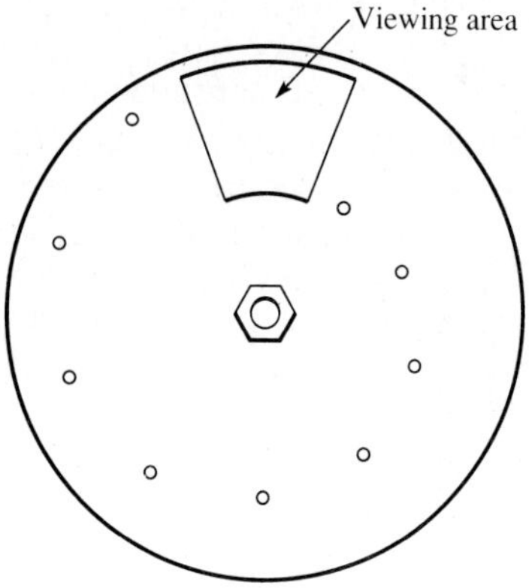

Viewing area

nize the scanning. The utility companies tried to maintain 60 Hz, but there were no unified power grids, so synchronization was always slow and laborious and often was impossible. To quote Percy Maxim of experimental radio station W1AW, ". . . for about half a second, I actually had a picture. It flickered and it was fuzzy and foggy, and about the time I was wondering why they picked on a cow to televise, it suddenly dawned on me that it was a man's face I was looking at. Then I lost synchronization and my man disappeared into a maze of badly intoxicated lines. . . ."

In 1929 Dr. V. K. Zworykin of the Westinghouse Research Laboratories demonstrated a television transmitter based on a mechanical scanning system like the Nipkow disk, but the disk at the receiver was replaced with a cathode-ray tube, much like the one used in oscilloscopes. It wasn't until 1931 that Isaac Schoenberg, with England's Electrical and Musical Industries (EMI), set up a TV research group and replaced the mechanical Nipkow disk at the studio with an electronic scan system using a vacuum tube. The broadcasts were experimental until 1939, when NBC's W2XBS televised President Roosevelt's opening of the World's Fair in New York.

On July 1, 1941, NBC's WNBT and CBS's WCBW were licensed as the first commercial television stations in the United States. The FCC authorization provided for an upgrade in picture-quality definition by adopting a 525-line standard for the picture and FM for the audio portion of the telecasts (replacing AM). American involvement in World War II caused a complete standstill in development of television, including experimental broadcasts. Television didn't really get under way again until 1948. At the end of the year, there were 36 stations on the air and 70 more were under construction, with an estimated 1 million television receivers sold.

It wasn't until 1953 that the Federal Communications Commission accepted the proposal developed by RCA and the National Television System Committee (NTSC) of a color system that would be compatible with the existing black-and-white system. The system assured that a color receiver would receive a color broadcast in color; a black-and-white receiver would receive a color broadcast without any degradation of the picture quality, only it would be in black and white. Reversed compatibility was also met, meaning that a color set could also receive black-and-white signals without degradation.

The system released from development in 1953 is still used in television. During this time, great improvements have been made in cameras, recording media, and receivers to give a clearer and higher-quality picture. Other enhancements, such as stereo sound and closed-caption capabilities, have been added to the system, but the basic system has not changed.

High-definition television (HDTV) is the next step in the development of television. This involves a system with nearly 1125 lines and an increased bandwidth. The picture is wider (more panoramic). It is a 3-unit-high by 5-unit-wide picture, as compared with the older 3-to-4 picture ratio. Other forms of high-definition television are currently used with computer graphics and in medical and industrial applications; they are more commonly referred to as high-*resolution* systems rather than high-*definition* systems.

Our current picture system was designed so that proper viewing distance is approximately five times the diagonal measurement of the screen. A 60-in. wide-screen television receiver should be viewed from a distance of at least 25 ft; otherwise, the scan lines and the imperfections of the picture in terms of sharpness and color resolution will be revealed. With high-resolution systems, much closer inspection of the screen will be permissible before resolution limitations are encountered.

16–2 BROADCAST STANDARDS

There are basically three systems worldwide for transmitting television signals. Most systems use amplitude modulation (AM) to broadcast picture information and frequency modulation (FM) for audio. Color information is encoded on the picture signal differently in each system.

The U.S. system was developed by the National Television Standard Committee and is known as the NTSC system. It has some problems with the way color is encoded because phase error is color-hue error. These errors have caused the NTSC system to be nicknamed "never twice the same color."

The color system developed in Europe is very much like our system except that the phase of the color signal on each alternate line is changed in a manner that cancels errors. This phase alternate by line system is better known as PAL. It has been nicknamed "perfection at last."

The French developed another system for encoding the color. The NTSC and the PAL system both depend on the phase of the color signal (referenced to a signal called the *burst reference*) to decode the color signal. All color signals have the same frequency, but a specific color is represented by shifting the phase of the color signal by a unique delay. This is known as phase modulation (PM). The French system encodes the color as FM. Each color is a unique frequency in the spectrum. Their system is called the SECAM system and has been nicknamed "something essentially contrary to the American method."

16–3 THE SCAN SYSTEM

Consider the problem of passing puzzle pieces to another person, one at a time, for that person to assemble into a complete picture as rapidly as possible. If the pieces were sent in random order, the puzzle might be put together in a random fashion. If, instead, the pieces were numbered on the back (encoded), it would be a simple matter to place them in the correct location on a grid; however, a considerable time would be required to encode and decode the pieces.

A better technique than encoding and decoding the pixels is to send the pixels in an agreed order; this is the idea behind virtually all television systems. The image is sent by a method called raster scanning. **Scanning** is a method of moving the electron beam in the receiver to form a series of horizontal lines that cover the entire picture area of the screen. In the absence of a picture, a white

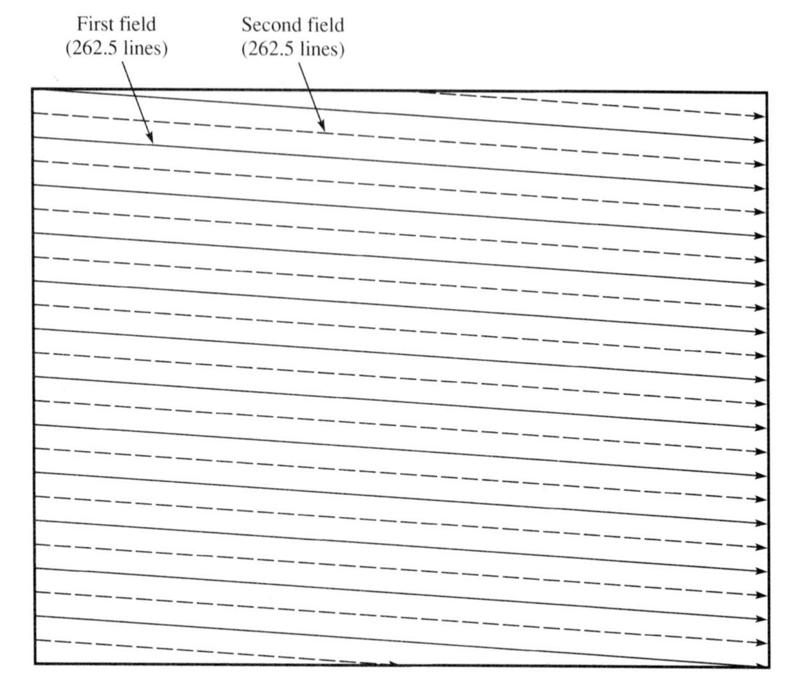

FIGURE 16–4
Raster scanning. The electron beam on a CRT first follows the solid lines (field 1) and then the dotted lines (field 2). There is a total of 525 lines in a complete picture.

area, called the **raster,** can be observed on the screen. In television, the odd-numbered horizontal lines, called *field* 1, are **interlaced** in a pattern with the even numbered lines, called *field* 2, to produce a **frame,** or single picture. This idea, illustrated in Figure 16–4, is used by all of the world's TV systems to prevent objectionable flicker. The NTSC system has about 220 pixels in a line and 525 lines in a frame.[1]

To keep the transmitter and receiver in step, it is necessary for the transmitter to send additional information in the form of *synchronizing,* or *sync,* pulses. **Horizontal synchronizing pulses** are rectangular pulses that are used to start each line of video at the proper time. Other pulses, called **vertical synchronizing pulses,** are used to start each new field of the picture. To keep picture and sync information separate, the sync pulses have the opposite polarity of the picture elements; picture elements have positive polarity and sync pulses have negative polarity. The difference between the vertical and horizontal sync pulses is that the vertical sync pulse is about 16 times as long as the horizontal sync pulse.

16–4 THE VIDEO WAVEFORM

The video signal is an amplitude-modulated waveform, meaning that the dc component of the signal directly affects the luminance of the picture. Newer video formats include component and digital video, but they are beyond the scope of

[1] High-resolution systems have over 1000 lines of video with over 600 pixels in each line.

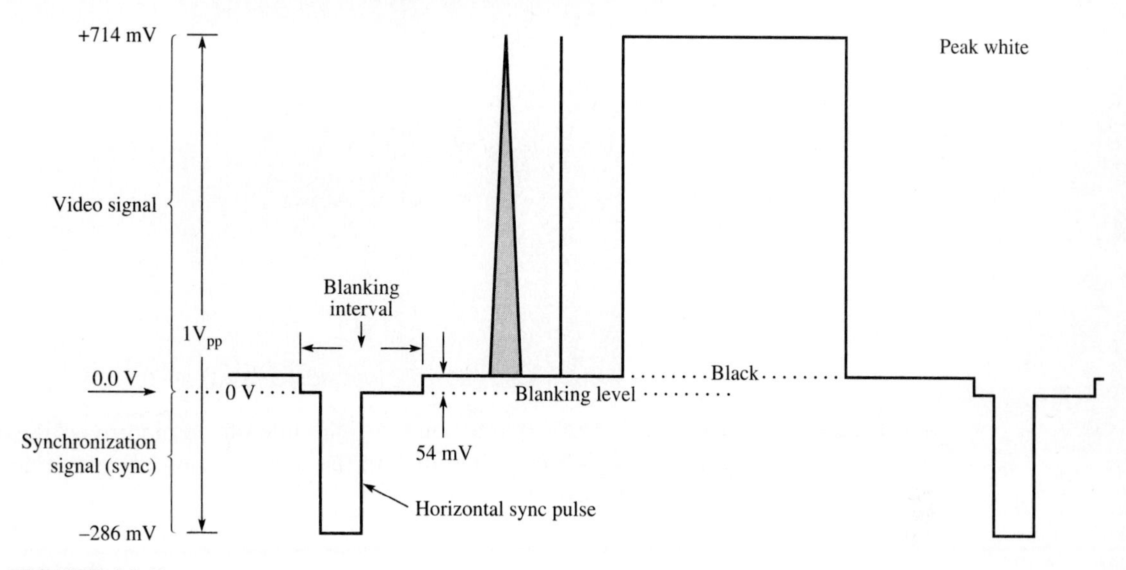

FIGURE 16–5
The composite video waveform, one line.

this text. We deal solely with an **analog composite signal,** the most common form of video, and methods for measuring this signal.

As previously mentioned, the analog composite signal is bipolar (having two polarities). The positive portion contains the video information, whereas the negative portion contains the synchronizing pulses. It is called *composite* because it has the video information as well as the synchronizing information. The synchronizing pulses don't change as a function of the picture content. They are always present, consistent, and stable.

The picture information is constantly changing with changes in the scenes. Figure 16–5 shows one line of the video signal. If the dc level of the signal is at 0.0 V (or 54 mV, as defined by older standards), then the picture transmitted is black. Likewise, if the signal is at maximum amplitude (714 mV), then the transmitted signal is white. All the shades of gray lie between the white and black video levels.

Figure 16–6 shows one line of video that is a RAMP signal. For the one line shown, the signal starts at the left-hand side of the waveform at 0.0 mV (black)

FIGURE 16–6
Ramp signal.

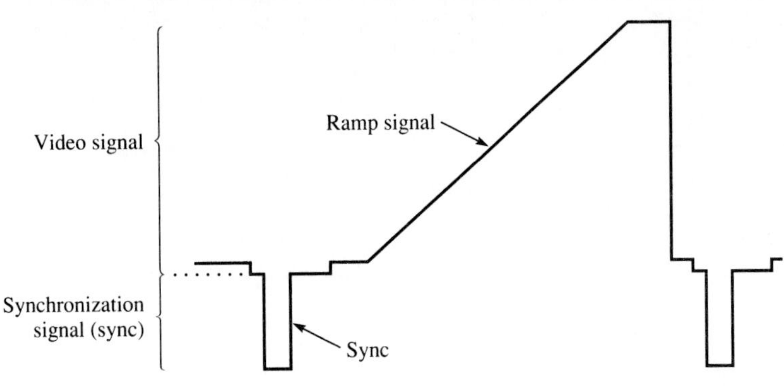

FIGURE 16–7
Actual representation of the ramp signal on a picture monitor.

and progresses linearly to 714 mV (white) on the right side of the picture. If this line is repeated again and again for all the lines in the picture, then the resulting picture is as illustrated in Figure 16–7.

EXAMPLE 16–1 Draw the waveforms needed for Figure 16–8(a) and Figure 16–9(a). Draw the waveform at lines *A*, *B*, and *C*, as indicated. Assume the white is at full level and the black it at blanking.

FIGURE 16–8

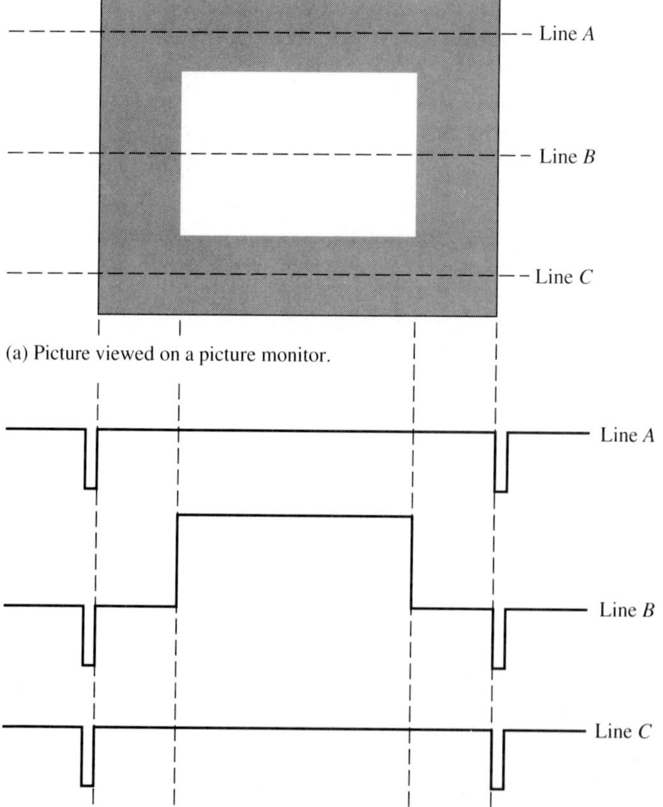

FIGURE 16–9

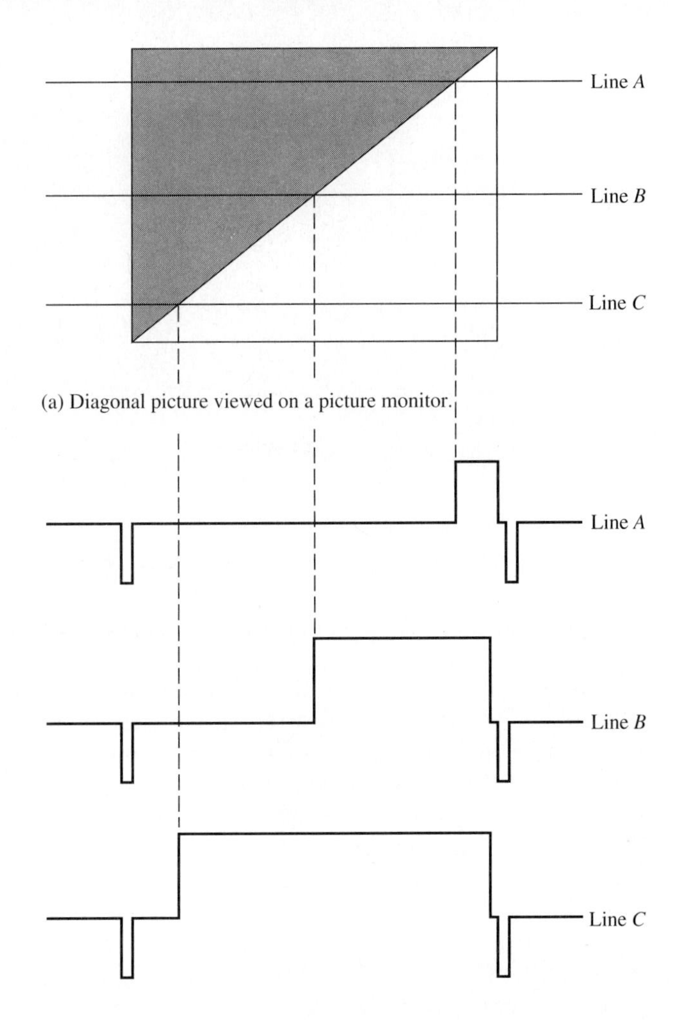

(a) Diagonal picture viewed on a picture monitor.

(b) Same picture viewed on a diagonal waveform monitor.

SOLUTION See Figures 16–8(b) and 16–9(b).

As previously mentioned, the vertical synchronizing pulse is very similar to the horizontal synchronizing pulse except that it is 16 times longer. It stays negative for three horizontal lines, returning to zero briefly to keep the horizontal sync section phase-locked. This is illustrated in Figure 16–10.

16–5 PRINCIPLES OF COLOR

Understanding the color signal starts with an awareness of how the eye sees color. This section deals with primary color and the blending of colors and how the

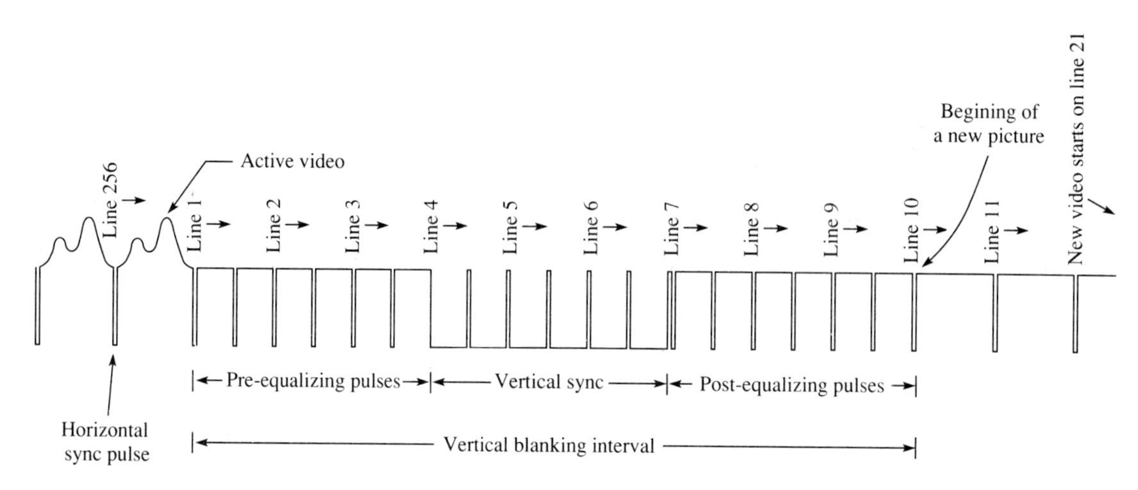

FIGURE 16–10
Vertical blanking/sync interval.

NTSC group of engineers used that information to establish the methods of color transmission.

There are two primary color systems. One has magenta, yellow, and cyan as the primary colors, and the other uses red, green, and blue. In either system, all other colors can be reproduced by mixing proportions of the primary colors together. The system that uses magenta, yellow, and cyan for primary colors is known as the **subtractive** system. The observed colors are a result of subtracting a range of colors through filtering or absorption. If you look at a sunset, it appears to be orange and yellow because the colors in the blue range have effectively been removed from the original white light. The cover of a green book looks green because all other colors are absorbed by the pigment in the ink, allowing just green to be reflected to the observer. The subtractive color system starts with white and removes color to obtain the final color.

Artists frequently refer to the subtractive primary colors (for paints) as red, yellow, and blue. These still represent subtractive primary colors, but red replaces magenta and blue replaces cyan. The light spectrum is a *band* of color and a slight shift of the primary colors (shifting from magenta to red as a primary) just means that color pigments need to be mixed in different proportions to yield the same results.

The system that uses red, green, and blue as the primary colors is the **additive** system. The observed colors are a result of *adding* colors. Colors add when they are combined from the *source* of light rather than from *reflected* or filtered light. Figure 16–11 illustrates this concept. Three light sources—three lamps that are red, green, and blue—are mixed. Notice that yellow, for example, is made up of red and green mixed together. The screen does not absorb (subtract) any of the light; it reflects it all. The blending of the green and red colors together takes place on the retina of the eye and is perceived as yellow by the brain. The additive color system starts with black and adds color to obtain the final result.

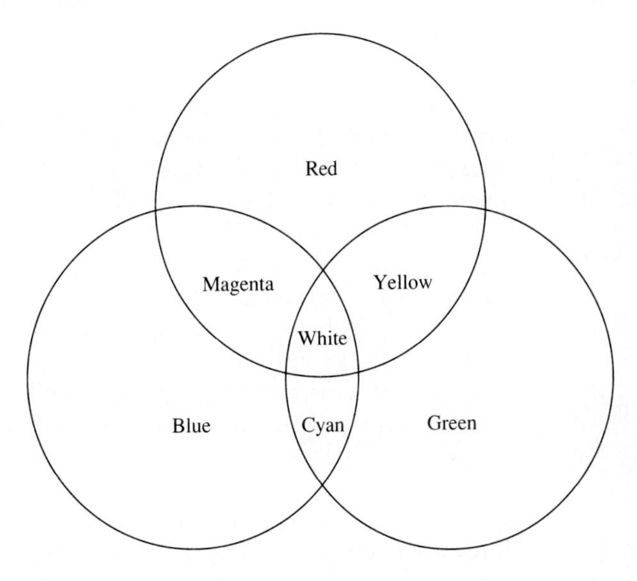

FIGURE 16–11
Additive color mixing.

Colors developed on the face of the color picture tube are produced by an electron beam striking a specific colored phosphor dot. Different phosphors emit different colors of light. Because the face of the tube is the *source* of the light, the primary colors used in television are the additive primaries: red, green, and blue.

When an electron collides with a phosphorous atom, the energy of the collision is absorbed by the outer-valance electrons. Those outer electrons are moved to a higher-energy band and reside there for a period of time; then the outer electron returns to its original energy band in steps, as shown in Figure 16–12. As it moves back, it gives off energy in the form of light. We see the emission of photons; different types of phosphors produce different colors of light.

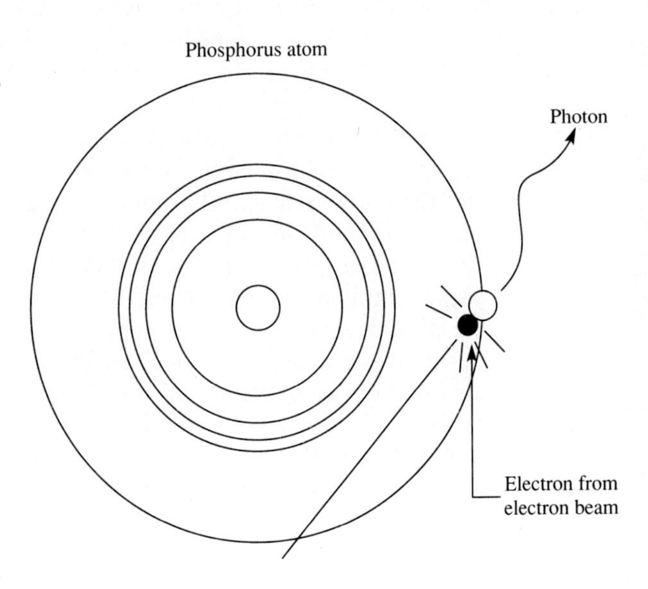

FIGURE 16–12
Emission of light due to electron collision.

The picture tube in the television system is very similar to the CRT explained in Section 8–2. The main difference is that the electron beam in a television picture tube is deflected electromagnetically instead of electrostatically. An electromagnet called the deflection yoke surrounds the neck of the picture tube. The reason for using electromagnetic deflection is that the deflection angles are much greater than those obtained with electrostatic deflection. Typical electrostatic deflections are 15°, whereas magnetic deflections for consumer television receivers are often as large as 55° (±55° from center, or 110° total angular deflection). Television picture tubes measuring 19 in. diagonally would have to be nearly 5 ft long if electrostatic deflection were used but are only about 10 in. long with magnetic deflection. The deflection rates for the picture tube are not as great as that needed for the oscilloscope because its highest scan frequency is only about 15 kHz, whereas the deflection rates for the oscilloscope are in the megahertz range; the faster sweep rates of the oscilloscope are not needed in broadcast television.

The other major difference between a television's CRT and an oscilloscope's CRT involves the type and color of phosphorous. The persistence of the phosphorous for each has been optimized for that particular application. As you might expect, black-and-white television's phosphorous is white. It emits the full spectrum of light when struck by the electron beam. If a particular section of the scene is to be black, then the electron beam is simply cut off as the beam passes. More electrons in the beam will cause the scene to be brighter. The quantity of electrons in the beam is controlled by the amplitude of the video signal described earlier. The video is connected to a **control grid** inside the CRT. The control grid acts like an electron valve, adjusting the flow of the electron beam. As the grid is made more negative compared to the cathode (the source of the free electrons), fewer electrons are in the beam. There is a voltage at which the control grid will repel all the free electrons back to the source, preventing any beam current. This condition is known as **cutoff;** the scene is black during cutoff.

A color picture tube has three color guns, as compared to a black-and-white tube, which has only one. Electrically, the three guns are identical; they differ from each other only by their physical placement. Some color picture tubes have the three guns arranged in a pattern 120° apart so that they are all the same distance from the screen. To understand the importance of their placement, it is helpful to understand how a pinhole camera operates. A color picture tube uses a similar principle in directing electron beams instead of light beams.

A basic pinhole camera is illustrated in Figure 16–13(a). Light from an image passes through the pinhole and travels in a straight line toward the screen. Since the pinhole directs the light from the source to only one point on the screen, an upside-down image is formed, as shown. If the source consists of separate red, green, and blue lights, an image of each light will appear in a unique location, as shown in Figure 16–13(b). Each image is separated from the others because the surface containing the pinhole has shadowed the other sources by placing a mask in the path. For this reason, the surface containing the pinhole is called a **shadow mask.**

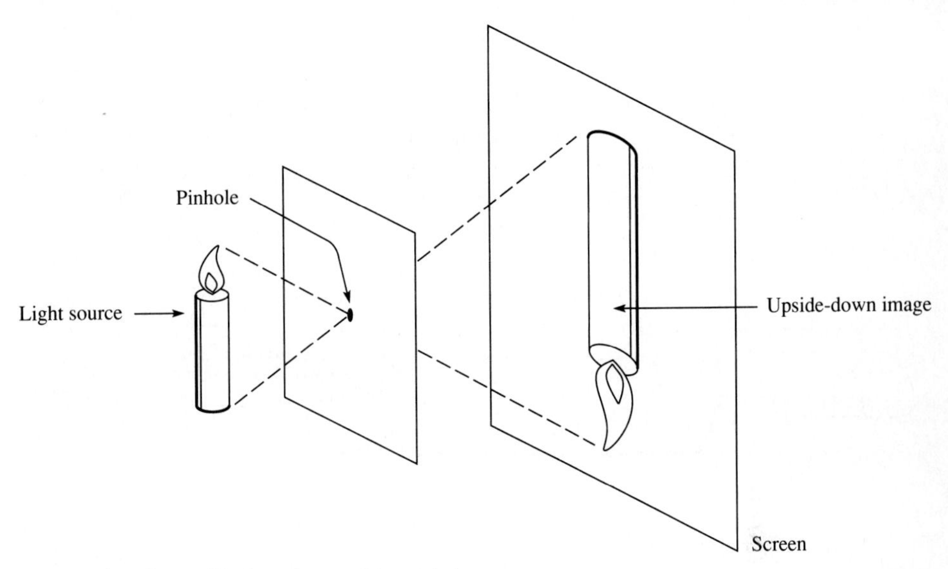

(a) Formation of an upside-down image with a pinhole camera.

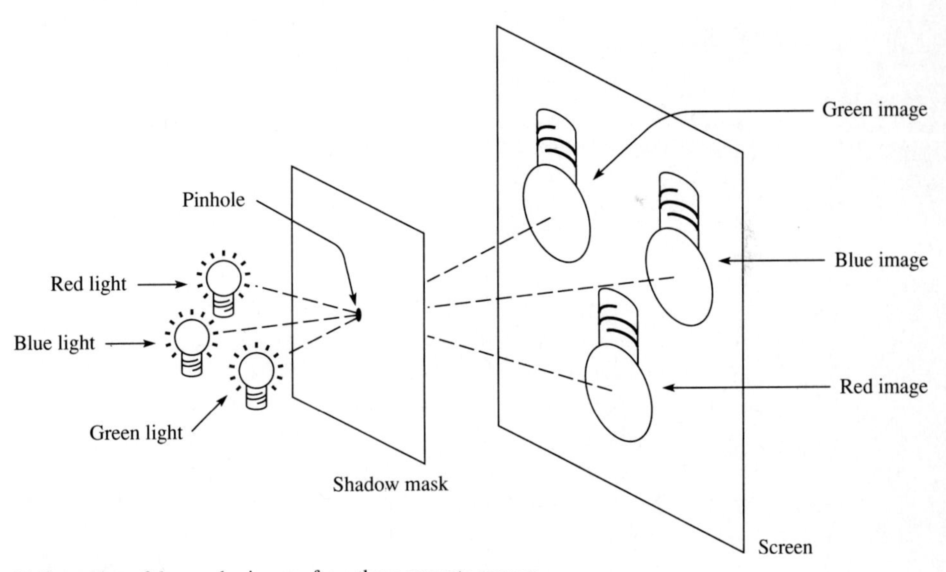

(b) Formation of three color images from three separate sources.

FIGURE 16–13
Pinhole cameras.

If the red, green, and blue lights were replaced with electron guns and the images on the screen were replaced with red, green, and blue phosphorous dots, you can see that the gun in the ''red'' position could send electrons only to the red phosphorous. The green and blue phosphorous dots are shadowed by the mask,

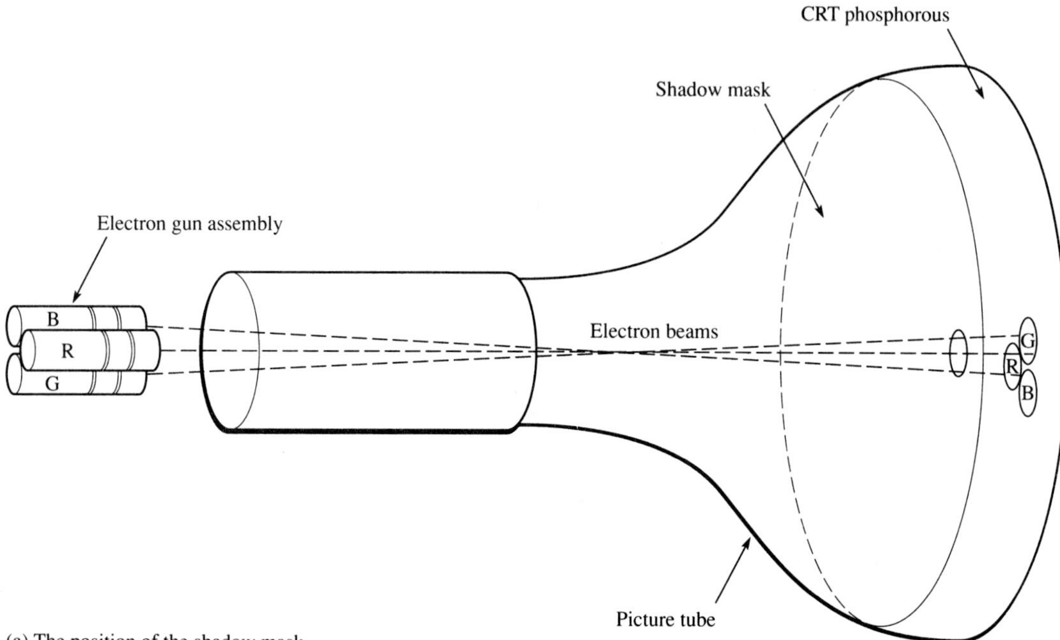

(a) The position of the shadow mask.

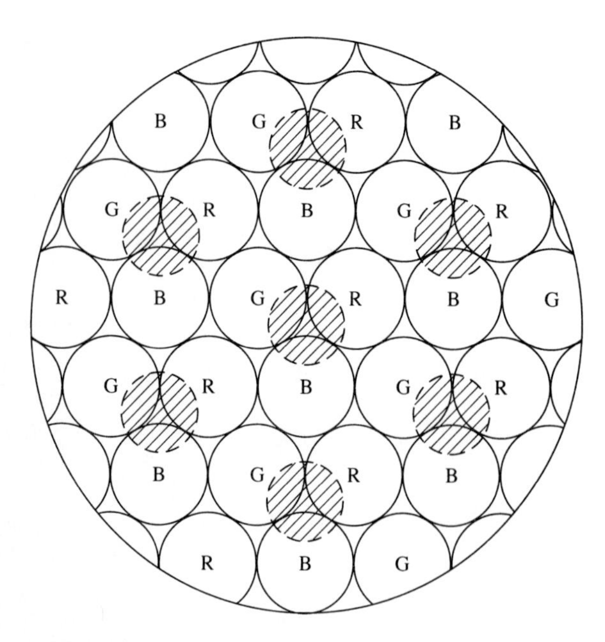

(b) The shaded areas represent the holes in the shadow mask. The red, green, and blue phosphorous dots are on the inner face of the color tube.

FIGURE 16–14
Color picture tube.

preventing them from being activated by the ''red'' gun. A similar situation exists for the other dots.

The entire front surface of the picture tube is covered with red, green, and blue phosphorous dots in discrete and unique locations, as shown in Figure 16–14(a). Just behind the front of the screen is a shadow mask containing thousands of holes, each of which is in line with the center of a trio of the color phosphorous dots. In the neck of the picture tube are three identical electron gun assemblies; their beams scan the face of the screen. The beam from the red gun can strike only the red phosphorous dots (masked from the others by the shadow mask), the beam from the blue gun strikes only the blue dots, and the green beam strikes only the green dots. If the picture is red, then only the red gun will produce an electron beam. For a yellow scene, both the red and green guns will be activated. For white, all three guns will be on at the same time with the same amount of beam intensity.

Figure 16–14(b) shows the relation of the hole (shaded area) in the shadow mask to that of the trio of color phosphorous dots. Each shadow mask hole (or color trio) represents a pixel. You can see how far apart these pixel groupings are by looking at the distance between shaded holes. Holes that are farther apart cause the horizontal and vertical resolution of the picture tube to be reduced. Other techniques have been developed that will improve the resolution by moving the colored phosphorous dots closer together and by arranging them into different patterns.

16–6 COLOR BROADCAST

The waveform in Figure 16–15(a) is a representation of one line of video, whose level varies between black and white; it is referred to as the **luminance signal.** The level shown is observed on the television screen as gray. It controls the *brightness* of a black-and-white picture by controlling the electron-beam current in the picture tube. The waveform in Figure 16–15(b) contains an additional sinusoidal signal called the **chrominance signal.** The chrominance signal has a frequency of 3.58 MHz (3.579549 MHz, to be exact). It is this signal, superimposed on the luminance signal, that contains the color information.

A waveform called a **burst reference** is present at the beginning of each line of active video in a color transmission. This signal is a ''burst'' of 8 to 11 cycles of the 3.58 MHz reference signal produced by the television studio. It does not change with the content of the picture. It is fixed in location, phase, amplitude, and frequency. Its purpose is to provide a reference signal for a phase-locked oscillator in the color television receiver (used for decoding the color picture). This is the same signal (described in Section 3–4) that can be used as a frequency standard.

The phase and amplitude of the 3.58 MHz waveform during the active portion of the picture vary with picture content. The phase difference between the picture waveform and the reference burst is decoded by the television receiver as the **hue** of the picture. The hue is the actual tint of the picture—that is, the reds,

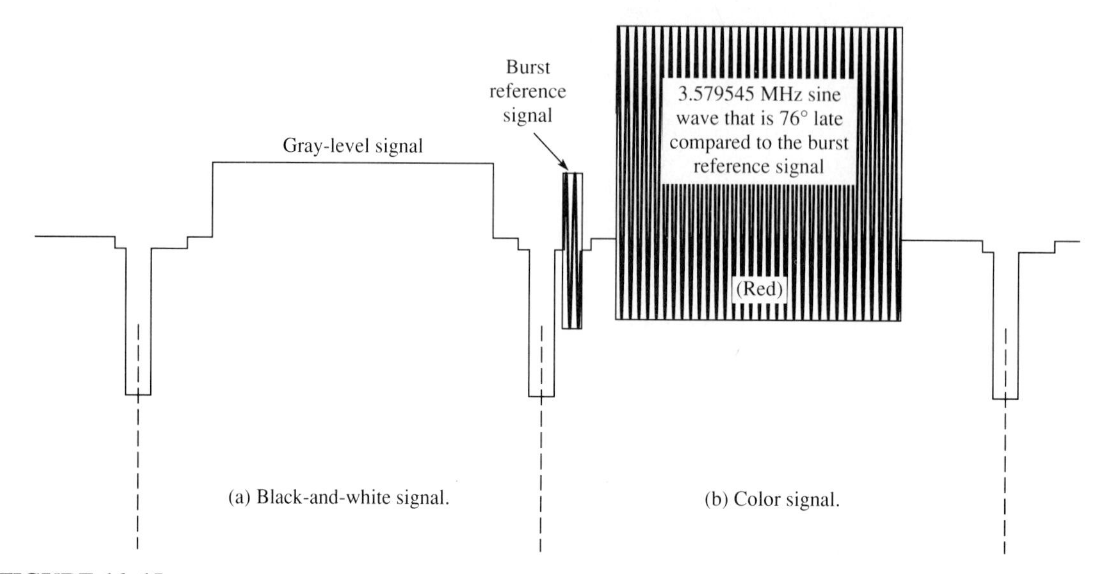

FIGURE 16–15
"Gray" level waveform and "red" signal waveform.

greens, yellows, violets, and other colors. The peak-to-peak amplitude of the waveform produces the **brilliance** of the color in the picture; pastel colors have low color brilliance (small peak-to-peak amplitude), whereas intense colors are often vivid, with high color brilliance (greater peak-to-peak amplitude). Table 16–1 is a table of phases and amplitudes of the 3.58 MHz signal for different colors using a standard test signal called **color bars.**

We can see the phase and amplitude of these signals on a polar coordinate system. The signals are plotted as vectors as illustrated in Figure 16–16. Notice that the horizontal and vertical axes are labeled $R - Y$ and $B - Y$, respectively. Any color can be constructed as a single vector on this polar graph. Later, we will discuss the vectorscope, which is a specialized measurement instrument that displays color phase and gain in the same manner.

TABLE 16–1

COLOR HUE	PEAK-TO-PEAK AMPLITUDE (mV)	PHASE RELATIVE TO BURST REFERENCE (DEGREES)
Red	908	104
Green	848	241
Blue	640	347
Yellow	640	167
Cyan	908	284
Magenta	848	61
Burst	286	180 (Reference)

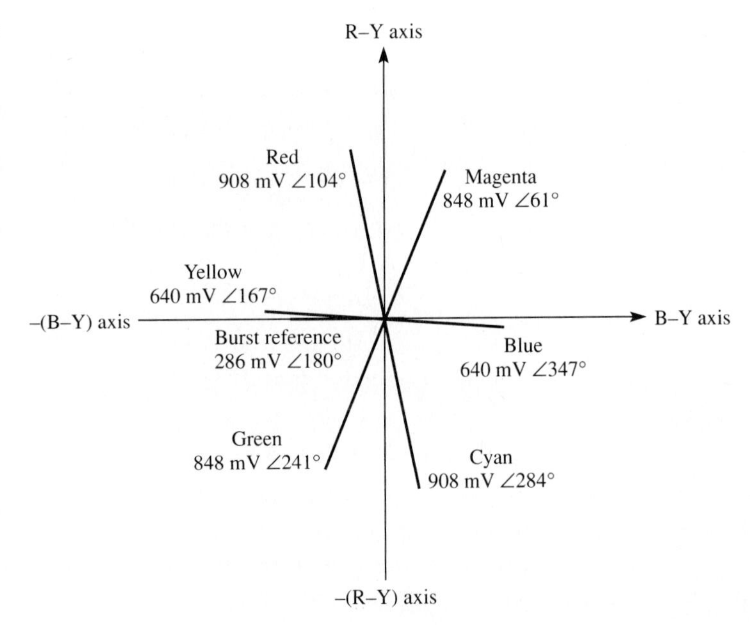

FIGURE 16–16
Vector display.

In order for color transmissions to be compatible with the existing black-and-white system, the luminance signal is sent. As mentioned before, the chrominance is modulated on the luminance as a 3.58 MHz sinusoidal wave. A black-and-white receiver ignores this 3.58 MHz color subcarrier. When color is transmitted, the compatible black-and-white signal is fabricated as

EQUATION 16–1

$$V_Y = 0.3V_R + 0.59V_G + 0.11V_B$$

where V_Y = the luminance signal used by black-and-white receivers, V
V_R = the voltage from the red color gun at the camera, V
V_G = the voltage from the green color gun at the camera, V
V_B = the voltage from the blue color gun at the camera, V

Each of the colors are mixed in this ratio so there will be the proper contrast between the colors when viewed with a black-and-white system; that is, blue will *look* darker on a black-and-white receiver than green, even though they may have the same lumens of brightness. Our eye responds much the same way; blue looks darker to us than green. A color receiver must also be able to receive a black-and-white transmission without impairment. To further complicate matters, the color receiver must be able to receive a color transmission and disregard the luminance terms (the luminance signal is needed only for black-and-white receivers). To have compatibility in either direction, from the black-and-white standard and the color standard, the color signal is sent as a **color difference** signal. The color differences are $R - Y$, $G - Y$, and $B - Y$, where Y is the luminance term and R (red), G (green), and B (blue) are the color terms. The color receiver adds the luminance (Y) term to the color difference terms to extract the color information, as shown:

$$R = (R - Y) + Y$$
$$G = (G - Y) + Y$$
$$B = (B - Y) + Y$$

If the broadcast is black and white, then the color difference terms are each zero, leaving

$$R = 0 + Y$$
$$G = 0 + Y$$
$$B = 0 + Y$$

Therefore

$$R = G = B = Y$$

If the three color guns equal each other in intensity, then the picture will be black and white, with shades of gray in between.

Example 16–2 calculates the $R - Y$ and $B - Y$ values (the actual transmitted color difference signals) for the color red.

EXAMPLE 16–2 Given that

$$\text{White} = V_Y = 0.59V_G + 0.30V_R + 0.11V_B$$

where V_Y = luminance voltage, V
V_G = green signal voltage, V
V_R = red signal voltage, V
V_B = blue signal voltage, V

(a) Find the phase and amplitude for the red vector at maximum saturation.
(b) Find the luminance value for red.

SOLUTION (a)

EQUATION 16–2

$$V_R - V_Y = V_R - (0.59V_G + 0.30V_R + 0.11V_B)$$
$$V_R - V_Y = -0.59V_G + 0.7V_R - 0.11V_B$$

EQUATION 16–3

$$V_B - V_Y = V_B - (0.59V_G + 0.30V_R + 0.11V_B)$$
$$V_B - V_Y = -0.59V_G - 0.30V_R + 0.89V_B$$

The FCC (Federal Communications Commission) requires that the $R - Y$ axis be reduced in amplitude to 87.7% and the $B - Y$ axis be reduced to 49.3% of their original values. This has to do with some problems with overmodulating the transmitter.

EQUATION 16–4

$$R - Y = 0.877(V_R - V_Y)$$
$$R - Y = -0.517V_G + 0.614V_R - 0.096V_B$$

EQUATION 16–5

$$B - Y = 0.493(V_B - V_Y)$$
$$B - Y = -0.291V_G - 0.1479V_R + 0.4388V_B$$

Only red is activated; blue and green are both off. Therefore, $V_G = V_R = 0.0$. Equations 16–4 and 16–5 reduce as follows:

EQUATION 16–6

$$R - Y = -0.517(0) + 0.614(V_R) - 0.096(0)$$
$$R - Y = 0.614(V_R)$$

EQUATION 16–7

$$B - Y = -0.291(0) - 0.1479(V_R) + 0.4388(0)$$
$$B - Y = -0.1479(V_R)$$

Luminance at full white is 714 mV. Each of the chroma terms (V_R, V_B, and V_G) are 714 mV when they, too, are at maximum. The equations then become

$$R - Y = 0.614(0.714) = 438 \text{ mV peak} \quad \text{or} \quad 876 \text{ mV}_{pp}$$
$$B - Y = -0.1479(0.714) = -105.6 \text{ mV peak} \quad \text{or} \quad -211.2 \text{ mV}_{pp}$$

See Figure 16–17.

FIGURE 16–17

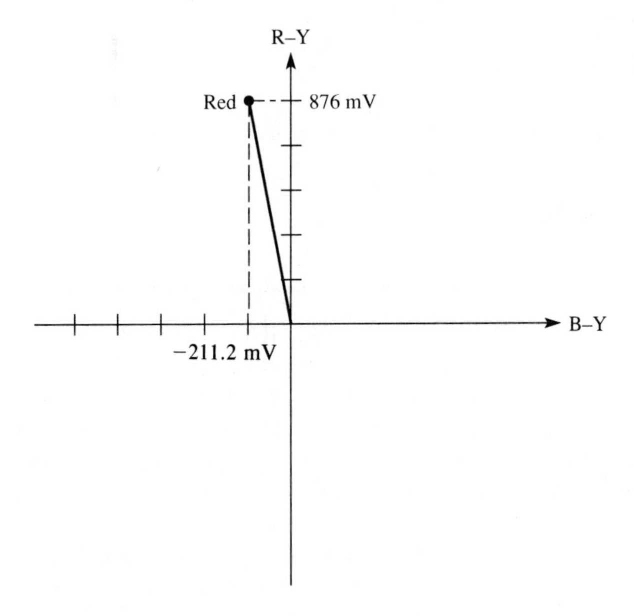

The chroma amplitude is

$$SC_R = \sqrt{(R - Y)^2 + (B - Y)^2}$$
$$= \sqrt{(876)^2 + (-211.2)^2}$$
$$= 901 \text{ mV}_{pp}$$

where SC = subcarrier. The chroma phase is

$$\phi_R = 180° - \arctan \frac{R - Y}{B - Y}$$
$$= 180° - \arctan \frac{876}{211.2}$$
$$= 180° - 76.4°$$
$$= 103.6°$$

(b) The luminance value is

$$V_Y = 0.59V_G + 0.30V_R + 0.11V_B$$
$$= 0.59(0) + 0.30(0.714) + 0.11(0)$$
$$= 214 \text{ mV}$$

Note that the values given for white, $0.59V_G$, $0.30V_R$, and $0.11V_B$, are approximate, as are the figures for the reduction factors. Therefore, the values calculated will not precisely match the values given in Table 16–1.

The solution is shown in Figure 16–18.

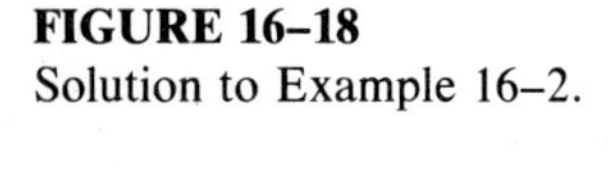

FIGURE 16–18
Solution to Example 16–2.

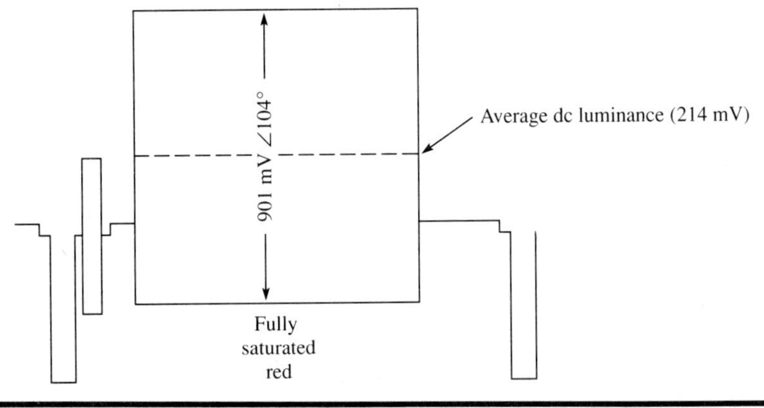

Before we discuss instruments, there are two terms used in video that can be very confusing: differential phase and differential gain. You should not confuse differential phase or differential gain with the word *differential* as used with differential amplifiers. With video, the term differential means the difference between a *black* picture level and a *white* picture level. The *phase* and *gain* refer to the phase and peak-to-peak amplitude of the color signal, respectively. Therefore, **differential phase** is the change in color *phase* (a distortion) between the black and the white levels. **Differential gain** is the change in color amplitude between the black and the white levels. Differential phase and differential gain have nothing to do with balanced inputs (as in op-amps). It's just the difference in the picture brightness and how it may distort the color.

16–7 THE WAVEFORM MONITOR

The **waveform monitor,** as shown in Figure 16–19, is a special-purpose oscilloscope. It is generally not as broadband as general-purpose oscilloscopes; its band-

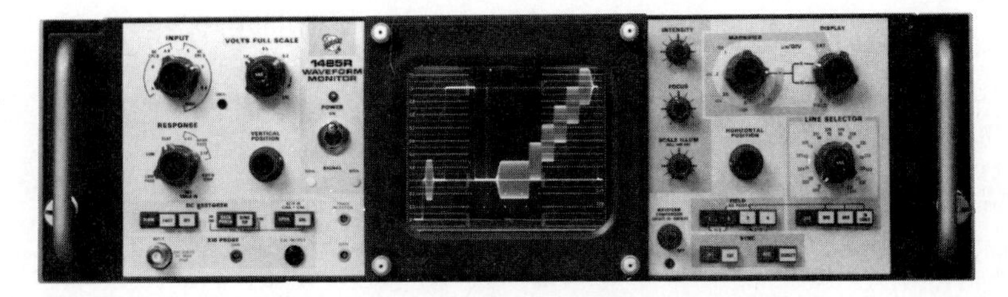

FIGURE 16–19
Waveform monitor (copyright Tektronix, Inc. Used with permission).

width is from dc to about only 20 MHz. But the frequency response (overall gain at specific frequencies) is flatter than the oscilloscope; that is, there is less variation in peak-to-peak voltage from dc to 20 MHz of the waveform monitor than an oscilloscope over this same frequency range. In addition, waveform monitors have special triggering circuits that are not common on oscilloscopes. These circuits aid in selecting which scan lines you wish to view, as test signals reside on certain lines. Also the graticule (discussed later) is marked specifically for video measurements.

CONNECTING THE WAVEFORM MONITOR

Because of their high frequency and noise-shielding requirements, video signals are transmitted over coax cable. The ends of the cable use BNC-type bayonet connectors, as shown in Figure 16–20. This cable is a transmission line with a characteristic impedance of 75 Ω. All video test equipment is designed to be connected with BNC connectors. Cables are driven with a video-driver circuit like the one illustrated in Figure 16–21(a). Its output impedance is as close to 0 Ω as possible. A precision 75 Ω resistor is used in series with the connection to the video cable. At the end of the video cable is another precision termination resistor of 75 Ω. This method is used to drive and terminate the coax cable with the same impedance as the cable itself, which avoids reflections and standing waves.

Many products have **loop through** connectors. This means that the chassis has two BNC jacks that are joined internally; the input to the instrument is high impedance. This allows the signal to be routed to several pieces of equipment via

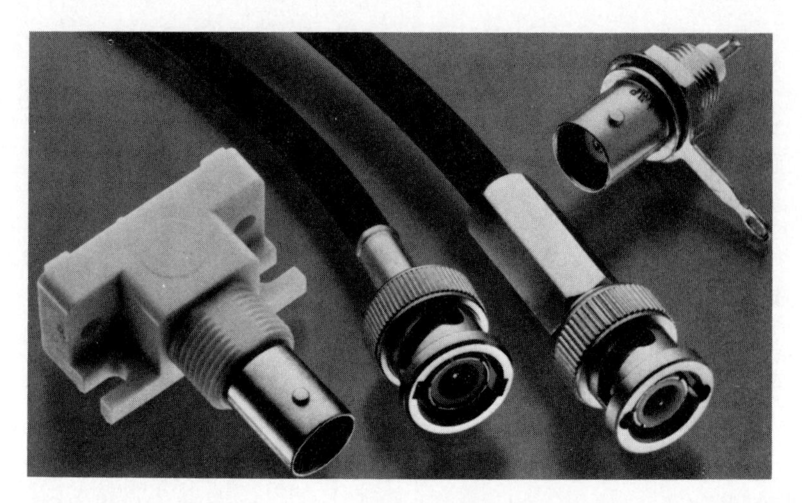

FIGURE 16–20
BNC connectors. (Courtesy of AMP, Inc.).

their loop-through connectors. The last piece of equipment in the line has a termination resistor connected, as illustrated in Figure 16–21(b). The terminating resistor is nothing more than a BNC connector containing a 75 Ω resistor between the center conductor and the shield.

Normal amplitudes of full-white signals (such as the pulse and bar signal that will be described) are 1.0 V_{pp} from the bottom of *sync* (the synchronizing pulse) to the top of the white portion of the signal. If you observe a signal that is twice as large as this after you have connected your equipment, don't panic! Chances are that you have *not* terminated the coax. The driving impedance and the termination impedance act as a −6 dB attenuator (a voltage divider of one-half). The source voltage is 2.0 V_{pp}. Leaving the termination resistor off permits the full source

FIGURE 16–21

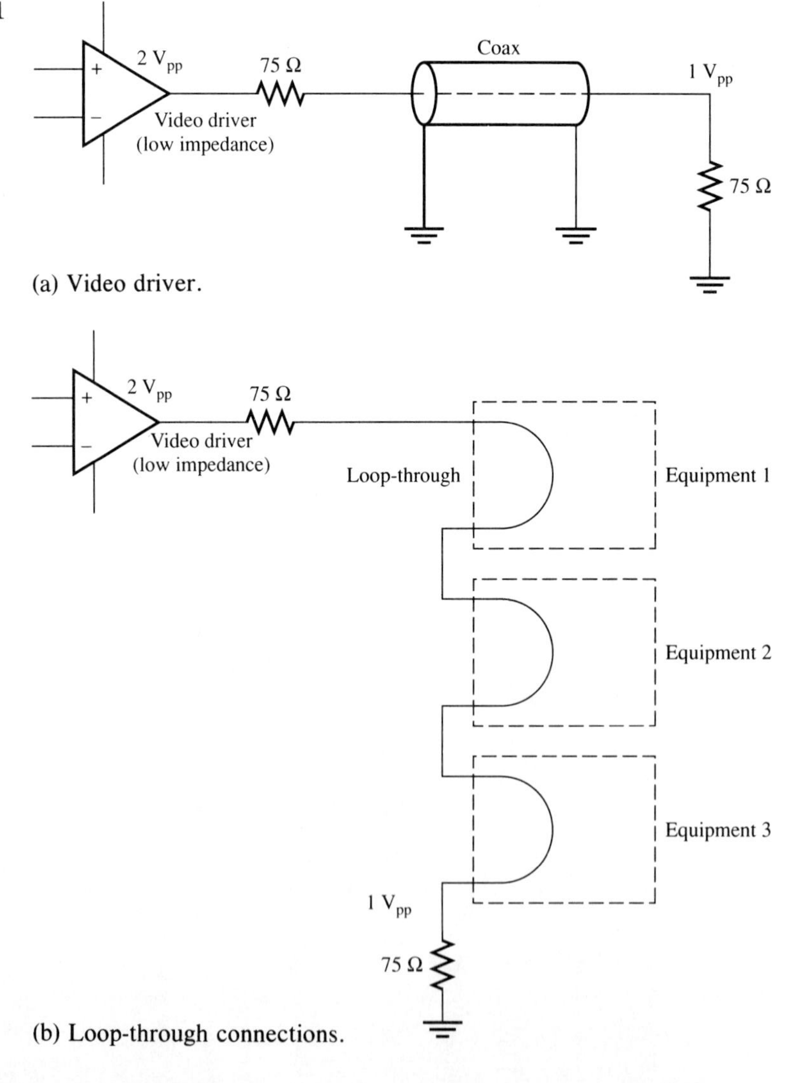

(a) Video driver.

(b) Loop-through connections.

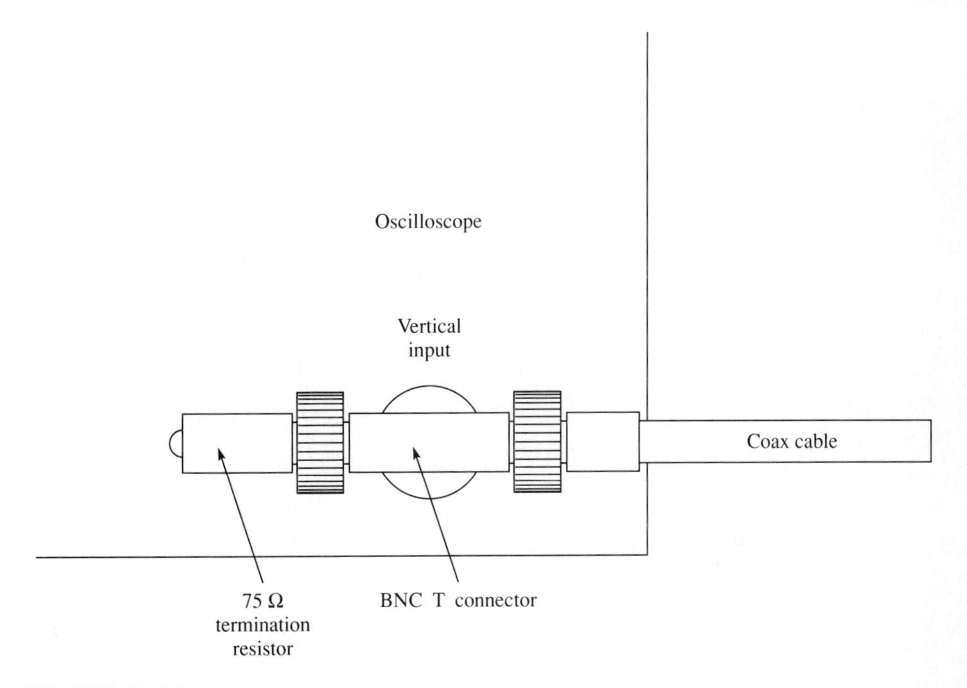

FIGURE 16–22
T connector to oscilloscope.

voltage to be applied to the high-impedance input of the waveform monitor without attenuation. Most oscilloscopes don't have this loop-through capability, so when connecting to an oscilloscope, you need to improvise a termination using a 75 Ω, 0.1% resistor, as shown in Figure 16–22.

THE GRATICULE

The waveform monitor's graticule is shown in Figure 16–23. Television engineering practice does not generally measure video in terms of absolute voltages; instead, measurements are made in relative terms to the specifications of the signal. For example, instead of stating a signal is 700 mV when it should be 714 mV, an engineer will indicate that the signal is 2% low. The graticule is designed to make this type of measurement easy. It is divided into 100 equal units, marked by small horizontal hash marks on a vertical scale. Each mark is equal to 1%. Full luminance (peak white signal) is considered to be 100%. The blanking is set to 0% (sync is below this 0% mark). Sync is not considered to be part of the active video and therefore is not included in the signal measurement. Other portions of the graticule are explained as they are encountered.

PULSE AND BAR TEST SIGNALS

Calibrating video equipment or testing video equipment for compliance to published specifications needs to be very quantitative and not subjective. That is,

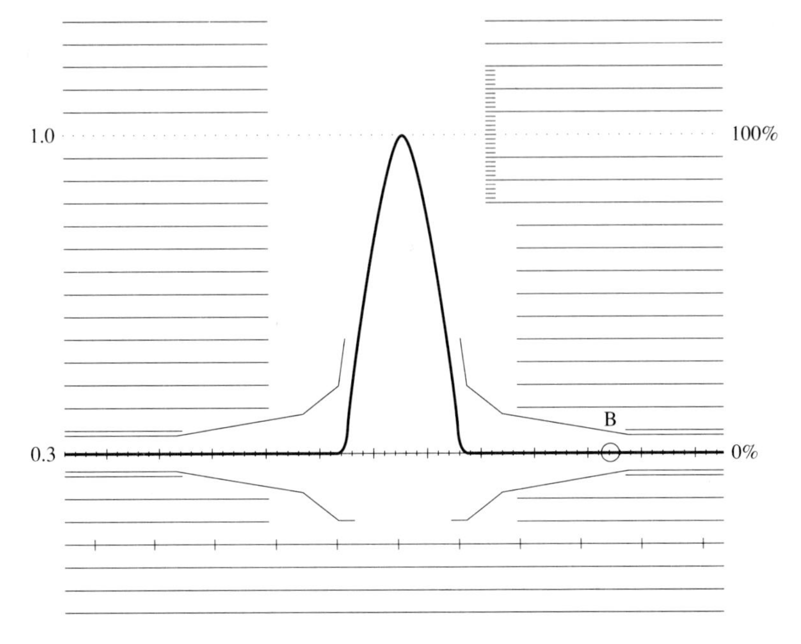

FIGURE 16–23
Graticule B for pulse-shape measurements.

measurements need to be very precise, measurable, and repeatable. For example, suppose the signal from a TV camera is routed through some equipment and viewed on a television monitor. If the picture doesn't look good, then you might wonder what caused the problem. Is the poor picture the fault of the camera, the equipment in the signal path, or the monitor, or is it a combination of problems? By using calibrated test signals and precision instrumentation, the problem can be pinpointed and corrected. If the equipment can pass these calibrated test signals accurately, then ordinary video will also pass through unimpaired. One of the most popular test signals used in conjunction with the waveform monitor is composed of the **12.5T pulse, 2T pulse,** and **bar test waveforms.** This test signal is shown in Figure 16–24.

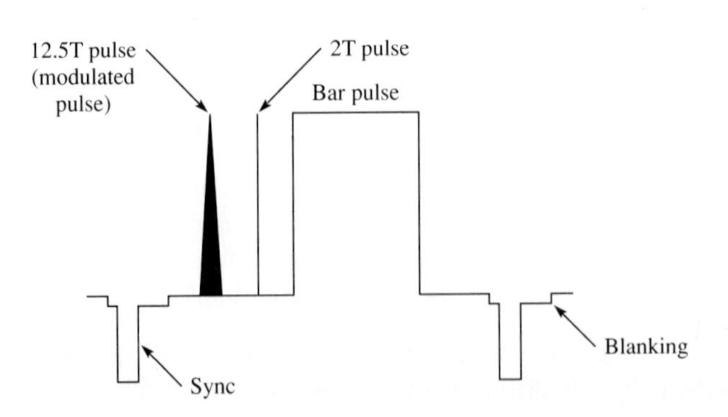

FIGURE 16–24
Pulse and bar test signal.

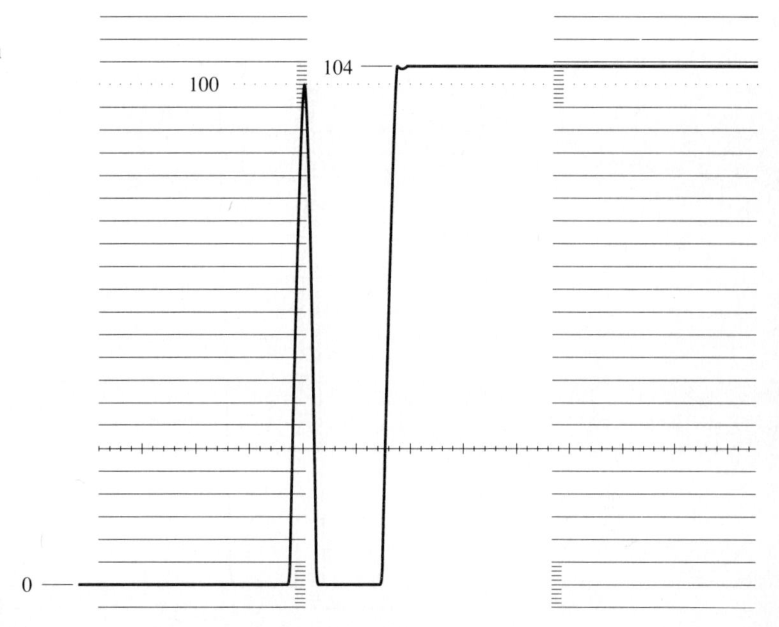

FIGURE 16–25
Waveform and graticule with 4.0% amplitude error.

MEASURING dc LEVEL AND GAIN

An important attribute of any video circuit is its ability to pass a video signal without shifting the dc level or changing the amplitude. If the dc level is shifted up, for example, black will become dark gray. If the gain is not unity, then the levels of brightness will be different than the original picture.

Dc levels are rather simple to measure, and on most video equipment there is a dc adjustment that can be used to correct any offsets. The portion indicated in Figure 16–24 as blanking is to be set to 0.0 V ±15 mV. A general-purpose oscilloscope can be used to measure and set this level as needed. Be sure to set the oscilloscope or waveform monitor's input coupling to dc, not ac.

The *bar* portion of the test video is 18 μs in duration and is specified as 714 mV above blanking (100% peak white). Figure 16–25 is an example of the pulse and bar signal positioned correctly on the graticule of a waveform monitor. Using an oscilloscope, you can measure the actual signal through the unit under test and convert the result into a percentage, as shown in the following example.

EXAMPLE 16–3 Assume a video signal from a unit under test is 723 mV at the output and is 714 mV (proper level) at the input. What is the gain error?

SOLUTION The gain error is

$$\frac{721 \text{ mV} - 714 \text{ mV}}{714 \text{ mV}} \times 100 \approx 1\%$$

The waveform monitor can give a direct reading from the graticule. Figure 16–26 shows the reading is 1% too high.

FIGURE 16–26
Bar gain at 101%.

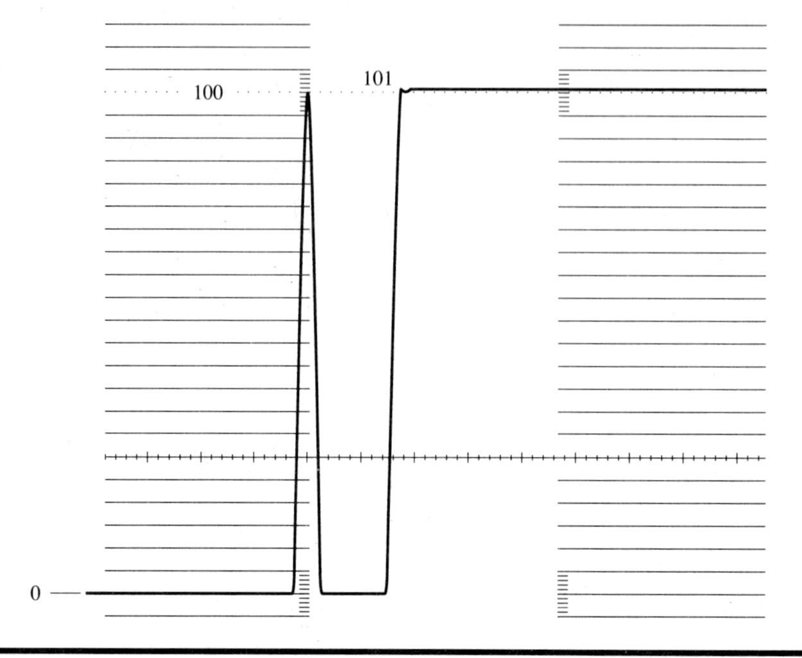

The following procedure can be used to make a dc level or gain test:

1. Connect the waveform monitor or oscilloscope to the unit under test, as shown in Figure 16–27, using a 75 Ω coax cable, and terminate the cable.
2. Select the GROUND input to the measurement equipment and move the trace to a reference point. The waveform monitor is illustrated.
3. Select the input to the measurement equipment to dc and measure the difference of the signal's blanking level to the reference. If the unit under test is to be adjusted, then adjust the blanking level until it is on the reference of the graticule. Refer to Figure 16–5 for the location of the blanking.
4. Measure the amplitude of the bar signal referenced to blanking. With an oscilloscope, convert the measured amplitude to a percentage. With a waveform monitor, this can be done directly. If the unit under test is to be adjusted, then adjust the gain of the unit under test so the amplitude of the signal is to specifications (100% peak white).

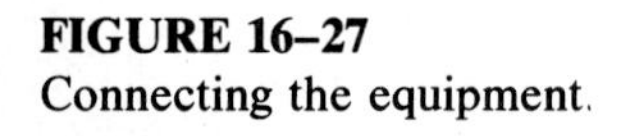

FIGURE 16–27
Connecting the equipment.

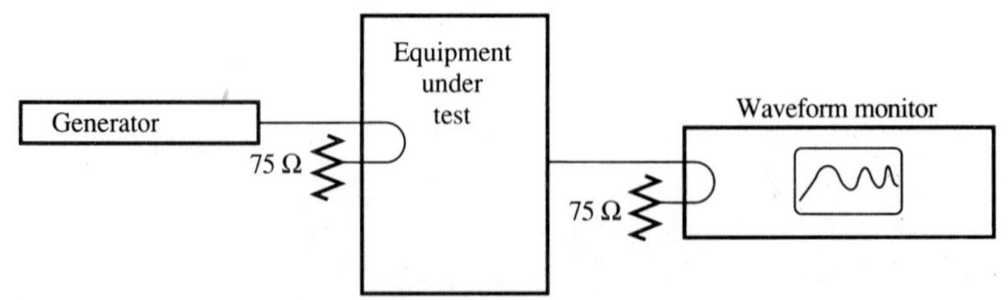

BAR TILT

The top of the bar (in the test waveform) should be as flat as possible. *Bar tilt* is the tilting of the bar, either up or down, as the name implies. Tilt usually occurs as a result of improper ac coupling, but it can also be due to hot transistor junctions, heated from the applied video's dc. The junction temperature can increase as the luminance level increases; the gain of the transistor likewise increases, causing the bar signal to tilt upward. You might expect that tilt due to ac coupling would be more prevalent at the lower vertical scan frequency of the picture, and this would be the case if it were not for a **dc restorer** or clamp circuit. A dc restorer circuit restores the signal to a reference at the beginning of each line. Any errors from line to line are *clamped* or *restored* to the reference level (keeping blanking at 0 V). AC coupling problems that persist after clamping or dc restoring are at the higher horizontal scan frequencies and are minimal due to the shorter time constants of the signal. The principle is illustrated in Figure 16–28. If a step response is applied to an ac coupled circuit, the instantaneous voltage diminishes at an exponential rate defined by the equation

EQUATION 16–8

$$v_t = V_p(e^{-t/RC})$$

FIGURE 16–28
Step response of *RC* coupled circuit.

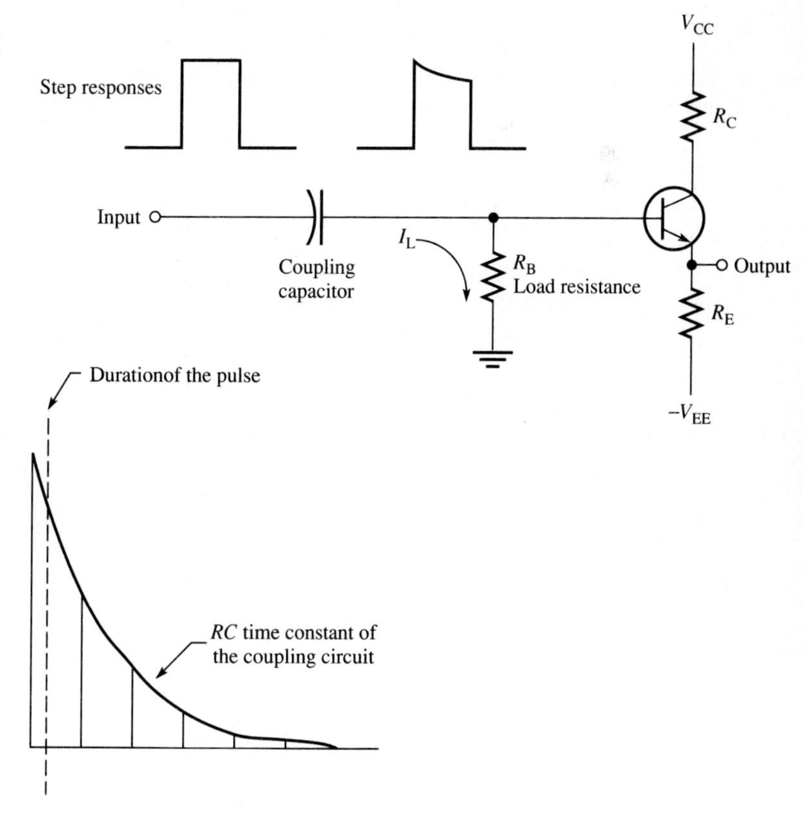

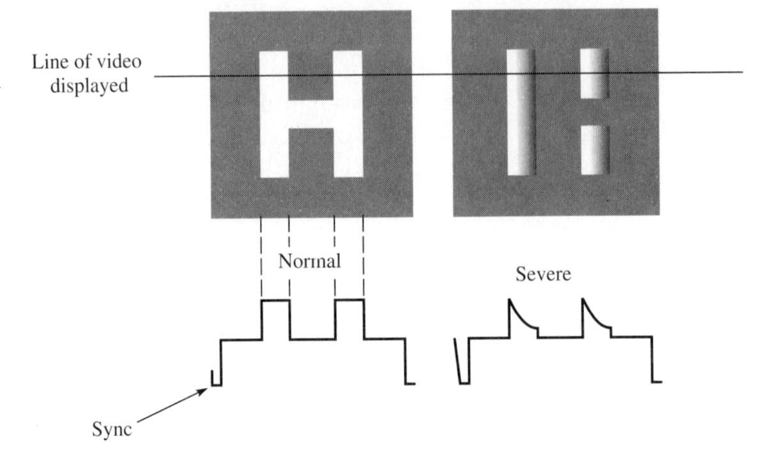

FIGURE 16–29
Example of severe tilt (poor low frequency response) on a picture monitor.

where v_t = instantaneous voltage on the resistor, V
V_p = peak voltage of the step, V
t = time, s
RC = time constant, s

If the RC time constant is *much* longer than the duration of the pulse, then the slope will be small and perhaps negligible. This type of distortion is exaggerated in Figure 16–29 for a television picture. The letter *H* is not consistent in the bright portion due to tilt distortion. Tilt distortion is measured using either an oscilloscope or the waveform monitor.

USING AN OSCILLOSCOPE TO MEASURE BAR TILT

The following is a procedure for measuring the bar tilt (line tilt) with an oscilloscope. Refer to Figure 16–30.

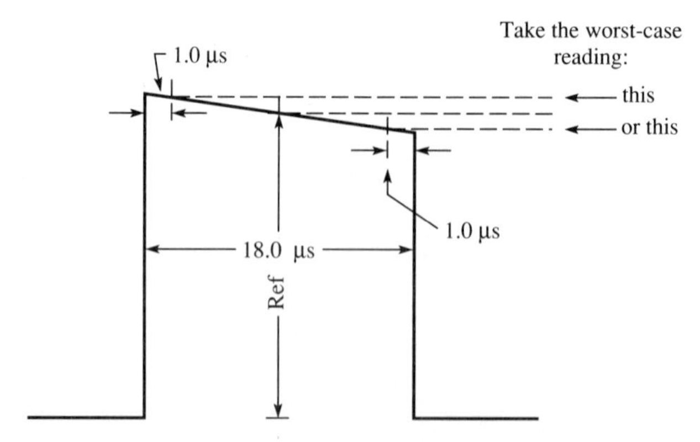

FIGURE 16–30
Measurement of bar tilt using an oscilloscope.

1. Locate the center of the bar.
2. Measure the amplitude of the center of the bar with respect to the blanking level. Record this level as V_{ref}.
3. Measure the amplitude of the waveform from the blanking level 1 μs after the leading edge of the bar and again at 1 μs before the falling edge of the bar.
4. Determine which of the two measurements taken in Step 3 is further from the value measured in Step 2. Record this worse-case level as V_W. Calculate the error as a K factor from

EQUATION 16–9

$$\%K_B = \frac{V_{ref} - V_W}{V_{ref}} \times 100\%$$

where $\%K_B$ = K factor of the bar, %
V_{ref} = reference taken at the center of the bar, V
V_W = worst-case reading taken 1.0 μs from either edge of the bar, V

A K factor is usually defined as simply a *ratio* of a measured parameter to a reference value. But sometimes a scaling factor is included in the equation (as in the pulse-to-bar measurement section, which follows). This scaling factor is needed because the K factor of one parameter is to have the same *perceived impairment* to the picture quality as the K factor of another parameter. That is, a K factor of 10 on the bar measurement should look as *bad* (different, perhaps, but still just as bad) as a K factor of 10 for the pulse-to-bar measurement.

USING A WAVEFORM MONITOR TO MEASURE BAR TILT

As mentioned before, the waveform monitor is designed to simplify video measurements. Figure 16–31 shows a portion of the graticule for measuring bar tilt.

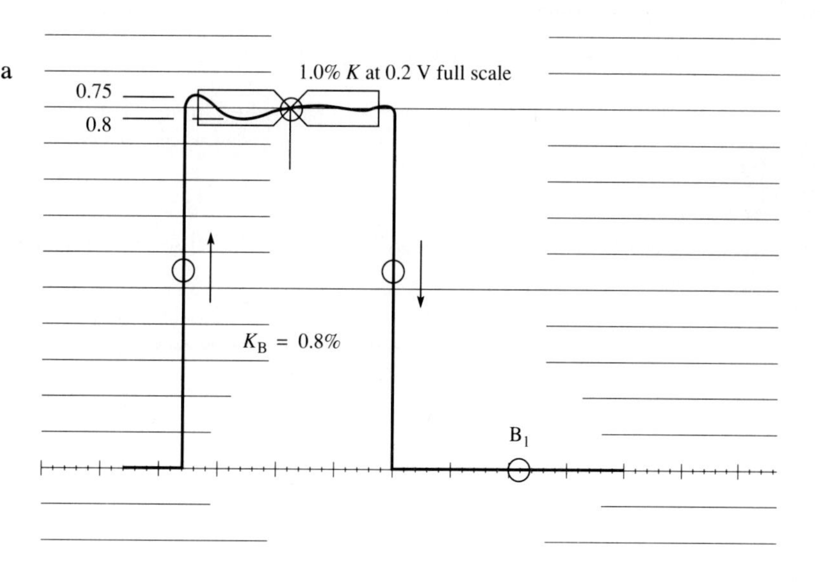

FIGURE 16–31
Bar tilt measured with a waveform monitor.

The **bow tie** portion of the graticule is a limit line. Anything within this limit line (vertical displacements only) is said to be within $\pm 5\% K$. The precision of the measurement can be improved by increasing the gain of the waveform monitor's vertical amplifier. The display shown is with the vertical amplifier set to 1.00 V full scale. If the gain is increased to 0.5 V full scale, the limit lines will measure $2.5\% K$. If the volt-full-scale reading is selected to 0.2 V, then the limit lines will be $1.0\% K$. The limit lines are *not* intended to allow the operator to interpolate measurements within the limit lines. They are intended only for go/no-go tests, such as "the bar response is equal to or better than $2.5\% K$."

Notice that the graticule shows two arrows, one that indicates the leading edge of the signal (up arrow) and the other, the trailing edge of the signal (down arrow). Some waveform monitors just have the arrows and others have the circle, as shown. If the edges of the signal are aligned with leading- and trailing-edge markers (by using the HORIZONTAL POSITION control), then the bow tie limits are automatically set 1.0 μs from the edges of the test signal. Any overshoot or undershoot outside of the right- and left-hand edges of the limit box are disregarded.

The procedure for measuring tilt with the waveform monitor is as follows:

1. Select the range switch for 1.0 V full scale.
2. Position the video waveform using the VERTICAL POSITION control so that the base line (blanking) goes through the zero reference line at B_1.
3. Adjust the HORIZONTAL POSITION so the leading and trailing edges of the signal go through the leading- and trailing-edge markers.
4. Adjust the VERTICAL GAIN VERNIER to place the top of the waveform just in the center of the bow tie crossover point.
5. If no portion of the waveform exceeds the limit lines, the video measurements are said to have K bar tilt less than 5.0%.
6. If better accuracy is required, use Table 16–2 to expand the resolution of the limit lines.

PULSE RESPONSE

Pulses that have equal rise and fall times have a fundamental frequency given by

EQUATION 16–10

$$f = \frac{1}{2t_w}$$

where t_w = the pulse width at the 50% points of the pulse, s

TABLE 16–2

RANGE SWITCH SETTING (VOLTS FULL SCALE)	TOLERANCE LIMITS ON THE GRATICULE
1.0 V	$5.0\% K$
0.5 V	$2.5\% K$
0.2 V	$1.0\% K$

For television signal measurements, there are two types of pulses. One is the T pulse, where the upper cutoff frequency is near 8 MHz. A more common pulse is the 2T pulse, which is closer to the 4.2 MHz bandwidth limit of the NTSC system. The 2T pulse should pass through most video equipment, but the T pulse may be greatly attenuated. The waveform of Figure 16–32 is the shape of a $\sin^2 x$ function used both for the T and 2T test signals.

The Fourier transform (see Figure 1–12) would show that the T pulse consists of a series of frequencies up to the 4 MHz limit. These frequencies must be delayed by the same amount as they pass through a video path within the equipment. That is, the lowest frequency must propagate at the same rate as the highest frequency; otherwise they will come out of the equipment's amplifiers at different times. The frequencies making up the pulse should also experience the same gain characteristics. The lowest frequency should have the same gain or attenuation as the highest frequency.

The PULSE graticule on the waveform monitor is useful for measuring group delay distortions. Again, the graticule is not an absolute measurement tool but a go/no-go guide.

The procedure for measuring group delay using the pulse waveform and the pulse graticule is as follows:

1. Turn on the DC RESTORER and set it to BACK PORCH and FAST.
2. Set the horizontal sweep magnification to 25× (0.2 ms/division).
3. Set the volts full scale for 1.0 V.
4. Adjust the HORIZONTAL POSITION to place the pulse waveform so it is centered within the graticule, as shown in Figure 16–33.
5. Adjust the VERTICAL POSITION to place the blanking level through the B on the reference line.
6. Adjust the VERTICAL GAIN so the top of the pulse is on the 100% line.
7. If no portion of the waveform exceeds the limit lines, the video measurements are said to have *K* pulse-to-bar equal to or better than 5.0%.

FIGURE 16–32
Fundamental frequency of a pulse waveform.

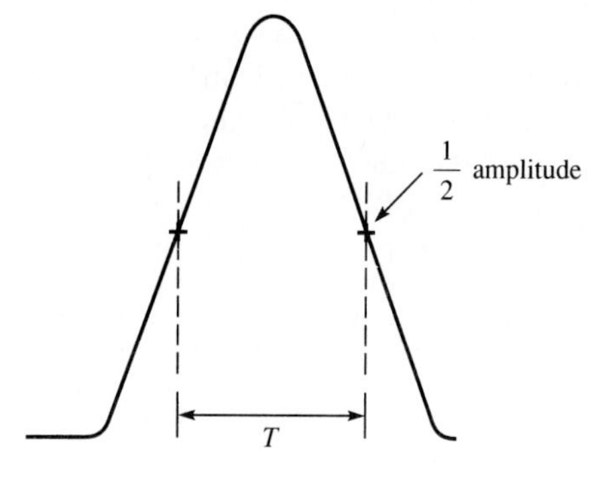

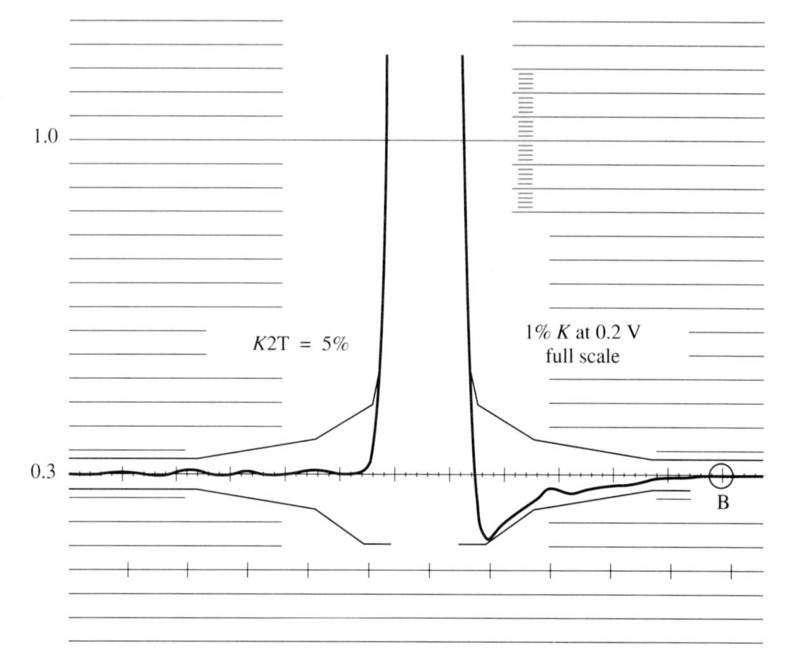

FIGURE 16–33
Measuring K_P, pulse response, with a waveform monitor.

Better accuracy can be obtained by adjusting the INPUT GAIN on the waveform monitor, as given in Table 16–3.

PULSE-TO-BAR MEASUREMENTS

To complete a quick check on the high-frequency response, compare the amplitude of the pulse signal to that of the lower-frequency bar signal. This check is done after the dc gain of the bar signal has been set and the response checks of the bar signal (bar tilt tests) are verified as being within specifications. There has been some debate concerning the merit of this performance check. There are more precise ways to check the high-frequency response. One such method uses a network analyzer, but these methods require much more sophisticated and expensive equipment. The pulse-to-bar method is simple, yet provides satisfactory results for most applications.

The K factor for the pulse to bar is a ratio (K_{PB}), but a scaling factor of 4 is included in the equation. The pulse-to-bar ratio must be four times larger to have a picture "look" as bad as with the other K factors. In this case, the comparison is more of an art than a pure science. We make precise measurements for repeatabil-

TABLE 16–3

RANGE SWITCH SETTING (VOLTS FULL SCALE)	TOLERANCE LIMITS ON THE GRATICULE
1.0 V	5.0%K
0.5 V	2.5%K
0.2 V	1.0%K

FIGURE 16–34
Pulse-to-bar measurement.

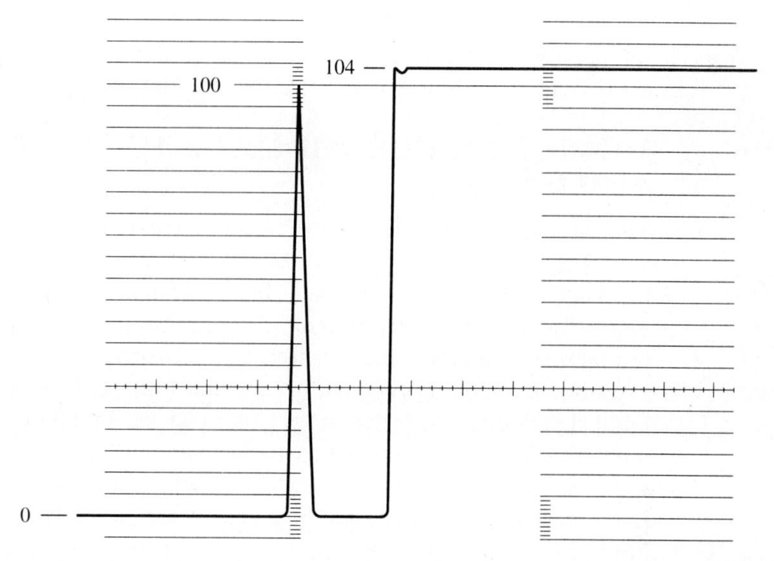

ity of the results, but the actual qualifying parameters were established by a consensus of the members of the NTSC committee as to what "looks" good or bad.

In making pulse-to-bar ratio measurements, it is conventional to use pulse amplitude as reference and then measure the amplitude of the center of the bar as in Equation 16–11.

EQUATION 16–11

$$K_{PB} = \frac{|B - P|}{4}$$

where K_{PB} = K factor of the pulse to the bar, $\%K$
B = amplitude at the center of the bar waveform, %
P = amplitude of pulse signal, %

The procedure for measuring pulse-to-bar ratio is as follows:

1. Adjust the vertical gain on the waveform monitor so the amplitude of the pulse is 100%, as shown in Figure 16–34.
2. Measure the amplitude of the bar in percentage (%) directly from the graticule.
3. Calculate K_{PB} from Equation 16–11.

EXAMPLE 16–4 A student has measured the amplitude of the bar to be 4% higher than the amplitude of the pulse, as shown in Figure 16–34. What is the K factor for this measurement?

SOLUTION

$$K_{PB} = \frac{|B - P|}{4}$$

$$= \frac{4\%}{4} = 1.0\%K$$

Notice that the ratio between these two signals is 4%, but the K factor is 1%. It has only one-fourth the visible effect on the picture compared to other K-factor measurements, such as the K factors for the bar (K_B).

MEASURING CHROMINANCE-TO-LUMINANCE DELAY AND GAIN

The last measurement to be discussed using a waveform monitor is the comparison of the propagation delays of the low-frequency luminance signal to the high-frequency chrominance signal. This is referred to as *chrominance-to-luminance delay*. In television equipment, the chrominance signal is often split from the luminance signal for special signal processing, as shown in Figure 16–35. An example is encoding the red, green, and blue signals as a polar- (phase-) modulated signal for color processing. Color processing takes extra time to accomplish, so the luminance signal needs to be delayed by an equal amount in order for the two signals to be in time with each other. In addition, the relative gain (gain balance) between the two must remain constant; this is referred to as chrominance-to-luminance gain.

Severe chrominance-to-luminance delay could occur if there were no added delay to the luminance path of the circuit to compensate for the processing delays in the chrominance path. The luminance would arrive before the chrominance signal. Since we are dealing with a pixel-by-pixel, line-by-line scan system, a luminance signal that occurs "early" will be placed on the screen further to the left. For example, if a picture of a woman who has bright red lips is shown and if there is significant delay of the chrominance path, then the coloring of her lips would start further to the right (from the viewer's perspective) than the luminance; the red would start halfway through her lips and would end on her cheek.

To measure the chrominance-to-luminance delay, a signal was developed, as shown in Figure 16–36. This test signal is very sensitive to both delay imbalance and gain imbalance. The test signal comprises two simultaneous signals added together. One signal is a simple $\sin^2 x$-shaped pulse, which is 12.5 times wider than a single cycle of the color subcarrier, and the other signal is the chroma subcarrier (3.58 MHz), neatly packaged into a $\sin^2 x$ envelope of the same width. Each signal has a peak positive excursion of half the amplitude of the total pulse signal. The two signals are added together, linearly, to form the test signal (refer to Figure 16–37). The positive amplitude swings of the modulated sine wave are added to

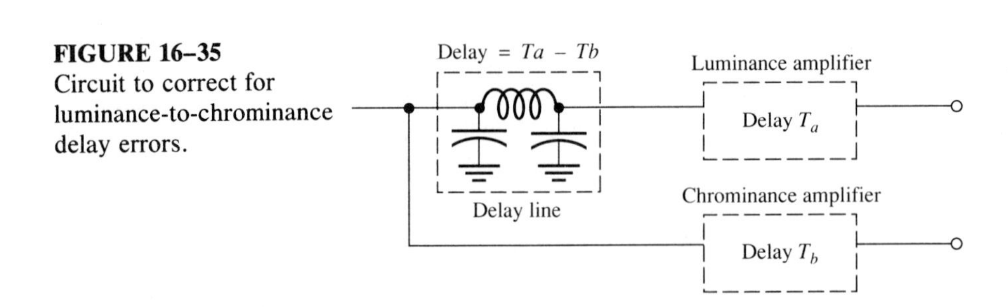

FIGURE 16–35
Circuit to correct for luminance-to-chrominance delay errors.

FIGURE 16–36
The modulated 12.5T pulse in (b) is formed by adding a luminance pulse to a chrominance pulse in (a).

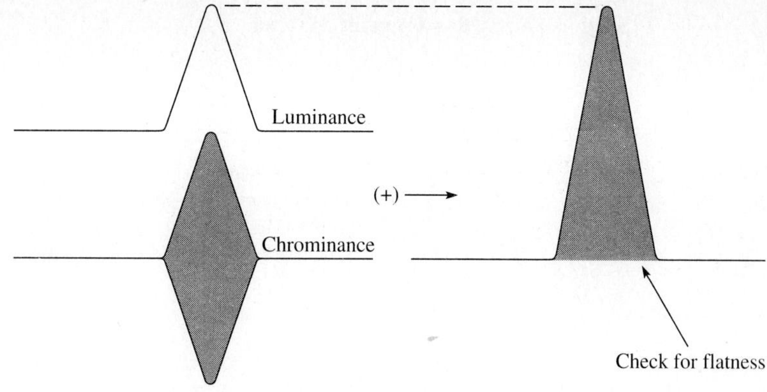

FIGURE 16–37
Formation of the modulated 12.5T pulse.

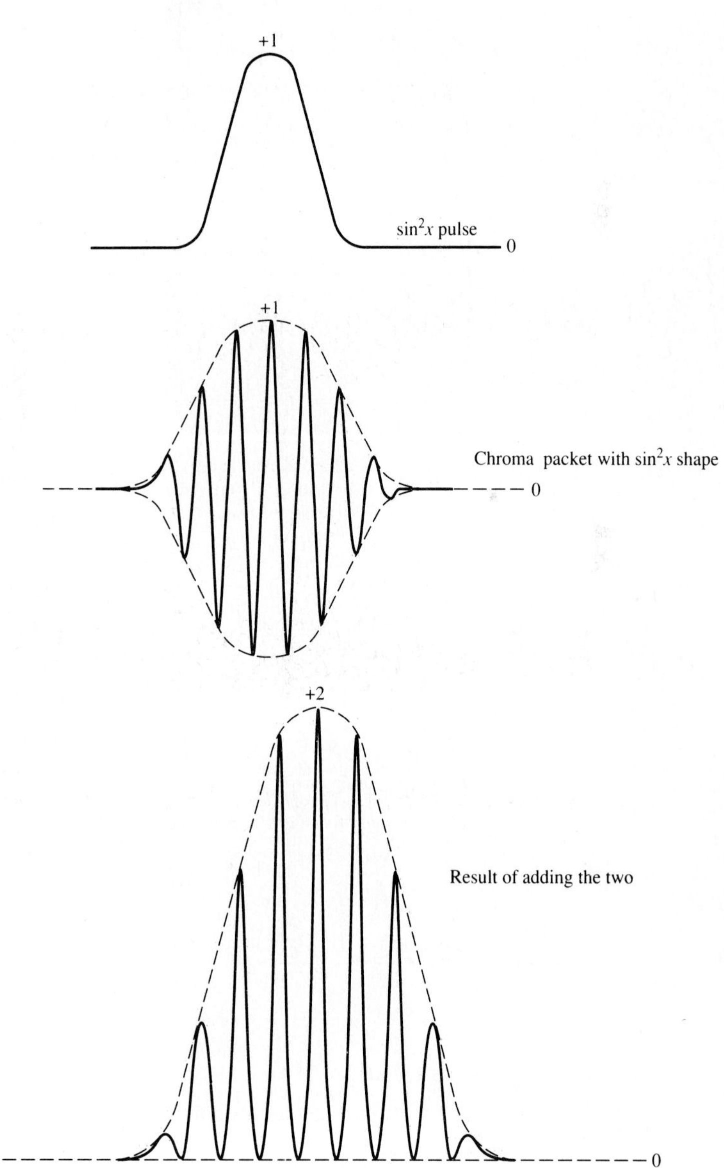

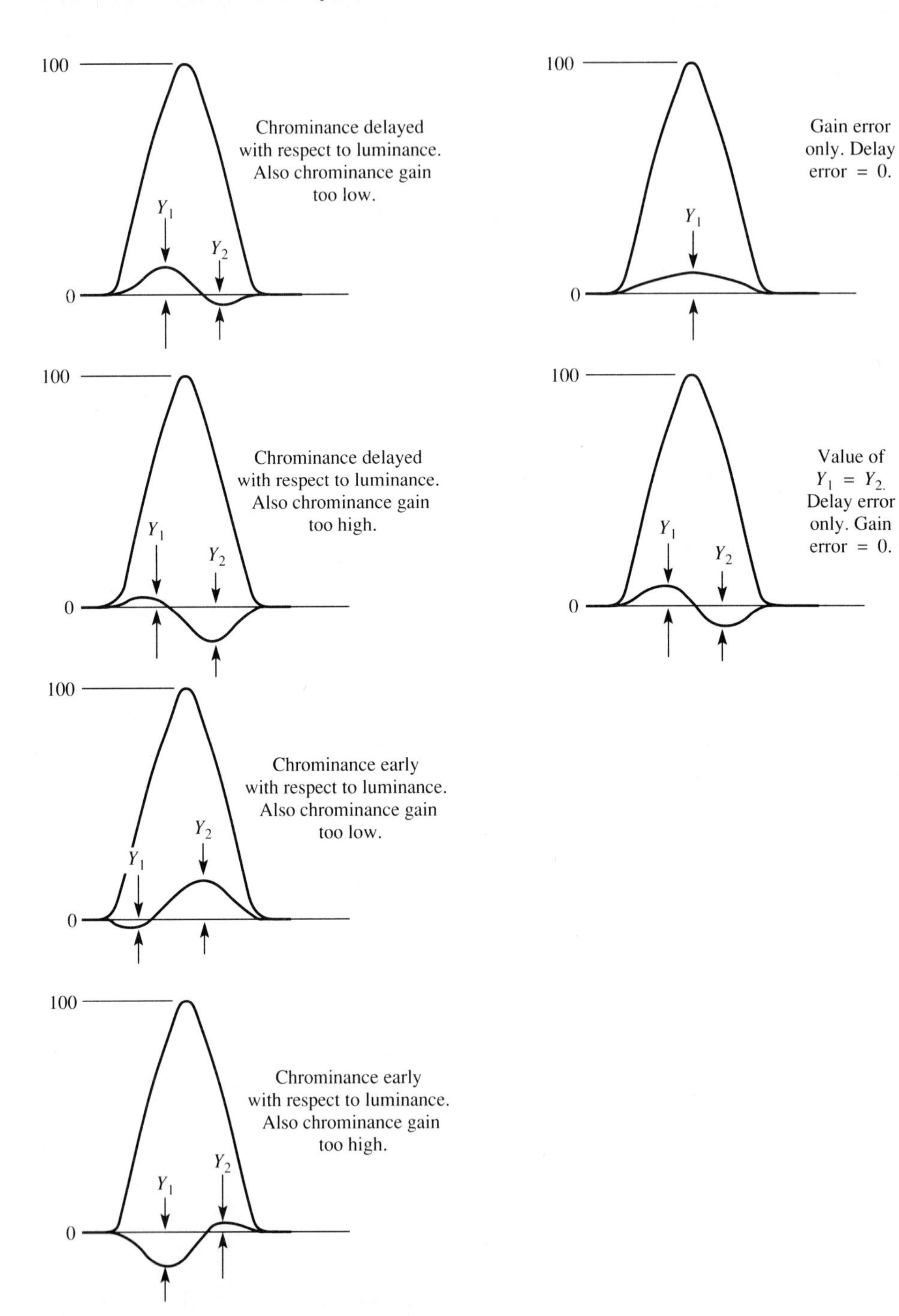

FIGURE 16–38
Examples of distortion using a modulated 12.5T pulse test signal.

the $\sin^2 x$ pulse (at any given time); the sum is a total of twice that of each considered separately. The negative peak excursions of the sine wave are added to the pulse; the result is 0 V, or a flat bottom on the composite waveform.

Differences in delay and gain are most noticeable when viewing the flatness at the bottom of the test signal. If the chroma gain is too great compared to the luminance signal, then the bottom will bulge downward. Conversely, too little chroma gain will cause the bottom to be concave upward. If the chroma is early or late in timing as compared to the luminance signal, then the bottom of the composite pulse will be S-shaped. Figure 16–38 illustrates different combinations of the chrominance-to-luminance gain and delay inequalities.

A nomograph is given for the purpose of calculating the gain and delay relationships, using a modulated 12.5T pulse, as shown in Figure 16–39. There are two axes on the graph. The vertical axis is titled "negative values of Y_1 or Y_2" and the horizontal axis is "positive values of Y_1 or Y_2." These values are the *units* by which the bottom of the pulse is distorted from flatness. One unit is equal to 1% of the total height of the pulse. A waveform monitor is used to make the measurements; VERTICAL POSITION is used to place blanking on the zero axis and VERTICAL GAIN is adjusted to place the top of the test signal at the 100% graticule.

The use of the nomograph is explained by an example. Suppose that the test signal on the waveform monitor curves up by 2% (a positive value) and there is no downward, or convex, distortion. You locate +2 along the horizontal axis of the nomograph and 0 along the vertical axis. The intersection of the two points gives 0 ns of delay error and about −0.3 dB of gain error. The chroma level is down by 0.3 dB compared to the luminance level.

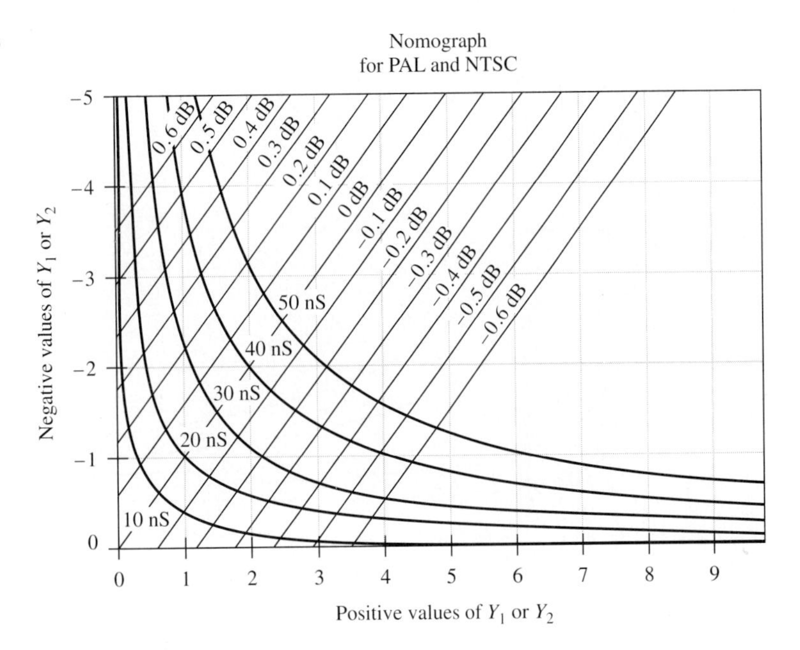

FIGURE 16–39
Nomograph.

EXAMPLE 16–5 A student uses a modulated 12.5T pulse test signal and measures the video waveform in Figure 16–40(a) and records that the bottom has first a 2% rise (positive value) and then a 3% fall (negative value). What is the gain and phase imbalance?

FIGURE 16–40

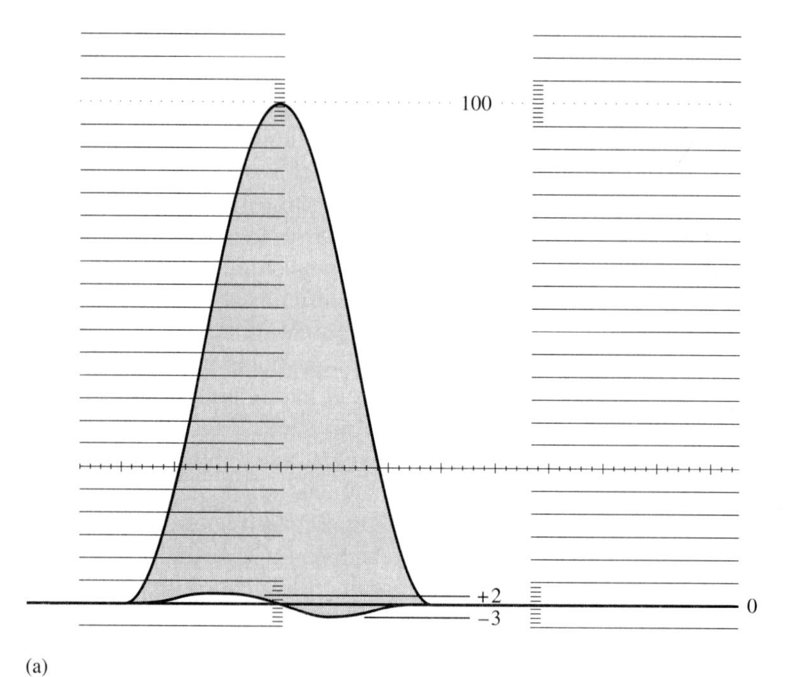

(a)

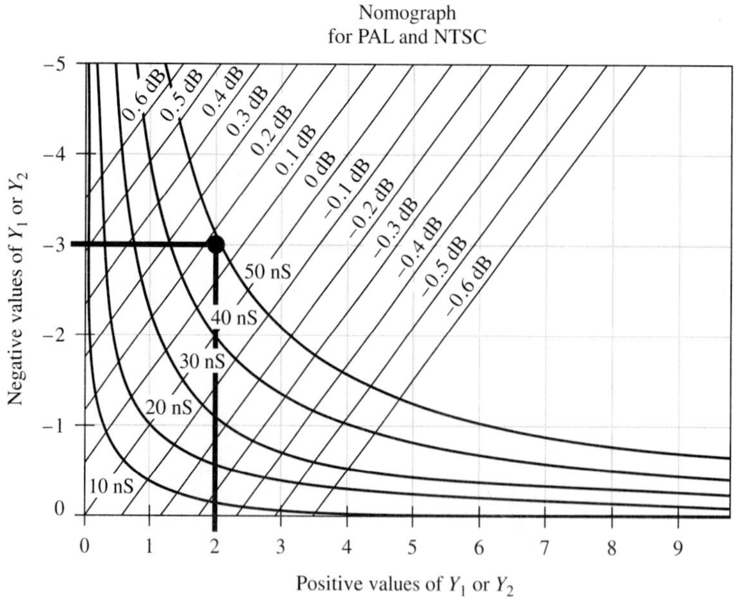

Nomograph
for PAL and NTSC

0.6 dB
0.5 dB
0.4 dB
0.3 dB
0.2 dB
0.1 dB
0 dB
−0.1 dB
−0.2 dB
−0.3 dB
−0.4 dB
−0.5 dB
−0.6 dB
50 nS
40 nS
30 nS
20 nS
10 nS
−5
−4
−3
−2
−1
0
Negative values of Y_1 or Y_2
0
1
2
3
4
5
6
7
8
9
Positive values of Y_1 or Y_2

(b)

SOLUTION As shown in Figure 16–40(b), the horizontal axis is for the positive values of the distortion. Draw a vertical line from the 2% mark. The vertical axis is for negative values in the distortion. Draw a horizontal line from the 3% mark. Read a delay of 49 ns and a gain of about +0.18 dB where the two lines intersect.

The same waveform can be observed with an oscilloscope. The scope's reading is valid as long as the distortion of the flatness is calculated as a ratio to the total amplitude of the pulse from blanking level to peak positive; then the nomograph can be used.

To make this measurement using a waveform monitor, follow these steps:

1. Turn on the dc restorer or clamp.
2. Set the response to FLAT.
3. Set the VOLTS FULL SCALE to 1.0 V.
4. Adjust the VERTICAL CENTERING and the VERTICAL GAIN to position blanking at 0% and the peak of the pulse on the 100% line of the graticule, as shown in Figure 16–41.
5. Move the HORIZONTAL POSITION as needed to place the positive and negative distortions in base line with the 1% markings of the graticule.
6. Use the nomograph to determine the errors in luminance to chrominance gain and delay, if any. Figure 16–41 has no delay error but has −0.45 dB of gain error. The chroma is too low.

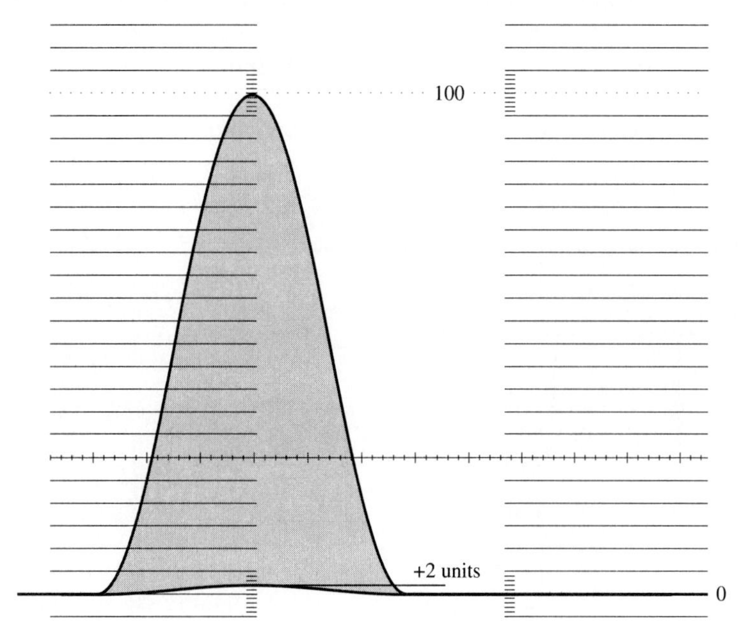

FIGURE 16–41
Chrominance amplitude is −0.45 dB relative to the luminance amplitude.

16–8 THE VECTORSCOPE

Recall that the color signal is transmitted by the 3.58 MHz subcarrier signal. The **vectorscope** is a specific measurement tool used to measure absolute gain and relative phase (relative to an input reference) of this signal. A Tektronix 1780 video measurement set that contains a vectorscope is shown on the front cover. The phase of the 3.58 MHz color signal as compared to a reference signal (burst reference) gives unique hues to the picture and the peak-to-peak amplitude gives unique color brilliance to the picture. The vectorscope displays the amplitude and phase of the 3.58 color signal as a *vector* rather than as the traditional oscilloscope display. For a point of reference, it should be noted that the eye can perceive color hues that differ by as little as 3° or 4° in phase; each degree is (1/3.58 MHz)/360° = 0.776 ns.

CONNECTING THE VECTORSCOPE

The vectorscope decodes the composite encoded color signal into a Cartesian coordinate system and displays color signals as vectors. The CRT is the same as an oscilloscope CRT; the beam is deflected electrostatically. With no input, the beam is centered. Deflection along the horizontal axis is a result of decoding the color signal into $\pm(B - Y)$ values. Deflection along the vertical axis is a result of decoding the color signal into $\pm(R - Y)$ values.

In order to decode the input signal, the vectorscope uses an internal 3.58 MHz oscillator, as shown in Figure 16–42. This oscillator is a phase-locked-loop oscillator with two phase adjustments. One phase adjustment is a coarse phase control that is continuously variable over 360° and the other is a high-precision phase control limited to ±10°. Reference to the oscillator can come from the input signal itself, which can be used for measuring differential phase or differential gain, or it can be externally referenced. The phase must be externally referenced if

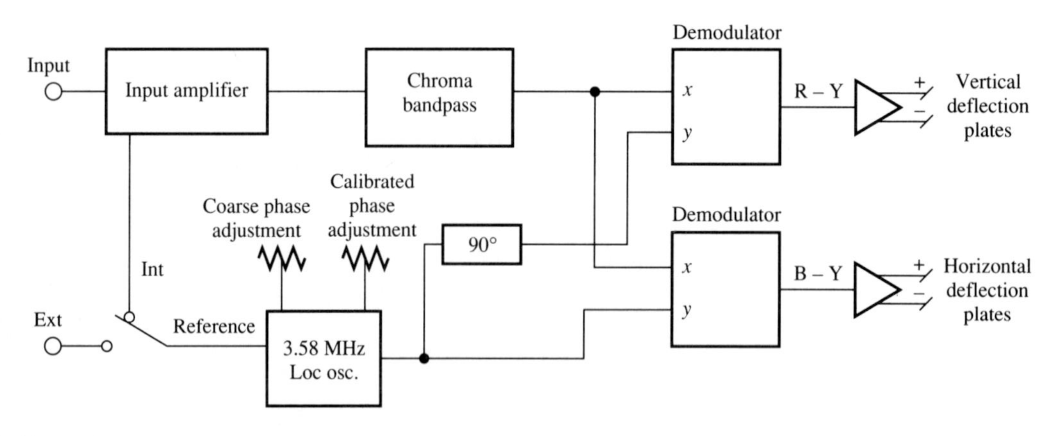

FIGURE 16–42
Basic block diagram of a vectorscope.

measuring delay. It is a good idea to connect the vectorscope to an external reference for all measurements to avoid mistakes.

VECTORSCOPE GRATICULE

There are two types of displays on the vectorscope. In the VECTOR mode, colors are displayed as vectors. In the DIFF PHASE or DIFF GAIN mode, color phase differences or color gain differences (respectively) are displayed as positive and negative excursions from a horizontally flat line along the center of the CRT.

In the vector display (Figure 16–43), each color signal is represented as a vector whose length represents the amplitude of the color signal; the angle with respect to the reference represents phase of the sinusoidal wave (color hue). The reference can be either an internal or an external reference. Normally the reference (burst signal) is also displayed.

Figure 16–44 shows the front graticule of the vectorscope. Notice the slightly arched rectangular boxes placed around the screen. Each box represents the correct location of a particular-color vector if a calibrated color bar signal is supplied as an input (the colors are yellow, cyan, green, magenta, red, and blue). The **color bar signal,** shown in Figure 16–45 and on the front cover, is an electronically generated signal that produces a series of colored bars, which have precise amplitudes and phase relations to the burst reference. If this signal can pass through the video system with negligible distortion, then any other picture can be sent through the video system without noticeable distortion. The waveform of the color bar signal is shown in Figure 16–46. The color-bar test signal is displayed in the photo on the front cover (the Tektronix 1780)—the vector display on the left screen and the waveform display of the same signal on the right screen.

There are two standards for color bar signals. One has the amplitudes given in Table 16–1 and is referred to as **100% color bars.** The other standard is referred to as **75% color bars.** The 75% bar amplitudes are 75% of the peak-to-peak values shown in the table. The 100% bars are most often used to test video equipment in the laboratory because they give the worst-case scenario; however, 100% bars can overdrive a broadcast transmitter and are not legal for on-air testing, so the 75% color bar signal is used for testing a video system while it is actively broadcasting.

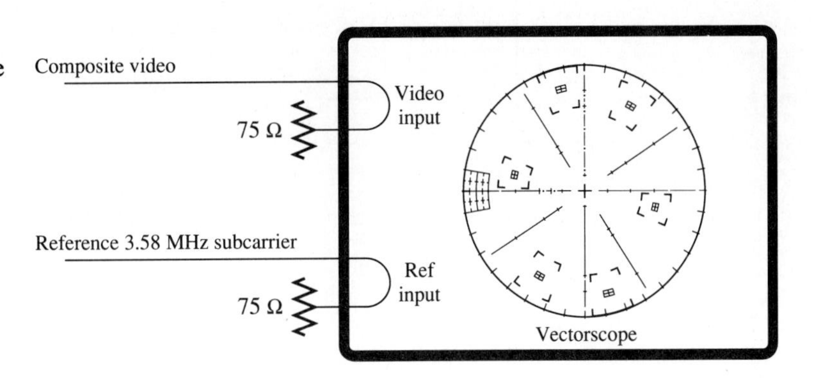

FIGURE 16–43
Connections to the vectorscope.

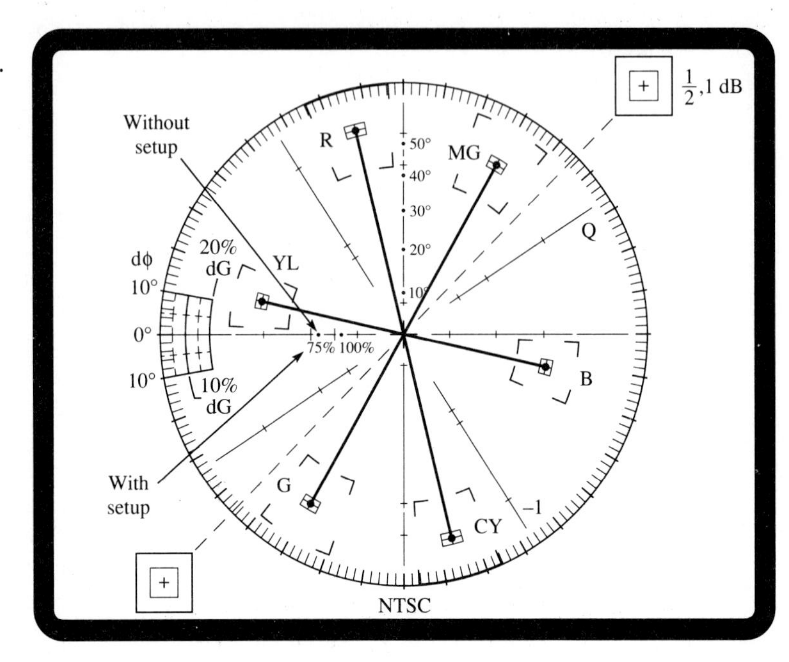

FIGURE 16–44
Vector scope's front graticule.

FIGURE 16–45
Standard color bars.

White	Yellow	Cyan	Green	Magenta	Red	Blue	Black

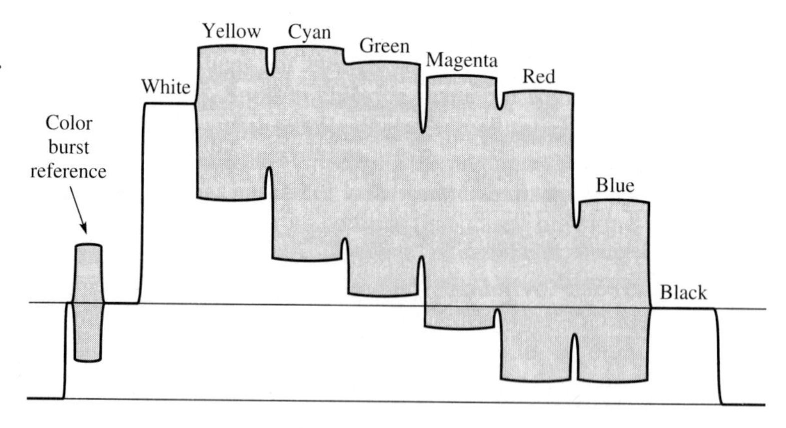

FIGURE 16–46
Waveform of color bars.

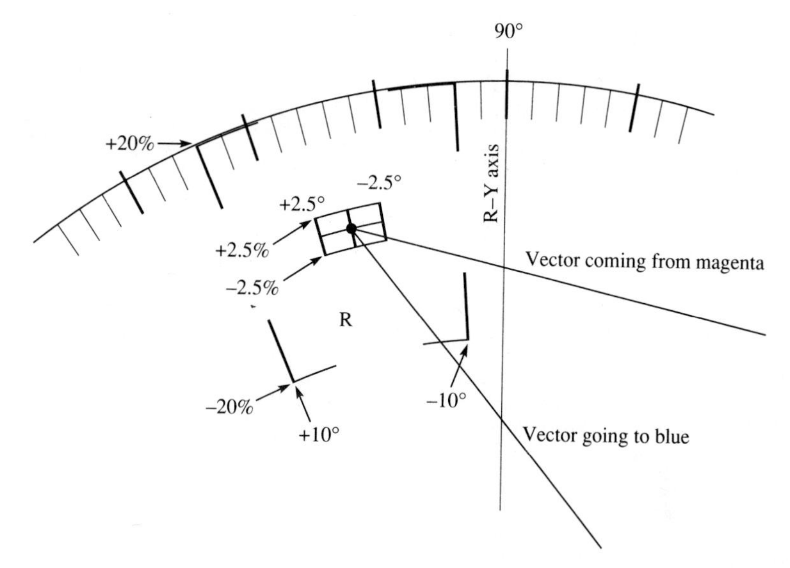

FIGURE 16–47
Red color box expanded.

Figure 16–47 shows the red box expanded and the vector dot shown within the box. Each of the color vectors should be clean round dots that land precisely in the center of their respective boxes. There will be faint lines between the dots, which are of no consequence; for example, there is a faint line between the red and blue vectors. That line exists because the picture (moving across a line of color bars) changes from red to blue at one point (see Figure 16–45). The red vector then shifts from the red position to the blue position, leaving this faint—but visible—unblanked line. The importance of the test is the placement of the dot within the square.

MEASURING DIFFERENTIAL GAIN AND DIFFERENTIAL PHASE

Recall that differential gain and differential phase are the measurements of the effect that luminance has on the gain and phase of the chrominance signal. As the luminance changes to different levels, say from black to white or visa versa, it should not affect the chrominance. To repeat, the term differential, when used in this context, has nothing to do with differential mode amplifiers but simply means the difference in luminance levels.

The color bar signal is a very useful signal to use in conjunction with the vectorscope. It can show several types of distortion at once. It can show differential gain, differential phase, timing errors, and slew rate problems.

Two common types of distortions are differential phase and differential gain. Each color bar in the standard calibrated color bar test signal has a unique amplitude and phase (vector length and angle). In addition, each color bar has a unique dc level, as shown in Figure 16–46. This dc level is the luminance value that is added to the 3.58 MHz chroma signal. This places each color bar on a dc ''pedes-

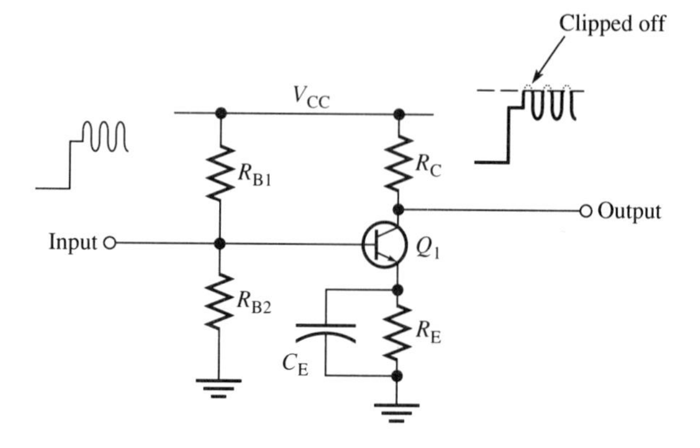

FIGURE 16–48
Improper bias causing chroma to clip, producing differential gain problems.

tal.'' The *luminance* level can affect how an amplifier passes the *color* signal by changing the dc bias on that stage. This can affect the gain through the stage, causing *differential gain error,* or it can affect the delay through the stage, causing *differential phase error*.

Two examples follow that show how circuit problems affect the color signal. The circuit in Figure 16–48 shows a transistor stage that is biased such that the transistor will start to saturate and clip the top of the signal as the luminance goes toward white. This clipping action attenuates the gain of the yellow and cyan color bars. It causes the yellow and cyan bars to fall short of their color boxes, as shown in Figure 16–49; however, the other color vectors hit the center of their perspective boxes. Although the signal has the proper hue, the intensity of the bars is incorrect.

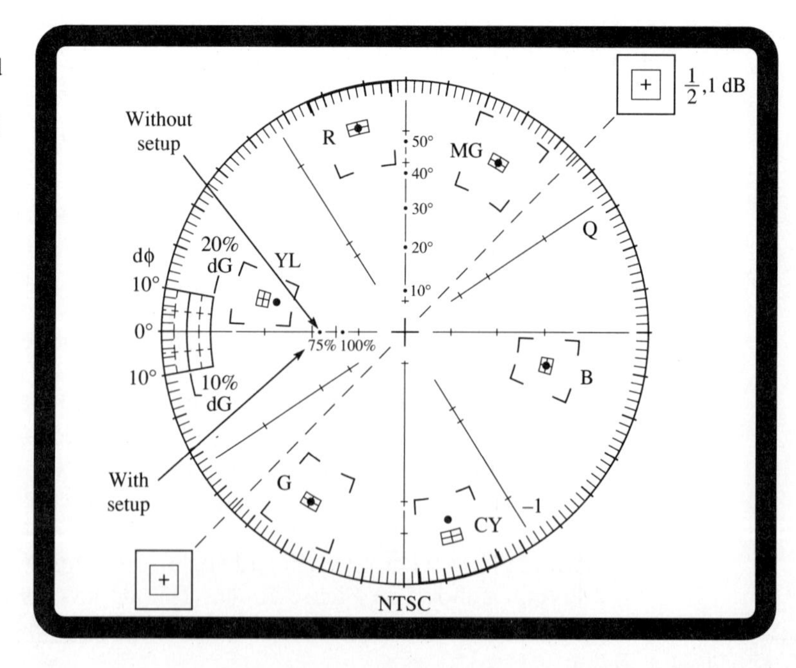

FIGURE 16–49
Yellow and cyan are affected by differential gain error. Other colors are not affected in this example.

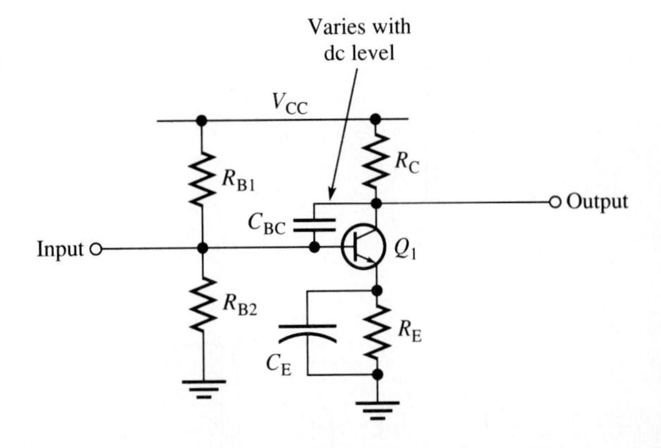

FIGURE 16–50
Large C_{BC}, causing the phase to vary with changes in input dc.

The circuit of Figure 16–50 shows a different problem. The transistor stage has a form of phase shift that is dependent on the luminance level of the input signal. The reversed-biased base-to-collector junction acts like a variable capacitor, forming an *RC* phase-shift network. In this case, the yellow and cyan color vectors have the proper amplitude, but they miss the box along a radial arc, as shown in Figure 16–51. The other color vectors hit the center of their respective boxes. Although the intensity of the color is correct, the hue of these bars is wrong.

Because the most common distortions that pertain to the color signal are differential gain and differential phase, a special test signal has been developed to measure the effects that the dc component (luminance) has on the chroma signal. This signal is called the *modulated ramp signal* (Figure 16–52(a)). It comprises a

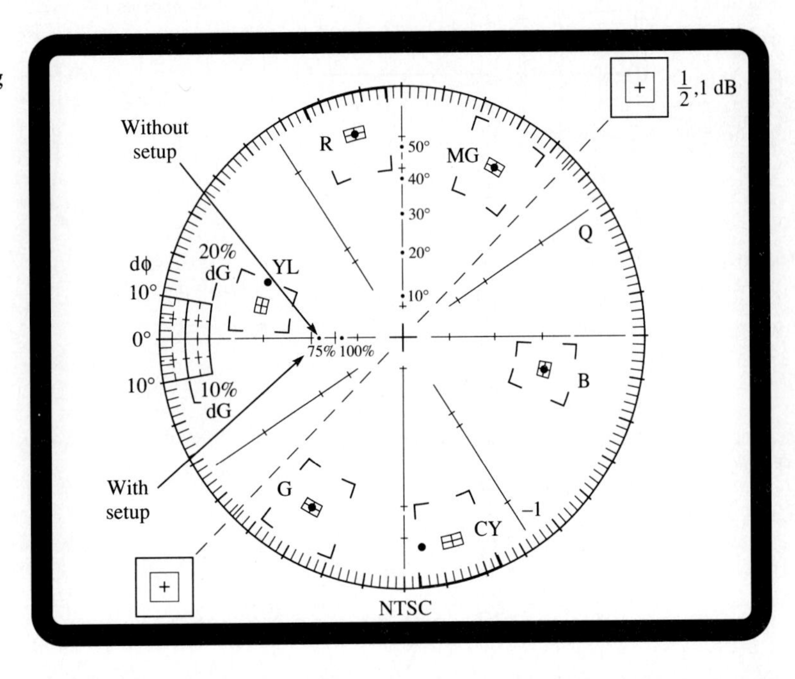

FIGURE 16–51
Differential phase affecting only yellow and cyan.

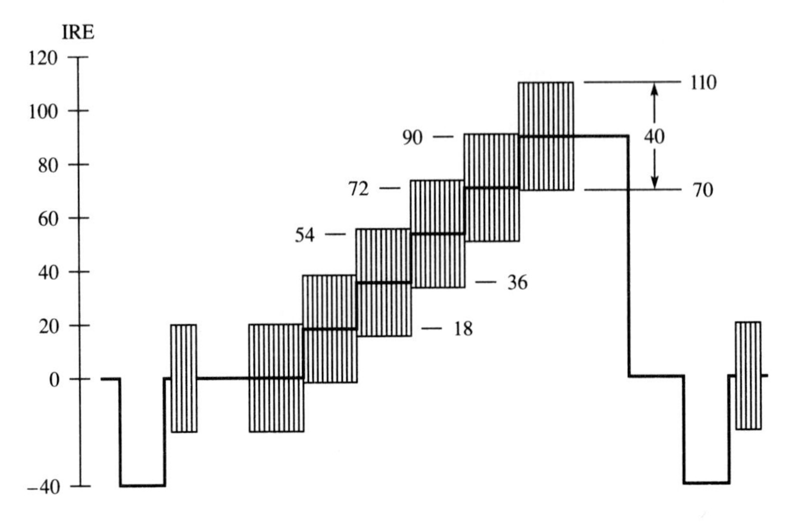

(a) Five-step ramp signal.

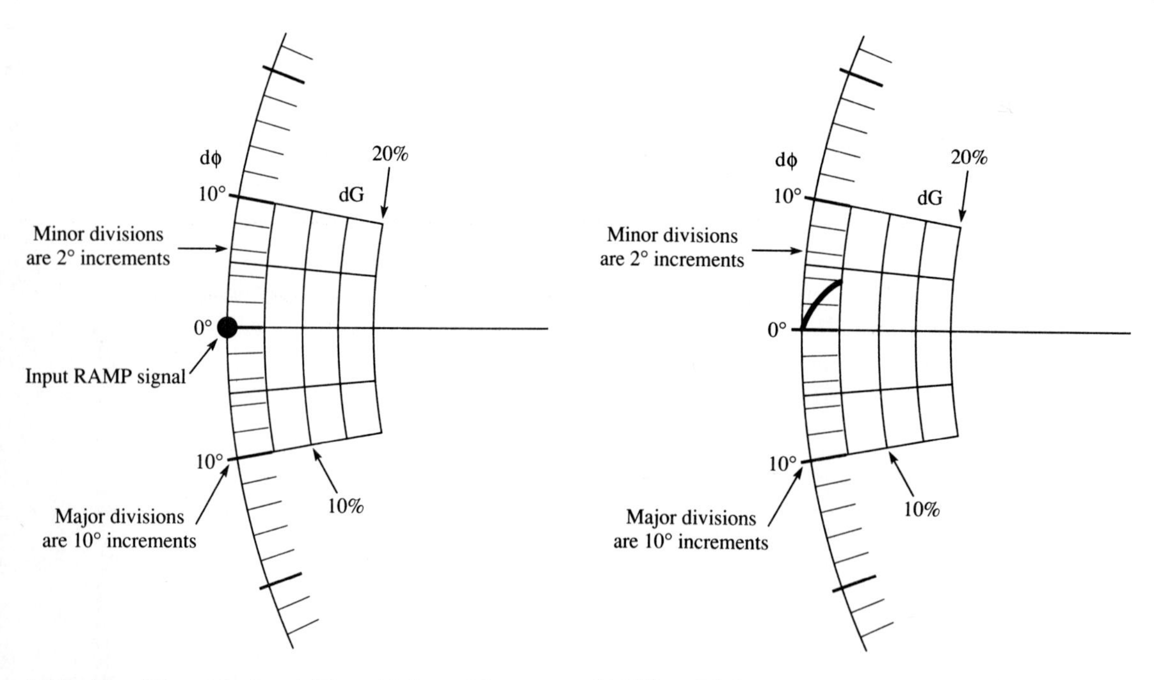

(b) Checking differential gain and differential phase of the test signal.

(c) Differential phase error of 4° and differential gain error of 5%.

FIGURE 16–52

ramp going from 0.0 V (black signal) at the left side of the picture to 0.714 V (full-peak white) at the right side of the picture. It repeats line after line over the entire picture. A chroma signal is modulated on the dc ramp and is constant in phase and gain with respect to the burst signal. The vector display of this signal is a single bright dot, as shown in Figure 16–52(b); dim lines to it from the center of the vector display will also be present but are of no consequence. Distortions of differential gain and differential phase would cause this dot to be a line or arc, as shown in Figure 16–52(c).

The following steps give the procedure for measuring the differential gain and differential phase with a vectorscope:

1. Connect an external 3.58 MHz subcarrier source to the vectorscope's reference input.
2. Connect the input of the vectorscope directly to a source of a modulated ramp test signal.
3. Adjust the PHASE and GAIN of the vectorscope to place the vector on the outer graticule of the vectorscope and on the 180° axis, as shown in Figure 16–52(b). Assure that the input signal is clear of distortions.
4. Connect the modulated ramp test signal to the input of the UUT and the vectorscope to the output of the UUT.
5. Readjust the PHASE and GAIN of the vectorscope to place the vector on the outer graticule of the vectorscope (see Figure 16–52(c)).
6. Use the grid on the CRT of the vectorscope to determine the differential phase and differential gain of the UUT.

EXAMPLE 16–6 A technician is asked to measure the differential gain and differential phase of a particular circuit. The vectorscope is connected to the circuit under test as described previously. The output of the circuit is shown on the vectorscope display in Figure 16–52(c). What are the differential gain and phase error?

SOLUTION The distortion can be directly read on the vectorscope as a differential gain error of 5% and a differential phase error of 4°.

Differential gain and differential phase have some interesting effects on the picture, if they are severe. Differential phase is the easiest to visualize. Suppose a camera has 45° of differential phase error and is pointed at the face of a person standing in bright sunlight. If the HUE control on the picture monitor (TV set) is set so the cheeks are the proper hue, then the forehead (in brighter sunlight) will experience a differential phase shift in one direction, causing a greenish tint to the skin. The shadow under the chin (a darker scene) will cause a phase shift in the other direction, causing a purple cast to the shadow. You would be unable to obtain proper hues for all levels of light intensity.

The effects of differential gain are a little more subtle. If someone in the shade is wearing a bright red shirt and moves into the sunlight, the red washes out to a pink color.

THE EFFECTS OF SLEW-RATE LIMITING

Another type of distortion is caused by *slew-rate limitation*. Slew rate is the ability of an output signal of a circuit to swing from one level to another as fast as the input signal commands it. If a circuit is slew-rate limited, then the output cannot keep pace with the input. This can cause gain compression as well as a phase shift of the output signal, as shown in Figure 16–53. A greater input signal swing will cause a larger effect on the output. The larger-amplitude signals such as cyan, green, and red will experience the greatest distortion, as shown in Figure 16–54.

MEASURING TIMING (DELAY)

Every circuit has some delay through it. In a color television studio, matching delays are important to maintain proper color timing between sources. There often is more than one camera in a studio as well as other video sources, such as video tape decks. If the sources are not *timed* with respect to each other, then there *could* be a hue shift *or timing glitch* when switching between video sources. This can cause video recorders to lose sync lock momentarily. Timing the sources means that the leading edge of the vertical sync pulses, the leading edge of the horizontal sync pulses, the phase of the burst reference signal, and the phase of the chroma signal are all coincident between the various video sources.

In order to set these source paths to a high degree of precision, a reference signal, such as color bars, is sent through each of the video paths. One video source (path) is used for a reference; the other video paths are selected and adjusted, if necessary, to match to the reference path. An oscilloscope or waveform monitor (externally triggered) is used to set the timing so the leading edge of vertical sync and the leading edge of horizontal sync of the various paths match

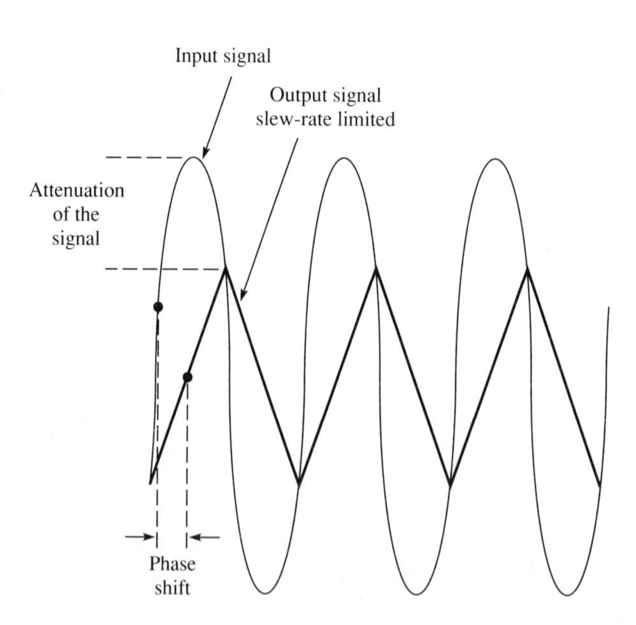

FIGURE 16–53
Severe slew-rate-limited waveform.

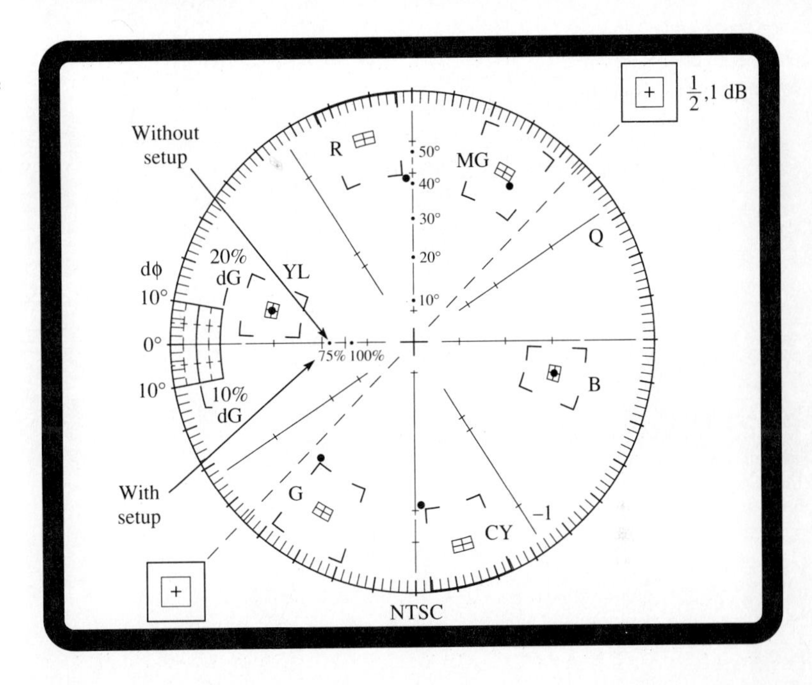

FIGURE 16–54
Vectors affected by slew-rate limiting.

that of the reference path. A vectorscope is then used to match the phase of the burst reference signal and the phase of the color bar signal. Once this has been done, a studio is said to be in *color time*. Figure 16–55 is an example of how the equipment is connected for setting the timing between video paths.

The following procedure shows how to check for color burst reference time.

1. Select a reference path and connect the vectorscope to it. Connect a 3.58 MHz subcarrier source to the vectorscope and select EXTERNAL REFERENCE on the vectorscope.
2. Adjust the GAIN and the PHASE of the vectorscope so that the burst vector is placed on the outer graticule of the screen, as shown in Figure 16–56. Note that this is the reference signal.
3. Connect a second source via its source coax cable to the vectorscope's input.

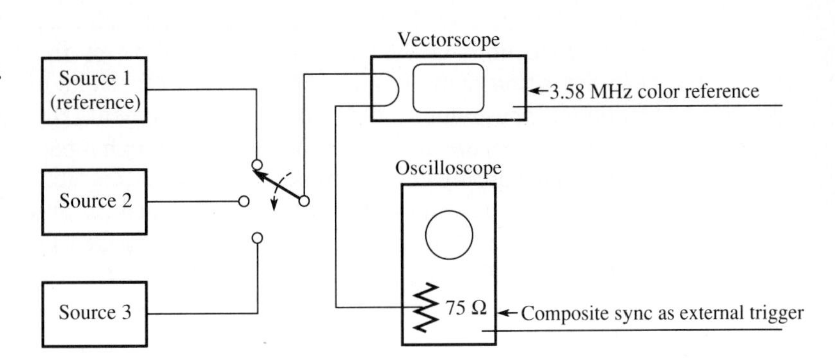

FIGURE 16–55
Timing sources together.

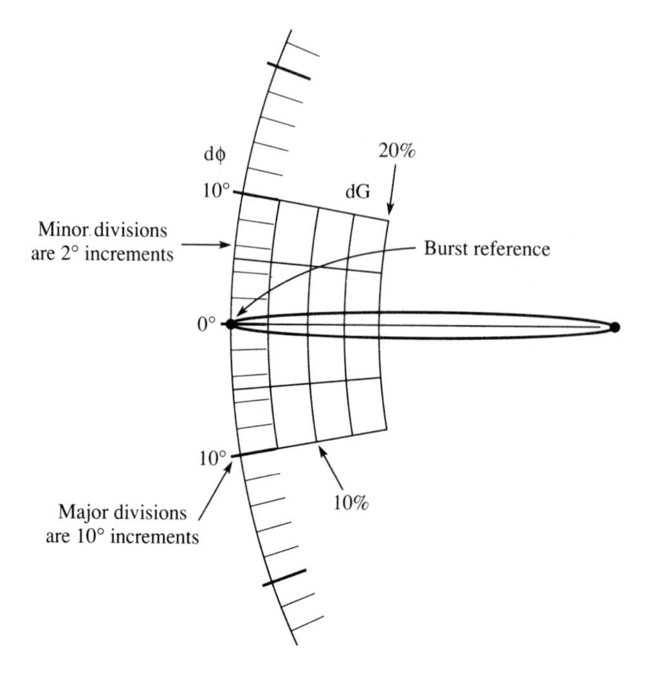

FIGURE 16–56
Setting burst reference.

4. Note the phase difference. Measure and record this difference as an error, or adjust the *color phase compensation* of the second source to match this signal to that of the reference (Step 2). The color phase compensation is a variable delay amplifier used in most video equipment for the purpose of timing signals.

DIFFERENTIAL GAIN AND DIFFERENTIAL PHASE DISPLAY MODES

Up to this point, timing, differential gain, and differential phase have all been measured using the vectorscope in the vector display mode. But as you saw in Figure 16–52(b), marks on the circular graticule are 2° apart for measuring phase and the gain marks are in steps of 5%. The size of the dot produced on the screen of the vectorscope is large enough to limit the precision of the measurement. The reduced precision in the vectorscope display is normally satisfactory for measuring the distortions in the communication link from the studio to the transmitter; these distortions can often be 2% to 5% in differential gain and several degrees in differential phase due to the microwave equipment. However, in the studio environment, where the signal may be passed through a particular piece of equipment several times during editing, these errors are intolerable. The signal quality needs to be much better and the test equipment must be able to measure signals with greater precision than with the vectorscope mode. The differential gain (DIFF GAIN) and differential phase (DIFF PHASE) modes of the vectorscope were developed to improve the precision of the differential measurements.

DIFFERENTIAL GAIN DISPLAY MODE

The differential gain display mode uses the modulated ramp as a video source, as shown in Figure 16–52(a). The amplitude of the chroma signal is constant as the luminance (dc level) is changed from black to white. The differential gain mode of display is much like that of an oscilloscope because it is a single line traced from left to right. The left side represents the beginning of a horizontal line (the black, or bottom, of the ramp signal) and the right side represents the end of a horizontal line (the white, or top, of the ramp signal). If the amplitude of the chroma subcarrier signal is constant from the black level to the white level, then the displayed trace is a flat horizontal line in the center of the CRT. If there is any deviation in the amplitude of the subcarrier, then this line will deviate a measurable amount. A reduction in the peak-to-peak amplitude of the subcarrier signal will cause the horizontal line to go downward, and an increase in the peak-to-peak amplitude causes an upward shift of the line.

To make a differential gain measurement, you must first adjust the gain of the vectorscope to be "in range." The gain window used by the differential gain mode is very narrow and sensitive; therefore, the amplitude of the signal going into this section has to fall within this window. This is done by selecting the vector mode of display and adjusting the GAIN of the vectorscope so that the vector is placed on the outer graticule, as shown in Figure 16–57. Then select DIFF GAIN mode on the vectorscope. The display will change from the vector display to the single line display as shown in Figure 16–58. Detailed instructions follow:

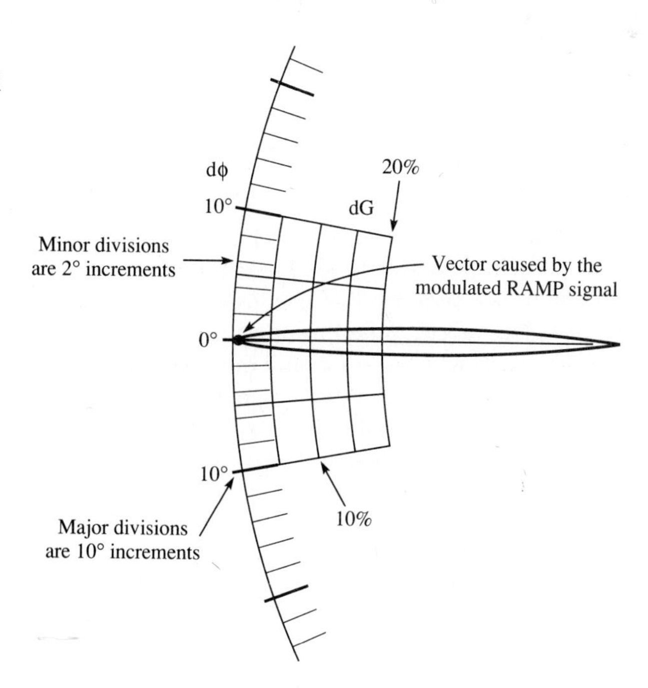

FIGURE 16–57
Initial setting for differential phase and differential gain measurements.

FIGURE 16–58
Differential gain error of 2%.

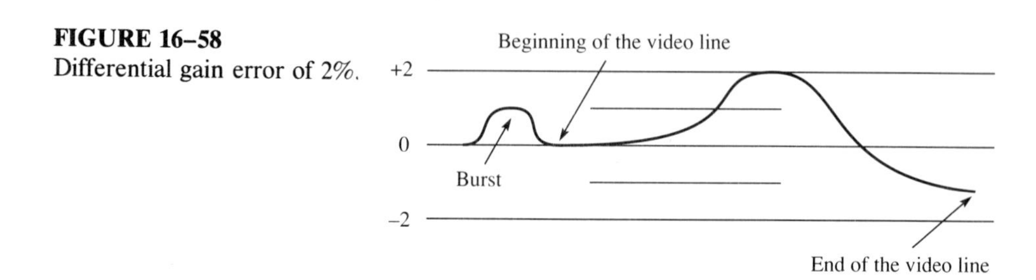

1. Connect a 3.58 MHz subcarrier source to the REFERENCE INPUT of the vectorscope or switch to INTERNAL REFERENCE.
2. Connect a source of modulated ramp to the equipment to be measured and terminate the coax with 75 Ω.
3. Connect the output of the UUT to the vectorscope and terminate the coax with 75 Ω.
4. Select the VECTOR mode of display and adjust the GAIN to place the vector representing the ramp signal on the outer graticule, as in Figure 16–57.
5. Select the DIFF GAIN mode and calibrate the left-hand side of the display as the reference gain by placing it on the 0% line using the VERTICAL POSITION control. See Figure 16–58.
6. Measure the maximum deviation from the 0% line against the scale as differential gain.

The very first portion of the display is the burst reference signal and should be ignored. Because this measurement depends on how flat this trace is, it is important that the TRACE ROTATION be checked before this measurement is made. This is easily done by selecting *Y* mode (luminance mode) without any input signal. Adjust TRACE ROTATION for a horizontal trace. Also check the input test signal (modulated ramp) to be sure that it is free of distortion by connecting the test signal directly to the vectorscope.

Figure 16–57 shows the proper setting of the vector on the outer graticule. There may be two vectors visible. One vector will be brighter than the other vector. The dim vector is the burst reference vector and should be ignored. The burst reference vector is dim because there are only 8 cycles of burst reference per line of video, as compared to more than 50 cycles of the modulated ramp.

The vector can be set at any radial position (any degree) on the graticule, but it *must* be on the outer graticule (gain). It is suggested that while the gain is being set on the outer graticule, the phase be set to the reference radial position, as shown in Figure 16–57 using the PHASE control because this will save time when setting up to measure differential phase in the next section.

The differential gain error measurement previously described has a special subgraticule illuminated (Figure 16–58). When DIFF GAIN is selected, a set of lamps that illuminates another graticule in front of the CRT is turned on.

Some vectorscopes have better display capabilities than the one just mentioned. They have a **double-line display** switch, which permits up to 0.1% display resolution. The double-line display is explained in the next section.

DIFFERENTIAL PHASE DISPLAY MODE

The differential phase mode uses the same modulated ramp signal that was used in the differential gain measurement (see Figure 16–52(a)). This measurement assures that the ramp signal stays constant in phase as the dc level of the signal changes from black to white.

There are two modes of display: single line and double line. The single-line mode is similar to the differential gain display. It is better to use the double-line mode of display as higher resolution can be obtained. The DOUBLE-LINE MODE of operation mirrors the displayed line about the horizontal axis. Adjustment of PHASE control or the CALIBRATED PHASE control will cause these two lines to converge or diverge.

A procedure to measure differential phase is as follows:

1. Connect a 3.58 MHz subcarrier to the REFERENCE INPUT to the vectorscope.
2. Connect a source of modulated ramp to the input of the UUT and terminate the coax with 75 Ω.
3. Connect the output of the unit under test to the vectorscope and terminate the coax with 75 Ω.
4. Select VECTORSCOPE mode of display. Adjust the CALIBRATED PHASE CONTROL to indicate 0.0° (see Figure 16–59). Adjust the PHASE to place the vector representing the ramp signal on the reference phase axis, as in Figure 16–57.
5. Select DIFF PHASE mode and adjust PHASE so the double lines converge at left-hand side of the display. See Figure 16–60.
6. Adjust the CALIBRATED PHASE CONTROL until the largest deviation of the display (most often it is at the full luminance or the right side of the display) has the double lines converged, as shown in Figure 16–61.
7. Measure differential phase by reading the value directly on the CALIBRATED PHASE CONTROL scale.

FIGURE 16–59
Calibrated phase control (copyright Tektronix, Inc. Used with permission).

FIGURE 16–60
Setting a reference point.

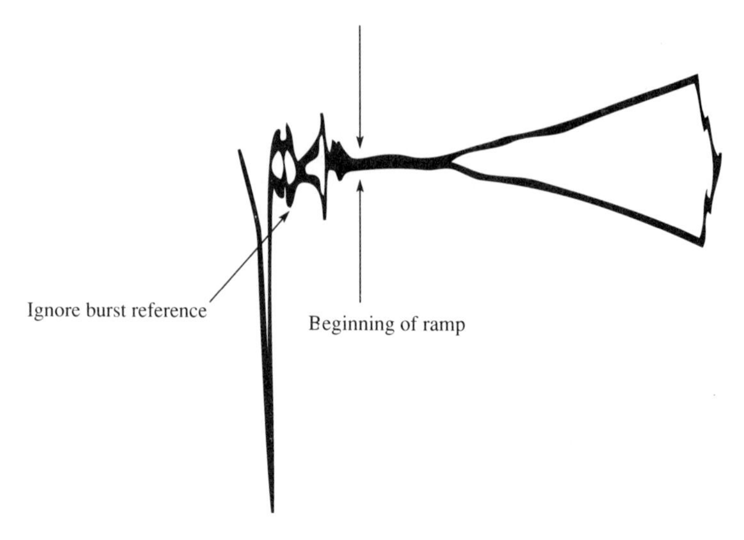

FIGURE 16–61
Setting for a measurement.

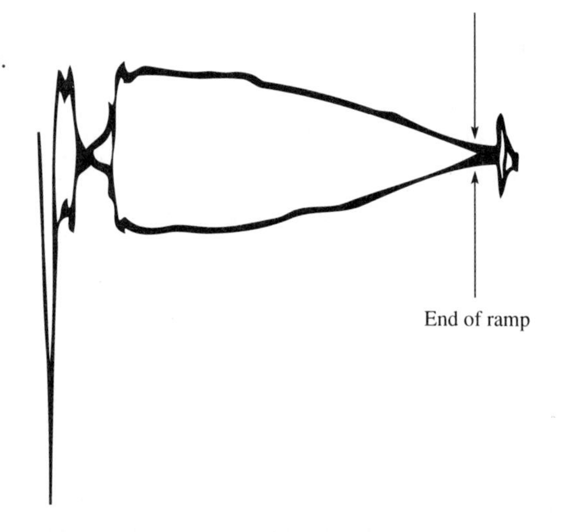

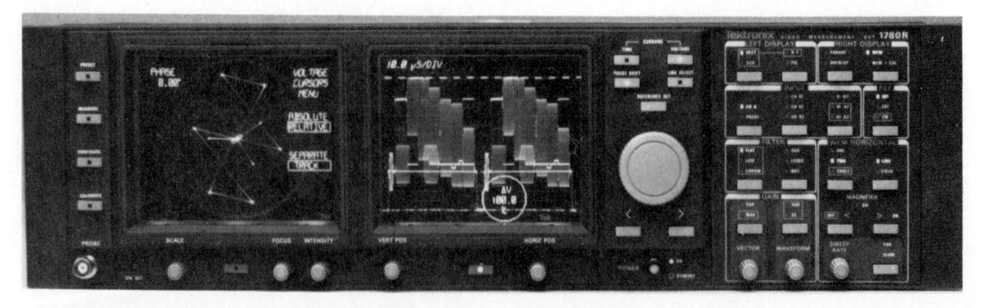

FIGURE 16–62
Tektronix 1780-series combination vectorscope, waveform monitor, and picture monitor (copyright Tektronix, Inc. Used with permission).

FIGURE 16–63
Tektronix VM 700 (copyright Tektronix, Inc. Used with permission).

Vectorscopes that have a double-line display for the differential gain measurement also have a CALIBRATED GAIN CONTROL similar to the calibrated phase, but it is calibrated in gain rather than in phase. Figure 16–62 shows the Tektronix 1780, which has a double-line display for differential gain as well as differential phase.

New video instrumentation comes in the form of data acquisition and measurement systems, as shown in Figure 16–63. By pressing the "measurement menu" button on the instrument panel, the CRT display shows several choices of measurement parameters. The screen is touch-sensitive; by touching the appropriate selection, the measurement is taken automatically.

SUMMARY

1. Special-application instruments may be needed for video measurements if general-purpose equipment is either not accurate enough or is too difficult to configure for frequently measured parameters. The waveform monitor and vectorscope are examples of special-application instruments used to measure the video signal.
2. There are three basic types of television broadcast media. They are NTSC (National Television System Committee) used in the USA, the PAL (Phase Alternate by Line) system used in much of Europe, and the SECAM (French development) system used in France and Russia. The systems are *not* compatible with each other.
3. The scan system used in the USA is a 525-line interlaced system. The frame rate is 30 frames per second. The scan system used in Europe (PAL) is a 625-line interlaced system at 25 frames per second.
4. The color signal is encoded by a sine wave at 3.58 MHz superimposed on the black-and-white signal. The phase of this sine-wave signal compared to the reference burst signal determines the hue. The ampli-

tude of the sine wave determines the color saturation.
5. The waveform monitor is a special-application oscilloscope. It has special triggering for viewing any selected line of video. It has clamping and dc restoration circuits in its input amplifier to correct for hum and long-time distortions of the video waveform. It also has filtering to remove the chrominance, leaving just the luminance, or vice versa. Its graticule is designed to make video measurements easily and accurately.
6. The vectorscope is a special-application instrument for displaying the chrominance amplitude and phase as a vector. It is useful for measuring very subtle distortions.

QUESTIONS AND PROBLEMS

1. What are the synchronizing pulses and what is their purpose?
2. The horizontal and vertical sync pulses are different than video signals. What distinguishes the sync from video?
3. Define a pixel.
4. In Figure 16–64, draw the video waveform for lines 1, 2, 3, and 4. Assume that black is 0 V and white is 714 mV.

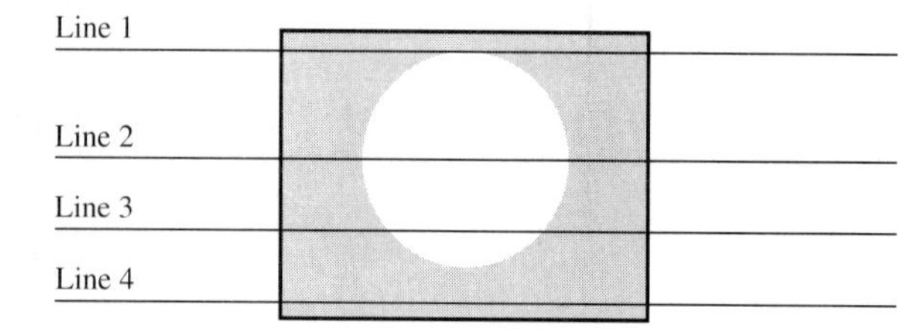

FIGURE 16–64

5. There are two sets of primary colors: One is associated with paints; the other is associated with light sources. What are the sets of primary colors and why do they differ?
6. At the beginning of each line of active video is a signal called the burst reference. What is the purpose of this reference signal?
7. (a) Find the gain and amplitude of the green vector at maximum saturation.
   (b) Find the luminance value for the green vector.
8. Define differential phase. Give an example of severe differential phase. That is, what would severe differential phase "look" like on a television screen?
9. Define differential gain. Give an example of severe differential gain. That is, what would severe differential gain "look" like on a television screen?
10. The output (source) impedance of a piece of video equipment is 75 Ω. When terminated with 75 Ω at the end of a coax cable (load), its nominal peak-to-peak signal voltage is 1.0 V.
    (a) What is the voltage if the termination is left off?
    (b) What is the voltage if it is terminated with 50 Ω?
    (c) Calculate the percent error if terminated with 50 Ω.
11. The video cable connecting several pieces of equipment together is looped through one piece to another, as shown in Figure 16–21. On one of the pieces of equipment, a switch was accidentally thrown that terminated the coax with 75 Ω; there is a termination at the end of the cable, on the last piece of equipment that also terminated the coax with 75 Ω. This condition is known as "double terminated."
    (a) Calculate the peak-to-peak voltage with double termination assuming that with single termination, the peak-to-peak voltage would be 1.0 V.

(b) What is the percent error with double termination?

12. A video signal is to be 714 mV above ground with sync being 286 mV below ground. The signal measures 735.42 mV above ground and sync stays at 286 mV.
    (a) What is the percent gain error of the signal?
    (b) What is the gain error if the sync increased, as well, to 294.58 mV?
13. The pulse-to-bar levels show 6% difference. What is the K factor for this error?
14. A circuit is ac coupled, as shown in Figure 16–65. What is the expected % K_B (tilt) of this signal?

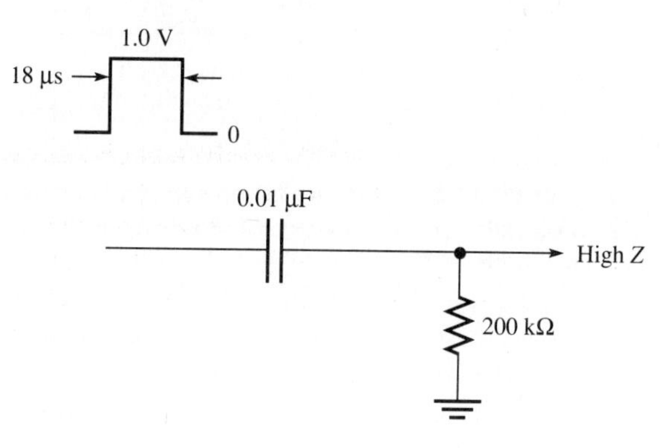

FIGURE 16–65

15. Luminance-to-chrominance delay and gain were measured and the waveform in Figure 16–66 was observed. What is the error in terms of gain and delay?

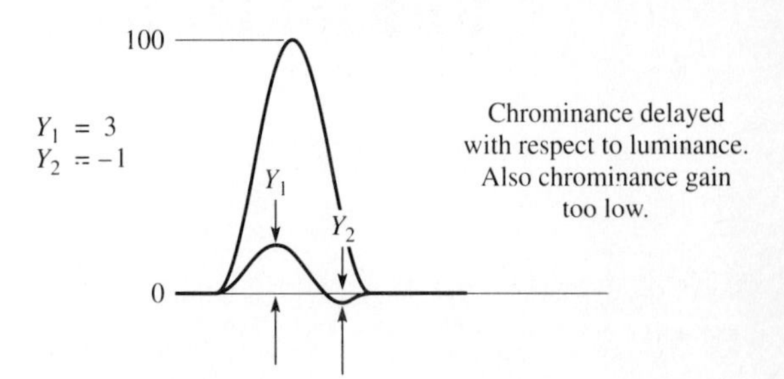

FIGURE 16–66

16. As the luminance of a signal goes from black to white, the circuit causes a differential phase error of −10° delay and a differential gain error of −20%. Show what this would look like on a vectorscope if modulated ramp (ramp with color) is used as a source signal and the red chroma box is used to make the measurement.
17. The grass is to look green and the sky is to look blue. What would their colors be if burst reference were 90° late? Refer to Figure 16–44.

Chapter 17

Biomedical Instruments

Richard D. Bliss

OBJECTIVES

Biomedical instruments are used to analyze the function of living organisms. Even the simplest organisms are complex compared with the machines made by humans, and thus the analysis of the function of organisms requires a variety of methods. In fact, biomedical instrumentation is a world of its own, encompassing hundreds of different specialized devices, transducers, techniques, and measurement units. Books much larger than this one have been written about the subject, but the field has really grown much too large for a single volume. For this reason, it would be impossible to present a complete overview of all of bioinstrumentation in a single chapter. Instead, we concentrate on some basic principles and instruments in order to give you a feeling for the field. We start with some principles of biology and *biochemistry* (the chemistry of life) and describe some important instruments based on these ideas. Then we develop more sophisticated biomedical principles, introducing new instruments as we go. Our development culminates in a discussion of the electrocardiograph, a widely used device that is perhaps the most "biomedical" of all clinical instruments. The chapter concludes with a discussion of safety in relation to biomedical instruments.

When you complete this chapter, you should be able to

1. Define and explain basic biomedical terms.
2. Describe the operation of instruments that measure solute concentration and the principles upon which these instruments are based.
3. Describe the properties of electrolyte solutions and the instruments used to measure ion concentration.
4. Define pH and describe the operation of pH meters.
5. Describe the construction and function of electrodes used to make biochemical and biomedical measurements.
6. Describe the characteristics of selectively permeable membranes and the origin of membrane potentials.
7. Describe the origin and nature of bioelectric potentials in nerve and muscle cells.
8. Describe the function and operation of instruments, particularly the electrocardiograph, that measure bioelectric potentials.
9. Explain the basic safety features of biomedical instruments and the reasons for these safety precautions.

HISTORICAL NOTE

In 1903 Willem Einthoven (1860–1927), a Dutch physiologist, perfected the *string galvanometer,* consisting of a silvered thread stretched between the poles of a strong magnet. Current from the circuit under investigation was passed through the thread. The instrument was read by observing the deflection of the thread in the magnetic field through a microscope. The magnified image of the thread could also be projected on a screen. Einthoven's instrument had both the sensitivity and the frequency response needed to record the tiny currents produced in the human body by the contraction of the heart. The device, when used this way, was the world's first electrocardiograph. Einthoven later worked out the principles and basic methods for electrical measurements of the heart. His pioneering work earned him the 1924 Nobel Prize for medicine and physiology. Einthoven's invention, updated and modernized over many decades, is today one of medicine's most important diagnostic tools.

17–1 BIOLOGICAL AND BIOCHEMICAL BACKGROUND

All organisms, including humans, are composed of cells. Cells are the fundamental structural and functional units of life. They are quite variable in size and shape, depending on their specialization, but all cells have certain basic features in common. Figure 17–1 is a drawing of a generalized cell. The outer boundary is the thin *cell membrane,* or *plasmalemma,* that is permeable only to water and certain other molecules. This membrane controls what substances enter or leave the cell, and without this function cells quickly die. The material inside the plasmalemma, the "stuff" of cells, is the *cytoplasm,* in which the chemical processes of life, called *metabolism,* occur. One of the most important processes of metabolism is

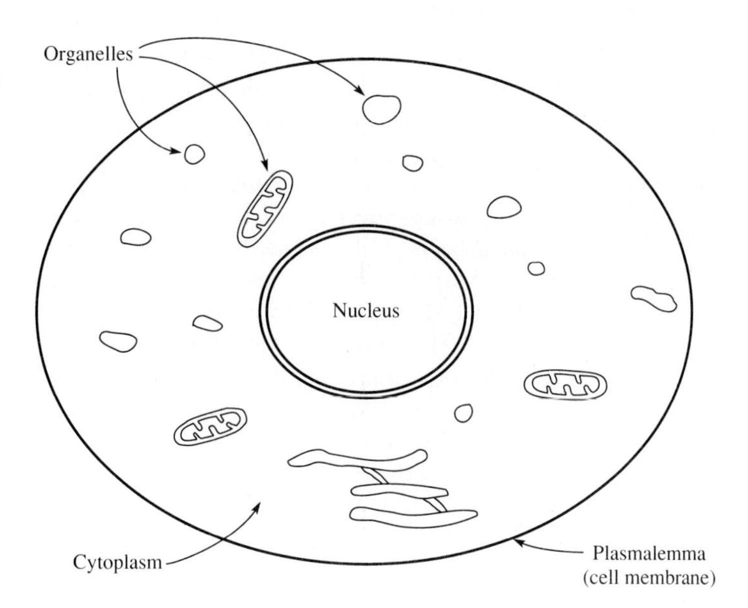

FIGURE 17–1
Schematic of a living cell.

the breakdown of food molecules that provide the cell the energy needed to keep going. All cells require energy for their activities. Embedded in the cytoplasm are a number of *organelles* (little organs), in which a variety of special processes occur. The largest organelle is the nucleus, which contains the chromosomes. Chromosomes are repositories of information that basically instruct the cell how to conduct the processes of life.

Anatomy is the study of the structure of organisms, but the word may also mean the structure itself. One may speak of the *anatomy,* or structure, of different kinds of organisms (for example, plant anatomy or human anatomy). The capabilities and function of any organism depend on the details of its anatomy. *Physiology* is the study of the function of organisms or of the parts of organisms. For example, human physiology is the study of the function of the human body, whereas *neurophysiology* is the study of the function of the nervous system. Sometimes the term physiology is used as a synonym for the term function. Naturally, the physiology of an organism is closely linked to its anatomy.

Malfunction of the structures that make up the organism usually leads to some sort of disease condition. *Pathology* is the study of disease and the anatomical and physiological changes in the body caused by disease. Malfunction may be due to the activities of a *pathogenic* (harmful) microorganism, to some kind of injury (*trauma*), or to breakdown of the structure involved (*degenerative* disease). In order to prescribe the correct *therapy* (treatment that promotes healing or recovery), the physician must *diagnose* (identify) the disease. Physicians can frequently make a diagnosis based on the examination of the anatomy of a patient. Many of the most important and sophisticated of biomedical instruments are designed to permit the examination of body structures. X-ray machines are an example of this kind of instrument. An important feature of instruments used to examine anatomy is that they produce an image, rather than data in the form of numbers or charts.

Not all physiological problems can be detected by examination of body structures. The human body is primarily a chemical machine, and changes in the chemical function don't always show up as a change in anatomical structure. For example, headaches are easily perceived by the sufferer, but a physician cannot detect them by anatomical examination alone (unless the headache is due to injury). Physiological measurements frequently depend on determining some chemical or **electrochemical** (involving both chemical and electrical phenomena) parameter of the body. Most of the "lab tests" ordered by physicians are basically chemical tests. Many biomedical instruments are designed to measure such chemical parameters. Examples of important and frequently used chemical laboratory instruments are pH meters and spectrophotometers. The most commonly used electrochemical instrument is the **electrocardiograph** (commonly abbreviated EKG, from the German spelling), which is used to assess heart function.

One of the difficulties that physicians face is the natural variability of human body structure and function. In fact, all species of organisms show considerable variability in normal function from one individual to the next. Consequently, there can be no exact definition of "normal" characteristics—only a range of values for healthy individuals. In fact, the normal values for such human characteristics as body weight and blood pressure are the average values for healthy people. Data gathered from real organisms will reflect the variability of nature and will likely show some fluctuation in values over time. That is, characteristics measured in the same organism and employing the same equipment may fluctuate as time passes. Physicians must rely in part on experience to tell whether an individual deviation from average is due to the variability of nature or due to disease.

The fundamental processes of life are basically chemical processes that occur in the watery interior of the body. To measure these processes, a number of instruments have been developed to determine various properties of *solutions*. A **solution** is a homogeneous mixture of a *solute* and a liquid *solvent,* such as water. The solute can be a solid, such as sugar, or a liquid (alcohol, for example) or a gas (such as the dissolved oxygen that fish breathe).

Before proceeding with the characteristics of solutions, it will be useful to review some basic principles of matter. Recall that all matter is composed of *atoms* and that atoms are made up of three kinds of elementary particles. Figure 17–2 is a drawing of the basic structure of atoms. The dense core, or *nucleus* of the atom contains two types of elementary particles, positively charged *protons* and electrically neutral *neutrons*. Surrounding the nucleus is a cloudlike cluster of negatively charged electrons. The number of electrons is equal to the number of protons, so that individual atoms have no net electric charge. Atoms of different chemical elements differ in the number of protons and electrons. That is, *atomic number* (number of protons) is characteristic of each element. Atoms that have the same number of protons (and electrons) but differ in the number of neutrons are *isotopes* of a single element. An alphabetic symbol and a name are used to

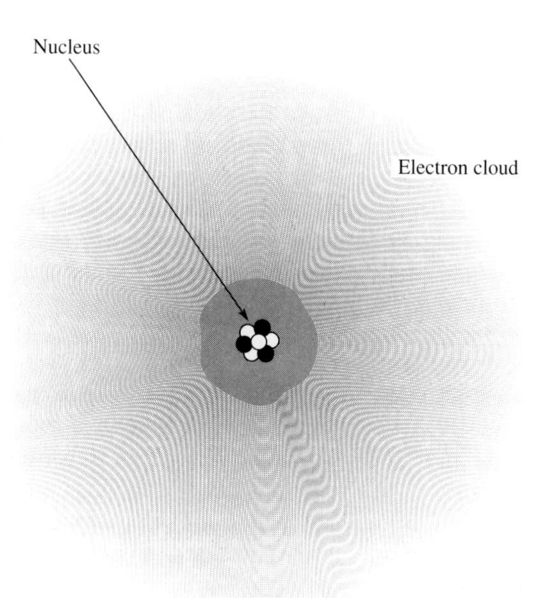

FIGURE 17–2
Structure of an atom.

TABLE 17–1

ELEMENT	SYMBOL	ATOMIC NUMBER	ATOMIC WEIGHT
Hydrogen	H	1	1
Carbon	C	6	12
Nitrogen	N	7	14
Oxygen	O	8	16
Sodium	Na	11	23
Magnesium	Mg	12	24
Chlorine	Cl	17	35
Potassium	K	19	39
Calcium	Ca	20	40

identify the different elements. The *atomic weight*—or, more properly, the *atomic mass*—is approximately equal to the number of protons and neutrons, since electrons have only a tiny fraction of the mass of the nuclear particles. The unit of mass is the *atomic mass unit* (amu) or *dalton*. Table 17–1 is a list of some of the elements that are important in biomedicine, along with their symbols, atomic numbers, and atomic masses.

Molecules are composed of two or more atoms bound together. The *molecular formula* gives the number and kind of atoms in the molecules of a given substance. For example, the molecular formula for water is H_2O, indicating that water molecules are composed of two hydrogen and one oxygen atom. The *molecular weight* (or *molecular mass*) is the sum of the atomic weights of all the atoms comprising the molecule. The molecular weight of water is 18 daltons, calculated as follows:

$$\text{Atomic weight of hydrogen} = 1$$
$$\text{Atomic weight of oxygen} = 16$$

Therefore, the weight of two hydrogen atoms and one oxygen atom is

$$2(1) + 1(16) = 18$$

17–2 SOLUTIONS

Water is the most abundant substance in living organisms. The human body, for example, is about 70% water. Water might be called the solvent of life, because virtually all chemical reactions in organisms occur in water-based solutions. However, the rate at which these reactions occur depends on the solute *concentration*. The term **concentration** means the amount of solute per unit volume of solution (such as grams of solute per liter of solution). A useful way of expressing chemical amounts is in *moles*. One mole of a substance is equal to the molecular weight of that substance expressed in grams. For example, the molecular weight of water is 18 and that of table sugar is 342; thus, 1 mole of water is 18 g and 1 mole of sugar is

342 g. One mole of any material contains a fixed number of molecules of that substance. The number of *molecules* per *mole* of all substances is always the same, 6.023×10^{23}, a constant called *Avogadro's number*. Concentration is usually expressed in units of moles per liter, or as *molar* concentration, abbreviated M (for instance, a 0.5M solution of sugar contains 1/2 mole of sugar, or 171 g, per liter of solution). Any solution contains the same number of solute molecules per unit volume as any other solution *of the same concentration,* regardless of the nature of the solute. For example, a 0.1M sugar solution contains the same number of sugar molecules *per cubic centimeter of solution* as 1 cm^3 of 0.1M ammonia solution contains ammonia molecules.

17–3 SPECTROPHOTOMETERS

The concentration of molecules has a direct effect on the rate at which chemical reactions occur. Consequently, the physiology of living organisms is also affected by the concentration of a wide variety of molecules. For example, the amount of sugar and salt in the blood is tightly controlled by the body because many physiological functions would be unbalanced by high or low concentrations of these materials. As you might expect, instruments that measure concentration are of great value in biomedical applications. **Spectrophotometers** are instruments that measure concentration on the basis of the ability of many substances to absorb light in direct proportion to their concentration. In order to understand this interaction, it will be useful to review the nature of light as electromagnetic radiation. Recall that the speed of electromagnetic radiation in free space is 3.0×10^8 m/s and that wavelength is the speed divided by frequency. The formula is

EQUATION 17–1

$$\lambda = \frac{c}{f}$$

where λ = wavelength, m
c = the speed of light, m/s
f = the frequency, Hz

The characteristics of electromagnetic radiation depend on the wavelength. Figure 13–30 illustrated the electromagnetic spectrum. For convenience, it is reproduced here as Figure 17–3. Visible light occurs in a very narrow band of wavelengths, from about 700 nm to about 400 nm. The human eye's receptor cells respond to the wavelength of light, which we perceive as color. The illustration on the back cover of the text shows the visible part of the electromagnetic spectrum in full color and the colors that the eye sees in response to various wavelengths. White light, such as sunlight, contains all wavelengths in roughly equal amounts.

The wavelength governs what happens when light interacts with matter. Many substances absorb some wavelengths but reflect others. A human observer perceives only the reflected light and sees the material as having the color of the reflected wavelengths. For example, plants appear green in sunlight because the leaves absorb long and short wavelengths but reflect the middle wavelengths that

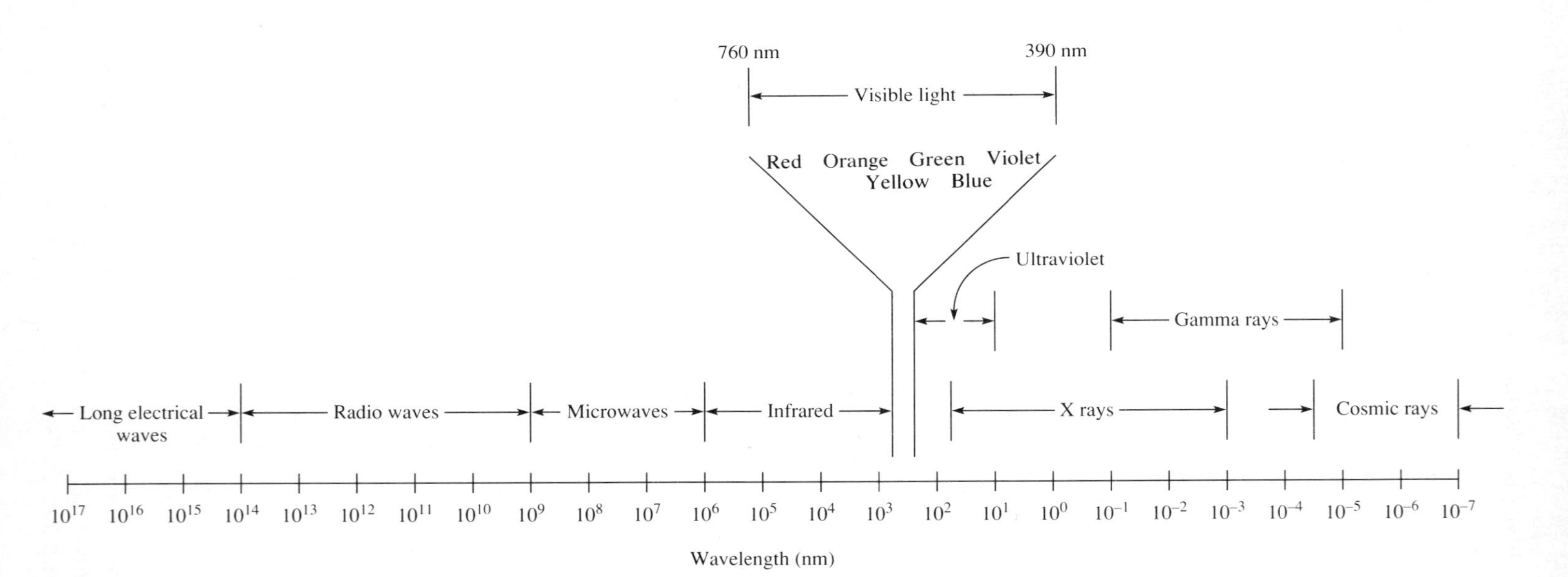

FIGURE 17–3
The electromagnetic spectrum.

the eye sees as green. Substances that absorb all wavelengths appear black to the eye, and those that reflect all wavelengths are white. Transparent substances may also selectively absorb or transmit different wavelengths. Beer, for example, transmits yellow wavelengths but absorbs those corresponding to other colors, whereas water transmits all wavelengths equally. It is possible to measure the amount of light of different wavelengths transmitted through transparent or translucent substances or solutions. If the amount transmitted is plotted against wavelength, the resulting plot is called an **absorption spectrum.** Figure 17–4 shows the absorption spectrum of a solution of chlorophyll, the green pigment responsible for photosynthesis in plants. Wavelengths corresponding to peaks in the absorption spectrum are called *absorption maxima*. Chlorophyll has absorption maxima at 410, 430, 578, 615, and 662 nm, but it absorbs light of wavelength 430 nm most strongly.

Solutions that appear colored contain solutes that absorb some of the wavelengths of white light. The more concentrated the colored solute, the more rapidly these wavelengths will be attenuated as light passes through the solution. The relationship between concentration and light absorption is given by *Beer's law*:

$$I = I_o e^{-k_a C}$$

This equation can be rearranged in logarithmic form:

EQUATION 17–2

$$\ln \frac{I}{I_o} = -k_a C$$

where I = the intensity of the light leaving the solution, lx
I_o = the intensity of the incident light beam, lx
C = the concentration of the absorbing solute, moles/L or M
k_a = a constant of proportionality, L/mole

The minus sign indicates that the light intensity decreases as concentration increases. This relationship can be used as the basis for measuring concentration of the solute if the incident and transmitted light intensities can be measured and if k_a

FIGURE 17–4
Absorption spectrum of chlorophyll, the green pigment in plants.

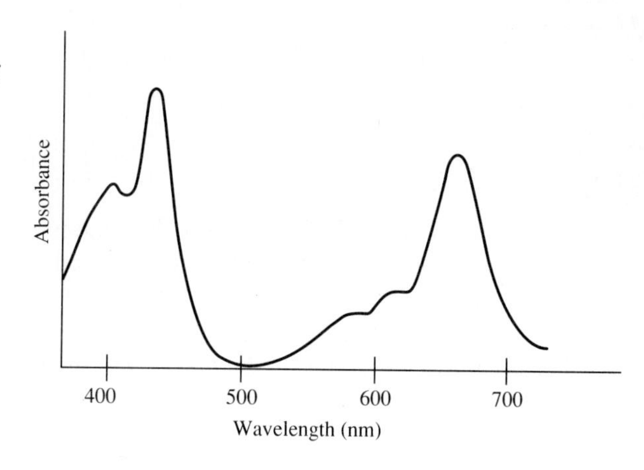

is known. However, the value of k_a will depend on the distance the light travels through the solution. The longer the light path, the more absorbing material it must pass through and the more the light will be absorbed. Therefore, light attenuation depends not only on concentration but also on the thickness of the solution sample being measured. The equation that gives the attenuation of light as a function of light path distance is called *Lambert's law*:

EQUATION 17–3

$$\ln \frac{I}{I_o} = -k_b d$$

where d = the distance light travels through the solution, cm

k_b = another constant, cm^{-1}

The minus sign indicates that light intensity decreases as the length of the light path increases. Combining Equations 17–2 and 17–3 gives the *Beer-Lambert law*:

$$\ln \frac{I}{I_o} = -k_c \mathrm{C}\mathrm{d}$$

which is customarily rearranged and written using base 10 logarithms

EQUATION 17–4

$$\log \frac{I_o}{I} = \frac{k_c}{2.303} \mathrm{C}\mathrm{d} = e\mathrm{C}\mathrm{d}$$

where e = a constant called the *extinction coefficient*

When the concentration, C, is expressed in moles per liter (molar concentration), then e is called the *molar extinction coefficient*. The value of e is greatest for wavelengths that are most strongly absorbed—that is, at the largest absorption maximum. The value of the extinction coefficient therefore depends on both the absorbing solute and the wavelength of light used to make the measurement. The term $\log(I_o/I)$ is a dimensionless quantity called *optical density,* or *absorbance (A)*. Spectrophotometers are usually designed to read out absorbance directly, since this allows easy calculation of concentration using Equation 17–4. Substituting the definition for absorbance into Equation 17–4 gives

EQUATION 17–5

$$A = eCd$$

Another common measure of light absorption is *percent transmission* ($\%T$), given by

EQUATION 17–6

$$\%T = 100 \times \frac{I}{I_o}$$

Figure 17–5 illustrates the fundamental features of a spectrophotometer, an instrument used for measuring concentrations by light absorption. The basic elements are a light source, a transparent *cuvette* that holds a sample of the solution under test, a wavelength selector that controls the wavelength of the light used to make the measurement, and a photodetector. The output of the photodetector is amplified and its intensity is displayed on a meter. The meter is usually calibrated to read both absorbance and percent transmission. Before use, the instrument is

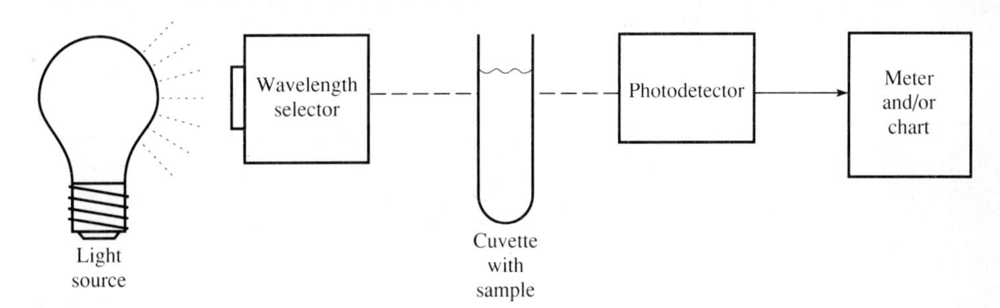

FIGURE 17–5
Block diagram of a spectrophotometer.

calibrated to 100% transmission (zero absorbance) by inserting a cuvette containing the pure solvent without light-absorbing solute. The light transmitted by the solvent and cuvette alone is equivalent to I_0 in Equations 17–2 through 17–6.

Although the block diagram of a spectrophotometer seems simple, each of the components must be carefully designed and selected for the instrument to have good performance. The problems with and the selection criteria for these components are discussed in the following paragraphs.

The ideal light source would have constant output intensity at all wavelengths, something that no real light source can achieve. Incandescent lamps produce much higher intensity at long wavelengths than at the blue (short-wavelength) end of the visible spectrum. However, tungsten lamps are advantageous since they are inexpensive and do not require special power supplies. They can be made to produce increased output at short wavelengths by operating them at high temperature (above their normal power rating), but this greatly shortens the life of the bulb. High-pressure gas-discharge tubes produce much greater intensities at short wavelengths and are more efficient than incandescent bulbs. These tubes are filled with a mixture of xenon gas and mercury vapor and have a pair of electrodes through which a current passes that heats and ionizes these gases. The "on" resistance of the arc is only 1 Ω or less, and consequently these tubes require sophisticated dc power supplies that can supply high current at very low voltage.

The wavelength selector is basically a bandpass filter for light. The ideal filter would have a very narrow passband so that the measurement wavelength could be precisely selected. Attenuation of wavelengths outside the passband should be infinite and wavelengths within the passband should be unaffected. However, real filters are subject to the same nonideal behavior as their electronic counterparts.

The simplest filters are made of colored glass or colored gelatin sheets. These may be inserted into the light path either before or after the sample cuvette. Such filters are inexpensive, have low passband attenuation, and are available in a variety of different colors (passbands). However, they are fairly broadband, and a different one is required for each wavelength used for measurement. Spectrophotometers that are designed for filters of this type are called **colorimeters.**

More sophisticated *interference* filters have recently come into common use. **Interference filters** consist of two reflecting surfaces separated by a transparent

FIGURE 17–6
An interference filter.

Reflecting surfaces
Glass substrate
Light
nL
n = an integer
L = wavelength

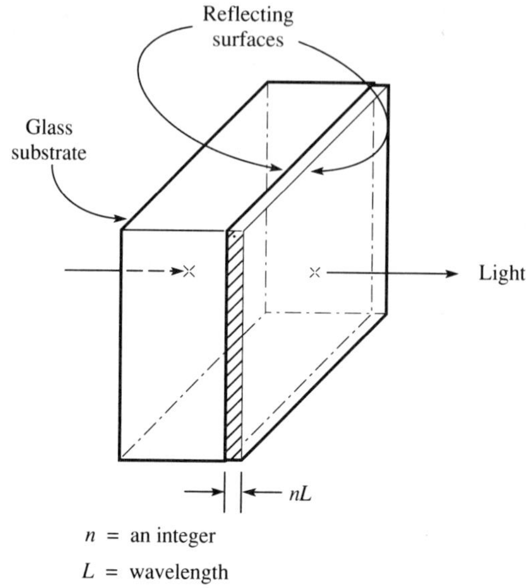

layer whose thickness is an integral number of wavelengths. These are made by vacuum deposition on a glass substrate. Figure 17–6 is a drawing of the structure of an interference filter. The only colors that pass are those that can interfere constructively—that is, those that have wavelengths that are integral fractions (1/2, 1/3, 1/4, etc.) of the distance separating the reflecting surfaces. Interference filters have much narrower passbands than colored glass filters, but they are more expensive. Like glass filters, a different one is required for each wavelength of interest.

A variable interference filter can be made by forming the transparent layer between the reflecting surfaces into a wedge shape, as shown in Figure 17–7.

FIGURE 17–7
A variable interference filter.

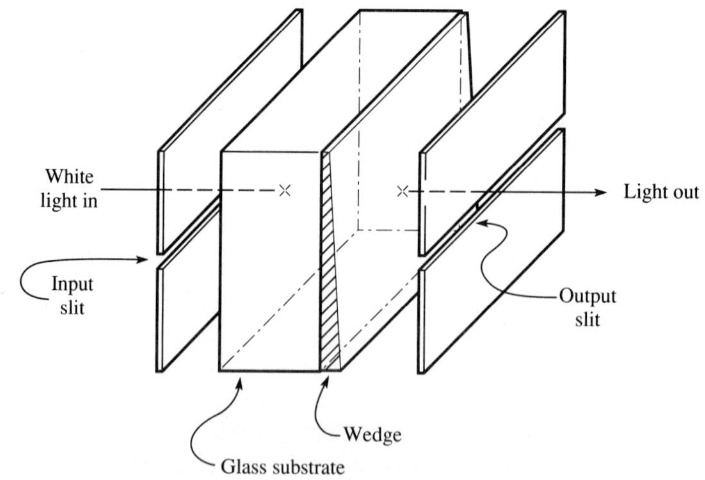

Since the passband wavelength is determined by the separation between the reflecting surfaces, a wedge-shaped filter will pass different wavelengths at different points along the wedge. An adjustable *monochromator* can be made by arranging the filter so that it can be shifted perpendicularly to the light beam to bring different zones of the wedge into the light path. (A monochromator is an adjustable wavelength selector.)

Another type of monochromator that is commonly used in spectrophotometers is made using prisms or diffraction gratings that disperse light in a spectrum. The structure of this kind of monochromator is shown in Figure 17–8. White light enters the monochromator through a slit and is focused on the prism or grating as a narrow beam. The angle of reflection from the grating (or refraction from a prism) depends on the wavelength. Therefore, the light is reflected as an expanding fan-shaped beam of different colors, looking much like a rainbow. The angle of reflection can be adjusted by physically rotating the prism or grating to concentrate different colors, or wavelengths, on an exit slit at the output of the monochromator. The mechanism that rotates the prism or grating can be calibrated to display the selected wavelength.

The cuvettes used for spectrophotometers must be made of material that is transparent to all wavelengths that might be used. Glass is transparent to visible wavelengths but is opaque to ultraviolet. Spectrophotometers that are capable of operating in the ultraviolet part of the spectrum require the use of quartz cuvettes. All cuvettes are carefully made so that the light path is of precise length, usually exactly 1 cm. Special cuvettes that have longer light paths for greater sensitivity are available, as are short path cuvettes that are intended for very small sample volumes.

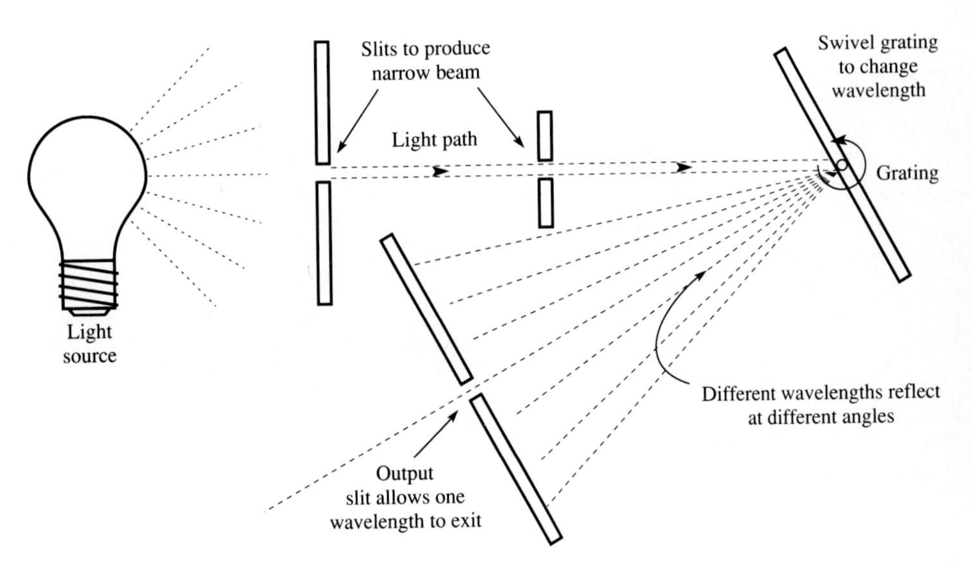

FIGURE 17–8
Grating monochromator.

The ideal photodetector for use in a spectrophotometer would be equally sensitive to all wavelengths for which the instrument is designed. This ideal is difficult to achieve in practice, since virtually all light transducers have an *action spectrum* that is not uniform across the range of wavelengths normally used in spectrophotometers. The simplest photodetector is the **phototube,** shown schematically in Figure 17–9. Light strikes the photocathode, which is made of a material that readily emits electrons in response to the energy of the absorbed photons of light. (This is called the *photoelectric effect*.) These electrons are attracted to the positively charged anode, and the anode current is linearly proportional to the intensity of the incident light.

The disadvantage of phototubes is that they are relatively insensitive. However, the more complex **photomultiplier** (PM) tubes, which operate on similar principles, can be made so sensitive that they can detect single photons of light. The structure and bias circuitry of a PM tube are shown in Figure 17–10. Light striking the photocathode causes the emission of electrons, as it does in the phototube. These electrons are accelerated toward the first of a series of electrodes called *dynodes* that act as amplifiers. Electrons striking the first dynode cause the secondary emission of additional electrons, which are accelerated toward the second dynode. These electrons again cause the secondary emission of electrons that are swept away to the next dynode, where the process continues. The number of electrons liberated by secondary emission is about twice the number of electrons striking each dynode. Therefore, the number of electrons that impinge on the anode is 2^n times the number emitted from the photocathode, where n is the number of dynodes. Typical PM tubes may have ten dynodes, making them 2^{10}, or 1024, times more sensitive than phototubes.

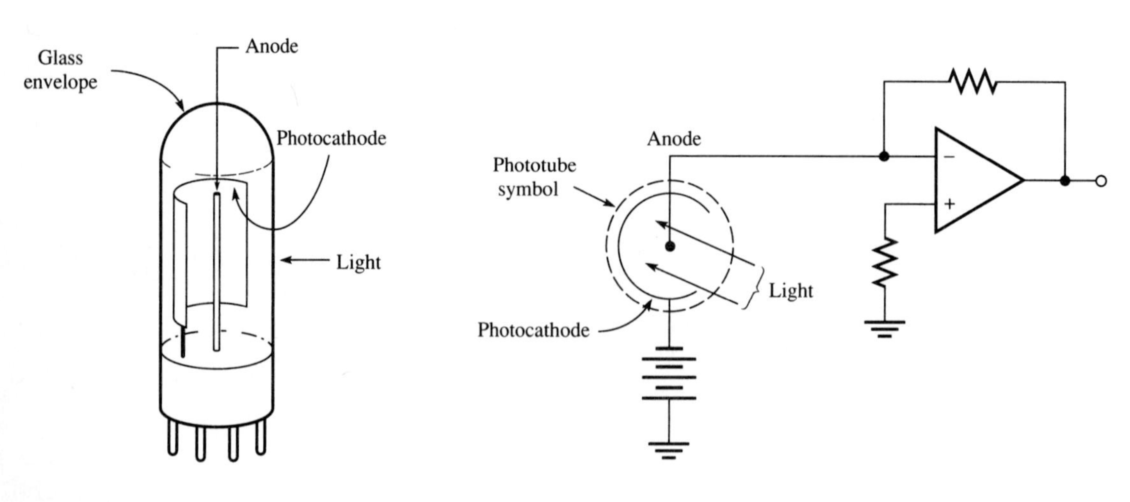

(a) Construction of a phototube.

(b) Typical phototube circuit.

FIGURE 17–9
The phototube.

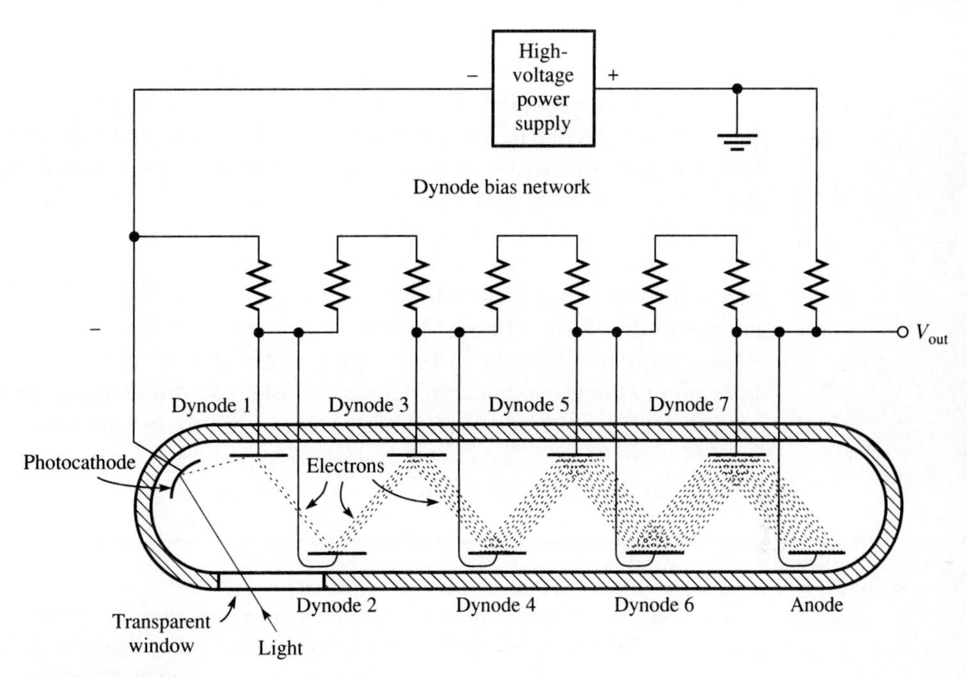

FIGURE 17–10
A photomultiplier.

Solid-state light transducers are becoming more common in spectrophotometer applications. The *photodiode* is made as a *pn* or PIN junction using special semiconductor materials. The energy of light striking the diode causes the formation of hole-electron pairs. These produce a current that is proportional to the intensity of the light incident on the junction. Photodiodes are normally operated in the reverse-biased mode, in which case the current caused by light appears as an increase in reverse leakage. However, an exceptionally linear detector can be made using a photodiode as an unbiased current source, as shown in Figure 17–11. The very small photocurrent is amplified by an operational amplifier connected as

FIGURE 17–11
Photodiode circuits.

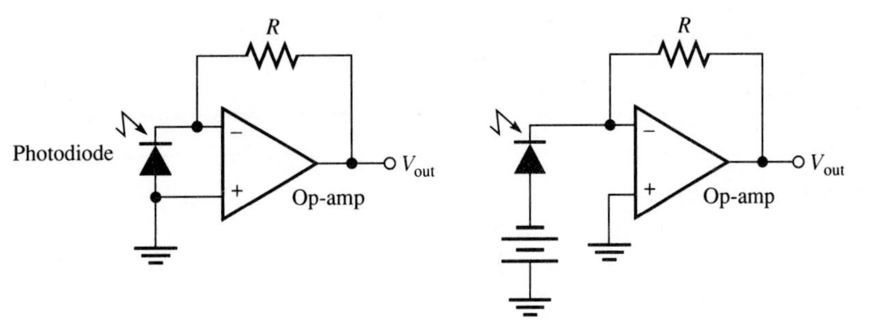

a current amplifier. The gain is set by a single resistor in the feedback path. The output voltage is proportional to light intensity over several orders of magnitude.

Spectrophotometers are used to measure the concentration of a very wide range of biologically important substances. Tests that determine the amount of a particular kind of substance present in a sample are called *assays*. Many materials that do not normally absorb light can be made to do so by treating them with chemicals that react to produce colored products. A classic example is the *biuret* assay for protein. Copper sulfate is added to a solution to be tested. The copper ions react with protein molecules to produce a violet-colored complex. The amount of protein can be assayed by adding copper sulfate to the sample and then measuring its absorbance at a wavelength of 550 nm. The assay is calibrated by determining the absorbance of standard protein solutions made up to known concentration. This permits the calculation of e, the extinction coefficient, in Equations 17–4 and 17–5. The concentration of an unknown sample can then be calculated using the same equation.

EXAMPLE 17–1 A series of protein samples of known concentration is tested in order to calibrate a biuret assay. Three specimens containing unknown concentrations of protein are tested at the same time. The measured absorbances (optical densities) are given in Table 17–2. Determine the value of the extinction coefficient and the protein concentration of the three specimens. Note that the units of concentration are grams per liter rather than moles per liter. This is because most biological specimens contain a mixture of several kinds of protein molecules, and the molecular weights are usually unknown.

TABLE 17–2

PROTEIN	ABSORBANCE (OPTICAL DENSITY)
0.0 g/L	0.00
1.0	0.07
2.0	0.14
3.0	0.22
4.0	0.27
5.0	0.36
6.0	0.41
Specimen A	0.13
Specimen B	0.25
Specimen C	0.38

SOLUTION Equation 17–5 is a linear equation; that is, it has the form $y = mx + b$, the algebraic equation for a straight line. Recall that m is the slope of the line and b is the vertical intercept at the point where x is zero. For our purposes, A is equivalent to y, the dependent variable, and C is equivalent to x, the independent variable. The intercept, equivalent to b, is 0, indicating that a plot of Equation 17–5 will pass through the origin. The slope, equivalent to m, is $(e \times d)$, the extinction coefficient times the length of the light path, d. Let's assume that d is the customary 1 cm, or 0.01 m. In order to calculate e, we must determine the slope of the best straight-line fit to the calibration data. This can be done either by graphing the data or by calculating the slope using linear regression (see Section 2–7). We choose the former and leave the reader the interesting exercise of calculating the slope by linear regression.

FIGURE 17–12

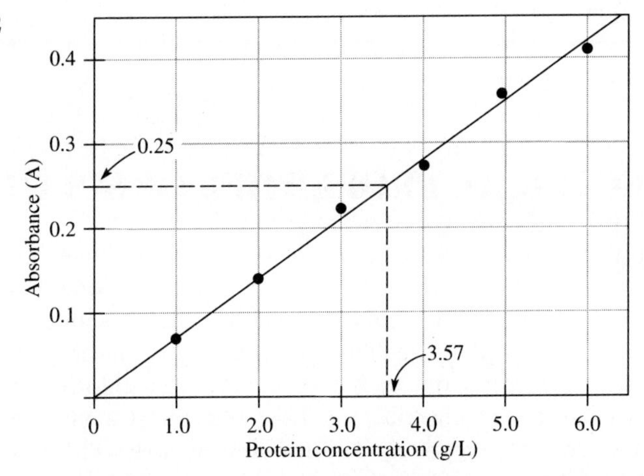

Figure 17–12 is a graph of the calibration data and a straight line that appears to fit the data points closely. The slope is the change in A (ΔA) divided by the change in C (ΔC). By inspection

$$\frac{\Delta A}{\Delta C} = \frac{0.42}{6.0 \text{ g/L}} = 7.00 \times 10^{-2} \text{ L/g}$$

The extinction coefficient can be calculated as follows:

$$ed = \text{slope}$$

$$e = \frac{\text{slope}}{d} = \frac{7.00 \times 10^{-2} \text{ L/g}}{0.01 \text{ m}} = 7.00 \text{ L/m·g}$$

Now that e is known, Equation 17–5 can be used to calculate the protein concentration of the three specimens. Solving for C gives

$$C = \frac{A}{ed} = \frac{A}{0.01e} = 100 \times \frac{A}{e}$$

For specimen A

$$C_{\text{A}} = \frac{100 \times 0.13}{7.00} = 1.86 \text{ g/L}$$

Similar calculations yield the results for specimens B and C.

$$C_{\text{B}} = \frac{100 \times 0.25}{7.00} = 3.57 \text{ g/L}$$

$$C_{\text{C}} = \frac{100 \times 0.38}{7.00} = 5.43 \text{ g/L}$$

These concentrations could actually have been determined from the calibration graph as follows: Locate the specimen absorbance on the vertical axis. Draw a horizontal line from that point to intersect the calibration line. Now draw a vertical line from the point of intersection down to the horizontal axis and read the concentration at the point where the vertical line crosses the axis. This method is illustrated on the graph for specimen B.

17–4 THE ELECTRICAL PROPERTIES OF SOLUTIONS

The molecules of many substances *dissociate* into *ions* when they dissolve in water. Ions are molecules or atoms that have gained or lost one or more electrons, causing them to be electrically charged. Table salt, or sodium chloride, is an example. The elements sodium and chlorine can react with each other to form ions by transferring an electron from each sodium atom to a chlorine atom. (The ionic form of chlorine is called chloride.) A salt solution is a mixture of positively charged sodium ions and negatively charged chloride ions. The dissociation of salt is shown by the following chemical equation:

$$\underset{\text{Sodium chloride}}{NaCl} \rightarrow \underset{\text{Sodium ion}}{Na^+} + \underset{\text{Chloride ion}}{Cl^-}$$

The superscripts indicate that the sodium ion carries one unit of positive charge, due to its having lost one electron. The chloride ion carries a unit of negative charge since it has one extra electron. Since the two kinds of ions have opposite charge, they are strongly attracted to each other and therefore remain mixed together in solution, preserving an overall electrical neutrality. The strong attraction keeps the ions stuck together when the salt solution dries out, causing the formation of salt crystals.

Many ionic substances, or **electrolytes,** have important physiological functions in living organisms. Sodium chloride turns out to be one of the most important, and most common, of the biological electrolytes. However, potassium ions also play important roles, as do *divalent* ions such as calcium and magnesium. A divalent ion is one that has gained or lost *two* electrons and therefore has two units of electrical charge. Ions such as sodium that carry only one unit of charge are said to be *monovalent.*

Ions cause solutions to be electrically conductive. The reason for this is that in solution, the charge carriers are the ions themselves, not electrons and holes as in solids. When a current is passed between electrodes through a solution of ions, the positively charged ions move toward the negative electrode, or cathode, whereas the negative ions migrate toward the positive anode. Positive ions are called *cations* because they are attracted to the cathode. Similarly, negative ions are called *anions.*

17–5 CONDUCTIVITY METERS

The conductivity of the solution depends on the concentration of ions present. Pure water is a poor conductor, but seawater is an excellent conductor thanks to its high concentration of salt and other electrolytes. It is possible to measure the concentration of ions in solution by measuring the conductivity. In fact, conductivity measurements are commonly used as a measure of water *salinity* (salt content). Salinity is an important determinant of water quality, especially for water used in the irrigation of crops.

The basic design of a **conductivity meter** is shown in Figure 17–13. The transducers of the system are the two electrodes, which convert conventional current from the instrument to ionic current in solution and vice versa. The electrodes are usually arranged on a probe in such a way that their spacing and contact surface area is held constant. An alternating current (typically, 1 kHz) is usually used for the measurement since this reduces some of the problems associated with the use of metal electrodes. As you will see in the next section, an electrode is not just a contact with the solution.

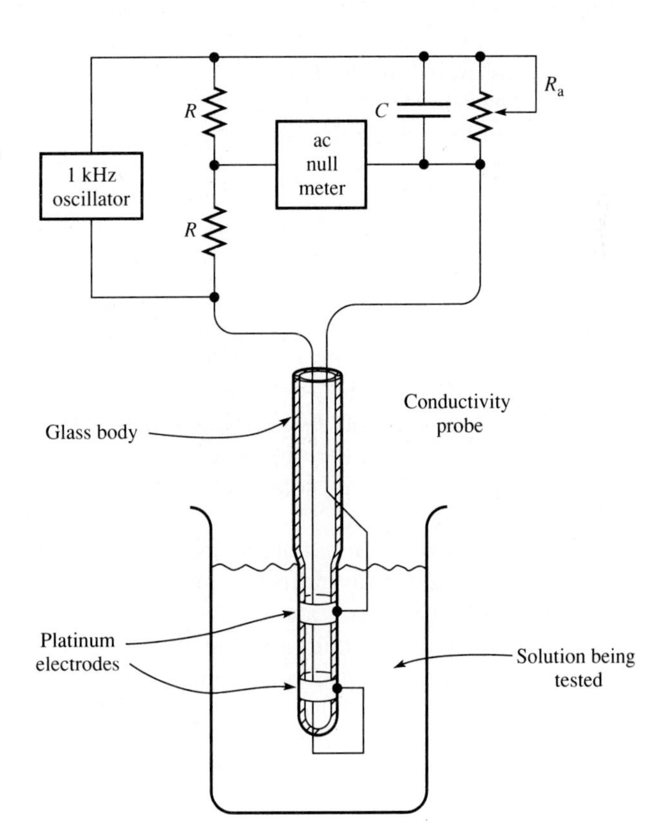

FIGURE 17–13
Conductivity bridge. When R_a is adjusted to give a null reading, the probe resistance is equal to R_a. Conductivity is the reciprocal of R_a.

17–6 ELECTRODES

Electrodes are the electrical connectors to solutions. Actually, they should be regarded as transducers and not merely as connectors, since they must convert the flow of electrons in the lead wires to the flow of ions that constitute current in a solution. Electrodes are used in a variety of biomedical applications, including connections to the body for such instruments as the electrocardiograph. Since the human body is more than 70% water, the potentials and currents that are part of the body's physiological processes are those of solutions rather than of conventional electronics. Electrodes are, therefore, an important part of biomedical instrumentation.

It may seem that all that is required to make an electrode is to insert a wire into the solution to which we wish to make a connection. However, such simple electrodes produce troublesome electrical potentials that vary depending on the solution and on the metal used. When a metal electrode is inserted into a solution of ions, several phenomena occur to produce a chemical and electrical disturbance near the metal. First, the metal attracts some kinds of ions more strongly than others, either cations or anions, forming a layer of charge at the metal surface. Ions of opposite charge are attracted to this layer; therefore, a second, oppositely charged, layer forms just outside the first. Second, some of the metal atoms dissolve, forming positively charged cations, leaving the electrons they have given up behind them in the electrode metal. These effects cause a double layer of charge to form at the metal surface. This double layer acts like a charged capacitor in series with the electrode, causing it to have an electrical potential that is different from that of the solution. This potential is called the **half-cell potential** of the electrode. The magnitude of the half-cell potential depends on the composition and concentration of the solution and on the kind of metal in the electrode.

Measuring the half-cell potential is more difficult than it may seem. A connection can easily be made to the electrode, but to measure the voltage across the electrode we would need to make another connection to the solution, and that can be done only with another electrode. For this reason we can measure the difference only between the half-cell potentials of two electrodes. (A simple battery cell can be made by inserting two dissimilar pieces of metal in an electrolyte solution. The battery voltage is the difference between the half-cell potentials.) The measurement problem can be solved by using a standard electrode, against which all others can be measured. The hydrogen electrode, formed by bubbling hydrogen gas over the surface of a sintered platinum electrode, is the accepted standard. Half-cell potentials for different metals measured under standard conditions against a hydrogen electrode are shown in Table 17–3.

The half-cell potentials given in Table 17–3 apply only to standard measuring conditions. The actual half-cell potential will depend on the solution in which the electrode is immersed. Further, the actual potential is susceptible to considerable drift, or low-frequency noise.

When a current passes through a pair of electrodes, the situation is changed because of the transduction to ionic current that occurs at the electrode surface. Cations, attracted to the negative cathode, pick up electrons at the cathode sur-

TABLE 17–3

METAL	HALF-CELL POTENTIAL
Aluminum	−1.662 V
Zinc	−0.763
Iron	−0.44
Hydrogen	0.0
Copper	0.521
Silver	0.799
Platinum	1.2
Gold	1.69

face, becoming uncharged atoms deposited on its surface. Anions give up their excess electrons at the anode, also becoming uncharged atoms in the process. Accordingly, electrons are liberated at the anode and absorbed at the cathode, or, looking at it another way, electrons flow into the solution at the cathode and out at the anode. The equivalent circuit for electrodes carrying current is shown in Figure 17–14.

R_s is the resistance of the solution, sometimes called the *bulk resistance*. The equivalent capacitor C is due to the double layer of surface charge described earlier. The equivalent resistance R_t is due to the transduction of current at the electrode surface and is quite different for different metals. The value of R_t also depends on the electrode's surface area; the larger the surface, the smaller R_t will be. Electrodes are classified on the basis of the magnitude of R_t. *Nonpolarizable* electrodes have small R_t values and are used for most biomedical applications.

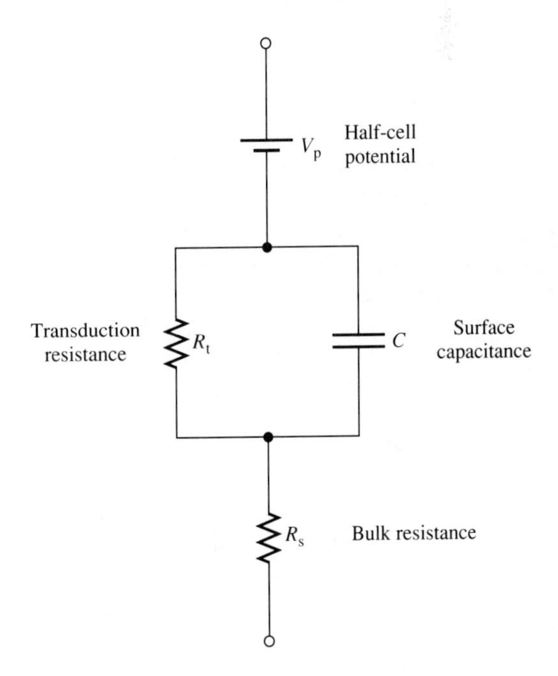

FIGURE 17–14
Equivalent circuit for electrodes carrying current.

However, *polarizable* electrodes, having very large equivalent resistances, are used in a number of special applications.

The ideal nonpolarizable electrode would have an equivalent resistance of 0 and a half-cell potential that does not drift with time. Two electrodes that fit these criteria fairly well are the *calomel* electrode and the *silver–silver chloride* electrode. The commonly used silver–silver chloride electrodes are made of silver wire that has been plated with silver chloride. The calomel electrode is more stable but is more complex and delicate in use. A simplified version of the calomel electrode is shown in Figure 17–15. Electrical connection to the electrode is made via a wire dipped into liquid mercury that is in contact with mercurous chloride, also called calomel. The calomel is in contact with a saturated solution of potassium chloride (KCl). **Saturated** means that the concentration is as high as it is possible to make it. A saturated solution is made by adding more solid potassium chloride than can dissolve, guaranteeing that as much dissolves as possible. The potassium chloride makes electrical contact with the solution. In Figure 17–15, this is shown as occurring through a siphon tube containing KCl solution gelled with agar, which is similar to gelatin but chemically more stable. In practice, such an electrode would be very awkward to use. A more refined commercial version is diagramed in Figure 17–16.

Electrodes used to make contact with the human body are subject to the same complexities as those used in solutions. Since the skin is usually dry, contact is made with a conductive cream or jelly between the electrode and the patient's skin. In other words, the electrode is in contact with a solution (the jelly), which is, in turn, in contact with the patient. The original electrodes used to measure biopotentials of the body were vessels of salt water into which the patient placed his or her hands. Einthoven (see the historical note) used such electrodes during his pioneering work on the EKG. Pictures of his early machines show miserable-looking patients with both hands and a foot immersed in buckets of saline solution. *Plate electrodes* replaced the uncomfortable buckets, but these are sensitive to movement by the patient, which can cause variation in the thickness of the layer of contact jelly. This causes the electrode potential and impedance to be

FIGURE 17–15
Simple calomel reference electrode.

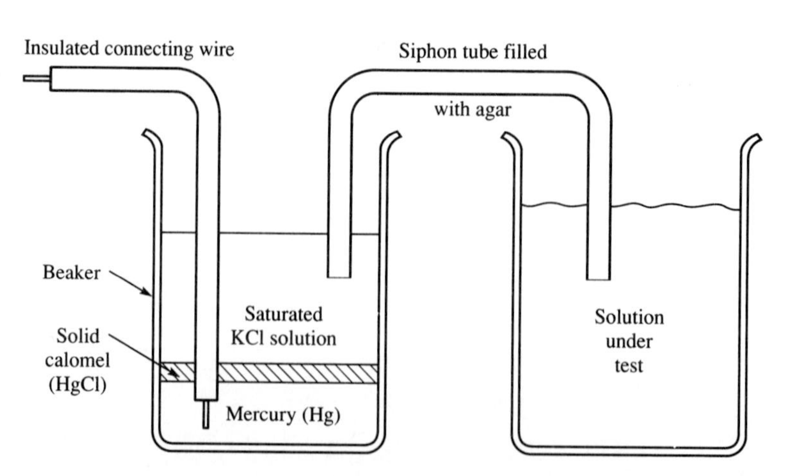

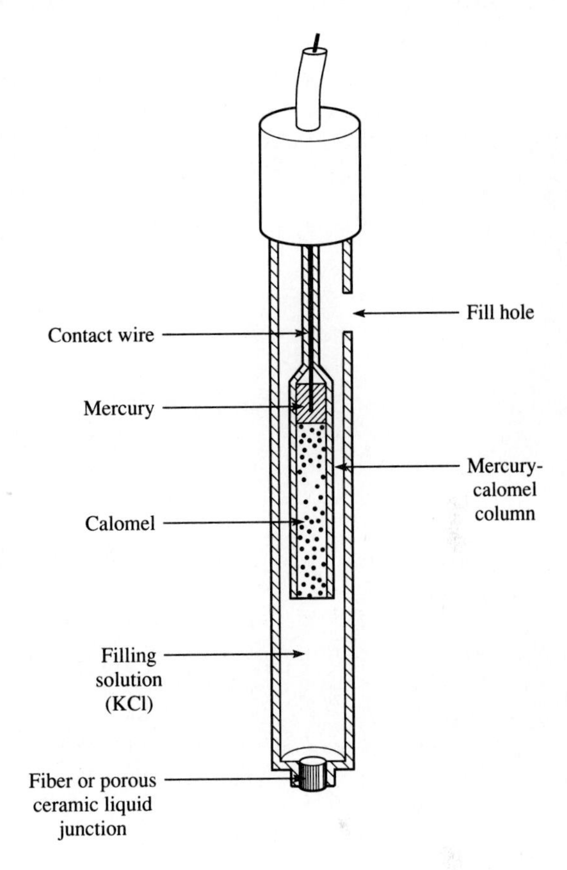

FIGURE 17–16
Practical calomel reference electrode.

variable. Modern *floating electrodes* eliminate this movement artifact by avoiding metal-to-skin contact. Figure 17–17 is a diagram of one kind of floating electrode. They are held in place by double-sticky tape rings that adhere to both the skin and the electrode.

Microelectrodes are used to make contact with individual cells. Much of what we have learned about the functioning of nerves has come from the use of microelectrodes implanted into single nerve cells. The construction of a glass microelectrode is illustrated in Figure 17–18. The body of the electrode is made from a piece of glass capillary tubing that has been heated and then drawn to a fine tip. The bore of the tip can be as small as 0.1 μm, too small for even tiny bacteria to pass through. The barrel of the electrode is filled with an electrolyte solution,

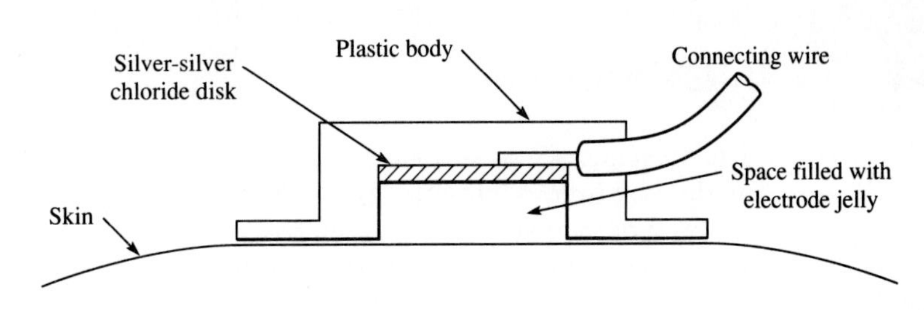

FIGURE 17–17
"Floating" skin electrode.

FIGURE 17–18
Glass microelectrode.

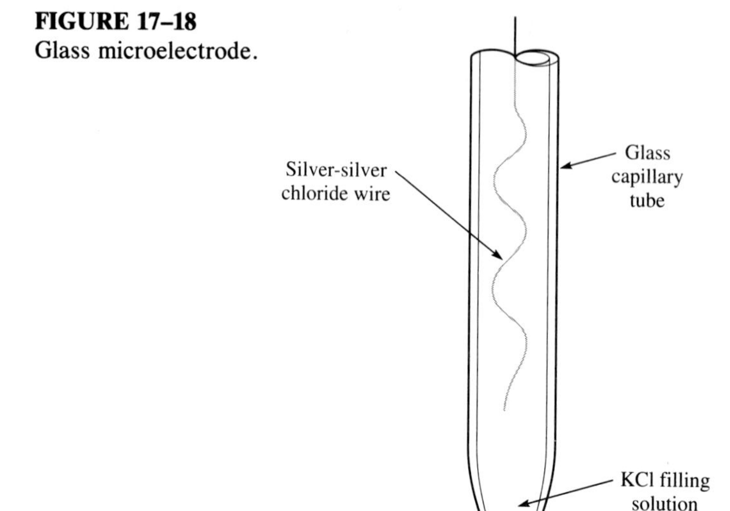

usually potassium chloride. A silver–silver chloride wire is inserted and into the barrel to make the electrical connection. The current transducer is therefore large compared with the size of the tip. Contact to the cell interior is made by puncturing the plasmalemma (cell membrane) with the small diameter tip. Because the tip bore is so small, these microelectrodes have very high dc resistance (10 to 300 MΩ).

17–7 MEMBRANES

The presence of a membrane in a solution greatly modifies the electrical properties of the solution. Moreover, the normal function of living cells depends on the function of their surrounding membrane, the plasmalemma. Therefore, it is necessary to have some knowledge of the properties of membranes and their interaction with solutions in order to understand what biomedical electrodes and the instruments to which they are connected are measuring.

Membranes are classified on the basis of the kinds of molecules that can pass through, or *permeate,* them. *Impermeable* membranes permit no molecules or ions to permeate. Plastic films are impermeable membranes. They act as electrical barriers, or insulators, in solutions, since they block the movement of ions. *Permeable* membranes, such as filter papers, allow all molecules and ions to permeate. They can have interesting electrical effects when they are placed between solutions of different composition and concentration. However, the most important membranes from a biomedical perspective are *selectively permeable,* allowing certain kinds of ions or molecules to permeate but not others. Biological cell membranes, for example, are quite permeable to water but are much less permeable to ions and are impermeable to large molecules.

To understand the electrical effects of selectively permeable membranes, imagine a system composed of two chambers separated by a membrane that allows only potassium ions to permeate. Figure 17–19 illustrates the situation.

Suppose that the left chamber is filled with $0.1M$ potassium chloride ($0.1M$ KCl) and the right chamber contains only water. Therefore, there is a concentration gradient across the membrane that will favor the diffusion of potassium ions through the membrane into the water chamber. However, since potassium ions carry electrical charge, their movement will cause the right chamber to become positively charged with respect to the left chamber. The positive charge in the right compartment can't be balanced by chloride ions, since they cannot cross the membrane. In other words, a voltage difference, called the *membrane potential,*

FIGURE 17–19
Membrane potential.

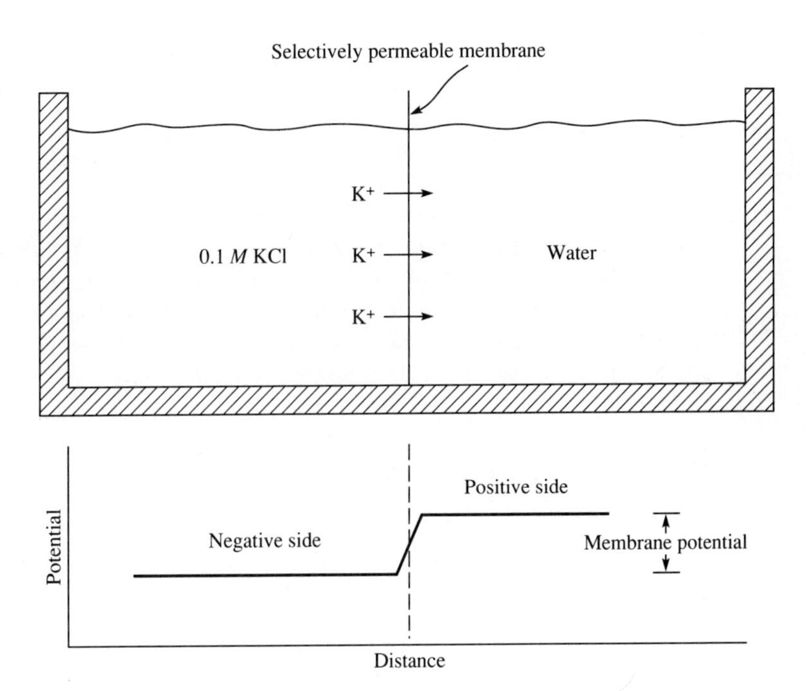

will occur across the membrane. This voltage difference is oriented so that it will favor the movement of positively charged potassium ions back through the membrane toward the negative left compartment. Eventually, the rate of diffusion due to the concentration gradient is balanced by the rate of return movement caused by the membrane potential. Thus an equilibrium is established, along with a constant membrane potential. The magnitude of the membrane potential will depend on the ratio of potassium concentrations across the membrane at equilibrium. The formula used to calculate the membrane potential is the famous *Nernst equation.*

EQUATION 17–7

$$P = -\frac{RT}{zF} \ln \frac{C_1}{C_2} = \text{membrane potential, V}$$

where R = the universal gas constant, 8.314 $JK^{-1}mole^{-1}$

T = the absolute temperature, K

z = the number of electrons gained or lost by the permeative ion

F = the Faraday constant, 9.648×10^4 C/mole

C_1 = the permeative ion concentration on side 1, mole/L (or M)

C_2 = the permeative ion concentration on side 2, mole/L (or M)

The minus sign indicates that the first compartment becomes negative as the positive potassium ions diffuse out. R, z, and F are constants. Usually the temperature is also a constant (room temperature, or 293 K) for most biomedical situations in which the Nernst equation is useful. It is more convenient to use common (base 10) logarithms instead of natural logarithms, as in Equation 17–7. With these considerations in mind, Equation 17–7 can be simplified for *monovalent cations* to

EQUATION 17–8

$$P = -58 \log \frac{C_1}{C_2} = \text{membrane potential, mV}$$

Equation 17–8 allows us to calculate the membrane potential at room temperature for any membrane that is permeable to only one kind of ion. It turns out that such *ion-selective membranes* are not merely a theoretical concept but can actually be made. It is possible to use such membranes and the Nernst equation to devise instrument electrodes that can be used to measure the concentration of specific ions. These are called *potentiometric sensors.*

17–8 POTENTIOMETRIC SENSORS

The construction of potentiometric probes is diagramed in Figure 17–20. The body of the probe is a glass or plastic cylinder filled with a conductive solution, such as potassium chloride. Joined to the end of the cylinder is an ion-selective membrane. A silver–silver chloride wire electrode is used to make contact with

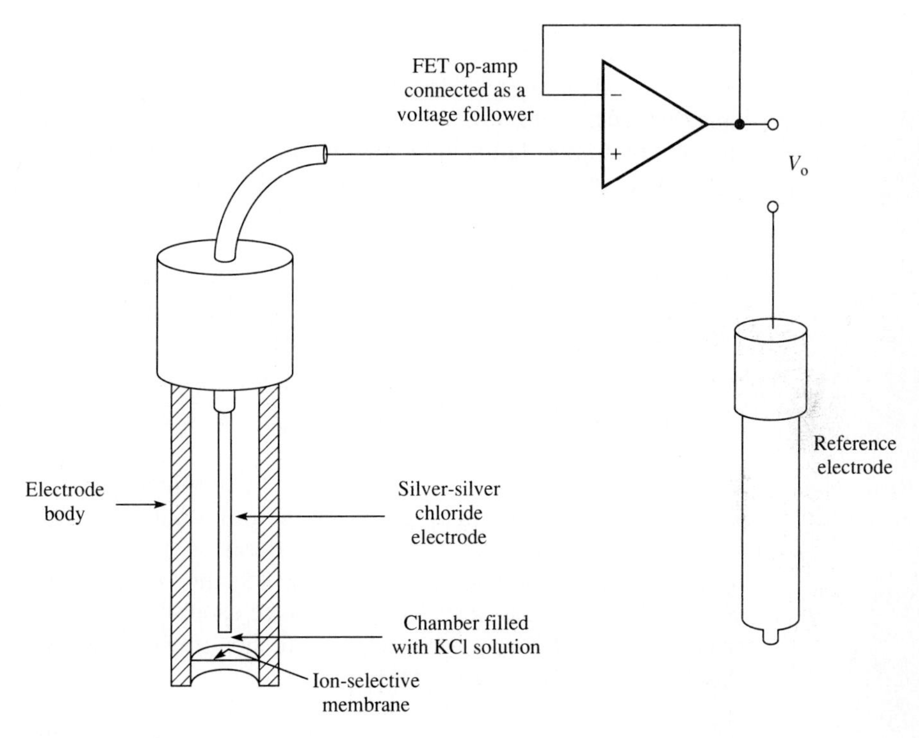

FIGURE 17–20
Ion-selective electrode.

the filling solution. The output of the probe is connected to a suitable amplifier. A *reference electrode,* such as silver–silver chloride, completes the circuit to the solution to be measured. The impedance of the ion-selective electrode depends on the permeability of the ion selective membrane, but it is usually very high, requiring the use of high-input-impedance amplifiers, such as FET-input voltage followers.

To measure the ion concentration in a sample solution, the electrodes are first calibrated by measuring the potential when the electrodes are immersed in solutions of known concentration. The calibration concentrations are normally chosen to bracket the sample concentration. Once the electrode response is known, a calibration chart can be made to relate measured potential to concentration. This is similar to the method of using a chart to relate absorbance to concentration discussed in Example 17–1. Alternatively, the amplifier gain and offset can be adjusted so that the output is a constant function of concentration.

The major difficulty with ion selective probes is interference from other kinds of ions to which the membrane is weakly permeable. The presence of such ions causes the output potential to deviate from that predicted by the Nernst equation. The magnitude of this deviation depends on the conditions and on the concentration of the interfering ions.

17–9 THE MEANING OF pH

The most commonly used ion-selective electrodes are used to measure pH. Biological processes are profoundly influenced by pH, and its determination is, therefore, of great importance in biomedical measurements. The term **pH** refers to the degree of acidity or alkalinity of a solution. More specifically, it is a measure of the concentration of hydrogen ions (H^+) in a solution. *Acid* solutions have a high concentration of hydrogen ions, whereas solutions that have low hydrogen ion concentration are said to be *alkaline,* or *basic*. All *aqueous* (water-based) solutions have a pH, since water molecules can dissociate to form a hydrogen ion and a *hydroxide* ion, as shown by this chemical equation:

$$\underset{\text{Water}}{H_2O} \rightleftharpoons \underset{\text{Hydrogen ion}}{H^+} + \underset{\text{Hydroxide ion}}{OH^-}$$

Arrows pointing in both directions indicate that this process is reversible. In pure water, the dissociation is balanced by the re-formation of water molecules from hydrogen and hydroxide ions. Chemists describe this situation by saying that an *equilibrium* exists, in which the rate of dissociation of water molecules is equal to the rate that molecules are re-formed. Since each water molecule that dissociates produces both a hydrogen and a hydroxide ion, the concentration of these ions is equal in pure water and is $10^{-7}M$.

Acids are substances that add hydrogen ions in a solution. Hydrochloric acid is secreted in the human stomach, which causes the hydrogen ion concentration of the stomach to increase. Hydrochloric acid in solution dissociates as shown in the following equation:

$$\underset{\text{Hydrochloric acid}}{HCl} \rightarrow \underset{\text{Hydrogen ion}}{H^+} + \underset{\text{Chloride ion}}{Cl^-}$$

Hydrochloric acid adds more hydrogen ions to the solution in the stomach than would be present by the dissociation of water molecules.

Bases are substances that decrease the hydrogen ion concentration in a solution. Bases work by either combining with hydrogen ions or by releasing hydroxide ions which, in turn, combine with hydrogen ions to form water. Magnesium hydroxide is a base that is sometimes used as an *antacid* to relieve an overly acid upset stomach.

$$\underset{\text{Magnesium hydroxide}}{Mg(OH)_2} \rightleftharpoons \underset{\text{Magnesium ion}}{Mg^{2+}} + \underset{\text{Hydroxide ions}}{2\ OH^-}$$

The hydroxide ions released by the dissociation of magnesium hydroxide combine with hydrogen ions in the stomach to form water, thereby reducing hydrogen ion concentration.

It turns out that the *product* of hydrogen ion concentration and hydroxide ion concentration is constant. This *ion product constant* is equal to 10^{-14} for

aqueous solutions. For water

$$[H^+] = 10^{-7}M$$
$$[OH^-] = 10^{-7}M$$

and

$$[H^+] \times [OH^-] = (10^{-7}) \times (10^{-7}) = 10^{-14}$$

The brackets indicate concentration. If an acid is added, causing the $[H^+]$ to go up, then the $[OH^-]$ must decrease accordingly. For example, if the hydrogen ion concentration goes up to 10^{-5}, the hydroxide ion concentration must decrease to 10^{-9} so that the product of concentrations remains constant at 10^{-14}.

The pH scale is used universally to indicate hydrogen ion concentration. The pH of a solution is defined as the negative logarithm of the hydrogen ion concentration:

EQUATION 17–9

$$pH = -\log[H^+]$$

In pure water, the pH is $-\log(10^{-7}) = 7$. For the example in the previous paragraph, the pH is $-\log(10^{-5}) = 5$. The hydrogen ion concentration in a human stomach is about $10^{-2}M$, so the pH is 2.

The pH scale ranges from 0 to 14. To be strictly correct, we should say that the 0–14 range applies only to dilute solutions. However, virtually all solutions of biomedical importance qualify as dilute. *Neutral* solutions are those having pH's close to that of pure water, or pH 7. Acidic solutions have pH between 0 and 7, whereas basic, or alkaline, solutions have pH greater than 7. Notice that low pH corresponds to high hydrogen ion concentration and vice versa.

Living organisms and their physiological processes are profoundly affected by changes in pH. For example, normal human blood pH is 7.4. A change, plus or minus, of less than one-half pH unit would unbalance the chemistry of the body so much that it would be lethal. Blood pH is tightly regulated by the body's physiological control mechanisms and by the presence of *buffers* in the blood. Buffers are substances that resist change in pH because they can reversibly combine with hydrogen ions. At high pH, the buffer dissociates, releasing its hydrogen ions. At high hydrogen ion concentration (low pH), the buffer combines with hydrogen ions, effectively removing them from solution. There are many substances that can act as buffers, and each works at its own pH range. The main buffer in blood is carbonic acid, H_2CO_3. It dissociates in solution as follows:

$$\underset{\text{Carbonic acid}}{H_2CO_3} \rightleftharpoons \underset{\text{Bicarbonate ion}}{HCO_3^-} + H^+$$

Carbonic acid is called a weak acid because it does not dissociate completely except at high pH. As pH increases, dissociation releases hydrogen ions to the blood. At low blood pH, excess hydrogen ions combine with bicarbonate ions, reforming carbonic acid and removing the hydrogens from the blood.

17–10 pH METERS

The usual sensor for instrumental pH measurements is an ion-selective electrode. The ion-selective membrane is a thin bulb of a special glass that is selectively permeable to hydrogen ions. These electrodes are sometimes called *glass electrodes* because of their construction. The term can be misleading, since chemists have devised glasses that are selectively permeable to a variety of other ions, such as sodium and potassium. These special glasses are also used in the construction of ion-selective electrodes. The construction of a pH glass electrode is shown in Figure 17–21(a).

As usual for electrode measurements, a reference electrode is needed to complete the connections to the solution to be measured. Both calomel and silver–silver chloride electrodes are used for this purpose. However, the most popular modern pH sensors are *combination electrodes,* in which the reference and sensing electrodes are combined in a single compact unit. Figure 17–21(b) is a

FIGURE 17–21
pH electrodes.

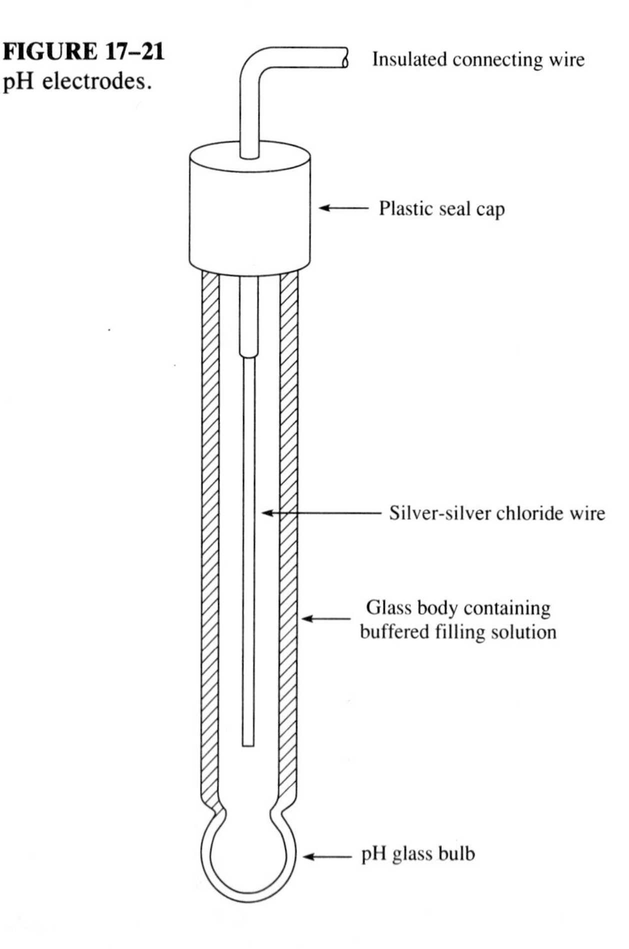

(a) Simple pH electrode.

FIGURE 17–21
(*continued*)

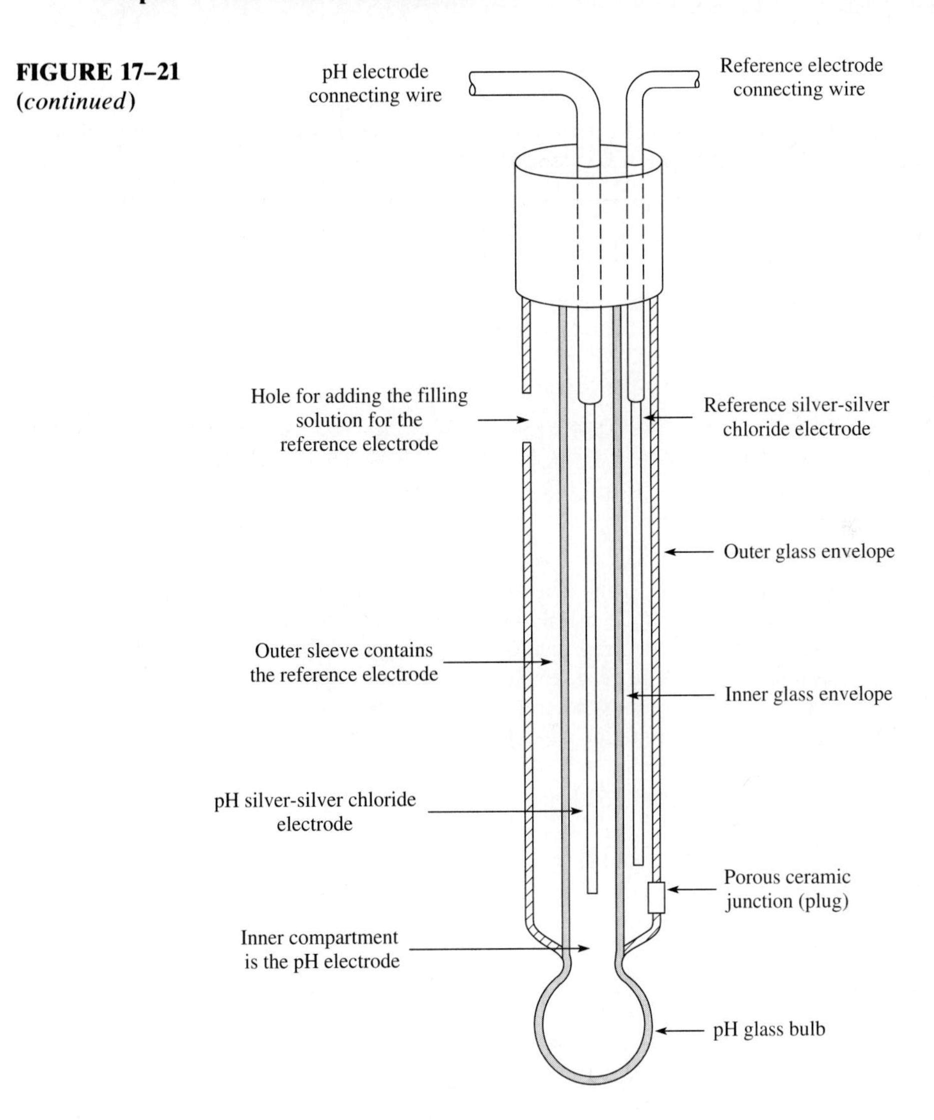

(b) Combination pH electrode.

diagram of one version of the combination electrode. The pH-sensing electrode is built into the center, where it connects to the glass membrane at the bottom. The reference electrode occupies the compartment that surrounds the sensing electrode. In it, a silver–silver chloride electrode makes contact with a potassium chloride filling solution. Ionic contact from the filling solution to the sample occurs through the ceramic junction. The junction acts as a nonselective membrane that passes ionic current but keeps the filling solution from running out. Nevertheless, the solution very slowly leaks into the sample solution and has to be replaced from time to time (every few months or so).

EXAMPLE 17–2 Suppose that a urine sample has a hydrogen ion concentration of $2.31 \times 10^{-5}M$. What is its hydroxide ion concentration? What is its pH? What would be the output voltage from a pH electrode immersed in this specimen?

SOLUTION Since urine is mostly water, the ion product constant can be used to calculate hydroxide ion concentration:

$$[H^+] \times [OH^-] = 10^{-14}M^2$$

or

$$[OH^-] = \frac{10^{-14}}{2.31 \times 10^{-5}M} = 4.33 \times 10^{-10}M$$

Equation 17–9 can be used to calculate pH:

$$pH = -\log[H^+] = -\log(2.31 \times 10^{-5}) = 4.64$$

Incidentally, normal urine pH ranges between 4.5 and 8.0.

The pH electrode membrane potential can be calculated from the Nernst equation (17–8). However, we need to know the hydrogen ion concentration of the electrode filling solution. It cannot be zero, since the solution is aqueous. Let's assume that it is a potassium chloride solution of neutral pH. Therefore, its hydrogen ion concentration is $10^{-7}M$. Substituting into Equation 17–8 gives

$$P = -58 \log \frac{2.31 \times 10^{-5}}{10^{-7}M} = -58 \log(231) = -137.1 \text{ mV}$$

The membrane potential is not the same as the electrode output voltage, since we have not considered the half-cell potentials in either the sensing or reference electrodes. However, these should be constants that are normally balanced out with a dc offset control on the pH meter.

The output impedance of glass electrodes is typically between 50 and 500 MΩ. They require high-input-impedance amplifiers for the metering circuitry and careful shielding and guarding of the connecting wires. (Guarding was discussed in Section 7–8.) However, the membrane potential is directly proportional to pH, since pH is a logarithmic function of hydrogen ion concentration. (Remember that the Nernst equation, (17–7) or (17–8), gives the membrane potential in terms of the logarithms of the ion concentrations.) A pH meter is principally just a millivolt meter with a high input impedance and a readout marked in pH units (see Figure 17–22). However, a gain adjustment, marked "temperature compensation," is needed, since the membrane potential varies with temperature (see Equation 17–7). A voltage offset control (marked "calibrate") is also needed to compensate for drift in either the reference or sensing electrode half-cell voltages. Frequent calibration of the meter and its electrodes against standard buffer solutions is

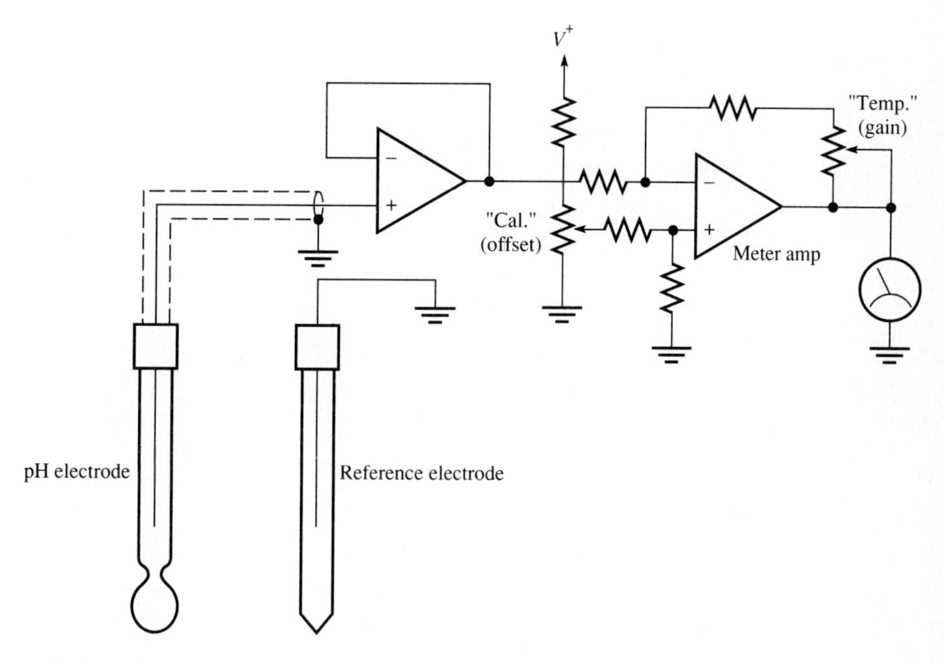

FIGURE 17–22
Circuitry of a pH meter.

essential for even reasonable accuracy. For critical measurements, the instrument should be calibrated just before making the actual measurement.

EXAMPLE 17–3 Suppose that a lab technician checks a pH meter and determines that the reading is 7.4 when the electrode is placed in a buffer solution known to have a pH of 7.00. The meter reads 4.7 when the electrode is immersed in a buffer solution having pH 4.00. What should be done to calibrate the meter?

SOLUTION Before deciding what to do, it is useful to examine the effect of the meter gain and offset controls on the instrument transfer characteristic. Figure 17–23(a) is a graph of the output voltage versus pH, showing the effect of changing the offset (''calibrate'') control. The slope of the line doesn't change, but the output voltage shifts by the same amount for all pHs. Figure 17–23(b) shows the effect of the gain (''temperature-compensation'') control. The slope of the line ''pivots'' around the pH 7 reading. (This is the pH of the electrode filling solution.) Therefore, it would be best to calibrate the instrument first at pH 7.00 by immersing the electrode in the proper buffer and then adjusting the offset to set the pivot point at the correct reading. The electrode then can be placed in a pH 4.00 buffer and the meter adjusted to read correctly using the gain control. The instrument should now be perfectly calibrated. In practice, the pivot point is rarely exactly at pH 7, so it may be necessary to repeat the procedure two or three times to refine the calibration.

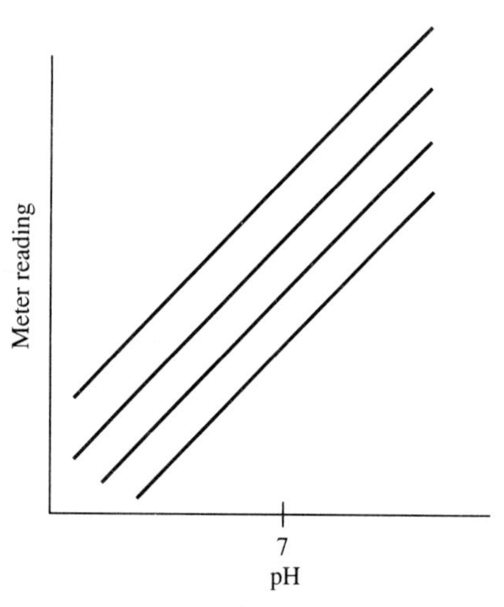

(a) Effect of changing the pH meter calibrate control on meter reading versus pH.

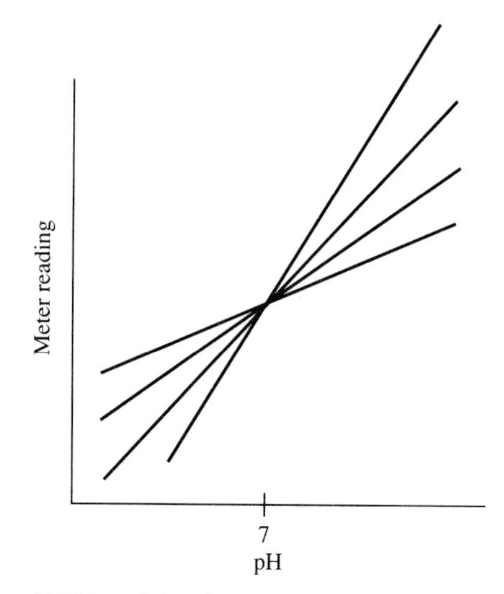

(b) Effect of changing the pH meter temperature control on meter reading versus pH.

FIGURE 17–23

17–11 BIOPOTENTIALS

You may have guessed from our discussion about membranes and the fact that living cells are encompassed in a selectively permeable membrane that there is a voltage difference between the inside and outside of cells. Virtually all living cells have membrane potentials. In most cases, the physiological significance of these potentials is not known. However, for nerve cells the membrane potential plays a clear role in the transmission of signals by nerves. The membrane potential is also important in the function of muscle cells, which are controlled by signals carried by nerves. The working of nerves and muscles in the human body produces potentials that can be detected using electrodes placed on the body surface. Several different instruments are used to measure these potentials as means of analyzing the state of health of the nervous system and of various muscles.

To understand the origin of these potentials, it is useful to take a brief tour of the nervous system and its special cells. The brain and the spinal cord comprise the *central nervous system* (CNS). While these organs perform a variety of complex functions, their basic mission is the processing of sensory information and the coordinating of a response. The nerves that connect the CNS to the rest of the body are part of the *peripheral nervous system* (PNS). Signals sent out from the CNS are transmitted to muscles via *motor nerves,* and signals are received through *sensory nerves,* both part of the peripheral nervous system. Nerves are made up of *neurons,* cells that actually carry the signal, plus some support cells. The neurons of the various parts of the total system are structurally somewhat

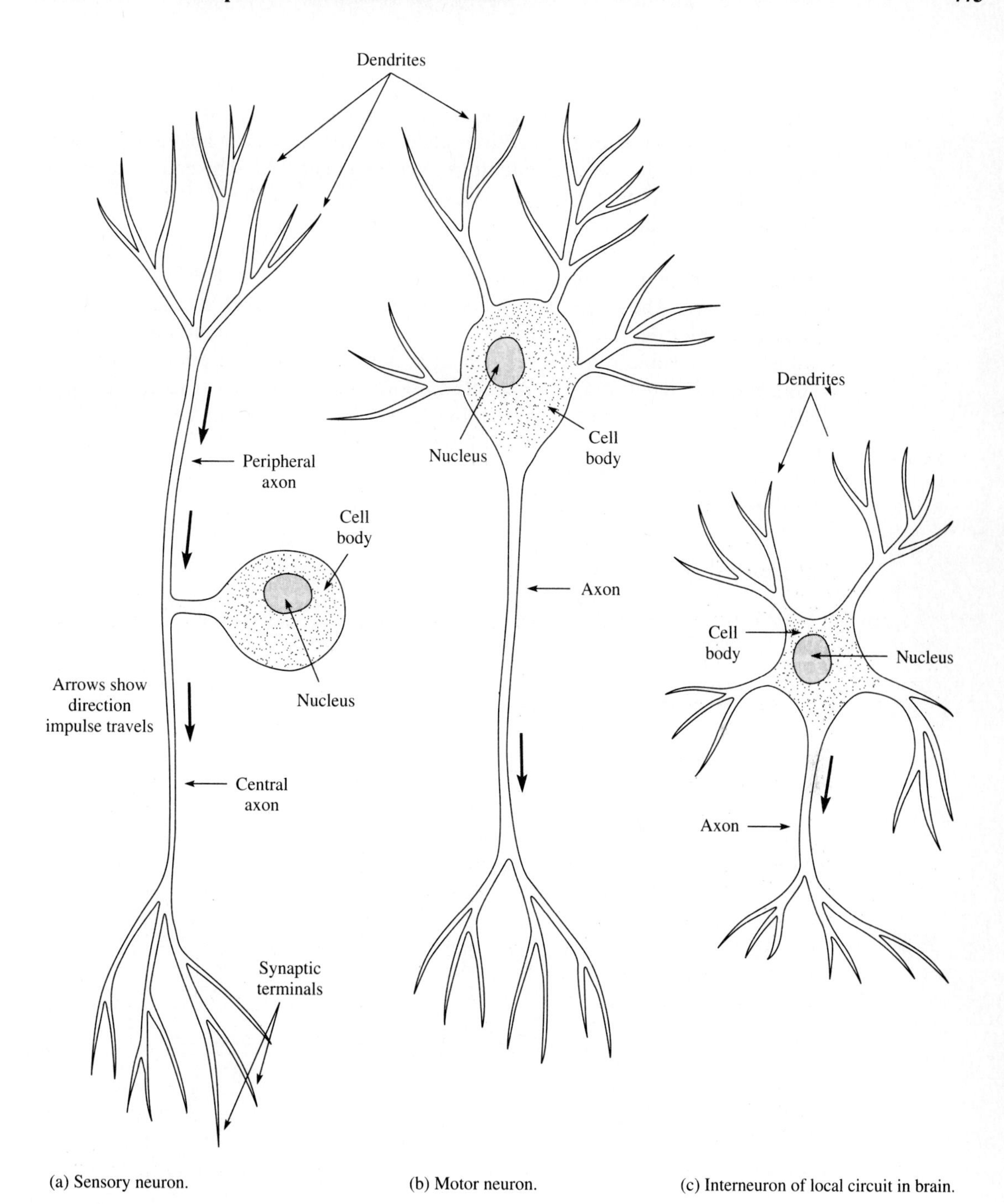

(a) Sensory neuron. (b) Motor neuron. (c) Interneuron of local circuit in brain.

FIGURE 17–24
Different types of neurons.

different from each other, depending on the details of the signal handling they perform (Figure 17–24). All transmit signals in the form of pulses, or *action potentials,* that propagate along the length of the cell. Neurons of the peripheral nervous system have long projections called *axons,* which are analogous to wires because they conduct action potentials over long distances. The cause and propagation of action potentials were originally discovered through the study of the electrical properties of axons.

Figure 17–25 is a diagram of an axon segment. The axon is a tube surrounded by body fluids that contain a number of electrolytes, but sodium chloride is the most abundant. The *axoplasm,* inside the axon, contains a high concentration of potassium chloride. The membrane (the plasmalemma) surrounding the axon is almost impermeable to chloride ions. It is permeable to both potassium and sodium, but it is much more permeable to potassium during its resting (no signal) state. In electrical terms, the membrane has high potassium conductance but behaves like an insulator to ion currents of sodium or chloride. The Nernst equation can be called back into service to calculate the axon resting potential, using the information in Table 17–4.

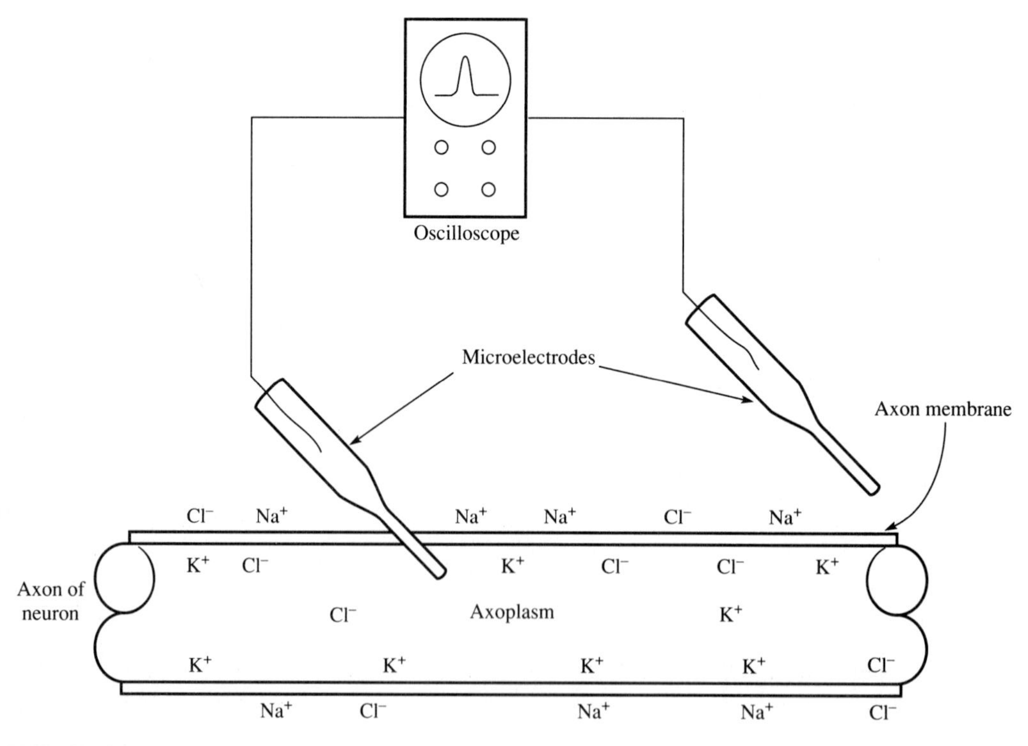

FIGURE 17–25
Axon segment.

TABLE 17–4

LOCATION	CONCENTRATION SODIUM [Na]	POTASSIUM [K]
Axoplasm	0.05*M*	0.40*M*
Exterior	0.44*M*	0.02*M*

Since potassium is the permeative ion, the Nernst equation (Equation 17–8) gives

$$P = -58 \log \frac{C_1}{C_2} = -58 \log \frac{0.4}{0.02} = -75 \text{ mV}$$

The minus sign indicates that the inside of the axon is negative with respect to the outside. The Nernst equation could also be used to calculate what the membrane potential would be if sodium permeability became the dominant conductance.

$$P = -58 \log \frac{C_1}{C_2} = -58 \log \frac{0.5}{0.44} = 55 \text{ mV}$$

Notice that the membrane potential would be of opposite polarity if sodium were the dominant charge conductor.

The membrane potential can be measured using a pair of glass microelectrodes. One is placed outside the axon and the second punctures the membrane to contact the axoplasm (Figure 17–25). The separation of charge across the axon membrane makes it equivalent to a charged capacitor. The total capacitance is quite respectable, in spite of the small size of the cell, since the membrane is extremely thin (7 to 9 nm).

Microelectrodes can also sense the changes that occur during an action potential. Nerve cell membranes, unlike those of most cells, are *excitable*; that is, a stimulus can cause dramatic changes in permeability. When the neuron is stimulated, the membrane becomes more permeable to sodium. This causes the membrane potential to *depolarize,* or become less negative, as sodium permeability begins to dominate over potassium permeability. In other words, current due to the sodium ions charges the membrane capacitance. The increasing potential stimulates further increases in sodium permeability, which in turn causes additional increase in membrane potential. This positive feedback situation results in rapid depolarization of the axon and reversal of the membrane potential polarity (see Figure 17–26). The potential reaches a maximum of about 35 mV, at which point the sodium conductance diminishes and potassium permeability increases. The membrane potential then swings negative, eventually dropping below—but then returning to—the resting potential. The number of ions that traverse the membrane during an action potential is a tiny fraction of those present. The ion current need be only enough to change the voltage on the membrane capacitance by a little over 100 mV.

Once the action potential begins, the positive feedback effect on sodium permeability guarantees that the process will undergo the complete cycle. The

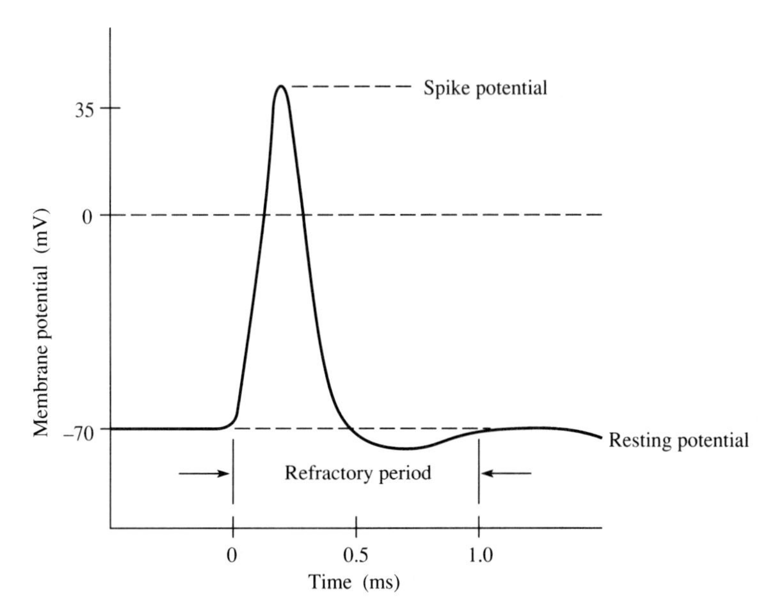

FIGURE 17–26
Action potential in nerve axon.

action potential is often described as an "all-or-none" phenomenon. The membrane is incapable of being restimulated until some time after the action potential is finished. That is, there is a *refractory period* during which a second action potential cannot start. When an action potential occurs in one area of the membrane, it triggers the same events in adjacent areas. The action potential therefore spreads from the original point of stimulus. However, as it propagates along the axon, the membrane just behind it is momentarily in the refractory state, whereas the membrane just ahead is susceptible to stimulation. Consequently, the action potential normally travels *away* from the point of the original stimulus.

Since all action potentials are identical, no information is conveyed by the magnitude of the voltage swing. The intensity of the nerve signals is controlled by the frequency of action potentials. In the absence of stimulation, nerve cells may have an action potentials repetition rate of a fraction of 1 Hz, but at maximum stimulation the rate can be as high as 1 kHz.

The electrical activity of neurons creates potentials in the body that could, in principle, be measured at the body surface. The signal from an individual neuron is too small to be measured that way. However, the total electrical activity of the brain can be picked up using skin-surface electrodes (such as the "floating" electrode described earlier). These signals are normally displayed on an oscilloscope or strip-chart recorder. Such a recording is called an **electroencephalogram,** or EEG for short. The voltages at the electrodes are very small, usually less than 50 μV, requiring that careful attention be given to noise reduction.

A normal brain produces a variety of signals, depending on brain activity. Much of the time there is no regular pattern apparent in the EEG, only a squiggly line. Under certain circumstances, regular patterns or brain waves do appear (Figure 17–27). *Alpha* waves, having a frequency between 8 and 13 Hz, occur when the person is in a quiet, relaxed, resting state. *Beta* waves have very low

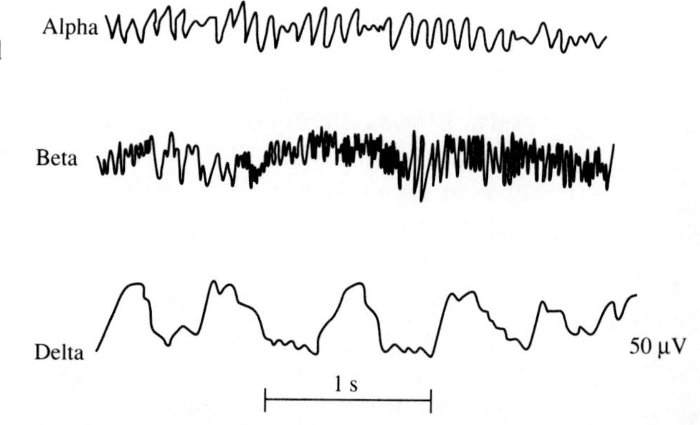

FIGURE 17–27
Different types of normal electroencephalographic waves.

amplitude and frequencies in the range 14 to 25 Hz. They occur during times of activation or tension. Low-frequency *delta* waves occur in infants and in adults during deep sleep. Delta waves at inappropriate times and certain other periodic waves indicate specific brain disorders. One of the principal uses of the EEG is in diagnosing *epilepsy,* a disease characterized by uncontrolled, excessive electrical activity of the brain.

Skin-surface electrodes can also be used to detect the electrical activity of muscles. Muscles are composed of bundles of tiny fibers that are specialized, fused cells containing contractile proteins. Muscles exert force only during contraction. Muscles are stimulated to contract by motor neurons coming in contact at the middle of the individual fibers. Stimulation of muscle fibers initiates an action potential that quickly propagates over the membrane surrounding each fiber. This action potential is very similar to the neuron axon action potential described earlier. Each action potential evokes a brief contraction, called a *twitch,* from the muscle fiber. A train of action potentials produces a series of twitches. If the frequency of action potentials is high enough, the summation of the twitches will be a smooth contraction. Because muscles are composed of large numbers of fibers that are coordinately activated, they produce potentials that can be easily measured at the body surface. A recording of the electrical activity of muscles is called an **electromyogram,** or EMG. Clinically, EMGs are used to diagnose abnormalities of muscle excitation.

17–12 THE HEART

The muscle most commonly examined using electrical measurements is the heart. A recording of the electrical activity of the heart is called an *electrocardiogram,* or EKG, and the instrument that produces it is an **electrocardiograph.** A trained cardiologist can tell a great deal about the physiology and pathology of the heart by reading its EKG. Because heart function is so vital to life and because heart disease is common, the electrocardiograph is one of the most frequently used clinical instruments. However, in order to learn how the instrument functions, it is necessary to learn a little about the anatomy and physiology of the heart and the circulatory system.

The circulatory system (Figure 17–28) is the major transport system of the body. Blood delivers oxygen carried from the lungs to all the body's tissues and there picks up carbon dioxide (CO_2), the waste product of respiration. The blood also carries nutrients, water, hormones, and many other materials throughout the body and it collects the waste products of metabolism and delivers them to the kidneys for excretion.

The circulatory system is actually two systems, each having its own pump to move the blood. You might guess from this statement that the heart is actually two separate pumps in one organ. The right side of the heart pumps blood through the lungs (*pulmonary circulation*), where oxygen is picked up and CO_2 is excreted. Oxygenated blood returns from the lungs to the left side of the heart. The left heart pumps blood to all parts of the body except the lungs (*systemic circulation*). Blood returning from systemic circulation enters the right side of the heart, ready to make another trip to the lungs.

Each side of the heart (Figure 17–29) is a two-stage, or two-chambered, pump. Blood entering the heart passes into the first chamber, or *atrium*. The walls

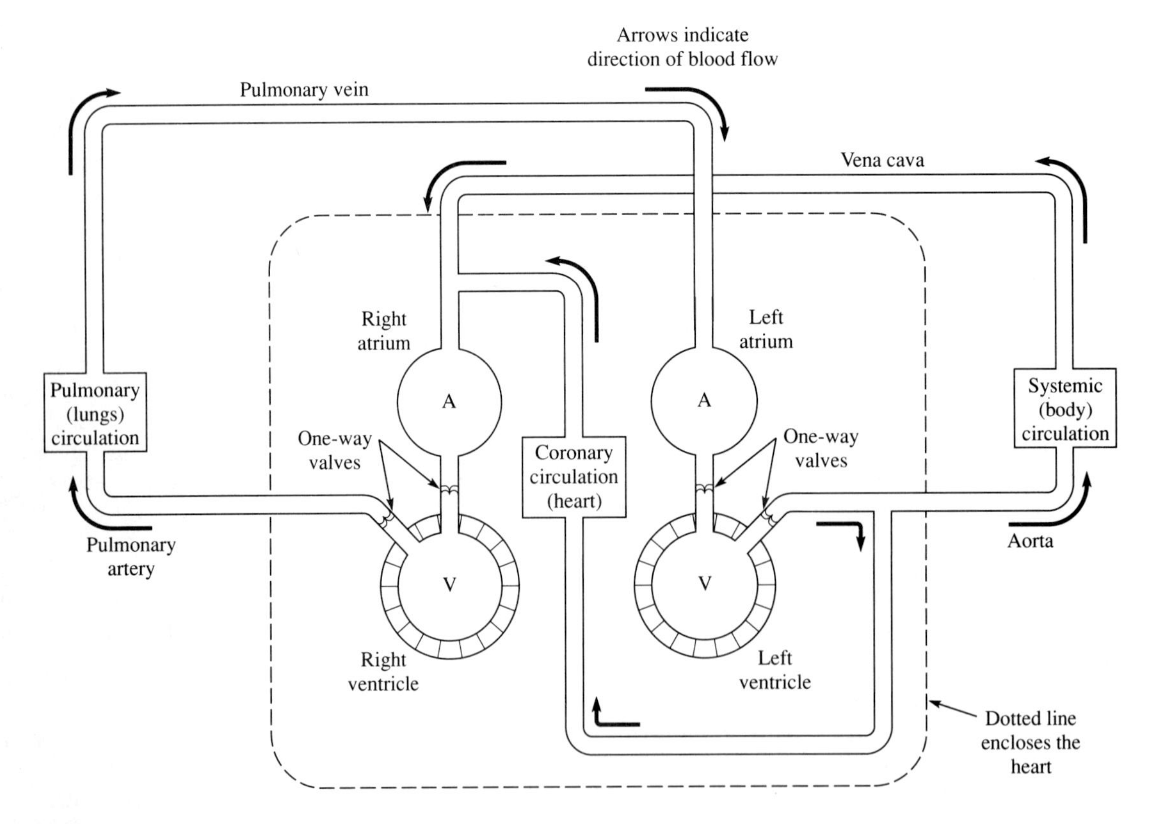

FIGURE 17–28
Schematic of the circulatory system.

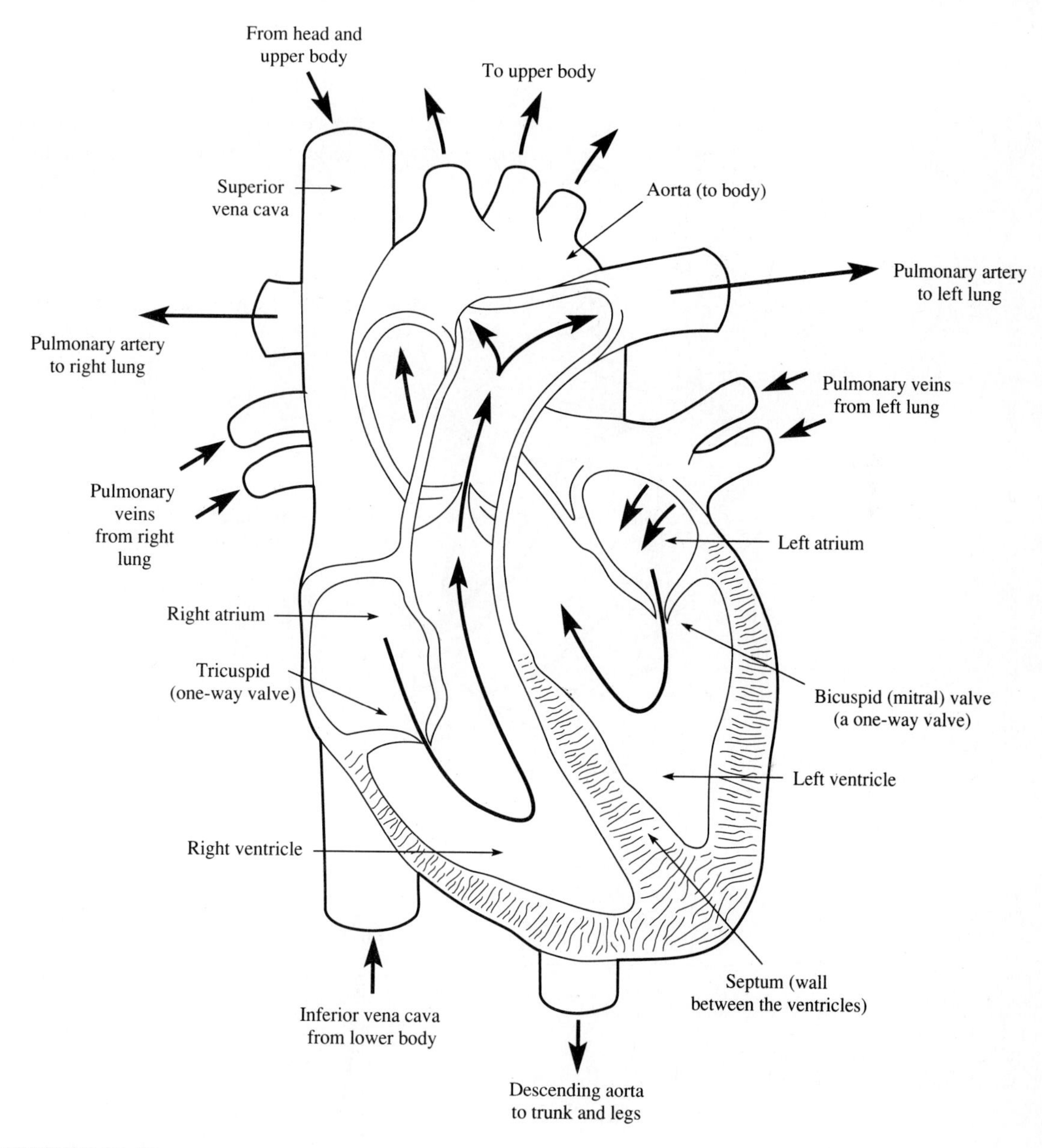

FIGURE 17–29
Anatomy of the human heart. Arrows show direction of blood flow.

of the atrium are composed of specialized *cardiac muscle* that contracts to squeeze blood out of the atrium into the second chamber, the *ventricle*. The thick-walled ventricles are responsible for the main pumping action of the heart. When the muscular ventricles contract, blood is forced out of the heart into the arteries. The period of contraction is called *systole,* and maximum blood pressure (*systolic*

FIGURE 17–30

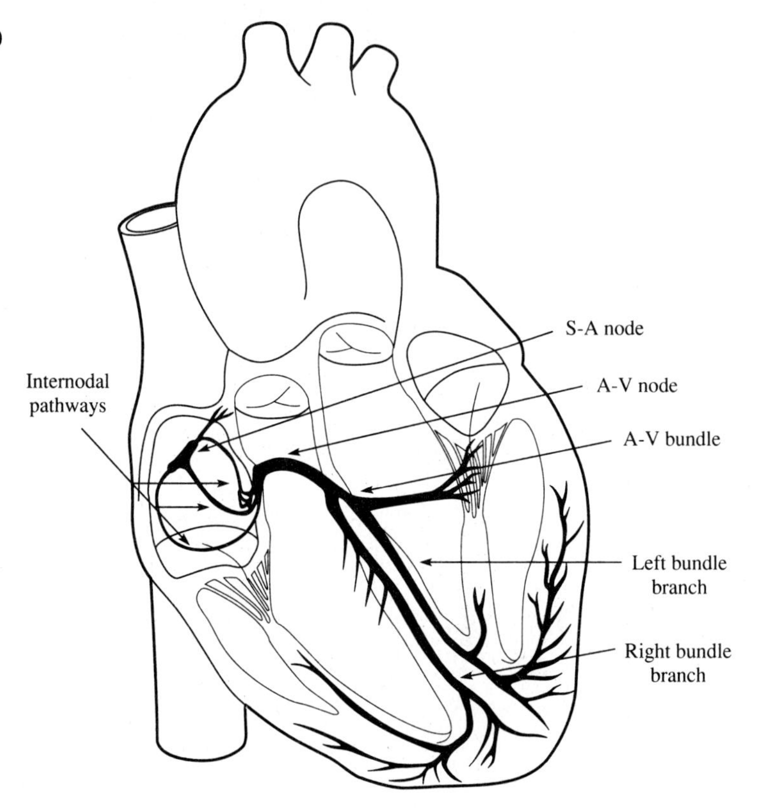

(a) Conduction of pacemaker signals in the heart.

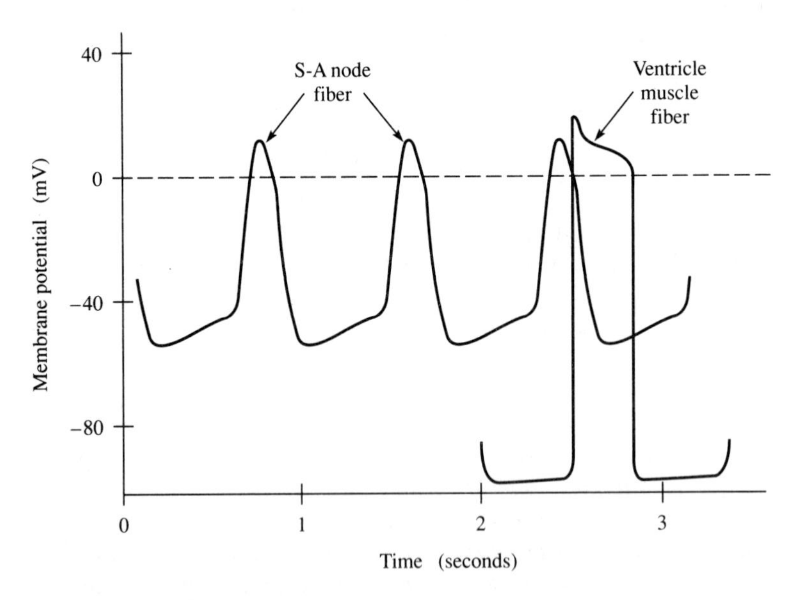

(b) Action potentials in the heart.

pressure) occurs at this time. Each ventricle is equipped with one-way valves to make sure that blood is not pumped backward into the atria. There are also one-way valves (the *semilunar valves*) at the output of the ventricles to prevent blood from surging back from the arteries into the ventricles when contraction ends.

The timing of heart contractions is important. The atria must contract before the ventricles, because it is their job to make sure that the ventricles are filled for efficient pumping. After contraction, both atria and ventricles must relax for a while to allow the heart to refill with blood. The period of relaxation is called *diastole*. During diastole, blood pressure falls to its minimum value (*diastolic pressure*) just before contraction begins anew. Contractions of the heart muscle are initiated and controlled by signals (action potentials) originating in the heart's *pacemaker*. The pacemaker is a small bundle of specialized muscle fibers, called the *sinoatrial node* (or SA node), located in the wall of the right atrium (see Figure 17–30). These fibers act as synchronized oscillators, spontaneously (without outside stimulation) generating action potentials at a fixed rate. These action potentials are similar to those described earlier for neurons, except that the depolarization pulse lasts much longer (about 300 ms). These pulses stimulate action potentials in the atrial muscles, causing them to contract for the 300 ms duration of the stimulating pulse. Meanwhile, the signal from the SA node is relayed by special muscle fibers to the *atrioventricular node* (AV node), where the signal is delayed due to very slow propagation. The delayed signal emerges into the *Bundle of His* (frequently called the *A-V bundle* for short) located in the wall between the ventricles. The A-V bundle splits into the left and right *bundle branches,* composed of *Purkinje fibers* that carry the signal to the muscle of the ventricles. The "signal" is still in the form of broad action potentials, because it is carried entirely by muscle fibers and not by nerves. The Purkinje fibers carry the signal down to the point of the heart and from there fan out to the muscle of the ventricle. The ventricles contract about 150 ms after the atria and the duration of contraction is about 300 ms, the same as the action potential depolarization pulse.

17–13 THE ELECTROCARDIOGRAPH

The potentials produced during the *cardiac cycle* cause ionic currents in the tissues surrounding the heart (remember that the watery interior of the body is an excellent conductor). The chest area is rather like a complex resistive network through which these currents flow. It is possible to measure the resulting voltages in the network using skin-surface electrodes. Of course, the signal will look slightly different for different electrode locations. Consequently, EKG electrodes are placed at standard locations that are easily found (the wrists and ankles) and give repeatable results. Figure 17–31 shows the conventional arrangement of electrodes and the voltages that they pick up. The arms and left leg connect to *Einthoven's triangle,* drawn around the heart to show the approximate vicinity of the conductive paths in the body fluids. The EKG signal is measured as the differences among these electrodes, but the right leg is frequently grounded. Physicians use the term *lead* for these differential signals. This is often confusing to

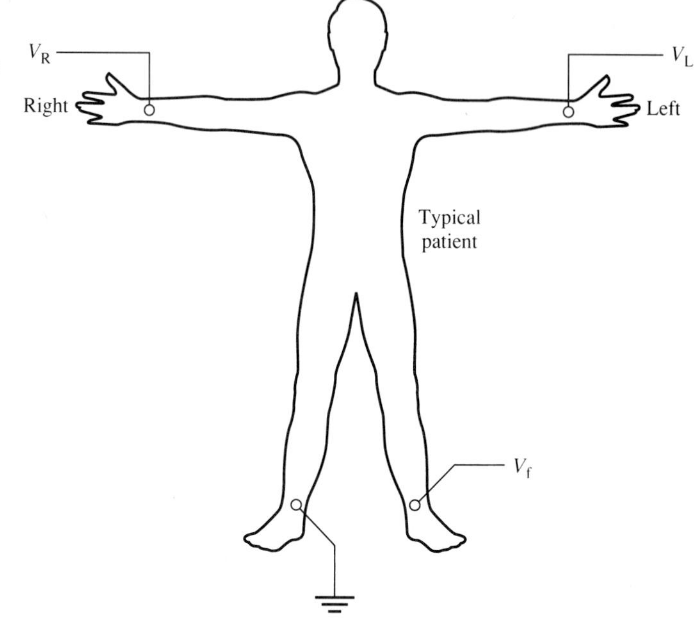

FIGURE 17–31
Placement of the EKG electrodes and the symbols for their voltages.

people trained in electronics, who think of "leads" as being the connecting wires. The medical definition of the three *bipolar limb leads* is

EQUATION 17–10

$$\begin{aligned} \text{Lead I} &= V_L - V_R = V_I \\ \text{Lead II} &= V_F - V_R = V_{II} \\ \text{Lead III} &= V_F - V_L = V_{III} \end{aligned}$$

Each lead can be thought of as the output of a differential amplifier. For example, for lead I the noninverting input is connected to the electrode on the left wrist and the inverting input is connected to the right wrist.

The lead voltages are usually displayed on a strip-chart recorder or a slow-sweep oscilloscope. The normal signal is a rather complex waveform having a number of characteristic components. These components are given letter names, starting with p and going alphabetically through t. Figure 17–32 shows what a normal (healthy) electrocardiogram looks like.

A useful mathematical relationship among the lead voltages is *Einthoven's law*:

EQUATION 17–11

$$V_I + V_{III} = V_{II}$$

This equation can be derived from the definitions of the bipolar limb leads. It allows any one of the instantaneous lead voltages to be calculated if the remaining two are known.

The EKG traces contain a surprising amount of information about how the heart is functioning. The repetition frequency of the waveform gives the *heart*

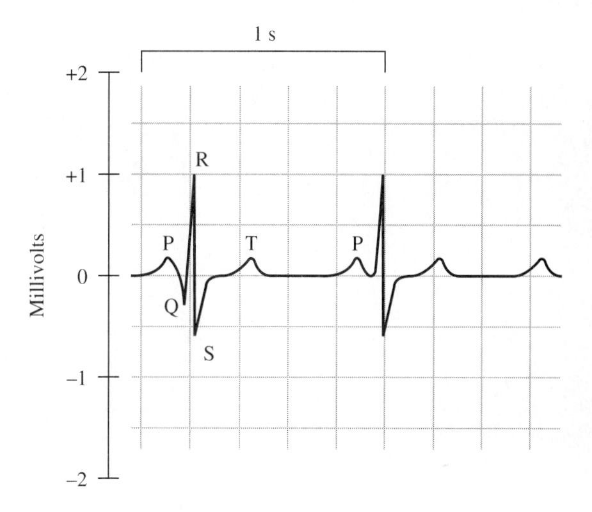

FIGURE 17–32
A normal electrocardiogram.

rate, which is normally about 72 beats (contractions) per minute, or 833 ms per contraction. The P wave is caused by contraction of the atria, and the *QRS complex* is due to contraction of the ventricles. Therefore, the time interval between the P wave and the Q wave is the delay between these contractions. Ventricle contraction lasts from the beginning of the Q wave to the T wave. The T wave is due to *repolarization* (decreasing membrane potential) in the ventricles following contraction. The *Q-T interval* is normally about 350 ms. Changes in the shape of the waveforms indicate a malfunction of the heart, either in transmission of the pacemaker signal or in contraction of the muscle tissue. In fact, most of the serious abnormalities of the heart can be detected by analysis of the EKG waveforms.

Different lead arrangements are sometimes used in order to extract as much information as possible from the electrocardiogram. Figure 17–33 shows the connections for *augmented unipolar limb leads*. The electrode placement is the same as for the bipolar limb leads described earlier, but two of the electrodes are connected through a symmetrical voltage divider to the inverting input of a differential amplifier. The third electrode is connected to the noninverting input. There are three possible leads connected this way:

EQUATION 17–12

$$aV_L = V_L - \tfrac{1}{2}(V_R + V_F)$$
$$aV_R = V_R - \tfrac{1}{2}(V_L + V_F)$$
$$aV_F = V_F - \tfrac{1}{2}(V_R + V_L)$$

A more modern but more complex electrode arrangement is shown in Figure 17–34. A series of six electrodes is placed on the chest over the heart. The six leads, $V_1, V_2, V_3, \ldots$, are the differences between each of these electrodes and the average voltage of the three limb electrodes. These "chest leads," or *precordial leads,* can be used to detect small abnormalities in the heart musculature, which is just beneath the chest wall. A physician wanting a complete check of a

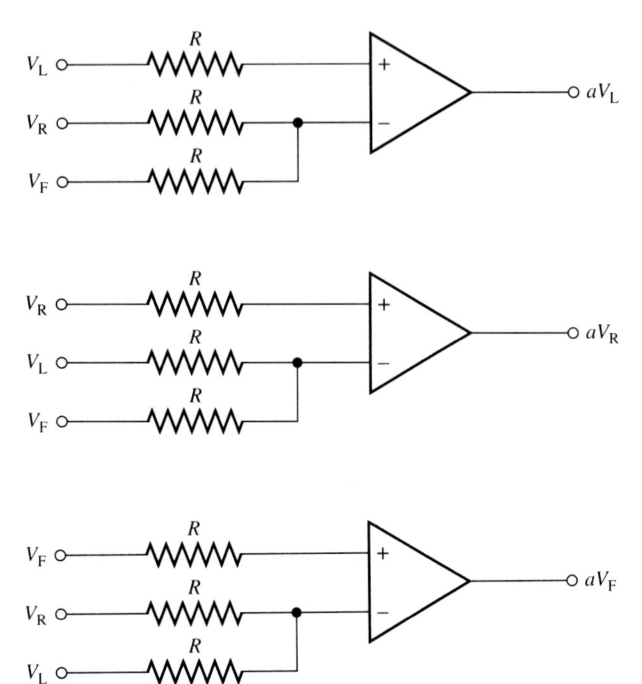

FIGURE 17–33
Connections for augmented unipolar limb leads.

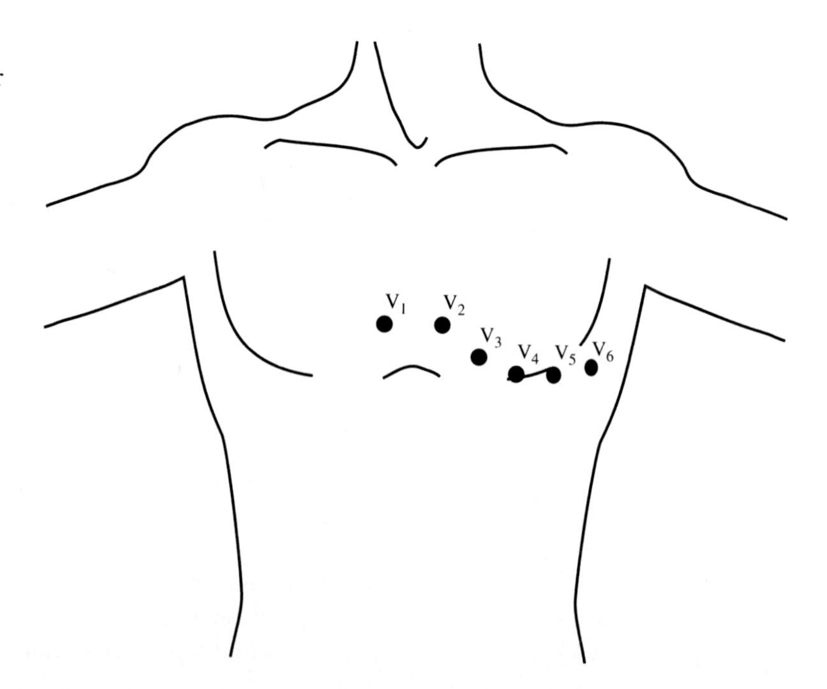

FIGURE 17–34
Placement of electrodes for precordial, or chest, EKG leads.

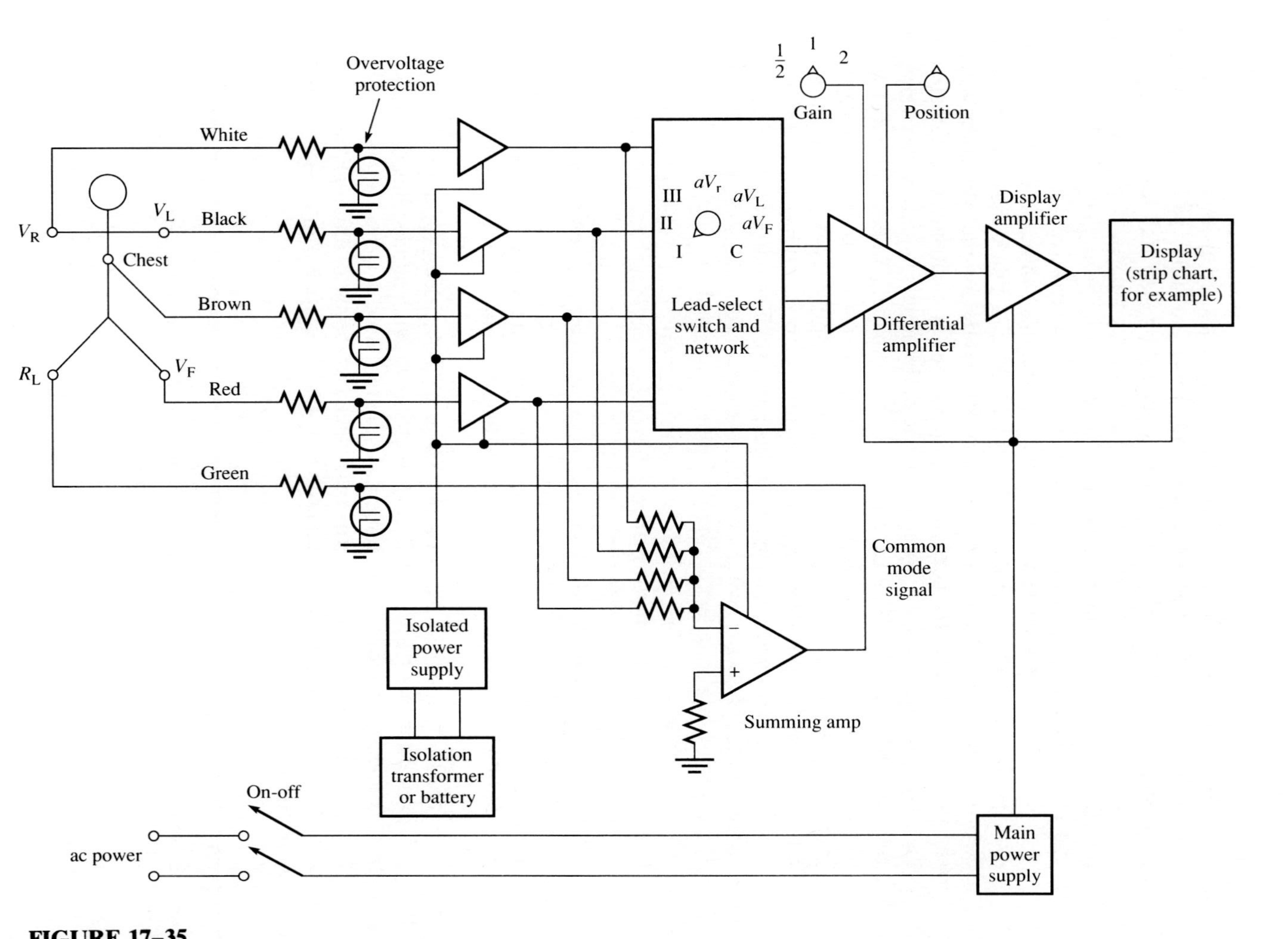

FIGURE 17–35
Block diagram of an electrocardiograph.

patient's heart will want to see recordings of all 12 leads, called, logically enough, a *twelve-lead electrocardiogram*.

Any time that potentials in the body are measured using skin surface electrodes, there is likely to be a problem with noise and 60 Hz pickup. The peak-to-peak amplitude to the QRS complex in standard bipolar limb leads is only about one millivolt. Careful circuit practice is needed, in particular to avoid 60 Hz interference. This usually requires that the patient be grounded (via the right leg) and that lead amplifiers with very high common-mode rejection be used.

Now that you understand some of the principles behind the EKG, it's time to look at the complete instrument. Figure 17–35 is a block diagram of a simple electrocardiograph. There is a standard color code for the electrode wires, as shown on the figure. In modern instruments, each electrode is connected to a high-input-impedance *buffer* amplifier. The buffer inputs usually have overvoltage circuits, such as neon bulbs, to protect the amplifiers in case an electroshock device called a *defibrillator* has to be used on the patient to restore normal heart rhythm. The buffer amplifiers, or at least the input stages, are powered by a separate, isolated power supply in order to reduce shock hazard to the patient. Sometimes a small battery is used for this purpose. The outputs of the buffers are connected to a selector switch and resistor network that derive the lead voltages. The switch can also select a 1.0 mV calibration signal.

The output of the switch and lead network is a differential signal that is amplified by a high-performance differential amplifier. The main requirement is very high CMRR, since noise and pickup from the patient are usually equally present on all the electrode voltages. Sometimes a common-mode signal is derived by combining the outputs of the buffer amplifiers using a resistive summing network and an op-amp. The common-mode signal may be used to reduce common-mode noise through a feedback arrangement to the patient's right leg.

The differential amplifier output drives a display device, traditionally a strip-chart recorder that traces the EKG waveform with a heated stylus on heat-sensitive paper. Some modern instruments use alternative displays, such as CRTs or LCD panels. The customary strip chart speed is 25 mm/s, although faster speeds can be switch-selected. The standard vertical sensitivity is 1 mV per centimeter of deflection.

EXAMPLE 17–4 Examine the EKG traces in Figure 17–36 and interpret their meaning.

SOLUTION The trace in Figure 17–36(a) appears at first glance to be a normal EKG except for the inverted T waves. Since the strip chart moves at 25 mm/s, each of the small (1 mm) squares corresponds to 40 ms. Therefore, the time between the first and second beat is about 820 ms, corresponding to about 73 beats per minute, well within normal range. The interval between the P and R waves is about 140 ms and the interval between the Q and T waves is about 260 ms, both normal. The third beat, however, is an *ectopic* (out-of-place) beat, since it occurs much sooner than 820 ms. Extra or missed beats occur occasionally in healthy hearts; only if ectopic beats occur regularly would this be cause for concern.

FIGURE 17–36
Sample EKG traces.

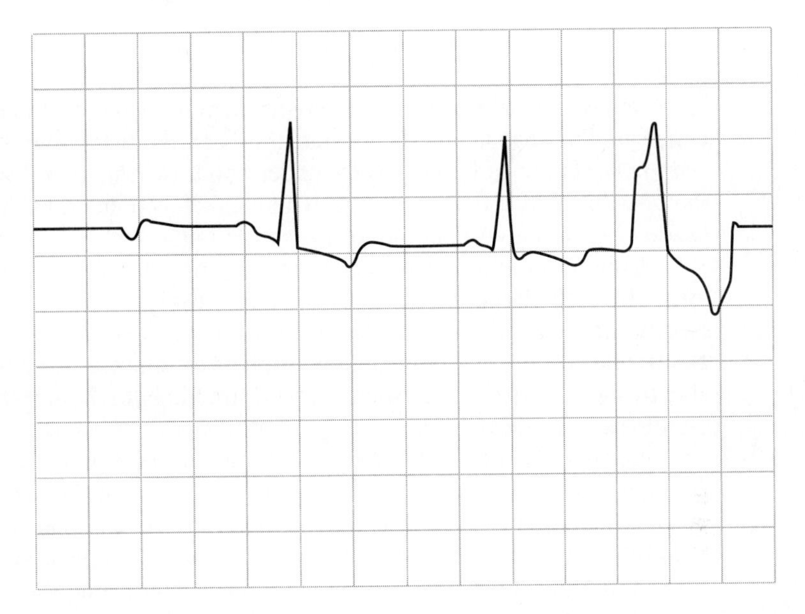

(a)

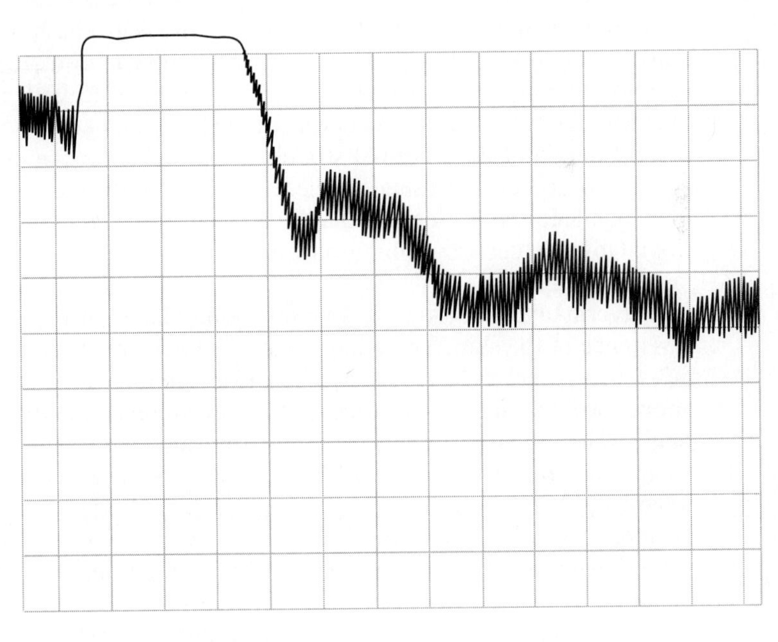

(b)

The trace in Figure 17–36(b) is definitely not normal, but the problem is most likely instrumental. The signal appears to be intermittent and noisy. This is probably due to faulty patient connections. The electrodes may be dry, the connectors may be loose, or the connecting wires may be damaged. Worn or damaged wires and connectors are the single most common biomedical instrumentation problem.

17–14 SAFETY IN BIOINSTRUMENTATION

In recent years, the safety of bioinstrumentation has become a matter of great concern. The reason for this is that the hazards of low-current electrical shock (*microshock*) are now more fully understood. Safety, therefore, means reducing the possibility of electrical shock, where shock is defined as the accidental occurrence of externally applied currents in the body.

You are now in a position to understand why electric currents in the body might be so hazardous. Both nerves and muscles respond to and function with electrochemical stimuli. The heart, as we have seen, is especially responsive to electrical stimuli. As you might expect, nerves, muscles, and heart are all responsive to shock currents, but such currents are likely to disrupt the normal functioning of these systems.

Because the possibility of shock is an ever-present danger in our electricity-dependent society, several studies have been made of the effects of shock on the body. The amount of current that is dangerous depends on where it is applied to the body. Current that passes through the chest is especially risky, for two reasons. First, current passing through the heart can excite the muscle fibers of the ventricles into uncoordinated contractions that don't pump blood effectively. This condition, called *ventricular fibrillation,* is fatal unless treated immediately. Ironically, the treatment is a brief but massive shock to the heart that causes all the muscle fibers to contract at once, thereby resynchronizing their activity. The second hazard of current across the chest has to do with its effect on skeletal muscles. As in the heart, electrical stimulation will cause the muscles to contract independently of their normal stimuli. Large currents can cause *tetany,* or maximal, uncontrolled contraction. These spasms can lead to paralysis of the muscles responsible for breathing, the diaphragm and the *intercostal* muscles (between the ribs). Obviously, such a condition would cause asphyxiation, followed by death in just a few minutes.

Normally, the body is fairly safe from electrical shock because of the high resistance of the skin. The outer layer of human skin is composed of dead, dried-out cells that are good insulators. Dry skin can have a contact resistance of up to about 1 MΩ, so accidental contact with low-voltage circuitry rarely causes even a tingle. The situation is very different for someone connected to an instrument such as an EKG via electrodes. The whole purpose of electrodes is to make low-impedance contact with the body. Furthermore, EKG electrodes are deliberately arranged to make optimum contact with the heart. Any EKG circuit failure that placed even low voltage on the electrodes could amount to a lethal risk to the patient.

Fortunately, much can be done to reduce the risk. A simple precaution is to insert current-limiting resistors in series with the electrodes. In this era of widespread use of FET-input op-amps to buffer the electrodes, these series resistances usually cause no compromise in instrument performance. Another way of reducing shock hazard is to use *isolation amplifiers,* described in Section 5–6. In these devices the input circuitry is completely isolated from the output. All signals are coupled to the output by optoisolators. The input circuitry can even be battery powered, which has the advantage of limiting the maximum current and voltage to which the patient could be exposed in even the most catastrophic circuit failure.

There is another shock hazard associated with the use of all modern electrical equipment. Government regulations now require grounding the metal chassis, or case, of electrical equipment, including biomedical instruments. Grounding is required for safety reasons, but under some circumstances a grounded system can increase, rather than reduce, shock hazard. To understand why, it is necessary to review the fundamentals of power distribution. Figure 17–37 is a schematic of a simple power-distribution system of the type used in the United States today. Power is delivered via a transformer that has a center-tapped secondary. The center tap is grounded—that is, physically connected to the earth via a water pipe or metal stake. Normal voltage across the transformer terminals is 220 V ac. The usual 110 V ac outlets are connected to one side of the transformer secondary and to the grounded, or *neutral*, center-tap. The connecting wires are color-coded. Black, the color of death, signifies the dangerous *hot* wire, and white indicates the benign neutral. A third, green-colored wire connects the ground wall contact to ground independently of neutral. Modern three-wire equipment is grounded through this independent ground. Therefore, any fault that occurs between the hot line and the equipment chassis will cause a short, which, in turn, will trip the circuit breaker. This removes the danger of shock and is a clear warning that something is wrong. What could be safer than that?

The problem is that an environment where everything is grounded allows people to make easy contact with one of the two connections to the body required to cause a shock. A "grounded" human who comes into accidental contact with

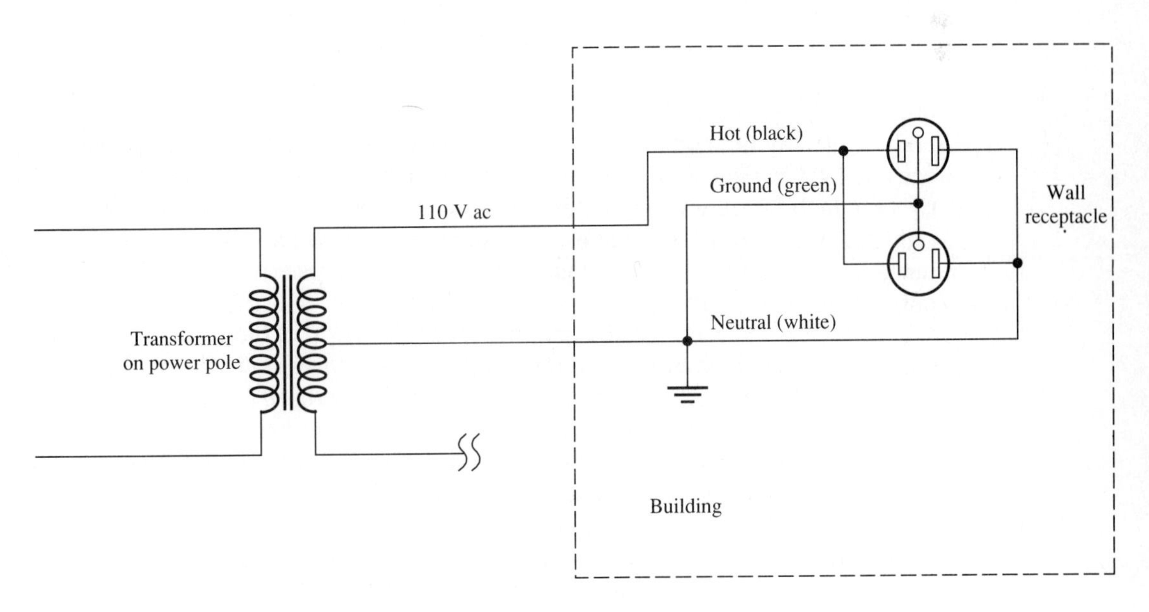

FIGURE 17–37
Modern power distribution to wall plugs.

the hot lead will be at risk. For example, consider the following unusual but possible sequence of events. Suppose that the ground wire in the power cord of an instrument becomes frayed and fails. Abuse and neglect of power cords is hardly unusual, so this is not an improbable event. Furthermore, loss of the ground is likely to go unnoticed, because the instrument will continue to function normally. Next, suppose that a fault occurs, putting the hot side of the power line in contact with the instrument case. No short will trip the circuit breaker and the instrument may still function normally. However, anyone who touches the case and any grounded object (another instrument, for example) is—literally—in for a nasty shock.

The shock will be worse than nasty for someone connected to an EKG or similar instrument. Recall that the right-leg electrode is frequently grounded. It is advantageous to ground the patient because this greatly reduces the common-mode noise and hum pickup on the other electrodes. Therefore, some means of improving the safety of the ground connection is warranted. Fortunately, there are several ways to do this.

A common safety precaution is the use of *ground fault interrupters* (described in Section 12–7). A second, even simpler safety measure is the use of *double insulation* for electrical equipment. The circuitry is enclosed in a grounded metal chassis, but the chassis is insulated by enclosing it in a nonconducting box, such as a plastic case. All controls are also insulated. Fortunately, plastic instrument cases have become quite fashionable, so they are both stylish and functional.

The risk to the patient connected to the EKG still has not been completely abolished, since the right leg is still connected to ground. It would be nice if the advantage of grounding could be retained without the risk. There is a simple circuit technique that can accomplish this. Figure 17–38 is a schematic of the *driven-leg* virtual ground. (Virtual ground is discussed in Section 5–6.) The circuit is essentially a voltage follower with its input connected to ground and the body plus a high-value resistance in the feedback path. The feedback connection to the body is usually the common-mode signal derived by summing the electrode voltages, shown in Figure 17–35. Since noise and 60 Hz hum are coupled to the body by very high impedance paths, only a very small drive current is required to the right leg to cancel these troublesome common-mode signals. The feedback resistance can be made large enough (1 MΩ or more) that the risk of dangerous currents is very small.

This concludes our brief exploration into the world of biomedical instrumentation. There is much more than this, but we wanted to avoid making the limited space available just a boring catalog of all the gizmos used in modern medicine. Instead, we thought it would be better to present the principles behind a few important biomedical instruments so that you might not only understand the electronics but also get some insight into the biological, chemical, and physical phenomena that govern the design and application of this equipment. Indeed, the field of biomedical instrumentation is both especially challenging and especially interesting because of its multidisciplinary nature. It is also especially satisfying to be a part of it because of its humanitarian purpose.

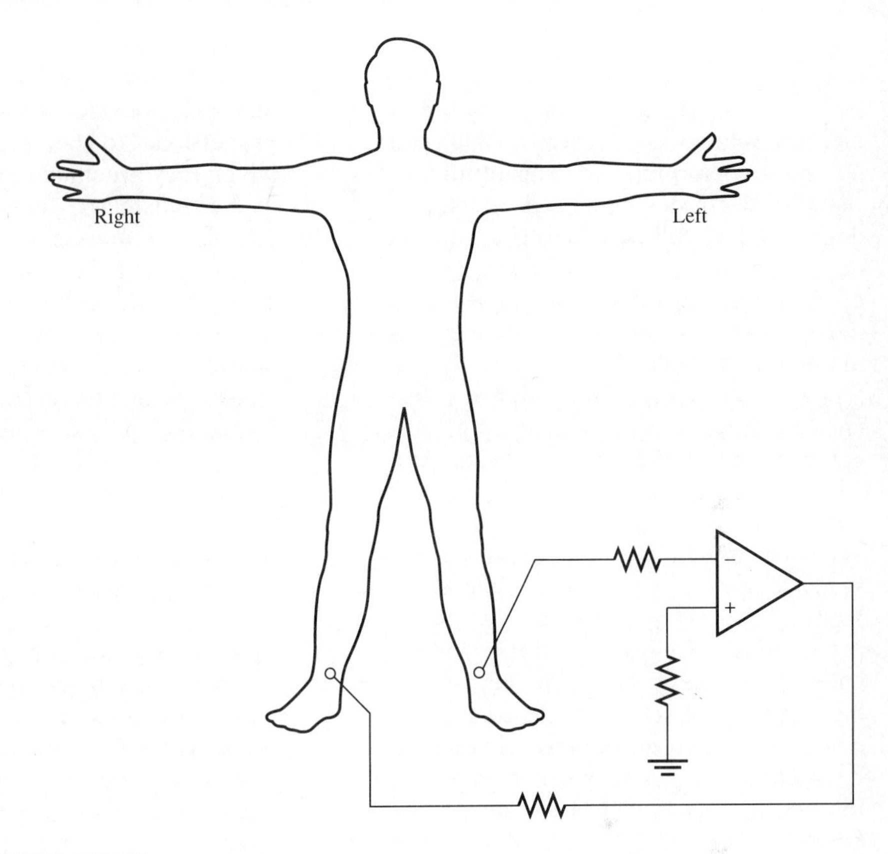

FIGURE 17–38
Driven-leg virtual ground. The amplifier inverting input may be connected to the EKG common-mode signal instead of the left-leg electrode.

SUMMARY

1. Cells are the functional units of organisms in which basic metabolic processes of life occur. Biomedical instruments are used to measure the processes occurring in organisms, particularly humans.
2. All matter is composed of atoms. Most substances are made up of molecules, groups of atoms linked together in specific arrangements by chemical bonds. The molecular weight is the sum of the atomic weights of all the atoms comprising the molecule.
3. Solutions are homogeneous mixtures of a solute dissolved in a solvent. Concentration is the amount of solute per unit volume of solution. Many solutes are colored or can be made colored for assay purposes. Colored substances absorb some visible light wavelengths but reflect or transmit others.
4. Spectrophotometers can be used to determine solute concentration. They measure the attenuation of a beam of light passing through a specimen.

5. Ions make solutions electrically conductive and are the charge carriers in solution. Conductivity meters may be used to measure solution conductance, which can be used to determine ion concentration.
6. Metallic electrodes immersed in solution develop a half-cell potential that depends on the electrode metal and solute, among other factors. Electrodes are transducers that convert electron flow in the metal to ion flow in the solution.
7. Silver–silver chloride and calomel reference electrodes have nearly constant half-cell potentials and are used to make electrical contact to solutions.
8. Selectively permeable membranes develop a membrane potential because some kinds of ions can permeate while other kinds do not. The membrane potential is proportional to the concentration difference of the permeative ion across the membrane. Ion-selective electrodes are made using such membranes. These can be used to determine ion concentration through the magnitude of the membrane potential developed in a solution under test.
9. pH is the negative log of hydrogen ion concentration. Special ion-selective glass electrodes develop a membrane potential proportional to the pH of the solution in which they are immersed. These are used as the transducers for pH meters.
10. Nerve and muscle cells develop membrane potentials that change due to changes in the cell membrane permeability to potassium and sodium ions. These action potentials carry nerve signals or cause muscles to contract.
11. Nerve and muscle potentials cause currents in the body that can be detected using electrodes at the body surface. Electrocardiographs chart the voltages produced by the heart over time. These records are useful in diagnosing heart disease.
12. Modern electrical equipment chassis and cases are grounded through a separate wire from the ac neutral. Special precautions and instrument circuitry are required to reduce shock hazard to patients connected via bioelectrodes to this equipment.

QUESTIONS AND PROBLEMS

1. Define the following.
   (a) Absorbance
   (b) Absorption spectrum
   (c) Action potential
   (d) Assay
   (e) Atrium
   (f) Beer-Lambert law
   (g) Dalton
   (h) Depolarization
   (i) Diastole
   (j) Electrocardiogram
   (k) Electroencephalogram
   (l) Electrolyte
   (m) Electromyogram
   (n) Extinction coefficient
   (o) Half-cell potential
   (p) Ion product constant
   (q) Ion selective membrane
   (r) Membrane potential
   (s) Mole
   (t) Monochromator
   (u) Nernst equation
   (v) pH
   (w) Refractory period
   (x) Selectively permeable membrane
   (y) Sinoatrial node
   (z) Systole

2. What is the difference between an atom and a molecule?
3. The molecular weight of sodium chloride (table salt) is 58.4. How much would be required to make 0.6 L of a 0.25*M* solution? How much would be needed for 50 mL of 0.05*M* solution?
4. What is the difference between a colorimeter and a spectrophotometer?
5. Describe how optical interference filters work.
6. Design a spectrophotometer (in block diagram form) capable of automatically measuring the absorption spectrum of a sample.
7. Most beers contain protein. Describe a method for determining the protein content of a beer sample. Explain how you would avoid problems caused by the beer bubbles and the normal beer color.
8. Pyridoxal hydrochloride (vitamin B6) has an absorption maximum at 293 nm in water solution. Suppose that three solutions of unknown concentration are measured using a spectrophotometer. A series of solutions of known concentration is measured at the same time. The results are shown in the accompanying table. Calculate the molar extinction coefficient and determine the concentration in the three unknown solutions. Notice the concentration is in moles/liter rather than grams/liter. (Assume *d* is 1 cm.)

CONCENTRATION	ABSORBANCE
0.0*M*	0.0
0.003	0.23
0.006	0.45
0.009	0.68
0.012	0.92
0.015	1.13
Unknown 1	0.076
Unknown 2	0.375
Unknown 3	0.83

9. Explain why solutions that contain ionic substances are electrically conductive.
10. Describe how a conductivity meter works.
11. Suppose that you are responsible for keeping the salinity of wastewater discharged from a pickle plant environmentally safe. Design a simple conductivity tester that will flash an alarm if the discharge exceeds a preset salinity level.
12. Lemon juice has a pH of about 2.3. What is its hydrogen ion concentration? The hydrogen ion concentration of cow's milk is 3.5×10^{-7}. What is the pH of milk?
13. Explain why pH meters require a reference electrode. Why won't a piece of bare wire work as a reference electrode?
14. Suppose that a new pH meter has just been unpacked. Describe how the instrument should be calibrated.
15. A common service problem with pH meters is caused by negligent users allowing the electrodes to dry out, causing the porous ceramic junction to become clogged with dried filling solution. What would be a simple way to repair a dried electrode?
16. Describe how nerve cells transmit signals along their length.
17. Explain what happens when the cell membrane (plasmalemma) depolarizes.
18. Suppose that bacteria have internal potassium chloride concentrations of 0.15*M* and that the bacterial plasmalemma is permeable only to potassium ions. What would the membrane potential be if these organisms are placed in a 0.05*M* potassium chloride solution?
19. (a) Suppose that human egg cells are permeable to potassium but not other ions. If they have a membrane potential of −40 mV when they are placed in an *in vitro* fertilization chamber containing 0.05*M* potassium chloride solution, what is the potassium concentration inside the egg?
    (b) After fertilization, the membrane potential normally increases to about +25 mV. Explain what happens to the

egg cell plasmalemma during fertilization.

20. Suppose that you are given the job of checking an EKG's power and ground connections to make sure that they are undamaged and functioning properly. Describe a procedure for doing this. How would you modify this procedure for double-insulated instruments?

21. Suppose that an EKG recording is made while the patient is exercising, doing sit-ups. The trace moves up and down in synchrony with the patient's movements. Explain what is happening to cause this. How could the instrument be modified to eliminate or reduce this problem?

Appendix A

Derivations

DERIVATION OF FREQUENCY AND GAIN FOR PHASE-SHIFT OSCILLATOR

The phase-shift oscillator uses an *RC* network to shift the phase of the output by 180° and an inverting amplifier to shift the phase another 180°, as required by the Barkhausen criterion. There is only one frequency at which the network will introduce the required 180° shift. At this frequency, the feedback network introduces significant attenuation that must be made up for by the gain of the amplifier. The frequency at which the network shifts the phase by 180° and the attenuation of the network at this frequency will be found in this derivation.

Usually, the feedback network is composed of three sections of equal resistors and capacitors. This circuit is frequently misunderstood; at first glance it would appear that each section of the feedback network shifts the phase by 60°. This is not the case, as this simplified approach ignores the loading effects of the following sections on the earlier sections. To account for these loading effects, the network is analyzed by writing a series of loop equations and solving for the current in each loop. This technique is shown in detail because of its general application to network analysis.

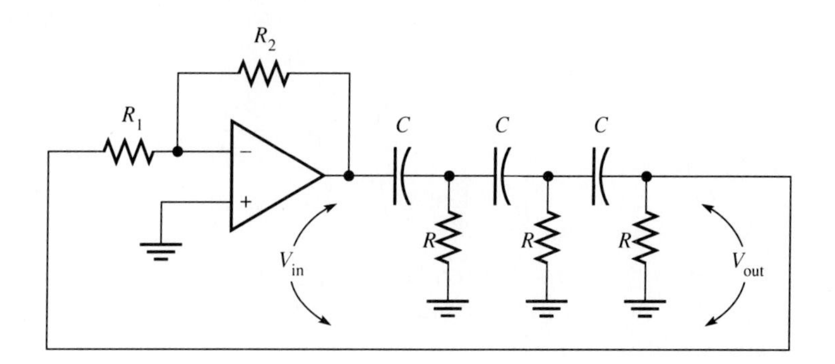

Write three loop equations for the *RC* network shown here:

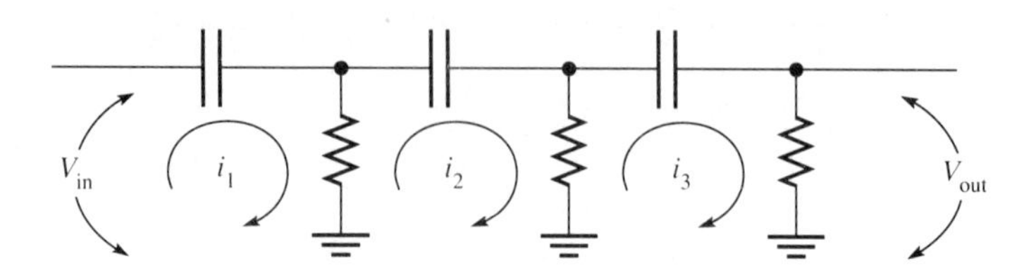

$$\begin{aligned}
&\text{Loop 1:} \quad -V_{\text{in}} + i_1(-jX_C) + (i_1 - i_2)R = 0 \\
&\text{Loop 2:} \quad (i_2 - i_1)R + i_2(-jX_C) + (i_2 - i_3)R = 0 \\
&\text{Loop 3:} \quad (i_3 - i_2)R + i_3(-jX_C) + i_3R = 0
\end{aligned}$$

Rewriting these equations in standard form gives

$$\begin{array}{llll}
i_1(R - jX_C) & + i_2(-R) & & = V_{\text{in}} \\
i_1(-R) & + i_2(2R - jX_C) & + i_3(-R) & = 0 \\
 & + i_2(-R) & + i_3(2R - jX_C) & = 0
\end{array}$$

The determinant is

$$\begin{aligned}
D &= \begin{vmatrix} R - jX_C & -R & 0 \\ -R & 2R - jX_C & -R \\ 0 & -R & 2R - jX_C \end{vmatrix} \\
&= (R - jX_C)(2R - jX_C)^2 - (R^2)(R - jX_C) - (R^2)(2R - jX_C) \\
&= 4R^3 - j4R^2X_C - RX_C^2 - j4R^2X_C - 4RX_C^2 + jX_C^3 - R^3 + jR^2X_C \\
&\quad - 2R^3 + R^2X_C \\
&= (R^3 - 5RX_C^2) + j(X_C^3 - 6R^2X_C)
\end{aligned}$$

Solving for i_3 results in

$$i_3 = \frac{\begin{vmatrix} R - jX_C & -R & -V_{\text{in}} \\ -R & 2R - jX_C & 0 \\ 0 & -R & 0 \end{vmatrix}}{D} = \frac{R^2V_{\text{in}}}{D}$$

$$V_{\text{out}} = i_3R = \frac{R_2V_{\text{in}}}{D}R = \frac{R^3V_{\text{in}}}{D}$$

$$\beta = \frac{V_{\text{out}}}{V_{\text{in}}} = \frac{R^3V_{\text{in}}}{DV_{\text{in}}} = \frac{R^3}{(R^3 - 5RX_C) + j(X_C^3 - 6R^2X_C)}$$

To obey the Barkhausen criterion, β must have a 180° phase shift. This means β is a real number (no imaginary component). Therefore, $(X_C^3 - 6R^2X_C)$ must be equal to zero.

$$X_C^3 - 6R^2X_C = 0 \qquad X_C^3 = 6R^2X_C \qquad X_C = \sqrt{6}R$$

but $X_C = 1/2\pi fC$. Therefore

$$\frac{1}{2\pi fC} = \sqrt{6}R$$

Solving for f gives

$$f = \frac{1}{2\pi\sqrt{6}RC} \qquad \text{oscillation frequency}$$

The real part of the expression for β gives the attenuation of the feedback network. That is

$$|\beta| = \frac{R^3}{R^3 - 5RX_C^2}$$

Let $X_C = 1/2\pi fC$ and $f = 1/2\pi\sqrt{6}RC$. Then

$$|\beta| = \frac{R^3}{R^3 - 5R\left(\dfrac{1}{2\pi fC}\right)^2} = \frac{R^3}{R^3 - 5R\left(\dfrac{2\pi\sqrt{6}RC}{2\pi C}\right)^2} = \frac{R^3}{R^3 - 30R^3}$$

$$|\beta| = \frac{1}{1 - 30} = \frac{1}{-29}$$

The feedback network attenuates the signal by 29 times and inverts it (minus sign).

DERIVATION OF $B = 0.35/t_r$

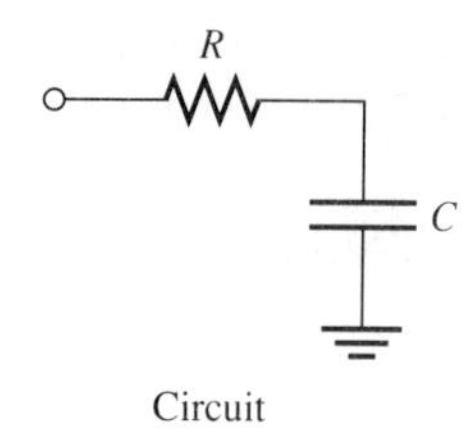

Circuit

$$B = \text{bandwidth, Hz}$$
$$t_r = \text{rise time, s}$$

The equation indicates that, for a series RC circuit, the bandwidth can be found if the rise time is known and vice versa. For instruments such as the oscilloscope, this is useful for determining the limits of the instrument for measuring rise or fall time with a given bandwidth instrument.

The equation for charging or discharging a capacitor is

$$v_t = (V_f - V_i)(1 - e^{-t/RC}) + V_i$$

where v_t = instantaneous voltage on C at time t, V
V_f = final voltage on C if the capacitor is allowed to fully charge or discharge, V
V_i = initial voltage on C, V

Assume the capacitor starts at 0 V and charges to V_f. The time t_1 to the 10% level is

$$0.10V_f = V_f(1 - e^{-t_1/RC})$$
$$0.10 = 1 - e^{-t_1/RC}$$
$$0.9 = e^{-t_1/RC}$$
$$\ln 0.9 = -t_1/RC$$
$$t_1 = 0.105RC \qquad (10\%)$$

The time t_2 to the 90% level is

$$0.90V_f = V_f(1 - e^{-t_2/RC})$$
$$0.90 = 1 - e^{-t_2/RC}$$
$$0.1 = e^{-t_2/RC}$$
$$\ln 0.1 = -t_2/RC$$
$$t_2 = 2.303RC \qquad (90\%)$$

The rise time is

EQUATION A–1

$$t_r = t_2 - t_1$$
$$= 2.303RC - 0.105RC$$
$$= 2.198RC$$
$$RC = \frac{t_r}{2.198}$$

The bandwidth of the same RC circuit is based on its response to a sinusoidal input. The lower frequency is zero; therefore, the bandwidth is equal to the cutoff frequency.

At the cutoff frequency

$$R = X_C$$
$$= \frac{1}{2\pi f_{co} C}$$

where f_{co} = cutoff frequency, Hz

By substitution

$$R = \frac{1}{2\pi BC}$$

Rearranging gives

$$B = \frac{1}{2\pi RC}$$

From Equation A–1

$$B = \frac{1}{2\pi\left(\frac{t_r}{2.198}\right)}$$
$$= \frac{0.35}{t_r}$$

GLOSSARY

absolute pressure. Pressure measured with reference to zero pressure.

absorption spectrum. The plot of light intensity versus wavelength that is transmitted by a transparent substance.

ac bridge. An instrument or circuit network that uses a bridge arrangement of resistive and reactive components and is excited by an ac source; it is used to determine the magnitude of an unknown resistive or reactive component or transducer.

accuracy. The degree to which a measured value represents the true or accepted value of a quantity.

ac resistance. The resistance of a device to a small ac signal at a designated point on the *IV* curve for the device; it is found by dividing a small change in voltage by the corresponding change in current.

acquisition clock. In a logic analyzer, a series of pulses that determine the sample points; it can be an internal (asynchronous) clock or an external (synchronous) clock that is provided by the system under test.

acquisition memory. In a logic analyzer, the random access memory that is used to store the data obtained during a cycle.

active device. A device that requires a source of power to function.

active filter. A frequency selective network that include active devices (transistors or op-amps) in their construction.

active probe. A probe that contains a very high impedance, low-capacitance amplifier to provide low loading to the test circuit.

active transducer. A transducer that requires an external source of power to function.

adder. A combinational logic circuit that performs binary addition, normally including carry-in and carry-out capability.

aliasing error. A problem that occurs when a time-dependent function is sampled at too low a rate, causing high-frequency signals to appear as lower frequency signals.

alternate. A method of displaying two signals on the screen of a dual trace oscilloscope by electronically switching between them in an alternating pattern after each has completed a sweep. The alternate mode is best for displaying relatively fast signals.

ammeter. An instrument for determining the magnitude of electrical current.

ampere (A). The fundamental unit for electrical current in the SI system; it is equal to 1 C/s.

amplifier. A circuit that increases the voltage or current of an input signal.

amplitude modulation (AM). A means of encoding a low-frequency signal onto a high-frequency carrier by causing the output amplitude to vary in a manner determined by the signal.

analog signal. A signal that can take on a continuous range of values between certain limits.

analog storage oscilloscope. An oscilloscope that is capable of retaining an image of the input waveform within the CRT for an extended time.

analog-to-digital (A/D) converter. A circuit that changes an analog input signal into an output signal that can take on only discrete values.

analyzer. A class of instruments that can measure several different variables.

aquadag. A conductive graphite coating placed on the inside of a CRT to provide a return path for electrons from the electron beam.

arbitrary function generator. A waveform generator that allows the user to create custom waveforms or select from a collection of stock waveforms.

arithmetic logic unit. A combinational logic circuit that can perform a variety of arithmetic and logic operations in accordance with a binary instruction.

arithmetic mean. See *mean.*

asynchronous sampling. In a logic analyzer, a sampling mode in which the sample clock is unrelated to the test system clock.

audio frequency (AF). Those frequencies in the range of human hearing; normally considered to be from 15 Hz to 20,000 Hz.

automatic test equipment (ATE). A group of interconnected instruments that can be controlled by a computer or a controller to perform parametric or functional tests on components, circuits, or systems.

autoranging. A switching circuit for a multirange instrument that selects the proper range for a measurement, automatically.

bandwidth. The frequency range between a low and high frequency in which the performance of an instrument or circuit is specified; typically, it is the frequencies at which the response is attenuated by 3 dB from the average response in the passband.

bandwidth selectivity. For a spectrum analyzer, the ratio of the 60 dB bandwidth to the 3 dB bandwidth for a given IF filter; it describes the shape of the IF response.

base. The number of symbols used in a weighted counting system.

baseband signal. The lowest-frequency band occupied by a modulating signal before it modulates the carrier.

baseline. (1) The amplitude level from which a pulse appears to originate. (2) On an oscilloscope, the line that is observed when no signal is present.

base unit. See *fundamental unit.*

bed-of-nails. An array of spring-loaded pins, often used in conjunction with a vacuum fixture, by which automatic test equipment can probe nodes for stimulus and measurement purposes.

Bessel function. A solution to a specific type of differential equation. They are used in instrumentation to specify the bandwidth of FM signals.

bioinstruments. A class of instruments used to measure biological parameters.

bipolar transistor. A three-terminal semiconductor device that uses both minority and majority carriers to function; it can be used as an amplifier, detector, or switch.

black level. The level of video corresponding to the cutting off of the electron beam in a CRT.

blanking. A negative pulse applied to a CRT to turn off the beam during retrace or other times.

block diagram. A diagram that shows the functional relationship and signal flow in a circuit without showing the individual components.

bolometer. A sensor that is used to determine microwave or infrared power by converting it into a resistance change.

boundary scan. An automatic testing method whereby shift registers are placed within a digital integrated circuit between each pin and the internal logic; the register can be loaded with test vectors from a separate input. The method provides access points for ATE where the bed-of-nails approach may be limited.

Bourdon tube. A pressure sensor made from a metal tube that expands or contracts under pressure; it is made into a spiral, helix, or C shape and is closed at one end.

bridge. See *ac bridge, Wheatstone bridge.*

burst. A mode on waveform generators that programs a set number of cycles of the output following a trigger or command. Also see *color reference burst.*

cable. Two or more conductors mechanically assembled into a common jacket that form a transmission path for electrical signals.

calibration. The process of comparing an instrument to a standard or other instrument of known accuracy in order to bring it into substantial agreement with the established standard.

calibration curve. A line connecting data points taken as a calibration record.

calibration record. Data taken during a calibration.

calibrator. An instrument that produces a reference level voltage or current that can be used as a working standard for calibrating other instruments.

candela (Cd). The fundamental unit for luminous intensity in the SI system.

carrier. (1) A continuous frequency capable of being modulated. (2) The high frequency electromagnetic signal that is the basic transmitting frequency of a broadcast station.

cathode-ray tube (CRT). An electron beam tube with a phosphorous screen that is used to show an image when struck by a narrow beam of electrons emitted from a hot cathode.

celsius. A temperature scale based on the freezing and boiling point of pure water at standard pressure, defined as 0°C and 100°C, respectively.

channel. A signal path for an instrument; for example, a dual-channel oscilloscope has two independent signal paths.

characteristic impedance. An ac quantity that is the ratio of the signal voltage to the signal current in a transmission line. It is determined by the physical diameter, spacing, and dielectric material of the line and is independent of the length of the line.

chart recorder. A graphic recorder that draws a permanent plot of an input variable as a function of time, normally using a moving stylus on rolled or circular paper.

chop. A method of displaying two signals on the screen of a dual-trace oscilloscope by very rapidly switching electronically between them as they both move across the screen. The chop mode is best for displaying relatively slow signals.

chrominance signal. In the NTSC television system, the 3.58 MHz signal that carries the picture color information. Color hue is transmitted as a phase change with respect to the color reference burst; color intensity is transmitted by the amplitude of this signal.

clamp. See *dc restorer*.

clipping. The removal of a portion of a signal above or below a threshold.

clock. A device that generates periodic timing pulses; in a logic analyzer, the clock determines the sample rate.

closed-loop gain. The gain of an amplifier, including the effects of negative feedback.

coaxial cable. A broadband transmission line formed by two concentric conductors separated by an insulating material and covered with an insulating material.

collector characteristics. A set of curves for a bipolar transistor that represent the collector current as a function of the collector-emitter voltage and the base current.

color-bar generator. An instrument that generates a video test signal in the form of colored bars for the purpose of aligning and testing television and other video equipment.

color difference signals. The three principle color signals and the luminance signal for color television sent as each color minus the luminance signal; they are named the R-Y, G-Y, and B-Y signals.

colorimeter. A form of spectrometer that uses color filters to select wavelengths of light for measurement of a sample.

combinational logic. A logic circuit in which the output is solely determined by the inputs.

common-mode interference. A form of noise that occurs when noise voltages are induced in both signal conductors of a transmission line by the same amount with respect to a common ground.

common-mode rejection ratio (CMRR). For a differential amplifier, the ratio of the differential gain to the common-mode gain, usually expressed as a logarithmic ratio.

common-mode signal. A signal that is applied with the same amplitude and phase to both inputs of a differential amplifier and ground.

comparator. A circuit that determines if a voltage is greater or less than a reference voltage.

component analyzer. An instrument that can perform failure analysis using dynamic tests on capacitors, inductors, and other components.

composite video. The color television signal that contains all the picture information (horizontal and vertical blanking, luminance, and chrominance) as well as the synchronization signals.

concentration. The amount of solute per unit volume of solution.

conductivity meter. An instrument that is used to determine the ion concentration of a solution by submerging two electrodes in the solution and observing current in the solution.

contact noise. A form of pink noise generated in switches and other contacts that are made from dissimilar materials.

continuous signal. A signal that changes smoothly, without interruption.
controller. A device on the IEEE-488 bus that is responsible for information management.
coordinated universal time. Astronomical time based on the average period of the earth's rotation and broadcast by the NIST stations WWV and WWVH.
correlation coefficient. A dimensionless number that is a qualitative measure of goodness of fit of a line to experimental data. A value of −1 represents a perfect negative linear fit; a value of +1 represents a perfect positive linear fit.
coulomb (C). A quantity of charge represented by the amount that passes a point in 1 s by a direct current of 1 A.
counter. A logic circuit that follows a predetermined sequence of states.
crest factor. The ratio of the peak value of a waveform to its rms value.
cross talk. A form of interference noise that occurs when an unwanted signal from one circuit is coupled to a separate circuit.
current tracer. A hand-held instrument, typically used in conjunction with a logic pulser, that can detect pulsating current in a circuit by responding to the changes in the magnetic field.
cursor. (1) A horizontal or vertical line that marks a particular location on the display of an oscilloscope, logic analyzer or other instrument. (2) A highlighted region on the display of a CRT that indicates the location where the next character from a keyboard will be entered.
curve tracer. An instrument that shows the response and characteristic curves of bipolar and field-effect transistors as well as other circuit components.
cutoff frequency. The frequency that represents the transition between the passband and attenuation band of a device; for an *RC* circuit, it represents the frequency at which the reactance of the capacitor is equal to the resistance (also called the critical frequency or half-power frequency).
cycle. The complete sequence of values that a periodic signal exhibits before repeating.
d'Arsonval meter. See *permanent-magnet moving coil meter*.
data. Collected information.
data-acquisition system. A system for monitoring, displaying, and/or storing data from one or more transducers. Typically, the system is controlled by a small computer that is dedicated to an application.
data analysis. The process of drawing meaningful results from data by manipulating it or performing computations on it.
dc restorer. A circuit used in video systems to restore the dc level of the signal after the signal has been processed; also called a clamp.
decade. A factor of 10.
decade counter. A circuit designed to count in base ten.
decibel. A logarithmic ratio for measuring relative signal voltage or power levels; it is 10 times the logarithm of the power ratio.
decoder. A combinational logic circuit that asserts one of several outputs depending on the code that appears on the input.
delayed sweep. For oscilloscopes, a precise time interval that can be inserted between the trigger time and the start of the sweep.
demultiplexer. A circuit that separates multiplexed signals. Multiplexed signals are multiple signals that are combined onto a single channel.
derived units. A measurement unit formed by combining fundamental and/or supplementary units; any unit that is not a fundamental unit is a derived unit.
descriptive statistics. The mathematical study dealing with the gathering and recording of data including graphs, charts, and tables.
detent. A latched position on an instrument control that represents the calibrated position for the control; the control tends to resist moving out of the detent position.
deviation. In a set of measurements, the amount a given value differs from the arithmetic mean.
dielectric absorption. An effect in a capacitor preventing the total discharge of the capacitor due to internal dipoles remaining in a polarized state even when a direct short is placed across the terminals.
differential amplifier. A two-input amplifier in which the output is proportional to the difference between the inputs.
differential gain. In television systems, a distortion that causes a change in the amplitude of the chrominance signal due to a change in amplitude of the luminance signal.

differential-mode signal. A signal that is applied between both inputs of a balanced differential amplifier; also called a normal-mode signal.
differential phase. In television systems, a distortion that causes a change in the phase of the chrominance signal due to a change in amplitude of the luminance signal.
differential pressure. The pressure difference between two points.
digital multimeter (DMM). An electronic test instrument designed to measure voltage, current, or resistance and display the results as a number.
digital pattern generator (DPG). An instrument that can generate complex digital patterns for testing; they are sometimes built into a logic analyzer.
digital signal. A signal that can take on a limited number of discrete values.
digital storage oscilloscope (DSO). An instrument that converts a waveform into a series of discrete, time-ordered samples that are stored in memory as a record. One or more records can be viewed on the display.
digital-to-analog (D/A) converter. A circuit that changes a numerical input signal into an analog output signal proportional to the input.
digital voltmeter (DVM). An electronic test instrument designed to measure voltage and display the measurement as a number.
discrete signal. A signal that changes in definite steps, or in a quantized manner. The term discrete can be applied to either the time or amplitude characteristics of a signal.
display memory. In a logic analyzer, the random access memory that is used to store data for viewing.
display section. One of the principal blocks of an oscilloscope that includes the CRT and controls for adjusting a sharp presentation with the proper intensity.
distortion analyzer. See *harmonic distortion analyzer*.
distributed parameters. Resistance, capacitance, or inductance that is spread over a long region such as a cable.
domain. The set of values assigned to the independent variable.
dual beam. An oscilloscope with two independent electron beams.
dual trace. An oscilloscope with one electron beam that is electronically switched to show two signals.
duty cycle. The ratio of the pulse width to the period for a repetitive pulse waveform; it may be expressed as a fraction or as a percentage.
dynamic behavior. For a transducer, a specification of how the transducer responds to a changing input.
dynamic range. The difference between the minimum and maximum signal to which a given system can respond with a stated degree of accuracy.
dynamic resistance. See *ac resistance*.
dynamometer wattmeter. A meter that measures true power by sensing both voltage and current simultaneously; also called an electrodynamometer wattmeter.
effective value. The root-mean-square of a waveform; it is the value of an ac waveform that produces the same heating value as a corresponding dc value. It can be found by finding the average of the squares of a series of values, and taking the square root of the result.
electrical noise. An unwanted electrical signal.
electrocardiograph (EKG, ECG). A graphic recorder for observing the electrical activity of the heart using electrodes on the surface of the body.
electrochemical. Processes that involve chemical change produced by electricity and vice versa; many physiological measurements involve these electrochemical changes.
electrode. (1) Any terminal to which a voltage is deliberately applied. (2) A conductor that is used to transfer charge between a solution and an external circuit.
electrodynamometer. A meter, similar to the d'Arsonval meter movement except that it uses an electromagnet in place of the permanent magnet; it is used in wattmeters (see *dynamometer wattmeter*).
electroencephalogram (EEG). A graphic recorder for observing the electrical activity of the brain using electrodes on the surface of the scalp.
electrolyte. A substance that forms ions in solution, thus forming an electrical conductor.
electromagnetic interference (EMI or RFI). Interference derived from waves that have both

a magnetic- and electric-field component; formerly called radio-frequency interference.

electromagnetic spectrum. The entire range of frequencies covering electromagnetic energy.

electrometer. A multifunctional meter that has an extremely high input impedance (typically above 100 TΩ) for measuring current, voltage, and resistance in very high impedance circuits.

electromyogram (EMG). A graphic recorder for observing the electrical response of the muscles during stimulation.

electrostatic shielding. An enclosure or baffle made from a conductive material and grounded at one point to help block electrostatic fields.

empirical data. Information taken directly from an experiment without processing; also called raw data.

empirical equation. An equation based on measurements rather than theory.

encoder. A combinational logic circuit that converts a character or message into a coded form at the output.

endpoints. The minimum and maximum values of a transducer's range.

engineering notation. Power-of-10 notation in which the power is always a multiple of 3.

envelope. The amplitude limits of an electrical signal.

envelope mode. A recording mode for digital storage oscilloscopes that stores two samples of the data for every clock pulse. The samples represent the smallest and largest value of the data over multiple sweeps.

equivalent time sampling. A method of collecting high-resolution discrete data from a periodic waveform by digitizing several cycles of the waveform. The waveform is reconstructed by ordering the samples with respect to a common starting point.

error. The difference between the true or best accepted value of some quantity and the measured value.

error voltage. The voltage difference between the input and output in a servo system; in a self-balancing recorder, it is the voltage difference between the input and that derived from the position of the stylus.

excitation. A source of energy for transducer operation; it is specified as a nominal voltage and current or range of voltage/current.

external clock. On a logic analyzer, a series of pulses that are normally derived from the system under test, causing the input sample time to be synchronized to the system.

external trigger. A mode of starting an action by deriving the trigger from a source that normally is synchronous to the system under test.

extrapolation. Finding corresponding values of x or y beyond measured data points.

extrinsic noise. Noise that is picked up in a circuit due to an external cause.

farad (F). The unit for capacitance; 1 F is the capacitance of a capacitor that has 1 V between the terminals when it is charged by 1 C.

feedback. The return of all or part of the output signal to the input. Feedback can be either positive (in phase) or negative (out of phase). Negative feedback is widely applied to circuits to stabilize the gain and other characteristics.

fiber-optic cable. A transmission medium that passes light along a thin optical material made from glass, plastic, or other transparent material. Fiber-optic cable may contain one or many optical fibers.

field. One part of a video picture containing either the odd or even scanning lines; two fields make up a complete picture.

field-effect transistor. A transistor whose conductivity is controlled by the action of an electric field applied to the gate; the electric field changes the thickness of the depletion layer between the source and drain.

filter. A frequency selective network that passes certain frequencies in a region called the passband and rejects others in a region called the stopband.

flat cable. A transmission line composed of side-by-side conductors that are insulated from each other and bound together; also called ribbon cable.

floating. A condition where earth ground is not connected to an instrument's or circuit's reference voltage; this produces a common-mode voltage between the instrument or circuit and earth ground.

flood gun. An electron gun in an analog storage CRT that sends a cloud of low-energy electrons toward the phosphorous screen, causing the image previously written by the write gun to intensify.

fluorescence. The property of a phosphor to emit light when struck by electrons.

form factor. The ratio of the rms to average value of sinusoidal waveform for a half-cycle.

Fourier analyzer. A digital instrument that takes data in the time domain and transforms it, using the Fourier transform, into the frequency domain for viewing.

Fourier series. A mathematical expression that enables any periodic function to be expressed as a dc term plus sine and cosine terms that are integer multiples of the fundamental frequency.

Fourier transform. An algorithm for breaking a continuous, periodic, time-dependent signal into its sinusoidal components or vice versa.

four-wire resistance measurement. A measuring technique for small resistances in which two wires carry a known current to the resistance and two wires are used to sense the voltage produced.

frame. A complete video picture, including two fields of interlaced scanning lines.

frequency counter. An instrument used to measure the frequency of an unknown by counting a number of cycles over a precisely defined amount of time.

frequency deviation. The peak difference between the instantaneous frequency and the carrier frequency of an FM signal.

frequency distribution. A table of data partitioned into a set of values with equal intervals between the partitions. A plot of a frequency distribution produces a bar chart.

frequency response. A table or graph of the magnitude of the output of an instrument, circuit, or device as a function of frequency.

frequency marker. On a spectrum or network analyzer, a small intensified spot on the display that corresponds to a known frequency.

frequency modulation (FM). A means of encoding a low-frequency signal onto a high-frequency carrier by causing the output frequency to vary in a manner determined by the signal.

frequency range. The span of frequencies over which an instrument's performance is specified.

frequency resolution. A measure of the ability of a spectrum analyzer to distinguish two signals that have nearly the same frequency; it is specified by the resolution bandwidth.

frequency response. A description of the amplitude or phase characteristic of a system as a function of frequency.

frequency synthesizer. A stable source of different frequencies that produces the output digitally from a fixed reference oscillator. The output can be a sine, square, triangle, or other waveform.

functional test. A test to determine if the unit under test is operating properly.

function generator. A signal source that can use a free-running oscillator to produce sinusoidal, square, sawtooth, and other periodic waveforms that can be varied in frequency and amplitude.

fundamental. The lowest single-frequency component of a periodic waveform.

fundamental unit. A basic unit of a measurement system that normally can be defined independently of all other units. For the SI system, there are seven fundamental units. (An exception in the SI system is the fundamental electrical unit, the ampere, which does use another fundamental unit, the second, in its definition.)

gage factor. For a strain gage, a constant of proportionality that represents the fractional change in resistance for a given strain.

gage pressure. A pressure that is measured with respect to that of the atmosphere.

galvonometer. A sensitive mechanical ammeter with a centered needle that indicates electrical current in either direction. It is useful as a null indicator for a bridge.

galvonometer recorder. A chart recorder that uses a galvonometer movement to control the movement of a stylus.

Gaussian distribution. See *normal distribution.*

general-purpose interface bus. (GPIB); see *IEEE-488 bus.*

glitch. A very fast, unwanted voltage transient. D/A converters can produce a glitch when the input digital word changes by a large amount, causing the analog output to slew past the voltage representing the new word.

glitch capture. The ability of a logic analyzer to detect and display a signal that crosses the logic threshold twice between successive clock cycles.

go/no-go test. A test method that distinguishes only acceptable and unacceptable performance; see *functional test.*

graph. A pictorial representation of the relationship between two or more variables.

graphic recorders. An instrument that draws a graph on paper of the input signal as it is being measured.

grass. Random noise that is observed on an instrument near the baseline.

graticule. A calibrated screen in front of or etched onto a CRT used on oscilloscopes, spectrum analyzers, and other instruments.

ground bus. A large conductor that is used for a common ground return for a system.

ground-fault. A condition that causes a potential difference between the enclosure of an electrical appliance or instrument and the ground.

ground-fault interrupter (GFI). A sensitive detector that quickly interrupts the ac power line in the event that current in the hot and neutral wires is not the same.

ground loop. More than one path for ground current in an instrumentation system; it can produce noise or oscillations in a measuring system.

guarding. (1) A noise-reduction technique that uses an extra shield (called a guard shield) on cables. The guard shield is normally connected to a low-impedance terminal of the source, causing common-mode noise flow around the input resistance of the device. (2) In ATE, guarding is a method of blocking currents in parallel paths from entering the measurement node.

guide pin. A pin used in automatic test fixtures to provide precise mechanical alignment of a circuit board with the fixture.

half-cell potential. The separation of the potential of an electrochemical cell into two halves to account for the separate reactions at each electrode. The algebraic sum of the two half-cell potentials gives the overall potential of the cell.

Hall effect. A transverse voltage produced in a current-carrying conductor when the conductor is in the presence of a magnetic field. The magnetic field is perpendicular to the current-carrying conductor, and the Hall-effect voltage is perpendicular to both.

harmonic distortion. A distortion due to the presence of unwanted multiples of a fundamental frequency in the output caused by nonlinearities.

harmonic distortion analyzer. An instrument that can measure the amount of harmonic distortion present in a signal by using filters to separate the fundamental and harmonics.

henry (H). The unit of inductance; it is the inductance of a closed circuit when a potential difference of 1 V is produced by a current that changes by 1 A/s.

heterodyne converter. An analog mixer that mixes an incoming signal with a high frequency to lower the frequency to a range that counter circuits can handle.

heterodyne spectrum analyzer. A type of spectrum analyzer in which the input signal is mixed with a high sweep frequency and the resulting difference signal is passed through fixed bandpass filters to determine the spectrum.

heterodyning. The process of mixing two frequencies in a nonlinear mixer causing the output to contain the sum and difference frequency.

hierarchy. In automatic testing, the logical organization of a test sequence in order to prevent the masking of problems or damaging the unit under test from the wrong sequence.

holdoff. An oscilloscope control that is used to prevent retriggering the sweep until a time has elapsed set by the control; it is useful to obtain stable triggering on complex waveforms.

horizontal section. One of the principal blocks of an oscilloscope that provides the time-base (a linear ramp) to the horizontal deflection plates.

horizontal synchronization pulse. A signal contained in the composite video television signal used to start the electron beam at a new line in the picture.

hue. A particular color corresponding to one region of the spectrum, such as red, yellow, or blue but not white or black.

hum. Interference in an audio system due to power line disturbances; the interference frequency is generally 60 Hz or 120 Hz.

hysteresis error. The maximum difference between consecutive measurements for the same quantity when the measured point is approached each time from a different direction for full-scale traverses.

IEEE-488 bus. An interconnection standard for instruments developed by Hewlett-Packard and used by a number of manufacturers; it is used to convey control and measurement data between instruments and one or more controllers. Also known as the HPIB (Hewlett-Packard interface bus) and the GPIB (general-purpose interface bus).

illuminance. The intensity of luminous flux on a surface per unit area, measured in lumens per square meter (photometric system) and watts per square meter (radiometric system).

impedance bridge. An ac bridge circuit for determining the resistive and reactive components of impedance.

impulse. A pulse that has zero width and infinite amplitude. In practice, a pulse that approximates this ideal and is used to characterize the transient response of a system.

incident wave. A wave that travels from the source to the load.

input bias current. The average of two bias currents for the differential inputs of an operational amplifier.

input offset current. The difference between two bias currents for the differential inputs of an operational amplifier.

input offset voltage. The dc voltage that must be applied between the input terminals of an operational amplifier to obtain zero output voltage.

input resistance. The ratio of the input voltage to the input current of an amplifier under stated conditions.

input threshold. The smallest detectable value of a quantity by a measurement system starting near the zero value of the variable.

input voltage range. The range of common-mode signal that can be applied to the input terminals for which the amplifier is operational.

instrument. A device or system that detects a physical quantity, converts it into a measurable quantity, processes it, and displays, records, or transmits the information for the user. The user may be a human or a machine.

instrumentation amplifier (IA). A differential-amplifier that has a very high common-mode rejection ratio, high input impedance, and fixed closed-loop gain. It is available in IC form and is widely used for measuring low-level signals from transducers.

integrated circuit (IC). A highly miniaturized circuit consisting of resistors, capacitors, diodes, and transistors fabricated together as a single package. Monolithic ICs have all parts on a single substrate; hybrid ICs use a ceramic base to interconnect the components.

intelligent instruments. Instruments that contain a built-in microprocessor that can process data, perform self-diagnostics, do self-calibration, or perform related functions.

interference. (1) Noise that tends to obscure a useful signal. (2) The interaction of two waves of the same frequency in which they can reinforce or cancel each other, depending on their phase.

interference filters. (1) A narrow-band optical filter that consists of two reflecting layers separated by a transparent layer. (2) An active or passive circuit that blocks or bypasses unwanted signals.

interlacing. In video, the process of placing odd-numbered lines in one field and even numbered lines in another; each successive line in the picture belongs to a different field.

intermediate frequency (IF). A sum or difference frequency produced after mixing an incoming frequency with an internal oscillator.

intermodulation distortion. A distortion that occurs when two signals interact in a nonlinear circuit, producing a new signal.

international system of units (SI). A system of measurement units based on the meter, kilogram, second, ampere, candela, kelvin, and mole as base units from which all other units are derived.

International Temperature Scale. A temperature standard in which specific values are assigned to certain melting and boiling points to enable laboratories to use these points for calibration.

interpolation. Calculating corresponding values of x or y between measured data points.

intrinsic noise. Noise that is generated from within a circuit as opposed to noise generated from an external source.

intrinsic standard. A measurement standard that can be realized directly from a definition rather than a physical artifact.

irradiance. The intensity of electromagnetic flux on a surface per unit area, usually measured in watts per square meter or milliwatts per square centimeter.

isolation amplifier. An amplifier designed to provide a very high impedance path between a circuit that follows it and the circuit that precedes it; a common application is to isolate electrodes connected to a patient from instruments.

jitter. (1) A variation in the starting time, duration, or amplitude of a pulse between successive cycles. (2) In video, an unsteady television picture usually attributed to improper synchronization.

Johnson noise. See *thermal noise*.

Kelvin (K). The fundamental unit for temperature in the SI system. The scale is based on absolute values, where all temperatures (other than zero) have positive values, and each unit is defined as 1/273.16 of the triple point of water.

Kelvin bridge. (Also called a Kelvin double bridge); a bridge circuit used to determine very small resistances. It uses an extra set of ratio arms to offset the effects of lead and contact resistance.

***K* factor.** A ratio of a measured parameter to a reference value; sometimes a scaling factor is included in the defining equation for a particular *K* factor.

kilogram (kg). The fundamental unit for mass in the SI system, defined by a prototype standard.

large-signal voltage gain. The ratio of the output voltage, when it swings over a specified range with a specified load, to the differential input voltage.

leakage current. Undesired current that passes through a capacitor or insulator.

least squares. A method of fitting a straight line to data points by minimizing the square of the vertical deviations from the line.

level. (1) Signal amplitude in relationship to a standard or as indicated on a standard measuring device. (2) The ratio of a measured value to a reference value, usually in decibels.

library. In automatic testing, a computer file containing test patterns (stimulus) information and the expected output (responses).

linear regression. Finding the best-fit straight line to a set of data.

linear variable differential transformer (LVDT). A displacement transducer that uses a moving magnetic core inside a specially wound transformer; the output is a voltage proportional to the displacement.

Lissajous figures. A pattern formed on a CRT due to the interaction of sinusoidal waveforms on both the horizontal and vertical axis.

listener. A device on the IEEE-488 bus that receives data.

load cell. A force-measuring transducer based on the deformation of a metal element which is instrumented with strain gages. The ideal load cell is characterized by a linear relationship between strain and force.

loading error. The error in a quantity that is being measured due to introducing a measuring instrument or transducer.

load regulation. The change in output voltage for a change in the load current at a constant temperature.

logarithmic amplifier. An amplifier in which the output is proportional to the logarithm of the input.

logarithmic scale. A scale in which each increment represents a fixed ratio; logarithmic scales are commonly used in spectrum analyzers.

logic analyzer. A multichannel digital storage instrument with flexible triggering modes and display formats; it is used to troubleshoot both hardware and software problems.

logic probe. A simple hand-held instrument that indicates if a logic level is high or low or if pulses are present.

logic pulser. A hand-held instrument that sources very short duration pulses into a circuit from which the power has been removed.

loop through. A method of connecting instruments in series with coaxial cable; the signal is picked off by a high-impedance input on each instrument and passed to the next instrument in the chain. The last instrument is terminated in the characteristic impedance of the cable.

luminance signal. A video signal that carries the picture-brightness information.

magnetic recorder. An instrument that uses magnetic tape or disks for storing analog or digital information.

magnitude comparator. A logic circuit that compares the size of two binary words and indicates if one word is larger than, smaller than, or equal to the other.

manufacturing defects analyzer (MDA). An automatic test system that tests completed circuit board assemblies by comparing impedance measurements at nodes on the test circuit board with a known good circuit board.

Maxwell bridge. An ac bridge that uses known

resistors and capacitors to measure the inductance of low-Q inductors.

mean. The average of a set of values; it is determined by dividing the sum of a set of values by the number of values in the set.

measurand. A physical quantity or condition that is measured. Typical measurands are temperature, pressure, frequency, and power.

measurement. Process of associating a number with a quantity by comparison of the quantity with a standard.

measurement system. A set of interconnected components that detects, transmits and processes a physical quantity and may display, control, or record the quantity.

megger. A portable instrument, equipped with a hand crank, that measures very high resistances by delivering a high voltage to the resistance and measuring the resulting current.

menu. A list of options that are available for control of an instrument.

meter (m). The fundamental unit for length in the SI system. It is defined as the length of a path traveled by light in a vacuum during a time interval of 1/299,792,458 s.

microchannel plate. A very thin glass plate with millions of tiny holes that are treated internally to promote the production of electrons. The plate is mounted on the inner surface of a CRT and greatly increases the writing rate of the CRT.

microwave frequency counter. An instrument that converts microwave frequencies to those suitable for counting by digital circuits and displays the microwave frequency.

minterm. A logic term formed as a product that contains all of the input variables, either with or without an overbar on each variable.

mixed sweep. An oscilloscope mode in which two different sweep rates are shown at the same time.

mixing. The process of combining two or more signals in a nonlinear circuit to produce the original signals as well as the sum and difference frequencies and their harmonics.

modulation index. The ratio of the peak variation of a modulating waveform to the maximum designed variation; for FM, it is the ratio of the frequency deviation (of the carrier) to the modulating frequency.

mole. The fundamental unit for quantity in the SI system. A mole is the amount of substance that contains as many elementary entities as there are atoms in 0.012 kg of carbon.

multiplex. The process of combining more than one signal onto a single channel of information.

nanovoltmeter. A sensitive voltmeter optimized to measure very small voltages from low impedance sources.

National Institute of Standards and Technology (NIST). The national organization in the United States charged with maintenance of measurement standards and doing research that improves understanding of measurements and scientific work in related fields (formerly called NBS).

network analyzer. An instrument that integrates a stimulus instrument and a measurement instrument together to make accurate measurements of magnitude, phase, complex impedance, or other characteristic of a network.

node. A junction in a circuit where two or more components are connected.

noise. An unwanted voltage or current that does not convey useful information.

noise equivalent temperature. A way of expressing noise by stating an equivalent temperature in a resistor that will produce the same noise.

noise factor. A means of expressing the added contribution of an amplifier to the signal-to-noise ratio; it is found by dividing the signal-to-noise ratio at the input by the signal-to-noise ratio at the output.

noise figure. The decibel ratio of the noise factor.

noise floor. The limiting lower level that signals can be discerned on an instrument; it is from noise produced from within the instrument. The noise floor on a spectrum analyzer is seen as a "grassy" level that appears even with no signal.

noise margin. For digital circuits, the difference between the input and output voltage specification. It is specified for both the logic high level and the logic low level.

nonlinearity. The deviation from a reference straight line for a transducer or instrument.

normal distribution. A distribution of data points characterized by a symmetrical bell-shaped curve for which the deviations have an equal probability of being above or below the true value. The normal distribution is based on the assumption that errors are strictly random, not systematic.

normal-mode signal. See *differential-mode signal.*

normal mode voltage. A voltage applied between an instrument's high and low input terminals.

Norton's theorem. A circuit simplification viewed from the output terminals in which a two-terminal, linear network is replaced with a single current source and a parallel impedance. The current source is the same as the output short circuit current and the parallel impedance is the impedance seen at the output terminals when the sources are replaced by their internal resistance. The impedance is the same as the Thevenin impedance.

NTSC. A color television standard developed in the United States by the National Television Standard Committee and used by a number of countries.

null. A condition of balance in a circuit where the output is zero.

Nyquist frequency. One-half the sampling rate of sampled data; above this frequency sample data cannot be reconstructed without distortion.

Nyquist theorem. A theorem that states that the sampling rate of an analog signal must be at least twice the frequency of the highest frequency component of the signal.

offset. (pulse) The algebraic difference between the amplitude of the baseline and the amplitude of a reference level (usually ground).

ohm. The unit for resistance; it is the resistance of a conductor when a potential difference of 1 V produces a current of 1 A in the conductor.

ohmmeter. A test instrument to measure resistance.

open-loop gain. The gain of an amplifier without feedback.

operational amplifier. A very high gain IC amplifier that uses a differential input. It is a basic building block for many linear and nonlinear circuits.

optical pyrometer. See *pyrometer.*

optoisolator. A device that provides electrical isolation between two circuits by converting the signal to light with an LED from one circuit, transmitting the light to a photocell, and converting it back to an electrical signal for the second circuit.

oscillator. An instrument or circuit that generates a known frequency determined by the circuit components.

oscilloscope. An instrument that produces a graph of voltage versus time on a cathode-ray tube.

overshoot. A peak distortion that occurs immediately after a major transition of a pulse; it is expressed as a percentage of the pulse amplitude.

parallax. A reading error that occurs with analog meters when the observer's eye is not exactly in line with the pointer and the scale. It can also be observed when a graticule overlay is used in front of a CRT.

parametric test. A test in which the measured quantity is assigned a value, as opposed to a functional or go/no-go test.

passive device. A device that cannot amplify and does not require any input other than a signal to function. Examples include resistors, capacitors, and inductors.

passive filter. A frequency-selective network made only from resistors, capacitors, and inductors.

passive transducer. A transducer that has no source of power other than the signal.

peak-to-peak. The difference between the upper and lower limits of a signal.

perceptual aliasing. The appearance that sampled data is a different frequency than the actual frequency because of the eye's tendency to connect close dots, even when they are not closest in time.

period. The minimum interval between points on a waveform that repeat.

periodic. A waveform that repeats at a fixed interval.

permanent-magnet moving coil meter (PMMC). An analog meter movement that indicates current with a pointer against a scale. The PMMC movement, also called the d'Arsonval movement, balances the force created in a fine coil against a spring.

persistence. The length of time a phosphorescent material continues to glow after the excitation has been removed.

pH. The negative logarithm of the hydrogen ion concentration of a solution; it is a measure of the acidity or alkalinity of a solution.

phase noise. Small unwanted shifts in the frequency of a sine wave due to short-term instability. Phase noise shows up on a spectrum analyzer as a broadening of the fundamental line.

pH meter. A meter that uses a specialized ion-selective electrode to convert the hydrogen ion concentration into an electrical signal.

phosphor. A crystalline compound coated on the inside faceplate of a CRT that emits visible light when bombarded by electrons from the electron beam.

phosphorescence. The property of a material to emit light after the source of excitation has been removed.

photodetector. A transducer that converts radiant energy into electrical energy; examples include photodiodes, phototubes, phototransistors, and photomultipliers.

photodiode. A light-sensitive reverse-biased *pn* junction in which the current varies linearly with the incident light.

photometry. The measurement of light that is visible to the human eye; this includes quantities such as luminance, luminance intensity, color, and spectral distribution.

photomultiplier tube. A light-sensitive vacuum tube containing a photocathode, a series of dynodes, and an output electrode; electrons emitted from the photocathode are amplified by secondary emission of electrons from the dynodes before being collected by the output electrode.

phototransistor. A light-sensitive bipolar transistor with a transparent window in the base-collector region that produces collector current proportional to the incident light.

phototube. A light-sensitive vacuum tube containing a sensitive cathode and collecting electrode; the output current is proportional to the intensity of the incident light.

photovoltaic cell. A cell composed of two dissimilar materials that generates a voltage when exposed to light.

picoammeter. A very sensitive ammeter optimized for measuring very small currents (in the range of 0.1 pA).

pink noise. Noise in which the power is inversely proportional to the frequency over a specified range; the spectrum rolls off at a constant 10 dB per decade.

Poisson's ratio. The ratio of the strain along the length of a uniform, homogeneous bar to the strain across its width when a stress is applied to the length of the bar; it is a dimensionless constant that is a function of the elastic property of the material.

precision. The degree to which a number of measured values agree; it is a measure of repeatability.

preprocessor. A data pod for a logic analyzer that is designed to disassemble (take apart) instruction codes that are executed by a system and convert them to assembly language mnemonics.

prescaler. A digital frequency divider used to divide the input frequency to a counter by some known amount.

preselector. On a spectrum analyzer, a narrow-band tracking filter located before the first mixer.

preshoot. A baseline distortion that occurs immediately before a major transition of a pulse; it is expressed as a percentage of the pulse amplitude.

pretrigger capture. Refers to acquiring data that occurs before a trigger event. The trigger is used to complete the ongoing storage process. Pretrigger capture is used by digital storage oscilloscopes and logic analyzers.

primary electrons. Electrons directly emitted from a surface such as a hot cathode or from an electron gun.

primary standard. A measurement standard that does not require any other reference for calibration. Primary standards are maintained and controlled by national standards laboratories such as the National Institute of Standards and Technology in the United States.

probe. A device used to couple a signal from a circuit under test to an instrument.

processed data. Experimental data that has been analyzed or manipulated.

programmable gain amplifier. An amplifier commonly used in data acquisition systems to set the input signal to a level that allows the full range of the A/D converter.

propagation delay time. The time required for a signal to travel between two given points.
prototype standard. A measurement standard realized from a physical artifact.
pulse. A voltage (or current) that begins at some level, changes rapidly to a new level, remains at the new level for a limited time, and returns to the original level.
pulse amplitude. The algebraic difference between the maximum excursion and the baseline of a pulse.
pulse generator. An instrument that produces pulses with fast rise times over a wide frequency range and with varying duty cycles. Most allow the user to control other pulse parameters such as amplitude, dc offset, triggering and polarity.
pyrometer. A temperature transducer that does not make direct contact with the body to be measured; it operates by measuring the radiated heat or sensing optical properties.
qualifier. A set of digital conditions that must be met before triggering a logic analyzer, oscilloscope, or other instrument. In a logic analyzer, the qualifier can be put on the acquisition clock as a gate that allows storage of the input data only if preset conditions are true.
quantitative data. Information that has numbers associated with the magnitude or intensity.
quantization. The process of assigning numeric values to sampled analog quantities.
quantization error. The difference between a sampled analog quantity and the digital number that represents the quantity.
quantize. To express a value in multiples of a fixed basic quantity.
radio frequency (RF). Those frequencies in the electromagnetic spectrum that are approximately from 10 kHz to 100 GHz.
radiometry. The measurement of the total optical spectrum; this includes quantities such as illuminance, irradiance, and spectral distribution.
radix. See *base*.
random errors. Measurement errors that tend to vary about the true value by chance.
range. The set of values between the upper and lower extremes that an instrument or transducer is designed to measure; it is normally specified by giving the upper and lower limits.
ratio measurement. A measurement in which a signal is compared to an external reference.
raw data. See *empirical data*.
real time. The processing of information as it is measured.
real-time sampling. Sampling in which all samples are collected sequentially, in a single acquisition.
real-time spectrum analyzer. A spectrum analyzer that uses a large number of filters to acquire the entire spectrum at the same time; it can show changes in a spectrum as they occur.
recording instrument. An instrument that retains a record of the input data as it is being measured.
record length. For a digital oscilloscope or logic analyzer, the number of sampled points in an acquisition.
reference levels. For a spectrum analyzer, the signal level that corresponds to the top graticule line.
reference memory. In a logic analyzer or a digital oscilloscope, the random-access memory that is used to store data obtained during a prior cycle; generally it is used for comparison with current data.
reflected wave. A wave that travels from the load to the source.
relative measurement. A measurement made by comparing it to a similar quantity, typically a standard. An example of a relative measurement is the dBm, which compares a power level to 1 mW.
repeatability. The agreement of a group of measurements taken by the same instrument under the same conditions.
resistance temperature detector (RTD). A temperature transducer constructed from fine wire or metal film that changes resistance in a known manner as a function of temperature.
resolution. The magnitude of the smallest detectable change in a measured quantity.
resolution bandwidth (RBW). For a spectrum analyzer, a measure of the width of the IF filter at some decibel level relative to the peak response (typically −3 dB or −6 dB); it determines the frequency resolution of the analyzer.
response instruments. Instruments that convert a quantity to be measured into a usable form for interpretation.

response time. The time required for the output of an instrument, system, or element to reach a specified percentage of the final value in response to an input change.

response-time error. Error associated with the inability of an instrument, system, or element to follow changes in the input.

ribbon cable. A flat package of wires placed side by side and fastened together by an insulating material.

ringing. A damped oscillation that appears immediately following a major transition.

ripple. An unwanted ac component of the power line, or harmonic of the power line, superimposed on a dc voltage.

rise time. The time required for a pulse to increase between two prescribed percentages of the final level, typically 10% and 90%.

root-mean-square (rms). A value assigned to a waveform found by taking the square root of the arithmetic mean of a series of squares; it is equivalent to the dc heating value of the waveform.

sample. One or more quantities selected at random from a group.

sample-and-hold. A circuit used in the process of converting an analog signal to a digital signal. The sample-and-hold circuit keeps the signal at a fixed level while it is quantized.

sample rate. The frequency at which periodic samples are taken. For digital storage oscilloscopes, the sample rate is given in megasamples per second (MSa/s); sample rate can refer to the frequency that an analog signal is encoded into a digital signal.

sampling. The process of converting a signal to a discrete-time signal.

saturated. A solution containing the maximum amount of solute that can be dissolved under given conditions.

Schmitt trigger. A comparator circuit with two trip points because of hysteresis; it is immune to noise.

second. The fundamental unit for time in the SI system.

secondary electrons. Electrons that are emitted from a surface due to bombardment by an incident (primary) electron.

secondary standard. A measurement standard that must be periodically compared to a primary standard.

Seebeck effect. The small thermionic voltage that appears between the leads of two dissimilar wires that are joined together; the voltage arises from the difference in density of free electrons in the wires and depends on the temperature. If the circuit is completed, and one junction is at a different temperature than the other, a current will flow.

self-balancing recorder. A type of chart recorder that uses a slide wire potentiometer to provide a voltage proportional to the position of the stylus; a servo motor is driven by the difference between the stylus position and the input signal.

sense leads. A pair of leads that is used to detect a condition; they carry very little current in order to avoid a voltage drop in the leads.

sensitivity. (1) The ratio of the output response to the input cause under static conditions. (2) For a radio receiver, the amplitude of an input radio carrier signal (in microvolts) that causes a given attenuation of noise with respect to the carrier (usually −20 dB).

sequential logic. A logic circuit in which the output is determined by the inputs and whatever has occurred previously. Sequential logic circuits contain combinational logic and memory.

settling time. The total time required for the output to remain within a specified error band of the final value after the application of a step input voltage. For pulses, it is measured from the 90% point on the rising edge.

setups. A set of instrument parameters that can be stored in a nonvolatile memory for quickly configuring an instrument for a particular measurement.

shield. A metal enclosure surrounding a circuit or wire to reduce interference or leakage current; shields are normally grounded.

shift register. A series of flip-flops connected together such that the output of one feeds the input of the next.

shot noise. A form of white noise generated by charge carriers crossing a semiconductor junction; it is also present in vacuum tubes due to random variations in number and energy of electrons leaving a heated cathode.

shunt resistor. A resistor placed in parallel with an ammeter to increase its full-scale range.
SI. See *international system of units*.
sideband. A signal component that is found above or below the carrier frequency; it is produced as a result of mixing a carrier and modulating waveform together in a nonlinear manner or it can be formed as a distortion product.
signal. Any physical quantity that varies with time (or other independent variable) and carries useful information.
signal generator. An instrument that produces high-frequency sine waves and modulated sine waves.
signal ground. The reference point in a circuit that acts as a return path for signal currents.
signal-to-noise ratio. The ratio of the amplitude of a signal to the amplitude of the noise at the same point, usually expressed in decibels.
signature analyzer. An instrument that checks for faults in a digital system by using a cyclic redundancy check calculator; the instrument compares expected values with those observed on a unit under test to help locate faults.
significant digits. Those digits in a measured quantity known to be correct; zeros used only to position the decimal point are not considered significant.
simple harmonic motion. The motion that occurs when a body is subjected to a force that is directly proportional to the displacement from a fixed point. An example is the motion of a pendulum.
sine-squared pulse. A video test signal used principally to evaluate differential gain and phase.
sine-wave generator. An instrument that generates an alternating frequency with very little harmonic content; the waveform has the shape of the trigonometric sine function.
single sweep. An operating mode used on oscilloscopes and spectrum analyzers in which the sweep control must be reset after each sweep; useful in photography of the display.
skin effect. The tendency for current density to be highest on the surface of a conductor and to decrease within the conductor; it is more pronounced at higher frequencies.
slew rate. The maximum rate of change of the output voltage when supplying rated output current when a step function is applied to the inputs, measured in V/μs.
solution. A homogeneous mixture of a solute and a liquid solvent; the solute can be a solid, liquid, or gas.
source-measurement unit. An instrument that can simultaneously source voltage and measure current or source current and measure voltage; it consists of an electrometer, DMM, voltage source and current source in one.
span. The difference between the maximum and minimum values on the scale of an instrument.
span shift. A change in calibration of a device characterized by a change in the slope of the measurement span; for transducers, it is generally caused by a temperature gradient.
spectral analysis. The process of dissecting a time-domain signal into its frequency distribution.
spectral purity. A measure of the short term stability of a signal; it is generally specified for communication systems in terms of the single sideband noise in a 1 Hz bandwidth as a function of the offset from the carrier frequency.
spectrophotometer. An instrument that measures the absorption of light as a function of wavelength through a sample; typically it is used to determine the concentration of a sample.
spectrum. (1) A plot of the amplitude of a waveform as a function of frequency. (2) The range of frequencies that compose a region of electromagnetic radiation.
spectrum analyzer. An instrument that measures the amplitude of a waveform as a function of frequency and displays them on a CRT; generally it is used to display the power spectrum. The polarity of the signal is ignored.
stability. The property of a measured quantity to be free of random variations.
stability error. The gradual change of a transducer or instrument due to physical, environmental, or other factor; it is specified in terms of a percentage change from full-scale output.
standard. A reference suitable for establishing the value of a physical quantity.
standard deviation. A measure of the scatter of a set of data; it is the root-mean-square of the deviations from the average.
state. One of the possible outputs of a logic circuit.

state diagram. A pictorial representation describing the sequence of a sequential logic circuit.

state table. In a logic analyzer, a tabular listing of digital data obtained during an acquisition cycle.

statistics. The branch of mathematics concerned with collecting, processing, and analyzing data (called descriptive statistics) and drawing conclusions from it (called statistical inference).

step generator. A circuit that produces a stair-step output of constant current or voltage at each step; it is used in a transistor curve tracer to produce a family of curves.

stimulus instruments. Instruments that are sources of signals.

strain. The elastic deformation of an object when it experiences a force; it is found by dividing the change in length of an object to its undistorted length.

strain gage. A transducer that consists of a thin metallic conductor or semiconductor that changes resistance slightly as it is stretched or compressed by strain.

stress. The force per unit area within a solid object; the force can be either a tension or compression force.

supplementary units. One of the two units in the SI system specially designated to measure angles. The units are the radian and the steradian.

surface-mounted device (SMD). A device that is installed (normally by soldering) onto the surface of a circuit board such that no part of it passes through the board.

sweep. A linear ramp signal that is applied to the horizontal plates of a CRT to move the beam across the display.

sweep oscillator. An instrument that generates a sine wave output that varies at a cyclic rate between two selected frequencies; also called a swept-frequency generator.

swept-frequency generator. See *sweep oscillator.*

synthesized function generator. A waveform generator that employs a crystal reference oscillator to generate highly accurate frequencies by using digital synthesis from the reference oscillator.

system. A combination of assemblies connected as a functional group.

systematic errors. Measurement errors that cause the result to be consistently above or below the true value; they are essentially constant with repeated measurements.

talker. A device on the IEEE-488 bus that sends data.

test vectors. A set of conditions for testing a digital device or system; the values are typically stored in a table.

thermal noise. Noise generated by the random movement of electrons in a resistance due to temperature; also called Johnson noise.

thermionic emission. The process of liberating electrons from a cathode by heating it.

thermistor. A sensitive temperature transducer made from a sintered mixture of a transition metal oxide that decreases resistance as a function of temperature.

thermocouple. A temperature sensor made by joining two dissimilar wires; it produces a voltage across the junction that is proportional to the temperature of the junction.

thermometer. A temperature transducer that makes direct contact with the body to be measured.

Thevenin's theorem. A circuit simplification viewed from the output terminals in which a two-terminal, linear network is replaced with a single voltage source and series impedance. The voltage source is the same as the output open circuit voltage and the series impedance is the impedance seen at the output terminals when the sources are replaced by their internal resistance.

threshold. (1) The smallest value of a voltage, current or other parameter that can be detected. (2) A voltage level that is used for comparison; logic thresholds are set on instruments such as logic analyzers and logic probes.

timebase. A sawtooth oscillator whose output is used as a time reference.

tooling hole. A reference hole in a circuit board for accepting a guide pin from an automatic test fixture to provide precise mechanical alignment of the circuit board with the fixture.

total harmonic distortion. The ratio of the rms value of all harmonics to the rms value of the fundamental when a pure sine wave is used as the input.

traceability. The calibration records that lead

from an instrument through the chain of standards and records leading to a national standards laboratory (in the United States, NIST).

transducer. A device that converts a physical quantity from one form to another.

transfer function. The ratio of the output signal of a system or subsystem to the input signal.

transfer impedance. The impedance determined by finding the ratio of a voltage at one point to a dependent current at another point.

transfer oscillator. An instrument that uses a phase-lock loop to lock an input signal with a known harmonic of an internal variable frequency oscillator; the variable oscillator's frequency can then be counted by a conventional frequency counter.

transient response. The time response of a system to an abrupt input change.

transitional sampling. In a logic analyzer, a sampling mode in which the sample clock is assigned a time value; the clock value and input signal are stored only when the input signal changes. Data is "time-stamped" for reconstruction.

transmission line. A conductive path, waveguide, or other system to transfer electrical signals from one point to another.

traveling-wave tube. A specialized vacuum tube used at microwave frequencies as an amplifier or oscillator. Traveling-wave tubes have been used to make specialized, very high frequency oscilloscopes.

trigger. A short pulse used to initiate an operation.

trigger probe. A probe that can generate a trigger pulse when certain predetermined conditions have been met in a digital circuit.

trigger section. One of the principal blocks of an oscilloscope that samples the input signal and provides a synchronizing pulse to the horizontal section.

truth table. A tabular listing of all possible inputs along with the respective output for a combinational logic element or circuit.

twisted-pair. A transmission line composed of two insulated wires twisted together; frequently the wires are contained within a common covering or shield.

undershoot. A baseline distortion that occurs immediately after a major transition of a pulse; it is expressed as a percentage of the pulse amplitude.

unit under test (UUT). A device or circuit assembly that is tested for parameters (parametric test) or functionality.

universal electronic counter. A multifunctional instrument that can measure frequency, period, time intervals, and do event counting; it operates by counting a number events that occur in a precisely controlled time interval.

vector impedance meter. An instrument that indicates the impedance of an unknown quantity at a stated frequency by measuring the voltage and current in the unknown and indicating the ratio.

vector scope. An instrument for displaying the color phase and gain of an encoded television signal.

vertical section. One of the principal blocks of an oscilloscope that attenuates or amplifies the input signal to provide the proper voltage to the vertical deflection plates.

vertical synchronization pulse. A signal contained in the composite video television signal used to start a new frame of the picture.

vestigal sideband. A form of amplitude modulation in which one of the sidebands, the carrier, and only part of the other sideband are transmitted.

video. The television signal; usually it refers to the bandwidth and spectral location of the signal; usually from 60 Hz to several megahertz.

video filter. A post-detection low-pass filter used in spectrum analyzers to remove noise from the displayed spectrum.

virtual ground. A junction at the inverting input of a linear operational amplifier circuit that assumes a potential near ground due to negative feedback when the noninverting input is connected to ground; it is a ground for voltage but not for current.

volt. The unit for potential difference; it is the difference in voltage across a conductor that carries a current of 1 A when the power dissipated is 1 W.

voltmeter. An instrument for measuring voltage; voltmeters can be designated for either direct or alternating voltage.

VOM. Volt-ohm-milliammeter; a portable test

instrument using an analog meter that measures voltage, resistance, or current.

vu meter. An indicator that incorporates a milliammeter calibrated to indicate the volume level of complex signals such as speech.

watt (W). The unit of power; 1 W is the power dissipated in a conductor when a current of 1 A develops a voltage of 1 V.

wattmeter. An instrument that indicates the power in a load by measuring the voltage and current.

waveform monitor. A specialized oscilloscope optimized for video measurements.

weber. The unit of magnetic flux; it is the flux which, linking a 1-turn coil, will produce a potential difference of 1 V when the flux is reduced to zero in 1 s.

weighted number system. A counting system in which the position of the symbols determines the value of the number.

Weston cell. A chemical cell that is used as a voltage standard; a bank of Weston cells was formerly used by the NIST as a primary standard.

Wheatstone bridge. A circuit containing four resistive arms used to detect a very small resistance change in one of the arms. The Wheatstone bridge is used to detect changes in resistance transducers and do precision resistance measurements.

white noise. Random noise in which the power is constant per hertz across a specified band.

word. A group of binary bits considered as an entity.

word recognition. The matching of a logic word; in a logic analyzer, the word recognition probe is used to sense logic for comparison with one previously stored in memory.

working standard. A standard used for routine calibration and certification. Working standards require periodic calibration and certification.

write gun. An electron gun in a storage CRT that sends a highly collimated beam of electrons toward a phosphorous screen.

writing rate. An indication of the ability of a CRT to give off light; a high writing rate is necessary to show rapidly changing waveforms.

WWV and WWVB. Radio stations operated by the NIST that provide high-quality carrier frequencies, audio tones, and time standards.

X-Y recorder. A type of graphic recorder that plots one variable against another on a rectangular sheet of paper; they are most suited to applications where time is not one of the variables.

Young's modulus. Stress divided by strain; it is a constant that describes one of the elastic properties of a noncrystalline material (also called the modulus of elasticity).

z axis. An oscilloscope input that controls the intensity of the electron beam.

zero beat. A process of comparing two frequencies by mixing them in a nonlinear mixer, creating the difference frequency. If the two frequencies are exactly the same, the difference frequency is zero and disappears.

zero offset. The output of an instrument or circuit when there is no input (input terminals shorted).

zero shift. A change in the output of a device with no input; for transducers, it is generally caused by a temperature shift that causes dimensional change or offset current change in the electronics.

INDEX